Joaquim Marques de Sá   Luís A. Alexandre
Włodzisław Duch   Danilo Mandic (Eds.)

Artificial Neural Networks – ICANN 2007

17th International Conference
Porto, Portugal, September 9-13, 2007
Proceedings, Part II

Springer

Volume Editors

Joaquim Marques de Sá
University of Porto, Faculty of Engineering
Rua Dr. Roberto Frias, 4200-465 Porto, Portugal
E-mail: jmsa@fe.up.pt

Luís A. Alexandre
University of Beira Interior, Dept. of Informatics
and
IT-Networks and Multimedia Group
Rua Marquês d'Ávila e Bolama, 6201-001 Covilhã, Portugal
E-mail: lfbaa@di.ubi.pt

Włodzisław Duch
Nicolaus Copernicus University, Dept. of Informatics
ul. Grudziadzka 5, 87-100 Torun, Poland
E-mail: wduch@is.umk.pl

Danilo Mandic
Imperial College, Communication and Signal Processing Research Group
Dept. of Electrical and Electronic Engineering
Exhibition Road, London, SW7 2BT, UK
E-mail: d.mandic@imperial.ac.uk

Library of Congress Control Number: 2007934292

CR Subject Classification (1998): F.1, I.2, I.5, I.4, G.3, J.3, C.2.1, C.1.3

LNCS Sublibrary: SL 1 – Theoretical Computer Science and General Issues

ISSN 0302-9743
ISBN-10 3-540-74693-5 Springer Berlin Heidelberg New York
ISBN-13 978-3-540-74693-5 Springer Berlin Heidelberg New York

Springer is a part of Springer Science+Business Media

springer.com

Printed in Germany

Typesetting: Camera-ready by author, data conversion by Scientific Publishing Services, Chennai, India
Printed on acid-free paper SPIN: 12116386 06/3180 5 4 3 2 1 0

Preface

This book includes the proceedings of the International Conference on Artificial Neural Networks (ICANN 2007) held during September 9–13, 2007 in Porto, Portugal, with tutorials being presented on September 9, the main conference taking place during September 10-12 and accompanying workshops held on September 13, 2007. The ICANN conference is organized annually by the European Neural Network Society in co-operation with the International Neural Network Society, the Japanese Neural Network Society, and the IEEE Computational Intelligence Society. It is the premier European event covering all topics related to neural networks and cognitive systems. The ICANN series of conferences was initiated in 1991 and soon became the major European gathering for experts in these fields. In 2007 the ICANN conference was organized by the Biomedical Engineering Institute (INEB - Instituto de Engenharia Biomédica), Porto, Portugal, with the collaboration of the University of Beira Interior (UBI - Universidade da Beira Interior), Covilhã, Portugal and ISEP, Polytechnic Engineering School, Porto, Portugal. From 376 papers submitted to the conference, 197 papers were selected for publication and presentation, following a blind peer-review process involving the Program Chairs and International Program Committee; 27 papers were presented in oral special sessions; 123 papers were presented in oral regular sessions; 47 papers were presented in poster sessions. The quality of the papers received was very high; as a consequence, it was not possible to accept and include in the conference program many papers of good quality. A variety of topics constituted the focus of paper submissions. In regular sessions, papers addressed the following topics: computational neuroscience and neurocognitive studies, applications in biomedicine and bioinformatics, spiking neural networks, data clustering, signal and times series processing, learning theory, advances in neural network learning methods, advances in neural network architectures, data analysis, neural dynamics and complex systems, ensemble learning, self-organization, robotics and control, pattern recognition, text mining and Internet applications, vision and image processing. Special sessions, organized by distinguished researchers, focused on significant aspects of current research, namely: emotion and attention, understanding and creating cognitive systems, temporal synchronization and nonlinear dynamics in neural networks, complex-valued neural networks. Papers presented in poster sessions were organized in the following topics: real-world applications, signal and time series processing, advances in neural network architectures, advances in neural network training, meta learning, independent component analysis, graphs, evolutionary computing, estimation, spatial and spatio-temporal learning. Prominent lecturers gave six keynote speeches at the conference. Moreover, well-known researchers presented seven tutorials on state-of-the-art topics. Four post-conference workshops, entitled "Cognitive Systems", "Neural Networks in Biomedical Engineering and

Bioinformatics", "What It Means to Communicate" and "Neural Networks of the Future?", concluded the focus of ICANN 2007 on the state-of-the-art research on neural networks and intelligent technologies. An in-depth discussion was held on the prospects and future developments both in theory and practice in those important topics. We would like to thank all the members of the local committee for their contribution to the organization of ICANN 2007. A special thanks to Alexandra Oliveira whose dedication and work quality were a major guarantee of the success of ICANN 2007. We also wish to thank Alfred Hofmann and the LNCS team from Springer for their help and collaboration in the publication of the ICANN 2007 proceedings.

July 2007

Joaquim Marques de Sá
Luís A. Alexandre

Organization

General Chair

Joaquim Marques de Sá
University of Porto, Portugal

Co-chair

Luís A. Alexandre
University of Beira Interior, Portugal

Program Chairs

Włodzisław Duch
Torun, Poland & Singapore, ENNS President

Danilo Mandic
Imperial College London, UK

Honorary Chair

John G. Taylor
Kings College, London, UK

Program Committee

Alessandro Sperduti, University of Padova, Italy
Alessandro Villa, University of Grenoble, France
Amir Hussain, University of Stirling, UK
Andreas Nuernberger, University of Magdeburg, Germany
Andreas Stafylopatis, NTUA, Greece
Andrzej Cichocki, RIKEN Brain Sci. Inst., JP
Bruno Apolloni, University of Milan, Italy
David Miller, University of Pennsylvania, USA
Dragan Obradovic, Siemens Corp. Res., Germany
Erkki Oja, Helsinki University, Finland
Erol Gelenbe, Imperial College London, UK
Hojjat Adeli, Ohio State University, USA
Jacek Mandziuk, Warsaw University, Poland

João Luís Rosa, Catholic University Campinas, Brazil
Jose Dorronsoro, Universidad Aut. de Madrid, Spain
José Príncipe, University of Florida, USA
Jürgen Schmidhuber, TU Munich, DE - IDSIA, Switzerland
Lefteri Tsoukalas, Purdue University, USA
Marios Polycarpou, University of Cyprus, Cyprus
Mark Embrechts, Rensselaer Inst., USA
Michel Verleysen, University of Louvain-la-Neuve, Belgium
Nikola Kasabov, Auckland University, New Zealand
Okyay Kaynak, Bogazici University, Turkey
Olli Simula, Helsinki University, Finland
Peter Andras, University of Newcastle, UK
Péter Érdi, HU & Kalamazoo College, USA
Stan Gielen, University of Nijmegen, The Netherlands
Stefan Wermter, University of Sunderland, UK
Stefanos Kolias, NTUA, Greece
Steve Gunn, University of Southampton, UK
Thomas Martinetz, University of Luebeck, Germany

Local Organizing Committee

Alexandra Oliveira, INEB
Ana Maria Tomé, Aveiro University
Bernardete Ribeiro, Coimbra University
Carlos Soares, Porto University
Daniel Carrilho, Health Techn. -IPP
Fernando Sereno, ESE-IPP
Helena Brás Silva, ISEP-IPP
Hugo Proença, UBI
Jorge Santos, ISEP-IPP
Lígia Carvalho, INEB
Luís Silva, INEB
Paulo Cortez, Minho University
Paulo Fazendeiro, UBI
Petia Georgieva, Aveiro University

Reviewers

Abe	Shigeo	Kobe University
Agell	Núria	Ramon Llull University
Aiolli	Fabio	Pisa University
Alexandre	Frederic	INRIA Lorraine/LORIA-CNRS
Alexandre	Luís	University of Beira Interior
Alhoniemi	Esa	Turku University
Andras	Peter	University of Newcastle

Anguita	Davide	Genoa University
Angulo-Bahon	Cecilio	Technical University of Catalonia
Apolloni	Bruno	University of Milan
Archambeau	Cédric	Université Catholique de Louvain
Arenas	Jerónimo	Universidad Carlos III de Madrid
Atencia	Miguel	Universidad de Málaga
Avrithis	Yannis	National Technical University of Athens
Barbosa	Jorge	University of Porto
Bermejo	Sergi	Universitat Politècnica da Catalunya
Bianchini	Monica	Università di Siena
L Boni	Andrea	University of Trento
Bourlard	Herve	IDIAP Research Institute
Boyer	Domingo	University of Córdoba
Cabestany	Joan	Universitat Politécnica da Catalunya
Colla	Valentina	Scuola Sup. Sant'Anna Pisa
Corchado	Emilio	Universidad de Burgos
Cornford	Dan	Aston University
Corona	Francesco	Helsinki University of Technology
Correia	Miguel	University of Porto
Cortez	Paulo	University of Minho
Crook	Nigel	Oxford Brookes University
Dorronsoro	José	Universidad Autónoma de Madrid
Dounias	Georgios	University of the Aegean
Duch	Wlodzislaw	Nicolaus Copernicus University
Duro	Richard	University of Coruña
Embrechts	Mark	Rensselaer Polytechnic Institute
Érdi	Péter	Henry Luce Foundation
Fazendeiro	Paulo	University of Beira Interior
Franco	Leonardo	Universidad de Málaga
Francois	Damien	Université Catholique de Louvain
Fyfe	Colin	University of Paisley
Garcia-Pedrajas	Nicolas	University of Córdoba
Georgieva	Petia	University of Aveiro
Gielen	Stan	University of Nijmegen
Giudice	Paolo	Istituto Nazionale di Fisica Nucleare
Gonzalez	Ana	Universidad Autónoma de Madrid
Gosselin	Bernard	Faculté Polytechnique de Mons
Grana	Manuel	University Pais Vasco
Gunn	Steve	University of Southampton
Hammer	Barbara	University of Osnabrueck
Heidemann	Gunther	Bielefeld University

Hollmen	Jaakko	Technical University of Helsinki
Honkela	Antti	Helsinki University of Technology
Hoyer	Patrik	Helsinki Institute for Information Technology
Igel	Christian	Ruhr-Universitaet Bochum
Indiveri	Giacomo	UNI-ETH Zurich
Jin	Yaochu	Honda Research Institute Europe
Jutten	Christian	LIS-INPG
Kaban	Ata	University of Birmingham
Kaiser	Marcus	Newcastle University
Karhunen	Juha	Helsinki University of Technology
Karpouzis	Kostas	ICCS-NTUA
Kasderidis	Stathis	Institute of Computer Science - FORTH
Kaynak	Okyay	Bogazici University
Kim	DaeEun	Max Planck Institute for Psychological Research
Kollias	Stefanos	National Technical University of Athens
Koroutchev	Kostadin	Universidad Autónoma de Madrid
Kounoudes	Tasos	SignalGeneriX
Laaksonen	Jorma	Technical University of Helsinki
Lagos	Francisco	Universidad de Málaga
Lang	Elmar	Universität Regensburg
Lansky	Petr	Academy of Sciences of the Czech Republic
Larochelle	Hugo	University of Montréal
Leclercq	Edouard	Université du Havre
Leiviskä	Kauko	University of Oulu
Lendasse	Amaury	Helsinki University of Technology
Likas	Aristidis	University of Ioannina
Loizou	Christos	Intercollege, Limassol Campus
Magoulas	George	Birkbeck College, University of London
Mandic	Danilo	Imperial College London, UK
Mandziuk	Jacek	Warsaw University of Technology
Marques de Sá	Joaquim	University of Porto
Martinetz	Thomas	University of Luebeck
Martinez	Dominique	LORIA
Masulli	Francesco	Polo Universitario di La Spezia G. Marco
Micheli	Alessio	University of Pisa
Moreno	Juan	Universidad Politécnica de Cataluña
Muresan	Raul	SC. NIVIS SRL
Müller	Klaus-Robert	University of Potsdam

Nakayama	Minoru	CRADLE
Navía-Vázquez	Ángel	Universidad Carlos III de Madrid
Neskovic	Predrag	Brown University
Nikolopoulos	Konstantinos	Lancaster University Management School
Nürnberger	Andreas	Otto-von-Guericke Universität Magdeburg
Obradovic	Dragan	Siemens AG
Oja	Erkki	Helsinki University of Technology
Osowski	Stanislaw	Warsaw University of Technology
Parra	Xavier	Technical University of Catalonia
Patan	Krzysztof	University of Zielona Góra
Paugam-Moisy	Helene	Institut des Sciences Cognitives
Peters	Gabriele	Universitaet Dortmund
Peterson	Leif	Dept. of Public Health of the Methodist Hospital
Petrosino	Alfredo	University of Naples "Parthenope"
Polani	Daniel	University of Hertfordshire
Porrmann	Mario	Heinz Nixdorf Institute
Príncipe	José	CNEL - University of Florida
Proença	Hugo	University of Beira Interior
Puzenat	Didier	Université Antilles-Guyane
Reyes	Jose	Universidade da Coruña
Ribeiro	Bernardete	University of Coimbra
Rocha	Miguel	University of Minho
Rosa	João	PUC-Campinas
Rospars	Jean-Pierre	INRA - Laboratoire de Biométrie
Rossi	Fabrice	INRIA Rocquencourt
Ruiz	Francisco	Universitat Politécnica de Catalunya
Sandoval	Francisco	Universidad de Málaga
Santos	Jorge	Instituto Superior de Engenharia do Porto
Scheper	Tjeerd	Oxford Brookes University
Schmidhuber	Jürgen	TU Munich, DE - IDSIA
Schwenker	Friedhelm	University of Ulm
Sereno	Fernando	Escola Superior de Educação do Porto
Serrano	Eduardo	Universidad Autónoma de Madrid
Silva	Luís	INEB - Instituto de Engenharia Biomédica
Simula	Olli	Helsinki University of Technology
Stafylopatis	Andreas	National Technical University of Athens
Steil	Jochen	University of Bielefeld
Suárez	Alberto	Universidad Autónoma de Madrid

Suykens	Johan	Katholieke Universiteit Leuven
Thomaidis	Nikos	University of the Aegean
Tomé	Ana Maria	University of Aveiro
Touzet	Claude	Université de Provence /CNRS
Trentin	Edmondo	Universitá di Siena
Tsakonas	Athanasios	University of the Aegean
Varona	Pablo	Universidad Autónoma de Madrid
Verleysen	Michel	Université Catholique de Louvain
L Vigário	Ricardo	Helsinki University of Technology
Villa	Alessandro	Université de Lausanne
Villmann	Thomas	Clinic for Psychotherapy
Vinciarelli	Alessandro	IDIAP Research Institute
Wennekers	Thomas	University of Plymouth
Wermter	Stefan	University of Sunderland
Wersing	Heiko	Honda Research Institute Europe GmbH
Wyns	Bart	Ghent University
Yearwood	John	University of Ballarat
Zervakis	Michalis	Technical University of Crete

Sponsors

ENNS - European Neural Networks Society
INNS - International Neural Networks Society
JNNS - Japanese Neural Networks Society
IEEE Computational Intelligence Society
EURASIP - European Association for Signal and Image Processing

INEB - Instituto de Engenharia Biomédica, Portugal
UBI - Universidade da Beira Interior, Portugal
ISEP - Instituto Superior de Engenharia do Porto, Portugal
UP - Reitoria da Universidade do Porto, Portugal
DEEC - Departamento de Engenharia Electrotécnica e de Computadores, UP
IPP - Instituto Politécnico do Porto

FCT - Fundação para a Ciência e Tecnologia
FLAD - Fundação Luso-Americana para o Desenvolvimento
Fundação Calouste Gulbenkian

Microsoft Research Cambridge Lab
PT - Portugal Telecom

Table of Contents – Part II

Computational Neuroscience, Neurocognitive Studies

Applications in Biomedicine and Bioinformatics

Pattern Recognition

Data Clustering

Self-organization

Text Mining and Internet Applications

Signal and Times Series Processing

Vision and Image Processing

Robotics, Control

Real World Applications

Independent Component Analysis

Graphs

Emotion and Attention: Empirical Findings Neural Models (Special Session)

Understanding and Creating Cognitive Systems (Special Session)

Table of Contents – Part I

Learning Theory

Advances in Neural Network Learning Methods

Ensemble Learning

Spiking Neural Networks

Advances in Neural Network Architectures

Neural Dynamics and Complex Systems

Data Analysis

Estimation

Spatial and Spatio-Temporal Learning

Evolutionary Computing

Meta Learning, Agents Learning

Complex-Valued Neural Networks (Special Session)

Temporal Synchronization and Nonlinear Dynamics in Neural Networks (Special Session)

A Marker-Based Model for the Ontogenesis of Routing Circuits

Philipp Wolfrum[1] and Christoph von der Malsburg[1,2]

[1] Frankfurt Institute for Advanced Studies, D-60438 Frankfurt am Main, Germany
[2] Computer Science Dept., University of Southern California, LA 90089-2520, USA

Abstract. We present a model for the ontogenesis of information routing architectures in the brain based on chemical markers guiding axon growth. The model produces all-to-all connectivity between given populations of input and output nodes using a minimum of cortical resources (links and intermediate nodes). The resulting structures are similar to architectures proposed in the literature, but with interesting qualitative differences making them biologically more plausible.

Keywords: Information routing, shifter circuits, dynamic links, visual cortex.

1 Introduction

An important part of brain function is the routing of information between different areas. The routes along which information flows cannot be static, but must be adaptable to the current requirements. The most prominent example for this necessity is visual attention, where a certain mechanism ensures that only a selected portion of the visual input reaches higher visual "target areas" like inferotemporal cortex (IT). Other examples in which information routing may be very useful include pitch-invariant recognition of melodies, or our ability to combine arbitrary words into grammatically correct sentences. Such abilities require routing structures that provide physical connections between all locations in a certain input region and all locations of a target area.

The necessity for dynamic information routing was appreciated early on [1], and models for its use in object recognition [2] and for frame-of-reference transforms [3] have been put forward. All-to-all routing between large cortical areas has to happen via intermediate stages to be biologically plausible (see problem definition in Sect. 2 and [4] for a detailed discussion). Several architectures for such a multi-stage routing have been proposed, like Shifter Circuits [5], the SCAN model [6], or the minimal architecture of [4]. What has been missing so far are models explaining the ontogenetic development of routing structures in the brain.

2 Routing Structures

Let us pose the information routing problem as follows:

- Given are an input layer and an output layer both consisting of N feature nodes (or simply *nodes*). We are looking for a routing network that establishes all-to-all connectivity between those layers.

J. Marques de Sá et al. (Eds.): ICANN 2007, Part II, LNCS 4669, pp. 1–8, 2007.

- The routing happens via $K - 1$ intermediate layers of N nodes each.
- Nodes of adjacent feature layers can be connected by *links*. For connecting the $K + 1$ feature layers, K stages of links are required, every stage containing N^2 potential links.

Several anatomically plausible architectures have been proposed that meet these requirements. The most prominent one is the so called *Shifter Circuit* [5]. While Shifter Circuits implement a redundant connectivity between input and output, in [4] we propose an architecture that provides full connectivity while requiring the minimally possible number of feature nodes and links. A one-dimensional version of this connectivity is shown in Fig. 1.

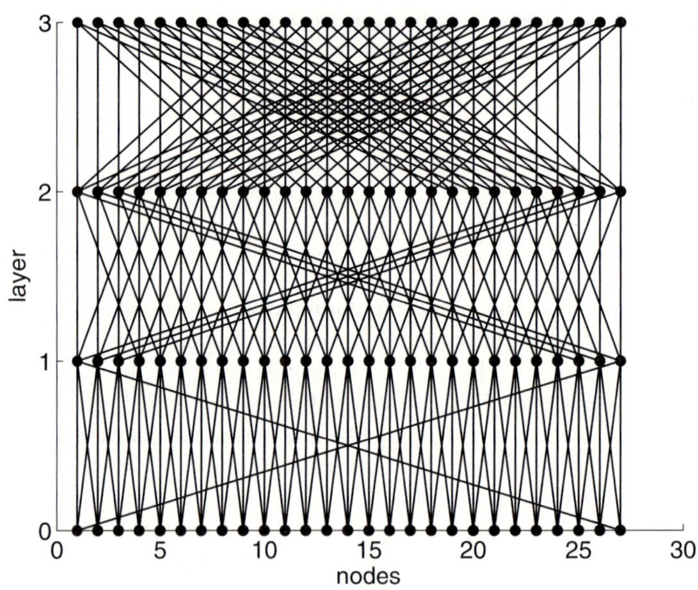

Fig. 1. Routing architecture from [4]. The $N = 27$ nodes of the input layer 0 are connected to all 27 nodes of output layer 3 via 2 intermediate layers and $K = 3$ stages of links. Note that this connectivity requires "wrap-around" links, i.e. links between one end of the presynaptic layer and the opposite end of the postsynaptic layer.

3 Ontogenetic Dynamics

How can ontogeny produce such routing circuits in the brain? Especially the large gaps necessary between links on higher stages are difficult to explain with traditional learning rules. Here we will investigate whether chemical markers can help forming such structures.

It is well known that axonal growth follows chemical gradients [7]. It was hypothesized early that chemical markers could help forming the point-to-point retinotopic mapping that exists between retina and the tectum [8], and Willshaw and von der Malsburg [9] presented a model that realizes the retinotectal mapping on the basis of a limited number of chemical markers present in the retina. Recent studies [10] have shown that the mechanisms by which axons detect these gradients are much more sensitive than previously assumed, allowing the question whether even more complicated patterns than the retinotectal map can arise from chemical marker interaction.

We here present a model that explains the development of routing structures based on chemical markers. For simplicity and ease of visualization we restrict ourselves here to the case of one-dimensional feature layers. Let $C_{i,j}^k$ denote the strength of the link between node i in layer k to node j of layer $k+1$. $C_{i,j}^k$ can vary between 0 and 1, with 0 representing an absent link and 1 a fully grown one. We will refer to all links of one stage by the $N \times N$ matrix C^k. When we make a statement that refers to the links of all K stages we will leave out the superscript k (this also applies for other variables introduced below).

We describe the growth of the links not directly in terms of C but of an unbounded variable U, which codes for the real links via the sigmoid function

$$C = \frac{1}{1+e^{-sU}}, \tag{1}$$

where s defines the steepness of the sigmoid. We let U start out at a homogeneous negative value with some noise added (see Sect. 4 for a discussion of robustness to noise), so that all links C are initially close to 0. The growth of U then follows the differential equation

$$\dot{U} = F^{\text{norm}} \times F^{\text{marker}} \times F^{\text{top}}, \tag{2}$$

where $\times$ denotes elementwise multiplication. The three terms have the roles of restraining local growth of connections (F^{norm}), keeping similarity of chemical markers on both sides of a link low (F^{marker}), and introducing topologic interactions (F^{top}). Thanks to the multiplicative combination, no "tuning" of the relative contributions of the terms is required; the mechanism works for different network sizes without need for adjusting many parameters.

The term

$$F_{i,j}^{\text{norm}} = d - \sum_{\tilde{j}} C_{i,\tilde{j}} \tag{3}$$

is a factor that tends to keep the sum of all efferent links from any position i close to a desired value d. Once the combined link strengths exceed d, F^{norm} turns negative, thus letting the respective link shrink.

The term F^{marker} makes a link's change sensitive to the similarity of chemical markers in the two nodes it connects. These markers are channeled from the input layer to higher levels by the very connectivity C whose growth in turn they influence. We assume each node of the input layer to contain a different type of chemical marker t_i (for a discussion of the plausibility of this and possible alternatives, see Sect. 5). In matrix notation this means that the marker distribution in layer 0 is the identity matrix, $M^0 = I_{N\times N}$, with the marker types on the 1st and the node location on the 2nd dimension. Markers are then transported to higher layers via the existing links C:

$$M^{k+1} = M^k C^k. \tag{4}$$

To calculate F^{marker}, we first define a similarity term

$$F^{\text{sim},k}_{i,j} = \sum_t M^k_{t,i}(M^{k+1}_{t,j} - C^k_{i,j} M^k_{t,i}), \tag{5}$$

which is the similarity (dot product) of the marker vector on the presynaptic side with that portion of the marker vector on the postsynaptic side that was not carried there by the link itself. Therefore, the similarity term signals to the link how well the routes between the part of input space it "sees" and its target node are already being served by other links (see Fig. 2). The role of F^{marker} is to let a link grow only if its similarity term is not too large. We therefore set

$$F^{\text{marker}}_{i,j} = 1 - H(F^{\text{sim}}_{i,j} - \alpha), \tag{6}$$

with $H(\cdot)$ denoting the Heaviside function and a fixed parameter α.

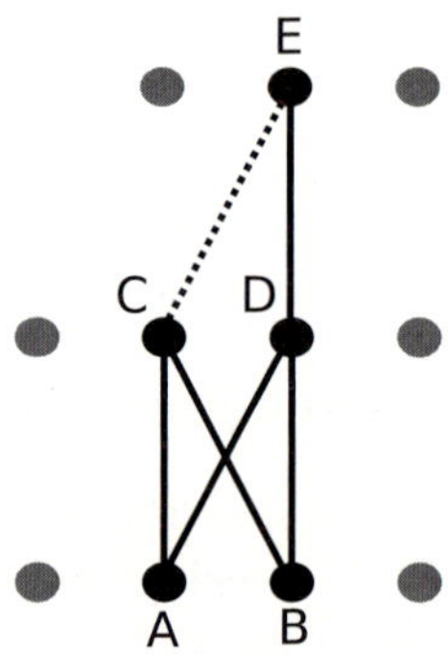

Fig. 2. Role of the similarity term. Already well-established links (solid lines) carry markers from input nodes A and B to intermediate nodes C and D, and from D to E. Therefore, a weak link C-E (dotted line) finds a marker distribution at its target E that is similar to the one at its origin C. This similarity keeps it from growing. Functionally, this mechanism prevents formation of redundant alternative routes between two points.

The term

$$F^{\text{top}}_{i,j} = \beta(C_{i-1,j-1} + C_{i+1,j+1}) + G_{i,j} \tag{7}$$

combines two different topological influences, their relative strength weighted by the parameter β. The first part adds cooperation between parallel neighboring links. The second term G favors the growth of links to the corresponding position in the next layer (i.e. $i = j$) over links to faraway positions. We assume it here to be a bounded hyperbolic function of position difference of the two end nodes: $G_{i,j} = \frac{\gamma}{|i-j|+\gamma}$, with γ defining the steepness (see Fig. 3). G is necessary to tell the ontogenetic mechanism how to align the coordinate systems of the layers it is connecting. A possible way of implementing this term is to first allow development of a point-to-point mapping (e.g. through the mechanism presented in [9]), which then serves as a guidance for the growth of a routing connectivity.

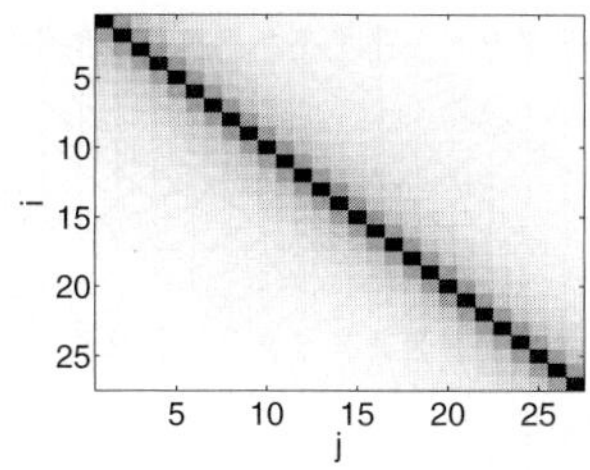

Fig. 3. The term G helps to align the coordinate systems of subsequent layers by favoring links between nodes with corresponding positions (middle diagonal) over links between distant nodes. $\gamma = 0.6$ like in the simulations of Sect. 4.

4 Results

We chose to investigate the growth of networks containing $K = 3$ link stages. Equation (2) was integrated using the Euler method and the following parameter settings: $s = 30$ (steepness of sigmoid), $\alpha = 0.5$ (threshold for marker similarity), $\beta = 0.6$ (strength of neighbor interaction), $\gamma = 0.6$ (steepness of the hyperbolic term G). A delayed onset of growth at higher stages improves the final results. We chose a delay of 15% and 30% of overall simulation time for the middle and the highest stage, respectively.

First, we assumed $d = 3$ as target number of links per node (see (3)). With $K = 3$ link stages, this means that $N = d^K = 27$ input and output nodes can be connected. The network resulting from the ontogenetic mechanism for these parameter settings is shown in Fig. 4. Note how the distance between links increases from 1 to 3 to 9 from bottom to top, thus producing non-redundant full connectivity. We can see in Fig. 4(a) that the resulting network differs qualitatively from the manually produced one of Fig. 1: There are no wrap-around links (i.e. links from a node on one side of the feature layer to the opposite side of the next layer). Instead, these links appear on the other side of the central link (cf. Fig. 4(b)). Interestingly, this new structure produces the same perfect all-to-all connectivity as the one arising from theoretical considerations in [4], while being biologically more plausible.

The mechanism can also grow routing structures between larger feature layers. For this we only have to adjust the target number of links per node d, without changing any of the other parameters. Fig. 5 shows simulation results for $d = 5$, i.e. $N = d^3 = 125$ nodes per layer. We see that qualitatively the resulting structure is similar to the one obtained for $d = 3$, except that now each node makes 5 connections to the next layer, with appropriate spacings of 1, 5, and 25 nodes.

However, we also see that the structure in Fig. 5 is not as clean as the one in Fig. 4, with several links not going to the "correct" targets. This results in an overall input-output connectivity that is not perfectly homogeneous, i.e. some input-output pairs are connected by two different routes, while others have no connection. For the structure shown in Fig. 5, the strengths of the input-output connections have mean value and standard deviation of $\mu \approx 1$ and $\sigma \approx 0.15$.

The reason for the uneven final structure lies in the noise that was introduced to the initial link strengths: We chose the initial values of U randomly from the interval

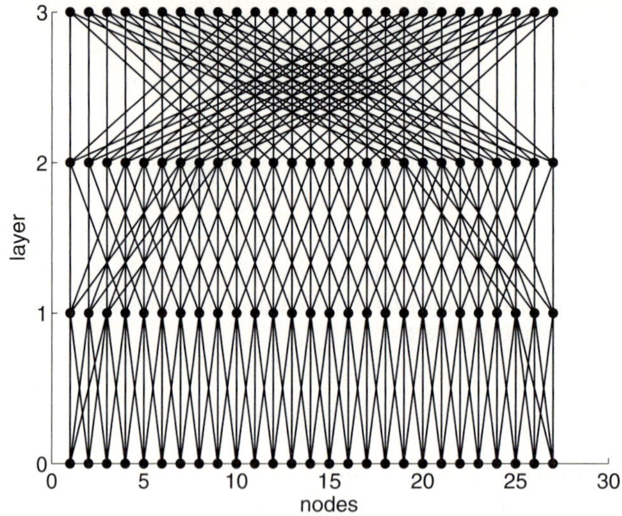

(a) Connection structure of the full network.

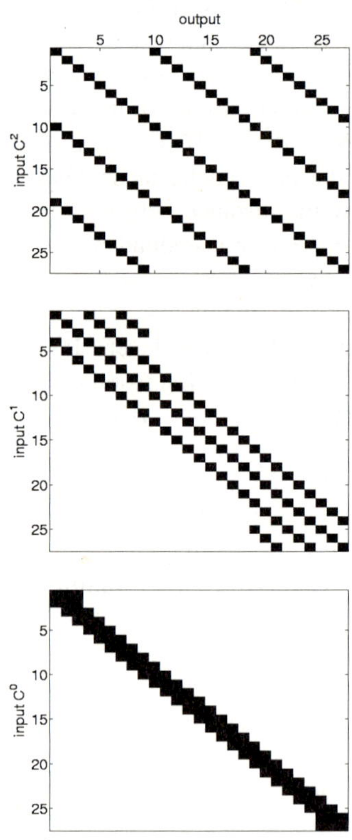

(b) Matrices C^k of the full network of (a) shown separately.

Fig. 4. Results for $N = 27$ nodes per layer and a target number of links $d = 3$

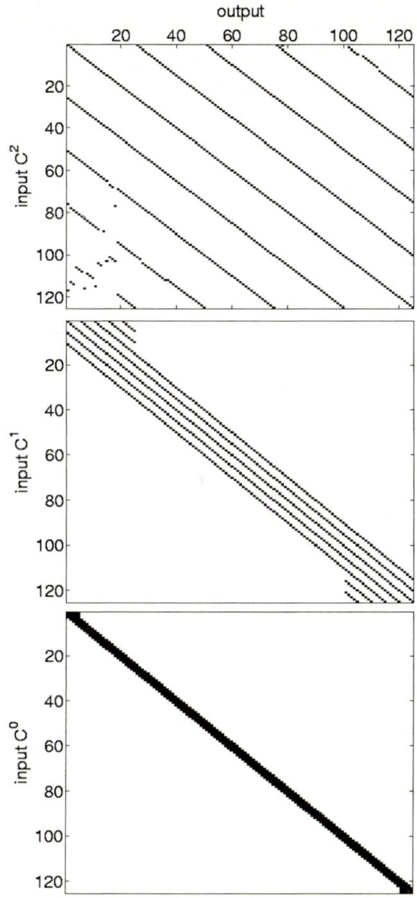

Fig. 5. Resulting connection matrices C^k for $N = 125$ nodes per layer and a target number of links $d = 5$. The initial values of U contained 10% of additive noise.

$[-16.5.. - 15]$, which means that they contain 10% of additive uniformly distributed noise. Further simulations have shown that the mechanism generally results in a flawless connectivity only if the initial conditions contain less than $\approx 5\%$ of noise. The growth of smaller networks is far less sensitive to noise: For $N = 27$, up to 20% of additive noise in the initial conditions practically always results in the correct final connectivity.

5 Conclusion

One assumption crucial for the model as presented here is that there be a unique chemical marker for every input node. This is very unlikely to be the case in the brain. Future work will address the question how the N different chemical markers can be replaced by a small number of marker gradients spread evenly over the input layer. Also, it is possible to replace the unique markers assumed here by stochastic, uncorrelated signals produced by the different input nodes. Since these signals are orthogonal, their linear superpositions arising at higher layers could still be decomposed and, most notably, the scalar product of such superpositions would give exactly the same similarity term as assumed in the model here. This would yield an activity-based instead of a marker-based mechanism following the same mathematical model.

We have presented a neurally plausible ontogenetic mechanism modeling the formation of routing circuits in the brain. The mechanism requires only signals that are available locally at source and/or target of the respective connection. While the mechanism may be important for understanding the development of certain wiring structures of the brain, it may also turn out to have technological applications like the automatic wiring of computer networks.

Acknowledgments

This work was supported by the European Union through project FP6-2005-015803 ("Daisy") and by the Hertie Foundation.

References

1. Pitts, W., McCulloch, W.S.: How we know universals: the perception of auditory and visual forms. Bulletin of Mathematical Biophysics 9, 127–147 (1947)
2. Lades, M., Vorbrüggen, J., Buhmann, J., Lange, J., von der Malsburg, C., Würtz, R., Konen, W.: Distortion invariant object recognition in the dynamic link architecture. IEEE Transactions on computers 42, 300–311 (1993)
3. Weber, C., Wermter, S.: A Self-Organizing Map of Sigma-Pi Units. Neurocomputing 70(13-15), 2552–2560 (2007)
4. Wolfrum, P., von der Malsburg, C.: What is the optimal architecture for visual information routing? Neural Computation (2007) (in press)
5. Olshausen, B.A., Anderson, C.H., van Essen, D.C.: A neurobiological model of visual attention and invariant pattern recognition based on dynamic routing of information. Journal of Neuroscience 13(11), 4700–4719 (1993)
6. Postma, E., van den Herik, H., Hudson, P.: SCAN: A Scalable Model of Attentional Selection. Neural Netw 10(6), 993–1015 (1997)
7. Yamamoto, N., Tamada, A., Murakami, F.: Wiring of the brain by a range of guidance cues. Prog Neurobiol 68(6), 393–407 (2002)
8. Sperry, R W: Chemoaffinity in the orderly growth of nerve fiber patterns and connections. Proc Natl Acad Sci U.S.A 50, 703–710 (1963)
9. Willshaw, D.J., von der Malsburg, C.: A marker induction mechanism for the establishment of ordered neural mappings: its application to the retinotectal problem. Philos Trans R. Soc. Lond B. Biol Sci. 287(1021), 203–243 (1979)
10. Rosoff, W.J., Urbach, J.S., Esrick, M.A., McAllister, R.G., Richards, L.J., Goodhill, G.J.: A new chemotaxis assay shows the extreme sensitivity of axons to molecular gradients. Nature Neuroscience 7, 678–682 (2004)

A Neural Network for the Analysis of Multisensory Integration in the Superior Colliculus

Cristiano Cuppini[1], Elisa Magosso[1], Andrea Serino[2], Giuseppe Di Pellegrino[2], and Mauro Ursino[1]

[1] Departement of Electronics, Computer Science and Systems - Univ. of Bologna, Italy
[2] Departement of Psychology – Univ. of Bologna, Italy
{ccuppini,emagosso,mursino}@deis.unibo.it,
{andrea.serino,g.dipellegrino}@unibo.it

Abstract. It is well documented that superior colliculus (SC) neurons integrate stimuli of different modalities (e.g., visual and auditory). In this work, a mathematical model of the integrative response of SC neurons is presented, to gain a deeper insight into the possible mechanisms implicated. The model includes two unimodal areas (auditory and visual, respectively) sending information to a third area (in the SC) responsible for multisensory integration. Each neuron is represented via a sigmoidal relationship and a first-order dynamic. Neurons in the same area interact via lateral synapses. Simulations show that the model can mimic various responses to different combinations of stimuli: i) an increase in the neuron response in presence of multisensory stimulation, ii) the inverse effectiveness principle; iii) the existence of within- and cross-modality suppression between spatially disparate stimuli. The model suggests that non linearities in neural responses and synaptic connections can explain several aspects of multisensory integration.

Keywords: Superior colliculus, multimodal integration, inverse effectiveness, cross-modality and within-modality suppression.

1 Introduction

Integration of stimuli from different sensory modalities (visual, auditive, tactile) plays a fundamental role in the correct perception of the external world and in determining the suitable behaviour of individuals towards external events [1]. The presence of multisensory neurons, able to integrate different sensory modalities into a complex response, is well documented in various structures of the mammalian brain outside the primary sensory areas [1]. An important locus of multisensory interaction is a layered midbrain structure, the superior colliculus (SC). Many neurons in the deep layers of the SC receive converging visual, auditory and somatosensory afferents from various subcortical and extraprimary cortical sources [2]. Responses of such neurons to a combination of modality-specific stimuli differ significantly from those evoked by any of their unisensory inputs in a way that substantially facilitates the role of the SC in controlling attentive and orientation behaviour [3].

J. Marques de Sá et al. (Eds.): ICANN 2007, Part II, LNCS 4669, pp. 9–18, 2007.

Many studies have been carried out to characterize how multisensory neurons in the SC integrate their unimodal inputs, and a consistent amount of physiological data has been gathered on their response to a variety of stimuli [4] [5] [6] [7]. These studies, which record single-unit activity in anesthetized animals, show that the responses of multimodal neurons in the SC are characterized by significant non-linear phenomena, which make their qualitative analysis extremely difficult without the help of mathematical quantitative tools. First, a multisensory SC neuron has multiple receptive fields (RFs), one for each of its sensory modalities. When two different sensory stimuli (e.g., auditory and visual) are present at close spatial proximity (as it occurs when they derive from the same event), their combination is typically synergistic producing a neuron's response which is significantly greater than that evoked by the most effective of the two unimodal inputs individually (*multisensory enhancement*) [1] [5] [8]. On the other hand, when the two stimuli are presented at different locations (i.e. they likely derive from different events) two alternative results can be observed: either no interaction occurs or the neuron's response to the within-field stimulus is considerably depressed (*multisensory depression*) [4] [9].

Multisensory enhancement is accompanied by another well known integrative principle called *inverse effectiveness*: combinations of weakly effective stimuli produce proportionally greater multisensory enhancement than more effective stimuli [1] [10]. Inverse effectiveness has functional sense in behavioural situations: the probability to detect a weak stimulus benefits more from multisensory enhancement than a high-intensity stimulus which is easily detected by a single modality alone [3] [11].

Despite the great number of experimental results on multisensory SC response which has been gathered in recent years, we are not aware of mathematical models and neural networks able to encompass these data into a coherent theoretical structure. It is reasonable to expect that the properties of multi-modal integration do not only depend on neuronal individual characteristics, but above all reflect the organization of the circuitry that processes unimodal stimuli and conveys these stimuli toward multisensory neurons.

A fundamental contribution to identify the mechanisms of multisensory integration in SC can be obtained with the use of neural networks and computer simulations. These models can be of value not only to provide putative explanation for existing data, but also to suggest new experiments and to provide some rules for artificial recognition systems.

The aim of this work is to develop an original neural network model, based on neurobiologically plausible mechanisms, able to reproduce and explain in-vivo results on multisensory integration in the SC. The model includes three neural networks, which communicate via synaptic connections. Two of them are unimodal and represent neurons coding visual and auditory stimuli, respectively; a downstream network, representing multimodal neurons in the SC, receives information from the upstream networks via feedforward synapses and integrates these information to produce the final response. Furthermore, neurons in each network are interconnected via lateral synapses. By adopting the previous structure and by using a single set of parameters, the model is able to reproduce a cluster of within- and cross-modality interactions in accordance with experimental data in the literature.

2 Method

In this section we will describe the general structure of the model and we will discuss parameter assignment, on the basis of existing data in the literature.

2.1 General Model Structure

- The model is composed of 3 areas (see Figure 1). Elements of each area are organized in *NxM* dimension matrices, so that the structure keeps a spatial and geometrical similarity with the external world: neurons of each area respond only to stimuli coming from a limited zone of the space. Neurons normally are in a silent state (or exhibit just a mild basal activity) and can be activated if stimulated by a sufficiently high input. Furthermore, each neuron exhibits a sigmoidal relationship (with lower threshold and upper saturation) and a first order dynamics (with a given time constant). The 2 upstream areas are unimodal, and respond to auditory and visual stimuli, respectively. A third downstream area represents neurons in the Superior Colliculus responsible for multisensory integration. These three areas have a topological organisation, i.e., proximal neurons respond to stimuli in proximal position of the space.

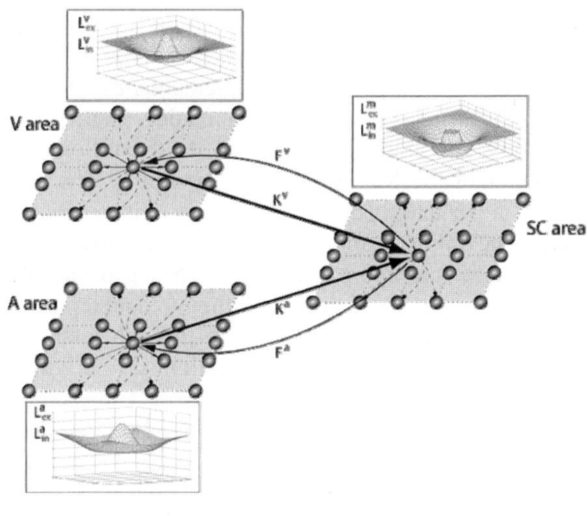

Fig. 1. Schematic diagram describing the general structure of the network. Each grey circle represents a neuron. Neurons are organized into 3 distinct areas of 40x40 elements. Each neuron of these areas (*V: visual*, *A: auditory* and *SC: multimodal in the superior colliculus*) is connected with other elements in the same area via lateral excitatory and inhibitory intra-area synapses (*arrows* L_{ex} and L_{in} within the area). Neurons of the unimodal areas send feed-forward excitatory inter-area synapses to multimodal neurons in the superior colliculus area located in the same position (*arrows K*). Multimodal neurons, in turn, send excitatory feed-back inter-area connections to neurons of the unisensory areas (*arrows F*) (see text for details).

- Each element of the unisensory areas has its own receptive field (RF) that can be partially superimposed on that of the other elements of the same area. The elements of the same unisensory area interact via lateral synapses, which can be both excitatory and inhibitory. These synapses are arranged according to a Mexican hat disposition (i.e., a circular excitatory region surrounded by a larger inhibitory annulus).
- The elements of the multisensory area in the Superior Colliculus receive inputs from the two neurons in the upstream areas (visual and auditory) located in the same spatial position. Moreover, elements in the SC are connected by long range lateral synapses.
- The multimodal neurons in the SC send back a feedback excitatory input to the unimodal neurons located in the same spatial position, i.e., detection of a multimodal stimulus may help reinforcement of the unisensory stimuli in the upstream areas.

2.2 Mathematical Description

In the following, quantities which refer to neurons in the auditory, visual or multisensory areas will be denoted with the superscripts a, v and m, respectively. The spatial position of individual neurons will be described by the subscripts ij.

The Receptive Fields of Unisensory Areas. In the present version we assume that each area is composed by 40x40 neurons (i.e.: N=40; M=40), to reduce the computational complexity of the computer implementation. Neurons in each area differ in the position of their receptive fields by 2.25 deg. Hence, each area covers 90 deg in the visual, acoustic or multisensory space. In the following, we will denote with x^i and y^j the center of the RF of a generic neuron ij. The receptive field (say $R^s_{ij}(x, y)$) of neuron ij in the unisensory area s ($s = a, v$) is described with a gaussian function. The standard deviation of this function has been given so that the receptive fields of the visual neurons are approximately 10-15 deg in diameter, and those of acoustic neurons approximately 20-25 deg, according to data reported in literature [5]. The amplitudes of the Gaussian functions are set to 1, to establish a scale for the strength of the inputs generated by the external stimuli. According to the previous description of the RF, an external stimulus excites not only the neuron centered in that zone, but also the proximal neurons whose receptive fields cover such position.

The Activity in the Unisensory Areas. Unisensory neurons are stimulated not only from external inputs, but also by connections with other elements in the same area and by a feedback connection from multisensory neurons in the downstream layer. Hence, the overall input for neuron in position ij can be written as follows:

$$u^s_{ij}(t) = r^s_{ij}(t) + l^s_{ij}(t) + f^s_{ij}(t) \ ; \qquad s = a, v. \tag{1}$$

r^s_{ij} represents the input that reaches the neuron ij in presence of a sensory stimulus; this is computed as the inner product of the stimulus and the RF. The term l^s_{ij} is the input coming from connections with other neurons in the same area. Synapses representing these connections are symmetrical and arranged according to a "Mexican hat" function. Parameters which establish the extension and the strength of lateral synapses in the unimodal areas have been assigned to simultaneously satisfy several criteria: i) the presence of an external stimulus produces an activation bubble of neurons which approximately coincide with the dimension of the receptive field; ii) according to data reported in [4] we assumed that the surrounding inhibitory area is much larger than the activation bubble; iii) inhibition strength must be strong enough to avoid instability; iv) stimulating the suppressive region with a second stimulus can induce within-modality suppression greater than 50%. To avoid undesired border effects, synapses have been realized by a circular structure so that every neuron of each area receives the same number of side connections. Finally, f^s_{ij} is the input to unisensory neurons induced by the feedback from the Superior Colliculus. Such connections exclusively link neurons placed in the same *ij*-position in the Colliculus and the unisensory area. Finally, neuron activity is computed from its inputs, through a static sigmoidal relationship and a first-order dynamic. This is described via the following differential equation:

$$\tau_s \cdot \frac{d}{dt} x^s_{ij}(t) = -x^s_{ij}(t) + \varphi\left(u^s_{ij}(t)\right). \tag{2}$$

The time constants, τ_s, which determines the speed of the answer to the stimulus, agrees with values (a few milliseconds) normally used in deterministic mean-field equations [12]. φ represents a static sigmoidal relationship, described by the following equation

$$\varphi\left(u^s(t)\right) = \frac{1}{1 + e^{-\left(u^s(t) - \vartheta^s\right) p^s}}. \tag{3}$$

where ϑ^s defines the threshold and p^s sets the slope at the central point. These 2 parameters have been assigned to have negligible neuron activity in basal condition (i.e., when the input is zero), and to have a reliable transition from silence to saturation in response to unimodal and cross-modal inputs. Such function identifies 3 regions of work, depending on the intensity of the input: the under-threshold behaviour of a neuron, an approximately linear region, and a saturation region. According to the previous equation, the maximal neuron activity is conventionally set at 1 (i.e., all neuron activities are normalized to the maximum).

The Activity in the Multisensory Area. Neurons in this area are stimulated by the activities of the neurons in the two unisensory areas located in the same *ij*-position. This choice has been adopted since, according to experimental data, the auditory and visual RFs of a multisensory neuron are in spatial register [4], i.e., they represent similar regions in space. Furthermore, neurons in the superior colliculus also receive lateral synapses from other elements in the same area.

We assumed that synapses in the multisensory area have a Mexican hat disposition, but they join only spatially distant neurons. This disposition of synapses has been adopted since data in the literature suggest the absence of within-modality integration of proximal stimuli, and cross-modal suppression between distal stimuli. Hence, the overall input, (say u^m_{ij}) to a multisensory neuron is computed as the sum of two elements: a feedforward term from upstream unimodal areas and lateral feedback from distal neurons in the same area. Then, the activity of a multisensory neuron is computed from its input by using equations similar to Eqs. (2) and (3).

3 Results

The steady state responses of an SC neuron vs. the magnitude of the input stimulus are reported in Fig. 2, for an auditory (pointed line), a visual (dashed line) and a multisensory (continuous line) stimulation. These responses have been obtained by using either a single stimulus (auditory or visual), of increasing strength, or two paired stimuli (visual + auditory) located at the centre of the RF. From these curves, the dynamical range (defined according to the literature [6] as the difference in neuron activity at saturation and at threshold) can easily been computed. Furthermore, by way of comparison the sum of the two unisensory responses is also presented in the same figure. Two aspects of these curves are of interest: first, the dynamical range to multisensory stimulation is much greater than the dynamical range to a single stimulus. Second, the neuron exhibits a superadditive behaviour (i.e. the response to a multisensory stimulus is greater than the sum of the two unimodal responses) at low values of the input stimuli (just above threshold), while the behaviour tends to become simply additive (i.e. the multisensory response is equal to the sum of the unisensory responses) at high stimulation levels (close to saturation). In order to quantitatively evaluate the multisensory integration we computed the so-called "interactive index" [6]. This is a measure of the augmentation of the response induced by two stimuli of different modality compared with a single stimulus, and is defined as follows:

$$\text{Interactive Index} = \left[\frac{Mr - Ur_{\max}}{Ur_{\max}}\right] \cdot 100. \tag{4}$$

where Mr (multisensory response) is the response evoked by the combined-modality stimulus, and Ur_{max} (unisensory response) is the response evoked by the most

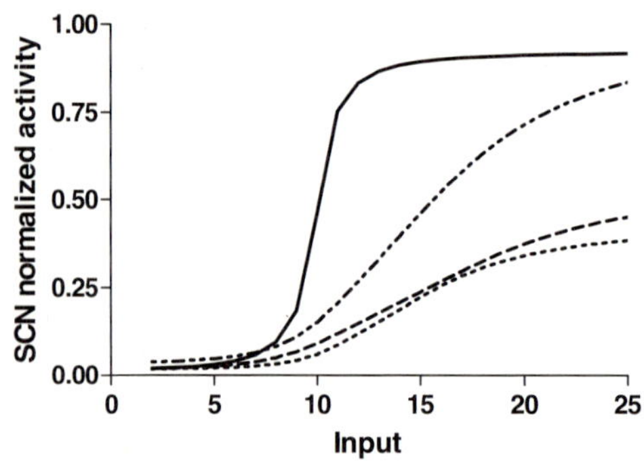

Fig. 2. Analysis of the response of multimodal neuron to unimodal and crossmodal stimuli. The responses were assessed stimulating the model with an acoustic (.......), a visual (— — —) and two paired multisensory (———) stimuli with increasing intensity. By way of comparison, the sum of the two unisensory responses (— · ·) is also presented in this figure. The stimulus was presented at the center of the RF of the observed SC neuron.

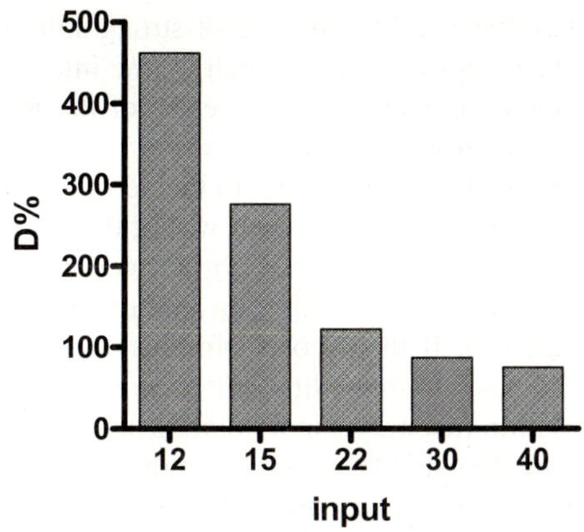

Fig. 3. Analysis of the interactive index vs. the intensity of the multimodal input, that underlines the phenomenon of the Enhancement and the inverse effectiveness principle. Interactive index (*D* %) is computed as the per cent increase of the multisensory response compared to the maximum unisensory response.

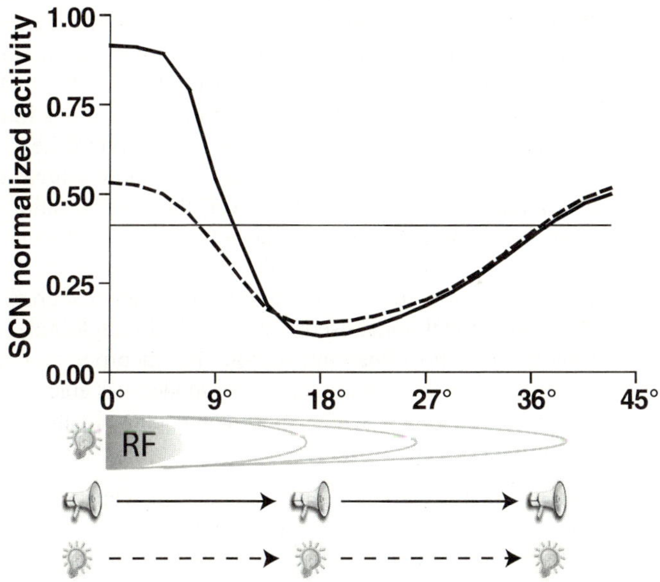

Fig. 4. Effect of the distance between two stimuli on the integrative response of the multisensory SC neurons. A first visual stimulus was located at the center of the RF of the observed neuron. A second stimulus, either of the same modality (*dashed line*) or of a different modality (*continuous line*), is progressively moved from the center of the RF to the periphery. The distance between the two stimuli is shown in the x-axis.

effective unisensory stimulus. Fig. 3 displays the interactive index computed at different values of the input stimuli. According to the principle of inverse effectiveness, this index decreases from more than 500% in case of small stimuli (just above the threshold, input = 8 - 12) down to 75% in case of strong stimuli. Finally, Fig. 4 analyzes the role of the distance between two stimuli on the integrated response. In these simulations a visual stimulus is located at the center of the RF, and either a second visual stimulus (within-modality interaction) or a second auditory stimulus (cross-modality interaction) is moved from the center to the periphery. Results confirm that a second stimulus of a different modality located within the receptive field causes significant cross-modal enhancement, whereas within-modality enhancement is mild (i.e., a second stimulus of the same modality, located inside the RF does not evoke a significantly greater response). If the second stimulus is moved away from the RF, one can observe significant within-modality suppression as well as significant cross-modality suppression. Within modality suppression is strong in both modalities (auditory and visual) leading to almost 70% reduction in the SC response. The suppressive regions are quite large (25-30 deg) in accordance with physiological data [4].

4 Discussion

The present work was designed to elucidate possible neural mechanisms involved in multisensory integration in the Superior Colliculus. To this end, we developed a simple neural circuit which encompasses several mechanisms, still maintaining a moderate level of complexity. Actually, the model aspires to represent a good compromise between completeness, on one hand, and conceptual (and computational) simplicity on the other. The basic idea of this model is that multimodal neurons in the Superior Colliculus receive their inputs from two upstream unimodal areas, i.e., one area devoted to a topological organisation of visual stimuli and another area devoted to a topological organisation of auditory stimuli. However, the exact location of these areas is not established in our model, i.e., we did not look for a definite anatomical counterpart. Experimental data suggest that multisensory neurons are created by the convergence of modality-specific afferents coming from different sources [2]. For the sake of simplicity, in this model somatosensory stimuli are neglected, i.e., we consider only the problem of audio-visual integration. By incorporating the previous mechanisms, and using a single set of parameters, the model was able to make several predictions, which can be compared with experimental data. In the following, the main simulation results are critically commented:

i) *Inverse effectiveness* – As it is evident in Figs. 2 and 3, the capacity of multisensory neurons to integrate cross-modal stimuli strongly depends on the intensity of the input. In the present work the facilitatory interaction has been quantified using the interactive index, which relates the multisensory response to the larger of the two unisensory responses. This index is affected by the intensity of the stimuli, and exhibits a significant decrease if stimulus intensity is progressively raised. This behaviour, which is known as "inverse effectiveness", is a consequence of the non-linear characteristic of neurons, and can be explained looking at the position of the working point on the sigmoidal relationship after application of the more effective input. First, let us consider the case in which, after application of the more effective stimulus, the SC neuron works in the lower portion of its sigmoidal relationship, close to the threshold. Then, application of a second stimulus may move the working point into the linear portion of the curve, thus causing a disproportionate increase in the response compared with that evoked by the first input (superadditivity, enhancement greater than 100%). By contrast, if the neuron works in the central (quasi-linear) region, the effect of a second stimulus is simply additive. Finally, if the upper saturation region is approached one can have sub-additivity, since a second stimulus can induce only a minor increase in neuron activity. The last case is not simulated in this work since, with the present value of feedforward synapses, a single stimulus cannot move the working point close to the upper saturation region. Sub-additivity, however, can be mimicked by increasing the feedforward synapses.

ii) *Dynamic range* – The multisensory dynamic range of multimodal neurons is greater than the unisensory dynamical range [6]. This signifies that the maximal response evoked by a combination of auditory and visual stimuli in close spatial and temporal register is greater than the maximal response evoked by a single stimulus of either modality [6]. Such a property is explained in our model by the presence of two sigmoidal relationships, disposed in a series arrangement. Let us consider a single stimulus and progressively increase its intensity: in our model, the maximal response in the SC (see Fig. 2) is determined by the upper saturation of neurons in the upstream uni-modal area, and by the strength of the feedforward synapses linking this unimodal neuron to the downstream (multimodal) neuron. This input does not lead multimodal neurons to saturation. Consequently, if we apply a combination of a visual and an auditory stimulus and progressively increase their intensity (multisensory dynamic range), the downstream multimodal neuron can be driven closer to its upper saturation and exhibits a greater response.

iii) *Cross-modality vs. within modality integration* – According to the literature [1] in our model a combination of two cross-modal stimuli within the RF results in significant enhancement of the SC response, but the same effect is not visible when the two stimuli are presented as within-modality pairs. A second within-modality stimulus applied within the RF causes just a mild enhancement (Fig. 4). This result is the consequence of the absence of lateral excitation between multi-modal neurons.

iv) *Spatial relationship between two (within-modal or cross-modal) stimuli* – In agreement with experimental data [4], our model shows that, when the spatial distance between two stimuli is increased, the integration performed by multimodal neurons changes from enhancement to suppression. In the present model the suppressive effect is evident both using within-modality and cross-modality stimuli. Similar exempla are reported in [4].

References

1. Stein, B.E., Meredith, M.A.: The Merging of the Senses. The MIT Press, Cambridge, MA (1993)
2. Wallace, M.T., Meredith, M.A., Stein, B.E.: Converging Influences from Visual, Auditory and Somatosensory Cortices onto Output Neurons of the Superior Colliculus. J. Neurophysiol. 69, 1797–1809 (1993)
3. Stein, B.E., Meredith, M.A., Huneycutt, W.S., McDade, L.: Behavioral Indices of Multisensory Integration: Orientation of Visual Cues is Affected by Auditory Stimuli. J. Cogn. Neurosci. 1, 12–24 (1989)

4. Kadunce, D.C., Vaughan, J.W., Wallace, M.T., Benedek, G., Stein, B.E.: Mechanisms of Within- and Cross-Modality Suppression in the Superior Colliculus. J. Neurophysiol. 78, 2834–2847 (1997)
5. Kadunce, D.C., Vaughan, J.W., Wallace, M.T., Stein, B.E.: The Influence of Visual and Auditory Receptive Field Organization on Multisensory Integration in the Superior Colliculus. Exp. Brain Res. 139, 303–310 (2001)
6. Perrault, T.J., Jr., V.J.W., Stein, B.E., Wallace, M.T.: Superior Colliculus Neurons Use Distinct Operation Modes in the Integration of Multisensory Stimuli. J. Neurophysiol. 93, 2575–2586 (2005)
7. Stanford, T.R., Quessy, S., Stein, B.E.: Evaluating the Operations Underlying Multisensory Integration in the Cat Superior Colliculus. J. Neurosci. 25, 6499–6508 (2005)
8. Perrault jr., T.J., Vaughan, J.W., Stein, B.E., Wallace, M.T.: Neuron-Specific Response Characteristics Predict the Magnitude of Multisensory Integration. J. Neurophysiol. 90, 4022–4026 (2003)
9. Meredith, M.A., Stein, B.E.: Spatial Factors Determine the Activity of Multisensory Neurons in Cat Superior Colliculus. Brain Res. 365, 350–354 (1986)
10. Meredith, M.A., Stein, B.E.: Visual, Auditory, and Somatosensory Convergence on Cells in Superior Colliculus Results in Multisensory Integration. J. Neurophysiol. 56, 640–662 (1986)
11. Jiang, W., Jiang, H., Stein, B.E.: Two Corticotectal Areas Facilitate Multisensory Orientation Behavior. J. Cogn. Neurosci. 14, 1240–1255 (2002)
12. Ben-Yishai, R., Bar, O., Sompolinsky, H.: Theory of Orientation tuning in Visual Cortex. Proc. Natl. Acad. Sci. USA 92, 3844–3848 (1995)

Neurotransmitter Fields

Douglas S. Greer

General Manifolds
dsgreer@gmanif.com

Abstract. Neurotransmitter fields differ from neural fields in the underlying principle that the state variables are not the neuron action potentials, but the chemical concentration of neurotransmitters in the extracellular space. The dendritic arbor of a new electro-chemical neuron model performs a computation on the surrounding field of neurotransmitters. These fields may represent quantities such as position, force, momentum, or energy. Any computation performed by a neural network has a direct analog to a neurotransmitter field computation. While models that use action potentials as state variables may form associations using matrix operations on a large vector of neuron outputs, the neurotransmitter state model makes it possible for a small number of neurons, even a single neuron, to establish an association between an arbitrary pattern in the input field and an arbitrary output pattern. A single layer of neurons, in effect, performs the computation of a two-layer neural network.

Keywords: Computational neuroscience, signal processing, pattern recognition, mathematical models, natural intelligence, neural fields.

1 Introduction

Neural fields have been studied for a long time [1]–[4] and comprehensively reviewed by several authors [5], [6]. This approach models the behavior of a large number of neurons by taking the continuum limit of discrete neural networks where the continuous state variables are a function in space representing the mean firing rates.

The distinction between neural fields and neurotransmitter fields is the physical quantity under consideration. Neural fields attempt to model the spatial distribution of mean neuron-firing rates as real-valued function, while neurotransmitter fields model the concentration of neurotransmitters in the extracellular space as a real-valued function. The localization of neurotransmitters to the space within the synaptic cleft is seen as an evolutionary adaptation that limits diffusion and increases the efficiency.

In order to develop the theory, we put forth a single proposition: the neurotransmitter cloud hypothesis. Empirical evidence and deductive arguments are provided which support this proposition, but verification will require further investigation and analysis. Acceptance of the hypotheses, like including an additional mathematical axiom, allows us to explore a new computational model that characterizes the electro-chemical properties of the neuron.

J. Marques de Sá et al. (Eds.): ICANN 2007, Part II, LNCS 4669, pp. 19–28, 2007.

1.1 Evolution of the Nervous System

Although the evolution of the senses and the central nervous system was a complex process that occurred over an extended time interval [7], we can attempt to understand some of the general constraints that may have influenced its development. One of these constraints was the need to evaluate the current state of the body and its immediate environment. This required the creation of internal representations that could be equated with physical quantities defined over the continuous variables of space, time, and frequency. These quantities included mass (external world), position (location of the body surface), energy (visible light and sound vibrations), and force (pressure on the body surface and the tension on the muscle cross-sections).

In the standard neural network model, a synapse is characterized mathematically by a single real-valued weight representing the effect one neuron has on another. The products of the weights times the activation values of the input neurons are summed, and a nonlinear transfer function is applied to the result [8]. This model describes electrical and chemical synapses uniformly, that is, by a single real value. Examining the difference between electrical and chemical synapses, we note that electrical synapses, which may have a weighted response proportional to the number of ion channels connecting the pre- and postsynaptic neuron, are more than ten times faster. They are also more efficient, since they do not require the metabolism of neurotransmitters, or the mechanics of chemical signaling. However, chemical synapses are found almost exclusively throughout the central nervous systems of vertebrates. This raises the question: Given a time interval of several hundred million years, and the wide range of species involved, why has nature consistently retained the cumbersome chemical synapses and not replaced them with electrical synapses?

We note that neurotransmitters, the core component of chemical synapses, are actually located outside the neuron cell walls in the extracellular space. Moreover, the chemical signaling often occurs in multiple-synapse boutons such as the one shown in Fig. 1. Within these complex synapses, which connect the axons and dendritic spines of many adjacent neurons, the density of neurotransmitter is equal to the sum of the contributions from each of the individual axons.

Another constraint during the course of evolution was the limited amount of processing power available. Solutions that required more than a very small number of neurons were not feasible. In addition, the space within the organism that could be devoted to representing physical quantities was limited, so small, compact representations were preferable.

If we leave the confines of the standard neuron model and consider the density of neurotransmitters as the state variables, we discover a number of advantages. The first is higher resolution; billions of small molecules can fit in the space occupied by a single neuron. The second is energy consumption; the concentration of neurotransmitters, like the concentration of ink on a sheet of paper, is passive and can store information indefinitely without expending energy. In contrast, action potentials require the continuous expenditure of energy in order to maintain state. Another advantage is that a very high-resolution representation can be maintained with only a few processing elements. For example, the terminal arbor of a single neuron that encodes a joint angle can support a high-resolution model describing the location of the surface of the limb in space. This collection of related concepts results in the following conjecture.

The Neurotransmitter Cloud Hypothesis. When multicellular fauna first appeared, organisms began to represent quantities such as mass, force, energy and position by the chemical concentration of identifiable molecules in the extracellular space. The basic principles of operation developed during this period still govern the central nervous system today.

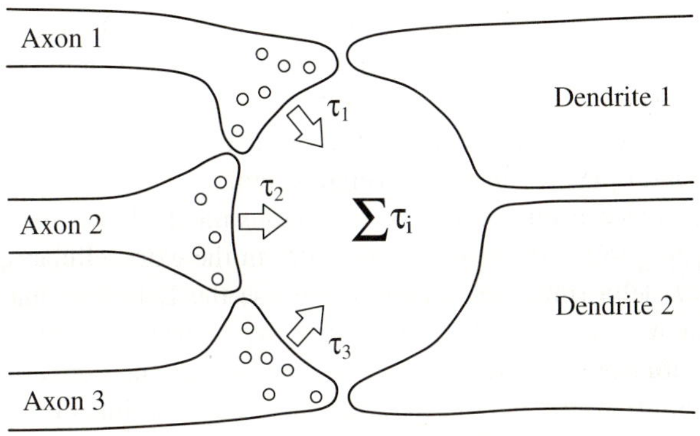

Fig. 1. This idealized view of multi-synapse boutons shows how the concentration of neurotransmitter perceived by multiple dendrites is the summation of that produced by three separate axon terminals. The summation occurs in the extracellular space and is separated from the intracellular summation by the nonlinear responses of the cell membranes.

The basic laws of physics are based on quantities defined in space, time, and frequency, which can be internally represented by the chemical concentration of neurotransmitters in three-dimensional space.

Neurotransmitter clouds in early metazoa would have suffered from two problems: chemical diffusion of the molecules and chemical inertia due to the large amounts of neurotransmitter required to fill in the extracellular space. As a result, evolutionary adaptation would have favored neural structures where the neurotransmitters remained confined to the small regions in the synaptic clefts between the pre- and postsynaptic neurons.

In order to visualize how a computation can be performed on a neurotransmitter cloud, imagine the dendritic arbor of a neuron as a leafless tree with its branches inside of the cloud. The surface of the tree is "painted" with a shade of gray that corresponds to its sensitivity to a particular neurotransmitter. When multiplied by the actual concentration of neurotransmitter present in the extracellular space, and integrated over a region of space that contains the dendritic tree, the result is a first-order approximation of the neuron's response. We can mathematically represent the "shade of gray" that corresponds to the sensitivity of a neuron's dendritic arbor in physical space by a function $\mu(x,y,z)$.

2 Neurotransmitter Field Theory

The mathematical formulation of neurotransmitter fields subsumes the functionality of the neural networks. That is, for every neural network, there exists a corresponding neurotransmitter field computation that generates an identical result.

2.1 Inner Products

The standard projection neural network calculation is based on the inner product of two vectors, a vector of input values, and a weight vector. Hilbert spaces generalize the inner product operation to continuous domains by replacing the summation of the products of the vector coefficients, with the integral of the product of two functions [9]. One of these two functions, $\mu(x,y,z)$, is used to represent the sensitivity of the dendritic arbor and is analogous to the weight vector.

Let H be the space occupied by the neurotransmitter cloud, and let $h(x,y,z)$ be a field corresponding to the density of transmitter in the extracellular space. To permit the use of Dirac delta (impulse) functions we use the Lebesgue integral and define $\mu(x,y,z)$ as a signed measure [9], [10]. Using the Lebesgue integral, instead of the conventional Riemann integral, allows us to model neurons that are able to discriminate neurotransmitter concentration at a single point, but may also exhibit sensitivity over entire regions. We conceptually model the operation of a neuron as an abstract *Processing Element* (PE).

The dendritic arbor computation of the PE, which is analogous to the vector inner product, is defined by the integral of h and with respect to μ

$$\text{response} = \int_H h(x,y,z)\, d\mu(x,y,z)\,. \tag{1}$$

To demonstrate why a neurotransmitter field calculation subsumes the functionality of the standard neural network model, we examine the computation performed by a single-layer network with a single output node. For an m-dimensional input vector $\mathbf{u} = (u_1, u_2, \ldots, u_m)^{\mathrm{T}}$, a weight vector $\mathbf{w} = (w_1, w_2, \ldots, w_m)^{\mathrm{T}}$, and a transfer function σ, the output v of a single-layer projection neural network is given by

$$v = \sigma(\mathbf{w}^{\mathrm{T}}\mathbf{u}) = \sigma\left(\sum_{k=1}^{m} w_k u_k\right). \tag{2}$$

To construct an analogous neurotransmitter field computation, identify the input vector $\mathbf{u}$ with any set of m distinct points $\{(x_k, y_k, z_k);\ 1 \le k \le m\}$ in H, and let the input vector coefficients $u_k = h(x_k, y_k, z_k)$ be defined by a function $h \in \mathbf{L}^2(H)$. Let $\{\delta_k\}$ be the set of three-dimensional Dirac delta functions (product measures) defined by

$$\delta_k = \delta(x - x_k)\delta(y - y_k)\delta(z - z_k)\,. \tag{3}$$

For a single PE, let the transfer function σ be the same as the one used for the neural network. Setting

$$\mu = \sum_{k=1}^{m} w_k \delta_k \tag{4}$$

we have

$$\sigma\left(\int_H h\,d\mu\right) = \sigma\left(\int_H h(x,y,z)\left(\sum_{k=1}^{m} w_k\, d\delta_k\right)\right) = \sigma\left(\sum_{k=1}^{m} w_k h(x_k, y_k, z_k)\right) = v \tag{5}$$

Thus, a PE with the measure μ performs the same calculation as the single-layer neural network. The biological meaning of the measure μ defined above is a mathematical model of a neuron with m points (idealized synapses) each with sensitivity w_k and located at the spatial positions (x_k, y_k, z_k) on the dendritic surface. Since the calculations may be carried out by cells other than neurons, we use the term *computational manifold* to refer to the generalization of discrete neural networks to continuous spaces.

2.2 Computational Manifolds

In addition to the dendritic arbor, each neuron also has an axonal tree, or terminal arbor, which releases neurotransmitter into the extracellular space. Let $\tau(x,y,z)$ denote the function that quantitatively describes the output of a neuron in terms of the spatial distribution of chemical transmitter it generates.

We use the index i to enumerate the set of neurons $\{N_i\}$. Each PE, N_i, has a unique dendritic arbor μ_i and a unique terminal arbor τ_i. Mathematically the neurotransmitter "clouds" are three-dimensional manifolds which we illustrate diagrammatically as rectangular blocks such as the input manifold H and the output manifold G shown in Fig. 2. To distinguish between the input and output spaces, we substitute the parameters (ξ,η,ζ) for (x,y,z) in the input manifold H.

Each processing element, N_i, such as the one shown in Fig. 2 consists of a *receptor measure*, $\mu_i(\xi,\eta,\zeta)$, a nonlinear *cell-body-transfer function*, σ, and a *transmitter function* $\tau_i(x,y,z)$. The receptor measure μ_i models the shape and sensitivity the dendritic arbor in the input manifold H, while transmitter function τ_i models the signal distribution in the terminal arbor and the concomitant release of neurotransmitters into the output manifold G.

The inherent nonlinear relationship between the concentration of neurotransmitter in the extracellular space and the gating of the ion channels on the dendritic surface is characterized by the *dendritic-cell-membrane-transfer function* χ_d. At some point, increasing the concentration of neurotransmitter has a diminishing effect on the ion channels. Consequently, χ_d is nonlinear. Similarly, the *axonal-cell-membrane-transfer function* χ_a, characterizes intrinsic nonlinear response corresponding to the release of neurotransmitters by the axons terminals as a function of the neuron firing rate. The two transfer functions, χ_d and χ_a, as well as the cell-body-transfer function σ are analogous to a sigmoid transfer function, such as $1/(1+\exp(-x))$ or the hyperbolic tangent function, used in neural networks. We model the spatial variations in the responses of each processing element N_i using μ_i and τ_i and assume that χ_d and χ_a, are fixed functions of a single real variable which are uniform throughout all cells.

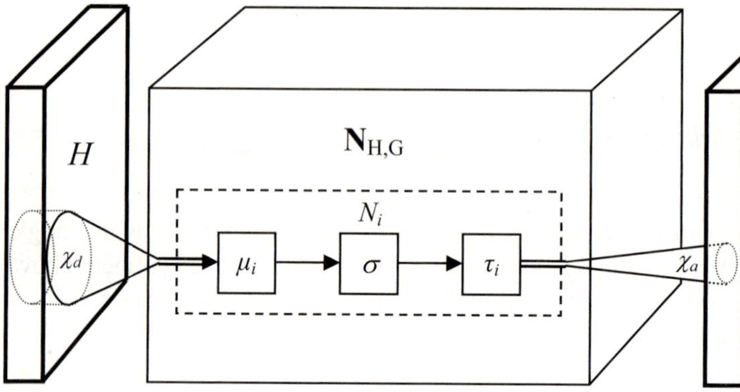

Fig. 2. The processing element N_i models the operation of a single neuron. The receptor measure μ_i converts the continuous distribution of neurotransmitter in the input manifold H to a single real value, while the transmitter function τ_i converts a single real value to a continuous distribution of neurotransmitter in the output manifold G. The operation σ models the nonlinear response of the cell to the dendritic inputs. The nonlinear response of the dendrite-cell membrane and the axon-cell membrane are represented by χ_d and χ_a respectively.

The transformation from a discrete real value back to a continuous field results from scaling the output of the nonlinear transfer function σ by the transmitter function $\tau_i(x, y, z)$. Taking into account the cell-membrane transfer functions and summing over all of the PEs gives the complete output function g.

$$g(x, y, z) = \sum_i \chi_a \left(\sigma \left(\int_H \chi_d \left(h(\xi, \eta, \zeta) \right) d\mu_i(\xi, \eta, \zeta) \right) \cdot \tau_i(x, y, z) \right) \tag{6}$$

The receptor measures and the transmitter functions perform the complementary operations of converting back and forth between fields defined on continuous manifolds and discrete real values.

2.3 Basis Functions

The continuous version of a projection neural network defined by (6) can be extended by generalizing the notion of radial basis functions [11] to computational manifolds. For discrete neural networks, a set of pattern vectors $\{\mathbf{u}_\alpha\}$ and a radial basis function θ form the discriminate functions $\theta(\|\mathbf{u} - \mathbf{u}_\alpha\|)$. The real-valued function $\theta(x)$ has its maximum at the origin and the properties $\theta(x) > 0$ and $\theta(x) \to 0$ as $|x| \to \infty$. Typically, $\theta(x)$ is the Gaussian, $\exp(-x^2 / 2\sigma^2)$, or a similar function.

To construct the analogous continuous basis functions, we replace the discrete pattern vectors $\mathbf{u}_\alpha$ with a continuous field ρ_α. Each of the functions $\rho_\alpha(\xi,\eta,\zeta)$ represents a "pattern" density defined on the input manifold. If we wish, we can associate a particular "target" function $g_\alpha(x,y,z)$ in the output manifold with each input pattern ρ_α. Assuming that there are several PEs available for each pattern, we assign a particular pattern to each N_i which we label ρ_i.

The equation corresponding to a basis-function neural network can be obtained by substituting either $\theta(\chi_d(h) - \chi_d(\rho_i))$ or the less complex $\theta(h - \rho_i)$ for $\chi_d(h)$ in (6)

$$g(x,y,z) = \sum_i \chi_a\left(\sigma\left(\int_H \theta(h-\rho_i)\,d\mu_i\right)\cdot\tau_i(x,y,z)\right) \tag{7}$$

where we have omitted the variables of integration (ξ,η,ζ) for h, ρ, and μ.

Each processing element now has an additional property ρ_i, which represents the pattern to which it is the most sensitive. For each PE, the integral inside (7) is maximum when $h = \rho_i$ over the region of integration. This in turn maximizes the coefficient for the transmitter function τ_i. The sum of the transmitter functions $\{\tau_i\}$ associated with a particular input pattern ρ_α can then be defined to approximate the desired target function g_α, thereby creating the required associations.

The measures μ_i in (7) can identify the regions where the pattern ρ_i is the most sensitive. For example, we can imagine photographs of two different animals that appear very similar except for a few key features. The photographs, representing two patterns ρ_1 and ρ_2, are approximately equal, but the measures can be trained so that their value where the patterns are the same is small, but in the key regions where the patterns differ, they have much larger values. In this way, even though the two image patterns are almost the same, the output functions g_α that result from the integrals in Equation (7) could be very different.

2.4 Computational Equivalence

While models that use action potentials as state variables can form associations by using matrix operations on a large vector of neuron outputs, equation (7) shows the neurotransmitter state model makes it possible for a small number of neurons, even a single neuron, to establish an association between an arbitrary input pattern $\rho_\alpha(\xi,\eta,\zeta)$ and an arbitrary output pattern $g_\alpha(x,y,z)$.

A two-layer discrete neural network and a continuous computational manifold are shown in Fig. 3. As we have seen, the measures $\{\mu_i\}$ in the computational manifolds can replace the weights $\{w_k\}$ in the neural network; the corresponding summation takes place inside the cell. Since the transmitter functions $\{\tau_i\}$ can extend over a large area, even the entire output manifold, many different processing elements may contribute to the concentration of neurotransmitter at any particular point (x,y,z). Consequently, the summations in (6) and (7) are equivalent to the summations in a neural network where the weights correspond to the values of the transmitter functions at a given point. This summation takes place outside the cell, as illustrated in Fig. 1.

Both the integrals with respect to the measures μ_i, and the summations over the transmitter functions τ_i, in effect perform operations analogous to the inner product with a weight vector in a single-layer neural network. Consequently, together they perform an operation analogous to a two-layer neural network.

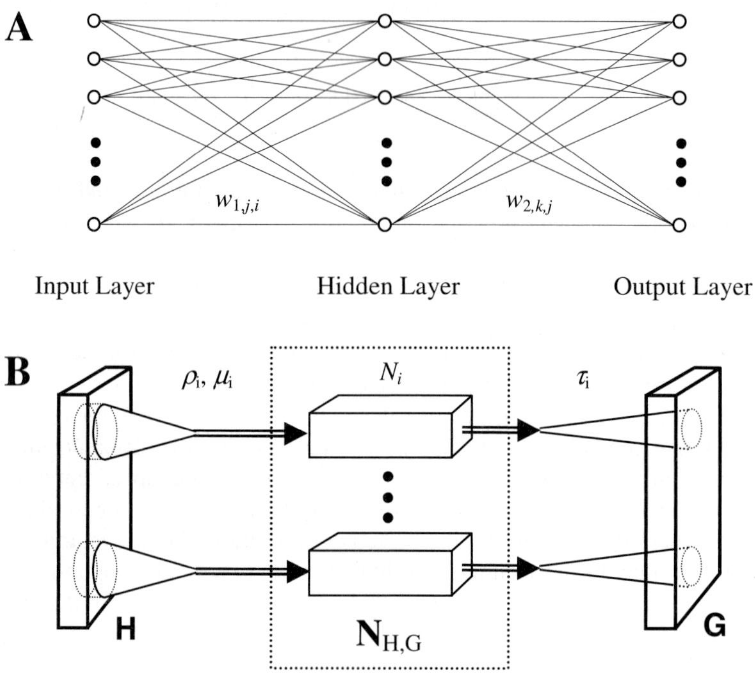

Fig. 3. A neural network (*A*) transforms discrete vectors, while a computational manifold (*B*) transforms continuous fields. Neurons are points in the function space $\mathbf{N}_{H,G}$. Since there are effectively two summations, one in the intracellular space and one in the extracellular space, the receptor measure μ, together with the transmitter function τ, allow a single layer of neurons to perform the equivalent computation of a two-layer neural network.

The collection of transmitter functions and receptor measures that comprise the synapses within a single neurotransmitter cloud can also be viewed as a two-layer neural network. In this formulation, a two-layer back propagation algorithm now takes place between the pre- and postsynaptic neurons, inside a single manifold, with the errors propagating back from the receptor measures to the transmitter functions.

Computation manifolds are useful for describing a wide range of cognitive operations [12]. In particular, the architecture outlined in Fig. 3, with processing elements incorporating the patterns densities defined by (7), is well suited for generating stable, recursive associations on spectral manifolds [13].

2.5 Function Spaces

The nodes of the neural network shown in Fig. 3A are partitioned into the input layer, the hidden layer, and the output layer. In the computational manifold model, the input layer is analogous to the input manifold *H*, and the output layer is analogous to the output manifold *G*. Both *H* and *G* represent the continuous distribution of neurotransmitters in physical space. The "hidden" layer is the space $\mathbf{N}_{H,G}$, which equals the Cartesian product of two function spaces: the space of all possible (receptor) measures on *H*, and the space all possible output (transmitter) functions on *G*. The individual neurons N_i are points in this infinite-dimensional product space.

When samples of a continuous function defined on a high-dimensional space are arranged in a lower dimensional space, the samples will in general appear to be discontinuous. Consequently, when a collection of processing elements, $\{N_i\}$, representing samples taken from the infinite-dimensional function space $\mathbf{N}_{H,Q}$ are arranged in three-dimensional physical space, the outputs will seem discontinuous. The resulting firing rates may appear to be stochastic when in fact they are deterministic. Moreover, realistic neural field models that attempt to describe the observed firing rates of large groups of neurons as a continuous function in physical space will be difficult or impossible to create.

Equations (6) and (7) express the computations of a neuron that is sensitive to a single neurotransmitter. Given the number of different chemicals that act as neurotransmitters, both inhibitory and excitatory, we clearly need to extend the model to account for their effects. If we have *n* different chemicals of interest in the extracellular space, we can model their concentration at each point as vector $\mathbf{h}(x,y,z) = (h_1(x,y,z),\ h_2(x,y,z),\ \ldots\ ,\ h_n(x,y,z))$. Any nonlinear interactions between the various neurotransmitters in the dendritic arbor will require the appropriate modifications to the integral equations on the input manifold.

3 Neuroglia

In the central nervous system of vertebrates, there are 10 to 50 times more glial cells than neurons [14]. Astrocytes, the most common type of neuroglia, are receptive to potassium ions and take up neurotransmitters in synaptic zones. Glial cells have also been shown to release neurotransmitters.

Unlike neurons, glial cells do not generate action potentials. Consequently, if state is encoded in the firing of neurons, glia are relegated to a support role. However, in a neurotransmitter-centric model, glia can take a central position along side neurons. They may participate in both short-term and long-term memory as well as computations. However, since they lack action potentials, glial cells transmit the results of their computations more slowly.

4 Conclusion

In the standard neural network model, the state variables are the neuron action potentials, and a synapse corresponds to a single weight that represents the effect the presynaptic neuron has on the postsynaptic neuron. In the neurotransmitter field model, the state variables are the concentrations of neurotransmitters in the extracellular space. In this formulation, a single layer of neurons is able to perform the computation of a two-layer neural network. One set of weights corresponds to the sensitivity of the dendritic arbor and the second set of weights corresponds to the amount of neurotransmitter released by the terminal arbor. The second summation occurs on the neurotransmitters in the extracellular space and remains separated from the intracellular summation by the nonlinear responses of the cell membranes.

Compared to a neuron action-potential model, a neurotransmitter-centric model presents a broader and more comprehensive view of natural intelligence. It allows the chemical reactions that take place in many types of cells, including neuroglia, to be incorporated into a general framework of memory and computation.

Acknowledgments. The author would like to thank Professor M. Tuceryan for his encouragement and helpful comments. This document contains material that is patent pending by General Manifolds LLC, http://www.gmanif.com/ip.

References

1. Beurle, R.L.: Properties of a Mass of Cells Capable of Regenerating Pulses. Philosophical Trans. of the Royal Society. Series B 240, 55–94 (1956)
2. Griffith, J.S.: A Field Theory of Neural Nets: I: Derivation of Field Equations. Bulletin of Mathematical Biophysics 25, 111–120 (1963)
3. Griffith, J.S.: A Field Theory of Neural Nets: II. Properties of the Field Equations. Bulletin of Mathematical Biophysics 27, 187–195 (1965)
4. Amari, S.: Dynamics of Pattern Formation in the Lateral-Inhibition Type Neural Fields. Biological Cybernetics 27, 77–87 (1977)
5. Ermentrout, B.: Neural Networks as Spatio-temporal Pattern-Forming Systems. Reports on Progress in Physics 61, 353–430 (1998)
6. Vogels, T.P., Rajan, K., Abbott, L.F.: Neural Network Dynamics. Annual Rev. Neuroscience 28, 357–376 (2005)
7. Sarnat, H.B., Netsky, M.G.: Evolution of the Nervous System, 2nd edn. Oxford (1981)
8. Widrow, B., Hoff, M.E.: Adaptive Switching Circuits. WESCON Convention Record, IRE, New York, pp. 96-104 (1960)
9. Royden, H.L.: Real Analysis, 3rd edn. Prentice-Hall, Englewood Cliffs (1988)
10. Kolmogorov, A.N., Fomin, S.V.: Introductory Real Analysis. Dover, New York (1970)
11. Bishop, C.M.: Neural Networks for Pattern Recognition. Oxford University Press, Oxford, UK (1995)
12. Greer, D.S.: A Unified System of Computational Manifolds. Tech. Rep. TR-CIS-0602-03, Dept. of Comp. and Info. Sci., IUPUI, Indianapolis (2003)
13. Greer, D.S.: An Image Association Model of the Brodmann Areas. In: Proc. 6th IEEE International Conf. on Cognitive Informatics (in press) (2007)
14. Kandel, E.R., Schwartz, J.H., Jessell, T.M.: Principles of Neural Science, 4th edn. McGraw-Hill, New York (2000)

SimBa: A Fuzzy Similarity-Based Modelling Framework for Large-Scale Cerebral Networks

Julien Erny[1,2,3], Josette Pastor[1,2], and Henri Prade[3]

[1] Inserm, U825, Toulouse, F-31000, France
[2] Université Toulouse III Paul Sabatier, Toulouse, F-31000, France
[3] CNRS, IRIT, Toulouse, F-31000, France
{julien.erny,josette.pastor}@toulouse.inserm.fr,
henri.prade@irit.fr

Abstract. Motivated by a better understanding of cerebral information processing, a lot of work has been done recently in bringing together connectionist numerical models and symbolic cognitive frameworks, allowing for a better modelling of some cerebral mechanisms. However, a gap still exists between models that describe functionally small neural populations and cognitive architectures that are used to predict cerebral activity. The model presented here tries to fill partly this gap. It uses existing knowledge of the brain structure to describe neuroimaging data in terms of interacting functional units. Its merits rely on an explicit handling of neural populations proximity in the brain, relating it to similarity between the pieces of information processed.

1 Introduction

Activation studies, where subjects are asked to perform a specific task while data of their brain functioning are collected through functional neuroimaging, have shown that sensorimotor or cognitive functions are the offspring of the activity of large-scale networks of anatomically connected cerebral areas [1]. However, knowing the cerebral substratum of a cognitive function is necessary, although not sufficient, to be able to make the accurate prognosis of the clinical aftermath of a lesion or the precise assessment of a rehabSect.ilitation procedure. The main point is to interpret functional neuroimaging data as the result of cerebral information processing, which can be tricky, even in the case of a basic function such as categorisation [2]. This is worsen by the fact that neuroimaging data are very indirect measures of the neuronal activity, e.g. the whole brain electrical activity "seen" by each electrode in EEG, or the haemodynamic response to neuronal energy demand provided by fMRI. Our long-term goal is to be able to predict the cognitive behaviour from neuroimaging data, which are an indirect evidence of the real, unknown, activity of the cerebral substratum. Currently, three main, and somehow independent, approaches tackle partly the problem. Statistical methods focus on the analysis of neuroimaging data that they relate loosely to cognitive functions through simplified (e.g. "additive") task models [3]. The other two approaches relate cerebral activity to cognitive functions. Top-down modelling relies usually on a functional decomposition of the large-scale networks components. The coarseness of the decomposition depends on how strongly the models make use of high-level symbolic tools [4,5]. Computational

J. Marques de Sá et al. (Eds.): ICANN 2007, Part II, LNCS 4669, pp. 29–38, 2007.

neuroscience is based on the idea that the function performed by a single region emerges from the activity of the neural network underpinning it [6]. The problem with this latter approach is that even in such an accurate model of a functional unit, it is likely that it will not inform us on how information is processed, due to the massive distribution of information representation and processing throughout the whole network. Related to the issue of explaining neuroimaging data, is the problem of information representation. Mukamel et al. [7] have shown that neuroimaging activation patterns are strongly correlated to the firing rates of the neurons. However when integrating activity over a neural population, the spatial activation pattern is lost. Moreover the widespread topical organisation of cerebral regions [8] and the possible modulation of the neural sensitivity to stimuli [9] are in favour of a symbolic component of information that we think is related to these activation patterns.

To address this challenge, we propose a preliminary model, called SimBa, that takes into account both overall populations firing rates (as a numerical component) and spatial activation patterns (as a fuzzy symbolic component). We hope that it will eventually help to interpret the cerebral networks revealed by clinical functional neuroimaging in terms of cognitive functions.

Section 2 presents the SimBa modelling framework itself while Sect.3 briefly presents two applications illustrating each components of the model. Section 4 discusses some other related models before concluding in Sect.5.

2 Presentation of SimBa

This section presents a model based on the causal connectivity paradigm that represents large-scale cerebral networks as neuroanatomy-based networks of functional units. Processed information is both numerical and symbolic and interactions between symbolic and numerical aspects occur inside each functional component.

2.1 Causal Connectivity and Information Representation

The causal connectivity approach [10] characterises the information processing that occurs in one functional unit along both symbolic and numerical aspects. This approach inspired probabilistic [11] and information's similarity-based [12] models (though the latter contains also probabilistic aspects). The main idea is to consider each functional unit as an *information processor* and the connections between them (i.e. axon bundles linking together neural populations) as *information transmitters*. The information itself is represented as two-dimensional data: i) a numerical component, called *magnitude*, stands for the overall activation of the neuronal population that processed this piece of information (thus allowing comparisons with neuroimaging data), and ii) a symbolic component, called *type*, that qualifies the pattern of the firing neurons in the population. Note that we do not manipulate rule-generated symbols that would then have to be grounded, but we rather give a symbolic label to a pattern of activation. While the model describes low-level functions, like sensory processing, those labels can be given a semantic meaning based on the topic organisation of the primary cortex areas [8]. Information transmission has two modes since the numerical part is propagated using a dynamic Bayesian formalism quite similar as in [11], while the symbolic part is just passed along the numerical part, without modifications between the processors. We describe now how information is processed by the functional units, where the two components of information will really interact.

2.2 Pattern Categorisation

A functional unit is characterised by the function it performs on the incoming information. This function depends on the role of this population in the network (whether it is inhibition, categorisation, etc.). In this we do not depart from the traditional view of causal connectivity [10]. As for the symbolic component, it will act as a way to modify this action. It is well known that the primary cortex is functionally organised according to the stimuli it receives [8]. For example, the auditory cortex presents a tonotopy, meaning anatomically localised populations will have a receptive field centred on a small interval of sound frequencies. Moreover, two overlapping populations will have overlapping receptive fields. This kind of topic organisation appears in every primary cortices and in fact can be generalised this way : **similar stimuli trigger similar cerebral activation**, i.e. activation of spatially-close neuronal populations. This suggests that similarity between stimuli can be represented in terms of spatial proximity of neural population responding to them, this without restriction on the modality of the stimuli (i.e. auditory, visual, etc.). From experimental observations [8] comes the main hypothesis motivating this model, that this is true not only in the primary cortex but also in the rest of the brain. Let us see in the next section how this will be practically used in the model.

2.3 System Workflow and Formalism

Patterns and fuzzy sets. The symbolic component of information is represented by a discrete fuzzy set, the core of which being the symbolic label of the population that produced this piece of information and the support being composed of the symbolic labels of populations known to be close to the producing one (see Fig.1). The single symbol in the core of a fuzzy set characterises the set and is called its *centre*. When processing incoming information, a functional unit compares its symbolic component (i.e. its type) with the receptive field of the unit. This receptive field is represented as a set of fuzzy set prototypes. This allows an easy comparison with the incoming type in terms of similarity.

Notation conventions. Time is discretised, with a time step Δt of usually 1 ms (this is sufficient regarding the time resolution of neural populations). An information is a couple (M, T) where M is the magnitude and T, the type. T is a fuzzy set defined on the discrete domain $\mathbb{D}_T$. When considering a functional unit, $(M_{in}^{(i)}, T_{in}^{(i)})$ is the incoming information on input i and (M_{out}, T_{out}) the outgoing information.

Spatial integration. Since a functional unit can receive several inputs, spatial integration is needed. This is where excitatory or inhibitory signals will be handled differently and where the different modalities will be combined either linearly or not depending of the role of the unit. For the magnitude, the combination uses classical addition and multiplication, whereas for the type, classical fuzzy operators are used [13]. For example,

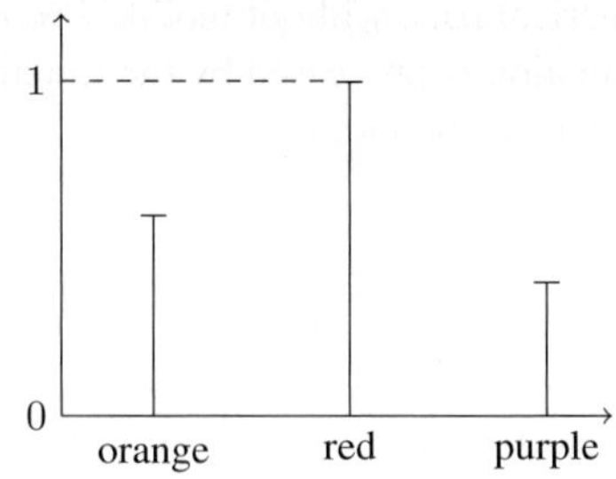

Fig. 1. A fuzzy set centred on the colour "red". If "orange" belongs to the core of another set, the patterns that get activated by "red" and "orange" are similar, meaning that they share neurons.

the following operators can be defined: (i) An operator AND ($\wedge$) used to aggregate two sets defined on two distinct domains : let A and B be two fuzzy sets defined on $\mathbb{D}_A$ and $\mathbb{D}_B$, then $\forall x = (x_A, x_B) \in \mathbb{D}_A \times \mathbb{D}_B, (A \wedge B)(x) = \min(A(x_A), B(x_B))$. (ii) An operator OR ($\vee$) used to aggregate two sets defined on the same domains : let A_1 and A_2 be defined on $\mathbb{D}_A$, then $\forall x \in \mathbb{D}_A, (A_1 \vee A_2)(x) = max(A_1(x), A_2(x))$. (iii) A function one used to make a set neutral for $\wedge$ and absorbent for $\vee$: let A be defined on $\mathbb{D}_A$, then $\forall x \in \mathbb{D}_A, one(A)(x) = 1$. (iv) A function $zero$ used to make a set neutral for $\vee$ and absorbent for $\wedge$: let A be defined on $\mathbb{D}_A$, then $\forall x \in \mathbb{D}_A, zero(A)(x) = 0$. The use of these two later functions will not be illustrated for the sake of brevity, however they play a noticeable role in the treatment of inhibition. Two aggregating functions are constructed using these operators, as shown below (since all variables refer to the same time step, it is omitted):

$$(\tilde{T}, \tilde{M}) = (f_{spat}(T_{in}^{(1)}, \cdots, T_{in}^{(n)}), g_{spat}(M_{in}^{(1)}, \cdots, M_{in}^{(n)}, u)) \tag{1}$$

where u is a random variable that stands for numerical errors and non-modelled influences.

Temporal integration. To get the kind of graded response that is expected from a stimulated neural population, a temporal integration is necessary. Moreover, by comparing the new incoming information with the previously processed one, *habituation* can be simulated. This is a property of neural populations that can be described as, when a stimulus is presented repeatedly, the overall activity of the population will decrease over time due to a lowering of activation thresholds and to fewer neurons being recruited [14].

Magnitude. The magnitude $M(t)$ got after temporal integration is the combination of $M(t - \Delta t)$, discounted by a discrepancy factor (i.e. a forgetting factor), and $\tilde{M}(t)$ given by (1). $\tilde{M}(t)$ is also discounted by a factor that represents both the compatibility between $T(t - \Delta t)$ and $\tilde{T}(t)$ and the fact that $\tilde{M}(t)$ is presented during only one time step. This is shown in (2).

$$M(t) = k_R.\alpha(t).\tilde{M}(t) + k_L.M(t - \Delta t) \tag{2}$$

where $\alpha(t) = \max \min_{x \in \mathbb{D}_T \cap \mathbb{D}_{\tilde{T}}} (T(t-\Delta t)(x), \tilde{T}(t)(x))$ is fuzzy-set consistency [13]. k_R and k_L can be related to respectively a response time τ_R ($k_R = 1 - e^{-\frac{3.\Delta t}{\tau_R}}$) and a relaxation time τ_L ($k_L = e^{-\frac{3.\Delta t}{\tau_L}}$) in a transfer function. Note also that the magnitude is bounded if $\tilde{M}$ is bounded : if $\forall t, \alpha(t) = 1$ (its maximum value), then $M_{max} = \tilde{M}_{max}.\frac{k_R}{1-k_L}$.

Type. Temporally integrating the type means to construct a new type based on the type at $t - \Delta t$ and on the new incoming one $\tilde{T}(t)$. On the one hand, (i) the importance that the latter should have in the combination depends on its consistency with the former (the more consistent it is, the more impact it has) but on the other hand, (ii) a repeatedly inconsistent incoming information should, after some time, trigger a shift of priority and make the incoming information paramount. Also, (iii) $\tilde{T}(t)$ being representative of only one time step should have much less influence on the combination than $T(t - \Delta t)$ which accounts for all the previous time steps. Finally, (iv) note that a high $\tilde{M}$ is associated with a change in the information, hence the update should be more efficient. (iii) suggests the use of a weighted disjunctive combination, while (i) and (ii) suggest a prioritised combination with a priority depending on the evolution of the consistency between T and $\tilde{T}$. The expression of a consistency-driven prioritised disjunction can be found in [15], here modified into an additive/multiplicative context. We define $\mathcal{I}$ as an inconsistency indicator. If x_T is the centre of a fuzzy set T, then:

if $\mathcal{I} < threshold,\ \forall x \in \mathbb{D} \setminus \{x_{T(t-\Delta t)}\}$,

$$T(t)(x) = T(t - \Delta t)(x) + [\tilde{T}(t)(x) - T(t - \Delta t)(x)].\alpha.s(\tilde{M}) \tag{3}$$

$$\mathcal{I} = \mathcal{I} + \tilde{T}(t)(x_T) - T(t - \Delta t)(x_T)].\alpha.s(\tilde{M}) \tag{4}$$

else $\forall x \in \mathbb{D} \setminus \{x_{\tilde{T}(t)}\}$,

$$T(t)(x) = \tilde{T}(t)(x) + [T(t - \Delta t)(x) - \tilde{T}(t)(x)].\alpha.s(\tilde{M}) \tag{5}$$

$$\mathcal{I} = 0 \tag{6}$$

Function s is increasing with the magnitude $\tilde{M}$, and should tend toward 0 (resp. 1) when $\tilde{M}$ tends toward 0 (resp. its maximum). These equations may look fairly complicated to understand and Fig. 2 illustrates their behaviour on an example. The idea is quite simple though : at each time step t, $T(t - \Delta t)$ is modified slightly in the direction of $\tilde{T}(t)$. If the centre of $T(t - \Delta t)$ is to be modified, $\mathcal{I}$ is incremented by the same value instead. Whenever $\mathcal{I}$ goes above a certain threshold, $\tilde{T}(t)$ becomes dominant, the priority is shifted and $\mathcal{I}$ is set to 0 for another cycle to begin.

Once the inputs are integrated both spatially and temporally, the result is compared to the set of prototypes that represents the population receptive field.

Comparison and decision. The receptive field of a functional population is composed of a set of prototypes $\{P_i\}_{i \in [1, \cdots, p]}$ defined on the same domain than the incoming information. Each of these P_i is associated with an output type E_i defined on $\mathbb{D}^{out}$. The centre of this pattern is the outgoing type characterised by the activation of the prototype P_i, while the other elements of the support are output types that are similar (i.e.

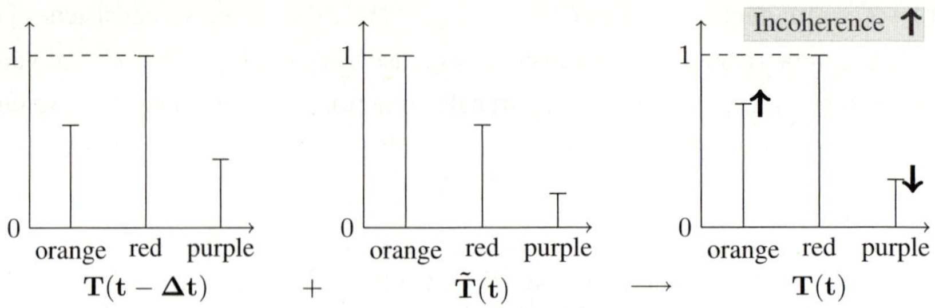

Fig. 2. The fuzzy set $T(t - \Delta t)$ is modified by $\tilde{T}(t)$, meaning that the values for "orange" and "purple" are moved in the direction of $\tilde{T}(t)$, while the value for "red" is unchanged to keep the result normalised. However, to reflect the inconsistency between the two types, the incoherence value is increased.

types whose associated prototypes share neurons, see Fig.1). By comparing T with each prototype, we can determine which patterns of neurons get activated by the incoming information. Compatibility a_i between T and P_i is $a_i = \max_{x \in \mathbb{D}} \min(T(x), P_i(x))$ (time is omitted since all variables are taken at the same time). Using then the similarities contained in the $\{E_i\}_{i \in [1,\cdots,p]}$, we can determine how overlapping patterns can influence each other activation. Let b_i be the total activation of E_i, namelly,

$$b_i = \sum_{k=1}^{p} E_k(x_i).\alpha_k \tag{7}$$

Once the activation of all patterns is known, the "winner-takes-all" principle is applied as a decision process, this is to account to the widespread lateral local inhibitions. The type of the output is then E_{max} such that $b_{max} = \max_{i \in [1,\cdots,p]}(b_i)$. Meanwhile, the output magnitude M_{out} is being calculated in the following way:

$$M_{out} = g^{(1)}_{out}(\beta_{max}).g^{(2)}_{out}(M, v) \tag{8}$$

where $g^{(1)}_{out}$ is a function from $\mathbb{R}$ to $[0, 1]$ which ensures that a badly recognised type will generate a low output magnitude, and where $g^{(2)}_{out}$ is a function defined from $\mathbb{R}$ to $\mathbb{R}$ which depends on the function of the unit in the network. v is a random variable modelling the stochastic nature of the neuronal signal.

This new model is now illustrated on two small scale applications.

3 Applications

The first application is designed to illustrate how the handling of similarity between patterns in our model can be used to account for a well-known perceptive illusion : the McGurk effect [16]. The second application focuses on using magnitude processing to reproduce synthetic neuroimaging results coming from an experimental study [17].

3.1 McGurk Effect

The McGurk effect is an illusion affecting language perception where a mismatch between a visual cue (lips articulating a phoneme) and an auditory cue (an actual phoneme) strongly modifies auditory perception. For example, *hearing* the sound [ba] while *seing* someone say [ga] results in the illusory perception of [da]. For this model, we use the articulatory theory of language that characterises a phoneme by the way we produce it. Table 1 represents a simplified French phonologic system. The second line of the table, and the fact that only the articulation locus can be considered a visual feature, hint towards a similarity-based interpolation between [b] and [g] to perceive [d]. Consider a phoneme discriminatory component, that takes as its inputs an auditory articulation mode M_A, an auditory articulation locus L_A and visual locus L_V. The spatial integration of these inputs is a linear pondered combination of L_A and L_V (L_V is given slightly more importance in the combination than L_A for the locus is primarily a visual cue), that is then non-linearly combined with M_A. The output domain is constituted by the different consonants. Similarity between the different output elements is also encoded (e.g. [d] is similar to both [b] and [g]) and the similarities between the different articulation modes and loci are transported by the inputs. While L_A and L_V are congruent in saying the locus is "labial", there is no problem, the combination will activate the prototype associated to [b] straightforwardly. However, if L_A and L_S carry different pieces of information (say "labial" and "velar"), then both prototypes associated to [b] and [g] will be activated, along with the prototype associated with [d] by means of the similarity between "dental" and "labial" and between "dental" and "velar". When then applying (7), providing [d] is similar enough to [b] and [g], [d] can easily be the most activated pattern and win. This is an informal description of how our model, which in several respects is close to the one used in [12], can yield the expected output and thus simulate the McGurk effect. Moreover it is important to notice that here the similarities have not to be set by the programmer but could rather be learned by the system (see Sect.4).

Table 1. Consonants in French phonologic system

locus \ mode	labial	dental	velar
plosive voiceless	p	t	k
plosive	b	d	g
nasal	m	n	

3.2 Visual Primary Cortex Response to a Simple Stimulus

The aim of this application is to replicate simulation results obtained by Pastor et al. [10] in modelling data coming from a PET (Positron Emission Tomography) study by Fox and Raichle [17]. In this study, a visual stimulus was repeatedly presented to participants while a PET camera was recording visual primary cortex activity. Different frequencies for stimuli presentation were used. The results (an increase of activity along with the frequency, then a decrease) let the authors of [17] with different possible explanations. By modelling the functional network involved, Pastor et al. [10] managed to support one of this hypothesis. This is only a partial illustration of our model since the symbolic part of information is actually unnecessary here. Its aim is to illustrate that SimBa also keeps the same modelling power compared to homologous models with respect to magnitude. We have used the same functional network than in [10] (see Fig.3). The $Total$ node is used to sum over time all activities coming from In_c. This can be compared to the activity detected by the PET camera, provided it is scaled appropriately. The scaled output of $Total$ (given by (8)) is compared to activation data from [17] and to simulation results from [10] in Fig.4.

It lacks though a real large-scale application that will make use of its similarity-based categorisation. This is discussed in the next section.

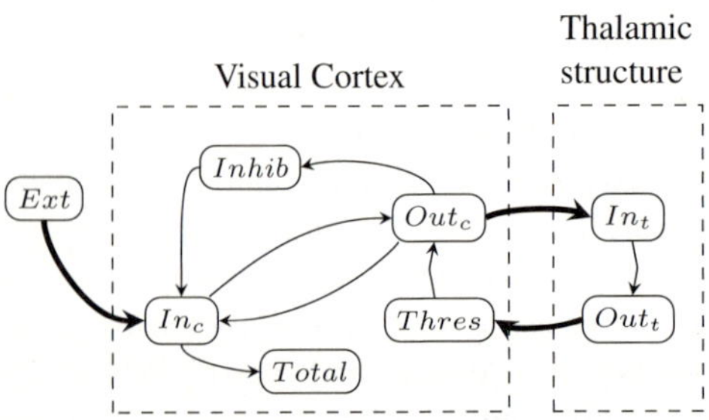

Fig. 3. A pre-processed information from Ext is passed to the input gate In_c. Out_c stands for the output gate. A cortico-thalamic loop allows dynamic threshold, hence habituation. Local inhibitions are modelled with $Inhib$ and the influence of Out_c on In_c represents a refractory period. $Total$ sums all activation in In_c to relate it to PET results.

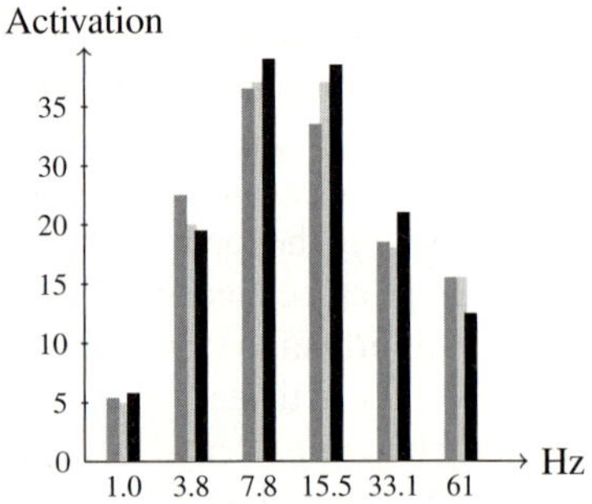

Fig. 4. Results of SimBa simulation (black on the right) are plotted for the different frequencies, along with the results appearing in [17] (gray on the left) and in [10] (light gray in the middle)

4 Related Works and Discussion

This section discusses the model in the light of other related approaches. There are several recent models that try to model cognitive cerebral functions while relating them to their structure in the brain. For example, Randall O'Reilly [18] is interested in modelling high-level cognitive function in a biologically plausible way. Although quite similar in the general principles underlying his approach to cerebral modelling, and letting aside the differences regarding the studied systems (high-level functions [19], automatised perceptions for SimBa), SimBa relate neural activity to interacting simple functional primitives (thanks to similarity-based model), while O'Reilly's models precisely describes the functional roles of small neural populations. The neural blackboard architecture proposed in [20] seems also to tackle the same issues that SimBa, but the difference lies in that the former tries to solve some cognitive fundamental problems (such as the binding and grounding problems) using neural-like computation, while Simba starts from the neural architecture to explain the cerebral information processing. We can also mention the ACT-R cognitive model that has been used quite recently to predict fMRI results [5], though in this case the modelling framework is not motivated by knowledge of the cerebral structure but was designed long ago as a cognitive architecture. SimBa is more strongly related to [12] but we claim that the model proposed here involves a much smaller number of parameters for obtaining results that are as good for the McGurk effect (the second illustration is not considered by [12]). In short, SimBa is not intented as a concurrent to any of these approaches, but more as providing a complementary angle of work, dealing with problems like similarity between patterns that have not been thoroughly explored yet, while including functional models that can describe neuroimaging data as cognitive processes.

SimBa still lacks learning ability, a property quite necessary to model cerebral networks. It can be incorporated by adjoining some simple mechanisms to the current framework. The learning can be resumed in an auto-organisation of the prototypes in every node according to incoming information. Several simple processes are involved to manage the set of prototypes : (i) introduction of a new prototype when the current information was unknown, (ii) fusion of prototypes that are too similar, (iii) forgetting of prototypes not used anymore. In addition to that, a way to learn similarities between output symbols is necessary. This relies on the observation that when two different prototypes are activated by an incoming information they must be somehow similar. Hence (iv) the similarity of their associated outputs is increased accordingly. Conversely, (v) when a prototype is activated alone, the similarity of its output with other outputs is decreased. Balancing these five mechanisms, the set of prototypes is build gradually according to the incoming information.

5 Conclusion

A framework for modelling large-scale networks as they appear in neuroimaging studies has been presented. Information is being represented as a numerical/symbolic couple. The numerical component relates to the integrated firing rates of neural populations while the symbolic one relates to the spatial configuration of the firing neurons in the same populations. The ability of the model to describe the links between the proximity of neural populations and the similarity of the information they process has been demonstrated on a simple application. Although some important features, like learning abilities, are still missing, as well as a large-scale modelling of real experimental data that could validate more strongly the approach, this model has interesting aspects and promising behaviour, in particular when it comes to cerebral categorisation.

References

1. Mesulam, M.: Large-scale neurocognitive networks and distributed processing for attention, language, and memory. Ann. Neurol. 28, 597–613 (1990)
2. Pernet, C., Schyns, P.G., Démonet, J.F.: Specific, selective or preferential: Comments on category specificity in neuroimaging. NeuroImage 35, 991–997 (2007)

3. Friston, K.J.: Functional and effective connectivity in neuroimaging: A synthesis. Human Brain Mapping 2, 56–78 (1994)
4. Sun, R., Alexandre, F. (eds.): Connectionist-Symbolic Integration: From Unified to Hybrid Approaches, Mahwah, NJ, USA. Lawrence Erlbaum Associates, Inc. (1997)
5. Anderson, J.R., Qin, Y., Jung, K.J., Carter, C.S.: Information-processing modules and their relative modality specificity. Cogn. Psych. 54, 185–217 (2007)
6. Dayan, P., Abbot, L.: Theoretical Neuroscience: computational and mathematical modeling of neural systems. MIT Press, Cambridge (2005)
7. Mukamel, R., Gelbard, H., Arieli, A., Hasson, U., Fried, I., Malach, R.: Coupling between neuronal firing, field potentials, and fmri in human auditory cortex. Science 309, 951–954 (2005)
8. Alexander, G., Delong, M., Crutcher, M.: Do cortical and basal ganglionic motor area use "motor programs" to control movement? Behav. Brain Sci 15, 656–665 (1992)
9. Tootell, R.B.H., Hadjikhani, N.K., Mendola, J.D., Marett, S., Dale, A.: From retinotopy to recognition: fmri in human visual cortex. Trends Cogn. Sci. 2(5), 174–182 (1998)
10. Pastor, J., Lafon, M., Trave-Massuyes, L., Demonet, J.F., Doyon, B., Celsis, P.: Information processing in large-scale cerebral networks: the causal connectivity approach. Biol Cybern 82(1), 49–59 (2000)
11. Labatut, V., Pastor, J., Ruff, S., Demonet, J.F., Celsis, P.: Cerebral modeling and dynamic bayesian networks. Artificial Intelligence in Medicine 30(2), 119–139 (2004)
12. Erny, J., Pastor, J., Prade, H.: A similarity and fuzzy logic-based approach to cerebral categorisation. In: Brewka, G., Coradeschi, S., Perini, A., Traverso, P. (eds.) Proc. of the 17th Euro. Conf. on Artificial Intelligence (ECAI'06), Riva del Garda, Italy, pp. 21–25. IOS Press, Amsterdam, Trento, Italy (2006), `http://www.iospress.nl/`
13. Zadeh, L.A.: Pruf–a meaning representation language for natural languages. International Journal of Man-Machine Studies 10(4), 395–460 (1978)
14. Miller, E.K., Li, L., Desimone, R.: A neural mechanism for working and recognition memory in inferior temporal cortex. Science 254(5036), 1377–1379 (1991)
15. Dubois, D., Prade, H., Yager, R.: Merging fuzzy information. In: Bezdek, J., Dubois, D., Prade, H. (eds.) Fuzzy sets in Approximate Reasoning and Information Systems. The Handbooks of Fuzzy Sets, Boston, Mass, USA, Kluwer, Dordrecht (1999)
16. McGurck, H., MacDonald, J.: Hearing lips and seeing voices. Nature 264, 246–248 (1976)
17. Fox, P., Raichle, M.: Stimulus rate dependence of regional brain blood flow in human striate cortex, demonstrated by positron emission tomography. J. Neurophy. 51, 1109–1120 (1984)
18. O'Reilly, R.C.: Biologically based computational models of high-level cognition. Science 314(5796), 91–94 (2006)
19. Herd, S.A., Banich, M.T., O'Reilly, R.C.: Neural mechanisms of cognitive control: An integrative model of stroop task performance and fmri data. J. Cogn. Neurosci. 18(1), 22–32 (2006)
20. van der Velde, F., de Kamps, M.: Neural blackboard architectures of combinatorial structures in cognition. The Behavioral and brain sciences 29, 37–108 (2006)

A Direct Measurement of Internal Model Learning Rates in a Visuomotor Tracking Task

Abraham K. Ishihara[1], Johan van Doornik[2], and Terence D. Sanger[2]

[1] Department of Aeronautics and Astronautics 4035, Stanford University, Stanford CA 94305-4035, U.S.A.
[2] Div. Child Neurology and Movement Disorders Stanford University Medical Center 300 Pasteur, room A345 Stanford, CA 94305-5235 USA
Tel.: (650)736-2154; Fax: 725-7459

Abstract. We investigate human motor learning in an unknown environment using a force measurement as the input to a computer controlled plant. We propose to use the Feedback Error Learning (FEL) framework to model the overt behavior of motor response to unexpected changes in plant parameters. This framework assumes a specific feedforward and feedback structure. The feedforward component predicts the required motor commands given the reference trajectory, and the feedback component stabilizes the system in case of imprecise estimates and initial conditions. To estimate the feedback gain, we employ a novel technique in which we probe the stability properties of the system by artificially inducing a time delay in the sensory feedback pathway. By altering the pole location of the plant during a sinusoidal tracking task, a feedforward learning bandwidth was computed for each subject which measures the ability to adaptively track time-varying changes in the plant dynamics. Lastly, we use the learning bandwidth to compute a learning rate with respect to the FEL model. This learning rate reflects the ability of the subjects' internal model to adapt to changes in an unknown environment.

1 Introduction

In this note, we will be concerned with the macroscopic level of brain-motor control during a visuo-motor tracking task, in which we will estimate the two macroscopic parameters: *feedback gain and learning rate.* The critical assumptions that we make are as follows: 1) *the central nervous system (CNS) utilizes internal models in the control of movement*, 2) *the CNS realizes a feedback system with constant gains*, and 3) *neuronal plasticity is a fundamental mechanism that allows that adaptive behavior of internal models.*

The first assumption is a highly debated topic in the field of motor control. Nevetheless the internal model concept is gaining ground as the results of various experimental and theoretical results (see [1]). Roughly speaking, internal models are 'neural mechanisms that can mimic the input/output characteristics, or their inverses, of the motor apparatus '[1]. Consistent with these assumptions is the Feedback Error Learning (FEL) framework originally proposed by Kawato [2] as a model for lateral cerebellum. This model is shown in Fig. 1.

J. Marques de Sá et al. (Eds.): ICANN 2007, Part II, LNCS 4669, pp. 39–48, 2007.

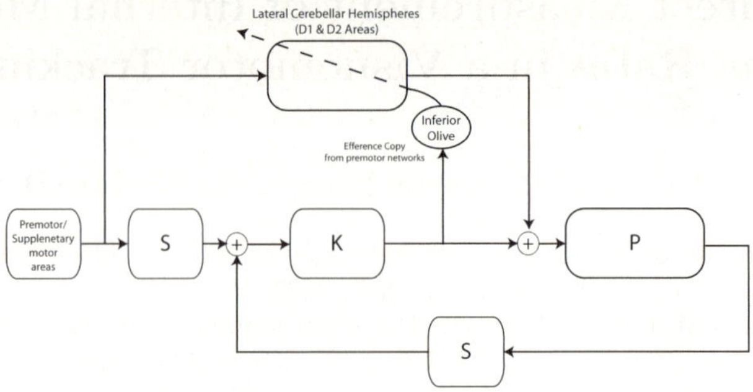

Fig. 1. Feedback Error Learning Block Diagram. The desired trajectory is generated in planning areas of cortex such as supplementary motor area (SMA), and premotor area (PMA). This command is then relayed to lateral cerebellum (D1 and D2 areas. See [3]) via internal capsule and pontine nuclei in brainstem.

The Task: We choose visuomotor tracking in an unknown, time-varying environment as the appropriate venue to explore the macroscopic parameters which we hope may characterize human motor behavior.

The FEL algorithm in [2] is:

$$\dot{W} = \Gamma \phi(q_d) \left(u_{fb}\right)^T \tag{1}$$

where W is a weight matrix, $\phi(q_d)$ is the basis function network, u_{fb} is the feedback control signal, and Γ is the learning rate.

We propose to use the FEL framework to model the motor response of human subjects during a tracking task. Our goal is to estimate each subject's learning rate, Γ, associated with the training of the feedforward component under FEL. To accomplish this goal, we first estimate the feedback gain, K, by artificially inducing time delays in the sensory feedback pathway, and observe the resulting motor response. Then, by varying the plant parameter sinusoidally, we compute the performance error and estimate the learning rate.

2 Materials and Methods

5 adults (age 22-30 years; two males and three females) with no known motor disorders participated in this study. Each subject gave written informed consent after the consent forms and study protocol were approved by the Stanford University Institutional Review Board.

2.1 Experimental Setup

Subjects were comfortably seated in a Biodex^{TM} chair in front of a large computer screen. Shoulder, chest, and leg straps were applied to restrict upper body

Fig. 2. Example of the feedback display during the baseline task

motion. The preferred arm was placed in a custom built device with the purpose of fixating the upper arm in a plane perpendicular to the floor, and with the elbow in a 90 degrees angle, such that the hand would point upwards. The device was setup individually for each subject according to his or her arm length and shoulder height. The lower arm was placed in a cup 2 inches below the wrist and tightened by a strap. A force sensor (Interface, Inc. 1500ASK-50 load cell) placed between the cup and the device allowed for a measurement of force between device and the subject's arm, which could be converted in a differential torque measure. The output of the force sensor was sampled at 1 kHz using a commercially available digital to analog interface (CED Technologies Inc., Manchester, UK) in connection with custom written software in Microsoft Visual C++. The filtered signal was then sent over a local area network using the Win32 named pipes protocol to another computer that simulated the plant and displayed the sensory feedback. The graphics were programmed in OPENGL. To guarantee a constant framerate, three separate threads were used for data collection, graphics display, and experiment control. This was important since we required precise control over the artificially induced delay in the sensory feedback pathway. All signals were stored for off-line analysis.

The measured signal was displayed on the screen by a torus. The subjects were instructed to keep the torus within the path at all times. The path was displayed by thin three dimensional rectangles that resembled a maze with right angled turns. The horizontal position of the torus moved at a constant velocity, while the vertical position of the torus was determined by the output of the plant. The plant was a first order linear, time-invariant system. An example of the feedback display is shown in Fig. 2.

Experimental Procedure
The experiment was divided into three separate parts discussed below.

1. *Baseline*: The first part entailed the learning of the baseline plant. It consisted of at least three trials through the path as depicted in Fig. 2.
2. *Probing the Feedback Gain*: This part consisted of eight separate trials in the same path used for baseline. During each trial, an artificial sensory feedback delay between 50 ms and 400 ms was introduced. The delays were arranged in a pseudo-random order. Additionally, the delay started at a time between 0 and 15 seconds on each trial. The rational for introducing time delays in explained in the next section.
3. *Probing Internal Model Learning*: The last part of the experiment consisted of tracking a sinusoidal desired trajectory. This consisted of six separate trials. During each trial, the pole of the plant was varied according to $a_0(t) = 2(0.5 + 0.4\sin(2\pi f_i t))$, where $f_0 = 0$, $f_1 = 0.05$, $f_2 = 0.1$, $f_3 = 0.15$, $f_4 = 0.2$, $f_5 = 0.4$. A typical performance observed will be shown in the following section.

2.2 Probing the Feedback Gain

We now discuss the second part of the experiment in more detail. We will assume that since the baseline trials were performed successfully, the subject has learned an adequate inverse internal model of the plant. Consider an nth order linear time invariant plant with fixed feedback delays in position (T_p) and velocity (T_v). Let

$$P(s) = \frac{b_m s^m + \cdots + b_0}{s^n + a_{n-1}s^{n-1} + \cdots + a_0} \tag{2}$$

be a minimum phase, strictly stable plant, that is, the poles and zeros are in open left half complex plane.

The constant feedback controller is given by: $K = \mathrm{diag}(K_p; K_v)$. It can be shown that the closed loop transfer function is given by:

$$H(s) = \frac{K_v b_m s^{m+1} + (K_p b_m + K_v b_{m-1})s^m + \cdots + (K_p b_1 + K_v b_0)s + K_p b_0}{(s^n + a_{n-1}s^{n-1} + \cdots + a_0) + (K_p e^{-sT_p} + sK_v e^{-sT_v})(b_m s^m + \cdots + b_0)} + \frac{s^n + a_{n-1}s^{n-1} + \cdots + a_0}{(s^n + a_{n-1}s^{n-1} + \cdots + a_0) + (K_p e^{-sT_p} + sK_v e^{-sT_v})(b_m s^m + \cdots + b_0)} \tag{3}$$

In the experiment, we simplify the analysis by considering the case when $n = 1$, and $K_v = 0$. The closed loop poles corresponding to the homogenous solution, are the zeros of $Z(s) = s + a_0 + K_p b_0 e^{-sT_p}$. If the delay, $T_p = 0$, then the system has a closed loop pole at $-a_0 < 0$. For $T_p > 0$, there are an infinite number of closed loop poles. It is well known that time delays can introduce instability by shifting the poles to the right. Indeed, in this case, as T_p increases, the poles begin to cross the $j\omega$ axis two poles at a time. We will be interested in the first pair of poles to reach the $j\omega$ axis as this will result in resonance of the closed loop system. This resonant frequency, ω_c, will depend on the open loop parameters, the feedback gain, and the time delay.

Consider poles on the imaginary axis at $s = i\omega$ for $\omega \in \mathcal{R}$. Thus, ω must satisfy: $Z(i\omega) = i\omega + a_0 + K_p b_0 e^{-i\omega T_p} = 0$. At $\omega = \omega_c$, we have: $k \sin T_p\omega_c = \omega_c$ and $k \cos T_p\omega_c = -a_0$ for $T_p\omega_c \in \left[\frac{\pi}{2} + 2n\pi, \pi + 2n\pi\right]$ where $n = 0, 1, 2, \cdots$ and $k = K_p b_0$. If given a critical frequency, ω_c, and a_0, then we can determine the corresponding position feedback time delay, T_p, and feedback gain, K_p (assuming b_0) is known. Thus, our approach to determine the feedback gain is to randomly introduce time delays, and determine the critical frequency of oscillation.

2.3 Probing Internal Model Learning

Having determined the feedback gain, K, we proceed to determine the learning rate. We begin by considering in detail FEL applied to the first order linear plant. We will assume that the pole of the system is unknown. The open loop dynamics are given by: $\dot{y} + a_0 y = b_0 u$. The ideal inverse dynamics satisfy: $u^*_{ff} = \frac{\dot{r}}{b_0} + H(r)a_0$, where $H(r) = r/b_0$. We assume that the actual feedforward command is given by: $u_{ff} = \frac{\dot{r}}{b_0} + H(r)\hat{a}_0$, where $\hat{a}_0$ is an estimate of a_0. Applying the FEL algorithm (1), the update rule is given by

$$\dot{\hat{a}}_0 = \gamma H(r) K e \tag{4}$$

Since we do not know the subjects' learning rate, γ, we cannot directly solve for $\hat{a}_0$. However, conside the relation: $u = u_{fb} + u_{ff} = Ke + \frac{\dot{r}}{b_0} + H(r)\hat{a}_0$. Solving for $\hat{a}_0$, we get

$$\hat{a}_0 = \frac{u - Ke - \dot{r}/b_0}{H(r)} \tag{5}$$

Notice that we have all the neccesary signals to compute $\hat{a}_0$. We measure the signal, u, the input to the plant. Since we determined K in the previous step, we can compute the feedback term, Ke. We also know $\dot{r}/b_0$ and $H(r)$, since we implement b_0 and r.

Computing the Learning Bandwidth and Learning Rate: The learning rate, γ, in (4) reflects the ability of the feedforward component to track parameter changes in the plant. A high learning rate will allow the feedforward component to track plant parameters that change rapidly, however, will be more susceptible to noise. A lower learning rate will reject parameter variation due to noise, but may be unable to track fast parameter changes in the plant.

Thus, as an indirect measure of learning rate, we can observe the the ability of the estimated parameters to track the time-varying plant parameters. In our analysis, we will assume no noise is present in any signal loop. Since our goal is to analyze the performance of the parameter estimate (output) to time-varying changes in the plant parameters (input), we view the closed loop system as the input-output system depicted in Fig. 3.

As the system depicted in Fig. (3). is a fairly complicated linear, time-varying system, we approximate it by the frequency domain relation as

$$\hat{A}_0(j(\omega) - T(j\omega)A_0(j\omega) \tag{6}$$

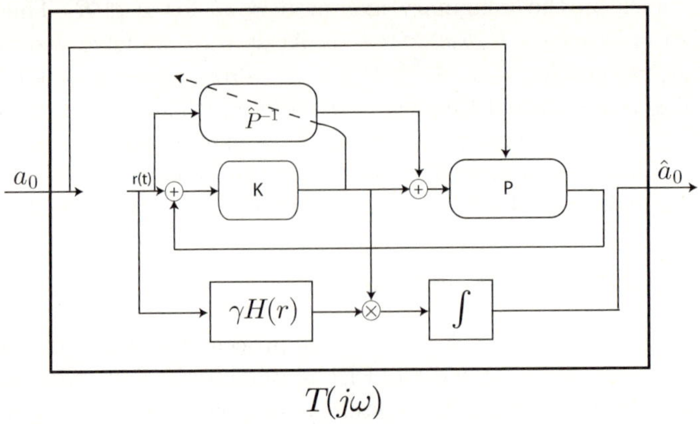

Fig. 3. In this figure, we recast the original framework to emphasize that we are viewing the plant parameter variation, $a_0(t)$, as the input to a linear, time-varying system where the parameter estimate, $\hat{a}_0(t)$, is considered as the output

That is, $T(j\omega)$ is the approximate bode plot of the closed loop system considering $a_0(t)$ as the input, and $\hat{a}_0(t)$ as the output. To compute $T(j\omega)$, we use sinusoidal inputs. Since the input is a sinusoid, the output will be given by $|\hat{A}_0(j\omega_i)| = |T(j\omega_i)|$, and thus we can construct a magnitude bode diagram. The procedure is summarized as follows:

1. Select the frequency, f_i, for the input sinusoid $a_0(t)$ in the system given in Fig. 3.
2. Compute the parameter estimate, $\hat{a}_0(t)$ using (5).
3. Compute the FFT of $\hat{a}_0(t)$ and evaluate the peak magnitude at f_i Hz.
4. Fit the Bode plot to $e^{-\frac{f}{\lambda}}$, where λ is defined to be the learning bandwidth.

There is a clear relationship between the learning bandwidth, λ, and learning rate, γ. The simulations suggests that a high learning rate results in a high learning bandwidth. Unfortunately, a closed form expression relating these two parameters is not easily determined for a general reference trajectory[1]. Nevertheless, we may compute numerically the learning rate as a function of learning bandwidth for the specific parameters implemented[2] in the experiment, namely $a_0(t), b_0$ and $r(t)$.

3 Experimental Results

To illustrate the methods, we show and discuss the input-output data of a typical subject during the different phases of the experiment.

[1] Asymptotic methods may be applicable when there exists a sufficient time-scale separation between the learning rate, plant dynamics, and reference trajectory.

[2] Note that the learning rate - learning bandwidth relationship also depends on the value of K which is determined, not implemented, by the methods described in section 2.2.

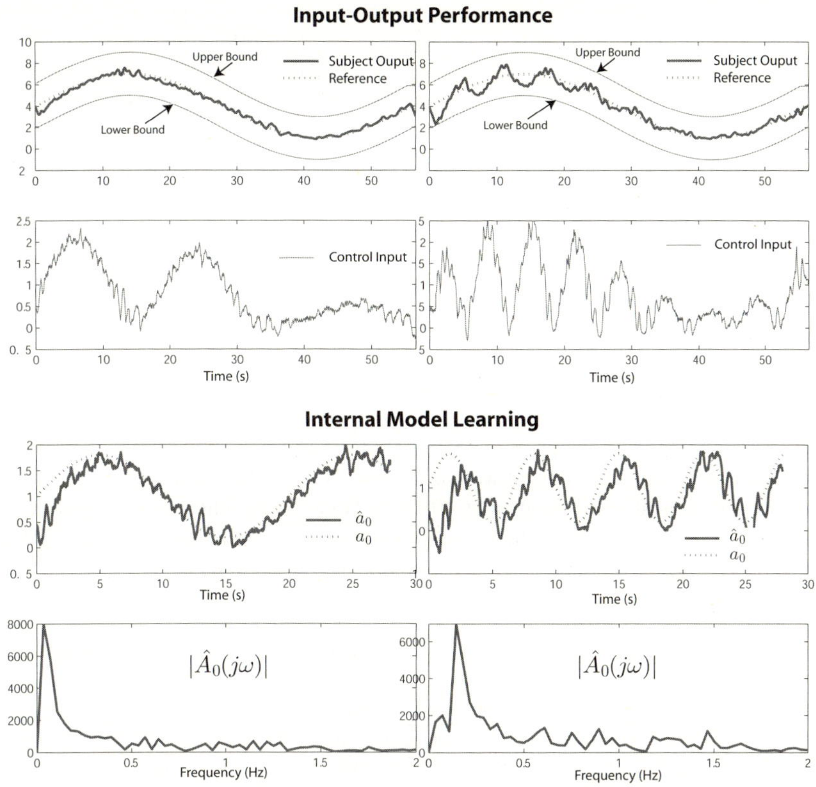

Fig. 4. The top left figure shows the input-output data for the case $f_i = 0.05$. In the top right figure, we show the subjects input-output data at $f_i = 0.15$ Hz. In the bottom left figure, we plot equation (5) using our estimate of the feedback gain. The plant pole was varied at $f_i = 0.05$ Hz. In the top subplot, we show the true parameter, a_0, in dotted, and the subject's estimated parameter, $\hat{a}_0$, in solid. In the bottom subplot, we compute the magnitude of the FTT of $\hat{a}_0$. We observe a peak magnitude at 0.05 Hz. In the bottom right figure, we plot equation (5) for the case, $f_i = 0.15$ Hz. In this case, we observe a decrease in parameter tracking performance. In the bottom subplot, a magnitude peak at 0.15 Hz is observed. However, notice that there are more additional frequencies present than in the previous case (bottom left). Also note that the peak amplitude at $f_i = 0.15$ Hz is lower than the peak amplitude at $f_i = 0.05$ Hz on bottom left.

Probing the Feedback Gain: For this subject, we found the critical time delay to be 300 ms corresponding to a frequency of instability of 0.625 Hz. At this delay, the observed frequency corresponded to a feedback gain of $K = 0.43$.

Probing the Learning Bandwidth: In Fig. 4., we show the subjects input-output data and parameter estimates at $f_i = 0.05$ and $f_i = 0.15$ Hz. The top left figure shows the input-output data for the case $f_i = 0.05$. The dotted line indicates the desired trajectory, while the solid line indicates the subject's performance.

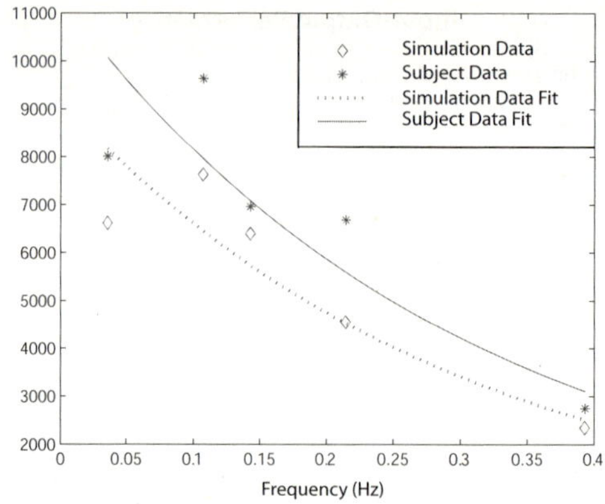

Fig. 5. Subject Bode Plot and Simulation Bode Plot Comparison: We observe a relatively good agreement between the two curves. Notice the similarity in the relative magnitude changes at each frequency. Subject Bode Plot: The peak magnitudes are indicated by stars. In solid, we fit the exponential, $Ce^{-\frac{f}{\lambda}}$, to determine the subjects learning bandwidth, λ. In this case, λ was found to be 0.30401.

Table 1. Table of feedback gain, learning bandwidth, and learning rate

Table of Results			
Subject	Feedback Gain	Learning Bandwidth	Learning Rate
A	0.43	0.304	0.642
B	0.43	0.3711	0.835
C	0.64	0.6542	1.15
D	0.73	0.73	1.2145
E	0.67	2.733	2.06

We also indicate the upper and lower bounds where points were deducted if the trajectory exceeded these bounds. Notice, that nearly perfect tracking is achieved while an oscillatory component of about 0.05 Hz is observed in the control signal. In the top right figure, we show the subjects input-output data at $f_i = 0.15$ Hz. In this case, we observe an increase in errors in the output as the subject is not able to compensate as well for the increase in frequency of the pole variation. The subjects control input reflect this variation, but also appears to contain frequencies which depend on the the increase in performance error. In the bottom left figure, we plot equation (5) using our estimate of the feedback gain. The plant pole was varied at $f_i = 0.05$ Hz. In the top subplot, we show the true parameter, a_0, in dotted, and the subject's estimated parameter, $\hat{a}_0$, in solid. In the bottom subplot, we compute the magnitude of the FTT of $\hat{a}_0$. We observe a peak magnitude at 0.05 Hz. In the bottom right figure, we plot equation (5) for the case, $f_i = 0.15$ Hz. In this case, we observe a decrease in parameter tracking performance. In the bottom subplot, a magnitude peak at 0.15 Hz is observed.

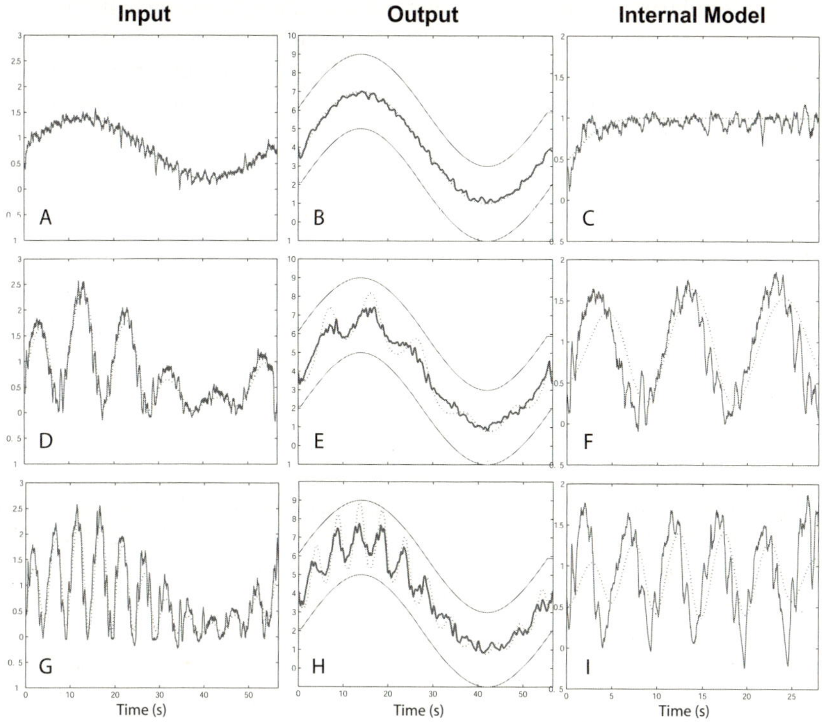

Fig. 6. Comparision of Subject A data with the FEL model based on the estimated parameters: In the left column, we plot the true input in solid and the simulated input in dotted. In the middle column, we plot the true output in solid, and the simulated output in dotted. In the far right column, we plot the subject's estimated parameter in solid using equation (5), and the parameter estimate obtained via simulation using equation (4) in dotted. The rows, (A-C), (D-F), and (G-I) correspond to pole frequency of 0 Hz, 0.1 Hz, and 0.2 Hz, respectively.

However, notice that there are more additional frequencies present than in the previous case (bottom left). Also note that the peak amplitude at $f_i = 0.15$ Hz is lower than the peak amplitude at $f_i = 0.05$ Hz on bottome left.

The Bode plot of $T(j\omega)$ in equation (6) is shown in Fig. 5. It was constructed by taking the peak magnitudes of $|\hat{A}_0(j\omega)|$ at the corresponding input frequencies, f_i. The peak magnitudes are indicated by stars. In solid, we fit the exponential, $Ce^{-\frac{f}{\lambda}}$, to determine the subjects learning bandwidth, λ. In this case, λ was found to be 0.30401. Having determines the subjects learning bandwidth, λ, and feedback gain, K, we can now numerically compute the feedforward learning rate. Given the specific learning bandwidth, γ is determined by linear interpolation between the data points. In this case, we found $\gamma = 0.64$.

Having determined all the required parameters in the FEL model, we can now go back and simulate equation (4) using the subject parameters K and γ. In Fig. 5., we compare the simulation Bode plot, and the subject's Bode plot.

Notice the similarity in the relative magnitude changes at each frequency. In Fig. 6., we compare the input-output data as well as the parameter estimates. That is, we compare the real subject data with the simulation based on the measured parameters determined previously. In the left column, we plot the subject's input to the plant, and the estimated input based on the measured learning rate and feedback gain. The true input is plotted in solid while the simulated input is plotted in dotted. In the middle column, we plot the true output in solid, and the simulated output in dotted. In the far right column, we plot the subject's estimated parameter in solid, and the parameter estimate obtained via simulation in dotted. Subplots (A-C) correspond to $f_i = 0$ Hz, (D-F) correspond to $f_i = 0.1$ Hz, and (G-I) correspond to $f_i = 0.2$ Hz.

4 Conclusion

In this study, we characterized feedforward motor learning by probing the human neuro-controller in an unknown time-varying environment. We assumed a simple yet powerful adaptive model known as Feedback Error Learning. We utilized a novel time delay analysis technique to estimate the subjects' feedback gains during a step tracking task. With the estimated feedback gains, we were able to construct a learning bandwidth by probing the system at various frequencies. An approximate Bode plot was constructed and used to determine the bandwidth. The learning rate was then numerically computed from the learning bandwidth. To validate our results, we simulated the FEL model with the estimated parameters compared the results with the subject's actual performance.

References

1. Kawato, M.: Internal models for motor control and trajectory planning. Current Opinion in Neurobiology 9, 718–727 (1999)
2. Kawato, M., Furukawa, K., Suzuki, R.: A Hierarchical Neural-Network Model for Control and Learning of Voluntary Movement. Biological Cybernetics 57, 169–185 (1987)
3. Ito, M.: Mechanisms of motor learning in the cerebellum. Brain Res. 886, 237–245 (2000)

Spatial and Temporal Selectivity of Hippocampal CA3 and Its Contribution to Sequence Disambiguation

Toshikazu Samura[1], Motonobu Hattori[2], and Shun Ishizaki[1]

[1] Graduate School of Media and Governance, Keio University
5322 Endo, Fujisawa-shi, Kanagawa, 252–8520, Japan
{samura,ishizaki}@sfc.keio.ac.jp
[2] Interdisciplinary Graduate School of Medicine and Engineering,
University of Yamanashi
4–3–11 Takeda, Kofu-shi, Yamanashi, 400–8511, Japan
m-hattori@yamanashi.ac.jp

Abstract. Many episodes are acquired in the hippocampus. An episode is expressed by a sequence of elements that are perceived in an event. Episodes are associated each other by events that contain information shared among the episodes. Sequences must be recalled individually, even if the sequences are overlapped at some representations. Therefore, sequence disambiguation is an essential function to dissociate overlapped sequences. In this study, we especially focus on the location-dependencies of the STDP effects on synaptic summation and the expression of AMPA receptor. We firstly show that the hippocampal CA3 is divided into two regions in which one region has spatial selectivity and the other has temporal selectivity. Moreover, we confirm that the divided CA3 could generate a code for sequence disambiguation in computer simulations. Consequently, we suggest that the CA3 can be divided into two regions characterized by their selectivity, and the divided CA3 contributes to sequence disambiguation.

1 Introduction

Eichenbaum suggested that daily episodes are memorized as a relational network in the hippocampus [1]. In the relational network, an episode is expressed by a sequence of elements that are perceived in an event. Episodes are associated each other by events that contain information shared among the episodes. For example, let the hippocampus compose a simple relational network from these two episodes, one is composed of events: A, B and C (A→B→C), the other is composed of D, B and E (D→B→E), where the event B associates the two episodes (C and E). Then, it is difficult to decide which pattern (C or E) should be retrieved from event B. Such ambiguity of sequences becomes a problem for retrieval. Thus, sequence disambiguation is an essential function for retrieving original episodes.

J. Marques de Sá et al. (Eds.): ICANN 2007, Part II, LNCS 4669, pp. 49–58, 2007.

In the hippocampus, the CA3 region has unique recursive axons that are called recurrent collaterals (RCs). Because of its uniqueness, many researchers focused on it and proposed many computational models of the CA3. Samura et al. suggested that the CA3 can be divided into autoassociative and heteroassociative memory [2] and the functional division of the CA3 contributes to sequence disambiguation [3]. They derived the hypothesis from location-dependencies of RCs and Spike-Timing Dependent Plasticity (STDP), which is a rule of changing synaptic weights. The former is that the projection of RCs differs according to the subregional location of a neuron (CA3a, b and c) [4]. The later is that the profiles of STDP differ depending on dendritic location of a synapse [5].

In addition to the above location-dependencies, we also incorporate two location-dependencies into our study. The first one is that the effect of STDP on synaptic summation differs according to the dendritic location of a synapse [6]. The second one is that the expression of AMPA receptor (AMPAR),which mediates fast synaptic transmission, also shows the dendritic location-dependency [7]. The new location-dependencies affect the selectivity of a neuron to inputs. In view of all location-dependencies, neurons, which have spatial or temporal selectivity to inputs, concentrate in the specific subregions of the CA3. Consequently, we suggest that the CA3 can be divided into two regions where there are spatial or temporal selectivity, rather than autoassociative and heteroassociative memory. Moreover, we show that the divided CA3 contributes to sequence disambiguation by computer simulations.

2 Anatomical and Physiological Backgrounds of Hippocampus

2.1 Structure of Hippocampal CA3

The hippocampus is divided into three regions: Dentate Gyrus (DG), CA3 and CA1. The CA3 is segmented into three subregions: CA3a (nearer CA1), CA3b and CA3c (nearer DG) (Fig.1(a)). The CA3 connects with DG and Entorhinal Cortex (EC) that works as an interface between the cortex and the hippocampus. DG connects to all CA3 subregions and EC connects to only CA3a and CA3b (Fig.1(a))[8].

2.2 Subregional Location-Dependencies

The CA3 neurons are connected recursively to other neurons by RCs. Fig.1(b) shows the relationship between the location of a neuron and the projection of its RCs [4]. First, the RCs of CA3c neurons are limited to the area surrounding them. Second, the RCs of CA3b neurons are widely spread. Projections onto CA3c become more temporal locations than their sources. Conversely, those onto CA3a become more septal locations. Finally, the RCs of CA3a neurons are limited to CA3a and CA3b. Projections onto CA3b become more temporal locations. In addition to the location-dependency of the projection, the dendritic locations of RCs depend on the relative positions between pre- and postsynaptic

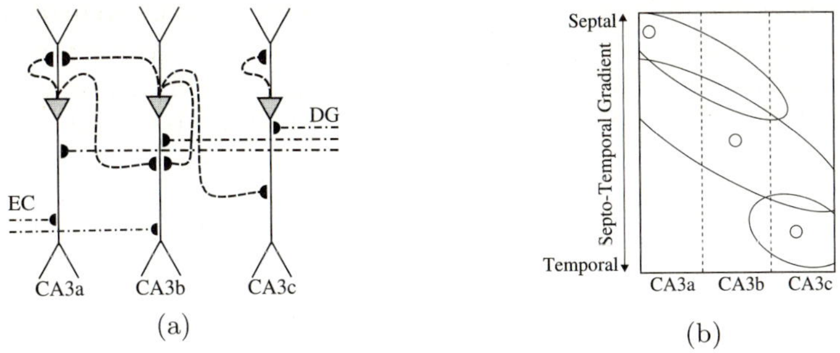

Fig. 1. Structure of the CA3. (a) Connections within the CA3 and its dendritic locations (inverse triangle : soma, forked line: dendrite, dashed line: RCs, chain line: connections from external regions). (b) Projections of RCs (circle: source neuron, ellipse: projection of circled neuron in it).

neurons [4]. As shown Fig.1(a), CA3a and CA3b neurons tend to receive RCs near a soma, while CA3c neurons tend to receive them remote from a soma. However, they receive RCs from CA3c near a soma.

2.3 Dendritic Location-Dependencies

In the hippocampus, STDP was observed as a rule of changing synaptic weights. STDP determines the magnitude of a synaptic change and its polarity (LTP:long-term potentiation or LTD:long-term depression) according to an interval between pre- and postsynaptic spikes. Additionally, recent study suggested that STDP turns asymmetric profile to symmetric one depending on the density of inhibitory interneurons [5]. Symmetric profile STDP (SSTDP) was observed from a high-density area near a soma, while asymmetric profile STDP (ASTDP) was observed from a low-density area remote from a soma. Therefore, the change of STDP profile correlates with a synaptic location in a dendrite. In other words, STDP shows the dendritic location-dependency.

Furthermore, STDP(in fact LTP or LTD) location-dependently modulates synaptic summation. Generally, synaptic summation is divided into two types. The one is spatial summation under which a neuron can fire when inputs arrive coincidentally. The other is temporal summation under which a neuron can fire without simultaneous inputs. Xu et al. suggested that distal dendrite enhances their activity when inputs arrive within a narrow time window ($<$ 5ms) after LTP induction. While proximal dendrite enhances their activity when inputs arrive within a long time window ($<$ 20ms) after that [6]. Consequently, after repeat of LTP, distal dendrite shows spatial summation and proximal one shows temporal summation.

Moreover, the expression of AMPAR also shows dendritic location-dependency and supports the dendritic location-dependency of synaptic summation. It was suggested that AMPAR expression becomes higher in distal stratum radiatum than proximal one, but the expression of NMDA receptor (NMDAR) show no location-dependency [7]. These receptors mediate EPSP. However, they differ in the time constant of EPSP. The EPSP time constant of AMPAR is shorter than the one of NMDAR. Then, the time constant of EPSP relates to the summation type of a synapse. A short time constant fits to spatial summation and a long time one fits to temporal summation. Thus, the predominance of AMPAR in distal dendrite means that distal dendrite suits for spatial summation, while the inferior of AMPAR in proximal dendrite suits for temporal summation. Additionally, the time constant of AMPAR and NMDAR, especially their rising time of EPSP (AMPAR: $\sim$ 5ms, NMDAR: 8 $\sim$ 20ms) [9] is similar to the above time window. Therefore, these findings are consistent with the dendritic location-dependency of synaptic summation.

3 Spatial and Temporal Selectivity in Hippocampal CA3

Firstly, the type of synaptic summation affects temporal tendency to fire a neuron. Under the temporal summation, neurons can fire when they receive inputs within a long time windows. Conversely, under the spatial summation, neurons can fire when they receive inputs within a short time window.

Moreover, the profiles of STDP also affects tendency to fire a neuron. Under SSTDP, simultaneous firing leads to potentiation, and time lag leads to depression. Neurons firing simultaneously are mapped onto synaptic weights as a firing pattern of a network. Thus, when a postsynaptic neuron receives inputs from neurons that compose the same firing pattern as the postsynaptic neuron, the postsynaptic one is likely to be activated regardless of the firing order of these neurons. In contrast, ASTDP potentiates synapses when postsynaptic neurons fire after presynaptic firings. Conversely, if their firing orders are reversed, synapses between them are depressed. As a result, synaptic weights reflect the order of firing. Therefore, when a neuron receives inputs from presynaptic neurons through potentiated synapses in the memorized order, the neuron is likely to be activated.

Here, we integrate the location-dependencies of the CA3. In CA3a and CA3b where neurons receive RCs at proximal dendrites, temporal summation and SSTDP coexist. Thus, neurons in CA3a and CA3b can fire when they receives inputs from the memorized set of neurons regardless of the coincidence and order of inputs. These regions are sensitive only to spatial information of inputs.While spatial summation and ASTDP coexist in CA3c where neurons receive RCs at distal dendrites, neurons in CA3c can fire when they coincidentally receives inputs from the memorized set of neurons in the memorized order. This region is sensitive to temporal information of inputs. Consequently, the CA3 region is divided into two regions where there are the spatial or temporal selectivity to inputs.

4 Hippocampal CA3 Model

4.1 Neuron Model

We revised a simple model that was suggested by Izhikevich [10] and the proposed hippocampal CA3 model consists of the neuron models. The following equation shows the membrane potential of the ith CA3 neuron at time t.

$$C\frac{dv_i(t)}{dt} = k\big(v_i(t) - v_{\mathrm{r}}\big)\big(v_i(t) - v_{\mathrm{t}}\big) - u_i(t) + I_i(t), \tag{1}$$

where C is the membrane capacitance, v_{r} is the resting membrane potential, v_{t} is the instantaneous threshold potential, u is the recovery current, and I denotes the sum of excitatory postsynaptic potential (EPSP) evoked by inputs. The recovery current of the ith neuron at time t is defined as follow:

$$\frac{du_i(t)}{dt} = a\{b(v_i - v_{\mathrm{r}}) - u_i(t)\}, \tag{2}$$

where a is the recovery time constant, b is the effect of u on v. When the membrane potential exceeds the threshold v^{peak}, the cell fires and the membrane potential is reset to c and the recovery current is also reset to $u_i(t) + d$, where d means the total amount of outward minus inward currents. The total EPSP of the ith neuron at time t is defined as follow:

$$I_i(t) = E_i^{\mathrm{DG \cdot EC}}(t) + E_i^{\mathrm{CA3}}(t), \tag{3}$$

where, $E_i^{\mathrm{DG \cdot EC}}(t)$ and $E_i^{\mathrm{CA3}}(t)$ are the sum of EPSP evoked by inputs during a period from the last spike timing of the ith neuron to present time t. They are calculated as follows:

$$E_i^{\mathrm{DG \cdot EC}}(t) = \Sigma_k w^{\mathrm{DG \cdot EC}} \varepsilon(t - t^k), \tag{4}$$

where $w^{\mathrm{DG \cdot EC}}$ is the synaptic weight from DG or EC to CA3, $\varepsilon(t - t^k)$ denotes the present amplitude of EPSP evoked by the kth spike during the period, t^k is the spike timing of the kth spike.

$$E_i^{\mathrm{CA3}}(t) = \Sigma_j \Sigma_k w_{ij}(t_j^k + \delta_{ij}) \varepsilon(t - t_j^k - \delta_{ij}), \tag{5}$$

where $w_{ij}(t_j^k + \delta_{ij})$ means the synaptic weight of RC between the ith and the jth neuron when the kth spike of the jth neuron at time t_j^k arrived at the ith neuron with axonal delay δ_{ij}, $\varepsilon(t - t_j^k - \delta_{ij})$ is the present amplitude of EPSP evoked by the kth spike of the jth neuron during the period. The following equation shows the amplitude of single EPSP at elapsed time t' since a spike arrived at a neuron,

$$\varepsilon(t') = \frac{\alpha}{\tau} t' \exp\left(\frac{-t'}{\tau}\right), \tag{6}$$

where α is the amplitude of EPSP and τ is the time constant of EPSP. These parameters differ according to the receptor type of a synapse.

4.2 Synaptic Formation

First of all, as shown in Fig.2(a), we consider CA3 as two dimensional map and defined x-axis and y-axis as CA3a→CA3c direction (the maximum value is W) and septal→temporal direction (the maximum value is H) respectively. We assume that neurons are located on each x–y integral coordinate. Then, the existing probability of a synapse between presynaptic neuron (x_i, y_i) and postsynaptic one (x_j, y_j) is given by equation below.

$$P = \exp\left(\frac{\{(y_j - y_i)Cos\theta + (x_j - x_i)Sin\theta\}^2}{\iota R(x_i)} + \frac{\{(x_j - x_i)Cos\theta + (y_j - y_i)Sin\theta\}^2}{\kappa R(x_i)} \right), \tag{7}$$

where $R(x)$ is the projection range defined according to x-coordinate of presynaptic neuron. It is given as follow:

$$R(x) = \lambda_{\min} + \lambda_{\max}\left(\frac{1}{1+\exp(x - 1.5W)}\right), \tag{8}$$

where $\lambda_{\min}$ is the minimum range and $\lambda_{\max}$ is the maximum range.

As shown Fig.2, a neuron projects its RCs to other neurons (Fig.2(c)) on the basis of the existing probability (Fig.2(b)). Then, the synaptic weights are randomly set as $0 < w_{ij} \leq w_{\text{init}}$. Moreover, the axonal delay and the receptor type of each synapse are defined. The axonal delay of a synapse between presynaptic neuron (x_i, y_i) and postsynaptic one (x_j, y_j) is calculated as follow:

$$\delta_{ij} = 1.0 + \delta_{\max}\sqrt{\frac{(x_i - x_j)^2 + (y_i - y_j)^2}{H^2 + W^2}}, \tag{9}$$

where $\delta_{\max}$ is the maximum delay. The receptor type of a synapse is defined according to the dendritic location of a synapse. In this study, we simply suppose that AMPAR is predominant in distal dendrite and NMDAR is predominant in proximal one. Thus, the receptor type of RCs is set as NMDAR (τ_{NMDA}) in the CA3a ($x_j \leq W/3$) and CA3b ($W/3 < x_j \leq W/1.5$). Conversely, the receptor type of RCs is set as AMPAR (τ_{AMPA}) in the CA3c ($W/1.5 < x_j$), but the receptor type of CA3c-CA3c connections is set as NMDAR (τ_{NMDA}).

4.3 Learning Rules

Each synaptic weight of RCs is changed by ASTDP or SSTDP. Spike interval Δt between the ith postsynaptic neuron and the jth presynaptic neuron is given by

$$\Delta t_{ij} = (T_i - T_j) - \eta, \tag{10}$$

where T_i and T_j denote the spike time of the ith postsynaptic neuron and that of the jth presynaptic one, respectively. η is defined in consideration of the activity of receptors that underlies STDP.

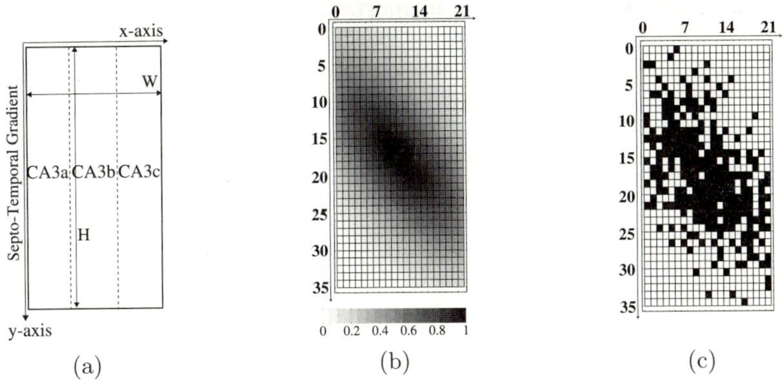

Fig. 2. Formation of RCs. (a) Definition of x–y axis. (b) Existing probability of sypase from the neuron (11,18) on the map (21×35) (*cell: synapse, gray scale: the probability*). (c) Existence of synapses from the neuron (11,18) (*black cell: existence of synapse*).

In this study, we employ the *semi-nearest-neighbor* manner for pairing spikes [11]. That is, for each presynaptic spike, we consider only one preceding postsynaptic spike and ignore all earlier spikes. All postsynaptic spikes subsequent to the presynaptic spike are also considered. For each pre-/postsynaptic spike pair, the synaptic weight is updated as follows:

$$\Delta w_{ij} = \beta \left\{ 1.0 - \gamma \left(0.12 \Delta t_{ij}\right)^2 \right\} \exp\left(\frac{-\left(0.12 \Delta t_{ij}\right)^2}{2} \right), \tag{11}$$

$$w_{ij}\left(t + \Delta t\right) = w_{ij}\left(t\right) + \Delta w_{ij}, \tag{12}$$

where β is the maximum modification width, γ shows the time constant of STDP. In CA3c, a synapse is updated between the ith postsynaptic neuron and the jth presynaptic one by ASTDP, the constant is defined by

$$\gamma = \begin{cases} 0.01 & \Delta t_{ij} \geq 0 \\ 0.65 & \Delta t_{ij} < 0. \end{cases} \tag{13}$$

While a synapse is updated by SSTDP in CA3a and CA3b, the constant is always set to 0.65. In this study, if the total synaptic weight of a neuron exceeds $w_{\max}$, the neuron suspends potentiation of its synapses. After the suspension, if the total synaptic weight of the neuron falls below $w_{\max}$, potentiation is resumed. Conversely, a synaptic weight becomes less than $w_{\min}$, it is set to $w_{\min}$.

4.4 Learning Phase and Retrieving Phase

We defined two phases (learning and retrieving) for this model. The model learns the input sequences by changing synaptic weights during learning phase. Then, the model suspends calculating equation (5) for memorizing smoothly. On the other hands, the model retrieves memorized patterns from inputs during retrieving phase. The connections from DG to CA3 contribute to memorization, while those from EC to the CA3 are required for retrieval and EC connects only to CA3a and CA3b [12]. Thus, during this period, inputs are limited to them. Furthermore synaptic modification is suspended. During both phases, input sequences are applied to the model by the conventional input procedure [3].

5 Computer Simulations

5.1 Conditions

In this simulation, we set parameters as shown in Table 1 and constructed the proposed model from 735 neuron models. Next, we defined 9 fixed patterns (A–I) and random patterns (*). Each fixed pattern was represented by the activation of 60 neurons and there were no overlap among them. The random patterns were represented by the activation of 5% neurons selected randomly. Using above patterns, we defined two overlapped sequences (sequence I and II). The sequence I was *→A→B→C→D→E→* and the other was *→F→G→C→H→I→*. Each sequence was applied to the model three times. Then the model memorized them (learning phase). Following the memorization, we applied a part of each sequence:*→A→B→C or *→F→G→C into the model (retrieving phase). Then, we confirmed that the model could discriminate between the pattern C of the sequence I and the pattern C of the sequence II by using a difference between the sequences. For the confirmation, we evaluated a similarity given by the direction cosine between the model output and a fixed pattern in each subregion.

Table 1. Parameters for the simulation

W	21	H	35	k	1.75	v_r	-55.0	v_t	-40.0	C	80
a	0.021	b	-1.7	c	-38.0	d	190.0	v^{peak}	10.0	w_{min}	1.0×10^{-7}
w_{max}	7.1	$w^{\mathrm{DG \cdot EC}}$	6.0	θ	0.25π	ι	30	κ	6	λ_{min}	0.5
λ_{max}	2.5	η_{CA3a}	0	η_{CA3b}	0	η_{CA3c}	10	δ_{max}	10	β	0.08
α_{NMDA}	66.0	τ_{NMDA}	5.0	α_{AMPA}	72.5	τ_{AMPA}	1.5				

5.2 Results

Figs. 3(a) and (b) show the similarity in each sequence. At the beginning of the cycle, random pattern was inputted to CA3a and CA3b. After that, next patterns were applied to them every 10 unit times in the order of each sequence. Although each pattern was applied only once in the cycle, as shown in these figures, they showed periodic activation of fixed patterns in CA3a and CA3b. Then, we compared the similarity of CA3c output in two sequences. As shown in the Fig. 3(a), when pattern C was applied to the model, CA3c outputted pattern D. When pattern C of sequence II was applied, CA3c outputted pattern H (Fig. 3(b)). This means that the proposed model generated different activities

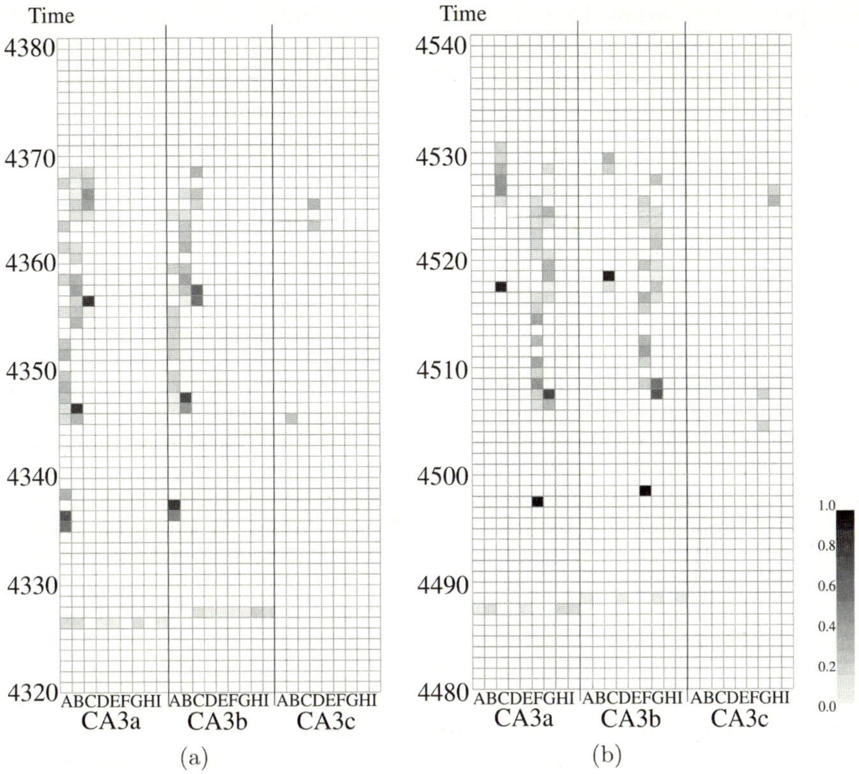

Fig. 3. Similarity in each subregion. Gray level of each cell means similarity between a retrieved pattern and a source pattern. (a) The similarity in the sequence I. (b) The similarity in the sequence II.

according to the differences between the sequences in spite of the same pattern C. The differences become a code for sequence disambiguation.

6 Conclusion

In this paper, we have focused on the location-dependencies elucidated from the anatomical findings and the physiological findings. On the basis of the findings, we have firstly suggested that CA3a and CA3b show temporal summation with SSTDP, which fits to spatial selectivity, while CA3c shows spatial summation with ASTDP, which fits to temporal selectivity. Consequently, we have suggested that the CA3 is divided into two regions characterized by their selectivity. Moreover, we have shown that the divided CA3 could generate a code for sequence disambiguation in the computer simulation. In the proposed model, previously inputted patterns were periodically retrieved in CA3a and CA3b. Thus, the model could buffer the differences between sequences. Then the information in the buffer was transmitted to CA3c through the connections, which are sensitive to temporal information. Therefore, the difference in the buffer caused the difference of retrieved pattern in CA3c. In other words, a code which dissociates same pattern in sequences was generated by the divided hippocampal CA3 model according to the difference of previous inputs. Consequently, we have suggested that the hippocampal CA3 is divided into two regions in which one region has spatial selectivity and the other has temporal selectivity and the divided CA3 contributes to sequence disambiguation.

References

1. Eichenbaum, H.: Hippocampus: Cognitive Processes and Neural Representations that Underlie Declarative Memory. Neuron 44, 109–120 (2004)
2. Samura, T., Hattori, M., Ishizaki, S.: Autoassociative and Heteroassociative Hippocampal CA3 Model Based on Physiological Findings. Abstracts of BrainIT2006, 62 (2006)
3. Samura, T., Hattori, M., Ishizaki, S.: Sequence Disambiguation by Functionally Divided Hippocampal CA3 Model. In: King, I., Wang, J., Chan, L., Wang, D. (eds.) ICONIP 2006. LNCS, vol. 4232, pp. 117–126. Springer, Heidelberg (2006)
4. Ishizuka, N., Weber, J., Amaral, D.G.: Organization of Intrahippocampal Projections Originating From CA3 Pyramidal Cells in the Rat. J. Comp. Neurol. 295, 580–623 (1990)
5. Tsukada, M., Aihara, T., Kobayashi, Y., Shimazaki, H.: Spatial Analysis of Spike-Timing-Dependent LTP and LTD in the CA1 Area of Hippocampal Slices Using Optical Imaging. Hippocampus 15, 104–109 (2005)
6. Xu, N., Ye, C., Poot, M., Zhang, X.: Coincidence Detection of Synaptic Inputs Is Facilitated at the Distal Dendrites after Long-Term Potentiation Induction. J. Neurosci. 26, 3002–3009 (2006)
7. Nicholson, D.A., Trana, R., Katz, Y., Kath, W.L., Spruston, N., Genisman, Y.: Distance-Dependent Differences in Synapse Number and AMPA Receptor Expression in Hippocampus CA1 Pyramidal Neurons. Neuron 50, 431–442 (2006)
8. Ishizuka, N., Maxwell, W., Amaral, D.G.: A Quantitative Analysis of the Dendritic Organization of Pyramidal Cells in the Rat Hippocampus. J. Comp. Neurol. 362, 17–45 (1995)
9. Hestrin, S., Nicoll, R.A., Perkel, D.J., Sah, P.: Analysis of Excitatory Synaptic Action in Pyramidal Cells Using Whole-Cell Recording from Rat Hipocampal Slices. J. Physiol. 422, 203–205 (1990)
10. Izhikevich, E.M.: Dynamical Systems in Neuroscience: The Geometry of Excitability and Bursting. The MIT press, Cambridge (2007)
11. Izhikevich, E.M., Desai, N.S.: Relating STDP to BCM. Neural Comput. 15, 1511–1523 (2003)
12. Treves, A., Rolls, E.T.: Computational Constraints Suggest the Need for Two Distinct Input Systems to the Hippocampal CA3 Network Hippocampus 2, 189–200 (1992)

Lateral and Elastic Interactions: Deriving One Form from Another

Valery Tereshko

School of Computing, University of Paisley, Paisley PA1 2BE, Scotland
valery.tereshko@paisley.ac.uk
http://cis.paisley.ac.uk/tere-ci0/

Abstract. Lateral and elastic interactions are known to build a topology in different systems. We demonstrate how the models with weak lateral interactions can be reduced to the models with corresponding weak elastic interactions. Namely, the batch version of soft topology-preserving map can be rigorously reduced to the elastic net. Owing to the latter, both models produce similar behaviour when applied to the TSP. Unlike, the incremental (online) version of soft topology-preserving map is reduced to the cortical map only in the limit of low temperature, which makes their behaviours different when applied to the ocular dominance formation.

1 Introduction

Competitive learning neural nets that utilize lateral interactions to perform a mapping from the stimulus space to the response space with preserving neighbourhood relations are called topology-preserving maps [1]. Well-known example of the above is Kohonen's self-organizing map that became a standard unsupervised learning algorithm [2].

It is known that elastic synaptic interactions can forge the topology as well. An elastic net was first applied to solve the travelling salesman problem (TSP) [3]. Another application of elastic synaptic interactions is the preservation of topology in cortical mappings [4, 5, 6].

We already demonstrated the benefits of using both lateral and elastic interactions for controlling the receptive field patterns [5,6]. In [7], we considered the model utilizing only lateral interactions, which, unlike elastic ones, are biologically plausible, and applied it to the problems previously solved only with elastic interactions. We proved that cortical map and elastic net can be derived from the incremental (online) and batch soft topology-preserving map respectively [8]. Applied to the TSP, the equivalence of the batch topology-preserving map and the elastic net was demonstrated.

In this paper, we consider the relations between lateral and elastic interactions further. First, we derive the free energy function for an unsupervised net of stochastic neurons with lateral interactions. The temperature incorporated in this function serves as control parameter in the annealing schedule. Then, we

J. Marques de Sá et al. (Eds.): ICANN 2007, Part II, LNCS 4669, pp. 59–68, 2007.

consider the incremental and batch modes of learning resulting in corresponding versions of soft topology-preserving mapping. The mapping utilizes only weak lateral interactions that can be, therefore, approximated by the nearest-neighbour ones. Considering the weight vector of a neuron as a "particle" moving in the space-time of imposed patterns, we decompose this particle trajectory over these patterns. Using the decomposition for incremental and batch modes of soft topology-preserving map, we derive the cortical map and the elastic net respectively. We show that the batch version of soft topology-preserving map is rigourously reduced to the corresponding elastic net. Unlike, the incremental version of soft topology-preserving map is reduced to the cortical map only in the low temperature limit. We tested the models on the relevant to them tasks: the TSP and the development of visual cortex topology, namely the formation of retinotopy and ocular dominance. The difference in derivation of the latter systems results into the difference in their behaviour: the batch soft topology-preserving map and the elastic net produce similar outputs whereas the incremental soft topology-preserving map and the cortical map behave differently.

2 Topology-Preserving Maps

We consider a one-dimensional net of n stochastic neurons trained by N patterns. The energy of this net, for a given stimulus, is

$$E_i(\mu) = \frac{1}{2}\sum_{j=1}^{n} h_{ij}|\boldsymbol{x}_\mu - \boldsymbol{w}_j|^2, \tag{1}$$

where $\boldsymbol{x}_\mu$ is a given sample pattern, $\boldsymbol{w}_j$ are the weight vectors, and $h(i,j)$ is the neighbourhood function.

Throughout, we consider weak, quickly decaying in the space, lateral interactions. The latter give us the opportunity to consider nearest-neighbour interactions only:

$$h_{ij} = \begin{cases} 1, & i = j; \\ \gamma, & |i-j| = 1; \\ 0, & |i-j| \geqslant 2, \end{cases} \tag{2}$$

with $0 < \gamma < 1$.

Instead of the "hard" assignment of Kohonen's original algorithm with an unique winner, we assume a "soft" assignment where every i-th neuron is assigned to a given μ-th pattern with a probability $p_i(\mu)$; $\sum_i p_i(\mu) = 1$ [9,10,5,6,7].

The assignment probabilities minimizing free energy of the system (that is a composite of the averaged energy and thermal noise energy) are found to be

$$p_i(\mu) = \frac{\mathrm{e}^{-\beta E_i}}{\sum_{k=1}^{n} \mathrm{e}^{-\beta E_k}}, \tag{3}$$

which gives the minimal free energy [9, 10, 5, 6, 7]

$$F(\mu) = -\frac{1}{\beta}\ln\Big(\sum_{i=1}^{n} \mathrm{e}^{-\beta E_i}\Big). \tag{4}$$

Incremental (online) learning strategies are derived through the steepest descent minimization of function (4). The dynamics follows the free energy gradient, which result in the soft topology-preserving mapping [5, 6, 7]:

$$\Delta \boldsymbol{w}_j = -\eta \frac{\partial F}{\partial \boldsymbol{w}_j} = \eta \sum_{i=1}^{n} p_i(\mu) h_{ij} (\boldsymbol{x}^{\mu} - \boldsymbol{w}_j). \tag{5}$$

Soft mapping is based on soft competition, which allows all neurons to adjust their weights with probabilities proportional to their topographic distortion. This makes the weights move more gradually to the presented patterns. The strength of the competition is adjusted by a temperature. The underlying mechanism, deterministic annealing, is derived from statistical physics: it mimics an ordering process during a system cooling. At high temperatures, the competition is weak and the original energy landscape is smoothed by the noise, which helps to eliminate local minima at the beginning of the ordering phase. On reducing the temperature, the competition becomes stronger, the smoothing effect gradually disappears, and the free energy landscape resembles the original one.

At low temperatures ($\beta \to \infty$), equation (5) reduces to Kohonen's map with only nearest-neighbour interactions:

$$\Delta \boldsymbol{w}_j = -\eta \frac{\partial F}{\partial \boldsymbol{w}_j} = \eta h_{jj^*} (\boldsymbol{x}^{\mu} - \boldsymbol{w}_j), \tag{6}$$

where j^* is the winning unit.

The batch learning mode, when the updating rule is averaged over the set of training patterns before changing the weights, gives the following free energy:

$$\langle F \rangle = -\frac{1}{\beta N} \sum_{\mu=1}^{N} \ln \Big(\sum_{i=1}^{n} \mathrm{e}^{-\beta E_i} \Big). \tag{7}$$

Minimization of energy (7) results in the batch version of soft topology-preserving map:

$$\Delta \boldsymbol{w}_j = -\eta \frac{\partial \langle F \rangle}{\partial \boldsymbol{w}_j} = \frac{\eta}{N} \sum_{\mu=1}^{N} \sum_{i=1}^{n} p_i(\mu) h_{ij} (\boldsymbol{x}_{\mu} - \boldsymbol{w}_j), \tag{8}$$

where η is the learning rate.

At low temperatures ($\beta \to \infty$), (8) reduces to the batch mode of the Kohonen map. Goodhill applied the latter model with the special lateral interaction function to modelling the formation of topography and ocular dominance in the visual cortex [11].

3 Cortical Maps

Neural receptive fields of visual systems are ordered. The projections from retina to optic tectum (in lower vertebrates), and from retina to lateral geniculate nucleus, then to primary visual cortex (in mammals) are topographic. The latter means that neighbouring point in the retina are mapped to neighbouring points in the cortex (tectum). The development of such continuous topographic mapping is called retinotopy. This order guarantees improvement of the recognition abilities and, hence, facilitates the species survival. Indeed, without ordering slight external or internal (neural) noise can results in absolutely unpredictable (and possibly completely wrong) outcome, whereas the ordered receptive field guarantees the recognition of a prototype that is, perhaps, not exactly the same, but very similar to the stimulus. The development of neural receptive fields in a way that they mimic stimulus distribution and become ordered is, thus, biologically meaningful. Together with retinotopy, ocular dominance and orientation preference are developed.

Mammalian primary visual cortex is naturally binocularly innervated. During development of many, though not all, mammalian species, each part of the visual cortex becomes more densely innervated by one eye and less densely innervated by the other. Eventually, so-called ocular dominance stripes, that are reminiscent of the zebra stripe pattern, are developed. Moreover, exact details of the stripes (their shape, spacing of the pattern, etc) are determined dynamically during development rather than by genetics.

During development the visual cortex cells become largely respond to some preferred orientations. Like ocular dominance, orientation selectivity forms its own pattern: cells with the same orientation preference group to the same domain.

The idea of cortex as a dimension-reducing map from high-dimensional stimulus space to its two-dimensional surface has proved to be fruitful [12, 4]. The backward projection of each position on the cortex sheet to the position in stimulus space is a convenient way to consider cortex self-organization — the way in which it fills stimulus space defines the receptive field properties. Performing such a mapping induces two conflicting tendencies: (i) the cortical surface should pass through the representative points in stimulus space; (ii) the area of the sheet should be kept a minimum. This ensures the formation of smooth receptive fields and, hence, the minimal "wiring" interconnecting the cortical cells, which, in turn, ensures the closeness of the cortical cells representing similar stimuli. The stripes and patches seen within cortical areas have been argued to be adaptations that allow the efficient wiring by such structures [13].

In cortical mappings, the topological order usually develops by elastic synaptic interactions [4]. Let us derive cortical map from a topology-preserving map.

Taking the Taylor series expansion (in power of γ) results in

$$F = -\frac{1}{\beta}\ln\sum_{i=1}^{n}\exp\big(-\frac{\beta}{2}|\boldsymbol{x}_\mu - \boldsymbol{w}_i|^2\big) + \frac{\gamma}{2}\sum_{i=2}^{n-1} p_i(\mu)\big(|\boldsymbol{x}_\mu - \boldsymbol{w}_{i-1}|^2 + |\boldsymbol{x}_\mu - \boldsymbol{w}_{i+1}|^2\big). \quad (9)$$

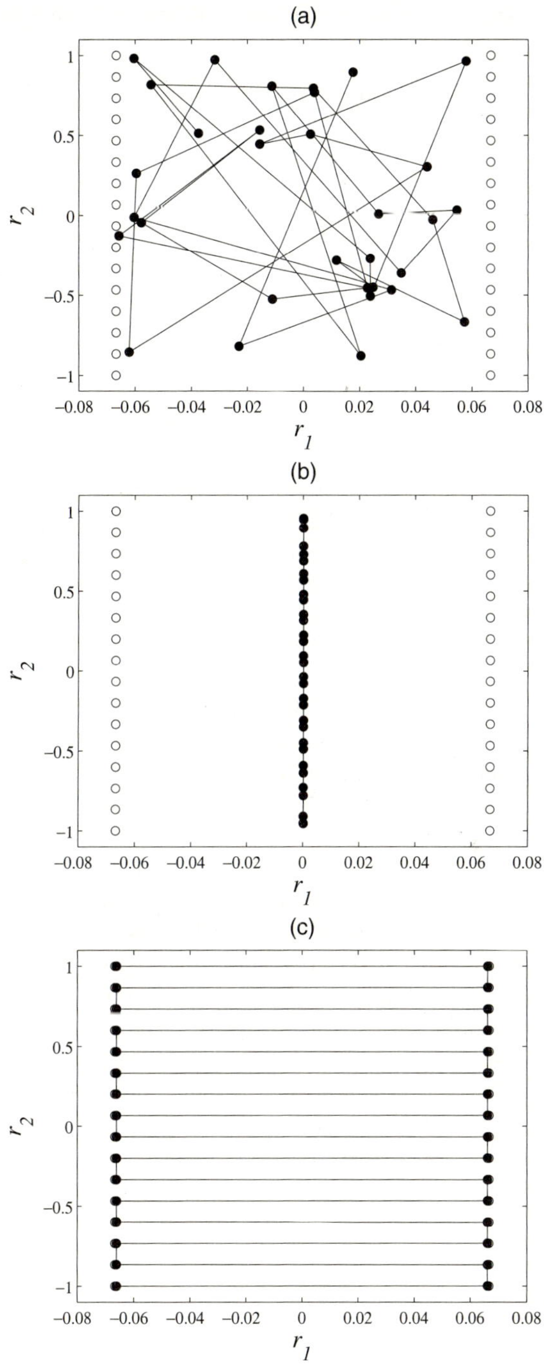

Fig. 1. Weight vector distribution of the cortical chain: (a) initial, (b) and (c) after applying mapping (12) with $\beta_f = 200$ and 1000 respectively (see other details in the text). Stimuli and weight vectors are marked by open and filled circles respectively.

Consider the weight vector as a "particle" moving in space-time $\boldsymbol{x}$ and decompose this particle trajectory:

$$\boldsymbol{w}_j = p_j(\nu)\boldsymbol{x}_\nu. \tag{10}$$

Applying decomposition (10) to free energy (9) and taking the low temperature limit yield

$$F = -\frac{1}{\beta} \ln \sum_{i=1}^{n} \exp\left(-\frac{\beta}{2}|\boldsymbol{x}_\mu - \boldsymbol{w}_i|^2\right) + \frac{\gamma}{2} \sum_{i=1}^{n-1} |\boldsymbol{w}_{i+1} - \boldsymbol{w}_i|^2. \tag{11}$$

Minimization of free energy function (11) results in the cortical mapping:

$$\Delta \boldsymbol{w}_j = -\eta \frac{\partial F}{\partial \boldsymbol{w}_j} = \eta\Big(\tilde{p}_j(\mu)(\boldsymbol{x}^\mu - \boldsymbol{w}_j) + \gamma(\boldsymbol{w}_{j+1} - 2\boldsymbol{w}_j + \boldsymbol{w}_{j-1})\Big), \tag{12}$$

where

$$\tilde{p}_j(\mu) = \frac{\exp(-\frac{\beta}{2}|\boldsymbol{x}^\mu - \boldsymbol{w}_j|^2)}{\sum_{k=1}^{n} \exp(-\frac{\beta}{2}|\boldsymbol{x}^\mu - \boldsymbol{w}_k|^2)} \tag{13}$$

is the reduction of $p_j(\mu)$ to the case of lateral-free interactions.

We apply mapping (12) to modelling the development of retinotopy and ocular dominance. Throughout, the training is cyclic with fixed sequence, i.e. before learning starts a particular order of pattern presentation is fixed. The simulations are performed for 32 cortical neurons with initial uniform random distribution of the weight vectors within $[-0.0667, 0.0667] \times [-1, 1]$ rectangular (Fig. 1(a)). The stimuli are placed regularly within the two columns of 16 units each at the left and right boundary of the rectangular, which represent left and right "eye" respectively. The ratio of the separation units between retinae to the separation of neighbouring units within a retina defines the correlation of retinal units, which is ≈ 1. The lateral interactions are allowed to decrease linearly with time: $\gamma = \gamma_0(1 - t/T)$ with $\gamma_0 = 0.03$ and $t = 0, .., T$. The learning rate linearly decreases too: $\eta = \eta_0(1 - t/T)$ with $\eta_0 = 1$. Let us look at evolution of the weight vector distribution when the inverse temperature increases from $\beta_0 = 4$ to different values of β_f in steps of 0.01. For β_f exceeding some threshold but remaining relatively small, the weight vectors become distributed on the line exactly between the left and right eye. For larger β_f, the clusters consisting two weight vectors are formed on this line (Fig. 1(b)). Increasing β_f further leads to breaking the spatial symmetry with one weight vector in a cluster moving toward the left eye and another weight vector moving toward the right eye. Thus, the retinotopy and ocular dominance simultaneously formed (Fig. 1(c)).

Unlike the considered cortical map, incremental soft topology-preserving mapping (5) doesn't produce satisfactory results being applied to this problem.

4 Elastic Nets

The elastic net is based on elastic, diffusion-type, interactions [3]. This algorithm works like an rubber ring: it gradually drags points on the ring towards the "cities" and an elastic force keeps neighbouring points close to one another.

Earlier, Simic showed the relationship between the Hopfied network and the elastic net: it derived the latter from Hopfield's objective function for the TSP [14,15]. Let us show how to derive the elastic net from the batch version of soft topology-preserving map.

Taking the Taylor series expansion (in power of γ) results in

$$\langle F \rangle = -\frac{1}{\beta N} \sum_{\mu=1}^{N} \ln \sum_{i=1}^{n} \exp\left(-\frac{\beta}{2}|\boldsymbol{x}_\mu - \boldsymbol{w}_i|^2\right) + \frac{\gamma}{2N} \sum_{i=2}^{n-1} \sum_{\mu=1}^{N} p_i(\mu)\left(|\boldsymbol{x}_\mu - \boldsymbol{w}_{i-1}|^2 + |\boldsymbol{x}_\mu - \boldsymbol{w}_{i+1}|^2\right). \quad (14)$$

Consider the weight vector as a "particle" moving in space-time $\boldsymbol{x}$ and decompose this particle trajectory:

$$\boldsymbol{w}_j = \langle \boldsymbol{x}(j) \rangle = \sum_{\nu=1}^{N} p_j(\nu)\boldsymbol{x}_\nu, \quad (15)$$

where $\langle \boldsymbol{x}(j) \rangle$ are the expected position of the particle at time j.

Applying decomposition (15) to free energy (14) yields

$$\langle F \rangle = -\frac{1}{\beta N} \sum_{\mu=1}^{N} \ln \sum_{i=1}^{n} \exp\left(-\frac{\beta}{2}|\boldsymbol{x}_\mu - \boldsymbol{w}_i|^2\right) + \frac{\gamma}{2N} \sum_{i=1}^{n-1} |\boldsymbol{w}_{i+1} - \boldsymbol{w}_i|^2. \quad (16)$$

Minimization of free energy function (16) results in the elastic net algorithm:

$$\Delta \boldsymbol{w}_j = -\eta \frac{\partial F}{\partial \boldsymbol{w}_j} = \frac{\eta}{N}\left(\sum_{\mu=1}^{N} \tilde{p}_j(\mu)(\boldsymbol{x}^\mu - \boldsymbol{w}_j) + \gamma(\boldsymbol{w}_{j+1} - 2\boldsymbol{w}_j + \boldsymbol{w}_{j-1})\right). \quad (17)$$

Defining $\beta \equiv \frac{1}{\sigma^2}$ with the Gaussian distribution width σ, energy (16) takes the exact form of the Durbin-Willshow elastic net energy [3]. Shrinking the distribution width is, thus, equivalent to decreasing the system temperature.

Let us demonstrate how different algorithms work for the TSP. The simulations are performed for 64 "cities" that are sites on a 8×8 regular square. The elastic ring has 128 points. The training is cyclic with fixed sequence. The inverse temperature β increases from 2 to 200 in steps of 0.01. The learning rate is linearly decreasing function of time, i.e. $\hat{\eta} = \frac{1}{N}\eta = \hat{\eta}_0(1 - t/T)$ with $\hat{\eta}_0 = 1$ and $t = 0, .., T$. Let us take $\gamma = 0.06$. Initially, the weight vectors are distributed equidistantly on the unit radius circle (Fig. 2(a)). The batch topology-preserving map (8) applied to the task produces one of the possible optimal tours (Fig. 2(b)). For the elastic net (17)), its elastic strength is allowed to decrease with time passing: $\gamma = \gamma_0(1-t/T)$ with $\gamma_0 = 0.06$, which can provide a finer pattern than one for fixed γ [1]. Fig. 2(c) demonstrates the result. Thus, both lateral and elastic interactions produce not the same but equally optimal tours for the given task.

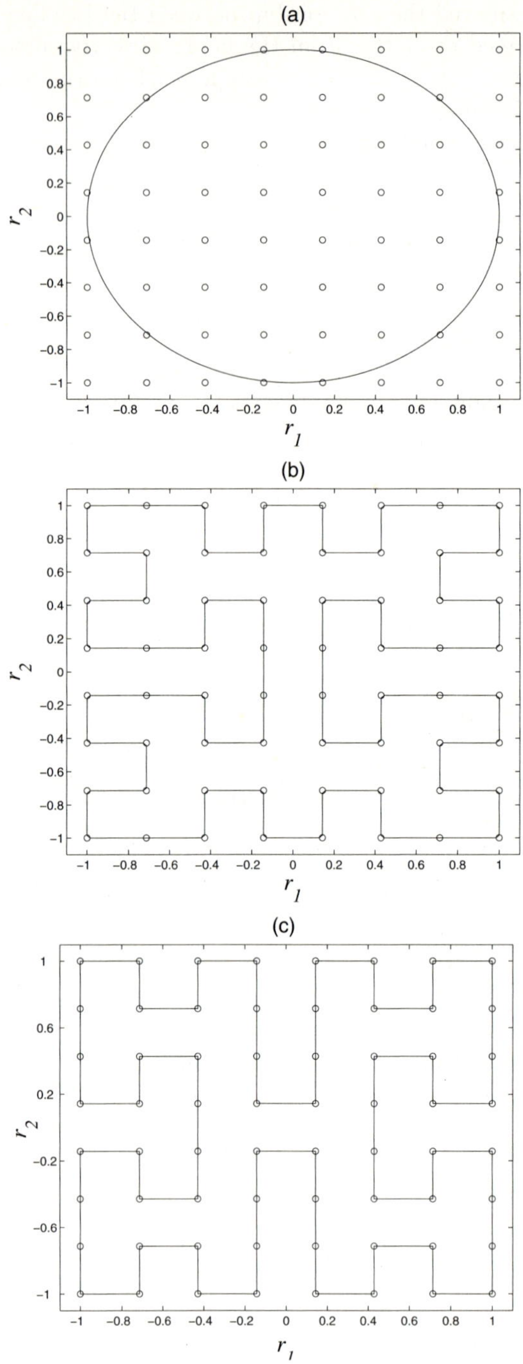

Fig. 2. Weight vector distribution: (a) initial, (b) and (c) after applying learning rules (8) and (17) respectively (see details in the text)

5 Conclusion

Researchers always paid attention to the similarity of neural patterns produced by lateral and elastic interactions, but not providing any rigorous proof of such relationship [1, 16, 17, 18, 12]. This paper is aimed to provide this proof.

First, we demonstrated that the weak lateral interactions can be transformed into the elastic ones. As result, the cortical map and the elastic net are derived from the incremental and the batch soft topology-preserving maps respectively. The temperature of the above maps is transformed into the Gaussian variance of the cortical map and the elastic net. This fact elucidates indirect incorporation of soft competition and deterministic annealing into the cortical map and the elastic net, which makes them to be very powerful neurocomputational models.

Second, we analyzed the relevant models by applying them to the same task. Application of the incremental soft topology-preserving map and the cortical map to the development of the visual cortex revels their differences. Indeed, these models are equivalent only in the low temperature limit. As known, at the beginning of learning process when the temperature is high, the state trajectory is very sensitive to any changes in the system and can take, therefore, any possible direction.

Unlike the above, the batch soft topology-preserving map and the elastic net are proved to be equivalent for all temperatures. As results, both models are appeared to be equally successful in solving the TSP.

References

1. Hertz, J., Krogh, A., Palmer, R.G.: Introduction to the Theory of Neural Computation. Addison-Wesley, Reading, London, UK (1991)
2. Kohonen, T.: Self-Organized Formation of Topologically Correct Feature Maps. Biological Cybernetics 43, 59–69 (1982)
3. Durbin, R., Willshaw, D.: An Analogue Approach to the Travelling Salesman Problem Using an Elastic Net Method. Nature 326, 689–691 (1987)
4. Durbin, R., Mitchison, G.: A Dimension Reduction Framework for Understanding Cortical Maps. Nature 343, 644–647 (1990)
5. Tereshko, V.: Topology-Preserving Elastic Nets. In: Mira, J.M., Prieto, A.G. (eds.) IWANN 2001. LNCS, vol. 2084, pp. 551–557. Springer, Heidelberg (2001)
6. Tereshko, V., Allinson, N.M.: Combining Lateral and Elastic Interactions: Topology-Preserving Elastic Nets. Neural Processing Letters 15, 213–223 (2002)
7. Tereshko, V.: Modelling Topology Preservation in a Cortex by Controlled Deformation of Its Energy Landscape. Neurocomputing 44-46, 667–672 (2002)
8. Tereshko, V.: Deriving cortical maps and elastic nets from topology-preserving maps. In: Cabestany, J., Priet, A., Sandoval, F. (eds.) IWANN 2005. LNCS, vol. 3512, pp. 326–332. Springer, Heidelberg (2005)
9. Graepel, T., Burger, M., Obermayer, K.: Phase Transitions in Stochastic Self-Organizing Maps, Phys. Rev. E. 56, 3876–3890 (1997)
10. Heskes, T.: Energy Functions for Self-Organizing Maps. In: Oja, E., Kaski, S. (eds.) Kohonen Maps, pp. 303–315. Elsevier, Amsterdam (1999)

11. Goodhill, G.J.: Correlations, Competition, and Optimality: Modelling the Development of Topography and Ocular Dominance. Cognitive Science Research Paper 226. University of Sussex, Brighton (1992)
12. Swindale, N.V.: The Development of Topography in the Visual Cortex: a Review of Models. Comput. Neur. Syst. 7, 161–247 (1996)
13. Mitchison, G.: Neuronal Branching Patterns and the Economy of Cortical Wiring. Proc. Roy. Soc. B. 245, 151–158 (1991)
14. Simic, P.D.: Statistical Mechanics as the Underlying Theory of "Elastic" and "Neural" Optimisations. Network: Comput. Neur. Syst. 1, 89–103 (1990)
15. van den, Berg, J.: Neural Relaxation Dynamics: Mathematics and Physics of Recurrent Neural Networks with Applications in the Field of Combinatorial Optimization. PhD Thesis. Erasmus University, Rotterdam (1996)
16. Folk, R., Kartashov, A.: A simple elastic model for self-organizing topological mappings. Network. Computation in Neural Systems 5, 369–387 (1994)
17. Wong, Y.: A Comparative study of the Kohonen self-organizing map and the elastic net. In: Hanson, S.J., Petsche, T., Kearns, M., Rivest, R.L. (eds.) Computational Learning Theory and Natural Learning Systems, vol. 2, pp. 401–413. MIT, Cambridge (1994)
18. Mitchison, G.: A Type of duality between self-organizing maps and minimal wiring. Neural Computation 7, 25–35 (1995)

A Survey on Use of Soft Computing Methods in Medicine

Ahmet Yardimci

Akdeniz University, Vocational School of Technical Sciences
Department of Industrial Automation, 07059, Antalya, Turkey
yardimci@akdeniz.edu.tr

Abstract. The objective of this paper is to introduce briefly the various soft computing methodologies and to present various applications in medicine. The scope is to demonstrate the possibilities of applying soft computing to medicine related problems. The recent published knowledge about use of soft computing in medicine is observed from the literature surveyed and reviewed. This study detects which methodology or methodologies of soft computing are used frequently together to solve the special problems of medicine. According to database searches, the rates of preference of soft computing methodologies in medicine are found as 70% of fuzzy logic-neural networks, 27% of neural networks-genetic algorithms and 3% of fuzzy logic-genetic algorithms in our study results. So far, fuzzy logic-neural networks methodology was significantly used in clinical science of medicine. On the other hand neural networks-genetic algorithms and fuzzy logic-genetic algorithms methodologies were mostly preferred by basic science of medicine. The study showed that there is undeniable interest in studying soft computing methodologies in genetics, physiology, radiology, cardiology, and neurology disciplines.

Keywords: Soft computing, Fuzzy-Neural systems, Fuzzy-Genetic algorithms, Neural-Genetic Algorithms, Probabilistic reasoning.

1 Introduction

Although computers were already used in medicine and the early medical systems appeared at about the same time as the seminal article by Zadeh - almost four decades ago - there was little communication between these research fields for many years [1,4]. But for the last two decades the situation has changed. A major transformation has occurred in the field of knowledge engineering and also medicine has been affected by this transformation. Many researchers had a bold vision of the way knowledge engineering would revolutionize medicine, and push the frontiers of technology forward. There are now numerous systems that use fuzzy logic (FL), neural networks (NNs), genetic algorithm (GA), and other techniques in approximate reasoning.

Many of the early efforts to apply artificial intelligence to medical reasoning problems have primarily used rule-based systems [5]. Until the late 1980s, the practice of building knowledge-intensive systems was viewed uniformly as "extracting" rules from application experts, and putting those rules into an expert-system shell. Such programs are typically easy to create, because their knowledge is catalogued in the form of *if-then* rules. Developers built systems rule by rule, attempting to mimic with their rule bases the

J. Marques de Sá et al. (Eds.): ICANN 2007, Part II, LNCS 4669, pp. 69–79, 2007.

problem-solving behaviours that application experts seemed to display. In relatively well-constrained domain such programs show skilled behaviour. But this "knowledge mining" view has many weaknesses, as Clancey [6] showed in his analysis of the MYCIN (Knowledge-based Medical Expert System) system. Firstly, it is generally impossible to elicit from professionals in a given application area an adequate set of rules for all but trivial tasks. This elicitation problem has been called the "knowledge acquisition bottleneck," a mournful phrase that has been repeated so often in literature that it almost has become trite. In real-life situations, there is considerable degradation of performance due to both presence of ambiguity and incomplete information as well as inadequate modeling of the diseases by the rules. It is difficult to construct automatic systems to provide classification or pattern recognition tools or help specialists make a decision. Second, the resulting knowledge- based systems are generally difficult to maintain: Adding one more rule can change the behaviour dramatically. Other conventional methods like Bayes classifier and flow charts are also unable to deal with most complex clinical decision making problems. The choice of a method to solve this problem depends on the nature of problem like classification, automatic diagnosis, decision support. But usually it is not possible to solve the problem completely by using just one methodology. It is necessary to use different methodologies together in various combinations which are chosen appropriate to the nature of the problem. At this point the importance of soft computing (SC) methodologies is to come out.

This paper surveys the use of SC in medicine based on searches in medical data base MEDLINE. The complementarities of FL, NNs and Probabilistic Reasoning (PR) have an important consequence: in many cases a problem can be solved most effectively by using FL, NN and PR in combination rather than exclusively. This is also one of our purposes in this paper: to present SC methodologies available to represent and manage imperfect knowledge in medicine.

2 Literature Review

We have already mentioned that one of the important goals of this paper is to survey the use of SC in medicine. Searches are based on MEDLINE medical and engineering database. The keywords used to search were based on the logical linguistic pattern;

1. "fuzzy logic *and* neural networks *and* biomedical *or* medicine"
2. "fuzzy logic *and* genetic algorithm *and* biomedical *or* medicine"
3. "neural networks *and* genetic algorithm *and* biomedical *or* medicine"
4. "fuzzy logic *and* neural networks *and* genetic algorithm *and* biomedical *or* medicine".

These linguistic patterns are suitable to find publications which contain SC methodologies. By using these patterns we also classify the publications in accordance with methodologies combinations. The search results show us popularity and applicability of the methodology combinations. To compare these results to the use of singular methodologies in studies, "fuzzy logic *and* biomedical *or* medicine", "neural network *and* biomedical *or* medicine" and "genetic algorithms *and* biomedical *or* medicine" logical linguistic patterns are also searched in MEDLINE and results are showed in a separate graphics.

Observing the former studies which survey on the use of artificial intelligence (AI) methodologies in medicine, the database search was restricted for the last decade and especially the last five years need to be highlighted. The articles of Abbod et al.[7,8] and Mahfouf et al.[9] have a good coverage for the use of fuzzy logic, smart and adaptive engineering systems in medicine until the year 2001. All papers survey the use of fuzzy logic and adaptive systems in diagnosis, therapy and imaging areas of medicine.

3 The Use of Soft Computing in Medicine

Medicine is a diverse field. It consists of various specialized sub-branches. Roughly we can divide it into four broad fields as follows: basic science, diagnostic science, clinical science and surgical science. Each of these fields can be further sub-classified. In the following sections, a brief description is given of key contributions which soft computing methodologies have made in each of the sub topics which have been identified in the literature search. Table 1 shows the separately distribution of soft computing studies as FL-NN, NN-GA and FL-GA, on medicine field. Please note that this table is prepared by considering only the last five years 2001-2006 studies.

3.1 Basic Science

According to MEDLINE database search results the use of SC methodologies in basic science of medicine is significantly increasing. Basic science is very suitable to all SC methodologies. For example, in biochemistry field there is a variety of phenomena with many complex chemical reactions, in which many genes and proteins affect transcription or enzyme activity of others. It is difficult to analyze and estimate many of these phenomena using conventional mathematical models. So, NNs, Fuzzy NNs, and the NN-GAs, have been applied to analysis in a variety of research fields. Especially biochemistry, biostatistics, genetics, physiology and pharmacology branches have applied to use of SC methodologies. Biochemistry, cytology, histology and pathology are the other branches which have applied to SC in their studies.

FL-NN Applications in Basic Science

The literature search results on the use of FL-NN methodologies in basic science of medicine take us to sub-branches like as cytology, physiology, genetics and biostatistics. In cytology, Ma et al. [10] have developed an application for cell slice image segmentation by using modern and traditional image segmentation technology. Because of the complex structure of cell and cell slice image it is always difficult to generally segmentate any kind of biological cell slice image. The study achieved to obtain good results for morphological image segmentation, which includes edge detection and regional segmentation, wavelet transform, by using fuzzy algorithms and artificial neural networks.

When we look at physiology branch studies, we can easily say that the complexity of biological signals to push the researchers to use FL-NN systems to solve the physiological problems and get acceptable results. Das et al. [11] proves this determination with their study. Because of the complexity of signals and in order to improve the reliability of the recognition of diagnostic system they have preferred to use hybrid fuzzy logic neural network methodology for recognition of swallow acceleration signals from artifacts. They train two fuzzy logic-committee networks (FCN); FCN-I and FCN-II. While the first one was used to recognize dysphagic swallow from artifacts, the second was used to recognize normal swallow from artifacts. Their evaluation results revealed that FCN correctly identified artifacts and swallows. Also they highlighted at the end of the study that the use of hybrid intelligent system consisting FL and NN provides a reliable tool for recognition and classification of biological signals. Catto et al. [12] and Futschik et al. [13] have used to FL-NN combining methodology to predict cancer tissue from gene expression data. Because of the poor accuracy of statistical analysis to prediction of tumor behaviors they preferred to use FL-NN methodology in their genetic science based studies. Knowledge-based neurocomputing was used by Futschik to contribute fuzzy rules which point to genes that are strongly associated with specific types of cancer.

NN-GA Applications in Basic Science

This category is the most preferred one for basic sciences disciplines. Biochemistry, biostatistics, genetics, histology, pathology, pharmacology and physiology are the disciplines which have some NN-GA applications.

Agatonovic-Kustrin et al. [14-16] have done some studies on pharmacology by using NN-GA 9combining methodologies. Agatonovic-Kustrin [15] developed a simple model for prediction of corneal permeability of structurally different drugs as a function of calculated molecular descriptors using artificial NNs. They used a set of 45 compounds with experimentally derived values to describe corneal permeability (log C). A genetic algorithm was used to select a subset of descriptors that best describe corneal permeability coefficient log C and a supervised network with radial basis transfer function was used to correlate calculated molecular descriptors with experimentally derived measure of corneal permeability. Their developed model was useful for the rapid prediction of the corneal permeability of candidate drugs based on molecular structure.

Most drugs are excreted into breast milk to some extent and are bioavailable to the infant. The ability to predict the approximate amount of drug that might be present in milk from the drug structure is very useful in the clinical setting. This mission is studied by Agatonovic-Kustrin et al. [16]. They used GA and NN for to simplify and upgrade their previously developed model for prediction of the milk to plasma (M/P) concentration ratio, given only the molecular structure of the drug. As mentioned befo9re in their previous study GA was used for a same aim, to select a subset of the descriptors that best describe the drug transfer into breast milk and NN to correlate selected descriptors with them M/P ratio and developed a quantitative structure-activity relationships (QSAR). The averaged literature M/P values were used as the artificial neural networks's (ANN) output and calculated molecular descriptors as the inputs. Before each training run, data sets were split randomly into three separate groups and both weights and biases were initialized with random values. As a result unlike previously reported models, this developed model does not require experimental parameters and useful prediction of M/P ratio of new drugs and reduce the need for actual compound synthesis and M/P ratio measurements.

Rask et al. [17] used GA to design NN structure in their automatic error reduction study for the real time dynamic biomechanical model of the human elbow joint with NN-GA combining methodology. They achieve the result of the GA networks reduced the error standard deviation across all subjects. Ogihara et al. [18] developed an anatomically and physiologically based neuro-musculo-skeletal model using NNs by optimized GA, to emulate the actual neuro-control mechanism of human bipedal locomotion. These two studies were given as a sample to application of NN-GA combining methodologies on physiology science.

3.2 Diagnostic Science

Diagnostic science mainly includes clinical laboratory sciences and radiology sciences. Database search results show that the almost SC application studies done in radiology especially for interventional radiology. Interventional radiology is concerned with using imaging of the human body, usually from CT, ultrasound, or fluoroscopy, to do biopsies, place certain tubes, and perform intravascular procedures.

If we compare to prefer and use of SC methodologies we clearly see that FL-NN applications is the first and nearly unique one for diagnostic science area. There are some applications which were done with NN-GA but these are very few when compare them to FL-NN applications.

Image segmentation is one of the most important steps leading to the analysis of digital images, its main goal being to divide an image into parts that have a strong correlation with objects or areas of the real world. Image segmentation is an indispensable process in the visualization of human tissues, particularly during clinical analysis of magnetic resonance (MR) images. But, MR images always contain a significant amount of noise caused by operator performance, equipment, and the environment, which can lead to serious inaccuracies with segmentation. Shen et al. [19], Meyer-Baese et al. [20] and Wismuller et al. [21] have used FL-NN methodology in their recent studies to solve magnetic resonance imaging (MRI) problems. These studies are also good examples to use FL-NN and NN-GA methodologies.

3.3 Clinical Disciplines

MEDLINE database search results showed that the clinical sciences are the most popular and suitable area for the SC methodology applications in medicine. So far, 60% of SC methodology applications were done for clinical science disciplines. Although the studies shows a regular dispersion to all sub-branches of clinical science, according to search results evaluation it is obvious that the cardiology, neurology, critical care medicine, anesthesiology and physical medicine and rehabilitation are the most preferred disciplines. When we compare to preference of SC methodologies we clearly see that the FL-NN methodology is significantly the most preferred one. This result also shows parallelism with a result of the preference of SC methodologies in medicine. Another important result related to the use of SC in the clinical science is there is not any study that has used FL-GA methodology so far.

FL-NN Studies in Clinical Science

Anesthesiology. Anesthesia is defined as the loss of sensation resulting from pharmacological depression of nerve function or from neurological dysfunction [8]. There are some good examples of the use of adaptive systems for controlling blood pressure, analgesia, paralysis, unconsciousness and septic shock in the field of anesthesia. Zhang et al. [22] and Allen and Smith [23] have studied modeling and controlling depth of anesthesia (DOA). In anesthesia, two approaches are being considered for measuring the DOA: indirect or direct [8]. The indirect method is achieved by monitoring clinical signs of anesthesia such as blood pressure and heart rate which are affected by the infused drugs. In contrast, the direct method measures anesthesia move directly from the nerves or the brain, such as in the muscle relaxation and evoked responses of the brain [8]. Allen and Smith have investigated utility of the auditory evoked potential (AEP) as a feedback signal for the automatic closed-loop control of general anesthesia using FL-NN methodology. A simple back-propagation NN was trained for AEP and its output used to FL infusion controller for the administration of anesthetic drugs.

Cardiology. Cardiology is concerned with the heart and cardiovascular system and their diseases. The majority of work done on the utilization of adaptive systems has been to adaptive pacemakers. Recently, Ubeyli and Guler [24] have introduced a successful tool which is an adaptive neuro-fuzzy inference system for detection of internal carotid artery stenosis and occlusion. The internal carotid arterial Doppler signals recorded from 130 subjects and internal carotid artery conditions were detected by three adaptive neuro-fuzzy inference system (ANFIS) classifiers. In spite of spectral analysis of the Doppler signals produced information concerning the blood flow in the arteries, NNs may offer a potentially superior method of Doppler signal analysis to the spectral analysis methods. The predictions of the tree ANFIS classifier were combined by the fourth ANFIS classifier. Study results showed that accuracy rates of the ANFIS model were found to be higher than that of the stand-alone NN model and indicate that the proposed ANFIS model has some potential in detecting internal carotid artery stenosis and occlusion.

The other literature works done successfully in cardiology discipline are; Kashihara et al. [25] have studied an automated drug infusion system using FL-NN methodology to control mean arterial pressure (MAP) in acute hypotension. Shyu et al. [26] proposed a method for detecting ventricular premature contraction (VPC) from the Holter system using wavelet transform and fuzzy neural network (FNN). Serhatlioglu et al. [27] have investigated the effects of diabetes mellitus on carotid artery by using a neurofuzzy system.

Critical Care Medicine. Critical care is concerned with the therapy of patients with serious and life-threatening disease or injury. Intensive care medicine employs invasive diagnostic techniques and temporary replacement of organ functions by technical means. Critical care applications are close to anesthesia and pulmunology in their medical function. Blood pressure and respiration regulation, electroencephalogram (EEG) monitoring and pain relief are the main application areas of critical care medicine.

Artificial ventilation of the lungs is one of the the major components of intensive care therapy. The aim is to deliver oxygen to the tissues and to remove carbon dioxide when the patient's lungs are not able to function adequately [28]. The clinicians in the critical care unit adjust the various ventilator settings in order to achieve a reasonable level of oxygenation in the blood. Clinicians make these decisions based upon knowledge of the pathophysiology of the lungs and the patient's condition and the past medical history [28]. Kwok et al. [28] have developed an ANFIS and multilayer perceptron (MLP) in rule-base derivation for ventilator control and tested it with closed-loop simulations. The developed ANFIS model was a Sugeno-type fuzzy inference system which had three inputs and one output. The consequent parts were constants. Firstly clustering was applied to the training data using MATLAB toolbox. A hybrid algorithm (gradient descent and the least-square estimation algorithms) was used for training. Using the same data training dataset they developed a feed-forward MLP. It has same inputs and output as for the ANFIS model. The performance of these two models was compared to that of FAVeM which is a previously developed fuzzy advisor for ventilator management [29]. As a result the use of adaptive neuro-fuzzy systems can facilitate the modeling of the clinicians' knowledge in the development of intelligent advisors for intensive care ventilators. Both the ANFIS and MLP were shown to be able to model clinicians' decision-making accurately. However, the ANFIS is more interpretable than the MLP.

Neurology. Neurology is concerned with the diagnosis and treatment of nervous systems (central, peripheral, autonomic, neuromuscular junction and muscle) diseases.

Most of the neurology studies were focused on sleep analysis, EEG and electromyogram (EMG) analysis subjects. Using nonlinear adaptive fuzzy approximator (NAFA), Zhang et al. [30] achieved to provide efficient nonlinear separation of single-sweep evoked potentials (EPs), which allows for quantitative examination of the cross-trial variability of clinical EPs. The NAFA is characterized by concise representation of structured knowledge, fast learning capability, as well as universal approximation property. It was applied to forecast the non-stationary EEG time-series and to estimate single-sweep EPs.

In an unusual study, Palaniappan and co-workers [31] used neural network architecture for incremental supervised learning of analog multidimensional maps (fuzzy ARTMAP) and NNs for to design a new brain-computer interface (BCI). They aim was to classify the best three of five available mental tasks for each subject using power spectral density (PSD) values of EEG signals. They tested the system with ten experiments; employing different triplets of mental tasks for each subject. Their findings showed that the average BCI- fuzzy ARTMAP outputs for four subjects gave less than 6% of error using the best triplets of mental tasks identified from the classification performances of fuzzy ARTMAP. This clearly implies that the BCI-fuzzy ARTMAP can be successfully used with a tri-state switching device.

4 Results and Discussion

An overview of different SC techniques is presented in this paper along with the review of important medicine applications. The proficiency of SC techniques has been explored in almost every field of medicine. Based on this study future developments of SC technology in medicine can be tentatively forecast. Table 1 summary the number of applications of SC methodologies in medicine on a yearly basis. FL, NN and GA based search results are given here to help reader understand the situation and compare the number of cited SC based papers in medicine.

Table 1. The number of applications of SC methodologies in medicine on a yearly basis

	Publication year							Total
	1995-1999	2000	2001	2002	2003	2004	2005	
*FL	184	41	81	44	45	58	42	495
NN	641	160	171	172	192	239	194	1769
GA	43	20	18	17	14	28	29	169
FL-NN	29	6	23	13	14	8	8	101
NN-GA	17	2	5	5	6	6	5	46
FL-GA	3	1	-	1	1	1	-	7
FL-NN-GA	1	-	-	1	1	1	-	4

* FL:Fuzzy logic, NN:Neural networks, GA:Genetic algorithms, FL-NN:Fuzzy logic-Neural networks, NN-GA:Neural networks-Genetic Algorithms, FL-GA:Fuzzylogic-Genetic Algorithms, FL-NN-GA: Fuzzy logic-Neural networks-Genetic algorithms.

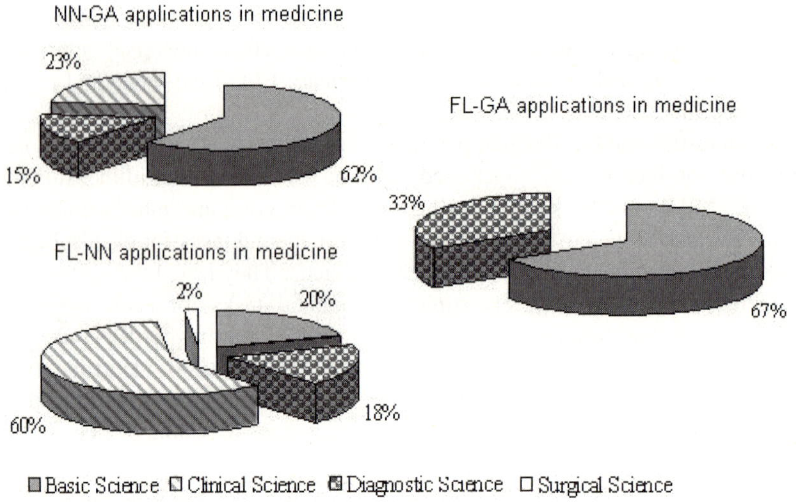

Fig. 1. Comparison of the use of SC methods in medicine and sub-branches of medicine

Especially radiology and neurology disciplines were used SC in classification and diagnosis studies. It should be noted that radiology, imaging and diagnosis studies are always related to other disciplines of medicine such as neurology, dermatology, pulmonology and oncology. To prevent to repeat one study in two areas and to find to correct group for study, all publications examined carefully.

The preference of SC methodologies in medicine is illustrated in Figure 1. The mostly used methodology is FL-NN 70% then NN-GA 27% and FL-GA 3%. As far

FL-NN methodology is significantly used in clinical science of medicine. 60% of cited FL-NN studies are related clinical science. On the other hand NN-GA and FL-GA methodologies were mostly preferred by basic science of medicine.

The main findings of the study are:

a. In genetics, physiology, interventional radiology, anesthesiology, cardiology, and neurology disciplines there are undeniable interest in studying SC methodologies. It proves to be very fruitful to study SC in these disciplines.
b. In the field of clinical laboratory science and surgical science, there are no specific applications to date.
c. SC methodologies give birth to new ideas in neighboring disciplines in medicine.

The last point which we got from search results is that the SC term is not used well enough as a keyword in studies. Hybrid systems, combining systems, fuzzy-neural, fuzzy GAs, neural GAs terms are mostly preferred instead of SC.

There is a growing interest in SC tools in medicine, which are used to handle imprecision and uncertainty, and to build flexibility and context adaptability into intelligent systems. It is obvious that the SC methodologies will be most preferred tools in medicine in the near future with its flexible information processing capability for handling real life ambiguous situations. A number of SC methods and theirs applications in medicine have been described in this paper. This paper can be used as a guide for future studies. The situation of present studies and virgin sub-branches of medicine may help researchers to orientate their study areas and to choose methodologies for their studies.

Acknowledgments. The author wishes to thank Akdeniz University Scientific Researches Projects Management Unit for their financial support to this work.

References

1. Zadeh, L.A.: "Fuzzy Sets". Information and Control 8, 338–353 (1965)
2. Zadeh, L.A.: Outline of a New Approach to the Analysis of Complex Systems and Decision Processes. IEEE Transactions on Systems, Man, and Cybernetics, V. SMC 3(1), 28–44 (1973)
3. Zadeh, L.A.: Possibility Theory and Soft Data Analysis (University of California, Berkeley, Memorandum UCB/ERL M79/66, 1979)
4. Akay, M., Cohen, M., Hudson, D.: Fuzzy sets in life science. Fuzzy Sets and Systems 90, 219–224 (1997)
5. Szolovits, P., Patil, R., Schwartz, W.: Artificial intelligence in medical diagnosis. Ann. Internal Med. 108, 80–87 (1998)
6. WJ, C.: The epistemology of a rule-based expert system - a framework for explanation Artificial Intelligence, V. Artificial Intelligence 20, 215–251 (1983)
7. Abbod, M.F., Diedrich, G.K., Linkens, D.A., Mahfouf, M.: Survey of utilization of fuzzy technology in Medicine and Healthcare. Fuzzy Sets and Systems 120, 331–349 (2001)
8. Abbod, M.F., Linkens, D.A., Mahfouf, M., Dounias, G.: Survey on the use of smart and adaptive engineering systems in medicine. Artificial Intelligence in Medicine 26, 179–209 (2002)

9. Mahfouf, M., Abbod, M.F., Linkens, D.A.: A survey of fuzzy logic monitoring and control utilization in medicine. Artificial Intelligence in Medicine 21(1-3), 27–42 (2001)
10. Ma, Y., Dai, R., Li, L., Wu, C.: The state and development of cell image segmentation technology. Journal of Biomedical 19(3), 487–492 (2002)
11. Das, A., Reddy, N.P., Narayanan, J.: Hybrid fuzzy logic committee neural networks for recognition of swallow acceleration signals. Computer Methods and Programs in Biomedicine 64(2), 87–99 (2001)
12. Catto, J.W.F., Linkens, D.A., Abbod, M.F., Chen, M., Burton, J.L., Feeley, K.M., Hamdy, F.C.: Artificial intelligence in predicting bladder cancer outcome: a comparison of neuro-fuzzy modeling and artificial neural networks. Clinical Cancer Research: An Official Journal of The American Association for Cancer Research 9(11), 4172–4177 (2003)
13. Futschik, M., Reeve, A., Kasabov, N.: Evolving connectionist systems for knowledge discovery from gene expression data of cancer tissue. Artificial Intelligence In Medicine 28(2), 165–189 (2003)
14. Agatonovic-Kustrin, S., Beresford, R., Yusof, A.: Theoretically-derived molecular descriptors important in human intestinal absorption. Journal of Pharmaceutical and Biomedical Analysis 25(2), 227–237 (2001)
15. Agatonovic-Kustrin, S., Evans, A., Alany, R.G.: Prediction of corneal permeability using artificial neural networks. Die Pharmazie 58(10), 725–729 (2003)
16. Agatonovic-Kustrin, S., Ling, L., Tham, S., Alany, R.: Molecular descriptors that influence the amount of drugs transfer into human breast milk. Journal of Pharmaceutical and Biomedical Analysis 29(1-2), 103–119 (2002)
17. Rask, J.M., Gonzalez, R.V., Barr, R.E.: Genetically-designed neural networks for error reduction in an optimized biomechanical model of the human elbow joint complex. Computer Methods in Biomechanics and Biomedical Engineering 7(1), 43–50 (2004)
18. Ogihara, N., Yamazaki, N.: Generation of human bipedal locomotion by a bio-mimetic neuro-musculo-skeletal model. Biological Cybernetics 84(1), 1–11 (2001)
19. Shen, S., Sandham, W., Granat, M., Sterr, A.: MRI fuzzy segmentation of brain tissue using neighborhood attraction with neural-network optimization. IEEE Transactions on Information Technology In. Biomedicine 9(3), 459–467 (2005)
20. Meyer-Baese, A., Wismueller, A., Lange, O.: Comparison of two exploratory data analysis methods for fMRI: unsupervised clustering versus independent component analysis. IEEE Transactions on Information Technology in Biomedicine 8(3), 387–398 (2004)
21. Wismuller, A., Meyer-Baese, A., Lange, O., Auer, D., Reiser, M.F., Sumners, D.W.: Model-free functional MRI analysis based on unsupervised clustering. Journal of Biomedical Informatics 37(1), 10–18 (2004)
22. Zhang, X.S., Huang, J.W., Roy, R.J.: Modeling for neuromonitoring depth of anesthesia. Critical Reviews in Biomedical Engineering 30(1-3), 131–173 (2002)
23. Allen, R., Smith, D.: Neuro-fuzzy closed-loop control of depth of anesthesia. Artificial Intelligence in Medicine 21(1-3), 185–191 (2001)
24. Ubeyli, E.D., Guler, I.: Adaptive neuro-fuzzy inference systems for analysis of internal carotid arterial Doppler signals. Computers in Biology and Medicine 35(8), 687–702 (2005)
25. Kashihara, K., Kawada, T., Uemura, K., Sugimachi, M., Sunagawa, K.: Adaptive predictive control of arterial blood pressure based on a neural network during acute hypotension. Annals of Biomedical Engineering 32(10), 1365–1383 (2004)
26. Wu, L.Y., Hu, Y.H.: Using wavelet transform and fuzzy neural network for VPC detection from the Holter ECG. IEEE Transactions on Bio-Medical Engineering 51(7), 1269–1273 (2004)

27. Serhatlioglu, S., Bozgeyik, Z., Ozkan, V., Hardalac, F., Guler, I.: Neurofuzzy classification of the effect of diabetes mellitus on carotid artery. Journal of Medical Systems 27(5), 457–464 (2003)
28. Kwok, H., Linkens, D., Mahfouf, M., Mills, G.: Rule-base derivation for intensive care ventilator control using ANFIS. Artificial Intelligence in Medicine 29(3), 185–201 (2003)
29. Goode, K.M., Linkens, D.A., Bourne, P.R., Cundill, J.G.: Development of a fuzzy rule-based advisor for the maintenance of mechanically ventilated patients in ICU: a model-based approach. Biomedical Engineering, Applications Basis Communications 10, 236–246 (1998)
30. Zhang, J.H., Bohme, J.F., Zeng, Y.J.: A nonlinear adaptive fuzzy approximator technique with its application to prediction of non-stationary EEG dynamics and estimation of single-sweep evoked potentials. Technology And Health Care: Official Journal of The European Society for Engineering and Medicine 13(1), 1–21 (2005)
31. Palaniappan, R., Paramesran, R., Nishida, S., Saiwaki, N.: A new brain-computer interface design using fuzzy ARTMAP. IEEE Transactions on Neural Systems and Rehabilitation Engineering 10(3), 140–148 (2002)

Exploiting Blind Matrix Decomposition Techniques to Identify Diagnostic Marker Genes

Reinhard Schachtner[1], Dominik Lutter[1,2], Fabian J. Theis[1], Elmar W. Lang[1], Ana Maria Tomé[3], and Gerd Schmitz[2]

[1] Institute of Biophysics, University of Regensburg, 93040 Regensburg, Germany
elmar.lang@biologie.uni-regensburg.de
[2] Institute of Clinical Medicine, University Hospital Regensburg, 93040 Regensburg, Germany
[3] DETI, IEETA Group, University of Aveiro, P-3810-193 Aveiro, Portugal

Abstract. Exploratory matrix factorization methods like ICA and LNMF are applied to identify marker genes and classify gene expression data sets into different categories for diagnostic purposes or group genes into functional categories for further investigation of related regulatory pathways. Gene expression levels of either human breast cancer (HBC) cell lines [5] mediating bone metastasis or cell lines from Niemann Pick C patients monitoring monocyte - macrophage differentiation are considered.

1 Introduction

The *transcriptom* comprises all cellular units and molecules needed to read out the genetic information encoded in the DNA. Among others, the level of messenger RNA (mRNA), specific to each gene, depends on environmental stimuli or the internal state of the cell and represents the gene expression profile (GEP) of the cell. High-throughput genome-wide measurements of gene transcript levels have become available with the recent development of microarray technology [1]. Microarray data sets are characterized by many variables (the GEPs) on only few observations (environmental conditions). Traditionally two strategies exist to analyze such data sets: a) *Supervised approaches* can identify gene expression patterns, called features, specific to each class but also classify new samples. b) *Unsupervised approaches* like PCA [3], ICA or NMF [2] represent exploratory matrix decomposition techniques for microarray analysis. Both approaches can be joined to build classifiers which allow to classify GEPs into different classes. We apply PCA, ICA and NMF to two well-characterized microarray data sets to identify marker genes and classify the data sets according to the diagnostic classes they represent.

2 The Data Sets

2.1 Breast Cancer Cell Lines - Bone Metastasis

The data set was taken from the supplemental data to [5]. The study investigated the ability of human breast cancer (BC) cells (MDA-MB-231 cell line) to form

J. Marques de Sá et al. (Eds.): ICANN 2007, Part II, LNCS 4669, pp. 80–89, 2007.

bone metastasis. Data set 1 comprised 14 samples; experiments 1-8 showed weak (7 and 8 mild) metastasis ability, while experiments 9-14 were highly active. Data set 2 consists of 11 experiments, 5 among them of high and 6 showing weak metastasis ability. Both data sets carry measured expression levels of 22283 genes using the Affymetrix U133a chip. For each measurement, the flags A(absent) or P(present) are provided. All genes showing more than 40% absent calls in one of the two data sets were removed. The remaining data sets contained the same 10529 genes. The authors published a list of 16 potential marker genes, 14 of which were still contained in the reduced data set.

2.2 Monocyte - Macrophage

For the monocyte - macrophage (MoMa) data set the gene-chip results from three different experiments were combined. In each experiment human peripheral blood monocytes were isolated from healthy donors (experiment 1 and 2) and from donors with Niemann-Pick type C disease (experiment 3). Monocytes were differentiated to macrophages for 4 days in the presence of M-CSF (50 ng/ml,R&D Systems). Differentiation was confirmed by phase contrast microscopy. Gene expression profiles were determined using Affymetrix HG-U133A (experiment 1 and 2) and HG-U133plus2.0 (experiment 3) Gene Chips covering 22215 probe sets and about 18400 transcripts (HG-U133A). Probe sets only covered by HG-U133plus2.0 array were excluded from further analysis. In experiment 1 pooled RNA was used for hybridization whereas in experiment 2 and 3 RNA from single donors were used. The final data set consisted of seven monocyte and seven macrophage expression profiles and contained 22215 probe sets. After filtering out probe sets which had at least one absent call 5969 probe sets remained for further analysis.

3 Data Analysis

The gene expression profiles are represented by an $(N \times M)$ data matrix $\mathbf{X} = [\mathbf{x}_1 \cdots \mathbf{x}_M]$ with each column $\mathbf{x}_m$ representing the expression levels of all genes in one of the M experiments conducted. Note that the data matrix is non-square with $N \approx 10^3 \cdot M$ typically. This renders a transposition of the data matrix necessary when techniques like PCA and ICA are applied. Hence ICA follows the data model $\mathbf{X}^T = \mathbf{AS}$. Thus in the data matrix each row represents the expression profile of all genes within one experiment, the rows of $\mathbf{S}$ contain the nearly independent component expression profiles, called expression modes, and the columns of $\mathbf{A}$ the corresponding basis vectors. In this study the JADE-algorithm [4] was used throughout, though with the natural gradient and the fastICA algorithm equivalent results were obtained. With NMF, a decomposition is sought according to $\mathbf{X} = \mathbf{WH}$ which is not unique, of course, and needs further specification. The columns of $\mathbf{W}$ are usually called *metagenes* and the rows of $\mathbf{H}$ are called *meta-experiments*. The localNMF (LNMF) algorithm [7] was applied in this study.

3.1 ICA - Analysis

We propose a method based on basic properties of the matrix decomposition model as well as on available diagnostic information to build a classifier. ICA essentially seeks a decomposition $\mathbf{X}^T = \mathbf{AS}$ of the data matrix. Column $\mathbf{a}_m$ of $\mathbf{A}$ can be associated with *expression mode* $\mathbf{s}_m$, representing the m-th row of $\mathbf{S}$. The m-th row of the matrix $\mathbf{A}$ contains the weights with which the $k \leq M$ expression levels of each of the N genes, forming the columns of $\mathbf{S}$, contribute to the m-th observed expression profile. Hence a concise analysis of matrix $\mathbf{A}$ hopefully provides insights into the structure of the data set.

Each microarray data set investigated here represents at least two different diagnostic classes. If the M expression profiles of $\mathbf{X}^T$ are grouped together according to their class labels, this assignment is also valid for the rows of $\mathbf{A}$. Suppose one of the independent *expression modes* $\mathbf{s}_m$ is characteristic of a putative cellular gene regulation process, which is related to the difference between the classes. Then in all experiments, this characteristic profile should only contribute substantially to experiments of one class and less so to the experiments of the other class (or vice versa). Since the m-th column of $\mathbf{A}$ contains the weights with which $\mathbf{s}_m$ contributes to all observations, this column should show large/small entries according to their class labels. In contrast to the method used by [6], the clinical diagnosis of the experiments is taken into account. The strategy concentrates on the identification of a column of $\mathbf{A}$, which shows a class specific signature. The *expression mode* related to that column is assumed to provide a good candidate for further class specific analysis. Informative columns were identified using the correlation of each column vector of $\mathbf{A}$ with a design vector $\mathbf{d}$ whose m-th entry is $d_m = \pm 1$, according to the class label of experiment $\mathbf{x}_i$.

3.2 Local NMF - Analysis

With NMF, each column of $\mathbf{X}$ comprises the expression profile resulting from one experiment. After applying the LNMF- algorithm [7], at least one column of $\mathbf{W}$, called a *metagene* is expected to be characteristic of a regulatory process, which is related to the class specific signature of the experiments. Its contribution to the observed expression profiles is contained in a corresponding row of matrix $\mathbf{H}$, called a *meta-experiment.* Once an informative meta-experiment is identified, further analysis can be focussed on the genes contained in the corresponding *metagene.* As before all experiments are labeled according to known diagnostics. The correlation coefficients $c(\mathbf{h}^j, \mathbf{d})$ between every *meta-experiment* $\mathbf{h}^j$ and $\mathbf{d}$ are then computed. Empirically, $|c| > 0.9$ signifies a satisfactory similarity between a *meta-experiment* and the design vector. The number of extracted basis components k, i.e. the *metagenes*, controls the structure of $\mathbf{W}$ and $\mathbf{H}$. For several decompositions $\mathbf{X} = \mathbf{WH}$ using different numbers k of *metagenes*, the rows of $\mathbf{H}$ are studied with respect to their correlation with the design vector. A *metagene* is considered **informative** only if **all** entries of the corresponding *meta-experiment* which belong to class 1 are smaller than **all** other entries of that *meta-experiment* (or vice versa). After 5000 iterations, the cost function of the LNMF algorithm did not show noticeable changes with any of the data sets investigated. For $k = 2, \ldots, 49$, ten separate simulations were carried out and only the simulation showing the smallest reconstruction error was stored. Further matrix decompositions with $k = 50, \ldots, 400$ *metagenes* were examined. In the latter case, three simulations were performed only for each k.

4 Results

4.1 Breast Cancer Data Set

In order to test the classification algorithms for diagnostic purposes we first selected the set of expression profiles from bone metastasis mediating breast cancer cell lines provided by [5].

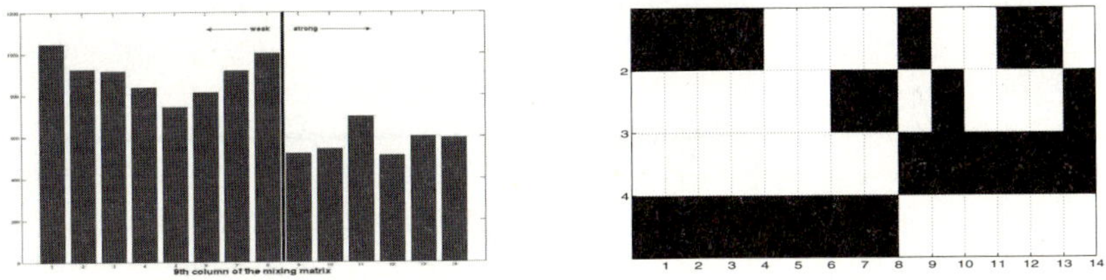

Fig. 1. *Left*: Entries of the 9-th column of matrix **A** as estimated with JADE, *Right*: Matrix **H** using $k = 4$ as estimated with LNMF. Row 3 and 4 show a clear separation between columns $1, \ldots, 8$ and columns $9, \ldots, 14$.

ICA Analysis. The analysis of the 14×14 matrix **A** identified one column with a correlation coefficient of 0.89 (see Fig. 1). Hence $\mathbf{s}_9$ should contain genes which provide diagnostic markers for the metastasis forming ability of the cell lines considered. In [5], a list of 16 putatively informative genes is provided. As shown in Table 1 the expression levels (taken from **S**) across all M experiments of many of these genes exhibit a high correlation with the design vector **d** indicating a rather high single discriminative power. Many of the genes belong to the most negatively expressed genes of *expression mode* 9.

An even more revealing picture appears if one divides componentwise the rows of the data matrix by the weighted row of the informative *expression mode*. The resulting diagram marks genes which contribute most to the observation. Many of the genes listed by [5] stick out as informative here (see Fig. 2).

NMF Analysis. The same data set was also analyzed using the LNMF algorithm. The decomposition is very robust and highly accurate. Considering the correlation between any row of matrix **H** and the design vector **d**, the decompositions into $k = 4, 20, 45, 47$ and $k = 120, 230, 400$ *metagenes* are suggested as being most informative. Considering the case $k = 4$, two of the four rows of the 4×14 matrix **H** show excellent correlations to the design vector (weak/strong),

Table 1. The correlation $|c|$ of the gene vector $\mathbf{s}_n$ with the design vector $\mathbf{d}$ for the 16 genes suggested by [5]. *number* denotes the column index in the data set $\mathbf{X}$, *gene name* denotes the affymetrix-ids, — genes missing in the reduced data set.

number	affymetrix-id	gene name	c-value	number	affymetrix-id	gene name	c-value
3611	204749-at	NAP1L3	-0.96	3694	204948-s-at	FST	-0.89
1586	201859-at	PRG1	-0.95	10480	222162-s-at	ADAMTS1	-0.86
5007	209101-at	CTGF	-0.94	6133	211919-s-at	CXCR4	-0.81
4311	207345-at	FST	-0.93	4233	206926-s-at	IL11	-0.57
1585	201858-s-at	PRG1	-0.92	3469	204475-at	MMP1	-0.47
4529	208378-x-at	FGF5	-0.92	4232	206924-at	IL11	-0.43
5532	209949-at	NCF2	-0.92	—	210310-s-at	FGF5	—
860	201041-s-at	DUSP1	-0.89	—	209201-x-at	CXCR4	—

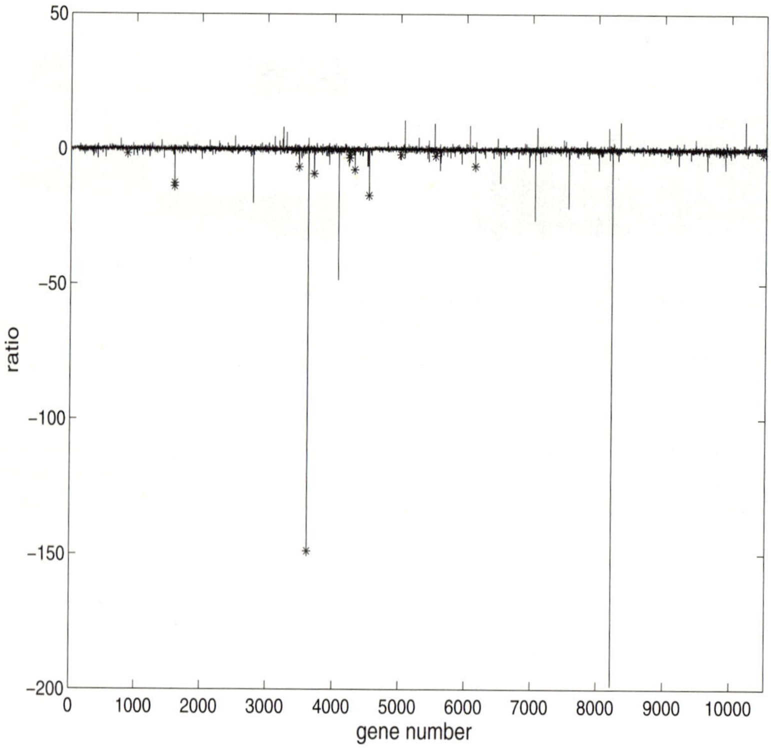

Fig. 2. Componentwise ratio of row $\mathbf{x}_9$ with $a_{n9}\mathbf{s}_9$. The genes of [5] are marked with crosses.

see Fig. 1, with coefficients $c(\text{row } 3, \mathbf{d}) = -0.91$ and $c(\text{row } 4, \mathbf{d}) = 0.91$, respectively. Thus, a decomposition in a comparatively small set of *metagenes* perfectly displays the diagnostic structure of the breast cancer data set. For the sake of comparison, a decomposition into $k = 20$ metagenes revealed four informative *meta-experiments* and their related *metagenes*. A comparison of the ten most expressed genes in each of the four identified metagenes shows, that 5 genes were also identified in case of $k = 4$, while 7 genes were also identified with ICA and 9 genes were identified with a SVM [8] approach as well. These genes are spread over all four *metagenes*.

4.2 Monocyte - Macrophage Data Set

4.3 ICA Analysis

Using a decomposition into $k = M = 14$ independent expression modes, column $\mathbf{a}_7$ of the resulting mixing matrix $\mathbf{A}$ showed a moderately strong correlation $|c| = 0.7$ with the design vector $\mathbf{d}_1$ with components $d_i = -1, i = 1, \ldots, 7$ and $d_i = 1, i = 8, \ldots, 14$ to discriminate GEPs taken from monocytes from those taken from macrophages. Column 1 showed a correlation coefficient $|c| = -0.95$ with design vector $\mathbf{d}_2$ with components $d_i = 1, i = 1, \ldots, 4$ and $i = 8, \ldots, 11$, while $d_i = -1, i = 5, \ldots, 7$ and $12, \ldots, 14$. These signatures are shown in Fig. 3. The signature of column 7 is not very clear cut. Hence a systematic investigation

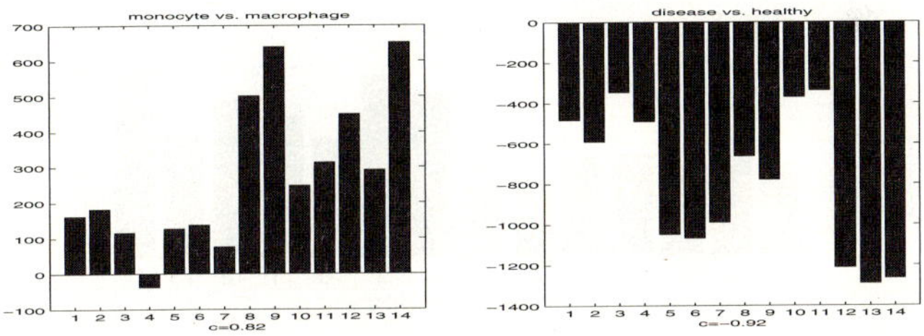

Fig. 3. Signature of column 7 (left) related with the discrimination monocyte vs macrophage (case 1) and column 1 (right) related with the discrimination healthy vs diseased (case 2) for $k = M = 14$ extracted *expression modes*

Table 2. Ten most strongly expressed genes in source 3 related with case 2 and in source 6 related with case 1; * detected by ICA, $k = 3$, ** detected by ICA, $k = 8$

	3neg		3pos		6neg		6pos	
	gene	loki-id	gene	loki-id	gene	loki-id	gene	loki-id
1	4546	OAZ1*	5834	3GM2A	752	NFKBIA	530	GPNMB*
2	4257	C6orf62	5863	STAB1*	1780	S100A9	1914	MMP9
3	2592	ARPC2*	3901	GM2A	1495	IL8*	304	CTSB
4	3552	RPL7*	4490	HLA-A	2191	FCN1	1485	FUCA1*
5	1634	S100A4	4482	SOD2	1525	S100A8*	958	LIPA
6	4686	ITM2B	752	NFKBIA	5675	CSPG2	160	CD63
7	619	ARHGDIB	1495	IL8*	1392	TNFAIP3	788	LAMP1
8	4588	TMSB4X*	2892	HLA-B	464	DUSP1	2601	TFRC
9	1973	ALOX5AP	3237	SAT	965	PRG1	572	CSTB
10	2750	HLA-DRA*	332	PSAP*	2176	FPR1	3855	K-ALPHA-1

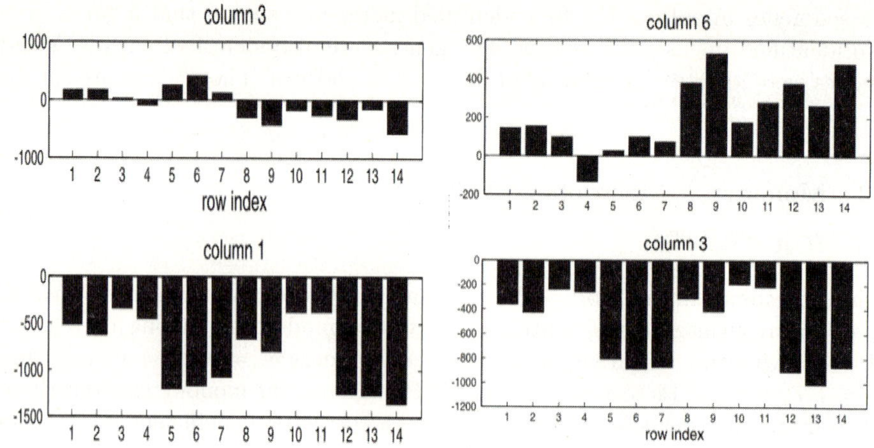

Fig. 4. Signatures of columns 1 & 3 (left) for $k = 3$ and columns 3 & 6 (right) for $k = 8$ related to cases 1 (monocyte vs macrophage) and 2 (healthy vs diseased)

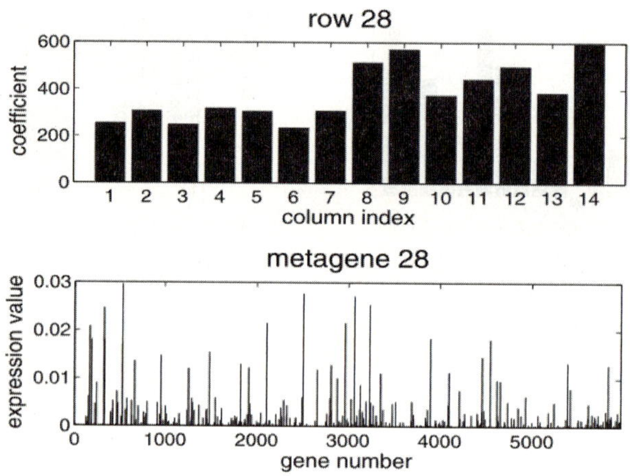

Fig. 5. Signature of row 28 of $\mathbf{H}_{k=29}$ (*meta-experiment* 28) and corresponding column $k = 28$ of $\mathbf{W}_{k=29}$ (*metagene* 28)

of the structure of the mixing matrices was carried out while increasing the extracted number of *expression modes* from $k = 2, \ldots, 14$. The resulting maximal correlation coefficients $|c(k)|$ showed little variation in both cases with average values $\langle |c_1| \rangle = 0.79$ and $\langle |c_1| \rangle = 0.94$. The maxima occur at $k = 3$ in case 1 and at $k = 8$ in case 2. The corresponding column signatures are also shown in Fig. 3. A list of the 10 most strongly expressed genes in each case is given in Table 2. Note that the dimension reduction can be done during the whitening step of the JADE algorithm. The information loss is not critical in any case as the first three principal components cover 96, 1% of the variance.

4.4 LNMF Analysis

Monocyte vs. Macrophage. A LNMF analysis was also performed on the 14 experiments of the MoMa data set. Again the number k of extracted *metagenes* was varied systematically to identify an optimal decomposition of the $N \times M$ data matrix $\mathbf{X}$. For every k, the correlation coefficients between the *meta-experiments* and the design vectors $\mathbf{d}_i, i = 1, 2$ were computed. For $k > 100$, several *meta-experiments* showing small entries for all monocytes < entries for the macrophage experiments, indicated by a correlation of $c > 0.9$. Up to $k = 90$ mostly one significant meta-experiment was observed, for $k > 90$, except $k = 120, 170$ and 190, at least two significant meta-experiments were detected. Rows of $\mathbf{H}$ related to the reverse case of macrophage < monocyte do not appear at a comparable level of correlation to the design vector. Figure 3 exhibits the signature of row 28 of $\mathbf{H}_{k=29}$ and the related *metagene*.

Healthy vs. Diseased. In this case, the number of *meta-experiments* with a strong correlation with the design vector reflecting over-expressed genes in

Table 3. The 10 most expressed genes in *metagene* 28; * detected by ICA, $k = 3$, ** detected by ICA, $k = 8$

metagene 28					
k	gene	id	k	gene-nr.	loki-id
1	530	*GPNMB**	6	2968	*HLA-DRB1
2	2511	*HLA-DRB1	7	2109	HLA-DRB1
3	3068	*CD74	8	170	GRN
4	3237	* SAT	9	3901	GM2A
5	327	*PSAP	10	4550	GRN

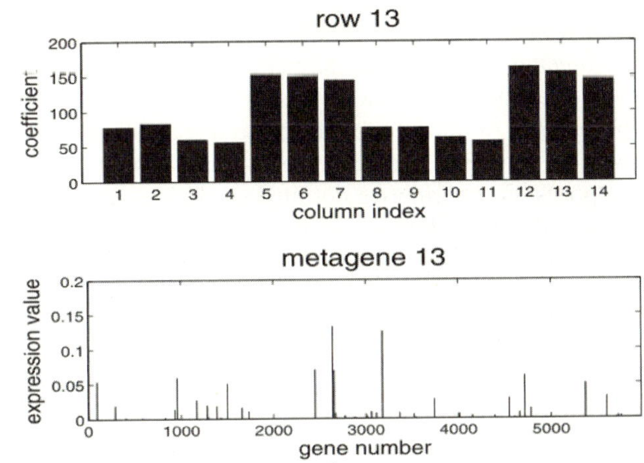

Fig. 6. Signature of *meta-experiment* $\mathbf{h}_{13}$ and corresponding *metagene* $\mathbf{w}_{13}$

Table 4. The 10 most expressed genes in *metagene* 13

metagene 13					
k	gene	loki-id	k	gene	loki-id
1	2641	SAP18	6	954	NEDD8
2	3179	—	7	95	RPS25
3	2450	RAB1A	8	1501	ATP6V1C1
4	2657	SUMO1	9	5368	CHMP5
5	4713	RAB31	10	5594	TSPYL1

case of cell lines taken from Niemann Pick C patients increases nearly linearly with increasing k. In case of under-expressed *metagenes* related to the disease, only a few significant meta-experiments appear for $k > 60$. As an example, a decomposition in $k = 370$ metagenes is considered. *Meta-experiment* 13 yields $c = -0.98$ with respect to the separation between the classes "healthy" and "diseased" (see figure 6). The 10 most strongly expressed genes in *metagene* 13 which qualify as marker genes for the discrimination between healthy subjects and Niemann Pick C patients are listed in Table 4.

5 Conclusion

The application of matrix decomposition techniques like ICA and NMF to microarray data explores the possibility to extract features like statistically independent *expression modes* or strictly positive and sparsely encoded *metagenes* which might offer a more favorable and intuitive interpretation of the underlying regulatory processes. Combined with a design function, reflecting the experimental protocol, biomedical knowledge is incorporated into the data analysis task which allows to construct a classifier for diagnostic purposes based on a global analysis of the whole data set rather than a statistical analysis based on single gene properties. This global analysis is based on the columns (ICA) or rows (NMF) of a matrix which contains the weights with which the underlying *expression modes* or *metagenes* contribute to any given observation in response to an applied environmental stimulus. If the signature of these column or row vectors matches the experimental design vector, the related *expression mode* or *metagene* contains genes with a high discriminative power which can be used as biomarkers for diagnostic purposes. Furthermore a detailed statistical analysis of these informative genes combined with a data bank search for their functional annotations might reveal underlying gene regulatory networks and can help elucidate the processes at the roots of the disease investigated. In any case knowledge of such marker genes allows to construct a simple and cheap chip for diagnostic purposes.

Acknowledgement

Support by grants from Siemens AG, Munich and the DFG (Graduate College 638) and the DAAD (PPP Hispano - Alemana) is gratefully acknowledged.

References

1. Baldi, P., Hatfield, W.: DNA Microarrays and Gene Epression. Cambridge University Press, Cambridge (2002)
2. Cichocki, A., Amari, S.-I.: Adaptive Blind Signal and Image Processing. John Wiley & Sons, Chichester (2002)
3. Diamantaras, K.I., Kung, S.Y.: Principal Component Neural Networks, Theory and Applications. Wiley, Chichester (1996)
4. Souloumiac, A., Cardoso, J.-F.: Blind Beamformimg for non-Gaussian signals. IEEE Proc. 140, 362–370 (1993)
5. Kang, Y., Siegel, P.M., Shu, A., Drobnjak, M., Kakonen, S.M., Cordón, C., Guise, T.A., Massagué, J.: A multigenic program mdeiating breast cancer metastasis to bone. Cancer Cell 3, 537–549 (2003)
6. Lee, S.-I., Batzoglou, S.: Application of independent component analysis to microarrays. Genome Biology, 4:R76.1–R76.21 (2003)
7. Zhang, H.J., Cheng, Q., Li, S.Z., Hou, X.W.: Learning spatially localized, parts-based representation. In: Proc. IEEE Conf. Computer Vision and Pattern recognition, Kauai, Hawaii, USA (2001)
8. Schachtner, R.: Machine Learning Approaches to the Analysis of Microarray Data. Diploma Thesis, University of Regensburg (2006)

Neural Network Approach for Mass Spectrometry Prediction by Peptide Prototyping

Alexandra Scherbart[1], Wiebke Timm[1,2], Sebastian Böcker[3], and Tim W. Nattkemper[1]

[1] Applied Neuroinformatics Group, Faculty of Technology, Bielefeld University, Postfach 10 01 31, 33501 Bielefeld
{ascherba,wtimm,tnattkem}@techfak.uni-bielefeld.de
[2] Intl. NRW Graduate School of Bioinformatics and Genome Research, Bielefeld University
[3] Bioinformatics Group, Jena University
boecker@minet.uni-jena.de

Abstract. In todays bioinformatics, Mass spectrometry (MS) is the key technique for the identification of proteins. A prediction of spectrum peak intensities from pre computed molecular features would pave the way to better understanding of spectrometry data and improved spectrum evaluation. We propose a neural network architecture of Local Linear Map (LLM)-type for peptide prototyping and learning locally tuned regression functions for peak intensity prediction in MALDI-TOF mass spectra. We obtain results comparable to those obtained by ν-Support Vector Regression and show how the LLM learning architecture provides a basis for peptide feature profiling and visualisation.

1 Introduction

In todays bioinformatics, Mass spectrometry (MS) is the key technique for the identification of proteins. Matrix-assisted laser desorption ionization (MALDI) is one of the most often used technique for the analysis of whole cell proteomes in high-throughput experiments. There are different applications of MALDI-MS where the prediction of peak heights (referred to as intensities) in the spectrum are needed for further improvements: Protein identification is commonly done by comparing the peak's masses from a spectrum – the so called protein mass finger print (PMF) – to theoretical PMFs in a data base, generating a score for each comparison. Different tools are available for this purpose. For an overview see [SCB05]. These tools rarely use peak intensities, because there is no model to calculate the theoretical PMFs directly. The use of peak intensities could improve the reliability of protein identification without lowering the error rate, as was shown by Elias *et al.* for tandem MS[EGK$^+$04].

Another application of MALDI where peak intensities are important is quantitative proteomics, where proteins in a complex sample are quantified or protein abundances across different samples are compared. For the prediction of MALDI PMF there has been one study so far by Gay *et al.* who applied different regression and classification algorithms[GBHA02]. Tang *et al.* used multi-layer neural

J. Marques de Sá et al. (Eds.): ICANN 2007, Part II, LNCS 4669, pp. 90–99, 2007.

networks to predict peptide detectabilities (i.e. the frequency with which peaks occur in spectra) in LC/MS ion trap spectra[TAA+06] which is a related problem.

An algorithmic approach for peak intensity prediction is a non-trivial task because of several obstacles: The extraction of PMF from spectra is a signal processing task which can not be done perfectly. Data from this domain is always very noisy and contains errors introduced by preprocessing steps in the wet lab as well as in signal processing. Misidentifications may even lead to wrong sequences. Intensity values can be distorted due to the unknown scale of spectra. It is nearly impossible to come by a large enough data set from real proteins where the content is known, i.e. there is no perfect gold standard, because of the not reproducible and non-unique peptide/intensity relation.

To overcome these obstacles to predict peak intensities in MALDI-TOF spectra based on a pre-selected training set of peptide/peak intensity pairs, a method is needed, that is able to (a) determine peptide profiles and (b) learn locally tuned regression functions for peak intensity prediction, since the peak intensity functions may vary fundamentally for two different peptide profiles. For this purpose we consider an artificial neural net architecture of Local Linear Map(LLM)-type, since it combines unsupervised (a) and supervised (b) learning principles.

The LLM-architecture is well suited for this task due to its transparency. It is simple to implement, can cope with large data sets, is easy adaptable to new data by a slight deviation of the parameters without loss of information. Other than for example support vector regression (SVR) it can be used for data mining once adapted in a straight forward manner, as demonstrated in this work. We propose a combination of unsupervised and supervised learning architecture with comparable results in predicting the peaks intensities to ν-Support Vector Regression (SVR) [TBTN06]. The mixture of linear experts derives implicit models for characterizing peptides and feature analysis as an unsupervised learning task. The second step consists of supervised adaptation of the neural network and prediction of peaks intensities.

2 Materials and Methods

2.1 Data

In this study we use one dataset **A** of peptides of MALDI mass spectra. It consists of 66 spectra of 29 different proteins, with 16 of these proteins being present in multiple spectra. Peak extraction steps include soft filtering, baseline correction, peak picking and isotopic deconvolution in the corresponding raw spectra. The resulting list of peaks is matched against masses derived from a theoretical tryptic digestion. These steps for **A** result in 857 matched peaks corresponding to 415 different peptides.

For preprocessing, normalization of the intensities is necessary, because spectra do not have the same scale. For a MALDI spectrum the exact amount of protein sample that leads to is not known, nor is it possible to scale spectra belonging to the same protein by the same amount. There is no peptide that connects the scales of spectra.

The normalized intensities are therefore computed following two different heuristics, in order to use values from different spectra together. In the remainder, the intensities refer to scaling the original intensity by "mean corrected ion current":

$$I_M = \frac{I_i^{\text{orig}}}{\sum_{i=1}^{N} I_i^{\text{orig}}} \quad \text{with} \quad I_i^{\text{orig}} = I_i - B_i - N_i, i = 1, \ldots, N$$

where B_i is the baseline value and N_i is the noise determined in the denoising step. The original matched peaks intensity I_i^{orig} after denoising and baseline correction is divided by the sum of all N values in the whole spectrum yielding the scaled intensity I_M. Subsequently, the natural logarithm of the intensities is applied to compute the final intensity output values.

2.2 Feature Sets

One of the most important questions in conjunction with finding a model for predicting peak intensities is the representation of the peptides. A suitable feature space is the precondition for success of any machine learning method.

We combine different properties of peptides to represent features of a peptide. Amino acid frequencies, typically used in bioinformatics, in conjunction with chemical features of the peptides are used to create the heuristically selected 18-dimensional feature space. It is built by different types of characterization we assume to be relevant for MALDI ionization and additional features that are chosen in an ad hoc feature forward selection.

Most of the peptides in the data set occur multiple times in different spectra with different intensity values. Due to limitation of training data, we eliminate outliers (potential noisy peptides) by mapping each peptide to one unique value, the α-trimmed mean of all intensities per distinct peptide with $\alpha = 50\%$. The α-trimmed mean is defined as the mean of the center 50% of an ordered list. In the case of less than 4 peptides in the list a simple mean is taken.

2.3 Local Linear Map / Mixture of Linear Experts

The task of mass spectrometry prediction and peptide prototyping corresponds to the task of unsupervised clustering as well as classification and supervised prediction. The problem can be stated as follows:

Given a training set $\Gamma = \{(\mathbf{x}, \mathbf{y})_i, i = 1, \ldots, N\}$, consisting of input-output pairs: peptide patterns $\mathbf{x}_i$, which are elements of feature space $\mathcal{X} = \mathbb{R}^t$, and real-valued outputs, i.e. intensities, $\mathbf{y}_i \in \mathbb{R}$. One promising approach would be to find a set of clusters and prototypes representing the data points best according to the statistical properties of the data provided. After assigning every input point to a prototype, a prediction of a real-valued output Y has to be done.

For determining peptide prototypes and learning into the mapping the output, i.e. intensity space, we propose to use the Local Linear Map (LLM)-architecture. The LLM combines unsupervised vector quantisation algorithm for computing a voronoi tessellation of the input space $\mathcal{X}$ with supervised techniques for feature classification.

The artificial neural net (ANN) of Local Linear Map-type [Rit91] was originally motivated by the Self-Organising Map by Kohonen [Koh82] and has been shown to be a valuable tool for the fast learning of non-linear mappings $\mathcal{C} : \mathbb{R}^{d_{in}} \to \mathbb{R}^{d_{out}}$.

A LLM consists of n_l nodes $\mathbf{v}_i, i = 1, \ldots, n_l$. Each node consists of a triple

$$\mathbf{v}_i = (\mathbf{w}_i^{in}, \mathbf{w}_i^{out}, \mathbf{A}_i).$$

The vectors $\mathbf{w}_i^{in} \in \mathbb{R}^{d_{in}}$ are used to build prototype vectors adapting to the statistical properties of the input data $\mathbf{x}_\xi \in \mathbb{R}^{d_{in}}$ provided. The vectors $\mathbf{w}_i^{out} \in \mathbb{R}^{d_{out}}$ approximate the distribution of the target values $\mathbf{y}_\xi \in \mathbb{R}^{d_{out}}$. The matrices $\mathbf{A}_i \in \mathbb{R}^{d_{in} \times d_{out}}$ are locally trained mappings from the input to the output space.

Vector quantisation - In the unsupervised training phase the n_l prototype vectors are adapted following Neural-Gas (NG) learning rule. The distance between input pattern $\mathbf{x}_\xi$ and prototype $\mathbf{v}_i$ yields a ranking among the neural gas neurons. The learning procedure changes the weights according to the ranking of the prototypes $r_i(\mathbf{x}, l)$, such that

$$\Delta \mathbf{w}_i^{in} = \epsilon^{in} \cdot h_\sigma(r_i(\mathbf{x}, l)) \cdot (\mathbf{x}_\xi - \mathbf{w}_i^{in}) = \epsilon^{in} \cdot \exp\left(-\frac{r_i(\mathbf{x}, l)}{\lambda}\right)(\mathbf{x}_\xi - \mathbf{w}_i^{in}).$$

After adapting the prototypes, each of the input vectors $\mathbf{x}$ can be associated with its closest prototype as a winner-takes-all (WTA) rule:

$$\mathbf{w}_\kappa = \arg\min_{\mathbf{w}} \left\{ \|\mathbf{x} - \mathbf{w}_i^{in}\| \right\}. \tag{1}$$

Training of local mappings from input to the output space is performed in the second step. Subsequently to unsupervised adaptation and tessellation of the input space $\mathcal{X}$, a local expert is assigned to each of the n_l voronoi cells.

The mapping of an arbitrary input vector $\mathbf{x}$ to an output $\mathcal{C}(X)$ is computed by

$$\mathcal{C}(\mathbf{x}) = \mathbf{w}_\kappa^{out} + \mathbf{A}_\kappa (\mathbf{x} - \mathbf{w}_\kappa^{in}) \tag{2}$$

by the corresponding local expert $\mathbf{w}_\kappa$. The weights $\mathbf{w}_i^{out}$ and the linear map $\mathbf{A}_i$ are changed iteratively applying the learning rules:

$$\Delta \mathbf{w}_i^{out} = \epsilon^{out} \cdot h_\sigma(r_i(\mathbf{x}, l)) \cdot (\mathbf{y}_\xi - \mathcal{C}(\mathbf{x}_\xi))$$
$$\Delta \mathbf{A}_i = \epsilon^{\mathbf{A}} \cdot h_\sigma(r_i(\mathbf{x}, l)) \cdot (\mathbf{y}_\xi - \mathcal{C}(\mathbf{x}_\xi)) \cdot \frac{(\mathbf{x}_\xi - \mathbf{w}_i^{in})^T}{\|\mathbf{x}_\xi - \mathbf{w}_i^{in}\|^2}$$

2.4 Evaluation

About 10% of the centered and normalized data are used for validation and put aside. The remaining dataset is used to train the LLM and to find the best parameter set using 10-fold Cross-Validation (CV). So, the remaining dataset is split into 10 portions and one set is used for testing performance of the selected model. It was ensured that peptides from one spectrum as well as peptides occurring in more than one spectrum are found in only one of the portions.

Model Selection: Grid search over the parameter space $P = (n_l, \epsilon^A)$ is performed to determine optimal parameters for learning. The remaining learning parameters for the LLM were set to initial values $\epsilon^{out} = 0.3, \epsilon^{in} = 0.5$ and $\lambda = 10$ decreasing over time. A 10-fold-CV is done for each parameter set. For every point in the parameter space the prediction accuracy for every training/test set is determined by squared pearson-correlation coefficient r^2 and root mean square error (RMSE) of the test set. The choice of the best parameter set is made by the best mean r^2 over all 10 test sets while training the learning algorithm.

Model Assessment: The final model with the optimal parameters is chosen. To validate its prediction (generalization) error on new data, the validation set is used, which has not taken part in training.

3 Results

The following results are evaluated for two subsets of the entire data set **A**. The first set, denoted by $\mathbf{A_{TM}}$, contains peptides mapped by α-trimmed mean (2.2), whereas the second set, $\mathbf{A_{Dup}}$, contains all peptides including these ones occurring in multiple spectra. In the remainder, we refer to them as duplicates. The validation set of $\mathbf{A_{TM}}$ consists of 44 items and for $\mathbf{A_{Dup}}$ of 73 items. One advantage of the ANN is, that is capable of dealing with duplicates straightforward, which means data points $(\mathbf{x}_i, \mathbf{y}_i), (\mathbf{x}_j, \mathbf{y}_j)$ in input space, $\exists i, j$, s. t. $\mathbf{x}_i = \mathbf{x}_j : \mathbf{y}_i \neq \mathbf{y}_j$ in contrast to many other techniques, as for example, Support Vector Regression (SVR).

3.1 Peptide Prototyping

A display of the prototype vectors resulting from the neural gas training allows a profiling of peptides. In the following Fig. **??** a resulting parallel coordinates plot used for multivariate data plotting for six prototypes in case of $\mathbf{A_{Dup}}$ is shown. A set of parallel axes are drawn for each feature, where for each variable the range of values (from min to max, from bottom to top respectively) covered by the prototypes is shown.

The correlation of input space reflects in the prototype distribution. We can see that some of the features ('VASM830103','WILM950102','FINA770101', 'ARGP820102') show a correlation to the mass, while no such tendency can be observed for the other features. 'OOBM850104' (measure of non-bonded energy), 'ROBB760107' (information measure) and 'M' (no. of methionine) show the least similarity to any other feature. If we look at prototype 0 and 4, it can be seen that they cover contrary outmost areas in data space in almost all features except the three mentioned ones. Another thing to be noted is that prototypes 2 and 4 are near to each other for almost all features, except for the three mentioned features and arginin_count as well as 'GB500' (the last two being highly

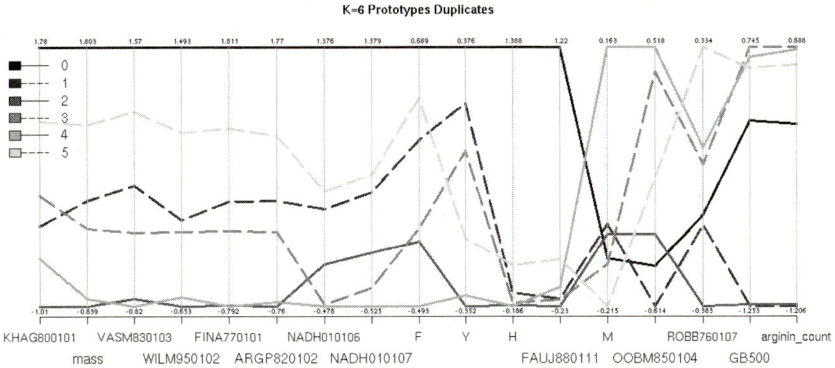

Fig. 1. Parallel coordinates plot for six prototypes in case of $\mathbf{A_{Dup}}$. For every feature the range of values covered by the prototypes is shown.

correlated), where they split up to almost the extremes. Similar behaviour can be observed for the prototypes 1 and 3.

The six prototypes take three to five levels for each feature, two or more prototypes sharing the same region. OOBM850104 and KHAG800101 (kerr constant increment) show the most even distribution of prototypes.

Thus the prototypes share ranges in certain features and split up for others, achieving a nice spread in data space: For a datapoint that is similar to two prototypes sharing their space for a set of features, other features decide which prototype it is assigned to.

Having observed these features being special, we can take a closer look at the data points' class assignment and there values these features (image not shown). There is no visible correlation between any of 'M','OOBM850104', 'ROBB760107' and the class label except for the extreme ranges of the latter. For 'GB500' there is a split between data of the classes 0, 1, 2 and 3, 4, 5 which corresponds well to the prototypes' distribution for this feature. If this is due to other features slightly correlated to each other or if they really carry no information with respect to the target values has to be the subject of further studies.

3.2 Predicting Peaks Intensities

We compare the prediction capabilities of the LLM for the two data sets $\mathbf{A_{TM}}$ and $\mathbf{A_{Dup}}$. The evaluation is done as described in 2.4. We perform a grid search over all parameters and the parameter set is chosen yielding the best mean r^2 of training/test sets. For the exact results of r^2 and RMSE see Table 1.

3.3 Penalizing Single Occurring Peptides

Peptides carrying not much information or being noisy should not be regarded in the learning process or at least should be less weighted. At about 50% of the peptides are represented only once in the data. We refer to them as "singles". In fact, their intensities show a wide spread distribution and may be not valuable

Table 1. Comparison of capabilities of the LLM in predicting intensities for the studied datasets $\mathbf{A_{TM}}$ and $\mathbf{A_{Dup}}$. The evaluation was done for $K = 3, 6$ and 10 prototypes. The resulting performances are given by the best r^2, the mean (r^2) and σ^2(r^2) of all 10 test sets and the corresponding error RMSE, as well as for the validation set.

	Test				*Valid*	
Dataset	r^2	ave(r^2)	σ^2(r^2)	RMSE	r^2	RMSE
$\mathbf{A_{TM}, K = 3}$	0.6773	0.4341	0.1811	1.0359	0.1611	1.6028
$\mathbf{A_{TM}, K = 6}$	0.6318	0.4391	0.1421	0.8619	0.1148	1.7938
$\mathbf{A_{TM}, K = 10}$	0.6003	0.3845	0.1881	0.9241	0.157	1.7447
$\mathbf{A_{Dup}, K = 3}$	0.647	0.433	0.1462	0.8237	0.1972	1.4516
$\mathbf{A_{Dup}, K = 6}$	0.6168	0.4286	0.1356	0.8222	0.2195	1.4294
$\mathbf{A_{Dup}, K = 10}$	0.5659	0.3599	0.1549	0.8872	0.137	1.5169

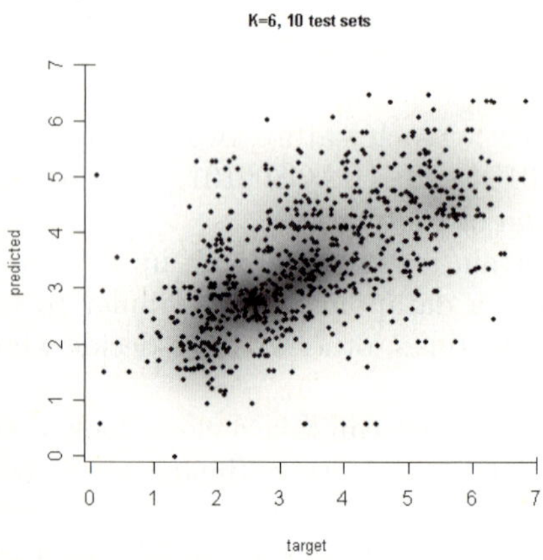

(a)

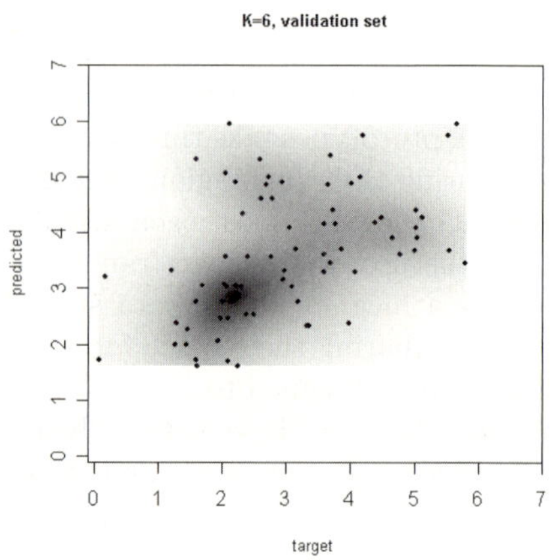

(b)

Fig. 2. Scatterplots of target vs. predicted values for a penalized prediction of $\mathbf{A_{Dup}}$ with $K = 6$ prototypes. The gray background results from a 2d density estimation of plotted points. The plot (a) is given by the resulting mapped test sets, (b) is given by the corresponding validation set. The mapped points show a wide spread and many are horizontally aligned due to the mapping of the same prototype to the same output value.

regarding signal-to-noise-ratio. It seems to be reasonable to take the duplicates more into account for the learning process. We introduce an additional factor to the learning algorithm

$$\epsilon^p = \epsilon \cdot f(peptide_{occ}), \quad \text{where} \quad \epsilon \in \left\{\epsilon^{in}, \epsilon^{out}, \epsilon^{A}\right\}$$

such that single peptides are penalized and peptides occurring more than once are more heavily weighted, ensuring, that not the whole learning process itself is slowed down. The best evaluated function of peptide occurrence was $f(x) = -\exp(-x/0.3)^{0.25} + 1.2$. Moreover, the best parameters found by evaluation were those, that took penalizing of single peptides into account.

4 Discussion

Our results show that the new LLM-approach combining data mining and supervised learning yields similar results in prediction accuracy to our first approach utilizing ν-SVR [TBTN06].

In Fig. 3 the results of the prediction accuracy for different number of prototypes and for the studied data sets $\mathbf{A_{TM}}$ and $\mathbf{A_{Dup}}$ are summed up. The mean performance of the 10 test sets is compared to the performance of the validation set in terms of r^2 and RMSE.

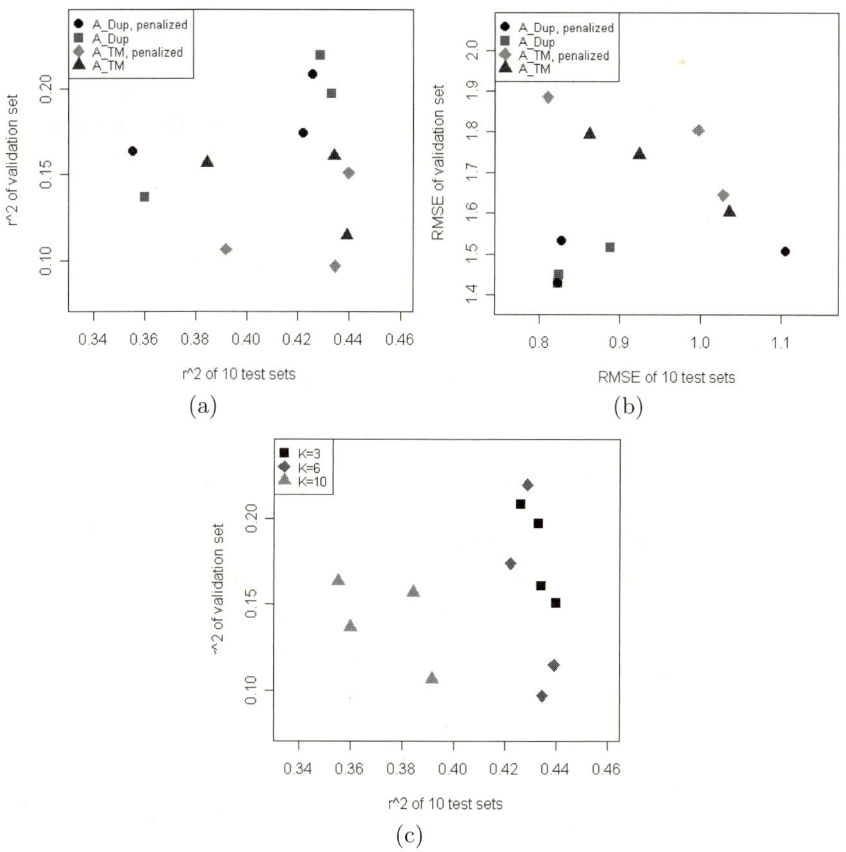

Fig. 3. Results of the prediction accuracy measured by (a) r^2 and (b)RMSE for the evaluated data sets $\mathbf{A_{TM}}$ and $\mathbf{A_{Dup}}$ taking into account the differences in penalizing single peptides. For every evaluation the results of the mean performance of the test sets is plotted against the performance of the validation set. The prediction accuracy of $\mathbf{A_{Dup}}$ yield (a) a higher correlation as well as (b) lower error compared to $\mathbf{A_{TM}}$, whereas the prediction accuracy of the penalized prediction in comparison to normal is not worse. (c) For the plot a discrimination according to the number of prototypes is done. The best results are found for $K = 3$ and for $K = 6$ in both cases of data sets.

From our results it is clear that peak intensities can be characterized and predicted by the use of the heuristically selected feature set with high prediction accuracy. It can be observed that considering the entire data set including duplicates instead of a α-trimmed mean mapped data set yields higher correlations and a better generalization performance. Moreover, an alternative mapping for the α-trimmed mean by the prototypes additionally incorporating the statistical properties of the input data is yielded by the LLM.

The investigation of the treatment of single occurring peptides, which amount 50% of the entire data, did not issue a precise statement. Therefore a comparison to data sets with more multiple spectra per protein would be helpful.

The visual inspection of the prototypes reveals that the peptides can be grouped around a set of approximately 6 profiles. Those seem to have individual mappings to peak intensity which can be discussed with biochemical experts.

The prediction capability and generalisation performance of our proposed learning architecture are determined by r^2 and RMSE of the validation set. For this set just the first 10% of the entire data set are taken. The fact that some test sets performance is worse than the performance of the chosen validation set, can be explained by the static choice of the set. The different portions of the entire data set yield a wide spread of correlation, resulting in high standard deviation of r^2 (see Table 1) over all the portions. There are test sets that seem significantly worse in prediction performance over all training sets. There exists a positive correlation to the number of test set examples.

5 Conclusion

We propose an algorithmic approach for peak intensity prediction in MALDI-TOF spectra. The proposed model for peptide prototyping and prediction of peak intensities with the architecture of Local Linear Map-type has been shown to be a valuable neural network tool for these tasks combining unsupervised and supervised learning architecture. The LLM includes determining peptide profiles in the data set and the mixture of linear expert are able to learn locally tuned regression functions for peak intensity prediction. The heuristically selected feature space is a good choice as the characteristics of peptides are reflected. Some features do not contribute to the assignment of data points to one of the prototypes. If this is due to the number of other features slightly correlated to each other or if they really carry no information with respect to the target values has to be the subject of further studies. The experiments with the considered data set have demonstrated the capabilities of the proposed neural net.

References

[EGK+04] Elias, J.E., Gibbons, F.D., King, O.D., Roth, F.P., Gygi, S.P.: Intensity-based protein identification by machine learning from a library of tandem mass spectra. Nat Biotechnol 22(2), 214–219 (2004)

[GBHA02] Gay, S., Binz, P.-A., Hochstrasser, D.F., Appel, R.D.: Peptide mass fingerprinting peak intensity prediction: extracting knowledge from spectra. Proteomics 2(10), 1374–1391 (2002)

[Koh82] Kohonen, T.: Self-organized formation of topologically correct feature maps. Biological Cybernetics 43, 59–69 (1982)

[Rit91] Ritter, H.: Learning with the self-organizing map. In: Kohonen, T., et al. (eds.) Artificial Neural Networks, pp. 379–384. Elsevier Science Publishers, Amsterdam (1991)

[SCB05] Shadforth, I., Crowther, D., Bessant, C.: Protein and peptide identification algorithms using MS for use in high-throughput, automated pipelines. Proteomics 5(16), 4082–4095 (2005)

[TAA+06] Tang, H., Arnold, R.J., Alves, P., Xun, Z., Clemmer, D.E., Novotny, M.V., Reilly, J.P., Radivojac, P.: A computational approach toward label-free protein quantification using predicted peptide detectability. Bioinformatics 22(14), 481–488 (2006)

[TBTN06] Timm, W., Böcker, S., Twellmann, T., Nattkemper, T.W.: Peak intensity prediction for pmf mass spectra using support vector regression. In: Proc. of the 7th International FLINS Conference on Applied Artificial Intelligence (2006)

Identifying Binding Sites in Sequential Genomic Data

Mark Robinson, Cristina González Castellano, Rod Adams,
Neil Davey, and Yi Sun

Science and Technology Research Institute
Univesrity of Hertfordshire, UK
{M.Robinson,R.G.Adams,Y.2.Sun,N.Davey}@herts.ac.uk,
c.gonzalezcastellano@yahoo.es

Abstract. The identification of *cis*-regulatory binding sites in DNA is a difficult problem in computational biology. To obtain a full understanding of the complex machinery embodied in genetic regulatory networks it is necessary to know both the identity of the regulatory transcription factors together with the location of their binding sites in the genome. We show that using an SVM together with data sampling, to integrate the results of individual algorithms specialised for the prediction of binding site locations, can produce significant improvements upon the original algorithms. These results make more tractable the expensive experimental procedure of actually verifying the predictions.

Keywords: Computational Biology, Support Vector Machine, Imbalanced data, Sampling, Transcription Factor Binding Sites.

1 Introduction

Binding site prediction is both biologically important and computationally interesting. One aspect that is challenging is the imbalanced nature of the data and that has allowed us to explore some powerful techniques to address this issue. In addition the nature of the problem allows domain specific heuristics to be applied to the classification problem. Specifically we can remove some of the final predicted binding sites as not being biologically plausible.

Computational predictions are invaluable for deciphering the regulatory control of individual genes and by extension aiding in the automated construction of the genetic regulatory networks to which these genes contribute. Improving the quality of computational methods for predicting the location of transcription factor binding sites (*TFBS*) is therefore an important research goal. Currently, experimental methods for characterising the binding sites found in regulatory sequences are both costly and time consuming. Computational predictions are therefore often used to guide experimental techniques. Larger scale studies, reconstructing the regulatory networks for entire systems or genomes, are therefore particularly reliant on computational predictions, there being few alternatives available.

Computational prediction of *cis*-regulatory binding sites is widely acknowledged as a difficult task [1]. Binding sites are notoriously variable from instance to instance and they can be located considerable distances from the gene being regulated in higher

J. Marques de Sá et al. (Eds.): ICANN 2007, Part II, LNCS 4669, pp. 100–109, 2007.

eukaryotes. Many algorithmic approaches are inherently constrained with respect to the range of binding sites that they can be expected to reliably predict. For example, co-regulatory algorithms would only be expected to successfully find binding sites common to a set of co-expressed promoters, not any unique binding sites that might also be present. Scanning algorithms are likewise limited by the quality of the position weight matrices available for the organism being studied. Given the differing aims of these algorithms it is reasonable to suppose that an efficient method for integrating predictions from these diverse strategies should increase the range of detectable binding sites. Furthermore, an efficient integration strategy may be able to use multiple sources of information to remove many false positive predictions, while also strengthening our confidence about many true positive predictions. The use of algorithmic predictions prone to high rates of false positive is particularly costly to experimental biologists using the predictions to guide experiments. High rates of false positive predictions also limits the utility of prediction algorithms for their use in network reconstruction. Reduction of the false positive rates is therefore a high priority.

In this paper we show how the algorithmic predictions can be combined so that a Support Vector Machine (SVM) can perform a new prediction that significantly improves on the performance of any one of the individual algorithms. Moreover we show how the number of false positive predictions can be reduced by around 80%.

2 Background

The use of a non-linear classification algorithm for the purposes of integrating the predictions from a set of *cis*-regulatory binding site prediction algorithms is explored in this paper. This is achieved by first running a set of established prediction algorithms, chosen to represent a range of different algorithmic strategies, on a set of annotated promoter sequences. Subsequently, an SVM is trained to classify individual sequence positions as a component of either a binding site or the background sequence. The set of predictions from the original algorithms, appropriately sampled to account for the imbalanced nature of the data set, and labeled with experimental annotations is used for the training inputs.

Table 1. The 12 Prediction Algorithms used

Strategy	Algorithm
Scanning algorithms	Fuzznuc MotifScanner [2] Ahab [3]
Statistical algorithms	PARS Dream (2 versions) [4] Verbumculus [5]
Co-regulatory algorithms	MEME [6] AlignACE [7] Sampler
Evolutionary algorithms	SeqComp [8] Footprinter [9]

A wide range of binding site prediction algorithms were used in this study. They were selected to represent the full range of computational approaches to the binding site prediction problem. The algorithms chosen were either reported in the literature or were developed in-house or by our collaborators in the case of PARS, Dream and Sampler. Table 1 lists the algorithms used along with references. Where possible, parameter settings for the algorithms were taken from the literature, if not available, default settings were used.

3 Description of the Data

Experimentally annotated sequences were used in this study. The yeast, *S.cerevisiae* was selected for the model organism; the use of this particularly well studied model organism ensures that the annotations available are among the most complete available. 112 annotated promoter sequences were extracted from the *S.cerevisiae* promoter database [10] for training and testing the algorithms. For each promoter, 500 base-pairs (*bp*) of sequence taken immediately upstream from the transcriptional start site was considered sufficient to typically allow full regulatory characterisation in yeast [10]. In cases where annotated binding sites lay outside of this range, then the range was expanded accordingly. Likewise, where a 500 bp upstream region would overlap a coding region then it was truncated accordingly. Further details about how the data was obtained can be found in [11].

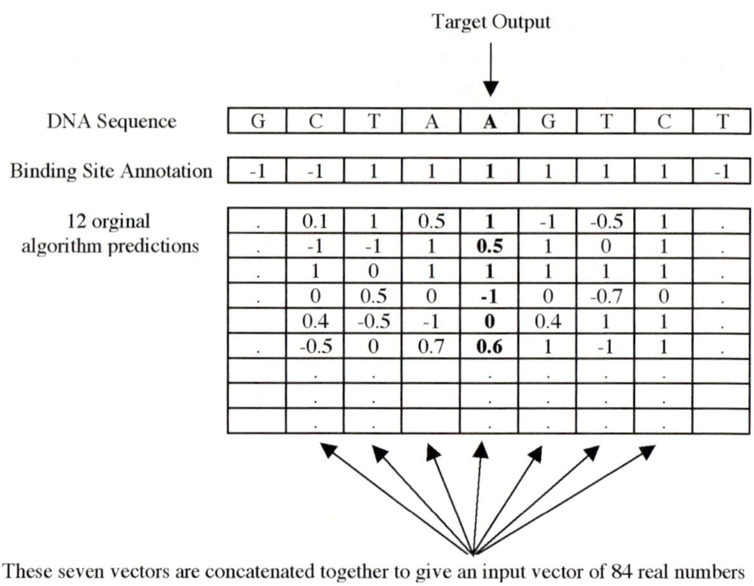

Fig. 1. The formation of the windowed data. The 12 predictions from the original algorithms for the target site are concatenated with the predictions from the 3 sites on either side. This gives an input vector of 12 by 7 real numbers. The corresponding label of this vector is the annotation of the central nucleotide.

Predictions made by the original algorithms across the dataset were placed into a matrix consisting of 67,782 12-ary real valued vectors, each associated with a binary label indicating the presence or absence of an experimental annotation at that position, see Figure 1.

Each 12-ary vector represents the predictions from all 12 original algorithms for a particular position in the dataset. All predictions in the matrix were normalised as real values in the range [-1,1] with the value of 0 allocated to sequence positions where an algorithm was unable to be run. Additionally, we contextualize the training and test datasets to ensure that the classification algorithms have data on contiguous binding site predictions. This is achieved by windowing the vectors within each of the 112 annotated promoter sequences. We use a window size of 7, providing contextual information for 3 bp either side of the position of interest.

Additionally this procedure carries the considerable benefit of eliminating a large number of repeated or inconsistent vectors which are found to be present in the data and would otherwise pose a significant obstacle to the training of the classifiers.

A number of statistics summarising the dataset are shown in Table 2.

Table 2. Summary of the data used

Total number of sequences	112
Total sequence length	67782 bp
Average sequence length	605 bp
Average number of TFBS sites per sequences	3.6
Average TFBS width	13.2 bp
Total number of TFBS sites	400
Number of unique TFBS sites	69
TFBS density in total dataset	7.8%

4 Performance Metrics

As approximately 8% of the dataset (see Table 2) is annotated as being a part of a binding site, this dataset is imbalanced. If the algorithms are to be evaluated in a useful manner simple error rates are inappropriate, it is therefore necessary to use other metrics. Several common performance metrics, such as *Recall* (also known as *Sensitivity*), *Precision*, *False Positive rate (FP-Rate)* and *F-Score*, can be defined using a confusion matrix (see Table 3) of the classification results. *Precision* describes the proportion of predictions that are accurate; *Recall* describes the proportion of binding site positions that are accurately predicted; *FP-Rate* describes the proportion of the actual negatives that are falsely predicted as positive; and the *F-Score* is the weighted harmonic mean of *Precision* and *Recall*. There is typically a trade off between *Precision* and *Recall*, making the *F-Score* particularly useful as it incorporates both measures. In this study, the weighting factor, β, was set to 1 giving equal weighting to both *Precision* and *Recall*. It is worth noting that for all these metrics a higher value represents improved performance with the solitary exception of *FP-rate* for which a lower value is preferable.

Table 3. The definition of performance measures

	Predicted Negatives	**Predicted Positives**
Actual Negatives	True Negatives - *TN*	False Positives - *FP*
Actual Positives	False Negatives - *FN*	True Positives - *TP*

$$Recall = \frac{TP}{TP + FN} \qquad Precision = \frac{TP}{TP + FP}$$

$$FP\_Rate = \frac{FP}{FP + TN} \qquad F\_Score = \frac{(1 + \beta^2)Recall \times Precision}{\beta^2 Recall + Precision}$$

5 Techniques for Learning Imbalanced Datasets

Without addressing the imbalance of the two classes in the data, classifiers will produce negligible true positive predictions. This is due to the fact that predicting that every base-pair is not part of a binding site will give high accuracy, being correct 92% of the time (with no false positives). However such a predictor is obviously worthless.

In this paper we address the problem of our imbalanced data in two ways: firstly by using data based sampling techniques [12, 13] and secondly by using different SVM error costs for the two classes [14].

5.1 Sampling Techniques

One way to address imbalance is simply to change the relative frequencies of the two classes by under sampling the majority class and over sampling the minority class. Under sampling the majority class can be done by just randomly selecting a subset of the class. Over sampling the minority class is not so simple and here we use the Synthetic Minority Oversampling Technique (*SMOTE*) [12]. For each member of the minority class its nearest neighbours in the same class are identified and new instances are created, placed randomly between the instance and its neighbours. In the first experiment the number of items in the minority class was first doubled and the number of randomly selected majority class members was then set to ensure that the final ratio of minority to majority class was 0.5. This value was selected using 5-fold cross validation experiments.

5.2 Different SVM Error Costs

In the standard SVM the primal Lagrangian that is minimized is:

$$L_p = \frac{\|\mathbf{w}^2\|}{2} + C\sum_{i=1}^{n}\xi_i - \sum_{i=1}^{n}\alpha_i[y_i(\mathbf{w}.\mathbf{x}_i + b) - 1 + \xi_i] - \sum_{i=1}^{n} r_i\xi_i$$

$$\text{subject to: } 0 \le \alpha_i \le C \text{ and } \sum_{i=1}^{n}\alpha_i y_i = 0$$

Here C represents the trade-off between the empirical error, ξ, and the margin. The problem here is that both the majority and minority classes use the same value for C, which as pointed out by Akbani et al [15] will probably leave the decision boundary too near the minority class. Veropoulos et al [14] suggest that having a different C value for the two classes may be useful. They suggest that the primal Lagrangian is modified to:

$$L_p = \frac{\|\mathbf{w}^2\|}{2} + C^+ \sum_{\{i|y_i=+1\}} \xi_i + C^- \sum_{\{i|y_i=-1\}} \xi_i - \sum_{i=1}^{n} \alpha_i [y_i(\mathbf{w}.\mathbf{x}_i + b) - 1 + \xi_i] - \sum_{i=1}^{n} r_i \xi_i$$

$$\text{subject to: } 0 \le \alpha_i \le C^+ \text{ if } y_i = +1, \quad 0 \le \alpha_i \le C^- \text{ if } y_i = -1 \quad \text{and} \sum_{i=1}^{n} \alpha_i y_i = 0$$

Here the trade-off coefficient C is split into C^+ and C^- for the two classes, allowing the decision boundary to be influenced by different trade-offs for each class. Thus the decision boundary can be moved away from the minority class by lowering C^+ with respect to C^-.

Akbani et al. [15] argue that using this technique should improve the position of the decision boundary but will not address the fact that it may be misshapen due to the relative lack of information about the distribution of the minority class. So they suggest that the minority class should also be over-sampled, using SMOTE, to produce a method they call SMOTE with Different Costs (SDC). This is one of the techniques we evaluate here.

6 Biologically Constrained Post-processing

One important concern when applying classifier algorithms to the output of many binding site prediction algorithms is that the classifier decisions could result in biologically unfeasible results. The original algorithms only predict reasonable, contiguous sets of base pairs as constituting complete binding sites. However when combined in our meta-classifier each base pair is predicted independently of the neighbouring base pairs, and it is therefore possible to get lots of short predicted binding sites of length one or two base pairs.

In this and a previous study, it was observed that many of the predictions made by the classifiers were highly fragmented and too small to correspond to biological binding sites. It was not clear whether these fragmented predictions were merely artifacts or whether they were accurate but overly conservative. Therefore, predictions with a length smaller than a threshold value were removed and the effect on the performance measures observed. It was found that removal of the fragmented predictions had a considerable positive effect on the performance measures, most notably for Precision and that an optimal value for the threshold is 6 bp. Interestingly, this value corresponds roughly to the lower limit of biologically observed binding site lengths which are typically in the range 5-30 bp in length.

7 Results

Before presenting the main results we should point out that predicting binding sites accurately is extremely difficult. The performance of the best individual original algorithm (Fuzznuc) is:

	Predicted Negatives	Predicted Positives
Actual Negatives	*TN*= 83%	*FP* = 10%
Actual Positives	*FN* = 4%	*TP* = 3%

Here we can see over three times as many false positives as true positives. This makes the predictions almost useless to a biologist as most of the suggested binding sites will need expensive experimental validation and most will not be useful. Therefore the key aim of our combined classifier is to significantly reduce the number of false positives given by the original algorithms.

7.1 Results Using Sampling

As described above the imbalanced nature of the data must be addressed. First the data is divided into a training set and test set, in the ratio 2 to 1. This gives a training set of 32,615 84-ary vectors and a test set of 16,739 vectors.

In the results here the majority class in the training set is reduced, by random sampling, from 30,038 vectors to 9,222 and the minority class was increased from 2,577 vectors to 4,611 vectors using the SMOTE algorithm. Therefore the ratio of the majority class to the minority class is reduced from approximately 12 : 1 to 2 : 1. Other ratios were tried but this appears to give good results. The test set was not altered at all.

As described earlier an SVM with Gaussian kernel was used as the trainable classifier, and to find good settings for the two free parameters of the model, C and γ standard 5-fold cross validation was used. After good values for the parameters were found ($C = 1000$, $\gamma = 0.001$), the test set was presented and the results are as follows:

	Recall	Precision	F-Score	FP-Rate
Best Original Algorithm	0.400	0.222	0.285	0.106
Combined Classifier - Sampling	0.305	0.371	0.334	0.044

The first notable feature of this result is that the combined classifier has produced a weaker Recall than the best original algorithm. This is because it is giving fewer positive predictions, but it has a much higher precision. Of particular significance is that the FP-Rate is relatively low at 0.04, so that only 4% of the actual non-binding sites are predicted incorrectly. However this is still too large a figure to make the classifier useful to biologists. So we turn to our second Combined Classifier using *SDC*.

7.2 Results Using *SDC*

First the minority class was over-sampled using SMOTE. The size of the minority class was tripled to 7731 vectors so that the ratio of majority to the minority class was now about 4 : 1. Once again 5-fold cross validation was used to find appropriate values for the three free parameters of the SVM with different costs, namely C^+, C^- and γ. The best values found were: $C^+ = 680$, $C^- = 1320$ and $\gamma = 0.0001$.

	Recall	Precision	F-Score	FP-RATE
Best Original Algorithm	0.400	0.222	0.285	0.106
Combined Classifier - Sampling	0.305	0.371	0.334	0.044
SDC	0.283	0.375	0.324	0.036

This method has produced a good classifier, but it is not much better than the classifier using a straightforward SVM and sampling. However the FP-Rate has been further reduced to 0.036.

7.3 Results After Post-processing

Finally we investigate how the results can be further improved by removing those predictions of base-pairs being part of a binding site that are not biologically plausible. As described earlier we find that removing predictions that are not part of a contiguous predicted binding site of at least six nucleotides gives an optimal result. So here we take the predictions of the SDC algorithm and remove all those that do not meet this criterion.

	Recall	Precision	F-Score	FP-Rate
Best Original Algorithm	0.400	0.222	0.285	0.106
Combined Classifier - Sampling	0.305	0.371	0.334	0.044
SDC	0.283	0.375	0.324	0.036
SDC + Post-Processing	0.264	0.517	0.350	0.021

This produces our best result by some way. The Precision of the prediction has been increased to 0.517 and the FP-Rate is now down to just 2%.

To see how this has come about Figure 2 shows a fragment of the genome with the original algorithmic predictions, the SVM predictions, the result of post-processing the SVM predictions and the actual annotation. It can be seen that for this fragment the removal of the implausible predictions eliminates almost all the false positive predictions.

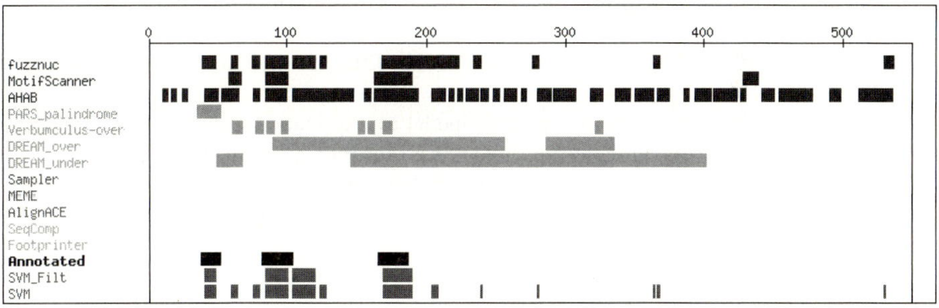

Fig. 2. A fragment of the genome with the 12 original predictions, the actual annotations in black. The last row shows the predictions of the SVM and above it the effect of removing unrealistically short predictions.

8 Discussion

The identification of regions in a sequence of DNA that are regulatory binding sites is a very difficult problem. Individually the original prediction algorithms are inaccurate and consequently produce many false positive predictions. Our results show that by combining the predictions of the original algorithms we can make a significant improvement from their individual results. This suggests that the predictions that they produce are complementary, perhaps giving information about different parts of the genome. The only problem of our approach is that the combined predictor can indicate implausibly short binding sites. However we have shown that by simply rejecting these binding sites, using a length threshold, gives a very low rate of false positive predictions. This is exactly the result that we wanted: false positives are very undesirable in this particular domain.

On the technical issue of dealing with the highly imbalanced data we found that both sampling of the two classes and using the SDC algorithm gave similar results, with both methods dealing well with our data.

References

[1] Tompa, M., Li, N., Bailey, T.L., Church, G.M., De Moor, B., Eskin, E., Favorov, A.V., Frith, M.C., Fu, Y., Kent, W.J., Makeev, V.J., Mironov, A.A., Noble, W.S., Pavesi, G., Pesole, G., Regnier, M., Simonis, N., Sinha, S., Thijs, G., van Helden, J., Vandenbogaert, M., Weng, Z., Workman, C., Ye, C., Zhu, Z.: Assessing computational tools for the discovery of transcription factor binding sites. Nat Biotechnol 23, 137–144 (2005)

[2] Thijs, G., Lescot, M., Marchal, K., Rombauts, S., De Moor, B., Rouze, P., Moreau, Y.: A higher-order background model improves the detection of promoter regulatory elements by Gibbs sampling. Bioinformatics 17, 1113–1122 (2001)

[3] Rajewsky, N., Vergassola, M., Gaul, U., Siggia, E.D.: Computational detection of genomic cis-regulatory modules applied to body patterning in the early Drosophila embryo. BMC Bioinformatics 3, 30 (2002)

[4] Abnizova, I., Rust, A.G., Robinson, M., Te Boekhorst, R., Gilks, W.R.: Transcription binding site prediction using Markov models. J Bioinform Comput Biol 4, 425–441 (2006)

[5] Apostolico, A., Bock, M.E., Lonardi, S., Xu, X.: Efficient detection of unusual words. J Comput Biol 7, 71–94 (2000)

[6] Bailey, T.L., Elkan, C.: Fitting a mixture model by expectation maximization to discover motifs in biopolymers. Proc Int Conf Intell Syst Mol Biol 2, 28–36 (1994)

[7] Hughes, T.R., Marton, M.J., Jones, A.R., Roberts, C.J., Stoughton, R., Armour, C.D., Bennett, H.A., Coffey, E., Dai, H., He, Y.D., Kidd, M.J., King, A.M., Meyer, M.R., Slade, D., Lum, P.Y., Stepaniants, S.B., Shoemaker, D.D., Gachotte, D., Chakraburtty, K., Simon, J., Bard, M., Friend, S.H.: Functional discovery via a compendium of expression profiles. Cell 102, 109–126 (2000)

[8] Brown, C.T., Rust, A.G., Clarke, P.J., Pan, Z., Schilstra, M.J., De Buysscher, T., Griffin, G., Wold, B.J., Cameron, R.A., Davidson, E.H., Bolouri, H.: New computational approaches for analysis of cis-regulatory networks. Dev Biol 246, 86–102 (2002)

[9] Blanchette, M., Tompa, M.: FootPrinter: A program designed for phylogenetic footprinting. Nucleic Acids Res 31, 3840–3842 (2003)

[10] Zhu, J., Zhang, M.Q.: SCPD: a promoter database of the yeast Saccharomyces cerevisiae. Bioinformatics 15, 607–611 (1999)

[11] Robinson, M., Sun, Y., Boekhorst, R.T., Kaye, P., Adams, R., Davey, N., Rust, A.G.: Improving computational predictions of cis-regulatory binding sites, Pac ymp Biocomput, pp. 391-402 (2006)

[12] Chawla, N.V., Bowyer, K.W., Hall, L.O., Kegelmeyer, W.P.: SMOTE: Synthetic minority over-sampling Technique. Journal of Artificial Intelligence Research 16, 321–357 (2002)

[13] Radivojac, P., Chawla, N.V., Dunker, A.K., Obradovic, Z.: Classification and knowledge discovery in protein databases. J Biomed Inform 37, 224–239 (2004)

[14] Veropoulos, K., Cristianini, N., Campbell, C.: Controlling the Sensitivity of Support Vector Machines. In: Proceedings of the International Joint Conference on Artificial Intelligence (IJCAI99), Stockholm (1999)

[15] Akbani, R., Kwek, S., Japkowicz, N.: Applying Support Vector Machines to Imbalanced Datase. In: Boulicaut, J.-F., Esposito, F., Giannotti, F., Pedreschi, D. (eds.) ECML 2004. LNCS (LNAI), vol. 3201, pp. 39–50. Springer, Heidelberg (2004)

On the Combination of Dissimilarities for Gene Expression Data Analysis

Ángela Blanco, Manuel Martín-Merino[1], and Javier De Las Rivas[2]

[1] Universidad Pontificia de Salamanca
C/Compañía 5, 37002, Salamanca, Spain
ablancogo@upsa.es, mmartinmac@upsa.es
[2] Cancer Research Center (CIC-IBMCC, CSIC/USAL)
Salamanca, Spain
jrivas@usal.es

Abstract. DNA Microarray technology allows us to monitor the expression level of thousands of genes simultaneously. This technique has become a relevant tool to identify different types of cancer.

Several machine learning techniques such as the Support Vector Machines (SVM) have been proposed to this aim. However, common SVM algorithms are based on Euclidean distances which do not reflect accurately the proximities among the sample profiles. The SVM has been extended to work with non-Euclidean dissimilarities. However, no dissimilarity can be considered superior to the others because each one reflects different features of the data.

In this paper, we propose to combine several Support Vector Machines that are based on different dissimilarities to improve the performance of classifiers based on a single measure. The experimental results suggest that our method reduces the misclassification errors of classifiers based on a single dissimilarity and a widely used combination strategy such as Bagging.

1 Introduction

DNA Microarray technology allows us to monitor the expression levels of thousands of genes simultaneously across a collection of related samples. This technology has been applied particularly to the prediction of different types of cancer with encouraging results [11].

A large variety of machine learning techniques have been proposed to this aim such as Support Vector Machines (SVM) [9] or k Nearest Neighbors [8]. However the algorithms considered in the literature rely frequently on the use of the Euclidean distance that fails often to reflect accurately the proximities among the sample profiles [7,15,17]. This increases the false negative errors (cancerous samples misclassified) although they are prohibitively expensive in our application. The SVM can be extended to work with non-euclidean dissimilarities [20]. In spite of this, no dissimilarity can be considered superior to the others

J. Marques de Sá et al. (Eds.): ICANN 2007, Part II, LNCS 4669, pp. 110–119, 2007.

because each one reflects just different features of the data and so a different set of patterns is misclassified.

In this paper, we try to reduce particularly the false negative errors by combining classifiers based on different dissimilarities. Several authors have pointed out that combining non-optimal classifiers can help to reduce particularly the variance of the predictor [16,21]. In order to achieve this goal, different versions of the classifier are usually built by sampling the patterns or the features [4]. Nevertheless, in our application, this kind of resampling techniques reduce the size of the training set. This may increase the bias of individual classifiers and the error of the combination [21]. To overcome this problem, the diversity among classifiers is induced in this paper by considering dissimilarities that reflect different features of the data. To this aim, the dissimilarities are first embedded into an Euclidean space where a SVM is adjusted for each measure. Next, the classifiers are aggregated using a voting strategy [16]. The method proposed has been applied to the prediction of different types of cancer using the gene expression levels with remarkable results.

This paper is organized as follows. Section 2 introduces the dissimilarities considered to build the ensemble of classifiers. Section 3 presents the method to combine classifiers based on dissimilarities. Section 4 illustrates the performance of the algorithm in the challenging problem of gene expression data analysis. Finally, section 5 gets conclusions and outlines future research trends.

2 The Problem of Distances for Gene Expression Data Analysis

An important step in the design of a classifier is the choice of a proper dissimilarity that reflects the proximities among the objects. However, the choice of a good dissimilarity is not an easy task. Each measure reflects different features of the data and the classifiers induced by the dissimilarities misclassify frequently different patterns. Therefore no dissimilarity can be considered optimal.

In this section, we comment shortly the main differences among several dissimilarities proposed to evaluate the proximity between celular samples considering the gene expression levels. For a deeper description and definitions see [7,15,10].

The Euclidean distance evaluates if the gene expression levels differ significantly across different samples. When the experimental conditions change from one sample to another the cosine dissimilarity is an interesting alternative. This measure will become small when the ratio between the gene expression levels is similar for the two samples considered. It differs significantly from the Euclidean distance when the data is not normalized.

The correlation measure evaluates if the expression levels of genes change similarly in both samples. Correlation based measures tend to group together samples whose expression levels are linearly related. The correlation differs significantly from the cosine if the means of the sample profiles are not zero. This measure is sensitive to outliers. The Spearman rank dissimilarity is less sensitive to outliers because it computes a correlation between the ranks of the gene expression levels. An alternative measure that helps to overcome the problem of outliers is the Kendall-τ index. The Kendall's τ is related to the Mutual Information probabilistic measure [10]. Finally, the Kullback-Leibler divergence evaluates the distance between the probability distribution of the gene expression levels for the samples. The Kullback-Leibler divergence is asymmetric and should be symmetrized using the following equation: $\delta_{ij}^{(s)} = (\delta_{ij} + \delta_{ji})/2$.

Due to the large number of genes, the sample profiles are codified in high dimensional and noisy spaces. In this case, the dissimilarities mentioned above are affected by the 'curse of dimensionality' [1,18]. Hence, most of the dissimilarities become almost constant and the differences among dissimilarities are lost [14]. To avoid this problem, it is recommended to reduce aggressively the number of features before computing the dissimilarities.

3 Combining Classifiers Based on Dissimilarities

In this section, we explain how the SVM can be extended to work directly from a dissimilarity measure. Next, the ensemble of classifiers based on multiple dissimilarities is presented. Finally we comment briefly the related work.

The SVM is a powerful machine learning technique that is able to deal with high dimensional and noisy data [22]. In spite of this, the original SVM algorithm is not able to work directly from a dissimilarity matrix. To overcome this problem, we follow the approach of [20]. First, the dissimilarities are embedded into an Euclidean space such that the inter-pattern distances reflect approximately the original dissimilarity matrix. Next, the test points are embedded via a linear algebra operation and finally the SVM is trained and evaluated. We comment briefly the mathematical details.

Let $D \in \mathbb{R}^{n \times n}$ be the dissimilarity matrix made up of the object proximities for the training set. A configuration in a low dimensional Euclidean space can be found via a metric multidimensional scaling algorithm (MDS) [5] such that the original dissimilarities are approximately preserved. Let $X = [\boldsymbol{x}_1 \dots \boldsymbol{x}_n]^T \in \mathbb{R}^{n \times p}$ be the matrix of the object coordinates for the training patterns. Define $B = XX^T$ as the matrix of inner products which is related to the dissimilarity matrix via the following equation:

$$B = -\frac{1}{2} J D^{(2)} J \,, \tag{1}$$

where $J = I - \frac{1}{n}\mathbf{1}\mathbf{1}^T \in \mathbb{R}^{n \times n}$ is the centering matrix, I is the identity matrix and $D^{(2)} = (\delta_{ij}^2)$ is the matrix of the square dissimilarities for the training patterns. If B is positive semi-definite, the object coordinates in the low dimensional Euclidean space $\mathbb{R}^k$ can be found through a singular value decomposition [5,12]:

$$X_k = V_k \Lambda_k^{1/2} \,, \tag{2}$$

where $V_k \in \mathbb{R}^{n \times k}$ is an orthogonal matrix with columns the first k eigen-vectors of XX^T and $\Lambda_k = diag(\lambda_1 \dots \lambda_k) \in \mathbb{R}^{k \times k}$ is a diagonal matrix with λ_i the i-th eigenvalue. Several dissimilarities introduced in section 2 generate inner product matrices B non semi-definite positive. Fortunately, the negative values are small in our application and therefore can be neglected [5] without losing relevant information about the data.

Once the training patterns have been embedded into a low dimensional Euclidean space, the test pattern can be added to this space via a linear projection [20]. Next we comment briefly the derivation.

Let $X_k \in \mathbb{R}^{n\times k}$ be the object configuration for the training patterns in $\mathbb{R}^k$ and $X_n = [\boldsymbol{x}_1 \ldots \boldsymbol{x}_s]^T \in \mathbb{R}^{s\times k}$ the matrix of the object coordinates sought for the test patterns. Let $D_n^{(2)} \in \mathbb{R}^{s\times n}$ be the matrix of the square dissimilarities between the s test patterns and the n training patterns that have been already projected. The matrix $B_n \in \mathbb{R}^{s\times n}$ of inner products among the test and training patterns can be found as:

$$B_n = -\frac{1}{2}(D_n^{(2)} J - U D^{(2)} J)\,, \tag{3}$$

where $J \in \mathbb{R}^{n\times n}$ is the centering matrix and $U = \frac{1}{n}\mathbf{1}^T\mathbf{1} \in \mathbb{R}^{s\times n}$. The derivation of equation (3) is detailed in [20]. Since the matrix of inner products verifies

$$B_n = X_n X_k^T \tag{4}$$

then, X_n can be found as the least mean-square error solution to (4), that is:

$$X_n = B_n X_k (X_k^T X_k)^{-1}\,, \tag{5}$$

Given that $X_k^T X_k = \Lambda_k$ and considering that $X_k = V_k \Lambda_k^{1/2}$ the coordinates for the test points can be obtained as:

$$X_n = B_n V_k \Lambda_k^{-1/2}\,, \tag{6}$$

which can be easily evaluated through simple linear algebraic operations.

Next we introduce the method proposed to combine classifiers based on different dissimilarities.

Our method is based on the evidence that different dissimilarities reflect different features of the dataset (see section 2). Therefore, classifiers based on different measures will not misclassify the same patterns. Figure 1 shows for instance that bold patterns are assigned to the wrong class by only one classifier but using a voting strategy the patterns will be assigned to the right class.

Hence, our combination algorithm proceeds as follows: First, a set of dissimilarities are computed. Each dissimilarity is embedded into an Euclidean space using equation (2). Next, we train a SVM for each dissimilarity computed. Thus, the misclassification errors will change from one classifier to another. So the combination of classifiers by a voting strategy will help to reduce the misclassification errors. Finally, the test points are embedded in the Euclidean space induced by the training patterns using equation (6).

A related technique to combine classifiers is the Bagging [4,2]. This method generates a diversity of classifiers that are trained using several bootstrap samples. Next, the classifiers are aggregated using a voting strategy. Nevertheless

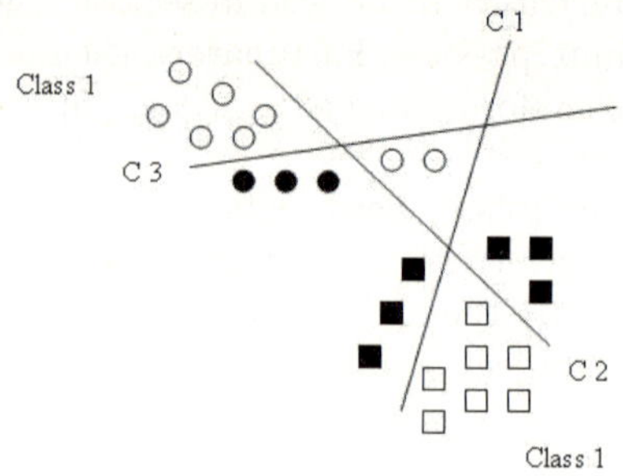

Fig. 1. Aggregation of classifiers using a voting strategy. Bold patterns are misclassified by a single hyperplane but not by the combination.

there are three important differences between Bagging and the method proposed in this section.

First, our method generates the diversity of classifiers by considering different dissimilarities and thus using the whole sample. Bagging trains each classifier using around 63% of the training set. In our application the size of the training set is very small and neglecting part of the patterns may increase the bias of each classifier. It has been suggested in the literature that Bagging doesn't help to reduce the bias [21] and so, the aggregation of classifiers will hardly reduce the misclassification error.

A second advantage of our method is that it is able to work directly with a dissimilarity matrix.

Finally, the combination of several dissimilarities avoids the problem of choosing a particular dissimilarity for the application we are dealing with. This is a difficult and time consuming task.

Notice that the algorithm proposed earlier can be easily applied to other classifiers such as the k-nearest neighbor algorithm that are based on distances. In this case, our method can be applied in a straightforward way because the dissimilarities should not be embedded in an Euclidean space.

4 Experimental Results

In this section, the ensemble of classifiers proposed is applied to the identification of cancerous tissues using Microarray gene expression data.

Two benchmark gene expression datasets have been considered. The first one consisted of 72 bone marrow samples (47 ALL and 25 AML) obtained from acute leukemia patients at the time of diagnosis [12]. The RNA from marrow mononuclear cells was hybridized to high-density oligonucleotide microarrays produced by Affymetrix and containing 6817 genes. The second dataset consisted of 49 samples from breast tumors [23], 25 classified as positive to estrogen receptors (ER+) and 24 negative to estrogen receptors (ER-). Those positive to estrogen receptors have a better clinical outcome and require a different treatment.

The RNA of breast cancer cells were hybridized to high-density oligonucleotide microarrays produced by Affymetrix and containing 7129 genes.

Due to the large number of genes, samples are codified in a high dimensional and noisy space. Therefore, the dissimilarities are affected by the 'curse of dimensionality' and the correlation among them becomes large [18]. To avoid this problem and to increase the diversity among dissimilarities we have reduced aggressively the number of genes using the standard F-statistic [10]. The dissimilarities have been computed without normalizing the variables because as we have mentioned in section 2 this operation may increase the correlation among them.

The algorithm chosen to train the Support Vector Machines is C-SVM. The C regularization parameter has been set up by ten fold-crossvalidation [19,3]. We have considered linear kernels in all the experiments because the small size of the training set in our application favors the overfitting of the data. Consequently error rates are smaller for linear kernels than for non linear ones.

Regarding to the ensemble of classifiers, an important issue is the dimensionality in which the dissimilarities are embedded. To this aim, a metric Multidimensional Scaling algorithm is first run. The number of eigenvectors considered is determined by the curve of eigenvalues.

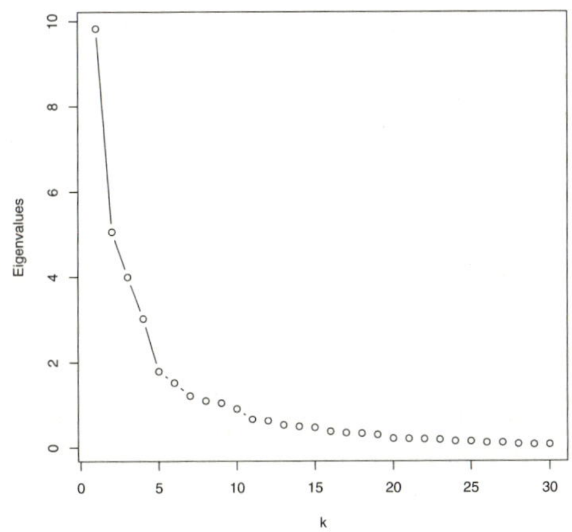

Fig. 2. Eigenvalues for the metric MDS with the χ^2 distance

Figure 2 shows the eigenvalues for the breast cancer data and the χ^2 dissimilarity. The first eleven eigenvalues account for 85% of the variance. Therefore, they preserve the main structure of the data.

The classifiers have been evaluated from two different points of view: on the one hand we have computed the misclassification errors. But in our application, false negative and false positives errors have unequal relevance. For instance, in breast cancer, false negative errors corresponds to tumors positive to estrogen receptors that have been classified as negative to estrogen receptors. This will lead a wrong treatment with very dangerous consequences to the patient. Therefore, false negative errors are much more important than false positive errors.

Table 1. Experimental results for the ensemble of SVM classifiers. Classifiers based solely on a single dissimilarity and Bagging have been taken as reference.

	% Error		% False negative	
Method	Breast	Leukemia	Breast	Leukemia
Euclidean	10.2%	6.9%	4%	6.94%
Cosine	14.2%	1.38%	4%	1.38%
Correlation	14.2%	2.7%	6.1%	2.7%
χ^2	12.2%	1.38%	4%	1.38%
Manhattan	12.2%	5.5%	4%	4.16%
Spearman	16.3%	8.3%	6.1%	5.5%
Kendall-Tau	18.3%	8.3%	6.1%	5.5%
Kullback-Leibler	16.3%	30.5%	12.2%	19.4%
Bagging	6.1%	1.38%	2%	1.28%
Random genes	4.2%	4.16%	2.04%	4.16%
Combination	8.1%	1.38%	2%	1.38%

The estimation of errors for a small sample size is not an easy task. In this paper, we have applied ten-fold cross-validation which gives good experimental results for this kind of problems [19].

The combination strategy proposed in this paper has been also applied to the k-nearest neighbor classifier. An important parameter in this algorithm is the number of neighbors which has been estimated by cross-validation.

Table 1 shows the experimental results for the ensemble of classifiers using the SVM. The method proposed has been compared with Bagging introduced in section 3 and a variant of Bagging that generates the classifiers by sampling the genes. Finally, the classifiers based on a single dissimilarity have been taken as a reference.

From the analysis of table 1, the following conclusions can be drawn:

- The error for the Euclidean distance depends on the dataset considered, breast cancer or leukemia. For instance, the misclassification error and false negative error are larger for Leukemia. On the other hand, the combination of dissimilarities improves significantly the Euclidean distance which is usually considered by most of SVM algorithms.
- The algorithm based on the combination of dissimilarities improves the best single dissimilarity which is χ^2. Notice that for breast cancer false negative errors are significantly reduced.
- The combination of dissimilarities performs similarly to Bagging sampling the patterns. However,we remark that our method is able to work directly from the dissimilarity matrix.

Table 2 shows the experimental results for the ensemble of k-NNs. The primary conclusions are the following:

- The combination of dissimilarities improves the best classifier based on a single dissimilarity and particularly for Leukemia data.

– The Euclidean distance performs very poorly and it is significantly improved by the combination of dissimilarities.
– The combination of dissimilarities outperforms clearly the Bagging algorithm.
– Bagging errors are larger for k-nn classifiers than for SVM classifiers. This can be justified because SVM is more robust when a subset of patterns is neglected due to bootstrap sampling. The combination of dissimilarities does not suffer from this drawback.

Table 2. Experimental results for the ensemble of k-NN classifiers. Classifiers based solely on a single dissimilarity and Bagging have been taken as reference.

	% Error		% False negative	
Method	Breast	Leukemia	Breast	Leukemia
Euclidean	14.2 %	6.94%	6.1%	4.1%
Cosine	16.3 %	2.77%	8.1%	2.77%
Correlation	14.2 %	4.16%	8.1 %	4.16%
χ^2	10.2%	2.77%	2.0%	2.77 %
Manhattan	8.1 %	2.7%	2.0%	2.7%
Spearman	10.2 %	2.77 %	4.0 %	2.77%
Kendall-tau	8.1 %	2.77 %	2.0 %	2.77 %
Kullback	51 %	76 %	46.9 %	11.1 %
Bagging	14.2 %	6.9 %	6.1 %	6.9 %
Combination	8.1 %	1.38 %	2.0 %	1.38 %

5 Conclusions and Future Research Trends

In this paper, we have proposed an ensemble of classifiers based on a diversity of dissimilarities. Our approach aims to reduce the misclassification error of classifiers based solely on a single distance working directly from a dissimilarity matrix. The algorithm has been applied to the classification of cancerous tissues using gene expression data.

The experimental results suggest that the method proposed improves both, misclassification errors and false negative errors of classifiers based on a single dissimilarity. We also report that for dissimilarity based classifiers such as k-NN our method improves significantly other combination strategies such as Bagging.

As future research trends, we will try to increase the diversity of classifiers by random sampling the patterns for each dissimilarity.

Acknowledgment

This work has been partially supported by the Junta de Castilla y León grant PON05B06.

References

1. Aggarwal, C.C.: Re-designing distance functions and distance-based applications for high dimensional applications. In: Proc. of the ACM International Conference on Management of Data and Symposium on Principles of Database Systems (SIGMOD-PODS), March 2001, vol. 1, pp. 13–18 (2001)
2. Bauer, E., Kohavi, R.: An empirical comparison of voting classification algorithms: Bagging, boosting, and variants. Machine Learning 36, 105–139 (1999)
3. Braga-Neto, U., Dougherty, E.: Is cross-validation valid for small-sample microarray classification? Bioinformatics 20(3), 374–380 (2004)
4. Breiman, L.: Bagging predictors. Machine Learning 24, 123–140 (1996)
5. Cox, T., Cox, M.: Multidimensional Scaling, 2nd ed., New York: Chapman & Hall/CRC Press (2001)
6. Cristianini, N., Shawe-Taylor, J.: An Introduction to Support Vector Machines and Other Kernel-Based Learning Methods. Cambridge University Press, Cambridge (2000)
7. Drãghici, S.: Data Analysis Tools for DNA Microarrays. Chapman & Hall/CRC Press, New York (2003)
8. Dudoit, S., Fridlyand, J., Speed, T.: Comparison of discrimination methods for the classification of tumors using gene expression data. Journal of the American Statistical Association 97, 77–87 (2002)
9. Furey, T., Cristianini, N., Duffy, N., Bednarski, D., Schummer, M., Haussler, D.: Support vector machine classification and validation of cancer tissue samples using microarray expression data. Bioinformatics 16(10), 906–914 (2000)
10. Gentleman, R., Carey, V., Huber, W., Irizarry, R., Dudoit, S.: Bioinformatics and Computational Biology Solutions Using R and Bioconductor. Springer, Heidelberg (2006)
11. Golub, T., Slonim, D., Tamayo, P., Huard, C., Gaasenbeek, M., Mesirov, J., Coller, H., Loh, M., Downing, J., Caligiuri, M., Bloomfield, C., Lander, E.: Molecular classification of cancer: Class discovery and class prediction by gene expression monitoring. Science 286(15), 531–537 (1999)
12. Golub, G.H., Loan, C.F.V.: Matrix Computations, 3rd edn. Johns Hopkins university press, Baltimore, Maryland, USA (1996)
13. Guyon, I., Weston, J., Barnhill, S., Vapnik, V.: Gene selection for cancer classification using support vector machines. Machine Learning 46, 389–422 (2002)
14. Hinneburg, C.C.A.A., Keim, D.A.: What is the nearest neighbor in high dimensional spaces? In: Proc. of the International Conference on Database Theory (ICDT), pp. 506–515. Morgan Kaufmann, Cairo, Egypt (2000)
15. Jiang, D., Tang, C., Zhang, A.: Cluster analysis for gene expression data: A survey. IEEE Transactions on Knowledge and Data Engineering 16(11) (2004)
16. Kittler, J., Hatef, M., Duin, R., Matas, J.: On combining classifiers. IEEE Transactions on Neural Networks 20(3), 228–239 (1998)
17. Martín-Merino, M., Muñoz, A.: Self organizing map and sammon mapping for asymmetric proximities. Neurocomputing 63, 171–192 (2005)
18. Martín-Merino, M., Muñoz, A.: A new sammon algorithm for sparse data visualization. In: International Conference on Pattern Recognition (ICPR), August 2004, pp. 477–481. IEEE Press, Orlando, Florida, USA (2004)
19. Molinaro, A., Simon, R., Pfeiffer, R.: Prediction error estimation: a comparison of resampling methods. Bioinformatics 21(15), 3301–3307 (2005)

20. Pekalska, E., Paclick, P., Duin, R.: A generalized kernel approach to dissimilarity-based classification. Journal of Machine Learning Research 2, 175–211 (2001)
21. Valentini, G., Dietterich, T.: Bias-variance analysis of support vector machines for the development of svm-based ensemble methods. Journal of Machine Learning Research 5, 725–775 (2004)
22. Vapnik, V.: Statistical Learning Theory. John Wiley & Sons, New York (1998)
23. West, M., Blanchette, C., Dressman, H., Huang, E., Ishida, S., Spang, R., Zuzan, H., Olson, J., Marks, J., Nevins, J.: Predicting the clinical status of human breast cancer by using gene expression profiles. PNAS 98(20) (2001)

A Locally Recurrent Globally Feed-Forward Fuzzy Neural Network for Processing Lung Sounds

Paris A. Mastorocostas, Dimitris N. Varsamis, Costas A. Mastorocostas, and Costas S. Hilas

Dept. of Informatics and Communications, Technological Educational Institute of Serres, Serres GR-621 24, Greece
mast@teiser.gr

Abstract. This paper presents a locally recurrent globally feedforward fuzzy neural network, with internal feedback, that performs the task of separation of lung sounds, obtained from patients with pulmonary pathology. The filter is a novel generalized Takagi-Sugeno-Kang fuzzy model, where the consequent parts of the fuzzy rules are Block-Diagonal Recurrent Neural Networks. Extensive experimental results, regarding the lung sound category of squawks, are given, and a performance comparison with a series of other fuzzy and neural filters is conducted, underlining the separation capabilities of the proposed filter.

1 Introduction

Pathological discontinuous adventitious sounds (DAS) are strongly related to pulmonary dysfunction. Their clinical use for the interpretation of respiratory malfunction depends on their efficient and objective separation from vesicular sounds (VS). In order to achieve this kind of separation, the non-stationarity of DAS must be taken into account. A number of nonlinear processing methods have been used in the past, with the wavelet transform-based stationary-non-stationary (WTST-NST) filter [1] providing the most accurate separation results. However, this method could not be easily implemented in real-time analysis of lung sounds.

During the last years computational intelligence models, such as neural and fuzzy-neural networks have been proposed, providing encouraging results. In particular, The Orthogonal Least Squares-based Fuzzy Filter has been suggested in [2] for real-time separation of lung sounds. A recurrent neural filter has been reported in [3], based on the Block-Diagonal Recurrent Neural Network (BDRNN) [4], which is a simplified form of the fully recurrent network, with no interlinks among neurons in the hidden layer. As shown in [3], due to its internal dynamics, the BDRNN filter is capable of performing efficient real-time separation of lung sounds.

It is common knowledge that the fuzzy neural networks combine the benefits of fuzzy systems and those of neural networks. In particular, fuzzy inference provides an efficient way of handling imprecision and uncertainty while neural learning permits determining the model parameters. Stemming from this fact and the modeling capabilities of the BDRNN mentioned above, in this work a recurrent fuzzy neural network is proposed, for real-time separation of DAS from VS. The novelty of the proposed model lies in the consequent parts of the fuzzy rules, which are small block-diagonal recurrent neural networks, thus introducing dynamics to the overall network.

J. Marques de Sá et al. (Eds.): ICANN 2007, Part II, LNCS 4669, pp. 120–128, 2007.

Extensive experimental results are given and a comparative analysis with previous works is conducted, highlighting the efficiency of the proposed filter.

2 The Dynamic Block-Diagonal Fuzzy Neural Network

The suggested dynamic block-diagonal fuzzy neural network (DBD-FNN) comprises r Takagi-Sugeno-Kang rules, [5], of the following form:

$$IF\ \boldsymbol{u}(k)\ is\ \mathcal{A}^{(l)}\ \ THEN\ \ g_l(k) = BDRNN_l\left(\boldsymbol{u}(k)\right),\ l = 1, ..., r, \tag{1}$$

where $\mathcal{A}^{(l)}$ is the fuzzy region in the premise part and the sub-model $BDRNN_l$ is a block-diagonal recurrent neural network that implements the consequent part of the *l-th* rule. For the sake of simplicity, a multiple–input – single–output model is used. Based on the structural characteristics of the TSK model, it can be divided into three major parts: the premise, the consequent and the defuzzification part.

At each time instant k, the premise part is fed with the process variables $u_1(k), ..., u_m(k)$, which are used for defining the fuzzy operating regions. The firing strengths of the rules are calculated by the following static function:

$$\mu_l(k) = \prod_{j=1}^{m} \exp[-\frac{1}{2} \cdot \frac{(u_j(k) - m_{lj}(k))^2}{(\sigma_{lj}(k))^2}],\quad l = 1,...,r, \tag{2}$$

where $\boldsymbol{m}_l = \{m_{lj}, j = 1,...,m\}$ and $\boldsymbol{\sigma}_l = \{\sigma_{lj}, j = 1,...,m\}$, $l = 1,...,r$, are the premise part parameters.

The consequent parts of the model are dynamic, including the r sub-models of the rules. Each sub-model $BDRNN_l$ is a block-diagonal recurrent neural network, which is a two-layer network, with the output layer being static and the hidden layer being dynamic. The hidden layer consists of pairs of neurons (blocks); there are feedback connections between the neurons of each pair, introducing dynamics to the network. Therefore, the overall model is a locally recurrent globally feedforward fuzzy neural network [6].

The operation of the BDRNN with m inputs, r outputs and N neurons at the hidden layer is described by the following set of state equations:

$$x_{2i-1}^{(l)}(k) = f_a[\sum_{j=1}^{m} b_{2i-1,j}^{(l)} \cdot u_j(k) + w_{1,i}^{(l)} \cdot x_{2i-1}^{(l)}(k-1) + w_{2,i}^{(l)} \cdot x_{2i}^{(l)}(k-1)], \tag{3}$$

$$x_{2i}^{(l)}(k) = f_a[\sum_{j=1}^{m} b_{2i,j}^{(l)} \cdot u_j(k) - w_{2,i}^{(l)} \cdot x_{2i-1}^{(l)}(k-1) + w_{1,i}^{(l)} \cdot x_{2i}^{(l)}(k-1)], \tag{4}$$

$$g_j(k) = f_b[\sum_{j=1}^{N} c_{lj} \cdot x_j^{(l)}(k)] = f_b[\sum_{j=1}^{N/2} c_{l,2j-1} \cdot x_{2j-1}^{(l)}(k) + \sum_{j=1}^{N/2} c_{l,2j} \cdot x_{2j}^{(l)}(k)], \tag{5}$$

$l = 1,...,r$, $i = 1,..., N/2$

where

(i) f_a and f_b are the neuron activation functions of the hidden and the output layer, respectively. In the following, the activation functions are both chosen to be the sigmoid function.

(ii) $\boldsymbol{x}^{(l)}(k)=[x_i^{(l)}(k)]$ is a N-element vector, comprising the outputs of the hidden layer of the *l-th* fuzzy rule. In particular, $x_i^{(l)}(k)$ is the output of the *i-th* hidden neuron at time *k*.

(iii) $B=\left[b_{ij}^{(l)}\right]$ and $C=\left[c_{lj}\right]$ are $r\times N\times m$ and $r\times N$ input and output weight matrices, respectively.

(iv) $w_{1,i}^{(l)}, w_{2,i}^{(l)}$ are the rules' feedback weights, that form the block diagonal feedback matrices, $W^{(l)}$. The *scaled orthogonal* form is employed in this work, [4], where the feedback matrices are described by the following formula:

$$W^{(l)}=\begin{bmatrix} w_{1,i}^{(l)} & w_{2,i}^{(l)} \\ -w_{2,i}^{(l)} & w_{1,i}^{(l)} \end{bmatrix} \quad i=1,2,\ldots,N/2 . \tag{6}$$

The output of the model at time *k*, *y(k)*, is determined using the weighted average defuzzification method:

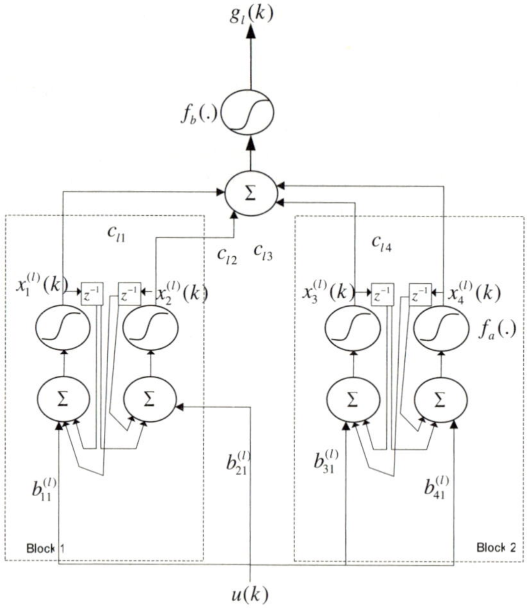

Fig. 1. Configuration of the consequent part of the fuzzy rules

$$y(k) = \frac{\sum_{l=1}^{r} \mu_l(k) \cdot g_l(k)}{\sum_{l=1}^{r} \mu_l(k)}. \tag{7}$$

The configuration of the proposed BDRNN consequent part is presented in Fig. 1, where, for the sake of simplicity, a single–input – single–output BDRNN with two blocks of neurons is shown.

The DBD-FNN is trained by use of the Dynamic Resilient Backpropagation algorithm, which was developed in [7] and constitutes a powerful first order training algorithm for batchwise learning, [8]. Only minor modifications are made, such that the method takes into consideration the special features of the DBD-FNN, requiring calculation of the error gradients for the feedback weights using the notion of ordered derivatives, [9]. Since the scope of the paper is to highlight the model's operation rather than the selected training algorithm, a detailed overview of RPROP can be found in [7].

3 The DBD-FNN Filter

The DBD-FNN estimates both the non-stationary (DAS) and the stationary (VS) parts of the input signal. The network operates in parallel mode and is fed with the input signal $u(k)$, which is the normalized zero-mean recorded lung sound. As a result, the outputs of the filter are estimations of the DAS (y_{NST}) and the VS (y_{ST}). The configuration of the proposed filter is shown in Fig. 2.

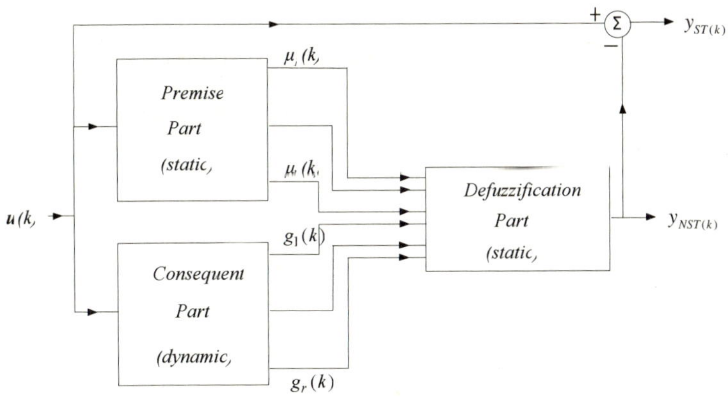

Fig. 2. Configuration of the DBD-FNN filter

The same pre-classified lung sound signals used in [1, 2, 3], i.e. ten cases, are used as model generation sets. The lung sounds are divided to three categories [1]: the coarse crackles (CC), the fine crackles (FC) and the squawks (SQ). The case of squawks is examined in the present work. The sounds have been drawn from an international sound database, [10]. The data set has been obtained by digitizing sections of 15 sec of the signals from the lung sound database by a 12-Bit Analog-to-Digital (A/D) converter at a sampling rate of 2.5 kHz, divided into successive records of 1024 or 2048 samples each, with zero mean value and normalized. Then, all these records have been processed by the WTST-NST filter in order to obtain an accurate estimation of their stationary and non-stationary parts. Therefore, the non-stationary and stationary outputs of the WTST-NST filter are considered to be the desired ones.

Several DBD-FNNs with different structural characteristics are examined. Additionally, various combinations of the learning parameters are tested. For each case, 100 trials are conducted with random initial weight values and the results are averaged. Selection of the model and the parameter combination is based on the criteria of *(a)* effective separation of DAS from VS and *(b)* a moderate complexity of the resulting model. The selected structural characteristics are given in Table 1. Training lasts for 1000 epochs and the selected learning parameters are given in Table 2.

Table 1. Characteristics of the DBD-FNN structure

Number of rules	4
Number of blocks	2
Coefficient of the sigmoid function, *a*	2
Overlapping coefficient between initial membership functions, [0,1]	0.4

Table 2. RPROP learning parameters

Premise part				Consequent part			
n^+	n^-	Δ_{min}	Δ_0	n^+	n^-	Δ_{min}	Δ_0
1.05	0.95	1E-4	0.02	1.2	0.9	1E-4	0.1

4 Experimental Results and Comparative Analysis

In this section, the results obtained using the DBD-FNN to the case of squawks are presented. The processed records were selected so that the main structural morphology of the DAS would be clearly encountered. The efficiency of the filter is tested using for evaluation the same cases of squawks that were used in [1, 2, 3].

The results obtained using the DBD-FNN filter are presented in Fig. 3 and Fig. 4, where "noisy" recorded vesicular sounds and separated DAS are depicted in parts (a) and (b), respectively. The position of waves identified visually (by a physician) as squawks were marked with arrowheads, in order to be compared with non-stationary signals separated automatically by the filter. Part (c) hosts the stationary filter's output that corresponds to the vesicular sound. In this way, the performance of the proposed filter in separating the non-stationary parts of breath sounds was verified, according to the evaluation procedure for the WTST-NST, OLS-FF and BDRNN models.

A burst of short squawks, which are included in a time section from a patient with allergic alveolitis (C5-2048 samples, 0.8192 sec) is displayed in Fig. 3(a). Despite the large concentration of squawks in this time section, their characteristics are clearly identified in the non-stationary output, as shown in Fig. 3(b). The pure vesicular

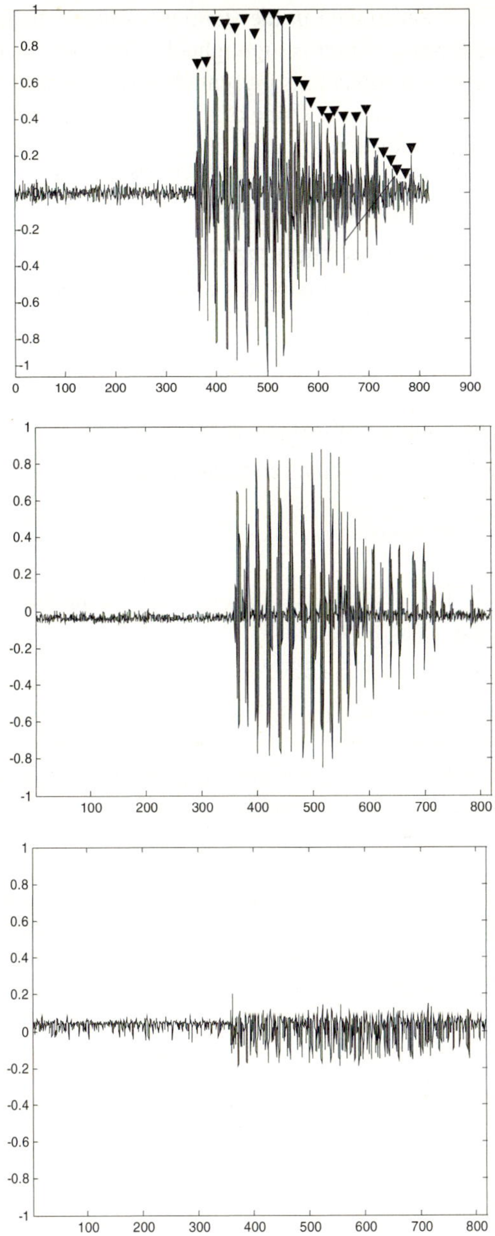

Fig. 3. (a) A time section of 0.8192 sec of squawks recorded from a patient with allergic alveolitis (case C5), considered as input. The arrowheads indicate waves, which have been visually identified as squawks. (b) The non-stationary output of the DBD-FNN filter. (c) The stationary output of the DBD-FNN filter.

sounds are accurately reconstructed in the stationary output, as depicted in Fig. 3(c). The efficiency of the proposed filter is highlighted in Fig. 4 as well. A time section of 10 msec, corresponding to a squawk, is presented. It is clear that the model has effectively detected the existence and the shape of the squawk, reproducing its morphology and location without any distortion.

The evaluation of the performance of the DBD-FNN filter is based on qualitative and quantitative measures introduced in [1]:

(i) Auditory inspection of the DBD-FNN's stationary output. The effect of the DBD-FNN on input breath sounds was tested in a qualitative manner by listening to its stationary outputs after digital-to-analog (D/A) conversion.

(ii) The rate of detectability: $D_R = \frac{N_E}{N_R} \cdot 100\%$

where N_E is the number of estimated DAS and N_R is the number of visually recognized DAS by a physician (considered as the true number of DAS in the input signal).

(iii) The root mean squared error:

$RMSE = \sqrt{\frac{1}{k_f}\sum_{k=1}^{k_f}\left[y(k)-y_d(k)\right]^2}$, where $y(k)$ is the DBD-FNN output of the k-*th* sample, $y_d(k)$ is the respective actual output and k_f is the number of samples. It should be noted that the values of the RMSE do not always represent good separation results since they do not focus on the particular signal details a physician is interested in. They are only intended to provide an indication of the quality of achieving the desired input-output relationships, given the evaluation by the first two criteria.

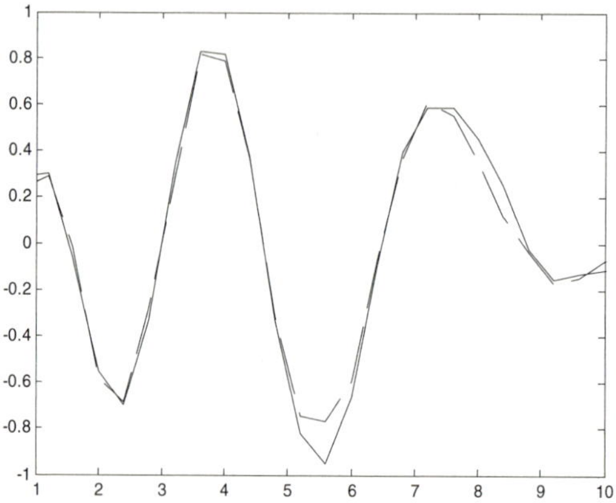

Fig. 4. A section of 10 msec that contains a squawk (solid line), together with the estimation of the non-stationary part by the DBD-FNN filter (dashed line)

The results of the quantitative evaluation are presented in Table 3 for each case. These results show that the proposed filter produces a perfect separation, since the average rate of detectability is 100%. Additionally, during the qualitative testing of the DBD-FNN filter, by listening to its stationary outputs after Digital-to-Analog (D/A) conversion, the sounds were practically not heard, confirming a high quality separation performance by the DBD-FNN filter.

Table 3. Performance of the DBD-FNN

Case	N	N_E/N_R	D_R (%)
C1	1024	2/2	100
C2	1024	4/4	100
C3	1024	5/5	100
C4	1024	6/6	100
C5	2048	26/26	100

N: Number of samples
N_R: Number of visually recognized DAS by a physician
N_E: Number of estimated DAS using the DBD-FNN filter
D_R: Rate of detectability of the DBD-FNN filter

It should be mentioned that, since the proposed filter requires only four neurons in the hidden layer of the consequent parts of the fuzzy rules, the operations of the DBD-FNN (multiplications, additions and look-up table operations) can be delivered within the sampling period (0.4 msec), ensuring its real-time operation and, consequently, improving the procedure of clinical screening of DAS.

In order to conduct a comparative analysis with other filters reported in literature, the DBD-FNN filter's performance is evaluated with regard to the WTST-NST [1], the OLS-FF [2] and the DBRNN [3]. All four models are applied to the same cases of patients and the results are shown in Table 4.

Table 4. Comparative analysis

	Average D_R *(%)*	*Average RMSE, non-stationary part*	*Average RMSE, Stationary part*
WTST-NST	100	Not defined	Not defined
OLS-FF	96.36	0.0711	0.0679
BDRNN	100	0.0650	0.0633
DBD-FNN	100	0.0588	0.0590

As shown in Table 4, the DBD-FNN filter exhibits a similar separation performance compared to the WTST-NST and the BDRNN, with respect to the average rate of detectability. Moreover, it outperforms all its competing rivals and particularly the static fuzzy filter OLS-FF with respect to the RMSE, leading to the conclusion that the recurrent model tracks effectively the dynamics of the non-stationary signal. Compared to the other recurrent model, the DBD-FNN attains a lower RMSE while requiring 50% less training time than the BDRNN does, for the same model complexity (72 weights). This result can be attributed to the enhanced modeling capabilities of the local modeling approach that the TSK fuzzy systems adopt. Furthermore, as discussed in the previous subsection, the proposed filter satisfies the real-time implementation issue, a fact that constitutes a clear advantage over the WTST-NST model since it improves the procedure of clinical screening of DAS.

5 Conclusion

A new recurrent fuzzy filter has been implemented for real-time separation of lung sounds. The filter model is based on the classic TSK fuzzy model, with the consequent parts of the fuzzy rules consist of small block-diagonal recurrent neural networks. The scheme has been evaluated on pre-classified DAS, belonging to the sound category of squawks, selected from an international lung sound database. From the experiments and the comparative analysis with other separation schemes, it is concluded that the DBD-FNN filter performs very efficiently in separating the DAS from VS despite the differences in their structural character, and is capable of performing real-time separation. Hence, the proposed filter can be used as an objective method for real-time DAS analysis.

Acknowledgement

The Project is co-funded by the European Social Fund and National Resources – (EPEAEK–II) ARXIMHDHS.

References

1. Hadjileontiadis, L.J., Panas, S.M.: Separation of Discontinuous Adventitious Sounds from Vesicular Sounds Using a Wavelet-Based Filter. IEEE Trans. Biomedical Eng. 44, 1269–1281 (1997)
2. Mastorocostas, P.A., Tolias, Y.A., Theocharis, J.B., Hadjileontiadis, L.J., Panas, S.M.: An Orthogonal Least Squares-Based Fuzzy Filter for Real-Time Analysis of Lung Sounds. IEEE Trans. Biomedical Eng. 47, 1165–1176 (2000)
3. Mastorocostas, P.A., Theocharis, J.B.: A Stable Learning Method for Block-Diagonal Recurrent Neural Networks: Application to the Analysis of Lung Sounds. IEEE Trans. Syst., Man, and Cybern. – Part B 36, 242–254 (2006)
4. Sivakumar, S.C., Robertson, W., Phillips, W.J.: On-Line Stabilization of Block-Diagonal Recurrent Neural Networks. IEEE Trans. Neural Networks 10, 167–175 (1999)
5. Takagi, T., Sugeno, M.: Fuzzy Identification of Systems and its applications to modeling and Control. IEEE Trans. Syst., Man, and Cybern. 15, 116–132 (1985)
6. Tsoi, A.C., Back, A.D.: Locally Recurrent Globally Feedforward Networks: A Critical Review of Architectures. IEEE Trans. Neural Networks 5, 229–239 (1994)
7. Riedmiller, M., Braun, H.: A Direct Adaptive Method for Faster Backpropagation Learning: The RPROP Algorithm. In: Proc. IEEE Int. Joint Conf. on Neural Networks, pp. 586–591 (1993)
8. Igel, C., Husken, M.: Empirical Evaluation of the Improved RPROP Learning Algorithms. Neurocomputing 50, 105–123 (2003)
9. Piche, S.: Steepest Descent Algorithms for Neural Network Controllers and Filters. IEEE Trans. Neural Networks 5, 198–221 (1994)
10. Kraman, S.S.: Lung Sounds: An Introduction to the Interpretation of the Auscultatory Finding. Northbrook, IL: Amer. College of Chest Physicians, audio tape (1993)

Learning Temporally Stable Representations from Natural Sounds: Temporal Stability as a General Objective Underlying Sensory Processing

Armin Duff[1,2], Reto Wyss[3], and Paul F.M.J. Verschure[2,4]

[1] Institute of Neuroinformatics, UNI - ETH Zürich, Winterthurerstrasse 190, CH-8057 Zürich, Switzerland
duffar@ini.phys.ethz.ch
[2] SPECS, IUA, Technology Department, Universitat Pompeu Fabra, Ocata 1, E-08003 Barcelona, Spain
[3] CSEM Centre Suisse d'Electronique et de Microtechnique SA, Untere Gründlistrasse 1, CH-6055 Alpnach-Dorf, Switzerland
[4] ICREA Institució Catalana de Recerca i Estudis Avançats, Passeig Lluís Companys 23, E-08010 Barcelona, Spain

Abstract. In order to understand the general principles along which sensory processing is organized, several recent studies optimized particular coding objectives on natural inputs for different modalities. The homogeneity of neocortex indicates that a sensitive objective should be able to explain response properties of different sensory modalities. The temporal stability objective was successfully applied to somatosensory and visual processing. We investigate if this objective can also be applied to auditory processing and serves as a general optimization objective for sensory processing. In case of audition, this translates to a set of non-linear complex filters optimized for temporal stability on natural sounds. We show that following this approach we can develop filters that are localized in frequency and time and extract the frequency content of the sound wave. A subset of these filters respond invariant to the phase of the sound. A comparison of the tuning of these filters to the tuning of cat auditory nerves shows a close match. This suggests that temporal stability can be seen as a general objective describing somatosensory, visual and auditory processing.

1 Introduction

The human neocortex is to a great extent homogeneous throughout all its areas [1, 2]. This suggests that it should be possible to describe its structure and dynamics with general concepts and models. This line of thinking appears most prominent in experimental and theoretical work that proposes a "canonical microcircuit" as a basic computational unit [3]. Such a general view does imply that any relevant computational neuronal model should explain response properties of neurons of different areas of the neocortex. Thus, one would expect that

J. Marques de Sá et al. (Eds.): ICANN 2007, Part II, LNCS 4669, pp. 129–138, 2007.

also in sensory processing, the same model can be applied to replicate response properties of different modalities.

It has long been assumed, that sensory systems adapt to the statistical properties of their input leading to the general approach of explaining receptive field properties of sensory systems by optimizing a particular coding objective for natural stimuli [4]. In a series of theoretical studies, several general coding principles have been exploited for different sensory modalities. In visual processing, learning sparse codes from natural images leads to simple-cell like receptive fields as found in primary visual cortex [5]. An extension of this approach to a multi-layer network enabled to learn contour coding in natural images [6]. Similarly, optimizing for temporal stability can replicate properties of V1 simple cells [7], but also invariant representations similar to V1 complex cells [8, 9, 10, 11, 12], color selective cells [13] and viewpoint invariant object representations [14, 15]. Recently it has been shown that the optimization of a multi layered network for temporal stability combined with local memory can account for a complete visual hierarchy, including place fields, by processing a continuous natural input stream generated by a mobile robot [16]. In a different approach, a hierarchical model was optimized based on a MAX-like operation for object recognition [17]. Another objective function, predictability, was proposed to yield self-emergent symbols in an optimization process [18]. In auditory processing, it was shown that optimizing a set of filters for efficiency can explain the formation of adequate auditory filters [19, 20, 21]. Further, optimizing a spectrographic representation of speech for sparseness leads to similar spectro-temporal receptive fields (STRF) as observed in primary auditory cortex [22]. Temporal stability optimization is not restricted to a visual input stream but was also successfully applied to preprocess data for somatosensory discrimination of texture [23]. This diversity of approaches to explain sensory processing contradicts several theoretical and anatomical studies suggesting that the same computational strategy is likely to be involved in processing information from different sensory modalities [1, 2, 24, 25, 26, 27].

In this study we investigate whether temporal stability may be an appropriate objective for the auditory domain and serve as a general objective to replicate neural responses in somatosensory, visual and auditory sensory processing. Auditory processing begins when the cochlea transforms sound energy into electrical signals and passes them to the auditory nerves. Ignoring the nonlinearities and amplification features of the cochlea and the primary auditory system, the response properties of auditory nerves can be described by a set of filters, with different frequency tunings, forming a spectro-temporal representation of the sound [28]. It is not clear how this representation is tuned to the statistics of the sound environment. In order to get a filter set adapted to the input statistics we optimize a set of complex filters to show a maximally stable response across time for natural sounds. Following this approach we do not describe the details of how cochlea and auditory nerve realize the spectro-temporal separation but reveal the general principle around which sensory processing is organized. While different types of filters emerge, we find that the tuning of the filters is in accordance with experimental data from cat physiological data. This suggests that the response properties of auditory nerve fibers can be explained in terms of temporal stability optimization. Thus the somatosensory, visual and auditory sensory processing can be explained within the same framework of temporal stability.

2 Methods

2.1 Input

The natural sounds used to learn the stable representations consist of different words in 6 distinct languages, spoken by three male and three female native speakers. The sound samples are re-sampled to 22050 Hz from the language illustrations accompanying the handbook of the International Phonetic Association (IPA) [29]. In addition to the raw sound wave we investigate band-passed sound waves. We consider four different band-pass filters with characteristic frequency bands of 1 Hz - 324 Hz, 324 Hz - 1050 Hz, 1050 Hz - 3402 Hz and 3402 Hz - 11025 Hz. According to the logarithmic organization of the cochlea we increased the bandwidth of the different filters logarithmically.

2.2 Filters

We optimized a set of complex filters with an analysis window covering 5.8 ms of the raw sound wave which corresponds to 128 data points. At each iteration of the optimization the sound wave is shifted through this window by the time interval τ. Thus, by changing τ we can change the overlap of subsequent analysis windows.

The activity (A_i) of the filter i is given by the absolute value of the scalar product between the input $\boldsymbol{I}$ and the *complex* filter function $\boldsymbol{h_i} \in \mathbb{C}^{128}$. Real and imaginary values can change independently and thus each filter is defined by 256 parameters.

$$A_i = abs(\boldsymbol{h}_i \cdot \boldsymbol{I}) \tag{1}$$

The absolute value function implies that the filters are not linear. Mathematically they are equivalent to the energy model proposed by Adelson and Bergen [30] and applied in different studies [9, 16, 23].

2.3 Optimization

In the present study we optimized a goal function which contains two terms such that the total objective Ψ is given by:

$$\Psi = (1-\gamma)\Psi_{stab} + \gamma\Psi_{decor} \tag{2}$$

where γ is used to balance between the relative contribution of the two objectives. The first part is the temporal stability objective Ψ_{stab} given by:

$$\Psi_{stab} = -\sum_i \frac{\langle \dot{A}_i^2 \rangle_t}{var_t(A_i)} \tag{3}$$

A_i is the activity of the filter and the sum is over all filters i. $\dot{A}$ denotes the discrete temporal differentiation given by:

$$\dot{A}_i = A_i(t) - A_i(t - \tau) \tag{4}$$

where τ is the time interval by witch the analysis window is shifted each iteration. The floating average at time t, $\langle\ .\ \rangle_t$ is calculated iteratively with a time constant $\zeta = 500\ ms$ and defined by:

$$\langle A_i \rangle_t = (1 - \frac{1}{\zeta})\langle A_i \rangle_{t-1} + \frac{1}{\zeta} A_i(t) \tag{5}$$

The temporal derivative in equation 3 is divided by the variance in order to avoid the trivial solution where all the parameters of the filter are zero. The variance is computed using:

$$var_t(A_i) = \langle A_i^2 \rangle_t - \langle A_i \rangle_t^2 \tag{6}$$

As the filters should have different receptive fields, collectively representing the statistics of the input, each of them must encode different information. Therefore, the second term, i.e the decorrelation objective Ψ_{decor} (7), is introduced to augment the statistical independence of the filters.

$$\Psi_{decor} = -\sum_{j \neq i} (\rho_{ij}(A_i, A_j))^2 \tag{7}$$

$$\rho_{ij}(A_i, A_j) = \frac{\langle A_i A_j \rangle_t - \langle A_i \rangle_t \langle A_j \rangle_t}{\sqrt{var_t(A_i) var_t(A_j)}} \qquad (correlation) \tag{8}$$

The filter functions h_i change following an on-line learning algorithm along the gradient in order to maximize the total objective function Ψ.

2.4 Time and Frequency Analysis

To characterize each of the filters we extracted the center frequency (CF), the spectral bandwidth (BW), the quality factor (Q), the temporal extent (TE) and the relative shift ϕ as key characteristics. The CF of the filter corresponds to the maximum of the power spectrum of the filter function. This is the frequency for which the filter is most sensitive. The BW is the width of the frequency response measured at 10 dB down from the peak at the CF. Q corresponds to the sharpness of the filter and is defined by the CF divided by the BW. The subscript in Q_{10dB} indicates the level at which the BW is measured. The TE of a filter is defined as the width that is used to cover 90 % of the filter power. ϕ is given by the relative phase shift of the real and the imaginary part of the filter and is calculated with respect to the CF of the filter.

3 Results

We optimize 64 filters on the raw sound wave using an update interval τ between 0.1 - 2.8 ms in steps of $\approx$ 0.1 ms for different simulations. Each filter is defined by its 256 parameters $h_i \in \mathbb{C}^{128}$. Most of the resulting filters are sinusoidal,

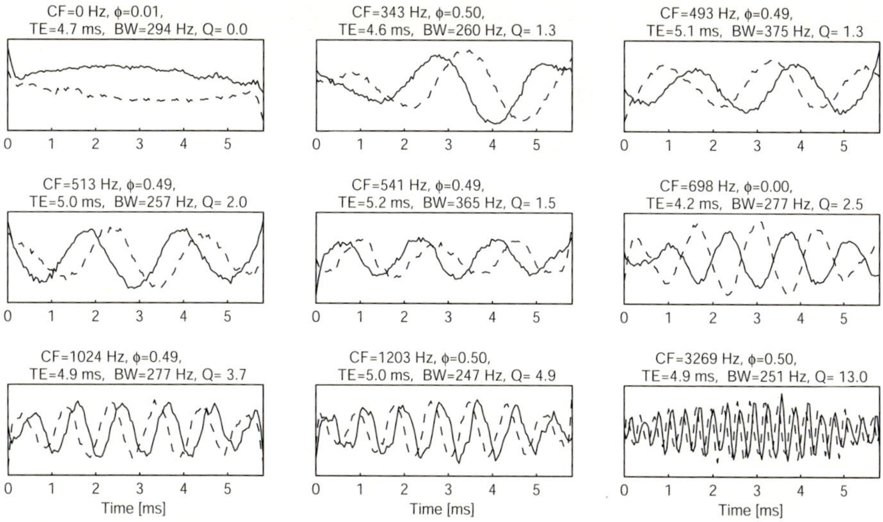

Fig. 1. Auditory filters after temporal stability optimization using an update interval τ of 0.7 ms. The plot shows a representative subset of the total population of 64 filters. The filters are plotted in the time domain where the solid line corresponds to the real part of the filter and the dashed line to the imaginary part. On top of each waveform the key characteristics of the filter are indicated: center frequency (CF), relative shift (ϕ), temporal extent (TE), spectral bandwidth (BW) and quality factor (Q).

amplitude modulated, and cover the whole window width (Figure 1). The filters have a single peak frequency tuning such that their CF is well defined.

The investigation of the distribution of the CFs, shows that it matches the Power Spectrum Density (PSD) of the sound (Figure 2). The higher the PSD in a frequency interval, the higher the number of filters with a CF in this interval.

A part of the nonlinear sinusoidal filters ($\approx$ 60%) exhibit a relative shift of $\approx \pi/2$ between their real and imaginary components that conforms to a filter selective for a particular frequency while being invariant to the phase of the sound wave. Other filters have no relative shift and therefore do depend on the phase.

As pointed out above, the distribution of the CFs is correlated with the PSD such that we only obtained filters within the lower part of the frequency spectrum. To enable the system to form filters with CFs in other frequency ranges we band-passed the signal of the speech ensemble before optimizing. For each band-passed signal we optimized 16 filters. The filter set that emerged covers a frequency range of 120 - 5500 Hz. Frequencies higher than 5500 Hz are not covered as the main energy of the band-passed signal with the highest pass range, lies between 3400 Hz and 5500 Hz. Preliminary experiments showed that higher CFs can be obtained by applying higher band-pass (data not shown).

To get an impression of the distribution and coverage of the filter sets in time and frequency, we considered the extent of the filters on the time-frequency plane. (Figure 3 A). The filter set possesses both, temporally localized and non-

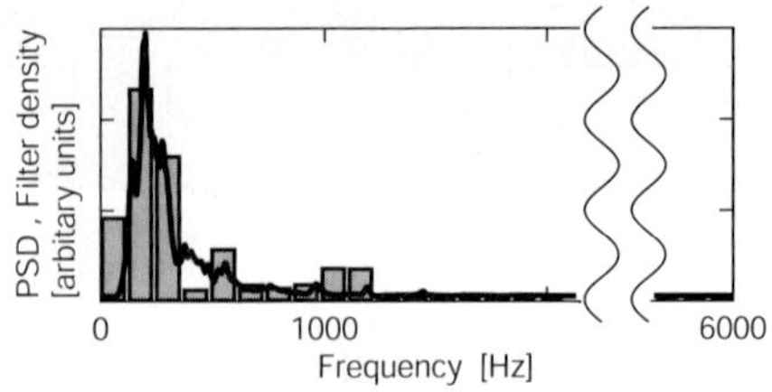

Fig. 2. Distribution of the CFs (bars) and the PSD of the signal (line). The bars indicate the relative density of filters for the corresponding frequency range. The PSD is superimposed where the energy density is given in an arbitrary scale.

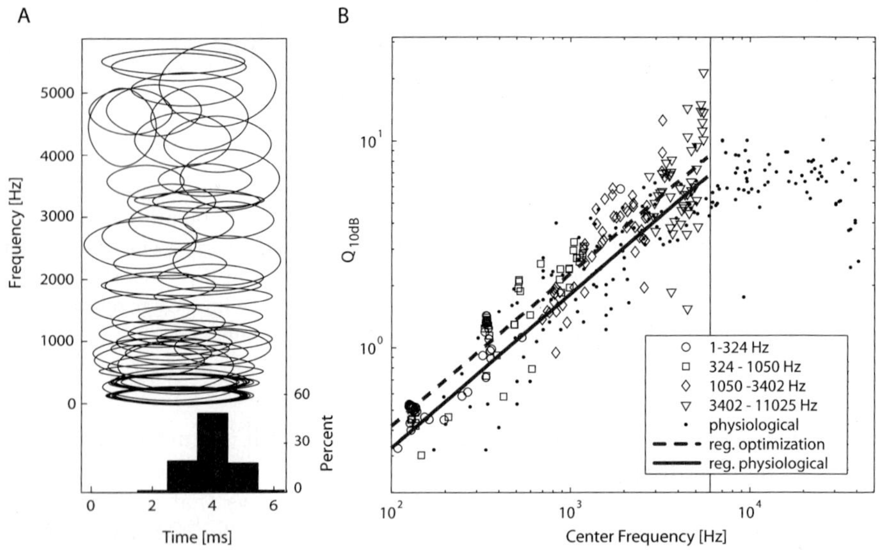

Fig. 3. Characteristics of the filter sets for the band-passed signals. **A** Tiling the time frequency plane. The extent of each filter is represented by an ellipse. The height of the ellipse corresponds to the frequency bandwidth and the width to the temporal extent of the filter. The histogram at the bottom indicates the distribution of the temporal extent for each filter set. **B** Q_{10dB} of the optimized filters compared to the Q_{10dB} measured from cat auditory nerve fibers. For comparison the regression line for the optimized filters and the physiological data are superimposed. Notice that the regression for the physiological data only includes the data points with a frequency lower than 5500 Hz. The physiological data is replotted from [28].

localized filters. Corresponding to the time frequency uncertainty relation, filters with narrower temporal extent (TE) have a broader frequency extent (BW). Further, one can observe that the TE decreases for higher frequencies.

In order to validate the emerged filters against physiological data we compare the tuning of the filters to cat auditory nerve fibers [28]. The sharpness (Q) of the optimized filters is consistent with the sharpness measured for cat auditory nerve fibers (Figure 3 B). Linear regression on the physiological data in a range of $1-5500\ Hz$ i.e. the range covered by the filters that emerged from temporal stability optimization, yields a stiffness of 0.744 $[0.675, 0.831]$ and an offset of -4.542 $[-5.176, -3.908]$. For the optimized filter set linear regression yields a stiffness of 0.736 $[0.703, 0.769]$ and an offset of -4.259 $[-4.487, -4.031]$. The numbers in brackets correspond to the 95% confidence interval. Thus, both the stiffness and the offset of the two curves are very close. The deviation for the stiffness is 1% and for the offset 6%. The confidence interval for the optimized filters lie within the confidence intervals of the physiological data. This suggests that some of the response properties of the auditory nerve fibers can be well explained in terms of temporal stability optimization.

4 Discussion

In this study we investigated if temporal stability is a general objective underlying sensory processing and can be applied in the auditory domain. We have shown that temporal stability optimization in the auditory domain leads to filters that extract the frequency content of speech. Due to a relative shift of $\pi/2$ between the real and imaginary parts of the complex filters, some filters have an invariant response to the phase of the sound wave. Further we found that the distribution of the CFs is related to the PSD such that frequency bands with high energy have a high filter density. Our approach rendered filters which show a quantitative match to the filter properties of the cat auditory nerve. This suggests that the physiological properties of this part of the auditory pathway can be explained in terms of temporal stability optimization.

The design of filters involves an inevitable trade-off between time resolution and frequency resolution [31]. To get precise information about the frequency content one has to integrate over a characteristic time length of the sound signal, leading to a decrease in temporal precision. In other words, it is not possible to design a filter that captures both the frequency and the timing of a sound with arbitrary accuracy. However, in order to be able to process natural sounds it is often important to have information about both frequency and timing. Which spectro-temporal representation is optimal depends on a number of factors such as the importance of the information available, the biological or computational constraints and the statistics of the sound. A common mis-characterization of the peripheral auditory system is that it performs a short time Fourier transform (STFT) or a kind of wavelet decomposition. For the STFT, the bandwidth of the signals is approximately constant whereas for a wavelet representation the sharpness of the filters remains constant for all frequencies. Neither of the two properties are observed experimentally [28]. Instead, similar to the filters found in our approach, the sharpness of the auditory nerve fibers follows a sub-linear power law (Figure 3 B). Therefore, an optimal set of filters for the analysis of speech must exhibit both, Fourier and wavelet characteristics.

This study is complementary to the work of Lewicki and Smith [19, 20, 21] who optimized a set of real-valued linear filters, for sparseness on different sound ensembles including speech. While all the filter responses resulting from Lewicki's approach must vary with the phase of the incoming signal due to linearity, we have found that a part of the nonlinear complex filters are phase invariant. Similarly, some of the auditory nerve fibers at the cochlea do code the phase of the signal, which is important to determine sound location in the early auditory pathway. At higher levels of processing, this phase information is lost, suggesting that phase invariance constitutes a first step towards sound classification [32]. A further difference to the work of Lewicki can be found in the distribution of the CF of the filters. Optimizing for sparseness leads to a distribution of CFs covering the whole frequency range up to the Nyquist frequency. Optimizing for temporal stability, however, results in a distribution which is correlated to the PSD and therefore confined to reagions with high energy. The question that arises is to what extent the features that carry the main energy also carry the relevant information. For speech, the slowly varying features are the vowels whereas the consonants vary much faster and have higher frequencies. Therefore, temporal stability optimization tends to extract the information that is contained in the vowels but mostly ignores the consonants whereas sparseness mainly extracts features with higher frequencies. However, for word discrimination both vowels and consonants are important. We have shown that we can account for the whole range of frequencies by band-passing the signal before optimizing for temporal stability. Given the acoustic properties of the cochlea, such band-passing is likely to happen already at a mechanical level at the basilar membrane [32]. The subsequent levels of auditory information processing would therefore already be supplied with an appropriately band-pass filtered signal.

In order to validate our results against experimental data, we compared the learned filters with cat physiological data. Primary auditory cortex (A1) neurons are characterized by Spectro-Temporal Receptive Fields (STRF) [33, 34, 35]. As our filters do not include a temporal component their activity is only dependent on the spectral content of the sound and thus a direct comparison is not possible. We could however compare the characteristics of the optimized filters with the characteristics of the cat auditory nerve fibers. For this comparison it is not clear to what extent human speech is an adequate auditory stimulus for cats or whether animal vocalization would be more appropriate. Speech contains a mixture of various auditory features which are present in different classes of natural sounds [19]. Thus, given that any animal is exposed to a mixture of environmental sounds and vocalization, speech constitutes a good compromise. We have found, that the sharpness of the filters for the band-passed speech ensemble matched the sharpness measured for cat auditory nerve fibers. Hence some of the response properties of auditory nerve fibers can be explained in terms of optimizing a set of complex filters for temporal stability. Therefore, temporal stability can be considered a general objective underlying sensory processing.

References

[1] Sur, M., Rubenstein, J.L.R.: Patterning and plasticity of the cerebral cortex. Science 310(5749), 805–810 (2005)

[2] Horng, S.H., Sur, M.: Visual activity and cortical rewiring: activity-dependent plasticity of cortical networks. Prog. Brain Res. 157, 3–11 (2006)

[3] Douglas, R., Martin, K.: Neocortex. In: Shepherd, G. (ed.) The Synaptic Organization of the Brain, pp. 459–509. Oxford University Press, New York (1998)

[4] Simoncelli, E.P., Olshausen, B.A.: Natural image statistics and neural representation. Annu. Rev. Neurosci. 24, 1193–1216 (2001)

[5] Olshausen, B., Field, D.: Emergence of simple-cell receptive field properties by learning a sparse code for natural images. NatureEmergence of simple-cell receptive field properties by learning a sparse code for natural images 381(6583), 607–609 (1996)

[6] Hoyer, P.O., Hyvärinen, A.: A multi-layer sparse coding network learns contour coding from natural images. Vision. Res. 42(12), 1593–1605 (2002)

[7] Hurri, J., Hyvärinen, A.: Simple-cell-like receptive fields maximize temporal coherence in natural video. Neural Comput 15(3), 663–691 (2003)

[8] Berkes, P., Wiskott, L.: Slow feature analysis yields a rich repertoire of complex cell properties. J Vis. 5(6), 579–602 (2005)

[9] Körding, K.P., Kayser, C., Einhäuser, W., König, P.: How are complex cell properties adapted to the statistics of natural stimuli? J. Neurophysiol. 91(1), 206–212 (2004)

[10] Hashimoto, W.: Quadratic forms in natural images. Network 14(4), 765–788 (2003)

[11] Einhäuser, W., Kayser, C., König, P., Körding, K.P.: Learning the invariance properties of complex cells from their responses to natural stimuli. Eur. J. Neurosci. 15(3), 475–486 (2002)

[12] Kayser, C., Einhäuser, W., Dümmer., O., König, P., Körding, K.P.: Extracting slow subspaces from natural videos leads to complex cells. In: Dorffner, G., Bischof, H., Hornik, K. (eds.) ICANN 2001. LNCS, vol. 2130, pp. 1075–1080. Springer, Heidelberg (2001)

[13] Einhäuser, W., Kayser, C., Körding, K., König, P.: Learning distinct and complementary feature-selectivities from natural colour videos. Journal of Neuroscience-Learning distinct and complementary feature-selectivities from natural colour videos 21, 43–52 (2003)

[14] Stringer, S.M., Rolls, E.T.: Invariant object recognition in the visual system with novel views of 3D objects. Neural Comput. 14(11), 2585–2596 (2002)

[15] Einhäuser, W., Hipp, J., Eggert, J., Körner, E., König, P.: Learning viewpoint invariant object representations using a temporal coherence principle. Biol. Cybern. 93(1), 79–90 (2005)

[16] Wyss, R., König, P., Verschure, P.F.M.J.: A Model of the Ventral Visual System Based on Temporal Stability and Local Memory. PLoS Biol. 4(5), e120 (2006)

[17] Riesenhuber, R., Poggio, T.: Hierarchical models of object recognition in cortex. Nature Neuroscience 2, 1019–1025 (1999)

[18] König, P., Krüger, N.: Symbols as Self-emergent Entities in an Optimization Process of Feature Extraction and Predictions. Biol. Cybern. 94(4), 325–334 (2006)

[19] Lewicki, M.S.: Efficient coding of natural sounds. Nat. Neurosci. 5(4), 356–363 (2002)

[20] Smith, E., Lewicki, M.S.: Efficient coding of time-relative structure using spikes. Neural. Comput. 17(1), 19–45 (2005)

[21] Smith, E.C., Lewicki, M.S.: Efficient auditory coding. Nature 439(7079), 978–982 (2006)
[22] Klein, D., König, P., Körding, K.: Sparse spectrotemporal coding of sounds. Eurasip JaspSparse spectrotemporal coding of sounds 3, 659–667 (2003)
[23] Hipp, J., Einhäuser, W., Conradt, J., König, P.: Learning of somatosensory representations for texture discrimination using a temporal coherence principle. Network 16(2-3), 223–238 (2005)
[24] Shamma, S.: On the role of space and time in auditory processing. Trends Cogn. Sci. 5(8), 340–348 (2001)
[25] Sur, M., Leamey, C.A.: Development and plasticity of cortical areas and networks. Nature Reviews Neuroscience 2, 251–262 (2001)
[26] Dennis, D., O'Leary, M.: Do cortical areas emerge from a protocortex? Trends in Neuroscience 12(10), 400–406 (1989)
[27] Sur, M., Garraghty, P., Roe, A.: Experimentally induced visual projections into auditory thalamus and cortex. Science 242, 1437–1441 (1988)
[28] Rhode, W.S., Smith, P.H.: Characteristics of tone-pip response patterns in relationship to spontaneous rate in cat auditory nerve fibers. Hearing Research 18, 159–168 (1985)
[29] International Phonetic Association, ed.: Handbook of the International Phonetic Association. Cambridge University Press, Cambridge, UK, (1999) Available at: `http://web.uvic.ca/ling/resources/ipa/handbook.htm`
[30] Adelson, E.H., Bergen, J.R.: Spatiotemporal energy models for the perception of motion. J Opt. Soc. Am. A 2(2), 284–299 (1985)
[31] Rioul., O., Vetterli, M.: Wavelets and signal processing. IEEE Signal Processing Magazine 8, 14–38 (1991)
[32] Hudspeth, A.J.: Hearing. In: Kandel, E.R., Schwartz, J.H., Jessell, T.M. (eds.) Principles of Neural Science, 4th edn., pp. 590–613. McGraw-Hill, New York (2000)
[33] Depireux, D.A., Simon, J.Z., Klein, D.J., Shamma, S.A.: Spectro-temporal response field characterization with dynamic ripples in ferret primary auditory cortex. J Neurophysiol 85(3), 1220–1234 (2001)
[34] Klein, D.J., Simon, J.Z., Depireux, D.A., Shamma, S.A.: Stimulus-invariant processing and spectrotemporal reverse correlation in primary auditory cortex. J Comput. Neurosci. 20(2), 111–136 (2006)
[35] Theunissen, F.E., Woolley, S.M.N., Hsu, A., Fremouw, T.: Methods for the analysis of auditory processing in the brain. Ann. N Y Acad. Sci. 1016, 187–207 (2004)

Comparing Methods for Multi-class Probabilities in Medical Decision Making Using LS-SVMs and Kernel Logistic Regression

Ben Van Calster[1], Jan Luts[1], Johan A.K. Suykens[1], George Condous[2], Tom Bourne[3], Dirk Timmerman[4], and Sabine Van Huffel[1]

[1] Department of Electrical Engineering (ESAT-SISTA), Katholieke Universiteit Leuven, Kasteelpark Arenberg 10, B-3001, Leuven, Belgium
[2] Nepean Hospital, University of Sydney, Sydney, Australia
[3] St Georges Hospital Medical School, London, UK
[4] University Hospitals K.U.Leuven, Leuven, Belgium

Abstract. In this paper we compare thirteen different methods to obtain multi-class probability estimates in view of two medical case studies. The basic classification method used to implement all methods are least squares support vector machine (LS-SVM) classifiers. Results indicate that multi-class kernel logistic regression performs very well, together with a method based on ensembles of nested dichotomies. Also, a Bayesian LS-SVM method imposing sparseness performed very well for methods that combine binary probabilities into multi-class probabilities.

1 Introduction

This paper focuses on two issues: multi-class classification and probabilistic outputs. Multi-class classification is less straightforward than binary classification, in particular for margin-based classifiers such as support vector machines (SVMs). Methods for multi-class tasks are based on the combination of binary classifiers or on 'all-at-once' classification. Binary classifiers are typically constructed by contrasting all pairs of classes (1-vs-1) or by contrasting each class with all other classes combined (1-vs-All), or by other techniques such as error-correcting output coding [1].

When using binary classifiers, results are frequently combined using a voting strategy. Often, however, one is interested in class probabilities. These are important because they give information about the uncertainty of class membership as opposed to black-and-white class predictions. In medical decision making, uncertainty information can influence the optimal treatment of patients.

In this paper, we compare many methods to obtain multi-class probability estimates using two medium sized medical data sets dealing with pregnancies of unknown location and ovarian tumors. For both conditions, early probabilistic predictions are needed for optimizing patient care and its financial implications. All-at-once, 1-vs-1, and 1-vs-All methods were used. Section 2 describes these methods. Section 3 describes the data and the analysis methodology. The results

J. Marques de Sá et al. (Eds.): ICANN 2007, Part II, LNCS 4669, pp. 139–148, 2007.

are reported and discussed in Section 4. We assume data sets with data $(\boldsymbol{x}_n, t_n)$, $n = 1, \ldots, N$, with $\boldsymbol{x}_n \in \mathbb{R}^q$ and t_n is one of k target classes ($k > 2$).

2 Obtaining Multi-class Probability Estimates

2.1 Least Squares Support Vector Machine Classifiers

The basic classification method used in this paper is the least squares support vector machine (LS-SVM) classifier [2],[3]. This variant of standard SVMs can be obtained by solving a linear system rather than a quadratic programming problem. This adjustment results, at least for standard LS-SVMs, in a lack of sparseness. However, methods to impose sparseness exist. The LS-SVM model formulation in the primal space is

$$\min_{\boldsymbol{w},b,e} \left(\frac{1}{2}\boldsymbol{w}^T\boldsymbol{w} + \frac{1}{2}\gamma \sum_{n=1}^{N} e_n^2 \right), \text{ such that } y_n \left[\boldsymbol{w}^T\varphi(\boldsymbol{x}_n) + b \right] = 1 - e_n \,, \qquad (1)$$

for the classifier $y(x) = \text{sign}[\boldsymbol{w}^T\varphi(\boldsymbol{x}) + b]$, where y_n is a binary class indicator (encoded as -1 vs $+1$), $\boldsymbol{w}$ is a parameter vector, b is a bias term, e_n is the error variable, and γ is a hyperparameter to control the amount of regularization. The mapping $\varphi : \mathbb{R}^q \rightarrow \mathbb{R}^r$ maps the input space into a high-dimensional feature space. By taking the Lagrangian for this problem, the classifier can be reformulated in the dual space as $y(x) = \text{sign}\left[\sum_{n=1}^{N} \alpha_n y_n K(\boldsymbol{x}, \boldsymbol{x}_n) + b\right]$, where $\alpha_1, \ldots, \alpha_N$ are the support values. In this way we implicitly work in the feature space by applying a positive definite kernel $K(\boldsymbol{x}, \boldsymbol{z}) = \varphi(\boldsymbol{x})^T\varphi(\boldsymbol{z})$. We used the radial basis function (RBF) kernel function throughout this paper. This kernel has parameter σ denoting the kernel width. Due to the 2-norm in the cost function, typically sparseness is lost.

2.2 Creating Probabilistic Output from Binary (LS-)SVMs

We implemented four techniques to get probabilistic outputs from the binary LS-SVM classifiers: a Bayesian LS-SVM [4] with and without sparseness induction, and a transformation of the LS-SVM output using either an improvement on Platt's method [5],[6] or isotonic regression [7]. For non-Bayesian LS-SVMs, the hyperparameters γ and σ have to be tuned using the training data. To do this, we applied the leave-one-out method suggested in [8] with the predicted residual sum-of-squares (PRESS) statistic.

Bayesian LS-SVM (bay; bays). MacKay's evidence procedure was used for the Bayesian LS-SVM [4],[9]. Here, two hyperparameters μ and ζ are used for the slightly modified cost function $0.5\mu\boldsymbol{w}^T\boldsymbol{w} + 0.5\zeta\sum_{n=1}^{N} e_n^2$ such that $\gamma = \zeta/\mu$. At the first level of the Bayesian procedure, the prior distribution for $\boldsymbol{w}$ and b is set to be multivariate normal. The prior is multiplied by the likelihood function to obtain the posterior. Maximization of the posterior (which is approximated by a normal distribution) yields the 'most probable' values for $\boldsymbol{w}$ and b, $\boldsymbol{w}_{\text{MP}}$ and

b_{MP}. Since the prior distribution corresponds to the regularization term $\boldsymbol{w}^T\boldsymbol{w}$ and the likelihood to the sum of the squared errors, such maximization is similar to solving an LS-SVM. On the second level, μ_{MP} and ζ_{MP} are obtained by using a uniform prior on $\log(\mu)$ and $\log(\zeta)$. At the third level, σ is updated. We denote this algorithm by `bay`. We also applied it with a sparseness procedure (`bays`). Since for LS-SVMs $\alpha_n = \gamma e_n$, support values can be negative for easy cases. Sparseness can be imposed by repeatedly pruning training data with negative support values until none are left [10].

LS-SVM + Platt (`plt`). Based on (LS-)SVM output, Platt [5] estimates the class probability $P(y=1|\boldsymbol{x}) = 1/\big(1+\exp(Af+B)\big)$, where the LS-SVM latent variable $f = \sum_{n=1}^{N} \alpha_n y_n K(\boldsymbol{x}, \boldsymbol{x}_n) + b$. Training data f values were used to find A and B using maximum likelihood estimation. These f values were derived using 5-fold cross-validation (5CV) on the training data in order to obtain less biased values. We used an improved implementation of Platt's method [6].

LS-SVM + Isotonic Regression (`iso`). Zadrozny and Elkan [7] propose the use of isotonic regression to obtain probabilities based on (LS-)SVM output (f). This nonparametric regression technique maps f to probabilities using a monotonically increasing function. The isotonic regression method used is that of pooled adjacent violators [11]. The training cases are ranked with respect to f. The corresponding class membership (coded as 0 versus 1 instead of -1 versus $+1$) is taken as the initial estimated probability of belonging to class $y = +1$. If these initial probabilities are entirely separated in the ranking (i.e., they are isotonic), they are taken as the final estimates. If the initial estimates are not isotonic for two cases, both probabilities are averaged to serve as the new estimates. This is repeated until the estimated probabilities are isotonic. Training data f values used for the isotonic regression were again based on 5CV. When a new f value fell between two training data f values with different estimated probabilities, we estimated the probability for the new output value by means of simple linear interpolation.

2.3 Methods to Combine Binary Probabilities

Let us denote the probability of a case to belong to the ith of k classes as $p_i = P(t=i|\boldsymbol{x})$ (estimated by $\hat{p}_i$), and the probability of a case to belong to the ith class conditional on membership of class i or j as $p_{ij} = P(t=i|t \in \{i,j\}, \boldsymbol{x})$, (estimated by $\hat{p}_{ij}$). To combine binary probabilities into multi-class probabilities we used ten methods that we will shortly describe.

Refregier and Vallet (`rvall`; `rvone`) [12]. Taking two classes i and j, one starts from the relation $\hat{p}_{ij}/\hat{p}_{ji} \approx p_i/p_j$ to estimate the multi-class probabilities using any $k-1$ binary classifiers. Because $\sum_{i=1}^{k} p_i = 1$ the multi-class probabilities are estimated by solving a linear system. We implemented this method once by averaging the multi-class probabilities for all subsets of $k-1$ binary classifiers (`rvall`) and once by randomly selecting one such subset (`rvall`).

Ensembles of Nested Dichotomies (`endall`; `endone`) [13]. This method obtains multi-class probabilities by advancing through a tree of binary classifiers. At the top level, all classes are divided into two subgroups for which a binary classifier is trained. Subgroups that consist of more than one class are again divided into two groups until all classes form a subgroup of their own. The tree then consists of k leaves. The multi-class probability p_i is obtained by multiplying all binary probabilities involving p_i. We implemented this method once by averaging the multi-class probabilities of all trees (`endall`) and once by randomly selecting one tree (`endone`). Of course, the number of different trees for k classes $T(k)$ grows exponentially with k such that `endall` is impractical for large classes: $T(4) = 15$ but $T(5) = 105$.

Price *et al.* (`pkpd`) [14]. This method uses the probability of a union of events, and by stating that $\hat{p}_{ij} \approx p_{ij} = p_i/(p_i + p_j)$, one ends up with $p_i \approx 1/\left[\sum_{j=1,i\neq j}^{k}(1/\hat{p}_{ij}) - (k-2)\right]$.

1-versus-All Normalization (`1van`). This method simply normalizes all obtained 1-versus-All probabilities such that they sum to 1.

Pairwise Coupling (`pc-ht`; `pc-wu1`; `pc-wu2`). The pairwise coupling method was introduced by Hastie and Tibshirani [15]. They estimate p_i by maximizing the negative weighted Kullback-Leibler distance ℓ between p_{ij} and $\hat{p}_{ij}$. After setting the derivative of ℓ with respect to p_i to zero, they aim to estimate the multi-class probabilities such that

$$\sum_{j=1,i\neq j}^{k} n_{ij} p_{ij} = \sum_{j=1,i\neq j}^{k} n_{ij} \hat{p}_{ij} \text{ with } \sum_{i=1}^{k} p_i = 1,\ p_i > 0 . \tag{2}$$

An iterative procedure is used for this. This method is called `pc-ht`.

Wu *et al.* [16] derived two other pairwise coupling methods, based on the assumption that the multi-class data are balanced such that weighting is not necessary. They suggest to solve

$$p_i \sum_{j=1,i\neq j}^{k} \frac{p_i + p_j}{k-1} \hat{p}_{ij},\ \forall i, \text{ with } \sum_{i=1}^{k} p_i = 1,\ p_i \geq 0 , \tag{3}$$

which can be done by a linear system. We call this method `pc-wu1`. These authors also suggest a second approach

$$\min_{p_1,\ldots,p_k} \sum_{i=1}^{k} \sum_{j=1,i\neq j}^{k} \left(\hat{p}_{ji} p_i - \hat{p}_{ij} p_j\right)^2 \text{ with } \sum_{i=1}^{k} p_i = 1,\ p_i \geq 0 , \tag{4}$$

which is again solved by a linear system. Note that the expression in (4) is similar to the basis of Refregier and Vallet's method [12]. We call this method `pc-wu2`.

Huang *et al.* (`gbt1va`) [17]. These authors proposed methods based on the generalized Bradley-Terry model [17]. The Bradley-Terry model estimates the probability that one object is preferred over another (in sports this can be the probability that one player beats another). In multi-class classification based on binary classifiers, $\hat{p}_{ij}$ can be seen as the probability that class i beats class j. Hastie and Tibshirani's pairwise coupling method [15] is formally similar to a Bradley-Terry model, but note that the different $\hat{p}_{ij}$ values are not independent. In the generalized Bradley-Terry model the players (classes) are repeatedly divided into two teams playing a series of games. The model is used to estimate the players' individual skill. We implemented the method where teams are created using the 1-versus-All strategy.

2.4 Duan *et al.*'s Softmax Methods (`sm1va`; `sm1v1`)

Duan *et al.* [18] propose to apply softmax functions to the outcomes of 1-versus-All or 1-versus-1 (LS-)SVMs to directly estimate multi-class probabilities. For the 1-versus-All strategy, the multi-class probabilities are computed as

$$\hat{p}_i = \exp(a_i f_i + b_i) \Big/ \left(\sum_{j=1}^{k} \exp(a_j f_j + b_j) \right), \tag{5}$$

where f_i is the LS-SVM output for the 1-versus-All classifier for class i, and a_i $(i = 1, \ldots, k)$ and b_i $(i = 1, \ldots, k)$ are parameters to be tuned by minimizing a negative log-likelihood function on the training data. For the 1-versus-1 strategy, the probabilities are computed analogously using the LS-SVM outputs for all 1-versus-1 classifiers [18]. Regularized versions were also suggested in [18] but it did not perform better in their own evaluation. The training data's 1-versus-All or 1-versus-1 LS-SVM outputs were obtained by 5CV on the training data.

2.5 Multi-class Kernel Logistic Regression (`mklr`)

Finally, an all-at-once method was used, being a weighted LS-SVM based version of multi-class kernel logistic regression [19],[20]. In standard multi-class logistic regression, the kth of k classes is taken as the reference class and $k-1$ binary models are simultaneously fitted in which each class is contrasted with the reference. This leads to the multi-class probabilities

$$p_i = \exp\left(\boldsymbol{\beta}_i^T \boldsymbol{x}\right) \Big/ \left(1 + \sum_{j=1}^{k-1} \exp\left(\boldsymbol{\beta}_j^T \boldsymbol{x}\right) \right). \tag{6}$$

The regularized model parameters are found by optimizing the penalized log-likelihood function using iteratively regularized re-weighted least squares. This can be changed into an iteratively re-weighted LS-SVM algorithm where probabilities are computed as in (6) using $\varphi(\boldsymbol{x})$ instead of $\boldsymbol{x}$.The hyperparameters γ and σ were tuned using 10CV.

3 Experimental Setup

3.1 Data

Two real world medical datasets were used in this paper. The first data set contains 508 women with a pregnancy of unknown location (PUL), collected at St Georges Hospital in London (UK). PUL is said to occur if a woman has a positive pregnancy test but no physical signs of a pregnancy are found either intra- or extra-uterine. A PUL can turn out to be any of three outcomes ($k = 3$): a failing PUL (283 cases, 56%), a normal intra-uterine pregnancy (IUP) (183, 36%), or an ectopic pregnancy (42, 8%). The last class the most difficult one, because it is small does not seem to be well separated from both other groups. The inputs used to predict the outcome ($q = 5$) are the ratio of the hCG levels at presentation and 48 hours later, the logarithm of the average hCG level, the logarithm of the average progesterone level, vaginal bleeding, and the mother's age.

The second data set contains 754 patients with at least one ovarian tumor (in case of bilateral tumors only the worst tumor is included), collected by the International Ovarian Tumor Analysis (IOTA) study group in nine centers across Europe [21]. The type of ovarian tumor is one of the following four ($k = 4$): benign (563 cases, 75%), primary invasive (121, 16%), borderline (40, 5%), and metastatic (30, 4%). Again, the classes are clearly unbalanced. This time, the borderline class will probably be the most difficult one since it is small and contains 'borderline malignant' tumors. The inputs used to predict the outcome ($q = 9$) are the woman's age, the maximal diameter of the lesion and of the largest solid component, and the presence of i) a personal history of ovarian cancer, ii) blood flow within the papillary projections, iii) irregular internal cyst walls, iv) ascites, v) an entirely solid tumor, vi) bilateral tumors.

3.2 Performance Evaluation

The performance of all methods was evaluated using stratified 5CV [22]. The data were split up in 5 folds of equal size with equal target class distributions. Each fold served once as test set to test the methods after they were applied to the other four folds that served as the training set. This resulted in five test set results for each method. This 5CV procedure was repeated 20 times [22] such that we obtained 100 test set results. All inputs were standardized to zero mean and unit variance separately in each of the 100 training sets. The inputs in the accompanying test sets were transformed with the same formulas used for their training set colleagues.

The test set results were summarized by the area under the receiver operating characteristic (ROC) curve (AUC) [23] and by the average entropy error (log loss). The AUC indicates how well a classifier has separated two classes. Perfect classification yields an AUC of 1 while an AUC of 0.5 reflects random classification. Even though multi-class AUC measures exist, we preferred to consider a separate AUC for each class. The average entropy is computed as the average of each case's -log($\hat{p}_i$) for its true class. The higher the estimated class probability of each case's true class, the lower the average entropy. The AUC and entropy results of the 100 test set evaluations were summarized by their median and interquartile range (IQR).

Due to the complex experimental setup, we adopted a heuristic strategy to compare methods. Based on [24], we used average ranks (AR). When comparing the ten methods to combine binary class probabilities, their performance was ranked for each of the four classification algorithms and these ranks were averaged. The opposite was done when comparing classification algorithms. This method summarized all obtained results well.

4 Results and Discussion

Because the first dataset has three classes and the second has four, we have AUC information on seven classes. To reduce the output, we averaged the k AUCs for each dataset. Lack of space precludes showing the full AUC results. The results for the PUL and IOTA data are shown in Table 1. Even though differences between methods were often not huge, the best probability combination methods were `endall` and `pc-wu1`. `pc-wu2` and `gbt1va` also performed very well. `gbt1va`'s AR of 5.0 on the IOTA dataset is caused by degraded performance for the two small classes when applying `gbt1va` to `bay`. Otherwise, this method performed excellent. The methods `pkpd`, `endone`, and `rvone` were outperformed.

Table 1. Average AUC for the different methods shown as the median over all 100 test set results after 20 stratified 5-fold CV runs (IQR between brackets)

PUL	bays	bay	plt	iso	AR	sm1va	sm1v1	mklr
endall	.966 (.015)	.959 (.024)	.961 (.020)	.961 (.024)	1.5	.963 (.024)	.960 (.025)	.964 (.018)
gbt1va	.966 (.016)	.960 (.025)	.960 (.023)	.960 (.025)	2.1			
pc-wu1	.964 (.015)	.946 (.038)	.960 (.023)	.959 (.027)	4.6			
pc-wu2	.964 (.015)	.946 (.040)	.960 (.024)	.958 (.026)	5.1			
pc-ht	.964 (.014)	.945 (.037)	.959 (.024)	.959 (.025)	5.9			
1van	.963 (.020)	.952 (.032)	.958 (.026)	.958 (.030)	6.0			
endone	.962 (.019)	.948 (.035)	.960 (.027)	.953 (.031)	6.3			
rvall	.966 (.013)	.946 (.037)	.954 (.031)	.951 (.030)	6.5			
pkpd	.964 (.016)	.945 (.053)	.958 (.025)	.956 (.028)	7.1			
rvone	.962 (.016)	.929 (.048)	.948 (.031)	.942 (.037)	9.9			
AR	1.0	3.9	2.3	2.9				
IOTA	bays	bay	plt	iso	AR	sm1va	sm1v1	mklr
endall	.881 (.029)	.859 (.037)	.869 (.027)	.868 (.029)	1.0	.862 (.032)	.861 (.034)	.883 (.027)
pc-wu1	.878 (.029)	.856 (.024)	.862 (.031)	.858 (.031)	2.5			
pc-wu2	.878 (.027)	.853 (.026)	.857 (.031)	.854 (.032)	4.1			
rvall	.877 (.029)	.857 (.026)	.857 (.034)	.851 (.032)	4.6			
gbt1va	.878 (.028)	.761 (.067)	.858 (.035)	.855 (.032)	5.0			
1van	.875 (.029)	.840 (.035)	.859 (.039)	.847 (.034)	5.8			
pc-ht	.865 (.039)	.852 (.037)	.850 (.048)	.853 (.026)	6.5			
pkpd	.874 (.029)	.847 (.030)	.849 (.036)	.845 (.033)	7.3			
endone	.873 (.032)	.835 (.038)	.837 (.055)	.831 (.031)	8.8			
rvone	.858 (.046)	.836 (.045)	.798 (.107)	.790 (.072)	9.5			
AR	1.0	3.4	2.4	3.3				

Differences in average AUC were reduced by the 'easier' classes. Bigger differences arose for the difficult classes ectopic pregnancy (PUL), borderline (IOTA), and metastatic (IOTA). Similar conclusions could be drawn when looking at these classes only: endall performed best, followed by pc-wu1, pc-wu2, and gbt1va.

The entropy results are shown in Table 2. This leads to a similar observation: endall, gbt1va, and pc-wu1 performed best among the probability combination methods. pkpd, endone, rvone, and 1van performed worst.

With respect to Duan *et al.*'s softmax methods [18], we can see that their performance did not differ substantially, yet sm1va always had the advantage over sm1v1. Also, their performance was similar to that of the best probability combination methods. When looking at AUC performance, however, these methods seemed to perform less than the probability combination methods applied to bays.

Table 2. Entropy for the different methods shown as the median over all 100 test set results after 20 stratified 5-fold CV runs (IQR between brackets)

PUL	bays	bay	plt	iso	AR	sm1va	sm1v1	mklr
endall	.298 (.104)	.378 (.169)	.286 (.056)	.288 (.095)	2.0	.273 (.070)	.288 (.071)	.276 (.063)
gbt1va	.307 (.115)	.402 (.170)	.287 (.062)	.310 (.154)	3.5			
pc-wu1	.289 (.098)	.421 (.229)	.293 (.056)	.302 (.107)	3.9			
pc-ht	.291 (.092)	.430 (.225)	.292 (.058)	.299 (.085)	3.9			
pc-wu2	.293 (.100)	.427 (.260)	.292 (.054)	.312 (.115)	4.6			
rvall	.288 (.097)	.404 (.204)	.302 (.065)	.318 (.122)	4.8			
endone	.320 (.107)	.469 (.295)	.291 (.065)	.356 (.187)	7.0			
1van	.310 (.118)	.441 (.216)	.298 (.062)	.323 (.156)	7.3			
pkpd	.324 (.123)	.508 (.337)	.293 (.060)	.377 (.206)	8.6			
rvone	.313 (.129)	.561 (.419)	.308 (.067)	.412 (.224)	9.5			
AR	1.8	4.0	1.3	2.9				
IOTA	bays	bay	plt	iso	AR	sm1va	sm1v1	mklr
endall	.495 (.063)	.653 (.099)	.512 (.039)	.497 (.048)	1.0	.504 (.052)	.510 (.054)	.492 (.055)
pc-wu2	.503 (.069)	.847 (.159)	.527 (.039)	.510 (.053)	2.8			
pc-wu1	.502 (.065)	.834 (.149)	.536 (.035)	.512 (.047)	3.3			
gbt1va	.546 (.117)	.877 (.239)	.517 (.046)	.542 (.110)	5.5			
pc-ht	.530 (.074)	.873 (.134)	.545 (.045)	.518 (.042)	5.8			
rvall	.506 (.066)	.874 (.144)	.559 (.038)	.529 (.054)	5.8			
endone	.526 (.101)	1.18 (.530)	.543 (.061)	.576 (.115)	6.8			
pkpd	.526 (.082)	1.20 (.217)	.534 (.059)	.603 (.176)	7.0			
1van	.563 (.146)	1.26 (.239)	.531 (.044)	.553 (.116)	7.3			
rvone	.568 (.127)	1.41 (.664)	.612 (.182)	.764 (.331)	10.0			
AR	1.5	4.0	2.3	2.2				

Interestingly, mklr had excellent performance throughout. It was similar to endall applied to bays, the best approach involving a probability combination method.

The AR results for the four algorithms bays, bay, plt, and iso reveal that a Bayesian LS-SVM with sparseness was better than one without, that bays was better than plt and iso, and that bay is worse than plt and iso. Differences between plt and iso were typically in favor of plt. Differences in performance between probability combination methods were often smaller when bays or plt were used.

Overall, on the present data sets the best methods are endall and mklr. While the latter is an all-at-once method, the former is an ensemble method for which many binary models have to be fit (25 when $k = 4$). This makes mklr more attractive. When using 1-versus-1 classifiers as the basis for obtaining multi-class probabilities, `pc-wu1` (preferably using `bays`) or `sm1v1` are good methods. When 1-versus-All classifiers are used, `sm1va` or `gbt1va` (preferably using `bays`) can be used. Out of the four implemented binary classification algorithms, `bays` seems a good choice to obtain probabilistic outputs. Using isotonic regression instead of Platt's method to obtain probabilities did not result in a clear benefit.

Acknowledgments. We thank Vanya Van Belle, Diana Sima, and Peter Karsmakers for constructive discussions. Research supported by Research Council KUL: GOA-AMBioRICS, CoE EF/05/006 Optimization in Engineering; Belgian Federal Science Policy Office IUAP P6/04 ('Dynamical systems, control and optimization', 2007-2011); EU: BIOPATTERN (FP6-2002-IST 508803), ETUMOUR (FP6-2002- LIFESCIHEALTH 503094).

References

1. Allwein, E.L., Schapire, R.E., Singer, Y.: Reducing multiclass to binary: a unifying approach for margin classifiers. J. Mach. Learn. Res. 1, 113–141 (2000)
2. Suykens, J.A.K., Vandewalle, J.: Least squares support vector machine classifiers. Neural Process. Lett. 9, 293–300 (1999)
3. Suykens, J.A.K., Van Gestel, T., De Brabanter, J., De Moor, B., Vandewalle, J.: Least squares support vector machines. World Scientific, Singapore (2002)
4. Van Gestel, T., Suykens, J.A.K., Lanckriet, G., Lambrechts, A., De Moor, B., Vandewalle, J.: Bayesian framework for least-squares support vector machine classifiers, Gaussian processes, and kernel Fisher discriminant analysis. Neural Comput. 14, 1115–1147 (2002)
5. Platt, J.: Probabilistic outputs for support vector machines and comparison to regularized likelihood methods. In: Smola, A.J., Bartlett, P.L., Scholkopf, B., Schuurmans, D. (eds.) Advances in Large Margin Classifiers, pp. 61–74. MIT Press, Cambridge (2000)
6. Lin, H.-T., Lin, C.-J., Weng, R.C.: A note on Platt's probabilistic outputs for support vector machines. Technical Report, Department of Computer Science, National Taiwan University (2003)
7. Zadrozny, B., Elkan, C.: Transforming classifier scores into accurate multiclass probability estimates. In: Proc. 8th ACM SIGKDD International Conference on Knowledge Discovery and Data Mining, pp. 694–699 (2002)
8. Cawley, G.C.: Leave-one-out cross-validation based model selection criteria for weighted LS-SVMs. In: Proc. 19th International Joint Conference on Neural Networks, pp. 2970–2977 (2006)
9. MacKay, D.J.C.: Probable networks and plausible predictions - a review of practical Bayesian methods for supervised neural networks. Netw.-Comput. Neural. Syst. 6, 469–505 (1995)
10. Lu, C., Van Gestel, T., Suykens, J.A.K., Van Huffel, S., Vergote, I., Timmerman, D.: Preoperative prediction of malignancy of ovarian tumors using least squares support vector machines. Artif. Intell. Med. 28, 281–306 (2003)
11. Ayer, M., Brunk, H., Ewing, G., Reid, W., Silverman, E.: An empirical distribution function for sampling with incomplete information. Ann. Math. Stat. 26, 641–647 (1955)

12. Refregier, P., Vallet, F.: Probabilistic approach for multiclass classification with neural networks. In: Proc. International Conference on Artificial Networks, pp. 1003–1007 (1991)
13. Frank, E., Kramer, S.: Ensembles of nested dichotomies for multi-class problems. In: Proc. 21st International Conference on Machine Learning, 39 (2004)
14. Price, D., Knerr, S., Personnaz, L., Dreyfus, G.: Pairwise neural network classifiers with probabilistic outputs. In: Tesauro, G., Touretzky, D.S., Leen, T.K. (eds.) Neural Information Processing Systems, vol. 7, pp. 1109–1116. MIT Press, Cambridge (1995)
15. Hastie, T., Tibshirani, R.: Classification by pairwise coupling. Ann. Stat. 26, 451–471 (1998)
16. Wu, T.-F., Lin, C.-J., Weng, R.C.: Probability estimates for multi-class classification by pairwise coupling. J. Mach. Learn. Res. 5, 975–1005 (2004)
17. Huang, T.-K., Weng, R.C., Lin, C.-J.: Generalized Bradley-Terry models and multi-class probability estimates. J. Mach. Learn. Res. 7, 85–115 (2006)
18. Duan, K., Keerthi, S.S., Chu, W., Shevade, S.K., Poo, A.N.: Multi-category classification by soft-max combination of binary classifiers. In: Windeatt, T., Roli, F. (eds.) MCS 2003. LNCS, vol. 2709, pp. 125–134. Springer, Heidelberg (2003)
19. Karsmakers, P., Pelckmans, K., Suykens, J.A.K.: Multi-class kernel logistic regression: a fixed-size implementation. Accepted for presentation at the 20th International Joint Conference on Neural Networks (2007)
20. Zhu, J., Hastie, T.: Kernel logistic regression and the import vector machine. J. Comput. Graph. Stat., 185–205 (2005)
21. Timmerman, D., Valentin, L., Bourne, T.H., Collins, W.P., Verrelst, H., Vergote, I.: Terms, definitions and measurements to describe the sonographic features of adnexal tumors: a consensus statement from the International Ovarian Tumor Analysis (IOTA) group. Ultrasound. Obstet. Gynecol. 16, 500–505 (2000)
22. Kohavi, R.: A study of cross-validation and bootstrap for accuracy estimation and model selection. In: Proc. 14th International Joint Conference on Artificial Intelligence, pp. 1137–1143 (1995)
23. Hanley, J.A., McNeil, B.J.: The meaning and use of the area under a receiver operating characteristic (ROC) curve. Radiology 143, 29–36 (1982)
24. Brazdil, P.B., Soares, C.: A comparison of ranking methods for classification algorithm selection. In: López de Mántaras, R., Plaza, E. (eds.) ECML 2000. LNCS (LNAI), vol. 1810, pp. 63–74. Springer, Heidelberg (2000)

Classifying EEG Data into Different Memory Loads Across Subjects

Liang Wu and Predrag Neskovic

Department of Physics and Institute for Brain and Neural Systems
Brown University, Providence, RI 02906, USA

Abstract. In this paper we consider the question of whether it is possible to classify n-back EEG data into different memory loads across subjects. To capture relevant information from the EEG signal we use three types of features: power spectrum, conditional entropy, and conditional mutual information. In order to reduce irrelevant and misleading features we use a feature selection method that maximizes mutual information between features and classes and minimizes redundancy among features. Using a selected group of features we show that all classifiers can successfully generalize to the new subject for bands 1-40Hz and 1-60Hz. The classification rates are statistically significant and the best classification rates, close to 90%, are obtained using conditional entropy features.

1 Introduction

Discovering subject-independent regularities in EEG signals associated with different types of cognitive processing is a very challenging problem which has been addressed in only few studies [1,2]. Among the obstacles to finding unique patterns of brain activity across subjects are: varying head shapes and sizes across subjects might result in different electrode positions, different subjects might use different strategies to solve a specific mental task resulting in different brain activations, and the cortical areas across subjects can be organized differently producing different patterns of activities.

The main objective of this paper is to construct a system that can classify n-back EEG signals into different memory loads across subjects. Our strategy in addressing this problem is to use highly informative and discriminative features in combination with machine learning algorithms. Specifically, in this work, we use newly proposed entropy-based features [3] that can capture non-linear spatial and temporal dependencies among electrodes. In order to reduce irrelevant and misleading features and improve generalization properties of the classifier, it often helps to reduce the number of features by selecting the most informative and discriminative features. In recent years, numerous algorithms for feature selection have been proposed [4,2], and mutual information has been advanced as an important measure of feature relevance [5,6]. In this work we use a feature selection method that maximizes mutual information between features and classes and minimizes redundancy among features [5,6]. We show that using a

J. Marques de Sá et al. (Eds.): ICANN 2007, Part II, LNCS 4669, pp. 149–158, 2007.

selected group of features our classifiers can easily generalize to new subjects with classification rates close to 90%.

2 Data Acquisition and Feature Extraction

The data set consists of EEG samples from 6 subjects while they performed an n-back memory task [3]. In this task, individuals were shown computerized visual displays consisting of a single target (a white circle) presented in 1 of 12 possible positions, each of which is on 1 of 12 equidistant radii of an imaginary circular array from the center of the monitor. For each trial, the subject decided whether the current target (a position of the white circle) was in the same or different spatial location as in the example presented n-times before.

Electrical activity (EEG) was recorded from 62 electrodes using 10-20 International System. We use 90 trials per subject and the total length of each trial is 2.2 seconds. In order to obtain better feature estimates, in this work we classify EEG data using 5 trials long segments. This means that the number of sampling points per segment is around 5,625, the length of a segment is 11 seconds, and the number of examples per subject is 18.

Feature Extraction. We use the following 3 types of features: power spectrum (F1), conditional entropy (F2), and conditional mutual information (F3). The power spectrum features measure the total power, P_i, for each channel i for a given frequency band. The number of F1 features is 62 (one per electrode)

$$F_1 = \{P_i\}_{i=1}^{62}. \tag{1}$$

In contrast to power spectrum features, the entropy-based features capture both linear and non-linear dependences. Furthermore, the conditional entropy and mutual information features also capture dynamical properties of the EEG signals [3]. In calculating F2 and F3 features we use the following equations

$$F_2 = \{H_i = H(X_i^t | X_i^{t-1})\}_{i=1}^{62}, \tag{2}$$

$$F_3 = \{I_{ij} = I(X_i^t; X_j^t | X_i^{t-1}, X_j^{t-1})\}_{i,j=1}^{62}, i \neq j, \tag{3}$$

where H_i denotes the entropy of the ith electrode and I_{ij} measures the conditional entropy between the ith and j^{th} electrodes. With symbol X_i^t we denote the output of the ith electrode at time t, and with X_j^{t-1} we denote the output of the jth electrode at time $t-1$. The total number of F2 features is 62, and the number of F3 features is 1,891. The entropy-based features, F2 and F3, are calculated following the procedure described in [3].

3 Feature Selection

Recently, we have demonstrated [3] that n-back EEG data can be successfully classified, using F1-F3 features, into different tasks with classification rates that exceed 90%. However, in that single-subject study the classifiers were tested on the same subject on which they were trained. Using the same features as reported in [3] the classification rates are close to chance if one tries to generalize from a group of subjects (used for training) to a new subject (used for testing).

In order to improve generalization properties of a classifier, it often helps to reduce the number of features by selecting the most informative and discriminative ones. A standard criterion used for selecting features is to maximize the statistical dependence between the features and the class distribution variable C [5,6]. A convenient measure for quantifying this dependence is mutual information, $I(f_1, ..., f_m; C)$, where f_i denotes the ith feature and C denotes a class variable. The feature selection method that uses this measure is called Mutual Information Feature Selection (MIFS). However, in order to calculate mutual information (MI), one has to estimate multivariate densities $p(f_1, ..., f_m)$ and $p(f_1, ..., f_m, C)$ which is very difficult when the number of samples is small. For this reason, a number of approaches have been proposed so that the MI is calculated incrementally, one feature at a time [5,6]. As a consequence of treating each feature independently, it is possible that some of the selected features carry similar information. To overcome this problem, an additional term that minimizes the redundancy between features is added to the objective function. In this work, we use one such approach, MIFS-U, that has been proposed in [5].

In the following, we briefly outline the procedure for selecting and ranking the features. We denote with F the initial set of all the features and with S the final set of selected features. Initially, S is empty while F has n features. To add the first feature to the set S, compute $I(C; f_i)$ for $\forall f_i \in F$ and select the feature f_i that maximizes the $I(C; f_i)$; set $F \leftarrow F \setminus \{f_i\}$ and $S \leftarrow \{f_i\}$. Then, using a greedy selection approach, repeat the following until the desired number of features is reached: a) for all pairs of features (f_s, f_i), compute $I(f_s; f_i)$, where $f_s \in S$ and $f_i \in F$, and b) choose the feature $f_i \in F$ that maximizes the following expression

$$I(C; f_i|f_s) \approx I(C; f_i) - \beta \sum_{f_s} \frac{I(C; f_s)}{H(f_s)} I(f_i; f_s), \tag{4}$$

where $H(f_s)$ is the entropy of the feature f_s and β is the parameter that controls the tradeoff between the class-feature dependence and redundancy among the features (feature-feature dependence). If we set $\beta = 0$, then the algorithm ignores the redundancy and selects the features only based on the amount of the MI between classes and features. By increasing β the algorithm excludes redundant features more efficiently. In our implementation we set $\beta = 1$ as suggested in [5].

The advantage of the greedy search algorithm is that it is very fast since it always considers one feature at a time. Unfortunately, the order in which the features are added to the set S does not represent an optimal ranking of the features (as a function of features' importance for classification). To improve the ranking of the features from the set S, we construct random subsets, each containing a few features from S, and then train a classifier using each subset separately. The ranking of each feature is obtained by comparing the average performance of the feature for its inclusion and exclusion in the input feature subset. Our procedure is similar to the Taguchi algorithm [5] but instead of orthogonal arrays we choose random arrays.

Once the selected features are ranked according to their importance for classification, the next step is to choose m features from the top of the list as the optimal feature set. A natural approach for selecting this number would be to train a classifier using a different number of top features and then select as the optimal number of features the one that gives the highest classification rate. Unfortunately, the optimal number obtained from the training set does not have to produce the highest rate on the testing set.

In this paper, we use two different methods for evaluating a selected feature set. In one method, we use a cross validation approach to select an optimal number of features and in the other method we evaluate the whole feature set by estimating the maximal performance on a new subject. In both methods we train a classifier and rank the features using exclusively the training subjects.

The Nested Cross Validation (NCV) Method. In this method, we partition the data repeatedly into three sets: a training set, an optimizing set, and a final test set. The data from the new subject are partitioned into 2 disjoint sets: an optimizing set and a test set. The optimal *number* of features is chosen by maximizing the classification rate of the optimizing set. The final test set is used only to evaluate a classifier and therefore provides an unbiased estimate of the accuracy.

The Maximal Accuracy (MA) Method. Assuming that the number of the selected features is m, we construct m different classifiers, C_k, where the kth classifier uses top k features and $k = 1, ..., m$. The final performance is estimated by recording the maximal accuracy for each fold (subject) and then averaged over the folds. This procedure can provide biased estimates (as previously noticed in [7]). The reason is that classification rates are usually different for different classifiers with different numbers of features even if we use random data. Some classification rates will be higher than others by chance alone. Therefore, the classification rate of the best performance could overestimate the true performance.

Both of the above methods have advantages and limitations. The advantage of the NCV approach is that it provides an unbiased procedure for selecting the number of features. However, in order to accurately estimate the number of features and the performance, one should have a sufficient number of examples in both the optimizing and the test sets. When the amount of data is limited, as is the case in our situation, both estimates can be skewed.

The MA method, on the other hand, utilizes all the test data and estimates the best possible performance regardless of the number of features used. However, the estimate can be highly biased. Fortunately, the estimates do not necessarily have to include bias or the bias can be very small as we will show in the next section.

4 Statistical Analysis of the MA Method

In this section we provide a theoretical analysis that can help us to understand the reasons why the estimates obtained using the MA method can be highly biased. We also provide a way to evaluate the statistical significance of the classification accuracy obtained with the MA method. We are testing the hypothesis that the expected value of the classification rate is not obtained by chance against the null hypothesis that the expected value of classification rate is obtained by chance. To do that, we need to establish a baseline, the probability that the correct results can be obtained by chance. Using the baseline value and the standard deviation, we then calculate the p-value in order to determine the statistical significance of our classification rates.

Baseline Estimation from a Theoretical Analysis. Let us assume that we want to classify n test samples from 2 classes, and that a classifier assigns the label to a test sample by chance. Then the probability that a sample is classified correctly is 1/2. If we use a random variable Y_i to indicate the classification outcome, i.e. $Y_i = 1$ to denote the correct classification and $Y_i = 0$ to denote the incorrect classification, then the classification rate (the proportion of labels that are correctly-assigned) of the samples from the classifier is $X = \frac{\sum_i^n Y_i}{n}$. From the law of large numbers, the classification rate asymptotically approaches normal distribution $\mathcal{N}(1/2, \frac{1}{4n})$. If we try m independent classifiers, each of which uses a different number of features, the expectation of the maximal classification rate of these m classifiers is the expectation of the maximal number out of m i.i.d. samples $\{X_i\}_{i=1}^m$ from $\mathcal{N}(1/2, \frac{1}{4n})$. In order to calculate the expected value, we have to know the cumulative probability function of the maximal classification rate $X^* = \max(X_1, ..., X_m)$, which can be easily calculated as

$$\begin{aligned} F(x) &= \Pr(X^* \leq x) \\ &= \Pr(\max(X_1, ..., X_m) \leq x) \\ &= \Pr(X_1 \leq x) \cdots \Pr(X_m \leq x) \\ &= \Phi^m(\sqrt{4n}(x - 1/2)), \end{aligned}$$

where $\Phi(\cdot)$ denotes the cumulative function of the standard normal distribution. From the cumulative function, in addition to the expectation value, one can also calculate the standard deviation of X^* and the p-value

$$\mathrm{E}[X^*] = \int xF'(x)dx, \tag{5}$$

$$\mathrm{Std}[X^*] = (\int (x - \mathrm{E}[X^*])^2 F'(x)dx)^{1/2}, \tag{6}$$

$$p = \Phi(\frac{-|r - \mathrm{E}[X^*]|}{\sqrt{\mathrm{Var}[X^*]}}), \tag{7}$$

where r is the classification rate that we obtain from our classifiers. To illustrate the relationship between the expectations (variances), and the number of classifiers (m) and samples (n), in Table 1 we show the results of numerical simulations using a few different values for m and n. One can see that the fewer test samples and the more classifiers one has, the larger the possibility that we get good results simply by chance. When there is only one classifier $m = 1$, the expected classification rate is 0.5, which means that the bias is zero. On the other hand, if one takes the maximum over 10 classifiers, $m = 10$, the likelihood of obtaining high rates by chance cannot be ignored even if one has a large pool of samples, e.g., $n = 144$.

Table 1. Expectation values and standard deviation for different values of m and n

E[X^*](Std[X^*])	n=18	n=36	n=72	n=144
m=1	.50 (.12)	.50 (.08)	.50 (.06)	.50 (.04)
m=10	.68 (.07)	.63 (.05)	.59 (.03)	.56 (.02)
m=20	.72 (.06)	.66 (.04)	.61 (.03)	.58 (.02)

Baseline Estimation by Shuffling the Test Samples. So far, we have considered random data and independent classifiers that assign a label to a test sample by chance. In our case, classifiers with different numbers of features are not independent since we always choose the features from the top of the ranking list and therefore classifiers share some common features. That means that we cannot just use the estimates from Table 1 but have to estimate the baseline (probability that correct results can be obtained by chance) for a given data set. We should note that the dependence among classifiers actually reduces bias. In the limit, if the classifiers are completely dependent on one another, one effectively has only one classifier and the bias is zero - the baseline is at chance. In the other limit, for an infinite number of independent classifiers, one can get 100% accuracy simply by chance. To get a more realistic estimation of the baseline, we shuffle the data by assigning a label to each sample with probability 1/2. We then calculate the classification rates from our classifiers using the MC method on shuffled data.

5 Results

The raw data is processed using the surface Laplacian [8] and filtered into 4 different bands: 1-20Hz, 1-40Hz, 1-60Hz, and 1-80Hz. Classification is done for each band separately in order to evaluate the importance of different frequency ranges on performance. Within each band, we then extract the power spectrum (F_1), conditional entropy (F_2), and mutual information (F_3) features and form three feature sets. For each feature set we select a subset of features and then perform classification using the NCV and the MA method.

Feature selection is done in two steps: filtering the initial set of extracted features into a smaller set, and ranking the smaller set of features according to their importance for classification. The filtering is done using a MIFS-U method [5] that maximizes mutual information between features and classes and minimizes redundancy among features. We use 10 bins to discretize feature values when calculating mutual information between features and classes as well as mutual information between features. When filtering F_1 and F_2 features, we start with 62 original features and reduce their number to 20 potentially relevant features. When filtering F_3 features, we start with a much larger set of 1,891 features, and end up with 30 features. To rank the features, we generate 30 random sets for F_1 and F_2 features, and 40 random sets for F_3. We then train a Naive Bayes (NB) classifier with Gaussian likelihood on each random set of features and use the procedure described in [5] to assign relevance to each feature.

We use three different architectures to classify test samples: nearest neighbor (NN) classifier [9], Naive Bayes classifier and support vector machines (SVM) [10]. The number of classes is two; the task is to associate a given EEG segment with either the 0-back or the 3-bask task. We train each of the classifiers using a leave one out procedure: to train a classifier, select and rank features we use five subjects. The tests are done on the subject that is left out. The number of training examples is 180 and the number of testing examples is 36 (18 examples from the 0-back task and 18 examples from the 3-back task).

To evaluate selected features, we use the NCV and the MA method. In addition, we estimate the baseline of the MA method when tested on shuffled data. The results are illustrated in Figures 1- 3 when using power spectrum, conditional entropy, and conditional mutual information features respectively.

The NCV Method. In this method, we partition the data from a new subject (test samples) into two sets: the optimizing set and the test set. We use the optimizing set only to determine the optimal *number* of features, k, from the set

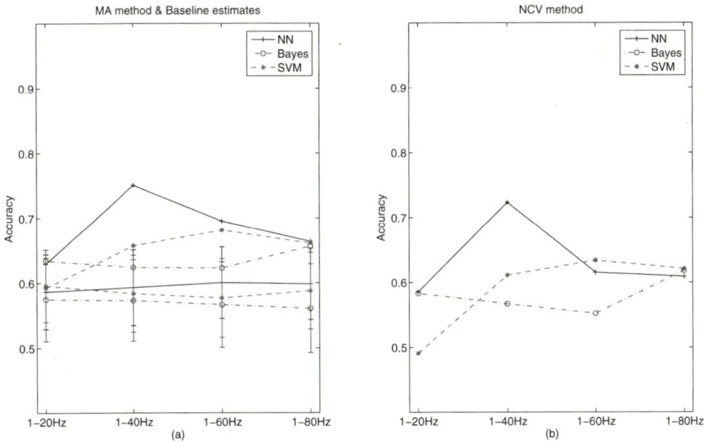

Fig. 1. The average classification rates across subjects and for different frequency bands using power spectrum features. Classification rates using: (a) the Maximal Accuracy (MA) method (top plots) and the baselines (bottom plots), (b) the Nested Cross Validation (NCV) method.

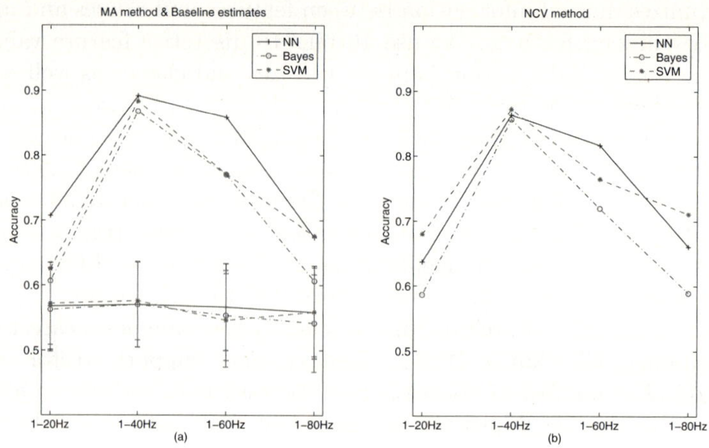

Fig. 2. The average classification rates across subjects and for different frequency bands using conditional entropy features. Classification rates using: (a) the MA method (top plots) and the baselines (bottom plots), (b) the NCV method.

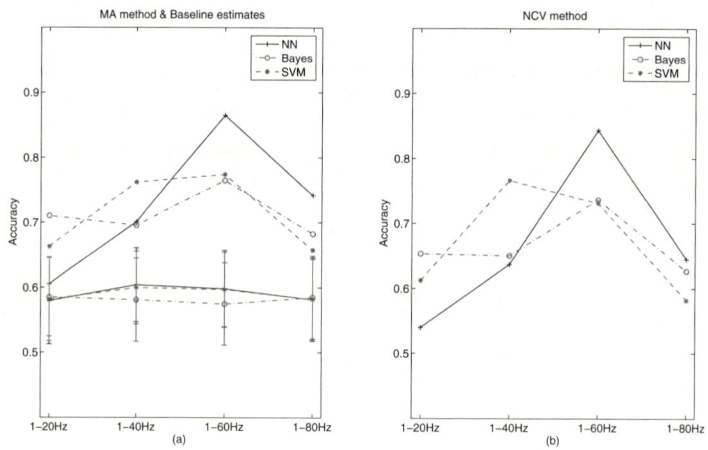

Fig. 3. The average classification rates across subjects and for different frequency bands using conditional mutual information features. Classification rates using: (a) the MA method (top plots) and the baselines (bottom plots), (2) the NCV method.

of already selected features. Note that all the classifiers (for all the architectures) are trained using different numbers of features on training data only and by selecting the number of features we also select one of the classifiers, for each of the three architectures (NN, NB, and SVM). Then, from the test data we extract the same k features and test the classifiers. In order to test the classifiers on all available data from a new subject, we divide the test samples into 3 folds. Each time, one fold of samples is taken as optimizing set and the other 2 folds as a test set. After trying all folds as optimizing sets, we calculate the average classification rate as shown in Figures 1(b)- 3(b).

The MA Method. Instead of selecting one classifier (for each architecture) and a specific number of features, in this method we train 10 different classifiers on 5 subjects and use all the classifiers on the test subject. Each classifier uses a different number of features (the kth classifier uses top k features). We record the accuracy of the best classifier (for each architecture) and then choose another group of 5 training and 1 testing subjects. The average over 6 folds is illustrated in Figures 1(a)- 3(a), top plots.

Estimating the Baseline. To estimate the baseline, we randomize the test samples to form 100 pools of random data in such a way that each sample is assigned a label by chance. We then apply the MA method to the randomized data to obtain the baseline for each frequency range. The results are illustrated in Figures 1(a)- 3(a), bottom plots. We should contrast these values to the baseline that one can get by chance when using 10 independent classifiers where each classifier assigns a label by chance. This baseline can be obtained from the Table 1 using $m = 10$ (since in MA method we use 1-10 top features), and $n = 36$ (since the number of test examples from each task is 18). For these two numbers, the expectation value is $\mathrm{E}(X) = 0.63$ which is higher compared to the baseline of our classifiers. It is clear that the strong dependence among the classifiers significantly reduces the possibility of getting high classification rates by chance. A relatively small baseline also means that the estimates obtained from the MA method are not very biased. Another support for this statement comes from the fact that the estimates obtained from the (unbiased) NCV method are very close to those obtained using the MA method. The classification rates of all classifiers are statistically significant for bands 1-40Hz and 1-60Hz with $p < 1e - 10$.

6 Summary

In this paper we consider the question of whether it is possible to classify n-back EEG data into different memory loads across subjects. Specifically, we consider whether a classifier that is trained on a group of subjects can generalize to a new subject and correctly classify an EEG segment into 0-back and 3-back tasks.

To capture relevant information from the EEG signal we use three types of features: power spectrum, conditional entropy, and conditional mutual information. In order to reduce irrelevant and misleading features we use a feature selection method that maximizes mutual information between features and classes and minimizes redundancy among features. We evaluate the selected features using the NCV and the MA method with three different types of classifiers. Using statistical analysis, we show that in situations where the number of classifiers is large and the number of testing samples is small, the MA can produce biased estimates. However, we demonstrate that a strong dependence among the classifiers can significantly alleviate this problem and result in less biased estimates.

From our tests we conclude that all classifiers can successfully generalize to the new subject for bands 1-40Hz and 1-60Hz and that classification rates of all classifiers are statistically significant. The best classification rates, close to 90%, are obtained using conditional entropy features. Furthermore, we show that both the conditional mutual information and the conditional entropy features significantly outperform the power spectrum features for bands 1-40Hz and 1-60Hz. In contrast to the results reported in [3], the performance sharply decreases for frequencies above 60Hz which is in accordance with physiological findings.

Acknowledgments. This work was supported in part by the Research Seed Fund Award from Brown University. The authors thank Elena Festa and William Heindel for providing the data and Leon N Cooper for helpful conversations.

References

1. Kaper, M., Ritter, H.: Generalizing to new subjects in brain-computer interfacing. In: Engineering in Medicine and Biology Society, pp. 4363–4366 (2004)
2. Schroder, M., Lal, T.N., Hinterberger, T., Bogdan, M., Hill, N.J., Birbaumer, N., Rosenstiel, W., Scholkopf, B.: Robust EEG channel selection across subjects for brain-computer interfaces. Eurasip Journal on Applied Signal Processing 2005(19), 3103–3112 (2005)
3. Wu, L., Neskovic, P.: Feature extraction for eeg classification: representing electrode outputs as a markov stochastic process. In: European Symposium on Artificial Neural Networks, pp. 567–572 (2007)
4. Guyon, I., Elisseeff, A.: An introduction to variable and feature selection. Journal of Machine Learning Research 3, 1157–1182 (2003)
5. Kwak, N., Choi, C.H.: Input feature selection for classification problems. IEEE Transaction on Neural Networks 13(1), 143–159 (2002)
6. Peng, H., Long, F., Ding, C.: Feature selection based on mutual information: criteria of max-dependency max-relevance, and min-redundancy. IEEE Trans. Pattern Anal. Mach. Intell. 27(8), 1226–1238 (2005)
7. Mitchell, T.M., Hutchinson, R., Niculescu, R.S., Pereira, F., Wang, X., Just, M., Newman, S.: Learning to decode cognitive states from brain images. Machine Learning 57(1-2), 145–175 (2004)
8. Hjorth, B.: An on-line transformation of EEG scalp potentials into orthogonal source derivations. Electroencephalog. Clin. Neurophysiol. 39(5), 526–530 (1975)
9. Cover, T.M., Hart, P.E.: Nearest neighbor pattern classification. IEEE Transactions on Information Theory 13(1), 21–27 (1967)
10. Christopher, J., Burges, C.: A tutorial on support vector machines for pattern recognition. Data Mining and Knowledge Discovery 2(2), 121–167 (1998)

Information Theoretic Derivations for Causality Detection: Application to Human Gait

Gert Van Dijck[1], Jo Van Vaerenbergh[2], and Marc M. Van Hulle[3]

[1,3] Computational Neuroscience Research Group, Laboratorium voor Neuro- en Psychofysiologie, K.U. Leuven, B-3000 Leuven, Belgium
[2] Laboratorium voor Neurorehabilitatie, Arteveldehogeschool, B-9000 Gent, Belgium
gert.vandijck@mtm.kuleuven.be, jo.vanvaerenbergh@cmat.be
marc@neuro.kuleuven.ac.be

Abstract. As a causality criterion we propose the conditional relative entropy. The relationship with information theoretic functionals mutual information and entropy is established. The conditional relative entropy criterion is compared with 3 well-established techniques for causality detection: 'Sims', 'Geweke-Meese-Dent' and 'Granger'. It is shown that the conditional relative entropy, as opposed to these 3 criteria, is sensitive to0. non-linear causal relationships. All results are illustrated on real-world time series of human gait.

Keywords: Conditional relative entropy, human gait, mutual information, surrogate testing, VARMA-GARCH.

1 Introduction

We address the problem whether one time series serves as an input, a driving system, to another time series, a driven system. The traditional techniques consist of modeling the time series by means of vector autoregressive models, such as in the 'Sims', 'Geweke-Meese-Dent' and 'Granger' criteria [3]. In this article a more general approach is taken by means of the conditional relative entropy (CRE). We derive a theorem in order to establish the link between CRE and conditional mutual information (CMI). This link allows to compute the CRE by means of available mutual information and entropy estimators.

A comparison between the CMI and the three mentioned criteria on real-case gait time series shows that it performs equally well in the linear regime and outperforms these techniques in the non-linear regime.

The paper is structured as follows. In section 2, the criterion is derived; this allows to appreciate conditional mutual information as a formal approach. Secondly, in section 3, we explain how the criterion can be computed based on finite sample estimators of entropy or mutual information. Section 4 considers an application of the criterion on real-case human gait time series. Finally, we end with conclusions.

2 The Conditional Relative Entropy Framework

As a formal criterion to causality detection we put forward that the distribution of a current observation of one time series, given its past observations, is statistically

J. Marques de Sá et al. (Eds.): ICANN 2007, Part II, LNCS 4669, pp. 159–168, 2007.

dependent on the previous observations of a second times series. This is a definition in accordance with Granger [4], except that here the full distribution on the current observation is considered, rather than the expected value. More formally, suppose we dispose of 2 time series: **Y**(t): Y(t), Y(t-1), …and **X**(t): X(t), X(t-1),… Suppose that we want to assess whether **X**(t) causally explains **Y**(t). For a particular observation of past samples of **Y**(t): y(t-1), …y(t-n1) and past samples of **X**(t): x(t-1), …x(t-n2) we need to assess whether the distribution of the most recent variable Y(t) of **Y**(t) is equal if we also observe the past of **X**(t):

$$\begin{aligned}&P(Y(t) \mid y(t-1),\dots,y(t-n_1)) = \\ &P(Y(t) \mid y(t-1),\dots,y(t-n_1),x(t-1),\dots,x(t-n_2))\end{aligned}. \tag{1}$$

For ease of notation we abbreviate the past observation of **Y**(t) from t-n_1 up to t-1 by $\mathbf{y}_{t-n_1}^{t-1}$ and a past observation of **X**(t) from t-n_2 up to t-1 by $\mathbf{x}_{t-n_2}^{t-1}$. Note that we use capitals to denote variables and bold style to denote series. An information theoretic approach to verify whether two distributions are equal is obtained by means of the Kullback-Leibler divergence:

$$\begin{aligned}&KL\left(P(Y(t) \mid \mathbf{y}_{t-n_1}^{t-1},\mathbf{x}_{t-n_2}^{t-1}) \,\|\, P\left(Y(t) \mid \mathbf{y}_{t-n_1}^{t-1}\right)\right) = \\ &\sum_{y(t)} P(y(t) \mid \mathbf{y}_{t-n_1}^{t-1},\mathbf{x}_{t-n_2}^{t-1}) \ln\left(\frac{P(y(t) \mid \mathbf{y}_{t-n_1}^{t-1},\mathbf{x}_{t-n_2}^{t-1})}{P\left(y(t) \mid \mathbf{y}_{t-n_1}^{t-1}\right)}\right)\end{aligned}. \tag{2}$$

However, conditioning on a particular past observation is overly simple and we need to average (2) over the prior of observing $\mathbf{x}_{t-n_2}^{t-1}$ and $\mathbf{y}_{t-n_1}^{t-1}$: $p(\mathbf{y}_{t-n_1}^{t-1},\mathbf{x}_{t-n_2}^{t-1})$:

$$\begin{aligned}&= \sum_{\mathbf{y}_{t-n_1}^{t-1},\mathbf{x}_{t-n_2}^{t-1}} p(\mathbf{y}_{t-n_1}^{t-1},\mathbf{x}_{t-n_2}^{t-1}) \sum_{y(t)} P(y(t) \mid \mathbf{y}_{t-n_1}^{t-1},\mathbf{x}_{t-n_2}^{t-1}) \\ &\ln\left(\frac{P(y(t) \mid \mathbf{y}_{t-n_1}^{t-1},\mathbf{x}_{t-n_2}^{t-1})}{P\left(y(t) \mid \mathbf{y}_{t-n_1}^{t-1}\right)}\right)\end{aligned}. \tag{3}$$

According to [2], see page 24, this is the definition of the conditional relative entropy:

$$= KL\left(P(Y(t) \mid \mathbf{Y}_{t-n_1}^{t-1},\mathbf{X}_{t-n_2}^{t-1}) \,\|\, P\left(Y(t) \mid \mathbf{Y}_{t-n_1}^{t-1}\right)\right). \tag{4}$$

Using the definition of conditional probability, (3) can be written similarly as:

$$= \sum_{y(t),\mathbf{y}_{t-n_1}^{t-1},\mathbf{x}_{t-n_2}^{t-1}} p(y(t),\mathbf{y}_{t-n_1}^{t-1},\mathbf{x}_{t-n_2}^{t-1}) \ln\left(\frac{P(y(t) \mid \mathbf{y}_{t-n_1}^{t-1},\mathbf{x}_{t-n_2}^{t-1})}{P\left(y(t) \mid \mathbf{y}_{t-n_1}^{t-1}\right)}\right). \tag{5}$$

From the information inequality theorem, see theorem 2.6.3 in [2], it is well known that the Kullback-Leibler divergence is always larger or equal to 0. The same holds for the conditional relative entropy, this is the second corollary of the information inequality in [2]:

$$KL\left(P(Y(t) \mid \mathbf{Y}_{t-n_1}^{t-1}, \mathbf{X}_{t-n_2}^{t-1}) \,\|\, P\left(Y(t) \mid \mathbf{Y}_{t-n_1}^{t-1}\right)\right) \geq 0 . \tag{6}$$

The expression in (6) will only be 0 if and only if for all y(t), $\mathbf{y}_{t-n_1}^{t-1}$ and $\mathbf{x}_{t-n_2}^{t-1}$ the equality in (1) holds. Hence, expression (6) allows to verify whether observation of $\mathbf{x}_{t-n_2}^{t-1}$ changes knowledge about y(t). The problem of directly computing (6) is that the required distributions in (1) as well as the prior $p(\mathbf{y}_{t-n_1}^{t-1}, \mathbf{x}_{t-n_2}^{t-1})$ are not known. On the other hand, estimators of the entropy and mutual information see e.g. [11], have gained a lot of attention in the literature. Therefore, if we are able to establish a relationship between conditional relative entropy and entropy or mutual information, the expression in (4) could be estimated from the mutual entropy or mutual information estimators. This relationship is established in following theorem.

2.1 Relationship Conditional Relative Entropy and Mutual Information

Theorem 1. The conditional relative entropy in (4) can be expressed as conditional mutual information:

$$KL\left(P(Y(t) \mid \mathbf{Y}_{t-n_1}^{t-1}, \mathbf{X}_{t-n_2}^{t-1}) \,\|\, P\left(Y(t) \mid \mathbf{Y}_{t-n_1}^{t-1}\right)\right) = MI(Y(t); \mathbf{X}_{t-n_2}^{t-1} \mid \mathbf{Y}_{t-n_1}^{t-1}) . \tag{7}$$

Proof. A sketch of the proof is based on the definition of conditional probabilities and mutual information. Starting from the definition in (5):

$$= \sum_{y(t), \mathbf{y}_{t-n_1}^{t-1}, \mathbf{x}_{t-n_2}^{t-1}} p(y(t), \mathbf{y}_{t-n_1}^{t-1}, \mathbf{x}_{t-n_2}^{t-1}) \ln \left(\frac{P(y(t) \mid \mathbf{y}_{t-n_1}^{t-1}, \mathbf{x}_{t-n_2}^{t-1})}{P\left(y(t) \mid \mathbf{y}_{t-n_1}^{t-1}\right)} \right) . \tag{8}$$

Multiplying both the numerator and the denominator within the logarithm with $p(y(t), \mathbf{y}_{t-n_1}^{t-1}, \mathbf{x}_{t-n_2}^{t-1})$ and using the convention that 0ln(0) = 0, this yields:

$$= \sum_{y(t), \mathbf{y}_{t-n_1}^{t-1}, \mathbf{x}_{t-n_2}^{t-1}} p(y(t), \mathbf{y}_{t-n_1}^{t-1}, \mathbf{x}_{t-n_2}^{t-1}) \ln \frac{p(y(t) \mid \mathbf{y}_{t-n_1}^{t-1}, \mathbf{x}_{t-n_2}^{t-1}) p(\mathbf{y}_{t-n_1}^{t-1}, \mathbf{x}_{t-n_2}^{t-1})}{p(y(t) \mid \mathbf{y}_{t-n_1}^{t-1}) P(\mathbf{y}_{t-n_1}^{t-1}, \mathbf{x}_{t-n_2}^{t-1})} . \tag{9}$$

Applying the definition of conditional probability, $p(\mathbf{y}_{t-n_1}^{t-1}, \mathbf{x}_{t-n_2}^{t-1}) = p(\mathbf{x}_{t-n_2}^{t-1} \mid \mathbf{y}_{t-n_1}^{t-1})$ $p(\mathbf{y}_{t-n_1}^{t-1})$, (9) becomes:

$$= \sum_{y(t), \mathbf{y}_{t-n_1}^{t-1}, \mathbf{x}_{t-n_2}^{t-1}} p(y(t), \mathbf{y}_{t-n_1}^{t-1}, \mathbf{x}_{t-n_2}^{t-1}) \ln \frac{p(y(t) \mid \mathbf{y}_{t-n_1}^{t-1}, \mathbf{x}_{t-n_2}^{t-1}) p(\mathbf{x}_{t-n_2}^{t-1} \mid \mathbf{y}_{t-n_1}^{t-1}) p(\mathbf{y}_{t-n_1}^{t-1})}{p(y(t) \mid \mathbf{y}_{t-n_1}^{t-1}) p(\mathbf{x}_{t-n_2}^{t-1} \mid \mathbf{y}_{t-n_1}^{t-1}) p(\mathbf{y}_{t-n_1}^{t-1})} . \tag{10}$$

Applying the definition of conditional probability, $p(y(t) \mid \mathbf{y}_{t-n_1}^{t-1}, \mathbf{x}_{t-n_2}^{t-1})$ $p(\mathbf{x}_{t-n_2}^{t-1} \mid \mathbf{y}_{t-n_1}^{t-1}) = p(y(t), \mathbf{x}_{t-n_2}^{t-1} \mid \mathbf{y}_{t-n_1}^{t-1})$, and deleting $p(\mathbf{y}_{t-n_1}^{t-1})$ from both the numerator and the denominator (10) yields:

$$= \sum_{y(t),\mathbf{y}_{t-n_1}^{t-1},\mathbf{x}_{t-n_2}^{t-1}} p(y(t),\mathbf{y}_{t-n_1}^{t-1},\mathbf{x}_{t-n_2}^{t-1}) \ln \frac{p(y(t),\mathbf{x}_{t-n_2}^{t-1} \mid \mathbf{y}_{t-n_1}^{t-1})}{p(y(t) \mid \mathbf{y}_{t-n_1}^{t-1}) p(\mathbf{x}_{t-n_2}^{t-1} \mid \mathbf{y}_{t-n_1}^{t-1})} . \tag{11}$$

The expression in (11) is equal to the conditional mutual information:

$$= MI(Y(t);\mathbf{X}_{t-n_2}^{t-1} \mid \mathbf{Y}_{t-n_1}^{t-1}) . \tag{12}$$

This completes the proof.

2.2 Techniques for Causality Detection

'Sims', 'Geweke-Meese-Dent' and 'Granger' criteria for causality detection are well-established criteria for causality detection, see reference [3].

In [10] and a reference therein, concepts are defined which use (coarse-grained) conditional mutual information in a heuristic manner. However, a formal derivation to justify conditional mutual information as the result of using the conditional relative entropy framework is lacking in [10].

Recently, causality in variance has been used to detect higher-order causal dependencies between times series, see e.g. [5]. In section 4, we will use a similar criterion as theorem 1 in [5] in order to verify whether the time series contain causal dependencies beyond those that can be assessed by Granger causality. Depending on the conditions, this theorem allows to assess whether there is 'causality in variance' or a 'causality linear in variance'. We explain the conditions for these causalities in more detail in section 4, after the VARMA-GARCH has been introduced as an appropriate model for gait time series.

3 Computation of Conditional Relative Entropy

An intuitive interpretation of theorem 1 is stated as follows: $\mathbf{X}_{t-n_2}^{t-1}$ does not causally explain Y(t), if it is conditionally independent of $\mathbf{X}_{t-n_2}^{t-1}$ given its past, i.e. $\mathbf{Y}_{t-n_1}^{t-1}$. Stated differently nothing can be learnt about Y(t) from $\mathbf{X}_{t-n_2}^{t-1}$ when $\mathbf{Y}_{t-n_1}^{t-1}$ is known or $\mathbf{X}_{t-n_2}^{t-1}$ is not redundant regarding Y(t) conditioned on $\mathbf{Y}_{t-n_2}^{t-1}$.

The mutual information from theorem 1 needs to be expressed in terms of unconditional mutual information, because mutual information and entropy estimators are most often provided unconditionally.

The following result can easily be shown by applying the *chain rule for information*:

$$MI(Y(t);\mathbf{X}_{t-n_2}^{t-1} \mid \mathbf{Y}_{t-n_1}^{t-1}) = MI(Y_t;\mathbf{Y}_{t-n_1}^{t-1},\mathbf{X}_{t-n_2}^{t-1}) - MI(Y_t;\mathbf{Y}_{t-n_1}^{t-1}) . \tag{13}$$

It is also easily shown that this can be expressed in terms of the unconditional entropies:

$$\begin{aligned} & MI(Y(t);\mathbf{X}_{t-n_2}^{t-1} \mid \mathbf{Y}_{t-n_1}^{t-1}) \\ &= MI(Y_t;\mathbf{Y}_{t-n_1}^{t-1},\mathbf{X}_{t-n_2}^{t-1}) - MI(Y_t;\mathbf{Y}_{t-n_1}^{t-1}) \\ &= H(Y(t)) - H(Y(t) \mid \mathbf{Y}_{t-n_1}^{t-1},\mathbf{X}_{t-n_2}^{t-1}) - \left(H(Y(t)) - H(Y(t) \mid \mathbf{Y}_{t-n_1}^{t-1})\right) \\ &= H(Y(t),\mathbf{Y}_{t-n_1}^{t-1}) - H(\mathbf{Y}_{t-n_1}^{t-1}) - \left(H(Y(t),\mathbf{Y}_{t-n_1}^{t-1},\mathbf{X}_{t-n_2}^{t-1}) - H(\mathbf{Y}_{t-n_1}^{t-1},\mathbf{X}_{t-n_2}^{t-1})\right) \end{aligned} \quad . \tag{14}$$

This last formulation of the conditional mutual information in terms of unconditional entropies will be used for the experiments.

The entropies are estimated by means of a recent k-nearest neighbor approach to entropy estimation, from [8]. The estimated entropy $\hat{H}(\mathbf{Y}_{t-n_1}^{t-1},\mathbf{X}_{t-n_2}^{t-1})$ is computed as follows:

$$\begin{aligned} & \hat{H}(\mathbf{Y}_{t-n_1}^{t-1},\mathbf{X}_{t-n_2}^{t-1}) = \\ & -\psi(k) + \psi(N) + \log\left(c_{d_x} c_{d_y}\right) + \frac{d_X + d_Y}{N} \sum_{i=1}^{N} \log \varepsilon(i) \end{aligned} \quad . \tag{15}$$

Where c_{d_x} en c_{d_y} are respectively the volumes of the d_x- and d_y-dimensional unit balls, d_x and d_y are respectively the dimensionality of $\mathbf{x}_{t-n_2}^{t-1}$ and $\mathbf{y}_{t-n_1}^{t-1}$, i.e. n_2 and n_1 respectively, and $\varepsilon(i)$ is twice the distance from $\left(\mathbf{x}_{t-n_2}^{t-1},\mathbf{y}_{t-n_1}^{t-1}\right)$ to its k-nearest neighbor. Finally, $\psi(.)$ is the digamma function. In the experiments we set 'k' equal to 8. We use the Euclidean distance here.

4 Experiments

The time series used in the experimental section of the paper are measured from the Tekscan FScan system. The time series are sampled at a sampling rate of 100 Hz. The forces exerted on the soles of the foot are measured from this system. An example of the global force/mass exerted on each foot is shown in figure 1.

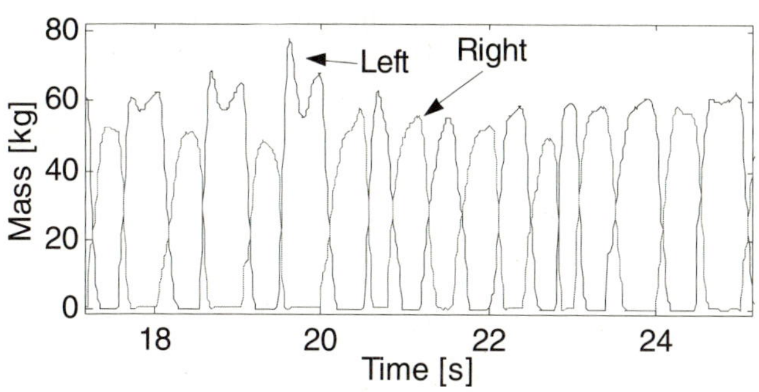

Fig. 1. Trace of the total mass on the left foot and the right foot

This figure should be interpreted as follows: when the foot is lifted from the ground the total force exerted on a foot becomes equal to 0 during the period when there is no contact with the ground. When a foot strikes the ground, the force gradually builds up until a maximum is achieved. Also note that 2 maxima may be achieved. This is due to the landing of a foot followed by the pushing off a foot prior to a swing of the foot.

In figure 2, the residual times series are shown after a 20th order VAR model has been fitted to the original gait time series. The original gait time series are shown as well, after the mean is removed and after scaling. This allows locating the residuals in the original time series. It is clear that the variance of the residuals is time dependent: the variance of the residual decreases when the corresponding foot does not touch the ground. Moreover, it can be observed that the variance of the residual in one time series is not only depending on the past variance of the same time series, but also on the past variance of the other time series. A high variance in 1 time series seems more likely when both the past variance of the same time series is high and the past variance of the other time series is low.

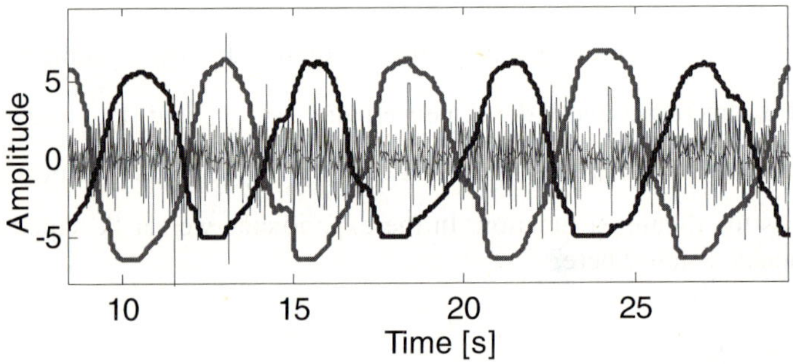

Fig. 2. The variance of the error changes with time

4.1 A Model for Gait Time Series

Gait characteristics, such as stride-to-stride variability, stride length, and gait speed have been studied before, see e.g. [6] for a review. The focus in gait literature so far has been on stride-to-stride fluctuations which exhibit fractal behavior. It has been shown that this fractal behavior changes with age and for certain pathologies [7]. Modeling of gait time series on a short time scale, taken here as approximately 30 s, has received so far little attention. Such models, however, may be important in the prediction of freezing of gait or falling [1].

The initial observations made in section 4 motivate to model the time series by means of a VARMA-GARCH model [9]. The VARMA (Vector Autoregressive Moving Average) part is able to capture the time varying mean behavior (oscillating behavior). The GARCH (Generalized Autoregressive Conditional Heteroscedasticity) part of the model is able to model the time varying variance. A VAR-GARCH formulation of the model can be written as follows:

$$\begin{bmatrix} X(t) \\ Y(t) \end{bmatrix} = \sum_{i=1}^{p} \mathbf{A}_i \begin{bmatrix} X(t-i) \\ Y(t-i) \end{bmatrix} + \mathbf{U}(t),\ \mathbf{U}(t) = \begin{bmatrix} U_x(t) \\ U_y(t) \end{bmatrix}, \\ \mathbf{U}(t) = \mathbf{\Sigma}_{t|t-1}^{1/2} \begin{bmatrix} \varepsilon_x(t) \\ \varepsilon_y(t) \end{bmatrix},\ \begin{bmatrix} \varepsilon_x(t) \\ \varepsilon_y(t) \end{bmatrix} \sim i.i.d(\mathbf{0}, \mathbf{I}_2) \tag{16}$$

$$vech(\Sigma_{t|t-1}) = \mathbf{C} + \sum_{j=1}^{q} \mathbf{\Gamma}_j vech(\mathbf{U}(t-j)\mathbf{U}'(t-j)) + \sum_{j=1}^{r} \mathbf{G}_j vech(\mathbf{\Sigma}_{t-j|t-j-1}). \tag{17}$$

Here, $\mathbf{A}_i$ are the matrices that express the dependencies of the time varying mean on past observations, and $U_x(t)$ and $U_y(t)$ are error variables with time varying covariance matrix $\mathbf{\Sigma}_{t|t-1}^{1/2}$. The subscript of the covariance matrix denotes the covariance matrix at time t, based upon all past information available up to time t-1. The variables $\varepsilon_x(t)$ and $\varepsilon_y(t)$ are i.i.d., i.e. independent and identically distributed, with zero mean and as covariance matrix $\mathbf{I}_2$, i.e. the identity matrix of size 2. The model in (17) for the covariance matrix is the VECH model of the GARCH formulation. In (17) VECH is used as an operator which takes a symmetric matrix as input and returns the lower triangular matrix as a column vector. The matrices $\mathbf{C}$, $\mathbf{\Gamma}_j$, and $\mathbf{G}_j$ express respectively the constant covariance matrix, dependency on past residuals and previous covariance matrices. In practice, q and r are taken equal to 1, due to the explosion in parameters in (17), e.g. for q = 1, r = 1, the total number of parameters is already equal to 21.

4.2 Validation

The gait time series can be used as ground-truth experimental time series, because forces exerted on the left foot will determine future forces on the right foot and the other way round. Hence, the left foot causally explains the right foot and the other way round.

Validation in the Linear Regime. Here we test 'Sims', Geweke-Meese-Dent' and 'Granger' criteria and the CMI on the original time series. The linear interactions in the signal prevail, considering the small residuals that are left after fitting a VAR(20) model, where 20 refers to time lags 1 up to 20, as in figure 2. It can be expected that the 3 mentioned techniques are able to capture the causalities within the original time series.

In total 20 experimental time series are measured from 7 different patients suffering from Parkinson disease. In table I, we test the performance of all criteria, this both for causality from the left to the right foot and the other way round. All criteria correctly identify the causal dependencies in all 20 time series, except for Sims' method which rejects causality for 1 patient from the right to the left foot.

All methods were tested at the $\alpha = 0.05$ significance level. For the traditional techniques the significance is computed from an F-test. However, for the conditional mutual information the distribution is not known in general. Hereto, we applied surrogate testing: blocks of n_2 samples from $\mathbf{X}(t)$ are randomly permutated against $\mathbf{Y}(t)$. This technique removes the causal dependencies between $\mathbf{X}(t)$ and $\mathbf{Y}(t)$. The last expression in (14) is then computed under this null hypothesis of no causal dependencies. This process has been repeated 1000 times. Then computing (14) without performing permutations and comparing with the 950'th order statistic under the null-hypothesis allows to assess the statistical significance at the $\alpha = 0.05$ level. Such a procedure is generally known as surrogate testing [12].

Table 1. Comparison between: 'Sims', 'Geweke-Meese-Dent', 'Granger' and 'Conditional Mutual Information' tests for causality detection

	Sims	Geweke-Meese-Dent	Granger	Conditional Mutual Information
Left to right	100 %	100 %	100 %	100 %
Right to left	95.0 %	100 %	100 %	100 %

As can be seen from formula (14) n_1 and n_2 are parameters for the conditional mutual information criterion. We set $n_1 = 20$ and $n_2 = 10$. Other settings are tried as well, but this setting was not critical. In the case of the 3 other tests, similar parameters need to be set, these were taken similar as in the conditional mutual information criterion.

Validation in the non-Linear Regime. The non-linear regime is obtained by first fitting a VAR(20) model to the original time series and considering the residual time series, as obtained in figure 2. On these residual time series we apply again the CMI criterion as in the linear regime and apply surrogate testing.

On the other hand we fit the VECH model from (17) to the residuals. Then causality towards X(t) can be observed from the coefficients in $\mathbf{\Gamma}_1$. The coefficients γ_{12} and γ_{13} reveal information that propagates into X(t) and that originates partly from the other time series by means of $U_x(t-1)U_y(t-1)$ or fully by means of $U_y(t-1)$ respectively. Causality towards Y(t) can be observed by the coefficients γ_{32} and γ_{31}. These coefficients reveal the information that propagates into Y(t) and that originates partly from the other time series by means of $U_x(t-1)U_y(t-1)$ or fully by means of $U_x(t-1)$ respectively. Hence, if at least one of the coefficients γ_{12} or γ_{13} is different from 0, there is a causal relationship from Y(t-1) towards X(t). On the other hand if at least one of the coefficients γ_{32} or γ_{31} is different from 0, there is a causal relationship from X(t-1) towards Y(t). Note that the causality can be bi-directional. The same holds for the coefficients in $\mathbf{G}_1$ of the VECH model in (17). Hence, only 1 of the coefficients γ_{12}, γ_{13}, g_{12} or g_{13} needs to be different from 0 to have a causality towards X(t). Similarly, only 1 of the coefficients γ_{31}, γ_{32}, g_{31} or g_{32} needs to be different from 0 to have a causality towards Y(t). This is in fact a test similar to theorem 1 in [5] in order to have 'causality in variance' or 'causality linear in variance'. We used the $\alpha =$ 0.05 significance level to test whether coefficients are significantly different from 0, after the GARCH model has been fitted to the residuals.

If a causal relationship is detected by the GARCH model this should also be the case for the for CMI criterion. In the case the GARCH model does not reveal a causal relationship, it is still possible that a causal relationship exists. The GARCH model can only detect these relationships for second order dependencies. We propose following measure to quantify the correspondence between GARCH and the CMI:

$$perf = \frac{\#\left(CMI = yes\ and\ GARCH = yes\right)}{\#\left(CMI = yes\ or\ GARCH = yes\right)} \times 100\% . \tag{18}$$

Hence, this is the ratio of the number of times both CMI and GARCH detect causality and the number of times at least one of the techniques detects causality. The performance measure of (18) is computed for the Sims, Geweke-Meese-Dent and Granger criteria in combination with the GARCH criterion as well. This performance measure is computed both for causality from the left foot to the right foot and the other way round.

In table 2 we show performances of the different criteria. As can be expected the 'Sims', 'Geweke-Meese-Dent', and 'Granger' perform very poorly. Once the linear causal dependencies are removed, these techniques cannot reveal any other dependencies. On the other hand CMI and GARCH causality detection have a higher correspondence, affirming that CMI is indeed able to capture non-linear causal relationships.

Table 2. Comparison between: 'Sims', 'Geweke-Meese-Dent', 'Granger' and 'Conditional Mutual Information' tests for causality detection in the non-linear regime

	Sims	Geweke-Meese-Dent	Granger	Conditional Mutual Information
Left to right	0 %	0 %	0 %	58.82 %
Right to left	0 %	0 %	0 %	66.67 %

5 Conclusion

The contribution in this paper is four-fold. Firstly, an information theoretic framework for causality detection is motivated from the conditional relative entropy (CRE) criterion. The CRE is formulated in terms of the information theoretic functionals: mutual information and entropy. The CRE allows to motivate conditional mutual information more thoroughly as opposed to previous research.

Secondly, it is shown that a VARMA-GARCH model is a plausible model for gait time series. Thirdly, experiments on real-case gait time series have shown that conditional mutual information performs as well as the well established Granger and Geweke-Meese-Dent criteria for causality detection in order to affirm the linear causal relationships between the left and the right foot. Finally, it is shown that conditional mutual information outperforms these techniques when non-linear causal dependencies exist.

Acknowledgements. The first author is supported by the Institute for the Promotion of Innovation through Science and Technology in Flanders (IWT Vlaanderen).

References

1. Bloem, B.R., Hausdorff, J.M., Visser, J.E., Giladi, N.: Falls and Freezing of Gait in Parkinson's Disease. A Review of Two Interconnected, Episodic Phenomena. Movement Disorders 19, 871–884 (2004)
2. Cover, T.M., Thomas, J.A.: Elements of Information Theory, 2nd edn. John Wiley & Sons, Hoboken, New Jersey (2006)
3. Geweke, J., Meese, R., Dent, W.: Comparing Alternative Tests of Causality in Temporal Systems. Journal of Econometrics 21, 161–194 (1983)
4. Granger, C.W.J.: Testing for Causality: a Personal Viewpoint. Journal of Econometrics Dynamic and Control 2, 329–352 (1980)
5. Hafner, C.M., Herwartz, H.: Testing for Causality in Variance using Multivariate GARCH Models, Economics Working Paper, no 2004-03, available at http://econpapers.repec.org/paper/zbwcauewp/1690.htm
6. Hausdorff, J.M.: Gait Variability: Methods, Modeling and Meaning. Journal of NeuroEngineering and Rehabilitation 2, 1–9 (2005)
7. Hausdorff, J.M., et al.: When Human Walking Becomes Random Walking: Fractal Analysis and Modeling of Gait Rhythm Fluctuations. Physica A 302, 138–147 (2001)
8. Kraskov, A., Stögbauer, H., Grassberger, P.: Estimating Mutual Information. Physical Review E 69, 1–16 (2004)
9. Ling, S., McAleer, M.: Asymptotic Theory for a Vector ARMA-GARCH Model. Econometric Theory 19, 280–310 (2003)
10. Paluš, M.: Synchronization as Adjustment of Information Rates: Detection from Bivariate Time Series. Physical Review E 63, 1–6 (2001)
11. Paninski, L.: Estimation of Entropy and Mutual Information. Neural Computation 15, 1191–1253 (2003)
12. Theiler, J., Eubank, S., Longtin, A., Galdrikian, B., Farmer, J.D.: Testing for Nonlinearity in Time Series: The Method of Surrogate Data. Physica D 58, 77–94 (1992)

Template Matching for Large Transformations

Julian Eggert[1], Chen Zhang[2], and Edgar Körner[1]

[1] Honda Research Institute Europe GmbH
Carl-Legien-Strasse 30, 63073 Offenbach/Main, Germany
[2] Darmstadt University of Technology, Institute of Automatic Control, Control Theory and Robotics Lab, 64283 Darmstadt, Germany

Abstract. Finding a template image in another larger image is a problem that has applications in many vision research areas such as models for object detection and tracking. The main problem here is that under real-world conditions the searched image usually is a deformed version of the template, so that these deformations have to be taken into account by the matching procedure. A common way to do this is by minimizing the difference between the template and patches of the search image assuming that the template can undergo 2D affine transformations. A popular differential algorithm for achieving this has been proposed by Lucas and Kanade [1], with the disadvantage that it works only for small transformations. Here we investigate the transformation properties of a differential template matching approach by using resolution pyramids in combination with transformation pyramids, and show how we can do template matching under large-scale transformations, with simulation results indicating that the scale and rotation ranges can be doubled using a 3 stage pyramid.

1 Introduction

Image registration using template matching, either directly on a pixel image or on an array of images that result from an appropriate preprocessing step on an image, is a fundamental step that serves as basis for many vision algorithms. The most straightforward way is to take the patch containing the template, overlay it onto the search image at all desired transformations (e.g. positions, rotations, etc.), and calculate a matching score that indicates how well the transformed template matches with the search image for each particular transformation.

In visual object classification and detection, this is the case for connectionist models that use nonlinearities alternated with correlation-based patch template matching with feature-sets in a weight-sharing architecture [2,3,4]. The weight sharing activity calculation basically corresponds to a feature search at all positions of the input image. More complex transformations of the templates (features) are usually not considered or included explicitly by building all transformation variants of a basis feature.

In template-based tracking, the picture is similar, with the difference that we can restrict the search to those templates and transformations that are likely to occur according to the tracker state predictions. A third field of research where

J. Marques de Sá et al. (Eds.): ICANN 2007, Part II, LNCS 4669, pp. 169–179, 2007.

template matching is important is motion processing (see e.g. [5]). Here, the task is to find patch-to-patch correspondences between two images from consecutive timesteps, in order to extract a displacement field that indicates how the different patches move from one timestep to the next.

The main problem for template matching is the number of transformations that have to be checked in order for the procedure to cope with deformations. The appearance of real-world objects undergo severe changes as the objects e.g. translate, rotate, come closer or rotate in 3D. Some of these transformations can (and have to) by covered by the template matching procedure, while others like true 3D appearance changes can only be captured as approximations and only if the transformations remain sufficiently small. This is the case e.g. for rotations in depth and approximately planar objects, whose transformation can then be approximated by a projective transformation.

A popular approach is to introduce 2D transformations into the matching process, in particular 2D affine transformations covering rotation, scaling and shearing in addition to translation. That is, we then search the best match between a search image and a template subject to its tansformations. In order to achieve this, an extensively large number of transformed templates has to be compared with the search image, corresponding to all possible combination of transformations.

Three things can be done to alleviate the costs of matching under transformations. Firstly, not all transformations are equally important and interdependent, so that we can sometimes search for separate transformation dimensions independently. As a second point, we can assume small differences between template and search image (introduced by small transformations), linearize and try to calculate the template match for small transformations computationally more effectively. And third, we can introduce search strategies for the transformation parameters, e.g. by sampling the transformation parameter space first coarsely to get a hint on the transformation range and then refining the search.

In this paper, we combine points 1, 2 and 3 by introducing a resolution pyramid in combination with a transformation pyramid that allows to estimate affine transformations for the image matching problem over a broad range of parameters, calculating first the coarse transformations and refining them in the successive stages of the pyramid. Although pyramidal approaches have been proposed already a number of times in the vision systems community, here we address explicitly the question of transformation pyramids for template matching, analysing the potential of such methods for large scale transformations in combination with Lucas-Kanade type 2D affine matching methods.

In the next section, we sketch the architecture of the approach. To this end, we first summarize a popular differential approach for 2D affine template matching and then show how we utilize it in a transformation pyramid. In the third section, we show in simulations how the approach performs for large-scale transformations. We show exemplar results of valid transformation ranges (since these depend also on intrinsic object characteristics and generally cannot be formulated for arbitrary objects).

2 Approach

2.1 Template Matching with 2D Affine Transformations

A popular approach for template search under small transformations that works well for the affine case has been introduced by Lucas and Kanade [1] and used for many extensions like tracking of objects by means of point features and the appearance transformations of the image patches around these points [6].

To compare two images we start with image points $\mathbf{x} = (x, y)^T$ a template $T(\mathbf{x})$ and its "mask" resp. window of validity $M(\mathbf{x})$ (either binary or continuous, but zero outside of the region of validity of the template) on one hand, and the search image $I(\mathbf{x})$ with its warping transformation $\mathbf{W}(\mathbf{x})$ on the other hand. For this paper, we restrict $\mathbf{W}(\mathbf{x})$ to be a linear transformation composed of an affine transformation matrix $\mathbf{A}$ and a translation vector $\mathbf{d}$, so that an image position $\mathbf{x}$ transforms according to

$$\mathbf{x} \rightarrow \mathbf{W}(\mathbf{x}) = \mathbf{A}\mathbf{x} + \mathbf{d} \; . \tag{1}$$

The target of the template match is to find the parameters $\mathbf{A}$ and $\mathbf{d}$ which minimize a functional

$$F = \int [I(\mathbf{A}\mathbf{x} + \mathbf{d}) - T(\mathbf{x})]^2 M(\mathbf{x}) d\mathbf{x} \; , \tag{2}$$

that is, which leads to the best Euclidean match between template and image under consideration of the geometrical transformation eq. 1 of the search image $I(\mathbf{x})$.

The reason that we transform the search image $I(\mathbf{x})$ and not the template $T(\mathbf{x})$ is that we consider the window $M(\mathbf{x})$ to be attached to the template. If we then transform the template, we would have to transform the window as well, which makes the derivation more complicated (nevertheless, this is a matter of interpretation of eq. 2 since template and search image are exchangeable). For tracking applications, or if we are interested in keeping the template fixed, the inverse transformation from the template to the search image can be easily calculated according to $\mathbf{A}^{-1}\mathbf{x} - \mathbf{A}^{-1}\mathbf{d}$ (e.g., if we want to say: "The pattern in the search image corresponds to the template rotated by ... degrees").

For small affine transformations it makes sense to write $\mathbf{A} = \mathbf{1} + \mathbf{D}$ (small deviation $\mathbf{D}$ from the unity matrix $\mathbf{1}$), with the deformation matrix

$$\mathbf{D} = \begin{pmatrix} d_1 & d_3 \\ d_2 & d_4 \end{pmatrix} \tag{3}$$

now completing the transformation parameters together with the displacement vector

$$\mathbf{d} = \begin{pmatrix} d_5 \\ d_6 \end{pmatrix} . \tag{4}$$

The 6 transformation parameters (4 for the deformation matrix $\mathbf{D}$ and 2 for the displacement vector $\mathbf{d}$) can be collected in a vector

$$\mathbf{z} = (d_1, d_2, d_3, d_4, d_5, d_6)^T \; . \tag{5}$$

A (local) minimization of the functional $F(\mathbf{z})$ can then be achieved by using gradient-descent. Nevertheless, since the warping of the image is the expensive step, it is desirable to avoid too many iterations during the gradient descent. To do few iterations, here we use the Newton method, setting $\nabla_{\mathbf{z}} F(\mathbf{z}) = 0$, with

$$\nabla_{\mathbf{z}} F(\mathbf{z}) = \int 2\left[I(\mathbf{Ax}+\mathbf{d}) - T(\mathbf{x})\right] \nabla_{\mathbf{z}} I(\mathbf{Ax}+\mathbf{d}) M(\mathbf{x}) d\mathbf{x} \tag{6}$$

and linearizing and solving the equation system repetitively for $\mathbf{z}$, as indicated in [7].

We assume smooth changes in the search image. Linearization with respect to $\mathbf{x}$ then yields

$$I(\mathbf{Ax}+\mathbf{d}) \approx I(\mathbf{x}) + \left[\nabla_{\mathbf{x}} I(\mathbf{x})\right]^T (\mathbf{Dx}+\mathbf{d}) \tag{7}$$

which we use to get

$$\nabla_{\mathbf{z}} I(\mathbf{Ax}+\mathbf{d}) \approx \left[\nabla_{\mathbf{x}} I(\mathbf{x})\right]^T \left[\nabla_{\mathbf{z}} (\mathbf{Dx}+\mathbf{d})\right] \tag{8}$$

with the $2 * 6$-matrix (for a 2-dimensional vector $\mathbf{v}$)

$$\nabla_{\mathbf{z}} \mathbf{v} = \begin{bmatrix} (\nabla_{\mathbf{z}} v_1)^T \\ (\nabla_{\mathbf{z}} v_2)^T \end{bmatrix} . \tag{9}$$

It is straightforward to calculate

$$\begin{aligned} \mathbf{g}(\mathbf{x}) &:= \left[\nabla_{\mathbf{x}} I(\mathbf{x})\right]^T \left[\nabla_{\mathbf{z}} (\mathbf{Dx}+\mathbf{d})\right] \\ &= \left[x\frac{\partial I(\mathbf{x})}{\partial x}, x\frac{\partial I(\mathbf{x})}{\partial y}, y\frac{\partial I(\mathbf{x})}{\partial x}, y\frac{\partial I(\mathbf{x})}{\partial y}, \frac{\partial I(\mathbf{x})}{\partial x}, \frac{\partial I(\mathbf{x})}{\partial y}\right]^T \end{aligned} \tag{10}$$

and setting $\nabla_{\mathbf{z}} F(\mathbf{z}) = 0$ and extracting $\mathbf{z}$, we arrive at the 6-dimensional linear equation system

$$\mathbf{Tz} = \mathbf{a} \tag{11}$$

with

$$\mathbf{T} = \int M(\mathbf{x}) \left\{\mathbf{g}(\mathbf{x})[\mathbf{g}(\mathbf{x})]^T\right\} d\mathbf{x} \quad \text{and} \tag{12}$$

$$\mathbf{a} = \int M(\mathbf{x}) \left[T(\mathbf{x}) - I(\mathbf{x})\right] \mathbf{g}(\mathbf{x}) d\mathbf{x} \tag{13}$$

that can be solved $\mathbf{z} = \mathbf{T}^{-1}\mathbf{a}$ by inverting $\mathbf{T}$.

2.2 Template Matching in Resolution and Transformation Pyramids

The problem of the method from section 2.1 is that it works well as long as the gradients introduced by eqs. 7 and 8 and incorporated into the method by eq. 10 provide sufficient information. For large transformation parameters, this may not be the case any more. A remedy then is to use coarser resolutions which smoothen the image, like in a Gaussian resolution pyramid, and combine them with a transformation pyramid that allows to calculate the total transformation as a concatenation of single differential transformations for each stage. (Transformation pyramids have been proposed repetitively in different contexts but mostly in combination with translational transformations, see e.g. [8] for an early proposal and [9] for an application of the same principle in a motion estimation system. Here we show in simulations to what extent they can be applied to the Lucas and Kanade type image matching procedures).

The idea is to use a pyramid with n levels (1: original, n: coarsest resolution, images and templates at different resolutions $I_i(\mathbf{x})$, $T_i(\mathbf{x})$), and apply the procedure on the coarsest level n with $I_n(\mathbf{x})$, $T_n(\mathbf{x})$ to get a first, rough estimation of the transformation parameters $\mathbf{A}_n$, $\mathbf{d}_n$, from which we get the first total transformation estimate $\mathbf{A}_{n-1}^{\text{tot}}$, $\mathbf{d}_{n-1}^{\text{tot}}$ (the indices are chosen indicating that this transformation is the currently best estimate for level $n-1$). Afterwards, we use the transformation parameters to warp the search image $I_{n-1}(\mathbf{x})$ in order to compare it with the template $T_{n-1}(\mathbf{x})$ for a refinement of the transformation parameters. As a result we then get $\mathbf{A}_{n-1}$, $\mathbf{d}_{n-1}$, which has to be composed with $\mathbf{A}_{n-1}^{\text{tot}}$, $\mathbf{d}_{n-1}^{\text{tot}}$ to get the improved estimate of the transformation parameters $\mathbf{A}_{n-2}^{\text{tot}}$, $\mathbf{d}_{n-2}^{\text{tot}}$, with

$$\mathbf{A}_{i-1}^{\text{tot}} := \mathbf{A}_i \, \mathbf{A}_i^{\text{tot}} \tag{14}$$

and

$$\mathbf{d}_{i-1}^{\text{tot}} := \mathbf{d}_i + \mathbf{d}_i^{\text{tot}} \; . \tag{15}$$

This has to be repeated until we arrive at the lower end of the pyramid which works on the images at original resolution. The expectation is that, since each stage of the pyramid already receives a search image that was moved closer to the template image (in terms of Euclidean match), the matching procedure from sec. 2.1 can be applied further to improve the transformation estimation, allowing to find image matches over a much broader range of parameters.

Figure 1 shows a schema of the resolution and transformation pyramid with the mentioned warping, transformation estimation and transformation concatenation steps. If we are e.g. in a tracking application, we usually do already have some initial estimate (from previous steps) of the overall transformation to start with, which we can include as $\mathbf{A}_n^{\text{tot}}$, $\mathbf{d}_n^{\text{tot}}$ to warp the search image $I_n(\mathbf{x})$ at the top of the pyramid. On the lower end of the pyramid, we get our total transformation estimate $\mathbf{A}_0^{\text{tot}}$, $\mathbf{d}_0^{\text{tot}}$. In the following examples, we used a Gaussian resolution pyramid with 3 levels, applied on images of $128x128$ pixel size, so that at each level of the pyramid the resolution halfened. The change in resolution has to be taken into accout in the calculation of the total translation since in eq. 15, $\mathbf{d}_{i-1}^{\text{tot}}$ and $\mathbf{d}_i^{\text{tot}}$ were assumed to operate on the same spatial scale. For different spatial scales, the translation vectors $\mathbf{d}$ have to be normalized, so that for our case of the Gaussian pyramid, we used $\mathbf{d}_{i-1}^{\text{tot}} := \mathbf{d}_i + 2\mathbf{d}_i^{\text{tot}}$.

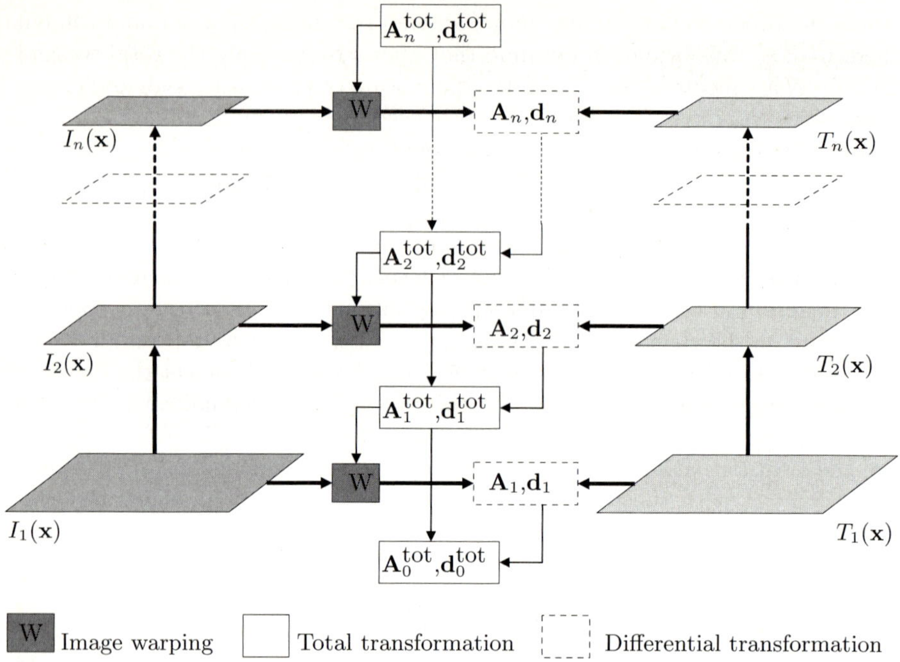

Fig. 1. Transformation and resolution pyramid for the estimation of large-scale transformations. On top of the pyramid coarse transformation estimates are calculated on coarse resolutions of the search and template images $I_n(\mathbf{x})$ and $T_n(\mathbf{x})$, which are used to warp the search image and refine the transformation estimate in the successively lower stages of the pyramid. $\mathbf{A}_n^{tot}, \mathbf{d}_n^{tot}$ is an initial transformation based on prior knowledge, $\mathbf{A}_0^{tot}, \mathbf{d}_0^{tot}$ is the overall transformation gained from the entire transformation pyramid.

3 Results for Large-Scale Transformations

The size of transformations that can be estimated with our method depend on the particular structure of the template and search images. Two types of properties are beneficial for the estimation of large transformations: 1. There has to be sufficient structure (so that there are pronounced minima in the functional eq. 2) and 2. the structure has to be sufficiently smooth so that gradients can drive the search towards a solution. Point 2 also implies that the minima are not too narrow, which is the case e.g. for templates with no pronounced autocorrelation lengths (in the simple case of a random pattern template created using white noise, there is basically no gradient information that can be used to find the minimum, if search image and template are more than one pixel offset).

To quantify the results of simulations with the presented algorithm, we took arbitrary single objects from the COIL[1] database. We chose not to average the results over a large image database (e.g., the entire COIL or more complex image

[1] Columbia Object Image Library.

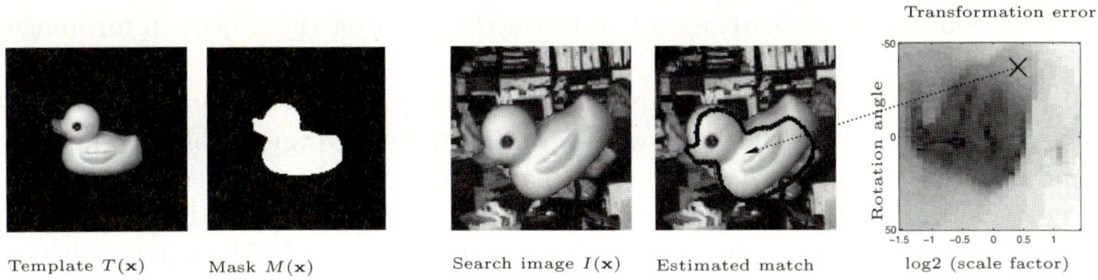

Fig. 2. Typical template, mask, (cluttered) search image and match result. The task is to find the transformation between the template and the rotated and scaled duck. On the 2 rightmost images, the match on the search image (indicated by a black contour) and the affine transformation error for different scaling factors and rotation angles are shown. The cross indicates the current target transformation (-40 degree rotation and scale factor 1.5) that was used to generate the search image. At the right, a logarithmic scale (of basis 10: e.g. -2 are errors in the 10^{-2}-range) indicates the transformation error (Euclidean distance between the gained and the true affine transformation matrix entries). Darker region correspond to good transformation matches, for the cross position the transformation estimation error is already considerable, so that the match is not perfect.

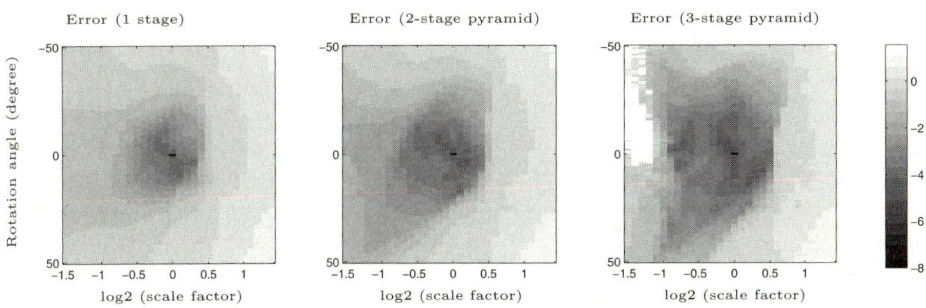

Fig. 3. Clean condition, Euclidean distance between real and computed affine transformation matrix for template matching pyramids consisting of 1, 2 and 3 stages, applied on the duck picture. The x and y-axes show the scaling factor (log 2. -1 meaning half size, +1 meaning double size) and the rotation angle of the template object in the search image. Darker regions of the graphs denote more accurate estimations of the affine transformation. Notice the increase in size of the dark region for an increasing number of stages, indicating that the pyramidal matching procedure is able to cope with scaling factors of nearly $0.5 - 2.0$ and rotation angles of about ± 20 degrees.

databases) because of the dependency on the intrinsic object properties explained above. Nevertheless, even from single objects the benefits of the method for large scale transformations can be evaluated.

We used 2 different paradigms: In the clean condition, we searched for a match between the original and the transformed version of the object and in the clutter condition, we searched a transformed version of the object in an image with clutter.

In the following two figures, we evaluated the affine transformation matrix error. We generated search images that were a rotated and scaled version of the template and then run the method with a transformation pyramid of $1-3$ layers. The resulting transformation matrix at each of the layers was then compared to the "real" transformation matrix, calculating the Euclidean distance of all its entries, shown in fig. 3. For easier visualization, we restricted the plots to affine transformations that are a combination of homogeneous scaling and rotation. It can be seen that the transformation pyramid increases the validity regions substantially.

To get a better quantitative idea of the estimation errors and to see how the error distributes among scaling and rotation parameters, we extracted from the affine transformation results a scaling parameter λ and a rotation angle α that approximate $\mathbf{A}$ according to

$$\mathbf{A} = \begin{pmatrix} a_1 & a_3 \\ a_2 & a_4 \end{pmatrix} \approx \begin{pmatrix} \lambda & 0 \\ 0 & \lambda \end{pmatrix} \begin{pmatrix} \cos(\alpha) & -\sin(\alpha) \\ \sin(\alpha) & \cos(\alpha) \end{pmatrix} . \tag{16}$$

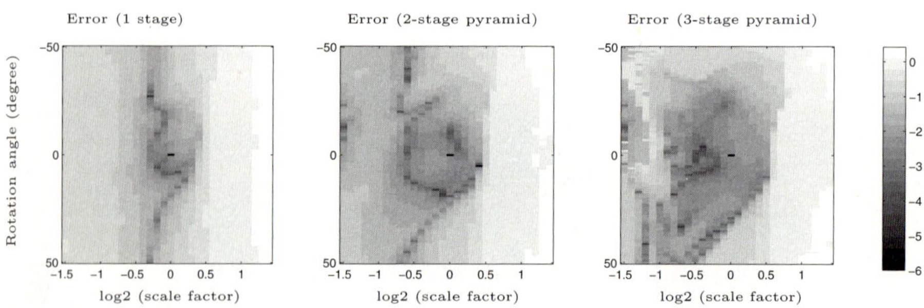

Fig. 4. Clean condition, scaling error for a template matching pyramid consisting of 1, 2 and 3 stages. The x and y-axes again show the scaling factor and the rotation angle of the template objects in the search image.

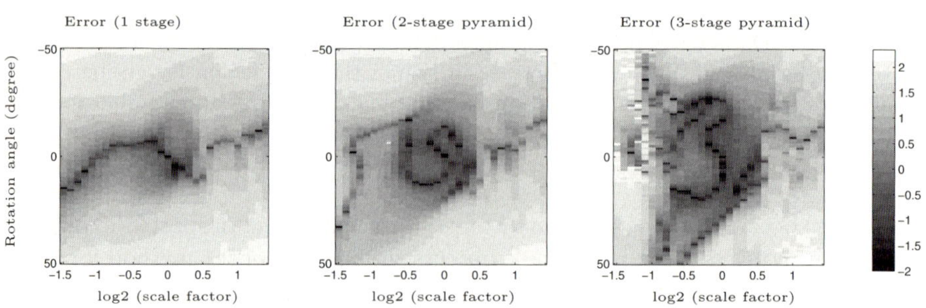

Fig. 5. Clean condition, rotation error for a template matching pyramid consisting of 1, 2 and 3 stages. The x and y-axes again show the scaling factor and the rotation angle of the template objects in the search image.

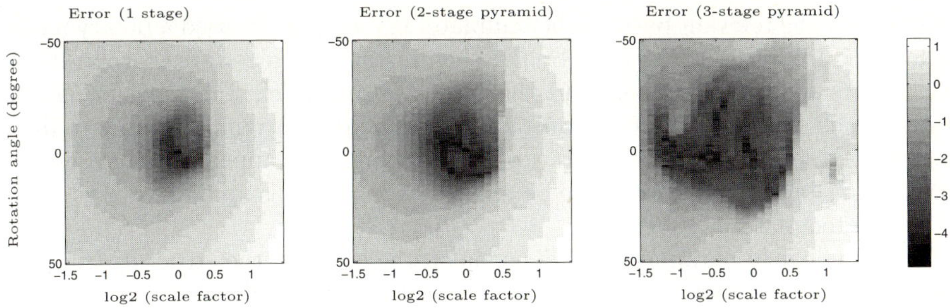

Fig. 6. Clutter condition, affine transformation error for a template matching pyramid consisting of 1, 2 and 3 stages. The x and y-axes again show the scaling factor and the rotation angle of the template objects in the search image. Although the absolute error is larger than for the clean case (fig. 6), the increase of the region of valid transformation estimations is even larger when going from 1 level to a 3-level pyramid.

Afterwards, we calculated the absolute difference between the real scale and orientation (used to generate the transformed objects for the search image) and the results after applying the transformation pyramid. Figs. 4 and 5 show the results for scaling and rotation separately.

Figure 6 shows the results for a search image that contains a transformed template object on a cluttered background. It can be seen that the gain (in terms of a larger valid region where transformation parameters are accurately estimated) is even larger than in the clean condition from fig. 3.

4 Conclusion

The translational parameters play a special role in the transformation estimation process. In many cases, the estimation of the full system eq. 11 leads to spurious or false minima, specially if the initial translational mismatch between template and object in the search image is large [7]. Then it is beneficial to separate the estimation of the translational parameters $\mathbf{d}$ from the estimation of the affine transformation parameters $\mathbf{A}$. This is something that we observed for the pyramid method at all levels. We therefore estimated first the translation at each level, and afterwards the affine transformation.

In the simulations we found that for the affine parameter estimation using a pyramid with 3 levels, a single iteration of the Newton method from section 2.1 at each level already suffices to produce good matching results. The costly part of the method, the warping of the search images at each pyramid level (see fig. 1), therefore occurs only 3 times (and only once for the full resolution image).

Since the Euclidean match in the functional eq. 2 is sensitive to differences over the entire region of the mask $\mathbf{M}(\mathbf{x})$, it is important that the mask matches covers only the relevant parts of the template. In our case, we used a binary mask calculated from the object itself by thresholding against a zero background (see fig. 2). As soon as the mask and the template mismatch, the transformation estimations degrade. Therefore, in tasks which require a template update, like e.g. when a real object is being tracked which changes its appearance beyong 2D affine transformations, care has to be taken that the mask is updated consistently with the template.

The number of pyramid levels depends on the size and the structure/texture of the template, since fine details are lost with increasingly coarse resolution. In our case, the 3-level pyramid provided to be a good choice; with more levels the results degraded since the highest level then did not have sufficient clues to estimate its transformation correctly. One way to estimate the number of needed pyramid levels without a fully extensive check is to regard the surface-averaged reconstruction error of the highest hypothetical pyramid level only (the level with the coarsest resolution), starting from the lowest possible resolution, and increasing step-by-step the number of pyramid levels, searching for a minimum.

All shown plots were gained using the same image (see fig. 2), which provided a good example with sufficiently detailed form but not too much resolution, so that it still presented a challenge for the template matching process. For other objects of the COIL database (which all have a comparable size), the gained results were very similar in terms of gain of applicable transformation range.

Summarizing, we have shown that transformation pyramids considerably extend the range of transformations that can be covered by template-matching procedures of the Lucas and Kanade type. Transformation ranges for objects increased from approx. $[0.65 - 1.4]$ to $[0.5 - 1.75]$ for the scaling factor and from $[-15, 15]$ to nearly $[-30, 30]$ degrees for the rotation angle, providing good results for situations with cluttered background.

References

1. Lucas, B.D., Kanade, T.: An iterative image registration technique with an application to stereo vision. In: International Joint Conference on Artificial Intelligence, pp. 674–679 (1981)
2. Wersing, H., Körner, E.: Learning optimized features for hierarchical models of invariant recognition. Neural Computation 15(7), 1559–1588 (2003)
3. Riesenhuber, M., Poggio, T.: Hierarchical models of object recognition in cortex. Nature Neuroscience 2(11), 1019–1025 (1999)
4. Fukushima, K.: Neocognitron: A self-organizing neural network model for a mechanism of pattern recognition unaffected by shift in position. Biological Cybernetics 39, 139–202 (1980)
5. Willert, V., Eggert, J., Adamy, J., Körner, E.: Non-gaussian velocity distributions integrated over space, time and scales. IEEE Transactions on Systems, Man and Cybernetics B (2006)
6. Tomasi, C., Kanade, T.: Detection and tracking of point features. Technical report, Carnegie Mellon University (1991)

7. Shi, J., Tomasi, C.: Good features to track. In: IEEE Conference on Computer Vision and Pattern Recognition (CVPR'94), Seattle (1994)
8. Bergen, J.R., Anandan, P., Hanna, K.J., Hingorani, R.: Hierarchical model-based motion estimation. In: Sandini, G. (ed.) ECCV 1992. LNCS, vol. 588, pp. 237–252. Springer, Heidelberg (1992)
9. Eggert, J., Willert, V., Körner, E.: Building a motion resolution pyramid by combining velocity distributions. In: Rasmussen, C.E., Bülthoff, H.H., Schölkopf, B., Giese, M.A. (eds.) Pattern Recognition. LNCS, vol. 3175, pp. 310–317. Springer, Heidelberg (2004)

Fuzzy Classifiers Based on Kernel Discriminant Analysis

Ryota Hosokawa and Shigeo Abe

Graduate School of Engineering
Kobe University
Rokkodai, Nada, Kobe, Japan
abe@kobe-u.ac.jp
http://www2.eedept.kobe-u.ac.jp/~abe

Abstract. In this paper, we discuss fuzzy classifiers based on Kernel Discriminant Analysis (KDA) for two-class problems. In our method, first we employ KDA to the given training data and calculate the component that maximally separates two classes in the feature space. Then, in the one-dimensional space obtained by KDA, we generate fuzzy rules with one-dimensional membership functions and tune the slopes and bias terms based on the same training algorithm as that of linear SVMs. Through the computer experiments for two-class problems, we show that the performance of the proposed classifier is comparable to that of SVMs, and we can easily and visually analyze its behavior using the degrees of membership functions.

1 Introduction

Support Vector Machines (SVMs) [1,2] are known to be a classifier with high generalization ability because in SVMs, the input space is mapped into a high-dimensional feature space to enhance linear separability, and the optimal separating hyperplane is determined with the maximum margin in the feature space.

Inspired by the success of SVMs, many linear techniques are extended to nonlinear forms using a mapping to a high-dimensional feature space. These techniques are called kernel-based methods. Kernel least squares [3], Kernel Principal Component Analysis (KPCA) [4], and Kernel Discriminant Analysis (KDA) [5] are the examples of such methods. Above all, because the component obtained by KDA maximally separates two classes in the feature space, it is suitable for pattern classification problems.

One of the disadvantages of SVMs, however, is that it is difficult to analyze their behavior because the input space is mapped into a high-dimensional, in some cases infinite, feature space. There are several approaches to tackle this problem. In [6,7], SVMs are visualized by using geometric methods or combining some linear classifiers. Another approach defines fuzzy classifiers in the feature space [8,9]. In [8], to improve generalization ability the membership functions are tuned so that the recognition rate of the training data is maximized.

J. Marques de Sá et al. (Eds.): ICANN 2007, Part II, LNCS 4669, pp. 180–189, 2007.

In this paper, we discuss fuzzy classifiers based on KDA for two-class problems which overcome the disadvantage of SVMs and show high performance comparable to SVMs. In our method, first we employ KDA to the given training data and calculate the component that maximally separates two classes in the feature space. Because this one-dimensional component well separates two classes, it is suitable for our aim to construct the powerful classifiers whose behavior is easily analyzable. Then, in the one-dimensional space obtained by KDA, we define a one-dimensional membership function [10] for each class based on the Euclidean distance from each class center.

To improve the performance of our fuzzy classifiers we need to tune fuzzy rules, in other words, we need to tune the slope and the bias term of each membership function. In our method, we show this tuning procedure can be performed based on the same training algorithm as that of linear SVMs by using the two-dimensional feature space whose axes are the distances from the centers of membership functions.

In Section 2 we summarize KDA, and in Section 3 we describe how to construct the proposed fuzzy classifiers. In Section 4, we show that the tuning procedure of membership functions reduces to the same training algorithm as that of linear SVMs, and in Section 5 we evaluate the proposed method using two-class data sets [11].

2 Kernel Discriminant Analysis

In this section we summarize KDA, which calculates the component that maximally separates two classes in the feature space in two-class problems.

Let the sets of m-dimensional data belong to class i ($i=1,2$) be $\{\mathbf{x}_1^i, \ldots, \mathbf{x}_{M_i}^i\}$, where M_i is the number of data belonging to class i, and data $\mathbf{x}$ be mapped into l-dimensional feature space by the mapping function $\mathbf{g}(\mathbf{x})$. The aim of KDA is to find the l-dimensional vector $\mathbf{w}$, whose direction maximally separates the two classes in the feature space.

The projection of $\mathbf{g}(\mathbf{x})$ on $\mathbf{w}$ is obtained by $\mathbf{w}^T\mathbf{g}(\mathbf{x})/\|\mathbf{w}\|$. In the following we assume that $\|\mathbf{w}\| = 1$. We find such $\mathbf{w}$ that maximizes the distance between the class centers, and minimizes the variances of the projected data.

The square difference of each class center of the projected data, d_c^2, is calculated as

$$d_c^2 = \mathbf{w}^T(\mathbf{c}_1 - \mathbf{c}_2)(\mathbf{c}_1 - \mathbf{c}_2)^T\mathbf{w}, \tag{1}$$

where $\mathbf{c}_i$ are the centers of class i data, obtained as follows:

$$\mathbf{c}_i = \frac{1}{M_i}\sum_{j=1}^{M_i}\mathbf{g}(\mathbf{x}_j^i) \quad \text{for} \quad i = 1, 2. \tag{2}$$

We define

$$Q_B = (\mathbf{c}_1 - \mathbf{c}_2)(\mathbf{c}_1 - \mathbf{c}_2)^T \tag{3}$$

and call Q_B the *between-class scatter matrix*.

Here, in order to simplify matters, we redefine the data sets as $\{\mathbf{x}_1^1, \ldots, \mathbf{x}_{M_1}^1, \mathbf{x}_1^2, \ldots, \mathbf{x}_{M_2}^2\} = \{\mathbf{x}_1, \ldots, \mathbf{x}_M\}$, where M is the number of all data and $M = M_1 + M_2$. In the following if there is no confusion, we use both representations for data sets.

The variance of all the projected data, s^2, is

$$s^2 = \mathbf{w}^T Q_T \mathbf{w}, \tag{4}$$

where

$$Q_T = \frac{1}{M}(\mathbf{g}(\mathbf{x}_1), \ldots, \mathbf{g}(\mathbf{x}_M))(I_M - \mathbf{1}_M) \begin{pmatrix} \mathbf{g}^T(\mathbf{x}_1) \\ \vdots \\ \mathbf{g}^T(\mathbf{x}_M) \end{pmatrix}. \tag{5}$$

Here, I_M is the $M \times M$ unit matrix and $\mathbf{1}_M$ is the $M \times M$ matrix with all elements being $1/M$. We call this matrix, Q_T, *total scatter matrix.*

Now, in KDA, we maximize the following criterion:

$$J(\mathbf{w}) = \frac{d_c^2}{s^2} = \frac{\mathbf{w}^T Q_B \mathbf{w}}{\mathbf{w}^T Q_T \mathbf{w}}, \tag{6}$$

but since $\mathbf{w}$, Q_B, and Q_T are defined in the feature space, we cannot calculate them explicitly. Here we need to use kernel tricks. Any solution $\mathbf{w}$ in the feature space can be written as an expansion of the form

$$\mathbf{w} = (\mathbf{g}(\mathbf{x}_1), \ldots, \mathbf{g}(\mathbf{x}_M))\boldsymbol{\alpha}, \tag{7}$$

where $\boldsymbol{\alpha} = (\alpha_1, \ldots, \alpha_M)^T$ and $\alpha_1, \ldots, \alpha_M$ are scalars. Substituting (7) into (6), we can rewrite the KDA criterion, J, as

$$J(\boldsymbol{\alpha}) = \frac{\boldsymbol{\alpha}^T K_B \boldsymbol{\alpha}}{\boldsymbol{\alpha}^T K_T \boldsymbol{\alpha}}, \tag{8}$$

where

$$K_B = (\mathbf{k}_{B_1} - \mathbf{k}_{B_2})(\mathbf{k}_{B_1} - \mathbf{k}_{B_2})^T, \tag{9}$$

$$\mathbf{k}_{B_i} = \begin{pmatrix} \frac{1}{M_i} \sum_{j=1}^{M_i} H(\mathbf{x}_1, \mathbf{x}_j^i) \\ \cdots \\ \frac{1}{M_i} \sum_{j=1}^{M_i} H(\mathbf{x}_M, \mathbf{x}_j^i) \end{pmatrix} \quad \text{for } \; i = 1, 2, \tag{10}$$

$$K_T = \frac{1}{M} \begin{pmatrix} H(\mathbf{x}_1, \mathbf{x}_1) \ldots H(\mathbf{x}_1, \mathbf{x}_M) \\ \cdots \\ H(\mathbf{x}_M, \mathbf{x}_1) \ldots H(\mathbf{x}_M, \mathbf{x}_M) \end{pmatrix} \times (I_M - \mathbf{1}_M) \begin{pmatrix} H(\mathbf{x}_1, \mathbf{x}_1) \ldots H(\mathbf{x}_1, \mathbf{x}_M) \\ \cdots \\ H(\mathbf{x}_M, \mathbf{x}_1) \ldots H(\mathbf{x}_M, \mathbf{x}_M) \end{pmatrix}^T. \tag{11}$$

Here $H(\mathbf{x}, \mathbf{x}') = \mathbf{g}^T(\mathbf{x})\mathbf{g}(\mathbf{x}')$ is a kernel function, and K_T is a positive semidefinite matrix. If K_T is positive definite, the solution of (8) is given by

$$\boldsymbol{\alpha} = K_T^{-1}(\mathbf{k}_{B_1} - \mathbf{k}_{B_2}). \tag{12}$$

If K_T is positive semidefinite, the inverse K_T^{-1} does not exist. One way to overcome singularity is to add positive values to the diagonal elements:

$$\boldsymbol{\alpha} = (K_T + \varepsilon I_M)^{-1}(\mathbf{k}_{B_1} - \mathbf{k}_{B_2}), \tag{13}$$

where ε is a small positive parameter.

From the assumption that $\|\mathbf{w}\| = 1$, we can calculate the projection of $\mathbf{g}(\mathbf{x})$ on $\mathbf{w}$, p, with kernel tricks as follows:

$$\begin{aligned} p &= \mathbf{w}^T \mathbf{g}(\mathbf{x}) \\ &= \mathbf{g}^T(\mathbf{x})\mathbf{w} \\ &= (\mathbf{g}^T(\mathbf{x})\mathbf{g}(\mathbf{x}_1), \ldots, \mathbf{g}^T(\mathbf{x})\mathbf{g}(\mathbf{x}_M))\boldsymbol{\alpha} \\ &= (H(\mathbf{x}, \mathbf{x}_1), \ldots, H(\mathbf{x}, \mathbf{x}_M))\boldsymbol{\alpha}. \end{aligned} \tag{14}$$

Using (14), training data $\{\mathbf{x}_1^i, \ldots, \mathbf{x}_{M_i}^i\}$ for class i are expressed by one-dimensional features $\{p_1^i, \ldots, p_{M_i}^i\}$, where $p_j^i = (H(\mathbf{x}_j^i, \mathbf{x}_1), \ldots, H(\mathbf{x}_j^i, \mathbf{x}_M))\boldsymbol{\alpha}$ for $j = 1, \ldots, M_i$. We call this one-dimensional space, obtained by (14), *KDA space*.

3 Classifier Architecture

3.1 Concept

We discuss fuzzy classifiers based on KDA for two-class problems. KDA is a powerful tool to obtain the one-dimensional feature that well separates two classes in the feature space. Hence, using KDA we can easily construct the powerful fuzzy classifier whose behavior is easily analyzable.

In the proposed method, using (14), we calculate the class i one-dimensional features. Then, in the KDA space, for each class we define the following fuzzy rules:

$$R_i : \text{ if } p \text{ is } \mu_i, \text{ then } \mathbf{x} \text{ belong to class } i, \tag{15}$$

where μ_i is the center of class i in the KDA space:

$$\mu_i = \frac{1}{M_i} \sum_{j=1}^{M_i} p_j^i \quad \text{for } i = 1, 2. \tag{16}$$

3.2 Definition of Membership Functions

In the KDA space, for the centers μ_i, we define one-dimensional membership functions $m_i(p)$ that define the degree to which p belongs to μ_i. Here, we consider the two types of $m_i(p)$, i.e., without a bias term and with a bias term.

Membership Functions Without Bias Terms. Based on the Euclidean distance $d_i(p)$ of p from the center μ_i, we define $m_i(p)$ as follows (see Fig.1):

$$m_i(p) = 1 - \frac{d_i(p)}{\beta_i}, \tag{17}$$

where $d_i(p)$ is the Euclidean distance from the center μ_i, $d_i(p) = |p - \mu_i|$, and β_i is a tuning parameter of the slope for $m_i(p)$. We allow negative values of $m_i(p)$ so that any point p can be classified into a definite class.

We calculate the degree of each membership function for input datum $\mathbf{x}$, $m_i(p)$, and classify it into the class whose membership is maximum. This is equivalent to finding the minimum Euclidean distance when β_i in (17) is equal to 1. If $m_i(p)$ is equal to 1, the input p is at the center of class i, μ_i.

Membership Functions with Bias Terms. In order to consider a more general form of membership functions, we add a bias term, b_i, to (17) as follows:

$$m_i(p) = 1 - \frac{d_i(p)}{\beta_i} - b_i. \tag{18}$$

Figure 2 shows membership functions with bias terms in the KDA space.

When we use membership functions given by (17) or (18), we need to determine the values of β_i or those of β_i and b_i. This process is called fuzzy rule tuning. This tuning procedure can be performed based on the same training algorithm as that of linear SVMs. This tuning method is described in Section 4.

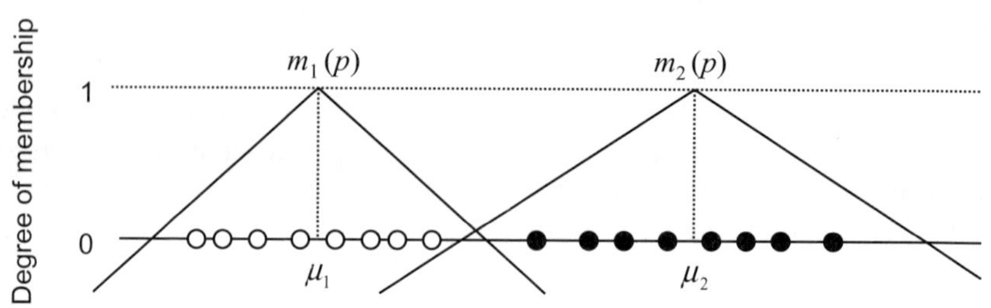

Fig. 1. Membership functions without bias terms in the KDA space

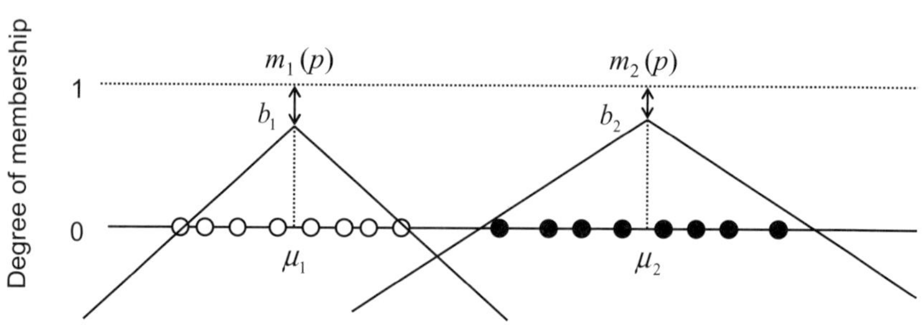

Fig. 2. Membership functions with a bias term in the KDA space

4 Fuzzy Rule Tuning

To tune membership functions of a fuzzy classifier, the steepest descent method is usually used. In [8], membership functions are tuned to maximize the recognition rate of the training data, by counting the numbers of correctly classified and misclassified data after tuning. Here, we consider using the same training algorithm as that of linear SVMs for fuzzy rule tuning of the classifier. In the following, we describe the tuning procedure.

First, we consider the membership function with a bias term given by (18). We define the function $L(p)$ as the difference of $m_1(p)$ and $m_2(p)$:

$$\begin{aligned} L(p) &= m_1(p) - m_2(p) \\ &= -\frac{d_1(p)}{\beta_1} + \frac{d_2(p)}{\beta_2} + (b_2 - b_1). \end{aligned} \tag{19}$$

Now we set $\boldsymbol{\beta} = (-\frac{1}{\beta_1}, \frac{1}{\beta_2})^T$, $\mathbf{d} = (d_1(p), d_2(p))^T$, and $b = b_2 - b_1$. Then (19) is rewritten as follows:

$$L(\mathbf{d}) = \boldsymbol{\beta}^T \mathbf{d} + b. \tag{20}$$

If the input datum p satisfies $L(\mathbf{d}) > 0$, the datum is classified into Class 1, and if $L(\mathbf{d}) < 0$, it is classified into Class 2. When $L(\mathbf{d}) = 0$, the datum is unclassifiable.

From the above discussion, (20) is equivalent to the decision function of a linear SVM in the two-dimensional space $(d_1(p), d_2(p))$. In the following, we call the two-dimensional space, $(d_1(p), d_2(p))$, *2-D tuning space*. If the point p is between μ_1 and μ_2 in the KDA space, the following equation is satisfied:

$$d_1(p) + d_2(p) = \mu \quad \text{for} \quad \mu_1 \le p \le \mu_2, \tag{21}$$

where μ is the Euclidean distance between μ_1 and μ_2. Namely, $\mu = |\mu_2 - \mu_1|$. Similarly, if point p is outside of the interval $[\mu_1, \mu_2]$, the following equation is satisfied:

$$|d_1(p) - d_2(p)| = \mu \quad \text{for} \quad p < \mu_1 \quad \text{or} \quad \mu_2 < p. \tag{22}$$

Hence, the relationship between $d_1(p)$ and $d_2(p)$ is described as follows:

$$d_2(p) = \begin{cases} -d_1(p) + \mu & \text{for} \quad \mu_1 \le p \le \mu_2, \\ d_1(p) \pm \mu & \text{for} \quad p < \mu_1 \text{ or } \mu_2 < p. \end{cases} \tag{23}$$

Therefore any point p in the KDA space is mapped on the reflexed line whose slopes are ± 1 in the 2-D tuning space, as shown in Fig. 3.

Classification using the decision function $L(\mathbf{d})$ in the 2-D tuning space is equivalent to that of fuzzy rules (15). Hence, given by training a linear SVM in the 2-D tuning space and calculating the weight vector $\boldsymbol{\beta}$ and bias term b, we can determine the parameters β_i and b_i. (But in this formulation, we cannot determine the values of b_1 and b_2 uniquely. However, because the classification boundary is invariant so long as $b = b_2 - b_1$ is constant, we can assume that either b_i is equal to 0).

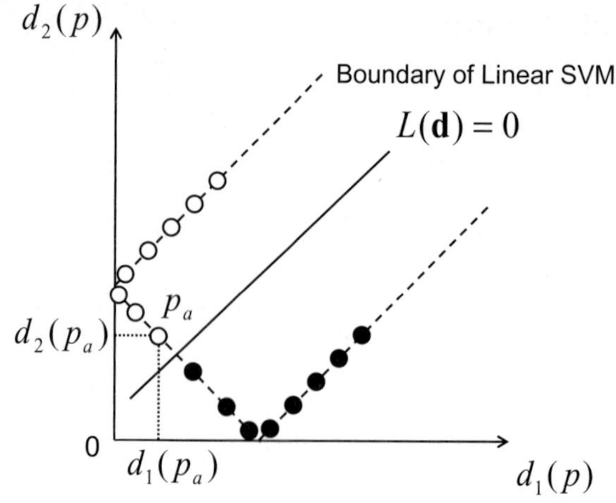

Fig. 3. 2-D tuning space

The decision function without a bias term in the 2-D tuning space is represented as

$$L(\mathbf{d}) = \boldsymbol{\beta}^T \mathbf{d}. \tag{24}$$

In this case, training a linear SVM without a bias term [12] in the 2-D tuning space, we can determine the parameters β_i. The decision boundary, $L(\mathbf{d}) = 0$, goes through the origin in the 2-D tuning space.

5 Performance Evaluation

We evaluated the proposed fuzzy classifier using two-class data sets used in [11]. As listed in Table 1, each problem has 100 training data sets and their corresponding test data sets.

Throughout the experiments, we used RBF kernels:

$$H(\mathbf{x}, \mathbf{x}') = \exp(-\gamma \|\mathbf{x} - \mathbf{x}'\|^2), \tag{25}$$

where γ is a positive parameter for controlling the radius, and we set $\varepsilon = 10^{-3}$ in (13) as used in [5].

We also show that we can easily analyze the behavior of the proposed fuzzy classifier by visualizing each class membership function.

5.1 Generalization Ability

We compared the generalization ability of the proposed fuzzy classifier and the SVM. In each method, we need to determine the value of kernel parameter γ and that of margin parameter C [1,2] of the SVM or the linear SVM used in the proposed classifier. In the experiments, we determine those parameters by fivefold cross-validation of the first five training data sets changing

Table 1. Benchmark data sets for two-class problems

Data	Inputs	Train	Test	Sets
Banana	2	400	4900	100
Heart	13	170	100	100
Ringnorm	20	400	7000	100
Thyroid	5	140	75	100
Twonorm	20	400	7000	100
Waveform	21	400	4600	100

Table 2. Parameter setting

Data	SVM		KDA-FC1		KDA-FC2	
	γ	C	γ	C	γ	C
Banana	15	100	15	50	15	1
Heart	0.1	50	1	100	0.5	5000
Ringnorm	15	1	0.5	8000	0.5	2000
Thyroid	15	100	15	50	10	500
Twonorm	0.5	1	0.1	10	0.5	1
Waveform	10	1	10	100	5	1

Table 3. Average recognition rates and standard deviations of the test data sets

Data	SVM	KDA-FC1	KDA-FC2
Banana	$\mathbf{89.3 \pm 0.5}$	88.1 ± 0.6	88.0 ± 0.6
Heart	83.7 ± 3.4	$\mathbf{83.8 \pm 3.3}$	83.4 ± 3.5
Ringnorm	97.8 ± 0.3	98.1 ± 0.2	$\mathbf{98.2 \pm 0.2}$
Thyroid	$\mathbf{96.1 \pm 2.1}$	95.7 ± 2.2	94.9 ± 2.2
Twonorm	$\mathbf{97.6 \pm 0.1}$	97.4 ± 0.3	96.1 ± 8.1
Waveform	90.0 ± 0.4	$\mathbf{90.4 \pm 0.3}$	90.1 ± 0.4

$\gamma = [0.1, 0.5, 1, 5, 10, 15]$ and $C = [1, 10, 50, 100, 500, 1000, 2000, 3000, 5000, 8000, 10000, 50000, 100000]$.

Table 2 lists the selected parameters by the above procedure. In Table 2, KDA-FC1 and KDA-FC2 denote the proposed fuzzy classifiers using membership functions without bias terms and with bias terms, respectively. For the optimal values of γ and C, we trained SVMs and the proposed two types of KDA-FCs for 100 training data sets and calculated the average recognition rates and the standard deviations for the test data sets. Table 3 shows the average recognition rates and the standard deviations for the two-class problems. The best result in the row is shown in boldface.

The KDA-FC1 showed the best performance for the heart and waveform data sets, and the KDA-FC1 performed better than the KDA-FC2 except for the ringnorm data sets. For the twonorm data set, the standard deviation of the KDA-FC2 is very large compared to those of other methods. Because the selected parameters by cross-validation were not suitable for several data sets among 100 sets, their performance degraded considerably. From this, we can say that the KDA-FC1 is more stable than the KDA-FC2. Except for the result of the KDA-FC2 for the twonorm data set, the performance of the KDA-FCs is comparable to that of SVMs.

5.2 Analysis of Classification

Here, we show that we can easily analyze the behavior of the KDA-FCs by visualizing each class membership function in the KDA space. Since KDA-FC1, without bias terms, performed better than KDA-FC2, here we analyze the behavior of KDA-FC1.

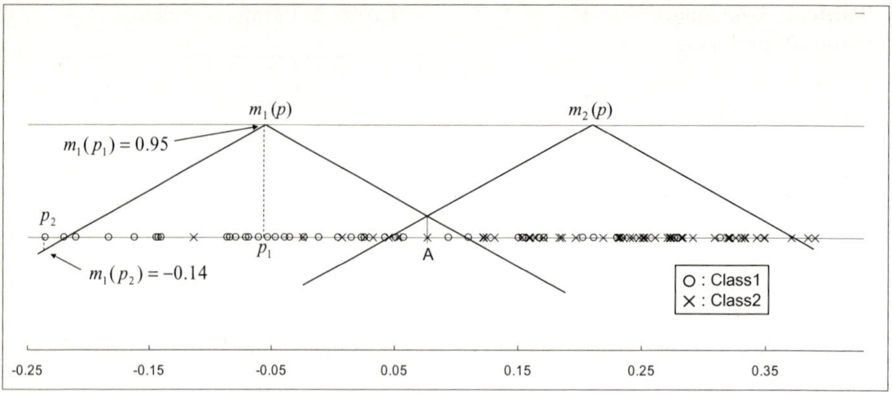

Fig. 4. Membership functions of the KDA-FC1 for the first heart test data set

Figure 4 shows the membership functions of the KDA-FC1 for the first heart test data set among 100 sets. In Fig. 4, $m_1(p)$ and $m_2(p)$ show the membership functions for Class 1 and Class 2, respectively, and their intersecting point A is a classification boundary in the KDA space. The datum p_1 belonging to Class 1 is correctly classified with $m_1(p_1) = 0.95$ and negative $m_2(p_1)$. Because $m_1(p_1)$ is almost 1 and $m_2(p_1)$ is negative, p_1 is considered to belong to Class 1 with high reliability. On the other hand, although the datum p_2 belonging to Class 1 is correctly classified, the degree of Class 1 membership is negative, i.e., $m_1(p_2) = -0.14$. Thus, the reliability of the classification result is considered to be low.

6 Conclusion

In this paper we discussed fuzzy classifiers based on KDA for two-class problems. In the proposed method, first we employ KDA to the given training data and calculate the component that maximally separates two classes in the feature space. Then, in the one-dimensional space obtained by KDA, we introduce a one-dimensional membership function for each class and generate classification rules. We showed that the tuning procedure of the membership functions reduced to the same training algorithm as that of linear SVMs.

From the computer experiments using benchmark data sets for two-class problems, we showed that the performance of the proposed fuzzy classifiers was comparable to that of SVMs. And by visualizing the membership functions in the KDA space, we showed that we could easily analyze the behavior of the proposed fuzzy classifiers.

References

1. Vapnik, V.N.: The Nature of Statistical Learning Theory. Springer, New York (1995)
2. Abe, S.: Support Vector Machines for Pattern Classification, Springer, London (2005)

3. Ruiz, A., López-de, T.P.E.: Nonlinear Kernel-Based Statistical Pattern Analysis. IEEE Transactions on Neural Networks 12(1), 16–32 (2001)
4. Schölkopf, B., Smola, A., Müller, K.R.: Nonlinear Component Analysis as a Kernel Eigenvalue Problem. Neural Computation 10, 1299–1319 (1998)
5. Mika, S., Rätsch, G., Weston, J., Schölkopf, B., Müller, K.-R.: Fisher Discriminant Analysis with Kernels. In: Hu, Y.-H., Larsen, J., Wilson, E., Douglas, S. (eds.) Neural Networks for Signal Processing. Proceedings of the 1999 IEEE Signal Processing Society Workshop, pp. 41–48 (1999)
6. Núñez, H., Angulo, C., Català, A.: Rule Extraction from Support Vector Machines. In: Proceedings of the Tenth European Symposium on Artificial Neural Networks (ESANN 2002), pp. 107–112 (2002)
7. Caragea, D., Cook, D., Honavar, V.: Towards Simple, Easy-to-Understand, yet Accurate Classifiers. In: Proceedings of the Third IEEE International Conference on Data Mining (ICDM 2003), pp. 497–500 (2003)
8. Kaieda, K., Abe, S.: KPCA-Based Training of a Kernel Fuzzy Classifier with Ellipsoidal Regions. International Journal of Approximate Reasoning 37(3), 145–253 (2004)
9. Abe, S.: Training of Kernel Fuzzy Classifiers by Dynamic Cluster Generation. In: Proceedings of the IEEE ICDM 2005 Workshop on Computational Intelligence in Data Mining (2005)
10. Abe, S.: Pattern Classification: Neuro-Fuzzy Methods and Their Comparison, Springer, London (2001)
11. Müller, K.-R., Mika, S., Rätsch, G., Tsuda, K., Schölkopf, B.: An Introduction to Kernel-Based Learning Algorithms. IEEE Transactions on Neural Networks, 181–201 (2001)
12. Vogt, M.: SMO Algorithms for Support Vector Machines without Bias Term. Technical report, Institute of Automatic Control, TU Darmstadt (2002)

An Efficient Search Strategy for Feature Selection Using Chow-Liu Trees

Erik Schaffernicht[1], Volker Stephan[2], and Horst-Michael Groß[1]

[1] Ilmenau Technical University
Department of Neuroinformatics and Cognitive Robotics, 98693 Ilmenau, Germany
[2] Powitec Intelligent Technologies GmbH, 45219 Essen-Kettwig, Germany
Erik.Schaffernicht@Tu-Ilmenau.de

Abstract. Within the taxonomy of feature extraction methods, recently the Wrapper approaches lost some popularity due to the associated computational burden, compared to Embedded or Filter methods. The dominating factor in terms of computational costs is the number of adaption cycles used to train the black box classifier or function approximator, e.g. a Multi Layer Perceptron. To keep a wrapper approach feasible, the number of adaption cycles has to be minimized, without increasing the risk of missing important feature subset combinations.

We propose a search strategy, that exploits the interesting properties of Chow-Liu trees to reduce the number of considered subsets significantly. Our approach restricts the candidate set of possible new features in a forward selection step to children from certain tree nodes. We compare our algorithm with some basic and well known approaches for feature subset selection. The results obtained demonstrate the efficiency and effectiveness of our method.

1 Introduction

If irrelevant features are used to adapt a Multi Layer Perceptron (MLP) to a certain task, the classifier has to handle a more complex decision surface. This leads to increased time requirements for a successful adaption, it may decrease the precision of the results and worsens the problem of overfitting. Therefore feature selection methods are applied to find and sort out the irrelevant features in the input data.

It is very common to characterize feature selection methods as "Filter", "Embedded" or "Wrapper" approaches (see [1] and [2]). Filter based approaches operate on the data to find intrinsic interrelations of the variables, prior to any application of a learning machine. On one hand, this includes data driven approaches like Principal Component Analysis or Non-Negative Matrix Factorization [3]. On the other hand, supervised methods are applied which investigate the correlation between the input data and the class labels or a target value. Examples are the linear correlation coefficient, Fisher discriminant analysis or information theoretic approaches.

Embedded methods use a specific learning machine, that is adapted with all data channels available. After the training process is complete, the importance

J. Marques de Sá et al. (Eds.): ICANN 2007, Part II, LNCS 4669, pp. 190–199, 2007.

of the inputs can be inferred from the structure of the resulting classifier. This includes e.g. weight pruning in neural network architectures with OBD [4], Bayes Neural Networks [5] or Random Forests [6].

The wrapper approach uses a learning machine, too, but the machine is arbitrary, since in this case it is considered a black box and the features selection algorithm wraps around the classifier, hence the name. A search strategy determines an interesting feature subset to train the learning machine. The resulting error rate is used as evaluation criterion in the subset search process. Since feature relevance and optimality with respect to the classification error rate is not always equivalent (as reported e.g. in [1]), it can be of advantage to use the same algorithm in the feature selection process and the classification task. The downside is, that this approach is prone to overfitting, a problem that has to be dealt with in additionally.

In a recent feature extraction competition (see results in [7]) the successful competitors used Embedded or Filter methods, while Wrappers were almost completely absent. In [7] the authors conclude, that Wrappers where omitted, not because of their capability, but their computational costs. Every time the search strategy determines a new candidate subset of features, the learning machine has to be adapted at least once, often even more, to produce reliable results. The time used to train the classifier is the dominating factor in terms of computational time. Since the used learning machine is considered a black box, it is not possible to optimize within the classifier without losing generality. Therefore, we aim to minimize the number of classifier evaluations imposed by the search algorithm without a significant increase of the risk of missing important feature subsets.

Our proposed method achieves this goal by constructing a Chow-Liu tree (CLT) [8] from the available data, see section 2. Then the obtained underlying tree structure is used by a forward search algorithm to create feature subsets. Through the inherent properties of the tree representation, the number of candidate subsets is considerably smaller, than e.g. in standard sequential forward selection methods [11]. As an positive side effect, possibly redundant features can be inferred directly from the CLT.

The main contribution of this work is the use of the CLT structure to minimize the number of evaluation steps. The paper is organized as follows. Section 2 explains the foundations of Chow-Liu trees, while the application of Chow-Liu trees in the context of feature selection is discussed in section 3. Some implication of the proposed method are discussed in section 4. Thereafter, we present some experimental results achieved with our method in comparison to other search strategies. Additionally, section 5 discusses related work of relevance, before we conclude in the final section.

2 Generation of Tree-Based Distributions

The basic idea of Chow-Liu trees (CLT) was presented in [8] and can be summarized as follows. In order to approximate a n-dimensional probability distribution, a first-order dependency tree with $n-1$ relationships is constructed. Within

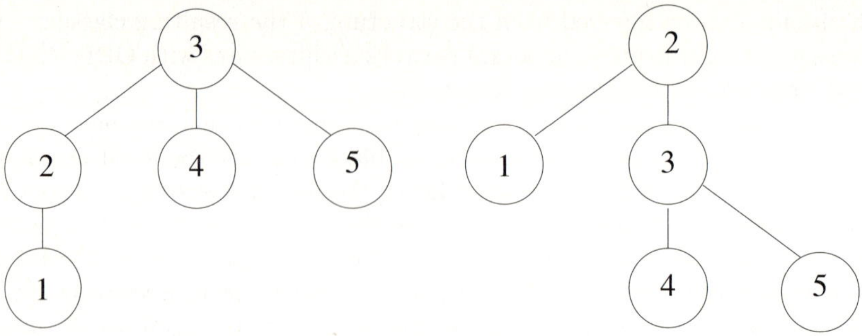

Fig. 1. Twice the same dependence tree with different root nodes. On the left the probability is expressed by $P(x) = P(x_3)P(x_4|x_3)P(x_5|x_3)P(x_2|x_3)P(x_1|x_2)$, on the right it is $P(x) = P(x_2)P(x_1|x_2)P(x_3|x_2)P(x_4|x_3)P(x_5|x_3)$.

the tree the underlying distribution is expressed as product of second-order distributions. The resulting representation can be used e.g. in pattern recognition tasks [8]. An example for a simple CLT is shown in figure 1. Please note that the choice of the root node is arbitrary. The algorithm to compute the tree structure minimizes the information difference between the original data and the dependency tree. It was shown, that the method is a maximum likelihood estimator for empirical data.

The problem of finding the optimal tree distribution is formulated as follow: Be $X = \{x^1, x^2, \dots, x^N\}$ the given samples data set in the features space F and we are looking for the tree T_{opt} that maximizes the log likelihood of the data:

$$T_{opt} = \arg\max_T \sum_{i=1}^{N} \log T(x^i) \tag{1}$$

The solution is obtained in three steps. The algorithm is outlined below:

Algorithm 1. CHOW-LIU-TREE(X)

Input: data set of observations X
Output: tree approximation T_{opt}

Determine the marginal distributions $P(x_i, x_j)$
Compute the mutual information matrix I
Compute the maximum-weight spanning tree T_{opt}

In the first part the pairwise mutual information I_{ij} between each pair of features $i, j \in F$ using the pairwise marginal distributions is computed:

$$I_{ij} = \sum_{x_i x_j} P(x_i, x_j) \log \frac{P(x_i, x_j)}{P(x_i)P(x_j)}, i \neq j \tag{2}$$

We used a histogram based approach to compute the pairwise marginal distributions and the mutual information, but kernel density estimation approaches are valid as well. For a more in depth discussion of different estimation methods, the interested reader is referred to [9].

The second part of the algorithm runs a maximum-weight spanning tree method on the mutual information matrix I, that is considered an adjacency matrix for this purpose. Beginning with the two nodes that have a maximum mutual information connection, further nodes are added with the next highest MI values. Any edges that would form a cycle are ignored. The resulting tree contains all nodes, while the sum of all weights of edges (which correspondes to the mutual information) in the tree is maximized. A modified Kruskal or Prim algorithm [10] can be applied for this task.

The obtained solution is non-ambiguous, if all the mutual information weights are different. Otherwise, if several weights are equal, the solution T_{opt} is possibly non-unique, but all alternatives still satisfy equation 1 and therefore this non-uniqueness property does not cause a problem.

In their work Chow and Liu showed, that the resulting dependence tree is indeed an optimal tree approximation of the underlying distribution.

3 Chow-Liu Trees for Feature Selection

We propose a supervised method for feature selection based on Chow-Liu trees. After further detailing our approach, we will discuss the benefits of using CLTs. We assume, that for each sample $x_i \in X$ we have a label $y_i \in Y$. We combine both information in a single matrix $Z = X \cup Y$, because for the purpose of constructing the tree, the labels are considered another input dimension. Then algorithm 1 is applied to compute the dependence tree. Each node of the tree now represents a feature or rather the label data.

Our algorithm uses the computed tree structure to guide the search process, that resembles the sequential forward selection strategy (SFS) (see chapter 4.3

Algorithm 2. SEQUENTIALFORWARDSELECTION(S, C, X, Y, E_S)

Input: data set of observations X, the corresponding labels Y, the current feature subset S, the candidate set of new features C, and the approximation error E_S for the subset S

Output: feature c_{best} to add to the feature subset

for $\forall c_i \in C$ **do**
  $E_i =$ TRAINCLASSIFIER($X, Y, S \cup c_i$)
end for
if $\exists E_i \in E; E_i + \varepsilon < E_S$ **then**
  $c_{best} = \arg\min_{c_i}(E)$
else
  $c_{best} = \emptyset$
end if

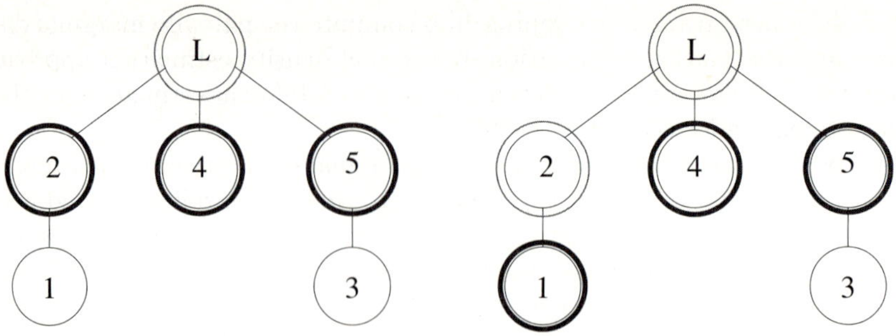

Fig. 2. On the left, the root (representing the label data) is the only member of the node set N, whose children form the candidate set. The SFS step is applied to the candidate set $C_1 = \{f_2, f_4, f_5\}$. The best improvement shall yield the inclusion of feature f_2, which is included in the feature set and is added to N (right). The new candidate set includes all children of f_2 $C_2 = \{f_1, f_4, f_5\}$.

in [7]). The basic SFS algorithm starts with an empty feature subset and adds one variable each step until a predefined number of features is reached, or the approximation result does not improve any further. For one step, each candidate is separatly added to the current subset and subsequently evaluated. The feature that induced the best improvement is included in the resulting subset. If the best new subsets improves more than a threshold ε, the new subset is returned, otherwise the algorithm terminates. A SFS step is shown in algorithm 2.

We consider the tree node, that represents the label data Y, as root instance of the dependence tree. All children of this node are treated as set of candidates C for a slightly modified SFS step. After this SFS step the chosen feature is added to the resulting feature subset. Besides this, the corresponding node in the tree is added to the set of nodes N, whose children are considered candidates for the next evaluation step. This is illustrated in figure 2.

Algorithm 3. FEATURE SELECTION WITH CLT(X, Y)

Input: data set of observations X and the corresponding labels Y
Output: feature subset S

$Z \leftarrow X \cup Y$
$T \leftarrow$ CHOW-LIU-TREE(Z)
$N \leftarrow t_y$ {start with the node corresponding to the label data}
$S \leftarrow \emptyset$ {start with empty feature subset}
repeat
  $C \leftarrow children(N)$ {all children of the current node set are candidates}
  $c \leftarrow$ SEQUENTIALFORWARDSELECTION(S, C, X, Y)
  $N \leftarrow N \cup c_{best} \cup c_{redundant}$ {add the best and possible redundant features to the search path}
  $S \leftarrow S \cup c_{best}$ {add the best node to feature set}
until $c_{best} = \emptyset$ AND $c_{redundant} = \emptyset$

The modification of the sequential forward selection is formed by the marking of features that do not improve the classification performance, since these nodes are either irrelevant or redundant features. The SFS step does not only return c_{best}, but any feature, whose inclusion did not improve the approximation performance more than threshold ε. In further candidate sets they are excluded, but added to the set of nodes N, so that their children in the tree are considered canidates in the next evaluation step.

The overall method is detailed in algorithm 3.

4 Discussion

In this section we will discuss the inherent advantages of using the computed tree structure to guide the search process. The Chow-Liu tree is constructed as maximum-weight spanning tree over the pairwise mutual information values of each feature and the labels. Let us assume, all features are statistically independent of each other, so there is no redundancy. A subset of these features contains information about the class label, so the mutual information between these ones and the label data is discriminatingly higher, than between any other pair. During the construction of the dependency tree, these features are connected to the label node, since this maximizes equation 1. All meaningful features are children of the root node. The labels that are irrelevant, are connected to any node, including the root node, with equal probability. Therefore using the CLT approach as filter method, stopping at this point and using only the children of the root node as features is not a good idea, because irrelevant inputs are possibly part of the children set.

Now consider adding redundant features f_1 and f_2 to the mix. The mutual information between one feature and the labels is the same as the joint mutual information between both features and the target value:

$$I(f_1, y) \approx I(f_1 \cup f_2, y) \tag{3}$$

They can be characterized by stating, that the mutual information between these feature f_1 and f_2 is greater than the mutual information between the features and the labels:

$$I(f_1, f_2) > \max(I(f_1, y), I(f_2, y)) \tag{4}$$

For this constellation of three nodes, the algorithm for constructing the maximum-weight spanning tree will always include the connection between the two features, since this maximizes the sum over the weights. Due to the tree structure, the root representing the labels can be connected to one of them only. This a plus in system identification tasks, because from the root's perspective, it is connected to the most informative feature and any features redundant to it are located in the same branch of the tree.

Any features that fulfill the condition of equation 4, but violate the redundancy condition 3, will be added to the same branch, even if they sustain new non-redundant information about the labels. Therefore the tree has to be searched down the branches. The search path has to include redundant features, because the informative feature is possibly connected to one.

From the sequential forward selection algorithm the CLT methods inherits the inability to detect any features that are useful only in combination with another (like the XOR problem [2]). The use of strategies to avoid this problem like sequential forward / backward floating selection (SFFS/SFBS) [11] are not very effective, because of the limited size of the candidate set. As a middle course we suggest using the proposed CLT method to construct a feature subset, that is used as starting point of a SBFS search afterwards. Preliminary results show that in this case the SBFS algorithm only performs very few steps, before it terminates as well.

Degenerated Trees

In the worst case, the tree is degenerated in such a way, that all nodes representing the input data are connected to the root node. All nodes contain information about the target and there is no significant dependency between the input channels. In this case the proposed method is reduced to the basic sequential forward selection method with the additional costs for tree construction, but such ill-conditioned input data indicates a problem that couldn't be solved by the means of feature selection methods. The maximum number of adaption cycles AC for the SFS-like subset search is given by

$$AC_{max} = \sum_{i=0}^{n_{sub}} (n_{all} - i) \, , n_{all} >= n_{sub}. \tag{5}$$

n_{all} is the number of all available features and n_{sub} is the number of features chosen for the final subset.

The other extreme case is a tree that has no splitting nodes, all features are lined up on a single path from the root to the only leaf. In terms of evaluation steps, this is optimal, since at each step the candidate set contains a single node only. So the minimum of adaption cycles is $AC_{min} = n_{all}$.

Both discussed cases are not common for real-world data and will occur in artificial datasets only. Typically, the obtained tree structures are somewhere in between the described extrema. The exact value depends on the underlying tree structure and the data interrelationship and is difficult to estimate. The average number of children per non-leaf node for the Spambase data set from UCI Machine Learning Repository [13] with 57 features is 2.48 with a variance of 5.72. For a number of different data sets ranging from 100 to 1000 features the average children per node is between 1.61 and 2.63, while the variance increased proportional to the amount of features. Hence for the average case, induced by the tree structure we conjecture an additional logarithmic dependency between the features and the number of adaption cycles, compared to AC_{min}.

5 Related Work and Experiments

The application of information theoretic measures like mutual information for feature selection was suggested before in several publications. The construction of classification and regression trees using the *Information Gain* criterion and their application in form of Random Forests [6] as Embedded method is an example.

A very similar idea to the Chow-Liu tree approach is the MIFS algorithm [12]. The MIFS criterion approximates the joint mutual information, between the features and the label data, which is difficult to estimate in high dimensional spaces. The method is a filter approach that evaluates the features with respect to the labels by computing the mutual information between those. Additionally the mutual information between the candidate and the previously chosen features in the subset is taken into account. The feature that maximizes the following term is added to the subset of features

$$\arg\max_{f_i}(I(f_i, y) - \beta \sum_{f_s \in S} I(f_i, f_s)). \tag{6}$$

The parameter β is used to balance the goals of maximizing relevance to the label and minimize the redundancy in the subset. Like the CLT method MIFS uses the pairwise relations of the features and the label data. The main difference is, that MIFS is used as data driven filter method, while the CLT approach is a wrapper using a black box classifier.

For a number of examples from the UCI repository [13] we compared the performance for the CLT, SFS and MIFS algorithms. As classifier we used a standard MLP with two hidden layers with 20 and 10 neurons respectively. After applying the feature extraction methods, the network was adapted using the selected features only. The balanced error rate

$$BER = \frac{1}{2}\left(\frac{false\ neg}{false\ neg + true\ pos} + \frac{false\ pos}{false\ pos + true\ neg}\right) \tag{7}$$

for the classification problems was calculated using 10-fold cross-validation. This measure accounts for any unbalanced class distributions. For comparison we adapted a network with all available features, too.

The stopping criteria for SFS and CLT were identical, see section 3. For the MIFS algorithm we introduced an additional random channel, independent from the labels and the rest of the features. The feature selection was stopped when the algorithm attempted to add this probe to the subset. In our tests we used $\beta = 0.15$.

The results for the experiments are shown in Table 1.

The MIFS algorithm shows a mixed performance. Given that it is a filter approach, MIFS does not have the advantage of using the classifier itself. Thus the information theoretic approach yields features, that are not optimal in every case for the training of the MLP. This seems to be the case for some examples (see the results for the Ionosphere data set), but not all data sets. The number

Table 1. The results for the different data sets and feature selection methods. The balanced error rate is given in percent. The number of chosen features and the number of evaluation steps are shown in parentheses.

Data set	Features f	Samples n	All	CLT	SFS	MIFS
			balanced error rate(features/evaluation steps)			
Ionosphere	34	351	20.08(34/-)	18.12(6/38)	18.47(3/130)	24.54(5/-)
Spambase	57	4601	13.81(57/-)	17.26(9/97)	17.39(8/477)	16.29(18/-)
GermanCredit	24	1000	41.70(24/-)	38.52(3/24)	39.06(4/110)	37.47(6/-)
Breast Cancer	30	569	13.78(30/-)	9.37(8/37)	13.44(4/140)	12.48(5/-)

of chosen features is higher, than for both wrapper approaches. Given its nature as filter approach, MIFS is the fastest algorithm considered.

CLT tends to produce smaller error rate results compared to the SFS algorithm, while the size of the feature set chosen by CLT is slightly higher. This observation seems to be somewhat counterintuitive, although both approaches act greedy when choosing the next feature, the difference is, that SFS does its selection on the global level, while CLT choses on a local level (the current candidate set). This can help avoiding local minima in the search, but comes at the cost of adding more features to the subset. The real advantage becomes clear if the number of evaluation steps is compared. The CLT algorithm performs only fractions of adaption cycles needed by the SFS method (given by equation 5).

6 Conclusion

We proposed a search strategy based on Chow-Liu trees for feature selection methods. The tree structure is utilized in a forward search strategy by restricting the candidate sets to the children of certain nodes in the tree. This way, some advantages of the information theoretic approach used to construct the tree are exploited.

This results in a significant reduction of performed evaluation steps in a wrapper feature selection strategy compared to standard methods like sequential forward selection, while retaining a similar performance error. Compared to the MIFS approach, a filter method using the mutual information in a similar way, the results for CLT based feature selection are slightly better, but in terms of computational costs the MIFS algorithm is cheaper. Within the domain of wrapper approaches the speed of the CLT based feature selection method is significant.

References

1. Kohavi, R., John, G.H.: Wrappers for feature subset selection. Artifical Intelligence 97, 273–324 (1997)
2. Guyon, I., Elisseeff, A.: An introduction to variable and feature selection. Journal of Machine Learning Research 3, 1157–1182 (2003)

3. Lee, D.D., Seung, H.S.: Algorithms for Non-negative Matrix Factorization. Advances in Neural Information Processing Systems, vol. 13, pp. 556–562. MIT Press, Cambridge, MA (2001)
4. LeCun, Y., Denker, J., Solla, S., Howard, R.E., Jackel, L.D.: Optimal Brain Damage. Advances in Neural Information Processing Systems, vol. 2. Morgan Kaufmann, San Francisco (1990)
5. Neal, R.M.: Bayesian Learning for Neural Networks. Springer, Heidelberg (1996)
6. Breiman, L.: Random Forests. Machine Learning 45, 5–32 (2001)
7. Guyon, I., Gunn, S., Nikravesh, M., Zadeh, L.: Feature Extraction: Foundations and Applications. Studies in fuzziness and soft computing, vol. 207. Springer, Heidelberg (2006)
8. Chow, C.K., Liu, C.N.: Approximating Discrete Probability Distributions with Dependence Trees. IEEE Transactions on Information Theory 14, 462–467 (1968)
9. Scott, D.W.: Multivariate density estimation: theory, practice, and visualization. John Wiley & Sons, New York (1992)
10. Cormen, T.H., Leierson, C.E., Rivest, R.L., Stein, C.: Introduction to Algorithms, 2nd edn. MIT Press, Cambridge, MA (2001)
11. Reunanen, J.: Search Strategies. In: Feature Extraction: Foundations and Applications. Studies in fuzziness and soft computing, ch. 4, vol. 207, Springer, Heidelberg (2006)
12. Battiti, R.: Using mutual information for selecting features in supervised neural net learning. IEEE Transactions on Neural Networks 5(4), 537–550 (1994)
13. Newman, D.J., Hettich, S., Blake, S.L., Merz, C.J.: UCI Repository of machine learning databases (1998), `http://www.ics.uci.edu/~mlearn/MLRepository.html`

Face Recognition Using Parzenfaces

Zhirong Yang and Jorma Laaksonen

Laboratory of Computer and Information Science*
Helsinki University of Technology
P.O. Box 5400, FI-02015 TKK, Espoo, Finland
{zhirong.yang,jorma.laaksonen}@tkk.fi

Abstract. A novel discriminant analysis method is presented for the face recognition problem. It has been recently shown that the predictive objectives based on Parzen estimation are advantageous for learning discriminative projections if the class distributions are complicated in the projected space. However, the existing algorithms based on Parzen estimators require expensive computation to obtain the gradient for optimization. We propose here an accelerating technique by reformulating the gradient and implement its computation by matrix products. Furthermore, we point out that regularization is necessary for high-dimensional face recognition problems. The discriminative objective is therefore extended by a smoothness constraint of facial images. Our Parzen Discriminant Analysis method can be trained much faster and achieve higher recognition accuracies than the compared algorithms in experiments on two popularly used face databases.

1 Introduction

Face Recognition (FR) is becoming an even more active research topic in the forthcoming years. The challenge of FR is at first induced by the high dimensionality of facial images. The problem is more challenging in presence of structured variations such as poses and expressions, which are difficult to be modeled and cause the data to distribute in a complicated manifolds. Therefore, the research in this field is not only useful for classifying faces, but also conducive to other high-dimensional pattern recognition problems.

A substantial amount of efforts has been devoted to the FR problem, among which Fisher's *Linear Discriminant Analysis* is widely used. Modeling each class by a single Gaussian distribution which shares a common covariance, LDA maximizes the Fisher criterion of between-class scatter over within-class scatter and can be solved by *Singular Value Decomposition* (SVD). The facial feature extraction by LDA is called *Fisherfaces* [1]. The Fisherface method is attractive for its simplicity, but the assumption of Gaussians with common variance heavily restricts its performance. Moreover, Fisherface requires preprocessing by *Principal*

* Supported by the Academy of Finland in the projects *Neural methods in information retrieval based on automatic content analysis and relevance feedback* and *Finnish Centre of Excellence in Adaptive Informatics Research.*

J. Marques de Sá et al. (Eds.): ICANN 2007, Part II, LNCS 4669, pp. 200–209, 2007.

Component Analysis (PCA) and the discriminative information may however be lost during the unsupervised dimensionality reduction. Later many variants of Fisherface such as [2] have been proposed. However, the Fisherface method and its variants make use of only the first- and second-order statistics of the class distributions while discarding the higher-order statistics.

Recently Goldberger et al. [3] proposed *Neighborhood Component Analysis* (NCA) which learns a linear transformation matrix by maximizing the summed likelihood of the labeled data. The probability density at each data point is estimated by using the neighbors in the transformed space, which turns out to be the Parzen estimation of the posterior of the class label. Peltonen and Kaski later proposed a very similar method called *Informative Discriminant Analysis* (IDA) [4], in which they instead employ log-likelihood, i.e. the information of predictive probability density. The likelihood formulation allows NCA and IDA to model very complicated class distributions. It was reported that these two methods outperform traditional discriminant analysis approaches in a number of low-dimensional supervised learning problems. However, the optimization of NCA or IDA requires the gradient of the Parzen-based objective, the computation of which is too expensive for most applications. To obtain an orthonormal transformation matrix, IDA employs a reparameterization based on Givens rotation, which even aggravates the computation and prevents its application to high-dimensional data. Peltonen et al. later proposed a modified version [5] to speed up the computation by using a small number Gaussian mixtures instead of the Parzen method. This nevertheless loses the advantage of nonparametric estimation. One has to insert additional EM iterations before computing the gradient, and how to select an appropriate number of Gaussians is unclear.

In this paper we point out that the computational burden of calculating the gradient in NCA and IDA can be significantly reduced by using matrix multiplication. Next, the Givens reparameterization in IDA can be replaced by geodesic updates in the Stiefel manifold, which further simplifies the optimization. Furthermore, we propose to regularize the projection matrix by employing a smoothness constraint. This is done by introducing an additional penalization term of local pixel variance. We name the new method as *Parzenface* when applying our discriminant analysis to the face recognition problem. The experiments on two public facial image databases, FERET [6] and ORL [7], demonstrate that our learning algorithm can achieve higher accuracy and run much faster than NCA and IDA.

2 Parzen Discriminant Analysis

2.1 Unregularized Objective

Consider a supervised data set which consists of pairs $(\mathbf{x}_j, c_j)$, $j = 1, \ldots, n$, where $\mathbf{x}_j \in \mathbb{R}^m$ is the primary data, and the auxiliary data c_j takes categorical values. We seek for an $m \times r$ orthonormal matrix $\mathbf{W}$ by which the primary data $\mathbf{x}_i$ are projected into a lower-dimensional space. The objective is to maximize the discriminative information, i.e. the sum of predictive log-likelihood

$$\mathcal{J}(\mathbf{W}) = \sum_{i=1}^{n} \log p(c_i|\mathbf{y}_i) = \sum_{i=1}^{n} \log \frac{p(\mathbf{y}_i|c_i)}{p(\mathbf{y}_i)} + \sum_{i=1}^{n} \log p(c_i) \tag{1}$$

in the projected space, where $\mathbf{y}_i = \mathbf{W}^T \mathbf{x}_i$.

If we estimate $p(\mathbf{y}_i|c_i)$ and $p(\mathbf{y}_i)$ by the Parzen window method, the objective becomes

$$\mathcal{J}(\mathbf{W}) = \sum_{i=1}^{n} \log \frac{\sum_{j:c_i=c_j} e_{ij}}{\sum_{j=1}^{n} e_{ij}} + \text{const} = \sum_{i=1}^{n} \mathcal{J}_i + \text{const}, \tag{2}$$

where $\mathcal{J}_i$ is the shorthand notation for $\log\left[\left(\sum_{j:c_i=c_j} e_{ij}\right) / \left(\sum_{j=1}^{n} e_{ij}\right)\right]$ and

$$e_{ij} = \begin{cases} \exp\left(-\frac{\|\mathbf{y}_i - \mathbf{y}_j\|^2}{2\sigma^2}\right) & \text{if } i \neq j \\ 0 & \text{if } i = j \end{cases} \tag{3}$$

with σ a positive parameter which controls the Gaussian window width.

2.2 Computing the Gradient

Our optimization algorithm is based on the gradient of $\mathcal{J}(\mathbf{W})$ with respect to $\mathbf{W}$

$$\nabla \equiv \nabla_{\mathbf{W}} \mathcal{J} = \sum_{i=1}^{n} \frac{\partial \mathcal{J}_i}{\partial \mathbf{W}} = \sum_{i=1}^{n} \sum_{j=1}^{n} \frac{\partial \mathcal{J}_i}{\partial \|\mathbf{y}_j - \mathbf{y}_i\|^2} \cdot \frac{\partial \|\mathbf{y}_j - \mathbf{y}_i\|^2}{\partial \mathbf{W}}. \tag{4}$$

Notice that the chain rule in the inner summation applies to the subscript j, i.e. treating $\mathbf{y}_j$ as an intermediate variable and $\mathbf{y}_i$ as a constant. Denote

$$G_{ij} \equiv \frac{\partial \mathcal{J}_i}{\partial \|\mathbf{y}_j - \mathbf{y}_i\|^2} \tag{5}$$

for notational simplicity. The gradient then becomes

$$\nabla = \sum_{i=1}^{n} \sum_{j=1}^{n} G_{ij} \left(\mathbf{x}_i - \mathbf{x}_j\right) \left(\mathbf{x}_i - \mathbf{x}_j\right)^T \mathbf{W} \tag{6}$$

Direct computation of ∇ by going through all the vector pairs is too expensive because it costs $O(n^2 m^2)$ time. However, careful examination of the formula reveals that the computation can be significantly reduced:

$$\sum_{i=1}^{n} \sum_{j=1}^{n} G_{ij} \left(\mathbf{x}_i - \mathbf{x}_j\right) \left(\mathbf{x}_i - \mathbf{x}_j\right)^T \tag{7}$$

$$= 2 \left(\sum_{i=1}^{n} \sum_{j=1}^{n} \mathbf{x}_i G_{ij} \mathbf{x}_i^T - \sum_{i=1}^{n} \sum_{j=1}^{n} \mathbf{x}_i G_{ij} \mathbf{x}_j^T \right) \tag{8}$$

$$= 2\left(\sum_{i=1}^{n} \mathbf{x}_i D_{ii} \mathbf{x}_i^T - \sum_{i=1}^{n}\sum_{j=1}^{n} \mathbf{x}_i G_{ij} \mathbf{x}_j^T\right) \tag{9}$$

$$= 2\left(\mathbf{X}\mathbf{D}\mathbf{X}^T - \mathbf{X}\mathbf{G}\mathbf{X}^T\right) \tag{10}$$

$$= 2\mathbf{X}(\mathbf{D} - \mathbf{G})\mathbf{X}^T, \tag{11}$$

where $\mathbf{X} = [\mathbf{x}_1, \mathbf{x}_2, \ldots, \mathbf{x}_n]$ and $\mathbf{D}$ is a diagonal matrix with $D_{ii} = \sum_{j=1}^{n} G_{ij}$. That is, the gradient can be computed by matrix operations as

$$\nabla = 2\mathbf{X}(\mathbf{D} - \mathbf{G})\mathbf{X}^T\mathbf{W}. \tag{12}$$

It is known that there exist fast algorithms that implement matrix multiplication in $O(\tau^q)$ time where $\tau = \min(m, n)$ and q a positive scalar less than 3 and towards 2 [8]. Many researchers believe that an optimal algorithm will run in essentially $O(\tau^2)$ time [9]. In practice, if the matrix multiplication is accelerated via the *Fast Fourier Transformation* (FFT), the computation of the gradient (12) can be accomplished in $O(\tau^2 \log \tau)$ time [8], which is already acceptable for most applications.

2.3 Geodesic Flows on the Stiefel Manifold

Orthonormality of the transformation matrix is preferred in feature extraction because it enforces the matrix to encode the intrinsic subspace in the most economic way. The orthonormality constraint also prevents the learning algorithm from falling into some trivial local minima. In addition, an orthonormal matrix as the learning result is convenient for us to compare the new method with many existing projective methods used in face recognition.

The set of $m \times r$ real orthonormal matrices forms a Stiefel manifold $\mathrm{St}(m, r)$. Given the gradient ∇ at $\mathbf{W}$, it has been shown [10] that the natural gradient in such a manifold is given by

$$\mathrm{grad}_{\mathbf{W}}^{\mathrm{St}(m,r)} \mathcal{J} = \nabla - \mathbf{W}\nabla^T\mathbf{W}, \tag{13}$$

and an approximated geodesic learning flow with the starting point $\mathbf{W}$ by

$$\mathbf{W}_{\text{new}} = \mathrm{expm}\left(t\left(\nabla\mathbf{W}^T - \mathbf{W}\nabla^T\right)\right)\mathbf{W}, \tag{14}$$

where expm represents the matrix exponential and t a usually small positive learning rate.

2.4 Regularization

An orthonormal matrix has $(m - r)r + r(r - 1)/2$ free parameters [11]. If this number is comparable to or larger than the number of samples n, the discriminant analysis problem probably becomes ill-posed. Unfortunately this is the case in face recognition, especially when the facial images are sampled in high resolutions. The learning objective must therefore be regularized.

However, simple L_2-norm used in e.g. *Support Vector Machines* is not suitable for penalization here because summing the squared entries of an $m \times r$ orthonormal matrix results in a constant r. One thus has to use some other regularization techniques.

Notice that each column of $\mathbf{W}$ acts as a linear filter and can be displayed like a *filter image*. It is a crucial observation we have made that many overfitting projection matrices have highly rough filter images. That is, local contrastive pixel groups dominate the filters, but they are too small to represent any relevant patterns for face recognition. This motivates us to adopt a penalization term $\mathrm{Tr}\left(\mathbf{W}^T \boldsymbol{\Omega} \mathbf{W}\right)$ [12] to emphasize the smoothness prior of images, where the constant matrix $\boldsymbol{\Omega}$ is constructed by

$$\Omega_{st} = \mathcal{N}(d(s,t);\rho). \tag{15}$$

Here $d(s,t)$ is the 2-D Euclidean distance of the locations s and t, and $\mathcal{N}$ the zero-mean normal distribution. The variance parameter ρ controls the neighborhood size and its value depends on the resolution of the facial images used. We find that $\rho \in (0.3, 0.8)$ works fine in our experiments with 32×32- and 23×28-sized images. It is not difficult to see that $\mathrm{Tr}(\mathbf{W}^T \boldsymbol{\Omega} \mathbf{W})$ is an approximated version of the Laplacian used in [12].

By attaching regularization term, we define the objective of *Parzen Discriminant Analysis* (PDA) to be the maximum of

$$\mathcal{J}_{\mathrm{PDA}}(\mathbf{W}) = \frac{1}{2}\sum_{i=1}^{n} \log \frac{\sum_{j:c_i=c_j} e_{ij}}{\sum_{j=1}^{n} e_{ij}} - \frac{1}{2}\lambda \mathrm{Tr}\left(\mathbf{W}^T \boldsymbol{\Omega} \mathbf{W}\right), \tag{16}$$

where λ is a positive parameter that controls the balance between discrimination and smoothness. The optimization of PDA is based on the gradient

$$\tilde{\nabla} = \mathbf{X}(\mathbf{D} - \mathbf{G})\mathbf{X}^T\mathbf{W} - \lambda \boldsymbol{\Omega} \mathbf{W}. \tag{17}$$

In the following experiments, we use the approximated geodesic update

$$\mathbf{W}_{\mathrm{new}} = \mathrm{expm}\left(t\left(\tilde{\nabla}\mathbf{W}^T - \mathbf{W}\tilde{\nabla}^T\right)\right)\mathbf{W}. \tag{18}$$

We name our new method *Parzenface* when the Parzen Discriminant Analysis is applied to the face recognition problem. The term Parzenface also refers to the extracted features of facial images by using PDA.

3 Connections to Previous Work

Fisherface is a combined method which applies Fisher's *Linear Discriminant Analysis* (LDA) on the results of *Principal Component Analysis* (PCA). Fisherface and its variants are attractive because they have closed-form solutions which can be obtained by (generalized) singular value decomposition. However, these methods model each subject class by a single Gaussian class, which heavily restricts their generalization in presence of different facial expressions, face poses and illumination conditions. In fact, these structural variabilities cause a subject class to stretch in a curved but non-Gaussian manifold.

Recently some unsupervised methods such as Laplacianfaces [13] have been proposed to unfold the structures of the face manifolds. Although it was reported that they have better recognition accuracy in some cases, the discriminative performance of these methods is naturally limited because they omit the supervised information.

There exist two gradient-based approaches that are closely related to our PDA method. *Neighborhood Component Analysis* (NCA) [3] learns a transformation matrix (not necessarily orthonormal) to maximize the *leave-one-out* (loo) performance of nearest neighbor classification. NCA measures the performance based on "soft" neighbor assignments in the transformed space, which is similar to the PDA objective except the logarithm function is dropped. However, without the logarithm function, NCA lacks the connection to the information theory. By contrast, PDA conforms the general assumption that the samples are *independently and identically distributed* (i.i.d.). Here the "independence" refers to the predictive version

$$p(\{c_i\}_{i=1}^n|\{\mathbf{y}_i\}_{i=1}^n) = \prod_{i=1}^n p(c_i|\mathbf{y}_i), \tag{19}$$

of which the maximization is equivalent to that of the PDA unregularized objective (1). In addition, the loss of orthogonality may cause NCA to fall in some trivial local optima, for example, all columns of the transformation matrix converging to a same vector.

Informative Discriminant Analysis (IDA) has the same objective as the unregularized one in Section 2.1. Peltonen and Kaski have shown that such predictive likelihood has asymptotic connection to the mutual information criterion and the learning metrics [4]. It has however been shown that the unregularized objective is prone to overfitting in high-dimensional cases [14], as will also be illustrated by experiment results in the next section. A significant drawback of IDA or NCA is their slow implementation of the gradient computing ((4) in [4] and (5) in [3]). The $O(n^2m^2)$ running time restricts IDA and NCA to only small-scale databases, which is nevertheless infeasible for face recognition. Another expensive computation in IDA is induced by the reparameterization by Givens rotation, which involves the trigonometric functions and further complicates the gradient computation. In contrast, Parzenface is a fast approach to compute the gradient by matrix multiplications and the updates within the Stiefel manifold naturally maintain the orthonormality.

4 Experiments

4.1 Data

We have compared PDA and five other methods on two databases of facial images. The first data set contains facial images collected under the FERET program [6]. 2409 frontal facial images (poses "fa" and "fb") of 867 subjects were stored in

Fig. 1. The sample images from (top) FERET and from (bottom) ORL databases

the database after face segmentation. In this work we obtained the coordinates of the eyes from the ground truth data of the collection, with which we calibrated the head rotation so that all faces are upright. Afterwards, all face boxes were normalized to the size of 32×32, with fixed locations for the left eye (26,9) and the right eye (7,9). The second data set comes from the ORL database [7] which includes 400 facial images of 40 subjects. There are 10 images taken in various poses and expressions from each subject. We resize the ORL images to the size of 23×28 without further normalization. Example images from FERET and ORL are displayed in Figure 1. We divide the images of each subject into two parts of equal size, the first half for training and the rest for testing. The whole training set is the union of the training part of all subjects, and so is the whole testing set.

4.2 Training Time

A major advantage of PDA (18) over its close cousins NCA [3] and IDA [4] is that PDA requires much less training time. We demonstrate this by running the compared algorithms on a Linux machine with 12GB RAM and two 64-bit 2.2GHz AMD Opteron processors.

We set the number of iterations for PDA and NCA to 10, and $10 \times n$ for IDA since IDA employs an online learning based on stochastic gradients. In this way all algorithms go through a same number of training samples. We repeated such training ten times and recorded the total time used in Table 1. It is easy to see that PDA significantly outperforms NCA and IDA in efficiency. PDA requires about 1/22 training time of NCA and 1/25 of IDA. The advantage is more obvious for the FERET database of larger scale, where PDA is almost 84 times and 100 times faster than NCA and IDA, respectively.

4.3 Visualizing the Filter Images

It is intuitive to inspect the elements in the trained projection matrix **W** before performing quantitative evaluation. If the projection matrix works well for the FR problem, it is expected to find some semantic connections between the filter images and our common prior knowledge about facial images.

Table 1. Training time of PDA and IDA on the facial image databases (in seconds)

database	PDA	IDA	NCA
FERET ($n = 1208, m = 1024, r = 10$)	3,733	362,412	313,811
ORL ($n = 200, m = 644, r = 10$)	1,502	40,136	32,914

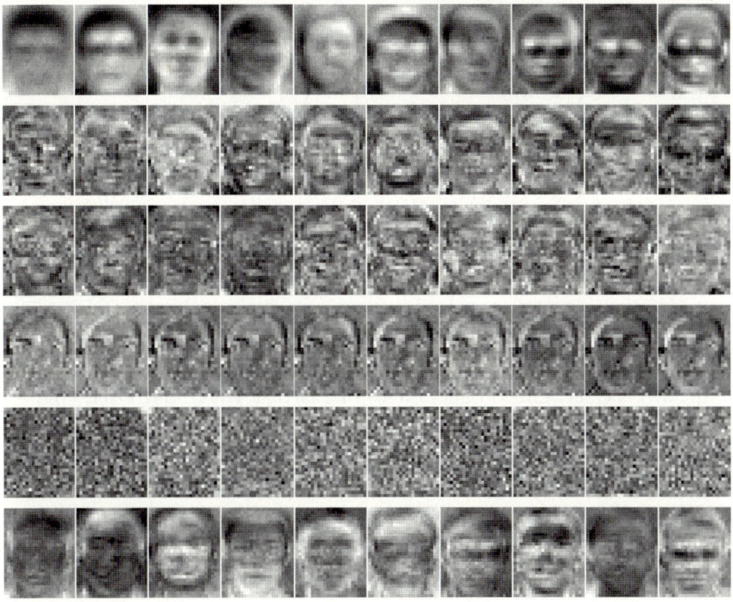

Fig. 2. The first ten filter images of the ORL database using (from top to bottom) Eigenface, Laplacianface, Fisherface, IDA and Parzenface

Figure 2 shows the first ten filter images of five compared methods, where the top two are unsupervised and the bottom four supervised. We only plot the ORL results due to space limit. The filter images of IDA contain almost random pixels and there seems no pattern related to faces. This is probably because IDA starts from the LDA projection matrix, but the latter suffers from data scarce in face recognition. Fisherface and Laplacianface are better than IDA since one can slightly perceive some contractive parts around or within the head-like boundary. These parts however are too small and scattered all over every filter image, which might cause overfitting of the projection matrix, e.g. being sensitive to small shifts and variation. The contrastive parts of the NCA basis mainly lie around the head, but these filters differ only in some tiny regions. This is probably caused by the removal of the orthogonal constraint in NCA. By contrast, Parzenface yields filter images that contain clearer facial semantics and are hence easier for interpretation. For example, the fourth filter image is likely related to the beard feature and the fifth may control the head shape. The filter images of Eigenface also comprise some facial parts like eyes and chins, which albeit are more blurred and may lead to underfitting for face recognition.

4.4 Face Recognition Accuracies

Classification of the testing faces is performed in the projected space by using the nearest neighbor classifier. The face recognition accuracies with r ranging from 10 to 70 are shown in Figure 3. Since the maximum output dimensionality

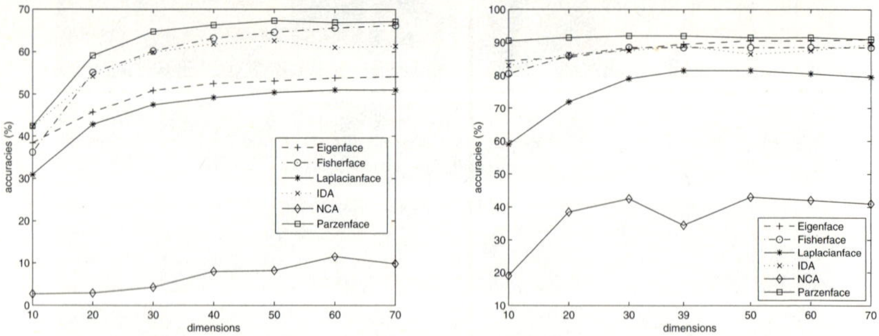

Fig. 3. Face recognition accuracies of FERET (left) and ORL (right)

of Fisherface is the number of classes minus one, i.e. 39 for the ORL database, we set a tick at 39 in the x-axis of the right plot for better comparison.

We found that NCA heavily suffers from the overfitting problem. Although it can achieve excellent classification accuracies for the training set, the NCA transformation matrix generalizes poorly to the testing data. Laplacianface performs the second worst. This is probably because it requires a large amount of data to build a reliable graph of locality, which is infeasible in our experiments. This drawback is more severe for the ORL database which contains facial images of different poses. The performance order of Eigenface, Fisherface and IDA depends on the database used. Fisherface is the best among these three for FERET while Eigenface is the best one for ORL.

The Parzenface learning can start from any orthonormal matrix. We set the initial matrix to be the one produced by its best opponent, i.e. Fisherface for FERET and Eigenface for ORL. The parameter σ in (3) and λ in (16) were obtained by cross-validation using the training set. The face recognition accuracies were then calculated by applying the trained Parzenface model to the testing set. From Figure 3 we can see that the face recognition accuracies using Parzenfaces are superior to all the other compared methods.

5 Conclusion

We have presented a new discriminant analysis method and applied it to the face recognition problem. The proposed Parzenface method overcomes two major drawbacks of existing gradient-based discriminant analysis methods by using information theory. Firstly the computation of the gradient can be greatly accelerated by using matrix multiplication instead of going through all the pairwise differences. Secondly we have proposed to employ the smoothness constraint of images to regularize the face recognition problem. The empirical study on two popular facial image databases shows that Parzenface requires much less training time than the IDA and NCA methods while achieving higher face recognition accuracies than the other compared methods.

In this paper we focused on feature extraction by linear projections, but our method can readily be generalized to the nonlinear version by using the kernel extension. Moreover, the optimization is not restricted to the gradient ascend flows. The convergence could be further improved by employing more advanced techniques, such as conjugate gradients, in Riemannian manifolds.

References

1. Belhumeur, P., Hespanha, J., Kriegman, D.: Eigenfaces vs. fisherfaces: recognition using class specific linear projection. IEEE Transactions on Pattern Analysis and Machine Intelligence 19(7), 711–720 (1997)
2. Howland, P., Wang, J., Park, H.: Solving the small sample size problem in face recognition using generalized discriminant analysis. Pattern Recognition 39(2), 277–287 (2006)
3. Goldberger, J., Roweis, S., Hinton, G., Salakhutdinov, R.: Neighbourhood components analysis. Advances in Neural Information Processing 17, 513–520 (2005)
4. Peltonen, J., Kaski, S.: Discriminative components of data. IEEE Transactions on Neural Networks 16(1), 68–83 (2005)
5. Peltonen, J., Goldberger, J., Kaski, S.: Fast discriminative component analysis for comparing examples. In: NIPS (2006)
6. Phillips, P.J., Moon, H., Rizvi, S.A., Rauss, P.J.: The FERET evaluation methodology for face recognition algorithms. IEEE Trans. Pattern Analysis and Machine Intelligence 22, 1090–1104 (2000)
7. Harter, F.S.A.: Parameterisation of a stochastic model for human face identification. In: Proceedings of the Second IEEE Workshop on Applications of Computer Vision, pp. 138–142 (1994)
8. Horn, R., Johnson, C.: Topics in Matrix Analysis. Cambridge (1994)
9. Robinson, S.: Toward an optimal algorithm for matrix multiplication. SIAM News 38(9) (2005)
10. Nishimori, Y., Akaho, S.: Learning algorithms utilizing quasi-geodesic flows on the Stiefel manifold. Neurocomputing 67, 106–135 (2005)
11. Edelman, A.: The geometry of algorithms with orthogonality constraints. SIAM J. Matrix Anal. Appl. 20(2), 303–353 (1998)
12. Hastie, T., Buja, A., Tibshirani, R.: Penalized discriminant analysis. The Annals of Statistics 23(1), 73–102 (1995)
13. He, X., Yan, S., Hu, Y., Niyogi, P., Zhang, H.J.: Face recognition using Laplacianfaces. IEEE Transactions on Pattern Analysis And Machine Intelligence 27, 328–340 (2005)
14. Yang, Z., Laaksonen, J.: Regularized neighborhood component analysis. In: Proceedings of 15th Scandinavian Conference on Image Analysis (SCIA), Aalborg, Denmark, pp. 253–262 (2007)

A Comparison of Features in Parts-Based Object Recognition Hierarchies

Stephan Hasler, Heiko Wersing, and Edgar Körner

Honda Research Institute Europe GmbH
D-63073 Offenbach/Germany
stephan.hasler@honda-ri.de

Abstract. Parts-based recognition has been suggested for generalizing from few training views in categorization scenarios. In this paper we present the results of a comparative investigation of different feature types with regard to their suitability for category discrimination. So patches of gray-scale images were compared with SIFT descriptors and patches from the high-level output of a feedforward hierarchy related to the ventral visual pathway. We discuss the conceptual differences, resulting performance and consequences for hierarchical models of visual recognition.

1 Introduction

The human brain employs different kinds of interrelated representations and processes to recognize objects, depending on the familiarity of the object and the required level of recognition, which is defined by the current task. There is evidence that for identifying highly familiar objects, like faces, holistic templates are used that emphasize the spatial layout of the object's parts but neglect details of the parts themselves. This holistic prototypical representation requires a lot of experience and coding capacity and therefore can not be used for all the objects in every day's life. A more compact representation can be obtained when handling objects as combinations of shared parts. There is various biological motivation for such a representation. The experiments of Tanaka [1] revealed that there are high-level areas in primates ventral visual pathway that predict the presence of a large set of features with intermediate complexity, generalizing over small variations and being invariant to retinotopical position and scale. The combinatorial use of those features was shown by Tsunoda [2]. He observed that complex objects simultaneously activate different spots in those areas and that this activation is caused by the constituent parts. A parts-based representation is especially efficient for storing and categorizing novel objects, because the largest variance in unseen views of an object can be expected in the position and arrangement of parts, while each part of an object will be visible under a large variety of 3D object transformations.

In computer vision literature there is a similar distinction into holistic and parts-based approaches, depending on how feature responses are aggregated over

J. Marques de Sá et al. (Eds.): ICANN 2007, Part II, LNCS 4669, pp. 210–219, 2007.

the image. Parts can be local features of any kind. The response of a part detector at different positions in an image means that the part might be present several times but not that the probability is higher that the part is present at all. So each peak in the multimodal response map is handled as a possible instance of the part. In contrary to this, holistic approaches contain a layer that simply accumulates the real-valued response of single features of the previous layer over the whole image. This is only comparable to the biological definition if the configurational information is kept.

Approaches with strong biological motivation are presented in [3,4]. Here hierarchies of feature layers are used, like in the ventral visual pathway, where they combine specificity and invariance of features. So there are cells that are either sensitive to a specific pattern of activation in lower layers, in this way increasing the feature's complexity, or that pool the responses of similar features, so generalizing over small variations. The output layer of the feedforward hierarchy proposed in [3] contains several topographically organized feature maps which are used directly by the final classifier. Following the above definition this is a holistic approach. The similar hierarchy of [4] employs in the highest feature layer a spatial max-pooling over each feature map in the previous layer, which makes it a parts-based approach. Multimodal response characteristics and the position of the parts are neglected.

Most other approaches work more directly on the images. Very typical holistic approaches apply histograms, so e.g. in [5] the responses to local features are simply summed and in [6] it is counted how often a response lies in a certain range. In other holistic methods the receptive fields of the features cover the whole image. So e.g. in [7] features obtained by principal component analysis (PCA) on gray-scale images were used to classify faces. These features, so called eigenfaces, show a very global activation and do not reflect parts of a face. In contrary to PCA other methods produce so called parts-based features like the nonnegative matrix factorization (NMF) proposed in [8] or a similar scheme proposed in [9] yielding more class-specific features. Although during training the receptive field of each feature covers the whole image, it learns to reconstruct a certain localized region that contains the same part in many training views (e.g. parts of normalized frontal views of faces). But usually those features are used in a holistic manner, meaning that they are extracted at a single position in the test image and in this way are only sensitive to the rigid constellation of parts that was present during training. This limits the possibilities to generalize over geometric transformations, which is especially a drawback when using few training examples in an unnormalized setting. Also the holistic approaches perform bad in the presence of clutter and occlusion and often require extensive preprocessing as localization and segmentation.

Other parts-based recognition approaches also use the maximum activation of each feature, like the highest layer in [4]. In [10] the features are fragments of gray-scale images. The response of a feature is binary and obtained by thresholding the maximum activation in the image. The approach selects features based on the maximization of mutual information for a single class. This yields fragments of intermediate complexity. An image is classified by comparing binary activation vectors to stored representatives in a nearest neighbor fashion. Other approaches make use of the position and treat each peak in the response map as possible part instance. In the scale invariant feature transform (SIFT) approach in [11] gradient-histograms are extracted for small patches around interesting points (see Fig. 1c). Each such patch descriptor is compared against a large repertoire of stored descriptors, where the best match votes for the presence of an object at a certain position, scale and rotation. The votes are combined using a Generalized Hough Transform and the maximally activated hypothesis is chosen. A similar scheme is proposed in [12]. Here image patches are used as features and the algorithm is capable to produce a segmentation mask for the object hypothesis that can be used for a further refinement process. In the bags of keypoints approaches, e.g. [13], it is counted how often parts are detected in an image. In contrast to holistic histogram-based approaches the presence of a part is the result of a strong local competition of parts. Therefore it is more a counting of symbol-type information than a summation over real-valued signal-type responses. Parts-based recognition can be used to localize and recognize objects at the same time and works well in the presence of clutter and occlusion.

In Sect. 2 we first comment on the task we want to solve and the nature of the features required for this. Then we describe the investigated feature types and our feature selection strategy. We give results for a categorization problem in Sect. 3 and present our conclusions in Sect. 4.

2 Analytic Features

To generalize from few training examples, parts-based recognition follows the notion that similar combinations of parts are specific for a certain category over a wide range of variations. In this work we investigate how suitable different feature types are for this purpose and which effort is needed in terms of the number of used features. As has been argued in [10], it is beneficial that a single part can be detected in many views of one category, while being absent in other categories. So we need a reasonable feature selection strategy that evaluates which and how many views of a certain category a feature can separate from other categories and, based on those results, choose the subset of features that in combination can describe the whole scenario best. For simple categories a single feature can separate many views and therefore only few features are necessary to represent the whole category. For categories with more variation more features have to be selected to cover the whole appearance. This dynamic distribution of resources is necessary to make best use of the limited number of features.

How well certain local descriptors can be re-detected under different image transformations, as scale, rotation and viewpoint changes, was investigated in [14]. Although this is a desired quality, it does not necessarily state something on the usefulness in object recognition tasks. To underline that the desired features should be meaningful, i.e. offer a compromise between specificity and generality at low costs, and to avoid confusion with approaches that learn parts-based features, we will use the term analytic features.

We decided to compare patches of gray-scale images, for their simplicity, SIFT descriptors, for their known invariance, and patches of the output of the feedforward hierarchy in [3], because of the biological background.

A SIFT descriptor as proposed in [11] describes a gray-scale patch of 16x16 pixels using a grid of 4x4 gradient-histograms (see Fig. 1c). Each histogram in the grid is made up of eight orientation bins. The magnitude of the gradient at a certain pixel is distributed in a bilinear fashion over the neighboring histograms (in general four), where the orientation of the gradient determines the bin. The gradient magnitudes are scaled with a Gaussian that is centered on the patch, in this way reducing the influence of border pixels. Prior to the calculation of the histogram grid a single histogram with a higher number of orientation bins is computed for the whole patch. The maximum activated bin in this histogram is used to normalize the rotation of the patch in advance. Finally the energy of the whole descriptor is normalized to obtain invariance to illumination. In contrast to [11], we do not extract SIFT descriptors at a small number of interesting keypoints, but for all locations where at least a minimum of structure is present. In this way only uniform, dark background is neglected and on the category scenario in Fig. 3 on average one third of all descriptors is kept. We reduce the number of descriptors for each image by applying a k-means algorithm with 200 components. A similar cluster step was also done in [15] to improve the generalization performance of the otherwise very specific SIFT descriptors.

For the gray-scale patches we decided to use the same patch size as for the SIFT approach and the influence of the pixels is also weighted with a Gaussian that is centered on the patch.

The feedforward hierarchy proposed in [3] is shown in Fig. 1a. The S1-layer computes the magnitudes of the response to four differently oriented gabor filters. This activation is pooled to a lower resolution in the C1-layer performing a local OR-operation. The 50 features used in S2 are trained as to efficiently reconstruct a large set of random 4x4x4 C1-patches from natural images and are therefore sensitive to local patterns in C1. Layer C2 performs a further pooling operation and is the output of the hierarchy. Columns of 2x2 pixels are cut from the C2-layer as shown in Fig. 1b and used as feature candidates. Because of the two pooling layers, which offer a small degree of invariance to translation, a column of 2x2 pixels in C2 corresponds roughly to a patch of 16x16 pixels in the gray-scale image.

We will refer to the parts-based approaches as GRAY-P, SIFT-P, and C2-P. For SIFT-P each image i is described by the $J = 4 \times 4 \times 8 = 128$ dimensional representatives of the 200 k-means clusters $\mathbf{p}_{in}$, $n = 1 \ldots 200$. For GRAY-P the $\mathbf{p}_{in}$ are the patches of image i at all distinct positions n ($J = 16 \times 16 = 256$). Similar to this for C2 each $\mathbf{p}_{in}$ is a column through the feature maps of image i at a distinct position n as shown in Fig. 1b ($J = 2 \times 2 \times 50 = 200$). The $\mathbf{p}_{in}$ show a large variety. Therefore we will use all $\mathbf{p}_{in}$ directly as feature candidates $\mathbf{w}_m$, where m is an index over all combinations of i and n, and select a subset

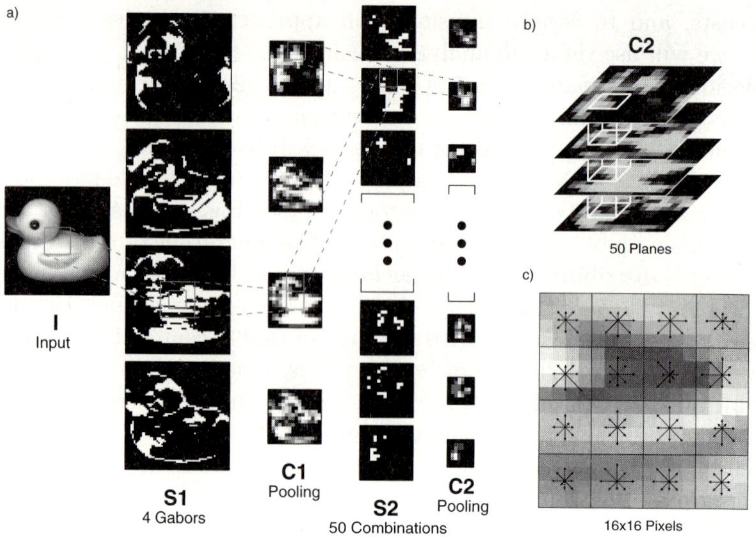

Fig. 1. a) Feedforward hierarchy in [3]. b) Columns of C2-layer are used as local features. c) SIFT descriptor [11] is grid of gradient histograms each with 8 orientations.

of those candidates with a strategy that is described later. The response r_{mi} of feature $\mathbf{w}_m$ on the image i is given by:

$$r_{mi} = \max_n \left(G(\mathbf{w}_m, \mathbf{p}_{in}) \right) . \tag{1}$$

For GRAY-P $G(\mathbf{w}_m, \mathbf{p}_{in}) = \frac{\sum_{j=1}^{J} h^j (w_m^j - \overline{w}_m)(p_{in}^j - \overline{p}_{in})}{\sqrt{\sum_j h^j (w_m^j - \overline{w}_m)^2 \sum_j h^j (p_{in}^j - \overline{p}_{in})^2}}$ is used which is the normalized cross-correlation, where $\overline{w}_m$ and $\overline{p}_{in}$ are the means of vector $\mathbf{w}_m$ and $\mathbf{p}_{in}$ respectively, and h^j is a weighting which decreases the influence of border pixels with a Gaussian. For C2-P the negative Euclidean distance $G(\mathbf{w}_m, \mathbf{p}_{in}) = -\sqrt{\sum_{j=1}^{J} (w_m^j - p_{in}^j)^2}$ shows better performance because of the sparseness in this layer. The similarity between SIFT descriptors is given by their dot product $G(\mathbf{w}_m, \mathbf{p}_{in}) = \sum_{j=1}^{J} w_m^j p_{in}^j$. The maximum activation per image is chosen as response and spatial information is neglected.

Reflecting the remarks on feature selection given above, we decided to use the following strategy: First we determine which views of a certain category each individual candidate feature $\mathbf{w}_m$ can separate. Therefore we compute the response r_{mi} for every training image with (1). Then the minimal threshold t_m is chosen that guarantees that all images with r_{mi} above or equal to t_m belong the same category (see Fig. 2):

$$t_m = \min \left\{ t | \forall_{\substack{i | r_{mi} \geq t \\ j | r_{mj} \geq t}} l_i = l_j \right\} . \tag{2}$$

Here l_i denotes the category label of image i. The images separated by the threshold is assigned a constant score $s_{mi} = k$ with respect to the feature $\mathbf{w}_m$.

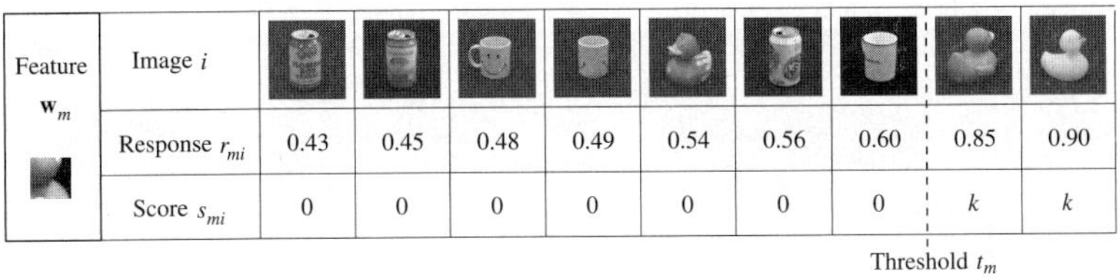

Fig. 2. Feature selection scheme. For visualization the images are sorted on their response r_{mi}. The threshold t_m separates views of a single category (here ducks) from all other images. To these views a score $s_{mi} = k$ is assigned.

When the scores s_{mi} are determined for the set of candidate features M an iterative process selects a given number of features by determining in each step the best candidate feature m with:

$$m = \arg\max_{m \in M} \left(\sum_i f \left(s_{mi} + \sum_{q \in Q} s_{qi} \right) \right) \tag{3}$$

and putting it from M into the set of already selected features Q. First ($Q = \emptyset$) the feature is selected that is detected in the most views of a certain category. Then successively the feature which causes the highest additional score is selected. The function $f(z)$ controls how effective a new feature can score for a single image. When using a Heaviside function only a single feature can score for an image. Here we use a Fermi function $f(z) = \frac{1}{1+e^{-z}}$ and set $k = 3$. In this way the feature gets only a high score for images that were not separated yet, and a much lower score for images in which features have already been detected.

3 Results

We tested the performance of the different feature types on the categorization scenario shown in Fig. 3. The gray-scale images have a resolution of 128x128 pixels and show centered objects on dark background. The objects belong to ten categories, where each category contains nine objects. Five objects per category are used for training and the remaining four for testing. Each object is represented by 30 views taken during a rotation around the vertical axis.

For each approach we ranked the candidate features from the complete set of training images with the introduced selection framework. The first 75 selected features for each approach are shown in Fig. 4a. The gray-scale patches contain a lot of similar parts under different orientations. For C2 less complex patches are selected that sometimes have only activation at the border or even seem to stem from the background. The SIFT patches show the largest variety.

For the different tests we then varied the number of used features and the number of training views that were used by a single layer perceptron (SLP), as the final classifier. So first for each training and test image a vector was

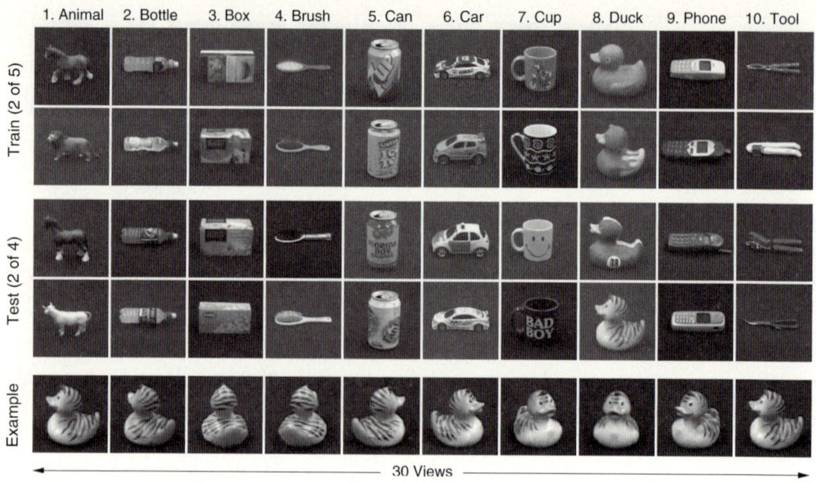

Fig. 3. Category scenario. Each category contains nine objects. Five are used for training and four for testing. Only two objects of both groups are shown here.

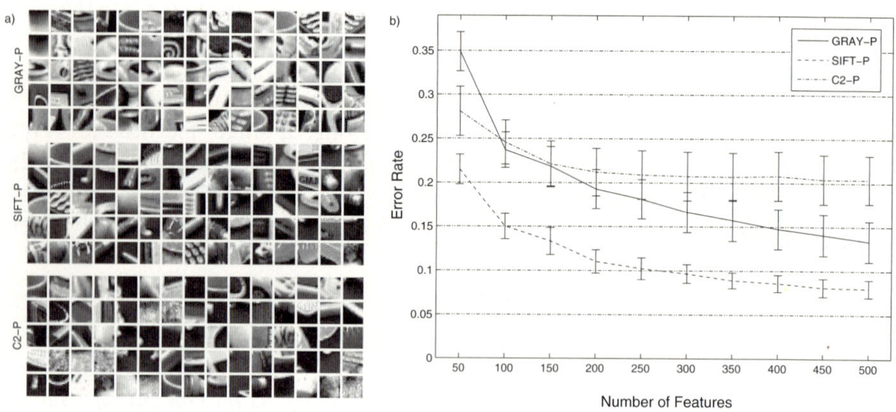

Fig. 4. a) First 75 top ranked features for parts-based approaches. For C2 the corresponding patch of the original gray-scale image and for SIFT the patch the descriptor of which is most similar to the selected k-means component is shown. b) Error rates depending on number of features for parts-based approaches.

calculated containing the responses of the selected features using (1). We let the SLP converge on the training vectors, and after this calculated the recognition performance on the complete set of test vectors. To increase both difficulty and objectivity we did not distribute the training examples equally over the single categories but repeated each test 50 times with random sets of training images and so obtained a mean performance together with a standard deviation.

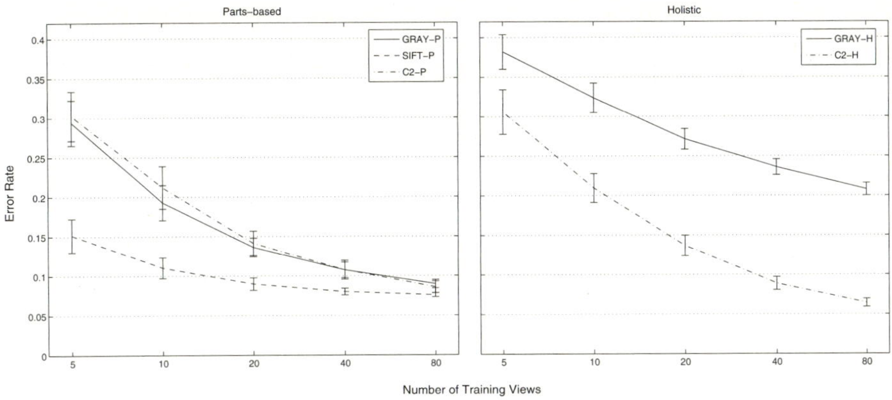

Fig. 5. Error rates depending on number of training views for different approaches

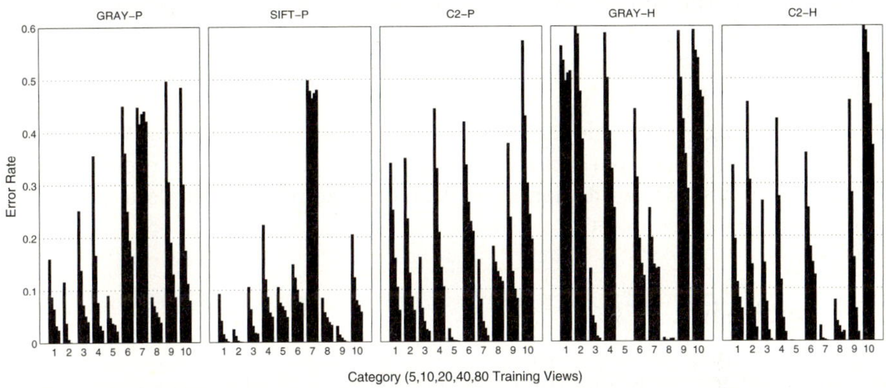

Fig. 6. Error rates of individual categories depending on number of training views

Fig. 4b shows how the classification performance depends on the number of selected features. For this test 10 random views were used per category for SLP training. SIFT-P outperforms C2-P and GRAY-P. For small numbers of features GRAY-P has the worst result, but shows the best improvement with increasing feature number, while the performance of C2-P saturates early. Maybe the variability that is gained via quantity helps to overcome the missing invariance of the very specific GRAY-P patches. C2-P patches make use of the invariance gained by the hierarchical processing from the beginning. But they maybe too general and only few qualitatively distinct features might exist.

Fig. 5 shows how the recognition performance depends on the number of views per category that were randomly chosen for the training of the SLPs. In this test we used 200 features for the parts-based approaches. On the right hand side of the figure we give also results of SLPs that were trained on the original gray-scale images (GRAY-H) and on the complete C2-activations (C2-H). For few training views SIFT-P is superior to the other approaches. C2-H is similar to C2-P and GRAY-P, and takes the lead when using a large number of views. GRAY-H shows the worst performance. More than the other approaches C2-H profits from an increase in the number of training views. This confirms the notion, that columns of C2-H, as used in C2-P, are invariant but too general. Although this is a drawback for C2-P, it helps C2-H together with position information to extrapolate well in the neighborhood of single views.

To provide reason for the shown differences Fig. 6 visualizes the same test with the mean error rates given for individual categories. SIFT-P especially works well for animals(1), bottles(2) and phones(9) but is outperformed by all other methods on cans(5) and cups(7), and by C2-H also on ducks(7). The performance of SIFT-P and GRAY-P on cups(7) is very poor and does not improve with more training views. The patches for SIFT-P and GRAY-P contain only few cup features but those are top-ranked and highly discriminative for the training images, but maybe too specific to generalize over the test images.

To conclude, the holistic approaches (C2-H, GRAY-H) are good for categories that do not vary much in shape during a rotation around the vertical axis, like cans(5) or cups (7). Also the results on ducks(8) are good because only the position of the head changes, while the body shape stays nearly unchanged. When the change of the global shape is more extreme during rotation SIFT-P performs better in comparison to the other approaches. This is especially true for categories where the rotation in depth looks like rotation in plane (bottle(2), brush(4), phone(9), tool(10)).

4 Conclusion

We evaluated the performance of different types of local feature when used in parts-based recognition. We showed that SIFT descriptors are good analytic features for most objects especially when the number of training views and the number of features is limited. The biological motivated feedforward hierarchy in [3] is powerful in holistic recognition with a sufficient number of training examples, but the patches from the output layer are too general and therefore show weak performance in parts-based recognition. This is interesting because also the calculation of a SIFT descriptor can be described as hierarchical processing: First features are used that extract the magnitudes for 8 different local gradient directions. Then a local winner takes all is applied over those features at each position. Each of the 16 histograms in the 4x4 grid integrates over each direction in a local neighborhood by summing the magnitudes (no non-linearity used as for pooling in [3,4]). Finally the SIFT descriptor stands for a more global activation pattern in the grid. Besides the normalization of rotation for SIFT, it would be interesting to investigate other reasons for the differences in performance in future work. This could be beneficial for both feature types.

The most related work in the direction of analytic features was done in [16], where Ullman introduced invariance over viewpoint in his fragments approach, or in the work of Dorko et al. in [15], where highly informative clusters of SIFT descriptors are used. Since both approaches have not been applied to scenarios with multiple categories, we hope that our comparative study provides further helpful inside into parts-based 3D object recognition.

References

1. Tanaka, K.: Inferotemporal Cortex And Object Vision. Annual Review of Neuroscience 19, 109–139 (1996)
2. Tsunoda, K., Yamane, Y., Nishizaki, M., Tanifuji, M.: Complex objects are represented in inferotemporal cortex by the combination of feature columns. Nature Neuroscience 4(8), 832–838 (2001)
3. Wersing, H., Körner, E.: Learning Optimized Features for Hierarchical Models of Invariant Object Recognition. Neural Computation 15(7), 1559–1588 (2003)
4. Serre, T., Wolf, L., Bileschi, S., Riesenhuber, M., Poggio, T.: Robust Object Recognition with Cortex-Like Mechanisms. IEEE Trans. Pattern Analysis and Machine Intelligence 29(3), 411–426 (2007)
5. Mel, B.W.: SEEMORE: Combining color, shape, and texture histogramming in a neurally inspired approach to visual object recognition. Neural Computation 9(4), 777–804 (1997)
6. Schiele, B., Crowley, J.L.: Object Recognition Using Multidimensional Receptive Field Histograms. In: European Conference on Computer Vision, pp. 1039–1046. Cambridge, UK (1996)
7. Turk, M., Pentland, A.: Eigenfaces for Recognition. Journal of Cognitive Neuroscience 3(1), 71–86 (1991)
8. Lee, D.D., Seung, H.S.: Learning the parts of objects by non-negative matrix factorization. Nature 401, 788–791 (1999)
9. Hasler, S., Wersing, H., Körner, E.: Class-specific Sparse Coding for Learning of Object Representations. In: Duch, W., Kacprzyk, J., Oja, E., Zadrożny, S. (eds.) ICANN 2005. LNCS, vol. 3696, pp. 475–480. Springer, Heidelberg (2005)
10. Ullman, S., Vidal-Naquet, M., Sali, E.: Visual features of intermediate complexity and their use in classification. Nature Neuroscience Vision Research 5(7), 682–687 (2002)
11. Lowe, D.G.: Distinctive Image Features from Scale-Invariant Keypoints. International Journal of Computer Vision 60(2), 91–110 (2004)
12. Leibe, B., Schiele, B.: Scale-Invariant Object Categorization Using a Scale-Adaptive Mean-Shift Search. In: Rasmussen, C.E., Bülthoff, H.H., Schölkopf, B., Giese, M.A. (eds.) Pattern Recognition. LNCS, vol. 3175, pp. 145–153. Springer, Heidelberg (2004)
13. Dance, C., Willamowski, J., Fan, L., Bray, C., Csurka, G.: Visual categorization with bags of keypoints. In: Pajdla, T., Matas, J(G.) (eds.) ECCV 2004. LNCS, vol. 3021, Springer, Heidelberg (2004)
14. Mikolajczyk, K., Schmid, C.: A Performance Evaluation of Local Descriptors. IEEE Trans. Pattern Analysis and Machine Intelligence 27(10), 1615–1630 (2005)
15. Dorko, G., Schmid, C.: Object Class Recognition Using Discriminative Local Features. In: INRIA (2005)
16. Ullman, S., Bart, E.: Recognition invariance obtained by extended and invariant features. Neural Networks 17(1), 833–848 (2004)

An Application of Recurrent Neural Networks to Discriminative Keyword Spotting

Santiago Fernández[1], Alex Graves[1], and Jürgen Schmidhuber[1,2]

[1] IDSIA, Galleria 2, 6928 Manno-Lugano, Switzerland
[2] TU Munich, Boltzmannstr. 3, 85748 Garching, Munich, Germany

Abstract. The goal of keyword spotting is to detect the presence of specific spoken words in unconstrained speech. The majority of keyword spotting systems are based on generative hidden Markov models and lack discriminative capabilities. However, discriminative keyword spotting systems are currently based on frame-level posterior probabilities of sub-word units. This paper presents a discriminative keyword spotting system based on recurrent neural networks only, that uses information from long time spans to estimate word-level posterior probabilities. In a keyword spotting task on a large database of unconstrained speech the system achieved a keyword spotting accuracy of 84.5 %.

1 Introduction

The goal of keyword spotting is to detect the presence of specific spoken words in (typically) unconstrained speech. Applications of keyword spotting include audio indexing, detection of command words in interactive environments and spoken password verification. In general, it is most useful in domains where a full speech recogniser is cumbersome and unnecessary, partly because less than perfect detection rates are still very satisfactory.

Nonetheless, the same mathematical framework used in the majority of speech recognisers also forms the basis for keyword spotting systems. The typical keyword spotting system consists of a set of hidden Markov models (HMM), one for each keyword plus one or more filler models [1]. The filler models characterize non-keyword events in the speech signal, such as other words, background noises and silence. The approach is generative and therefore finds the sequence of models most likely to have produced the observations. The output of the system is post-processed before deciding on the presence or absence of keywords in the utterance [2]. First, the confidence of the predictions is estimated. This is typically done by computing the ratio of likelihoods between keyword model hypotheses and filler model hypotheses. Finally, a threshold level is applied to the confidence measure in order to achieve a compromise between the number of true and false positives predicted by the system.

In general, a discriminative approach to keyword spotting is more suitable, as it allows discrimination between keyword and non-keyword events, and also among similar keywords. In addition, posterior probabilities can directly be

J. Marques de Sá et al. (Eds.): ICANN 2007, Part II, LNCS 4669, pp. 220–229, 2007.

used as an effective confidence measure. Recently, discriminative keyword spotting systems have been explored in various papers. The most common approach [3,4,5] uses artificial neural networks (ANN) in the framework of hybrid ANN/HMM systems [6,7]. In these systems, keywords are modelled by concatenating phoneme models. A phoneme classifier (typically a multi-layer perceptron) estimates frame-level posteriors. These local posterior estimates are accumulated over the duration of a segment and, typically, the result is normalised by the length of the segment. Another approach to discriminative keyword spotting uses kernel machines and large margin classifiers with a set of models and feature functions consistent with hybrid ANN/HMM systems [8].

The approach presented in this paper attempts to model full keywords in the sequence data stream, while previous discriminative keyword spotting systems are based on sub-word units. It is based on recurrent neural networks (RNNs) trained with the connectionist temporal classification (CTC) objective function. This objective function allows artificial neural networks to map unsegmented sequential data onto a sequence of labels [9,10]. The proposed system uses information over long time spans, thereby providing non-local estimates of the a posteriori probability of the presence of either keyword or non-keyword events in the speech signal.

There is a plethora of different keyword spotting systems, often tailored to meet the requirements of particular tasks. However, there is little consensus on how to define benchmark tasks [11]. We have opted to tackle a realistic keyword spotting task in a large database of spontaneous and unconstrained speech where an HMM-based speech recogniser achieves a word accuracy of only 65 %.

The remainder of the paper is structured as follows. Section 2 provides a description of the system. Keyword spotting experiments are presented in section 3 and the results are shown in section 3.4. Section 4 offers a discussion on differences between previous approaches and the one presented in this paper, and gives directions for future work. Final conclusions are given in section 5.

2 Method

2.1 Outline

The network architecture selected for the keyword spotting task is the bidirectional long short-term memory recurrent neural network (BLSTM), which has shown good performance in a series of speech tasks [12,13]. The network is trained with the connectionist temporal classification (CTC) objective function [9,10].

The input to the recurrent network is the data sequence of a speech utterance. The network has a soft-max output layer with as many output units as keywords to be detected, plus one output unit associated with non-keyword events. The target for the training algorithm is a list (which may be empty) of keywords in the order in which they appear in the input speech utterance. No further constraints are applied to the system. In particular, the segmentation of the speech signal is not required and various pronunciation variants of the same keyword can appear in the data set.

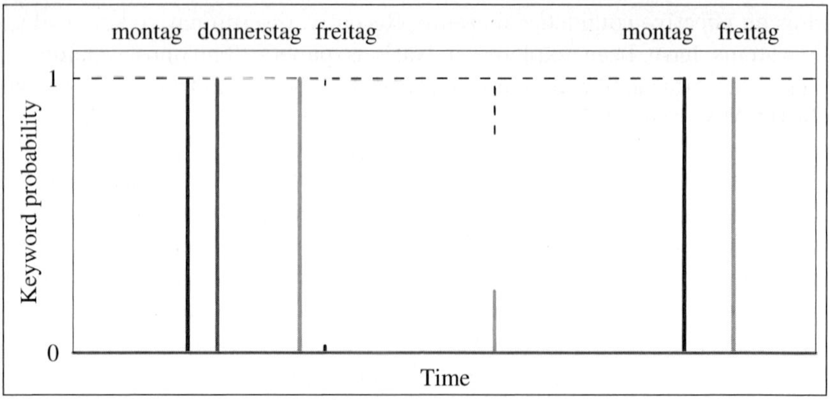

Fig. 1. Discriminant keyword spotting with BLSTM and CTC. When a keyword is detected the ANN produces a spike-like output with higher probability than the non-keyword output (dashed line).

Once the network has been trained, its output typically consists of a series of spikes corresponding to keyword events that have been detected in the input, separated by long periods of activation of the non-keyword output unit (see figure 1). The activation of every output unit at every time step is an estimate of the probability of detecting a particular keyword (or non-keyword event) at that time step.

2.2 BLSTM

The long short-term memory (LSTM) [14,15] is an RNN architecture designed to deal with long time-dependencies. It addresses the problem of the back-propagated error either blowing up or decaying exponentially for long time lags in conventional RNNs. The hidden layer of an LSTM network consists of a set of recurrently connected blocks containing one or more memory cells and three multiplicative units (the input, output and forget gates), which allow writing, reading or resetting of the information in the memory cell.

Bi-directional RNNs [16] address in an elegant way the need for delayed decisions in some sequential tasks such as speech processing. Data sequences are presented forwards and backwards to two separate recurrent networks, which are connected to the same output layer. Therefore, for every point in a given sequence, the network has complete sequential information about the points before and after it. An implementation of bi-directional LSTM (BLSTM) can be found in [12].

2.3 CTC

Connectionist temporal classification (CTC) [9,10] is an objective function to label unsegmented sequential data with RNNs. The basic idea behind CTC is to interpret the network outputs as a probability distribution over all possible label sequences, conditioned on the input data sequence. Given this distribution, an objective function can be derived that directly maximises the probability of the correct labelling. Since the objective function is differentiable, the network can be trained with standard back-propagation through time.

3 Experiments

3.1 Material

The experiments were carried out on a set of dialogues in German from the Verbmobil database [17]. The dialogues deal with the scheduling of date appointments for meetings. The database consist in spontaneous and unconstrained speech, with long silences and noises. The results provided along with the database for a baseline automatic speech recogniser give an approximate idea of the difficulty of processing this database: the HMM-based speech recogniser achieves a word accuracy of 65 % in an automatic, full transcription, of the test set [17].

The material used in the experiments corresponds to version 2.3 (March 2004) of the Verbmobil database (except for CD-ROM number 53.1 of the training set which was not available) [17]. It includes separate training, validation and test sets. The database is speaker independent. Speakers were distributed equally across sexes in all sets and every speaker appears in only one of the sets. The training set includes 748 speakers and 23975 dialogue turns for a total of 45.6 hours of speech. The validation set includes 48 speakers and 1222 dialogue turns for a total of 2.9 hours of speech. The test set includes 46 speakers and 1223 dialogue turns for a total of 2.5 hours of speech.

Due to the nature of the dialogues included in the database, we decided to use dates and places as keywords for the detection task. This ensures a relatively good coverage of keywords in the training, validation and test data sets. The twelve keywords chosen were:

> april, august, donnerstag, februar, frankfurt, freitag, hannover, januar, juli, juni, mittwoch, montag

Note that the database includes various pronunciation variants of some of these keywords (e.g. "montag" can end either with a /g/ or with a /k/). In addition, several keywords appear as sub-words, e.g. in plural form such as "montags" or as part of another word such as "ostermontag" (Easter Monday).

The orthographic transcription provided along with the database was examined for the presence in every utterance of one or more of the twelve keywords selected. Once a keyword was found, the begin and end times for the keyword were saved for evaluating the performance of the keyword spotting system at a later stage. These times were provided along with the database and correspond to the segmentation given by an automatic speech recognition system.

This procedure gave a total of 10469 keywords on the training set with an average of 1.7 % keywords per non-empty utterance (73.6 % of the utterances did not have any keyword); 663 keywords on the validation set with an average of 1.7 % keywords per non-empty utterance (68.7 % of the utterances did not have any keyword); and 620 keywords on the test set with an average of 1.8 % keywords per non-empty utterance (71.1 % of the utterances did not have any keyword).

The list of keywords (which may be empty) in the order in which they appear in every utterance, forms the target sequence for the network's training algorithm.

Finally, the speech data was characterised as a sequence of vectors of thirty nine coefficients, consisting of twelve Mel-frequency cepstral coefficients (MFCC) plus energy and first and second order derivatives of these magnitudes. The coefficients were computed every 10 ms over 25 ms-long windows. First, a Hamming window was applied; secondly, a mel-frequency filter bank of 26 channels was computed; and finally, MFCC coefficients were calculated with a 0.97 preemphasis coefficient. The input data was normalised to have zero mean and standard deviation one on the training set.

3.2 Difficulty with HMMs

We briefly illustrate the difficulty of performing keyword spotting with a generative HMM with as few constraints as those required by our system. In particular, no post-processing is applied. For these experiments every keyword was modelled as a left-to-right HMM with sixteen states and observation probabilities given by a mixture of Gaussians with diagonal covariance matrices. For the filler model we used an HMM of the same type with three states. The grammar allows one or more repetitions of any keyword or filler per utterance.

We experimented with: a) doubling the number of mixtures from two until a maximum of 128 and re-estimating the parameters twice after every step; b) the parameters of the HMMs with eight Gaussians were re-estimated for more than five hundred iterations, with an evaluation of the results carried out every ten iterations (henceforth, one step); and c) the first experiment was repeated with HMMs initialized using the reference segmentation and increasing the number of mixtures up to 256. The performance was evaluated on the validation set after every step with insertion penalties from zero to minus one hundred in steps of ten. For experiment b), the performance did not improve (it slightly decreased) after 350 iterations.

The results on the test set for all experiments, optimised on the validation set, showed negative accuracy due to the many false positives generated by the system: of the order of three thousand in all cases. Hence, the importance of a post-processing stage for HMMs that estimates the confidence of the predictions and a threshold level for balancing true and false positives.

3.3 Setup

The CTC-BLSTM network was trained with the input sequences and keyword targets from section 3.1. The network has, therefore, 39 input units and 13 units in the output soft-max layer. Both the forward and backward hidden recurrent layers consist of 128 LSTM memory blocks with one memory cell per block, peephole connections and forget gates. The input and output cell activation functions are a hyperbolic tangent. The gates use a logistic sigmoid function in the range [0,1]. The input layer is fully connected to the hidden layer and the hidden layer is fully connected to itself and to the output layer. The total number of weights in the network is 176,141.

Training was carried out with a learning rate of 10^{-4} and a momentum coefficient of 0.9. Weights were initialised with a Gaussian random distribution with mean zero and standard deviation 0.1. In addition, Gaussian noise with a standard deviation of 0.5 was added to the inputs during training to improve generalization. The network was tested every five epochs on the validation set and training finished once the error on the validation set stopped decreasing.

3.4 Results

The typical sequence of outputs for a trained network can be seen in figure 1. It consists in a sequence of spikes associated with detected keywords and separated by long periods during which the non-keyword output unit is the most active. The CTC algorithm allows the network to keep any output active for as long as necessary or as little as one time step only. To evaluate the results, at every time step the output with the highest activation is chosen. If an output is the most active for a period of time, we choose the highest activation during that period as the probability of detecting the keyword, and the time at which the highest activation occurs as the location of the detected keyword.

Spikes whose identity match a keyword in the speech signal and that appear within the boundaries of the keyword in the speech signal are counted as true positives, unless more than one spike is output in that period, in which case only one of them counts as a true positive and the rest count as false positives. Those spikes that appear out of the boundaries of the keyword in the speech signal are considered false positives. The accuracy of the system is given by the number of true positives minus the number of false positives divided by the number of keywords in the data set.

Four networks were trained on the same task with different initial random weights. Average accuracy over four runs was 84.5 % with a standard error of 1.2 %. The average probability of true positives was 0.98 with a standard error of 0.004. The average probability of false positives was 0.80 with an standard error of 0.01. As expected, given the discriminative nature of the algorithm, setting a posteriori a threshold level in between this probability levels did not increase performance significantly (84.8 % accuracy, with an standard error of 1.2 %). Table 1 shows the number of true and false positives by keyword for the network with the best performance. As shown in the table, the system discriminates almost perfectly between similar keywords such as "juni" and "juli".

As of now, the CTC training algorithm does not constrain the spikes to appear within the begin and end times of the speech pattern (it is assumed that a reference segmentation is not available). The algorithm trains the network to detect the correct sequence of keywords and in the right order for any utterance

Table 1. Results by keyword for the ANN with the best performance. Accuracy is the number of hits minus false positives divided by the actual number of keywords in the test set. Average accuracy over four runs is 84.5 % with a standard error of 1.2 %.

Keyword	# Hits	# FPs	# Actual	% Accuracy
april	27	2	32	78.12
august	29	1	34	82.35
donnerstag	55	6	56	87.50
februar	55	1	60	90.00
frankfurt	18	0	25	72.00
freitag	40	4	45	80.00
hannover	76	5	86	82.56
januar	35	4	38	81.58
juli	53	1	56	92.86
juni	63	2	66	92.42
mittwoch	36	1	39	89.74
montag	79	3	83	91.57
Overall	566	30	620	86.45

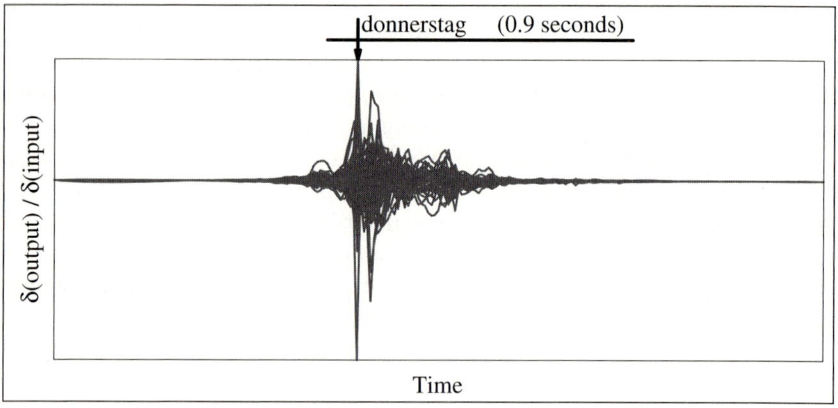

Fig. 2. The figure shows the dependency of the output unit associated with keyword "donnerstag" at the time step when this keyword is detected in figure 1, with respect to the network inputs at various time steps around the detection point (indicated by an arrow at the top of the figure). In addition, the extent (0.9 s) and location of the keyword in the speech signal is shown at the top of the figure. As can be seen, the probability of detecting the keyword depends on the inputs over a long time span, and decreases towards the end of the keyword which is the least discriminative part of it: "tag" (day).

in the training set. Probably this makes training slower, but it seems that it is not necessary to add this constraint to the training algorithm in order to achieve good performance: when the results are evaluated without using the segmentation, the average accuracy over four runs is 86.1 % with a standard error of 0.7 %. Of those keyword candidates appearing outside the boundaries in the speech signal, only one was more than 0.5 s outside the reference segment. This level of precision should be enough for tasks where high detection accuracy is more important.

Finally, figure 2 shows, for a successful detection, the dependency of the network output at the time when the keyword is detected with respect to the inputs at some previous and following time steps. As can be seen, the probability of detecting the keyword depends on the inputs over a long time span (of about one second, or a hundred network time steps) and, for the example shown, the dependency decreases towards the end of the keyword, which is the least discriminative part of it.

4 Discussion and Future Work

The main difference between our approach to discriminative keyword spotting and other discriminative systems is the capability of computing non-local posteriors for the keywords, i.e. the system uses information over long time spans to compute the probability of a keyword appearing in the speech signal. In addition to this, the a posteriori estimation of a threshold level to balance true and false positives is not required.

The algorithm makes very few assumptions about the domain, which facilitates the development of systems for detection tasks. Nonetheless, if very high precision in the location of spots is required, the system might benefit from adding to the algorithm the constraint that keywords must be detected within the boundaries established by the reference segmentation. However, not having this constraint greatly simplifies the preparation of training data.

To make the system scale, one solution consists in implementing smaller networks. For example, one for each keyword, or one for each set of similar keywords among which the system must discriminate. New networks can be added to the system at any time. Under the assumptions that, for different networks, the training data has similar characteristics and that the non-keyword output unit is associated to a model of similar characteristics, which is reasonable, the estimates of the a posteriori probabilities given by different networks can be compared and used to determine the presence or absence of keywords in the input data stream.

Another option to improve scalability consists in detecting sub-word units in the signal and use these outputs as inputs for one of the existing systems that builds keywords from sub-word units. The detection of sub-word units with the system proposed in this paper will benefit from the same features and ease of use described in this paper (see [9] for results on phoneme recognition with CTC-BLSTM). Besides this, it is possible to feed the sub-word level predictions to another CTC-BLSTM network with keyword-level outputs [10].

The choice of sub-word or word models depends on the task. In general, sub-word units are more practical for audio indexing tasks because the system must respond to searches for unknown keywords. However, in tasks such as detection of command words and spoken password verification the vocabulary is unlikely to change. Also, the benefits of word-level discrimination (e.g. decreased confusion among similar keywords) are most desirable in these cases, where security is paramount.

5 Conclusion

We have presented a word-level discriminative keyword spotting system that is fast, accurate and easy to use. The probability of detecting a keyword is computed using information from long time spans. These probabilities can be read directly from the outputs of a recurrent neural network. An a posteriori estimate of an acceptance threshold level is not required. Finally, the system makes few assumptions about the domain: only the input speech signal and a list of keywords (which may be empty) in the order in which they occur in the signal are necessary to train the system. The algorithm is general and can be used for any task requiring the detection of patterns in sequence data. In a keyword spotting task in a large database of unconstrained speech the system achieved a keyword spotting accuracy of 84.5 %.

Acknowledgments

This research was funded by SNF grant 200021-111968/1.

References

1. Wilpon, J.G., Rabiner, L.R., Lee, C., Goldman, E.R.: Automatic recognition of keywords in unconstrained speech using hidden Markov models. IEEE Transactions on Acoustics, Speech and Audio Processing 38(11), 1870–1878 (1990)
2. Benítez, M.C., Rubio, A., García, P., de la Torre, A.: Different confidence measures for word verification in speech recognition. Speech Communication 32, 79–94 (2000)
3. Silaghi, M.C., Bourlard, H.: Iterative posterior-based keyword spotting without filler models. In: Proceedings of IEEE Automatic Speech Recognition and Understanding Workshop, Keystone - Colorado (U.S.A.) (1999)
4. Silaghi, M.: Spotting subsequences matching an HMM using the average observation probability criteria with application to keyword spotting. In: Proceedings of the 20th National Conference on Artificial Intelligence and the 17th Conference on Innovative Applications of Artificial Intelligence, pp. 1118–1123. Pittsburgh - Pennsylvania (U.S.A.) (2005)
5. Ketabdar, H., Vepa, J., Bengio, S., Bourlard, H.: Posterior based keyword spotting with a priori thresholds. In: Proceedings of the 9th International Conference on Spoken Language Processing, Pittsburgh - Pennsylvania (U.S.A.) (2006)
6. Robinson, A.J.: An application of recurrent nets to phone probability estimation. IEEE Transactions on Neural Networks 5(2), 298–305 (1994)
7. Bourlard, H.A., Morgan, N.: Connectionist speech recognition: a hybrid approach. Kluwer Academic Publishers, Dordrecht (1994)

8. Keshet, J., Grangier, D., Bengio, S.: Discriminative keyword spotting. In: Workshop on Non-Linear Speech Processing NOLISP, Paris (France) (2007)
9. Graves, A., Fernández, S., Gomez, F., Schmidhuber, J.: Connectionist temporal classification: labelling unsegmented sequence data with recurrent neural nets. In: Proceedings of the 23rd International Conference on Machine Learning, Pittsburgh - Pennsylvania (U.S.A.) (2006)
10. Fernández, S., Graves, A., Schmidhuber, J.: Sequence labelling in structured domains with hierarchical recurrent neural networks. In: Proceedings of the 20th International Joint Conference on Artificial Intelligence, Hyderabad (India) (2007)
11. Silaghi, M.C., Vargiya, R.: A new evaluation criteria for keyword spotting techniques and a new algorithm. In: Proceedings of InterSpeech (2005)
12. Graves, A., Schmidhuber, J.: Framewise phoneme classification with bidirectional LSTM and other neural network architectures. Neural Networks 18(5–6), 602–610 (2005)
13. Graves, A., Fernández, S., Schmidhuber, J.: Bidirectional LSTM networks for improved phoneme classification and recognition. In: Duch, W., Kacprzyk, J., Oja, E., Zadrożny, S. (eds.) ICANN 2005. LNCS, vol. 3697, pp. 799–804. Springer, Heidelberg (2005)
14. Hochreiter, S., Schmidhuber, J.: Long short-term memory. Neural Computation 9(8), 1735–1780 (1997)
15. Gers, F., Schraudolph, N., Schmidhuber, J.: Learning precise timing with LSTM recurrent networks. Journal of Machine Learning Research 3, 115–143 (2002)
16. Schuster, M., Paliwal, K.K.: Bidirectional recurrent neural networks. IEEE Transactions on Signal Processing 45, 2673–2681 (1997)
17. Verbmobil: Database version 2.3 (2004), `http://www.phonetik.uni-muenchen.de/Forschung/Verbmobil/Verbmobil.html`

Spatiostructural Features for Recognition of Online Handwritten Characters in Devanagari and Tamil Scripts

H. Swethalakshmi, C. Chandra Sekhar, and V. Srinivasa Chakravarthy

Indian Institute of Technology Madras, Chennai, India

Abstract. The spatiostructural features proposed for recognition of online handwritten characters refer to offline-like features that convey information about both the positional and structural (shape) characteristics of the handwriting unit. This paper demonstrates the effectiveness of representing an online handwritten stroke using spatiostructural features, as indicated by its effect on the stroke classification accuracy by a Support Vector Machine (SVM) based classifier. The study has been done on two major Indian writing systems, Devanagari and Tamil. The importance of localization information of the structural features and handling of translational variance is studied using appropriate approaches to zoning the handwritten character.

1 Introduction

Handwritten character recognition (HCR) assumes considerable importance because of its applicability to pen-based interfaces and recognition of handwriting on scanned documents. Online and offline handwriting recognition entail different modes of input, representation, processing and recognition strategies. In this paper, we present studies on online handwriting data of two major Indian writing systems. Offline-like features extracted from temporal sequences of positional co-ordinates for the characters are used to represent the units of online handwriting.

Recognition of Indian script characters pose diverse challenges. Indian writing systems derived from the Brahmi script have syllabic character sets. Such a writing system is characterized as an *abugida*, and has distinct symbols for consonants, vowels and conjunct consonants as well as for the modifier symbols of the above units. The combination of such modifiers with specific consonant symbols often gives rise to a new symbol derived from the smaller units. The shape of character units is perceptually more complex than in many other writing systems and occasionally varies with the context of its co-occurrence with a vowel modifier unit. A character can be written using one or more strokes (a maximum of around 10 for Devanagari script). Here a stroke is defined as the trace of the pen-tip captured from a pen-down event till the following pen-lift event. Some characters have common constituent stroke units. Hence it involves less number of classes to represent the character set in Indian scripts when using strokes compared to using character units.

J. Marques de Sá et al. (Eds.): ICANN 2007, Part II, LNCS 4669, pp. 230–239, 2007.

Tamil script is used in southern states of India as well as in Sri Lanka and South East Asian countries. Tamil script consists of 33 basic consonant and vowel units. Characters representing syllabic combinations of the above units can be realized using 98 strokes, including dot strokes. Devanagari is the most widely used Indian writing system. The 45 Devanagari consonant and vowel characters, their modifiers and composite ligatures formed from these can be written using 123 strokes including dot strokes.

The next section gives an overview of approaches to online HCR. Section 3 describes the spatiostructural features for online handwriting. Section 4 describes different methods for extraction of spatiostructural features. Section 5 presents the results of studies on recognition of strokes.

2 Approaches to Online HCR

Online handwritten character recognition uses spatiotemporal features to represent the stroke unit, which involve the time sequence of sample values representing the information about the pen-tip. Features extracted can be positional, structural or statistical. Offline character recognition uses spatioluminance based features captured from the image of the character[11]. Since it is not possible to demarcate stroke boundaries in the images of characters, stroke level recognition cannot be performed. Feature extraction techniques applied to offline handwritten character data are presented in [12].

Zoning is a technique extensively used for extraction of features from offline data. Zoning refers to the partition of a character image into distinct regions called zones, which can be considered as sub-images, for feature extraction. Zoning strategies and feature extraction involving zoning for offline recognition are described in [12]. Online HCR approaches also considered zoning[9][13] for identification of the position of stroke points in a character with respect to predefined regions for a small set of strokes. These approaches use the encoded sequence of zones traversed to obtain the online features[3][6][7]. However, there has been no mention of structural feature extraction from zones in any of these works.

There have been numerous studies to identify cardinal points on a stroke[4][10]. Authors of [4] analyze the shape of handwritten characters to identify *shape points* that are less susceptible to perturbations and useful for description of characters. They also define a *valence* feature to describe a shape point. Based on an empirical study, a subset of such features with low *valence* have been identified to represent handwritten Tamil characters.

Various studies have presented observations that offline handwriting recognition offers less classification accuracy compared to online handwritten character recognition[2][11]. Online data for a character also offers significant reduction in space complexity compared to its offline counterpart. Since the two techniques view a character from different perspectives, there is a significant amount of complementarity in information across the representations, which can be used for increasing the recognition accuracy[14]. There have also been methods proposed for the extraction of online features from offline data[5].

3 Spatiostructural Features

Spatiostructural features characterize the positional and structural or shape features of a character. Localization of the structural features is achieved by considering regions called zones. Each structural feature characterizes the shape of a stroke segment. The shape features are comparable to offline features because these features can be extracted from the contour profiles of a stroke image. The shape features considered in our studies are as shown in Figure 1. The cusp feature is located around a point where the x and y derivatives of a stroke curve vanish simultaneously. The bump points, where the derivative with respect to either x or y vanishes, as well as stroke terminals have been quantized in 8 directions, where consecutive axes are oriented at a spacing of 45°.

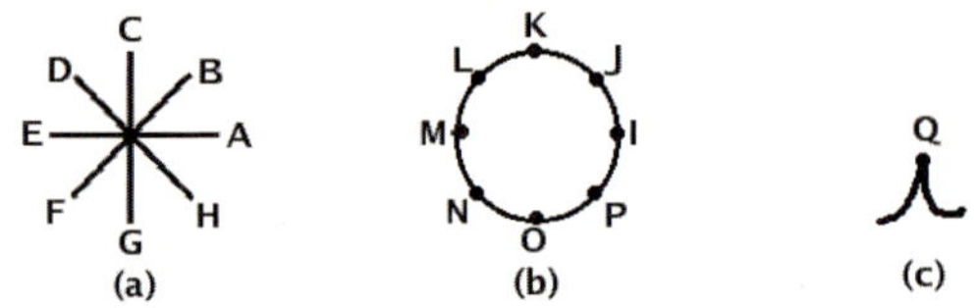

Fig. 1. The structural features that are extracted from the zones for representation of a stroke: (a) line terminals with directional information, (b) bumps with orientation information and (c) a cusp

A segment of online handwriting data corresponding to a structural feature is a short temporal sequence of two-dimensional positional coordinates. This is comparable to small spatially continuous segments that can be identified from the contour of the stroke image occurring in the same spatial region within the stroke bounds. The feature extraction module identifies the spatial regions in the data of a stroke and determines the frequency of occurrence of shape features in each zone. The objective of this work is to consider the spatiostructural features to represent the online handwritten character data and to study the effectiveness of these features extracted using different partitioning approaches or zoning strategies.

The main issues in the extraction of spatiostructural features are as follows:

- What is the optimal number of zones needed to represent a handwritten character in a writing system?
- What is the method of partitioning that will help in the identification of optimal zones?
- What are the features to be extracted from each zone?
- How are shape features occurring at zone boundaries handled?

It is not straightforward to determine the number and choice of zones for an effective representation. Most of the zoning techniques applied to offline recognition rely on empirical studies to identify the best choice of zoning strategy. The use of genetic algorithms[8] is suboptimal in terms of its time and algorithmic complexity. Carefully identified zone boundaries (using training data variations) may be more useful for printed or neatly written characters where variations in writing a character are not significant.

With reduction in zonal area, the spatial resolution is enhanced. However, a large number of zones or over-optimized zone boundaries may be inefficient in handling inherent translational or rotational variations of the structural features of a handwritten character. Significant positional information (local and global) may be available with an optimal choice of the minimum number of zones. Our approach involves empirical determination of the optimal zoning strategy. Features that occur in the boundary of a zone have been assigned membership to all zones sharing the boundary.

4 Zoning Strategies for the Extraction of Spatiostructural Features

Figure 2 illustrates the different zoning strategies for a Devanagari stroke. Structural features are extracted from the segments of a stroke occurring in each *zone*.

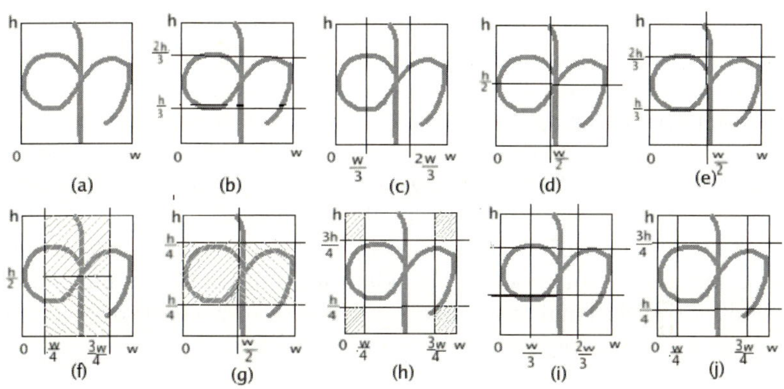

Fig. 2. Strategies for zoning the data of a handwritten stroke: (a) single zone; (b) 3 horizontal zones; (c) 3 vertical zones; (d) quadrant zoning; (e) 6 zones; (f),(g),(h) the 5 zone approach which involves overlapping and disjoint sub-regions; (i) 9-zones homogeneous partitioning; (j) 9-zones heterogeneous partitioning

1. *Single Zone Representation:* The entire region of the stroke boundary is considered to form one zone (Figure 2(a)). The study based on this representation presents the effectiveness of using the frequency of occurrence of structural features in a stroke for classification.
2. *3-Zone Representation – Horizontal Partitioning:* Here the bounding rectangle of the stroke is divided into three horizontal zones (Figure 2(b)). This representation uses the information of whether the shape based features identified in a stroke belong to the top, bottom or middle part of the stroke (when viewed as an image).

3. *3-Zone Representation – Vertical Partitioning:* Here the stroke area is divided into three homogeneous vertical zones. The approach identifies the structural features that occur in the 3 vertical blocks shown as in Figure 2(c). The maximum density of features mostly occurs in the middle zone.
4. *4-Zone Representation:* The stroke area has been divided into quadrants, similar to the division in the cartesian co-ordinate space (Figure 2(d)).
5. *6-Zone Representation:* This gives information of whether a structural feature is located in the top, middle or bottom, as well as whether it is in the left or the right half of the stroke area. This partitioning is shown in Figure 2(e).
6. *5-Zone Overlapping and Disjoint Subzone Partition:* Figures 2(f)(g) and (h) depict this zoning strategy, where the shaded regions represent the zones. Figures 2(f) and (g) show two zones each, whereas the shaded corner regions in Figure 2(h) all correspond to a single region. These 5 zones are equal in area. But the difference from the previous zoning strategies is that we consider overlap across zonal regions in Figures 2(f) and (g). Moreover, the zone depicted in Figure 2(h) has disjoint sub-regions. This representation is helpful in obtaining translational invariance for shape features within the spatial bounds of a stroke.
7. *9-Zone Homogeneous Partitioning:* In this case, the stroke area is divided into 9 equal-area zones as shown in Figure 2(i). The advantage in this case is a high level of localization of the structural feature information. This intuitive partitioning approach can be viewed as a combination of the two 3-zone partitioning strategies, where the availability of horizontal and vertical information helps in locating the grid where the feature is positioned.
8. *9-Zone Heterogeneous Partitioning:* Here, the zonal orientation is the same as in the previous case. But the zones are chosen to be of different sizes or areas. The reason for choosing this kind of zoning is to accommodate some amount of translational invariance, similar to the 5-zone strategy. The 9-zone heterogeneous partitioning is shown in Figure 2(j).

5 Studies on Stroke Recognition for Devanagari and Tamil Writing Systems

Studies have been conducted on stroke databases of two Indian writing systems, Devanagari and Tamil. The data used consists of 19398 training examples and 9080 test examples for Tamil, and 14269 training and 6929 test examples for Devanagari writing system. The training and test sets have no examples in common. The number of stroke classes used for the study is 93 for Tamil and 117 for Devanagari. Strokes having very small horizontal and vertical extent (like the dot strokes for Devanagari and Tamil) have not been included in the study. Their classification is subject to preclassification techniques based on their structural properties.

The preprocessing techniques are aimed at producing a uniform representation of the stroke, reducing variability across examples of the same class and reducing noise, while maintaining the structural characteristics of a stroke. Preprocessing of strokes involves position normalization, followed by upsampling, smoothing and size normalization preserving the stroke shape. The smoothing operation is achieved by convolving a Gaussian filter with the abscissae and ordinates. The values of parameters involved in preprocessing are experimentally determined.

The structural features have been extracted from a windowed sequence of 2-dimensional positional co-ordinates, with the window length empirically chosen. The zones are identified within the rectangular area enclosed by the bounding box of the stroke. The frequency of occurrence of shape features within each zone constitutes its zonal information. Feature extraction results in a fixed dimensional feature vector describing the stroke. The length of the feature vector is proportional to the number of regions considered because the zonal features are concatenated to obtain the stroke feature vector. The classification module used here is implemented using an SVM classifier. One-against-the-rest strategy is used for multiclass classification. A rule-based approach is applied to identify the character unit from the component strokes.

Table 1 shows the classification accuracy obtained using different partitioning strategies for extraction of spatiostructural features from Devanagari and Tamil strokes. A Gaussian kernel SVM has been used for classification, with an empirically determined value of width (σ) and regularization (C) parameter. The optimal value of σ is 20 and 10 for Devanagari and Tamil strokes respectively. The value of C is 100 for both scripts. The discussion of the results in each case follows.

Table 1. Classification accuracy using different zoning strategies for the extraction of spatiostructural features

Zoning Strategy			1-zone	3 Horizontal zones	3 Vertical zones	4-zones	5-zones	6-zones	9-zone Homogeneous	9-zone Heterogeneous
Classification Accuracy	Devanagari	1-best	26.54	43.86	65.12	76.35	79.97	87.17	89.54	90.86
		3-best	47.75	73.15	79.83	86.92	87.30	93.74	94.90	95.56
	Tamil	1-best	39.53	82.35	84.28	81.38	87.69	83.45	89.69	89.97
		3-best	60.80	90.12	91.75	87.42	93.63	89.54	94.87	95.63

The results of the study using *single zone representation* show that the frequency of occurrence of structural features without spatial information is insufficient for classification. Analysis of the confusion matrix and a study of the feature vectors of examples from confusable classes shows that there are groups of different looking strokes which have the same frequencies of structural shape features. Such a group of Tamil strokes is shown in Figure 3(a). It is the confusion among such groups of strokes that causes a high rate of misclassification.

The inclusion of minimal y-positional information with the *3-zone horizontal partitioning* leads to a significant increase in the recognition accuracy. Analysis of the structure of Tamil and Devanagari strokes (with a height of approximately

$\frac{1}{3}^{rd}$ the height of the character or more) reveals an inherent 3-zone structure for most of the stroke classes, with the relative position of the characteristic shape features maintained with respect to implicit horizontal reference lines (Figure 3(b)).

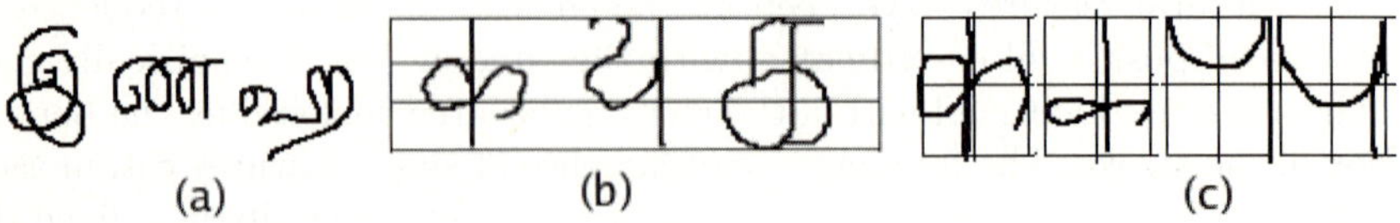

Fig. 3. (a) 3 Tamil strokes which have similar frequencies of occurrence of structural features for the single-zone approach; (b) inherent 3-zone structures in Devanagari and Tamil strokes; (c) 2 pairs of Devanagari strokes showing the translational variations of shape features that alter their zonal locations with respect to 4-zone strategy

With *3-zone vertical partitioning*, the performance obtained is significantly better than previous cases, implying that the discriminable information present in such a vertical partitioning is more compared to a corresponding horizontal partitioning. An implicit localization of structural features into 3 zones as obtained through vertical partitioning is more observable in Tamil strokes, which contributes to the better performance on Tamil data.

The classification accuracy corresponding to *quadrant zoning* is seen to be improved for Devanagari script, but it is less than the 3-zone partitioning for Tamil script. The lack of effectiveness of this kind of partitioning can be attributed to the translational variations of characteristic shape features within the stroke area. In some cases, as shown in Figure 3(c), minor shifts of the features can result in different zonal locations of shape features. These variations reduce discriminability when quadrant zoning is applied.

The *5-zone representation* is seen to have considerable advantage, especially for Tamil strokes, possibly due to the choice of zones such that a reasonable degree of translation invariance has been accommodated. The presence of shape features across overlapped zones gives a more robust measure of spatial information. The choice of zones with a focusing on the central regions of the character and the treatment of corners separately is another salient difference from the previous approaches. The corner features are less susceptible to translational variation because they are associated with line terminals and the bounding box of the stroke. Moreover, since there is observed to be no pair of strokes which have the same occurrence of corner structural features but which are differentiated by the location of the corner, we are able to consider all corner features as belonging to one (the fifth) region. The performance has largely improved with this zoning strategy for Tamil strokes.

Partitioning a stroke into *6 homogeneous spatial zones* increases spatial localization and provides x positional information in addition to the 3-zone horizontal partitioning approach. This additional information is observed to be more useful for Devanagari script.

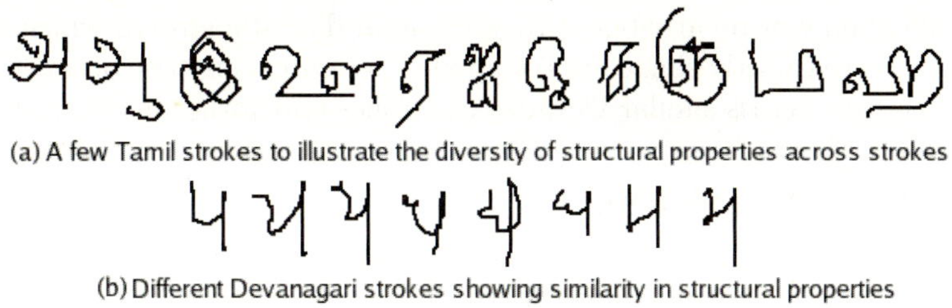

Fig. 4. Figure showing structural properties of Devanagari and Tamil strokes

The amount of spatial information available in the *9-zone representations* is seen represent Devanagari and Tamil strokes well. The length and sparsity of the feature vector have also increased compared to the previous approaches. The feature vector has a dimension of 153 with an average of 10 – 20 non-zero feature values, depending on the complexity of the shape of a stroke.

Heterogeneous 9-zone partitioning handles spatial displacement of structural features within the stroke (which occurs naturally due to variations in handwritten characters). The best observed recognition accuracy is 90.86% for Devanagari strokes and 90.77% for Tamil strokes, which corresponds to the 9-zone heterogeneous partitioning strategy with σ values for the Gaussian kernel chosen empirically as 20 and 15 respectively. Thus by this partitioning method, we are able to get a considerable improvement over the structural approach described in [1].

In the studies that have been presented, it is generally observed that the recognition accuracy for Tamil strokes is higher than that for Devanagari strokes, which can be explained based on the structural properties of Devanagari and Tamil strokes. Most of the Tamil characters (consonants, vowels, consonant-vowel or conjunct-consonant-vowel) are written as single strokes. The number of stroke classes considered for Tamil is 93, as compared to 117 for Devanagari. However, the distinctive structural features of Tamil strokes have more to add to discriminability than their number. There are well-defined shape differences even across similar Tamil strokes, which are not likely to be confused with noise or with variations in writing a stroke. Some typical Tamil strokes have been plotted in Figure 4(a).

Most Devanagari characters have a central vertical line with characterizing shape features occurring on the left or on both sides. For such characters, we can identify a nuclear stroke structure of which many other strokes can be realized as perturbations. Common ways of writing such Devanagari characters are as a single stroke or as 2 (or rarely 3) strokes. With 2 or more strokes, each unit represents a smaller part of the character and some units are seen to be repeated across characters. This kind of realization poses a problem when the perturbations intended for a different stroke are likely to be sources of confusion for variations from the "neat" way of writing the stroke. A number of Devanagari strokes which appear to be derived from perturbations of a nuclear stroke structure are illustrated in Figure 4(b).

Analysis of the confusion matrix also shows that most of the misclassified strokes are small in size have a simple structure. For Devanagari strokes, the misclassification can be attributed to very small diacritic strokes which have a simple structure (mostly single lines or arcs). The confusions arise between structural differences across similar Devanagari strokes and variations for a stroke.

5.1 Performance Comparison

The classification accuracy reported for a structural classifier built on Tamil strokes using online features is 82.80%[1]. The performance of the classifier using spatiostructural features is also compared with the performance using purely online spatiotemporal features (Table 2). The performance of spatiotemporal features is observed to be better as recognition accuracies with online features are expected to be higher[11]. Spatiostructural features can be observed to offer discriminability comparable to spatiotemporal features along with complementarity of information. They also possess writing direction invariance which is not present in a representation using spatiotemporal features.

Table 2. Comparison of the accuracy of spatiostructural features based classifier with positional coordinate sequence features, using SVM-based classifiers for both cases

Feature	Spatiostructural	Spatiotemporal
Devanagari (%)	90.86	95.14
Tamil (%)	90.77	92.03

6 Summary and Conclusion

The spatiostructural features that describe a handwritten stroke are extracted by the evaluation of the frequency of occurrence of structural features in zones defined within the bounding rectangle of the stroke. The empirical determination of optimal zoning strategies for Devanagari and Tamil strokes has been described. It has been observed that increasing the localization information of the features adds to discriminability in most cases. Incorporation of translational invariance for structural features is seen to enhance recognition accuracy and it has been studied with heterogeneous zoning approaches. The best of the zoning strategies identified in this study are seen to yield over 90% classification accuracy for both Devanagari and Tamil writing systems. The Tamil recognition system based on zoning has already been integrated into a handwriting interface used in real word applications. The offline-like features are observed to be more suited to strokes having larger curve lengths, especially complex strokes that represent conjunct consonants and composite consonant-vowel units which do not have such high classification accuracies with the online features based systems.

We must also keep in mind that the systems built with offline features are not expected to perform as well as those built with online features. The complementary information provided by such systems is useful for disambiguation of confusable strokes and for classifier ensembles. The advantage of spatiostructural features is that it gives a robust direction-invariant representation of the stroke, using offline-like features traditionally extracted from images. But the proposed method of extraction of shape features is done from online data and consequently, it is easier to implement, more accurate and free from the overhead associated with the storage and processing of offline character images.

Future work in this area is directed towards automating the identification of optimal zones for each dataset under study. The zoning strategy should not be too complex with complicated boundary information and zones that are far from regular. There should not be a large number of zones, either, because in all such cases, the complexity of processing and sparsity of the feature vector will be very high.

References

1. Aparna, K.H., Subramanian, V., Kasirajan, M., Prakash, G.V., Chakravarthy, V.S., Madhvanath, S.: Online handwriting recognition for Tamil. In: Proceedings of the Ninth International Workshop on Frontiers in Handwriting Recognition, pp. 438–443 (2004)
2. Brakensiek, A., Kosmala, A., Willett, D., Wang, W., Rigoll, G.: Performance evaluation of a new hybrid modeling technique for handwriting recognition using identical on-line and off-line data. In: Proceedings of the International Conference on Document Analysis and Recognition, pp. 446–449 (1999)
3. Brown, R.M.: On-line computer recognition of handprinted characters. IEEE Transactions on Electronics and Computers EC-13, 750–752 (1964)
4. Chakravarthy, V.S., Kompella, B.: The shape of handwritten characters. Pattern Recognition Letters 24(12), 1901–1913 (2003)
5. Viard-Gaudin, C., Lallican, P., Knerr, S.: Recognition-directed recovering of temporal information from handwriting images. Pattern Recognition Letters 26(16), 2537–2548 (2005)
6. Hall, R.E., Hulbert, L.N.: Machine recognition of symbols. U.S. Patent 3676848 (1972)
7. Hanaki, S., Yamazaki, T.: On-line recognition of handprinted Kanji characters. Pattern Recognition 12, 421–429 (1980)
8. Impedovo, S., Lucchese, M.G., Pirlo, G.: Optimal zoning design by genetic algorithms. IEEE Transactions on Systems, Man and Cybernetics Part A 36(5), 833–846 (2006)
9. Li, X., Plamondon, R., Parizeau, M.: Model based on-line handwritten digit recognition. In: Proceedings of the 14th International Conference on Pattern Recognition, 2004, vol. 2, pp. 1134–1136 (1998)
10. Li, X., Parizeau, M., Plamondon, R.: Segmentation and reconstruction of on-line handwritten scripts. Pattern Recognition 31(6) (1998)
11. Plamondon, R., Srihari, S.N.: Online and off-line handwriting recognition: a comprehensive survey. IEEE Trans. Pattern Anal. Mach. Intell. 22(1), 63–84 (2000)
12. Trier, O., Jain, A., Taxt, T.: Feature extraction methods for character recognition - a survey. Pattern Recognition 29, 641–662 (1996)
13. Verma, B., Lu, J., Ghosh, M., Ghosh, R.: A feature extraction technique for online handwriting recognition. In: Proceedings of the IEEE International Joint Conference on Neural Networks, vol. 2, pp. 1337–1341 (2004)
14. Vinciarelli, A., Perrone, M.: Combining online and offline handwriting recognition. In: Proceedings of the Seventh International Conference on Document Analysis and Recognition, pp. 844–848 (2003)

An Improved Version of the Wrapper Feature Selection Method Based on Functional Decomposition*

Noelia Sánchez-Maroño, Amparo Alonso-Betanzos, and Beatriz Pérez-Sánchez

University of A Coruña, Department of Computer Science, 15071 A Coruña, Spain
nsanchez@udc.es, ciamparo@udc.es, bperezs@udc.es

Abstract. This paper describes an improved version of a previously developed ANOVA and Functional Networks Feature Selection method. This wrapper feature selection method is based on a functional decomposition that grows exponentially as the number of features increases. Since exponential complexity limits the scope of application of the method, a new version is proposed that subdivides this functional decomposition and increases its complexity gradually. The improved version can be applied to a broader set of data. The performance of the improved version was tested against several real datasets. The results obtained are comparable, or better, to those obtained by other standard and innovative feature selection methods.

1 Introduction

Enhancing learning machine generalization often motivates feature selection. Feature selection reduces the number of original features by selecting a subset that still retains enough information to achieve a good performance. In general, feature selection approaches can be grouped into two categories [1]:

- Filter models that rely on the general characteristics of the training data to select certain features without the need for any learning algorithm.
- Wrapper models that require a predetermined learning algorithm for feature selection and that use this to evaluate and determine which features should be selected.

Although wrapper models tend to be more computationally expensive that filter models, since they tend to find features better suited to the predetermined learning algorithm, they typically result in better performance [2].

The AFN-FS (ANOVA and Functional Networks Feature Selection) is a wrapper method based on functional networks and an analysis of variance decomposition [3]. In several experiments this method produced accurate results while maintaining a reduced set of variables. However, its main disadvantage is exponential complexity, given a large number of features. This drawback limits the

* This work has been funded in part by Project PGIDT05TIC10502PR of the Xunta de Galicia.

J. Marques de Sá et al. (Eds.): ICANN 2007, Part II, LNCS 4669, pp. 240–249, 2007.

application of the method to datasets with a small number of features. This paper describes a modification of the AFN-FS method that overcomes this drawback. The new version of the method is applied to real-world classification datasets and its performance results are compared to those obtained by other innovative feature subset selection methods.

2 ANOVA and Functional Networks Feature Selection

2.1 A Brief Introduction to the AFN Method

The ANOVA and Functional Networks Feature Selection (AFN) method, as its name indicates, is based on ANOVA and Functional Networks [4], which estimates a function f in terms of n variables, $f(x_1, x_2, \ldots, x_n)$, by approximating its functional components.

According to Sobol [5], any square integrable function, $f(x_1, x_2, \ldots, x_n)$, can always be written as the sum of the 2^n orthogonal summands:

$$f(x_1, \ldots, x_n) = f_0 + \sum_{i=1}^{n} f_i(x_i) + \sum_{i=1}^{n-1} \sum_{j=i}^{n} f_{ij}(x_i, x_j) + \cdots + f_{12\ldots n}(x_1, x_2, \ldots, x_n).$$

This can be rewritten, in a simplified form, as:

$$f(x_1, \ldots, x_n) = f_0 + \sum_{\nu=1}^{2^n-1} f_\nu(\mathbf{x}_\nu) \tag{1}$$

where ν represents each possible subset formed with the variables $\{x_1, x_2, \ldots, x_n\}$ and where f_0 is a constant that corresponds to the function with no arguments.

Furthermore, if $f(x_1, x_2, \ldots, x_n)$ is square integrable, then each summand is also square integrable. Hence:

$$\int_0^1 \int_0^1 \cdots \int_0^1 f^2(x_1, \ldots, x_n) dx_1 dx_2 \ldots dx_n - f_0^2 = \sum_{\nu=1}^{2^n} \int_0^1 f_\nu^2(\mathbf{x}_\nu) d\mathbf{x}_\nu.$$

Denoting the left part of this equation as D, the variance, and each summand in the right part as D_ν:

$$D = \sum_{\nu=1}^{2^n-1} D_\nu.$$

The variance of the initial function can be obtained by summing the variance of the components. This enables global sensitivity indices to be assigned to the different functional components, adding up to one, such that:

$$GSI_\nu = \frac{D_\nu}{D} \qquad \nu - 1, 2, \ldots, 2^n - 1.$$

The AFN method approximates each functional component $f_\nu(\mathbf{x}_\nu)$ in (1) as:

$$f_\nu(\mathbf{x}_\nu) = \sum_{j=1}^{k_\nu} c_{\nu j} p_{\nu j}(\mathbf{x}_\nu), \tag{2}$$

where $c_{\nu j}$ are the parameters to be estimated and where p_ν is a set of orthonormalized basis functions. There are several alternative ways of choosing these functions [4]. One is to use one of the families of univariate orthogonal functions (for example, Legendre polynomials), form tensor products and select a subset.

The $c_{\nu j}$ parameters are learnt by solving an optimization problem:

$$Minimize \quad J = \sum_{s=1}^{m} \epsilon_s^2 = \sum_{s=1}^{m} (y_s - \hat{y}_s)^2, \tag{3}$$

where m is the number of available samples, y_s is the desired output for the sample s, and $\hat{y}_s$ is the estimated output obtained by:

$$\hat{y}_s = \hat{f}(x_{s1}, \ldots, x_{sn}) = f_0 + \sum_{\nu=1}^{2^n-1} \sum_{j=1}^{k_\nu} c_{\nu j} p_{\nu j}(\mathbf{x}_{s\nu}). \tag{4}$$

2.2 The AFN-FS Method

An extension of the AFN method was proposed as a feature selection method in [3]. This is a wrapper method that applies a backward selection search but discards several features in each step. The AFN method is used as an induction algorithm that guides the search process based on the following indices that are directly derived from the coefficients $c_{\nu j}$ in (4):

- Global sensitivity indices (GSI), obtained as the sums of the squares of the coefficients c_ν:

$$GSI_\nu = \sum_{j=1}^{k_\nu} c_{\nu j}^2 \quad \nu = 1, 2, \ldots, 2^n - 1. \tag{5}$$

- Total sensitivity indices (TSI), calculated for each variable x_i adding the overall global sensitivity index of each subset ν where the feature x_i is included:

$$TSI_i = \sum_{\nu=1}^{2^n-1} GSI_\nu \text{ such that } x_i \in \nu \quad i = 1, \ldots, n \tag{6}$$

It is important to emphasize that, while GSI denotes overall importance, in terms of the variance of each functional component in (1), TSI indicates the importance of each feature.

The AFN method was initially intended to solve regression problems [4]. However, as most of the feature selection studies [1,6,7] refer to classification problems, the mean squared error used for the optimization problem in (3) may not be the most suitable [8]. Hence, cross-entropy was selected for the cost function in order to improve results. For binary classification tasks, the optimization problem in (3) can thus be written as:

$$Minimize \quad J = -\sum_{s=1}^{m} y_s ln(\hat{y}_s) + (1 - y_s) ln(1 - \hat{y}_s). \tag{7}$$

This wrapper method requires an evaluation function, and, in this case, chosen was mean accuracy from a five-fold cross validation (as in [1]). The pseudo-code in Figure 1 illustrates the different steps in the AFN-FS method, the main points of which are discussed below.

As a backward selection method, the selection process starts with the complete set of features. An initial selection is done (lines 1-4) to ensure that some features are discarded (note that the main **repeat** – **until** loop in lines 6-26 of the pseudo-code may not discard any features). The term EVAL (first mentioned

Input : Complete set of features $\{\mathbf{x_1}, \mathbf{x_2}, \ldots, \mathbf{x_n}\}$ and the desired output $\mathbf{y}$
Output : Reduced set of features $\{\mathbf{x_{t1}}, \mathbf{x_{t2}}, \ldots, \mathbf{x_{tr_t}}\} \mid \mathbf{r_t} < \mathbf{n}$ and an estimate of the desired output $\hat{\mathbf{y}}$

1 $acc_{ini}, \mathbf{C_{ini}} \leftarrow$ EVAL$(\{\mathbf{x_1}, \mathbf{x_2}, \ldots, \mathbf{x_n}\})$
2 Calculate GSI in (5) and TSI in (6) for each set of parameters in $\mathbf{C_{ini}}$
3 Discard features according to the thresholds σ_{Tini}, σ_{Gini} and σ_{Kini}
4 Obtain a reduced set of features $S_{ini} = \{\mathbf{x_{ini1}}, \mathbf{x_{ini2}}, \ldots, \mathbf{x_{inir_{ini}}}\}$
5 $acc_{(t-1)}$, $\mathbf{C_{(t-1)}} \leftarrow$ EVAL(S_{ini})
6 **Repeat**
7 Calculate GSI in (5) and TSI in (6) for each set of parameters in $\mathbf{C_{(t-1)}}$
8 Discard features according to the thresholds σ_T, σ_G and σ_K
9 Obtain a reduced set of features $S_t = \{\mathbf{x_{t1}}, \mathbf{x_{t2}}, \ldots, \mathbf{x_{tr_t}}\}$
10 acc_t, $\mathbf{C_t} \leftarrow$ EVAL(S_t)
11 **If** $(acc_t < acc_{t-1})$ **then**
12 **Repeat**
13 Increase the number of coefficients $(ncoef_t)$ with the same set of features
14 $acc'_t, \mathbf{C'_t} \leftarrow$ EVAL$(\mathbf{x_{t1}}, \mathbf{x_{t2}}, \ldots, \mathbf{x_{tr_t}})$
15 **until** $(ncoef_t > ncoef_{t-1})$ or $(acc'_t \geq acc_{t-1})$
16 **If** $(acc'_t > acc_t)$ **then**
17 $acc_t \leftarrow acc'_t$
18 **Endif**
19 **Endif**
20 **If** $(acc_t < acc_{t-1})$ **then**
21 **Repeat**
22 Add a previously discarded feature, $S'_t = \{\mathbf{x_{t1}}, \mathbf{x_{t2}}, \ldots, \mathbf{x_{tr_t}}\} \cup \mathbf{x_{(t-1)d}}$
23 $acc'_t, \mathbf{C'_t} \leftarrow$ EVAL(S'_t)
24 **until** $(card(S'_t) \geq card(S_{(t-1)}))$ or $(acc'_t \geq acc_{t-1})$
25 **Endif**
26 **until** $(acc_t > acc_{t-1})$

Fig. 1. Pseudo-code for the AFN-FS algorithm. EVAL stands for the evaluation function and *card* is abbreviation for cardinality.

in line 1) refers to a complex process involving the induction algorithm and the AFN method (briefly explained below in Figure 2). Note that the **for** loop in Figure 2 is required because of the five-fold cross validation.

Input : Set of features $\{\mathbf{x_{t1}}, \mathbf{x_{t2}}, \ldots, \mathbf{x_{tr_t}}\} \mid \mathbf{r_t} \leq \mathbf{n}$ and the desired output $\mathbf{y}$
Output : Mean accuracy, *acc*, and a set of learnt parameters, C

27 C $\leftarrow \emptyset$
28 **for** $i \leftarrow 1$ **to** 5
29   Approximate the desired output using (4)
30   Solve the optimization problem in (7)
31   Obtain the learnt parameters (c_ν)
32   C $\leftarrow$ C $\cup \{c_\nu\}$
33   Calculate accuracy, acc_i
34 **endfor**
35 Calculate mean accuracy, *acc*

Fig. 2. Pseudo-code for the EVAL process

The accuracy in line 1 (acc_{ini}) and subsequent lines in Figure 1 is the mean accuracy from the five-fold cross-validation (for the sake of clarity, the standard notation was not used). However, the set ($\mathbf{C_{ini}}$) includes five different sets of parameters for each of which the corresponding GSI and TSI indices are calculated (line 2). The step in line 3 is subdivided into:

- Select those features whose TSI value is above an established threshold σ_{Tini} in a minimum of σ_{Kini} folds.
- Using the GSI value, determine if a feature is important in its own right or in combination with other features. Features or combinations of features under the threshold σ_{Gini} are eliminated.

Once several features are discarded, the selection process is repeated iteratively (lines 6-26) as long as the mean accuracy obtained in the present state (acc_t) is higher or equal than that obtained in the previous state (acc_{t-1}). The selection process (lines 7-10) has been explained above, so we will just focus on the steps in the method that overcome degradation in accuracy, that is, lines 11-25 (for further details, see [3]):

- Increase the number of coefficients. Reducing the number of features means decreasing the number of coefficients. Even with the right features, a reduced number of parameters may not achieve a good estimate. This step increases the number of coefficients, by considering more complex functions, in order to avoid the degradation of the estimated output.
- Include discarded variables. Several variables can be discarded in one step according to the global and total sensitivity indices. When a method becomes less accurate, this step enables thresholds (σ_G and σ_T) to be reestablished in order to reconsider some of the discarded variables.

3 Improving the AFN-FS Method

The AFN-FS method produces a satisfactory performance [3], and even outperforms other feature selection methods while reducing the number of selected features. It also has other advantages, such as allowing several variables to be discarded in one step and interpreting the relevance of each selected or discarded feature. However, it has one major drawback that limits its application, namely, the exponential complexity of the functional decomposition in (1). Therefore, the AFN-FS method can only be applied to datasets with a high ratio between the number of samples, m, and the number of features, n. Specifically, the relation should be $m > 2^n - 1$. This requirement is not feasible in real-world datasets. Therefore, we propose several modifications to the AFN-FS method that will improve its scope of application. The pseudo-code for the AFN-FS method, as illustrated in Figure 1, remains largely unchanged. Nevertheless, instead using the equation in (4), we designed an incremental approximation to estimate desired output.

3.1 Incremental Functional Decomposition

The main modification consists of handling functional decomposition *incrementally.* This means that the desired output is initially estimated using a simple approximation that does not include interactions between features. Thus, only the univariate components in (1) are considered and all the other components are 0:

$$\hat{y}_s = f_0 + \sum_{i=1}^{n} \sum_{j=1}^{k_i} c_{ij} p_j(x_{si}), \tag{8}$$

where s is a specific sample, n is the number of initial features and k_i is the set of functions employed to estimate the functional component i. In this case, only the feature x_i is considered in the functional component.

Note that since the feature selection process depends on the function learnt, the mean accuracy in (8) should have be high.

GSI and TSI are derived from the learnt parameters c_{ij} in the previous estimation. As no interactions between features are considered, both sets of indices are exactly equal. These indices point to the most relevant features, and using these, the bivariate components in the functional decomposition can be considered for estimating the desired output:

$$\hat{y}_s = f_0 + \sum_{i=1}^{r_t} \sum_{j=1}^{k_i} c_{tij} p_j(x_{sti}) + \sum_{i_1=1}^{r_t - 1} \sum_{i_2=i_1}^{r_t} \sum_{j=1}^{k_{(i_1,i_2)}} c_{(ti_1,ti_2)j} p_j(x_{sti_1}, x_{sti_2}),$$

where $\{x_{st1}, x_{st2}, \ldots, x_{str_t}\}$ is the subset of selected features in this step t.

Again, this estimation produces a set of TSI and GSI values that suggest a subset of features from the previous set. A new approximation is implemented using this subset of features and including the trivariate components of the functional decomposition. The same process continues, with the addition of new functional components increasing the complexity of the approximation. The selection process described in lines 1-4 and lines 5,7-8 in Figure 1 is thus subdivided into several substeps that gradually increase complexity.

3.2 Repeating the Initial Selection

Using the modification explained in the previous section, the initial subset of features is obtained by a simple approximation (see equation 8). Nevertheless, in the previous AFN-FS method, this initial selection is never reconsidered, although the next selections depend on it. Therefore, it is necessary to guarantee the adequacy of this first selection. Therefore, a new loop is included in the pseudo-code shown in Figure 1 that repeats the initial selection until the accuracy obtained with the subset of features is greater than a threshold (σ_{acc}). Thus, line 1 in Figure 1 changes to:

Repeat
  $acc_{ini}, \mathbf{C_{ini}} \leftarrow \text{EVAL}(\{\mathbf{x_1}, \mathbf{x_2}, \ldots, \mathbf{x_n}\})$
until $(acc_{ini} > \sigma_{acc})$

The goal of this first selection is to discard as many features as possible, but without negatively affecting mean accuracy.

3.3 Establishing the Thresholds

The AFN-FS method is very dependent on the different thresholds employed (σ_T, σ_G and σ_{acc}). In order to develop a fully automated process, an initial value for these thresholds had to be determined. This is no easy task, as thresholds tend to vary from one dataset to another.

For the threshold σ_T, the idea was to establish an initial value according to different metrics obtained from the set $\{TSI_{t1}, TSI_{t2}, \ldots, TSI_{tr_t}\}$, where r_t is the number of selected features in the step t of the process, $r_t <= n$. Several experiments were conducted using different combinations of the mean and the standard deviation for the set. The final value established was:

$$\sigma_T = \overline{TSI} - \frac{SD_{TSI}}{2}, \tag{9}$$

where $\overline{TSI}$ is the mean of the set $\{TSI_{t1}, TSI_{t2}, \ldots, TSI_{tr_t}\}$ and where SD is the standard deviation.

Although a similar process was implemented to assign σ_G, the high variability of the global sensitivity indices –which depend on the problem being solved– makes it very difficult to establish a starting value. However, several trials with different datasets indicated that using a value between $[0.01 - 0.02]$ led to a good result. What this means is that the features or combinations of features that represent more than 1% or 2% of the total variance need to be taken into account.

As for σ_{acc}, this was fixed as the value obtained after training the AFN method with the complete set of features.

4 Experimental Results

In order to evaluate the method described above, some real-world classification problems used in previous studies [1,6,7] were selected. All are binary classification problems that can be obtained from the UCI-Irvine repository [9], and only the first problem could be dealt with using the original AFN-FS method. Table 1 describes these problems briefly.

Table 1. Dataset description. Baseline accuracy: accuracy when the main class is selected.

Dataset	Features	Number of Samples	Baseline Accuracy
crx	15	699	55.51
wdbc	30	569	62.74
wpbc	32	198	76.26
Ionosphere	34	351	64.10

The AFN method estimates a function by approximating its functional components from a family of basis functions that has to be orthonormalized. In a first approach, the polynomial family was selected for the experiments, and the univariate polynomial functions $\{1, x_1, x_1^2, x_1^3\}$ were selected, leading to the following set of orthonormalized functions:

$$\begin{aligned}\{p_{1;1}(x_1), p_{1;2}(x_1), p_{1;3}(x_1)\} =& \{\sqrt{3}(2x_1 - 1), \sqrt{5}(6x_1^2 - 6x_1 + 1),\\ & \sqrt{7}(20x_1^3 - 30x_1^2 + 12x_1 - 1)\}.\end{aligned} \tag{10}$$

Tensor products with these functions were formed to obtained bivariate and trivariate functions. If we select polynomials of degree d, as univariate basis functions for the n-dimensional basis functions, the tensor product technique leads to polynomials of degree $d \times n$, which is too high. However, we can limit the degree of the corresponding n-multivariate basis to contain only polynomials of degree d_n or less. This was done with the dataset in our case, thus limiting d_n to 4 or 5, depending on the problem. Note that these bases are obtained independently of the dataset, which means that they are valid for all the datasets considered.

It is important to note that the experimental results used for the comparative study were extracted from different works. Kohavi and John [1] described a broad-based feature selection study mainly devoted to wrapper methods. These wrapper methods were obtained by combining different induction algorithms (Naive-Bayes, ID3 and C4.5) with standard search strategies (best-first-search and hill-climbing). An innovative method for feature subset selection based on neural networks and ant colony optimization (ANN-AC) is explained in [7]. Finally, Liu and Zhen use Support Vector Machines to develop an interesting method called FS-FSF (Filtered and Supported Sequential Forward Search) [6]. For a fair comparison, the training and test sets for our research were generated in the same way as in those studies. Ten-fold cross-validation was employed in [1] while, in the other works [6,7], 20% of the samples were randomly selected to construct the test set. The rest of the samples form a training set that was used for the feature selection process. Twenty different simulations were carried out for each dataset in [6]. Results are summarized in Table 2.

Table 2. Results for the improved AFN-FS method compared to results for other methods. Features: the mean number of features selected. Test acc: accuracy for test datasets. BFS: best-first-search. HC: hill-climbing. ANN-AC: artificial neural networks and ant colony optimization (see [7]).

Method	Dataset	Other-studies		Improved AFN-FS method	
		Features	Test acc	Features	Test acc
AFN-FS	crx	7.2	85.05	6.2	85.36
Naive-Bayes-BFS		5.9	86.23		
ID3-HC		2.9	85.65		
C4.5-BFS		7.7	85.80		
ANN-AC	wpbc	14	77.50	8	78.00
	wdbc	12	95.57	11	99.00
FS-SFS	ionosphere	10	92.00	19	92.37

As previously mentioned, only the *Crx* dataset could be evaluated with the previous AFN-FS method and the modified method, because of the impossibility of applying the previous AFN-FS method to the other datasets. Note that, with respect to the previous AFN-FS results, the mean accuracy for this dataset increases as the number of features is reduced. Furthermore, accuracy results are similar to those obtained by the other methods, although they indicate that more significant feature reduction can be achieved. In comparison with the ANN-AC method, the improved AFN-FS method performs better with fewer features. Finally, the improved AFN-FS method is slightly more accurate using more features than the FS-SFS method for the ionosphere dataset. It is important to note that this is a preliminary study and better performance results might be obtained using families of functions other than (10). Finally, regarding computational time, although the method requires the evaluation of different subsets of features, this evaluation is quite fast and the number of steps is reduced because several variables are discarded in each. For example, it took around 2.5 seconds to evaluate each possible subset of the *wdbc* dataset, and required 11 different steps to reach a solution, for a total time of 11×2.5 seconds.

5 Conclusion

This paper describes several modifications to the AFN-FS method that extend its scope of application and improve performance results. The modifications maintain the useful properties of the AFN-FS method, namely: a) several variables can be discarded in a single step, without the need to check all the possible subsets, and b) the importance of each feature is given in terms of variance, which allows results to be interpreted. The same modifications, however, also overcome the main limitation of the method, namely, its exponential complexity. It does this by subdividing the functional decomposition used to estimate a given function. In order to obtain a fully automated method that was less dependent on the data, we conducted several experiments to establish suitable threshold values. The experimental results indicate the adequacy of the method proposed. Note, however, that this is a work in progress, and that better results are likely to be obtained using other families of functions. Future research will address the application of the AFN-FS method to multi-class problems.

References

1. Kohavi, R., John, G.: Wrappers for feature subset selection. Artificial Intelligence, Special issue on relevance 97(1-2), 273–324 (1997)
2. Blum, A.L., Langley, P.: Selection of relevance features and examples in machine learning. Artificial Intelligence, Special issue on relevance 97(1-2), 245–271 (1997)
3. Sánchez-Maroño, N., Caamaño-Fernández, M., Castillo, E., Alonso-Betanzos, A.: Functional networks and analysis of variance for feature selection. In: Corchado, E., Yin, H., Botti, V., Fyfe, C. (eds.) IDEAL 2006. LNCS, vol. 4224, Springer, Heidelberg (2006)
4. Castillo, E., Sánchez-Maroño, N., Alonso-Betanzos, A., Castillo, M.: Functional network topology learning and sensitivity analysis based on anova decomposition. Neural Computation 19(1) (2007)
5. Sobol, I.M.: Global sensitivity indices for nonlinear mathematical models and their Monte Carlo estimates. Mathematics and Computers in Simulation 55, 271–280 (2001)
6. Liu, Y., Zheng, Y.F.: A novel feature selection method for support vector machines. Pattern Recognition 39, 1333–1345 (2006)
7. Sivagaminathan, R.K., Ramakrisham, S.: A hybrid approach for feature subset selection using neural networks and ant colony optimization. Experts systems with applications 33, 49–60 (2007)
8. Bishop, C.: Neural Networks for Patter Recognition. Oxford University Press, New York (1995)
9. Blake, C., Merz, C.: UCI repository of machine learning databases (1998), http://www.ics.uci.edu/mlearn/MLRepository.html

Parallel-Series Perceptrons for the Simultaneous Determination of Odor Classes and Concentrations

Gao Daqi[1,2], Sun Jianli[1], and Li Xiaoyan[1]

[1] Department of Computer Science, East China University of Science and Technology, Shanghai 200237, China
[2] State Key Laboratory of Bioreactor Engineering, East China University of Science and Technology, Shanghai 200237, China

Abstract. The simultaneous determination of odor classes and concentrations is solved by a kind of parallel-series perceptron models. Two groups of parallel single-output perceptrons are in series, and the former is responsible for classification, and the latter for location. The number of parallel perceptrons is equal to the number of odor classes. A multi-class learning problem is first decomposed into multiple two-class problems, and then solved by multiple parallel perceptrons, one by one. Each training subset is composed of the most necessary samples. And furthermore, some virtual samples are added to the weak side of any two-class learning subsets in order to arrive at a virtual balance. The experimental results for 4 kinds of fragrant materials show that the proposed parallel-series perceptrons with the electronic nose are effective.

Keywords: Parallel-series perceptrons, simultaneous determination, odor classes and concentrations, electronic nose.

1 Introduction

An electronic nose is an instrument which comprises an array of electronic chemical sensors with partial specificity and an appropriate pattern recognition model, capable of recognizing simple or complex odors [1-2]. This kind of technique has wide application prospects in such fields as foods, fragrances and flavors, cosmetics, etc [1-2]. At the present day, however, electronic noses are only used to determine the classes of odors [3]. From a long-term point of view, electronic noses will unavoidably be employed to simultaneously estimate the classes, concentrations or strengths of odors. In order to implement the task, we need both a superior electronic nose and an appropriate classification method.

Odors can be neither seen by eyes nor felt by hands. Not only that, it is well- known that the strengths or concentrations of odors are closely related to the environmental temperature. A material may be odorous when the environmental temperature is over a certain critical point; otherwise odorless when below. Therefore, a key point is how to make an electronic nose keep constant in spite of the changing environment. In addition, it is very important to avoid the operation effect given by different persons for the purpose of improving the reproducibility of electronic noses.

J. Marques de Sá et al. (Eds.): ICANN 2007, Part II, LNCS 4669, pp. 250–260, 2007.

In mathematics, the problem of how to simultaneously determinate the classes and strengths of odor samples is a multi-input multi-output (MIMO) function approximation problem if its physical meaning is left aside. There are the following several kinds of solutions. (A). Treat such a problem as a classification problem, one concentration for one class, and employ either an MIMO or multi-input single-output (MISO) classifiers [4-5] to solve it. (B). Look upon such a problem first as a classification problem and then as multiple MISO approximation problems [6]. (C). Regard such a problem as an approximation problem, and use multiple approximation model ensembles to solve it [3]. Because methods (A) and (C) are too complicated in structure and relatively low in predicted accuracy, this paper pays main attention to approach (B).

Now no people yet suspect the capability of perceptrons for solving the small-scale classification problems [4-5]. However, the structure optimization and learning algorithms of multi-layer perceptrons (MLPs) suitable for the small-scale problems cannot be simply generalized to the larger-scale ones. As the number of classes increases, the computational complexity will quickly reach an unmanageable proportion [4-5]. What is more serious is that an MLP becomes quite easy to be caught in some local points during learning when the number of categories is relatively many. Therefore, up to now it is still a challenging task for MLPs to solve the multi-class learning problems. Modular neural networks [7-9] and neural trees [10] thus emerge as the times require. And the class-modular networks are especially attractive because they are relatively simple in structure and quite effective for the large-set classification problems [7]. It is a pity that the modular networks unavoidably run up against the imbalanced training problems when an n-class learning task is decomposed into n two-class tasks. Some solutions aiming at boosting up the minority training data examples include prior duplication and snowball [8, 11], re-calculation for the weight updated direction [12], example filtering and pruning from the majority classes [13-14], re-sampling [15], etc. However, the existing methods often result in the huge increase of computational load and the decrease of classification accuracy [8-9].

This paper uses parallel-series perceptron models to simultaneously determine the classes and concentrations of odor samples. Our ideas are as follows. Such a problem is regard as multiple MISO classification problems and solved with multiple single-output perceptrons. After determining the category of one odor sample by an MISO MLP, we then employ another MISO MLP to estimate its concentration and strength. In order to attain the goal, this paper will study the task decomposition of large-scale learning sets, the virtual balance of unbalanced sample distribution, and the structure optimization and fast learning of parallel-series perceptrons.

2 A Practical Electronic Nose

See Ref. [3] for details, omitted here.

3 Parallel-Series MLPs and Task Decomposition

3.1 Parallel-Series Perceptrons

We use parallel-series MLPs to simultaneously determine the categories and concentrations of large-scale samples from n kinds of odors. At the learning stage, the problem is regarded first as an MIMO classification task and then as an MIMO function approximation task if its physical content is left aside. The MIMO classification task is accomplished by n single-output MLPs, and so is the following MIMO function approximation task. In that way, parallel-series MLPs come into being, shown as Fig. 1. According to the figure, two single-output MLPs in serial are on behalf of a specified odor, and the first is responsible for classification and the second for approximation. In order to determine their structures, each MLP for classification learns all the samples from the represented class and a small part from the neighboring, but each MLP for approximation only learns all the samples from the represented class. There is no need for the MLPs to learn the original large-scale dataset. At the stage of decision making, for a specified sample $\boldsymbol{x}$, all n MLP classifiers ought to give their respective predictions. The MLP classifier with the max output, say MLP classifier $j1$, determines the label of $\boldsymbol{x}$, and then, MLP $j2$ for approximation estimates the position of $\boldsymbol{x}$, shown as Fig. 2. In other words, we need to use n MLP classifiers to determine $\boldsymbol{x}$'s category according the max combination rule, but only a single MLP to estimate $\boldsymbol{x}$'s concentration.

3.2 Task Decomposition and Virtual Balance of Imbalanced Dataset

The discussion given in the section about the solutions of task decomposition and imbalanced dataset is only suitable to MLP classifiers, because there are not such problems for MLPs for approximation. Each MLP for approximation only learns the samples from the presented class, and the problem of data imbalance does not exist.

It is easy to decompose an n-class problem into n two-class ones. Naturally, n single-output MLP classifiers come into being, and each is responsible for forming the decision boundaries of its represented class. What is important is how to get rid of those distant futile patterns. Let us take the shaping of a 2-class training subset $\Xi^{(j)}=\{\boldsymbol{X}^{(j)}, \boldsymbol{X}^{(\sim j)}\}$ as an example to go into details on the formation of economic learning subsets, as shown in Fig. 3. The sample set $\boldsymbol{X}^{(j)}$ from class ω_j are expressed by dashed line in magenta; otherwise in other colors. When MLP classifier j is used to solve the 2-class problem $\{\omega_j, \sim\omega_j\}$, say to separate $\boldsymbol{X}^{(j)}$ from $\boldsymbol{X}^{(\sim j)}$, an evident fact is that a large number of distant samples from $\sim\omega_j$ may have no any use for the formation of decision boundaries of ω_j. In other words, besides all the samples from ω_j, it is enough to only let a small part of samples from $\sim\omega_j$ take part in training MLP classifier j. Though the samples from ω_j may distribute in irregular regions, we are always able to enclose all of them with an initial oblique ellipsoid Θ_{j0}. Of course, some samples from $\sim\omega_j$ will also be included in Θ_{j0} at the moment. Next we can expand Θ_{j0} to some suitable sphere. Our intention is like this. If within Θ_{j0} the ratio of number of samples from ω_j and $\sim\omega_j$ is small, which means that relatively many samples from $\sim\omega_j$ are included in, Θ_{j0} is enlarged a little less; otherwise a little more. Let $X\in R^{N\times m}$ be the original learning set, $N^{(j)}$, N_j and $N_{\sim j}$ the numbers of samples in $\Xi^{(j)}$, $\boldsymbol{X}^{(j)}$ and $\boldsymbol{X}^{(\sim j)}$, respectively, and $\boldsymbol{X}^{(j)}=(\boldsymbol{x}_1^{(j)}, \boldsymbol{x}_2^{(j)},\ldots,\boldsymbol{x}_{Nj}^{(j)})\in R^{Nj\times m}$ the sample matrix from ω_j. The detailed process forming the economic learning subset $\Xi^{(j)}$ is given as follows.

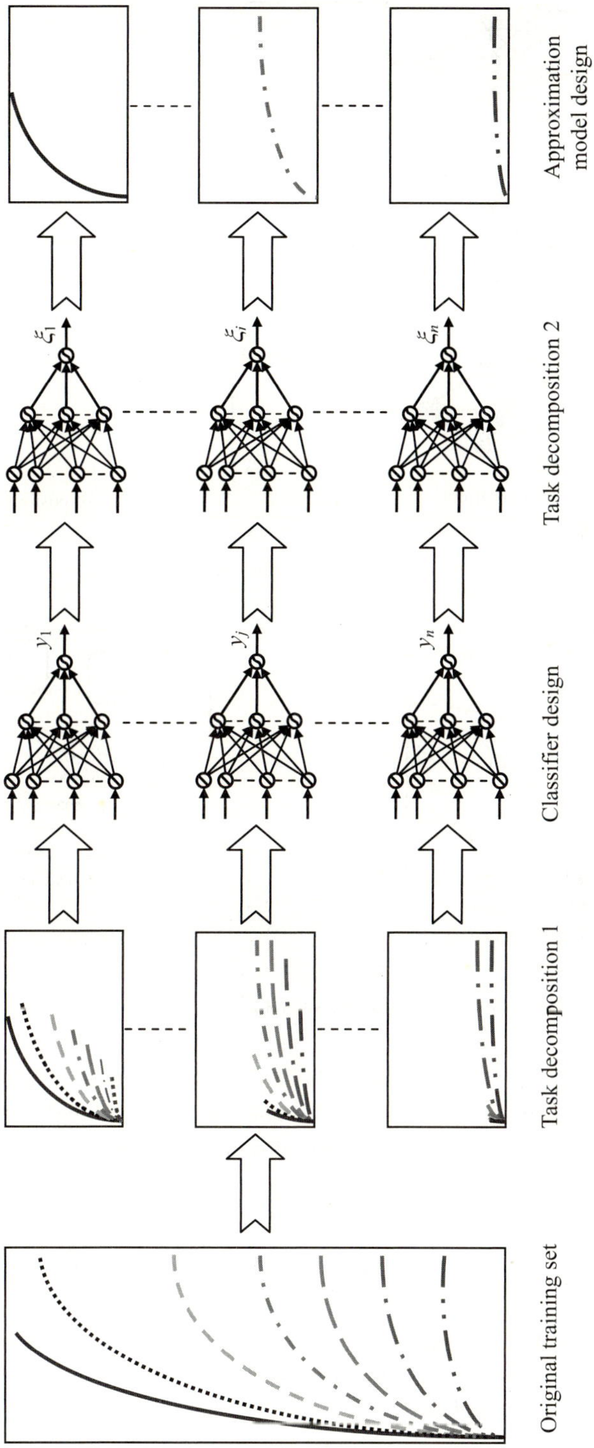

Fig. 1. Learning process of parallel-series perceptrons

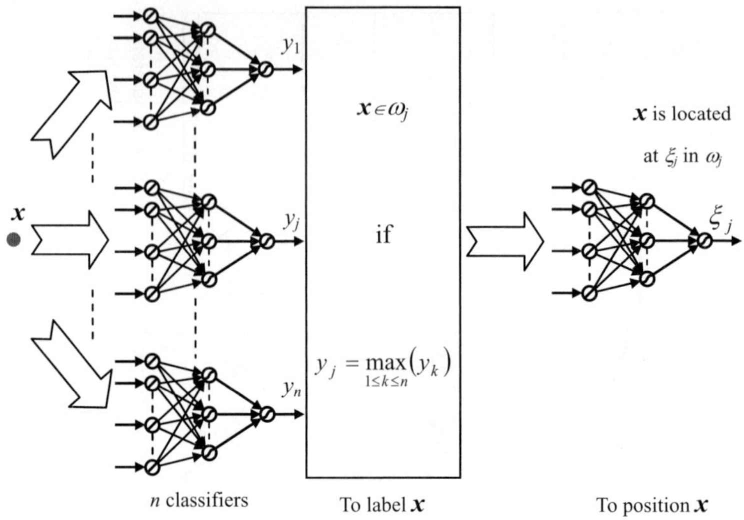

Fig. 2. Decision process of parallel-series perceptrons

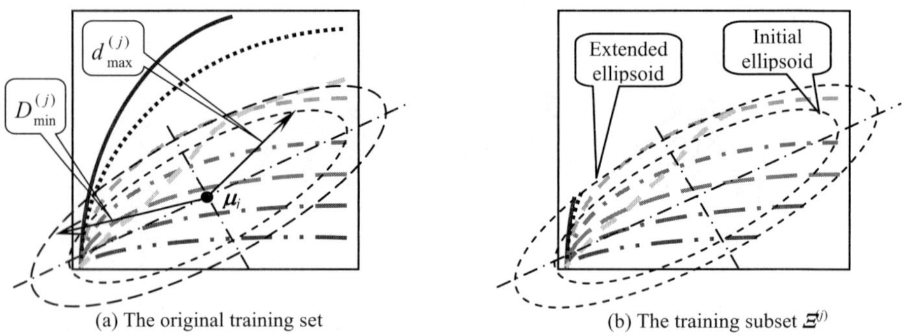

Fig. 3. Formation of the training subset $\boldsymbol{\Xi}^{(j)}$, '.....'–Samples from ω_j

(A) Draw an initial hyper-dimensional oblique ellipsoid Θ_{j0}, which center $\boldsymbol{\mu}_j$ and half axis directions coincide with those of $\boldsymbol{X}^{(j)}$, and which sizes of major and minor axes are determined by the max Mahalanobis distance $d_{\max}^{(j)}$ between all samples from ω_j and the center $\boldsymbol{\mu}_j$. Let $\boldsymbol{\Sigma}_j$ be the covariance matrix of $\boldsymbol{X}^{(j)}$, $d_{\max}^{(j)}$ is calculated by

$$d_{\max}^{(j)} = \max_{1\le p\le N_j} \sqrt{\left(\mathbf{x}_p^{(j)} - \boldsymbol{\mu}_j\right)\boldsymbol{\Sigma}_j^{-1}\left(\mathbf{x}_p^{(j)} - \boldsymbol{\mu}_j\right)^T} \tag{1}$$

If $\boldsymbol{\Sigma}_j$ is singular, $\boldsymbol{X}^{(j)}$ is added a normal noise matrix $0.01N(\boldsymbol{\theta}, \boldsymbol{I})$ with a mean vector $\boldsymbol{\theta}$ and a variance matrix $0.01\boldsymbol{I}$, and $\boldsymbol{\Sigma}_j$ is recalculated with $\boldsymbol{X}^{(j)}+0.01N(\boldsymbol{\theta}, \boldsymbol{I})$. If still singular, $\boldsymbol{\Sigma}_j$ has to be replaced by its main diagonal matrix.

(B) Calculate the initial number $N_{\sim j0}$ of samples included within Θ_{j0} and from $\sim\omega_j$. Supposed $\boldsymbol{x}_p \in \boldsymbol{X}$ and $\boldsymbol{x}_p \notin \boldsymbol{X}^{(j)}$, $N_{\sim j0}$ is calculated by the following loop:
$N_{\sim j0}=0$.

For $p=1:N\text{-}N_j$
  If $(\boldsymbol{x}_p - \boldsymbol{\mu}_j)\boldsymbol{\Sigma}_j^{-1}(\boldsymbol{x}_p - \boldsymbol{\mu}_j)^T \le (d_{\max}^{(j)})^2$
    $N_{\sim j0} \leftarrow N_{\sim j0}+1$.
End

(C) Determine the minimum radius of the extended ellipsoid Θ_j. Our intent is that the fewer the samples from $\sim\omega_j$ are included within Θ_{j0}, the larger the max radius of Θ_j is. The min Mahalanobis radius of Θ_j can be determined by

$$D_{\min}^{(j)} = (1 + N_j / N_{\sim j0})d_{\max}^{(j)} \tag{2}$$

If $N_{\sim j0}=0$, which really means that ω_j can be separated from $\sim\omega_j$ by an oblique ellipsoid, $D_{\min}^{(j)} = 10 d_{\max}^{(j)}$.

(D) Form the final learning subset $\Xi^{(j)}$, which consists of all the samples within Θ_j. Supposed $\boldsymbol{x}_p \in \boldsymbol{X}$ and $\boldsymbol{x}_p \notin \boldsymbol{X}^{(j)}$, $\Xi^{(j)}$ can be determined by the following loop
$\Xi^{(j)}(N_j)=\boldsymbol{X}^{(j)}$.
$k=N_j$.
For $p=1:N\text{-}N_j$
  If $(\boldsymbol{x}_p - \boldsymbol{\mu}_j)\boldsymbol{\Sigma}_j^{-1}(\boldsymbol{x}_p - \boldsymbol{\mu}_j)^T \le (D_{\min}^{(j)})^2$
    $\Xi^{(j)}(k+1) \leftarrow \Xi^{(j)}(k)+\boldsymbol{x}_p$,
    $k \leftarrow k+1$.
End
$N_{\sim j}=k$.
$N^{(j)}=N_j+N_{\sim j}$.

In brief, $\Xi^{(j)}=\{\boldsymbol{X}^{(j)}, \boldsymbol{X}^{(\sim j)}\} \in R^{(Nj+N\sim j)\times m}$ in the m-dimensional space is only made up of all samples $\boldsymbol{X}^{(j)}$ from ω_j and a small part of samples $\boldsymbol{X}^{(\sim j)}$ from $\sim\omega_j$ that are most neighboring to ω_j, see Fig. 3 for details. In case a new class joins in after all the economic learning subsets have come into being and all the corresponding MLP classifiers been trained, it is enough to only re-train a small part of MLPs with the changed subsets.

Without a doubt, if the number of samples in a two-class learning subset $\Xi^{(j)}=\{\boldsymbol{X}^{(j)}, \boldsymbol{X}^{(\sim j)}\}$ is equal each other, the final decision boundaries formed by MLP classifier j will be close to the central sections of margins between the two classes. However, the number of samples in two classes of $\Xi^{(j)}$ is often unequal when the one-versus-all decomposition method is employed. Under the situations, the final decision boundaries will unavoidably deviate from the central sections, which is quite unfavorable for the generalization improvement of MLP classifiers.

Let us illustrate the case with a simple example. Suppose two classes ω_1 and ω_2 are in the 1-dimensional space, only one sample is situated at point 0.0 in ω_1, and 10 samples all located at point 1.0 in ω_2. We take a single neuron with the sigmoid activation function (SAF) $f(\varphi)=(1+\exp(-\varphi))^{-1}$ to divide ω_1 and ω_2. When the decision equation is x-0.5=0, which is just the midpoint, the sum-of-squared error is $10\times(1-(1+exp(-0.5))^{-1})^2+(0-(1+exp(0.5))^{-1})^2=1.5679$. If the equation is x=0.0, which coincides across point 0.0, the sum-of-squared error is $10\times(1-(1+exp(-1))^{-1})^2+(0-(1+exp(0))^{-1})^2=0.9733$. The sum-of-squared error of the former is unexpectedly larger than that of the latter. How incredible it is! And it is just the learning result of an MLP using the back-propagation (*BP*) algorithm! As a result, the generalization performance of MLPs trained with the unbalanced training subsets will be quite poor.

According to the above analysis, we can infer that the final decision boundaries formed by MLP classifier j learning the imbalanced subset $\Xi^{(j)}=\{X^{(j)}, X^{(\sim j)}\}$ will be closer to $X^{(j)}$ because N_j is often several times smaller than $N_{\sim j}$. Our solution to the unbalanced problem existing in the two-class training subsets is to add some virtual samples to the smaller side $X^{(j)}$. Definitely speaking, the samples in $X^{(j)}$ are virtually enlarged $N_{\sim j}/N_j$ times. If $x_p^{(j)}$ is from $X^{(j)}$, the pth updated weight increment $\Delta w_p^{(j)}(\tau)$ is multiplied by an enlargement coefficient $N_{\sim j}/N_j$; otherwise $\Delta w_p^{(j)}(\tau)$ keeps unchanged. All one needs to do is only to add some judgment conditions in the programming. Therefore, the following judgments are added in the τth iteration loop:

$\Delta w^{(j)}(\tau)=0$.

For $p=1: N^{(j)}$,

If $x_p^{(j)} \in X^{(j)}$

$$\Delta \mathbf{w}^{(j)}(\tau) \leftarrow \Delta \mathbf{w}^{(j)}(\tau) - \left(\frac{N_{\sim j}}{N_j}\right)\frac{\partial E_j(\tau)}{\partial \mathbf{w}_p^{(j)}(\tau)} \tag{3}$$

Else

$$\Delta \mathbf{w}^{(j)}(\tau) \leftarrow \Delta \mathbf{w}^{(j)}(\tau) - \frac{\partial E_j(\tau)}{\partial \mathbf{w}_p^{(j)}(\tau)} \tag{4}$$

End

In that way, all is right! There is hardly any additionally computational burden for training MLP classifier j with the unbalanced subset $\Xi^{(j)}=\{X^{(j)}, X^{(\sim j)}\}$!

4 Experimental Results

The experiment aims at using an electronic nose of 16 semi-conducting metal-oxide gas sensors to simultaneously estimate the categories and concentrations of four kinds of fragrant materials, ethanol, ethyl acetate, ethyl caproate, and ethyl lactate. Each of fragrant materials is diluted with distilled water into 4 to 6 kinds of concentrations, and further, 50 liquid samples of each concentration are made up. In order to get the good repeatability, the headspace vapor of a liquid sample is measured only once. The measurement method and device were given in Ref [3]. Responses of sensors are limited in the range of 0.0 to 10.0V by hardware. Fig. 4 givens the principal component analysis (PCA) result for all the samples. According to the figure, there are 22 measured points of concentration in 4 kinds of odors. 40 of measures for each

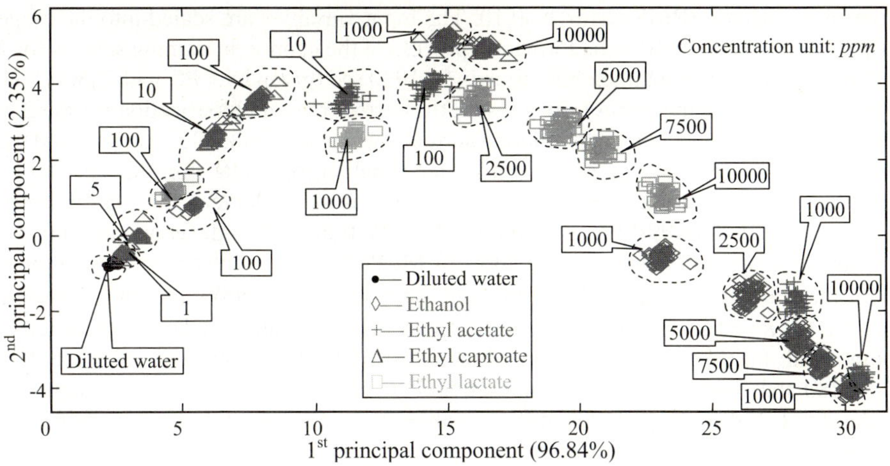

Fig. 4. Principal component analysis for the original dataset of 4 kinds of fragrant materials

Table 1. Training subsets and learning parameters of 4 MLP classifiers

Odor	Ethanol	Ethyl acetate	Ethyl caproate	Ethyl lactate
No. samples in ω_j, N_j	280	200	280	280
No. samples from $\sim\omega_j$, $N_{\sim j0}$	79	92	6	82
$N_j/N_{\sim j0}$	3.0380	2.1739	46.6667	3.4146
No. training samples	511	728	650	781
Max Mah. distance $d_{\max}^{(j)}$	7.9064	615	8.9822	6.7458
Weight enlargement factor	0.8250	2.0750	1.3214	1.7893
Iteration epochs	2520	3770	15250	9730
Learning time (*sec*)	3.7296	6.6953	28.7991	21.9690

concentration are used as a part of the training set, and the remains as a part of the test set. Consequently, there are 40×23=920 samples in the training set, and 10×23=230 in the test set. Here, diluted water is also regarded as a measure point, namely the original point. The 4 classes are nonlinearly separable each other, and the relationship between the sensor responses and the odor concentrations is probably nonlinear.

The number of input dimensions is 16 because there are 16 gas sensors in the array. Naturally, a compulsory decision rule comes into being that a sample is assigned to distilled water only on condition that all the 4 MLP classifiers say 'Yes'. In order words, a certain sample ought not to be assigned to distilled water as long as one or more MLP classifiers say 'No'. The structure and learning parameters of MLPs for classification are as follows. The numbers of hidden nodes are respectively h=8, the learning rate η=0.02, the max iteration step τ_{max}=20000, and the allowable root-mean- square (RMS) error ε^*=0.10. The input variables are scaled into the range [0, 6.0] [16]. Table 1 given the formation process of the economic training subsets for 4 MLP classifiers. The total learning time is 61.1930 sec (given by a PC with 2.6G CPU, 256M RAM, the same below). As a result, the correct rate of classification for the 230 test samples 100%. When every training subset is with the original 920 samples and the step of virtual balance is not adopted, the final correct rate for the test set is 227/230=98.70%. If the step of virtual balance is not adopted, the classification accuracy is 228/230=99.13%, even if the MLPs are trained with the economic subsets.

The structure and learning parameters of MLPs for approximation are as follows. h=5, the learning rate η=0.02, the max iteration step τ_{max}=15000, and the allowable RMS error ε^*=0.025. Since the measurement of distilled water is taken as the original point of all the curves, the numbers of samples for MLP for approximation are 280, 200, 280 and 280 in each training subset, and 70, 50, 70 and 70 in each test subset in order, see Fig. 4 for details. The input variables are scaled into the range [0, 6.0] [16], and the target outputs are expressed by

$$d_{p3}^{(j)} = \phi\left(C_p^{(j)}\right) = \left(1 + \log_{10} C_p^{(j)}\right) * 2.9/5.0 + 0.05 \tag{5}$$

Here, $C_p^{(j)}$ is the target concentration of the pth sample $\boldsymbol{x}_p^{(j)}$ in odor ω_j, which measurement unit is *ppm*. The concentration of distilled water is forced to be 0.1 *ppm*. The target values of MLP classifiers are thus scaled in the range of [0.05, 2.95].

All the 16-5-1 MLPs for approximation iterate 15000 epochs and take 7.11, 5.19, 7.11, and 7.11 *sec*, respectively. The RMS errors of predicted logarithmic concentrations of the test samples given by the 4 MLPs for their represented odors are 0.017, 0.027, 0.030, and 0.016 in order.

For the sake of comparison, we still consider the problem first an MIMO classification problem and then 4 MISO approximation problems, and employ either single-output MLPs with the different training subsets or support vector machines (SVMs) to solve it. For SVMs, the width factor γ=0.25, the insensitivity constant ε=0.005, and the capacity constant C=1000 [17]. Table 2 gives the learning and predicted results of different models and methods for simultaneously estimating the classes and concentrations of 4 kinds of fragrant materials. According to the table, the

Table 2. Learning and predicted results of MLPs and SVMs for the 4 kinds of fragrant materials

Classifiers and approximation model	No. samples in the training subset	Learning time (sec)	Virtual balance	Pred. acc. for the test set (%)	No. equivalent samples
From 4 MISO MLPs to 4 MISO MLPs	(511-781) / (200-280)	87.71	Yes	100	59.00
From 4 MISO MLPs to 4 MISO MLPs	(511-781) / (200-280)	87.71	No	99.13	59.00
From 4 MISO MLPs to 4 MISO MLPs	920 / (200-280)	191.34	No	98.70	59.00
From 4 MISO SVMs to 4 MISO SVMs	920 / (200-280)	103.66	No	99.57	733.13

proposed method and parallel-series perceptrons has advantages on comprehensive performance. However, the SVM models have to store a large number of support vectors, which is quite disadvantageous for solving the large-scale learning problems, in particular, the multi-output ones.

5 Conclusion

This paper regards the simultaneous determination of odor classes and strengths as a series classification and approximation problem. We first decompose such a problem into multiple MISO problems, and then employ multiple series MLPs to solve them one by one. An MLP is trained only a small part of samples from the represented class and the neighboring classes. The experimental results for simultaneously estimating the odor classes and strengths show that the proposed models are quite effective.

Acknowledgments

This work is supported by the National Science Foundation of China (NSFC) under Grant No. 60275017, 60373073, 60575027, the high-tech development program of china (863) under Grant No. 2006AA10Z315; the Doctor Unit Foundation of the Education Ministry of China under Grant No. 20060251013, Shanghai Science Foundation of China under Grant No. 06ZR14026, and the open project program of the State Key Laboratory of Bioreactor Engineering.

References

[1] Gardner, J.W., Bartlett, P.N.: Electronic noses–principles and applications. Oxford Press (1999)
[2] Pearce, T.C., Schiffman, S.S., Nagle, H.T., Gardner, J.W. (eds.): Hand of Machine Olfaction. Wiley-VCH Press, Chichester (2003)
[3] Daqi, G., Wei, C.: Simultaneous estimations of odor classes and concentrations using an electronic nose with function approximation model ensemble, Sensors and Actuators B, vol. 120(2), pp. 584–594 (2007)
[4] Duda, R.O., Hart, P.E., Stork, D.G.: Pattern classification. John Wiley & Sons, New York (2000)
[5] Bishop, C.M.: Neural networks for pattern recognition. Clarendon Press, Oxford (1995)
[6] Liobet, E., Brezmes, J., Vilanova, X., et al.: Qualitative and quantitative analysis of volatile organic compounds using transient and steady-state responses of a thick-film tin oxide gas sensor array. Sensors and Actuators B 41(1), 13–21 (1997)
[7] Oh, I.S., Suen, C.Y.: A class-modular feedforward neural network for handwriting recognition. pattern recognition 35(1), 229–244 (2002)
[8] Ou, G.B., Murphey, Y.L.: Multi-class pattern classification using neural networks. Pattern Recognition 40(1), 4–18 (2007)
[9] Anand, R., Mehrotra, K.G., Mohan, C.K., et al.: Efficient classification for multiclass problems using modular neural networks. IEEE Transactions on Neural Networks 6(1), 117–124 (1995)

[10] Song, H.H., Lee, S.W.: A self-organizing neural tree for large-set pattern classification. IEEE Transactions on Neural Networks 9(3), 369–380 (1998)
[11] Murphey, Y.L., Guo, H., Feldkamp, L.A.: Neural learning from imbalanced data. Applied Intelligence and Neural Networks Application 21(2), 117–128 (2004)
[12] Anand, R., Mehrotra, K., Mohan, C.K., Ranka, S.: An improved algorithm for neural network classification of imbalanced training sets. IEEE Trans. on Neural Networks 4(6), 962–969 (1993)
[13] Jamshid, D., Mustafa, K., Valdivieso, C.M.: A rule-based scheme for filtering examples from majority class in an imbalanced training set. In: Third International Conference on Machine Learning and Data Mining in Pattern Recognition, Leipzig, Germany. Lecture Notes in Artificial Intelligence (LNAI), vol. 2734, pp. 215–223 (2003)
[14] Cantador, I., Dorronsoro, J.R.: Parallel perceptions, activation margins and imbalanced training set pruning. In: Marques, J.S., Pérez de la Blanca, N., Pina, P. (eds.) IbPRIA 2005. LNCS, vol. 3523, pp. 43–50. Springer, Heidelberg (2005)
[15] Andrew, E., Taeho, J., Nathalie, J.: A multiple resampling method for learning from imbalanced data sets. Computational Intelligence 20(1), 18–36 (2004)
[16] Daqi, G., Genxing, Y.: Influences of variable scales and activation functions on the performances of multilayer feedforward neural networks. Pattern Recognition 36(4), 869–878 (2003)
[17] Vapnik, V.N.: The nature of statistical learning theory. Springer, New York (2000)

Probabilistic Video-Based Gesture Recognition Using Self-organizing Feature Maps

George Caridakis, Christos Pateritsas, Athanasios Drosopoulos, Andreas Stafylopatis, and Stefanos Kollias

School of Electrical and Computer Engineering, National Technical University of Athens, Politechnioupoli, Zographou, Greece
{gcari,ndroso}@image.ece.ntua.gr, pater@softlab.ece.ntua.gr, {andreas,stefanos}@cs.ece.ntua.gr

Abstract. Present work introduces a probabilistic recognition scheme for hand gestures. Self organizing feature maps are used to model spatiotemporal information extracted through image processing. Two models are built for each gesture category and, along with appropriate distance metrics, produce a validated classification mechanism that performs consistently during experi-ments on acted gestures video sequences.

Keywords: Hand tracking, gesture recognition, gesture classification, self organizing feature map, markov processes.

1 Introduction

Gesture recognition continuously receives abundant attention, especially throughout the research fields of sign language recognition, multimodal human computer interaction, cognitive systems and robotics. Renewed focus on interdisciplinary studies lead scientists to review and confront the questions raised when attempting to model and extract the information that a gesture conveys. Since hand gestures can be used for a wide variety of communicative purposes, classification becomes a significant problem, starting at the level of defining gesture taxonomy through psychological studies. Most commonly, gesturing behavior can be classified on a spectrum that ranges from highly structured languages (e.g. sign languages), through universal symbols, to natural and unconscious gesticulation [1]. Studies also show that gesture classification is, in general, a multimodal task that should make use of both hand movement trajectories and linguistic cues [2], [3].

In terms of computer vision, appropriate feature extraction and tracking is the focus of many researchers, in order to apply classification schemes for the hand trajectory and/or the hand shape. Depending on the scope of a study, approaches vary from multimodal interpretation (gesticulation, natural language, facial expression, domain knowledge, etc.) to gesture classification through a single modality. Present work deals with the classification of gestures from visual cues, focusing on robustness, performance and user independence. Aiming for naturalistic data, our intention is to localize and track hands, classifying their trajectories regardless of the hand shape.

An extensive review of several techniques is presented both in [4] and [5]. The first focuses mainly on SL recognition and classification issues, while examining closely

J. Marques de Sá et al. (Eds.): ICANN 2007, Part II, LNCS 4669, pp. 261–270, 2007.

hand localization and tracking, and on various feature extraction techniques related to automatic analysis of manual signing. In addition, it addresses the linguistic aspect of SL and non manual signals, along with methodologies to incorporate these in the SL recognition chain. On the other hand, Wu and Huang delve more into works related to hand modeling (shape analysis, kinematics chain and dynamics) and computer vision, and pattern recognition issues associated to hand localization and feature extraction from image sequences. Classification schemes involve several methods, depending on the features and the stages of the procedure. Methods used include neural networks and variants, hidden markov models and variants, principal component analysis, and numerous other machine learning methods or combinations (decision trees, template matching, etc.).

One of the most commonly proposed approaches involves feature extraction from the input signal and utilization of these features as input for a fine tuned HMM [6], [7]. In addition, variations of the previous group have been widely adopted [8], [9]. Other approaches employ alternate machine learning and artificial intelligence techniques such as recurrent fuzzy network [10], time delay neural network [11], finite state machines [12], Bayesian classifiers [13], etc. Finally, there have been several efforts combining more than one technique. Mantyla et al. [14] present a system for static gestures recognition using a self-organizing mapping scheme, while a hidden Markov model is used to recognize dynamic gestures. Black and Jepson [15] present an extension of the "condensation" algorithm, modeling gestures as temporal trajectories of the velocity of the tracked hands. Fang et al. [16] present an additional layer enhancing the HMM architecture with SOFM and improving their recognition rate by 5%, while introducing a fuzzy decision tree in an attempt to reduce the search space of recognized classes without loss of accuracy.

Present work introduces a novel approach for applying a combination of self organizing maps and markov models for gesture classification. The features extracted include the trajectory of the hand and the direction of motion in the various stages of the gesture. The classification scheme is based on the transformation of a gesture representation from a series of coordinates and movements to a symbolic form and on building probabilistic models using these transformed representations. Our study indicates that, although each of the two sets of features (trajectory and motion direction) can provide distinctive information in most cases, only an appropriate combination can result in robust and confident user independent gesture recognition.

2 System Overview

The steps of the introduced procedure, which is depicted in Fig. 1, begin with an image processing module. Taking into consideration computational cost and robustness, we employed an accurate, near real-time skin detection and tracking module [17] allowing a rate of around 12 fps (frames per second) on a usual PC configuration, which is adequate for continuous gesture tracking. The process involves the creation of moving skin masks and tracking their centroids to produce an estimate of the user's movements. The object correspondence heuristic makes it possible to individually track the hand segments correctly while the fusion of color and motion information eliminates any background noise or artifacts.

Following, each gesture instance is represented by a time series of points, representing the hand's location with respect to the head of the person performing the gesture. Consequently, a gesture G_i containing l points can de expressed as an ordered set of points:

$$G_i = \{(x_1, y_1), (x_2, y_2), ...(x_l, y_l)\}, \tag{1}$$

where l varies across different gesture instances. The system's input is a set of gestures D, assigned to c different categories.

The proposed modeling scheme is based on the transformation of a gesture representation from a series of coordinates and movements to a symbolic form which, in turn, is used to build the respective probabilistic models. The first transformation is based on the relative position of the hand during the gesture and is achieved using a self-organizing map model. Despite the fact that the map units are treated as symbols, the map's neighborhood function provides a distance metric between them, that is used during the classification of an unlabeled gesture. Additionally, this enables the use of the Levenshtein distance metric for the comparison between these sequences of symbols and the definition of a "mean" string of symbols representing e.g. the gestures included in a D_j set.

The second transformation is based on the optical flow of the gesture, aiming to describe the gesture's direction changes. The symbols generated from this transformation constitute the set of angles of the gesture's trajectory. This set is limited to quantized values that are treated as symbols in order to be used for the creation of an additional set of Markov models.

For the classification of an unlabeled gesture, the Markov models created from the first transformation play the primary role, while the models created from the second transformation are used for validation and decisions in cases of low confidence classification.

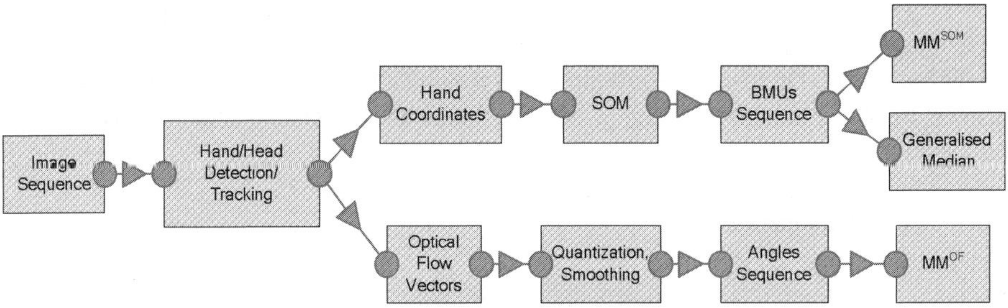

Fig. 1. Building gesture models from transformed gesture representations

3 Probabilistic Models of Hand Movement

The coordinates of all the points from all the gestures are used to train a hexagonal, two-dimensional grid SOM with the batch mode learning procedure. The points are fed to the map in an unordered form, inconsequently to the gesture instance they belong to and to their ranking position into the gesture. Following training, each point is assigned to the respective best matching unit (BMU) on the map, i.e. the unit of the map closer to the point in the input data space, according to the Euclidean distance of the two vectors. Thus, a gesture G_i can be transformed from a series of points to a series of map units.

$$T(G_i) = (u_1, u_2, ..., u_l), \text{ where } u_i = BMU(x_i, y_i) . \quad (2)$$

Function $BMU(x_i,y_i)$ returns the index of the best-matching unit for point (x_i,y_i) and $T(G_i)$ is the modified gesture representation. Given that u_i is the index of a map unit, this function can be is declared as BMU:R^2 ->S, where S is the set of the indices of all map units and can be treated as a set of symbols. In many cases, the u_i value of consequent points of a gesture remains the same since, although the continuous movement of the hand is represented by the distinct points, consequent points are generally close in the input data space. Replacing consequent equal values of u_i with a single value results in the following gesture definition,

$$G_i^{'} = N(T(G_i)) = \{u_1, u_2, ..., u_m,\} : m \le l, \forall t \in [2, l]\ u_t \ne u_{t-1}, \quad (3)$$

where N is a function that removes consecutive equal u_i values and $G_i^{'}$ is the transformed gesture instance. The transformation of the gestures with the use of the SOM can be considered a transformation of the continuous trail to a sequence of m discrete symbols, different for every gesture class, that define the finite states to build first order Markov chain models.

Such a model, for each of the categories in the gestures' data set, is created. The sequence of the u_i values into the transformed gestures $G_i^{'}$ of $D_j^{'}$ set, will be used for the calculation of the transition probabilities of the model MM_j^{som} describing the j category and for the determination of the values of the function π_j^{som}, which is the first state probability function of this model. The result is a set MM^{som} of c Markov models.

$$MM^{som} = \{MM_1^{som}, MM_2^{som}, ..., MM_c^{som}\} : D_i^{'} = \{G_1^{'}, G_2^{'}, ..., G_n^{'}\} \rightarrow MM_i^{som} \quad (4)$$

These models are used to evaluate a new unlabeled gesture in order to be classified in one of the c categories. Fig. 2 depicts the above described transformation for a gesture instance.

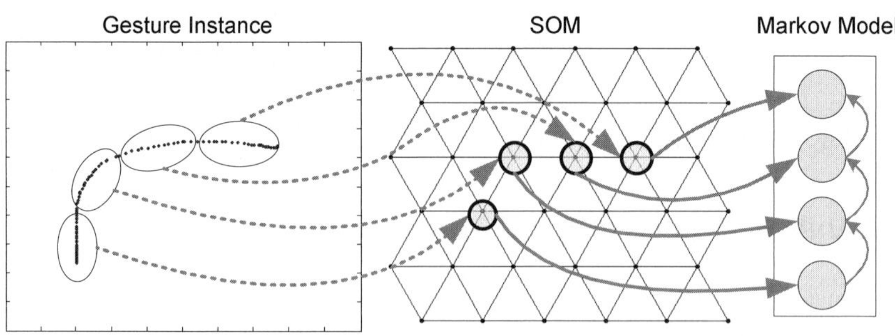

Fig. 2. Correspondence of gesture trajectory points to their respective BMUs on the SOM. These BMUs constitute the states of the Markov models.

With the purpose of providing a more descriptive representation of each gesture instance, an additional transformation is introduced, based on the optical flow of each gesture. This describes the different directions that the gesture trajectory presents instead of the spatial position of gesture points. In order to achieve such a representation, direction vectors are calculated from the consecutive gesture trajectory points. These angles are then quantized in 8 different symbolic values as depicted in Fig. 3. The segments of coordinates in Fig. 2 and Fig. 3 are considered to be a set of coordinates that belong to the same cluster (BMU and Quantized Angle for Fig. 2 and Fig. 3 respectively). In that sense, we define the transformation of a gesture instance G_i using the *OF* function as:

$$OF(G_i) = \{v_1, v_2, ..., v_m\} : v_i = W_r(Q(\arctan(\frac{y_i - y_{i-1}}{x_i - x_{i-1}}))), \quad (5)$$

where v_i are the quantized values, Q the quantization function and W_r a median function applied to the values of a fixed length window around the input value. The purpose of the later is to smooth the quantized values against possible instabilities of the hand during the gesture. Applying the transformation function along with function N (eq. 3) for the removal of the equal consecutive values we get

$$G_i'' = N(OF(G_i)) = \{v_1, v_2, ..., v_m\} \quad (6)$$

The v_i values define the states for a new set of Markov models MM^{of} that is built using the transformed set D_j''. The first state probability function π_j^{of} is also calculated using this set.

$$MM^{of} = \{MM_1^{of}, MM_2^{of}, ..., MM_c^{of}\} : D_i'' = \{G_1'', G_2'', ..., G_n''\} \rightarrow MM_i^{of} \quad (7)$$

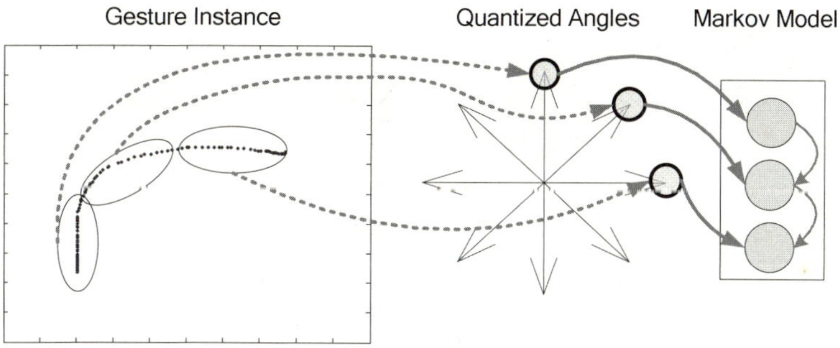

Fig. 3. Building a Markov model for a gesture's optical flow

4 Classification of an Unlabeled Gesture

The classification of an input gesture will be based on the two sets of Markov models (eqs. 4 & 7). Let G_k be a gesture instance of unknown category, and G_k' and G_k'' its transformed representations. Using the MM^{som} set of models, the probability of this gesture to belong in category j can be calculated as:

$$P(G_k^{'} \mid MM_j^{som}) = \frac{\sum_{i=1}^{m} S_i^{som}}{m} \tag{8}$$

The above equation averages the values S_i^{som}, which represent an evaluation factor for each u_i value of the $G_k^{'}$ transformed gesture with respect to the MM_j^{som} Markov model. These values are calculated as:

$$S_i^{som} = \max_z (NF_{u_i}^{som}(z) P(z \mid u_i, MM_j^{som})) \tag{9}$$

$$u_i = \arg\max_z (S_i^{som}), \tag{10}$$

where z is a variable that indexes the units of the trained map, $NF_{u_i}^{som}(z)$ is the distance of the unit z as defined by the self-organizing map Gaussian neighborhood function with the u_i unit as its center. In equation (9), the proximity between the state-unit z and the previous state-unit u_{t-1} of the gesture is multiplied with the probability of the transition from state-unit z to state-unit u_{t-1}. As the z variable varies across all the units of the map, this product will provide the unit that combines a considerable transition probability from the previous state with a small distance onto the map grid from the current state. This unit will also be used as the previous state in the next step as defined by equation (10). The initial values used in the sum derive from the following equations.

$$S_1^{som} = \max_z (NF_{u_1}^{som}(z) \pi_j^{som}(z)), u_1 = \arg\max_z (S_1^{som}) \tag{11}$$

Using the MM^{of} set of models, the probability of this gesture to belong in category j can be calculated as:

$$P(G_k^{''} \mid MM_j^{of}) = \frac{\sum_{i=1}^{m} S_i^{of}}{m} \tag{12}$$

The values S_i^{of} are calculated from the following equations:

$$S_i^{of} = \max_z (NF_{v_{i-1}}^{of}(z) P(z \mid v_{i-1}, MM_j^{of})), v_i = \arg\max_z (S_i^{of}), \tag{13}$$

where z is a variable that indexes the different states-directions and $NF_{u_i}^{of}(z)$ a distance function between these states. These equations implement a search similar to the previous search on the map grid, but in this case the search is performed among the different possible gesture directions. The initial values are calculated in a similar way from the following equations.

$$S_1^{of} = \max_z (NF_{v_1}^{of}(z) \pi_j^{of}(z)), v_1 = \arg\max_z (S_1^{of}) \tag{14}$$

In order to compare the length of the unknown gesture with the length of the gestures included in each $D_j^{'}$ set, a distance metric for the comparison of symbol strings is necessary. From each set D_j', a *Generalized Median* gesture is calculated. Let S be a set of symbol strings s_i. We can then define m as a string that consists of a combination of all or some of the symbols used in the set and which minimizes the following expression.

$$\sum_{s_i} L(s_i, m), \forall s_i \in S \tag{15}$$

where $L(,)$ denotes the Levenshtein distance, one of the most widely used string distance metric. If the search for string m is restricted to the members of the set then m is the *set median*. But if m is a hypothetical string and the search is not restricted then m is the *Generalized Median* of the set. Using the above definition we calculate the Levenshtein distance $L_{kj} = L(G_k' \mid M(D_j'))$ between G_k' and the *Generalized median* $M(D_j')$ of each D_j' set.

The category of the unknown gesture is primarily decided using the MM^{som} set of models. Subsequently, the category would be equal to:

$$\arg\max_j P(G_k' \mid MM_j^{som}) \tag{16}$$

In order for the category of the unknown gesture to be decided by the above equation the three following conditions must be fulfilled.

$$\max_j(P(G_k' \mid MM_j^{som})) \geq \alpha \tag{17}$$

$$\max_j(P(G_k' \mid MM_j^{som}) - 2^{nd} \max_j(P(G_k' \mid MM_j^{som}) \geq \beta \tag{18}$$

$$L_{k,\arg\max_j(P(G_k' \mid MM_j^{som})} \leq \gamma LM(\arg\max_j(P(G_k' \mid MM_j^{som})) \tag{19}$$

The two first conditions requires that the maximum probability calculated using position based models must exceed a threshold value a while the difference between the maximum probability and the second ranked ones must also exceed a threshold value β. These two values represent confidence thresholds. The last condition applied is that the Levenshtein distance between the gesture and the *Generalized Median* of the category with the maximum probability must be larger than the *LM* value of this category, multiplied by a user defined factor γ. This last comparison is made in order to assess the length of the unknown gesture with respect to the average length of the gestures of the category with the maximum probability. If one of these conditions is not fulfilled then the category of the unknown gesture is defined from a combination of values:

$$\arg\max_j(P(G_k' \mid MM_j^{som})P(G_k'' \mid MM_j^{of})\frac{1}{\frac{L_{kj}}{\|M(D_j)\|}}) \tag{20}$$

This classification rule combines the evaluation provided from both the MM^{som} and MM^{of} set of Markov models with the Levenshtein distance of the gesture and the *Generalized median* of the each category normalized by the length of the *Generalized median*.

5 Experimental Results

Fig. 4 shows a sample of frames from the input sequences that where used and indicative results of the hand localization and tracking module.

Fig. 4. Sample of input sequences with hand tracking results

Experiments were conducted, with the above dataset, in order to evaluate the recognition performance of the proposed method. When all the gesture instances are used for both training and testing, the recognition rate is 100%. To evaluate the generalization capabilities of the proposed method the 10-fold cross validation strategy was used. In this case the average recognition rate was 93%. presents in detail the recognition percentages of each category.

In order to compare the results of our system with one of the most commonly used approaches in the literature we employed an HMM based classifier [7], training one HMM per gesture class. We used continuous left-to-right models and a mixture of 3 Gaussian probability density functions. During the decoding of a gesture it was tested against all models and the one with the highest log-likelihood value was selected as the winner. The above described process produced an average recognition rate of 85%.

Table 1. Proposed method's recognition rate per gesture category (93% average)

Category	%	Category	%	Category	%	Category	%	Category	%
1	100	7	100	13	80	19	100	25	90
2	100	8	100	14	80	20	90	26	90
3	100	9	100	15	100	21	50	27	90
4	100	10	100	16	90	22	70	28	100
5	100	11	100	17	100	23	100	29	100
6	100	12	100	18	100	24	60	30	100

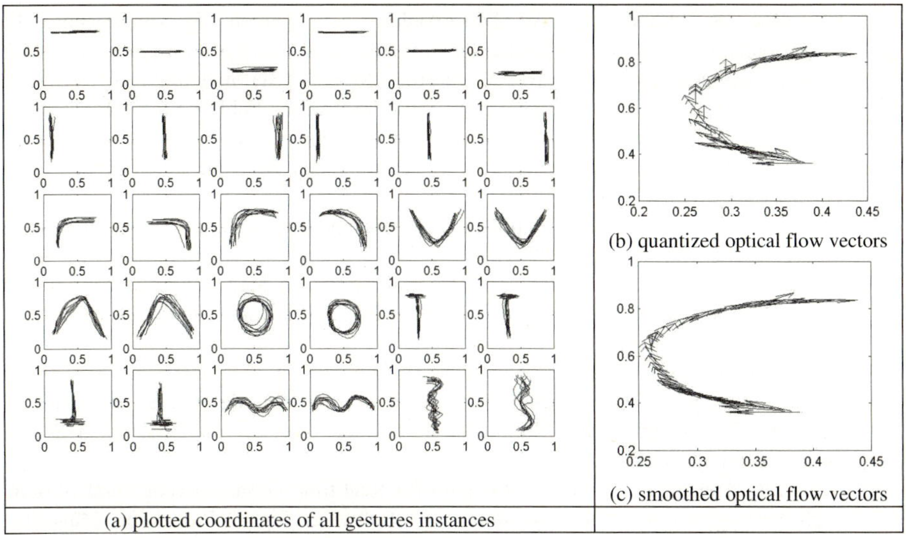

(a) plotted coordinates of all gestures instances

(b) quantized optical flow vectors

(c) smoothed optical flow vectors

Fig. 5. The gesture dataset

6 Conclusion

Present work introduced a novel modeling scheme for gesture recognition from hand trajectories. The system builds models for gesture categories utilizing SOMs that are trained with features extracted through image processing. Experimental results indicate that the system is capable of performing robustly while also evaluating its results. Intended experiments on alternate gesture corpora will be used to assess the capabilities of the system in a broader spectrum of gesture based interaction. Through further research, we intend to address the classification strategy for gestures that present low confidence results, i.e they belong to unknown categories, as well as the evaluation of the system's gesture prediction capabilities.

Acknowledgements

This work has been partially supported by the "EHG" ("Vision-Sound-Language") Programme of the Greek General Secretariat of Research and Technology and the EU in the framework of the Feelix-Growing (www.feelix-growing.org) STREP programme.

References

[1] Kendon, A.: Conducting Interaction. Cambridge, University Press, Cambridge (1990)

[2] Eisenstein, J., Davis, R.: Visual and Linguistic Information in Gesture Classification. In: Proceedings of the 6th International Conference on Multimodal Interfaces (ICMI '04), USA, October 13 - 15, 2004, pp. 113–120. ACM Press, New York (2004)

[3] Karpouzis, K., Raouzaiou, A., Drosopoulos, A., Ioannou, S., Balomenos, T., Tsapatsoulis, N., Kollias, S.: Facial expression and gesture analysis for emotionally-rich man-machine interaction. In: Sarris, N., Strintzis, M. (eds.) 3D Modeling and Animation: Synthesis and Analysis Techniques, pp. 175–200. Idea Group Publ., USA (2004)
[4] Ong, S.C.W., Ranganath, S.: Automatic Sign Language Analysis: a Survey and the Future beyond Lexical Meaning. Pattern Analysis and Machine Intelligence. IEEE Transactions 27(6), 873–891 (2005)
[5] Wu, Y., Huang, T.S.: Hand Modeling, Analysis and Recognition. Signal Processing Magazine, IEEE 18(3), 51–60 (2001)
[6] Starner, T., Weaver, J., Pentland, A.: Real-time American Sign Language Recognition Using Desk and Wearable Computer-based Video. IEEE Trans. Pattern Analysis and Machine Intelligence (1998)
[7] Balomenos, T., Raouzaiou, A., Ioannou, S., Drosopoulos, A., Karpouzis, K., Kollias, S.: Emotion Analysis in Man-Machine Interaction Systems. In: Bengio, S., Bourlard, H. (eds.) Machine Learning for Multimodal Interaction. LNCS, vol. 3361, pp. 318–328. Springer, Heidelberg (2004)
[8] Ozer, I.B., Tiehan, L., Wolf, W.: Design of a Real-time Gesture Recognition System. High Performance through Algorithms and Software. Signal Processing Magazine, IEEE 22(3), 57–64 (2005)
[9] Wilson, B.A.: Parametric Hidden Markov Models for Gesture Recognition. IEEE Trans. Pattern Analysis and Machine Intelligence 21(9) (1999)
[10] Juang, C.-F., Ku, K.C.: A Recurrent Fuzzy Network for Fuzzy Temporal Sequence Processing and Gesture Recognition. Systems, Man and Cybernetics, Part B. IEEE Transactions 35(4), 646–658 (2005)
[11] Yang, M.H., Ahuja, N., Tabb, M.: Extraction of 2D Motion Trajectories and its Application to Hand Gesture Recognition. Pattern Analysis and Machine Intelligence. IEEE Transactions 24(8), 1061–1074 (2002)
[12] Hong, P., Turk, M., Huang, T.S.: Gesture modeling and recognition using finite state machines. In: Proc. Fourth IEEE International Conference and Gesture Recognition, Grenoble, France (March 2000)
[13] Wong, S., Cipolla, R.: Continuous Gesture Recognition using a Sparse Bayesian Classifier. In: Proceedings of the 18th international Conference on Pattern Recognition - Volume 01 2006. ICPR, pp. 1084–1087. IEEE Computer Society Press, Washington, DC (2006)
[14] Mantyla, V.-M., Mantyjarvi, J., Seppanen, T., Tuulari, E.: Hand Gesture Recognition of a Mobile Device User. Multimedia and Expo, ICME 2000, IEEE International Conference 1, 281–284 (2000)
[15] Black, M.J., Jepson, A.D.: Recognizing Temporal Trajectories Using the Condensation Algorithm. In: Proceedings of the 3rd. international Conference on Face & Gesture Recognition FG, Washington, DC, IEEE Computer Society, Los Alamitos (1998)
[16] Fang, G., Gao, W., Zhao, D.: Large Vocabulary Sign Language Recognition based on Fuzzy Decision Trees. Systems, Man and Cybernetics, Part A. IEEE Transactions 34(3), 305–314 (2004)
[17] Martin, J.-C., Caridakis, G., Devillers, L., Karpouzis, K., Abrilian, S.: Manual annotation and automatic image processing of multimodal emotional behaviors: validating the annotation of TV interviews. Personal and Ubiquitous Computing, Special issue on Emerging Multimodal Interfaces. Springer, Heidelberg (2007)

Unbiased SVM Density Estimation with Application to Graphical Pattern Recognition

Edmondo Trentin and Ernesto Di Iorio

DII - Università di Siena, V. Roma 56, Siena, Italy
{trentin,diiorio}@dii.unisi.it

Abstract. Classification of structured data (i.e., data that are represented as graphs) is a topic of interest in the machine learning community. This paper presents a different, simple approach to the problem of structured pattern recognition, relying on the description of graphs in terms of algebraic binary relations. Maximum-a-posteriori decision rules over relations require the estimation of class-conditional probability density functions (pdf) defined on graphs. A nonparametric technique for the estimation of the pdfs is introduced, on the basis of a factorization of joint probabilities into individual densities that are modeled, in an unsupervised fashion, via Support Vector Machine (SVM). The SVM training is accomplished applying support vector regression on an unbiased variant of the Parzen Window. The behavior of the estimation algorithm is first demonstrated on a synthetic distribution. Finally, experiments of graph-structured image recognition from the Caltech Benchmark dataset are reported, showing a dramatic improvement over the results (available in the literature) yielded by state-of-the-art connectionist models for graph processing, namely recursive neural nets and graph neural nets.

1 Introduction

Learning from structured data, i.e. data that are represented as graphs, is a topic that has received a significant attention from the machine learning community. Classification of structured data is relevant to areas such as natural language processing, bioinformatics, structural pattern recognition, and the Web applications. Neural network models were proposed, such as recursive neural nets (RNN) [11] and graph neural nets (GNN) [7]. These architectures have the capability of unfolding over labeled graph, in a backpropagation-through-time fashion, recursively encoding information on the topology and on the label values into an internal activation state. In spite of their strong theoretical properties, RNNs and GNNs suffer from some drawbacks that might limit their real-world application: (i) RNNs can process only acyclic structures, which only seldom fit real-world scenarios; (ii) both RNNs and GNNs are complex machines, both from a formal and from a computational point of view; (iii) unconnected graphs cannot be processed; (iv) above all, they suffer from a drawback which they share with the classic recurrent neural nets, the "long term dependencies" problem [1]. In terms of graphical structures this problem takes the form of graphs with long shortest-paths between certain pairs of nodes, e.g. high trees. The paper

J. Marques de Sá et al. (Eds.): ICANN 2007, Part II, LNCS 4669, pp. 271–280, 2007.

looks at the problem of learning (discriminant functions) on structured domains from a different perspective and introduces a simple and effective attempt to overcome the above limitations. The proposed approach is suitable for directed or undirected, connected or unconnected, cyclic or acyclic graphs. Labels may be attached to nodes or edges as well. The approach simplifies significantly the formalism and may be implemented by means of ordinarily available software simulators. The idea is to describe the graph as an algebraic binary relation, i.e., as a subset of the Cartesian product in the definition domain of the graph. The class-posterior probabilities given the graph can then be reduced to a product (joint probability) of probability density functions (pdf) evaluated over the pairs in the relation. In order to apply the formalism, the problem of proper pdf estimation has to faced.

Estimating pdfs is one major topic in unsupervised learning and in supervised pattern recognition [5]. In fact, the right-hand-side of Bayes' theorem requires the knowledge of pdfs known as class-conditional probabilities [5]. Parametric techniques (e.g. maximum-likelihood) are not promising in the present scenario, since they rely on an arbitrary assumption on the form of the underlying, unknown distribution. Nonparametric techniques (e.g. k_n-nearest neighbors [5]) remove this assumption and attempt a direct estimation of the pdf from a data sample. The Parzen Window (PW) is one of the most popular approaches, relying on a combination of local window functions centered in the patterns of the training sample [5]. PW is limited in several respects, including: (1) the estimate is not expressed in a compact functional form, but it is a sum of as many local windows as the size of the sample; (2) the nature of the window functions may yield a fragmented model, which is basically "memory based" and (by definition) is prone to overfitting; (3) the whole training sample has to be kept in memory in order to compute the estimate of the pdf over any new (test) patterns, resulting in a high complexity of the technique in space and time; (4) the PW model heavily depends on the choice of an initial width of the local windows. For these reasons, artificial neural networks (ANN) could be considered instead, given the fact that they are nonparametric, "universal" models. Unfortunately, in spite of the popularity of ANNs probability estimation (e.g., class posterior probability modeling), only few and sub-optimal connectionist techniques for pdf estimation are presented in the literature. This is due to the fact that modeling a pdf (whose values are possibly in the range $(0, +\infty)$ and whose integral over the feature space has to be 1) via ANN is a difficult unsupervised task. Support vector machines (SVM) are a feasible alternative to ANNs for density estimation. SVMs were originally developed within the framework of Vapnik's machine learning theory as trainable non-parametric discriminant functions for supervised classification problems [12]. Supervised support vector regression (SVR) was then proposed for function approximation and regression problems, relying on the empirical risk minimization principle [10]. Although the problem of pdf estimation was considered by Vapnik since the early Nineties [12], only a few SVM approaches have been proposed. In [14], *ad hoc* training algorithms for SVM density estimation are introduced, relying either on the empirical cumulative distribution function or on the PW estimates. In [6] a kernel-based density estimator is used, relying on the idea of reducing the original data sample to a small subset, with the aim of reducing the computational cost of classic nonparametric methods.

This paper introduces a SVM algorithm for unsupervised, nonparametric density estimation that shares similarities with the technique described in [14]. The algorithm uses standard SVR (i.e., it may be implemented using any SVM simulation software which is suitable to SVR) and an unbiased version of the PW. It takes in input an unlabeled data sample, and returns a SVM which encapsulates the estimated pdf. The unbiased technique has a significant impact on the quality of the result. It overcomes the limitations of PW to a significant extent, and it may lead to better pdf models than previous SVM techniques for density estimation. Eventually, the model is used as a suitable paradigm for pdf estimation over graphs, within the proposed structured pattern classifier. The topics are addressed according to the following organization: the SVM density estimation algorithm is presented first (Section 2), followed by a demonstration (Section 3) featuring comparative evaluations on samples drawn from a mixture of Fisher-Tippett pdfs. The probabilistic classifier for graphical data is presented next (Section 4). Experiments on the Caltech Benchmark dataset are reported (Section 5), showing a dramatic improvement over the established results yielded by RNNs and GNNs. Conclusions are drawn in Section 6.

2 Unbiased Pdf Estimation Via SVM

Let us consider a pdf $p(\mathbf{x})$, defined over a real-valued, d-dimensional feature space. Again, let $\mathcal{T} = \{\mathbf{x}_1, \ldots, \mathbf{x}_n\}$ be an unsupervised sample of n patterns, identically and independently distributed (i.i.d.) according to $p(\mathbf{x})$. The PW estimate of $p(\mathbf{x}')$ from the sample $\mathcal{T}$ over a generic feature vector $\mathbf{x}'$ has the form $p(\mathbf{x}') \simeq \frac{1}{n}\sum_{i=1}^{n} \frac{1}{V_n}\varphi(\frac{\mathbf{x}'-\mathbf{x}_i}{h_n})$ [5], where $\varphi(.)$ is the window function having edge h_n, and $V_n = h_n^d$ is the corresponding volume. The edge and the volume are explicitly written as a function of n, since smaller regions around $\mathbf{x}'$ are considered as the sample size n increases, e.g. $h_n = h_1/\sqrt{n}$. The initial edge h_1 has to be chosen empirically, and it heavily affects the resulting model on finite samples. Asymptotic convergence of nonparametric models of this kind can be found in [5]. Let us now consider a SVM that we wish to train in order to learn a model $\tilde{p}(.)$ of the probability law $p(\mathbf{x})$ from the unsupervised dataset $\mathcal{T}$. The idea is to use the PW model as a target output for the SVM (as in [14]), and to apply standard SVR. An unbiased variant of this idea is proposed, according to the following unsupervised algorithm (expressed in pseudo-code):

1. Let $h_n = h_1/\sqrt{n}$, and $V_n = h_n^d$
2. For i=1 to n do /* loop over $\mathcal{T}$ */
2.1 Let $\mathcal{T}_i = \mathcal{T} \setminus \{\mathbf{x}_i\}$
2.2 Let $y_i = \frac{1}{n}\sum_{\mathbf{x}\in\mathcal{T}_i} \frac{1}{V_n}\varphi(\frac{\mathbf{x}_i-\mathbf{x}}{h_n})$ /* target output */
3. Let $\mathcal{S} = \{(\mathbf{x}_i, y_i) \mid i = 1, \ldots, n\}$ /* supervised training set */
4. Train the SVM via SVR over $\mathcal{S}$
5. Let $\tilde{p}(.)$ be equal to the SVM

Unsupervised selection of optimal parameters for the SVM kernels may be obtained applying the method described in [3]. Since the SVM output is assumed to be an estimate of a pdf, it must be non-negative. The algorithm does not prevent from negative values over certain regions of the feature space, e.g. regions that are not covered by the data sample. A viable solution is forcing negative outputs to zero once training is completed. In this respect, the introduction in the training set of a small fraction of synthetic patterns featuring a zero target in the regions adjacent to the support of the pdf may turn out to be effective (smoothing down the estimate of the pdf along its tails, too). The support may be obtained applying the technique proposed in [9]. Nevertheless, as in the popular k_n-nearest neighbor nonparametric pdf estimation [5], $\tilde{p}(.)$ is not necessarily a pdf (in general, the integral of $\tilde{p}(.)$ over the feature space does not equal 1). If the size of the data sample and/or the dimensionality of the feature space are large, the computational cost of the algorithm may become troublesome. The PW estimation time (step 2.2 of the algorithm) grows as n^2 with the sample size, the SVR slows down its convergence, and the search for adequate parameters for SVR becomes excessively demanding. In addition, if $p(\mathbf{x})$ is varying substantially over its support, the resulting number of support vectors may exceed a critic fraction of the original patterns (e.g., $70-80\%$ of the sample size). Under these circumstances, a variant of the algorithm may be used, by taking the Parzen window functions $\varphi(.)$ as the kernels for the SVM (several SVM software simulators feature the adoption of pre-computed kernels).

There is a major aspect of the algorithm that shall be clearly pointed out: the PW generation of target outputs (steps 2-2.2) is unbiased. Computation of the target for i-th input pattern $\mathbf{x}_i$ does not involve $\mathbf{x}_i$ in the underlying PW model. This is crucial in smoothing the local nature of PW. The target (i.e., estimated pdf value) over $\mathbf{x}_i$ is determined by the concentration of patterns in the sample (different from $\mathbf{x}_i$) that occur in the surroundings of $\mathbf{x}_i$. In particular, if an isolated pattern (i.e., an outlier) is considered, its exclusion from the PW model turns out to yield a close-to-zero target value. This phenomenon is evident along the tails of certain distributions (Section 3). It is seen that, in spite of its simplicity, the approach overcomes most of the PW limitations. The following Section shows that it may also turn out to be more accurate than other kernel-based methods for density estimation.

3 Demonstration

An illustrative estimation task is considered. Samples are randomly drawn from a mixture of 3 Fisher-Tippett distributions:

$$p(x) = \sum_{i=1}^{3} \frac{\Pi_i}{\beta_i} \exp\left(-\frac{x-\mu_i}{\beta_i}\right) \exp\left\{-\exp\left(-\frac{x-\mu_i}{\beta_i}\right)\right\} \tag{1}$$

where $\Pi_1 = 0.35$, $\Pi_2 = 0.5$, and $\Pi_3 = 0.15$ are the mixing parameters, and the component densities are identified by the locations $\mu_1 = 2.0$, $\mu_2 = 5.0$,

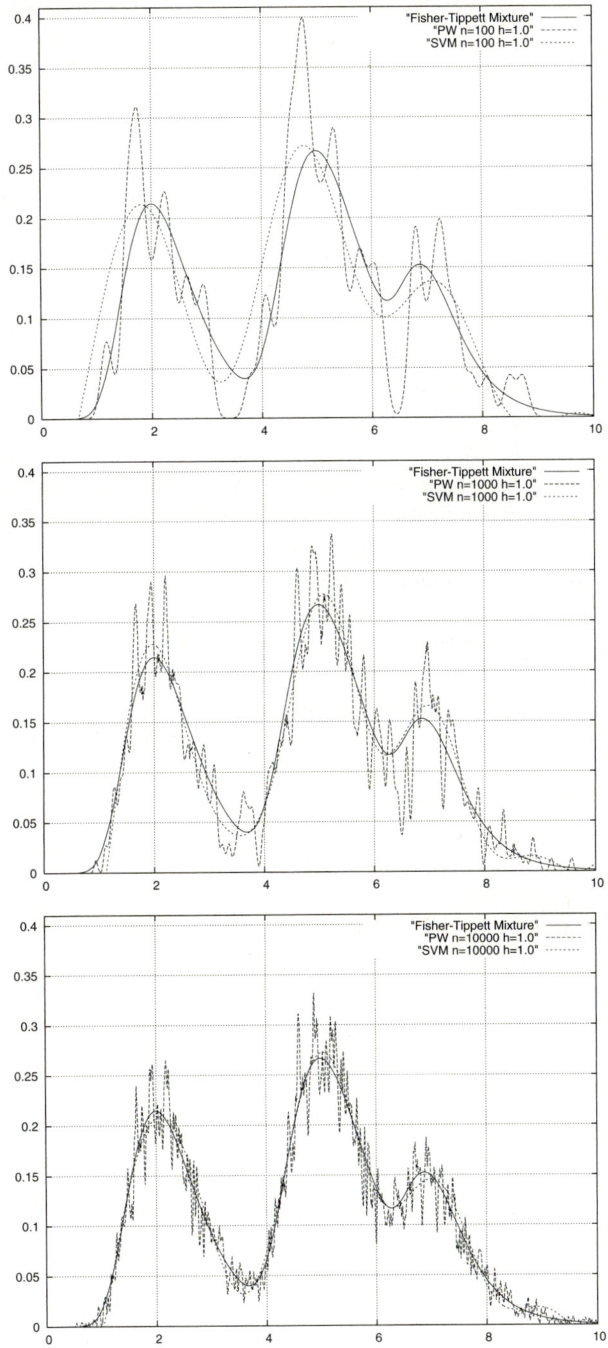

Fig. 1. Estimates for an increasing sample size: $n = 100$ (top), $n = 1000$ (mid), $n = 10000$ (bottom)

$\mu_3 = 7.0$, and by the scales $\beta_1 = 0.6$, $\beta_2 = 0.7$, $\beta_3 = 0.5$, respectively. Figure 1 shows the resulting (classic) PW and (unbiased) SVM models, estimated from data samples of increasing size ($n = 100$, $n = 1000$, and $n = 10000$) randomly drawn from the mixture. The estimates are plotted against the original pdf. A standard Normal window function was used in both the classic PW and in the proposed technique, with a standard initial edge $h_1 = 1.0$. SVMs with radial basis functions kernels are applied throughout the paper. As expected, both models improve as the size of the sample increases. The SVM estimate is closer to the reference pdf than the PW is, turning out to be also much smoother and less sensitive to variations in the number n of training patterns. The SVM estimates were forced to non-negative values by converting negative outputs to zero. The PW is significantly affected by the presence of individual training points (e.g., along the tails of the distribution) belonging to low-probability regions. The presence of local peaks violates the natural shape of the underlying pdf. It is worth noticing the difference between the SVM curve and the form of the PW model that, roughly speaking, provides the SVM with the target outputs during training. Note that, by definition, no unbiased PW can be applied at test time (no direct comparison w.r.t the SVM is possible), since the algorithm applies only to patterns that are included in the training set. Fig. 2 plots the estimates obtained using a fixed-size data sample of $n = 200$ random patterns from the mixture, as a function of the initial edge h_1, namely (from top-left to bottom-right) 0.5, 1.0, 1.5 and 2.0. The PW estimates are heavily affected by the value h_1, while the SVM model exhibits a stabler behavior.

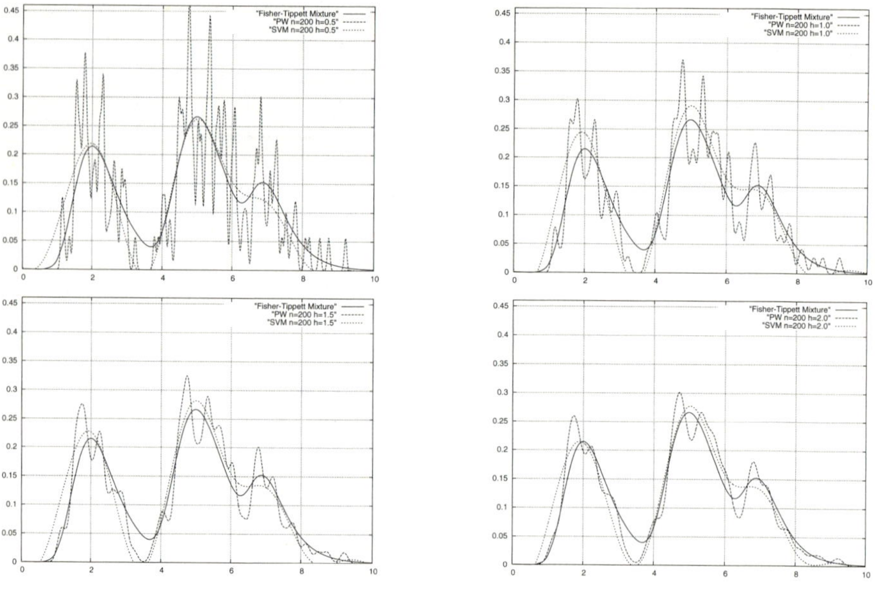

Fig. 2. Estimates for a fixed sample size ($n = 200$) and varying edge: $h_1 = 0.5$ (top left), $h_1 = 1.0$ (top right), $h_1 = 1.5$ (bottom left), $h_1 = 2.0$ (bottom right)

Table 1. Integrated squared errors for increasing sample size

	Parzen Window	*Weston* et al. *SVM*	*Unbiased PW SVM*
100 samples	1.324e-2	9.057e-03	6.219e-3
1000 samples	5.571e-3	1.403e-3	8.07e-4
10000 samples	2.188e-3	2.522e-4	1.403e-4

Finally, a quantitative comparison with a previous SVM density estimation technique ([14], section 1.11) is carried out. In [14] the evaluation is expressed in terms of integrated squared error (ISE), computed using Simpson's method over random samples drawn from a mixture of 2 Normal densities. The same criterion (i.e., ISE) is adopted in this paper, taking in consideration random samples from the Fisher-Tippet mixture and applying Simpson's method over the support of the pdf. Results are reported in Table 1 (for an increasing number of training patterns).

4 A Novel Framework for Graphical Pattern Recognition

A graph $\mathcal{G}$ is a pair $\mathcal{G} = (V, E)$ where V is an arbitrary set of nodes (or, vertices) over a given universe U, and $E = \{\mathbf{x}_j = (a_j, b_j) \mid a_j, b_J \in U, j = 1, \ldots, n\}$ is the set of edges. We consider directed as well as undirected, connected and unconnected finite graphs ($\mathcal{G}$ is undirected iff $(a, b) \in E \leftrightarrow (b, a) \in E$), either cyclic or acyclic. From an algebraic point of view, the graph is a binary relation over U, i.e. $\mathcal{G} \subseteq U \times U$. All the binary relations (graphs) involved in the learning problem at hand (both in training and test) are assumed to be defined over the same domain U. We rely on the assumption that the universe U is a (Lebesgue) measurable space, in order to ensure that probability measures can actually be defined. The measurability of finite graphs defined over measurable domains (and with measurable labels) like countable sets or real vectors is shown in [8].

Labels may be attached to vertices or edges, assuming they are defined over a measurable space. For the vertices, we consider a labeling $\mathcal{L}$ in the form of d-dimensional vectors associated with nodes, namely $\mathcal{L}(\mathcal{G}) = \{\ell(v) \mid v \in V, \ell(v) \in \mathcal{R}^d\}$. As regards the edge labels, for each $(a_j, b_j) \in E$ a label is allowed in the form $\ell_e(a_j, b_j) \in \mathcal{R}^{d=(a_j,b_j)e}$, where d_e is the dimensionality of the continuous label domain. Labels are accounted for by modifying the definition of $\mathbf{x}_j = (a_j, b_j) \in E$ slightly, taking $\mathbf{x}_j = (a_j, \ell(a_j), b_j, \ell(b_j)), \ell_e(a_j, b_j))$. Note that the present framework requires that the nodes in the graph are individual elements of a well-defined universe. Consequently, it does not explicitly cover scenarios in which the nodes act only as "placeholders" in the specific graphical representation of the data. If this is the case, and the actual input features are completely encapsulated within label vectors, the previous definitions may be replaced by $\mathbf{x}_j = (\ell(a_j), \ell(b_j))$ for each pair $(a_j, b_j) \in E$. This may turn out to be effective in practical applications, but it is mathematically justified only iff each label uniquely identifies the corresponding node. Examples of structures that fit the present framework are: semantic networks, e.g. whose nodes are words from a given dictionary; subgraphs of the World Wide Web, where nodes are extracted from the universe of possible URLs and node labels are a representation of the information contained in the web page; scene descriptions in syntactic pattern recognition, whenever nodes are extracted from the universe of terminal/nonterminal symbols.

Let $\omega_1, \ldots, \omega_c$ be a set of separate classes. We assume that each graph belongs to one of the c classes. The posterior probability of i-th class given the graph is $P(\omega_i \mid \{\mathbf{x}_1, \ldots, \mathbf{x}_n\})$, where each $\mathbf{x}_j = (a_j, b_j)$ is interpreted as a random vector whose characteristics and dimensionality depend on the nature of the universe U. The assumption of dealing with measurable universes allows the adoption of probabilistic measures, and applying Bayes' theorem [5] we can write:

$$P(\omega_i \mid \{\mathbf{x}_1, \ldots, \mathbf{x}_n\}) = \frac{p(\{\mathbf{x}_1, \ldots, \mathbf{x}_n\} \mid \omega_i) P(\omega_i)}{p(\{\mathbf{x}_1, \ldots, \mathbf{x}_n\})} \tag{2}$$

where $P(.)$ denotes a probability measure, and $p(.)$ denotes a probability density function (pdf). The quantity $p(\{\mathbf{x}_1, \ldots, \mathbf{x}_n\} \mid \omega_i)$ is a joint pdf that expresses the probabilistic distribution of the overall binary relation $\{\mathbf{x}_1, \ldots, \mathbf{x}_n\}$ over its domain according to the law $p(.)$. We assume that the pairs $\mathbf{x}_j, j = 1, \ldots, n$ (including the corresponding labels) are independently and identically distributed (iid) according to the class-conditional density $p(\mathbf{x}_j \mid \omega_i)$. Roughly speaking, $p(\mathbf{x}_j \mid \omega_i)$, encapsulates three different, yet joint probabilistic quantities, all of them conditioned on ω_i: (1) the likelihood of observing any given pair of nodes (edge), (2) the probability distribution of node labels, and (3) the pdf of edge labels. In so doing, the probability of having an edge between two vertices is modeled jointly with the statistical properties of the nodes and of their labels.

The iid assumption is in line with classical and state-of-the-art literature on statistical pattern recognition and on random graphs. For instance, in the ER random graph model [2] edges are iid according to a unique (e.g. uniform) probability distribution over the whole graph. In the *small worlds* paradigm [13] iid edges are inserted during the rewiring process that generates the graph. Similar arguments apply also to scale-free networks. It is noteworthy that the iid assumption does not imply any loss in terms of structural information. Once a graph is given, its structure is encapsulated within the binary relation, which does not depend on the probabilistic quantities involved in Eq. 2. Applying the iid assumption, Eq. 2 can be rewritten as:

$$P(\omega_i \mid \{\mathbf{x}_1, \ldots, \mathbf{x}_n\}) = \frac{\prod_{j=1}^{n} p(\mathbf{x}_j \mid \omega_i) P(\omega_i)}{\prod_{j=1}^{n} p(\mathbf{x}_j)} \tag{3}$$

where $p(\mathbf{x}_j) = \sum_{k=1}^{c} P(\omega_k) p(\mathbf{x}_j \mid \omega_k)$. Since the pairs $\mathbf{x}_j$ are extracted from a well-defined universe and the joint probabilities are invariant w.r.t. arbitrary permutations of their arguments, there is no "graph matching" problem in the present framework. Representing the graph as a relation implies looking at the structure as a whole. This is a major difference w.r.t. other techniques that require a visit of the graph in a specific order, and that are faced with the problem of possible infinite recursion over cyclic structures.

In order to apply Eq. 3 as a discriminant function for graphical pattern recognition, we need to estimate $P(\omega_i)$ and the class-conditional pdf $p(\mathbf{x}_j \mid \omega_i)$ for $i = 1, \ldots, c$ and $j = 1, \ldots, n$. If good estimates of these quantities are obtained, the maximum-a-posteriori decision rule expressed by Eq. 3 is expected to yield the minimum Bayesian risk (i.e., minimum probability of classification error) [5]. The quantity $P(\omega_i)$ can be estimated from the relative frequencies of classes over the training sample, as usual. The technique described in Section 2 is used to estimate $p(\mathbf{x}_j \mid \omega_i)$. It has to be applied for c times, once for each independent subsample of the data which belongs to a specific class. Since $\prod_{j=1}^{n} p(\mathbf{x}_j)$ does not depend on ω_i, it can be dropped from Eq. 3. Taking the logarithm we obtain an equivalent discriminant function $g_i(\{\mathbf{x}_1, \ldots, \mathbf{x}_n\}) = \sum_{j=1}^{n} \log\{p(\mathbf{x}_j \mid \omega_i)\} + \log\{P(\omega_i)\}$. In so doing, numerical stability is gained when dealing with joint probabilistic quantities over large graphs.

5 Experiments: Image Classification from the Caltech Benchmark Dataset

The proposed technique is compared with RNNs and GNNs on an image classification problem from the Caltech benchmark dataset. The experiment (as in [4]) is based on 4 classes, i.e. images of *bottles*, *camels*, *guitars*, and *houses*. For each class, a subset of 350 images was extracted from the Caltech dataset. Half of the images consists of positive examples of the class, while the others are negative examples, i.e. images randomly sampled from the other classes. The same data subsets as in [4] were used, each divided into training, validation and test sets (150, 50, and 150 images, respectively). Each image was represented as an undirected Region Adjacency Graph (RAG), obtained using the Mean Shift algorithm and the k-means color quantization procedure as in [4]. Since RNNs cannot deal with undirected graphs, application of the RNNs requires that the RAGs are transformed into directed acyclic graphs (DAG) *via* breadth-first visit and substitution of each undirected edge with a directed one. Each node of the RAG has a 23-dimensional vector label, while edge labels are 5-dimensional [4].

Results are expressed in terms of recognition accuracy on a class-by-class basis (see [4]). Average of the accuracies is reported in the last column of the Table 2. It is seen that, although simple, the present approach outperforms the RNNs, and it yields also a significant average improvement over the GNNs.

Table 2. Recognition accuracies on the Caltech Benchmark Dataset w.r.t. the results reported in [4]

Models	*Bottles*	*Camels*	*Guitars*	*Houses*	Avg.
Present approach	90.51	82.00	84.51	98.77	**88.94**
GNN	84.67	74.67	70.67	84.67	**77.84**
RNN	70.66	65.33	62.67	81.33	**69.33**

6 Conclusion

There are two contributions of the paper: (1) an unbiased and easy to implement technique for multivariate pdf estimation via SVM trained with standard SVR. The model is effective, much stabler than PW, and it may yield improvement over previous SVM algorithms for density estimation; (2) a novel and simple approach to structured pattern recognition. Its core aspect is to look at a graph as a binary relation, and to introduce a Bayesian classifier that involves the estimation of a joint pdf over the relation itself. The SVM estimation technique may be successfully used in this respect. Experimental comparison w.r.t. state-of-the-art connectionist models for structured data confirms that the framework, albeit not universal, turns out to be sound.

References

1. Bengio, Y., Simard, P., Frasconi, P.: Learning long-term dependencies with gradient descent is difficult (Special Issue on Recurrent Neural Networks, March 94). IEEE Transactions on Neural Networks 5(2), 157–166 (1994)
2. Bollobs, B.: Random Graphs, 2nd edn. Cambridge University Press, Cambridge, UK (2001)
3. Cassabaum, M.L., Waagen, D.E., Rodríguez, J.J., Schmitt, H.A.: Unsupervised optimization of support vector machine parameters. In: Sadjadi, F.A. (ed.) Proc. of SPIE, vol. 5426, pp. 316–325 (2004)
4. Di Massa, V., Monfardini, G., Sarti, L., Scarselli, F., Maggini, M., Gori, M.: A comparison between recursive neural networks and graph neural networks. World Congr. on Comp. Intelligence, 778–7825 (2006)
5. Duda, R.O., Hart, P.E.: Pattern Classification and Scene Analysis. Wiley, New York (1973)
6. Girolami, M., He, C.: Probability density estimation from optimally condensed data samples (2003)
7. Gori, M., Monfardini, G., Scarselli, F.: A new model for learning in graph domains. In: Proc. of IJCNN-05 (August 2005)
8. Hammer, B., Micheli, A., Sperduti, A.: Universal approximation capability of cascade correlation for structures. Neural Computation 17(5), 1109–1159 (2005)
9. Scholkopf, B., Platt, J.C., Shawe-Taylor, J.C., Smola, A.J., Williamson, R.C.: Estimating the support of a high-dimensional distribution. Neural Comput. 13(7), 1443–1471 (2001)
10. Smola, A.J., Scholkopf, B.: A tutorial on support vector regression. Statistics and Computing 14(3), 199–222 (2004)
11. Sperduti, A., Starita, A.: Supervised neural networks for the classification of structures. IEEE Transactions on Neural Networks 8(3), 714–735 (1997)
12. Vapnik, V.N.: The Nature of Statistical Learning Theory. Springer, New-York (1995)
13. Watts, D., Strogatz, S.: Collective dynamics of small world networks. Nature 393, 440–442 (1998)
14. Weston, J., Gammerman, A., Stitson, M., Vapnik, V., Vovk, V., Watkins, C.: Support vector density estimation. In: Scholkopf, B., Burges, C.J.C., Smola, A.J. (eds.) Advances in Kernel Methods — Support Vector Learning, Cambridge, MA, pp. 293–306. MIT Press, Cambridge (1999)

Neural Mechanisms for Mid-Level Optical Flow Pattern Detection

Stefan Ringbauer, Pierre Bayerl, and Heiko Neumann

Ulm University, Inst. for Neural Information Processing, Ulm, Germany

Abstract. This paper describes a new model for extracting large-field optical flow patterns to generate distributed representations of neural activation to control complex visual tasks such as 3D egomotion. The neural mechanisms draw upon experimental findings about the response properties and specificities of cells in areas V1, MT and MSTd along the dorsal pathway. Model V1 cells detect local motion estimates. Model MT cells in different pools are suggested to be selective to motion patterns integrating from V1 as well as to velocity gradients. Model MSTd cells considered here integrate MT gradient cells over a much larger spatial neighborhood to generate the observed pattern selectivity for expansion/contraction, rotation and spiral motion, providing the necessary input for spatial navigation mechanisms. Our model also incorporates feedback processing between areas V1-MT and MT-MSTd. We demonstrate that such a re-entry of context-related information helps to disambiguate and stabilize more localized processing along the primary motion pathway.

Keywords: Motion perception, optical flow, motion integration, motion gradient, feedback, navigation, area MT, area MSTd.

1 Introduction

The robust analysis of optical flow patterns is one of the basic computational tasks for steering egomotion of human and higher animals [9]. Self-motion induces typical patterns of optical flow on the retina which are analyzed by the visual system over several stages of hierarchically organized visual areas in the dorsal cortical pathway. The two cortical areas MT (medial temporal) and MSTd (dorsal medial superior temporal) are primarily concerned with the processing of optical flow. Cells in area MT represent proper features of optical flow information, e.g., motion direction and speed, in medium-size regions of the visual field [1]. MT cells project to area MSTd where cells have huge receptive fields and are tuned to specific patterns of optical flow. While [8] found evidence for a dominance of cells selective to radial and rotational flow (defining an orthogonal basis), the investigation of [10] found support for the existence of a continuum of flow sensitive cells with equally likely preference to spiral motion patterns.

Previous approaches to modeling optical flow extraction have been suggested which consider MSTd cells sensitive to large-field optical flow. For example, [4] investigated the development of optical flow sensitive cells using a two-layer backpropagation network. [11] studied optical flow processing by MSTd cells using a

J. Marques de Sá et al. (Eds.): ICANN 2007, Part II, LNCS 4669, pp. 281–290, 2007.

spatial retino-cortical mapping with log-polar magnification. In this model the output from direction selective MT cells were spatially integrated by MSTd cells with different selectivities that are fed forward to a heading map. The model by [17] was the first incorporating the functionality of speed gradient cells (at MT). Different pools of MSTd neurons spatially integrate motion responses and speed gradient responses separately to generate cells tuned to complex motion patterns.

Our model extends previous models drawing upon own previous investigations on recurrent feedforward-feedback processing for integration and disambiguation of motion signals. Here we propose a new model mechanism for extracting changes in velocity by employing cells with asymmetric receptive field profiles oriented along the motion direction. This extends previously proposed speed gradient mechanisms by computing velocity changes. MSTd in turn integrates information signaled by MT gradient cells to form large-field motion cells selective to complex motion patterns. We demonstrate that MSTd gradient cells show position-independence properties with respect to variations of the location of centers of motion. Furthermore, we also incorporate predictive feedback from MSTd to MT that stabilizes the extraction of noisy complex motion patterns even in the presence of moving objects.

2 Mid-Level Optical Flow Pattern Detection

2.1 Input Sequences and Initial Motion Detection

Our approach draws upon experimental evidence about the structure and function of the primary stages of the dorsal pathway in visual cortex. Initial motion detection is realized at the stage of the primary cortical area V1, while visual area MT focuses on the robust detection of salient motion components, tracking of localized features, and the disambiguation of locally ambiguous motion patterns that were caused, e.g., by the aperture problem [13]. Visual area MSTd is primarily concerned with the detection and representation of large whole field flow patterns, such as those that were generated by observer self-motion.

Local motion detection is accomplished by a Reichard-like correlation scheme with shunting inhibition between half-detectors tuned to opposite movement directions. These activities feed into a feedforward-feedback loop modeling V1-MT interaction [2]. In a nutshell, motion sensitive cells in model MT integrate activities from V1 direction selective cells by pooling same velocities over a spatial as well as directional neighborhood, essentially low-pass filtering the noisy motion estimates. The resulting activities are represented in a pool of motion cells in area MT. Pooled activities $a_{\mathbf{u}}^{MT}$ are subsequently fed back to enhance V1 cell activations that are compatible with the velocity represented in model MT. The resulting activations were also fed forward into a population of MT model cells that compute *changes* in the MT velocity field representation. This processing and subsequent integration of these activities by large-field model MSTd cells are at the primary focus of this contribution and are described in subsection 2.2.

The dynamics of each model area is described by a cascade model of linear-non-linear-non-linear (LNN) processing [12] derived from single-compartment neuron models with firing-rate activation dynamics. We utilize three-stages of (a) input filtering, (b) non-linear signal amplification (via feedback), and (c) shunting inhibition to realize activity normalization in a local neighborhood. This basic architecture has been utilized in previous models for static form as well as dynamic motion perception [14][2].

2.2 Detection of Motion Patterns and Gradients

The V1-MT motion detection scheme is extended in several ways, namely (a) a neural mechanism to compute changes in the dense velocity representation (motion gradients) at the level of the MT/MSTl complex, (b) a stage of integrating gradient signals over a large spatial neighborhood to generate cell responses in model area MSTd, and (c) incorporating feedback to deliver a re-entry signal for stabilizing the processing of motion gradients. Similar to the approach developed by Tsotsos and colleagues [17] we employ mechanisms to measure local changes in velocity generating a gradient representation. Experimental evidence for the existence of cells sensitive to velocity gradients has been presented by, e.g., [16], and the spatial pattern of surround weighting was studied in [18].

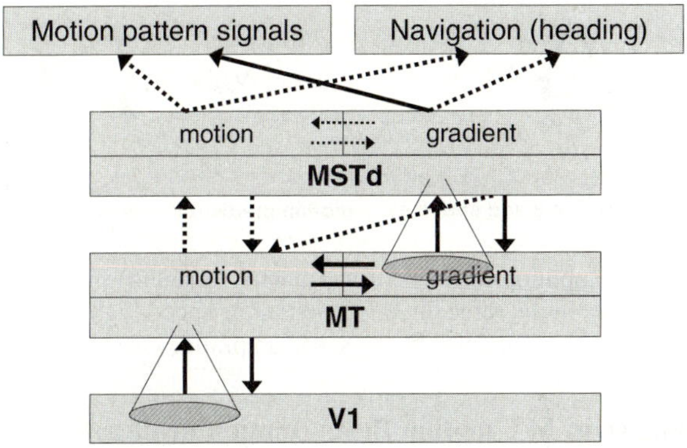

Fig. 1. General outline of processing stages and communication pathways of the model. Bold arrows denote information flow currently implemented in the model, dashed arrows depict those connectivities that are planned to be included in further model extensions. Representation of *motion* in MT and MSTd denotes a location-velocity space, <x,y,θ,s>, while the *gradient* representations denote a location-velocity-gradient space, <x,y,Δθ,Δs>.

The primary input to area MSTd is delivered via area MT where it is integrated over a large spatial neighborhood to generate the observed pattern specificity to large-field motion patterns. Physiological investigations have demonstrated that cells in area MSTd respond primarily to large field motion patterns having different selectivities, e.g., expanding or contracting radial motion, clockwise or counter-clockwise rotation, or linear superpositions that lead to spiral motion patterns [10]. A sub-population of cells also responds to pure translatory motion. MSTd activation in turn is fed back to cells in area MT to incorporate prediction and enhancement of noisy activation distributions. In all, we suggest a coherent architecture in which the bidirectional signal flow defines a key component of functionality to achieve robust cortical motion representation (Fig. 1). Size ratios between cells in the different model cortical areas are defined in accordance to experimental data [6, 10] and were set to V1:MT:MSTd = 1:5:25.

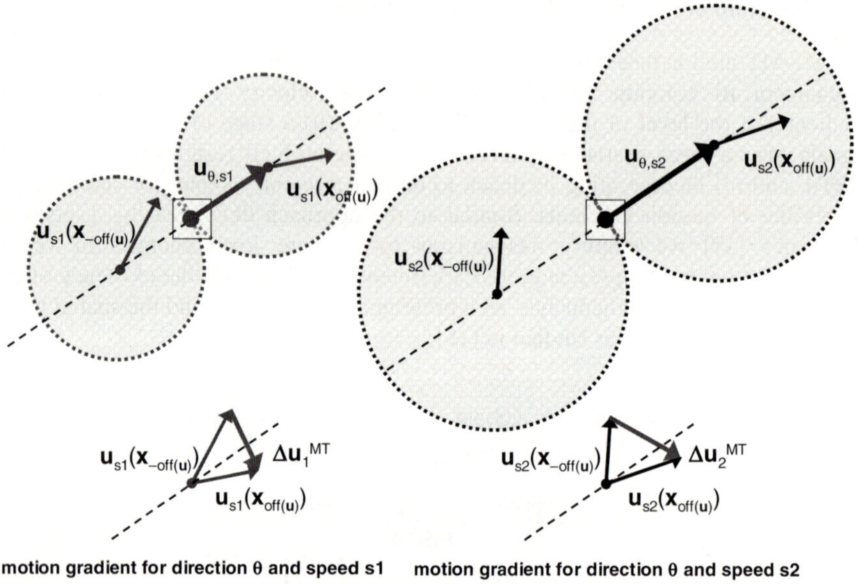

Fig. 2. Mechanism of computing differences in velocity (gradient) in model area MT for two different speed amplitudes in the same direction (left: slow speed, right: fast speed). The size of sub-field integration scales as a function of the speed amplitude (see text).

Detection of changes in MT motion fields (motion gradients). Motion is encoded by populations of cells sensitive to direction and speed, i.e., $\mathbf{u} = (\theta, s)$. In order to keep the computational efforts in reasonable bounds we employ a rank-order coding approach [15] in which spike sequences are generated algorithmically and represented in ordered lists. This allows highly efficient on-demand representation of motion in various directions and allowing the detection of arbitrary speed amplitudes (see [3] for details).

We propose a scheme to measure changes in the velocity field along local motion directions in MT, which are represented in a second pool of MT neurons, $a_{\Delta\mathbf{u}_\theta}^{MT}$, where $\Delta\mathbf{u} = (\Delta\theta, \Delta s)$ symbolizes changes in velocity. The population of motion sensitive cells is sampled at each location by two sub-fields that are spatially offset along the movement direction at the target cell. For a given motion along direction θ and speed $s = |\mathbf{u}|$, sampling locations of cells are at $\mathbf{x}_{-off(\mathbf{u})} = \mathbf{x} - \mathbf{u}_\theta$ and $\mathbf{x}_{off(\mathbf{u})} = \mathbf{x} + \mathbf{u}_\theta$, respectively ($\mathbf{x}$ denoting the spatial target location). The radius of the sampling kernels varies with the speed of the current velocity, i.e. $r = |\mathbf{u}|$, such that we get an increased spatial uncertainty with increasing speed (see Fig. 2). In each sub-field the velocities are weighted and summed to derive a population response for the velocity.

The differences between velocities of the populations from both offset locations are defined by the quantity[1]

$$\Delta\mathbf{u}^{MT}(\mathbf{x}) = \mathbf{u}(\mathbf{x}_{off(\mathbf{u})}) - \mathbf{u}(\mathbf{x}_{-off(\mathbf{u})}). \tag{1}$$

We specify cells to represent certain $\Delta\mathbf{u}$ which compute their activity utilizing the following non-linear mechanism[2],

$$a_{\Delta\mathbf{u}}^{MT}(\mathbf{x}) = a_{\mathbf{u}}^{MT}(\mathbf{x}) \cdot a_{\mathbf{u}}^{MT}(\mathbf{x}_{off(\mathbf{u})}) \cdot a_{\mathbf{u}}^{MT}(\mathbf{x}_{-off(\mathbf{u})}) \tag{2}$$

to generate a field of motion gradients. These activities are efficiently approximated similar as in [3]. Each local difference in activity is represented with respect to the direction of motion at the respective target location that defines a local gauge coordinate system (compare [17]). Activities of motion cells as well as of motion gradient cells are subsequently normalized by a process of divisive (shunting) center-surround competition in the space-feature domain. For simplicity, at the moment, we employ normalization over gradient cell activities at single locations, namely

$$a_{\Delta\mathbf{u}_\theta}^{MT}(\mathbf{x}) = a_{\Delta\mathbf{u}_\theta}^{MT}(\mathbf{x}) / \sum_{all\, a_{\Delta\mathbf{u}}(\mathbf{x})} a_{\Delta\mathbf{u}}^{MT}(\mathbf{x}). \tag{3}$$

The resulting activations are then fed forward to cells in model area MSTd.

MSTd gradient motion integration. Cells in model MSTd sum up the responses of corresponding MT gradient units over a large spatial neighborhood similar as in previous models [4][11][17]. The mechanism utilizes a convolution by a suitable kernel to weight activities in the spatial-feature domain

$$a_{\Delta\bar{\mathbf{u}}_\theta}^{MSTd}(\mathbf{x}) = \sum_{\mathbf{x}',\phi} a_{\Delta\mathbf{u}_\phi}^{MT}(\mathbf{x}') \cdot \Lambda_{\mathbf{x}\mathbf{x}'} \cdot \Psi_{\theta\phi} \tag{4}$$

with separable kernels Λ and Ψ for weighting in the spatial and the direction domain, respectively.

The spatial resolution of model area MSTd is down-sampled by a factor of 1:5. In the current version of our model we integrate responses of the MT gradient cells to be represented in a pool of motion cells, $a_{\Delta\bar{\mathbf{u}}_\theta}^{MSTd}$. Those cell responses resulting from non-zero directional differences, $\Delta\theta^{MT}(\mathbf{x}) \neq 0$, encode large-field motion patterns such as radial expansion/contraction, rotation, and spiral motion. For zero directional differences, $\Delta\theta^{MT}(\mathbf{x}) = 0$, and vanishing speed gradients, $\Delta s^{MT}(\mathbf{x}) = 0$, one gets cells with pure large-field translatory motion pattern selectivity (center of motion shifted to infinity).

---

[1] To simplify the necessary calculations we employed a vector notation for computing the velocity differences. In order to represent the signals in a neurally plausible scheme direction and speed components could be represented in separate sub-populations of cells with proper sampling resolution. These quantities could then enter into competitive interactions.

[2] We suggest taking the product of the three measures to calculate a response that denotes a likelihood of the presence of the particular change in velocity. The independence of the three measures is assumed.

Motion pattern prediction and modulatory MSTd-MT feedback. Integrated measures of velocity changes, in turn, provide a context for more localized measures at the earlier computational stages (model area MT in our case). In other words, MSTd activation serves as a predictor signal that is fed back to gradient responses in area MT. This feedback is proposed to be modulatory and excitatory such that activities of cells in MT selective for a particular gradient direction can be amplified by MSTd gradient activation for the same (or similar) gradient direction. The modulation is denoted by the scheme (with constant coefficient C)

$$a_{\Delta \mathbf{u}_\theta}^{MT} \cdot \left(1 + C \cdot a_{\Delta \bar{\mathbf{u}}_\theta}^{MSTd}\right). \tag{5}$$

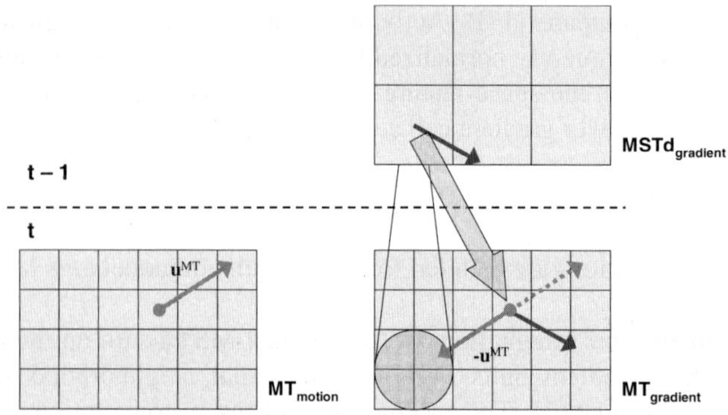

Fig. 3. Sketch of mechanism employed for model MSTd-MT feedback of velocity measures

The rationale is that MT activations gate the modulation signal so that existing gradient representations will be amplified by MSTd feedback or left unchanged in the case when no feedback signal exists. In case no feedforward activation has been computed at a given location feedback alone cannot generate new activities. This property stabilizes the network behavior in accordance with the no-strong loop hypothesis [7]. The feedback signal amplifies filtered feedforward activations at the second stage of the three-level cascade model briefly sketched above. The amplified responses subsequently undergo center-surround shunting competition (stage three of the cascade model). This realizes a biased competition since those activities in the competitive pool that were amplified now have a stronger bias and consequently reduce the activities of cells which have not received any feedback.

The feedback mechanism is predictive in the sense that the amplification is shifted spatially to a location that coheres with the target motion direction, $\mathbf{u}^{MT}(\mathbf{x})$, in the next frame of the sequence. In a nutshell, the prediction of a gradient measure (with amplitude and direction) utilized "votings" that were spatially shifted according to the velocity $\mathbf{u}^{MT}$ at the target location (see Fig. 3).

3 Simulation Results

The neural model has been tested on a variety of input motion sequences. In order to demonstrate its selectivity in the processing of large-field motion patterns and the response distribution of flow sensitive MSTd cells over time we first probe the network with a motion sequence of a simulated flight through a corridor with synthetically generated wall texture patterns (Fig. 4). Here the observer maneuvers such that a continuous sequence of steering commands lead to forward motion followed by left and right turns as well as reversing the spatial movement in between.

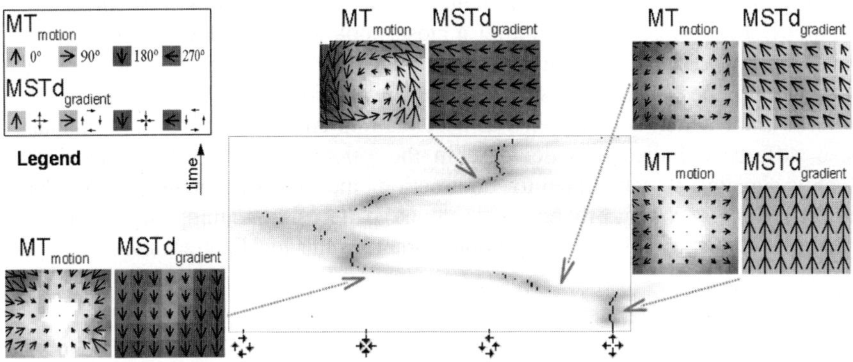

Fig. 4. Results of recurrent V1-MT-MSTd processing for an image sequence from a flight through a tunnel and respective maneuvers over time. The four different pairs of flow field patterns represent the equilibrated motion estimates at different times each showing MT motion responses and MSTd gradient cell responses, respectively (color coded directions represent a mapping of corresponding radial/rotation/spiral patterns shown in the legend top left), response trace of pattern selective MSTd gradient cells over time shown in center (see text).

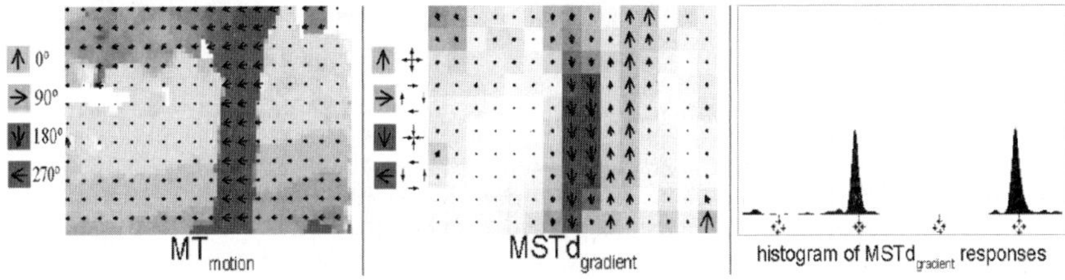

Fig. 5. Results of recurrent V1-MT-MSTd processing for the "Flower garden" sequence. Left: MT motion responses, center: MSTd gradient cell responses, right: histogram of MSTd gradient responses (see text for discussion).

Equilibrated MT motion responses are shown that were generated by V1-MT feedforward/feedback interaction at different times of the temporal sequence and the corresponding results of integrated velocity gradients at the stage of area MSTd (equilibrated response of MT-MSTd feedforward/feedback interaction). The response tuning of large-field motion pattern responses of a continuum of MSTd cells (selective to radial expansion/contraction, clockwise/counter-clockwise rotation, and spiral motion) is shown in the panel at the center. The display of responses of the motion pattern selective cells with different tunings demonstrates that the temporal course helps to infer the observer motion.

We also tested the model using the "Flower garden" sequence (Fig. 5). This specifically investigates the computational results for scenes containing mutually occluding objects thus generating locations of temporally disappearing structure as well as structure reappearing at locations that were uncovered. The key observations are that (1) dense motion is computed with speeds corresponding to different depths of the scenic objects (including speed gradient in vertical direction of the ground plane), and (2) that motion gradient cells detect occlusion boundaries by signaling opposing motion directions. The gradient directions indicated might be surprising at a first glance but can be explained after a closer look at the mechanisms involved. The signal for "expansion" motion along the right tree boundary is generated by the different velocities (increasing speed) in motion direction from background to the tree region in front which is interpreted as acceleration. At the opposite tree boundary the speed difference indicates a decrease in the velocity (indicative for deceleration) which is signaled by cells tuned to "contraction" motion. A prediction of the model is that MSTd cells when probed by stimulus patterns containing significant depth structure and thus motion parallax should signal similar local pattern motion.

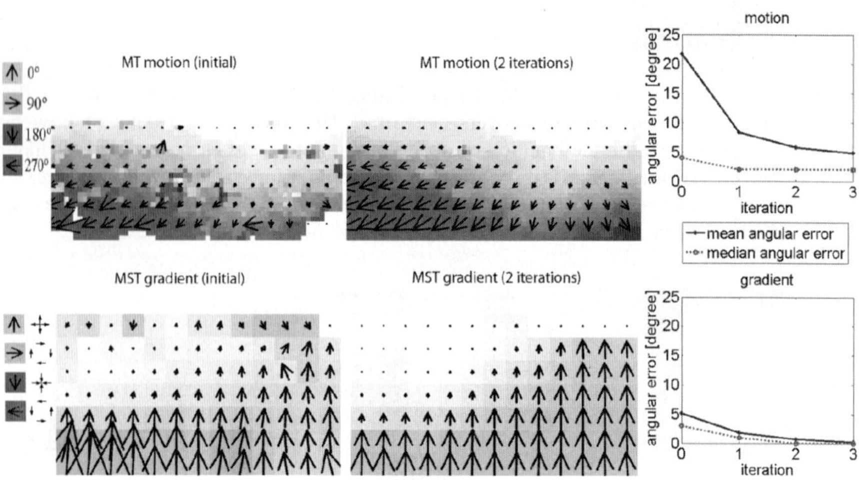

Fig. 6. Results of recurrent V1-MT-MSTd processing for ground plane motion with view direction offset from motion direction. Top: MT responses (initial response, after 2nd iteration), bottom: MSTd gradient cell responses, right panels: improvement of motion and gradient estimates (mean and median angular error) over time (see text).

Finally, we probed the model with the flow pattern generated by forward observer motion over a ground plane with a view direction 15 deg. offset to the left from the translation axis (Fig. 6). This investigates the model properties in navigation tasks with different centers of motion due to varying view directions. The perspective effect of flows on the ground plane poses a problem for the model proposed by Tsotsos and colleagues which employs speed gradient measures for flow pattern extraction [17]. As a consequence, the depth gradient for the ground plane leads to patches of different optical flow field interpretations, namely rotations to both sides of the center of motion, expansion below and above the center, and spiral motion segments in between. Our approach leads to a response pattern that is invariant against the surface projection. In addition, the study demonstrates how MT flow estimates as well as MSTd pattern representations were improved by the iterative feedforward-feedback interaction. MT motion responses are shown for the (noisy) initial estimates and after 2 iterations as well as the corresponding responses for MSTd gradient cells. The temporal course of angular error reduction is shown over 3 iterations (right panels).

4 Discussion and Conclusions

We propose a novel neural architecture of motion detection, integration and the extraction of large field optical flow patterns building upon evidence about cells at different stages along the dorsal pathway of primate visual cortex. Our model makes several new contributions in comparison with previous computational models. In particular, a previous model of early motion detection and integration was extended using the same type of basic mechanisms, namely feedforward integration, modulatory feedback, and shunting competition. Unlike many other models, with the notably exception of the architecture proposed by Tsotsos and coworkers [17], we propose a stage of making explicit velocity changes. Unlike the Tsotsos' model, we propose a speed sensitive scheme of difference filtering between sub-fields along the direction of motion at a target location. Our scheme can be considered as a more generalized approach that also allows detecting direction gradients (in addition to speed gradients) and measuring such velocity gradients in a gauge coordinate frame along the local direction of motion. This idea is reminiscent of the approach for dense texture flow field extraction that has been proposed by Zucker and coworkers [5]. Whereas, Zucker focuses on the long-range lategral connectivities for integration of oriented patterns in static form processing, we are concerned with measuring the direction changes in flow fields. For that reason, we suggested to employ a scheme that utilizes oriented receptive fields with excitatory and inhibitory sub-fields whose sizes scale with the speed of the motion patterns. Unlike the model proposed by Grossberg and colleagues [11] which utilizes feedforward integration only, we employ feedback and integrate velocity gradient information as well. The proposed scheme does not at the moment incorporate learning to automatically develop cells being selective for mid-level motion patterns, such as, e.g., [4].

As we have indicated in Fig. 1 that displays an overview of the computational architecture the extraction of large-field motion patterns serves as an input representation for further computations supporting different behavioral tasks. For example, the estimation of heading is useful for navigation in the spatial environment. In order to reliably extract the heading direction the complex motion field must be somehow decomposed into translatory and rotational flow-field components of different relative amounts. A mixture of expansion and rotation component in flow patterns occurs routinely during fixations of a target object while an observer is moving in a particular direction. Also scenic objects can move in certain directions independently of the observer and must be detected and segmented from the global flow field. We suggest that motion and motion gradient information provide complementary information since the extraction of global motion pattern signals requires invariance against, e.g., the focus of expansion. On the other hand, estimating the heading direction needs to gain information about the localization of the focus-of-expansion. Both type of information is robustly encoded in the motion and gradient signals at the level of model MSTd. Their proper combination remains a topic for further investigation.

Acknowledgments. This research has been supported in part by a grant from the European Union (EU FP6 IST Cognitive Systems Integrated project: Neural Decision-Making in Motion; project no. 027 198).

References

1. Albright, T.D.: Direction and orientation selectivity of neurons in visual area MT of the macaque. J. Neurophysiol. 52, 1106–1130 (1984)
2. Bayerl, P., Neumann, H.: Disambiguating visual motion through contextual feedback modulation. Neural Comp. 16, 2041–2066 (2004)
3. Bayerl, P., Neumann, H.: A fast biologically inspired algorithm for recurrent motion estimation. IEEE Trans. on PAMI 29, 246–260 (2007)
4. Beardsley, S.A., Vaina, L.M.: Computational modeling of optical flow sensitivity in MSTd neurons. Comp. Neural Syst. 9, 467–493 (1998)
5. Ben-Shahar, O., Zucker, S.: Geometrical computations explain projection patterns of long-range horizontal connections in visual cortex. Neural Comp. 16, 445–476 (2004)
6. Born, R.T., Bradley, D.C.: Structure and function of visual area MT. Ann. Rev. Neurosci. 28, 157–189 (2005)
7. Crick, F., Koch, C.: Constraints on cortical and thalamic projections: The no-strong loops hypothesis. Nature 391, 245–250 (1998)
8. Duffy, C.J., Wurtz, R.H.: Sensitivity of MST neurons to optic flow stimuli I. A continuum of response selectivity to large-field stimuli. Neurophysiol 65, 1329–1345 (1991)
9. Gibson, J.J.: The Ecological Approach to Visual Perception. LEA, Hillsdale, NJ (1986)
10. Graziano, M.S.A., Anderson, R.A., Snowden, R.: Tuning of MST neurons to spiral motions. J. Neurosci. 14, 54–67 (1994)
11. Grossberg, S., Mingolla, E., Pack, C.: A neural model of motion processing and visual navigation by cortical area MST. Cerebral Cortex 9, 878–895 (1999)
12. Herz, A.V.M., Gollisch, T., Machens, C.K., Jaeger, D.: Modeling single-neuron dynamics and computations: A balance of detail and abstraction. Science 314, 80–85 (2006)
13. Pack, C.C., Born, R.T.: Temporal dynamics of a neural solution to the aperture problem in cortical area MT. Nature 409, 1040–1042 (2001)
14. Thielscher, A., Neumann, H.: Neural mechanisms of cortico-cortical interaction in texture boundary detection: A modeling approach. Neuroscience 122, 921–939 (2003)
15. Thorpe, S., Delorme, A., Van Rullen, R.: Spike-based strategies for rapid processing. Neural Networks 14, 715–726 (2001)
16. Treue, S., Anderson, R.A.: Neural responses to velocity gradients in macaque cortical area MT. Vis. Neurosci. 13, 797–804 (1996)
17. Tsotsos, J.K., Liu, Y., Martinez-Trujillo, J.C., Pomplun, M., Simine, E., Zhou, K.: Attending to visual motion. Computer Vision and Image Understanding 100, 3–40 (2005)
18. Xiao, D.-K., Raiguel, S., Marcar, V., Orban, G.: The spatial distribution of the antagonistic surround of MT/V5 neurons. Cerebral Cortex 7, 662–677 (1997)

Split–Merge Incremental LEarning (SMILE) of Mixture Models

Konstantinos Blekas and Isaac E. Lagaris

Department of Computer Science, University of Ioannina, 45110 Ioannina, Greece
{kblekas,lagaris}@cs.uoi.gr

Abstract. In this article we present an incremental method for building a mixture model. Given the desired number of clusters $K \geq 2$, we start with a two-component mixture and we optimize the likelihood by repeatedly applying a *Split-Merge* operation. When an optimum is obtained, we add a new component to the model by splitting in two, a properly chosen cluster. This goes on until the number of components reaches a preset limiting value. We have performed numerical experiments on several data–sets and report a performance comparison with other rival methods.

1 Introduction

Clustering, apart from being on its own a challenging field of research, is useful to a wide spectrum of application areas, such as pattern recognition, machine learning, computer vision, bioinformatics, etc. The large interest of the scientific community for the problem of clustering is reflected by the growing appearance of related monographs [1],[2],[3],[4], journal articles and conferences. With the advent of the Internet and the World Wide Web, scientific data from a wide range of fields have become easily accessible. This convenience has further raised the interest and expanded the audience of clustering techniques. Clustering can be viewed as the identification of existing intrinsic groups in a set of unlabeled data. Associated methods are often based on intuitive approaches that rely on specific assumptions and on the particular characteristics of the data sets. This in turn implies that the corresponding algorithms depend crucially on some parameters that must be properly tuned anew for each problem.

A plethora of clustering approaches has been presented over the last years. Hierarchical methods are based on a tree structure over the data according to some similarity criteria. Methods based on partitioning, relocate iteratively the data points into clusters until the optimum position of some cluster representatives (e.g. centers) is found; the popular "K-means" algorithm for instance belongs to this category. On the other hand, model-based methods are closer to the natural data generation mechanism and assume a mixture of probability distributions, where each component corresponds to a different cluster[1],[3],[4]. In these methods the *Expectation-Maximization* (EM) algorithm [5] is the preferred framework for estimating the mixture parameters due both to its simplicity and flexibility. Moreover, mixture modeling provides a powerful and useful platform for capturing data with complex structure. A fundamental concern in applying

J. Marques de Sá et al. (Eds.): ICANN 2007, Part II, LNCS 4669, pp. 291–300, 2007.

the EM algorithm, is its strong dependence on the initialization of the model parameters. Improper initialization may lead to points corresponding to local (instead of global) maxima of the log-likelihood, a fact that in turn may weigh on the quality of the method's estimation capability. Attempts to circumvent this, using for example the K-means algorithm to initialize the mixture parameters, amounts to shifting the original problem to initializing the K-means. Recently, several methods have been presented, aiming to overcome the problem of poor initialization. They are all based on an incremental strategy for building a mixture model. In most cases these methods start from a single component and iteratively add new components to the mixture either by performing a split procedure [6], or by performing a combined scheme of global and local search over a pool of model candidates [7]. A similar in nature technique is to follow an entirely opposite route and start with several components that iteratively will be discarded [8]. An alternative strategy has been presented in [9] where a split-and-merge EM (SMEM) algorithm was proposed. Initially the SMEM method performs the usual EM algorithm to a K-order mixture model and an initial estimation of the parameters. At a second level, repeated split-merge operations are performed exhaustively among the K components of the mixture model that re-estimate the model parameters until a termination criterion is met.

The idea of the SMILE method is to start with a mixture model with $k = 2$ and then to apply a Split & Optimize, Merge & Optimize (SOMO) sequence of operations. If this leads to a model with higher likelihood we accept it and repeat the SOMO procedure. In the opposite case we choose the model created just after the Split & Optimize (SO) step, which corresponds to a mixture model with an additional component. This is continued up to a preset number of components. At that stage if the SOMO sequence does not produce a higher likelihood value, the algorithm concludes. We have tested SMILE on a suite of benchmarks, with both simulated and real data sets, taking in account a variety of cases, with promising results. Comparisons have been made with existing methods of similar nature. The quality of the solutions offered by each method is rated in terms of the associated log-likelihood value. An important test for SMILE is its application to image segmentation problems. Here we have considered data arising from MRI images and the results are quite encouraging.

The rest of the paper is organized as follows. In section 2 we present the mixture models and the EM algorithm for parameter estimation, in section 3 we present in detail our incremental scheme where we lay out an algorithmic description, while in section 4 we report results obtained by applying SMILE to several data sets. Our conclusions and a summary are included in section 5 along with some remarks and speculations.

2 Mixture Models

Given a set of N data points $A = \{x_i | x_i \in R^d,\ i = 1, \cdots, N\}$, the task of clustering is to find a number of K subsets $A_j \subset A$ with $j = 1, \cdots, K$, containing points with common properties. These subsets are called *clusters*. We consider here that the properties of a single cluster j, may be described implicitly via a probability distribution with parameters θ_j.

A mixture model is a linear combination of these cluster-distributions, e.g.:

$$f(x|\Theta_K) = \sum_{j=1}^{K} \pi_j p(x|\theta_j) \tag{1}$$

The parameters $0 < \pi_j \leq 1$ represent the mixing weights satisfying $\sum_{j=1}^{K} \pi_j = 1$, while $\Theta_K = \{\pi_j, \theta_j\}_{j=1}^{K}$ represents the vector of all unknown model parameters. Mixture models provide an efficient method for describing complex data sets. The parameters can be estimated by maximizing the log-likelihood, by using for example the EM algorithm [5]. EM performs a two-step iterative procedure: The E-step calculates the posterior probabilities:

$$z_{ij}^{(t)} = p(j|x_i, \theta_j^{(t)}) = \frac{\pi_j^{(t)} p(x_i|\theta_j^{(t)})}{\sum_{l=1}^{K} \pi_l^{(t)} p(x_i|\theta_l^{(t)})}\,, \tag{2}$$

while the M-step updates the model parameters by maximizing the complete log-likelihood function. If we assume multivariate Normal densities $\theta_j = \{\mu_j, \Sigma_j\}$ maximization yields the following updates:

$$\pi_j^{(t+1)} = \frac{\sum_{i=1}^{N} z_{ij}^{(t)}}{N}\,,\ \mu_j^{(t+1)} = \frac{\sum_{i=1}^{N} z_{ij}^{(t)} x_i}{\sum_{i=1}^{N} z_{ij}^{(t)}}\,,\ \Sigma_j^{(t+1)} = \frac{\sum_{i=1}^{N} z_{ij}^{(t)} (x_i - \mu_j^{(t+1)})(x_i - \mu_j^{(t+1)})^T}{\sum_{i=1}^{N} z_{ij}^{(t)}}\,. \tag{3}$$

3 *Split–Merge* Mixture Learning

In this section we describe in detail the proposed method. To begin with, we describe the operations used in the SOMO sequence.

3.1 The Split Operation

Suppose that the model currently contains $k \geq 2$ components (clusters). The selection of the cluster to be split is facilitated with one of the criteria below.

1. Maximum Entropy:

$$H(j) = -\int p(x|\theta_j) \log p(x|\theta_j) dx \tag{4}$$

2. Minimum Mean Local Log-likelihood:

$$L(j) = \frac{\sum_{i=1}^{N} p(j|x_i\theta_j)\log(p(x_i|\theta_j))}{\sum_{i=1}^{N} p(j|x_i,\theta_j)} \tag{5}$$

3. Maximum local *Kullback divergence*: (used also by SMEM [9])

$$J(j) = \int f(x|\Theta)\log\frac{f(x|\Theta)}{p(x|\theta_j)}dx \tag{6}$$

where the density $f(x|\Theta)$ represents an empirical distribution [9].

Suppose that cluster j^* is being selected for the split operation. Two clusters are then created labeled as j_1^* and j_2^*. Their parameters are initialized as follows:

$$\pi_{j_1^*} = \pi_{j_2^*} = \frac{\pi_{j^*}}{2} \ , \Sigma_{j_1^*} = \Sigma_{j_2^*} = \frac{\Sigma_{j^*}}{2} \tag{7}$$

$$\mu_{j_1^*} = \mu_{j^*} + \frac{\sqrt{\lambda_{max}}}{2}v_{max} \ , \mu_{j_2^*} = \mu_{j^*} - \frac{\sqrt{\lambda_{max}}}{2}v_{max} \ , \tag{8}$$

where λ_{max}, v_{max} are the maximum eigenvalue and its corresponding eigenvector of the covariance matrix Σ_{j^*} .

3.2 The Optimization Operation

Let $f(x|\Theta_k^*)$ be the mixture without the j^*-th component, i.e.:

$$f(x|\Theta_k^*) = f(x|\Theta_k) - \pi_{j^*}p(x|\theta_{j^*}) = \sum_{j=1,j\neq j^*}^{k} \pi_j p(x|\theta_j) \tag{9}$$

The resulting mixture after the split operation takes the following form:

$$f(x|\Theta_{k+1}) = f(x_i|\Theta_k^*) + (\pi_{j^*} - \alpha)p(x|\theta_{j_1^*}) + \alpha p(x|\theta_{j_2^*}) \tag{10}$$

with $0 \leq \alpha \leq \pi_{j^*}$. In this mixture, the first term is inherited from the original model, while the rest two, are the newly introduced components by the split operation. The first term remains intact, while the other two are to be adjusted so as to maximize the likelihood. This is facilitated by a partial application of the EM algorithm, that modifies only the new component parameters $\alpha, \theta_{j_1^*}, \theta_{j_2^*}$. The following updates are obtained:

At the E-step:

$$z_{ij_1^*}^{(t)} = \frac{(\pi_{j^*} - \alpha^{(t)})p(x_i|\theta_{j_1^*}^{(t)})}{f(x|\Theta_{k+1}^{(t)})}, \ z_{ij_2^*}^{(t)} = \frac{\alpha^{(t)}p(x_i|\theta_{j_2^*}^{(t)})}{f(x|\Theta_{k+1}^{(t)})} \ , \tag{11}$$

and at the M-step:

$$\alpha^{(t+1)} = \pi_{j^*}\frac{\sum_{i=1}^{N} z_{ij_2^*}^{(t)}}{\sum_{i=1}^{N} z_{ij_2^*}^{(t)} + z_{ij_1^*}^{(t)}}, \tag{12}$$

$$\mu_m^{(t+1)} = \frac{\sum_{i=1}^{N} z_{im}^{(t)} x_i}{\sum_{i=1}^{N} z_{im}^{(t)}}, \; \Sigma_m^{(t)} = \frac{\sum_{i=1}^{N} z_{im}^{(t)} (x_i - \mu_m^{(t+1)})(x_i - \mu_m^{(t)})^T}{\sum_{i=1}^{N} z_{im}^{(t)}}, \; m = \{j_1^*, j_2^*\} \tag{13}$$

After obtaining an optimum in the subspace of the newly introduced parameters, a full space optimization is performed, again by the EM algorithm (Eqs. 2, 3).

3.3 The Merge Operation

During this operation two clusters are fused into one. Two clusters $\{k_1, k_2\}$ are selected according to any of the criteria that follow.

1. Minimum Distribution Distance (*Symmetric Kullback Leibler*)

$$\int p(x|\theta_{k_1}) \log \frac{p(x|\theta_{k_1})}{p(x|\theta_{k_2})} dx + \int p(x|\theta_{k_2}) \log \frac{p(x|\theta_{k_2})}{p(x|\theta_{k_1})} dx \tag{14}$$

2. Maximum Distribution Overlap (used also in [9])

$$\sum_{i=1}^{N} p(k_1|x_i, \theta_{k_1}) p(k_2|x_i, \theta_{k_2}) \tag{15}$$

Let the resulting cluster be labeled by k. Its parameters are then initialized as:

$$\pi_k = \pi_{k_1} + \pi_{k_2} \;, \mu_k = \frac{\pi_{k_1} * \mu_{k_1} + \pi_{k_2} * \mu_{k_2}}{\pi_{k_1} + \pi_{k_2}} \;, \Sigma_k = \frac{\pi_{k_1} * \Sigma_{k_1} + \pi_{k_2} * \Sigma_{k_2}}{\pi_{k_1} + \pi_{k_2}} \tag{16}$$

The optimization step following the merge operation is in the same spirit as that of section 3.2, e.g. we perform partial EM steps, allowing only the new (merged) cluster parameters to vary. After obtaining an optimum in the subspace of the newly introduced parameters, a full space EM optimization is performed.

3.4 Description of the Method

Initially we construct a mixture with two components, i.e. $k = 2$. Denote by Θ_k^1 the mixture parameters, and by $L(\Theta_k^1)$ the corresponding value of the log-likelihood function. We perform in succession a split and an optimization operation, obtaining so a model Θ_{k+1}^m with $k+1$ components. Similarly in what follows, we perform a merge and an optimization operation, that creates a model again with k components. Let Θ_k^2 be the new mixture parameters after this split-merge operation and $L(\Theta_k^2)$ the corresponding log-likelihood. If $L(\Theta_k^2) > L(\Theta_k^1)$ then we update the k-order model to Θ_k^2 and we repeat the SOMO procedure. In the case where $L(\Theta_k^2) \leq L(\Theta_k^1)$, i.e. when the SOMO procedure fails to obtain a better value for the likelihood, we discard the last merge operation and update our model to Θ_{k+1}^m, which was obtained after the last SO operation, with $k+1$ components. The algorithm proceeds so, until we obtain a model with the prescribed number of components (K) and the SOMO iterations fail to provide further improvement to the likelihood.

We now can proceed and describe our method algorithmically.

- Start with $k = 2$
- while $k < K$
  1. Estimate the current log-likelihood L_1
  2. Perform SOMO operation:
     - Split: select a cluster j^* and divide it into two clusters j_1^* and j_2^*.
     - Optimization operation: Perform partial-EM and then full EM.
     - Merge: select two clusters k_1 and k_2 and merge them.
     - Optimization operation: Perform partial-EM and then full EM.
     - Estimate the log–likelihood L_2
  3. if $L_2 > L_1$ then:
     Accept the fused cluster. Set $L_1 \leftarrow L_2$ and go to step 2.

     else:
     Reject the last merge operation. Set $k \leftarrow k + 1$.
- endwhile

4 Experimental Results

We have performed several experiments to examine the effectiveness of the SMILE method. We have considered both real and simulated data sets of varying dimensionality. We compare against three incremental approaches, namely the Greedy EM method[1] [7], the Split and Merge EM (SMEM) [9], the MML-EM [2] [8], as well as with the simple K-means initialized EM. The initialization scheme in SMILE is predetermined in distinction to the contestant schemes that depend heavily on random numbers. Hence, in order to obtain a meaningful comparison, we performed 30 different runs for each data set with different seeds and kept records of the mean value and the standard deviation of the log-likelihood.

Experiments with simulated data sets

In Fig. 1 we give an example of the performance of our algorithm in a typical 2-dimensional data set that has been generated from a $K = 5$ Gaussian mixture. Note that ellipses were created by covariances. Step 0 shows the solution with one cluster, which in step 1 is split into two, with a log-likelihood estimation $L_1 = -1727$. Then, SMILE tests the optimality of this solution by performing a SOMO procedure (steps 2a, 2b), leading to a solution with $L_2 = -1973$. Since the SOMO fails ($L_2 < L_1$), we discard the last MO operation leading to a $K = 3$ mixture model and continue with the next SOMO process (steps 3a, 3b). In this case, this SOMO operation found a better solution $L_2 = -1272$ (step 3b) in comparison with the one $L_1 = -1537$ of step 2a. Therefore, we accept this updated $K = 3$ model and perform another SOMO operation (steps 4a,

[1] The software was downloaded from http://staff.science.uva.nl/~vlassis/software/
[2] The software was downloaded from http://www.lx.it.pt/~mtf/

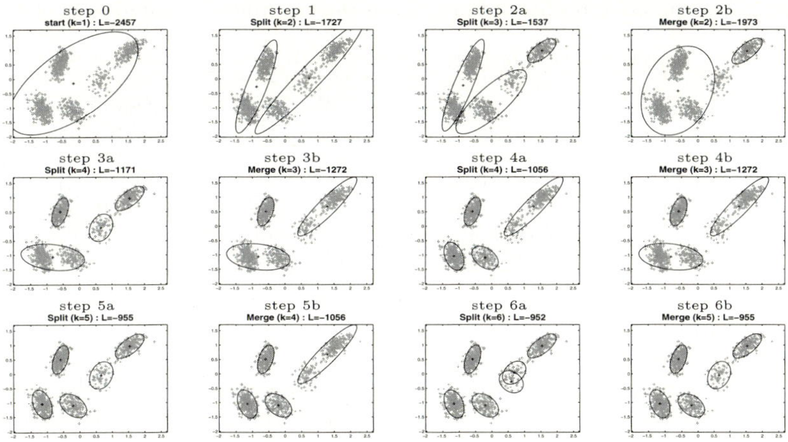

Fig. 1. Visualization of the SMILE steps on a typical data set. Each figure shows the current clusters and the corresponding value of the log-likelihood.

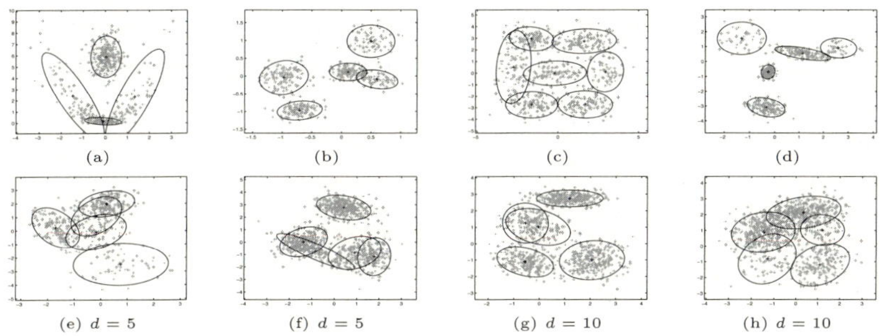

Fig. 2. Simulated data sets used during experiments. We give also the clustering solution obtained by our method, i.e. their centers and the elliptic shapes.

4b), which however fails to further improve the likelihood. Finally, two other SOMO calls are made that both fail before the final $K = 5$ solution is reached (step 5a).

Several experiments were conducted using simulated data sets created by sampling from Gaussian mixture models. Figure 2 illustrates eight (8) such data sets containing $N = 500$ points. The first four sets (a,b,c,d) are in 2-dimensions, while (e,f) (g,h) are in 5 and 10-dimensions respectively. The visualization for the sets with dimensionality 5 and 10, is performed by projecting on the plane spanned by the first two principal components. The clustering obtained by SMILE is displayed in Fig. 2. Table 1 summarizes the results obtained by the application of the five contestants to the above mentioned data sets. Note that SMILE has recovered the global maximum in all cases; from the rest, only the Greedy EM and SMEM methods yielded comparable results. For the data set of Fig.2c, SMILE was the only method that obtained the global solution.

Table 1. Comparative results obtained form the experiments in data sets of Fig. 2

Data set	*SMILE*	*Greedy EM*	*SMEM*	*MML-EM*	*K-means EM*
(a)	−2.82	−2.87(0.02)	−2.82(0.00)	−2.86(0.05)	−2.89(0.01)
(b)	−0.83	−0.83(0.01)	−0.85(0.02)	−0.98(0.13)	−0.86(0.05)
(c)	−3.92	−3.94(0.01)	−3.94(0.00)	−3.95(0.02)	−3.93(0.01)
(d)	−1.87	−1.87(0.00)	−1.89(0.04)	−2.00(0.17)	−1.99(0.19)
(e)	−4.54	−4.63(0.04)	−4.55(0.02)	−4.59(0.04)	−4.61(0.02)
(f)	−4.74	−4.74(0.00)	−4.75(0.03)	−4.80(0.07)	−4.85(0.14)
(g)	−7.19	−7.19(0.00)	−7.19(0.00)	−7.40(0.09)	−7.44(0.33)
(h)	−7.71	−7.71(0.00)	−7.75(0.16)	−7.82(0.03)	−7.83(0.23)

Experiments with real data sets

Additional experiments were made using real data sets. In particular, we have selected two widely used benchmarks. The first one is the CRAB data set of Ripley [2], that contains $N = 200$ data belonging to four clusters ($K = 4$). Original CRAB data are in five dimensions. Here we have also created a 2-dimensional data set by projecting the data on the plane defined by the second and third principal components. We have also considered the renowned Fisher-IRIS data set [10] with $N = 150$ points in $d = 4$ dimensions belonging to three clusters ($K = 3$). In Table 2 we summarize the results obtained by the 5 contestants. Note, that in the case of the original CRAB data set, SMILE was the only one that recovered the optimal solution.

Another experimental benchmark used is the Phoneme data set [10]. This is a collection of two-class five dimensional data points. In our study we have randomly selected a training set with $N = 2800$ and a test set with 2604 data points. Figure 3 illustrates the performance of each method by plotting the log-likelihood value versus the number of components $K = [2, 10]$, in both the training and

Table 2. Comparative results obtained from the CRAB and the IRIS data sets

Data set	*SMILE*	*Greedy EM*	*SMEM*	*MML-EM*	*K-means EM*
CRAB $d = 5$	−6.14	−6.35(0.14)	−6.35(0.12)	−6.86(0.01)	−6.60(0.19)
CRAB $d = 2$	−2.49	−2.50(0.01)	−2.50(0.00)	−2.55(0.06)	−2.52(0.06)
IRIS $d = 4$	−1.21	−1.23(0.02)	−1.23(0.04)	−1.25(0.04)	−1.28(0.09)

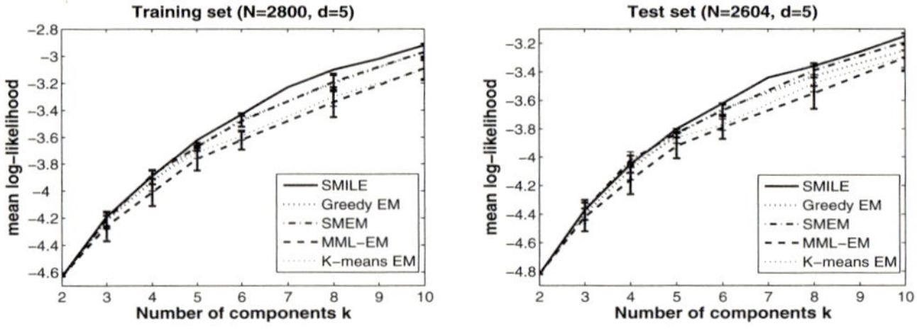

Fig. 3. Plots of mean log-likelihood objective function estimated by each method against number of components K to the Phoneme data set

the test sets. Observe that SMILE's curve is consistently above all others, both for the training and for the test set, implying superiority in performance and in generalization as well. Note that again, there were cases where SMILE was the only method that arrived at the global solution.

Application in image segmentation
In computer vision clustering finds application in image segmentation, i.e. the grouping of image pixels based on attributes such as their intensity and spatial location. We have tested SMILE to simulated brain MRI images available on the site BrainWeb [11], where we have reduced them into half of their original size (181×217). The segmentation of MRI mainly requires the classification of the brain into three types of tissue: (GM, WM, CSF). Since we are aware of the true class labels of the pixels we evaluate each method according to the computed total classification error. Figure 4 illustrates four such MRI images together with the segmentation result using a $K = 5$ Gaussian mixture, where in the reconstructed images every pixel assumes the intensity value of the cluster center that belongs. The overall classification error obtained from all the clustering methods to these images are presented at Table 3. It is obvious that our method achieves superior results for the tissue segmentation.

Original image

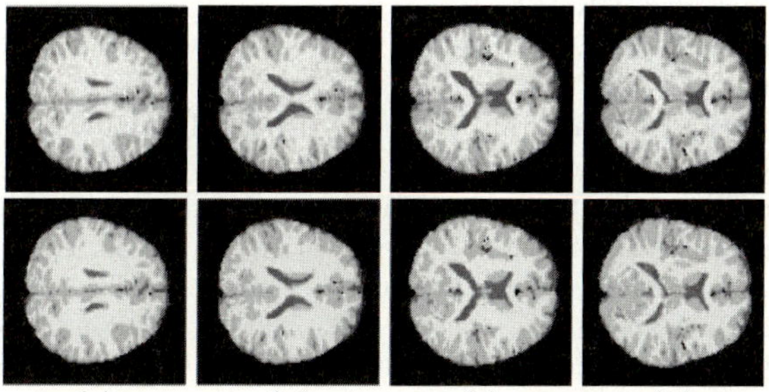

Segmentation result

(a) (b) (c) (d)

Fig. 4. Image segmentation results obtained by our method in four MRI images

Table 3. Percentage of misclassified pixels for the MRI of Fig.4 using $K = 5$ Gaussians

MRI image	*SMILE*	*Greedy EM*	*SMEM*	*MML-EM*	*K-means EM*
(a)	36.76	37.24(0.23)	37.12(0.00)	38.31(0.94)	37.84(0.01)
(b)	35.88	36.50(0.04)	36.69(0.00)	37.48(0.89)	36.57(0.08)
(c)	35.48	35.60(0.12)	36.18(0.30)	37.05(0.48)	36.21(0.29)
(d)	37.88	38.20(0.19)	37.98(0.00)	39.37(0.58)	38.90(0.32)

5 Conclusion

In this study we have presented SMILE, a new incremental mixture learning method based on successive split and merge operations. Starting from a two-component mixture, the method performs split-merge steps to improve the current solution maximizing the log-likelihood. SMILE has the advantage of not relying on good initial estimates, unlike the other rival methods studied in this article. The results of the comparative study are very promising. Several developments are possible that need further research. For example, consecutive multiple split operations followed by corresponding merge steps may lead to even better models. The persistent issue of discovering the optimal number of clusters in a data set may be examined in the framework of this method as well.

Acknowledgments

This work was partially supported by Interreg IIIA (Greece-Italy) Grant (No I2101005).

References

1. Fukunaga, K.: Introduction to Statistical Pattern Recognition. Academic Press, San Diego (1990)
2. Ripley, B.: Pattern Recognition and Neural Networks. Cambridge Univ. Press Inc, Cambridge, UK (1996)
3. Duda, R., Hart, P., Stork, D.: Pattern Classification. Wiley-Interscience, New York (2001)
4. Bishop, C.: Pattern Recognition and Machine Learning. Springer, Heidelberg (2006)
5. Dempster, A., Laird, N., Rubin, D.: Maximum Likelihood from incomplete data via the EM algorithm. J. Roy. Statist. Soc. B 39, 1–38 (1977)
6. Li, J., Barron, A.: Mixture density estimation. In: Advances in Neural Information Processing Systems, pp. 279–285. The MIT Press, Cambridge, MA (2000)
7. Vlassis, N., Likas, A.: A greedy EM algorithm for Gaussian mixture learning. Neural Processing Letters 15, 77–87 (2001)
8. Figueiredo, M., Jain, A.: Unsupervised learning of finite mixture models. IEEE Trans. Pattern Analysis and Machine Intelligence 24(3), 381–396 (2002)
9. Ueda, N., Nakano, R., Ghahramani, Z., Hinton, G.: SMEM algorithm for mixture models. Neural Computation 12(9), 2109–2128 (2000)
10. Merz, C., Murphy, P.: UCI repository of machine learning databases.Irvine, CA (1998), http://www.ics.uci.edu/~mlearn/MLRepository.html
11. Cocosco, C., Kollokian, V., Kwan, R.S., Evans, A.: BrainWeb: Online interface to a 3D MRI simulated brain database. NeuroImage 5(3), 2/4, S425 (1997)

Least-Mean-Square Training of Cluster-Weighted Modeling

I-Chun Lin and Cheng-Yuan Liou*

Department of Computer Science and Information Engineering
National Taiwan University
Republic of China
Supported by National Science Council
cyliou@csie.ntu.edu.tw

Abstract. Aside from the Expectation-Maximization (EM) algorithm, Least-Mean-Square (LMS) is devised to further train the model parameters as a complementary training algorithm for Cluster-Weighted Modeling (CWM). Due to different objective functions of EM and LMS, the training result of LMS can be used to reinitialize CWM's model parameters which provides an approach to mitigate local minimum problems.

1 Introduction

Foundations for data manifold have been set down for factorial components [6], oblique transformations [7], ICA [13], generalized adalines [14] [8]. They have been successfully applied in various temporal data analyses [5] [15]. Cluster-Weighted Modeling (CWM) was introduced by Neil Gershenfeld [2] as an elegant approach to approximate an arbitrary function and it can also be applied to temporal time-series data prediction, characterization and synthesis. It is derived from hierarchical mixture-of-experts type architectures [4]. The framework is based on density estimation around Gaussian kernels which contain simple local models describing the system dynamics of a data subspace. CWM then assembles the local models into a global model that can handle nonlinear and discontinuous data. CWM is trained by Expectation-Maximization (EM) [1] algorithm which converges quickly. The resulting model has transparent local structures and meaningful parameters, it allows one to identify and analyze data subspaces. Least-Mean-Square is a common learning approach for universal function approximators. In some cases one does not make full use of the density estimation carried out by CWM, but only the expected value of the output for a given input vector. Employing LMS learning to further train CWM's resulting model parameters provides a good training option. The LMS learning for cluster centers is similar to the supervised selection of centers in RBF networks [10]. Wettschereck and Dietterich [12] have compared the performance of (Gaussian) radial-basis function networks with unsupervised learning of the centers'

* Corresponding author.

J. Marques de Sá et al. (Eds.): ICANN 2007, Part II, LNCS 4669, pp. 301–310, 2007.

locations and supervised learning of the centers' locations. In their work the latter is able to exceed substantially the generalization performance of multilayer perceptrons while the former can not.

EM algorithm is prone to local minimum which sometimes can lead to very poor performance. The objective functions of EM and LMS are different, thus the local minimum that CWM confronts would not be the same local minimum in LMS learning. The training result of LMS learning can be used to reinitialize CWM's model parameters which provides an approach to mitigate local minimum problems.

Section 2 reviews the concept of CWM. LMS learning for CWM is introduced in section 3. Some experimental results are presented in section 4 and concluding remarks are presented in section 5.

2 Cluster-Weighted Modeling

We start with a set of discrete or real valued input features $\mathbf{x}$ and an discrete or real valued output target vector $\mathbf{y}$. The most general model infers the joint density $p(\mathbf{y}, \mathbf{x})$ of the data set, from which conditional quantities such as the expected $\mathbf{y}$ given $\mathbf{x}$, $\langle \mathbf{y} \mid \mathbf{x} \rangle$, and the expected covariance of $\mathbf{y}$ given $\mathbf{x}$, $\langle P_y \mid \mathbf{x} \rangle$ can be derived.

We expand this joint density with M clusters which contain an output distribution, and an input domain of influence, and an unconditioned cluster probability

$$p(\mathbf{y}, \mathbf{x}) = \sum_{m=1}^{M} p(\mathbf{y} \mid \mathbf{x}, c_m) p(\mathbf{x} \mid c_m) p(c_m) \tag{1}$$

The input distribution is taken to be Gaussian distribution,

$$p(\mathbf{x} \mid c_m) = \frac{\left| P_m^{-1} \right|^{1/2}}{(2\pi)^{D_x/2}} e^{-(\mathbf{x}-\mu_m)^T \cdot P_m^{-1} \cdot (\mathbf{x}-\mu_m)/2}, \tag{2}$$

where P_m is the cluster-weighted covariance matrix in the feature space and D_x is the dimension of input vectors.

The output distribution is taken to be

$$p(\mathbf{y} \mid \mathbf{x}, c_m) = \frac{\left| P_{m,y}^{-1} \right|^{1/2}}{(2\pi)^{D_y/2}} e^{-(\mathbf{y}-\mathbf{f}(\mathbf{x},\beta_m))^T \cdot P_{m,y}^{-1} \cdot (\mathbf{y}-\mathbf{f}(\mathbf{x},\beta_m))/2}, \tag{3}$$

where the mean value of the output Gaussian is replaced by the function $\mathbf{f}(\mathbf{x}, \beta_m)$ with unknown parameters β_m. D_y is the dimension of output vectors.

Consider the conditional forecast of the expected $\mathbf{y}$ given $\mathbf{x}$,

$$\langle \mathbf{y} \mid \mathbf{x} \rangle = \frac{\sum_{m=1}^{M} \mathbf{f}(\mathbf{x}, \beta_m) p(\mathbf{x} \mid c_m) p(c_m)}{\sum_{m=1}^{M} p(\mathbf{x} \mid c_m) p(c_m)} \tag{4}$$

The predicted $\mathbf{y}$ is a superposition of all the local functionals, where the weight of each contribution depends on the posterior probability that an input point was generated by a particular cluster.

Generally the local models are taken to be polynomial functions. CWM is trained by a variant of the Expectation-Maximization (EM) algorithm.

In the E-step, we estimate the posterior probability using the current model parameters

$$p(c_m \mid \mathbf{y}, \mathbf{x}) = \frac{p(\mathbf{y} \mid \mathbf{x}, c_m)p(\mathbf{x} \mid c_m)p(c_m)}{\sum_{l=1}^{M} p(\mathbf{y} \mid \mathbf{x}, c_l)p(\mathbf{x} \mid c_l)p(c_l)} \tag{5}$$

In the M-step, we assume that the current data distribution is correct and find the cluster parameters that maximize the likelihood of the data. The prior probability is updated with $p(c_m) = \frac{1}{N}\sum_{n=1}^{N} p(c_m \mid \mathbf{y}_n, \mathbf{x}_n)$ and the cluster centers are updated with

$$\mu_m = \frac{\sum_{n=1}^{N} \mathbf{x}_n p(c_m \mid \mathbf{y}_n, \mathbf{x}_n)}{\sum_{n=1}^{N} p(c_m \mid \mathbf{y}_n, \mathbf{x}_n)} \tag{6}$$

Define a cluster-weighted expectation of any function $\Theta(\mathbf{x})$,

$$\langle \theta(\mathbf{x}) \rangle_m = \frac{\sum_{n=1}^{N} \theta(\mathbf{x}_n) p(c_m \mid \mathbf{y}_n, \mathbf{x}_n)}{\sum_{n=1}^{N} p(c_m \mid \mathbf{y}_n, \mathbf{x}_n)} \tag{7}$$

which lets us update the cluster weighted covariance matrices, $[P_m]_{i,j} = \langle (x_i - \mu_i)(x_j - \mu_j) \rangle_m$.

The model parameters are found by taking the derivative of the logarithm of the total likelihood function with respect to parameters.

If we choose a local model that has linear coefficients with a bias term,

$$f(\mathbf{x}, \beta_m) = \sum_{i=1}^{D_x+1} \beta_{m,i} f_i(\mathbf{x}) \tag{8}$$

Take the derivative with respect to the $j-$th component of β_m, then this gives for the coefficients of the $m-$th cluster

$$0 = \langle [y - f(\mathbf{x}, \beta_m)] f_j(\mathbf{x}) \rangle_m = \langle y f_j(\mathbf{x}) \rangle_m - \sum_{i=1}^{I} \beta_{m,i} \langle f_j(\mathbf{x}) f_i(\mathbf{x}) \rangle_m \tag{9}$$

For every components of β_m, we can have its matrix form $\beta_m = B_m^{-1} \mathbf{a}_m$ with $[B_m]_{ij} = \langle f_i(\mathbf{x}) f_j(\mathbf{x}) \rangle_m$ and $[\mathbf{a}_m]_j = \langle y f_j(\mathbf{x}) \rangle_m$.

Finally the output covariance matrices associated with each model are estimated,

$$P_{y,m} = \left\langle [\mathbf{y} - \langle \mathbf{y} \mid \mathbf{x} \rangle]^2 \right\rangle_m = \left\langle [\mathbf{y} - \mathbf{f}(\mathbf{x}, \beta_m)][\mathbf{y} - \mathbf{f}(\mathbf{x}, \beta_m)]^T \right\rangle_m . \tag{10}$$

3 Least-Mean-Square Training of CWM

Minimizing squared-error cost function of CWM's training result to find another solution from that we can have a more precisely fitted function. And it also provides an alternative to find another solution when CWM is trapped in local minimum. In general, we wish to improve CWM's performance by applying LMS learning. In our experiments and derivations, we use separable Gaussians with off diagonal terms in the covariance matrices are zero for both input and output covariances.

Here we derive the formulas of the LMS learning for CWM. Consider the cost function E^N is the sum of squared errors over the entire data set

$$E^N = \frac{1}{2}\sum_{n=1}^{N}(y_n - < y \mid \mathbf{x}_n >)^2 \tag{11}$$

where y_n is the $n-th$ desired data and $< y \mid \mathbf{x}_n >$ is the expected value given the $n-th$ input vector $\mathbf{x}_n$. Consider the squared error produced by a single input-output pair

$$E^n = \frac{1}{2}(y_n - < y \mid \mathbf{x}_n >)^2 = \frac{1}{2}e_n^2 \tag{12}$$

where $e_n = y_n - < y \mid \mathbf{x}_n >$. And from the formula of $< y \mid \mathbf{x}_n >$,

$$< y \mid \mathbf{x}_n >= \frac{\sum_{m=1}^{M} f(\mathbf{x}_n, \beta_m) p(\mathbf{x}_n \mid c_m) p(c_m)}{\sum_{m=1}^{M} p(\mathbf{x}_n \mid c_m) p(c_m)} = \frac{A}{B} \tag{13}$$

Let A and B denote the nominator and the denominator respectly, we have the following derivations.

Take the partial derivative with respect to priors $p(c_m)$,

$$\frac{\partial E^n}{\partial p(c_m)} = -e_n p(\mathbf{x}_n \mid c_m)\left[\frac{f(\mathbf{x}_n, \beta_m)B - A}{B^2}\right] \tag{14}$$

In order to keep the probabilistic constraint of priors, we define

$$p(c_m) \equiv \frac{\exp(z_m)}{\sum_{l=1}^{M}\exp(z_l)} \tag{15}$$

With the softmax activation function we can make sure that the value of $p(c_m)$ is between 0 and 1. And their summation is equal to 1. We can initailize the value of $z_m$by assigning $z_m = \ln(p(c_m))$.

Hence, we must update z_m,

$$\frac{\partial E^n}{\partial z_m} = -e_n p(\mathbf{x}_n \mid c_m)\left[\frac{f(\mathbf{x}_n, \beta_m)B - A}{B^2}\right][p(c_m) - p(c_m)^2] \tag{16}$$

Take the partial derivative with respect to cluster mean

$$\frac{\partial E^n}{\partial \mu_{m,i}} = \frac{-e_n p(c_m) p(\mathbf{x}_n \mid c_m) \left(\frac{x_{n,i}-\mu_{m,i}}{\sigma^2_{m,i}}\right) [f(\mathbf{x}_n, \beta_m)B - A]}{B^2} \tag{17}$$

Take the partial derivative with respect to cluster standard deviation

$$\frac{\partial E^n}{\partial \sigma_{m,i}} = \frac{-e_n p(c_m) p(\mathbf{x}_n \mid c_m) \left[\frac{(x_{n,i}-\mu_{m,i})^2}{\sigma^3_{m,i}} - \frac{1}{\sigma_{m,i}}\right] [f(\mathbf{x}_n, \beta_m)B - A]}{B^2} \tag{18}$$

Take the partial derivative with respect to β_m,

$$\frac{\partial E^n}{\partial \beta_{m,i}} = -e_n \left[\frac{p(\mathbf{x}_n \mid c_m) p(c_m) f(\mathbf{x}_n)_i}{B}\right] \tag{19}$$

We first initialize the model parameters using CWM's training result, then we calculate the error and the gradients. We update the cluster centers using $\mu^{k+1}_{m,i} \leftarrow \mu^k_{m,i} - \eta \frac{\partial E^n}{\partial \mu^k_{m,i}}$, where η is the learning rate and k is the iteration number. Other parameters are updated in the similar way. In the cost function of LMS it does not involve the output covariance at all. But it is still desirable to estimate the log-likelihood value and the output covariances. Here we apply EM algorithm to estimate only the output covariance in every epoch. The learning rates can be different values between 0 and 1. Empirically, the learning rates we choose for our experiments are between 0.001˜0.06. To scale each feature of the input and output data to $[-1, 1]$ beforehand could avoid numerical errors while updating these parameters.

4 Experiments

4.1 Local Minimum

In this experiment we examine the four-clump data with known local minimum [11]. There are four clusters generated from a Gaussian distribution with standard deviation of 0.2 around its center. There are totally 2000 points, the true probability of data in the two dense clumps is 0.475 per clump, whereas the probability of data in two sparse clumps is 0.025 per clump. The centers of the two dense clumps are located at $[1, 1]^T$ and $[-1, -1]^T$ respectively. The centers of the two sparse clumps are located at $[1, -1]^T$ and $[-1, 1]^T$. In our experiment, the input space is 2 dimensional and the output of each cluster is a linear function of each cluster's input features. Figure 1 shows an example of local minimum. There are two clusters positioned near each other thereby sharing density for the same clump with center at around $[-1, -1]^T$, and the cluster center at around $[0, 0]^T$ tries to do the job of two clusters (modeling the two sparse clumps). We initialize CWM with this particular intial condition, after CWM converged we apply LMS learning. Here the learning rate for cluster centers is 0.06 and the

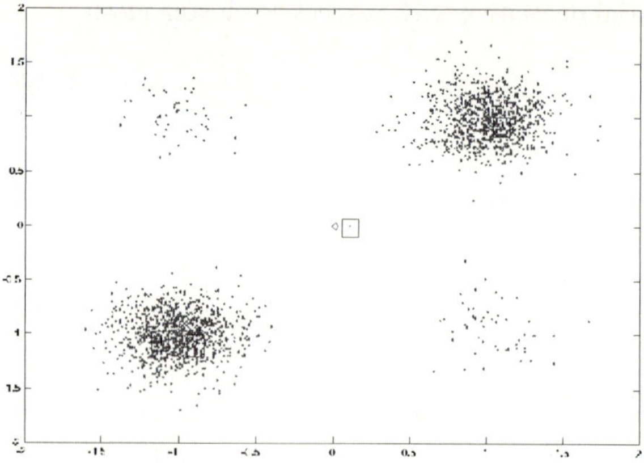

Fig. 1. The initial location of local minima of four cluster centers

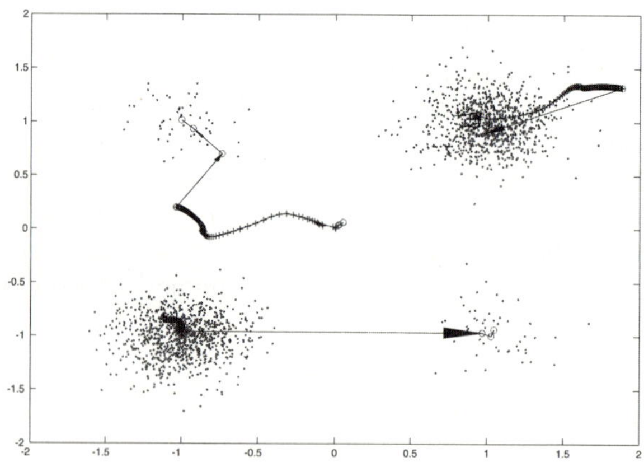

Fig. 2. The red circles represent the converged cluster centers of CWM. The black cross '+', represent the converged cluster centers of LMS-CWM. The arrows indicate the directions of each cluster shift.

other learning rates are all set to 0.01. Then we reinitialize CWM with LMS's training result. First we initialize k-means with LMS's resulting cluster centers, then enters the EM algorithm. After CWM takes over LMS's training result, it successfully escapes from the local minimum and the centers are all located at the correct locations. Figure 1 shows the initial centers of CWM and LMS-CWM. Figure 2 shows the resulting centers' locations after each training session of CWM and LMS-CWM.

4.2 Mackey-Glass Chaotic Time-Series Prediction

Mackey-Glass time-series is generated by the Mackey-Glass differential equation [9]:

$$\frac{dx(t)}{dt} = -0.1x(t) + 0.2\frac{x(t-17)}{1+x(t-17)^{10}} \tag{20}$$

1000 samples of Mackey-Glass time-series are generated. The first 500 points are used as the training set, and the last 500 points are used as the test set. CWM is used to predict the values of the time-series at point $x(t+85)$ from the earlier points $[x(t),\, x(t-6),\, x(t-12),\, x(t-18)]$. The delay time and the embedding dimension in this paper are followed by the convention of other literatures. There are some methods to decide the delay time and the embedding dimension [3].We use the last 100 points from training set as our validation set and decide the number of clusters which gives the minimal mean-square-error on validation set. Here we choose 30 clusters with local linear models. The learning curves of CWM and LMS-CWM are shown in Figure 3. Table 1 compares the Mean-Square-Error of their performances.

We can see that the prediction accuracy is much improved by further using LMS learning. The predicted time-series made by CWM and LMS-CWM are shown in Figure 4 and Figure 5 respectively. The prediction made by LMS-CWM is also much smoother.

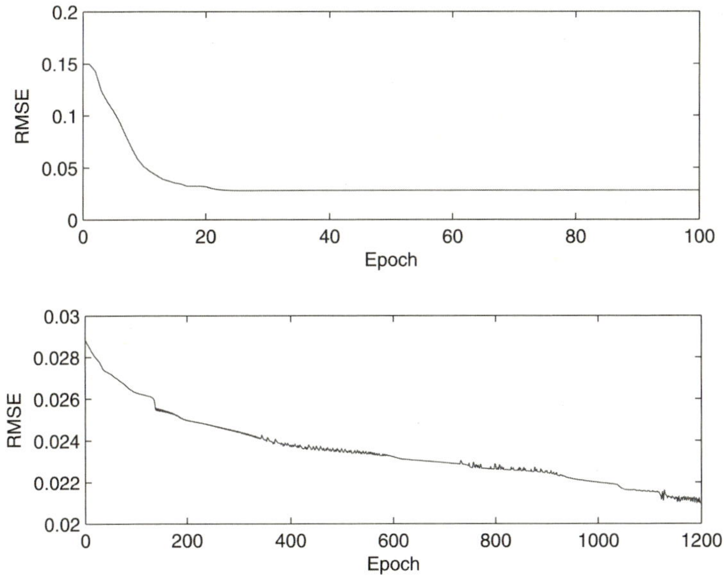

Fig. 3. The RMSE learning curves of CWM(top) and LMS-CWM(bottom)

Table 1. Mean-Square-Error of CWM and LMS CWM on training and test set

MSE	CWM	LMS-CWM
Training Set	0.0008027	0.0004293
Test Set	0.0006568	0.0004480

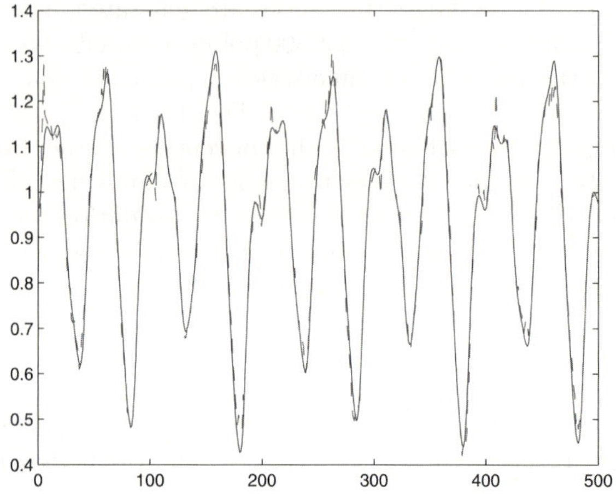

Fig. 4. The predicted time-series of CWM. The dashed line is CWM prediction, and the solid line is the system output.

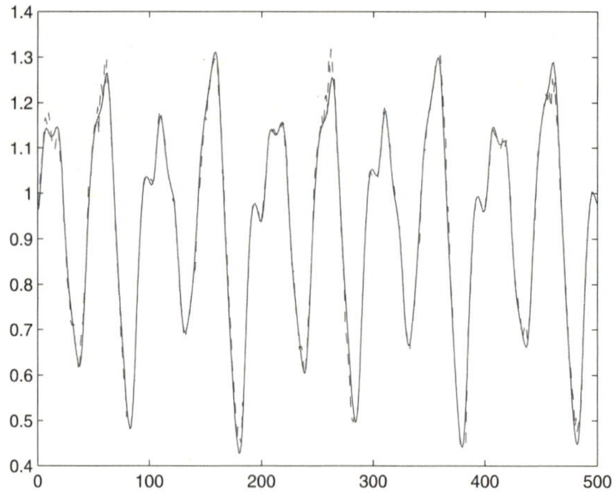

Fig. 5. The predicted time-series of LMS-CWM. The dashed line is LMS-CWM prediction, and the solid line is the system output.

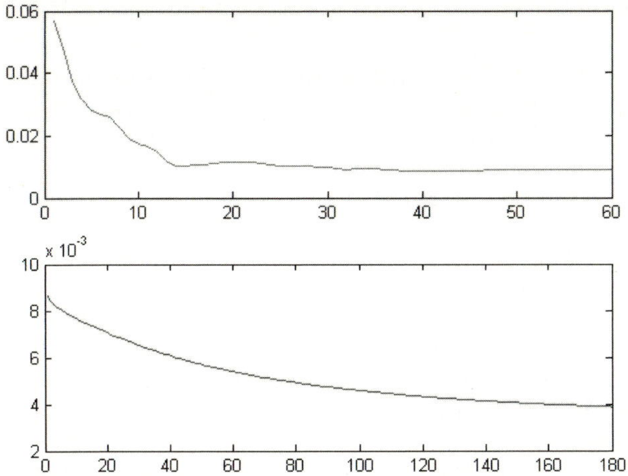

Fig. 6. The learning curves of CWM(top) and LMS-CWM(bottom) on Rössler data

4.3 Rössler Chaotic System

The equations of Rössler system is given as

$$\dot{x} = -(y + z); \dot{y} = x + .15y; \dot{z} = .2 + z(x - 10) \tag{21}$$

These equations are numerically integrated by fourth-order Runge-Kutta with a fixed time step of $\Delta t = .01$. 2000 points are generated. The state vector is reconstructed from sampling the time series of variable x as $\mathbf{z}(t) = [s(t), s(t+1), s(t+2)]^T$, where s is the time series generated by variable x. Then we model its evolution trajectories by fitting the fuction of input $\mathbf{x} = \mathbf{z}(t)$ and output $\mathbf{y} = \mathbf{z}(t+1)$. Here we use 6 clusters with quadratic local models. The learning curves of CWM and LMS-CWM are shown in Figure 6.

5 Conclusion

In this work, we develop a LMS learning algorithm for CWM as a complementary learning method. CWM provides a good initialization to LMS learning and we wish to make use of both the advantages of EM and LMS. After LMS learning, generally we can improve the prediction accuracy, but it may lose the benefit of data density estimation when the log-likelihood value decreases. Due to different objective functions of EM and LMS, the local minimum should be different. Therefore, we can use the resulting parameters of LMS to reinitialize parameters for CWM, and the parameters can be trained alternatively. This provides an approach that we can mitigate the local minimum problem presented in this work. The resulting model of LMS learning could be veiwed as a refinement of CWM if only prediction accuracy is our main concern. Regularization techniques shoud be investigated for the future work.

References

1. Dempster, A.P., Laird, N.M, Rubin, D.B.: Maximum likelihood from incomplete data via the em algorithm. Journal of the Royal Statistical Society 39, 1–38 (1977)
2. Gershenfeld, N., Schoner, B., Metois, E.: Cluster-weighted modeling for time-series analysis. Nature 397, 329–332 (1999)
3. Hegger, R., Kantz, H., Schreiber, T.: Practical implementation of nonlinear time series methods. The tisean package. Chaos 9, 413 (1999)
4. Jordan, M., Jacobs, R.: Hierarchical mixtures of experts and the em algorithm. Neural Computation 6, 181–214 (1994)
5. Liou, C.Y., Chang, C.C.: The determination of modal damping ratios and natural frequencies from bispectrum modeling. OCEANS 2, 548–553 (1987)
6. Liou, C.Y., Musicus, B.R.: Separable cross-entropy approach to power spectrum estimation. IEEE Transactions on Acoustics, Speech and Signal Processing 38, 105–113 (1990)
7. Liou, C.Y., Musicus, B.R.: Cross entropy approximation of structured covariance matrices. arXiv (August 2006), http://arxiv.org/PS_cache/cs/pdf/0608/0608121v1.pdf
8. Liou, C.Y., Wu, J.M.: Self-organization using Potts models. Neural Networks. 9(4), 671–684 (1996)
9. Mackey, M.C., Glass, L.: Oscillation and chaos in physiological control systems. Science 197, 716–723 (1997)
10. Poggio, T., Girosi, F.: Networks for approximation and learning. Proceedings of the IEEE 78, 1481–1497 (1990)
11. Prokhorov, D.V., Feldkamp, L.A., Feldkamp, T.M.: A new approach to cluster weighted modeling. Procedings of International Joint Conference Neural Networks 3, 1669–1674 (2001)
12. Wettschereck, D., Dietterich, T.: Improving the performance of radial basis function networks by learning center locations. In Advances in Neural Information Processing Systems 4, 1133–1140 (1992)
13. Wu, J.M., Chiu, S.J.: Independent component analysis using Potts models. IEEE Transactions on Neural Networks 12, 202–212 (2001)
14. Wu, J.M., Lin, Z.H., Hsu, P.H.: Function approximation using generalized adalines. IEEE Transactions on Neural Networks 17, 541–558 (2006)
15. Wu, J.M., Lu, C.Y., Liou, C.Y.: Independent component analysis of correlated neuronal responses in area MT, International Conference on Neural Information Processing, ICONIP, pp. 639–642 (2005)

Identifying the Underlying Hierarchical Structure of Clusters in Cluster Analysis*

Kazunori Iwata and Akira Hayashi

Graduate School of Information Sciences, Hiroshima City University,
Hiroshima, 731-3194, Japan
{kiwata,akira}@hiroshima-cu.ac.jp

Abstract. In this paper, we examine analysis of clusters of labeled samples to identify their underlying hierarchical structure. The key in this identification is to select a suitable measure of dissimilarity among clusters characterized by subpopulations of the samples. Accordingly, we introduce a dissimilarity measure suitable for measuring a hierarchical structure of subpopulations that fit the mixture model. Glass identification is used as a practical problem for hierarchical cluster analysis, in the experiments in this paper. In the experimental results, we exhibit the effectiveness of the introduced measure, compared to several others.

1 Introduction

Hierarchical cluster analysis (HCA) [1, Ch. 6][2, Ch. 14][3, Ch. 10] has developed as demand for data analysis increases. It is, in general, used to classify unlabeled samples in a database into hierarchical clusters, based on a dissimilarity or similarity measure between two samples. According to a resulting dendrogram, we classify the samples into several classes. This process might be called labeling samples with HCA. On the other hand, HCA can also be used to identify the underlying hierarchical structure of clusters. When used for this purpose, the hierarchical structure is derived from the clusters of labeled samples. This paper focuses on using HCA to identify a hierarchical structure. The key in this identification is to select a suitable measure of dissimilarity among clusters, each of which is characterized by a subpopulation of the samples. Accordingly, we introduce a dissimilarity measure that was recently proposed in [4]. This measure is used because it is appropriately defined to quantify dissimilarity among multiple subpopulations of a population based on the mixture model. Glass identification, that is identifying glasses by type, is used as a practical problem in the experiments in this paper. In the experiments, we examine the effectiveness of the measure in identifying the hierarchical structure of different types of glass and compare this to several other measures. The results show that the introduced measure is especially appropriate for identifying the hierarchical structure of

* This work was supported in part by Grant-in-Aid 18700157 and 18500116 for scientific research from the Ministry of Education, Culture, Sports, Science, and Technology, Japan.

J. Marques de Sá et al. (Eds.): ICANN 2007, Part II, LNCS 4669, pp. 311–320, 2007.

clusters, and that the measure yields an accurate dendrogram without producing chain effects.

This paper is organized as follows. In Section 2, we present some notations and introduce the measure to be used in HCA. We show the experimental results in Section 3. Finally, our conclusions are set out in Section 4.

2 Preliminaries

We assume that the population of samples is based on the mixture model that is commonly understood in the literature throughout this paper. Accordingly, in this section, we explain the mixture model and then introduce a dissimilarity measure that fits the population. This measure is used in HCA to identify a hierarchical structure.

We formulate the mixture model shown in Figure 1 as follows. The population of samples consists of M different subpopulations, each with a prior probability. The subpopulation for generating a sample varies at each time-step according to the prior probability for the choice of the subpopulation. At each time-step, one of the subpopulations is chosen with the prior probability, and a sample is drawn according to the probability distribution (PD) of the chosen subpopulation. To facilitate exposition, we number the subpopulations with a label number. The label number of each sample indicates the subpopulation from which it was generated. We use X to express a stochastic variable (SV) over an arbitrary sample space $\mathcal{X}$ and use X_i to denote X at time-step $i \in \mathbb{N}$. Let $\mathcal{L} \triangleq \{1, \ldots, M\}$ be the entire set of label numbers. Let $\mathcal{P}(\mathcal{L})$ be the set of the probability density functions (PDFs) of the subpopulation, that is,

$$\mathcal{P}(\mathcal{L}) \triangleq \{P_m | m \in \mathcal{L}\}, \tag{1}$$

where P_m denotes the PDF of the subpopulation m over $\mathcal{X}$. If the label number of a sample $x \in \mathcal{X}$ is $m \in \mathcal{L}$, then from here on it is expressed: $x \sim P_m$. For every time-step i and every $m \in \mathcal{L}$, we define the prior probability as

$$\omega(m) \triangleq \Pr(X_i \sim P_m). \tag{2}$$

This means that a subpopulation is chosen at each time-step to generate each sample independently and according to its prior probability ω. For simplicity, we assume that $\omega(m) > 0$ for every $m \in \mathcal{L}$. This is an underlying assumption throughout this paper. Hence, the following is always satisfied:

$$\sum_{m \in \mathcal{L}} \omega(m) = 1. \tag{3}$$

For any positive number n, let $\boldsymbol{X}_n = (X_1, X_2, \ldots, X_n)$ be SVs of the population. This is sometimes written as $\boldsymbol{X}$ for brevity when we do not need to indicate n explicitly. The expected value of any function $y(x)$ over $\mathcal{X}$ with respect to P_m is denoted by

$$E_{P_m}[y(x)] \triangleq \int_{x \in \mathcal{X}} P_m(x) y(x)\, dx, \tag{4}$$

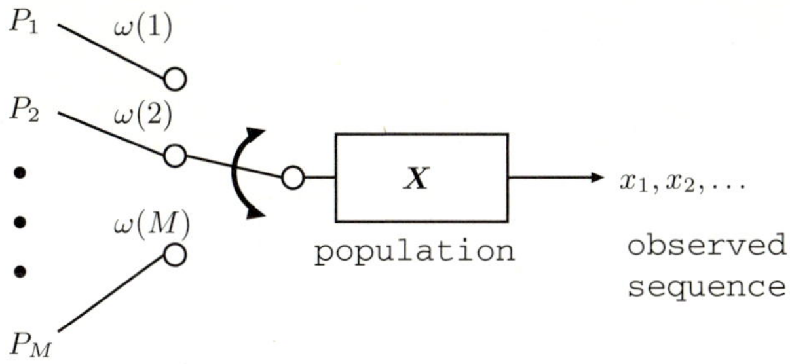

Fig. 1. The population based on the mixture model

for every $m \in \mathcal{L}$. We use I_C to denote an indicator function such that for any condition C,

$$I_C = \begin{cases} 1, & \text{if } C \text{ is true,} \\ 0, & \text{otherwise.} \end{cases} \tag{5}$$

Next, we introduce the following measure that was recently proposed in [4]. It is referred to as redundancy-based dissimilarity among PDs (RDSP). We propose employing this measure in HCA because it is appropriately defined to quantify dissimilarity among multiple subpopulations based on the mixture model.

Definition 1 (RDSP [4]). *For any subset $\underline{\mathcal{L}} \subseteq \mathcal{L}$, we define the squared dissimilarity measure among multiple PDFs $\mathcal{P}(\underline{\mathcal{L}})$ as*

$$\{RDS(\mathcal{P}(\underline{\mathcal{L}}))\}^2 \triangleq \sum_{m \in \underline{\mathcal{L}}} \lambda_{\underline{\mathcal{L}}}(m) \mathrm{D}(P_m \| Q_{\underline{\mathcal{L}}}), \tag{6}$$

where $\lambda_{\underline{\mathcal{L}}}$ is a normalized probability given by

$$\lambda_{\underline{\mathcal{L}}}(m) \triangleq \frac{\omega(m)}{\sum_{m \in \underline{\mathcal{L}}} \omega(m)}, \tag{7}$$

and $\mathrm{D}(P_m \| Q_{\underline{\mathcal{L}}})$ denotes the information divergence given by

$$\mathrm{D}(P_m \| Q_{\underline{\mathcal{L}}}) = E_{P_m}\left[\log \frac{P_m(x)}{Q_{\underline{\mathcal{L}}}(x)}\right], \tag{8}$$

where the PDF $Q_{\underline{\mathcal{L}}}$, for any $x \in \mathcal{X}$ is:

$$Q_{\underline{\mathcal{L}}}(x) \triangleq \sum_{m \in \underline{\mathcal{L}}} \lambda_{\underline{\mathcal{L}}}(m) P_m(x). \tag{9}$$

From an information theory viewpoint, this squared RDSP represents the amount of information loss involved in regarding multiple PDFs $\mathcal{P}(\underline{\mathcal{L}})$ as merged $Q_{\underline{\mathcal{L}}}$ such that the population of the mixture model is drawn according to a single

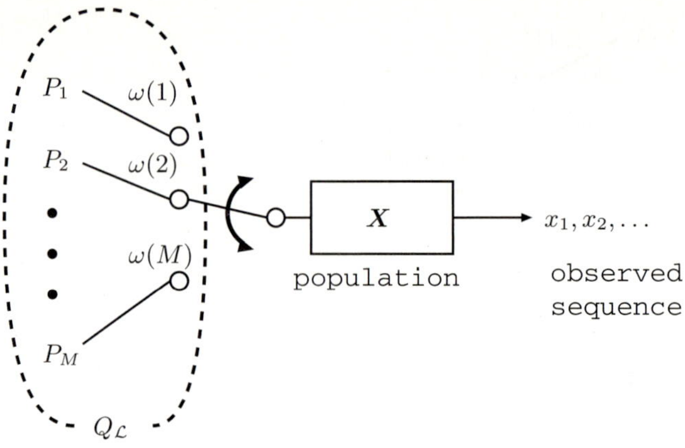

Fig. 2. The population drawn intuitively according to a single PDF $Q_{\underline{\mathcal{L}}}$ where $\mathcal{L} = \underline{\mathcal{L}}$

PDF $Q_{\underline{\mathcal{L}}}$ (see Figure 2). More intuitively, RDSP represents how multiple PDFs are placed over the sample space. Hence, it vanishes if and only if all the PDFs in $\mathcal{P}(\underline{\mathcal{L}})$ are equal. It was shown in [4] that when $|\underline{\mathcal{L}}| = 2$, $RDS(P_1, P_2)$ becomes the Jensen-Shannon divergence [5,6] which is a metric between P_1 and P_2.

3 Experiments

The experimental results discussed here are for real data processed using HCA for clusters of labeled samples. The experiments used a glass identification database from the UCI repository of machine learning databases [7]. Glass identification is a particularly interesting problem, as is mentioned in the database, as it is often of use in criminal investigations, because the glass left at the scene of a crime can be crucial evidence if it is correctly identified. The database consists of 214 labeled samples featuring nine dimensions (attributes). These nine dimensions are the refractive index and weight percents of sodium, magnesium, aluminum, silicon, potassium, calcium, barium, and iron in their corresponding oxides. The glass samples in the database are labeled according to coded type. Each sample is classified into one of six classes corresponding to the hierarchical structure shown in Figure 3. The six types of glass are classified broadly as window or non-window class. The non-window class includes containers (C), tableware (T), and headlamps (H). The window class is furthermore classified into float-processed versus non-float-processed class. The float-processed class includes float-processed building windows (FB) and float-processed vehicle windows (FV), and the non-float-processed class includes non-float-processed building windows (NB). We partitioned all samples in the database into eight clusters such that samples of each cluster have the same label, as shown in Table 1. The samples of the FB and NB types are separated into two clusters according to their exact ID numbers, because the number of samples for these two types is much larger than those of the others.

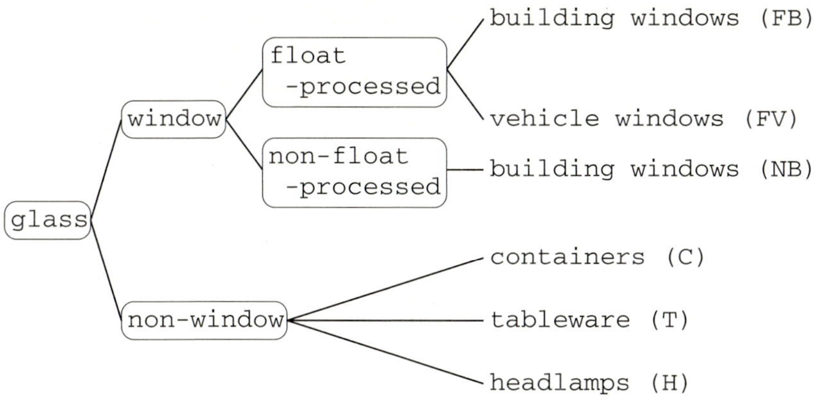

Fig. 3. The hierarchical structure of the types of glass

Table 1. Cluster number, glass type, number of samples, and ID number

cluster number ($m = 1, \ldots, 8$)	glass type (label code)	number of samples in each cluster	ID number in database (1–214)
1	FB	35	1–35
2	FB	35	36–70
3	NB	40	71–110
4	NB	36	111–146
5	FV	17	147–163
6	C	13	164–176
7	T	9	177–185
8	H	29	186–214

The goal of this clustering task is to find the underlying hierarchical structure of the types of glass from the clusters. We expect from Table 1 that the closest clusters are 1 and 2, or 3 and 4 at the first or the second step, since the two clusters in each pair are the same type. According to Figure 3, we expect clusters 1-2 and 5 to merge in a subsequent step, (where cluster 1-2 is the cluster produced by merging clusters 1 and 2,) because clusters 1, 2 and 5 are all float-processed. Moreover, we expect that clusters 1-2-5 and 3-4 will also merge, as cluster 1 to 5 all fall into the window class category.

We employed the following six methods with different measures to form hierarchical clusters. Of these six methods, the nearest neighbor and the group average methods are sometimes referred to as linkage methods [2, Ch. 14][3, Ch. 10], Ward's method [8,9] is a centroid-based method, the clustering evaluation function method [10] is a kernel-based method, and the symmetric information divergence and RDSP methods are clustering methods based on probabilistic-dependent measures.

NN Method: For the nearest neighbor (NN) method, the distance (dissimilarity) measure between any two (prime or merged) clusters is the Euclidean distance of the two closest samples in the different clusters. That is, for any subset $\underline{\mathcal{L}}' \subset \mathcal{L}$ and $\underline{\mathcal{L}}'' \subset \mathcal{L}$ such that $\underline{\mathcal{L}}' \cap \underline{\mathcal{L}}'' = \emptyset$, the dissimilarity is expressed by

$$\Delta_N(\underline{\mathcal{L}}', \underline{\mathcal{L}}'') = \min_{x' \in \boldsymbol{x}^{(\underline{\mathcal{L}}')},\ x'' \in \boldsymbol{x}^{(\underline{\mathcal{L}}'')}} \sqrt{(x' - x'')^T (x' - x'')}, \tag{10}$$

where T is the transposition of a vector, and for any subset $\underline{\mathcal{L}}$ the set $\boldsymbol{x}^{(\underline{\mathcal{L}})}$ of nine-dimensional samples means

$$\boldsymbol{x}^{(\underline{\mathcal{L}})} \triangleq \bigcup_{m \in \underline{\mathcal{L}}} \boldsymbol{x}^{(m)}, \tag{11}$$

where for a total number n of samples, $\boldsymbol{x}^{(m)}$ is defined as

$$\boldsymbol{x}^{(m)} \triangleq \{x_i \in \mathcal{X} \mid x_i \sim P_m, i = 1, \ldots, n\}. \tag{12}$$

In this experiment, $\boldsymbol{x}^{(m)}$ is the set of samples of cluster m in Table 1. The NN method always attempts to merge two clusters that minimize the distance measure at each step of the mergence.

GA Method: The group average (GA) method is well recognized in the literature to effectively measure the distance between two clusters. In this method, the distance measure between any two (prime or merged) clusters is the average distance between them. That is, for any subset $\underline{\mathcal{L}}' \subset \mathcal{L}$ and $\underline{\mathcal{L}}'' \subset \mathcal{L}$ such that $\underline{\mathcal{L}}' \cap \underline{\mathcal{L}}'' = \emptyset$, the distance between them is given by

$$\Delta_G(\underline{\mathcal{L}}', \underline{\mathcal{L}}'') = \frac{1}{n_{\underline{\mathcal{L}}'} n_{\underline{\mathcal{L}}''}} \sum_{x' \in \boldsymbol{x}^{(\underline{\mathcal{L}}')}} \sum_{x'' \in \boldsymbol{x}^{(\underline{\mathcal{L}}'')}} \sqrt{(x' - x'')^T (x' - x'')}, \tag{13}$$

where for any subset $\underline{\mathcal{L}}$, $n_{\underline{\mathcal{L}}}$ denotes

$$n_{\underline{\mathcal{L}}} \triangleq \sum_{m \in \underline{\mathcal{L}}} n_m, \tag{14}$$

where n_m is the number of elements in the set $\boldsymbol{x}^{(m)}$. The GA method always attempts to merge two clusters that minimize the distance measure at each step.

Ward's Method: In this method, the distance between any two (prime or merged) clusters is written as Ward's measure [8,9]. That is, for any subset $\underline{\mathcal{L}}' \subset \mathcal{L}$ and $\underline{\mathcal{L}}'' \subset \mathcal{L}$ such that $\underline{\mathcal{L}}' \cap \underline{\mathcal{L}}'' = \emptyset$, the distance between them is written as

$$\Delta_W(\underline{\mathcal{L}}', \underline{\mathcal{L}}'') = \frac{n_{\underline{\mathcal{L}}'} n_{\underline{\mathcal{L}}''}}{n_{\underline{\mathcal{L}}'} + n_{\underline{\mathcal{L}}''}} \left(\bar{x}^{(\underline{\mathcal{L}}')} - \bar{x}^{(\underline{\mathcal{L}}'')}\right)^T \left(\bar{x}^{(\underline{\mathcal{L}}')} - \bar{x}^{(\underline{\mathcal{L}}'')}\right), \tag{15}$$

where for any subset $\underline{\mathcal{L}}$, $\bar{x}^{(\underline{\mathcal{L}})}$ is given by

$$\bar{x}^{(\underline{\mathcal{L}})} \triangleq \frac{1}{n_{\underline{\mathcal{L}}}} \sum_{x \in \boldsymbol{x}^{(\underline{\mathcal{L}})}} x. \tag{16}$$

Similarly, Ward's method always attempts to merge two clusters that minimize the distance measure at each step.

CEF Method: Recently, an advanced similarity measure was proposed in [10] to evaluate the information potential between samples in multiple clusters. The information potential is expressed as Renyi's quadratic entropy and is based on a Gaussian kernel function. For any subset $\underline{\mathcal{L}}' \subset \mathcal{L}$ and $\underline{\mathcal{L}}'' \subset \mathcal{L}$ such that $\underline{\mathcal{L}}' \cap \underline{\mathcal{L}}'' = \emptyset$, the similarity measure is given by

$$CEF(\underline{\mathcal{L}}', \underline{\mathcal{L}}'') = \frac{1}{2n_{\underline{\mathcal{L}}'}n_{\underline{\mathcal{L}}''}} \sum_{x' \in \boldsymbol{x}(\underline{\mathcal{L}}')} \sum_{x'' \in \boldsymbol{x}(\underline{\mathcal{L}}'')} G(x' - x'', 2\sigma^2), \tag{17}$$

where for any $x \in \mathbb{R}^9$ and any $v^2 \in \mathbb{R}$, the Gaussian kernel function G is:

$$G(x, v^2) \triangleq \frac{1}{2\pi v} \exp\left(-\frac{x^T x}{2v^2}\right). \tag{18}$$

Equation (17) is referred to as the clustering evaluation Function (CEF). From here on when we employ the pseudo-distance measure [10, Eq. 16] based on the CEF we call it the CEF method. In this method, the pseudo-distance measure between any two (prime or merged) clusters is the kernel-based distance between them. For any subset $\underline{\mathcal{L}}' \subset \mathcal{L}$ and $\underline{\mathcal{L}}'' \subset \mathcal{L}$ such that $\underline{\mathcal{L}}' \cap \underline{\mathcal{L}}'' = \emptyset$, the pseudo-distance measure is given by

$$\Delta_C(\underline{\mathcal{L}}', \underline{\mathcal{L}}'') = -\log CEF(\underline{\mathcal{L}}', \underline{\mathcal{L}}''). \tag{19}$$

The CEF method always attempts to merge two clusters that minimize the distance measure at each step. The parameter σ was carefully tuned to ensure the method works well. As a result, we set $\sigma = 0.3$ in this experiment. We have seen that the distance between clusters is sensitive to the choice of the σ parameter of the Gaussian kernel function. In fact, if $\sigma = 0.5$, the resulting dendrogram is completely different from Figure 4(d).

SID Method: From here on when we employ to measure the distance between two clusters we call it the SID method. The symmetric information divergence measure, or SID, is one of the most well-known probabilistic-dependent measures in the field. For any subset $\underline{\mathcal{L}}' \subset \mathcal{L}$ and $\underline{\mathcal{L}}'' \subset \mathcal{L}$ such that $\underline{\mathcal{L}}' \cap \underline{\mathcal{L}}'' = \emptyset$, the SID between $Q_{\underline{\mathcal{L}}'}$ and $Q_{\underline{\mathcal{L}}''}$ is given by

$$\Delta_S(\underline{\mathcal{L}}', \underline{\mathcal{L}}'') = \mathrm{D}(Q_{\underline{\mathcal{L}}'} \| Q_{\underline{\mathcal{L}}''}) + \mathrm{D}(Q_{\underline{\mathcal{L}}''} \| Q_{\underline{\mathcal{L}}'}). \tag{20}$$

However, since the integration of information divergence creates serious computational complexity, this is generally difficult to compute directly. Hence, for the PDFs of the subpopulations, we use the approximated (empirical) SID designated by

$$\Delta_S(\underline{\mathcal{L}}', \underline{\mathcal{L}}'') \approx \frac{1}{n_{\underline{\mathcal{L}}'}} \left(\sum_{m \in \underline{\mathcal{L}}'} \left| \sum_{x \in \boldsymbol{x}(m)} \log \frac{Q_{\underline{\mathcal{L}}'}(x)}{Q_{\underline{\mathcal{L}}''}(x)} \right| \right) + \frac{1}{n_{\underline{\mathcal{L}}''}} \left(\sum_{m \in \underline{\mathcal{L}}''} \left| \sum_{x \in \boldsymbol{x}(m)} \log \frac{Q_{\underline{\mathcal{L}}''}(x)}{Q_{\underline{\mathcal{L}}'}(x)} \right| \right). \tag{21}$$

From the weak law of large numbers, we can readily show that (21) converges to (20) in probability as $n_{\underline{\mathcal{L}}'} \to \infty$ and $n_{\underline{\mathcal{L}}''} \to \infty$. Accordingly, the SID method always attempts to merge two clusters that minimize the distance measure given by (21) at each step.

RDSP Method: When the distance measure is produced by squared RDSP, we call it the RDSP method. As with the above methods, the RDSP method attempts to form hierarchical clusters. Since the RDSP is also expressed in terms of information divergence, we approximate the squared RDSP to avoid computing the integration. Whenever merging two clusters, for the PDFs of the subpopulations, the RDSP method always selects the two (prime or merged) clusters $\underline{\mathcal{L}}'$ and $\underline{\mathcal{L}}''$ that minimize the squared RDSP approximated by

$$\Delta_R(\underline{\mathcal{L}}', \underline{\mathcal{L}}'') = \left\{ RDS(\mathcal{P}(\underline{\mathcal{L}}' \cup \underline{\mathcal{L}}'')) \right\}^2, \tag{22}$$

$$\approx \frac{1}{n_{\underline{\mathcal{L}}'} + n_{\underline{\mathcal{L}}''}} \left\{ rds_{\mathcal{P}(\underline{\mathcal{L}}' \cup \underline{\mathcal{L}}'')}(x_1, \ldots, x_{n_{\underline{\mathcal{L}}'}+n_{\underline{\mathcal{L}}''}}) \right\}^2, \tag{23}$$

where for $i = 1, \ldots, n_{\underline{\mathcal{L}}'} + n_{\underline{\mathcal{L}}''}$, each sample is $x_i \in \boldsymbol{x}^{(\underline{\mathcal{L}}')} \cup \boldsymbol{x}^{(\underline{\mathcal{L}}'')}$ and

$$\left\{ rds_{\mathcal{P}(\underline{\mathcal{L}}' \cup \underline{\mathcal{L}}'')}(x_1, \ldots, x_{n_{\underline{\mathcal{L}}'}+n_{\underline{\mathcal{L}}''}}) \right\}^2 \triangleq \sum_{m \in \underline{\mathcal{L}}' \cup \underline{\mathcal{L}}''} \left| \sum_{i=1}^{n_{\underline{\mathcal{L}}'}+n_{\underline{\mathcal{L}}''}} I_{x_i \sim P_m} \log \frac{P_m(x_i)}{Q_{\underline{\mathcal{L}}}(x_i)} \right| . \tag{24}$$

Equation (24) asymptotically converges to the squared RDSP in probability as had been proved in [4]. In addition to this approximation, the RDSP deals with multiple clusters (rather than only two) alleviating computational complexity considerably. For example, when $\underline{\mathcal{L}}' = \{1, 2\}$ and $\underline{\mathcal{L}}'' = \{3, 4\}$ in (22), it is simply calculated using $\{RDS(P_1, P_2, P_3, P_4)\}^2$ rather than $\{RDS(Q_{1,2}, Q_{3,4})\}^2$ where $Q_{1,2}$ and $Q_{3,4}$ denote merged PDFs.

In this experiment, since the subpopulation PDFs to be used in the SID and RDSP methods are unknown, we assumed that the PDF P_m of each subpopulation, where $m = 1, \ldots, 8$, is a normal distribution of the nine dimensions and estimated the mean and covariance matrix of P_m from the set of samples of cluster m. For example, the mean and covariance matrix of P_1 were computed from samples whose ID numbers are 1–35 in Table 1. We also set $\exp(y) = 1.5^y$ for any $y \in \mathbb{R}$ in calculating the exponent of the PDF P_m of a normal distribution and the base of the logarithm in (21) and (24). This is done simply to scale up the tail of the PDF of a normal distribution to avoid overly small values. Note that this scaling does not essentially affect any resulting dendrogram.

Figure 4 shows the resulting dendrograms of the six methods for the eight clusters of samples shown in Table 1. The vertical axis in the figures represents the distance measure for each method, which was adopted as a minimum value at each step. The horizontal broken lines separate the clusters at proper levels of the hierarchical structure. We confirm from the horizontal broken lines that only the RDSP method successfully classified the clusters into float-processed and non-float-processed classes at a low level in the hierarchical structure, and

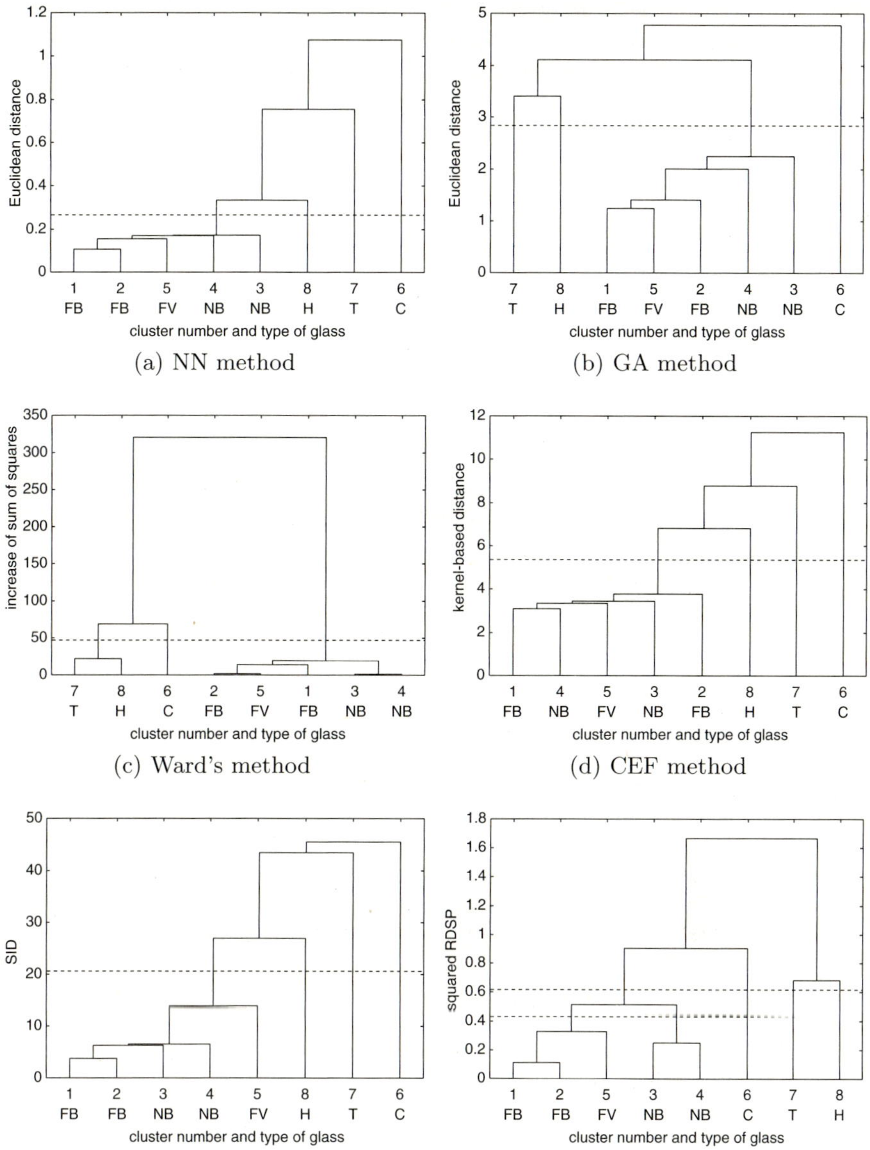

(a) NN method (b) GA method

(c) Ward's method (d) CEF method

(e) SID method (f) RDSP method

Fig. 4. Dendrograms for the eight clusters in Table 1

furthermore into the windows versus non-windows at a high level. From the horizontal broken lines, we see that although the other methods also provided appropriate classification into the window and non-window classes, they failed to classify the clusters into float-processed and non-float-processed classes. It follows that RDSP is especially appropriate for identifying the hierarchical structure of clusters. Moreover, Figure 4 shows that the NN, CEF, and SID methods produced a chain effect. In contrast, the RDSP method efficiently constructed an accurate dendrogram without producing a chain effect.

4 Conclusion

This paper has introduced RDSP to identify the underlying hierarchical structure of clusters in HCA. The experiments for the glass identification database confirmed that RDSP is especially appropriate for identifying the hierarchical structure of clusters, and that RDSP efficiently and accurately creates dendrograms without chain effects.

References

1. Anderberg, M.R.: Cluster Analysis for Applications. Probability and Mathematical Statistics, vol. 19. Academic Press, New York (1973)
2. Hastie, T., Tibshirani, R., Friedman, J.: The Elements of Statistical Learning: Data Mining, Inference, and Prediction. Springer series in statistics. Springer, New York (2001)
3. Webb, A.R.: Statistical Pattern Recognition, 2nd edn. John Wiley & Sons, New York (2002)
4. Iwata, K., Hayashi, A.: Theory of a probabilistic-dependence measure of dissimilarity among multiple clusters. In: Proceedings of the 16th International Conference on Artificial Neural Networks, Part II, Athens, Greece. LNCS, vol. 4132, pp. 311–320. Springer, Heidelberg (2006)
5. Topsøe, F.: Some inequalities for information divergence and related measures of discrimination. IEEE Transactions on Information Theory 46(4), 1602–1609 (2000)
6. Endres, D.M., Schindelin, J.E.: A new metric for probability distributions. IEEE Transactions on Information Theory 49(7), 1858–1860 (2003)
7. Newman, D.J., Hettich, S., Blake, C.L., Merz, C.J.: UCI repository of machine learning databases (1998), `http://www.ics.uci.edu/~mlearn/MLRepository.html`
8. Ward, J.H.: Hierarchical grouping to optimize an objective function. Journal of the American Statistical Association 58(301), 236–244 (1963)
9. Ward, J.H., Hook, M.E.: Application of an hierarchical grouping procedure to a problem of grouping profiles. Educational Psychological Measurement 23(1), 69–82 (1963)
10. Gokcay, E., Principle, J.C.: Information theoretic clustering. IEEE Transactions on Pattern Analysis and Machine Intelligence 24(2), 158–171 (2002)

Clustering Evaluation in Feature Space

Alissar Nasser, Pierre-Alexandre Hébert, and Denis Hamad

ULCO, LASL, 50 rue F. Buisson, BP 699, 62228 Calais, France

Abstract. Many clustering algorithms require some parameters that often are neither a priori known nor easy to estimate, like the number of classes. Measures of clustering quality can consequently be used to a posteriori estimate these values. This paper proposes such an index of clustering evaluation that deals with kernel methods like kernel-k-means. More precisely, it presents an extension of the well-known Davies & Bouldin's index. Kernel clustering methods are particularly relevant because of their ability to deal with initially non-linearly separable clusters. The interest of the following clustering evaluation is then to get around the issue of the not explicitly known data transformation of such kernel methods. Kernel Davies & Bouldin's index is finally used to a posteriori estimate the parameters of the kernel-k-means method applied on some toys datasets and Fisher's Iris dataset.

Keywords: Clustering evaluation, Kernel-k-means, Davies & Bouldin's index, Number of classes estimation, Kernel estimation.

1 Introduction

Clustering methods based upon a kernel feature space usually reach a good clustering accuracy, even with datasets whose clusters cannot be linearly separated. This non-linearly separation boundary may be indirectly obtained by transforming the input data space $x \in \Re^n$ into a high-dimensional space by applying a function ϕ, which may facilitate the discrimination [2]. This function is usually not explicitly known, but simply induced by the selected kernel. This fact does not trouble kernel clustering methods like kernel-k-means, nor other kernel methods like kernel-principal component analysis (kernel-PCA). All needed variables of these algorithms are indeed expressed as dot products that can then be computed as values of the kernel function, from the moment that this function fulfills Mercer's conditions.

A plethora of new kernel clustering methods have been proposed in recent years which give very impressive results with highly different data shapes. The most important among them include kernel-PCA which has been observed to perform clustering [4], and kernel-based clustering in feature space like kernel-k-means [1,8]. Whilst, two issues of clustering still remain: (1) estimating the parameters of the clustering method, like the number of clusters within the dataset, (2) evaluating the quality of the obtained clustering. Some heuristics exist to estimate the number of clusters [1,5]. But it appears as simple as pertinent to extend a classical evaluation method based upon the notion of compactness measure in order to deal with a kernel feature space.

J. Marques de Sá et al. (Eds.): ICANN 2007, Part II, LNCS 4669, pp. 321–330, 2007.

In this paper, we propose to calculate Davies & Bouldin's index in the feature space using the kernel trick. This allows to evaluate the result of the kernel clustering methods, and consequently to estimate the adequacy of their parameters like the number of clusters and the parameters of the kernel.

The paper is organized as follows: Section 2 describes the kernel-based clustering method that we decided to deal with: the kernel-k-means. Section 3 presents DB index in both input and feature space. Section 4 shows some experimental results where DB index is used to determine the parameters of the kernel-k-means.

2 From k-Means Algorithm to Its Kernel Extension

Classical k-means algorithm [6] aims at clustering the dataset into a predefined number k of groups by minimizing a formal objective mean-squared-error (MSE) function. It is an easy iterative method whose main calculation consists in obtaining the distances from points to the centers of clusters.

2.1 K-Means in Input Space

Let X be a dataset of points $\{x_i, i = 1 : n\}$, where $x_i \in \Re^n$, and let k be the predefined number of clusters. K-means algorithm consists in the following iterative method:

1. Initialize the k centers $c_1, \ldots, c_k$ of clusters $C_1, \ldots, C_k$.
2. Assign each point x_i in X to its nearest center by computing the distances from x_i to all centers: $\left\|x_i - c_j\right\|^2$ for each j in $\{1,..,k\}$. This induces a new partition $\{C_1, \ldots, C_k\}$.
3. Update the centers $c_1, \ldots, c_k$: they are defined as the centers of gravity of the corresponding clusters $C_1, \ldots, C_k$.
4. Go to step 2 until the partition $\{C_1, \ldots, C_k\}$ is stabilized.

The k-means algorithm is significantly sensitive to the initial clusters centers. Relevant centers can be obtained by a random selection in the dataset X. Moreover, the algorithm should better be run multiple times with different initializations. The objective function MSE minimized by the algorithm can then be used to determine the best initialization: $MSE = \frac{1}{n}\sum_{j=1}^{k}\sum_{x_i \in C_j}\left\|x_i - c_j\right\|^2$.

2.2 k-Means Algorithm in Feature Space: Kernel-k-Means

The k-means method gives poor results when the dataset is composed of non-linearly separable clusters. A way to encompass this problem is to perform k-means in a kernel-determined feature space using the kernel trick [8]. If the kernel K is adequate and well-tuned, then the algorithm becomes able to recognize non-Gaussian or non-convex shaped clusters in input space.

The objective function MSE applied in the feature space corresponding to a function ϕ is defined as follows [7]:

$$MSE^{\phi} = \frac{1}{n}\sum_{j=1}^{k}\sum_{x_i \in C_j^{\phi}} \left\| \phi(x_i) - c_j^{\phi} \right\|^2, \tag{1}$$

where $c_j^{\phi} = \frac{1}{\left|C_j^{\phi}\right|}\sum_{x_i \in C_j^{\phi}} \phi(x_i)$ denotes the center of gravity of C_j^{ϕ}, the cluster j in the feature space defined by Function ϕ. This function does no need to be explicitly known: according to the kernel trick, the calculation of any dot products in the feature space $\langle \phi(x), \phi(y) \rangle,\quad (x,y) \in X^2$ can be done without any explicit definition of ϕ. It is rather induced by the choice of a positive semi-definite kernel function: $K: X \times X \rightarrow \Re$, that explicitly defines the dot product between the points in X: $K(x,y) = \langle \phi(x), \phi(y) \rangle$. The reason why k-means algorithm can easily be extended in the feature space is that its computation only needs dot products: indeed, as well points $x_i \in X$ or centers c_j^{ϕ} do not need to be computed, and the only necessary variables are the following Euclidian distances:

$$\left\| \phi(x_i) - c_j^{\phi} \right\|^2 = \phi(x_i).\phi(x_i)^T - \frac{2}{\left|C_j^{\phi}\right|}\sum_{a \in C_j^{\phi}} \phi(a).\phi(x_i)^T + \frac{1}{\left|C_j^{\phi}\right|^2}\sum_{a \in C_j^{\phi}}\sum_{b \in C_j^{\phi}} \phi(a).\phi(b)^T \tag{2}$$

$$\left\| \phi(x_i) - c_j^{\phi} \right\|^2 = K(x_i, x_i) - \frac{2}{\left|C_j^{\phi}\right|}\sum_{a \in C_j^{\phi}} K(a, x_i) + \frac{1}{\left|C_j^{\phi}\right|^2}\sum_{a \in C_j^{\phi}}\sum_{b \in C_j^{\phi}} K(a,b), \quad \forall\, x_i \subset X \tag{3}$$

Except these equations, the kernel-k-means algorithm is completely equivalent to the precedent one.

3 Davies and Bouldin's Index

Davies & Bouldin's index is a well-known cluster validity index [6,11] that is used in classical clustering because of its simplicity and its relevance. In particular, it may be preferred to MSE in k-means, because it is let as a ratio of distances that does not necessarily decrease when the number k of classes increases; it can then be used in order to evaluate the accuracy of the selected k value.

Clustering evaluation may be obtained by other ways like the visualization approaches, such as PCA, MDS, Sammon's maps [6]. More recently, learning techniques like Autoassociator networks, SOM, k-PCA have been developed to visualize high-dimensional data as two dimensional scatter plots [5,10,12]: the resulting representations allow a straightforward analysis of the inherent structure of clusters within the input data. Roberts & all [9] show that the minimization of a partition entropy or the maximization of a partition certainty may be used to estimate the right partitioning, by deciding the most adequate number of classes.

In this paper, we propose to use the well-known cluster validity Davies & Bouldin's index [6,11] in the feature space, as a way to evaluate the clustering quality of a given kernel clustering method. Clearly, any other clustering evaluation index could be used if and only if it can be written as a function of dot products $\langle \phi(x), \phi(y) \rangle$, $x, y \in X$.

3.1 Davies and Bouldin Index in the Input Space: DB Index

Let $\{C_1, C_2 \ldots C_k\}$ denotes a partition of the points in the set X obtained by a clustering method. The DB index for k clusters is defined as follows:

$$DB = \frac{1}{k} \sum_{j=1}^{k} R_j , \tag{4}$$

where $R_j = \max_{m \neq j} \left\{ \frac{\Delta(C_j) + \Delta(C_m)}{\delta(C_j, C_m)} \right\}$, with $\Delta(C_j)$ (resp. $\Delta(C_m)$) the intra-cluster dispersion of cluster C_j (resp. C_m), and $\delta(C_j, C_m)$ the inter-cluster distance between clusters C_j and C_m.

In the input space, the classical choice of the Euclidean distance induces:

$$\Delta(C_j) = \frac{1}{|C_j|} \sum_{x_i \in C_j} \left\| x_i - c_j \right\|^2 \text{ and } \delta(C_j, C_m) = \left\| c_j - c_m \right\|^2 .$$

DB index is a ratio, i.e. a variable without unit whose values can so be compared despite distinct numbers of classes. It decreases as clusters become more compact (lower $\Delta(C_.)$) and more distinctly separated (higher $\delta(C_., C_.)$). That is the reason why its minimal value may be used to indicate an adequate number of classes k.

We now propose to extend its computation to deal with kernel clustering methods.

3.2 DB Index in Feature Space: Kernel-DB

After the transformation of the set X into a feature space by Function ϕ, DB can still be easily computed. Indeed Euclidean intra-cluster dispersions and inter-clusters distances can be written as dot products of the ϕ-transformed points.

First, intra-cluster dispersion is computed as follows:

$$\Delta(C_j^{\phi}) = \frac{1}{\left|C_j^{\phi}\right|} \sum_{x_i \in C_j^{\phi}} \left\|\phi(x_i) - c_j^{\phi}\right\|^2; \tag{5}$$

substituting (3) in (5) we get:

$$\Delta(C_j^{\phi}) = \frac{1}{\left|C_j^{\phi}\right|} \sum_{x_i \in C_j^{\phi}} K(x_i, x_i) - \frac{2}{\left|C_j^{\phi}\right|^2} \sum_{x_i \in C_j^{\phi}} \sum_{a \in C_j^{\phi}} K(x_i, a) + \frac{1}{\left|C_j^{\phi}\right|^3} \sum_{x_i \in C_j^{\phi}} \sum_{a \in C_j^{\phi}} \sum_{b \in C_j^{\phi}} K(a, b); \tag{6}$$

yielding:

$$\Delta(C_j^{\phi}) = \frac{1}{\left|C_j^{\phi}\right|} \sum_{x_i \in C_j^{\phi}} K(x_i, x_i) - \frac{1}{\left|C_j^{\phi}\right|^2} \sum_{a \in C_j^{\phi}} \sum_{b \in C_j^{\phi}} K(a, b) \cdot \tag{7}$$

With the commonly used Gaussian kernel function $K_{G(\sigma)}(x, y) = e^{-\frac{\|x-y\|^2}{2\sigma^2}}$, equation (7) is simplified as follows:

$$\Delta(C_j^{G(\sigma)}) = 1 - \frac{1}{\left|C_j^{G(\sigma)}\right|^2} \sum_{x_i \in C_j^{G(\sigma)}} \sum_{a \in C_j^{G(\sigma)}} K_{G(\sigma)}(x_i, a) \cdot \tag{8}$$

Then inter-clusters distance between two centers c_j^{ϕ} and c_m^{ϕ} is defined in this way:

$$\begin{aligned}
\delta(C_j^{\phi}, C_m^{\phi}) &= \left\|c_j^{\phi} - c_m^{\phi}\right\|^2 = \left\| \frac{1}{\left|C_j^{\phi}\right|} \sum_{a \in C_j^{\phi}} \phi(a) - \frac{1}{\left|C_m^{\phi}\right|} \sum_{b \in C_m^{\phi}} \phi(b) \right\|^2 \\
\delta(C_j^{\phi}, C_m^{\phi}) &= \frac{1}{\left|C_j^{\phi}\right|^2} \sum_{a \in C_j^{\phi}} \sum_{c \in C_m^{\phi}} \phi(a).\phi(c)^T - \frac{2}{\left|C_j^{\phi}\right| \left|C_m^{\phi}\right|} \sum_{a \in C_j^{\phi}} \sum_{b \in C_m^{\phi}} \phi(a).\phi(b)^T + \frac{1}{\left|C_m^{\phi}\right|^2} \sum_{b \in C_m^{\phi}} \sum_{d \in C_m^{\phi}} \phi(b).\phi(d)^T \\
\delta(C_j^{\phi}, C_m^{\phi}) &= \frac{1}{\left|C_j^{\phi}\right|^2} \sum_{a \in C_j^{\phi}} \sum_{c \in C_m^{\phi}} K(a, c) - \frac{2}{\left|C_j^{\phi}\right| \left|C_m\right|} \sum_{a \in C_j^{\phi}} \sum_{b \in C_m^{\phi}} K(a, b) + \frac{1}{\left|C_m^{\phi}\right|^2} \sum_{b \in C_m^{\phi}} \sum_{d \in C_m^{\phi}} K(b, c).
\end{aligned} \tag{9}$$

Kernel DB index is then computed in the feature space according to the following algorithm:

1. For each cluster C_j^{ϕ} in the feature space, calculate the intra-cluster dispersion $\Delta(C_j^{\phi})$ using Equation (7) (or Equation (8) for Gaussian kernel).
2. Calculate the inter-clusters distances $\delta(C_j^{\phi}, C_m^{\phi})$ between all pairs of centers using Equation (9).

3. For each cluster C_j^{ϕ}, compute the corresponding value R_j according to Equation (4).
4. The value of kernel-DB obtained with k clusters is finally defined as the mean of the k values R_j, $\forall j \in \{1,\ldots,k\}$.

4 Experiments

The purpose of this section is to use kernel-DB as an estimation tool of the parameters of a kernel-clustering method. We focus on kernel-k-means, which is certainly one of the most used. Then, we decide to use the Gaussian kernel, because of its ability to make separable clusters that are initially compact and contiguous but not linearly-separable.

Gaussian kernel-k-means needs two parameters: the radius σ of the Gaussian kernel, and the number of classes k. We try to a posteriori estimate them conjointly, by considering that the better the choice of parameters, the better the quality of clustering i.e. the lower kernel-DB.

For both parameters k and σ, some ranges of values can reasonably be a priori set: in all following examples, we decide: $k \in \{2,\ldots,8\}$ and $\sigma \in [\sigma_{\min}, \sigma_{\max}]$, where $\sigma_{\min}$ and $\sigma_{\max}$ respectively denote the minimal and maximal Euclidian distances between points in the initial space. The range of σ is then discretized by selecting 60 values equally spaced in the interval; the 5 smallest values are then conveniently rejected as irrelevant, because $\sigma_{\min}$ is very close to 0 in our examples.

Kernel-DB results are presented on two graphs:

- Graph "Effect of the parameters on the clustering quality": for each $k \in \{2,\ldots,8\}$, a function kernel-DB is plotted in term of the variable $\sigma \in [\sigma_{\min}, \sigma_{\max}]$; these k-plots should help to tune both parameters;
- Graph "Kernel-DB and MSE$^{\Phi}$ indexes": kernel-DB is compared to the MSE index computed in the feature space (MSE$^{\Phi}$). The comparison is applied for a constant σ: indeed, MSE$^{\Phi}$ is not defined as a ratio, and simple distances can not reasonably be compared in different feature spaces.

The problem of the initializations of the kernel-k-means for each pair of parameters (k,σ) is solved by computing the algorithm 10 times with different initial cluster centers which are randomly selected in X. Only the minimal values of kernel-DB and MSE$^{\Phi}$ among the 10 initializations are kept in the graphs.

Finally, we represent for each example the classification obtained with the a priori known number k of classes (that should correspond to the value minimizing kernel-DB). It allows verifying the accuracy of the clustering in the input space.

4.1 Example 1: "Gaussian Square"

The dataset consists of 520 datapoints in a 2-dimensional space (Fig. 1): 4 Gaussian clusters are centered in the four corners of the square {(-1,-1),(1,-1),(1,1),(-1,1)}.

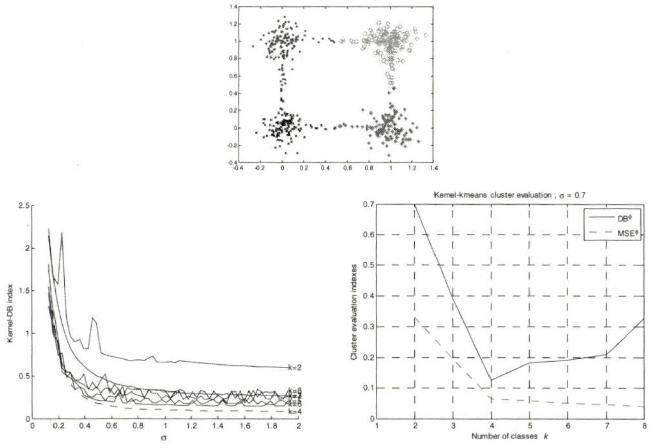

Fig. 1. "Gaussian square" classification σ=0.7 (up), Effect of the parameters on the clustering evaluation (left), Kernel-DB and MSE^{Φ} indexes (right)

Moreover, some points are randomly selected on the four segments of this square, according to a uniform probability density function.

We observe that the "best" classification meant by DB is achieved with the dot line *k=4* (Fig. 1, left and right). Moreover, this classification corresponds to our perception (Fig. 1, up). In this case, the kernel clustering is not sensitive to the sigma value: smaller or larger values of σ would give similar results. MSE^{Φ} can be thought as an easily interpretable index, but kernel-DB could be preferred because of its clear minimum.

4.2 Example 2: "2-Spheres"

The second example we investigate is composed of two spheres in 3D (Fig. 2). The dataset consists of 800 points defined as follows: 400 points are randomly selected (uniform pdf) within a shell defined by the two spheres of radius 0.015 and 0.08; the 400 other points are selected in the same way in the shell defined by the two spheres of radius 0.485 and 0.49 centered like the first sphere.

In this example (Fig. 2, up), the clusters are no more linearly-separable, but the kernel-clustering still gives good clusters. Without any knowledge of σ, the estimation of the value k is complex (Fig. 2, left): the dot line "*k=2*" reaches a minimal value, but only on the reduced interval $\sigma \in [0,0.25]$. Such a reduction could be obtained following Jenssen [3], by first selecting a plausible radius of the Gaussian kernel thanks to a Parzen window size selection procedure.

Then, we remark that if one of these parameters is already known or estimated then the other one may easily be assessed (Fig 2, left and right).

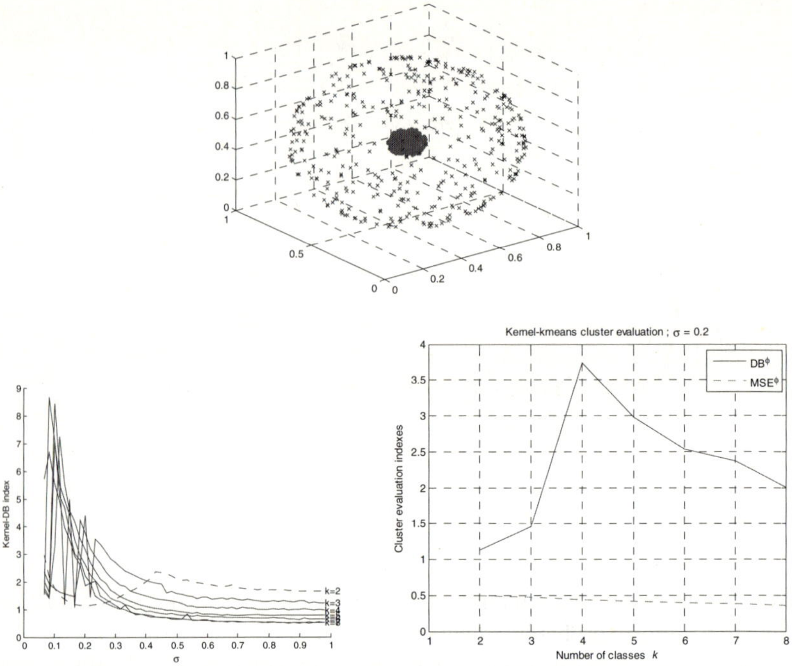

Fig. 2. "2-spheres" classification σ=0.2 (up), Effect of the parameters on the clustering evaluation (left), Kernel-DB and MSE^{Φ} indexes (right)

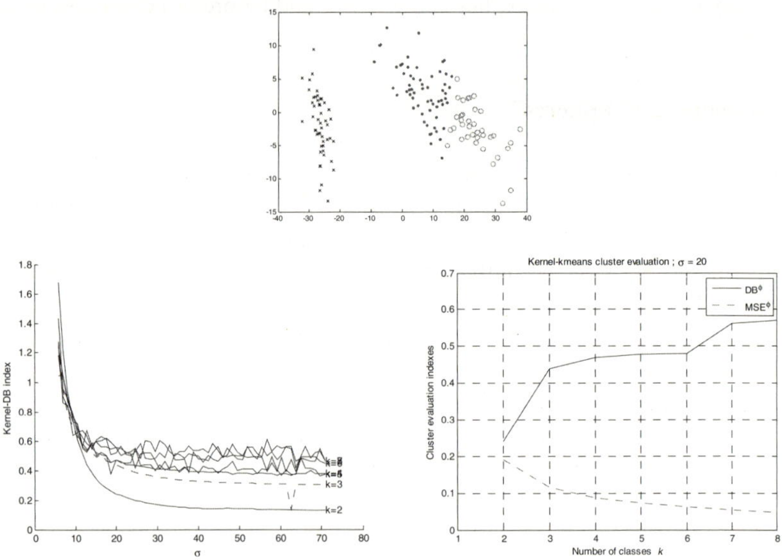

Fig. 3. "Fisher's iris" σ=20 classification plot on the 2 dimensional PCA (up), Effect of the parameters on the clustering evaluation (left), Kernel-DB and MSE^{Φ} indexes (right)

4.3 Example 3: Fisher's Iris

Iris dataset is formed with 150 points in R^4 that belong to 3 different classes [13].

This case is more difficult: the dot line "k=3" (Fig. 3, left) does not allow detecting 3 as the right number of classes, nor the kernel-DB function with $\sigma=20$ (Fig. 3, right). The assessed value seems rather to be 2, although 3 looks like the second most plausible value. Iris is indeed known as a complex case, because of the proximity between two classes (Fig. 3, up).

5 Conclusion

We presented an extension of the well-known Davies and Bouldin's index in the feature space using the kernel trick. We were motivated by the fact that most clustering algorithms need to evaluate their results, particularly in order to a posteriori estimate the number of clusters. The experiments on some small examples show that this kernel-DB index could be used as a helpful clustering evaluation tool for the kernel-clustering methods. The problem of conjointly estimating the whole set of parameters is clearly not achieved, but kernel-DB still gives good results when some ranges of parameters are beforehand reduced.

These results encourage the application of such kernel clustering evaluation indexes as tools to choose the kernel or to estimate its parameters as well as the number of classes. A future work would consist in making automatic the tuning of the parameters and in applying it on more interesting real datasets.

References

1. Girolami, M.: Mercer kernel-based clustering in feature space. IEEE Trans. on Neural Networks 13(3), 780–784 (2002)
2. Schölkopf, B., Smola, A.J.: Learning with Kernels: Support Vector Machines, Regularization, Optimization and Beyond, Massachusetts, London, England. MIT Press, Cambridge (2002)
3. Jenssen, R., et al.: Some Equivalences Between Kernel Methods and Information Theoretic Methods. Journal of VLSI Signal Processing 45, 49–65 (2006)
4. Christianini, N., Shawe-Taylore, J., Kandola, J.: Spectral kernel methods for clustering. Neural Information Processing Systems 14 (2002)
5. Su, M.-C., Chang, H.-T.: A new model of self-organizing neural networks and its application in data projection. IEEE trans. on Neural Network 12(1), 153–158 (2001)
6. Jain, A.K., Dubes, R.C.: Algorithms for Clustering Data. Prentice Hall, Englewood Cliffs (1988)
7. MacQueen, J.B.: Some Methods for classification and Analysis of Multivariate Observation. In: Proceedings of 5-th Berkeley Symposium on Mathematical Statistics and Probability, vol. 1, pp. 281–297. University of California Press, Berkeley (1967)
8. Shawe-Taylor, J., Cristianini, N.: Kernel Methods for Pattern Analysis. Cambridge University Press, Cambridge (2004)
9. Roberts, S.J., Everson, R., Rezek, I.: Maximum Certainty Data Partitioning. Pattern Recognition 33(5), 833–839 (2000)

10. Mao, J., Jain, A.K.: Artificial neural networks for features extraction and multivariate data projection. IEEE Trans. Neural Networks 6(2), 296–317 (1995)
11. Davies, D.L., Bouldin, D.W.: A cluster separation measure. IEEE Trans. Pattern Analysis Machine Intelligence 1(4), 224–227 (1979)
12. Nasser, A., Hamad, D., Nasr, C.: Kernel PCA as a Visualization Tools for Clusters Identifications. In: Kollias, S., Stafylopatis, A., Duch, W., Oja, E. (eds.) ICANN 2006. LNCS, vol. 4132, pp. 321–329. Springer, Heidelberg (2006)
13. Fisher, R.A.: The Use of Multiple Measurements in Taxonomic Problems. Annals of Eugenics 7, 179–188 (1936)

A Topology-Independent Similarity Measure for High-Dimensional Feature Spaces

Jochen Kerdels[1] and Gabriele Peters[2]

[1] DFKI - German Research Center for Artificial Intelligence, Robotics Lab, Robert Hooke Str. 5, D-28359 Bremen, Germany
[2] University of Dortmund, Department of Computer Science, Computer Graphics, Otto-Hahn-Str. 16, D-44221 Dortmund, Germany

Abstract. In the field of computer vision feature matching in high dimensional feature spaces is a commonly used technique for object recognition. One major problem is to find an adequate similarity measure for the particular feature space, as there is usually only little knowledge about the structure of that space. As a possible solution to this problem this paper presents a method to obtain a similarity measure suitable for the task of feature matching without the need for structural information of the particular feature space. As the described similarity measure is based on the topology of the feature space and the topology is generated by a growing neural gas, no knowledge about the particular structure of the feature space is needed. In addition, the used neural gas quantizes the feature vectors and thus reduces the amount of data which has to be stored and retrieved for the purpose of object recognition.

1 Introduction

In the field of computer vision objects are often represented by feature vectors describing local areas of them (e.g., [1,2,3]). These local descriptors often are vectors of high-dimensional feature spaces. To identify equal or similar objects, for example for the purpose of object recognition, feature matching techniques are common, and for these matching techniques similarity measures for the feature vectors are needed. One major problem when choosing the similarity measure is often the lack of knowledge about the structure of the feature space. For example the features in the SIFT feature space as described by Lowe [1] are not uniformly distributed. Using the Euclidean distance – as Lowe does – leads to the problem, that the direct distance between two features cannot be used as an absolute measure of their similarity. Accordingly, Lowe uses a workaround for this problem with the drawback that it requires each object to have at least one unique (i.e., identifying) feature. This is a general problem of non-uniform feature spaces. Wrongly presumed uniformity can result in a classification of unsimilar features as similar and vice versa. Some approaches try to improve the matching of features in the non-uniform feature space by using dimensionality reduction techniques such as Principal Component Analyses (PCA) [4]. For example Ke and Sukthankar showed in [5] that using PCA can improve

J. Marques de Sá et al. (Eds.): ICANN 2007, Part II, LNCS 4669, pp. 331–340, 2007.

the matching of features in SIFT space. But they still rely on the Euclidian distance and the proposed workaround of Lowe. This paper describes how to obtain a similarity measure which is suitable for the task of feature matching without knowledge of the particular structure of the high-dimensional feature space. Using a growing neural gas as described by Fritzke in [6], the topology of the feature space is first learned and then used as a basis for the similarity measure. The similarity between two feature vectors will incorporate the length of the shortest path between those two nodes of the neural gas the feature vectors are mapped on. Besides the ability to adapt to non-uniformly distributed feature spaces the neural gas also quantizes the feature vectors. On the one hand, this can be accompanied by a possible loss of information. But on the other hand, it also vastly reduces the amount of data which has to be stored and retrieved for feature matching purposes. In section 2 we recall the functionality of growing neural gas on which our similarity measure will be based. The proposed measure is derived in section 3, after which the description of the experiments (section 4), the summary of the results (section 5), and a conclusion (section 6) follow.

2 Growing Neural Gas Revisited

The growing neural gas (GNG) used for the similarity measure is described in detail by Fritzke in [6]. The GNG is similar to the self organizing maps of Kohonen [7]. In contrast to the self organizing maps the GNG does not have a fixed number of nodes (often also called "neurons" or "units"). It is subject to a data driven growing process which ends when a halting criterion (e.g., a minimal quantization error or a maximum number of nodes) is complied with. Figure 1 depicts the growing process of a GNG on a non-uniformly distributed, two-dimensional feature space Ω. The set $\mathcal{S}$ of nodes of the GNG is initialized with two nodes. Both nodes are associated with different random vectors $w \in \Omega$, called *reference vectors*. That are random positions in the high-dimensional feature space Ω. In addition, every node $s \in \mathcal{S}$ has an accumulated error E_s initialized with 0. The edges of the GNG, which connect the nodes, have an attribute "age". During the growing process this attribute makes it possible to detect edges which are not needed any more and thus to delete them. At first, the set of edges $\mathcal{C}, \mathcal{C} \subseteq \mathcal{S} \times \mathcal{S}$, between the nodes is empty. A new feature vector $\xi \in \Omega$ is processed as follows: Those two nodes s_1 and s_2 the reference vectors w_{s_1} and w_{s_2} of which are closest (in terms of Euclidean distance) to the feature vector ξ are selected. If there is no connecting edge between s_1 and s_2 in $\mathcal{C}$, an edge between s_1 and s_2 is added to $\mathcal{C}$. By setting the age of the edge to 0 the edge (if it already existed) is refreshed. The accumulated error E_{s_1} of the nearest node s_1 is increased by the square of the distance between the feature vector ξ and the reference vector w_{s_1}. Next, the reference vector of s_1 and all reference vectors of the direct neighbors of s_1 are adapted. The age is increased by one for those edges the endpoints of which have been moved in the previous step. If the age of an edge reaches a threshold $a_{\max}$, the edge will be removed. When this leads to an isolated node it is also removed. The value of $a_{\max}$ defines the

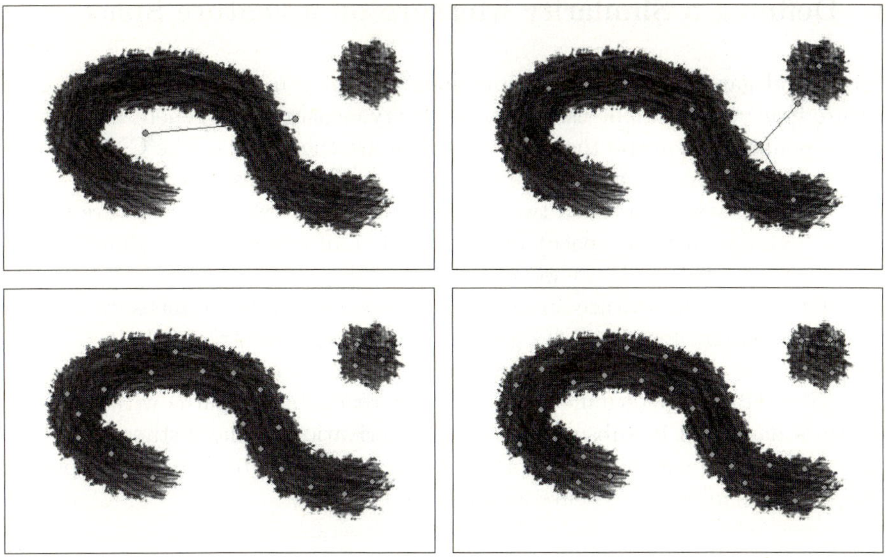

Fig. 1. A growing neural gas at different points in time. The dark shaded areas represent the values of a non-uniformly distributed, two-dimensional feature space Ω which should be characterized by the GNG.

stiffness of the generated topology. A small value leads to an unstable topology, a very high value leads to an only slow detachment of isolated areas in the feature space. A value of 100 for $a_{\max}$ turned out to be a good medium choice. After a number λ of input feature vectors (e.g., $\lambda = 300$) a new node r is added to the GNG if the halting criterion is not met yet. For this purpose two nodes of $\mathcal{S}$ are selected: first, the node q with the highest accumulated error E_q and secondly, that node f of all nodes adjacent to q that has the highest accumulated error E_f. The new node r is added to $\mathcal{S}$ and obtains a reference vector w_r which is the average between w_q and w_f. The accumulated error E_r of r is interpolated between the accumulated errors E_q and E_f, which were reduced by a fraction in a preceding step. Next, the set of edges $\mathcal{C}$ is extended by an edge between r and q and an edge between r and f. The edge between q and f is removed from $\mathcal{C}$. In a last step the accumulated errors of all nodes are reduced by a fraction β. This last step simulates a kind of aging on the accumulated errors, thus giving newer errors more weight and avoiding a build-up of small errors over time. With respect to the behaviour of the neural gas the parameter β influences how good the neural gas can adapt to fine structures in the feature space. Summarizing, the algorithm produces a graph with nodes explicitly linked to their closest neighbors. The graph is a subset of the Delaunay triangulation, a property we refer to later in subsection 3.3 as Delaunay property. Using the GNG to quantize vectors of high-dimensional feature spaces, the return value for an input feature vector $\xi \in \Omega$ could be the reference vector or just the the nearest node (which is the definition we will apply in subsection 3.3).

3 Defining a Similarity Measure on a Feature Space

The neural gas described in the previous section generates a topology of the feature space which can be used for a similarity measure. The generated topology is represented by a graph the nodes of which are the nodes of the GNG and the edges of which connect neighboring and thus similar nodes. Accordingly, we can describe the distance between two nodes (and later between two feature vectors of the high-dimensional space) by the number of edges on the shortest path between them. By doing so we utilize the ability of the neural gas to reflect the structure of the feature space. To develop our similarity measure we need a distance matrix for the GNG graph. This distance matrix is derived by the calculation of paths of length n, where n is the number of edges connecting two nodes. How to determine nodes that are reachable on paths with a distinct length is described in subsection 3.1. The derivation of the distance matrix is then described in subsection 3.2, after which we are able to define our topology-independent similarity measure in subsection 3.3.

3.1 Paths of Distinct Length

Once the growing neural gas has learned the topology of the feature space using N nodes, an $N \times N$ distance matrix D can be generated that contains for every node the shortest distance to all other nodes. The distance matrix D can be calculated using the fact that the adjacency matrix of a graph to the power of n codes for every node the subsequent nodes which are n edges away. This is explained in the following. First we give an example of a simple graph:

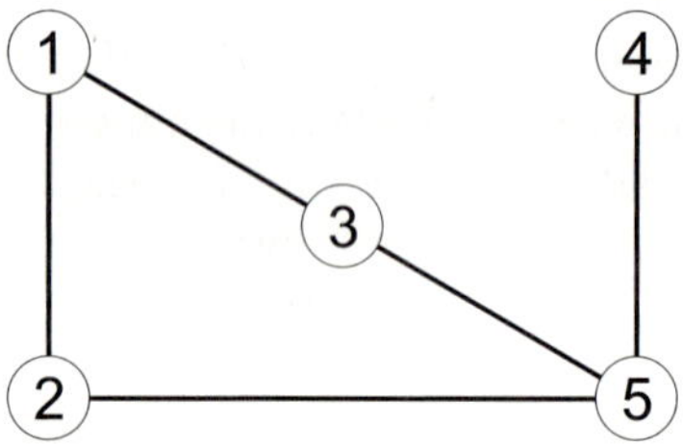

The adjacency matrix A of this graph describes in every column $\boldsymbol{a}_i$ the direct neighbors of node $i, i = 1, \ldots N$:

$$A = \begin{pmatrix} 0\,1\,1\,0\,0 \\ 1\,0\,0\,0\,1 \\ 1\,0\,0\,0\,1 \\ 0\,0\,0\,0\,1 \\ 0\,1\,1\,1\,0 \end{pmatrix}$$

The multiplication of a column vector $\boldsymbol{b} := (b_1, b_2, \ldots, b_N)^T$ and an $N \times N$ matrix M can be seen as a linear combination of the columns $\boldsymbol{m}_i$ of M:

$Mb = b_1 m_1 + b_2 m_2 + \ldots + b_N m_N$. The multiplication of two $N \times N$ matrices G and H can be done column by column as multiplication of columns h_i of H and matrix G:

$$GH = \begin{pmatrix} Gh_1 & Gh_2 & \ldots & Gh_N \end{pmatrix}.$$

Thus, the square $A' = AA$ of adjacency matrix A contains in every column a'_i a linear combination of the columns a_j of A:

$$A' = \begin{pmatrix} a'_1 & a'_2 & \ldots & a'_N \end{pmatrix} = \begin{pmatrix} Aa_1 & Aa_2 & \ldots & Aa_N \end{pmatrix}.$$

As every column a_i of adjacency matrix A describes the adjacent nodes of node i, every column a'_i of A' describes all adjacent nodes of the adjacent nodes of node i or in other words, it describes those nodes which are reachable from node i on paths of length 2. The values of the entries a'_{ij} of matrix A' describe, how many paths of length 2 between node i and node j exist. Accordingly, another multiplication of A' with A results in a matrix $A'' = A'A = AAA$ the columns a''_i of which describe the nodes that are reachable from node i on paths of length 3. In general, the adjacency matrix of a graph to the power of n codes the nodes that are connected via paths of length n.

3.2 Distance Matrix D

For the task of calculating the distance matrix D, i.e., the matrix that contains for every node the shortest distance to all other nodes, the precise values of the exponentiated adjacency matrix are not needed. The information whether or not there is a path of length n between two nodes is satisfactory. Thus it is sufficient to use boolean values 0 and 1 and to replace the addition by the disjunction and the multiplication by the conjunction when exponentiating the adjacency matrix. Then the distance matrix D is calculated as follows:

$$D = D_0 - \sum_{i=0}^{N-1} \bigvee_{j=0}^{i} A^j \qquad \mathit{with}$$

$$D_0 = \begin{pmatrix} N & \cdots & N \\ \vdots & \ddots & \vdots \\ N & \cdots & N \end{pmatrix}, \qquad A^0 = \begin{pmatrix} 1 & & 0 \\ & \ddots & \\ 0 & & 1 \end{pmatrix},$$

and $A^1 = A$ the adjacency matrix, $A^2 = A'$, $A^3 = A''$, etc. D is symmetric and its entries are either positive or zero, zero if and only if they are elements of the diagonal. The complete computation of distance matrix D requires N matrix multiplications. For each matrix multiplication N^2 matrix elements have to be computed, which requires N conjunctions and $N-1$ disjunctions for every element. Therefore the distance matrix D can be calculated in $\mathcal{O}\left(N^4\right)$. But the computing time can be further reduced to $\mathcal{O}\left(kN^3\right)$ if a maximum depth k for a path between two nodes is used instead of a complete computation of the distance matrix with depth N. As normally one is not interested in the similarity of features beyond a certain threshold, the exact distance between those very unsimilar features can be disregarded without posing too many restrictions to possible applications. Thus, for most applications at stake the computational complexity of $\mathcal{O}\left(N^3\right)$ of the proposed similarity measure compares to other all-pairs shortest path algorithms such as Floyd-Warshall [8].

3.3 Topology-Independent Similarity Measure *d*

Having defined the distance matrix D for the nodes of the GNG, we will now derive our topology-independent similarity measure d. (As we will define d as a pseudometric, we should properly speak about a *dissimilarity measure* rather than a *similarity measure*, but we will adhere to the more colloquial term.)

Let Ω be the high-dimensional feature space. Features ξ are represented as points in this metric space: $\xi \in \Omega$. Furthermore, let $\mathcal{S} := \{s_1, s_2, \ldots, s_N\}$ be the set of nodes of the GNG as introduced in section 2. The quantization operation induced by the growing neural gas is a mapping $Q : \Omega \to \mathcal{S}, Q(\xi) = s$, with s the node ξ is assigned to by the GNG. Now we first can define a metric $\widetilde{d}$ on $\mathcal{S}$:

$$\widetilde{d} : \mathcal{S} \times \mathcal{S} \to \mathrm{I\!R}, \qquad \widetilde{d}(s_i, s_j) := \mathtt{d}_{ij}$$

with $\mathtt{d}_{ij}$ the entries of the distance matrix D: $D = \left(\mathtt{d}_{ij}\right)_{i,j=1,\ldots,N}$. The metric axioms (i) non-negativity, (ii) identity of indiscernibles, (iii) symmetry, and (iv) triangle inequality obviously hold true for $\widetilde{d}$ because of the properties of D mentioned in subsection 3.2. Given $\widetilde{d}$, we can define a topology independent pseudometric d on Ω now:

$$d : \Omega \times \Omega \to \mathrm{I\!R}, \qquad d(\xi, \eta) := \widetilde{d}\left(Q(\xi), Q(\eta)\right).$$

As different feature vectors can be mapped on the same node of the GNG the second metric axiom is not necessarily fulfilled. This means d is a pseudometric only, i.e., only the following axioms hold true:

(i) $d(\xi, \eta) \geq 0$,
(iii) $d(\xi, \eta) = d(\eta, \xi)$,
(iv) $d(\xi, \eta) \leq d(\xi, \rho) + d(\rho, \eta)$.

Properties (i) and (iii) are induced by the corresponding properties of $\widetilde{d}$. Property (iv) holds true because of the Delaunay property of the growing neural gas mentioned in section 2. The concrete values of d depend on the granularity of the similarity measure. This granularity is determined by the halting criterion of the GNG. Thus, the precision of the similarity measure can be controlled by adjusting the halting criterion. Figure 2 shows an example of our similarity measure in a schematical way.

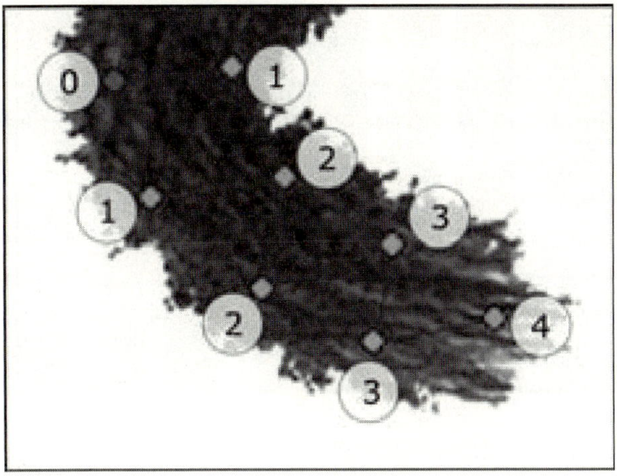

Fig. 2. Topology-independent similarity measure. The gray areas represent the feature vectors ξ of the high-dimensional space Ω. Some nodes of the growing neural gas, i.e., some elements of set $\mathcal{S}$, are depicted by red and green dots. The numbers they are labeled with are those values of the metric $\widetilde{d}$ which express the distance between the red node and each of the green nodes. For example, the shortest distance between the red node and the upper green node labeled with "2" is a path of 2 edges.

4 Experiments

We carried out our experiments on a database of 798 gray value images, a few of which are shown in figure 3. As features we consider patches of 18×18 pixels, thus our high-dimensional feature space Ω has 324 dimensions. These features are not optimal descriptors for the purpose of object recognition. Nevertheless, we chose them for the evaluation of the similarity measure because they can be evaluated more easily by visual inspection than more advanced feature descriptors such as the SIFT vectors, for which a visual interpretation is much harder. Per sample image we extracted about 250 feature vectors, the positions of which have

Fig. 3. Database. A selection of 798 sample images on which we carried out our experiments.

Fig. 4. Quantization of feature vectors by the growing neural gas. Each column stands for one node of the GNG, e.g., the first column represents $s_1 \in \mathcal{S}$, the second column $s_2 \in \mathcal{S}$, etc. Each entry (i.e., row) of a column shows one feature vector $\xi \in \Omega$ in form of a 18×18 gray value patch, which has been assigned to this node. In each case a column shows the last 10 feature vectors which have been mapped onto it. The features marked by a green frame are those which are the last assigned.

been determined with a KLT detector [9], resulting in a total of about 200,000 features. We ran the growing neural gas algorithm as described in section 2. After the processing of $\lambda = 300$ feature vectors we added one node to the GNG and stopped the growing procedure after it consisted of 300 neurons. (All of the 200,000 features have been used for the generation of the GNG according to the algorithm described in section 2. After 300 nodes have been incorporated into the gas not many changes of the topolopy of the GNG were caused by the remaining features. Thus, those remaining features contributed to the stabilization of the GNG only, rather than to its overall topoloy.) Figure 4 shows exemplarily how some feature vectors have been quantized.

5 Results

The final purpose of defining feature vectors and endowing their space with a suitable similarity measure, is the adequate encoding of the characteristics we intend to measure. In this case the application is an encoding of the visual characteristics of objects for purposes such as storage, classification, or recognition. Therefore, we have to analyze whether the features classified as being (mathematically) similar are also assessed by humans as being (visually) similar. In other words, the similarity (or difference) we determine by the proposed method must bear some correlation with the perceptual similarity (or difference) of two feature vectors. As Santini and Jain point out in [10], if our systems have to respond in an "intelligent" manner, they must use a similarity model resembling the perceptual similarity model of humans. Having these considerations in mind, we decided to assess the quality of the similarity measure d by a visual inspection

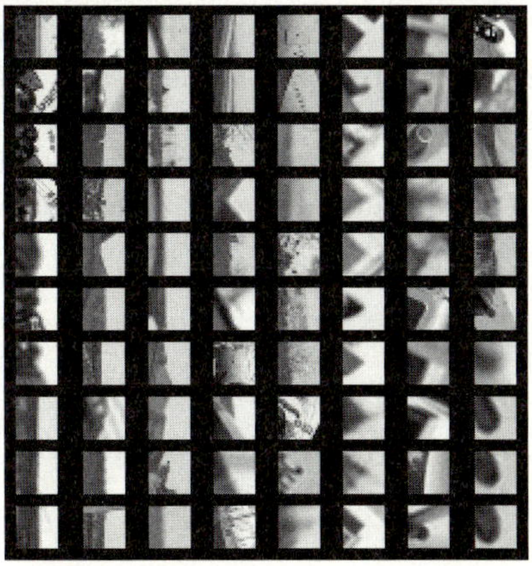

Fig. 5. Classification of feature vectors into 8 neighboring nodes of the growing neural gas. The columns represent the nodes. The entries of the columns are the last 10 feature vectors which were mapped onto them. The distance between the nodes is equivalent to the number of columns between them.

of the feature classification. Figure 5 shows again nodes of the growing neural gas, represented by 8 columns of the 10 last assigned feature vectors each. This time neighboring columns show *adjacent* nodes of the GNG, thus the number of columns between two nodes in the diagram is proportional to the distance between the nodes in the GNG. For example, the node represented by the first column and the node represented by the last column have a distance of 7 edges. We can summarize the results of the visual inspection as follows: Firstly, the similarity between features belonging to one node (i.e., features within one column) is, in general, larger than between features of different nodes. Secondly, one can observe a gradual decrease in similarity from the left to the right node for most of their assigned features. For example, the second column displays a larger overall similarity to the first column than the last column. Summarizing, one can say that the classification of features emerged from the proposed similarity measure corresponds to the assessment of the perceptual similarity by humans. Object recognition experiments remain to be done.

6 Conclusion

We considered the problem of endowing a feature space with an adequate similarity measure. Often researchers make unwarranted assumptions about the metric of the space. Usually it is assumed to be Euclidean. In this paper we presented a similarity measure for high-dimensional feature vectors which is independent from the actual structure of the feature space in the sense that no a priori knowledge on the topology of the feature space is necessary. The similarity measure is based on the advantageous distribution of the nodes in a growing neual gas. In addition, the use of a growing neural gas provides a quantization of the high-dimensional feature vectors. Despite a possible loss of information, this reduces the amount of data which has to be stored and searched for in further processing. The described similarity measure is particularly useful for object recognition tasks where an object is represented by a set of feature vectors, as it seems to correspond to human perceptual similarity assessment.

Acknowledgements

This research was funded by the German Research Association (DFG) under Grant PE 887/3-1.

References

1. Lowe, D.: Distinctive Image Features from Scale-Invariant Keypoints. International Journal of Computer Vision 60(2), 91–110 (2004)
2. Schmid, C., Mohr, R.: Local Grayvalue Invariants for Image Retrieval. IEEE Trans. on Pattern Analysis and Machine Intelligence 19(5), 530–535 (1997)
3. Kadir, T., Brady, M.: Saliency, Scale and Image Description. Int. J. of Computer Vision 45(2), 83–105 (2001)
4. Kessler, W.: Multivariate Datenanalyse. Wiley-VCH, Weinheim (2006)
5. Ke, Y., Sukthankar, R.: PCA-SIFT: A More Distinctive Representation for Local Image Descriptors. In: Proc. IEEE Conf. Comp. Vis. and Pat. Rec (CVPR) (2004)
6. Fritzke, B.: A Growing Neural Gas Network Learns Topologies. In: Advances in Neural Information Processing Systems, 7th edn., pp. 625–632. MIT Press, Redmond, Washington (1995)
7. Kohonen, T.: Self-Organizing Maps. Springer Series in Information Sciences, vol. 30. Springer, Berlin, Heidelberg, New York (2001)
8. Cormen, T.H., Leiserson, C.E., Rivest, R.L.: Introduction to Algorithms, pp. 558–565. MIT Press and McGraw-Hill, Cambridge (1990)
9. Tomasi, C., Kanade, T.: Detection and Tracking of Point Features. Carnegie Mellon University Technical Report CMU-CS-91-132 (1991)
10. Santini, S., Jain, R.: Similarity Measures. IEEE Transactions on Pattern Analysis and Machine Intelligence 21(9), 871–883 (1999)

Fuzzy Labeled Self-organizing Map with Kernel-Based Topographic Map Formation

Iván Machón González and Hilario López García

Universidad de Oviedo, Escuela Politécnica Superior de Ingeniería, Departamento de Ingeniería Eléctrica, Electrónica de Computadores y Sistemas, Edificio Departamental 2, Zona Oeste, Campus de Viesques s/n, 33204 Gijón (Asturies), Spain

Abstract. Fuzzy Labeled Self-Organizing Map is a semisupervised learning that allows the prototype vectors to be updated taking into account information related to the clusters of the data set. In this paper, this algorithm is extended to update individually the kernel radii according to Van Hulle's approach. A significant reduction of the mean quantization error of the numerical prototype vectors is expected.

1 Introduction

The main objective of this paper is to carry out a comparison of a Fuzzy Labeled Self-Organizing Map algorithm (FLSOM) with its modified version in terms of mean quantization and topographic errors. In this proposed version of FLSOM, the kernel radii of the neurons are individually updated, maximizing the mutual information between the input and output of the neuron and minimizing the mutual information between the outputs of the neurons. A significant reduction of the mean quantization error is expected due to this reform.

The paper contains brief descriptions of individual kernel radii updating in section 2 and FLSOM algorithm in section 3. The proposed modification of FLSOM is presented and tested in sections 4 and 5.

This work is part of the SensorControl project "Sensor-based online control of pickling lines", which is sponsored by ECSC whose agreement number is RFS-CR-04052. The contractors are Betriebsforschungsinstitut GmbH (BFI), Rasselstein GmbH (TKS-RA), Centro Sviluppo Materialia S.p.A. (CSM), Universidad de Oviedo (UniOvi) and Svenska Miljinstitutet AB(IVL). The main objective of the SensorControl project is the development of a sensor for acidic measurement and supervision techniques of coil surface defects for implementation in steel pickling plants.

2 Kernel-Based Topographic Map Formation

The SOM algorithm [6][7] has a kernel function centered on the winning neuron that is usually considered as a Gaussian. In this way, a neighborhood radius is created influencing to a high degree the neighbor neurons of the winning unit while the values of the prototype vectors are updated during the training. This

J. Marques de Sá et al. (Eds.): ICANN 2007, Part II, LNCS 4669, pp. 341–348, 2007.

kernel function is responsible for the SOM property of topology preservation. However, the quantization error can be improved modifying this kernel function by means of maximizing the mutual information between the input and output of the neuron and minimizing the correlations between the outputs of the neurons [11]. This can be achieved maximizing the differential entropy [2] whenever the kernel output distribution is uniform.

In [11] Van Hulle obtains the distribution of the squared Euclidean distances of the data vectors to the mean of the Gaussian kernel and its output is defined according to the cumulative of that distribution as an incomplete gamma distribution. The learning rules appear in equations (1) and (2). They are obtained from the entropy of the kernel output by means of derivation with respect to the prototype vector $\mathrm{w_i}$ and the kernel radius σ_i.

$$\Delta \mathrm{w_i} = \eta_\mathrm{w} \, \Lambda(i, i^*, \sigma_\Lambda) \, \frac{\mathrm{v} - \mathrm{w_i}}{\sigma_\mathrm{i}^2} \tag{1}$$

$$\Delta \sigma_\mathrm{i} = \eta_\sigma \, \Lambda(i, i^*, \sigma_\Lambda) \, \frac{1}{\sigma_\mathrm{i}} \left(\frac{\| \mathrm{v} - \mathrm{w_i} \|^2}{d \, \sigma_\mathrm{i}^2} - 1 \right), \qquad \forall i, \tag{2}$$

$$\Lambda(i, i^*, \sigma_\Lambda) = \exp\left(- \frac{\| \mathrm{r}_i - \mathrm{r}_{i^*} \|^2}{2\sigma_\Lambda^2} \right), \tag{3}$$

$$\sigma_\Lambda(t) = \sigma_{\Lambda 0} \, \exp\left(-2 \, \sigma_{\Lambda 0} \, \frac{t}{t_{\mathrm{max}}} \right), \tag{4}$$

where v is the input data vector, η_w is the learning rate for prototype vectors, η_σ is the learning rate for kernel radii, d is the number of dimensions in the input data space, t is the time step, t_{max} is the maximum number of time steps, r_i and r_{i^*} are the lattice coordinates of the updated neuron i and the winning neuron i^*, respectively, Λ is a monotonous decreasing function of the lattice distance from the winner (in this case is a Gaussian), σ_Λ and $\sigma_{\Lambda 0}$ are the neighborhood range, which is used as a neighborhood cooling term and its initial value, respectively. Term (3) is added to allow competitive-cooperative learning. The winning neuron and its neighborhood determine the values during the training giving the topological information. Moreover, the values of the kernel radii are updated individually using this term avoiding the statistical dependency between the outputs of the neurons.

3 FLSOM Algorithm

Using SOM for classification tasks is an important feature to identify the clusters of the input data space, their relationships between each other and with the process variables. A classical method of classification after training consists of carrying out an assignment of each prototype vector to a certain cluster obtaining a new component plane. In this case, the clustering process is composed of two phases [13]. In the first phase, the SOM network is trained whereas in the second phase a partitive clustering algorithm, e.g. K-means, is applied to obtain several clustering structures of several numbers of clusters since the optimum number of clusters is unknown. The optimum clustering structure is chosen by means of a clustering validation index. This procedure is useful when the number of clusters and the classification (fuzzy or crisp distribution) of each data vector are not known beforehand [8][9]. However, these methods have the drawback of not influencing the training since they happen after it and don't modify the values of the prototype vectors.

This can be corrected using FLSOM [14]. In this algorithm, the classification task influences the values of the prototype vectors and both of them take place at the same time during the training. In this way, FLSOM can be considered as a semisupervised algorithm. The training algorithm is based on an energy cost function of the SOM (E_{SOM}) proposed by Heskes in [4]. The cost function of equation (5) includes a term (E_{FL}) that represents the labeling or classification error of the prototype vectors $\mathrm{y_i}$ of the classification map with regard to the probabilistic vectors $\mathbf{x}$ supplied by the training data set.

$$E_{\mathrm{FLSOM}} = (1-\beta)\, E_{\mathrm{SOM}} + \beta\, E_{\mathrm{FL}}, \tag{5}$$

Each labeling data vector $\mathbf{x}$, which is associated with a numerical data vector $\mathbf{v}$, is assigned in a fuzzy way according to a probabilistic membership to the clusters of the data set. This stage is critical in the data preprocessing, but obviously the algorithm can be applied to crisp distributions. A Gaussian kernel in the input data space is included within term E_{FL} so that the prototype vectors $\mathbf{w}$ close to the data vectors $\mathbf{v}$ determine the classification task. It is formulated in equation (6).

$$g_\gamma(\mathbf{v}, \mathrm{w_i}) = \exp\left(-\frac{\| \mathrm{v} - \mathrm{w_i} \|^2}{2\gamma^2}\right), \tag{6}$$

The learning rules are obtained by means of derivation of energy cost function (5) with respect to the numerical and labeling prototype vectors, $\mathrm{w_i}$ and $\mathrm{y_i}$, respectively. These learning rules are expressed in equations (7) and (8) considering using the squared Euclidean metric in the algorithm.

$$\Delta \mathrm{y_i} = \alpha_l\, \beta\, g_\gamma(\mathrm{v}, \mathrm{w_i})\, (\mathrm{x} - \mathrm{y_i}), \tag{7}$$

$$\Delta \mathrm{w_i} = \alpha\,(1-\beta)\exp\left(-\frac{\| \mathrm{r}_i - \mathrm{r}_{i^*} \|^2}{2\sigma(t)^2}\right)(\mathrm{v} - \mathrm{w_i}) + \alpha_{\mathrm{w}}\, \beta\, \frac{1}{4\gamma^2}\,(\mathrm{v} - \mathrm{w_i})\ \| \mathrm{x} - \mathrm{y_i} \|^2 \tag{8}$$

4 Modification of the Algorithm

As stated above, FLSOM is formulated according to energy cost function (5). First term E_{SOM} corresponds to the typical approximation of the numerical prototype vectors $\mathbf{w}$ to the input data vectors $\mathbf{v}$. This prototype update is carried out by the gradient of E_{SOM} or $\frac{\partial E_{\mathrm{SOM}}}{\partial \mathrm{w_i}}$ that turns out the first term of learning rule (8) considering the squared Euclidean as metric to be used. It can be observed that this term corresponds to the basic sequential version of the SOM training algorithm where the numeric prototype vectors $\mathrm{w_i}$ are updated. In that version, the kernel radii take the same value instead of being individually updated as in the kernel-based topographic map formation proposed in [11]. This individual updating should imply an improvement of the mean quantization error of the trained map. In this way, the original version of the FLSOM has been extended to this individual kernel radius updating obtaining learning rule (9) to update the numerical prototype vectors $\mathrm{w_i}$.

$$\Delta \mathrm{w_i} = \alpha_{\mathrm{w}} (1 - \beta) \Lambda(i, i^*, \sigma_\Lambda) \frac{\mathrm{v} - \mathrm{w_i}}{\sigma_{\mathrm{i}}^2} + \alpha_{\mathrm{w}} \beta \frac{1}{4\gamma^2} (\mathrm{v} - \mathrm{w_i}) \parallel \mathrm{x} - \mathrm{y_i} \parallel^2 \quad (9)$$

The algorithm is completed with learning rules (2) and (7) to update kernel radius σ_{i} and labeling prototype vector $\mathrm{y_i}$, respectively.

Therefore, there are two mean quantization errors. The first is quantization error e_{qw} related to the approximation of numerical prototype vectors $\mathrm{w_i}$ to numerical data vectors $\mathbf{v}$ and the other is quantization error e_{ql} related to the approach of labeling prototype vectors $\mathrm{y_i}$ to labeling data vectors $\mathbf{x}$. Both errors can be weighted by parameter β of the cost function.

Two maps are obtained after training with this algorithm. One map contains the component planes that correspond to the numerical variables and it is related to topographic error e_{tw}. The other map represents the component planes of the clusters defined in the data set and it is associated with topographic error e_{tl}.

Several advantages are obtained by means of the use of this algorithm. The visualization of the clusters improves the data understanding. Moreover, the algorithm is a robust classifier since it uses fuzzy data sets allowing the use of contradictory data.

5 Experimental Testing

Four data sets were used to check the algorithm performance according to the mean quantization error and the topographic error. This topographic error measures the percentage of map units whose two BMUs are not adjacent in the output space [5]. The map size must be equal for both algorithms since the errors depend on it. Parameter β was chosen equal to 0.5 for both of them taking an intermediate weight of the quantization errors e_{qw} and e_{ql}.

In this case, the algorithms are being evaluated on the same data that is used for training. This is tolerable for errors e_{qw} and e_{tw} because both of them measure a kind of distortion which should basically be the same for both training and test data. Since the aim of the paper is to improve e_{qw}, that option has been chosen.

In FLSOM algorithm the radii $\sigma(t)$ decrease monotonically and linearly. The advisable minimum is equal to 1, since any value below produces maps which do not preserve the data topology. The training was carried out in two phases according to [12]: first with large neighborhood radius in phase 1, and then finetuning with small radius in phase 2.

Parameter γ represents the kernel radius in the input space and it was considered as a constant since it is not described in detail in [14]. It affects both quantization and topographic errors. Its value was chosen equal to 0.5 because it seems the most appropriate for a normalized input data distribution with zero mean and unitary variance.

Iris Data Set. The iris data set was extracted from UCI repository of machine learning databases [10]. It was considered as a crisp distribution and is composed of 150 instances, 4 numerical variables and 3 clusters. The trained map size was a lattice of 8 x 8. Regarding FLSOM, the number of epochs was equal to 150 for both training phases. The kernel radius $\sigma(t)$ was within interval [2,1] in phase 1 and it was equal to 1 in phase 2. The parameters for this algorithm were $\beta = 0.5$, $\alpha = 0.01$, $\alpha_l = 0.1$, $\gamma = 0.5$. The number of epochs was equal to 150 in relation to the proposed algorithm and its parameters were $\beta = 0.5$, $\alpha_w = 0.001$, $\alpha_l = 0.01$, $\gamma = 0.5$, $\eta_\sigma = 0.0001 \cdot \alpha_w$, $\sigma_{\Lambda 0} = 4$, t_{max}=101250.

Balance Scale Data Set. This data set was extracted from UCI repository of machine learning databases [10]. It was considered as a crisp distribution and it has 625 instances, 4 numerical variables and 3 clusters. The map size was 10 x 10. Regarding FLSOM, the number of epochs was equal to 30 for both training phases. The kernel radius $\sigma(t)$ was within interval [3,1] in phase 1 and it was equal to 1 in phase 2. The parameters for this algorithm were $\beta = 0.5$, α=0.01, $\alpha_l = 0.5$, $\gamma = 0.5$. The number of epochs was equal to 30 in relation to the proposed algorithm and its parameters were $\beta = 0.5$, $\alpha_w = 0.01$, $\alpha_l = 0.5$, $\gamma = 0.5$, $\eta_\sigma = 0.0001 \cdot \alpha_w$, $\sigma_{\Lambda 0} = 4$, t_{max}=84375.

Fuzzy Data Set. The fuzzy data set was formed as a fuzzy distribution of three clusters applying fuzzy c-means to a random data set composed of 200 samples of two dimensions. It was formulated in this way since it was difficult to find a correct fuzzy distribution in a known database. The trained map size was equal to 10 x 10. Regarding FLSOM, the number of epochs was equal to 150 for both training phases. The kernel radius $\sigma(t)$ was within interval [3,1] in phase 1 and it was equal to 1 in phase 2. The parameters for this algorithm were $\beta = 0.5$, $\alpha = 0.01$, $\alpha_l = 0.01$, $\gamma = 0.5$. The number of epochs was equal to 150 in relation to the proposed algorithm and its parameters were $\beta = 0.5$, $\alpha_w = 0.001$, $\alpha_l = 0.01$, $\gamma = 0.5$, $\eta_\sigma = 0.0001 \cdot \alpha_w$, $\sigma_{\Lambda 0} = 4$, t_{max}=135000.

SensorControl Data Set. The data from the SensorControl project was considered as a crisp distribution and is composed of 215 instances, 5 process variables and 3 clusters. The lattice of the trained map was 11 x 11. Regarding FLSOM, the number of epochs was equal to 30 for both training phases. The kernel radius $\sigma(t)$ was within interval [3,1] in phase 1 and it was equal to 1 in phase 2. The parameters for this algorithm were $\beta = 0.5$, $\alpha = 0.5$, $\alpha_l = 0.5$, $\gamma = 0.5$. The number of epochs was equal to 30 in relation to the proposed algorithm and its parameters were $\beta = 0.5$, $\alpha_w = 0.01$, $\alpha_l = 0.5$, $\gamma = 0.5$, $\eta_\sigma = 0.0001 \cdot \alpha_w$, $\sigma_{\Lambda 0} = 4$, t_{max}=29025.

Table 1. Description of columns according to their number in Fig. 1

No.	Description
1	e_{qw} or mean quantization error of numerical prototype vectors $\mathbf{w_i}$ using FLSOM
2	e_{qw} or mean quantization error of numerical prototype vectors $\mathbf{w_i}$ using FLSOM & Kernel-based
3	e_{tw} or topographic error of numerical prototype vectors $\mathbf{w_i}$ using FLSOM
4	e_{tw} or topographic error of numerical prototype vectors $\mathbf{w_i}$ using FLSOM & Kernel-based
5	e_{ql} or mean quantization error of labelling prototype vectors $\mathbf{y_i}$ using FLSOM
6	e_{ql} or mean quantization error of labelling prototype vectors $\mathbf{y_i}$ using FLSOM & Kernel-based
7	e_{tl} or topographic error of labelling prototype vectors $\mathbf{y_i}$ using FLSOM
8	e_{tl} or topographic error of labelling prototype vectors $\mathbf{y_i}$ using FLSOM & Kernel-based

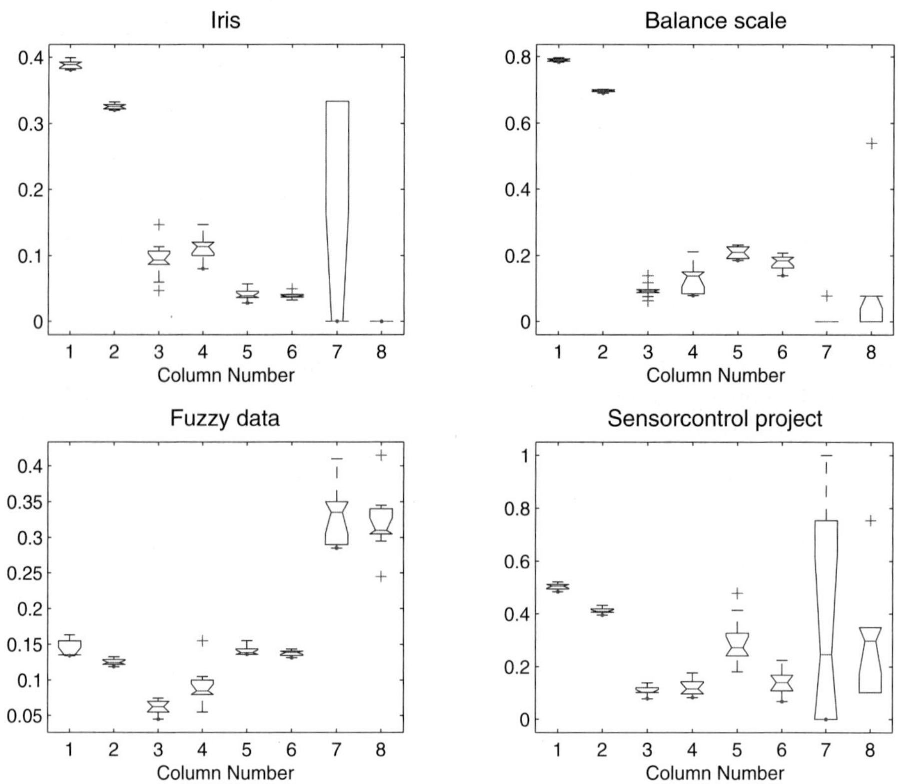

Fig. 1. Results for the four data sets

5.1 Results

In the proposed algorithm, the radii were initialized randomly from a uniform distribution [0, 0.1]. The weights or values of the prototype vectors ($\mathrm{w_i}$ and $\mathrm{y_i}$) were randomly initialized for both algorithms. Fifty maps were obtained for each algorithm and dataset. The errors mentioned in Table 1 were calculated for each map and their distributions are shown in Fig. 1. The boxes represent the typical distribution indicating the lower quartile, median, and upper quartile values. Notches estimate the medians for box to box comparison.

The results seem to prove that the mean quantization using the proposed algorithm is always less than the map trained using FLSOM, especially for e_{qw}. Also, the distribution variance of this type of error is quite narrow and less than obtained with FLSOM. The mean quantization error e_{ql} related to the classification task is slightly better than obtained with FLSOM. However, the topographic error of the numerical variables e_{tw} seems to be better in FLSOM. Although a similar value of e_{tw} could be achieved by the proposed algorithm since its variance allowed it to be obtained. It seems that the proposed algorithm reduces the mean quantization error at the expense of increasing its numbers of folds in the input space. However, the definition of topographic preservation depends on the chosen tool. Besides Kiviluoto's approach [5], other topology preservation indexes [1][3] carry out a more complete analysis, taking into account the neighborhood and the isometry that corresponds between the lattice neurons and the prototype vectors, but their computational costs are higher. Moreover, the original version of FLSOM uses twice the number of epochs.

6 Conclusion

FLSOM is a version of SOM algorithm where the prototype vectors are influenced by the labeling data vectors that define the clusters of the data set. This paper aims to improve the quantization error of this algorithm by means of individual kernel radius updating according to Van Hulle's approach, modifying the term that corresponds to numeric prototype vectors. The proposed algorithm was compared with the original one using four data sets to test both of them. The results seem to prove that the mean quantization error is reduced with the proposal of modification.

As future work, this proposed algorithm will be employed to train the data that belongs to the SensorControl project to obtain several models which will be selected with a suitable topographic error. This model will try to estimate the optimum line speed that minimizes the surface defects of the steel strip in the pickling line.

Acknowledgments

Our warmest thanks are expressed to the following for their financial support: The Commission of the European Communities, European Coal and Steel (ECSC), supporting SensorControl project "Sensor-based online control of pickling lines" with RFS-CR-04052 as agreement number.

References

1. Bauer, H.-U., Pawelzik, K.: Quantifying the neighbourhood preservation of self-organizing feature maps. IEEE Transactions on Neural Networks 3(4), 570–579 (1992)
2. Bell, A.J., Sejnowski, T.J.: An information-maximization approach to blind separation and blind deconvolution. Neural Computation 7, 1129–1159 (1995)
3. Bezdek, J.C., Pal, N.R.: An index of topology preservation for feature extraction. Pattern Recognition 28, 381–391 (1995)
4. Heskes, T.: Energy functions for self-organizing maps. In: Oja, E., Kaski, S. (eds.) Kohonen Maps, pp. 303–316. Elsevier, Amsterdam (1999)
5. Kiviluoto, K.: Topology preservation in self-organizing maps. In: Proceedings of the IEEE International Conference on Neural Networks, pp. 249–299 (1996)
6. Kohonen, T.: Self-organizing maps. Berlin: Springer, 3rd extended ed. 2001 (1995)
7. Kohonen, T., Oja, E., Simula, O., Visa, A., Kangas, J.: Engineering applications of the self-organizing map. Proceedings of the IEEE 84, 1358–1384 (1996)
8. López, H., Machón, I.: Self-organizing map and clustering for wastewater treatment monitoring. Engineering Applications of Artificial Intelligence 17(3), 215–225 (2004)
9. Machón, I., López, H.: End-point detection of the aerobic phase in a biological reactor using SOM and clustering algorithms. Engineering Applications of Artificial Intelligence 19(1), 19–28 (2006)
10. Newman, D.J., Hettich, S., Blake, C.L., Merz, C.J.: UCI repository of machine learning databases (1998)
11. Van Hulle, M.M.: Kernel-based topographic map formation achieved with an information-theoretic approach. Neural Networks 15, 1029–1039 (2002)
12. Vesanto, J., Alhoniemi, E., Himberg, J., Kiviluoto, K., Parviainen, J.: Self-Organizing Map for Data Mining in MATLAB: the SOM Toolbox. Simulation News Europe, 25–54 (1999)
13. Vesanto, J., Alhoniemi, E.: Clustering of the Self-Organizing Map. IEEE Transactions on Neural Networks 11(3), 586–600 (2000)
14. Villmann, T., Seiffert, U., Schleif, F.-M., Brss, C., Geweniger, T., Hammer, B.: Fuzzy Labeled Self-Organizing Map with Label-Ajusted Prototypes. Artificial neural Networks in Pattern Recognition 4087, 46–56 (2006)

Self-organizing Maps of Spiking Neurons with Reduced Precision of Correlated Firing

Francisco J. Veredas, Luis A. Martínez, and Héctor Mesa

Dpto. Lenguajes y Ciencias de la Computación,
Universidad de Málaga, Málaga 29071, Spain
fvn@lcc.uma.es

Abstract. Early studies on visual pathway circuitry demonstrated that synapses arrange to self-organize cortical orientation selectivity maps. It is still a debate how these maps are set up, so that diverse studies point to different directions to conclude about the main role played by feed-forward or intracortical recurrent connectivity. It is also a subject of discussion the way neurons communicate each other to transmit the information necessary to configure the circuits supporting the features of the central nervous system. Some studies claim for the necessity of a precise spike timing to provide effective neural codes. In this article we simulate networks consisting of three layers of integrate-and-fire neurons with feed-forward excitatory modifiable synapses that arrange to conform orientation selectivity maps. Features of receptive fields in these maps change when the precision of correlated firing decreases as an effect of increasing synaptic transmission jitters.

1 Introduction

Studies of cortical activity in primary visual cortex of cat demonstrated that cortical neurons were more effectively activated when a certain receptive field on the retina was stimulated. Many neurons in the cortex respond preferentially to bars of stimulation disposed on a concrete orientation on the visual space. Pioneering work of Hubel and Wiesel showed that the position of the receptive fields, the orientation selectivity and the ocular dominance present a global columnar organization in the cortex.

However, it is still a debate how cortical orientation selectivity is set up. While diverse physiological and theoretical studies support the role of feed-forward connectivity for orientation selectivity, other reputable works point to a larger influence of intracortical circuitry and inhibitory connections.

It is also an issue of controversy the way neurons communicate each other to conform the proper circuitry that robustly supports different features of the central nervous system. Much of the initial discussion focused on whether neurons use rate coding or temporal coding. While many works address the possibility of having a sparse neural code based on neural assemblies [2], synchronous fire chains [3], or oscillations [6], other studies claim for the existence of neural codes that require a precise inter-neural communication to be arranged [5].

J. Marques de Sá et al. (Eds.): ICANN 2007, Part II, LNCS 4669, pp. 349–358, 2007.

Across the visual pathway of mammals neural activity becomes more variable, the correlated activity among neurons becomes less precise and, furthermore, synaptic jitters (i.e. the variability of synaptic transmission latencies) increase. For example, in the developing rat neocortex, Markram *et al.* (1997) [4] measured fluctuations in excitatory post-synaptic potentials latency of 1.5 ms between layer 5 neurons which is a very large value in comparison with the sub-millisecond produced by retinogeniculate connections [1]. Increasing transmission jitters reduces significantly the precision of the correlated firing [9]. The question that arises now is how this reduction of the time accuracy of spike transmissions could affect to the organization of feature maps.

In this article we analyze the self-organization of receptive fields and orientation selectivity maps in feed-forward networks of integrate-and-fire (IAF) neurons with modifiable connections. Our results show how changes in the precision of the correlated firing among neurons do not prevent the emergence of orientation selectivity maps but affect the features of receptive fields. This means that reducing the exactitude of inter-neural communication does significantly influences the configuration of receptive fields in this sort of feed-forward networks. The way we modified this accuracy of the correlated activity was by introducing incremental physiologically plausible changes in the synaptic transmission jitters, i.e. increasing the variability of synaptic transmission delays.

We attack the problem by a spike-time dependent plasticity (STDP) approach, where modifiable excitatory synapses change during a learning process to converge towards a self-organizing structure that presents orientation sensitive receptive fields.

2 The Integrate-and-Fire Neuron Model

The traditional form of an IAF neuron consists of a first order differential equation (eq. 1) with a subthreshold integration domain [where the neuron integrates its inputs $I(t)$] and a threshold potential (not explicitly represented in the equations) for action potential generation [7].

$$C_m \frac{dV_m(t)}{dt} = I(t) - \frac{[V_m(t) - V_{rest}]}{R_m}, \quad (1)$$

where C_m is the neuronal membrane capacitance, V_m the membrane potential, R_m the membrane resistance, V_{rest} the resting potential and $I(t)$ the synaptic input current. For the purposes of this paper, $I(t)$ has been modeled as in eq. 2.

$$I(t) = \sum_j \omega_j g(\hat{t}_j)[E_{syn} - V_m(t)] + noise, \quad (2)$$

where ω_j represents the connection weight (or synaptic efficacy) between neuron j (presynaptic) and the current neuron (postsynaptic), $g(s)$ is the synaptic conductance, $\hat{t}_j$ is the time relative to the last spike of the presynaptic neuron j [also taking into account the existence of a synaptic delay (δ) and a jitter of transmission delay (ι), modelled as the mean and the standard deviation of a Gaussian distribution of synaptic latency, respectively], E_{syn} is the reversal potential or synaptic equilibrium and, finally, *noise* represents the background synaptic noise [following a Gaussian distribution with parameters (μ_n, σ_n)].

Synaptic conductance has been modeled (eq. 3) by the difference of two negative exponential functions with different time-constants, τ_r and τ_d, which affect the rise and decay times of the conductance curve, respectively.

$$g(t) = \hat{g}\left[e^{-1/\tau_r} - e^{-1/\tau_d}\right], \tag{3}$$

where $\hat{g}$ modulates the maximum synaptic conductance.

For the computational implementation of the IAF neuron model, equation 1 has been integrated by the *Backward Euler Integration Method*, and we chose a time resolution of $\Delta t = 0.1\,ms$ to ensure the stability of the system and to allow the introduction of all the physiological phenomena that could contribute to precise correlated firing and —as for transmission synaptic jitter— occur within a time order smaller than $1\,ms$.

To complete the neuron model, it was included an absolute refractory period (of $2.5\,ms$) and an after-hyperpolarization potential where the membrane potential decays to when an action potential occurs. This after-hyperpolarization potential is calculated as the 20% of the membrane potential once it exceeded the threshold potential: $V_m(t^f) = 1.2 \times V_m(t^f - \Delta t)$, where t^f is the spike time.

In computer simulations of the neuron model above, synaptic conductances have to be updated at each simulation step, which constitutes a crucial task from the point of view of computational efficiency. The basic problem consists of computing the overall conductance in a more efficient manner than the simple convolution of spike trains with conductance functions of all synapses at each simulation step.

Starting from conductance equation 3, the $\mathcal{Z}$-transformation of the discretization of $g(t)$ was used to get a recursive expression of $g[n]$ as a function of only a few previous terms. This recursive function allows us to state each value of the overall conductance at step n as a function of a reduced number of the previous values of the conductance and the presynaptic spike trains, which significantly increases the computational efficiency of the model [8].

3 Effects of Jitter on the Precision of Correlated Firing

As revealed by physiological and theoretical studies (see [9]), increasing the synaptic transmission jitters —defined as the variability of the synaptic delay— affects the precision of the correlated firing in a monosynaptic connection. When this correlated activity is measured by cross-correlation analysis of the activity of two monosynaptically connected neurons, the increasing of transmission jitters is evidenced by significant changes in the monosynaptic peak of the correlogram. For results in figure 1 two monosynaptically connected IAF neurons following the model above were simulated. The left graph of the figure shows the correlogram for a simulation with jitter of $\iota = 0.1\,ms$, while the jitter for the right

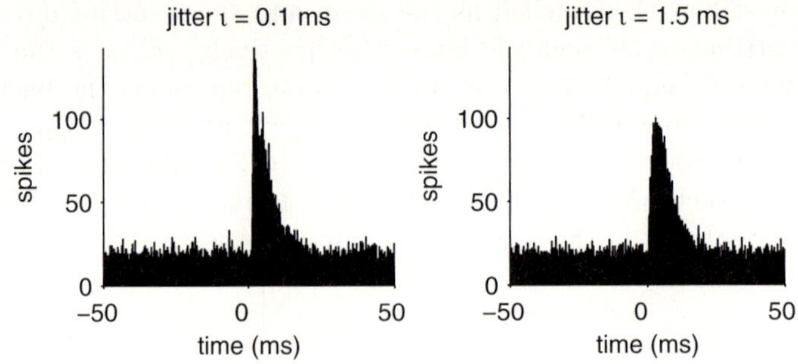

Fig. 1. Simulation of a monosynaptic connection with two different synaptic jitters. Left: jitter $\iota = 0.1\,ms$; position of maximum peak $= 2.1\,ms$; maximum peak amplitude $= 143\,spikes$; width at 50% of maximum peak amplitude $= 3.3\,ms$; witdth at 25% of maximum peak amplitude $= 7.4\,ms$. Right: jitter $\iota = 1.5\,ms$; position of maximum peak $= 3.1\,ms$; maximum peak amplitude $= 10\,spikes$; witdth at 50% of maximum peak amplitude $= 6.6\,ms$; witdth at 25% of maximum peak amplitude $= 10.6\,ms$.

graph of the figure was $1.5\,ms$. As can be visually appreciated, changes in the jitter affect the amplitude, position and width of the monosynaptic peak and, consequently, reveal changes in the precision of the correlated firing.

In this paper we present the results of simulation of feed-forward networks of excitatory IAF neurons with spike-timing plastic synapses and analyze the effects of changing synaptic jitters into the self-organization of receptive fields and their orientation selectivity.

4 STDP Learning Mechanisms

Modifiable synapses are modeled by a STDP rule (see equations 4 and 5). STDP rules model synaptic plasticity by considering into their formulation the precise timing of neuronal activity, so that synaptic efficacies are enhanced or reduced in a way that depends on the time distance of each post- and presynaptic spike. For the network architecture that we present in this paper, only excitatory synapses are considered, which are governed by equations 4 and 5.

$$W(t) = \begin{cases} 0 & if\ t < 0 \\ e^{-t/\tau_w} & if\ t \geq 0, \end{cases} \tag{4}$$

where τ_w is the learning time constant for synaptic modifications that depend on the distance between the post- and the presynaptic spikes (see figure 2). Synaptic depression by negative distances of post- and presynaptic spikes (left half of the learning window) has not been included in the model: weight reductions are only modulated by the occurrences of postsynaptic spikes (see equation 5).

$$\Delta\omega_j(t) = s(t)\left[\sum_{t_j^f<t} W\left(t - t_j^f\right) + \eta\right] + \zeta - \xi\omega_j(t), \tag{5}$$

where $s(t)$ is the activity of the postsynaptic neuron at the current time, with $s(t) = 1$ if the postsynaptic neuron fires, and $s(t) = 0$ if the postsynaptic neurons remains at rest; W is the STDP learning window (see figure 2); t_j^f it the time step f when the presynaptic neuron emitted an action potential; η is a learning rate constant that modulates the amplitude of incoming synaptic weights reductions due to the emission of each postsynaptic spike; ζ represents a sort of activity-independent synaptic enhancement: without activation, a synapse will slowly approach some non-zero efficacy; finally, ξ modulates the reduction of each synapse at a rate proportional to its current weight.

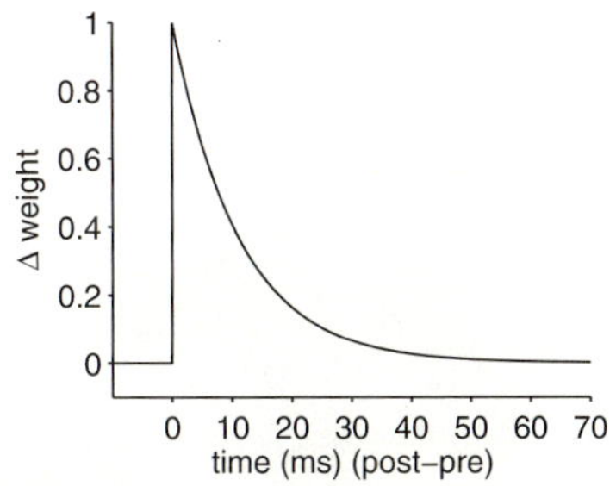

Fig. 2. Learning window $W(t)$ for excitatory synapses. The change in the synaptic weight depends on the difference between the post- and the presynaptic spikes.

5 Simulating Feed-Forward Networks with Different Jitters

The network architecture for simulations consists of three two-dimensional layers of 32×32 IAF neurons with feed-forward modifiable excitatory connections from input to output. There are not lateral connections. Receptive fields of neurons in intermediate and output layers consists of a 11×11 array of afferent synapses from the presynaptic layer, centered at the position of the postsynaptic neuron, assuming layers as circular matrixes for neighboring considerations. Synaptic weights are initialized as a function of the distance between each post- (i) and presynaptic (j) neuron: $w_{ij}(0) = \lambda e^{\sqrt{(x_i - x_j)^2 + (y_i - y_j)^2}/\sigma}$, where x and y are the row and column of the neuron position in the layer, and λ is a scale factor.

In table 1 default neuronal, synaptic and learning parameters are shown. Neurons in the input layer are stimulated by convolving two-dimensional Gaussian functions with a stationary noise signal that follows a Gaussian distribution [with parameters $\mathcal{N}(\mu_n, \sigma_n)$] (see figure 3). This convolution introduces an activity correlation in the network that allows the emergence of receptive fields and orientation selectivity maps in the absence of lateral connections, as it will be shown later.

Neurons in the intermediate and output layers receive, as a stimulation signal, a source of synaptic noise current which integrates with the presynaptic input

Table 1. Simulation parameters. A) Default physiological parameters for each IAF neuron. τ_m represents the membrane time-constant; V_{thres} is the threshold potential; ϑ is an after-hyperpolarization factor; ρ is an absolute refractory period. B) Synaptic and transmission parameters. C) Parameters for synaptic learning. (*i*, *ii* and *iii* represent input, intermediate and output layers, respectively).

A) Neuronal parameters

Δt	τ_m	C_m	V_{rest}	V_{thres}	ϑ	ρ
$0.1\,ms$	$1\,ms$	$0.1\,nF$	$-70\,mV$	$-40\,mV$	0.2	$2.5\,ms$

B) Synaptic parameters

δ	τ_r	τ_d	μ_n	σ_n
$1.0\,ms$	$4.0\,ms$	$1.0\,ms$	$2.7\,nA$	$0.5\,nA$

C) Learning parameters

τ_w^{i-ii}	τ_w^{ii-iii}	η	ζ	ξ
$11\,ms$	$5\,ms$	-0.57	$9.5 \cdot 10^{-4}$	$1.25 \cdot 10^{-6}$

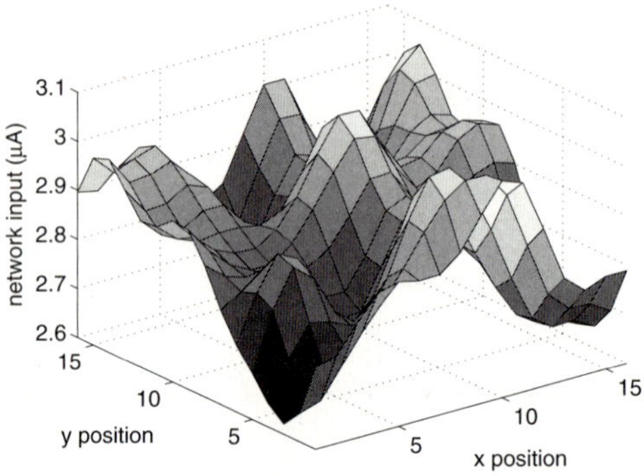

Fig. 3. Network input stimulation pattern: stationary signals following a Gaussian distribution $\mathcal{N}(2.7\,\mu A, 0.5\,\mu A)$ are convolved with a two-dimensional Gaussian function. X and Y axes represent position of neurons in the input layer.

from neurons in the preceding layer. Synaptic transmission delay has been modelled (see section 2) as a Gaussian distribution of synaptic latencies with a mean of $\delta = 0.1\,ms$ —the same for all simulations— and a standard deviation that represents synaptic jitter (ι).

In figure 4 results from one simulation of the feed-forward architecture above are shown. All synapses in the network have a same synaptic transmission jitter of $\iota = 0.1\,ms$. On the left panel of the figure, an orientation selectivity map for the output layer is represented which emerges from the self-organization of all synapses in the network, which are modifiable via the STDP rule of equations 4 and 5. On the center panel a smoothed version of the orientation selectivity map is shown. As can be observed at both panels, a typical arrangement of orientation selectivity bands and singularities emerges on the map. Finally, on the right panel of the figure, the receptive fields of the neurons in the output layer are visualized: a gray-level scale has been used to represent the synaptic weights (0: black; 1: white) at each 11×11 receptive field. As can also be visually appreciated at this panel, clusters of neurons with similar receptive fields additionally emerge with the self-organizing process.

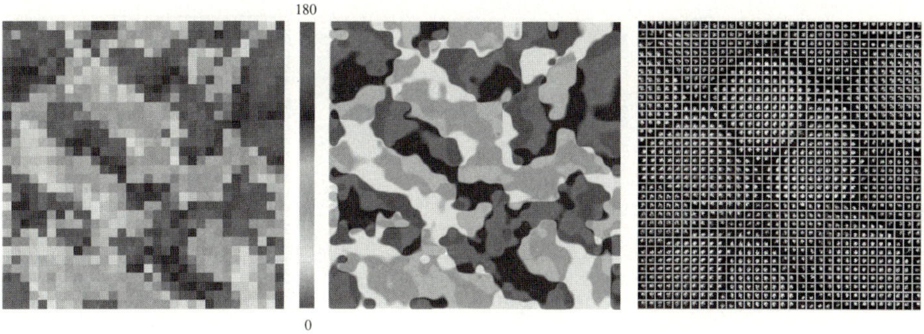

Fig. 4. Self-organization of feed-forward synapses from the intermediate layer to the output layer (simulation time = $50,000\,ms$). All synapses in the network had an uniform jitter of $\iota = 0.1\,ms$. Left panel shows the orientation selectivity map for the output layer: each colour square represents one neuron's orientation selectivity as it is indicated at the color scale bar on the right (in degrees). Center panel is a smoothed version of left panel. Right panel shows the receptive fields for the output layer: each gray-level square of $11 \times 11\,pixels$ represents one's cell incoming synaptic weights from 11×11 presynaptic neighboring neurons.

In figure 5 similar graphs as in figure 4 are shown but for one simulation with a larger synaptic jitter of $\iota = 1.5\,ms$ instead for the whole synaptic tree of the network. Moreover, in figure 6 results from a simulation of the network architecture with a non-uniform jitter are presented. For the simulation of this figure, jitter was modeled as a bi-dimensional inverse Gaussian function to have a minimun value of jitter at the central synapse at each receptive field in the network and also an increasing jitter when it is going from the center to the periphery of the receptive fields, as it has been formulated in the following equation:

$$J(x,y) \;=\; \Gamma - \varrho e^{-[([x-\chi/2]^2+[y-\Upsilon/2]^2)/\sigma]}, \tag{6}$$

where Γ is the maximum likely jitter, ϱ is a scale factor; x and y are the row and column of position of the presynaptic neuron, respectively; $\chi \times \Upsilon = 11 \times 11\,neurons$ is the size of the receptive field (matrix dimensions of incoming synaptic weights), and σ is the standard deviation of a Gaussian function.

At both figures (5 and 6) orientation selectivity maps and clusters of receptive fields still emerge, as for the smaller jitter values of figure 4, by synaptic self-organization, so that a significant reduction in the precision of the correlated firing introduced by increasing the jitters (see section 3) does not dramatically preclude the formation of receptive fields and orientation selectivity maps.

In table 2 we have gathered several measurements taken on the results from simulations of the network architecture above with different jitter arrangements. Receptive field clusters (i.e. grouping of receptive fields visually observable) have been identified on each panel (right panel of figures 4, 5 and 6) by image processing techniques based on region growing. A threshold level has been set to consider a receptive field as formed by those afferent synapses with an efficacy

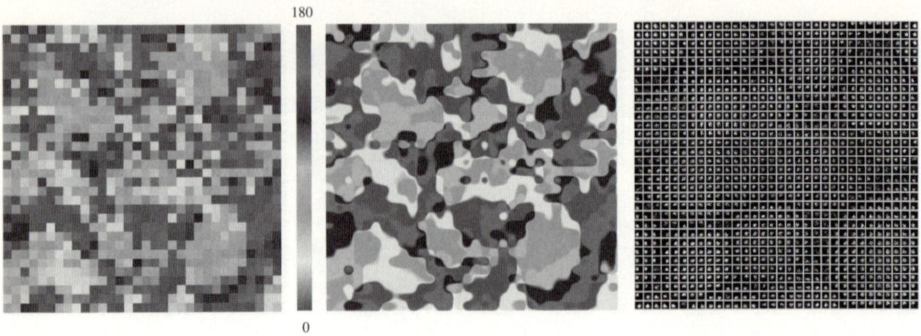

Fig. 5. Self-organization of synaptic feed-forward projections from the intermediate layer to the output layer, for a simulation of $50,000\,ms$. All synapses in the network had an uniform jitter of $\iota = 1.5\,ms$. See figure 4 for explanations and details.

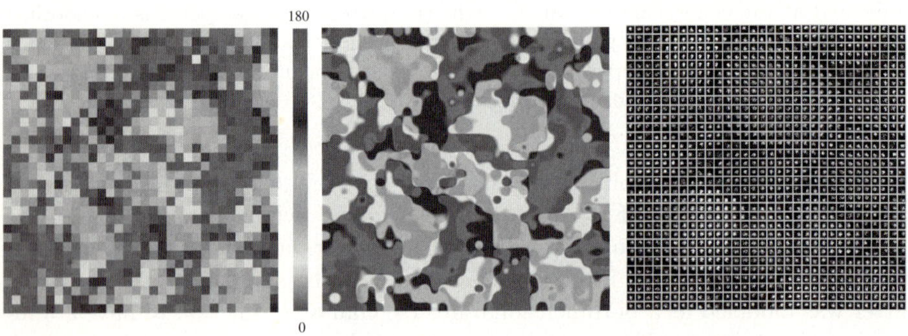

Fig. 6. Self-organization of synaptic feed-forward projections from the intermediate layer to the output layer, for a simulation of $50,000\,ms$. Jitter was modeled by an inverse Gaussian distribution with $\Gamma = 1.5\,ms$, $\varrho = 1.4\,ms$ and $\sigma = 9$ (see equation 6), giving a maximum likely jitter of $\iota =\approx 1.5\,ms$ at the periphery of each receptive field and a minimum jitter of $\iota = 0.1\,ms$ at its center. See figure 4 for details.

weight ≥ 0.5. Elongation measurements of clusters and receptive fields have been calculated as $1 - 4\pi \cdot area/perimeter^2$.

As revealed from the results shown in figures 4, 5 and 6, although increasing jitters (in an uniform or non-uniform manner) does not prevent the self-organization of receptive fields and orientation selectivity maps, significant differences can be found by closely inspecting the size and shape of the receptive fields and their clusters. Data from cluster measurements (see the upper table) throw a not very dissimilar number of clusters and neurons conforming them, but the high standard deviations blur the significance of those dissimilarities. On the other hand, receptive field measurements show significant differences in size and elongation of receptive fields: as uniform jitter increases, the size and the elongation of the receptive fields linearly decrease (with linear regression coefficients of $\gamma = 0.94$ and $\gamma = 0.85$, respectively). Moreover, as can be seen in table 2,

Table 2. Cluster and receptive field measurements from simulations of networks with different jitters. Each measurement was obtained as an average of three simulations with identical parameters (and $50,000\,ms$ of simulation time each). Four former columns show measures from simulations with an uniform jitter for the whole synapses in the net; two later columns show the results for synaptic jitters configured by an inverse Gaussian function ($\Gamma = 1.0\,ms$; $\varrho = 0.9\,ms$; see equation 6). Cluster elongation for all jitters (0: circle; 1: line): mean ≈ 0.4, std. dev. ≈ 0.1, median ≈ 0.4.

Cluster measurements (averaged for 3 simulations of $50,000\,ms$ each)

	$0.1\,ms$	$0.5\,ms$	$1.0\,ms$	$1.5\,ms$	$\sigma = 18$	$\sigma = 9$
#Clusters	12.7	15.3	14.3	16	13.6	14
Mean of #neurons	81	64	63.5	77.8	77.3	66.5
Std. dev. of #neurons	73.3	65.3	58.2	102.1	64.7	69.7
Median of #neurons	87	41.5	63.6	28	62.7	45.5

Receptive field measurements (averaged for 3 simulations of $50,000\,ms$ each)

	$0.1\,ms$	$0.5\,ms$	$1.0\,ms$	$1.5\,ms$	$\sigma = 18$	$\sigma = 9$
Mean of size (synapses)	19.6	17	15.3	12.5	14.8	14.3
Std. dev. of size (synapses)	4.2	3.9	4.5	4.1	4.3	4
Median of size (synapses)	19.7	17.3	15.6	13	15	14.3
Mean of elongation (0: circle; 1:line)	0.71	0.70	0.69	0.67	0.69	0.69
Std. dev. of elongation	0.16	0.16	0.16	0.17	0.17	0.17
Median of elongation	0.73	0.72	0.72	0.65	0.72	0.72

non-uniform jitter arrangements confirm this tendency that would seriously impede synaptic self-organization and formation of orientation selectivity maps for larger non-realistic jitter values. (Our results for uniform jitters larger than $1.5\,ms$ addressed to that direction, although they are not shown here).

Finally, we have not appreciated significant differences in statistics of the distributions of orientation selectivity in maps with different jitter arrangements for the physiological values of jitter managed in our simulations. Parameters of table 1 were empirically set to ensure convergence to significant network states. Changes in those parameters can drive the network to trivial or oscillating states.

6 Conclusion

In this paper we simulated feed-forward networks of IAF excitatory neurons with modifiable synapses —by a STDP learning rule— to study how the precision of the correlated activity affects the self-organization of receptive fields and orientation selectivity maps. For this purpose, we analyzed the effects of increasing the jitters of transmission delay to the final arrangement of neural receptive fields. Our main results show how by reducing the precision of the interneural correlated activity, the mean size and elongation of receptive fields at the output layer get linearly decreased.

If synaptic jitter values remain into realistic physiological ranges, these changes in the shape of receptive fields do not preclude the self-organization of orientation selectivity maps in this sort of excitatory feed-forward architecture. These results are in close agreement with works claiming for the unnecessary sharpness of time precision of interneural spike transmission for the conformation of some sort of emerging neural features as orientation selectivity. Moreover, although the increasing variability of neural responses along the visual pathway could affect the shape of receptive fields, as our results support, measured physiological magnitude of these variability seems not to be a determinant to prevent the emergence of self-organizing orientation selectivity maps.

References

1. Cleland, B., Dubin, M., Levick, W.: Sustained and transient neurones in the cat's retina and lateral geniculate nucleus. J. Pysiol. 217, 473–496 (1971)
2. Gerstein, G., Kirkland, K.: Neural assemblies: technical issues, analysis, and modeling. Neural Networks 14, 589–598 (2001)
3. Gewaltig, M.-O., Diesmann, M., Aertsen, A.: Propagation of cortical synfire activity: survival probability in single trials and stability in the mean. Neural Networks 14, 657–673 (2001)
4. Markram, H., Lubke, J., Frotscher, M., Roth, A., Sakmann, B.: Physiology and anatomy of synaptic connections between thick tufted pyramidal neurones in the developing rat neocortex. J. Physiol. 500, 409–440 (1997)
5. Reinagel, P., Reid, R.: Temporal coding of visual information in the thalamus. J. Neurosci. 20, 5392–5400 (2000)
6. Ritz, R., Sejnowski, T.J.: Synchronous oscillatory activity in sensory systems: new vistas on mechanisms. Curr. Opin. Neurobiol. 7, 536–546 (1997)
7. Stein, R.: The frequency of nerve action potentials generated by applied currents. Proc. Roy. Soc. Lond. B 167, 64–86 (1967)
8. Veredas, F., Mesa, H.: Optimized synaptic conductance model for integrate-and-fire neurons. In: 10th IASTED International Conference Artificial Intelligence and Soft Computing, Palma de Mallorca, Spain, pp. 97–102 (August 2006)
9. Veredas, F., Vico, F., Alonso, J.: Factors determining the precision of the correlated firing generated by a monosynaptic connection in the cat visual pathway. J. Physiol. 567(3), 1057–1078 (2005)

Visualising Class Distribution on Self-organising Maps

Rudolf Mayer, Taha Abdel Aziz, and Andreas Rauber

Institute for Software Technology and Interactive Systems
Vienna University of Technology
Favoritenstrasse 9-11/188, A-1040 Vienna, Austria
{mayer,rauber}@ifs.tuwien.ac.at, abdel_aziz_taha@hotmail.com

Abstract. The *Self-Organising Map* is a popular unsupervised neural network model which has been used successfully in various contexts for clustering data. Even though labelled data is not required for the training process, in many applications class labelling of some sort is available. A visualisation uncovering the distribution and arrangement of the classes over the map can help the user to gain a better understanding and analysis of the mapping created by the SOM, e.g. through comparing the results of the manual labelling and automatic arrangement. In this paper, we present such a visualisation technique, which smoothly colours a SOM according to the distribution and location of the given class labels. It allows the user to easier assess the quality of the manual labelling by highlighting outliers and border data close to different classes.

1 Introduction

The *Self-Organising Map* (SOM) [1] has been successfully used for clustering various kinds of data. It provides a topology-preserving mapping from a high-dimensional input space to a lower-dimensional, in most cases a two-dimensional, output space, which is an easily understandable representation. To reveal the cluster structures on the SOM, many visualisations techniques have been developed to help the user analyse and use the map.

In many applications, the data might already be assigned to classes, which can be utilised to compare the manual labelling with the automatic arrangement of the SOM. The distribution and arrangement of these classes on the map may reveal outliers, and 'conflicts' between the (manual) assignment and the clustering of the SOM - those data items are especially interesting to inspect, as they might denote a labelling of bad quality, or correlations between data items that were not obvious to the observer. In this paper, we propose a visualisation of the SOM based on class labels that helps the user in quickly and easily finding those interesting data points, exploiting the class information for unsupervised learning. We want to colour the map in continuous regions in such a way that the regions reflect the distribution and location of the classes over the map, similar as e.g. a political map does for countries.

J. Marques de Sá et al. (Eds.): ICANN 2007, Part II, LNCS 4669, pp. 359–368, 2007.

The remainder of this paper is organised as follows. Section 2 gives an overview of related work on the SOM and SOM-based visualisations. Section 3 presents the approach of colour flooding, while Section 4 describes some graph-based approaches. In Section 5 we demonstrate our methods in experiments, and Section 6 gives conclusions and an outlook on future work.

2 Self-organising Map

The SOM is a neural network model for unsupervised learning. It provides a mapping from a high-dimensional input space to a lower-dimensional, in most cases a two-dimensional, output space. This output space is in many applications organised as a rectangular grid of units, a representation that is easily understandable for users due to its analogy to 2-D maps. Each of the units on the map is assigned a *weight vector*, which is of the same dimensionality as the vectors in the input space. During the training process, randomly selected vectors from the input space are presented to the Self-Organising Map, and the unit with the most similar weight vector to this input vector is determined. The weight vector of this unit, and, to a lesser extent, of the neighbouring units, are adapted towards the input vector, i.e. their distance in the input space is reduced. As a result of this training process, the output space will be arranged in a way to represent the input space as closely as possible. An important property of the SOM is that the mapping it generates is topology preserving – elements which are located close to each other in the input space will also be closely located in the output space, while dissimilar patterns will be mapped on opposite regions of the map.

2.1 Self-organising Map Visualisations

The SOM itself does not explicitly assign data items to clusters, nor does it identify cluster boundaries, as opposed to other clustering methods. Thus, visualising the mapping created by the SOM is a key factor in supporting the user in the analysis process. A wealth of methods has been developed, mainly to visualise the cluster structures of the data.

Some visualisation techniques rely solely on the weight vectors of the units. Among them, the *unified distance matrix* (U-Matrix [2]) is a technique that shows the local cluster boundaries by depicting pair-wise distances of neighbouring prototype vectors. It is the most common method associated with SOMs and has been extended in numerous ways. The Gradient Field [3] has some similarities with the U-Matrix, but applies smoothing over a larger neighbourhood and uses a different style of representation. It plots a vector field on top of the lattice where each arrow points to its closest cluster centre.

Other visualisation techniques take into account the distribution of the data. The most simple ones are hit histograms, which show how many data samples are mapped to a unit, and labelling techniques, which plot the names and categories, provided they are available, of data samples onto the map. More sophisticated methods include Smoothed Data Histograms [4], which show the clustering structure by mapping each data sample to a number of map units, or graph-based methods [5], showing connections for units that are close to each other in the feature space. The P-Matrix [6] is another density-based approach that depicts the number of samples that lie within a sphere of a certain radius around the weight vectors.

Other techniques adjust the inter-unit distances in the SOM during the training process to more clearly separate the cluster boundaries [7].

Several advanced methods for colouring a SOM have been proposed. In [8] a method to assign colours to the SOM to visualise cluster structures is employed. The colours are not chosen randomly as in other applications, but in a way that differences perceived in the colours reflect the distances within the cluster structures as much as possible. The method is based on a non-linear projection of the SOM into the CIELab colour space. [9] employs a similar method: first, a clustering is applied on top of the SOM, then the map is coloured by projecting it to a sub-space in an RGB-cube.

If the input data mapped is (manually or automatically) assigned to classes, this information can be visualised on the SOM. This can help the user in identifying similarities between classes. She can spot outliers, which might be errors in the manual labelling, and data items close to other classes, which might be worth taking a closer look at. [10] e.g. uses Gabriel-graphs, a subgraph of the Delauny triangulation, on top of projection methods such as the SOM, to highlight isolated (all neighbours have different classes) and border (at least one neighbour has a different class) data. When it comes to displaying such information by colouring the SOM, a very simple approach to visualise the class distribution is by displaying a pie-chart for each node on the map. The pie-chart is split into n sectors, where n is the number of different classes of the data items mapped onto this unit. The sizes of the sectors are determined by the fraction each class contributes to the total number of data items mapped to the unit, and the sectors are coloured with the colour assigned to each class. However, this visualisation is not suited to allow the user to get a quick overview over the class distribution on the whole map.

3 Colour Flooding the SOM

One intuitive approach to visualise the class distribution over the map is to use the method of *colour flooding*. Each unit on the map is thereby first substituted by a coloured point. Then, the points iteratively spread their colour outwards. At each iteration, we can either add one more ring-shaped layer around the already coloured area of the point, or grow by one graphical unit (e.g. pixel). The spreading continues as long as the areas reached are not yet coloured by a different substitution point. The result is a coloured map showing the class distribution, as illustrated in Figure 1(a).

If there are more class labels per unit, we substitute the units by a single-coloured point per class. Next, we arrange the points so that they are aligned as

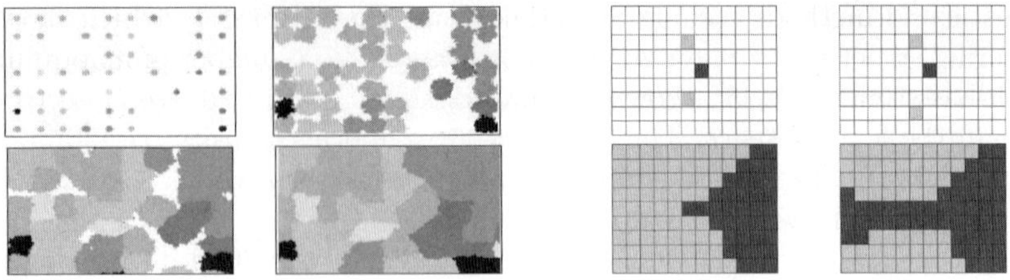

(a) Sequential spreading of colours (b) Instability effect

Fig. 1. Colour flooding

much as possible with points of the same colour on neighbouring units. Then, the flooding process works as described above.

Colour flooding seems to be a feasible approach and generates good visualisations, however, it has a serious disadvantage when it comes to the stability of the resulting colouring. A slightly different layout of the map, leading to a slightly different locations of substitution points, may result in a drastic change in the size and shape of the coloured regions. An illustration of this problem is given in Figure 1(b). The slightly different located lower yellow point results in a 'path' open for the red point to spread to the left edge of the map. Even without changing the position of colour points, a change in the order in which pixels are occupied (e.g. by random selection) can have similar effects. The problem seemed to be in the nature of the method, therefore we abandoned this approach.

4 Graph-Based Colouring

A more promising approach in terms of stability is to use a graph-based segmentation of the map. In this section we outline the used segmentation, Voronoi diagrams, and present different solutions for dealing with the multi-class problem.

4.1 Voronoi Diagrams

A Voronoi diagram [11] of a set of Points P, located on a plane, partitions this plane in exactly $n = |P|$ Voronoi regions, each being assigned to one point $p \in P$. We define some notations which we will use for describing our method:

- $R = \{r_1, r_2, ..r_n | n \geq 0\}$ is the set of all Voronoi regions on the SOM.
- $C = \{c_1, c_2, ..c_m | m \geq 0\}$ is the set of all classes that exist in the data set.
- $C(r_i) = \{c_{i1}, c_{i2}, ...\}$ is the set of classes present in the region r_i.
- $F(r_i, c_j) \in R^+$ is the contribution fraction which class c_j has from all the data items of region r_i.

A Voronoi region is defined as all the points x on the plane belonging to one region $R(p_i)$ that are closest to the point p_i:

$$R(p_i) = \{x : |x - p_i| <= |x - p_j|, \forall j \neq i\}. \quad (1)$$

In our case, the number of regions is equal to the number of units with at least one data item mapped onto. Units with no data items mapped onto them will be split and become parts of other regions.

If each region contains only data items from one class, they can be assigned the corresponding class colour. Since this is a rather unrealistic case, we have to find methods to solve the multi-class problem.

4.2 Basic Voronoi Region Segmentation

If each Voronoi region is thought of as a grid of pixels, we can generate a chessboard-like visualisation. Each class colour present in the region occupies its corresponding fraction of pixels, which are assigned using a uniform distribution function. An illustration of this approach can be seen in Figure 2(a).

Another method is similar to unit substitution in the colour flooding approach: each unit is substituted by n points, each of which is assigned to one of the n classes present. The points are located almost in the same position of the unit, and are arranged to be as closely as possible to neighbouring regions of the same colour. The Voronoi region is then further partitioned into smaller areas, but the global arrangement, i.e. the size and position of the other regions, is not changed. This is illustrated in Figure 2(b). There are still shortcomings with this approach, e.g. classes spanning over 3 neighbouring regions might not be connected well.

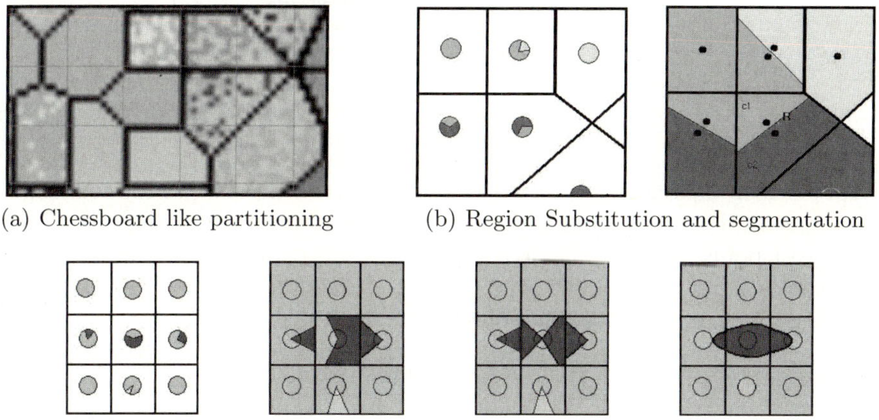

(a) Chessboard like partitioning

(b) Region Substitution and segmentation

(c) Angular Sector partitioning

Fig. 2. Different segmentation methods

Yet another technique is to segment the Voronoi regions angularly into sectors, similar to how pie charts are constructed. The angles are calculated according to the class contribution fraction on the unit, the orientation of the sectors according to the neighbouring classes. There are, however, still cases that cannot be solved satisfactorily. Consider the setting in Figure 2(c). The method described would produce a colouring as in the second image. The red area, however should be connected both to the left and right region, which could to some extent be achieved if we allow sectors to be split, as in the third image. An ideal colouring would be however smoother. It should also show isolated areas, as the yellow area in the example of Figure 2(c), as isolated areas in the middle of the region, rather than as a sector stretching to the edges of it.

4.3 Smoothed Voronoi Region Segmentation

In this section we present an approach to achieve a smooth and optimal segmentation of the regions based on an attractor function. We introduce the concept of *connection lines*, which are imaginary lines that connect the points of a region r_i with the middle of the edge to a neighbouring region r_j, if there is a class c which is present in both regions. This is illustrated in Figure 3. The function assigning class colours to pixels is called attractor function, as it is similar in effect to a magnet attracting objects. The function is applied to a region, a connection line segment, a class and a number n denoting the number of pixels to be assigned.

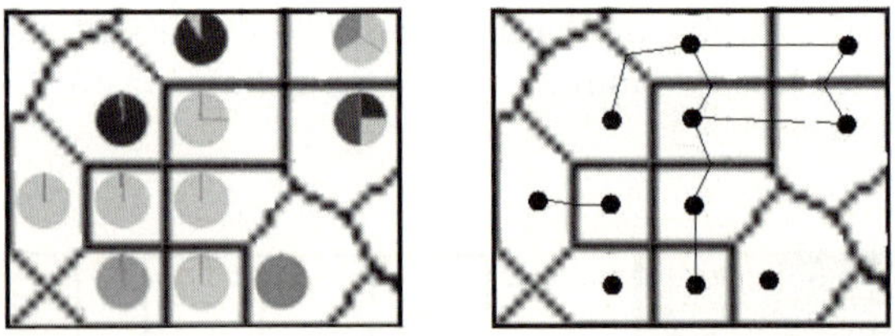

Fig. 3. Connection lines between neighbouring Voronoi regions

All pixels in the region are sorted according to their distance to the connection line segment. The distance d is defined as the distance between a point P and the closest point on a line segment $\overline{P_0P_1}$. After all pixels are sorted, the first n unassigned pixels nearest to the line segment are coloured by the given class. The following types of attractor functions can be identified for a class c in region r:

1. The class c is isolated: there is no neighbouring region $r_j \in r_n$ that contains data of the same class c, as shown in Figure 4(a). This is the simplest case, as we do not need to consider a cluster that extends to other regions. Thus we use a point attractor by adding a segment with length 0 (a point) in the location of the point of the region and assign the corresponding fraction $F(r, c)$ of pixels for the class c.
2. There is only one neighbouring region r_1 that contains the class c, as in Figure 4(b)). We want to attract the coloured pixels at the region border close to r_1. Thus we add a line segment from the site location to the middle point of the common edge e (the edge between r and r_1) and apply the attractor function on this segment with the corresponding fraction of pixels.

3. There are two neighbouring regions r_1 and r_2 that contain data items of the class c and that are themselves neighbours to each others (cf. Figure 4(c)), which implies that r,r_1,r_2 share a vertex v. In this case we use a point attractor by adding a line segment of length 0 to the point at the common vertex v. When all of the three regions concentrate the colour at this point, we have a coloured cloud extending over the three regions.
4. There is more than one neighbouring region, namely m regions $\{r_j, r_k, .., r_m\}$ that have the class c, but none of them is a neighbour of the other (see Figure 4(d)). We treat each of these regions similar as in second case, with the difference that we now have m line segments, one to each neighbouring region. Consequently, the number of pixels assigned to each line is not corresponding to $F(r, c)$, but rather $F(r, c)/m$.

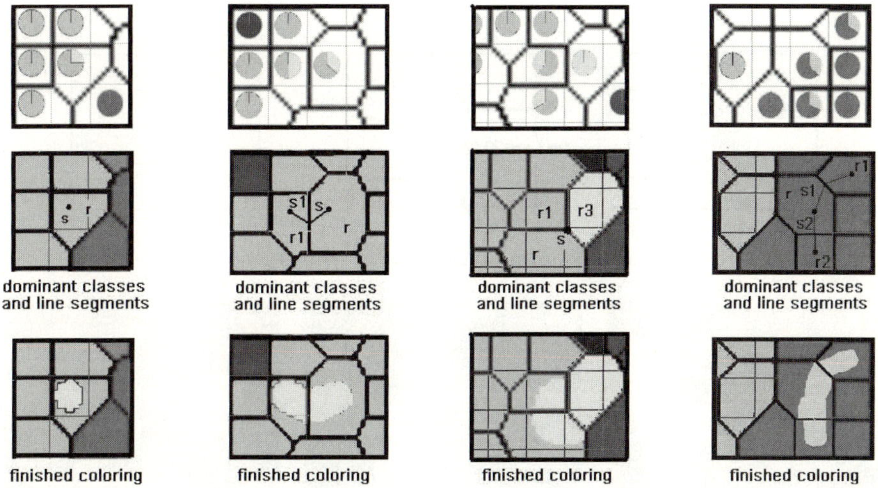

(a) (b) (c) (d)

Fig. 4. Types of Attractor Functions

Border smoothing by weighting the line segments. Although using the attractor function creates a smooth partitioning of the Voronoi regions itself, rough transitions can occur on the border of two regions, as illustrated in the left image in Figure 5(a). This is the case when either the contribution to the class c in the two regions is different, or the regions itself differ in size.

To solve this problem, we modify the above mentioned function which sorts the pixels according to their distances to a line segment. Two weights are assigned to the two ends of the line segment, and are used in measuring the distance. If p is the point to be measured and $\overline{p_1p_2}$ is the line segment, then the weighted distance is:

$$d_w(p, p_1, p_2) = d(p, p_1, p_2) + |\overline{pp_1}|(1/w_1)^2 + |\overline{pp_2}|(1/w_2)^2 \quad (2)$$

where $d(p, p_1, p_2)$ the normal (not weighted) distance and w_1, w_2 the weights for the line segment ends p_1 and p_2 respectively. As a result, the pixels tend to be concentrated more around the end point with the larger weight. For two neighbouring regions having the same class, where the contribution fraction for the class c are f_1 for region r_1 and f_2 for region r_2, we assign to the points of the line segments connecting the two regions the weights as follows: the points at the distant ends of the segment receive the weights $w1 = f_1$ and $w_2 = f_2$, and the common point on the edge of the Voronoi region is assigned a weight of the value $\frac{w_1+w_2}{2}$, as illustrated in the middle image of Figure 5(a). This implies that the pixel concentration on both sides of the common edge is the same for both regions regardless of the values of the contribution fractions. As a result, we get a smoothed border as in the right image of Figure 5(a).

(a) Smoothing region border transition

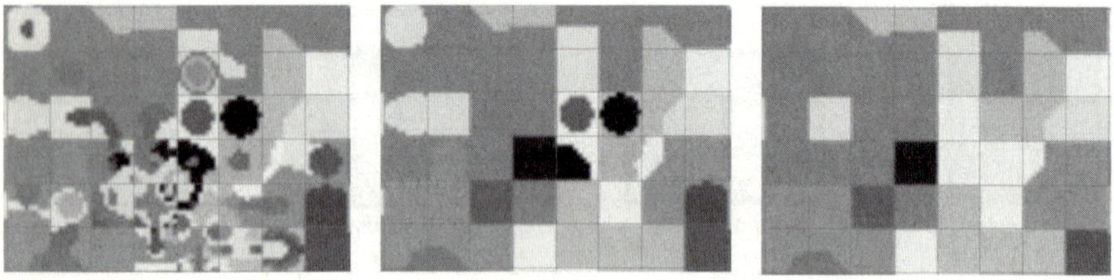

(b) Levels of granularity: threshold of 0%, 50% and 100% contribution fraction

Fig. 5. Improvements to the smooth segmentation method

Minimum class contribution threshold. Some regions have classes with a low contribution value – if one is interested in having a visualisation of a certain abstraction level, the class colouring might show too many unimportant details. This can simply be solved by introducing a threshold for the minimum contribution fraction a class must have of a unit in order to be shown in this unit. Depending on the application, it may be useful to not apply this threshold to the dominant class of the region, otherwise some regions might be left uncoloured. Figure 5(b) shows a comparison between three different thresholds of 0% (showing all), 50% and 100% (showing only the dominant class), respectively.

5 Experiments

For the experiments presented in this section, we used the BankSearch data set, a standard benchmark for clustering and classification methods. It consists of four main categories (Banking & Finance, Programming languages, Science and Sport), which are further divided into a total of eleven subcategories (cf. the class-legend in Figure 6). Each category contains 1.000 documents, thus totalling to 11.000 documents for the whole data set.

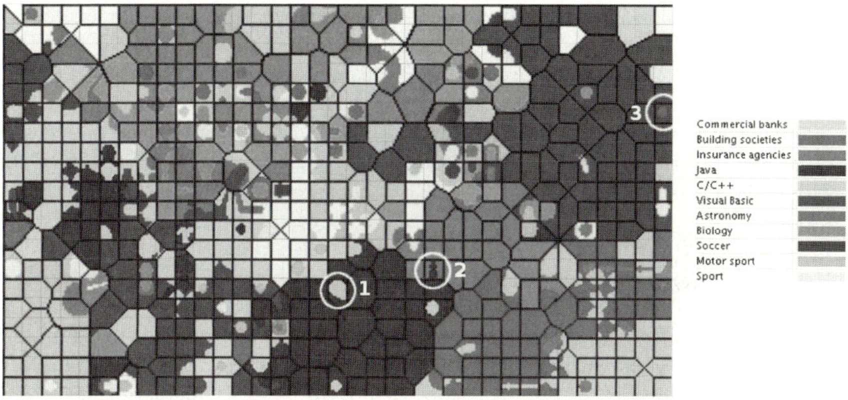

Fig. 6. Class Colouring applied on the Banksearch data set

Figure 6 depicts the trained SOM, using 20% as the minimum class fraction for the colouring. Some areas on the map are marked by a circle – these are sample regions where isolated or boundary data is found. The area marked with '1' holds documents from the *Java* and *Commercial banks* categories, the later however mainly describing a financial application implemented in Java. The area marked with '2' has two documents from *Astronomy* and a single *Java* document mapped onto. However, after inspecting it becomes clear that the 'Java' document is actually a wrongly labelled document from the Astronomy category. The *Soccer* category contains documents about soccer and other related team sports. Besides those, area '3' holds additionally some documents from the category *Building societies*, which all talk about a specific organisation becoming the sponsor of a rugby cup competition – the documents would therefore as well fit in the sports category. In all examples described, the visualisation assisted the user in quickly spotting the interesting areas.

6 Conclusions and Future Work

In this paper we presented several approaches for colouring a Self-organising Map according to the class labels attached to the input data items. The basic idea of using colour flooding was extended to a graph-based segmentation of the map using Voronoi regions. The visualisation provided by this method offers a smoothly coloured map that can assist the user in quickly discovering interesting data items on the map as outliers or other overlapping areas. As future work, we want to combine the chessboard segmentation with the attractor functions.

References

1. Kohonen, T.: Self-Organizing Maps. Springer Series in Information Sciences, vol. 30. Springer, Heidelberg (1995)
2. Ultsch, A., Siemon, H.P.: Kohonen's self-organizing feature maps for exploratory data analysis. In: Proc. of the Intl. Neural Network Conference (INNC'90), Paris, France, pp. 305–308. Kluwer Academic Publishers, Dordrecht (1990)
3. Pölzlbauer, G., Dittenbach, M., Rauber, A.: A visualization technique for self-organizing maps with vector fields to obtain the cluster structure at desired levels of detail. In: Proc. of the Intl. Joint Conference on Neural Networks (IJCNN'05), Montreal, Canada, pp. 1558–1563. IEEE Computer Society Press, Los Alamitos (2005)
4. Pampalk, E., Rauber, A., Merkl, D.: Using Smoothed Data Histograms for Cluster Visualization in Self-Organizing Maps. In: Dorronsoro, J.R. (ed.) ICANN 2002. LNCS, vol. 2415, pp. 871–876. Springer, Heidelberg (2002)
5. Pölzlbauer, G., Rauber, A., Dittenbach, M.: Advanced visualization techniques for self-organizing maps with graph-based methods. In: Wang, J., Liao, X.-F., Yi, Z. (eds.) ISNN 2005. LNCS, vol. 3497, pp. 75–80. Springer, Heidelberg (2005)
6. Ultsch, A.: Maps for the visualization of high-dimensional data spaces. In: Proc. of the Workshop on Self organizing Maps (WSOM'03), Kyushu, Japan, pp. 225–230 (2003)
7. Liao, G., Shi, T., Liu, S., Xuan, J.: A novel technique for data visualization based on som. In: Duch, W., Kacprzyk, J., Oja, E., Zadrożny, S. (eds.) ICANN 2005. LNCS, vol. 3697, pp. 421–426. Springer, Heidelberg (2005)
8. Kaski, S., Venna, J., Kohonen, T.: Coloring that reveals high-dimensional structures. In: Proceedings of the International Conference on Neural Information Processing (ICONIP'99), Perth, Australia. Volume II., Piscataway, NJ, IEEE Service Center, 729–734 (1999)
9. Himberg, J.: A SOM based cluster visualization and its application for false coloring. In: Proceedings of the IEEE-INNS-ENNS International Joint Conference on Neural Networks (IJCNN '00), Washington, DC, USA, pp. 587–592. IEEE Computer Society Press, Los Alamitos (2000)
10. Aupetit, M.: High-dimensional labeled data analysis with gabriel graphs. In: Proceedings of the European Symposium on Artificial Neural Networks (ESANN'03), Bruges, Belgium, pp. 21–26 (2003)
11. Aurenhammer, F.: Voronoi diagrams – a survey of a fundamental geometric data structure. ACM Computing Surveys 23(3), 345–405 (1991)

Self-organizing Maps with Refractory Period

Antonio Neme[1] and Victor Mireles[2]

[1] Universidad Autónoma de la Ciudad de México
neme@nolineal.org.mx
[2] Uiversidad Nacional Autónoma de México
syats.vm@gmail.com

Abstract. Self-organizing map (SOM) has been studied as a model of map formation in the brain cortex. Neurons in the cortex present a refractory period in which they are not able to be activated, restriction that should be included in the SOM if a better description is to be achieved. Altough several works have been presented in order to include this biological restriction to the SOM, they do not reflect biological plausibility. Here, we present a modification in the SOM that allows neurons to enter a refractory period (SOM-RP) if they are the best matching unit (BMU) or if they belong to its neighborhood. This refractory period is the same for all affected neurons, which contrasts with previous models. By including this biological restriction, SOM dynamics resembles in more detail behavior shown by the cortex, such as non-radial activity patterns and long distance influence, besides the refractory period. As a side effect, two error measures are lower in maps formed by SOM-RP than in those formed by SOM.

1 Introduction

The self-organizing map (SOM) is presented as a model of the self-organization of neural connections, which is translated in the ability of the algorithm to produce organization from disorder [1]. One of the main properties of the SOM is the ability to preserve in the output map those topographical relations present in input data [2], This property is achieved through a transformation of an incoming signal pattern of arbitrary dimension into a low-dimensional discrete map (usually one or two-dimensional) and to adaptively transform data in a topologically ordered fashion [3,2]. Each input data is mapped to a single neuron in the lattice, the one with the closest weight vector to the input vector, or best matching unit (BMU). The SOM preserves relationships during training through the learning equation, which establishes the effect each BMU has in any other neuron. Weight neurons are updated accordingly to:

$$w_n(t+1) = w_n(t) + \alpha_n(t)h_n(g,t)(x_i - w_n(t)) \quad (1)$$

Where $\alpha(t)$ is the learning rate at time t and $h_n(g,t)$ is the neighbourhood function from BMU neuron g to neuron n at time t. In general, the neighbourhood function decreases monotonically as a function of the distance from neuron g

J. Marques de Sá et al. (Eds.): ICANN 2007, Part II, LNCS 4669, pp. 369–378, 2007.

to neuron n. The SOM tries to preserve relationships of input data by starting with a large neighbourhood and reducing it during the course of training [2,4].

As pointed out by Ritter [3], SOM and related algorithms share the idea of using a deformable lattice to transform data similarities into spatial relationships. The lattice is deformed by applying learning equation (1).

Altough SOM has been widely applied in data visualization and clustering, it has also been studied as a model of the brain cortex. For example, in [5,6] it has been studied to understand the map formation in visual cortex, and in [7] as a model of brain maps from sensorial areas to cortical regions. However, SOM fails to reproduce the activity patterns present in the cortex [15,?], altough some variants, as the one proposed in [8], in which a modification in the kernel is considered to include surround inhibition, achieve the formation of pinwheel patterns similar to those observed in the visual cortex. In these models, however, influence from BMUs to neighbors is radial and symmetrical. In [18] non-radial patterns of activity from BMU to its neighbors are reported as a consequence of differentiated influence based on the relative frequency each BMU includes neurons as neighbors. In [17], a recursive rule allows non radial neighborhood adaptation as a consequence of the pulling from BMUs to the direct neighbors whereas these neighbors further pull their neighbors.

Concepts analogous to the refractory period have also been included in the SOM. For example, in [12] the BMU is relaxed in order to modify the magnification exponent. In [19] a traveling wave of activity induced in BMU allows its neirghbors to be more likely to be the next BMU, which can be interpreted as a refractory period for those neurons not included in the wavefront. In [20] an activation memory is defined for each neuron, in order to define the new active neuron, and a modification in the BMU selection mecanism is presented, so if the memory parameter is high, the previous winner neuron will win again unless another neuron matches very close the input data, which, again, may be seen as a refractory resctriction for those neurons with a low memory parameter.

In these modifications, the refractory time depends on the weight modification which may not be a biologically realistic behavior [14]. Here, we study the effects of including a refractory period that does not depend on this or any other quantity in the neurons for the SOM, the SOM-RP, and show that it is possible to obtain maps equivalent to those obtained with the SOM.

2 The Self-organizing Map with Refractory Period (SOM-RP)

Once an input vector is maped to a neuron, the later becomes a BMU and is activated, as well as its neighbors. Biologically, the activation of a given neuron is achieved trough the electrical opening of ion channels which, due to concentration gradients and charge differences between the out and inside of the cell, drive positive ions into the cell. This changes the potential (inside relative to outside) of the cell, from it's resting potential to the activity one [13].

For this change of potential to take place, a fair amount of positive ions must flow into the cell, thus reversing the concentration gradient of this ions for a moment. Since the diffusion force plays an important role in the activation process, this cannot take place until the concentration difference is reestablished, so when the proper channels open, positive ions will indeed flow into the neuron. The time in which the necessary concentration is not present is referred to as the absolute refractory period, referred here as refractory period [14,13]. While there also exists a relative refractory period in which the neuron can become active but needs a far greater stimulus to do so, we will not consider it in the present work.

In this work, we are interested in the SOM capabilities to form maps that resemble the input space distribution even when some of the neurons in the network are not able to learn for a given period of time, which affects the SOM dynamics as the weight folding may follow other routes.

We propose a modification that allows active neurons to enter a refractory period. By active neurons we mean BMU and those neurons within its neighborhood. The studied modifications are two: in the first one, each BMU and a subset of its neighbors enter a refractory period in which they are not able to be affected by any other BMU for a given time. Two variables are defined here: τ and d. τ is the refractory period and is defined as the number of input vectors for which an affected neuron is not able to learn trough eq. (1). τ does not depend on the weight update value or any other variable and is the same throughout the learning process and for all affected neurons. d is the radius of a hypersphere in the lattice centered at the BMU, which defines the set of neurons that will fall into refractory period. If d is greater than the actual neighborhood width, then d is set to that width.

In the second modification studied in this work, BMUs and all its neighbors present a maximum number of times they can be affected before entering the refractory period, named c. Once this maximum is achieved, they enter the refractory period and stay there for τ vectors. That is, c acts as a delay for the neurons to enter refractory period. Once again, τ is the same for all sleeping neurons.

3 Simulations and Results

In order to study SOM-RP dynamics, several thousand maps were formed for six data sets: spiral, random and unitary circumference (2-dimensional); iris (4-dimensional); Mexican elections (ME) (6-dimensional) and ionosphere (34-dimensional) data sets.

In order to verify self-organization in the SOM-RP two error measures were quantified and compared to the error measures present in the maps formed by eq (1). Altough there are several error measures for the maps obtained by SOM and there is no solid definition of the energy function [9,10,11], the topographic error (TE) as well as the error quantization (EQ) were the error measures quantified for the obtained maps, as they are good measures of the quality of topographic mapping and vector quantization. In order to test sensitivity and self-organization, several thousands of experiments were made for two lattice sizes, $N \times N$ ($N = 20$ and 30), as well as for the initial learning parameter $0 < \alpha(0) \leq 1$ and for the initial neighborhood size $1 < h_n(g, 0) \leq N$. Practically, for each

- learning set (circumference, spiral, random, iris, ionosphere, ME data),
- number of epochs (between 1 and 30),
- τ (between 0 and 25)
- d (between 0 and 25) (for modification 1)
- c (between 0 and 25) (for modification 2)

the initial learning parameter $\alpha(0)$ was chosen randomly from $(0, 1]$ as well as the initial neighborhood width was chosen from $[1, N]$. The final learning parameter was 0.0001 ($\alpha(r) = 0.0001$) whereas the final width was decreased to 0 by an exponential function. In both modifications, if $\tau = 0$ then SOM-RP is reduced to SOM.

Fig. 1 shows TE and EQ as a function of τ and d for the first modification, whereas fig. 2 presents TE and QE as a function of c and τ for the second modification. What is observed is that error measures are in general lower for SOM, corner (0,0), than for SOM-RP. However, it is important to notice that this is condensed information for several thousand maps, with different epoch numbers, as well as different neighborhood width and α values.

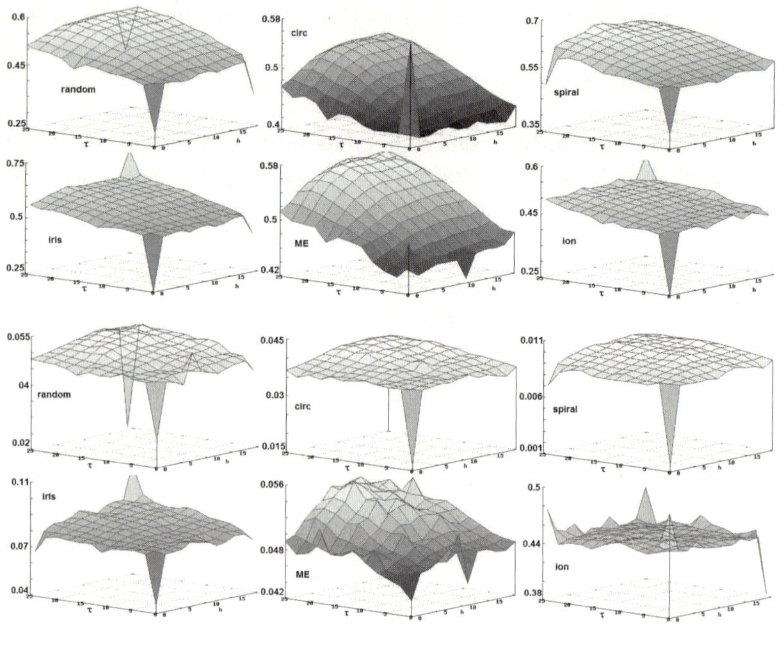

Fig. 1. TE (top) and EQ (bottom) for modification 1 for the six data sets

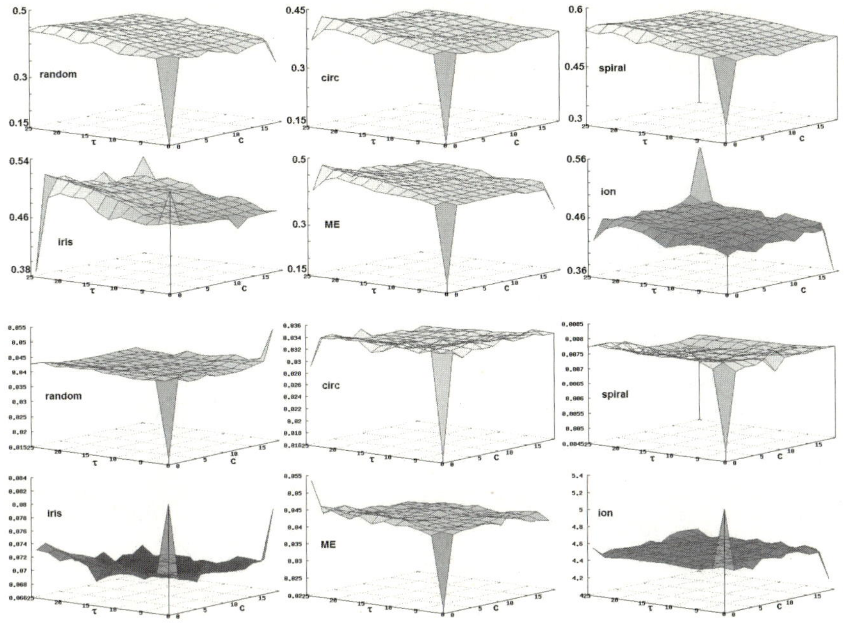

Fig. 2. TE (top) and EQ (bottom) for modification 2 for the six data sets

Altough TE and EQ are in general greater in SOM-RP than in SOM, let us consider, as shown in fig. 3, only the 5% of maps with the lowest TE values. Here maps formed by SOM-RP ($\tau > 0$) are much more frequent than those formed by SOM. In some data sets, none of the low-error maps were formed by SOM. While this does not justify the use of SOM-RP if one desires to optain a single map with low TE or EQ, it does suggest that, if the time and computational power are available to make several thousands of maps for the same data set, and choose the best one, it is a good idea to use SOM-RP, for that will yield maps with lower error meassures. If, how ever, a single or a couple of maps are to be performed on each data set, as argumented below, it is preferable to use the traditional SOM. This low error maps were obtained for several values of both τ and d (or c for modification 2) (see fig. 4).

On a 30x30 lattice we performed thousands of simulations of both SOM and SOM-RP and examined how frequent a given TE was achieved. The results are shown in fig 5.

As we can see, on all cases, the traditional SOM presents spikes in the frequency histogram which are further left than the spikes present in the SOM-RP frequency histogram. This means that the most likely TE of the SOM is lower than the most likely TE of the SOM-RP.

This drawback is of importance if a small number of maps are made, however, as can be seen above, if a large enough number is made and the map with lowest TE is chosen, then the SOM-RP is very likely to produce a map with significantly lower TE than SOM.

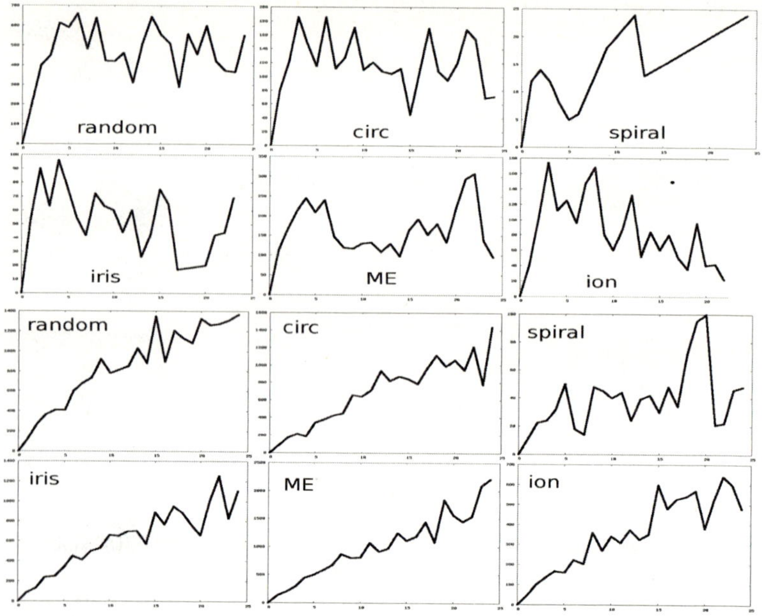

Fig. 3. The 5% of the maps with lowest TE are considered. τ is indicated in the x axis ($\tau = 0$ is the original SOM), while in the y axis is considered the number of maps for that τ that are in the group of the maps with very low errors. The first six figures are for modification 1 and the last six for modification 2.

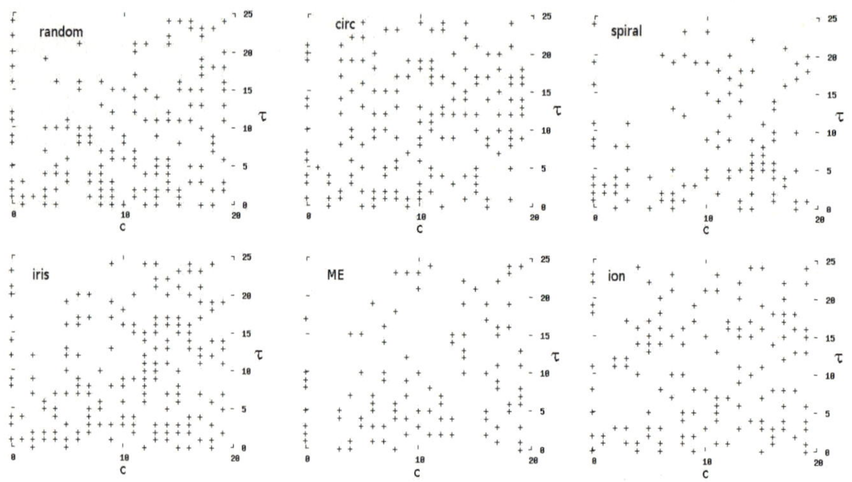

Fig. 4. τ and c for the 5% of the maps with the lowest TE for the six data sets and lattice of size 20x20

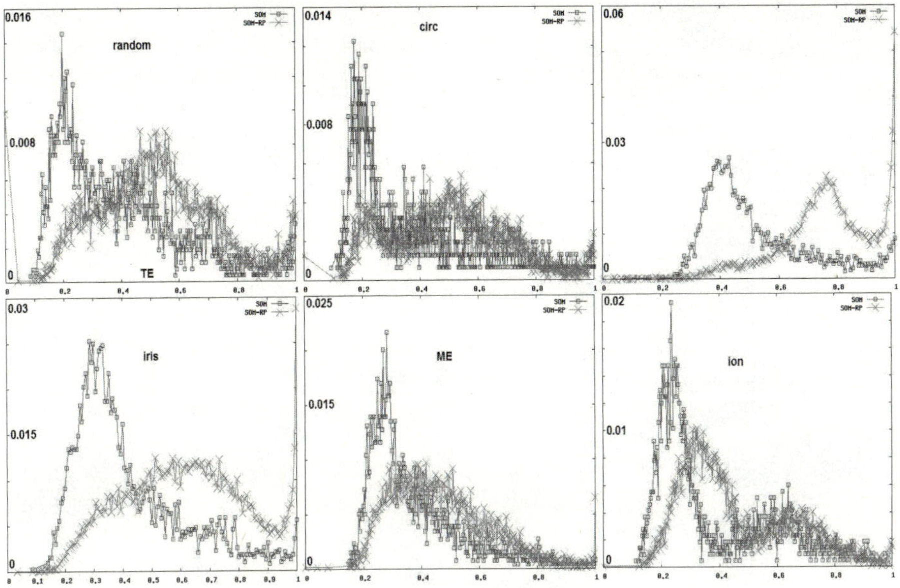

Fig. 5. Frequency of maps (y axis) for TE (x axis) for lattices of size 30x30 for SOM (squared) and SOM-RP (cross)

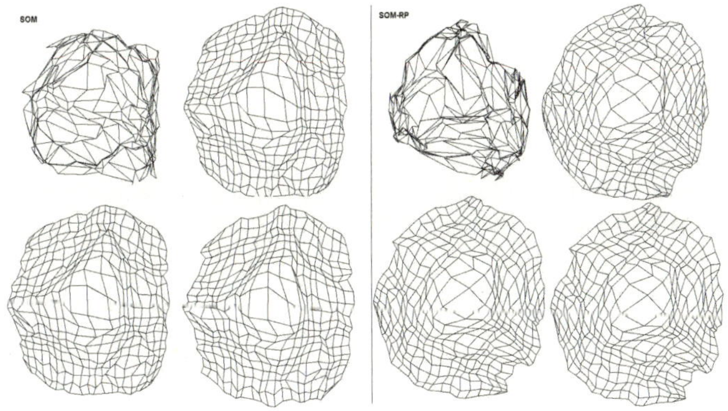

Fig. 6. Weight folding in the SOM (left) and in the SOM-RP (right) for $t = 2, 4, 6, 10$ for the ring data set (only included as an example for weight folding). It is observed that the SOM-RP presents, in general, smooth borders that fit the input vectors. SOM-RP parameters were $\tau = 2$ and $d = 2$.

Folding in the SOM-RP is affected as shown in fig. 6. It is observed that the SOM-RP approximates better the ring data set as it shows smooth borders which contrast with the borders in the SOM weights. This is a consequence of the refractory period in which neurons enter after being affected. When a neuron enters this period, some of its neighbors may be able to learn the new input.

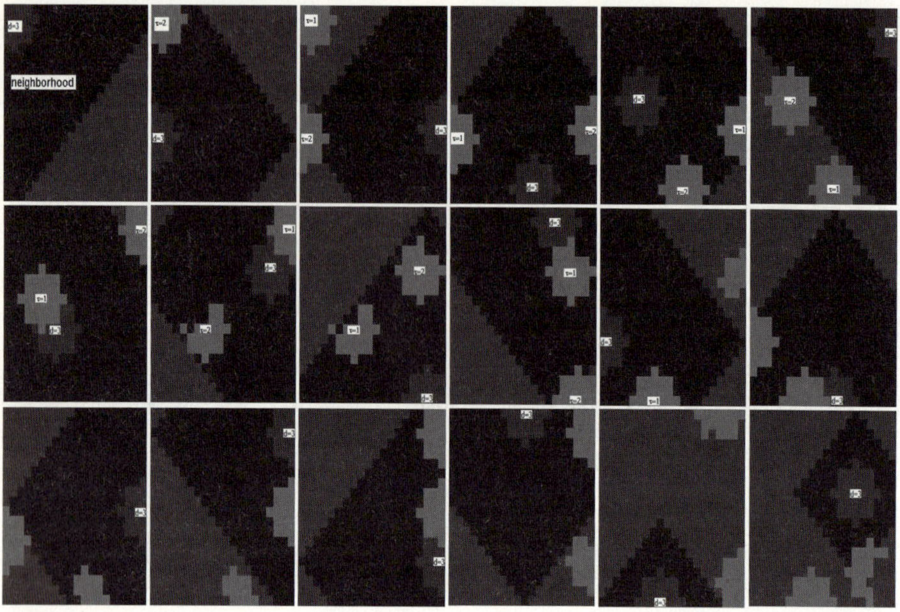

Fig. 7. Activity patterns in the SOM-RP for consecutive times. Non radial patterns of activity are formed because of the refractory period. Here, the parameters for the SOM-RP were $\tau = 2$ and $d = 3$ in a 20x20 lattice.

Thus, the network learning may be improved if some of the neurons do not take part in the process for some periods of time. It is also observed that the bottom-right corner is not properly folded, once again, as a consequence of the refractory period. The refractory period may help the network to properly fold, but, if τ is large enought and the area of influence is also large, then the folding may be disrupted.

The activity patterns in the SOM are radial and symetrical, which is different from the patterns formed in SOM-RP, as shown in fig 7. The BMU affects all its neighbors, but only a subset of them (defined by the d parameter, $d = 3$ in the example) will become inactive for τ input vectors ($\tau = 2$). Once τ input stimulus are maped, the refractory neurons become susceptible and may be affected by BMUs or even become BMU. Neurons in BMU's neighborhood might not modify their weight vector, as they might be in a refractory period. Closer neurons to BMU may be refractory, while farther neurons (also included in neirghborhood) may be affected. This long-distance effect has biological foundations [21].

4 Discusion and Conclusions

Map formation is possible when homogeneus refractory periods are included in neurons. We are interested in studying the properties of SOM when a refractory period is included in its neurons. Altough it is not possible to give a recipe for τ, d and h values, we have shown it is possible that the map folds properly to approximate input space, which is an important restriction to be included in the SOM if is to be studied as a more realistic model of the brain cortex.

In nature, refractory periods are important as they allow neurons to repolarize and become susceptible for further activation. In general, it had been identified as a restriction in neurons that should not be considered in artificial models, no matter they are supposed to explain the general aspects of self-organization in the brain. Here, we have incorporated a homogeneous refractory period and the results are, we believe, interesting, in the sense that are equivalent (in terms of errors) to those obtained by SOM, but achieved by following a different and more realistic route.

The fact that learning took place in the SOM-RP, reafirms the fact that the learning process is a distributed one, for modifications in 'small' regions of the lattice do not affect the overall behaviour of the map. This, of course, can be seen with many other modifications to the original SOM. It is however of interest, that the overall behaviour of the map is little affected by phenomena which are not instantaneous: for several values of τ the properties of the SOM are not dramatically altered, which suggests that the distribution of information and processing capabilities in the SOM are robust enough as to go arround lasting obstacles.

The existence of a refractory period drives the BMU arround the grid, forcing it to fall into non-recently visited sites. Since in general τ will be significantly smaller than the number of input vectors, this is translated into further spreading the neurons associated with each stimulus. This will in turn make the convergence of the mapping slower. This is not necesarly a drawback, for it also makes the distribution of the weight vectors more uniform in the input space, thus allowing a better mapping.

Non-symetrical activity patterns are present in the cortex, but the SOM fails to reproduce them. With the proposed SOM-RP, those patterns, as well as long-distance influence, are achieved.

References

1. Cottrell, M., Fort, J.C., Pagés, G.: Theoretical aspects of the SOM algorithm. Neurocomputing 21, 119–138 (1998)
2. Kohonen, T.: Self-Organizing maps, 3rd edn. Springer, Heidelberg (2000)
3. Ritter, H.: Self-Organizing Maps on non-euclidean Spaces Kohonen Maps. In: Oja, E., Kaski, S. (eds.), pp. 97–108 (1999)
4. Erwin, O., Schulten, K.: Self-organizing maps: Ordering, convergence properties and energy functions. Biol. Cyb. 67, 47–55 (1992)
5. Mayer, N., Herrmann, J., Geisel, T.: Retinotopy and spatial phase in topographic maps. Neurocomputing, 32–33, 447–452 (2000)
6. Bednar, J., Kelkar, A., Miikkulainen, R.: Scaling self-organizing maps to model large cortical networks. Neuroinformatics (2004)
7. Kohonen, T.: Self-organizing neural projections. Neural Networks 19, 723–733 (2006)

8. Liljeholm, M., Lin, A., Ozdzynski, P., Beatty, J.: Quantitative analysis of kernel properties in Kohonen's self-organizing map algorithm: Gaussian and difference of Gaussians neighborhoods. Neurocomputing, 44–46, 515–520 (2002)
9. Goodhill, G., Finch, S., Sejnowski, T.: Quantifying neighborhood preservation in topographic maps. 3rd. Joint Symp. in neural comp., pp. 61–82 (1996)
10. Kiviluoto, K.: Topology preservation in Self-Organizing maps. In: Proc. ICNN96, IEEE Int. Conf. on Neural Networks (1996)
11. Bauer, H., Herrmann, M., Villmann, T.: Neural maps and topographic vector quantization. Neural networks 12, 659–676 (1999)
12. Claussen, J.: Winner-relaxing self-organizing maps. Neural computation 17, 996–1009 (2006)
13. Thompson, R.: The brain: a neuroscience primer. Worth pub. (2000).
14. Koch, C.: Biophysics of computation and information processing in single neurons. Oxford University press, New York (1998)
15. Koulakov, A., Chklovsky, D.: Orientation preference patterns in mammalian visual cortex: a wire length minimization approach. Neuron 29, 519–527 (2001)
16. Goodhill, G., Cimponeriu, A.: Analysis of the elastic net model applied to the formation of ocular dominance and orientation columns. Network: comput. neural systems 11, 153–168 (2000)
17. Lee, J., Verleysen, M.: Self-organizing maps with recursive neighborhood adaption. Neural Networks 15, 993–1003 (2002)
18. Neme, A., Miramontes, P.A: parameter in the SOM learning rule that incorporates activation frequency. ICANN 1, 455–463 (2006)
19. Principe, J., Euliano, N., Garani, S.: Principles and networks for self-organization in space-time. Neural Networks 15, 1069–1083 (2002)
20. Chappell, G., Taylor, J.: The temporal Kohonen map. Neural Networks 6, 441–445 (1993)
21. Sporns, O., Chialvo, D., Kaiser, M., Hilgetag, C.: Organization, development and function of complex brain networks. Trens in cognitive sciences. 8, 418–425 (2004)

Improving the Correlation Hunting in a Large Quantity of SOM Component Planes

Classification of Agro-Ecological Variables Related with Productivity in the Sugar Cane Culture

Miguel A. Barreto S.[1,2,3] and Andrés Pérez-Uribe[2]

[1] Université de Lausanne, Hautes Etudes Commerciales (HEC), Institut des Systèmes d'Information (ISI)
Miguel-Arturo.Barreto-Sanz@heig-vd.ch

[2] University of Applied Sciences of Western Switzerland (HEIG-VD)(REDS)
andres.perez-uribe@heig-vd.ch

[3] Corporación BIOTEC

Abstract. A technique called component planes is commonly used to visualize variables behavior with Self-Organizing Maps (SOMs). Nevertheless, when the component planes are too many the visualization becomes difficult. A methodology has been developed to enhance the component planes analysis process. This methodology improves the correlation hunting in the component planes with a tree-structured cluster representation based on the SOM distance matrix. The methodology presented here was used in the classification of similar agro-ecological variables and productivity in the sugar cane culture. Analyzing the obtained groups it was possible to extract new knowledge about the variables more related with the highest productivities.

1 Introduction

A traditional technique to detect dependencies between variables is the use of scatter plots. In addition, when the variables are more than a pair, it is possible to generate a scatter plot matrix with several sub-plots where each variable is plotted against each other variable. However, in this technique the number of pairwise scatter plots increases quadratically with the number of variables [5]. This type of visualization is thus not practical in applications where the analysis of many variables is necessary.

Another visualization technique consists on using the so-called SOM components planes [6], the number of sub-plots grows linearly with the number of variables. In addition, this technique is able to cluster variables with similar behaviors. Every SOM component plane is formed by the values of the same component in each prototype vector. Therefore, they can be seen as a sliced version of the map [10]. After plotting all component planes, relations between variables can be easily observed. The task of organizing similar components planes in order to find correlating components is called correlation hunting [13].

J. Marques de Sá et al. (Eds.): ICANN 2007, Part II, LNCS 4669, pp. 379–388, 2007.

However, when the number of components is large it is difficult to determine which planes are similar to each other. Different techniques can be used to reorganize the component planes in order to aid the correlation hunting. The main idea is to place correlated components close to each other.

One of the most often used techniques in correlation hunting is the projection of the component planes on another plane. This projection can be done using, e.g. another SOM, as the work of Vesanto et Ahola. [13]. Another interesting approach was introduced by Sultan et al. [9] they presented a binary tree-structured vector quantization (BTSVQ) algorithm. The BTSVQ uses SOMs for visualization, and the partitive k-means clustering, to group similar component planes and organizing them into a binary tree structure. This hybrid algorithm is used to improve the process of data analysis and visualization of gene expression profiles.

The approach of Vesanto and Ahola [13] is adequate to organize the component planes, but it is an inefficient tool for visualization when the number of planes is large. It is difficult to clearly observe the relationships between the component planes due to the quantity of planes to show in a same space. Sultan's algorithm is more adequate to organize a large quantity of data. Its binary tree structure allows the analysis of groups of component planes at different levels. Nevertheless, the algorithm proposed to organize the SOM component planes make use of k-means as a clustering algorithm, and the SOM is only employed to show the data.

In this paper we present a methodology to enhance the visualization and analysis process of a large quantity of component planes. This methodology uses a SOM to project the component planes. This SOM is partitioned into in clusters with a technique based on the SOM distance matrix. A tree structure is generated from different clustering levels of the SOM, in order to clearly visualize the groups of component planes. The methodology presented here was used in the classification of similar agro-ecological variables and productivity in the sugar cane culture. Analyzing the obtained groups it was possible to extract new knowledge about the variables more related with the highest productivity.

This paper presents the following structure. In the next section the methodology is explained. Third section focuses on the application of the methodology to the sugar cane case. Finally, in section four conclusions and future extensions of this work are presented.

2 Methods

2.1 Self-organizing Maps

A Self-Organizing Map (SOM) [6] is composed of artificial neurons situated on a but regular low-dimensional grid. This grid can be in one, two or three dimensions, generally two are used. The neurons in the grid have rectangular or hexagonal form. Each neuron i represents an n-dimensional prototype vector $mi = [mi_1, \ldots, mi_s]$, where s is equal to the dimension of the input space. In the beginning of the training process the prototype vectors are initialized with random values. On each step of the training a data vector x from the input data is selected and presented to the SOM. The unit mc closest to x is located into the map, this winner unit is called the best-matching unit (BMU). The BMU and its neighboring prototype vectors on the grid are moved in the direction of the sample vector, $mi = mi + \alpha(t)h_{ci}(t)(x - mi)$ where $\alpha(t)$ is the learning rate and $h_{ci}(t)$ is a neighborhood kernel centered on the winner unit c. The learning rate and neighborhood kernel radius decrease monotonically with time. Through the iterative training, the SOM organizes the neurons so that neurons that represent similar vectors in the input space are located on the map in contiguous zones, trying to conserve the linear or nonlinear relations of the input space.

2.2 SOM Component Planes

SOM allows a straightforward visual inspection because the prototype vectors are organized according to their similarity in a low-dimensional grid. This feature is helpful when it is needed to handle large multidimensional vectors. A way to improve this inspection is by means of the component plane representation. A component plane (CP) is a projection of the same input variable from each vector prototype on a grid. For example, having the prototype vectors $m1, \ldots, mi$. The component plane which represents the first input variable will be formed by $CP1 = [m1_1, \ldots, mi_1]$ in general $CPs = [m1_s, \ldots, mi_s]$ where s is equal to the dimension of the input space. Hence, the number of component planes will be equal to the input space dimension. In addition, the component planes are visualized in an grid identical to that of the SOM. However, the difference between the component plane grid and the SOM grid is that on this new grid each neuron does not plot a prototype vector, instead it represents a component of this vector. Each component in the component plane grid conserves the same place that the prototype vector in the SOM grid. Finally, every component on the component plane is visualized by giving to each neuron a color according to the relative value of the respective component in that neuron. As a result, it is possible to obtain color maps of the component planes in order to compare them and look for relations between variables.

2.3 Correlation Hunting

The component planes analysis can be a tool for discovering relations between variables. Comparing the planes, it is possible to observe similar patterns in identical positions indicating correlation between the respective components. Even, local correlations can be found if two parameter planes resemble each other in some regions. The process to find these relationships is called correlation hunting. The expression correlation does not include just linear correlations, but also nonlinear and local or partial correlations between variables [13].

The correlation hunting can be realized manually or automatically. However, in many cases the manual analysis is difficult because usually the component planes are not ordered. In addition, the comparison becomes more difficult when the number of components increases. In order to overcome this drawback, it is possible to apply reorganization of the component planes such that similar component planes could be located close to each other [14]. To do this, the component planes can be projected on a plane. The projection could be done using, e.g., Sammon's mapping [8], CCA [3] or another SOM. In this paper SOM was used as projection technique. The projection process using SOM is the following:

1. Each component plane is transformed into a vector and then normalized to ignore different scales of the components.
2. The vectors are further processed by calculating a measure of distance between them.
3. The measure of distance between component planes i and j can be defined as the value of the correlation of each map position, formally $distCP(i,j) = mc*(CP_i, CP_j)$ where mc is a suitable measure of correlation, in this paper the Pearson correlation coefficient is used.
4. A covariance matrix is generated with the obtained distances.
5. The vectors of the covariance matrix are used as inputs to a new SOM.
6. Each component plane grid from the old SOM is projected to the new SOM.
7. This projection is realized locating in the place of the BMUs of the new SOM, the respective component planes grids from the old SOM. Hence, planes with high correlation are located near each other.

An advantage of using a SOM for component plane projection is that the placements of the component planes can be shown on a regular grid. In addition, an ordered presentation of similar components is automatically generated. A disadvantage is that the choice of grouping variables is left to the user. This task is complicated when the number of component planes is large.

2.4 Distance Matrix Based Clustering of the SOM

Having a projection of component planes in a new SOM, it is possible to use a method to cluster the new SOM in order to find component plane groups. For example, partitive (e.g., k-means) or agglomerative clustering algorithms (e.g., agglomerative hierarchical clustering) are used to cluster the prototype vectors [15]. Nevertheless, those approaches do not take into account the SOM neighborhoods. To cope with this drawback, a cluster distance function can be used to consider the neighborhoods into account. The U-matrix [11] had been used as an effective cluster distance function [16]. The U-matrix visualizes distances between each map unit and its neighbors, thus it is possible to visualize the SOM cluster structure. This method is usually applied to select clusters from the map by hand. This selection is normally subjective because it is based on the visual perception of each person. Vellido et al. [12] proposed an algorithm to do distance matrix based clustering automatically. In this algorithm, the U-matrix is used to identify cluster centers from the SOM. The rest of the map units are then assigned to the cluster whose center is closest. The algorithm is the following:

1. Compute the distance matrix local minima. This is done by finding the set of map units i for which:

$$f(m_i, N_i) \leq f(m_j, N_j), \forall j \in N_i \tag{1}$$

   where N_i denoted the set of neighboring map units of the map unit i and $f(m_i, N_i)$ is some function of the set of neighborhood distances $\|m_i - m_j\|$ $j \in N_i$, associated with map unit i. In the experiments, median distance was used. The set of local minima may have units which are neighbors of each other. Only one minimum from each such group is retained.
2. For the initialization, let each local minimum be one cluster: $C_i = m_i$. All other map units j are left unassigned.
3. Calculate distance $d(C_i, m_j)$ from each cluster Ci to (the cluster formed by) each unassigned map unit j.
4. Find the unassigned map unit with smallest distance and assign it to the corresponding cluster.

This algorithm provides an automatic discrimination of clusters which permits an easier exploration of similar component planes. Although, when the number or component planes is large is desirable an approach that permits to organize the component planes in a structure, and to analyze clusters at several levels of detail. Hence, the idea of considering super-clusters, consisting of several sub-clusters, making easier the analysis of the large quantity of planes.

2.5 Tree-Structured Component Planes Clusters Representation

In order to analyze the component planes clusters at several detail levels, it is possible to make a tree-structured representation of them. The Vellido's algorithm is used to obtain different partitioning levels of the clustering of the SOM in an attempt to achieve this goal. The Vellido's algorithm provides a partitioning of the map into a set of base clusters. The number of clusters is equal to the number of local minima on the U-matrix; allowing different levels of clustering. Regarding the equation 1 it is possible to observe that the local minima depends on the set of neighboring map units (Ni) from the map unit i. Hence, Ni depends on the amount of neighbors chosen to i. As a result, when the neighborhood is large, the number of local minima is small and therefore the number of clusters too. Varying the neighborhood size it is possible to obtain different cluster quantities, see figure 1.a. So, it is possible to find different cluster levels, and as a result, build a tree structure that will permit to have several levels of detail, see figure 1.b and figure 2.

An outline of the algorithm to generate the tree-structured component planes clusters representation is given below.

1. Tree-generation.
   (a) Calculate the SOM U-matrix used for the projection process.
   (b) Apply the Vellido's algorithm to the obtained U-Matrix. Use a neighborhood value of 1, and save the results (nodes and component planes clusters).

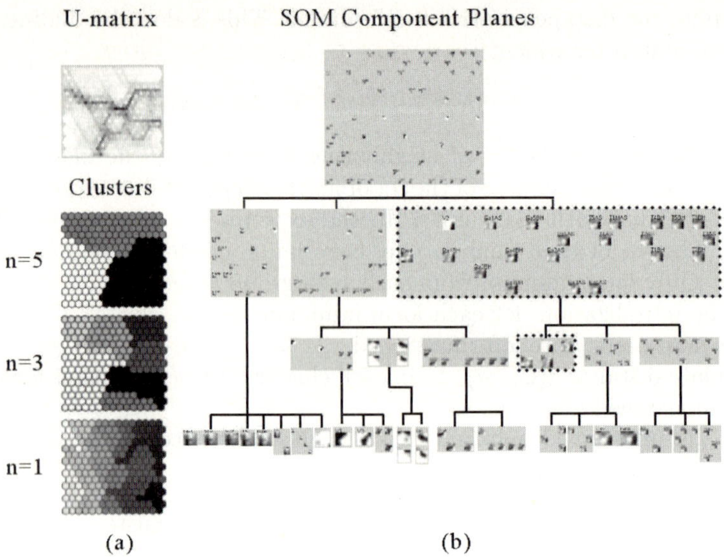

Fig. 1. Tree-structured component planes clusters representation. (a) Clusters obtained varying the neighborhood size (n) in Vellido's algorithm. (b) Clusters levels obtained.

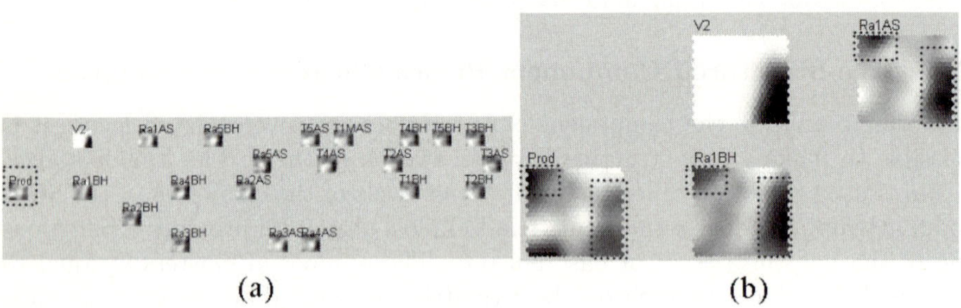

Fig. 2. Different levels of detail of the tree-structured component planes cluster representation. (a) When neighborhood is 3 ($n = 3$) temperature (T), radiation (Ra), sugar cane variety 2 (V2), and productivity (Prod) are in the same cluster. Productivity is shown on dotted line. (b) When neighborhood is 1 ($n = 1$) radiation of first month after seed (Ra1AS), radiation of first month before harvest (Ra1BH), sugar cane variety 2 (V2) and productivity (Prod) form a cluster. The local correlation between Ra1BH, Ra1AS and productivity is shown on dotted lines.

(c) Use the Vellido's algorithm to partition the map again increasing the neighborhood by 1.
  - If the cluster's quantity is equal to the previously obtained, repeat c.
  - If the cluster's quantity is different, save the results and repeat c.
  - Stop when the neighborhood value is the same to that of the maximal neighborhood value taken to train the SOM used for the projection.

2. Visualization.
   (a) Arrange the component planes at nodes of the tree generated in step 1.
   (b) Each group found with a specific neighborhood value is showed as a level of the tree and every cluster parent is connected to his children.
   (c) The cluster structure is visualized using the component planes already computed in the first SOM.

The aforementioned algorithm uses the projection of component planes in a SOM, and the distance matrix based clustering in a complementary fashion. Thus, a tree-structurated representation of the component planes clusters is built in order to improve the correlation hunting process. The highest confidence in the clustering result of this method is achieved when samples with visually similar component planes are placed in the same child of the clustering tree.

3 Case Study: Sugar Cane Culture

3.1 Problem Description

SOM has proved to be effective for the exploratory analysis of agro-ecologic data and has become a very useful technique in ecological modeling [7]. SOMs are recommended in cases when it is essential to extract features out of a complex data set [1]. Moreover, it is useful for generating easily comprehensible low-dimensional maps, improving the visualization and data interpretation [2,4]. For these reasons, methodologies based in SOM were selected as tools for exploring the data in this study case. The objective of this case study was to determinate which variables are more related to the productivity in the sugar cane culture in a specific region. The methodologies previously shown were used to classify zones with similar productivity, in order to find similar patterns of behavior. Finally, analyzing these patterns it was possible to acquire new knowledge about the relationship between the agro-ecological variables and productivity. A more detailed description of the problem is presented as following:

A plant is affected by diverse variables (e.g., climate, soil) during its life. These variables have different effects on the plant at different moments of its development (e.g., germination, flowering). Moreover, the combination and/or change of these variables in certain moments determines the development states of the plant. This mixture of factors finally determines the crop production. For example, in the sugar cane case, expert knowledge indicates that the most relevant periods are the beginning and the end of plant development. In the first months (after sowing) the vegetative structure is formed (e.g., leafs grows allowing the photosynthesis process), in this moment the water is very important to improve the development of the plant. During the last months (approximately thirteen months after sowing) the plant accumulates the major part of saccharose. In this moment not much water is essential because the plant is totally developed. These periods are the most important in the agricultural productivity. Accordingly, to determine how and when the variables affect the plant development would be very helpful to support decision making (e.g. in what moment to seed and/or to harvest in order to obtain a better productivity).

3.2 Classification of Agro-Ecological Variables Related with Productivity

The database used was provided by a sugar cane research center located in the region under study. The data base contains information collected during seven years (1999 to 2005). The agro-ecological variables used for this experiment are listed as follows. Climate variables are Temperature Average (T), Relative Humidity Average (RH), Radiation (Ra), and Precipitation (P). Soil variables are Order (Ord), Texture (Tex) and Depth (Dee). Topographic variables are Landscape (Ls) and Slope (Sl). Other variables are Water Balance (WB) and Variety (V). Finally, productivity (P) of each cultivated zone. As it was mentioned before, the most relevant periods in the sugar cane are the beginning and the end of plant development. Therefore, it is possible not using all the climate data set and to use only the data from $1,\ldots,x$ Months After Sowing (xAS) and $1,\ldots,x$ Months Before Harvest (xBH). In our case study $x = 5$ was used. Soil variables and Variety were ordered using a presence/absence coding, 0 represents presence and 1 absence. As a result, the vector which defines a cultivated zone (CZ) is compound of 54 variables, 1328 vectors were used representing each one the characteristics of a cultivated zone.

All the variables were scaled [-1,1] in order to allow their comparison in magnitude. Then, it was created a matrix with 1328 vectors CZs ($CZmatrix$) composed by 54 variables each one. Notice that the output of this sugar cane model is the productivity. Nevertheless, in this case the productivity was used as input in order to find the component planes related with. The $CZmatrix$ was used as input for a SOM with 400 neurons (20x20) and it was trained with the batch algorithm. With this SOM, it was possible to generate 54 component planes, one for each agro-ecological variable. These component planes were projected in a new SOM composed of 400 neurons (20x20) and it was trained with the batch algorithm. Finally, this last map was clustered and the clusters were organized. For this aim it was used the algorithm for tree-structured component planes clusters representation showed in the previous section. The results can be observed in figure 1.

Some interesting aspects can be found here. When $n = 3$ (where n is the neighborhood), it is possible to locate in a same cluster the temperature, radiation and production, each component plane with similar patterns, figure 1.b on dotted line and figure 2.a. In addition, it is possible to view when $n = 1$ that radiation of first month after seed, radiation of first month before harvest and productivity present similar patterns, figure 1.b on dotted line and figure 2.b. Thanks to the tree-structured representation, it is easier to group the clusters to facilitate the observation of local correlations. As an example, in the right side and in the top of left side of the component planes, figure 2.b, it was possible to see that when the productivity was high most of the values of Ra1BH and Ra1AS were high too (represented for a dark gray). Drawing the BMUs of the component planes productivity, Ra1BH and Ra1AS in a scatter plot, figure 3, it is possible to detect high values of productivity when there are high values of Ra1BH and Ra1AS.

As a conclusion, the radiation of the first month after seed and the radiation of the first month before harvest are more correlated with the productivity than the other variables. In addition, a local correlation is observed between a majority of high values of radiation and high productivity.

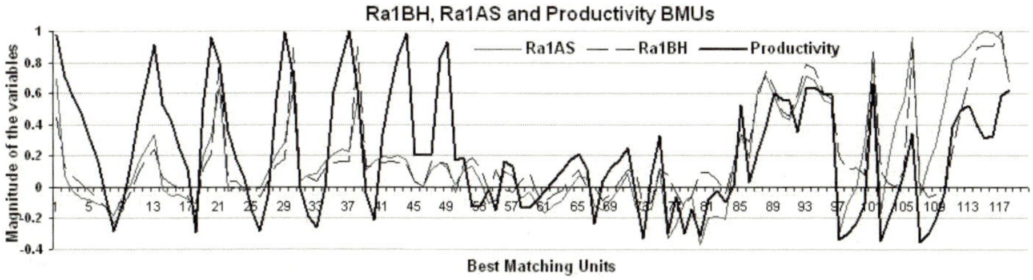

Fig. 3. BMUs of the component planes: productivity, radiation 1 month before harvest (Ra1BH) and radiation 1 month after seed (Ra1AS)

4 Conclusion

This paper has presented a methodology to enhance the component planes analysis process. This methodology improves the correlation hunting in the component planes with a tree-structured clusters representation based on the SOM distance matrix. This tree-structured representation permits the analysis of component planes clusters at several levels of detail. This methodology can be applied in cases where the number of component planes is very large, witch is quite often in agro-ecological modeling. As an case study, the methodology presented here was used in the classification of zones with similar agro-ecological conditions and productivity in the sugar cane culture. Analyzing the obtained groups of agro-ecological variables and cultivated zones it was possible to find a relationship between the radiation during the first month after seed, the first month before harvest, and high productivity. More analysis can be made in order to improve the decision support in the sugar cane culture based on the aforementioned methodology. This paper shows only a part of this work. Future work will be focus on the analysis of other patterns.

Although, the tree-structured is a good method to show clusters, it would also be desirable to obtain a measurement of similarity between clusters in each branch of the tree. Future work will be focused on the study of a similarity measurements to improve the tree-structured clusters representation.

Acknowledgements

This work is part of a cooperation project between BIOTEC, CIAT, CENICANA (Colombia) and HEIG-VD (Switzerland) named Precision agriculture and the construction of field-crop models for tropical fruits. The economical support is given by several institutions in Colombia (MADR, COLCIENCIAS, ACCI) and the State Secretariat for Education and Research (SER) in Switzerland.

References

1. Chon, T., Park, Y., Moon, K., Cha, Y.: Patternizing communities by using an artificial neural network. Ecological Modelling. 90, 69–78 (1996)
2. Chung, H., Hsieh, J., Chang, T.: Prediction of daily maximum ozone concentrations from meteorological conditions using a two-stage neural network. Atmospheric Research 81, 124–139 (2006)
3. Demartines, P., Hrault, J.: Curvilinear Component Analysis: a Self-Organizing Neural Network for Nonlinear Mapping on Data Sets. IEEE Transactions on Neural Network 8, 148–154 (1997)
4. Giraudel, J., Lek, S.: A comparison of self-organizing map algorithm and some conventional statistical methods for ecological community ordination. Ecological Modelling 146, 329–339 (2001)
5. Himberg, J.: Enhancing the SOM-based Data Visualization by Linking Different Data Projections. In: Proceedings of 1st International Symposium IDEAL'98, Intelligent Data Engineering and Learning–Perspectives on Financial Engineering and Data Mining, pp. 427–434 (1998)
6. Kohonen, T.: Self-Organizing Maps. Springer, Heidelberg (1997)
7. Liu, Y., Weisberg, H., He, R.: Sea surface temperature patterns on the West Florida Shelf using Growing Hierarchical Self-Organizing Maps. Journal of Atmospheric and Oceanic Technology 23, 325–338 (2006)
8. Sammon, J.A: Nonlinear Mapping for Data Structure Analysis. IEEE Transactions on Computers. 18, 401–409 (1969)
9. Sultan, M., Wigle, D., Cumbaa, C., Maziarz, M., Glasgow, J., Tsao, M., Jurisica, I.: Binary tree-structured vector quantization approach to clustering and visualizing microarray data. Bioinformatics 18, 111–119 (2002)
10. Tryba, V., Goser, K.: Self-Organizing Feature Maps for Process control in Chemistry. In: Proc. ICANN, Helsinki, pp. 847–852 (1991)
11. Ultsch, A., Siemon, P.: Kohonen's self organizing feature maps for exploratory data analysis. In: Proc. INNC'90, Int. Neural Network Conf, pp. 305–308 (1990)
12. Vellido, A., Lisboa, P., Meehan, K.: Segmentation of the on-line shopping market using neural networks. Expert Systems with Applications 17, 303–314 (1999)
13. Vesanto, J., Ahola, J.: Hunting for Correlations in Data Using the Self-Organizing Map. Proceeding of the International ICSC Congress on Computational Intelligence Methods and Applications, pp. 279–285 (1999)
14. Vesanto, J.: SOM-based data visualization methods. Intelligent Data Analysis. 3, 111–126 (1999)
15. Vesanto, J., Alhoniemi, E.: Clustering of the self-organizing map. IEEE Transactions on Neural Networks 11, 586–600 (2000)
16. Vesanto, J., Sulkava, M.: Distance Matrix Based Clustering of the Self-Organizing Map. In: Dorronsoro, J.R. (ed.) ICANN 2002. LNCS, vol. 2415, pp. 951–956. Springer, Heidelberg (2002)

A Dynamical Model for Receptive Field Self-organization in V1 Cortical Columns

Jörg Lücke

Gatsby Computational Neuroscience Unit, UCL, London WC1N 3AR, UK

Abstract. We present a dynamical model of processing and learning in the visual cortex, which reflects the anatomy of V1 cortical columns and properties of their neuronal receptive fields (RFs). The model is described by a set of coupled differential equations and learns by self-organizing the RFs of its computational units – sub-populations of excitatory neurons. If natural image patches are presented as input, self-organization results in Gabor-like RFs. In quantitative comparison with *in vivo* measurements, we find that these RFs capture statistical properties of V1 simple-cells that learning algorithms such as ICA and sparse coding fail to reproduce.

1 Introduction

Self-organizing systems are commonly used to study learning in biological networks and/or to learn from a set of presented inputs. Classical examples are systems that learn input categories [1] or systems that are learning the neighborhood relationship of the input data [2]. Based on recent results on the anatomical fine-structure of cortical columns [3], we show how self-organization can be used to extract basic constituents of the input. The presented bottom-up approach is based on earlier work on this subject [4,5] and shows the applicability of the approach to natural images. We find that the components extracted by the model have a higher degree of similarity with *in vivo* measurements of simple-cells than the classical algorithmic approaches of ICA [6,7] and sparse coding [8,9].

2 System Dynamics

Our model column consists of k neuron populations or *hidden units* $p_1, \ldots, p_k$ and N input units $y_1, \ldots, y_N$. The inputs $\tilde{I}_1, \ldots, \tilde{I}_k$ to the hidden units originate from external neural units $y_1, \ldots, y_N$ and influences the hidden units via afferent fibers $R_{\alpha j}$ with $I_\alpha = \sum_j R_{\alpha j} y_j$. We implement a feed-forward inhibition that ensures that the inputs $\tilde{I}_1, \ldots, \tilde{I}_k$ sum to zero: $\tilde{I}_\alpha = I_\alpha - \frac{1}{k}\sum_\beta I_\beta$. It follows that $\tilde{I}_\alpha = \sum_j R^{\mathrm{eff}}_{\alpha j} y_j$ with $R^{\mathrm{eff}}_{\alpha j} = R_{\alpha j} - \frac{1}{k}\sum_\beta R_{\beta j}$. $\mathbf{R}^{\mathrm{eff}}_1, \ldots, \mathbf{R}^{\mathrm{eff}}_k$ will be referred to as *effective RFs* or just *RFs* of the column. Fig. 1 visualizes the afferents and the internal connectivity of the system. Note that the dynamics is formulated on the population level and that there are different alternatives of its implementation. Fig. 1 is a visualization of entities that are relevant for the dynamics.

The hidden units of the system model sub-networks of excitatory neurons as found in cortical columns [3]. We use a polynomial approximation of such

J. Marques de Sá et al. (Eds.): ICANN 2007, Part II, LNCS 4669, pp. 389–398, 2007.

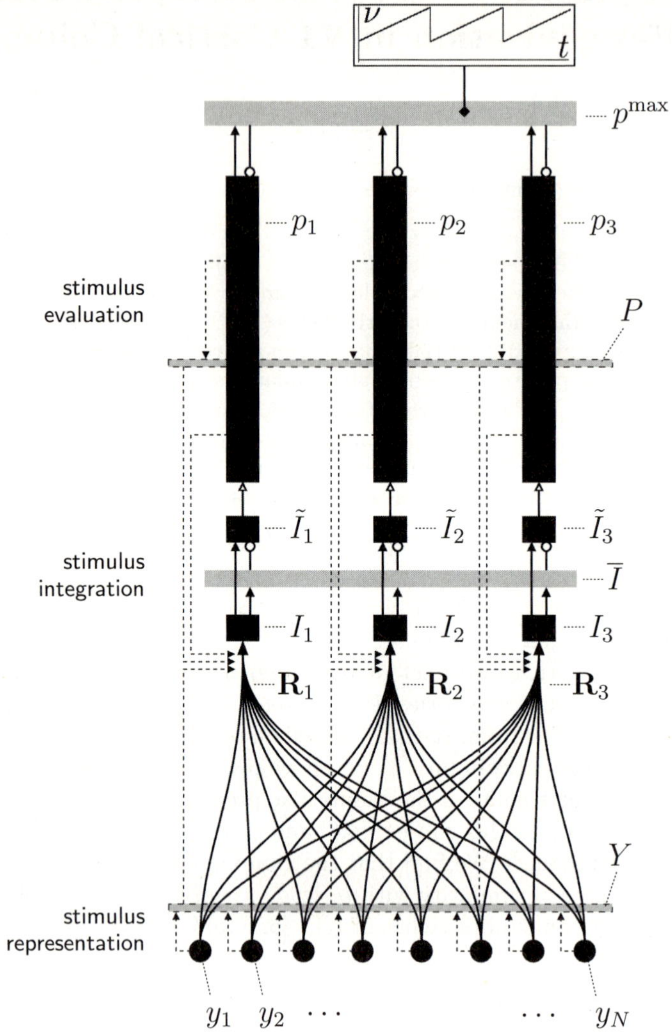

Fig. 1. Sketch of a cortical column with $k = 3$ sub-populations of excitatory neurons (visualized as black vertical bars). Input to the column originates from N external units y_1 to y_N. In a first processing stage the input is integrated. The inputs I_1 to I_3 are via feed-forward inhibition transformed to mean-free inputs $\tilde{I}_1$ to $\tilde{I}_3$. These inputs drive the self-excitatory sub-populations with activities p_1 to p_3. $\overline{I}$ is the mean input. Triangular arrow hats denote excitatory (solid) and mixed (hollow) influences, empty circles inhibitory influences. Lateral inhibition between the self-excitatory populations is modulated by the bifurcation parameter ν. Dashed lines indicate the influence of dynamic variables on the modification of the populations' receptive fields $\mathbf{R}_1$ to $\mathbf{R}_3$. P and Y are the sums of input unit activities and population activities, respectively. $p^{\max}$ is equal to the greatest population activity.

self-excitatory units as suggested in [10]. Together with a particular inhibitory coupling between the self-excitatory units the dynamics is given by:

$$\nu(t) = (\nu_{\max} - \nu_{\min})\tilde{t} + \nu_{\min}, \qquad \tilde{t} = \frac{1}{T}\mathrm{mod}(t,T), \tag{1}$$

$$\frac{d}{dt}p_\alpha = a\Big(\underbrace{p_\alpha^2}_{\text{self-excitation}} \underbrace{- \nu(t)\,p_\alpha \max_{\beta=1,\ldots,k}\{p_\beta\}}_{\text{lat. inhibition}} \underbrace{- p_\alpha^3}_{\text{self-inhibition}} \Big) \underbrace{+ \kappa \tilde{I}_\alpha}_{\text{input}} \underbrace{+ \sigma\, p_\alpha\, \eta_t}_{\text{noise}}, \tag{2}$$

where κ is the coupling to input $\tilde{I}_\alpha$ and where σ parameterizes multiplicative Gaussian white noise. Eqn. 1 represents a linear increase of the dynamics' bifurcation parameter ν. The time $\tilde{t}$ runs from 0 to 1 within the time interval $[0,T)$. After time $t = T$ a new input is presented and the increase of ν starts anew. We refer to such a cycle as ν-cycle. Dynamics (1) and (2) implement a particular kind of lateral competition between the hidden units that has proven to be advantageous for learning distributed input encodings. The parameter ν increases competition between the hidden units, and initially active units are deactivated during a ν-cycle. The dynamics of neural activity, (1) and (2), has been studied earlier [10] and represents an abstraction of a model that was based on sub-populations of explicitly modeled excitatory neurons [4].

The non-linear evaluation of an input pattern according to (1) and (2) couples into a dynamics of Hebbian-type synaptic plasticity given by:

$$\frac{d}{dt}R_{\alpha j} = \frac{\epsilon}{N}\left([p_\alpha]^+ y_j - [p_\alpha]^+ Y R_{\alpha j}\right) \;\text{ iff }\; P(t) < \chi\,, \tag{3}$$

where $P = \sum_{\alpha=1}^{k} p_\alpha$ and $Y = \sum_{j=1}^{N} y_j$ are overall-activities of hidden units and input units, respectively, and where $[p_\alpha]^+ = p_\alpha$ if $p_\alpha \geq 0$ and $[p_\alpha]^+ = 0$ otherwise. To learn with slowly increasing competition between the RFs[1] we change, after each ν-cycle, the threshold for learning χ and the maximal level of lateral inhibitory coupling $\nu_{\max}$:

$$\Delta\chi = -\lambda_\chi(\chi - a_\chi P) \quad \text{and} \quad \Delta\nu_{\max} = -\lambda_\nu(a_\nu - P), \tag{4}$$

where λ_χ and λ_ν are modification rates. The second equation increases $\nu_{\max}$ to counteract the effect of RFs that are increasingly specialized to the input.

Self-organization of an initially unstructured system is commonly characterized by: (A) random fluctuations that result in structural seeds, (B) a positive feed-back loop amplifying certain structures or modes, and (C) a negative feed-back loop that counteracts amplification and finally keeps the system in an active equilibrium. For our system we start in a state with all afferents initialized to the same value $R_{\alpha j} = \frac{1}{N}$. (A) After an input is presented, ν is increased and the noise in (2) breaks the symmetry among the activities p_α (compare [10]). Hidden units are deactivated until $P(t) < \chi$. During the remainder of the ν-cycle the RFs are modified according to their activities (3). This implies that just some

[1] Which is related but not identical to the competition between the hidden units.

or only one RF is significantly changed to become more similar to the presented input. (B) Hidden units with RFs similar to a certain type of input are likely to remain active if such an input is presented. Their RFs will therefore further specialize to this input type. (C) Lateral inhibition in (2) forces the system to specialize to different input patterns and the negative term in (3) prevents the afferents from growing infinitely.

For our dynamics the control of lateral inhibition represents the crucial part. If we learn *after* inhibition selects a single unit, i.e. as in winner-take-all (WTA) networks, a distributed encoding of presented input is not observed even if it consists of easily to identify combinations of basic components. The same applies if we learn *after* inhibition has selected, e.g., K units (K-WTA). In contrast, dynamics (3) modifies RFs *during* the process of deactivation of hidden units. Furthermore, learning favors input that results in a relatively sparse activation of the hidden layer (small $P(t)$). If we learn according to dynamics (1) to (4) and if the presented input consists of combinations of basic constituents, the system self-organizes its RFs to represent these constituents. Continuously increasing competition between the RFs (4) crucially helps in guiding the self-organization process. With increasing competition groups of initially similar RFs (we initialize with a relatively large χ) decay into ever smaller sub-groups. For a similar system, such a type of self-organization was therefore termed *hierarchical self-organization* in [11].

Equations (1) to (4) represent a dynamical model of a cortical column. Before applying the dynamics to input, we have to find a set of parameters that lets the system operate in the interesting non-linear regime between no competition and competition with WTA characteristics. Choosing parameters is straightforward and good results can be obtained for a large range of different parameters. To determine the particular set of parameters used in this paper[2] we have tuned the parameters using the so-called bars test [12] as a benchmark. Once the set of parameters is chosen, the system is extraordinarily robust with respect to different types of input.

3 Simulation Results

To model input to our dynamics that resembles input received by cortical columns in V1, we will use gray-level images patches as input. We expect the column model to self-organize its RFs such that the activities of hidden units can represent the input distributedly. Input is taken from 20 randomly selected images of the van Hateren database [7] that do not show man-made structures. We use difference of Gaussians (DoG) transformed versions of these images to emulate the preprocessing of visual input by the retina and the lateral geniculate nucleus (LGN). We used a standard deviation of $\sigma_+ = 1.0$ pixels for the positive part and $\sigma_- = 3.0$ pixels for the negative part, consistent with the biologically measured ratio [13] of $\frac{\sigma_+}{\sigma_-} \approx \frac{1}{3}$. From the 20 images we randomly selected patches of

[2] Eqn. 1: $\nu_{\min} = 0.4$, $T = 25$ms; Eqn. 2: $a = 200\text{ms}^{-1}$, $\kappa = 1.0\text{ms}^{-1}$, $\sigma = 0.01\text{ms}^{-1}$; Eqn. 3: $\epsilon = 0.02$; Eqn. 4: $\lambda_\chi = 5 \times 10^{-5}$, $a_\chi = 1.2$, $\lambda_\nu = 10^{-3}$, $a_\nu = 0.7$.

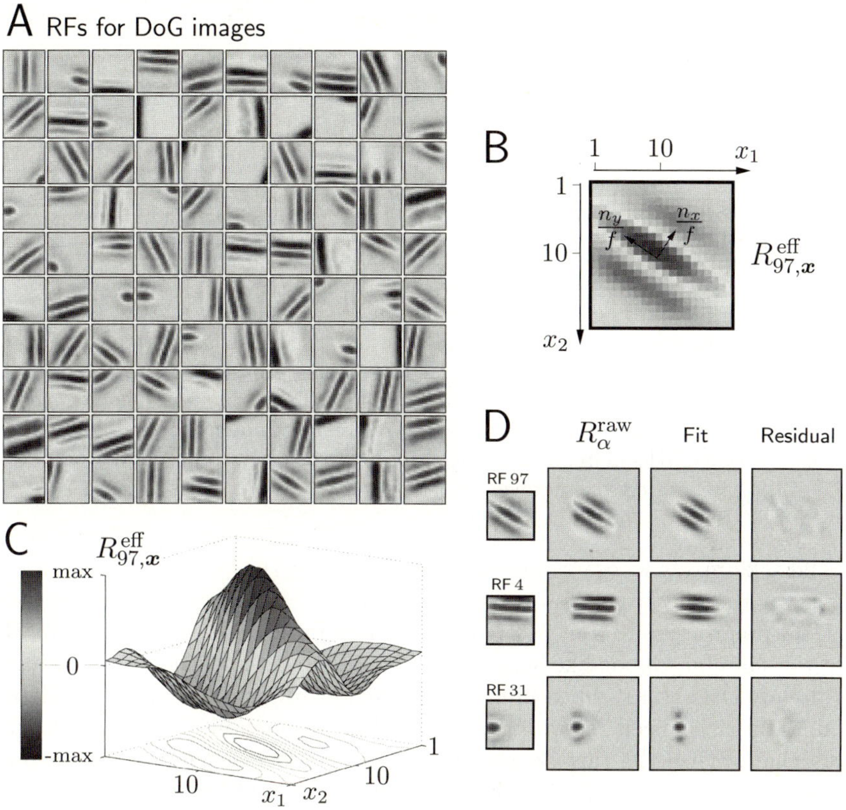

Fig. 2. RFs of the model column if natural images are used as input. **(A)** Effective RFs $\mathbf{R}^{\text{eff}}_{\alpha}$ after 2×10^6 ν-cycles if difference of Gaussians (DoG) filtered images are used as input. For each of $k = 100$ RFs, values are color coded to lie between dark blue (-max) and dark red (+max) where max is the maximal absolute value of the RF, $\text{max} = \max_j |R^{\text{eff}}_{\alpha j}|$, which ensures that $\mathbf{R}^{\text{eff}}_{\alpha} = 0$ is assigned to the same color (green) for all RFs. See **(C)** for the color coding scheme. **(B)** Effective RF $\alpha = 97$ of the simulation displayed in **(A)**. The index j is replaced by the two-dimensional vector $\boldsymbol{x}$. The same coding scheme as in **(A)** is used with $\text{max} = 1.67 \times 10^{-3}$ in this case. The small arrows in the center represent the principal axes of the Gaussian envelope after the RF was matched with a Gabor wavelet. The lengths of the arrows are the wavelet's standard deviations $\sigma_x = \frac{n_x}{f}$ and $\sigma_y = \frac{n_y}{f}$. The dimensionless entities n_x and n_y (f is the wavelet frequency) are used for further analysis (see Fig. 4). **(C)** The same RF as in **(B)** plotted in three dimensions to illustrate the color coding. **(D)** Three examples of matching the RFs with Gabors. RFs 4 and 31 illustrate artifacts of rectangular sampling and Gabor matching, respectively. Columns show original RFs (1st column), corresponding filters that act on the raw pixel images (2nd column), Gabor wavelet matches of the raw filters (3rd column), and differences (residuals) of raw filters and Gabor fits, which shows the error made by the wavelet approximation (4th column).

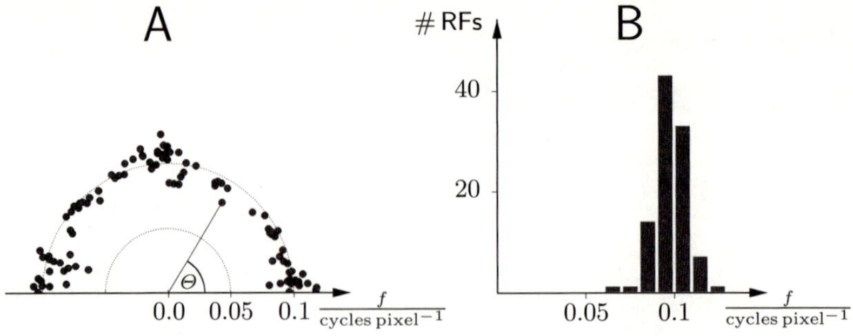

Fig. 3. Orientation and frequency distribution of RFs. (**A**) Distribution of frequency vs. angle for the RFs displayed in Fig. 2A after being transformed to filters on the raw pixel values and matched with Gabor wavelets. RFs are relatively homogeneously distributed in orientation space apart from a preference of horizontal and vertical RFs. (**B**) Distribution of RFs in frequency space. Receptive fields cluster around $f \approx 0.1 \frac{\text{cycles}}{\text{pixel}}$ and cover approximately one octave.

$N = 20 \times 20$ pixels whose values were scaled linearly to lie in the interval $[0, 1]$. In Fig. 2A the effective RFs of a system with $k = 100$ hidden units are shown for DoG preprocessed input after being trained for 2×10^6 ν-cycles[3]. The RFs have the familiar Gabor-like shape (see Fig. 2C) and represent filters acting on the already DoG transformed images. For comparison with physiological data as obtained in simple-cell recordings, we first have to compute the corresponding filters that act on the raw images. For DoG preprocessed input this amounts to a convolution of the RFs in Fig. 2A using the same DoG-kernel as for the preprocessing of the image. In Fig. 2D the resulting filters are shown for three examples. The filters are, in theory, infinitely large but are virtually zero for all pixels outside a central region of 40×40 pixels. For further analysis, and as is customary in the literature, we match the real filters using Gabor wavelet functions (see e.g. [14,15]). Matching works well in most cases but the artificial rectangular sampling can result in notable artifacts. The effect of these artifacts on the later analysis can be well understood, however, and they will be discussed below using RFs 4 and 31 in Fig. 2D.

An analysis of the parameters of the matched Gabors shows a relatively even distribution of RF positions (data not shown). Plotting orientation vs. frequency (Fig. 3A) shows a distribution similar to the ones obtained by using independent component analysis (ICA) [6,7] and sparse coding [8,9]. The orientation preferences are relatively evenly distributed apart from a stronger preference for

[3] To reduce undesirable boundary effects, image patches (20×20 pixels) are large compared to patch sizes used by other methods (e.g. 12×12 in ICA [6] and sparse coding [8,9]). Larger patch sizes exceed the already extensive computational resources required to simulate dynamics (1) to (4). Note that the range of preferred spatial frequencies is determined by the DoG preprocessing (Fig. 3B). Increasing this frequency with a smaller DoG kernel is not possible because the standard deviation of its positive part is already as small as one pixel.

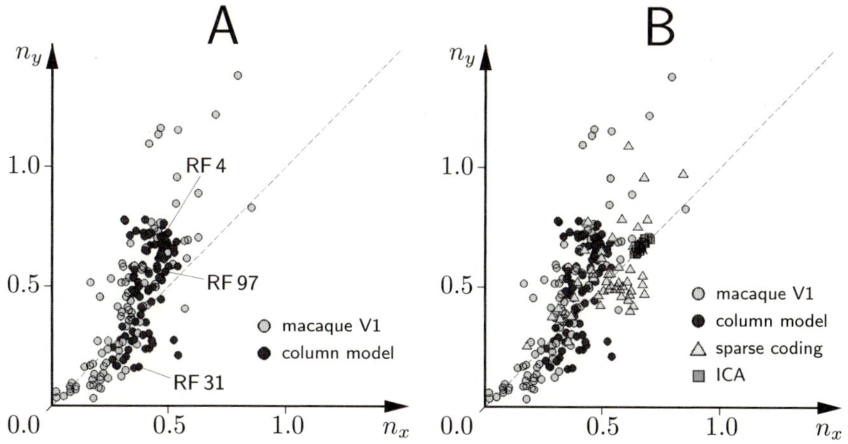

Fig. 4. Distributions of RF extensions parallel and orthogonal to the RFs' wave vector (compare Fig. 2B). n_x and n_y are parameters of wavelets that were matched to the RFs acting on the raw input, $\mathbf{R}^{\text{raw}}$ (compare Fig. 2D). n_x is the size of the Gaussian envelope in wave vector direction expressed in terms of the wave length of the wavelet and n_y is the size of the Gaussian envelope orthogonal to the wave vector. **(A)** Distribution in the n_x/n_y-plane of the $k = 100$ RFs displayed in Fig. 2A (blue) together with the distribution of macaque simple-cells (bright magenta) as measured *in vivo* [15]. The data points of the three RFs displayed in Fig. 2D are explicitly labeled. **(B)** The distributions displayed in **(A)** overlaid with the corresponding distributions of RFs obtained using sparse coding (yellow) as described in [9] and ICA (green) as described in [7]. Data are taken from [15]. The dashed diagonal line is the bisection line.

vertical and horizontal orientations. As for sparse coding and ICA, and unlike RFs measured *in vivo*, the RFs obtained in our model are clustered around a preferred frequency which is given by $f \approx 0.1\frac{\text{cycles}}{\text{pixel}}$ (see Fig. 3). In our case, the frequency distribution is largely explained by the bandpass properties of the DoG preprocessing, which prefers frequencies in the range of the obtained filters.

The most notable feature of the RFs in Fig. 2A is the variation in the widths and lengths of their Gaussian envelopes. In [15] an analysis of properties of the envelopes relative to the wavelets' frequency was suggested as a means for comparing the RFs of computational models with data. For quantitative comparison the RF parameters are plotted in the n_x/n_y-plane with $n_x = \sigma_x f$ and $n_y = \sigma_y f$ where f is the frequency of the matched wavelet and σ_x and σ_y are, respectively, the standard deviations of the Gaussian envelope in the direction of, and perpendicular to, the wave vector (compare Fig. 2B). Fig. 4A shows the distribution predicted by our model together with *in vivo* measurements of simple-cell RFs in macaque primary visual cortex [15]. Both distributions have similar variance and show the same correlation between n_x and n_y, with a preference for RFs to be elongated in the n_y-direction if they are distant from the origin. Note that the RF distribution in cat primary visual cortex also shows this property [14,15]. Two notable differences between the measured distribution and the RFs of the model are the absence of model RFs with values near the origin and with values far away from it.

The absence of RFs distant from the origin can be explained by the chosen data representation which restricts RFs to a rectangular patch size. As can be seen in Fig. 2D (2nd row), the Gabor match results in a Gaussian envelope that is artificially restricted in n_y-direction. Because of the range of preferred frequencies, Fig. 3, this prevents values of n_y from being larger than $n_y \approx 0.8$. The cluster of data points near $(0.5, 0.7)$ in Fig. 4A can therefore be interpreted as a consequence of rectangular patch sizes (compare RF 4). Less artificial or larger patches would move some of these points, e.g. RF 4, further away from the origin and presumably closer to the measured data.

RFs with values near to the origin are associated with what was called 'globular' RFs in [15], i.e., RFs with no or weak orientation preference. Although RFs with weak orientation preference are actually developed by our system (see, e.g., RF 31 in Fig. 2D), the plot in Fig. 4 does not show any points near the origin. The reason for this is that we use Gabor wavelets to match localized RFs whose positive and negative parts sum to zero. Even in the case of perfectly radial symmetric RFs, Gabor matching would break the symmetry to a preferred orientation (compare Fig. 2D, 3rd row). RFs measured *in vivo* have values near the origin in Fig. 4 because they are not subject to this effect. Many of the simple-cell RFs in [15] are therefore essentially matched by a Gaussian, which represents a degenerated wavelet with zero frequency.

Comparison of the distribution predicted by our model and distributions predicted by sparse coding [8,9] and ICA [6,7] are shown in Fig. 4B. Sparse coding seems to partly predict the measured n_y/n_x-correlation but shows an incorrect preference for RFs elongated in n_x-direction near $(0.5, 0.5)$. The distribution of RFs obtained using ICA is concentrated in a small region on the bisecting line near $(0.7, 0.7)$. ICA neither predicts the variety of Gaussian envelopes nor any preference for RF elongation in any direction.

In contrast to sparse coding and ICA, the distribution predicted by our model is in good agreement with *in vivo* measurements (see Fig. 4). It has to be clearly stated, however, that the results depend on various choices made in the experiment and the analysis. In particular these are: the DoG preprocessing, rectangular patches that restrict RFs, and Gabor matching which can artificially remove globular RFs. Likewise, different preprocessing and analysis techniques can affect the RF properties of other computational systems (see [15] for a discussion). Bearing in mind these various influences, it can nevertheless be stated that the dynamics described in this paper captures a property of simple-cell RFs that has not been reproduced by earlier systems. That is, the broad distribution of RFs in the n_y/n_x-space and their n_y/n_x-correlation with elongation of RFs in n_y-direction if they are distant from the origin (Fig. 4). For our RFs this property is already recognizable by considering the raw RF data as displayed in Fig. 2A.

4 Discussion

Motivated by cortical interconnectivity and earlier modeling work we have developed and analyzed a dynamical bottom-up model of a cortical column. The system models the first stages of columnar processing and self-organization of afferent fibers. Its dynamics is based on self-excitatory populations, feed-forward inhibition, and modulation of lateral inhibitory coupling (compare [3] and [16]). Anatomically the model predicts that two stages of processing (see Fig. 1) are required to enable the emergence of a distributed stimulus encoding. The integration stage has linear response properties and projects to an evaluation stage with non-linear responses of neurons. For V1 our model predicts that these two stages are a pre-requisite for the emergence of Gabor-like RFs. For natural images, no simple-cell-like RFs were obtained without feed-forward inhibition. With the use of feed-forward inhibition, self-organization generated a rich diversity of RFs if randomly selected and DoG filtered image patches were presented. Similar to properties of simple-cell RFs and classical models thereof the obtained RFs show sensitivity to different spatial orientations, frequencies, and locations. Furthermore, as analyzed by matching Gabor wavelets, the RFs show a specific variety in the extents of their Gaussian envelopes relative to their frequencies. This feature is consistent with *in vivo* measurements of simple-cell like RFs [14,15] (Fig. 4A) and has earlier not been reproduced to such an extend as reported here. Only very recently, in two functionally motivated approaches developed in parallel to this work, distributions of Gaussian envelopes comparable to the one presented here were reported. The resulting distributions in [17] are broader than *in vivo* measurements, however. In the model suggested in [18], which implements a form of sparse coding, RF distributions are obtained that contain globular RFs. These RFs can be matched by degenerated Gabor wavelets and correspond to points near the origin of an n_x/n_y-plot. Otherwise the RFs are reminiscent of those obtained by sparse coding (compare Fig. 4B). That is, they partly match the measured distributions well but show, distant from the origin, numerous RFs that are incorrectly elongate in n_x- instead of n_y-direction. Also note that the model in [18] was tuned to fit the data. The classical models for the emergence of simple-cell RFs [8,7,6], in particular ICA, do not accurately reproduce the experimental data (see Fig. 4B). Bottom-up models for the emergence of Gabor-like RFs include BCM [19] and CBA [20], which model learning based on single neurons. Quantitative comparison of the RFs obtained with these systems is difficult. To the knowledge of the author, no data about the variability of Gaussian envelops (as used in Fig. 4) is available for BCM; and RFs obtained with CBA do not seem localized enough for an analysis using Gabor matching.

To conclude, we have defined and simulated a dynamical model of a cortical column. The dynamics evaluates input using a balance between excitation and two forms of inhibitory interaction. If coupled to Hebbian synaptic plasticity, the dynamics induces a self-organization process of afferent fibers. If natural images are used as input, self-organization results in Gabor-like RFs of columnar subpopulations. These RFs match the variability of simple-cell RFs better than classical methods. The dynamics models important functions associated with cortical columns and represents, by directly combining cortical anatomy and the emergence of neuronal RFs, a coherent model of the first stages in columnar processing.

Acknowledgments. I gratefully acknowledge funding by the Gatsby Charitable Foundation and by the EU project FP6-2005-015803.

References

1. Carpenter, G.A., Grossberg, S.: Adaptive resonance theory. In: Arbib (ed.) The handbook of brain theory and neural networks, pp. 87–90. MIT Press, Redmond, Washington (2003)
2. Kohonen, T.: Self-Organizing Maps. Springer, Heidelberg (1995)
3. Yoshimura, Y., Dantzker, J.L., Callaway, E.M.: Excitatory cortical neurons form fine-scale functional networks. Nature 433, 868–873 (2005)
4. Lücke, J., von der Malsburg, C.: Rapid processing and unsupervised learning in a model of the cortical macrocolumn. Neural Computation 16, 501–533 (2004)
5. Lücke, J., Bouecke, J.D.: Dynamics of cortical columns – self-organization of receptive fields. In: Duch, W., Kacprzyk, J., Oja, E., Zadrożny, S. (eds.) ICANN 2005. LNCS, vol. 3696, pp. 31–37. Springer, Heidelberg (2005)
6. Bell, A.J., Sejnowski, T.J.: The independent components of natural scenes are edge filters. Vision Research 37(23), 3327–3338 (1997)
7. van Hateren, J.H., van der Schaaf, A.: Independent component filters of natural images compared with simple cells in primary visual cortex. Proceedings of the Royal Society of London B 265, 359–366 (1998)
8. Olshausen, B.A., Field, D.J.: Emergence of simple-cell receptive field properties by learning a sparse code for natural images. Nature 381, 607–609 (1996)
9. Olshausen, B.A.: Probabilistic Models of the Brain: Perception and Neural Function. In: chapter Sparse Codes and Spikes, pp. 257–272. MIT Press, Redmond, Washington (2002)
10. Lücke, J.: Dynamics of cortical columns – sensitive decision making. In: Duch, W., Kacprzyk, J., Oja, E., Zadrożny, S. (eds.) ICANN 2005. LNCS, vol. 3696, pp. 25–30. Springer, Heidelberg (2005)
11. Lücke, J.: Hierarchical self-organization of minicolumnar receptive fields. Neural Networks 17/ 8-9, 1377–1389 (2004)
12. Földiák, P.: Forming sparse representations by local anti-Hebbian learning. Biological Cybernetics 64, 165–170 (1990)
13. Somers, D., Nelson, S., Sur, M.: An emergent model of orientation selectivity in cat visual cortical simple cells. The Journal of Neuroscience 15, 5448–5465 (1995)
14. Jones, J.P., Palmer, L.A.: An evaluation of the two-dimensional Gabor filter model of simple receptive fields in cat striate cortex. Journal of Neurophysiology 58(6), 1233–1258 (1987)
15. Ringach, D.L.: Spatial structure and symmetry of simple-cell receptive fields in macaque primary visual cortex. Journal of Neurophysiology 88, 455–463 (2002)
16. Dantzker, J.L., Callaway, E.M.: Laminar sources of synaptic input to cortical inhib. interneurons and pyram. neurons. Nature Neuroscience 3, 701–707 (2000)
17. Osindero, S., Welling, M., Hinton, G.E.: Topographic product models applied to natural scene statistics. Neural Computation 18, 381–414 (2006)
18. Rehn, M., Sommer, F.T.: A network that uses few active neurones to code visual input predicts the diverse shapes of cortical receptive fields. Journal of Computational Neuroscience (2006)
19. Bienenstock, E.L., Cooper, L.N., Munro, P.W.: Theory for the development of neuron selectivity. The Journal of Neuroscience 2(1), 32–48 (1982)
20. Jerez, J., Atencia, M., Vico, F.J.: A Learning Rule to Model the Development of Orientation Selectivity in Visual Cortex. Neural Processing Letters 21, 1–20 (2005)

Meta-evolution Strategy to Focused Crawling on Semantic Web

Jason J. Jung[1], Geun-Sik Jo[1], and Seong-Won Yeo[2]

[1] School of Computer and Information Engineering, Inha University
253 Yonghyun-Dong, Nam-Gu, Incheon 402-751 Korea
j2jung@intelligent.pe.kr, gsjo@inha.ac.kr
[2] Siemens VDO Automotive AG, Korea
seongwon.yeo@siemens.com

Abstract. In this paper, we propose an evolutionary approach to deal with shortcomings on conventional focused crawling systems in semantic web environment. Thereby, meta-evolution strategy for collaboration among multiple crawlers has to be efficiently carried out. It is based on incremental aggregation of partial semantic structures extracted from web resources, which are in advance annotated with local ontologies. To do this, we employ similarity-based matching algorithm, so that fitness function is formulated by summing all possible semantic similarities. As a result, the best mapping condition (i.e., the fitness is maximized) is obtained for efficiently i) reconciling semantic conflicts between multiple crawlers, and ii) evolving semantic structures of web spaces over time.

1 Introduction

Since the first idea of focused web crawlers was introduced in [1], this work has been regarded as a potential solution to the problem of indexing the exponentially growing information on the web. *Focused crawling* is designed to only gather documents on a specific topic, thus reducing the amount of network traffic and downloading cost. Basically, focused crawling has to be able to analyze the behavior information about the users. Examples of such data are social connections [2], a set of bookmarks [3], and a set of URL sequences visited by users [4]. As depended on the characteristics of a given dataset, the proper methodologies such as heuristic searching, graph matching, time-series analysis, and semi-supervised learning should be chosen (sometimes, in hybrid manners).

Now, in this paper, we consider on the focused crawling process on semantic web space[1] in which resources are, in advance, annotated by referring to the local domain ontologies. Each user behavior while browsing these semantic web spaces can be enriched with semantic information (e.g., metadata) extracted from the corresponding resources. Thereby, the crawler can be expected to efficiently estimate the particular intention (or contexts) of users for assisting their information searching tasks. More importantly,

[1] In fact, we have firstly introduced a preliminary study of focused crawling on semantic space with an example of the product image files assumed to be annotated with the local domain ontologies of e-commerce business sites [5].

J. Marques de Sá et al. (Eds.): ICANN 2007, Part II, LNCS 4669, pp. 399–407, 2007.

multiple crawlers can share and integrate the pieces of partial information about a certain web space all together.

However, problem is that local domain ontologies are semantically heterogeneous with each other. Such heterogeneities are caused by lexical differences as well as structural mismatches [6]. Thereby, we propose an evolution strategy to cooperate to find out the optimal semantic structures of local ontologies. This crawling system is composed of a centralized meta-crawler, and multiple crawlers. Mainly, the meta-crawler can merge the semantic substructures discovered by multiple crawlers, and spread the refined (or crystallized) semantic information for the other crawlers. We want to note our contributions of this paper, as follows;

- Minimization of semantic conflicts between partial knowledge obtained from the crawlers, and
- Detection of newly changed semantic information in evolvable ontologies.

The remainder of this paper is as follows. In the following Sect. 2, we simply address the problem and preliminary notations for formalizing the problem. The evolution strategy will be explained in Sect. 3, and experimentation will be shown in Sect. 4. Finally, we will compare the proposed approach with the previous work in Sect. 5, and draw a conclusion in Sect. 6.

2 Problem Description

Given a set of candidate (or frontier) resources on semantic web space, we want user intention model to be comparable with annotations[2] of the resources.

Definition 1 (User intention). *User intention is represented as*

$$\mathcal{FU}_i = \{\langle f_\alpha, w_\alpha\rangle | f_\alpha \in \widetilde{\mathcal{FS}}_j, w_\alpha \in [0,1]\} \tag{1}$$

where each feature f_α (e.g., labels obtained from annotations) is assigned the weight value w_α between $[0,1]$, and initial value Δ is defined by the user (in this paper, as 0.5). $\widetilde{\mathcal{FS}}_j$ stands for a set of features of j-th semantic web space.

More importantly, the user intention model is changed, as he continues to browse. We note that it is based on i) adjusting feature weights with ii) merging features.

Property 1 (Adaptability). With resources selected by users over time, the user model $\mathcal{FU}_i^{(t)}$ is changed into

$$\mathcal{FU}_i^{(t+1)} = \mathcal{FU}_i^{(t)} + \{f_\beta | f_\beta \in \mathcal{FS}_j\}^{(t)}. \tag{2}$$

If f_β discovered at (t) is already existing in $\mathcal{FU}_i^{(t)}$, the corresponding weight value is reinforced (and the rest are discouraged) by

$$w_\alpha^{(t+1)} = w_\alpha^{(t)} \times \begin{cases} 1+\eta & \text{if } f_\beta = f_\alpha\ (\in \mathcal{FU}_i^{(t)}) \\ 1-\eta \times \frac{w_\alpha^{(t)}}{N_i^{(t)}-1} & \text{otherwise.} \end{cases} \tag{3}$$

[2] Resource annotation is represented as a set of triples, e.g., $\langle f_i, r_{ij}, f_j\rangle$. These features f_i and f_j can be regarded as classes in ontologies. For details, please refer to [6].

where η is the coefficient for the learning rate. $N_i^{(t)}$ is the size of $\mathcal{FU}_i^{(t)}$, i.e., the number of different features related to the corresponding user U. On the other hand, if feature f_β is new, it should be simply merged. We can assert these unmatched features at time t into $\mathcal{FU}_i$, because they can be possibly related to user intentions and preferences. Let the matched and unmatched features denoted as $\mathcal{F}_i^+$ and $\mathcal{F}_i^-$, respectively. (In this paper, they are simply obtained by string matching between a feature and a label from an annotation.) The matching ratio ρ^+ is defined as $\rho^+ = \frac{|\mathcal{F}_i^+|}{|\mathcal{FU}_i|}$ and we can trivially compute the unmatching ratio $\rho^- = 1 - \rho^+$.

Thus, the size of $\mathcal{FU}_i^{(t)}$ is expanded to

$$\mathcal{N}_i^{(t+1)} = \mathcal{N}_i^{(t)} + |\mathcal{F}_i^-| \tag{4}$$

where $|\mathcal{F}_i^-| \leq |\mathcal{FU}_i|$, and also, the weight values of $\mathcal{F}_i^-$ should be initialized by Δ. In next step, all the weight values in $\mathcal{FU}_i$ should be normalized by

$$w'_\alpha = \frac{w_\alpha}{\sum_{\langle f_\gamma, w_\gamma \rangle \in \mathcal{FU}_i} w_\gamma} \tag{5}$$

and then, $\mathcal{FU}_i$ is represented as $\{\langle f_\alpha, w'_\alpha \rangle | f_\alpha \in \mathcal{FS}_j, w'_\alpha \in [0,1]\}$. This can be regarded as the error reduction procedure caused by initialization of newly merged features.

Now, we want to briefly describe the problem caused by semantic heterogeneity. As browsing the resources under heterogeneous spaces, the user intention model $\mathcal{FU}_i$ is repeatedly expanded by $|\mathcal{F}_i^-|$ as shown in Equ. 4. It makes the existing major features converges to zero by subtracting $\Delta \cdot \left(\frac{|\mathcal{F}_i^-|}{\mathcal{N}_i + |\mathcal{F}_i^-|}\right)$ and much harder to discriminate the user intention. This problem is very similar to the so-called "curse of dimensionality" in machine learning community. In order to maintain the discrimination power, the crawlers need to keep the number of features constant ($|\mathcal{F}_i^-| = 0$) or, if possible, lower ($\mathcal{N}_i^{(t+1)} < \mathcal{N}_i^{(t)}$) by mapping semantic features (or substructures).

3 Evolution Strategy Discovering Semantic Structures

Once a crawler c_i of user u_i finds new feature $f_i^{(t)}$ from a semantic web space $\mathcal{FS}_j$, it has to share the feature with other crawlers via a centralized meta-crawler, as shown in Fig. 1. As introduced in [7], the meta-crawler needs to conduct *knitting* operation $\mathcal{K}(\mathcal{FS}_j, \bigcup_{c_i \in C} \langle f_\alpha, rel_{\alpha\beta}, f_\alpha \rangle)$, i.e., aggregating semantic substructures to find out the best mappings among all of the substructures.

3.1 Measuring Semantic Similarity

Beside the numerous relationships that can be found, new relationships between features can be inferred between $\mathcal{FS}_j$ and $\langle f_\alpha, rel_{\alpha\beta}, f_\alpha \rangle$. One particular relationship that will be interesting here is *similarity*. In order to obtain possible relationships between different semantic structures, semantic identification of the entity pairs (i.e., correspondences) is very important. As a matter of fact, most of the matching algorithms use some

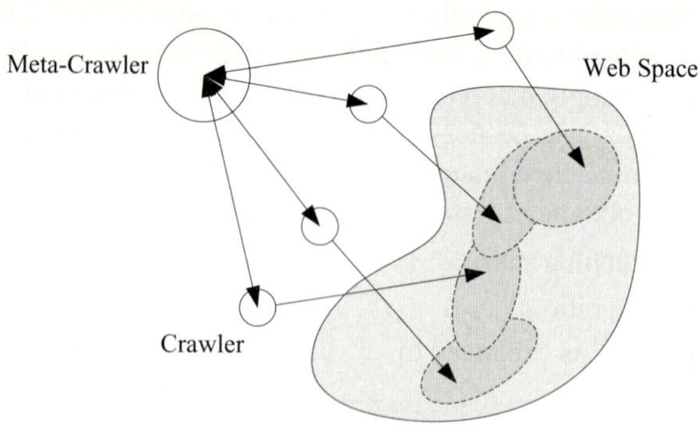

Fig. 1. Collaborative focused crawling architecture; The dotted circles are indicating the partial knowledge discovered by multiple crawlers

similarity measure (or distance) in order to match entities. Some distances can be established from the local features of entities. For instance, the name of entities can be the basis for matching them. Many techniques have been developed for comparing strings, based on their structures (like edit distance), their morphology (through lemmatization), their entry in lexicons (using WordNet). Some other distances, more in the spirit of network analysis, can be defined from the structure of the network. For instance, Euzenat and Valtchev [8] defines all possible similarities (e.g., Sim_C, Sim_R, Sim_A) between classes, relationships, attributes, and instances.

Given a pair of features from an annotation newly discovered from a web space by a crawler and semantic substructure already established in meta-crawler, semantic similarity measure $Sim_{\mathcal{F}}$ is assigned in $[0, 1]$.

Definition 2 (Semantic similarity). *Semantic similarity ($Sim_{\mathcal{F}}$) between f and f' is defined as*

$$Sim_{\mathcal{F}}(f, f') = \sum_{E \in \mathcal{N}(\mathcal{F})} \pi_E^{\mathcal{F}} MSim_Y(E(f), E(f')) \quad (6)$$

where $\mathcal{N}(\mathcal{F}) \subseteq \{E^1 \dots E^n\}$ is the set of all relationships in which features participate (for instance, subclass, instances, or attributes). The weights $\pi_E^{\mathcal{F}}$ are normalized (i.e., $\sum_{E \in \mathcal{N}(\mathcal{F})} \pi_E^{\mathcal{F}} = 1$).

If we consider feature labels (L) and three relationships in $\mathcal{N}(\mathcal{F})$, which are the superclass (E^{sup}), the subclass (E^{sub}) and the sibling feature (E^{sib}), Equ. 6 is rewritten as:

$$\begin{aligned} Sim_{\mathcal{F}}(f, f') = {} & \pi_L^{\mathcal{F}} sim_L(L(A_i), LF(B_j)) \\ & + \pi_{sup}^{\mathcal{F}} MSim_{\mathcal{F}}(E^{sup}(f), E^{sup}(f')) \\ & + \pi_{sub}^{\mathcal{F}} MSim_{\mathcal{F}}(E^{sub}(f), E^{sub}(f')) \\ & + \pi_{sib}^{\mathcal{F}} MSim_{\mathcal{F}}(E^{sib}(f), E^{sib}(f')). \end{aligned} \quad (7)$$

where the set functions $MSim_{\mathcal{F}}$ compute the similarity of two entity collections.

As a matter of fact, a distance between two set of features can be established by finding a maximal matching maximizing the summed similarity between the features:

$$MSim_{\mathcal{F}}(S,S') = \frac{\max(\sum_{\langle f,f'\rangle \in Pairing(S,S')} (Sim_{\mathcal{F}}(f,f'))}{\max(|S|,|S'|)}, \tag{8}$$

in which $Pairing$ provides a matching of the two set of classes. Methods like the Hungarian method allow to find directly the pairing which maximizes similarity. The OLA algorithm is an iterative algorithm that compute this similarity [8]. This measure is normalized because if $Sim_{\mathcal{F}}$ is normalized, the divisor is always greater or equal to the dividend.

Finally, given two semantic structure, we can measure the semantic similarity by using Equ. 8.

3.2 Meta-evolution Strategies

Meta-evolution strategy is a global combinatorial optimization method based on heuristics [9]. In this paper, the meta-crawler can find out the optimal state with the fitness function. Once the meta crawler can gather semantic substructures from multiple crawlers, it has to find out the best mapping between a set of the substructures and local ontology (we do not need to consider how much the onology is constructed). By using semantic similarity measurement given in Equ. 6, the fitness function is defined, as follows.

Definition 3 (Fitness). *Given a semantic web space $\tilde{\mathcal{FS}}_j$, let a meta-crawler be able to communicate with K crawlers $\{c_k|k \in [1,K]\}$. The fitness function is formulated as*

$$fitness^{(t)}_{\mathcal{FS}_j} = MSim_{\mathcal{F}}(\mathcal{FS}^{(t-1)}_j, \bigcup_{k=1}^{K^\star} \Delta\mathcal{FU}^{(t)}_k) \tag{9}$$

where $\Delta\mathcal{FU}^{(t)}_k = \mathcal{FU}^{(t)}_k - \mathcal{FU}^{(t-1)}_k$ is a semantic substructure sent by crawler c_k. $K^\star = \frac{K}{N_G}$, i.e., the number of crawlers in a group where N_G is the number of randomly generated groups.

Newly discovered features are collected (i.e., $\bigcup_{k=1}^{K^\star} \Delta\mathcal{FU}^{(t)}_i$), and they are compared with the semantic structure which is built until previous step $t-1$. The collecting process $\bigcup$ means generating population to be evolved for the combinatorial optimization. Here, at a certain time t, the population can be generated by applying feature merging scheme to a set of features discovered by multiple crawlers.

Definition 4 (Population). *The population is represented as a set of tuple $\mathcal{P}^{(t)} = \langle \mathcal{F}, \mathcal{R}, \mathcal{E} \rangle$ where*

- *$\mathcal{F}$ and $\mathcal{R}$ are finite sets whose elements are labels of features and relations, respectively, and*
- *$\mathcal{E}$ is a set of multiple relations among features, i.e., $\mathcal{E} \subseteq \mathcal{F} \times_{\mathcal{R}} \mathcal{F}$.*

As an example, when two feature triples

$$\langle \mathsf{price}, SubClass, \mathsf{shipping\_cost}\rangle \text{ and } \langle \mathsf{price}, SiblingClass, \mathsf{brand\_name}\rangle \tag{10}$$

are collected, the population is represented as

$$\langle \mathsf{shipping\_cost} \times_{SuperClass} \mathsf{price} \times_{SiblingClass} \mathsf{brand\_name}\rangle. \tag{11}$$

Thus, the meta-crawler in this paper should be simply carrying out the following evolution steps. It is very similar to the well-known *natural selection* process [10].

- **Reproduction (or Initialization).** Multiple crawlers are randomly organized into N_G populations. Each population is represented as $\mathcal{P}_k^{(t)} = \bigcup_{i=1}^{K_k^\star}\{\langle \mathcal{F}, \mathcal{R}, \mathcal{E}\rangle\}$ where a set of features merging the features discovered by crawlers. We take into account two cases of grouping scheme; the group sizes $K^\star$ (i.e., the number of crawlers) are i) identical, and ii) distinct.
- **Crossover.** The populations $\mathcal{P}_k^{(t)}$ are *randomly* split into two sub-populations (i.e., $\mathcal{P}_k^{(t)+} + \mathcal{P}_k^{(t)-}$), and intermixed with others (i.e., $\mathcal{P}_k^{(t+1)} = \mathcal{P}_k^{(t)+} + \mathcal{P}_{k'}^{(t)-}$).
- **Mutation.** With predefined mutation coefficient τ_M, *randomly* selected populations and features are mutated by $\mathcal{P}_k^{(t+1)}$. Simply, for this mutation, we change relationships between the selected features.
- **Selection.** Each population is evaluated by fitness function in Equ. 9. The only populations whose fitness is greater than threshold τ_S are selected.

Mainly, we emphasize two effects by exploiting this meta-evolution strategy to collaboration based on meta-crawler.

1. Reconciling conflicts: In [11], semantic conflicts are classified into three levels; instance level, concept level, and relation level. In this paper, we are focusing on conflicts in concept level.
2. Detecting changes: As monitoring the new populations containing semantic substructures during a certain time, we can realize semantic changes on the local ontologies.

4 Experimental Results and Discussion

In order to prove the performance of this system, we acquired and analyzed experimental results. We used Java Genetic Algorithms Package[3], and more importantly, for mapping the semantic substructures, Alignment API[4]. We evaluated the processes for estimating local ontologies and constructing meta-evolution strategy, by interacting multiple crawlers.

Five students were asked to browse the web spaces[5] to search for particular items. We generated 20 crawlers for each student. Fig. 2 shows three different cases of building

[3] JGAP, http://jgap.sourceforge.net/

[4] Alignment API, http://alignapi.gforge.inria.fr/

[5] This testing bed is obtained from [6].

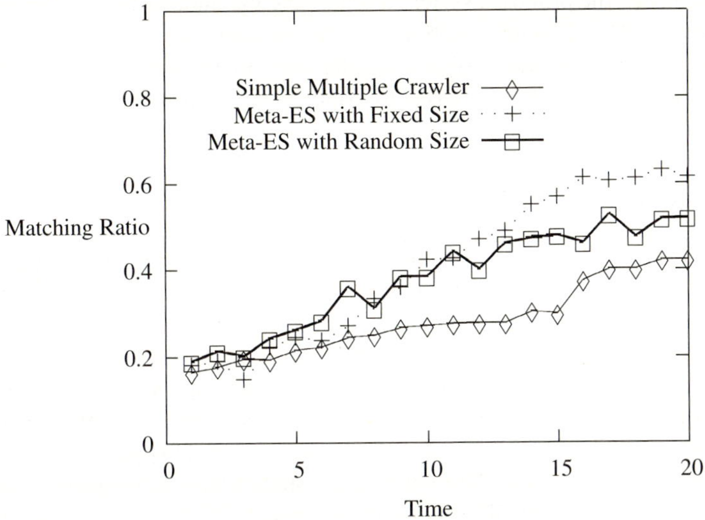

Fig. 2. Average matching ratio

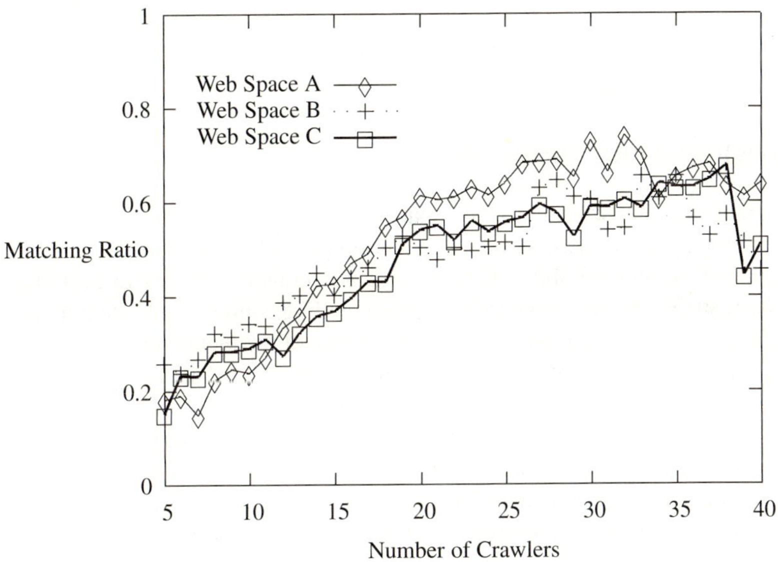

Fig. 3. Matching ratio for three web spaces

semantic structure of a given web space. The proposed methods (e.g., 'Meta-ES with Fixed/Random Size') using meta-evolution strategy have outperformed about 12% and 15% matching ratio, respectively. More particularly, we found out that the meta-crawler of which population size is static has shown the best performance.

Also, we tested the proposed system as changing the number of crawlers (from 5 to 40). As shown in Fig. 3, in three web spaces, the matching ratios are linearly in proportion to the number of crawlers. However, after over 30-34 crawlers, the performances were decreased. We think this is caused by conflicts between crawlers. It means that we need to find the optimal number of crawlers.

5 Related Work

There have been various studies to model and infer the context on user searching tasks in information retrieval (IR) systems. Especially, like our user intention modeling, some studies have proposed to model user contexts from implicit relevance feedback (Implicit RF) for efficient focused crawling. Kelly and Teevan [12] classified the contextual IR systems, with respect to what kind of user behaviors are observed. The interested behaviors are depended upon the goal of target applications. Internet surfing [13] is related to scrolling and clicking, while bookmark sharing system [14] has to focus on bookmarking, deleting, and reusing. Moreover, some studies have investigated to model the contents accessed to by users like binary voting model [15] and the proposed method.

In this paper, user's browsing pattern is the only behavior that we are interested. Also, the annotated semantic information, instead of content itself, is the main target of this study. Meanwhile, it also has involved in some related studies for building consensual knowledge base acquired from end-users. Such methods are CO_4 [16] and meta-level patterns [17]. In our work, the consensus are represented as a set of features most frequently applied to local ontologies.

6 Conclusion and Future Work

With semantically heterogeneous web spaces, we were motivated to overcome the information searching problem based on cooperative crawler framework. The aim of this paper is seamless interoperability for user's focused crawling with meta-evolution strategy. In particular, emergent semantics (e.g., reconciling conflicts and detecting changes) were efficiently recognized to improve performance of the meta-crawler.

Most importantly, we are expecting that this study is very appropriate to many applications (e.g., business process, information integration, and so on) in era of semantic web.

As future work, we, above all, plan to develop the *semantic transformer* for translating different ontology languages, because it was the main problem of merging operation. Also, after building consensus ontology, we want local ontologies to be clearly socialized based on extracting social features from relationships between ontologies [18], in order to construct society of ontologies.

References

1. De Bra, P.M.E., Post, R.D.J.: Information retrieval in the World-Wide Web: Making client-based searching feasible. Computer Networks and ISDN Systems 27(2), 183–192 (1994)
2. Chakrabarti, S., van den Berg, M., Dom, B.: Focused crawling: a new approach to topic-specific web resource discovery. Computer Networks 31(11-16), 1623–1640 (1999)

3. Jung, J.J.: Collaborative web browsing based on semantic extraction of user interests with bookmarks. Journal of Universal Computer Science 11(2), 213–228 (2005)
4. Liu, H., Milios, E., Janssen, J.: Probabilistic models for focused web crawling. In: Proceedings of the 6th annual ACM international workshop on Web information and data management (WIDM '04), pp. 16–22. ACM Press, New York (2004)
5. Jung, J.J., Lee, K.S., Park, S.B., Jo, G.S.: Efficient web browsing with semantic annotation: A case study of product images in e-commerce sites. IEICE Transactions on Information and Systems E88-D(5), 843–850 (2005)
6. Jung, J.J.: Exploiting semantic annotation to supporting user browsing on the web. Knowledge-Based Systems 20(4), 373–381 (2007)
7. Jung, J.J.: Ontological framework based on contextual mediation for collaborative information retrieval. Information Retrieval 10(1), 85–109 (2007)
8. Euzenat, J., Valtchev, P.: Similarity-based ontology alignment in OWL-Lite. In: Proceedings of the 16th European Conference on Artificial Intelligence, pp. 333–337 (2004)
9. Beyer, H.G., Schwefel, H.P.: Evolution strategies - a comprehensive introduction. Natural Computing 1(1), 3–52 (2002)
10. Cantú-Paz, E.: Order statistics and selection methods of evolutionary algorithms. Information Processing Letters 82(1), 15–22 (2002)
11. Nguyen, N.T.: Conflicts of ontologies - classification and consensus-based methods for resolving. In: Gabrys, B., Howlett, R.J., Jain, L.C. (eds.) KES 2006. LNCS (LNAI), vol. 4252, pp. 267–274. Springer, Heidelberg (2006)
12. Kelly, D., Teevan, J.: Implicit feedback for inferring user preference: a bibliography. SIGIR Forum 37(2), 18–28 (2003)
13. Claypool, M., Brown, D., Le, P., Waseda, M.: Inferring user interest. IEEE Internet Computing 5(6), 32–39 (2001)
14. Jung, J.J.: An application of collaborative web browsing based on ontology learning from user activities on the web. Computing and Informatics 23(4), 337–353 (2004)
15. White, R.W., Jose, J.M., Ruthven, I.: An implicit feedback approach for interactive information retrieval. Information Processing and Management 42(1), 166–190 (2006)
16. Euzenat, J.: Building consensual knowledge bases: context and architecture. In: Proceedings of the 2nd International Conference on Building and Sharing very large-scale Knowledge Bases (KBKS), pp. 143–155. IOS Press, Amsterdam (1995)
17. Kim, J.: Meta-level patterns for interactive knowledge capture. In: Proceedings of the 3rd International Conference on Knowledge Capture (K-CAP '05), pp. 207–208. ACM Press, New York (2005)
18. Kleinberg, J.M.: Authoritative sources in a hyperlinked environment. Journal of the ACM 46(5), 604–632 (1999)

Automated Text Categorization Based on Readability Fingerprints

Mark J. Embrechts[1], Jonathan Linton[2], Walter F. Bogaerts, Bram Heyns[3], and Paul Evangelista[4]

[1] DSES, Rensselaer Polytechnic Institute, Troy, NY 12180, USA
[2] Telfer School of Managment, University of Ottawa, Canada
[3] Science and Engineering Library, University of Leuven, B-3001 Leuven, Belgium
[4] Paul Evangelista, U.S. Military Academy, West Point, NY, USA

Abstract. This paper introduces the use of 15 different readability indices as a fingerprint that enables the classification of documents into different categories. While a classification based on such fingerprints alone is not necessarily superior to document categorization based on dedicated dictionaries per se, the document fingerprints can enhance the overall classification rate by applying proper data fusion techniques. For other applications text mining related applications such as language classification, the detection of plagiarism, or author identification, the accuracy of text categorization methods based on readability fingerprints can even exceed a dictionary-based approach. A novel addition to the readability indices is the addition of histograms based on the word length of all the dictionary words used in the text and a dictionary of the most common easy words in the English language.

1 Introduction

Document classification and Automated Text Categorization (ATC) refers to the classification of documents into one or more predefined classes or categories. Practical applications of document classification can range from spam filtering, author identification, document searching, homeland security, and fault diagnosis [9,12]. The vast amount of documents that can be retrieved online in digital form motivates the need for efficient automated document organization and classification systems [14].

A readability index is a metric that indicates how hard a text is to read. Often a readability index is expressed in grade level, originally indicating the grade level for which a particular book is appropriate. The literature cites a variety of readability indices often for different purposes such as the ease of automation, estimating the language complexity for military manuals and instructions, evaluating readability for readers with English as a second language and so on. The readability indices are often based on the average number of words per sentence and the averaged word complexity as expressed by the number of syllables or the word length. Many of the commonly used readability indices were introduced in the fifties, sixties, and seventies and ease of computerized evaluation was usually

J. Marques de Sá et al. (Eds.): ICANN 2007, Part II, LNCS 4669, pp. 408–416, 2007.

not considered while defining readability indices. This paper introduces the idea that a collection of different readability indices can be used as a characteristic fingerprint for a document that can bring additional value for document classification. While for many documents classification tasks the use of dedicated dictionary-based text encoding might lead to an inherently superior classification per se, even in these cases the generation readability fingerprints can be used in a data fusion procedure to further increase the correct classification rate.

This paper is organized as follows: section 2 provides a brief introduction to readability indices and describes the generation of a readability-based fingerprint for a document; section 3 shows the use of readability fingerprints for text categorization by language using an unsupervised self-organizing map; section 4 shows how the same technique can be used for author identification; section 5 compares binary supervised text categorization on the TechTC benchmark dataset from Gabrilovich and Markovitch at Technion [8].

2 Use of Readability Indices to Fingerprint Documents

A readability index indicates how easy a text is to understand for a particular target population of readers and documents. Readability indices have been defined for different target populations of readers such as students in the lower six grade levels, high school students and students with a university education, foreign students with English as a second language, military pilots, and soldiers. By the same token different readability indices have been defined for amongst others: computer manuals, military instruction manuals, and electronic documents. Most readability indices are based on heuristic formulas that take into account the average number of words per sentence, the average word complexity, and sometimes consider which fraction of words in a text can be considered as familiar words.

The comprehension of a document can be considered as is an interaction between the particular reader, whose possible prior knowledge of aspects of the content and the text features would influence the ease with which they access the text, as well as the aspects of the text itself, such as text size, layout, and font type. Readability can be assessed by conducting surveys and readability tests.

We developed software to automatically determine 15 commonly used readability indices for electronic documents. The software is a C-based scriptable shell that generates Perl routines for dedicated text analysis tasks. The software for estimating readability indices was first extensively benchmarked with several online programs listing different readability indices. As an example, Figure 1 shows text statistics and readability indices for the Notebooks of Leonardo.

In addition to the readability indices, the text fingerprints consist of 48 additional numbers that describe three 16-bin histograms: (a) a histogram for the wordlengths for the 3,000 Dale-Chall easy words [4] ; (b) a histogram for the wordlength with the easy words removed from the text; and (c) a wordlength histogram for all the words in the text. The Dale-Chall easy word list [4] contains

Description	Abbreviation	MetricValue	Interpretation	Comment	year	Introduced by
Number of characters without spaces	CHR	103011				
Number of syllables	SYL	310378				
Number of words	WRD	239016				
number of sentences	SEN	9761				
Estimated # of long words (>10 char)	(LONG)	14373				
Estimated # of complex words (3 syl)	(3SYL)	33057				
Estimated # of complex words (4 syl)	(4SYL)	11736				
Estimated # of fog words (3 syl*))	(3SYL*)	16624		Drop trivial ending syllables such as -ing, -ed, etc.		
Average # of characters per word	A_CHR_W	4.31				
Average # of syllables per word	A_SYL_W	1.30				
Average # of words per sentence	A_WRD_S	24.49				
MJE Gunning Fog Index	FOG_MJE	12.2				
Wordstem-based Gunning Fog Index	FOG_STM	12.58		Number of years of formal education necessary to easily read a text	1952	Robert Gunning
Gunning Fog Index	FOG3	15.33				
Gunning Fog Index	FOG4	11.76				
Coleman-Liau Readability Index	CLRI	8.36		U.S. grade level for grades 4 to college level	1975	Meri Coleman and T. L. Liau
Flesch-Kincaid Grade Level	FKGL	9.28		Grade Level, for U.S. Navy technical documents	1976	Rudolph Flesch and J. Kincaid
Automated Readability Index	ARI	11.11		U.S. grade level, originally for Air Force technical documents	1967	E. A. Smith and R. J. Senter
SMOG Index	SMOG	9.65		"Simple Measure of Gobbledygook" in # years of US education	1969	G. Harry Mclaughlin
FRES Score	FRES	72.12		Numerical score between [1..100] for [easy .. Difficult]	1948	Rudolph Flesch
FORCAST readability Grade Level	FORCAST	13.91		Grade level, U.S. Army technical Manuals and forms	1992	Thomas Sticht
Powers-Sumner-Kaerl Grade Level	PSK_GL	5.61		Kindergarten to 7th graders, US army technical manuels & froms	1958	Powers, Sumners, Kaerl
Laesbarhedsindex	LIX	49.37	difficult	[0..100] Works for any Western European language	1979	Björnsson, Hård, and Segerstad
Rate Index	RIX	6.21	grade 12	Can be translated into grade level, any european language	1981	Anderssen
Linsear Grade Level	LWR	15.63		Grade level, U.S. Air force technical Manuels		
McAlpine EFLAW Index	EFLAW	45.59	confusing	Devised for readers with English as a second language	2004	McAlpine
Dale-Chall RGS	DC_RGS	8.19		1975 revised list of 3000 familiar words, for all grade levels	1948/1975	Edgar Dale and Jeanne S. Chall
Easy word percentage	EW_PER	78.88		Based on list of 3000 familiar words	1975	Edgar Dale and Jeanne S. Chall
SPACH*:	SPACH1	3.98		Intended for grades 1-3; 1974 revised list of 1065 easy words	1974	
SPACH:	SPACH2	4.66		Same as above without wordstemming		
FRY_X= Avg syllables/100 words	FRY_X	129.86		Use chart to translate Fry indices in a grade level	1977	Edward Fry
FRY_Y = Avg sentences/100 words	FRY_Y	4.08				

Fig. 1. Summary of 24 readability related indices (highlighted in column 3) used in a text fingerprint for The Notebooks of Leonardo da Vinci [3]

3,000 familiar words that can be understood by more than eighty percent of fourth grade students. Words that are not on the Dale-Chall list can be interpreted as unfamiliar and are therefore more difficult to read. An example of two of these histograms is shown in Figure 2.

The Dale-Chall readability index is based on the fraction of Dale-Chall words in a text and was described in the now out-of-print 1948 book by Edgar Dale and Jeanne S. Chall entitled "A Formula for Predicting Readability." The original list initially contained 763 familiar words and was modified to 3,000 words in 1995. As an example, the formula for the Dale-Chall reading Grade Score (RGS) is given below:

$$RGS = 0.1579 \times DS + 0.0496 \times ASL + 3.6365$$

Where DS is the percentage of words in the text that are not in the list of 3000 familiar dale-Chall words, and ASL represents the average sentence length, in words per sentence. The Dale-Chall readability index is not the most widely used index, but is considered more accurate than more popular indices such as the Flesch Reading Ease score [7], or the Flesch-Kincaid Grade Level [10].

A nice comprehensive survey on the different readability indices can be found in a document written by William H. DuBay [6]. In this study we used a total of 72 fingerprints: (a) 24 fingerprints that are or are associated with several of the popular and less popular readability indices listed by DuBay; and (b) 48 fingerprints associated with the three 16-bin wordlength histograms. Most readability indices are based on simple metrics that have to be derived from the text such as the average number of characters per word, the average number of syllables per word, and the average number of words per sentence. Some

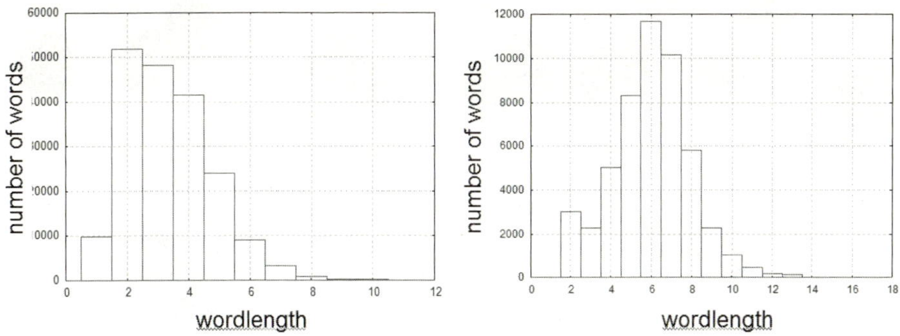

Fig. 2. Histograms for the wordlength of the familiar words (left hand side) and the wordlength of all the words with the familiar words removed for the notebooks of Leonardo da Vinci [3]

readability indices have in addition a dictionary of easy words used to determine the fraction of easy words in a text. Different readability indices have different rules for operations such as wordstemming, and counting syllables. It is word pointing out again here that most readability indices were defined without any consideration for the easy by which their calculation could be automated for processing digital documents.

3 Characterizing Text by Language with Readability Indices

Readability fingerprints were applied to encode 71 books in 5 different languages. The books were obtained as digital documents from the Project Gutenberg website [15] and consisted of 34 English books, 6 Latin books, 12 Dutch books, 7 German books, and 12 books in French. A typical Kohonen map [11] is reproduced in Figure 3.

Note that the different language categories are not equally represented, that some books are split up in several parts, other books are represented several times in different languages, and that certain authors (e.g. Shakespeare) have several books in the pool. The only preprocessing of the documents, after determining the readability fingerprints, was a straightforward normalization (i.e., center each variable and divide by the standard deviation). The purpose of this study is not to claim that fingerprinting text based on (mostly English language-based) readability fingerprints is the best way to classify text. Obviously a dictionary encoding with typical small frequent word dictionaries for the different languages would work as well if not better. Nevertheless we can definitely conclude that fingerprinting texts with readability indices deserves consideration, and can probably boost the performance of any language classifier if proper data fusion techniques are applied. An example of data fusion for increased prediction performance based on different representations of the data is presented

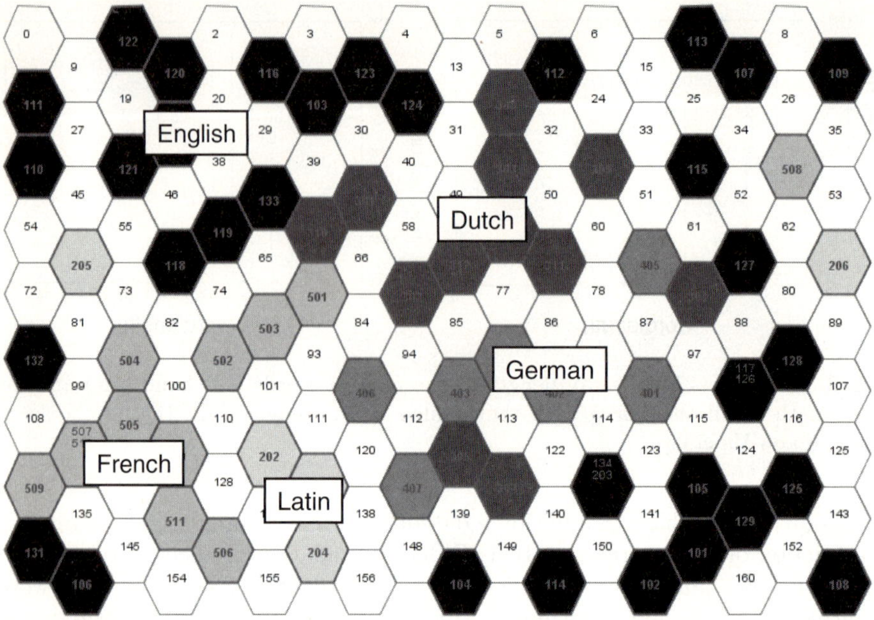

Fig. 3. Kohonen map (toroidal geometry) for 71 documents in 5 languages with 72 fingerprint features. Note that top and bottom and right-hand and left-hand side should wrap around.

in these proceedings in the paper by Chanjian Huang et al. A representation of the 71 books based on the first and the second principal components is shown in Figure 4. Different language regions can easily be delineated in this case there is just one document that does not fit the linear separations.

A spectral clustering [1] with a Gaussian kernel was fairly robust (i.e., similar results can be obtained within a wide range for the kernel Parzen window, and the specified number of clusters) and resulted in the confusion matrix reproduced in Table 1. The results for spectral clustering were slightly better, but otherwise similar to those obtained based on a K-means clustering. Note that the largest source for the error is that 4 Dutch texts are classified as German books. Considering that the four misclassified texts actually dated from the seventeenth and eighteenth century, where Dutch was a lot more similar to German as is currently the case, this is actually a very reasonable result. The other Dutch text that was misclassified as a Latin book is actually a medical text with an abundance of Latin terms in the text.

The experiments on book categorization by language were repeated in several different ways, but now rather than using the 72 fingerprints containing readability indices information and three 16-bin wordlength histograms, we either considered only the 48 variables associated with the word histograms, or eliminated

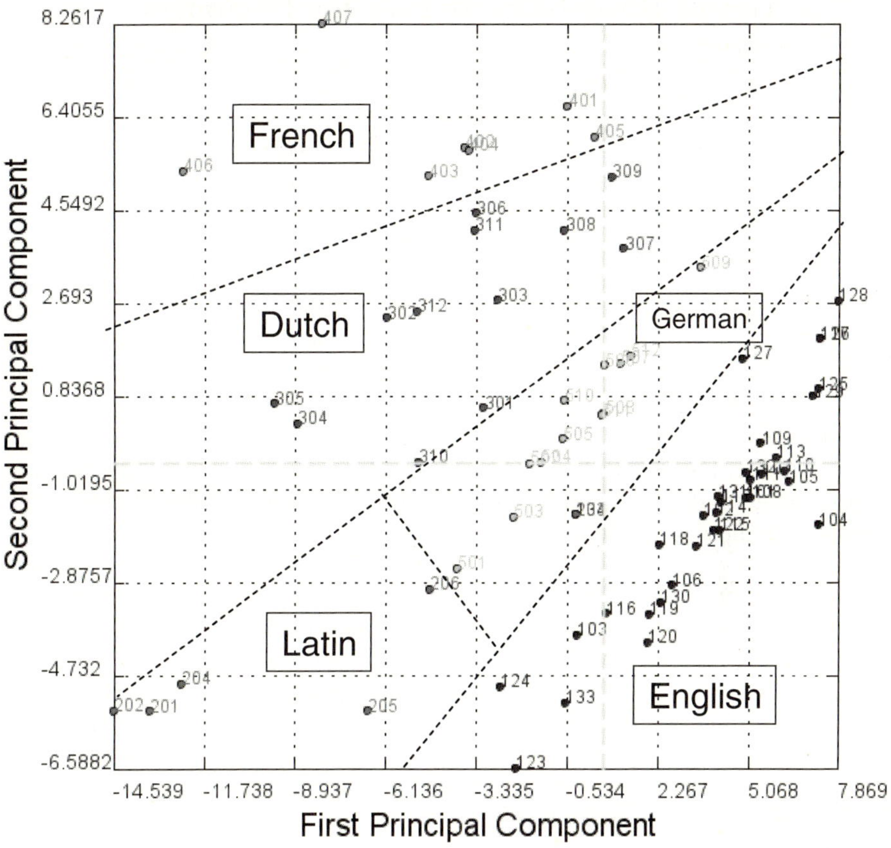

Fig. 4. Principal component representation of 71 books in 5 different languages

Table 1. Confusion matrix for spectral clustering of 71 texts in 5 different languages

Language	English	Dutch	German	French	Latin
English	34	0	0	0	0
Dutch	0	7	4	0	1
German	0	0	7	0	0
French	1	0	0	11	0
Latin	1	0	0	0	5

the 48 variables associated with the word histograms were eliminated from the fingerprints. In both cases the classification results were by far inferior to the classification based on the 72-variable fingerprints. Depending on the specific documents, and the specific text categorization task we consistently noticed that the wordlength histogram information is about as relevant as the variables that are immediately associated with the readability index information.

4 The Use of Readability Indices for Document Classification

Good text categorization benchmark datasets are the Reuters-21578 collection, the 20 Newsgroup OHSUMED and the TIPSTER corpus. For this study we choose a subset of Tech TC-100, which contains 100 binary classification datasets whose categorization difficulty (as measured by a baseline linear SVM classifier) is uniformly distributed between 0.6 and 0.92. The data can be downloaded from the Technion repository for text categorization datasets [Gabrilovich 2004]. The data were collected with an automated data acquisition methodology for datasets with desired properties as described by Davidov et al. [5]. Each dataset consists of binary categories with an average of 150-200 documents (depending on the specific test collection).

For this study we used a subset of Tech TC-100 consisting of 19 binary classification datasets: the first 10 binary datasets and the remaining datasets with the highest categorization difficulty. The datasets were encoded with 72 readability fingerprint indices (24 readability related indices, and three 16-bin wordlength histograms). In this case we selected the same number of documents as indicated in the study by Gabrilovich and Markovitch [8]: the documents had an equal number of instances for both categories, and only the largest documents were used in this study. We used partial-least squares (PLS) [16] and kernel partial least squares (K-PLS) [13] in a leave-one-out mode for doing the binary classifications. The results are summarized in Figure 5 and compared with results obtained with dictionary-based encoding by Gabrilovich and Markovitch.

Looking at Figure 5, it can be concluded that the classifications based on K-PLS fingerprints are comparable or superior in six cases to classifications based on linear SVMs and a dictionary-based encoding. It is actually expected

No	Cat_ID1	Cat_ID2	# Docs	PLS_finger	K-PLS_finger	SVM_100	C4.5_100	KNN_100	SVM_0.5	C4.5_0.5	KNN_2%
1	1622	42350	163	0.613	0.662	0.863	0.773	0.838	0.744	0.75	0.729
2	6920	8366	140	0.643	0.735	0.919	0.897	0.927	0.897	0.897	0.907
3	8308	8366	144	0.583	0.618	0.829	0.743	0.793	0.829	0.807	0.862
4	10341	10755	145	0.706	0.790	0.771	0.811	0.708	0.820	0.792	0.794
5	10341	14271	145	0.628	0.731	0.681	0.788	0.618	0.813	0.820	0.796
6	10341	14525	158	0.632	0.674	0.590	0.507	0.590	0.564	0.615	0.514
7	10341	61792	153	0.712	0.79	0.757	0.803	0.717	0.796	0.822	0.810
8	10341	186330	147	0.721	0.796	0.701	0.800	0.729	0.848	0.854	0.826
9	10341	194927	159	0.660	0.761	0.737	0.770	0.737	0.801	0.769	0.821
17	10385	312035	145	0.675	0.717	0.650	0.536	0.655	0.486	0.519	0.543
28	10567	46076	142	0.690	0.746	0.650	0.900	0.593	0.943	0.936	0.879
46	14630	20186	157	0.688	0.764	0.596	0.949	0.468	0.929	0.949	0.833
51	17360	20186	145	0.696	0.751	0.597	0.875	0.563	0.903	0.931	0.852
53	18479	20186	152	0.651	0.756	0.645	0.934	0.533	0.921	0.928	0.875
58	20186	61792	153	0.758	0.817	0.592	0.892	0.582	0.842	0.928	0.731
61	20673	312035	139	0.640	0.676	0.640	0.787	0.618	0.839	0.838	0.812
75	186330	46076	145	0.752	0.772	0.650	0.716	0.672	0.822	0.764	0.832
80	194915	20186	157	0.669	0.79	0.404	0.561	0.547	0.398	0.603	0.442
82	194915	194927	164	0.579	0.713	0.631	0.706	0.613	0.763	0.706	0.639

Fig. 5. Comparison of 20 binary text classifications for a subset of tech TC-100, using PLS and K-PLS on readability fingerprints and other methods based on dictionary encodings [8]

that dictionary-based encoding would be superior to readability fingerprints: the fact that this is not always the case is actually surprising and warrants further study. In the past we extensively compared K-PLS classification and regression on benchmark data with SVMs, and found that both methods yield comparable results when the same kernel is used [2]. In this case of the benchmarks in Figure 5, the SVMs are actually linear SVMS, but the K-PLS method still compares favorably with C4.5 or K Nearest Neighbors (KNN). Note also that the classification accuracy based on fingerprints would drastically decrease if the 48 wordlength histogram data were not considered in the text fingerprints (results not shown).

From these benchmark studies it can be concluded that for certain datasets and binary text categorization tasks readability fingerprints are an appropriate way to represent the data and that such a text encoding can possible boost the accuracy of a text categorization system by applying proper data fusion techniques.

5 Conclusion

This paper shows for the first time that the use of different readability indices can be utilized for document categorization. Two benchmark studies were conducted: a language classification experiment, and a comparison on a subset of the Tech TC-100 data. While a classification based on such fingerprints alone is not necessarily superior to document categorization based on dedicated dictionaries per se, the document fingerprints can enhance the overall classification rate by applying proper data fusion techniques.

References

1. Bach, F.R., Jordan, M.I.: Learning Spectral Clustering, with Application to Speech Separation. Journal of Machine Learning Research 7, 1963–2001 (2006)
2. Bennett, K.P., Embrechts, M.J.: An Optimization Perspective on Partial Least Squares. In: Suykens, J., Horvath, G. (eds.) Advances in Learning Theory: Methods, Models and Applications. NATO Science Series III: Computer & Systems Sciences, vol. 190, pp. 227–246. IOS Press, Amsterdam (2003)
3. da Vinci, L.: The Notebooks of Leonardo Da Vinci (translated by Jean-Paul Richter in 1880). Oxford University Press, New York (1952)
4. Dale, E., Chall, J.S.: A Formula for Predicting Readability: Instructions. Educational Research Bulletin 27(2), 11–20 (1948)
5. Davidov, D., Gabrilovich, E., Markovitch, S.: Parameterized Generation of Labeled Datasets for Text Categorization Based on a Hierarchical Directory. In: The 27th Annual International ACM SIGIR Conference, Sheffield, UK, July 2004, pp. 250–257 (2004)
6. William, H.: DuBay. Unlocking Language: The Classic Readability Studies. Accessed (31 March, 2007), http://www.impact-information.com/impactinfo/research/classics.pdf
7. Flesh, R.: A New Readability Yardstick. Journal of Applied Psychology 32, 221–223 (1948)

8. Gabrilovich, E., Markovitch, S.: Text Categorization with Many Redundant Features: Using Aggressive Feature Selection to Make SVMs Competitive with C 4.5. In: 21st International Conference on Machine Learning (ICML), Banff, Alberta, Canada, pp. 321–328 (2004)
9. Joachims, T., Sebastiani, F.: Special Issue on Automated Text Categorization. Journal on Intelligent Information Systems 2 (2002)
10. Kincaid, J.P., Fishburne, J.R.P., Rogers, R.L., Chissom, B.S.: Derivation of New Readability Formulas (Automated Readability Index, Fog Count and Flesch Reading Ease Formula) for Navy Enlisted Personnel. In: Naval Technical Training, U. S. Naval Air Station, Memphis, Research Branch Report 8-75, Millington, TN (1975)
11. Kohonen, T.: Self-Organizing Maps: Third Extended Edition. Springer Series in Information Sciences, vol. 30. Springer, Heidelberg (2001)
12. Lewis, D.D., Hayes, P.J.: Guest Editorial for the Special Issue on Text Categorization. ACM Transactions on Information Systems 12(3), 231 (1994)
13. Rosipal, R., Trejo, L.J.: Kernel Partial Least Squares Regression in Reproducing Kernel Hilbert Space. Journal of Machine Learning Research 2, 97–128 (2001)
14. Sebastiani, F.: Machine Learning in Automated Text Categorization. ACM Computing Surveys 34(1), 1–47 (2002)
15. Wikipedia, The Free Encyclopedia. Project Gutenberg. Accessed (31 March 2007), http://en.wikipedia.org/wiki/Project_Gutenberg
16. Wold, S., Sjöström, M., Erikson, L.: PLS Regression: A Basic Tool of Chemometrics. Chemometrics and Intelligent Laboratory Systems 58, 109–130 (2001)

Personalized Web Page Filtering Using a Hopfield Neural Network

Armando Marin[1], Juan Manuel Adán-Coello[2], João Luís Garcia Rosa[2], Carlos Miguel Tobar[2], and Ricardo Luís de Freitas[2]

[1] Senac Ribeirão-Preto, Ribeirão Preto, SP, Brazil
amarin@sp.senac.br
[2] PUC-Campinas, Cx. P. 317, CEP 13012-970, Campinas, SP, Brazil
{juan,joaoluis,tobar,rfreitas}@puc-campinas.edu.br

Abstract. The immense amount of unstructured information available on the Web poses increasing difficulties to fulfill users' needs. New tools are needed to automatically collect and filter information that meets users' demands. This paper presents the architecture of a personal information agent that mines web sources and retrieves documents according to users' interests. The agent operates in two modes: "generation of space of concepts" and "document filtering". A space of concepts for a domain is represented by a matrix of asymmetrical coefficients of similarity for each pair of relevant terms in the domain. This matrix is seen as a Hopfield neural network, used for document filtering, where terms represent neurons and the coefficients of similarity the weights of the links that connect the neurons. Experiments conducted to evaluate the approach show that it exhibits satisfactory effectiveness.

1 Introduction

With the expansion of the Web, users face increasing difficulties to fulfill their information needs. Contribute to this scenario the unstructured nature of the data stored on the web, mostly text documents formatted using the HTML language, and the dynamic nature of the Web that requires that users continually carry out new searches to check for new documents of interest. This situation has motivated the development of personal information filtering software agents that continuously search the web looking for documents of interest for their users.

Although the research on information software agents is recent, the idea of developing systems directed to fulfill information needs of specific users can be traced back to the late 1950'ies. In his *Business Intelligence System*, Luhn (1958) proposed that librarians create user profiles that should be used to produce lists of suggestions of new documents for each user. Requests of particular documents should be recorded and used to automatically update users' profiles. Luhn identified several key aspects of modern information filtering systems, although the technology available at that time severely restricted the implementation of his ideas.

Information needs change from user to user. Therefore, information filtering systems have to be personalized in order to fulfill individual user's interests, taking the

J. Marques de Sá et al. (Eds.): ICANN 2007, Part II, LNCS 4669, pp. 417–424, 2007.

role of personal assistants. Such a personalized information filtering system has to satisfy three requirements:

1. Specialization: the system selects the documents relevant to the user and discards the others;
2. Adaptation: information filtering is an iterative process done for large periods of time, during which user's interests change;
3. Exploration: the system should be able to explore new domains, in order to find something new potentially interesting to the user.

A number of different models and systems have been implemented for information retrieval and filtering. Typically, these systems consist of three main components: document representation, user's interests representation and algorithms used to match user's interests to documents representations (Salton, G.; McGill, 1997; Yan, T. W.; Garcia-Molina, 1999).

Hopfilter is a personal agent that mines web information sources and retrieves documents according to user's interests. This paper describes in detail the architecture of Hopfilter; a general overview can be found elsewhere (Adán-Coello, Tobar, Freitas, and Marin, 2007).

In general lines, Hopfilter works in the following way: initially, the user manually selects a document collection on an interesting subject. Through automatic indexation, the system generates the terms that better represent the content of the collection. With the use of statistical functions that calculate the co-occurrence of the terms in the collection, the system generates a similarity matrix containing terms and coefficients that indicate the degree of relationship between every pair of terms. This matrix of similarity is interpreted as a knowledge net or a space of concepts. The space of concepts represents a neural net with neurons (terms) interconnected through synaptic weights (relationship coefficients). The filtering process starts when a document to be filtered is used as an input pattern to the network. When the network converges to a state of equilibrium, the amount of active neurons indicates whether the document is compatible with the concepts stored in the memory of the net or not. If the document is considered relevant, the user is notified that the filtered document contains information of interest.

The generation of the space of concepts for filtering is done by an adaptation of the method used by Chen et al. (1994, 1998) to generate a list with the main ideas (concepts) raised during a session of an Electronic Brain Storming System.

The article is organized as follows. In section 2 the architecture of Hopfilter is presented. Section 3 presents the results of some experiments conducted to evaluate the agent. Section 4 closes the paper with some final remarks.

2 The Architecture of Hopfilter

The filtering agent is composed of the following modules: User interface (UI), Web interface (WI), Document Preprocessing (DPP), Automatic Indexing (AI), Generation of the Space of Concepts (GSC), and Artificial Neural Net (ANN). The UI and WI modules interface with the user and with the web, as their names suggest, and are not on the focus of this paper.

The filtering agent can operate in two modes: "generation of space of concepts" and "document filtering". The generation of space of concepts mode uses the modules DPP, AI and GSC. The document filtering mode involves the modules DPP, AI and ANN. For running in the "document filtering" mode, a space of concepts (SC) for the considered domain must already be available. The functioning of each mode of operation is briefly described below; each module will be described in details in the sequence.

To generate the space of concepts a collection of *n* documents selected by the user about a specific domain must be submitted to the agent. The preprocessing module removes the tags of the document, producing plaint text words. From the words identified, the module of automatic indexing generates the terms (indexes) that better represent the document collection and computes asymmetric coefficients of similarity for each pair of generated terms. The terms with their respective asymmetric coefficients of similarity are arranged into a matrix of similarity (space of concepts). The agent interprets the space of concepts as being a Hopfield artificial neural net (Hopfield, 1982), where terms correspond to neurons and asymmetric coefficients of similarity to neuron connections weights.

To filter a document, the agent initially identifies the words in the document, using the preprocessing module, and generates the terms that represent the document using the automatic indexing module. The artificial neural net module uses the terms generated by the indexing module to create a vector of the form $\mathbf{x} = \{ x_1, x_2..., x_n \}$, where n is equal to the number of terms in the space of concepts, and x_i, corresponds to the i'th term generated by the mechanism. x_i will be 1 if the i'th term is present in the space of concepts or 0 otherwise. This vector is used to activate the Hopfield neural net. Once activated, the net will search for a state of equilibrium. When such a state is found, the number of active neurons indicates the relevance of the filtered document.

2.1 Preprocessing

The preprocessing module receives a Web page and outputs plain text. HTML tags, punctuation marks, dates, numbers and several other symbols are removed from the input document.

2.2 Automatic Indexing

When a document is indexed the result is a list of terms, or indexes that represents the document content. Automatic indexing consists of three operations: removal of stopwords, work stemming, and term formation.

Removing stopwords. After identifying the words in the input document, using the preprocessing module, the words that are not relevant for characterizing the content of the document are removed. To assist in this process it is used a dictionary with around 46,000 words. Dictionary entries can be manually marked by the user as stopwords. By default, articles, pronouns, conjunctions, prepositions, adverbs, numerals, and linking and auxiliary verbs are not considered discriminant words, and as such treated as stopwords. All words in the input document that are not found in the dictionary are kept in a table for posterior analysis. If desired, these words could be included in the dictionary by the user, allowing its constant update.

Word stemming. The purpose of this step is to reduce the number of cognate words to be indexed. The algorithm implemented is an adaptation of the Lancaster Stemming Algorithm (Paice, 1990) for the Portuguese language that removes only the suffixes of the words.

Term formation. Adjacent words resulting from the previous step are used to form terms. A term can be formed by one, two or three words. For example, with the three words, "tax", "credit" and "exchange" it is possible to form up to six terms: "tax", "tax-credit", "tax-credit-exchange", "credit", "credit-exchange" and "exchange". For each term it is computed a term frequency, *tf*, that represents the number of times the term appears in the document. When the agent operates in the "generation of space of concepts" mode, it is also calculated the document frequency for the term, *df*, that represents the number of documents, in the collection of documents, where the term appears.

2.3 Generation of the Space of Concepts

The objective of this module is to calculate asymmetrical coefficients of similarity for each pair of terms, generating a matrix containing the terms and the respective coefficients, or degrees of relationship. This matrix, also called matrix of similarity, represents the space of concepts.

Term selection. Normally, the amount of terms generated by the module of automatic indexation is very high. To reduce the time needed to compute the coefficients of similarity, only the most important terms must be considered. Term selection is based on the importance of the terms in a document collection. We define the importance of term j, d_j, in a collection of n documents by the following equation

$$d_j = tf_j \times \log(\frac{n}{df_j} \times ts) \tag{1}$$

where tf_j is the frequency of term j in the document collection, df_j is the document frequency of term j in the document collection, ts is the term size in words, and n the amount of documents that compose the collection. ts is used to increase the weight of terms formed by two or three words, because those terms are more descriptive than single word terms.

The frequency of term j in a collection of documents, tf_j, is calculated by equation 2:

$$tf_j = \sum_{i=1}^{n} tf_{ij} \tag{2}$$

where tf_{ji} is the frequency of term j in document i.

Computing the asymmetric coefficients of similarity. The asymmetric coefficients of similarity for each pair of terms i and j are calculated by the asymmetric clustering function used by Chen et al. (1994) represented by equations 3 and 4 below.

$$acs_{jk} = \frac{\sum_{i=1}^{n} d_{ijk}}{\sum_{i=1}^{n} d_{ij}} \quad (3)$$

$$acs_{kj} = \frac{\sum_{i=1}^{n} d_{ijk}}{\sum_{i=1}^{n} d_{ik}} \quad (4)$$

Equation 3 computes the similarity from term j to term k and equation 4 from term k to term j; d_{ij} represents the weight of term j in document i; d_{ik} represents the weight of term k in document i, and d_{ijk} represents the weight of terms j and k in document i.

The factor d_{ij}, that represents the weight of term j in document i, is calculated by equation 5 below:

$$d_{ij} = tf_{ij} \times \log \frac{n}{df_j} \quad (5)$$

where tf_{ij} represents the number of occurrences of term j in document i, and df_j represents the number of documents in a collection of n documents in which the term occurs. The factor d_{ij} is proportional to the frequency of term j in document i and inversely proportional to the number of times term j occurs in the collection of n documents: a term that seldom appears in the documents has a high weight, while a term that occur in many documents has a low weight. Terms that occur quite frequently can be very general and as such have very little discriminant power.

The factor d_{ijk} indicates the relative importance of terms j and k in document i. It is computed by equation 6 below.

$$d_{ijk} = tf_{ijk} \times \log \frac{n}{df_{jk}} \quad (6)$$

where tf_{ijk} represents the number of simultaneous occurrence of terms j and k in document i, and d_{fjk} represents the number of documents, in a collection of n documents, in which terms j and k occur simultaneously.

The result of this step is a matrix of similarity or a measure of the distance between each term. The similarity matrix represents the space of concepts and can be seen as a neural network, where terms represent neurons and the coefficients of similarity the weights of the links that connect the neurons.

2.4 Hopfield Neural Network

This module is used to filter a document by means of the Hopfield neural network as discussed in section 2.3. Initially the document to be filtered is submitted to the preprocessing and automatic indexing modules, in such way that it will be represented by a set of terms or indexes, as done for generating the space of concepts. These terms will define the elements of an n dimension input vector to the net, for a space of concepts with n terms. The input vector will have the form $\mathbf{x} = \{x_1, x_2, ..., x_n\}$, where x_i will 1 if the i-th term of the space of concepts is present in the document to be filtered and 0 otherwise.

Initialization of the net. At time t = 0, each neuron of the net (representing a given term) will assume its corresponding value in vector **x**, as described by equation (7).

$$y_i(t) = x_i,\ 0 \leq i \leq n-1 \tag{7}$$

where $y_i(t)$ is the output of neuron i at time $t = 0$ and x_i (with value 0 or 1) is the input value to neuron i at time $t = 0$.

Activation and iteration. Network activation and iteration are done by the transfer function shown below.

$$y_j(t+1) = fs(net_j) = \frac{1}{1+\exp\left[\dfrac{-(net_j - \theta_1)}{\theta_0}\right]},\ 1 \leq j \leq n \tag{8}$$

where $fs(net_j)$ is the sigmoid function. θ_1 e θ_0 are empirically defined constants and net_j is computed by the following expression

$$net_j = \sum_{i=1}^{n} w_{ij} x_i(t) \tag{9}$$

where w_{ij} is the weight of the connection between neurons i and j, and $x_i(t)$ is the value of neuron i at time t.

The implemented artificial neural net executes the update of the neurons synchronously, that is, the values of all neurons are updated at the same time.

Convergence. The previous step is repeated until no significant alterations in the outputs of the neurons between two consecutive iterations are detected, as described by equation 10.

$$\sum_{j=0}^{n-1} [\mu_j(t+1) - \mu_j(t)]^2 \leq \mathrm{E} \tag{10}$$

where E, defined empirically, corresponds to the maximum allowed variation between the output values of the neurons in two consecutive iterations, indicating that the net should be stable.

When the net converges, the number of active neurons is counted. The higher the number of active neurons the higher the relevance of the document, according to the space of concepts stored in the memory of the net.

3 Experimental Evaluation

Hopfilter was evaluated using documents copied from the Brazilian Federal Revenue site (http://www.receita.fazenda.gov.br) to a local document base. A collection of 59 documents was selected and classified by an expert user. The space of concepts was constructed using 22 documents of the collection dealing with the subject "income tax". For the filtering experiment, 37 documents were chosen randomly, 17 of which were relevant to the subject "income tax", according to the classification made by the expert.

The agent effectiveness in filtering documents was measured using the *precision* and *recall* rates, widely used by the information retrieval community (Rijsbergen, 1979). The *precision* is calculated dividing the number of relevant documents retrieved by the total number of retrieved documents. The *recall* is calculated dividing the number of relevant documents retrieved by the number of relevant documents available in the document base.

Four parameters were decisive to produce satisfactory results: the number of neurons in the net, the energy E, the bias θ_1, and the curvature θ_0. For a net with 25 neurons, $\theta_0 = 0.01$, $\theta_1 = 0.7$, and $E = 0.025$, precision and recall rates were of 83.33% and 88.23%, respectively. A good performance when compared with similar systems (Vallim and Adán-Coello, 2003; Salton and McGill, 1997).

4 Concluding Remarks

The nature of natural language used in Web pages presents semantic characteristics that make it very difficult to construct mechanisms for automatic information retrieval, filtering and extraction. For example, synonymy (different words express the same idea) and polysemy (a word has several meanings) make it difficult to determine if a given page is relevant for the user's interests.

Despite the inherent difficulties to filter texts written in natural language, the experiments conducted so far suggest that concept spaces and associative memories are an effective option to represent document content. However, the selection of terms to compose the space of concepts is a decisive factor for a good performance of this approach. Terms very common in the considered domain have low descriptive power and can reduce substantially the filtering precision rate. The investigation of semiautomatic methods for identifying such and the use of ontologies to face the difficulties presented by synonymy and polysemy are among the planned future works for increasing the effectiveness of Hopfilter.

References

Adán-Coello, J.M., Tobar, C.M., Freitas, R.L., Marin, A.: Hopfilter: an agent for filtering Web pages based on the Hopfield artificial neural network model, 24th British National Conference on Databases (2007)

Chen, H., Hsu, P., Orwig, R., Hoopes, I., Nunamaker, J.F.: Automatic Concept Classification of Text from Electronic Meetings. Communications of the ACM 37(10), 56–73 (1994)

Chen, H., Zhang, Y., Houston, A.L.: Semantic Indexing and Searching Using a Hopfield Net. Journal of Information Science 24(1), 18–33 (1998)

Hopfield, J.J.: Neural Network and Physical Systems with Collective Computational Abilities. Proceedings of the National Academy of Science, USA 79(4), 2554–2558 (1982)

Paice, C.D.: Another Stemmer. SIGIR Forum. 24(3), 56–61 (1990)

Rijsbergen, C.J.: Information Retrieval. Butterworths, London (1979)

Salton, G., McGill, M.J.: The SMART and SIRE Experimental Retrieval Systems. In: Readings in Information Retrieval, pp. 381–399. Morgan Kaufmann Publishers, Pine Street, Sixth Floor. San Francisco (1997)

Vallim, M.S., Adán Coello, J.M.: An Agent for Web Information Dissemination Based on a Genetic Algorithm. In: Proceedings of the IEEE International Conference on Systems, Man and Cybernetics (October 5-8, 2003)

Yan, T.W., Garcia-Molina, H.: The SIFT Information Dissemination System. ACM Transactions on Database Systems 24(4), 529–565 (1999)

Luhn, H.P.: A business intelligence system. IBM Journal of Research and Development 2(4), 314–319 (1958)

Robust Text Classification Using a Hysteresis-Driven Extended SRN

Garen Arevian and Christo Panchev

University of Sunderland, School of Computing and Technology
St Peter Campus, St Peter's Way
Sunderland SR6 0DD, United Kingdom
{garen.arevian,christo.panchev}@sunderland.ac.uk

Abstract. Recurrent Neural Network (RNN) models have been shown to perform well on artificial grammars for sequential classification tasks over long-term time-dependencies. However, there is a distinct lack of the application of RNNs to real-world text classification tasks. This paper presents results on the capabilities of extended two-context layer SRN models (xRNN) applied to the classification of the Reuters-21578 corpus. The results show that the introduction of high levels of noise to sequences of words in titles, where noise is defined as the unimportant stopwords found in natural language text, is very robustly handled by the classifiers which maintain consistent levels of performance. Comparisons are made with SRN and MLP models, as well as other existing classifiers for the text classification task.

1 Introduction

This paper explores the capabilities of extended two-context layer simple recurrent neural network models (xRNN) [21,23,22] for classifying real-world *news titles* from the Reuters-21578 Corpus [13]. The more complex architectures of RNNs have been shown to have powerful computational capabilities [24,4] and stable behaviour [19,11].

However, many of these recurrent models have only been applied to artificially generated grammars and benchmarking datasets, but there is no equivalent research on real-world text classification (TC). The reasons for applying such neural network approaches to the domain of TC are many, including robust and adaptable behaviour in the face of noisy [8], new and ever-changing information such as that presented by the Internet. In addition, RNNs specifically have the inherent property of being able to encode and learn information that is dependent on learning memory-dependent, contextual and sequential information.

A Simple Recurrent Network (SRN) or Elman Network [6] can only embody network activation information from the *t-1* timestep during learning, the memory span is not sufficient for sequence learning that requires longer time-dependencies.

Recent work [19,9] on extended RNN architectures with more context layers and various layer-to-layer configurations has shown that the stability and

J. Marques de Sá et al. (Eds.): ICANN 2007, Part II, LNCS 4669, pp. 425–434, 2007.

computational power of these architectures is greater than with SRNs and a potential solution to the rapid decay of information during training in RNNs, often referred to as the "problem of vanishing gradients" [7].

2 Description of Recurrent Neural Model Used for Text Classification Task

The two hidden layer SRN architecture (xRNN) [21,23,22] used in this study is shown in Figure 1. The context layers take copies of previous hidden layer activations, and add them to respective current activations. In a conventional RNN, context layers essentially take one-to-one copies of activations. However, using the hysteresis decay function (discussed in subsection 2.2), the context layers are able to selectively maintain varying percentages of activations from previous states, giving an architecture that is potentially more finely tunable.

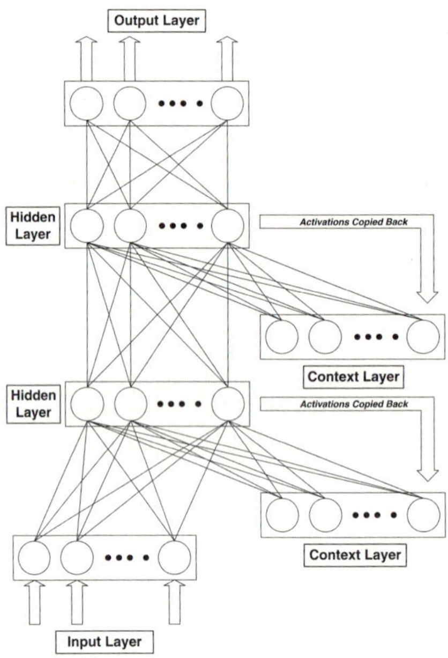

Fig. 1. An xRNN recurrent network, which is an Elman/SRN network with two hidden and two context layers instead of one of each. The arrows indicate the hidden unit activations that are copied to the respective context units.

2.1 Formal Definition of xRNN Model

The input to a hidden layer H_n is constrained by the underlying layer H_{n-1}, as well as the incremental context layer C_n. The activation of a unit $H_{ni}(t)$ at time t is computed on the basis of the weighted activation of the units in the previous layer $H_{(n-1)i}(t)$, and the units in the current context of this layer $C_{ni}(t)$, which are constrained by the logistic function f:

$$H_{ni}(t) = f(\sum_k w_{ki} H_{(n-1)i}(t) + \sum_l w_{li} C_{ni}(t)) \quad (1)$$

The units in the context layers sum the information using the equation:

$$C_{ni}(t) = (1 - \varphi_n) H_{ni}(t-1) + \varphi_n C_{ni}(t-1) \quad (2)$$

where $C_{ni}(t)$ is the activation of a unit in the context layer at time t. The self-recurrent connections are represented using the *hysteresis value* φ_n. The hysteresis value of the context layer C_{n-1} is lower than the hysteresis value of the next context layer C_n. This ensures that the context layers closer to the input layer will perform as memory that represents a more dynamic context. Having higher hysteresis values, the context layers closer to the output layer will incrementally build more stable sequential memory.

2.2 Hysteresis Function of the Context Layers

The hysteresis function [14], which is applied to the context layers of the xRNN model, is inspired from the mechanism in biological neurons that enables them to maintain their activation states even after the excitatory impulse disappears [12]. This parameter needs to be optimised for the architecture and the dataset being used, allowing the fine tuning of performance as required for each classification task and dataset.

Fine tuning the hysteresis values allows the xRNNs to superimpose and extend different time-dependencies, retaining contextual information over longer sequences for the classification task to be learned. For example, a value of 0.8 means that the context unit is retaining 80% of its current activation and therefore only 20% of new activation is copied to it.

3 Description of Text Corpus Used in Experiments

The Modified Apté ("ModApté") Split [1] of the Reuters-21578 Text Categorisation Test Collection corpus [13] was used as it has been an important, prevalent and standard benchmarking corpus in Information Retrieval (IR) and Machine Learning (ML).

In order to ensure that there were enough training samples for the networks, two pairs of related categories from this split were concatenated to give 8 categories, as opposed to the 10 normally used categories from this corpus subset. The corpus consists of ≈10,000 titles. For the training set, ≈1,000 news titles were used, the first 130 of each of the 8 categories; all the other ≈9,000 news titles were used in the test set.

3.1 Definition of Noise

Noise is generally defined in the field of IR as the insignificant, irrelevant words or so-called stopwords which are normally present in any natural language text. Stopwords have an average distribution in any standard English language corpus and do not normally contribute any information to classification tasks; examples are "and", "but" and "the".

4 Representation of Text Corpus: Deriving Word Vectors and Feature Reduction

With over 10,000 unique words in the Reuters-21578 corpus, techniques have to be used to transform the dimensionality or feature space of the data into a smaller feature set. The dimension reducing preprocessing step adopted represents the titles of the corpus in the form of "semantic vectors" [23,22], a variation on the well-known Vector Space Models [16] such as the TF-IDF (term frequency–inverse document frequency) representation.

The features are transformed into weighted vector representations for each word in such a way that the number of dimensions for each word vector is thus a function of the total number of classes in the corpus. Each word is given a unique decimal value ranging from 0 to 1 that associates it with a particular class, with high values for strong associations to a class, and small values for weak associations.

4.1 Semantic Vectors

The semantic vector preprocessing strategy represents words as vectors which have a likelihood of occurring in a particular semantic category; the advantage of this representation is that word vectors are independent of the number of examples present in each category:

$$v(w, c_i) = \frac{Normalized\ frequency\ of\ w\ in\ c_i}{\sum_{j}^{C} Normalized\ frequency\ of\ w\ in\ c_j} \tag{3}$$

$for\ j \in \{1, \cdots n\}$

where:

$$Normalized\ freq.\ of\ w\ in\ c_i = \frac{Freq.\ of\ w\ in\ c_i}{No.\ of\ titles\ in\ c_i} \tag{4}$$

The *normalised* frequency of appearance of a word w in a semantic category c_i (i.e. *the normalised category frequency*) is computed as a value $v(w, c_i)$ for each element of the semantic vector, divided by normalising the frequency of appearance of a word w in the corpus (i.e. *the normalised corpus frequency*). C is the total number of classes.

5 Performance Measures Used for Comparison of Experimental Results

An important step that must always be taken after TC experiments is to quantify the performance of the various approaches used in a way so as to derive useful and comparable representations of the results.

5.1 Recall and Precision

The standard measures of recall and precision (R-P) [15] used in IR and ML were chosen for evaluating the performances of the classifiers, to address the capabilities of xRNNs from an IR viewpoint. These two measures were then combined to give what is termed the F_{score} measure. If:

- α, number of correctly classified documents to a specific category
- β, number of incorrectly classified documents to a specific category
- γ, number of incorrectly rejected documents from a specific category

then Table 1 gives the formulae used to derive R-P measures used for presenting results in this paper.

Table 1. Efficiency measures used in information retrieval and machine learning to evaluate the performance of various classification approaches

IR	AI	
Recall	Sensitivity	$\frac{\alpha}{\alpha+\gamma}$
Precision	Predictive Value	$\frac{\alpha}{\alpha+\beta}$

5.2 F_{score} Measure

$$F_N = \frac{(1 + N^2) * precision * recall}{(precision + (N^2 * recall))}$$

For the purposes of comparison, results are represented as the weighted mean of precision and recall; this F_{score} is termed the F_1 measure where recall and precision are evenly weighted by setting $N = 1$.

6 Experimental Setup: Optimising xRNN Parameters

Well-defined methodologies must be adopted to ensure that good model generalisation is maintained and potential pitfalls avoided. The following optimisation steps were taken to ensure that the models and experiments were properly validated.

6.1 Determining the Optimal Number of Input, Output and Hidden Units

The input layer of the xRNN accepts a series of vectors, one for each word in the title, whilst the output layer represents the class to which the title belongs. Since there were 8 main categories for the training/testing, the input layer of the xRNN had 8 units for the 8 dimensions of the semantic vectors for each word. Furthermore, there were 8 output units that represented the 8 possible desired output preferences for the respective semantic category of the title.

For determining the optimal number of hidden units, the "forward selection" method was used until the best rate of convergence and a stopping criterion were found – this was set to 6 hidden units. The hidden layers had a one-to-one connection to their respective context layers, that is, the 6 hidden units were connected to 6 respective units in the context layer.

6.2 Cross-Validation and Determination of Optimal Hysteresis Values

Generalisation error was maintained at low levels by the adoption of a bootstrapping cross-validation approach, where subsamples of the data were used. Each subsample is a random sample with replacement from the full sample that allows the reuse of the dataset efficiently [3]. The networks were optimised by constantly resampling [20] during the initial optimisation experiments to derive the parameter values that gave the best R-P outputs and the lowest generalisation error on the training set. These settings were maintained throughout all the experiments unless otherwise stated or required.

The important derivation of the optimal hysteresis parameters for the xRNN network have been described in [2]. The experiments were performed by iterating through incremental hysteresis values in steps of 0.1, starting from $\{0.0, 0.0\}$ for the first (C_1) and second context (C_2) layers through to $\{1.0, 1.0\}$ respectively. The optimal settings adopted for C_1 and C_2 were set at 0.2 and 0.7 respectively. These values were taken as the benchmarks for all comparisons. All the experiments that follow were performed using the same conditions for learning rate (0.01), number of epochs (900) and momentum (0.6) using the standard backpropagation algorithm.

7 Experiments Performed and Results Obtained

The xRNN models were used to classify various configurations of the Reuters-21578 Corpus training/test sets to test several hypotheses. Each series of experiments was repeated a number of times and the results compared; the main appropriate [5] statistical test done for significance between different results was the *two-sample T-test*. Given differences in the results can be deemed statistically significant, unless otherwise stated.

7.1 Experimental Series I: Importance of Word Order and Ranking

This series of experiments used training and test sets of the Reuters-21578 Corpus with the original, reversed, randomised, ranked and reverse ranked word orders from the titles to ascertain whether sequential word ordering was important. The results are compared and presented in Table 2; in all the tables, the results are ranked according to the mean F_{score} measures for 50 separately initialised networks given in the final column.

Table 2. Summary of micro-averaged R-P classification performances on the five main orderings of the Reuters-21578 Corpus used to train/test the xRNNs

Version of corpus dataset used for the xRNN models	Highest performing network on dataset (%)			Mean performance for 50 networks (%)		
	Recall	Precision	F-Measure	Recall	Precision	F-Measure
Reversed Ranked	94.41	94.05	94.23	94.14	93.41	93.80
Fully Ranked	94.82	94.33	94.57	93.88	93.28	93.58
Randomised	93.64	93.26	93.45	92.72	92.12	92.42
Original Corpus	94.10	93.32	93.71	92.59	91.73	92.16
Reversed Original	93.72	92.96	93.34	92.26	91.39	91.83

The xRNNs performed best on the reverse ranked versions of the corpus, followed by the ranked version. However, unexpectedly, the xRNNs trained/tested on the randomised version of the corpus show a statistically significant improvement in their classification over the original word order of the corpus. The worst performance was obtained when the reversed version of the original corpus was used.

7.2 Experimental Series II: Removal of Stopwords, TFIDF Representation and Sensitivity Analysis Comparisons

The next set of experiments used the corpus stripped of all stopwords from the titles to show the degree of significance of noise. Another experiment used the classic TF-IDF representation of words. Then, a series of sensitivity analysis experiments were done, whilst keeping all parameters constant but using the three, two and single highest word vector values from each title of the corpus to train/test the xRNNs; these experiments would show whether there was a cut-off point for the number of words required to maintain classification performance. The results are compared and presented in Table 3.

The xRNNs trained/tested using only the three highest value ranked vectors from each title in the corpus showed the best classification results of this series of experimental comparisons. Compared to the original word order results as given in Table 2, using the three highest vectors is statistically better as is using a corpus with all the stopwords removed, but only by a very small amount. The gain from removing stopwords does not seem extremely important, at least not to the degree expected. Usefully, the xRNNs trained/tested using the TF-IDF

Table 3. Summary of R-P classification performances on the five other datasets derived from the Reuters-21578 Corpus used to train/test the xRNNs

Version of corpus dataset used for the xRNN models	Highest performing network on dataset (%)			Mean performance for 50 networks (%)		
	Recall	Precision	F-Measure	Recall	Precision	F-Measure
3 Highest Vectors	93.82	93.55	93.69	93.16	92.64	92.90
No Stopwords	94.00	93.04	93.52	92.72	91.90	92.31
2 Highest Vectors	93.31	92.59	92.95	92.47	91.88	92.17
TF-IDF	93.40	91.95	92.67	92.12	91.04	91.58
1 Highest Vector	91.29	91.26	91.28	90.94	90.84	90.90

vector representation produced significantly lower results (91.58%) when compared to the results using the original word order (92.16%) using the semantic vector representation, to give a significant advantage over TF-IDF.

7.3 Experimental Series III: Artificially Noisy Datasets

Experimental Series I demonstrated, unexpectedly, that the xRNNs were not relying on word ordering. In fact, randomising the word orders in titles tended to statistically *improve* classification performance by the xRNNs. Thus, this series of experiments using artificially augmented versions of the Reuters-21578 Corpus were deemed appropriate.

Three versions of the corpus were created. Titles were artificially made noisy by introducing controlled amounts of random stopwords by a factor of two, four and six times; this effectively increased each title in the original corpus from an average of 8 words to ≈16, ≈32, and ≈48 words respectively. Table 4 shows the results obtained.

Table 4. Summary of R-P classification performances on the three artificially noisy datasets derived from the Reuters-21578 Corpus by the introduction of stopwords

Version of corpus dataset used for the xRNN models	Highest performing network on dataset (%)			Mean performance for 10 networks (%)		
	Recall	Precision	F-Measure	Recall	Precision	F-Measure
Noise Factor 2	93.43	92.59	93.01	92.39	91.63	92.01
Noise Factor 4	93.09	92.54	92.81	91.28	90.37	90.82
Noise Factor 6	92.42	91.71	92.07	86.40	85.63	86.01

Although there is a statistically significant drop in the mean classification performance values using these artificially very noisy datasets, nevertheless, it can be seen that the xRNNs are able to maintain reasonably significant R-P performances even in the face of high levels of noise. However, mean performance begins to drop when the title lengths reach 48 words.

8 Conclusion

The experiments showed that for this task of classifying the Reuters-21578 Corpus, the xRNN models, interestingly, did not rely on the strict sequential ordering of the words in the titles. In fact, *randomising* titles *improved* classification – the networks seemed to be aggregating important contextual keyword information over long sequences instead to give $\approx$94.0% in terms of R-P performances. On the same original corpus, SRNs gave performances with an average value of 45.29% and 45.10% for recall and precision respectively, an unexpectedly low value, highlighting the importance of the hysteresis-driven xRNN models. The Multi-Layer Perceptron (MLP) models gave $\approx$90.0% for R-P using the highest keyword values from each title of the corpus. There was a very significant gain of $\approx$4% to using the xRNN model in both comparisons. The memory capacity for contextual information was being maintained by the xRNNs using the hysteresis feedback parameter even in the face of excessive levels of noise.

Although mean values over many networks tended to be lower with greater levels of noise, it was still possible to find, by careful model selection from a population of trained networks, xRNNs that maintained very high levels of R-P. Other results [2] have shown that carefully choosing the hysteresis decay values of the context layers is critical for maintaining consistent R-P performance. Slight deviations from the optimal derived values can cause a catastrophic drop in R-P performance; the hysteresis value settings are especially important for networks trained on increasing noise levels.

Comparisons with different ML approaches on this corpus are given in [18]. For example, SVMs [10] achieve $\approx$92.0% in their combined F-measure scores using micro-averaging for 10 categories. Using the KL-divergence approach to Naive Bayes [17], R-P performances of $\approx$90.0%-91.0% have also been achieved. Nevertheless, the results reported in this paper reinstate confidence in RNN approaches to TC, giving performances that are on par with the most powerful alternative ML techniques to TC.

References

1. Apté, C., Damerau, F.J., Weiss, S.M.: Automated learning of decision rules for text categorization. ACM Transactions on Information Systems 12(3), 233–251 (1994)
2. Arevian, G., Panchev, C.: Optimizing the hystereses of a two context layer rnn for text classification, Proceedings of the International Joint Conference on Neural Networks 2007 (20th IJCNN'07) (Orlando, Florida) (August 2007)
3. Baxt, W.G., White, H.: Bootstrapping confidence intervals for clinical input variable effects in a network trained to identify the presence of acute myocardial infarction. Neural Computation 7(3), 624–638 (1995)
4. Bodén, M., Jacobsson, H., Ziemke, T.: Evolving context-free language predictors. In: Whitley, L.D., Goldberg, D.E., Cantú-Paz, E., Spector, L., Parmee, I.C., Beyer, H.-G. (eds.) GECCO, pp. 1033–1040. Morgan Kaufmann, San Francisco (2000)
5. Demsar, J.: Statistical comparisons of classifiers over multiple data sets. Journal of Machine Learning Research 7, 1–30 (2006)

6. Elman, J.L.: Finding structure in time, Tech. Report CRL 8901. University of California, San Diego, CA (1988)
7. Frasconi, P., Gori, M., Soda, G.: Local feedback multilayered networks. Neural Computation 4(1), 120–130 (1992)
8. Giles, C.L., Lawrence, S., Tsoi, A.C.: Noisy time series prediction using a recurrent neural network and grammatical inference. Machine Learning 44(1/2), 161–183 (2001)
9. Huang, B.Q., Kechad, T.: A recurrent neural network recogniser for online recognition of handwritten symbols. In: Chen, C.-S., Filipe, J., Seruca, I., Cordeiro, J. (eds.) ICEIS, pp. 27–34 (2005)
10. Joachims, T.: Learning to classify text using support vector machines. Kluwer Academic Publishers, Boston (2002)
11. Kaikhah, K.: Text summarization using neural networks, WSEAS (2004)
12. Koch, C., Segev, I.: Methods in neuronal modelling. MIT Press, Cambridge, Massachusetts (1989)
13. Lewis, D.D.: Reuters- 21578 text categorization test collection (1997), `http://www.research.att.com/~lewis`
14. O'Reilly, R.C., Munakata, Y.: Computational explorations in cognitive neuroscience: Understanding the mind by simulating the brain. MIT Press, Redmond, Washington (2000)
15. Van Rijsbergen, C.J.: Information retrieval, Butterworths, London (1979)
16. Salton, G., McGill, M.J.: Introduction to modern information retrieval. McGraw-Hill, New York (1983)
17. Schneider, K.M.: A new feature selection score for multinomial naive bayes text classification based on KL-divergence (June 29, 2004)
18. Sebastiani, F.: Fabrizio Sebastiani, *Machine learning in automated text categorization.* ACM Comput. Surv. 34(1), 1–47 (2002)
19. Shi, X., Liang, Y., Xu, X.: An improved elman model and recurrent backpropagation control neural networks (in Chinese with English abstract, keywords, figures, tables, and references). Journal of Software 14(6), 1110–1119 (2003)
20. Weiss, S.M., Kulikowski, C.A.: Computer systems that learn, classification and prediction methods from statistics, neural networks, machine learning and expert systems, San Mateo, CA. Morgan Kaufmann, San Francisco (1991)
21. Wermter, S.: Hybrid connectionist natural language processing. Chapman and Hall, Thomson International, London, UK (1995)
22. Wermter, S., Arevian, G., Panchev, C.: Recurrent neural network learning for text routing. In: Proceedings of the International Conference on Artificial Neural Networks, Edinburgh, UK, pp. 898–903 (1999)
23. Wermter, S., Panchev, C., Arevian, G.: Hybrid neural plausibility networks for news agents. In: Proceedings of the National Conference on Artificial Intelligence, Orlando, USA, pp. 93–98 (1999)
24. Wiles, J., Elman, J.: Learning to count without a counter: A case study of dynamics and activation landscapes in recurrent networks (1995)

Semi-supervised Metrics for Textual Data Visualization

Ángela Blanco and Manuel Martín-Merino

Universidad Pontificia de Salamanca
C/Compañía 5, 37002, Salamanca, Spain
ablancogo@upsa.es, mmartinmac@upsa.es

Abstract. Multidimensional Scaling algorithms (MDS) are useful tools that help to discover high dimensional object relationships. They have been applied to a wide range of practical problems and particularly to the visualization of the semantic relations among documents or terms in textual databases.

The MDS algorithms proposed in the literature often suffer from a low discriminant power due to its unsupervised nature and to the 'curse of dimensionality'. Fortunately, textual databases provide frequently a manually created classification for a subset of documents that may help to overcome this problem.

In this paper we propose a semi-supervised version of the Torgerson MDS algorithm that takes advantage of this document classification to improve the discriminant power of the word maps generated. The algorithm has been applied to the visualization of term relationships. The experimental results show that the model proposed outperforms well known unsupervised alternatives.

1 Introduction

The Multidimensional Scaling Algorithms (MDS) are multivariate data analysis techniques that allow us to visualize high dimensional object relationships in an intuitive way. A large variety of algorithms have been proposed to this aim (see for instance [15,10,16]). In particular, the Torgerson MDS algorithm [10] has been applied to visualize the underlying structure of high dimensional data.

An interesting application of the Torgerson MDS algorithm is the visualization of the semantic relations among terms or documents [15,17] in text mining problems. This visual representation gives more information than just the hard classification of terms or documents and is particularly helpful for novel users [8]. However, the algorithms proposed in the literature often have a low discriminant power, that is, terms that belong to different topics overlap strongly in the word map. Therefore, the resulting maps are often useless to identify the different semantic groups in a given textual collection. This is mainly due to the unsupervised nature of the algorithm and to the 'curse of dimensionality' [6,17].

J. Marques de Sá et al. (Eds.): ICANN 2007, Part II, LNCS 4669, pp. 435–444, 2007.

The visualization of term relationships has been usually done in the literature by non-supervised techniques. Moreover, common semi-supervised algorithms [14] can not be applied because the categorization of a small subset of terms is a complex and time consuming task [1,19]. However, textual databases provide often a classification for a subset of documents [1] because this is easier for human experts. This suggests that the organization of terms into topics can be improved considering the available labels in the space of documents.

In this paper we present a modification of the Torgerson MDS algorithm that improves the visualization of the term relationships taking advantage of a categorization for a subset of documents. To this aim, rather than modifying the error function as is usually done by supervised clustering algorithms [20] we define a semi-supervised similarity that takes into account the document class labels. Next, the Torgerson MDS algorithm is applied to generate the word map considering this similarity matrix. Finally the new algorithm has been tested using a real textual collection and has been exhaustively evaluated through several objective functions.

This paper is organized as follows. In section 2 the Torgerson MDS algorithm is introduced. Section 3 presents the new semi-supervised MDS algorithm. In section 4 the algorithm is applied to the visualization of term relationships. Finally section 5 gets conclusions and outlines future research trends.

2 The Torgerson MDS Algorithm

Let $\boldsymbol{X}(n \times d)$ be a matrix of n objects represented in $\mathbb{R}^d$ and $\boldsymbol{D} = (\delta_{ij})$ the dissimilarity matrix made up of the object proximities. The Multidimensional Scaling algorithms look for an object configuration in a low dimensional space (usually two for visualization) in such a way that the inter-pattern Euclidean distances reflect approximately the original dissimilarity matrix.

A large variety of algorithms have been proposed in the literature. In this paper we have considered the Torgerson MDS algorithm [10] because it exhibits several properties interesting for text mining problems. First, the algorithm with the Euclidean distance is equivalent to a linear PCA [10] that can be solved efficiently through a linear algebraic operation such as the Singular Value Decomposition (SVD) [5]. Second, the optimization problem doesn't have local minima. Notice that many MDS algorithms such as Sammon or certain neural based techniques [16] rely on non-linear optimization methods that can get stuck in local minima. Finally, the Torgerson MDS algorithm can be considered with certain similarities equivalent to the Latent Semantic Indexing (LSI) [4] that has been succesfully applied in text mining problems.

Next we introduce briefly the Torgerson MDS algorithm. For a detailed explanation see [10].

Define the matrix $\boldsymbol{A}(n \times n)$ as $[\boldsymbol{A}]_{ij} = a_{ij} = -\frac{1}{2}\delta_{ij}^2$. The inner product matrix $\boldsymbol{B}$ can be obtained as:

$$\boldsymbol{B} = \boldsymbol{X}\boldsymbol{X}^T = \boldsymbol{H}\boldsymbol{A}\boldsymbol{H} \tag{1}$$

where $\boldsymbol{H}$ is a centering matrix defined as:

$$\boldsymbol{H} = \boldsymbol{I} - \frac{1}{n}\mathbf{1}\mathbf{1}^T \tag{2}$$

where $\mathbf{1} = [1\,1\ldots 1]^T$ is a column matrix of n ones and $I(n \times n)$ is the identity matrix.

The Torgerson MDS algorithm looks for a projection $W : \mathbb{R}^d \rightarrow \mathbb{R}^k$ to a lower dimensional space such that the Euclidean distances in $\mathbb{R}^k$ preserve as much as possible the original dissimilarities. The object coordinates that verify this condition are given [10] by:

$$\boldsymbol{X}_k = \boldsymbol{V}_k \boldsymbol{\Lambda}_k^{\frac{1}{2}}, \tag{3}$$

where $\boldsymbol{V}_k$ is the $n \times k$ orthonormal matrix whose columns are the kth first eigen vectors of $\boldsymbol{B}$ and $\boldsymbol{\Lambda}_k = diag(\lambda_1, \ldots, \lambda_k)$ is a diagonal matrix with λ_i the ith eigenvalue of $\boldsymbol{B}$.

The object coordinates in equation (3) can be obtained through a SVD. This operation is particularly efficient when only the first eigenvectors are needed [11] as it happens for visualization purposes.

3 A Semi-supervised MDS Algorithm

The word maps generated by the Torgerson MDS algorithm often suffer from a low discriminant power. The unsupervised nature of the algorithm favors the overlapping between different topics in the map. Moreover, due to the "curse of dimensionality" the words concentrate around the center map and the smaller distances become often meaningless [18,6].

In this section we explain how the categorization of a subset of documents by human experts can be exploited to improve the word maps generated by the MDS algorithm. The novelty of this problem relies in that we are trying to improve an unsupervised technique that works in the space of terms considering the available labels in the space of documents. To this aim rather than modifying the error function as is usually done by supervised clustering algorithms [20] we define a semi-supervised similarity that takes into account the class labels. This similarity will reflect both, the semantic classes of the textual collection and the term relationships inside each class. Once the semi-supervised dissimilarities are computed, the Torgerson MDS algorithm can be applied to generate a visual representation of the term relationships. Notice that our approach allow us to extend to the semi-supervised case any algorithm that works from a dissimilarity matrix.

Let t_i, t_j be two terms and $\{C_k\}_{k=1}^{c}$ the set of categories created by human experts. The association between terms and categories are usually evaluated in the Information Retrieval literature by the Mutual Information [23] defined as:

$$I(t_i; C_k) = \log \frac{p(t_i, C_k)}{p(t_i)p(C_k)}, \tag{4}$$

where $p(t_i, C_k)$ denotes the joint coocurrence probability of term t_i and class C_k. $p(t_i)$, $p(C_k)$ are the a priori probability of occurrence of term t_i and class C_k

respectively. The Mutual Information is able to capture non-linear relationships between terms and categories.

However, it has been pointed out in the literature [23] that the index (4) gives higher score to rare terms. To overcome this problem we have considered a weighted version of the previous index defined as

$$I'(t_i; C_k) = p(t_i, C_k) \log \frac{p(t_i, C_k)}{p(t_i)p(C_k)} . \tag{5}$$

This index reduces obviously the weight of the less frequent terms.

Now, we can define a similarity measure between terms considering the class labels. This measure will be referred as supervised similarity from now on. Obviously, this similarity should become large for terms that are related/unrelated with the same categories of documents. This suggests the following definition for the term similarity:

$$s_1(t_i, t_j) = \frac{\sum_k I'(t_i; C_k) I'(t_j; C_k)}{\sqrt{\sum_k (I'(t_i; C_k))^2} \sqrt{\sum_k (I'(t_j; C_k))^2}} . \tag{6}$$

The numerator of this similarity will become large for terms that are correlated with similar categories. Notice that the index (6) can be considered a cosine similarity between the vectors $I'(t_i; \cdot) = [I'(t_i; C_1), \ldots, I'(t_i; C_c)]$ and $I'(t_j; \cdot) = [I'(t_j; C_1), \ldots, I'(t_j; C_c)]$. This allow us to interpret the new similarity as a non-linear transformation to a feature space [22] where a cosine similarity is computed. Other dissimilarities can be considered in feature space but we have chosen the cosine because it has been widely used in the Information Retrieval literature. Finally the similarity (6) is translated and scaled so that it takes values in the interval $[0, 1]$.

The similarity defined above can be considered an average over all the categories. Next, we provide an alternative definition for the supervised similarity that considers only the class with higher score. It can be written as

$$s_2(t_i, t_j) = \max_k \{ \bar{I}(t_i; C_k) * \bar{I}(t_j; C_k) \} , \tag{7}$$

where $\bar{I}$ is a normalized Mutual Information defined as

$$\bar{I}(t_i; C_k) = \frac{I(t_i; C_k)}{\max_l \{ I(t_i; C_l) \}} . \tag{8}$$

This normalization factor guarantees that $s_2(t_i, t_i) = 1$. The similarity (7) will get large when both terms are strongly correlated with one of the classes.

The supervised measures proposed earlier will score high terms that are related with the same categories. However, for visualization purposes it is also interesting to reflect the semantic relations among the terms inside each class or among the main topics. This information is provided by unsupervised measures such as for instance the cosine. This justifies the definition of a semi-supervised similarity as a convex combination of a supervised and an unsupervised measure.

This similarity will reflect both, the semantic groups of the textual collection and the term relationships inside each topic. It is defined as follows:

$$s(t_i, t_j) = \lambda s_{sup}(t_i, t_j) + (1 - \lambda) s_{unsup}(t_i, t_j) , \tag{9}$$

where s_{sup} and s_{unsup} denote the supervised and unsupervised measures respectively. The parameter λ verifies $0 \leq \lambda \leq 1$. This parameter will determine if the resulting map reflects better the semantic classes of the textual collection (λ large) or the semantic relations among the terms (λ small).

The semi-supervised similarity (9) has an interesting property that is worth to mention. The standard deviation for the semi-supervised similarity histogram is 0.18 while for the cosine similarity is 0.04. Therefore, most of the similarities for the unsupervised measure are zero or close to zero. This suggests that cosine similarity is strongly degraded by the 'curse of dimensionality' [2]. On the other hand, the standard deviation for the semi-supervised dissimilarity is larger and thus, it is more robust to the 'curse of dimensionality'. Hence, any algorithm based on dissimilarities will perform better [18,6].

Finally the Torgerson MDS algorithm introduced in section 2 is applied to derive a visual representation of the semi-supervised similarities. To this aim, the similarity (9) must be transformed into a dissimilarity using for instance the following rule $\delta_{ij} = 1 - s_{ij}$ [10]. Then, the Torgerson MDS algorithm can be used to get an approximate representation of the data in a space of dimension $< n - 1$ where n is the sample size.

The semi-supervised MDS algorithm presented earlier assumes that the whole textual collection is categorized by human experts. However, it is very common in text mining problems that only a small percentage of the textual collection is labeled [1]. Hence, we have a small training set of categorized documents and a much larger test set of documents not labeled. In this case, it has been suggested in the literature that the test documents should be classified considering the clustering hypothesis [7]. That is, the classifier should not split test documents that belong to the same cluster. A large variety of techniques have been proposed in the literature based on this hypothesis. In this paper we have considered the Transductive Support Vector Machines (TSVM) [22] because they have been successfully applied to the categorization of document collections [12]. The Transductive SVM aims at finding a decision function that maximizes the margin of both, labeled and unlabeled patterns. This technique allow us to reduce significantly the misclassification error of the inductive SVM, particularly when the training set is very small [12].

Once the documents are classified using the TSVM, the semi-supervised similarity (9) is computed as in the supervised case. However, those terms that appear in less than five documents categorized by human experts or by the TSVM are considered unreliable and the similarity is computed in an unsupervised way ($\lambda = 0$).

4 Experimental Results

In this section we apply the proposed algorithms to the construction of word maps that visualize term semantic relationships. The textual collection considered, is made up of 2000 *scientific abstracts* retrieved from three commercial databases 'LISA', 'INSPEC' and 'Sociological Abstracts'. For each database a thesaurus created by human experts is available. Therefore, the thesaurus induces a classification of terms according to their semantic meaning. This will allow us to exhaustively check the term associations created by the map.

Assessing the performance of algorithms that generate word maps is not an easy task. In this paper the maps are evaluated from different viewpoints through several objective functions. This methodology guaranty the objectivity and validity of the experimental results.

The objective measures considered in this paper quantify the agreement between the semantic word classes induced by the map and the thesaurus. Therefore, once the objects have been mapped, they are grouped into topics with a clustering algorithm (for instance PAM [13]). Next we check if words assigned to the same cluster in the map are related in the thesaurus. To this aim we have considered the following objective measures:

The F measure [3] has been widely used by the Information Retrieval community and evaluates if words from the same class according to the thesaurus are clustered together. The entropy measure [23] evaluates the uncertainty for the classification of words from the same cluster. Small values suggest little overlapping among different topics in the map and are preferred. Finally the Mutual Information [21] is a nonlinear correlation measure between the word classification induced by the thesaurus and the word classification given by the clustering algorithm. This measure gives more weight to specific words and therefore provides valuable information about changes in the position of specific terms.

Table 1. Empirical evaluation of several semi-supervised visualization algorithms for a collection of scientific abstracts

	F	E	I
Torgerson MDS	0.46	0.55	0.17
Least square MDS	0.53	0.52	0.16
Torgerson MDS (Average)	0.69	0.43	0.27
Torgerson MDS (Maximum)	0.77	0.36	0.31
Least square MDS (Average)	0.70	0.42	0.27
Least square MDS (Maximum)	0.76	0.38	0.31

Table 1 shows the experimental results for the semi-supervised MDS algorithms proposed in this paper. Two unsupervised techniques have been considered as reference, the Torgerson MDS algorithm introduced in section 2 and a standard least square MDS algorithm [10]. For each technique, two semi-supervised similarities have been considered, the average (see equation (6)) and the maximum (see equation (7)). As unsupervised measure we have selected the cosine because it has been widely used by the information retrieval community [9] with reasonable good results. For other possible choices see [9].

The least square MDS has been always initialized by a PCA to avoid that the algorithm get stuck in a local minima. The λ parameter in the semi-supervised measures has been set up to 0.5 which achieves a good balance between structure preservation and topic separation in the map. From the analysis of table 1 the following conclusions can be drawn:

- The semi-supervised techniques improve significantly the word maps generated by the unsupervised ones. In particular, the semi-supervised Torgerson MDS (rows 3-4) reduces significantly the overlapping among the different topics in the map (E is significantly reduced). The Mutual Information is particularly improved which suggests that the overlapping among the specific terms that belong to different topics is reduced in the map. Finally, the F measure corroborates the superiority of the proposed algorithm.

  The least square MDS algorithm (rows 5-6) improves similarly the maps generated when the semi-supervised dissimilarities are considered. This suggests that many algorithms that work from a dissimilarity measure can benefit from the ideas presented in this paper.
- The maximum semi-supervised similarity gives always better results than the average. This can be explained because the maximum supervised similarity is defined considering only the class that is more correlated with the terms. This feature improves the separation of the topics in the map.

Finally figure 1 illustrates the performance of the semi-supervised Torgerson MDS algorithm from a qualitative point of view. For the sake of clarity only a subset of words that belong to three topics have been drawn. We report that

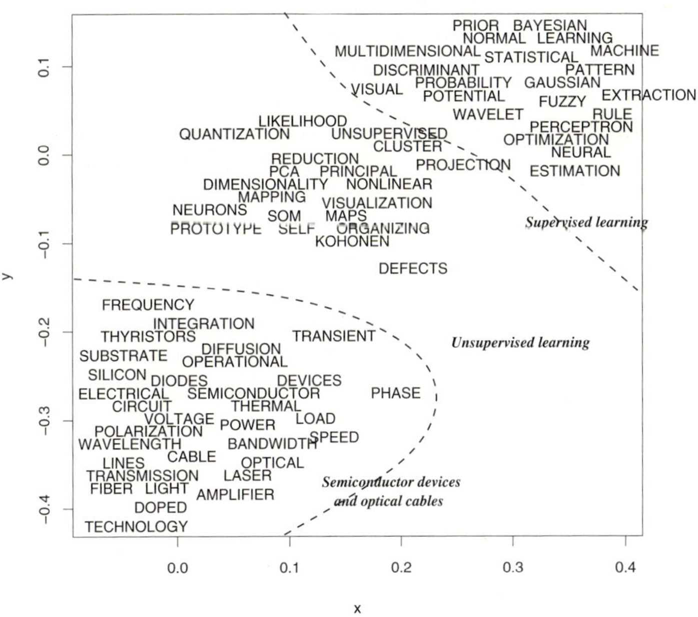

Fig. 1. Map generated for the semi-supervised Torgerson MDS algorithm with average similarity for a subset of words that belong to three topics with different overlapping

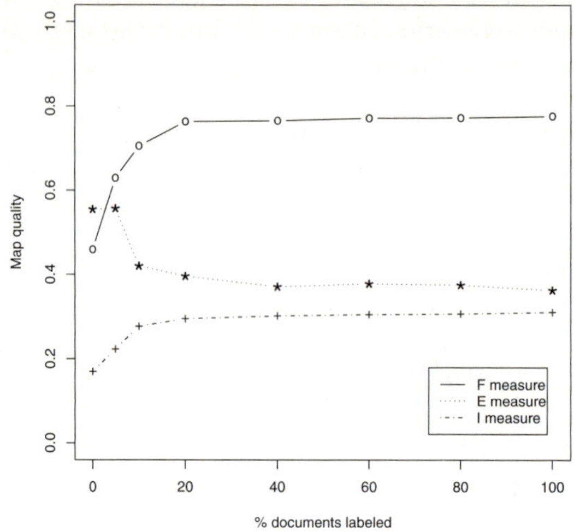

Fig. 2. Evaluation measures for the semi-supervised Torgerson MDS algorithm with average similarity when the percentage of documents labeled range from 0% to 100%

the term associations induced by the map are satisfactory and that the semantic topics can be easily identified in the map.

As we have mentioned earlier, in text mining applications only a small subset of documents is categorized by human experts. Therefore, from a practical point of view it is very important to evaluate the sensibility of the method proposed to the percentage of categorized documents. The Transductive SVM has been implemented using the SVM^{light} software. For the multicategory classification we have considered the 'one against one' approach. The regularization parameters C and C^* for training and test sets respectively have been set up to one. Finally, $\lambda = 0.5$ in the semi-supervised measures. The empirical results suggest that this value allow us to identify easily the semantic topics in the word maps and the term relationships.

Figure 2 shows the evaluation measures when the percentage of documents labeled range from 0% to 100%. According to this figure the quality of the word maps generated is similar whenever the percentage of documents labeled is larger than 10%. Moreover, with only 5% of documents categorized the performance is not significantly degraded. Finally, we report that the semi-supervised MDS algorithms with only 10% of documents categorized improve significantly the unsupervised counterparts (0% of documents labeled).

5 Conclusions and Future Research Trends

In this paper we have proposed a semi-supervised version of the Torgerson MDS algorithm for textual data analysis. The new model takes advantage of a categorization of a subset of documents to improve the discriminant power of the word maps generated. The algorithm proposed has been tested using a real textual collection and evaluated through several objective functions.

The experimental results suggest that the proposed algorithm improves significantly well known alternatives that rely solely on unsupervised measures. In particular the overlapping among different topics in the map is significantly reduced improving the discriminant power of the algorithms.

Future research will focus on the development of new semi-supervised clustering algorithms.

Acknowledgements. Financial support from Junta de Castilla y León grant PON05B06 is gratefully appreciated.

References

1. Aggarwal, C.C., Gates, S.C., Yu, P.S.: On Using Partial Supervision for Text Categorization. IEEE Transactions on Knowledge and Data Engineering 16(2), 245–255 (2004)
2. Aggarwal, C.C.: Re-designing distance functions and distance-based applications for high dimensional applications. In: Proc. of SIGMOD-PODS, vol. 1, pp. 13–18 (2001)
3. Baeza-Yates, R., Ribeiro-Neto, B.: Modern information retrieval, Wokingham, UK. Addison-Wesley, London, UK (1999)
4. Bartell, B.T., Cottrell, G.W., Belew, R.K.: Latent Semantic Indexing is an Optimal Special Case of Multidimensional Scaling. In: Proceedings of ACM SIGIR Conference, Copenhagen, pp. 161–167 (1992)
5. Berry, M.W., Drmac, Z., Jessup, E.R.: Matrices, vector spaces and information retrieval. SIAM review 41(2), 335–362 (1999)
6. Buja, A., Logan, B., Reeds, F., Shepp, R.: Inequalities and positive default functions arising from a problem in multidimensional scaling. Annals of Statistics 22, 406–438 (1994)
7. Chapelle, O., Weston, J., Schölkopf, B.: Cluster kernels for semi-supervised learning. In: Conference Neural Information Processing Systems, vol. 15 (2003)
8. Chen, H., Houston, A.L., Sewell, R.R., Schatz, B.R.: Internet browsing and searching: User evaluations of category map and concept space techniques. Journal of the American Society for Information Science (JASIS) 49(7), 582–603 (1998)
9. Chung, Y.M., Lee, J.Y.: A corpus-based approach to comparative evaluation of statistical term association measures. Journal of the American Society for Information Science and Technology 52(4), 283–296 (2001)
10. Cox, T.F., Cox, M.A.A.: Multidimensional scaling, 2nd edn. Chapman & Hall/CRC, USA (2001)
11. Golub, G.H., Van Loan, C.F.: Matrix Computations, 3rd edn. Johns Hopkins university press, Baltimore, Maryland, USA (1996)
12. Joachims, T.: Learning to Classify Text using Support Vector Machines. Methods, Theory and Algorithms. Kluwer Academic Publishers, Boston (2002)
13. Kaufman, L., Rousseeuw, P.J.: Finding groups in data. An introduction to cluster analysis. John Wiley & Sons, New York (1990)
14. Kothari, R., Jain, V.: Learning from Labeled and Unlabeled Data Using a Minimal Number of Queries. IEEE Transactions on Neural Networks 14(6), 1496–1505 (2003)

15. Lebart, L., Salem, A., Berry, L.: Exploring Textual Data, Netherlands. Kluwer Academic Publishers, Boston, MA (1998)
16. Mao, J., Jain, A.K.: Artificial neural networks for feature extraction and multivariate data projection. IEEE Transactions on Neural Networks 6(2) (March 1995)
17. Martín-Merino, M., Muñoz, A.: A New MDS Algorithm for Textual Data Analysis. In: Pal, N.R., Kasabov, N., Mudi, R.K., Pal, S., Parui, S.K. (eds.) ICONIP 2004. LNCS, vol. 3316, pp. 860–867. Springer, Heidelberg (2004)
18. Martín-Merino, M., Muñoz, A.: A New Sammon Algorithm for Sparse Data Visualization. In: International Conference on Pattern Recognition, Cambridge, vol. 1, pp. 477–481 (August, 2004)
19. Mladenié, D.: Turning Yahoo into an Automatic Web-Page Classifier. In: Proceedings 13th European Conference on Aritficial Intelligence, Brighton, pp. 473–474 (1998)
20. Pedrycz, W., Vukovich, G.: Fuzzy Clustering with Supervision. Pattern Recognition 37, 1339–1349 (2004)
21. Strehl, A., Ghosh, J., Mooney, R.: Impact of similarity measures on web-page clustering. In: Proceedings of the 17th National Conference on Artificial Intelligence: Workshop of Artificial Intelligence for Web Search, Austin, USA, pp. 58–64 (July 2000)
22. Vapnik, V.N.: Statistical Learning Theory. John Wiley & Sons, New York (1998)
23. Yang, Y., Pedersen, J.O.: A comparative study on feature selection in text categorization. In: Proc. of the 14th International Conference on Machine Learning, Nashville, Tennessee, USA, pp. 412–420 (July, 1997)

Topology Aware Internet Traffic Forecasting Using Neural Networks

Paulo Cortez[1], Miguel Rio[2], Pedro Sousa[3], and Miguel Rocha[3]

[1] Department of Information Systems/R&D Algoritmi Centre, University of Minho, 4800-058 Guimarães, Portugal
pcortez@dsi.uminho.pt
http://www.dsi.uminho.pt/~pcortez
[2] Department of Electronic and Electrical Engineering, University College London, Torrington Place, WC1E 7JE, London, UK
m.rio@ee.ucl.ac.uk
[3] Department of Informatics, University of Minho, 4710-059 Braga, Portugal
{pns,mrocha}@di.uminho.pt

Abstract. Forecasting Internet traffic is receiving an increasing attention from the computer networks domain. Indeed, by improving this task efficient traffic engineering and anomaly detection tools can be developed, leading to economic gains due to better resource management. This paper presents a Neural Network (NN) approach to predict TCP/IP traffic for all links of a backbone network, using both univariate and multivariate strategies. The former uses only past values of the forecasted link, while the latter is based on the neighbor links of the backbone topology. Several experiments were held by considering real-world data from the UK education and research network. Also, different time scales (e.g. every ten minutes and hourly) were analyzed. Overall, the proposed NN approach outperformed other forecasting methods (e.g. Holt-Winters).

Keywords: Link Mining, Multilayer Perceptrons, Multivariate Time Series, Network Monitoring, Traffic Engineering.

1 Introduction

Nowadays, more and more applications are migrating into TCP/IP networks (e.g. VoIP, IPTV). Hence, it is important to develop techniques to better understand and forecast the behavior of these networks. In effect, TCP/IP traffic prediction is gaining more attention from the computer networks community [18,12,1,2]. By improving this task, network providers can optimize resources, allowing a better quality of service. Also, traffic forecasting can help to detect anomalies (e.g. security attacks, viruses or an irregular amount of SPAM) by comparing the real traffic with the forecasts [8,7].

Often, TCP/IP traffic prediction is done intuitively by network administrators, with the help of marketing information on the future number of costumers and their behaviors [12]. Yet, this may not be suited for serious day-to-day network administration. Developments from the areas of Operational Research and Computer Science as lead to solid forecasting methods that replaced intuition based ones. In particular, the field of

J. Marques de Sá et al. (Eds.): ICANN 2007, Part II, LNCS 4669, pp. 445–454, 2007.

Time Series Forecasting (TSF), deals with the prediction of a chronologically ordered variable, where the goal is to model a complex system as a black-box, predicting its behavior based in historical data [10]. The TSF approaches can be divided into univariate and multivariate, depending if one or more variables are used. Multivariate methods are likely to produce better results, provided that the variables are correlated [14].

Several TSF methods have been proposed, such as the Holt-Winters [10] and Neural Networks (NN) [9,16,2]. Holt-Winters was developed for series with trended and seasonal factors and more recently a double seasonal version has been proposed [17]. In contrast with the conventional TSF methods (e.g. Holt-Winters), NNs can predict nonlinear series. In the past, several studies have proved the predictability of network traffic by using similar methods. For instance, the Holt-Winters was used in [8,6] and NNs have also been proposed [18,7,2].

Recently, there has been an increasing interest in Link Mining, which aims at the discovery of useful patterns in graph structured datasets [4]. For a given goal (e.g. prediction), the idea is to use models that learn from data extracted from correlated links. The Internet backbone, which is made up of core routers that transport data through countries or continents, is a fertile ground for Link Mining.

This work will use recent real-world data from the United Kingdom Education and Research Network (UKERNA) backbone. NNs will be used to predict the traffic for all 18 links of this backbone network, under univariate and multivariate approaches. The former is based on the previous traffic from the current link, while a heuristic rule is proposed for the latter, where the NNs are fed with data from current and the direct neighbor links. Furthermore, the predictions will be analyzed at different time scales (e.g. every ten minutes, hourly) and compared with other methods (e.g. Holt-Winters).

2 Time Series Data

The data collection was based in the Simple Network Management Protocol (SNMP), which quantifies the traffic passing through every network interface with reasonable accuracy [15]. SNMP is widely deployed by every Internet Service Provider/network and the collection of this data does not induce any extra traffic on the network. This work will analyze traffic data (in Mbit/s) from all links of the UK academic network backbone (UKERNA). This backbone contains a total of eight core routers and 18 links. Figure 1 plots the respective direct graph. The data was recorded into two datasets (every 10 minutes and every hour), between 12 AM of 14th June 2006 and 12 AM of 23th July 2006. The obtained multivariate series included 2 missing periods for the 10 minute data, which were replaced with a linear interpolation. The missing values are explained by the fact that the SNMP scripts are not 100% reliable, since the SNMP messages may be lost or the router may not reply on time. Yet, this occurs very rarely and it is statistically insignificant. The hourly multivariate series contains 936 observations for each link, while the 10 minute data encompasses a total of 5613 time records.

As an example, the hourly traffic of two neighbor links, London-Cosham (LC) and Cosham-Bristol (CB), is plotted in Figure 2. In the first case (LC), it is clear the influence of two seasonal components due to the the intraday and intraweek cycles. The weekly pattern is less visible in the second example (CB).

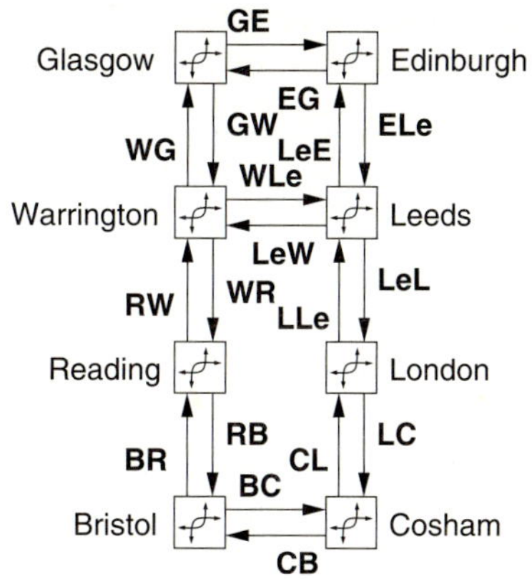

Fig. 1. The schematic of the UK academic Internet backbone

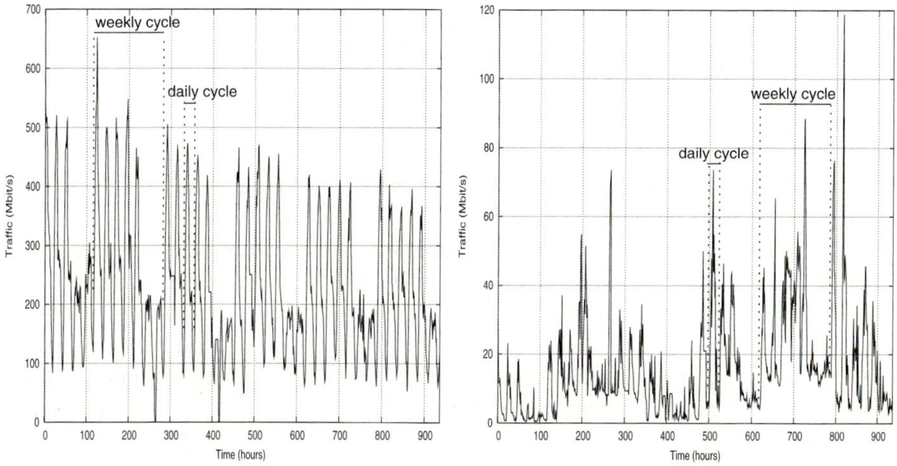

Fig. 2. The hourly IP traffic rate for the London-Cosham (left) and Cosham-Bristol (right) links

3 Forecasting Methods

A Time Series Forecasting (TSF) model assumes that past patterns will occur in the future. Let $y_t = (y_{1t}, \ldots, y_{kt})$ denote a multivariate series, where y_{ij} is the jth chronological observation on variable i and k is the number of distinct time variables ($r = 1$ when a univariate setting is used). Then [14]:

$$\begin{aligned} \widehat{y}_{pt} &= f(y_{1t-1}, \ldots, y_{1t-n}, \ldots, y_{rt-1}, \ldots, y_{rt-n}) \\ e_t &= y_{p,t} - \widehat{y}_{pt} \end{aligned} \quad (1)$$

where $\widehat{y}_{pt}$ denotes the estimated value for the pth variable and time t; f the underlying function of the forecasting model; and e_t is the error (or residual).

The overall performance of a model is evaluated by an global accuracy measure, namely the Root Mean Squared Error (RMSE) and Relative RMSE (RRMSE), given in the form [19]:

$$\begin{array}{l} RMSE = \sqrt{\sum_{i=P+1}^{P+N} e_i^2/N} \\ RRMSE = RMSE/RMSE_{\overline{y}_{pt}} \times 100\ (\%) \end{array} \quad (2)$$

where P is the present time; N is the number of forecasts; and $RMSE_{\overline{y}_{pt}}$ is the $RMSE$ given by the simple mean prediction. The last metric ($RRMSE$) will be adopted in this work, since it has the advantage of being scale independent, where 100% denotes an error similar to the mean predictor ($\overline{y}_{pt}$).

Due to the temporal nature of this domain, a sequential holdout will be adopted for the forecasting evaluation. Hence, the first $TR = 2/3$ of the series will be used to adjust (train) the forecasting models and the remaining last $1/3$ to evaluate (test) the forecasting accuracies. Also, an internal holdout procedure will be used for model selection, where the training data will be further divided into training ($2/3$ of TR) and validation sets (1/3 of TR). The former will be used to fit the candidate models, while the latter will be used to select the models with the lowest error ($RMSE$). After this selection phase, the final model is readjusted using all training data.

3.1 Neural Networks

Neural Networks (NNs) are innate candidates for forecasting due to their nonlinear and noise tolerance capabilities. Indeed, the use of NNs for TSF began in the late eighties with encouraging results and the field has been growing since [9,16,18,2].

The multilayer perceptron is the most popular NN used within the forecasting domain [9,16,18]. When adopting this architecture, TSF is achieved by using a *sliding time window*, in a combination also named *Time Lagged Feedforward Network* in the literature. A sliding window is defined by the set of time lags used to build a forecast. For instance, given the univariate time series 1,2,3,4,5,6 and sliding window $\{1, 2, 4\}$, the following training examples can be built: $1, 3, 4 \rightarrow 5$ and $2, 4, 5 \rightarrow 6$. In a multivariate setting, k sliding windows are used: $\{L_{11}, \ldots, L_{1W_1}\}, \ldots, \{L_{k1}, \ldots, L_{kW_k}\}$, where L_{ij} denotes a time lag for the ith variable.

In this work, a fully connected multilayer network with one hidden layer of H hidden nodes and bias connections will be adopted (Figure 3). The logistic activation function is applied on the hidden nodes and the output node uses a linear function [5]. The overall model is given in the form:

$$\widehat{y}_{p,t} = w_{o,0} + \sum_{i=I+1}^{I+H} f\left(\sum_{s=1}^{k} \sum_{r=1}^{W_s} y_{st-L_{sr}} w_{i,j}\right) \quad (3)$$

where $w_{d,s}$ is the weight from node s to d; (if $d = 0$ then it is a bias connection); $j \in \{1, \ldots, I\}$ is an input node; o is the output node; and f the logistic function ($\frac{1}{1+e^{-x}}$).

Before training, all variables are scaled with a zero mean and one standard deviation. Then, the initial NN weights are randomly set within $[-0.7, +0.7]$. Next, the training algorithm is applied and stopped when the error slope approaches zero or after a maximum of E epochs. Since the NN cost function is nonconvex (with multiple minima), NR runs are applied to each neural setup, being selected the NN with the lowest error [5]. After training, the NN outputs are rescaled to the original domain.

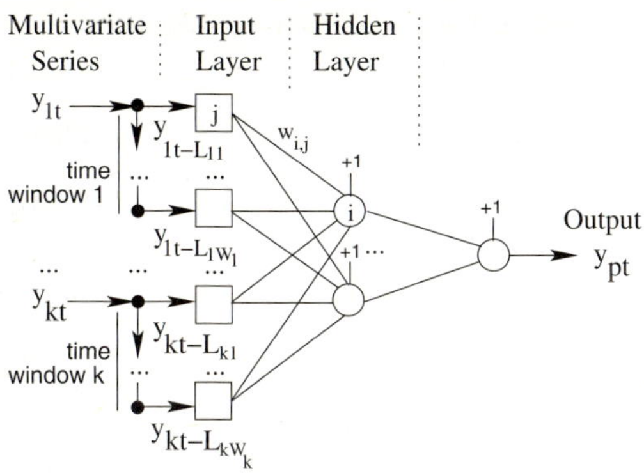

Fig. 3. The multilayer perceptron architecture for multivariate time series forecasting

Under this setting, the NN performance will depend on the number of hidden nodes (H), the selection of the k variables used in the multivariate model and the time window used for each variable. All these parameters can have a crucial effect in the forecasting performance. Feeding a NN with uncorrelated variables or time lags with affect the learning process due to the increase of noise. In addition, a network with few hidden nodes will have limited learning capabilities, while an excess of hidden nodes will lead to overfitting or generalization loss. Since the search space for these parameters is high, an heuristic procedure will be used for the model selection step (see Section 4).

3.2 Naive and Holt-Winters Methods

Two TSF methods will be used as a baseline comparison with the proposed NNs. The most common naive forecasting method is to predict the future as the present value. This setup will be termed NV1. Other possibility is to use a seasonal variant, where the forecast will be given by the observed value for the same period related to the previous daily (NVD) or weekly (NVW) cycles [17].

The *Holt-Winters* [10] is another important univariate forecasting technique from the family of Exponential Smoothing methods. The predictive model is based on some underlying patterns such as a trend or a seasonal cycle (K_1), which are distinguished from random noise by averaging the historical values. Its popularity is due to advantages such as the simplicity of use, the reduced computational demand and the accuracy of the forecasts, specially with seasonal series. More recently, this method has been extended to encompass two seasonal cycles (K_1 and K_2) [17].

4 Experiments and Results

All forecasting methods were implemented in the **R** environment, an open source and high-level programming language for data analysis [13]. The NNs were trained with the $E = 100$ epochs of the BFGS algorithm [11], from the family of quasi-Newton methods and available at the **nnet R** function, while the number of NN runs was set to $NR = 3$. The number of tested hidden nodes (H) was within the range $\{0,2,4,6\}$ [2].

Table 1. The best neural forecasting models

Link	Scale					
	10 minutes			1 hour		
	H	W_p	W_n	H	W_p	W_n
BR	6	{1,2,3,144,145}	–	0	{1,24,25,168,169}	–
BC	0	{1,2,3,4,5,6}	{1}	0	{1,24,25}	–
LC	0	{1,2,3,4,5,6}	{1,2,3,4,5,6}	0	{1,24,25,168,169}	{1,24,25}
LLe	0	{1,2,3,4,5,6}	–	0	{1,24,25,168,169}	–
WR	0	{1,2,3,4,5,6}	–	0	{1,24,25,168,169}	–
WL	0	{1,2,3,4,5,6}	–	0	{1,24,25,168,169}	–
WG	0	{1,2,3,144,145}	{1,2,3,144,145}	2	{1,24,25}	–
ELe	4	{1,2,3,72,73}	{1,2,3,4,5,6}	0	{1,24,25,168,169}	{1,168,169}
EG	0	{1,2,3,4,5,6}	{1,2,3,4,5,6}	0	{1,168,169}	{1,168,169}
RB	0	{1,2,3,144,145}	–	0	{1,24,25,168,169}	{1,24,25}
CB	0	{1,2,3,4,5,6}	{1,2,3,144,145}	0	{1}	{1,168,169}
CL	0	{1,2,3,4,5,6}	{1,2,3,144,145}	0	{1,168,169}	{1,24,25}
LeL	4	{1,2,3,144,145}	–	0	{1,24,25}	{1,168,169}
RW	4	{1,2,3,144,145}	{1,2,3,144,145}	0	{1,24,25,168,169}	{1}
LW	0	{1,2,3,4,5,6}	{1}	0	{1,24,25}	–
GW	0	{1,2,3,4,5,6}	{1}	2	{1,168,169}	–
LeE	0	{1,2,3,144,145}	–	0	{1,24,25,168,169}	{1}
GE	0	{1}	{1}	2	{1,24,25,168,169}	{1,24,25}

Two configurations are used for the variable selection. The first is the simple univariate model. The second multivariate setup will use topology information from the backbone (Figure 1), where the predicted traffic is based on the past values of the current link (p) plus the previous traffic observed in the closest neighbor links that are expected to influence the predicted link (p). For instance, the link Londom-Cosham (LC) presents only one direct neighbor (LeL)[1], while the connection Leeds-London (LeL) contains two (WLe and ELe). Several sliding windows were heuristically set for each series based on their characteristics. It should be noted that in previous univariate IP traffic forecasting work [2], this sliding window setup obtained high quality results. For the single variable model, the tested time window (W_p) was within the range $\{1\}$, $\{1,2,3,4,5,6\}$, $\{1,2,3,72,73\}$ and $\{1,2,3,144,145\}$ (10 minute scale); and $\{1\}$, $\{1,24,25\}$, $\{1,168,169\}$ and $\{1,24,25,168,169\}$ (hourly data). Under the multivariate setting, similar sliding windows were used for the target variable (p). Regarding the other variables, the same window (W_n) will be applied to all neighbor links. For these links, the tested windows were $\{1\}$, $\{1,2,3,4,5,6\}$ and $\{1,2,3,144,145\}$ (10 minute data) and $\{1\}$, $\{1,24,25\}$, $\{1,168,169\}$ (1 hour scale).

The forecasting neural models appear in Table 1. Interestingly, the multivariate neighborhood heuristic is the best option to forecast 11 (10 minute series) and 10 (hourly data) of the 18 links. In general, the multivariate model uses a similar or even higher number of time lags for the predicted variable p than the neighbor links (the

[1] The link CL is not considered, since its origin (Cosham) matches the LC connection destination.

Table 2. Comparison of the forecasting models ($RRMSE$ values, in percentage)

Link	10 minutes			1 hour		
	NV	HW	NN	NV	HW	NN
BR	7.1	4.8	**4.5**±0.0 U	35.3	25.2	**24.8**±0.0 U
BC	19.1	18.0	**16.2**±0.0 M	69.8	68.7	**63.2**±0.0 U
LC	7.5	5.4	**5.2**±0.0 M	37.1	27.8	**22.4**±0.0 M
LLe	6.7	4.0	**3.8**±0.0 U	35.6	25.2	**21.3**±0.0 U
WR	8.9	6.8	**6.4**±0.1 U	40.5	**34.0**	34.1±0.0 U
WL	12.9	**10.5**	**10.5**±0.0 U	59.1	69.0	**58.4**±0.0 U
WG	7.1	**4.6**	**4.6**±0.0 M	36.0	25.1	**24.8**±0.1 U
ELe	9.6	8.5	**8.3**±0.0 M	44.3	44.3	**40.6**±0.0 M
EG	13.1	10.8	**10.2**±0.0 M	**56.3**	67.5	57.5±0.0 M
RB	6.6	2.9	**2.6**±0.0 U	36.5	15.0	**14.6**±0.0 M
CB	13.4	11.1	**10.2**±0.0 M	57.8	58.0	**53.8**±0.0 M
CL	10.9	9.7	**8.8**±0.0 M	**42.3**	45.4	57.2±0.0 M
LeL	7.4	4.6	**4.4**±0.0 U	37.6	**31.7**	34.9±0.0 M
RW	6.9	4.0	**3.7**±0.0 M	36.5	**19.0**	19.5±0.0 M
LW	21.6	21.5	**18.8**±0.0 M	87.0	87.0	**80.8**±0.0 U
GW	9.3	7.4	**6.7**±0.0 M	41.9	**39.4**	41.7±0.1 U
LeE	7.5	4.6	**4.3**±0.0 U	38.7	30.1	**29.2**±0.0 M
GE	**11.5**	**11.5**	11.8±0.0 M	**54.8**	80.8	90.7±1.6 M
Mean	10.4	8.4	**7.8**	47.8	44.1	**42.8**

exception is link CB for the hourly series). Moreover, only seven models denote non-linearity ($H > 0$): BC, ELe, LeL and RW, for the 10 minute data; and WG, GW and GE for the hourly series. These results confirm the notion that real/short time Internet traffic can be modeled by small networks.

The three naive methods (NV1, NVD and NVW) were tested on the model selection step. For all cases, the best model was NV1, which will be adopted as the naive benchmark. Turning to the Holt-Winters (HW) models, the internal parameters were optimized using a 0.05 grid search for the best training error ($RMSE$), which is a common procedure within the forecasting field. For the hourly series, non seasonal, seasonal ($K_1 = 24$ or $K1 = 168$) and double seasonal variants ($K_1 = 24$ and $K_2 = 168$) were tested. Within the selection stage, the weekly seasonal variant ($K1 = 168$) presented the lowest errors. The exception were the links BC, CB, LW (non seasonal model) and LC (daily seasonal with $K1 = 24$). Regarding the 10 minute series, only non seasonal and daily seasonal ($K_1 = 144$) models were tested, since trended effects should be higher than seasonal components at this scale. In effect, the non seasonal version was the best option for all links except BC, CB and GE.

The forecasts with the selected models were performed on the test sets (with 1871 values for the 10 minute series and 312 elements for the hourly data). Table 2 shows the forecasting errors ($RRMSE$) for each method. Thirty runs were applied for the NNs and the results are shown as the mean with the respective 95% t-student confidence intervals. The type of forecasting model is also shown for the NN method: univariate

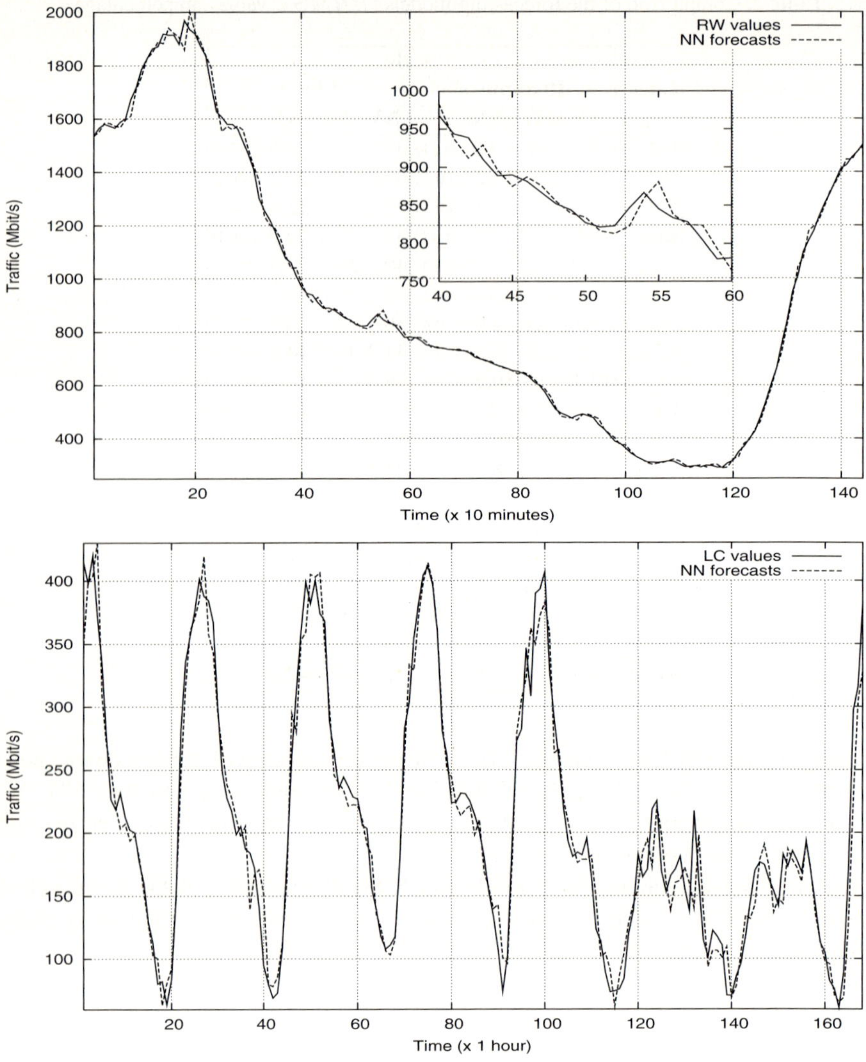

Fig. 4. Neural forecasts for the first day RW (top) and first week LC (bottom) links

(U) or Multivariate (M). Finally, the global performance is presented in the last row in terms of the mean error.

The analysis will start with the 10 minute data. As expected, the naive method gets the worst performance. The NV is only the best option for the last link (GE), presenting the highest mean error. The HW comes in second place. When compared with NV, the mean error decreases 2 percentage points. Moreover, it is the best method for 3 series (WL, WG and GE). Nevertheless, the proposed approach (NN) is clearly the best solution, outperforming (with statistical significance) other methods in 15 (of 18) links and presenting the lowest mean error. Also, it should be noted that the multivariate heuristic (M) is highly relevant, exceeding the NV/HW models in 10 of 11 cases.

The hourly scale is harder to predict, since the $RRMSE$ values are around five times higher than those obtained for the 10 minute series. NV is still the worst strategy, although it is now the best choice for 3 links (EG, CL and GE). Next comes the HW, which presents the lowest errors in 4 cases (WR, LeL, RW and GW). Again, the NNs obtain the best forecasts, presenting an overall performance 1.3/5.0 percentage points below the HW/NV errors and being the best option for 11 links. At this time scale, the multivariate model outperforms the other methods in half the cases (5 of 10).

For demonstrative purposes, Figure 4 presents the traffic forecasts for the first day of the 10 minute RW data (top) and the first week of the hourly LC series (bottom). In both cases, a high quality fit is achieved by the NN forecasts, which are close to the real values. Another relevant issue is related with the computational complexity. The proposed solution is very fast and can be used in real-time. For instance, with a Pentium Dual Core 3GHz processor, the thirty runs of the NN training and testing required only 15.8 (10 minute RW link data) and 3 (the hourly LC series) seconds.

5 Conclusion

In this work a Neural Network (NN) is proposed to forecast the Internet traffic for all 18 links of the UK academic network backbone. In particular, univariate and multivariate strategies were tested. The former used past data from the predicted link, while the latter used topology information, i.e. the direct neighbor links were also fed into the predictive model. Recent data, collected from the United Kingdom Education and Research Network (UKERNA), was analyzed using two forecasting types (or scales): real-time (every 10 minutes) and short-term (hourly values). Also, a comparison was made with two baseline benchmarks, the naive (NV) and Holt-Winters (HW) methods.

The NN multivariate strategy outperformed the univariate approach in 61% (real-time forecasts) and 56% (short-term predictions) of the links considered. Overall, the NN results are promising, with a global Relative Root Mean Square Error (RRMSE) of 7.8% (10 minute series) and 42.8% (hourly data). Indeed, the proposed NN solution produces the best forecasts, surpassing other methods in 83% (10 minute scale) and 61% (hourly series) of the cases. Moreover, the NNs are very fast and can be applied in real-time. Therefore, the proposed approach opens room for producing better traffic engineering tools and methods to detect anomalies in the traffic patterns. This can be achieved with minimal use of computation resources and without producing any extra traffic in the network, since a passive monitoring system was adopted.

In future work, the comparison will be extended to other forecasting techniques (e.g. ARMA models [3]); the proposed approach will be applyed to traffic demands of specific Internet applications (e.g. VoIP); and distinct forecasting horizons will be tested, i.e. from one to several lookaheads. For this last option, several models could be used, where each NN is trained for a specific n-ahead forecast. As an alternative, the one step-ahead forecasts could be used iteratively as inputs. Under a multivariate setting, this would require the simultaneous forecasting of the predicted and direct neighbor NNs.

References

1. Babiarz, R., Bedo, J.: Internet Traffic Mid-term Forecasting: A Pragmatic Approach Using Statistical Analysis Tools. Lecture Notes on Computer Science 3976, 111–121 (2006)
2. Cortez, P., Rio, M., Rocha, M., Sousa, P.: Internet Traffic Forecasting using Neural Networks. In: Proceedings of the IEEE 2006 International Joint Conference on Neural Networks, Vancouver, Canada, pp. 4942–4949 (2006)
3. Cortez, P., Rocha, M., Neves, J.: Evolving Time Series Forecasting ARMA Models. Journal of Heuristics 10(4), 415–429 (2004)
4. Getoor, L., Diehl, C.P.: Link Mining: A Survey. SIGKDD Explorations 7(2), 3–13 (2005)
5. Hastie, T., Tibshirani, R., Friedman, J.: The Elements of Statistical Learning: Data Mining, Inference, and Prediction, NY, USA. Springer, Heidelberg (2001)
6. He, Q., Dovrolis, C., Ammar, M.: On the Predictability of Large Transfer TCP Throughput. In: Proc. of SIGCOMM'05, Philadelphia, USA, ACM Press, New York (2005)
7. Jiang, J., Papavassiliou, S.: Detecting Network Attacks in the Internet via Statistical Network Traffic Normality Prediction. Journal of Network and Systems Management 12, 51–72 (2004)
8. Krishnamurthy, B., Sen, S., Zhang, Y., Chen, Y.: Sketch-based Change Detection: Methods, Evaluation, and Applications. In: Proc. of Internet Measurment Conference (IMC'03), Miami, USA, ACM Press, New York (2003)
9. Lapedes, A., Farber, R.: Non-Linear Signal Processing Using Neural Networks: Prediction and System Modelling. Tech. Rep. LA-UR-87-2662, Los Alamos National Laboratory, USA (1987)
10. Makridakis, S., Weelwright, S., Hyndman, R.: Forecasting: Methods and Applications. John Wiley & Sons, New York, USA (1998)
11. Moller, M.: A scaled conjugate gradient algorithm for fast supervised learning. Neural Networks 6(4), 525–533 (1993)
12. Papagiannaki, K., Taft, N., Zhang, Z., Diot, C.: Long-Term Forecasting of Internet Backbone Traffic. IEEE Trans. on Neural Networks 16(5), 1110–1124 (2005)
13. R Development Core Team R: A language and environment for statistical computing. R Foundation for Statistical Computing, Vienna, Austria, ISBN 3-900051-00-3 (2006), http://www.R-project.org
14. Reinsel, G.: Elements of Multivariate Time Series Analysis, 2nd edn. Springer, Heidelberg (2003)
15. Stallings, W.: SNMP, SNMPv2, SNMPv3 and RMON 1 and 2. Addison-Wesley, London, UK (1999)
16. Tang, Z., Fishwick, F.: Feed-forward Neural Nets as Models for Time Series Forecasting. ORSA Journal of Computing 5(4), 374–386 (1993)
17. Taylor, J., Menezes, L., McSharry, P.: A Comparison of Univariate Methods for Forecasting Electricity Demand Up to a Day Ahead. Int. Journal of Forecasting 21(1), 1–16 (2006)
18. Tong, H., Li, C., He, J.: Boosting Feed-Forward Neural Network for Internet Traffic Prediction. In: Proc. of the IEEE 3rd Int. Conf. on Machine Learning and Cybernetics, Shanghai, China, pp. 3129–3134 (August 2004)
19. Witten, I.H., Frank, E.: Data Mining: Practical Machine Learning Tools and Techniques with Java Implementations, 2nd edn. Morgan Kaufmann, San Francisco, CA (2005)

Boosting Algorithm to Improve a Voltage Waveform Classifier Based on Artificial Neural Network

Milde M.S. Lira[1], Ronaldo R.B. de Aquino[1], Aida A. Ferreira[2], Manoel A. Carvalho Jr[1], Otoni Nóbrega Neto[1], Gabriela S.M. Santos[1], and Carlos Alberto B.O. Lira[1]

[1] Federal University of Pernambuco, Acadêmico Helio Ramos s/n, Cidade Universitária, Cep: 50.740-530, Recife – PE, Brazil
[2] Federal Federal Center of Technologic Education of Pernambuco, Av. Prof. Luiz Freire, 500 Cidade Universitária, Cep: 50.740-540, Recife – PE, Brazil
milde@ufpe.br, rrba@ufpe.br, aidaaf@gmail.com, macj@ufpe.br, otoninobrega@hotmail.com, gabrielaseabra@hotmail.com, cabol@ufpe.br

Abstract. An ANN-based classifier for voltage wave disturbance was developed. Voltage signals captured on the power transmission system of CHESF, Federal Power Utility, were processed in two steps: by wavelet transform and principal component analysis. The classification was carried out using a combination of six MLPs with different architectures: five representing the first to fifth-level details, and one representing the fifth-level approximation. Network combination was formed using the boosting algorithm which weights a model's contribution by its performance rather than giving equal weight to all models. Experimental results with real data indicate that boosting is clearly an effective way to improve disturbance classification accuracy when compared with the simple average and the individual models.

Keywords: Artificial Neural Network, Power Quality Disturbance, Principal Component Analysis, Wavelet Transform.

1 Introduction

In recent years, the power quality has become an important issue for electricity companies, equipment manufacturer, and users. The main reason for this interest is the great proliferation of devices and microprocessors used in several electronic equipments employed in the industrial control systems, like as computers, high-speed drivers, and other no-linear loads. That is, any device that depends on a volatile memory chip for information storage is potentially at risk from power quality events.

To improve electric power quality, some electric utility companies are investing significantly in monitoring their power system in real time. By analyzing the data recorded in the monitoring system, it is possible for the engineers to diagnose the problems and recommend appropriate actions to mitigate their effect, besides evaluating the electric power quality provided.

This paper proposes a method that automatically analyzes voltage waveforms and classifies its type by means of Artificial Neural Network (ANN). This methodology

J. Marques de Sá et al. (Eds.): ICANN 2007, Part II, LNCS 4669, pp. 455–464, 2007.

analyzes the voltage signal samples by the Wavelet Transform (WT); then applies Principal Component Analysis (PCA) to the wavelet coefficients; and finally classifies different voltage disturbances using combination of ANNs.

Information about the wavelet coefficients of all decomposition level of the voltage signal was used by introducing the PCA which is useful to reduce the dimension of the input vectors. Then a neural classifier constructed from this data will certainly presents better performance in relation to other base that only uses the wavelet coefficients of the first decomposition level or the level of higher energy, observing that the wavelet coefficients of other levels that were discharged contain relevant information of the voltage signal.

The resilient backpropagation (RPROP) training algorithm was used to train the ANNs and the boosting learning method was applied to combine them.

A comparison between the individual models and the combined approach of the models using the simple average and the boosting method was performed.

This paper is organized as follows. Sections 2 and 3 introduce a description of the WT and PCA, respectively. Section 4 describes the preprocessing of the data. Section 5 discusses about the training and architecture of the ANNs. Section 6 provides experimental results of the proposed classifier involving multiple neural networks using the boosting algorithm and simple average. Finally, section 7 concludes with a summary of this paper.

2 Discrete Wavelet Transform

The main goal of the WT is to decompose the information contained in a signal into characteristics at different scales. Wavelet analysis overcomes the limitations of the Fourier methods by employing analyzing functions that are local both in time and frequency. The WT is well suited to wideband signals that are not periodic and may contain both sinusoidal and impulse components typical of fast power system transients [1].

Wavelet transform has been used successfully for different applications in areas of signal processing, and recently WT has been proposed as an analysis tool of transients in power systems [2].

An efficient way to implement the discrete wavelet transform (DWT) using filters was developed in 1988 by Mallat [3]. Besides the discretization of the time-frequency plane, the independent variable of the signal is also discretized. That is possible by a simple modification in the CWT mathematical notation, which will result in the following expression:

$$DWT(m,n) = \frac{1}{\sqrt{a_0^m}} \sum_{k=0}^{N-1} f[k]\psi^*\left(\frac{k - nb_0a_0^m}{a_0^m}\right). \tag{1}$$

where $\psi(k)$ is the mother wavelet. The variables m and n are integers that scale and dilate to generate wavelets. The scale index m indicates the wavelet's width, and the location index n gives its position. The $DWT(m,n)$ are the wavelet coefficients. Usually a_0 and b_0 are integers. The smallest integer step for the scale, a_0 =2, is known as dyadic scale, and the smallest integer step of translation is $b_0 = 1$. Thus, the dyadic wavelets became:

$$DWT(m,n) = 2^{-m/2} \sum_{k=0}^{N-1} f[k] \psi^* \left(2^{-m} k - n\right). \tag{2}$$

The dyadic wavelet is implemented by the pyramidal multiresolution decomposition technique. At first, the digitalized signal $c_0[n]$ is decomposed into its detailed $d_1[n]$ and approximation $c_1[n]$ versions, using the filters $g[n]$ and $h[n]$, respectively. The digital filter g[n] is a highpass filter. Therefore, the filtered signal $d_1[n]$ is a detailed version of the signal $c_0[n]$ and possesses high-frequency components (e.g. sharp variations in the power disturbance signal) by comparison to the approximation signal $c_1[n]$, because the filter $h[n]$ is a lowpass filter. The decomposition of the original signal $c_0[n]$ into $c_1[n]$ and d_1[n] represents the first level decomposition. Mathematically, they are defined as:

$$c_1[n] = \sum_k h[k-2n] c_0[k]. \tag{3}$$

$$d_1[n] = \sum_k g[k-2n] c_0[k]. \tag{4}$$

3 Principal Components Analysis

A common problem in statistical pattern recognition is that of feature selection or extraction. Feature extraction refers to a process whereby a data space is transformed into a feature space that, in theory, has exactly the same dimension as the original data space. However, transformation is designed in such a way that the data set may be represented by a reduced number of effective features and yet retains most of the intrinsic information content of the data. This process is known as dimensionality reduction. As we can see, the aim of this technique is to minimize information loss while maximizing reduction in dimensionality.

Principal Components Analysis (PCA) [4], also known as the Karhunen-Loève transformation in communication theory, is a quantitatively rigorous method for achieving this simplification. The method generates a new set of variables called principal components. Each principal component is a linear combination of the original variables. All the principal components are orthogonal to each other so there is no redundant information. The principal components as a whole form an orthogonal basis for the space of the data.

In order to perform PCA on the data, we represent the set of feature vectors by an n-dimensional wavelet coefficient vector (x):

$$\mathbf{x} = \left\langle x_1, x_2, \ldots, x_n \right\rangle. \tag{5}$$

where n is the vector size, and the *ith* variable in $\boldsymbol{x}$ (x_i) takes on values from the wavelet coefficient of the *ith* element in the wavelet coefficient vector. We now find a set of n-dimensional orthogonal unit vectors, ($\mathbf{u}_1$, $\mathbf{u}_2$, . . ., $\mathbf{u}_n$), to form an orthonormal basis for the n-dimensional feature space. We form projections of $\mathbf{x}$ (a_i) onto the set of unit vectors:

$$a_i = \mathbf{x}^T \mathbf{u}_i . \tag{6}$$

In doing so, we perform a coordinate transformation in the feature space, such that the unit vectors ($\mathbf{u}_1$, $\mathbf{u}_2$, . . ., $\mathbf{u}_n$) form the axes of the new coordinate system and transform the original random vector $\mathbf{x}$ into a new random vector $\mathbf{a}$ with respect to the new coordinate system:

$$\boldsymbol{a} = \langle a_1, a_2, \cdots, a_n \rangle . \tag{7}$$

In PCA, the choice of the unit vectors ($\mathbf{u}_1$, $\mathbf{u}_2$, . . ., $\mathbf{u}_n$) is such that the projections (a_i) are uncorrelated with each other. Moreover, if we denote the variance of a_i by λ_i, for i = 1, 2, …, n , then the following condition is satisfied:

$$\lambda_1 > \lambda_2 > \cdots > \lambda n . \tag{8}$$

In other words, the projections, a_i contain decreasing variance, these projections a_i are called the *principal components*. It can be showed [4] that the variance $(\lambda_1, \lambda_2, \cdots, \lambda n)$ corresponds to the eigenvalues of the data covariance matrix $\mathbf{R}$. In order to reduce the dimensionality of the feature space from n to p where $p < n$ while minimizing the loss in data variance, we form a reduced feature space by taking the first p dimensions with the largest variance. In this case, the reduced feature vectors of the documents are represented by the p dimensional random vector:

$$\boldsymbol{a}_p = \langle a_1, a_2, \cdots, a_p \rangle . \tag{9}$$

4 Data Preprocessing

The 60Hz voltage waveform captured by the Digital Fault Recorder (DFR) during the event at a sampling rate of 128 sample/cycle per 14 cycles, yielded 1792 samples. This signal is preprocessed in two steps:

First Step. The main objective at this stage is to extract the maximum information in the wavelet domain at different levels of resolution. In this regard, the signal is decomposed by mean of db6 wavelet [5]. The number of subbands (level of resolution), to be used for signal decomposition, is chosen in such a way that the signal at the fundamental frequency, is included in the middle of the lowest frequency subband, in order to limit the effects of fundamental spectral contents on the other subbands [6]. Consequently, as the DFR captures the 60Hz voltage waveform at 7,680Hz, the signal was decomposed at level 5 thus yielding a total of 1844 wavelet coefficients.

Second Step. The wavelet coefficients are submitted to a transformation to reduce the number of characteristics by mean of PCA, however still keeping most of the information content [7]. At this stage, the characteristics were reduced by taking the first few principal components, where the sum of the variance exceeded 90% of the total variance of the original data (wavelet coefficients), to guarantee the minimization of information loss while maximizing reduction in dimensionality. Table 1 shows the size of the wavelet coefficients and PCA vectors, and the sum of the total variance in the retained components with the largest variance.

Table 1. Size of the wavelet coefficients and PCA vectors

Wavelet coefficients	Size of the wavelet coefficients vector	Size of the PCA vector	Total variance (%)
cA5	66	5	97.3
cD5	66	35	92.1
cD4	122	50	90.6
cD3	233	85	90.5
cD2	456	100	90.4
cD1	901	195	90.3

cA- Approximation coefficient; cD- Detail coefficient

Preprocessing the voltage signal data provides the ANN for a faster and more efficient learning, thus these new data are now normalized and uncorrelated.

4.1 Development of Input/Output Data

In this work, five disturbance types are proposed to be classified, namely: Voltage Sag, Voltage Swell, Harmonic Distortion, Oscillatory Transients, and Supply Interruption. Therefore, it is necessary that the neural network accomplishes six classification classes, where the additional class corresponds to the case of no disturbance, i.e., normal operation.

Six knowledge bases were created, according to the level of resolution of the decomposed signal by the WT, as shown in Table 2.

The database to be preprocessed consists of 730 examples of each class, and the algorithm to accomplish this process was implemented in the MATLAB®. This program prepares the neural network training, validation and test set, and stores all the parameters (averages, standard deviations, and PCA transformation matrix) necessary to preprocess new voltage signal.

The Partitioning of Data. In accordance to the 10-fold cross validation technique with stratification, the examples should be divided into 10 stratified partitions.

The 4,380 examples were divided up into 20% for the test set, 20% for the validation set, and 60% for the training set.

Codification of the Neural Network Output. The classification is encoded using 1-of-m output representation for the m classes, where the k-th element of the desired output is indicated by 1 (one) if the input vector belongs to class k, otherwise it is indicated by 0 (zero).

Table 2. Composition of the Input Data

Filter Bandwidth (Hz)	Base					
	1	2	3	4	5	6
0 – 120	-	-	-	-	-	cA5
120 – 240	-	-	-	-	cD5	-
240 – 480	-	-	-	cD4	-	-
480 – 960	-	-	cD3	-	-	-
960 – 1,920	-	cD2	-	-	-	-
1,920 – 3,840	cD1	-	-	-	-	-

5 ANNs Architecture and Training

The resilient backpropagation (RPROP) training algorithm eliminates the harmful effect of having a small slope at the extreme ends of sigmoid squashing transfer functions. Only the sign of the derivative of the transfer function is used to determine the direction of the weight updates; the magnitude of the derivative has no effect on the weight update. RPROP is generally much faster than the standard backpropagation algorithm. It also has the remarkable property of requiring only a modest increase in memory requirements [8].

In order to find the best MLP network architecture having a single hidden layer, an algorithm was implemented in MATLAB®. To decide on the best node configuration in the hidden layer, ten experiments were carried out with random initialization of weights and varying number of hidden nodes.

All of the ANNs used have one input layer, one hidden layer, and one output layer. The nodes of the hidden and output layers used the sigmoid activation function. The maximum number of iterations for all of the trainings was set to 2500 epochs. The training stops if the early stopping flag implemented in MATLAB® occurs 20 times consecutively, or if the maximum number of epochs is reached, or if the error gradient reaches a minimum, or still if the error goal in the training set is met. The early stopping method has the objective of improving generalization of the neural networks. MATLAB® implements this technique, monitoring the error on the validation set during the training process.

The developed program reads the preprocessed data set and accomplishes the following steps:

1. Generates the 10 partitions;
2. Combines partitions in a random way to arrange the training, validation, and test set;
3. Takes the first number of hidden node in the range $[N_i, N_f]$ as defined by the user;
4. Initiates the weights and bias at random;
5. Trains the neural network until any stopping condition occurs;
6. Stores the MSE in the validation set and the parameter values of the ANN; re-initializes weights and bias, and trains again using the same architecture;

7. Compares the error in the current validation set with the previous one, and stores the parameter values of the ANN associated with the lowest error.
8. Increments the hidden node and goes to step 4, repeating this procedure until the number of hidden nodes has been achieved N_f;
9. When N_f is reached, takes the next partition and goes to step 4, repeating this procedure until the Partition 10 has been accomplished;

When the Partition 10 is reached, the parameter value of the architecture corresponding to the best ANN is stored and the program is finished.

The selection of the best architectures for each base was accomplished by varying the number of hidden node from 10 to 100 with an increment of 1. The best architecture which was chosen using the smallest mean square error (MSE) on the validation set and the MSE on the training, validation and test set are shown in Table 3.

Table 3. ANNs Arquitecture and Mean Square Error on Training, Validation, and Test Set

Base	ANNs Architecture	MSE		
		Training	Validation	Test
6	5-88-6	0.0280	0.0347	0.0364
5	35-95-6	0.0055	0.0215	0.0292
4	50-100-6	0.0116	0.0342	0.0431
3	85-86-6	0.0127	0.0373	0.0427
2	100-73-6	0.0190	0.0465	0.0512
1	195-47-6	0.0205	0.0481	0.0489

6 Results

6.1 Individual Neural Networks

To evaluate the network performance, new data was collected from the Transmission System of CHESF. Table 4 presents the number of misclassified examples by each ANN using this new test set.

As shown in Table 4, the two best neural networks are the 5 and 6, which were trained using the coefficients of the fifth-level detail (cD5) and approximation (cA5), respectively.

6.2 Networks Ensemble

An obvious approach to making decisions more reliable is to combine the outputs of different models. Several machine learning techniques do this by learning an ensemble of models and using them in combination: prominent among these are schemes called bagging, boosting, and stacking [9].

Table 4. Number of Examples Misclassified by each ANN

Disturbance	No.	ANN 1 (cD1)	ANN 2 (cD2)	ANN 3 (cD3)	ANN 4 (cD4)	ANN 5 (cD5)	ANN 6 (cA5)
Sag	52	23	26	20	23	8	6
Normal	88	19	18	11	7	2	12
Harmonic	27	8	5	5	0	4	4
Swell	42	0	0	0	0	0	0
Transient	49	5	3	5	5	1	2
Interruption	47	22	17	8	3	0	0
Total	305	77	69	49	38	15	24

Simple Average. A composite output was formed for each of the six output nodes by averaging the six individual networks outputs. In [10] this simple average is used but only two knowledge bases (all detail coefficients and all approximation coefficients) are used to train the ANNs. The final decision of the classifier corresponds to the output with the largest value (the winner-takes-all method). Table 5 presents the classification results of the ANNs ensemble.

Even though four of the six ANNs have presented very poor performance when working by itself, the combination of them applying the simple average improves the system significantly. It is important to point out that the quality of individual classifier in the ensemble has substantial implication for the overall classifier performance [11], [12].

Table 5. Performance of the Neural Networks Ensemble by Simple Average

Disturbance	No.	True	False	Percentage Error
Sag	52	47	5	9.61
Normal	88	86	2	2.27
Harmonic	27	27	0	0.00
Swell	42	42	0	0.00
Transient	49	48	1	2.04
Interruption	47	47	0	0.00
Total	305	297	8	2.62

Boosting Method. Metalearning algorithms take classifiers and turn them into more powerful learners (models). Some of these algorithms work for both classification and regression, depending on the base learner.

Ideally, the models complement one another, each being a specialist in a part of the domain where the other models don't perform very well. The boosting method for combining multiple models exploits this insight by explicitly seeking models that complement one another. Boosting uses voting (for classification) or averaging (for numeric prediction) to combine the output of individual models. It combines models of the same type – e.g. decision trees. It is interactive, i.e. each models is influenced by the performance of those built previously. It encourages new models to become experts for instances handled incorrectly by earlier ones. Boosting weights a model's contribution by its performance rather than giving equal weight to all models.

In order to combine the six ANNs developed we used the free software Weka [13] which provides implementations of learning algorithm that could be easily applied to the networks outputs.

By experiments, the attributes types of the base learner, which produced best results to solve the problem of power disturbance classification, were numeric to the inputs (outputs of the ANNs) and nominal to the output (desired output).

Table 6 shows the performance of the neural network ensemble by bagging, where the combination of all ANNs presented a percentage error of 1.97 against 2.62 presented by the simple average.

Table 6. Performance of the Neural Networks Ensemble by Bagging

Disturbance	No.	True	False	Percentage Error
Sag	52	48	4	7.69
Normal	88	87	1	1.14
Harmonic	27	26	1	3.70
Swell	42	42	0	0.00
Transient	49	49	0	0.00
Interruption	47	47	0	0.00
Total	305	299	6	1.97

7 Conclusion

In this paper, we proposed an ANN-based automatic classifier for power system disturbance waveforms. The real base was preprocessed by WT and PCA in order to be transformed into a suitable training data to the ANNs. To construct the classifier, single hidden layer MLP networks were trained using the RPROP algorithm. Six models were obtained after training the ANNs with six different bases. The six networks with the best performance were then selected. Next, they were combined by the simple average and bagging algorithm.

Comparing the outputs of individual neural networks and combined approaches using simple average and bagging, revealed how the combined approaches were able to outperform each component model used individually. The combined ANNs by the bagging algorithm provided a mean to improve performance of the system.

Even though four ANNs performed poorly when acting alone, the new system formed by combining models using the bagging algorithm was capable of overcoming this issue in a way that was superior to the simple average.

Acknowledgments. The authors would like to thank to CNPq and CAPES-Brazil for financial research support and CHESF for kindly offering the disturbance data presented in this paper.

References

1. Kim, C.H., Aggarwal, R.: Wavelet Transforms in Power Systems. Part1: General Introduction to the Wavelet Transforms. IEEE Power Engineering Journal 14, 81–87 (2000)
2. Ribeiro, P.F: Wavelet Transform: An Advanced Tool for Analyzing Non-Stationary Harmonic Distortions in Power Systems. In: Proceedings IEEE ICHPS VI, Bologna, Italy, pp. 365–369 (1994)
3. Mallat, S.: A Theory for Multiresolution Signal Decomposition: The Wavelet Representation. In: IEEE Trans. Pattern Analysis and Machine Intelligence, pp. 674–693 (1989)
4. Jolliffe, I.T.: Principal Component Analysis, New Work. Springer, Heidelberg (1986)
5. Arruda, E.F., Filho, O.D., Coury, D.V., Carneiro, A.A.F.M.: Um Estudo das Famílias Wavelets Aplicadas à Qualidade da Energia Elétrica, XVI - CBA - Congresso Brasileiro de Automática (2002)
6. Angrisani, L., Daponte, P., D'Apuzzo, M., Testa, A.: A Measurement Method base don the wavelet Transform for Power Quality Analysis. IEEE Trans on Power Delivery 13(4), 990–998 (1998)
7. Lam, S.L.Y., Lee, D.L.: Feature Reduction for Neural Network Based Text Categorization. In: 6th International Conference on Data-base Systems for Advanced Applications, pp. 195–202 (1999)
8. Riedmiller, M., Braun, H.: A Direct Adaptive Method for Faster Backpropagation Learning. The RPROP Algorithm, IEEE International Conference on Neural Networks 1, 586–591 (1993)
9. Witten, I.H., Frank, E.: Data Mining: Practical Machine Learning Tools and Techniques, 2nd edn., pp. 315–336. ELSEVIER, North-Holland, Amsterdam (2005)
10. Lira, M.M.S., Aquino, R.R.B., Ferreira, A.A., Carvalho Júnior, M.A., Lira, C.A.B.O.: Improving Disturbance Classification by Combining Multiple Artificial Neural Networks. In: IEEE World Congress on Computational Intelligence/ IJCNN, Vancouver-Canada. IEEE Xplore, 3436-3442 (2006)
11. Menezes, L.M., Bunn, D.W., Taylor, J.W.: Review of Guidelines for the Use of Combined Forecasts. European Journal of Operational Research 120(1), 190–204 (2000)
12. Hibon, M., Evgenious, T.: To Combine or not to Combine: Selecting among Forecasts and their combination. International Journal of Forecasting 21(1), 15–24 (2005)
13. Weka Software, University of Waikato - New Zealand. Available: http://www.cs.waikaoto.ac.nz/ml/weak

Classification of Temporal Data Based on Self-organizing Incremental Neural Network

Shogo Okada[1] and Osamu Hasegawa[2]

[1] Department of Computer Intelligence and Systems Science, Tokyo Institute of Technology
[2] Imaging Science and Engineering Laboratory, Tokyo Institute of Technology

Abstract. This paper presents an approach (SOINN-DTW) for recognition of temporal data that is based on Self-Organizing Incremental Neural Network (SOINN) and Dynamic Time Warping. Using SOINN's function that eliminates noise in the input data and represents topological structure of input data, SOINN-DTW method approximates output distribution of each state and is able to construct robust model for temporal data. SOINN-DTW method is the novel method that enhanced Stochastic Dynamic Time Warping Method (Nakagawa,1986). To confirm the effectiveness of SOINN-DTW method, we present an extensive set of experiments that show how our method outperforms HMM and Stochastic Dynamic Time Warping Method in classifying phone data and gesture data.

1 Introduction

Recognition and modeling of time-series data are fundamental techniques for gesture recognition and speech processing. Time-series data observed from human gestures and speech have noise and individual differences. Therefore, it is difficult to recognize and model these time-series data.

A common approach for time-series data is the use of the Hidden Markov Model (HMM), a powerful generative model that includes hidden state structure. In fact, HMM is used for speech recognition[1] and speaker adaptation [2] and speech synthesis[3]; it is standard technique for speech processing. The most successful cases of its application are in speech processing. Therefore, HMM is frequently used for gesture recognition. In [4][5], the HMM state which outputs a discrete value is used. In [6][7], the HMM state which outputs a continuous value is used according to the distribution of a mixture of Gaussians. A Hidden semi-Markov model (Segment model), which is an HMM with explicit state duration probability distributions, is proposed in [8]. In HMM, time-series data are modeled using a Markov model that has finite states. For example, phoneme data are often modeled using an HMM that has three states because it is difficult to estimate parameters of HMM that have too many states using finite training data. In general, timescales of data change according to the kind of data. For that reason, long sequential data might not be able to be modeled using an HMM that has finite states.

J. Marques de Sá et al. (Eds.): ICANN 2007, Part II, LNCS 4669, pp. 465–475, 2007.

On the other hand, Dynamic Time Warping (DTW) is capable of calculating summation of local distance, and calculating the global distance between two sequential data correctly. Nevertheless, using DTW, it is difficult to model sequential data containing individual differences.

With that background, Stochastic DTW[9], which uses advantages of both DTW and HMM, is proposed. In Stochastic DTW, probabilistic distance is used instead of Euclidean distance (local distance); a probabilistic path is used instead of a warping path. In addition, a time-series number of a template sequence corresponds to a number of state, and a Gaussian distribution is used for the output distribution. However, not all output distribution is approximated by a Gaussian distribution because the output distribution changes according to feature vector types and their number involved.

In this paper, we propose a method that can approximate the output distribution of each state in detail according to the kind of feature vector and its number. In the proposed model, the output distribution of state is approximated using a Self-Organizing Incremental Neural Network (SOINN), which can grow incrementally and accommodate input data of online non-stationary data distribution [10]. As [10] reported, the design of a two-layer neural network enables it to both represent the topological structure of unsupervised data and to report the reasonable number of clusters. It can eliminate noise in the input data both dynamically and efficiently. Using SOINN's function of representation of the topological structure, the topological structure of the output distribution is represented in a self-organizing manner. In addition, the time-series number of the template sequence corresponds to the number of state. We define this proposed model as SOINN-DTW.

The main contribution of this paper is introducing this approach, SOINN-DTW, which can approximate the output distribution of state in a detailed and flexible manner according to the kind of feature vector. Results show that this approach has advantages for the classification performance of time-series data. We present an extensive set of experiments that highlight how SOINN-DTW outperforms HMM and Stochastic DTW in classifying phone data and human motion data.

2 Approach

2.1 Overview of SOINN

The SOINN is a self-organized growing neural network. This network can grow incrementally and accommodate input data of online non-stationary data distribution [10]. As [10] reported, the design of a two-layer neural network enables it to represent the topological structure of unsupervised online data, report the reasonable number of clusters, and give typical prototype data of every cluster without prior conditions such as a suitable number of nodes or a good initial codebook. The SOINN implements two main functions for our purposes. One is to perform unsupervised clustering of online data. The other is to learn topology structures to represent unsupervised online data.

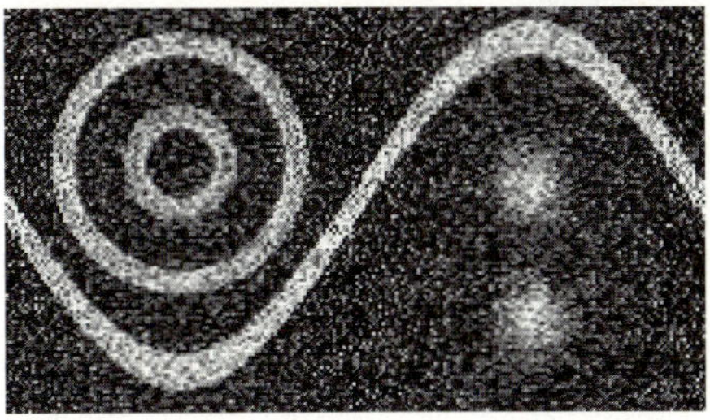

(a) This figure illustrates an artificial 2D dataset with noise as input to the SOINN. The dataset is separable into five parts: a sinusoid, two concentric circles, and two datasets that satisfy a 2D Gaussian distribution.

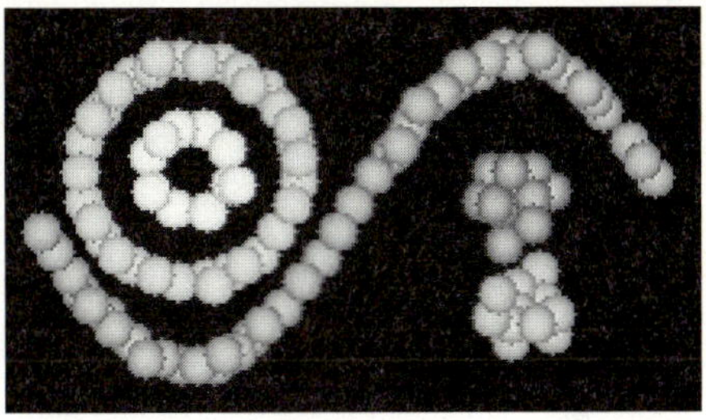

(b) This figure illustrates the SOINN output. We infer that five clusters exist in the artificial input data.

Fig. 1.

Two major benefits are gained from using SOINN: (1) It can eliminate noise in the input data both dynamically and efficiently. (2) It can represent the topological structure of unsupervised online data. As one instance, Fig. 1 (a) depicts raw input data to the SOINN. The output we obtained are shown in Fig. 1 (b). We can find that the noise in the input data was reduced considerably and that the output contained five clusters that approximate the raw data. We describe a brief outline of SOINN here. Details of SOINN are described elsewhere [10].

In SOINN-DTW, SOINN is used for approximation of the distribution of each state. Using two major benefits, the topological structure of each state's distribution is represented.

2.2 SOINN-DTW

In SOINN-DTW, the global distance between training data is calculated using DTW. In addition, a template model is constructed based on DTW. Let N be the number of training data which belong to category $\mathcal{C}$. Then we explain the construction procedure of the template model from N training data.

[STEP 1: Selection of standard data]
Standard data P^* of the template model are selected from among N training data belonging to category $\mathcal{C}$. Standard data P^* are determined using the following equation.

$$P^* = \arg\min_{P_m} \left\{ \sum_{n=1}^{N} D(P_m, P_n) \right\} \quad (\{P_n, P_m\} \in \mathcal{C}) \qquad (1)$$

In eq. (1),P_m, P_n denote training data which belong to category $\mathcal{C}$. In addition, $D(P_m, P_n)$ denotes the global distance in symmetric DTW, where T^* is the time length of standard data P^*.

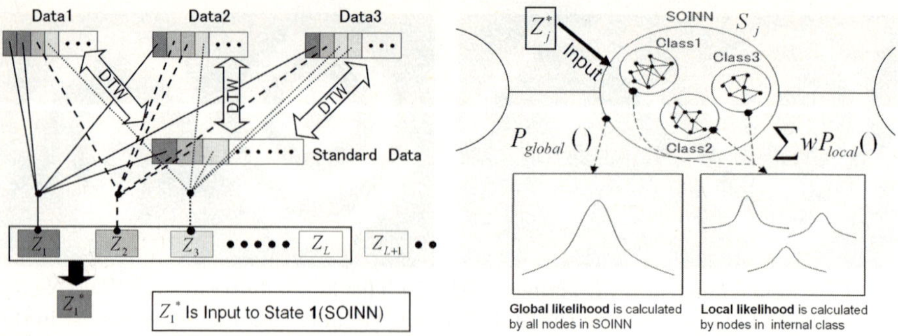

(a) Process of STEP 2. (After DTW, the optimal path between the standard data and training data is determined. Corresponding data in the optimal path are input into each SOINN.))

(b) Two kinds of probability distribution formed by the node set of SOINN

Fig. 2.

[STEP 2: Allocation of samples to each state]
Let p^*_j be a sample of standard data $\boldsymbol{P^*}$ at time j. Let p^n_i be a sample of training data $\boldsymbol{P_n}(n \in \mathcal{C})$ at time i. After DTW, the standard data $\boldsymbol{P^*}$ and the training data optimal (warping) path between p^*_j and p^n_i is determined as $i = w^n_j(j = 1, 2, \cdots, T^*)$, such that the global distance $D(P^*, P_n)$ is minimized. Sample p^n_i at time i is divided to each state (SOINN) j of the template model according to $i = w^n_j$. This allocation of samples is done from time 1 to time T^*.

[STEP 2] is executed for all training data (without $\boldsymbol{P^*}$). As a result, $N-1$ optimal path is determined as $w^n(n = 1, \cdots, N-1)$. The allocation of samples is also done according to w^n. We define the set of samples that is allocated to each state j (SOINN) as $\mathcal{Z}_j$.

Next, the set of samples $\mathcal{Z}_j$ is input to SOINN. The topological structure of the distribution generated by$\mathcal{Z}_j$ is represented after learning by SOINN. Samples in $\mathcal{Z}_j$ are scarce when training data are scarce. The learning performance of SOINN worsens if samples which are included in $\mathcal{Z}_j$ are scarce. A set of samples from $\mathcal{Z}_j$ to $\mathcal{Z}_{j+L-1}$ is input to SOINN (state j) to prevent the problem. The set of samples which is allocated to each state j (SOINN) is defined as ${\mathcal{Z}^*}_j = \{\mathcal{Z}_j, \mathcal{Z}_{j+1}, \cdots, \mathcal{Z}_{j+L-1}\}$ again, where L denotes the number of segments and is the parameter in SOINN-DTW. The method of setting L is described in Section 3.2. The state number is re-defined as $T^* - L - 1$ in SOINN-DTW. Fig. 2(a) represents the process of STEP 2.

[STEP 3: Learning by SOINN]
After ${\mathcal{Z}^*}_j$ is input into SOINN and learned by SOINN, the number of node sets (clusters) that are connected by edges is output by SOINN. We estimate the output distribution of SOINN (state j) from these node sets. The method of parameter estimation is described in Section 2.3.

2.3 Recognition of Input Data

By calculation of the global distance between the template model and input data, the category to which input data belongs is decided.

Recurrence Formula of SOINN-DTW: The global distance between the template model TM_c of category c and input data IP is calculated using the following recurrence formula.

$$Q(i,j) = \max \begin{cases} Q(i,j-1) + C(x_i, S_j) \\ Q(i-1,j-1) + 2C(x_i, S_j) \\ Q(i-1,j) + C(x_i, S_j) \end{cases} \tag{2}$$

In eq. (2), $C(x_i, S_j)$ denotes likelihood. The $C(x_i, S_j)$ is output from the state j (S_j) of the template model TM_c when the samples x_i of input data IP are input to S_j. In SOINN-DTW, DTW is performed such that the summation of likelihood is maximized. Results obtained using DTW show that summation of likelihood $\boldsymbol{E}(\boldsymbol{IP}, \boldsymbol{TM_c})$ is represented in the following equation:$\boldsymbol{E}(\boldsymbol{IP}, \boldsymbol{TM_c}) = Q(I_{IP}, J_c)\frac{2J_c}{(I_{IP}+J_c)}$

Here,I_{IP} denotes the time length of the input data IP, and J_c denotes the time length of the template model TM_c

Calculation of likelihood $\boldsymbol{C(x_i, S_j)}$: After learning by SOINN, the number of node sets (clusters) which are connected by edges is output by SOINN (fig.2(b)). We define one cluster that is output by SOINN as inner class. Likelihood $C(x_i, S_j)$ is calculated using the position vector of the node set in the inner class. For this study, we define two likelihoods (global likelihood and local likelihood) and define the sum of those two likelihoods as $C(x_i, S_j)$.

Global likelihood
All nodes in state S_j are used for calculation of global likelihood. The node set in state S_j is approximated by a Gaussian distribution. Here, the Gaussian distribution has a full-covariance matrix $\boldsymbol{\Sigma_j}$. We define the output probability from the Gaussian distribution as $P_{global}(\boldsymbol{x_i}|S_j)$ and define global likelihood as $\log(P_{global}(\boldsymbol{x_i}|S_j))$. In addition, $\boldsymbol{\mu}_j$ of $P_{global}(\boldsymbol{x_i}|S_j)$ is the average vector of all nodes in state S_j. $\boldsymbol{\Sigma_j}$ are calculated using maximum likelihood estimation.

Local likelihood
Local likelihood is calculated using nodes in inner classes. $Classes1$–3 in Fig. 2(b) represent inner classes. Here, nodes in an inner class are scarce. For that reason, the node set of inner classes is approximated not by a Gaussian distribution, which has a full-covariance matrix, but by a Gaussian kernel function.

Let U_{jk} be inner class k in SOINN (S_j); the output probability $P_{local}(\boldsymbol{x_i}|U_{jk})$ is estimated from all nodes in U_{jk} as follows.

$$P_{local}(\boldsymbol{x_i}|U_{jk}) = \frac{1}{(2\pi h_{jk}^2)^{M/2}} \exp\{-\frac{\|\boldsymbol{x_i} - \boldsymbol{x_{jk}}\|^2}{2h_{jk}^2}\} \tag{3}$$

In eq. (3), x_{jk} is the average vector of all nodes in U_{jk}. Width parameters h_{jk} are calculated using maximum likelihood estimation.

Using $P_{global}(\boldsymbol{x_i}|S_j)$ and $P_{local}(\boldsymbol{x_i}|U_{jk})$, likelihood $C(x_i, S_j)$ is represented as follows:

$$C(\boldsymbol{x}, S_j) = \alpha \log(\sum_{k}^{K} \omega_k P_{local}(\boldsymbol{x_i}|\boldsymbol{U_{jk}})) + (1-\alpha)\log(P_{global}(\boldsymbol{x_i}|\boldsymbol{S_j})) \quad (4)$$

where $\alpha = 0.5$ is in SOINN-DTW. Weight ω_k is $\omega_k = \frac{N_{jk}}{N_j^{all}}$ $(\sum_k^K w_k = 1)$. In eq. (4), N_{all} denotes the number of nodes in SOINN(S_j); K is the number of inner classes in SOINN.

3 Experiment

In this section, to evaluate the general modeling performance of SOINN-DTW, we performed experiments and used two datasets (gesture data and phoneme data) for experiments. In particular, we used moving images (gestures) that directly captured human motion, using no devices such as data gloves. For this experiment, we compared our SOINN-DTW to HMM and Stochastic DTW.

3.1 Comparative Method

Hidden Markov Model: The HMM that we used was the left-to-right model based on mixed Gaussian probabilities having a diagonal-covariance matrix for each state. The model parameters are learned from training data using the Baum-Welch algorithm. In addition, to improve the performance of estimation, the segmental k-means method is used for setting of initial parameters.

Stochastic DTW: An asymmetric recurrence formula, which is used in the Stochastic-DTW method, is as follows:

$$Q(i,j) = \max \begin{cases} Q(i-2, j-1) + \log P(\boldsymbol{a}_{i-1}|j) + \log P(\boldsymbol{a}_i|j) + \log P_{DP1}(j) \\ Q(i-1, j-1) + \log P(\boldsymbol{a}_i|j) + \log P_{DP2}(j) \\ Q(i-1, j-2) + \log P(\boldsymbol{a}_i|j) + \log P_{DP3}(j) \end{cases} \quad (5)$$

State probability $P(\boldsymbol{a}_i|j)$ and state transition probability $P_{DP1,2,3}(j)$ are calculated as shown in [9].

In Stochastic DTW, because the symmetric recurrence formula is used in SOINN-DTW, we used the asymmetric recurrence formula (eq. (5)) and the symmetric recurrence formula for calculation of global distance. Here, we used a recurrence formula that exchanged $C()$ for $P(\boldsymbol{a}_i|j)$ in eq. (2).

3.2 Parameter Setting

Through preliminary experimentation, we set parameters that SOINN had and parameters that SOINN-DTW had. Here, the content of preliminary experimentation was an isolated word recognition task of five classes. For the phoneme classification task and the gesture classification task, we used parameters that were set through preliminary experimentation.

Table 1. Conditions of the phoneme classification experiment

Classification object:	Phoneme data 39 classes (aa,ae,ah,ao,ax,ay,bcl,ch,dcl,dh,dx,eh,er, ey,f,gcl,h,ih,iy,jh,k,kcl,l,m,n,ng,ow,p,pcl,r,s,sh,t,tcl,uw,v,w,y,z)
Data set:	Total 3900 data (100 data per class)
Speaker:	1 male
Evaluation method:	10 experiments using cross-validation method (exchange training data and test data)
Dataset:	Number of data in 1 experiment (training data, test data)=(40,60),(80,20)
Signal Sampling:	25 ms Hamming window, 5 ms frame rate
Feature vector:	12-MFCCs, log-energy, 12-ΔMFCCs, Δlog-energy, 12-$\Delta\Delta$MFCCs, $\Delta\Delta$log-energy (Total 39 dimensions)
Segment number L:	$L = 6$(in training data 40), $L = 3$(in training data 80)

Table 2. Classification rate in the phone classification task [%] (TD40(TD80) denotes the number of training data)

Method	SO-DTW	ST-DTW(1)	ST-DTW(2)	HMM
TD40 [%]	56.36	30.81	51.71	47.69 (5S,2M)
TD80 [%]	62.55	33.83	55.46	51.90 (5S,4M)

Parameter of SOINN: In SOINN, to eliminate noise, parameters (λ,age_{dead}) were used for creation and deletion of edges. From results of preliminary experimentation, we set these parameters as $\lambda = 10000$, $age_{dead} = 30000$. SOINN also has parameters $c, \alpha_1, \alpha_2, \alpha_3, \beta, \gamma$. We set these parameters respectively as $c = 1, \alpha_1 = 1/6, \alpha_2 = 1/4, \alpha_3 = 1/4, \beta = 2/3, \gamma = 3/4$ as [10] reported.

Parameter of SOINN-DTW: Furthermore, SOINN-DTW has a parameter (number of segments) L. The number of samples was adjusted using that parameter L. When we set L as a large value, samples at different times were input into the same state; features of the time series were ignored. In contrast, when we set L as a small value, samples were scarce, and the learning performance of SOINN worsened.

Therefore, we set L according to feature dimension p and number of training data N. From results of preliminary experimentation, we set L as the minimum value at which $L \geq \frac{6p}{N}$ was filled.

3.3 Phoneme Classification

Experimental condition: We used the ked-TIMIT dataset[11]. Experimental conditions are shown in Table 1. With regard to HMM, we performed experiments (table 1) and searched for optimal parameters (number of state, number of mixture), such that HMM had the best classification performance.

Table 3. Condition of gesture classification experiment

Classification object:	7 gestures (fig. 3)
Data set:	Total 168 data (24 data per class)
Performer:	5 people
Evaluation method:	100 experiments using cross-validation method (exchange training data and test data)
Data set:	Number of data in one experiment (training data, test data)=(10,14)
Signal Sampling:	29 ms frame rate
Feature vector:	Self-correlation feature (Total 8 dimensions)
Segment number L:	$L = 5$

Table 4. Correct classification rate in the motion classification task [%]

Method	SO-DTW	ST-DTW(1)	ST-DTW(2)	HMM
[%]	98.70	97.50	96.53	93.71(S11)

Result of phoneme classification: After 10 experiments, the average classification rate is shown in Table 2. Here, [SO-DTW] denotes SOINN-DTW, [ST-DTW(1)] denotes Stochastic-DTW, which uses the asymmetric recurrence formula, [ST-DTW(2)] denotes Stochastic-DTW, which uses the symmetric recurrence formula. In Table 2, $(\cdot S, \cdot M)$ in the classification rate of HMM denotes the parameter of the situation in which HMM has the best classification performance. In addition, S is the number of states and M is the number of mixtures.

For HMM, we changed the state's number $N1$ $(1 \leq N1 \leq 13)$ and performed experiments. Results of experiments show that the classification rate is maximum in $3 \leq N1 \leq 7$. Next, in $3 \leq N1 \leq 7$, we changed the state's number $M1$ $(1 \leq M1 \leq 4)$ and performed experiments. Consequently, the classification rate is maximum, as shown in Table 2.

Regarding Stochastic-DTW, the performance of Stochastic-DTW using the symmetric recurrence formula was better than that of Stochastic-DTW using the asymmetric recurrence formula ([9]).

According to Table 2, the performance of the SOINN-DTW was clearly better than that of either Stochastic-DTW or HMM.

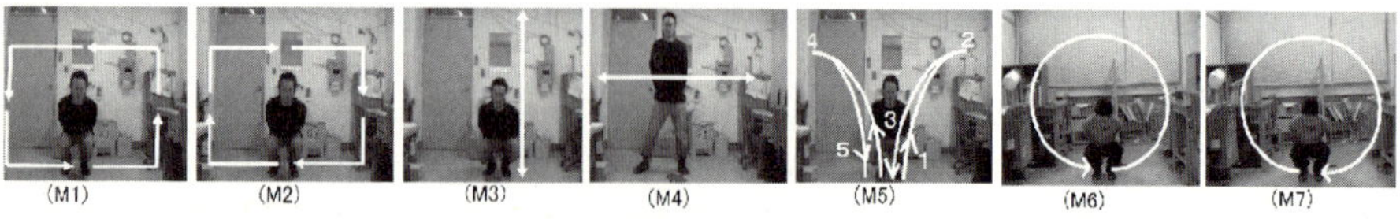

Fig. 3. Examples of moving images used for experiments

3.4 Gesture Classification

The dataset used for the experiment was created manually, extracting only the gesture part from the moving images. Extracted gestures are M1–M7 in Fig. 3. The time length of gestures are various: 110–440 frames.

Image processing: The image processing method that is applied in this experiment is represented below.

Step 1. After smoothing each frame of the input images, we calculated the difference between frames.

Step 2. We converted the RGB value to a luminance value, and converted color images to binary images.

Step 3. We extracted a self-correlation feature [12] for the time-series direction. A (3×3) mask was applied to calculate the correlation feature, giving a nine-dimensional real value vector. Here, the value of the center mask does not represent the feature of motion. And, we did not use the value of the center mask for the feature vector.

Experimental condition: The experimental conditions are as shown in Table 3. For HMM, we performed experiments (Table 3) and searched for the optimal number of states $N2$ $(1 \leq N2 \leq 15)$ such that HMM has the best classification performance.

Result of gesture classification: After 100 experiments, the average classification rate is shown in Table 4. According to Table 4, the performance of the SOINN-DTW was better than that of Stochastic-DTW and HMM.

4 Discussion

4.1 Comparison to Stochastic DTW and HMM

In SOINN-DTW, to implement a robust model for time-series data, The output distribution of state was approximated in detail by SOINN. In addition, we used a symmetric recurrence formula for calculation of the global distance.

Results of experiments show that SOINN-DTW markedly improved the classification performance over that of Stochastic-DTW. Using SOINN-DTW, it is possible to represent the topological structure of the output distribution in each state. In addition, because the number of states is determined as the time length of standard data, we need not set an optimal number of states preliminarily. On the other hand, when HMM is used for modeling of time-series data, we must set the number of states and the number of mixtures preliminarily. For this reason, we performed experiments and searched for optimal parameters that gave HMM the best classification performance.

Experiments show that the classification performance of SOINN-DTW was better than the best classification performance of HMM. Consequently, SOINN-DTW considerably improved the classification performance over that of HMM.

Table 5. Comparison of classification rates obtained using SOINN-DTW and classification rates obtained using methods that do not use SOINN

	Not using SOINN		Use SOINN	
Approach	[Approach.1]	[Approach.2]	SOINN-DTW	SOINN-DTW
[%]	58.41	58.86	62.55 ($\alpha = 0.5$)	63.22($\alpha = 0.45$)

4.2 Contribution to Classification by SOINN

In this section, we discuss how the SOINN performance contributes to classification performance. We compared SOINN-DTW to two methods that do not use SOINN and evaluated the contribution to classification performance of SOINN. The two methods that do not use SOINN are the following.

Approach 1. In Approach 1, $\mathcal{Z}^*{}_j$ were input into SOINN. From $\mathcal{Z}^*{}_j$, we calculated the Gaussian distribution $P(\boldsymbol{x_i}|S_j)$ using maximum likelihood estimation. We defined likelihood $C(x_i, S_j)$ as $C(x_i, S_j) = \log(P(\boldsymbol{x}|S_j))$ (only global likelihood).

Approach 2. For Approach 2, $\mathcal{Z}^*{}_j$ were input into SOINN. We defined likelihood $C(x_i, S_j)$ as $C(x_i, S_j) = \log(P_{global}(\boldsymbol{x_i}|S_j))$ (only global likelihood).

Experimental conditions are as shown in Table 1. We set the number of training data as 80 in the experiment. The average classification rates of SOINN-DTW, Approach 1 and Approach 2 are shown in Table 5. According to Table 5, the classification performance of SOINN-DTW was better than that of either Approach 1 or Approach 2.

In SOINN-DTW, the local likelihood obtained after learning by SOINN is used for calculation in DTW. However, in Approach 1 and Approach 2, the local likelihood is not used for calculation in DTW. The result suggests that the local likelihood obtained after learning by SOINN contributes to the classification performance.

Next, we discuss which likelihood (global and local) contributes to the classification performance. We performed experiments and searched for optimal α in eq. (4) such that SOINN-DTW had the best classification performance. Consequently, the best classification rate is obtained in $(\alpha) = 0.45$, which suggests that both the global likelihood and the local likelihood contribute to classification performance.

5 Conclusion

We proposed SOINN-DTW, which improved Stochastic DTW. We used two real datasets for experiments and performed experiments to evaluate the general modeling performance of SOINN-DTW in comparison to our SOINN-DTW method with HMM and Stochastic DTW. Results show that SOINN-DTW markedly improved the classification performance over that of HMM and Stochastic DTW.

Acknowledgment

This study was supported by Industrial Technology Research Grant Programin '04 from New Energy and Industrial Technology Development Organization (NEDO) of Japan.

References

[1] Rabiner, L.R.: A tutorial on hidden markov models and selected applications in speech recognition. In: Proc. IEEE, pp. 257–286 (1989)

[2] Kim, D.K., Kim, N.S.: Rapid online adaptation based on transformation space model evolution. IEEE Transactions on Speech and Audio Processing 13(2), 194–202 (2005)

[3] Donovan, R.E., Woodland, P.C.: A hidden markov-model-based trainable speech synthesizer. Computer Speech and Language 13, 223–241 (1999)

[4] Yamato, J., Ohya, J., Ishii, K.: Recognizing human action in time-sequential images using hidden markov models. In: Proc. IEEE International Conference on Computer Vision, pp. 379–387 (1992)

[5] Elgammal, A., Shet, V., Yacoob, Y., Davis, L.S.: Learning dynamics for exemplar-based gesture recognition. In: Proc. IEEE Conference on Computer Vision and Pattern Recognition, vol. 1, pp. 16–22 (2003)

[6] Wilson, A., Bobick, A.: Learning visual behavior for gesture analysis. In: Proc. IEEE International Symposium on Computer Vision. vol. 5A, Motion2 (1995)

[7] Hamdan, R., Heits, F., Thoraval, L.: Gesture localization and recognition using probabilistic visual learning. In: Proc. IEEE Conference on Computer Vision and Pattern Recognition(CVPR), pp.98–103 (1999)

[8] Ostendorf, M., Digalakis, V., Kimball, O.: From hmms to segment models: A unified view of stochastic modeling for speech recognition. IEEE Trans. Speech and Audio Process 4(5), 360–378 (1996)

[9] Nakagawa, S.: Speaker-independent consonant recognition in continuous speech by a stochastic dynamic time warping method. In: Proc. of 8th International Conference Pattern Recognition, vol. 70, pp. 925–928 (1986)

[10] Shen, F., Hasegawa, O.: An incremental network for on-line unsupervised classification and topology learning. Neural Networks 19(1), 90–106 (2006)

[11] University of edinburgh, cstr us ked timit, http://festvox.org/dbs/dbs_kdt

[12] Kobayashi, T., Otsu, N.: Action and simultaneous multiple persons identification using cubic higher-order local auto-correlation. In: Proc. International Conference on Pattern Recognition, vol. 19, pp. 741–744 (2004)

Estimating the Impact of Shocks with Artificial Neural Networks

Konstantinos Nikolopoulos[1], Nikolaos Bougioukos[2], Konstantinos Giannelos[2], and Vassilios Assimakopoulos[3]

[1] Decision Sciences and Operations Management, Manchester Business School, Booth St. West, Manchester M15 6PB, United Kingdom
kostas.nikolopoulos@mbs.ac.uk
[2] Forecasting Systems Unit, School of Electrical and Computer Engineering, National Technical University of Athens, 9 Iroon Polytechniou Str., 15773, Zografou Athens, Greece
nbougiou@softlab.ntua.gr, giannelos@softeng.org
[3] Special Secretary for Digital Planning, Ministry of Economy and Finance, 5-7 Nikis Str., 10180, Athens, Greece
vassim@mnec.gr

Abstract. Quantitative models are very successful forr extrapolating the basic trend-cycle component of time series. On the contrary time series models failed to handle adequately shocks or irregular events, that is non-periodic events such as oil crises, promotions, strikes, announcements, legislation etc. Forecasters usually prefer to use their own judgment in such problems. However their efficiency in such tasks is in doubt too and as a result the need of decision support tools in this procedure seem to be quite important. Forecasting with neural networks has been very popular across the Academia in the last decade. Estimating the impact of irregular events has been one of the most successful application areas. This study examines the relative performance of Artificial Neural Networks versus Multiple Linear Regression for estimating the impact of expected irregular future events.

Keywords: Shocks, Irregular events, Forecasting, Artificial Neural Networks.

1 Introduction

Historically, quantitative models have appeared to be very successful in extrapolating the basic trend-cycle component of time series [3], [6], [15], [17]. However time series models have been proved insufficient when it comes to handling shocks or irregular events, that is non-periodic events such as promotions, announcements, regulations, strikes, dramatic price increases due to oil crises etc. That kind of events, introduce an additive or multiplicative impact in the baseline series [2], [13]. Both Literature review and common practice suggest that forecasters use their own judgment for adjusting time series for irregular events, surprisingly not so successfully too [8], [9].

J. Marques de Sá et al. (Eds.): ICANN 2007, Part II, LNCS 4669, pp. 476–485, 2007.

Forecasting with neural networks has been very popular across the Academia in the last decade [1], [5], [11], [12], [19]. While batch forecasting was not one of the success areas [4, [17], estimation of irregular events' impact proved to be a very fertile one [14]. Lee and Yum have proposed a formal framework for judgmental adjustments in time series for irregular events using Artificial Neural Networks (ANN) in order to estimate the additive impact of irregular events. This framework has been enhanced by Nikolopoulos and Assimakopoulos in (2003) [18], so as to incorporate a variety of supportive models, such as Multiple Linear Regression (MLR) or Nearest Neighbors approaches.

2 Adjusting Time Series for Irregular Events

A shock or irregular event is a non-periodic event; an expert should be able to identify its appearance in a series, as well as the parameters that affected it (and potentially quantify them). Consequently the impact of such an event in the past could be quantified and removed from a series resulting in a filtered series – far more appropriate for extrapolation. On the other hand, the future appearance of irregular events could be known in advance, and if an expert could forecast or even know the values of the event's parameters, an appropriate model could estimate the potential impact on the baseline series. As baseline series is defined as a filtered series, including only trend and cycle components, that is the series excluding seasonality and irregular events [16].

Lee and Yum in (1998) [14], have defined all the necessary entities so as to describe, adjust for and forecast the impact of an irregular event . An expert in the field (someone possessing all the necessary domain knowledge for the time series under consideration) should:

- *Define Judgmental Factors:* i.e *promotions*
- *Identify Historical Judgmental Events:* an expert should be able to identify when exactly an irregular event has happened
- *Remove Impact of Judgmental Instances from Time-series Historical Data and create Filtered Time-series Data*

On the expectation of a future irregular event, an expert should adjust his forecasts. This is a very tricky procedure that introduces major amount of error in the forecasting process, thus need for automated system support is more than essential. The forecaster/expert should be able to:

- *Create Judgmental Case Base:* the expert should be aware of a future scheduled *promotion* so as to adjust baseline forecasts respectively.
- *Build a model for judgmental adjustment:* a model should be built so as to estimate the impact of the future events, based on the impacts of similar events in the past. There are very few studies on this topic; methods used from time to time include complex methods such as ANN, the Delphi method and AHP [7].

3 Forecasting the Impact of Future Irregular Events Using Neural Networks

An irregular event may be explained from more than one judgmental factor and parameters. Many and different, multiple or single events may appear in the historic data of a time series. The ideal objective would be to approximate the relationship between each factor's parameter and the impact it results in the historic baseline series. An assumption that such a non-linear function exists has to be made. This function takes as arguments all parameters values for a specific historic event, and gives as output the impact in the series under consideration. For the parameters/arguments that do not appear in a historic irregular event, value equal to zero is assigned. In this section, we will present the use of special case of Artificial Neural Networks, the Multilayer Perceptrons).

3.1 Multilayer Perceptrons

We employed a fully connected multilayer perceptron with two hidden layers and an output layer [10]. Similar networks have been utilized in order to forecast the impact of future events. In our case the size of the input signal is equal to the maximum number of factors' parameters explaining the historic irregular events, with each input representing one parameter. For example if promotions (with duration, type, budget and media used as exploratory parameters) and regulations (with date of announcement, date of application, type as exploratory parameters) describe past irregular events for a specific product, seven parameters are used in total thus the multilayer perceptron will have seven input nodes. The size of the output signal is equal to 1. In order to forecast the impact of a possible future event, the network is trained first with all historic data as the training set, using Back Propagation algorithm. Each historic event is considered to be one training example and the impact it has is considered to be the desired response of the network. Then the future event's parameters are fed into the network with the form of an input signal. For each parameter in the input signal that does not appear in the future event, value 0 is assigned. Signal flow through the network progresses in a forward direction, from left to right and on a layer-by-layer basis reaching the output node. The outcome is the predicted impact of the event.

3.2 ANN Training Methods

Three methods were applied for training ANN in our specific problem of estimating irregular events' impact in time series.

- *1st Method (Simple Training):* the fully connected model with 2 hidden layers and the number of neurons in each layer being equal to the input size is trained with Back Propagation algorithm once. Leave one out method is used if the size of the training sample is less than 100 and non-asymptotic mode is used on any other case. The final trained network is used to produce forecasts.
- *2nd Method (Repeated Training):* the fully connected model with 2 hidden layers and the number of neurons in each layer being equal to the input size is trained

with Back Propagation algorithm 10 times. Each time the Mean Square Error (MSE) over the training sample is calculated. The network with the smallest MSE error is used to produce the forecasts.

- *3rd Method. (Advanced training):* all different fully connected networks up to 2 hidden layers and no more neurons per hidden layer than the input size are trained 10 times each. The number of different networks is limited by the constraint that the second hidden layer is not allowed if the first hidden layer does not contain neurons equal in number with the input size. Each time MSE over the training sample is calculated. The network with the smallest MSE error is used to produce the forecasts. For example, if input size is 5 each of the different ANN networks is trained 10 times.

From these three methods the most appropriate, in terms of out of sample forecasting accuracy, for the problem under consideration was the 3rd one, and that is the one used in the following evaluation section.

4 Simulation

Each dataset is constructed from 15 single or multiple events – the 10 first as a learning set and the next 5 for out of sample accuracy - measured in terms of Mean Absolute Percentage Error (MAPE)

Factors-Parameters
Two Factors are considered, *promotions* and *strikes*:

- *Promotion* has positive effect and is described via three parameters
  - *Budget*, ranging from 50 –150, with step 10 (representing 1000€),
  - *Duration*, ranging from 1-14 (days),
  - *Media-used*, with values: 1 (Paper), 2 (Paper+Radio), 3 (Paper+Radio+TV)
- *Strike* with negative effect described via two parameters
  - *Percentage*, ranging from 20%-100%=0.2-1 with step 0.05 (representing % participation in the strike)
  - *Duration*, ranging from 1-7 (days)

The parameters are set in priority order, that is a promotion is defined either with only a, or a and b, or a, b and c. (not only b or b and c etc). As a result the input vector of the ANN networks is five.

Impact Formula
For the construction of the test events we consider the following formalisation:

Impact = f(Factors parameters) + Noise

For function *f*, two formulas are considered: *Linear* and *Non-linear*

- *Promotion - Linear : 0.7 x Budget + 5 x Duration + 20 x Media*
- *Strike – Linear : - 100 x Percentage – 2 x Duration*
- *Promotion - Non Linear : (0.5 x Budget) x ((Duration/14) + Media)*
- *Strike – Non Linear : - 50 x Percentage x ((Duration/7)+1)*

Noise
Two levels of Gaussian *noise* $N(\mu,\sigma^2)$ are considered: *Low* N(0,10) and *High* N(0,30)

Runs
We consider at least 10 different randomization seeds for the Noise distribution for each combination of the experiment factors (parameters involved, function formula, level of noise).

In total 360 datasets with 15 events each have been constructed, thus the forecasting accuracy is tested on 360x5=**1800** events (out of sample). In each dataset ANN is following a new learning process over 10 events (or even less if it uses an out of sample criterion) and produces forecasts for 5 events. The events are put in an order in each dataset. The first 10 are used for learning and the next 5 for testing.

ANN trained with the third proposed method are competed with classical MLR, with independent variable the relative factors' parameters and dependent variable the impact in the baseline time series (absolute value or alternatively signaled percentage change). The results are presented in the following tables 1 to 11:

Table 1. P3S2, All cases (MAPE)

P3S2	
MLR	ANN
43.3%	*42.6%*

Table 2. P3S2, Formula type and Noise level (MAPE)

P3S2_FL	
MLR	ANN
13.4%	29.3%
P3S2_FNL	
MLR	ANN
73.2%	**56.0%**
P3S2_NH	
MLR	ANN
47.1%	47.5%
P3S2_NL	
MLR	ANN
39.5%	**37.7%**

P3S2 is the most complex class involving two factors described by three and two parameters respectively. These two factors have also controversial impact, that is some times when applied simultaneously, might result in zero impact, fact that makes this problem even more difficult. Randomly distributed single and multiple events are used in this experiment. Thus models have to provide forecasts both for multiple and single expected future irregular events.

Table 3. P3S2, Combinations of formula type and noise level (MAPE)

P3S2_FL_NH	
MLR	ANN
13.4%	28.9%
P3S2_FL_NL	
MLR	ANN
13.3%	29.7%
P3S2_FNL_NH	
MLR	ANN
80.8%	**66.2%**
P3S2_FNL_NL	
MLR	ANN
65.7%	**45.8%**

In general ANN performs slightly better. ANN present clear advantage in non-linear relationships, while MLR present their advantage in the linear ones. The effect of noise creates problems in both approaches.

Table 4. P1S1, All cases (MAPE)

P1S1	
MLR	ANN
21.3%	30.7%

Table 5. P1S1, Noise level (MAPE)

P1S1_NH	
MLR	ANN
23.3%	32.9%
P1S1_NL	
MLR	ANN
9.3%	28.5%

Table 6. P2, All cases (MAPE)

P2	
MLR	ANN
26.5%	28.7%

P1S1 is the simplest case of those involving two judgmental factors. Since each factor is described via only one parameter, non-linear functions cannot be applied (as defined in the framework of this experiment). In general MLR performs better, without this result being differentiated according to the level of noise.

Table 7. P2, Formula type and Noise level (MAPE)

P2_FL	
MLR	ANN
4.6%	6.6%
P2_FNL	
MLR	ANN
48.4%	50.8%
P2_NH	
MLR	ANN
29.3%	32.7%
P2_NL	
MLR	ANN
23.7%	24.8%

Table 8. P2, Combinations of formula type and noise level (MAPE)

P2_FL_NH	
MLR	ANN
6.2%	7.5%
P2_FL_NL	
MLR	ANN
3.1%	5.8%
P2_FNL_NH	
MLR	ANN
52.3%	57.6%
P2_FNL_NL	
MLR	ANN
44.4%	**43.7%**

Table 9. P1, All cases (MAPE)

P1	
MLR	ANN
7.2%	8.5%

P2 involves only one judgmental factor described by two parameters. In general MLR performs slightly better. ANN present an advantage only in non-linear relationships with low level of noise. The effect of noise creates problems in both approaches boosting the error up to 50%.

Table 10. P1, Noise level (MAPE)

P1_NH	
MLR	ANN
10.1%	10.3%
P1_NL	
MLR	ANN
4.3%	6.7%

P1 is the simplest case involving only one judgmental factor described by only one parameter. As expected, MLR presents better results than ANN for all levels of noise. Also the level of error is smaller than any other case examined.

Table 11. All cases (MAPE)

	P3S2	P1S1	P2	P1
MLR	43.3%	**21.3%**	**26.5%**	**7.2%**
ANN	**42.6%**	30.7%	28.7%	8.5%

This final table proves, that as the problem under consideration becomes more difficult/complex, the level of error escalates accordingly. The only case where ANN presents better results - no matter what the formula type - is the most complex case of P3S2. The errors are climaxing regularly from 7.2% up to almost 43% for the most difficult to forecast case. The final level of error – over 40% - indicates as well the difficulty of the problem of forecasting the impact of irregular events.

5 Conclusion

This study examined the relative performance of Artificial Neural Networks versus Multiple Linear Regression for estimating the impact of expected irregular future events. There is strong evidence that no winner can be nominated at this point. The simpler the problem (more linear and lower dimension) the greater the advantage for regression approaches. The more complex the problem and ANN present critical advantage. Thus, a selective protocol would be ideal so as ANN can be used in certain cases and MLR in others.

In detail, the effect of following factors has been discussed:

- *Linearity:* the major finding of this study is that when nonlinearities exist in the underlying formula that describes the irregular event, ANN seem to be more appropriate.
- *Noise:* the level of noise just makes things worse in terms of forecasting accuracy, but it does not nominate a winner. So, ANN and MLR perform equally badly in greater levels of noise.
- *Complexity:* the more complex the irregular event (more factors X parameters needed to describe it) and ANN perform comparatively better. However in the simpler cases, that manager face all the time in every-day forecasting, MLR performs pretty adequately.

Future research should focus on constructing an expert model that either through a selective protocol or an intelligent combination could exploit the advantages of these two different approaches.

Acknowledgments. Dr K. Nikolopoulos has been supported for this research from the Manchester Business School Research Support Fund 2006/7 [Research Project - "Innovative Forecasting Methodology for Television Ratings" - £2000].

References

1. Aiken, M.: Using a neural network to forecast inflation. Industrial Management and Data Systems 99, 296–301 (1999)
2. Armstrong, J.S.: Principles of Forecasting: A handbook for researchers and practitioners. Kluwer Academic Publishers, Dordrecht (2001)
3. Assimakopoulos, V., Nikolopoulos, K.: The theta model: a decomposition approach to forecasting. International Journal of Forecasting 16, 521–530 (2000)
4. Balkin, S., Ord, K.: Automatic Neural Net Modelling for univariate time series. International Journal of Forecasting 16, 509–515 (2000)
5. Chu, C., Zhang, G.P.: A comparative study of linear and nonlinear models for aggregate retail sales forecasting. International Journal of Production Economics 86, 217–231 (2003)
6. Fildes, R., Hibon, M., Makridakis, S., Meade, N.: Generalising about univariate forecasting methods: further empirical evidence. International Journal of Forecasting 14, 339–358 (1998)
7. Flores, B.E., Olson, D.L., Wolfe, C.: Judgmental adjustment of forecasts: A comparison of methods. International Journal of Forecasting 7, 421–433 (1992)
8. Goodwin, P.: Improving the voluntary integration of statistical forecasts and judgment. International Journal of Forecasting 16, 85–99 (2000)
9. Goodwin, P.: Integrating management judgment and statistical methods to improve short-term forecasts. Omega 30, 127–135 (2002)
10. Haykin, S.: Neural Networks: A Comprehensive Foundation (International Edition), Pearson US Imports & PHIPEs. New York (1998)
11. Heravi, S., Osborn, D.R., Birchenhall, C.R.: Linear versus neural network forecasts for European industrial production series. International Journal of Forecasting 20, 435–446 (2004)

12. Kim, K.-J.: Financial time series forecasting using support vector machines. Neurocomputing 55, 307–319 (2003)
13. Lawrence, M., O'Connor, M.: Judgment or models: The importance of task differences. Omega 24, 245–254 (1996)
14. Lee, J.K., Yum, C.S.: Judgmental adjustment in time series forecasting using neural networks. Decision Support Systems 22, 135–154 (1998)
15. Makridakis, S., Andersen, A., Carbone, R., Fildes, R., Hibon, M., Lewandowski, R., Newton, J., Parzen, E., Winkler, R.: The forecasting accuracy of major time series methods. John Wiley & Sons, New York (1984)
16. Makridakis, S., Wheelwright, S., Hyndman, R.: Forecasting, Methods and Applications, 3rd edn., Wiley, New York (1998)
17. Makridakis, S., Hibon, M.: The M3-Competition: Results, conclusions and implications. International Journal of Forecasting 16, 451–476 (2000)
18. Nikolopoulos, K., Assimakopoulos, V.: Theta Intelligent Forecasting Information System. Industrial Management and Data Systems 103, 711–726 (2003)
19. Zhang, G., Patuwo, B.E., Hu, M.Y.: Forecasting with artificial neural networks. The state of the art. International Journal of Forecasting 14, 35–62 (1998)

Greedy KPCA in Biomedical Signal Processing

Ana Rita Teixeira[1], Ana Maria Tomé[1], and Elmar W. Lang[2]

[1] DETI/IEETA, Universidade de Aveiro, 3810-193 Aveiro, Portugal
ana@ieeta.pt
[2] Institute of Biophysics, University of Regensburg, D-93040 Regensburg, Germany
elmar.lang@biologie.uni-regensburg.de

Abstract. Biomedical signals are generally contaminated with artifacts and noise. In case artifacts dominate, the useful signal can easily be extracted with projective subspace techniques. Then, biomedical signals which often represent one dimensional time series, need to be transformed to multi-dimensional signal vectors for the latter techniques to be applicable. In this work we propose the application of a greedy kernel Principal Component Analysis(KPCA) which allows to decompose the multidimensional vectors into components, and we will show that the one related with the largest eigenvalues correspond to an high-amplitude artifact that can be subtracted from the original.

1 Introduction

In many biomedical signal processing applications a sensor signal is contaminated with artifact related signals as well as with noise signals of substantial amplitude. The former sometimes can be the most prominent signal component registered, while the latter is often assumed to be additive, white, normally distributed and non-correlated with the sensor signals. Often signal to noise ratios (SNR) are quite low. Hence to recover the signals of interest, the task is to remove both the artifact related signal components as well as the superimposed noise contributions. Projective subspace techniques can then be used favorably to get rid of most of the noise contributions to multidimensional signals. But many biomedical signals represent one dimensional time series. Clearly projective subspace techniques are not available for one dimensional time series to suppress noise contributions, hence time series analysis techniques often rely on embedding a one dimensional sensor signal in a high-dimensional space of time-delayed coordinates [6], [16]. As embedding is a non-linear signal manipulation [11] Kernel Principal Component Analysis (KPCA) should be a suitable technique. KPCA can simultaneously retain the non-linear structure of the data while denoising is achieved with better performance because the projections are accomplished in the higher-dimensional feature space. The KPCA method represents a projective subspace technique applied in feature space which is created by a non-linear transformation of the original data. In the feature space a linear principal component analysis is performed. Denoising is achieved by considering the projections related to the largest eigenvalues of the covariance matrix.

The mapping into feature space is avoided by using kernel functions which implicitly define a dot product in feature space computed using data in input space [10]. The kernel matrix (a dot product matrix) of the mapped data is easily computed and naturally its dimension depends on the size of the data set and the transformation. The

J. Marques de Sá et al. (Eds.): ICANN 2007, Part II, LNCS 4669, pp. 486–495, 2007.

entries $k(i, j)$ of the matrix depend on the corresponding data points and are computed according to the defined kernel function. The size of the kernel matrix represents a computational burden once its eigendecomposition must be achieved. In practice, the goal of projective subspace techniques is to describe the data with reduced dimensionality by extracting meaningful components while still retaining the structure of the raw data. Then only the projections on the directions corresponding to the most significant eigenvalues of the kernel (or covariance matrix) need to be computed. The exploitation of methods like the well-known Nyström method to achieve a low rank eigendecomposition is a strategy that has been considered in [4],[15]. Furthermore these techniques can also achieve a solution without the manipulation of the full matrix. We show how Nyström's method can be applied to KPCA leading to what is usually known as greedy KPCA. In this work we reformulate both KPCA and greedy KPCA under a unifying algebraic notation underlying the differences between both approaches. And we exploit such greedy techniques to extract high-amplitude artifacts from EEG recordings. The extracted artifact is then subtracted from the original recording thus obtaining a corrected version of the original signal.

EEGs are generally distorted by signals generated by eye movements, eye blinking, muscle activity, head movements, heart beats and line noise. Particularly, eye movements and blinking are major sources of EEG contamination. These ocular movement based signals (EOG) are of larger amplitude than cortical signals (EEG). As they propagate over the scalp, they are recorded in most EEG derivations. Especially the frontal channels often show prominent EOG artifacts which obscure the underlying EEG signals. The availability of digitalized EEG signals makes possible the application of more sophisticated techniques than simple linear filtering. More recently independent component analysis (ICA) [14], [9], blind source separation [8] or adaptive filtering techniques [7] have been discussed. The most recent works use independent component analysis: [9] used the INFOMAX algorithm [13], [17] applied the joint approximative diagonalization of eigen-matrices algorithm (JADE), in [8] an approximate joint diagonalization of time-delayed correlation matrices (SOBI) was used while in [14] the fast fixed point algorithm (FastICA) has been applied. In all the works except [13], the EOG channels are included in the processed data set. In this work we propose the application of a kernel technique to the multidimensional signal obtained by embedding a single univariate EEG signal in its time-delayed coordinates. And after reconstruction and reverting the embedding, the artifact is isolated. The corrected EEG is obtained subtracting the artifact from the original signal. Naturally, this method can be applied also in parallel to a subset of channels.

2 The Kernel Greedy Approaches to Decompose Univariate Signals

Time series analysis techniques often rely on embedding one dimensional sensor signals in the space of their time-delayed coordinates [11]. Embedding can be regarded as a mapping that transforms a one-dimensional time series $x = (x[0], x[1], ..., x[N-1])$ to a multidimensional sequence of $K = N - M + 1$ lagged vectors

$$\mathbf{x}_k = [x[k-1+M-1], \ldots, x[k-1]]^T, \ k = 1, \ldots, K \tag{1}$$

The lagged vectors $\mathbf{X} = [\mathbf{x}_1 \cdots \mathbf{x}_K]$ lie in a space of dimension M.

Kernel Principal Component Analysis (KPCA) relies on a non-linear mapping of given data to a higher dimensional space, called feature space. Without loosing generality, let's assume that the data set is centered and split into two parts yielding the mapped data set

$$\mathbf{\Phi} = [\phi(\mathbf{x}_1)\phi(\mathbf{x}_2)\ldots\phi(\mathbf{x}_R), \phi(\mathbf{x}_{R+1})\ldots\phi(\mathbf{x}_K)] = [\mathbf{\Phi}_R\ \mathbf{\Phi}_S] \tag{2}$$

In denoising applications, the first step of KPCA is to compute the projections of a mapped data set onto a feature subspace. Considering L eigenvectors (columns of $\mathbf{U}$) of a covariance matrix (a correlation matrix if the data is centered) corresponding to the L largest eigenvalues, the projections of the mapped data set $\mathbf{\Phi}$ are

$$\mathbf{Z} = \mathbf{U}^T\mathbf{\Phi} \tag{3}$$

The columns of the matrix $\mathbf{U}$ form a basis in feature space onto which the data set is projected. This basis can be written as a linear combination of the mapped data

$$\mathbf{U} = \mathbf{\Phi}_B\mathbf{A} \tag{4}$$

The matrix $\mathbf{A}$ is a matrix of coefficients and either $\mathbf{\Phi}_B = \mathbf{\Phi}$ (KPCA) or $\mathbf{\Phi}_B = \mathbf{\Phi}_R$ (greedy KPCA), representing a subset of the data set only. Note that the column j of $\mathbf{Z}$ depends on the dot products $\mathbf{\Phi}_B^T\phi(\mathbf{x}_j)$. However to avoid an explicit mapping into feature space, all data manipulations are achieved by dot products [10] and the kernel trick is applied. For instance, using an RBF kernel, the dot product between a vector $\phi(\mathbf{x}_i)$, with $i \in B$, and $\phi(\mathbf{x}_j)$ is computed using a kernel function that only depends on the input data

$$k(\mathbf{x}_i, \mathbf{x}_j) = exp(-\frac{\|\mathbf{x}_i - \mathbf{x}_j\|^2}{2\sigma^2}) \tag{5}$$

Finally, to recover the noise-reduced signal after denoising in feature space, the non-linear mapping must be reverted, i.e. the pre-image in input space of every signal, denoised and reconstructed in feature space, must be estimated. Denoising using KPCA thus comprises two steps after the computation of the projections in the feature space: a) the reconstruction in feature space and b) the estimation of the pre-image of the reconstructed point $\hat{\phi}(\mathbf{x}_j) = \mathbf{U}\mathbf{z}_j$, where $\mathbf{z}_j$ represents the projections of a noisy point $\mathbf{x}_j$. These two steps can be joined together by minimizing the Euclidian distance of the image $\phi(\mathbf{p})$ of a yet unknown point $\mathbf{p}$ from $\hat{\phi}(\mathbf{x}_j)$

$$\begin{aligned}\tilde{d}^{(2)} &= \|\phi(\mathbf{p}) - \hat{\phi}(\mathbf{x}_j)\|^2 \\ &= (\phi(\mathbf{p}) - \hat{\phi}(\mathbf{x}_j))^T(\phi(\mathbf{p}) - \hat{\phi}(\mathbf{x}_j))\end{aligned} \tag{6}$$

The central idea of the fixed-point method [10] consists in computing the unknown pre-image of a reconstructed point in the projected feature subspace by finding a $\mathbf{p}$ which minimizes that distance (eq.6). If an RBF kernel is considered, the iterative procedure is described by the following equation

$$\mathbf{p}_{t+1} = \frac{\mathbf{X}_B(\mathbf{g}\diamondsuit\mathbf{k}_{\mathbf{p}_t})}{\mathbf{g}^T\mathbf{k}_{\mathbf{p}_t}} \tag{7}$$

where $\diamondsuit$ represents a Hadamard product, $\mathbf{g} = \mathbf{A}\mathbf{z}_j$. The components of the vector $\mathbf{k}_{\mathbf{p_t}} = \mathbf{k}(\mathbf{X}_B, \mathbf{p}_t)$ are given by the dot products between $\phi(\mathbf{p}_t)$ and the images $\mathbf{\Phi}_B$ of the training subset $\mathbf{X}_B$. The algorithm must be initialized and $\mathbf{p}_0 \equiv \mathbf{x}_i$ is a valid choice [11]. The points $\mathbf{p}_k$ then form the columns of $\hat{\mathbf{X}}$, the noise-free multidimensional signal matrix in input space. The one-dimensional signal, $\hat{x}[n]$, is then obtained by reverting the embedding, i.e. by forming the signal with the mean of the values along each descendent diagonal of $\hat{\mathbf{X}}$ [12]. Note that if $\hat{x}[n]$ corresponds to the high amplitude artifact, then the corrected signal is computed as $y[n] = x[n] - \hat{x}[n]$.

2.1 Computing the Basis

The projections $\mathbf{Z}$ of the training set are also related with the eigenvectors of a matrix computed using only dot products ($\mathbf{K}$), the kernel matrix. Naturally the entries of the latter matrix can be easily computed using the kernel trick thus avoiding an explicit mapping of the data set. Furthermore, considering a singular value decomposition of the data set

$$\mathbf{\Phi} = \mathbf{U}\mathbf{D}^{1/2}\mathbf{V}^T \tag{8}$$

where D is a diagonal matrix with its non-zero eigenvalues of the kernel matrix (or of the covariance matrix) ordered according to $\lambda_1 > \lambda_2 > ... > \lambda_L ... > \lambda_R$ and $\mathbf{V}$ and $\mathbf{U}$ are the R eigenvectors of the kernel and covariance matrices, respectively. The data set can be approximated using an SVD decomposition with the L most significant singular values and the corresponding eigenvectors. Then substituting the SVD decomposition (eq.8) in eqn. (3) the L projections for each element of the data set read

$$\mathbf{Z} = \mathbf{D}^{1/2}\mathbf{V}^T \tag{9}$$

The two approaches, KPCA and greedy KPCA, respectively, arise from two distinct strategies to deal with the eigendecomposition of the kernel matrix ($\mathbf{K}$) of the data set. In KPCA, where the whole data set is used to compute the kernel matrix, matrix $\mathbf{A}$ is computed using the largest eigenvalues ($\mathbf{D}$) and corresponding eigenvectors ($\mathbf{V}$). And by manipulation of eqns. (3) and (9), the basis vector matrix is obtained as

$$\mathbf{U} = \mathbf{\Phi}\mathbf{V}\mathbf{D}^{-1/2} \tag{10}$$

In greedy KPCA a low-rank approximation of the kernel matrix is considered. This leads to the eigendecomposition of matrices with reduced size. Considering that the training set was divided into two subsets, the kernel matrix can be written in block notation [15],[4]

$$\mathbf{K} = \begin{bmatrix} \mathbf{K}_r & \mathbf{K}_{rs} \\ \mathbf{K}_{rs}^T & \mathbf{K}_s \end{bmatrix} \tag{11}$$

where $\mathbf{K}_r$ is the kernel matrix within subset $\mathbf{\Phi}_R$, $\mathbf{K}_{rs}$ is the kernel matrix between subset $\mathbf{\Phi}_R$ and $\mathbf{\Phi}_S$ and $\mathbf{K}_s$ is the kernel matrix within the subset $\mathbf{\Phi}_S$. The approximation is written using the upper blocks of the original matrix [15], [4]

$$\tilde{\mathbf{K}} = \begin{bmatrix} \mathbf{K}_r \\ \mathbf{K}_{rs}^T \end{bmatrix} \mathbf{K}_r^{-1} \begin{bmatrix} \mathbf{K}_r & \mathbf{K}_{rs} \end{bmatrix} \tag{12}$$

It can be shown that the lower block is approximated by $\mathbf{K}_s \approx \mathbf{K}_{rs}^T \mathbf{K}_r^{-1} \mathbf{K}_{rs}$. The R eigenvectors $\mathbf{V}$ corresponding to the R largest eigenvalues are then computed as

$$\mathbf{V}^T = \mathbf{H}^T \left[\mathbf{K}_r \; \mathbf{K}_{rs} \right] = \mathbf{H}^T \mathbf{\Phi}_R^T \left[\mathbf{\Phi}_R \; \mathbf{\Phi}_S \right] \tag{13}$$

where $\mathbf{H}$ can be computed following different strategies [15], [4] and [1]. The latter [1] refers to an incomplete Cholesky decomposition $\tilde{\mathbf{K}} = \mathbf{C}^T \mathbf{C}$, with

$$\mathbf{C} = \left[\mathbf{L} \; \mathbf{L}^{-T} \mathbf{K}_{rs} \right] \tag{14}$$

The matrix $\mathbf{L}$ is a triangular matrix corresponding to the complete Cholesky decomposition of $\mathbf{K}_r = \mathbf{L}^T \mathbf{L}$.

The $R \times R$ matrix $\mathbf{Q} = \mathbf{C}\mathbf{C}^T$ and its eigendecomposition $\mathbf{V}_q \mathbf{D} \mathbf{V}_q^T$ are used to obtain the low rank approximation of the kernel matrix. In particular the eigenvectors (see eq. 13) are computed with $\mathbf{H} = \mathbf{L}^{-1} \mathbf{V}_q \mathbf{D}^{-1/2}$. The eigenvectors are orthogonal ($\mathbf{V}^T \mathbf{V} = \mathbf{I}$) and consequently the data, in the feature space, is represented by non-correlated projections ($\mathbf{Z}$). The manipulation of equations (13), (9) and (3) gives the basis vector matrix

$$\mathbf{U} = \mathbf{\Phi}_R \mathbf{L}^{-1} \mathbf{V}_q \tag{15}$$

The eigenvectors in the matrix $\mathbf{V}_q$ should be placed according to their corresponding eigenvalues. The first column should have the eigenvector corresponding to the largest eigenvalue and so on. Furthermore the matrix should have $L < R$ columns to enable projections of the data onto the directions related to the L largest eigenvalues. Note that the pivoting index of the incomplete Cholesky decomposition [1] leads to the selection of $\mathbf{\Phi}_R$ within the training set. Further note that in any practical implementation the data is not centered in the feature space. However their centering can be achieved in matrix $\mathbf{C}$, i.e. , after the incomplete Cholesky decomposition of the kernel matrix.

2.2 Choosing Training and Testing Subsets

In the last section it was discussed that with an incomplete Cholesky decomposition it is possible to compute the parameters of the model using a training data set divided into two subsets. But the parameters usually depend on both subsets. However, there are approaches, where the parameters depend on only one of the subsets [15] whose elements are chosen randomly. In this application, we consider an hybrid approach which leads to the choice of three subsets of data. We start by splitting the data (with J vectors) into two data sets: the training set with K vectors and the testing set which contains the remaining data to be processed. In this application, we consider to form the training set with two strategies

- choosing the K vectors randomly.
- choosing the K vectors that correspond to a subsegment (with artifact) of the segment to be processed.

The subset R of the training set is chosen using a Cholesky decomposition performed with the symmetric pivoting algorithm [1]. The methodology is based on the minimization of the trace $tr(\mathbf{K}_s - \mathbf{K}_{rs}^T \mathbf{K}_r^{-1} \mathbf{K}_{rs})$ applied to iteratively update the subset R. So by identifying the maximum value of the trace operator (the pivot), an element of subset S is moved to the subset R and the matrix $\mathbf{K}_r$ increases its size while the others decrease. The process stops when the trace of the matrix corresponding to the actual approximation is less than a threshold [1], [5] and/or the matrix $\mathbf{K}_r$ is not well conditioned [2].

3 Numerical Simulations

The method described in the previous section will be applied to the removal of prominent EOG artifacts from EEG recordings. The corrected EEG is obtaining by subtracting the artifact from the original channel.

Two experiments will be discussed: one dealing only with the details of the application of the method to a single channel; the other application illustrates the use of the algorithm to ease visual inspection of critical segments. The algorithm is always applied to segments of $12s$ or $10s$ (typical window size on displays) duration. The multidimensional signal is obtained with an embedding into an $M = 11$ feature space. An RBF

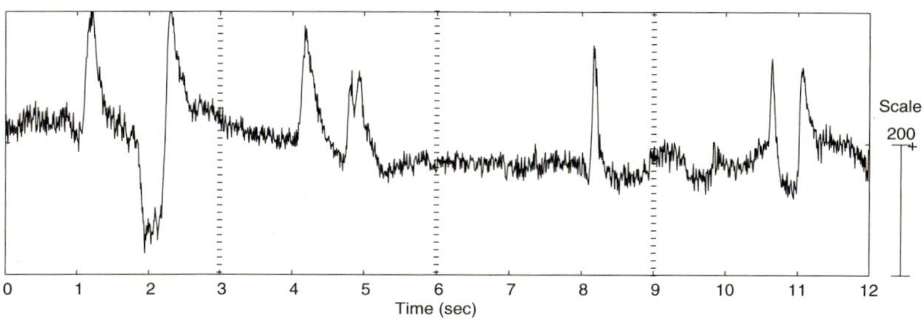

Fig. 1. Fp2 channel (with reference to Cz): all subsegments contain an EOG artifact

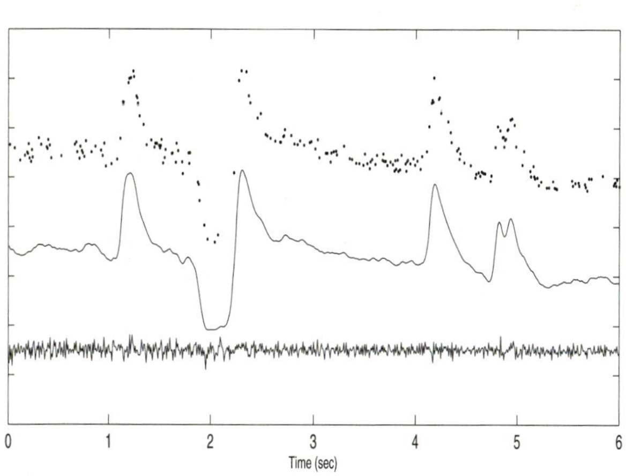

Fig. 2. Greedy KPCA algorithm in different steps: - *top* random selected training set, - *middle* extracted EOG signal and *bottom* - corrected EEG

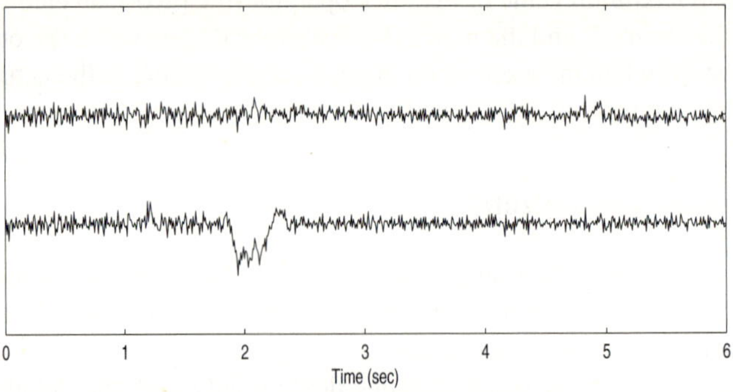

Fig. 3. Corrected EEG using as training set: *top* first subsegment of fig. 1, *bottom* the second segment of fig. 1

kernel is used with $\sigma = max_i(\|\mathbf{x}_i - \mathbf{x}_{mean}\|), i = 1, \ldots, J$, where $\mathbf{x}_{mean}$ is the mean of the data set.

Single Channel Analysis. A frontal (Fp2-Cz) EEG channel sampled at $128Hz$ was used. The segment of the signal containing high-amplitude EOG artifacts is shown in fig. 1). After embedding, the data set with $J = 1526$ vectors was split into testing and training sets. The training set is formed with roughly 25% of the data considering the following strategies

1. A random choice of $K = 381$ data vectors. Fig. 2 (plot on top) illustrates the random choice by plotting the first row of the training data vectors according to their time reference.
2. Using all vectors that correspond either to the first subsegment of the signal $(0-3s)$ or to the second subsegment $(3-6s)$.

The incomplete Cholesky decomposition algorithm is used with $R = 20$. For the three cases data were projected onto $L = 6$ basis vectors ($\mathbf{U}$) which correspond to the leveling off of the eigenspectrum of matrix $\mathbf{Q}$. Fig. 2 shows the result of the application of the algorithm when the training set is selected randomly. Fig. 3 shows the first $6s$ of the corrected EEG when the training set was formed by a subsegment of the signal. It is easily visible that when the training set do not have "examples" of the features to extract the algorithm does not work well. The strongly negative features present in the first subsegment are not represented in the second subsegment. Then when the training set is formed with vectors belonging to the 2nd segment, the artifact is not removed (see fig. 3- plot on bottom).

Multichannel Analysis. The algorithm is applied in parallel to recordings from 4 different EEG electrodes all of which contain a high amplitude EOG interference. The segments each have a duration of $10s$ and fig. 4 (on the left) shows a subsegment of $3s$. After embedding, the training set was formed with roughly 25% of the multidimensional data set. The incomplete Cholesky decomposition with $R = 20$ was used. In

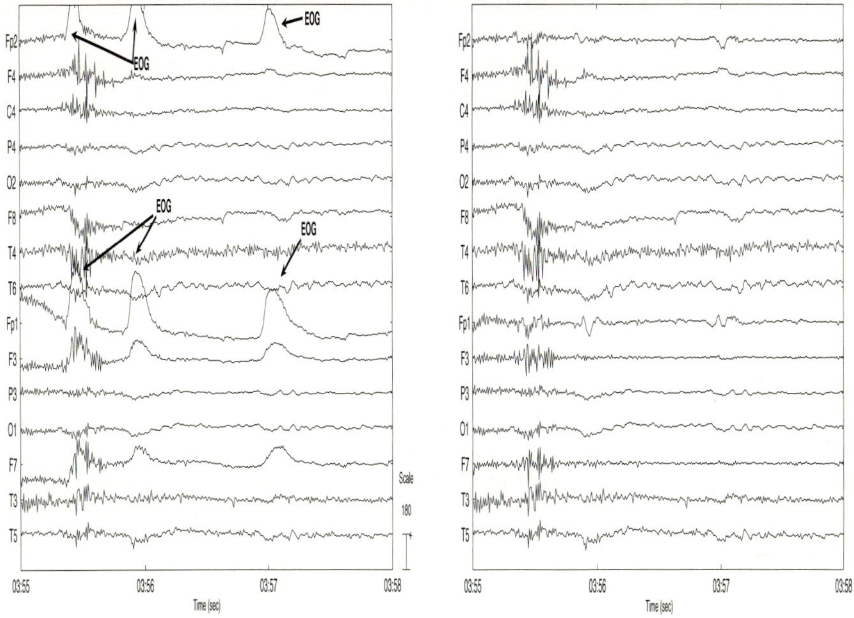

(a) Original EEG (b) Corrected EEG

Fig. 4. EEG signals placed according $10 - 20$ system with reference to Cz. Only $3s$ of the $10s$ are plotted. Channels processed: Fp2, Fp1, F3 and F7.

the feature space, the mapped multidimensional signal is projected onto L directions chosen according to the eigenspectrum of matrix $\mathbf{Q}$ (see table 1).

Note that the number of directions of the Fp channels is higher because the artifacts have a higher energy in these recordings. The results of the processing are plotted in fig. 4 and firmly corroborate our previous conclusions. Note that the algorithm has only one parameter (L), so it is very easy to do an automatic tool to provide on-line help in any visual inspection and analysis of the recordings.

Table 1. Multichannel Analysis. Number of directions to project data

	$Fp1$	$Fp2$	$F3$	$F7$
L	6	6	4	3

4 Concluding Remarks

In this work we present a new variant of greedy KPCA algorithms where the data is split into training and testing sets. The data set is projected onto basis vectors computed with the training set only. This strategy is important to decrease the computational burden of kernel principal component analysis methods when large data sets are involved. In the proposed variant the complexity of the algorithm is related with the splitting of the training set into two subsets. This procedure is achieved by computing the trace of a matrix in each step which involves the manipulation of the whole training set. In our application, we conclude experimentally that with 20 steps we form a subset with $R = 20$ elements that allows to estimate the $L < R$ basis vectors needed to project the data. In what concerns the artifact elimination, the proposed method needs the information contained within a single channel only, hence can be applied to each channel separately. Thus only channels which contain such artifacts need to be processed. Our preliminary results show good performance in removing artifacts like eye or head movements. In summary, the method achieves the separation of EEG signal recordings into two components: artifacts and undistorted EEG. Although this is ongoing work, we present a method that is intended to ease a visual inspection of the EEG recordings by an experienced clinician, hence might be useful in some critical segment analysis like the onset of ictal seizures. It is also to be noticed that despite the variety of methods applied, it is not possible to conclude about their performance once they use distinct databases, different measures and goals. The proposed method needs to be evaluated in a more quantitative and systematic approach, concerning for instance the spectral distortion in the important frequency ranges of EEG. A plug-in to EEGLAB platform [3] has been developed to introduce the method in the clinical routine and facilitate the comparison with other methods.

Acknowledgments

A.R. Teixeira received a PhD Scholarship (SFRH/BD/28404/2006) supported by the Portuguese Foundation for Science and Technology (FCT). This work was also supported by grants from DAAD and CRUP which is gratefully acknowledged. We are grateful to A. Martins da Silva and Hospital Geral de Santo António for providing the signals and the helpful discussions.

References

1. Bach, F.R., Jordan, M.I.: Kernel independent component analysis. Journal of Machine Learning Research 3, 1–48 (2002)
2. Cawley, G.C., Talbot, N.L.C.: Efficient formation of a basis in a kernel induced feature space. In: Verleysen, M. (ed.) European Symposium on Artificial Neural Networks, Bruges, Belgium, pp. 1–6 (2002)
3. Delorme, A., Makeig, S.: EEGLAB: an open source toolbox for analysis of single-trial EEG dynamics. Journal of Neuroscience Methods 134, 9–21 (2004)
4. Fowlkes, C., Belongie, S., Chung, F., Malik, J.: Spectral grouping using the nyström method. IEEE Transactions on Pattern Analysis and Machine Intelligence 26(2), 214–225 (2004)
5. Franc, V., Hlaváč, V.: Greedy algorithm for a training set reduction in the kernel methods. In: Petkov, N., Westenberg, M.A. (eds.) CAIP 2003. LNCS, vol. 2756, pp. 426–433. Springer, Heidelberg (2003)

6. Golyandina, N., Nekrutkin, V., Zhigljavsky, A.: Analysis of Time Series Structure: SSA and Related Techniques. Chapman & HALL/CRC (2001)
7. He, P., Wilson, G., Russel, C.: Removal of ocular artifacts from electroencephalogram by adpative filtering. Medical & Biological Engineering & Computing 42, 407–412 (2004)
8. Joyce, C.A., Gorodniysky, I.F., Kutas, M.: Automatic removal of eye movement and blink artifacts from EEG data using blind component separation. Psychophysiology 41, 313–325 (2004)
9. Jung, T.-P., Makeig, S., Humphries, C., Lee, T.-W., Mckeown, M.J., Iragui, V., Sejnowski, T.j.: Removing electroencephalographic artifacts by blind source separation. Psychophysiology 37, 163–178 (2000)
10. Müller, K.-R., Mika, S., Rätsch, G., Tsuda, K., Schölkopf, B.: An introduction to kernel-based algorithms. IEEE Transactions on Neural Networks 12(2), 181–202 (2001)
11. Teixeira, A.R., Tomé, A.M., Lang, E.W., Stadlthanner, K.: Nonlinear projective techniques to extract artifacts in biomedical signals. In: EUSIPCO2006, Florence, Italy (2006)
12. Teixeira, A.R., Tomé, A.M., Lang, E.W., Gruber, P., da Silva, A.M.: Automatic removal of high-amplitude artifacts from single-channnel electroencephalograms. Computer Methods and Programs in Biomedicine 83(2), 125–138 (2006)
13. Urrestarazu, E., Iriarte, J., Alegre, M., Valencia, M., Viteri, C., Artieda, J.: Independent component analysis removing artifacts in ictal recordings. Epilepsia 45(9), 1071–1078 (2004)
14. Vigário, R.N.: Extraction of ocular artefacts from EEG using independent component analysis. Electroencephalography and Clinical Neurophysiology 103, 395–404 (1997)
15. Williams, C.K.I., Seeger, M.: Using the nyström method to speed up kernel machines. In: Advances in Neural Information Processing Systems, pp. 682–688. MIT Press, Cambridge (2000)
16. You, C.H., Koh, S.N., Rahardja, S.: Signal subspace speech enhancement for audible noise reduction. In: ICASPP 2005, Philadelphia, USA, vol. I, pp. 145–148. IEEE, Los Alamitos (2005)
17. Zhou, W., Gotman, J.: Removing eye-movement artifacts from the EEG during the intracarotid amobarbital procedure. Epilepsia 46(3), 409–414 (2005)

The Use of Artificial Neural Networks in the Speech Understanding Model - SUM

Daniel Nehme Müller, Mozart Lemos de Siqueira, and Philippe O.A. Navaux

Federal University of Rio Grande do Sul
Porto Alegre, Rio Grande do Sul, Brazil
{danielnm,mozart,navaux}@inf.ufrgs.br

Abstract. Recent neurocognitive researches demonstrate how the natural processing of auditory sentences occurs. Nowadays, there is not an appropriate human-computer speech interaction, and this constitutes a computational challenge to be overtaked. In this direction, we propose a speech comprehension software architecture to represent the flow of this neurocognitive model. In this architecture, the first step is the speech signal processing to written words and prosody coding. Afterwards, this coding is used as input in syntactic and prosodic-semantic analyses. Both analyses are done concomitantly and their outputs are matched to verify the best result. The computational implementation applies wavelets transforms to speech signal codification and data prosodic extraction and connectionist models to syntactic parsing and prosodic-semantic mapping.

1 Introduction

The research of Spoken Language Understanding (SLU) software is derived from two joint technologies: Automatic Speech Recognition (ASR) and Natural Language Processing (NLP). These two technologies are complementary: the natural language can help in the speech recognition through information in syntax and semantics, and the speech recognition can improve the understanding of the language with contextual information, such as the intonation of the words (prosody)[1].

This work argues that it is possible to unify several computational systems to represent the speech understanding process. Thus, we propose a software architecture the SUM, a Speech Understanding Model, based on a neurocognitive model of auditory sentence (section 4). Through SUM, we searched a computational representation for speech signal codification, prosody, syntactic and semantic analysis. The SUM is illustrated in the figure 1.

Wavelet transform is used for the signal processing and prosodic codification. The wavelets codification is fully described in the section 4. The connectionist subsystems used in syntactic parsing and definition of semantic contexts are described in the sections 4. Finally, in section 5 we describe how the integration of the subsystems occurs, while in section 6, the results concerning the consequences of this work are presented.

J. Marques de Sá et al. (Eds.): ICANN 2007, Part II, LNCS 4669, pp. 496–505, 2007.

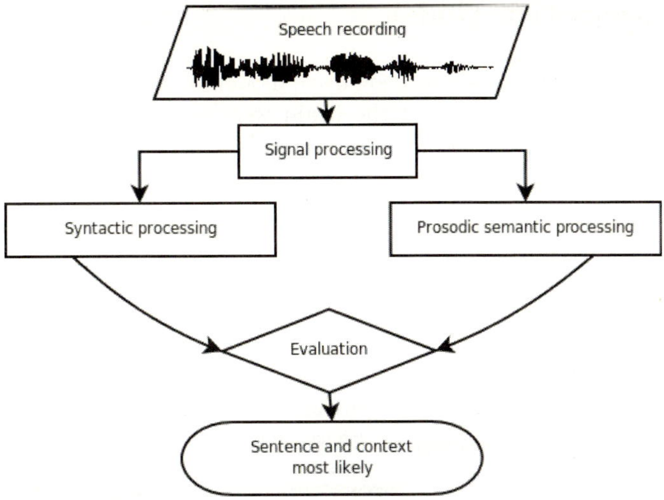

Fig. 1. The Speech Understanding Model - SUM

2 Related Work

The SLU is the first part of spoken dialog systems [2]. In our work, we apply two parts of SLU systems: speech recognizer and language parser. Moreover, to improve this process, we take as base recent neurocognitive researches that demonstrate the relevance of prosody to the natural processing of auditory. Thus, we present to follow some related works in speech recognizer, language parser and prosody. Wavelets transforms have been proposed as an alternative to traditional methods to perform speech analysis as MFCC (Mel Frequency Cepstral Coefficients), which applies Fourier transform. Some limitations of MFCC have been observed, such as easily noise corruption and a signal frame representation that can hold more than one phoneme [3]. Wavelets extract speech signal characteristics by sub-band division [4]. Software to determine semantic context has been the focus of researches, in spite of some work about handmade sentence labeling [5]. This handmade method is expensive and laborious, however, it permits a full and deep semantic analysis [6]. On the other hand, the shallow semantic analysis is fully statistical and more robust than deep analysis [6]. Some methods have been developed using shallow semantic analysis approach. All of them do a classification by distance clustering based on context determination by sentence structure. Neural networks are applied to thematic indexing of spoken documents [7]. Kompe [8] pointed out the importance of accent analysis to word recognition and focus in speech context determination. In this direction, Zhang et al. [9] used Kompe's concepts to create an automatic focus kernel extraction and to differentiate a pair of words with similar linguistic structure but with different meanings.

3 Neurocognitive Model

Angela Friederici proposes a neurocognitive model of auditory sentence processing that identifies which parts of the brain were activated at the time, given the different applied tests. She divided the processing of the auditory sentences in four large phases [10]. Indeed, the most recent research indicates that the prosody processing description must be added to the neurocognitive model [11]. This model is illustrated in figure 2.

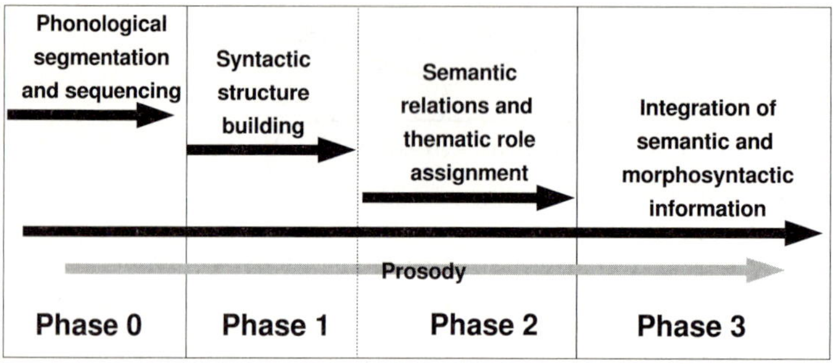

Fig. 2. Improved sequence of neurocognitive model of auditory sentence

In the first phase (*phase 0*), it is done an acoustic characteristic extraction and signal codification. In this process, the pitch is isolated in the right hemisphere and the affective signal (emotions) is distinguished from the linguistic signal. Thus, the pitch variation (affective signal) defines the prosodic characteristics, which determine the processing segmentation. The linguistic characteristics will be analyzed at the syntactic level by the left hemisphere of the brain during the second phase [10]. The second phase (*phase 1*) performs the syntactic analysis and it occurs only in the left hemisphere of the brain. The syntactic analysis is not affected by the prosody and the semantics, and it is processed according to an independent manner [12]. The syntactic evaluation process occurs where the structure sentence errors cause the need of corrections without semantic analysis [13]. This means that the syntactic structure errors must be corrected before semantic analysis. The semantic analysis is performed in the third phase (*phase 2*), where there is a query in the words category memory, which can be observed in tests where the sentences had been organized to produce conflicts related to category and gender [12]. The semantic analysis apparently *awaits* the syntactic analysis output in order to solve interpretation problems brought about mainly by the words categories contextualization. If the sentences are well structured, they will be evaluated by gender, category and semantic context of the involved words [10][13]. The fourth and last phase (*phase 3*) the integration among syntax, semantics and prosody, necessary to review problems not resolved in the previous phases takes place. The syntax structure correction is necessary

when there are lexical terms organization problems [10]. The syntax structure correction is necessary when lexical terms organization problems occur [10].

4 The Implementation of Speech Understanding Model - SUM

We have propose a software architecture based on the natural auditory sentence processing to represent Friederici's neurocognitive model [14]. From the four described phases in the neurocognitive model, we propose the architecture of SUM, as illustrated in figure1. In SUM, the first phase extracts from speech signal coefficients, which were obtained through mathematic transforms applied. These coefficients provide information about the speech signal and they are used in the subsequent phases. The second computational phase is the application of speech coefficients to carry out the syntactic parsing. In the third phase the coefficients are used to define semantic contexts. The linguistic and prosodic information embedded in the coefficients is used to verify the similarity with predefined semantic patterns. The fourth phase receives the analyses from second and third phases' outputs. In this phase, the most likely context is indicated as answer to each analyzed sentence.

Speech signal processing. In this work, wavelet multiresolution analysis is used to extract the characteristics of the speech signal. We divided these signal characteristics in prosodics and semantics. Wavelet permits wave decompositions in scale (dilation) and temporal displacement (translation). The scale enables the signal differentiation between frequency levels, whereas the translation defines the band wideness in analysis [15]. If we vary only the scale, the wavelets can work as filterbanks with multiresolution analysis [16]. This means that it is possible to obtain more details of signal in different frequency levels. The scale is defined by

$$\phi(x) = \sqrt{2}\sum_{k} h_k \phi(2x - k)$$

where h_k is a low-pass filter and k is the scale index. The mother-wavelet will be inserted in this scale by

$$\psi(x) = \sqrt{2}\sum_{k} g_k \psi(2x - k)$$

where g_k is a high-pass filter. These parameters permit high frequencies to be analyzed in narrow windows and low frequencies in wide windows.

In this SUM implementation we use the multiresolution analysis to speech signal codification. We define this process in two ways: phonetic and prosodic approaches (fig 3). The *phonetic coefficients* are obtained from a single decomposition of wavelet coefficients. The *prosodic coefficients* are extracted by wavelets from F0 variation (pitch).

We can extract the *spectral density* in each sub-band to phoneme identification [17] and calculate from the leaves of the wavelet multiresolution tree are the

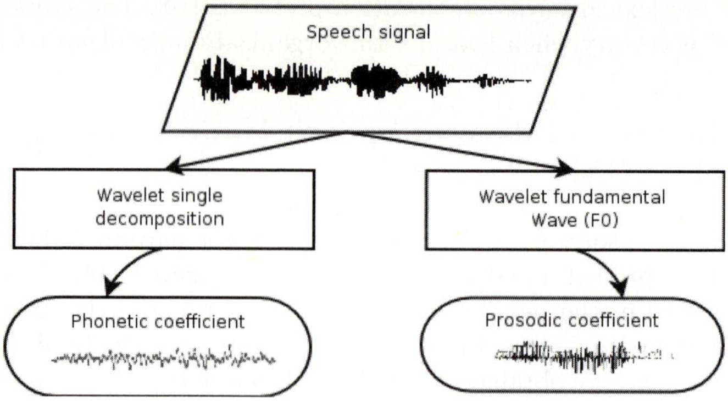

Fig. 3. Coefficients extraction from speech signal

phonetic coefficients. This coefficients are used as word patterns in connectionist systems to process language. The identification of prosodic characteristics is done through the F0 analysis. According to Kadambe, it is necessary to detect the wavelet maximum points to acquire information on the variations of the F0 speech, which correspond to the *glottal closure instants* (GCI) [18]. The *prosodic coefficients* are the variance of F0 detected by wavelet multiresolution analysis and will be applied to semantic and prosodic analysis.

Syntactic analysis. The syntactic analysis is processed by SARDSRN-RAAM system, developed by Mayberry and Miikkulainen [19]. The phonetic coefficients are trained by RAAM (Recursive Auto-Associative Memory) net, whose activation enables the sequencing of the words in the phrase by the SARDSRN-RAAM. In this training, the phonetic coefficients are associated to words. Afterward, the temporal sequence of the component words is initiated, and the sentence pattern presented in the input layer is distributed to the hidden layer and to SARDNET (Sequential Activation Retention and Decay Network). This net, also feeds the hidden layer that, in turn, transfer its codes to a context layer, characterizing the SRN (Simple Recurrent Network) in the SARDSRN-RAAM. In the end, the output layer generates a pattern sentence that is decoded by the RAAM net (see figure 4).

Prosodic-semantic analysis. The prosodic-semantic analysis system receives the phonetic and the prosodic coefficients from wavelet transform. The semantic processing is composed by four chained Growing Hierarchical Self-Organizing Maps (GHSOM) [20](see figure 5). In the first GHSOM net, the input is provided by prosodic coefficients from wavelet transform applied on the speech signal. In the second GHSOM net, the *phonetic map* organizes groups of words according to their linguistic structure and pronunciation. The net that forms the *prosodic-semantic map* uses the output information from the activated neuron in the phonetic map and the activation in the prosodic map. The training of this composition enables the semantic word clustering in the map. Finally, the last map

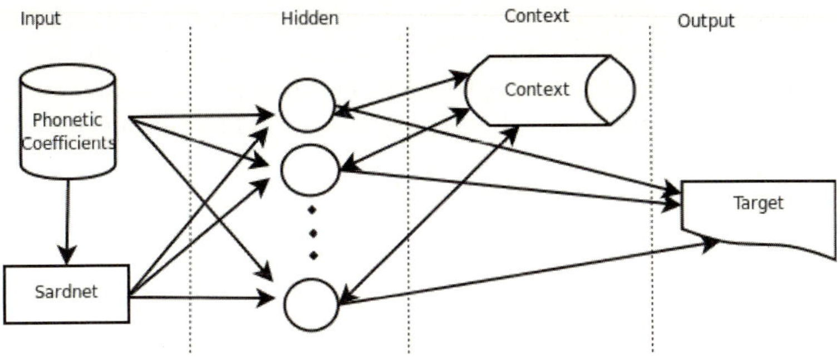

Fig. 4. SARDSRN schema

is responsible for grouping sentences that are informed by the user. The codification of each component word of the sentence supplied by the system user is made by the activation of the preceding maps. The composition of the output of semantic map for each word is the input of the *sentences map*. The recognition of speech patterns is performed by the sentences map, which indicates the most likely sentence.

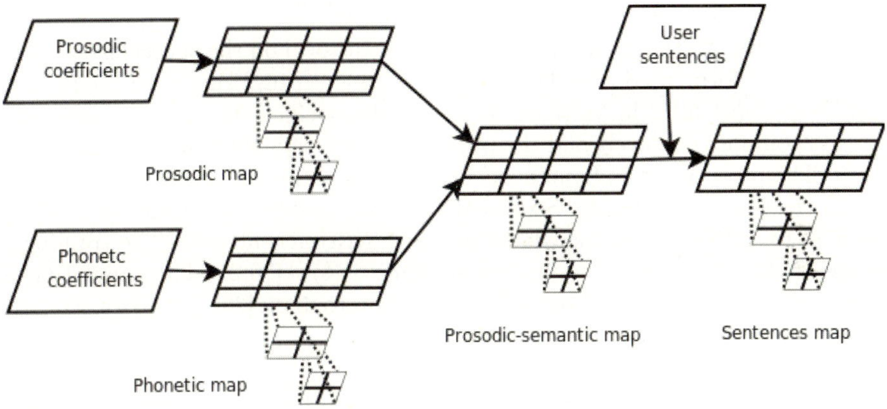

Fig. 5. Maps organization to phonetic, prosodic, semantic and sentences clustering

Evaluation. The sentences resulting from systems' recognitions are evaluated after syntactic and prosodic-semantic processing. The SARDSRN-RAAM system indicates an error rate in each sentence output and the semantic maps system points to the winner neuron in the sentences map. We elaborated an algorithm to perform this evaluation: if a syntactic error rate is lower than 0.5, then we must reject the syntactic structure; if the distance from trained patterns in the sentences map is higher than 2, we must reject the semantic context; if there is a failure just in the syntactic analysis, the system points out the best sentence of the prosocid-semantic analysis; if the failure occurs just in the semantics, the system returns the sentence found in the syntactic analysis; if both analyses fail, the sentence is considered incomprehensible; if both have a successful recognition, they are indicated to the user as being good results. The result of this algorithm is the output of our SUM implementation. For each spoken sentence, the system generates the most reasonable written sentence as result.

5 Simulation of the SUM

Training. We apply wavelet transforms to obtain patterns to train the neural networks. We record 8 sentences to extract 13 words segmented by hand spoken by three Brazilian Portuguese speakers. The records are made in Brazilian Portuguese, but here we present only the translation of the sentences. The spoken sentences trained were (with acronyms marked with small letters):

acronym	sentence	acronym	sentence
tbstc	the boy saw the cat	tdctc	the dog chased the cat
tcltg	the cat liked the girl	tcstd	the cat saw the dog
tdbtb	the dog bit the boy	tbbtg	the boy bit the girl
tgltd	the girl liked the dog	tgctb	the girl chased the boy

We used the Matlab to apply the wavelet transform on the speech signal. The wavelet transform used was Daubechies' wavelet transforms with eight filter coefficients (*db4* filter in Matlab). The phonetic coefficients were obtained with three decomposition levels and the prosodic coefficients with two decomposition levels. The prosodic and phonetic coefficients are obtained directly from wavelet transform (figure 6). In the SARDSRN-RAAM system, the order of training input is guided by a grammar definition with written sentences. The training is stopped when the error rate in each sentence is lower 0.01. The simulation of system maps was proceeded by GHSOM Matlab toolbox [20]. All GHSOM maps were chained as described in section 4.

Recognition. In recognition, the same wavelet codification process was realized. We select three spoken sentences with each verb of the lexicon. The sentences were spoken by a single speaker. The selected sentences to recognize were (with acronyms marked with capital letters):

acronym	sentence	acronym	sentence
TBSTG	the boy saw the girl	TCBTB	the cat bit the boy
TGSTC	the girl saw the cat	TDBTC	the dog bit the cat
TDSTB	the dog saw the boy	TGBTC	the girl bit the cat
TDLTG	the dog liked the girl	TBCTC	the boy chased the cat
TGLTB	the girl liked the boy	TDCTG	the dog chased the girl
TBLTC	the boy liked the cat	TCCTD	the cat chased the dog

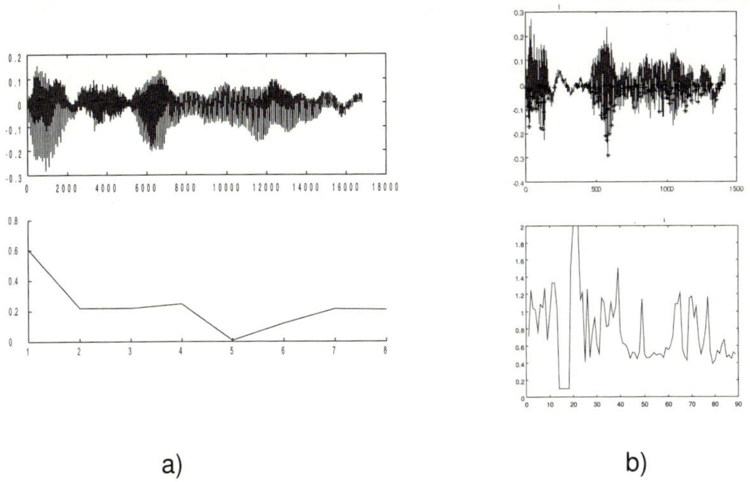

Fig. 6. a) original wave and phonetic coefficients and b) wavelet coefficients with zero-crossing marks and prosodic coefficients

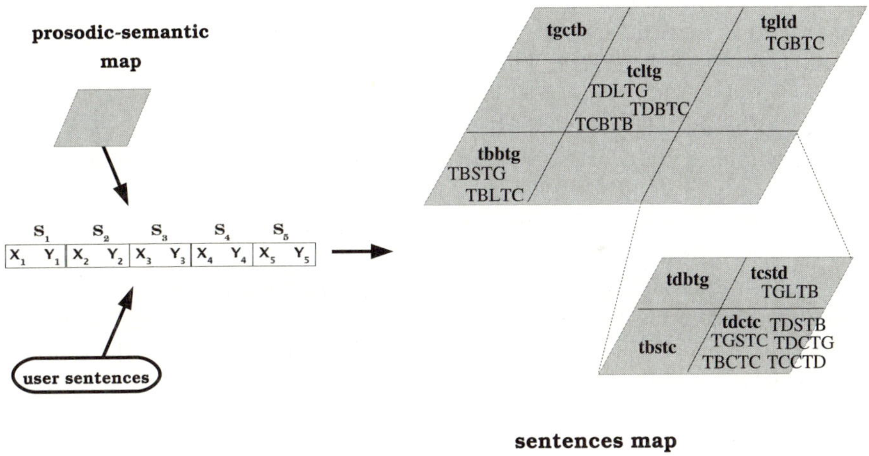

Fig. 7. The final sentences map

In the SARDSRN-RAAM recognition, for each sentence the system presents the best likely sentence. For example, the sentence TBSTG was recognized as sentence trained *tbbtg*, TGSTC as *tgctb*, TDSTB as *tdbtb*, TBLTC as *tbstc*, and so on. It is important to stand out that we recognized (by estimation) not known sentences. As illustration of the map estimation, the not trained sentence TDLTG match with trained sentence *tcltg*, TBCTC with *tdctc*, and so on. Each sentence corresponds to recognized neuron, as it is shown in figure 7, that it illustrates the resultant grouping of sentences in the sentences map. We chosed the sentences TDBTB (the dog bit the boy) and TCCTD (the cat chased the dog), that they are not in the set of training. In the first sentence, we got a prosodic-semantic match in *tcltg* (the cat liked the girl). The syntactic subsystem returned the same written sentence (the dog bit the boy), although not trained. In the second sentence, the syntactic subsystem presented the *wrong* sentence *the cat boy the dog*. The result of sentences map was a distance 0 from pattern *tdctc* (the dog chased the cat) (see figure 7). These two examples mean that, in the case of the first sentence, the syntactic subsystem got the correct sentence, but the maps system have a bad choice. In the case of the second sentence, the syntactic system fails, but the maps system returned the best likely sentence.

6 Conclusion

The SUM model is a software architecture to guide the computational implementation of the auditory sentence process. The proposal of implementation to SUM consists in the codification of the speech signal through coefficient wavelets.

The results obtained from wavelet transform allowed an appropriate speech signal and prosody codification for the use in the connexionist systems to syntactic and semantic representation. The resultant codification demonstrates that there is an interface between existent linguistic parsing connectionists systems to analyze text and the speech. This opens a new method to implementation of systems for written language. The use of artificial neural nets in the syntactic and prosodic-semantic processing was presented as a facilitator in the language modeling process. The training through examples provided by connectionists systems simplifies the work necessary to define grammars and contexts of the language.

The use of hierarchies of maps for definition of semantic contexts might use the prosodics information as guide for linguistic parsing. As analyzed, this notion has a biological inspiration in the presented neurocognitive model.

Finally, the computational prototype that demonstrates the processing of the SUM resulted in a complementing analysis system. Therefore, when the syntactic analysis does not offer reliability, it is possible to evaluate prosodic-semantic analysis, such as in human speech understanding.

References

1. Price, P.: Spoken language understanding. In: Cole, R.A., et al. (eds.) Survey of the State of the Art in Human Language Technology, Cambridge University Press, Stanford, Cambridge (1996)
2. Higashinaka, R., et al.: Incorporating discourse features into confidence scoring of intention recognition results in spoken dialogue systems. Speech Communication 48, 417–436 (2006)
3. Tufekci, Z., Gowdy, J.N.: Feature extraction using discrete wavelet transform for speech recognition. In: Proc. IEEE Southeastcon 2000, pp. 116–123 (2000)
4. Indrebo, K.M., et al.: Sub-banded reconstructed phase spaces for speech recognition. Speech Communication 48, 760–774 (2006)

5. Wang, Y.-Y., Acero, A.: Rapid development of spoken language understanding grammars. Speech Communication 48, 390–416 (2006)
6. Erdogan, H., et al.: Using semantic analysis to improve speech recognition performance. In: Computer Speech and Language, vol. 19, pp. 321–343. Elsevier (2005)
7. Kurimo, M.: Thematic indexing of spoken documents by using self-organizing maps. Speech Communication 38, 29–45 (2002)
8. Kompe, R.: Prosody in Speech Understanding Systems. Springer, Berlin (1997)
9. Zhang, T., Hasegawa-Johnson, M., Levinson, S.: A hybrid model for spontaneous speech understanding. In: Proceedings of the AAAI Workshop on Spoken Language Understanding, Pittsburgh, pp. 60–67 (2005)
10. Friederici, A.D., Alter, K.: Lateralization of auditory language functions: A dynamic dual pathway model. Brain and Language 89, 267–276 (2004)
11. Eckstein, K., Friederici, A.D.: Late interaction of syntactic and prosodic processes in sentence comprehension as revealed by erps. Cognitive Brain Research 25, 130–143 (2005)
12. Heim, S., et al.: Distributed cortical networks for syntax processing: Broca's area as the common denominator. Brain and Language 85, 402–408 (2003)
13. Rossi, S., et al.: When word category information encounters morphosyntax: An erp study. Neuroscience Letters 384, 228–233 (2005)
14. Müller, D.N., de Siqueira, M.L., Navaux, P.O.A.: A connectionist approach to speech understanding. In: Proceedings of 2006 International Joint Conference on Neural Networks - IJCNN'2006, pp. 7181–7188 (July, 2006)
15. Daubechies, I.: Ten lectures on wavelets, Siam (1992)
16. Mallat, S.G.: A theory for multiresolution signal decomposition: The wavelet representation. IEEE Trans. Pat. Anal. Mach. Intell. 11, 674–693 (1989)
17. Ricotti, L.P.: Multitapering and a wavelet variant of mfcc in speech recognition. IEEE Proc. Vision, Image and Signal Processing 152, 29–35 (2005)
18. Kadambe, S., Boudreaux-Bartels, G.F.: Application of the wavelet transform for pitch detection of speech signals. IEEE Trans. Information Theory 38, 917–924 (1992)
19. Mayberry III, M.R., Miikkulainen, R.: SARDSRN: a neural network shift-reduce parser. In: Proceedings of IJCAI-99, pp. 820–825. Kaufmann (1999)
20. Elias Chan, A.: Pampalk: Growing hierarchical self organising map (ghsom) toolbox: visualisations and enhancemen. In: Proceedings of the 9th International Conference on Neural Information Processing (ICONIP'02), vol. 5, pp. 2537–2541 (2002)

On Incorporating Seasonal Information on Recursive Time Series Predictors

Luis J. Herrera, Hector Pomares, Ignacio Rojas, Alberto Guilén, and G. Rubio

Computer Architecture and Computer Technology Department
University of Granada, 18071, Granada, Spain

Abstract. In time series prediction problems in which the current series presents a certain seasonality, the long term and short term prediction capabilities of a learned model can be improved by considering that seasonality as additional information within it. Kernel methods and specifically LS-SVM are receiving increasing attention in the last years thanks to many interesting properties; among them, these type of models can include any additional information by simply adding new variables to the problem, without increasing the computational cost. This work evaluates how including the seasonal information of a series in a designed recursive model might not only upgrade the performance of the predictor, but also allows to diminish the number of input variables needed to perform the modelling, thus being able to increase both the generalization and interpretability capabilities of the model.

1 Introduction

Time series forecasting is a challenge in many fields. There exist many techniques that are applied to the problem of predicting new values in univariate time series [1]. Neural Networks, Fuzzy Systems, Support Vector Machines, AR-ARMA models, etc., are among the paradigms that have been applied to that problem. It is a complex problem, that in general has several points in common with function approximation problems. A good analysis of the time series is necessary to properly tackle the prediction problem, normally treated as a modeling problem, in which new series values $\hat{x}(t+h)$ are predicted using previous values ($x(t-0)$, $x(t-1)$, ..., $x(t-\tau)$) in the general model

$$\hat{x}(t+h) = F\left(x\left(t-0\right),\; x\left(t-1\right),\ldots\; x\left(t-\tau\right)\right). \tag{1}$$

For example, when it is detected that the series presents a certain seasonality, this seasonality has to be included in the prediction model [10,11,13,14]. In kernel methods, including LS-SVM, this seasonality information can be included by simply adding new variables to the problem that represent this information, with the important advantage that they don't increase the computational cost [10,13].

This work emphasizes the importance of including the seasonal information of a series in a recursive model when dealing with a seasonal time series prediction

J. Marques de Sá et al. (Eds.): ICANN 2007, Part II, LNCS 4669, pp. 506–515, 2007.

problem. As it will be shown in the examples, including seasonal information might not only upgrade the performance of a recursive predictor, but also can be seen as a way to diminish the number of input variables needed to perform the modelling with an objective performance.

This work will make use of Least Squares Support Vector Machines, that are a powerful and convenient tool for time series prediction problems, as shown in several works [10,12]. The simulations will be performed in three well known time series, using recursive LS-SVM predictors.

The rest of the work is structured as follows. Section 2 briefly reviews the LS-SVM learning methodology for modelling problems, and addresses how to consider seasonal information within the kernel method. Section 3 reviews the concepts of recursive prediction and direct prediction for long term time series forecasting. Section 4 analyses the time series proposed for the simulations, and addresses the seasonality present in the series. Section 5 presents the simulations, and section 6 concludes the paper.

2 Least Squares Support Vector Machines

LS-SVMs are reformulations to standard Support Vector Machines (SVMs) that lead to solving linear Karush-Kuhn-Tucker systems [5]. LS-SVMs are regularized supervised approximators, closely related to regularization networks and Gaussian processes, but that additionally emphasize and exploit primal-dual interpretations from optimization theory [6].

The LS-SVM model [6] is defined in its primal weight space by

$$\hat{y} = \boldsymbol{\omega}^T \phi(\boldsymbol{x}) + b \tag{2}$$

where $\boldsymbol{\omega}^T$ and b are the parameters of the model, $\phi(\boldsymbol{x})$ is a function that maps the input space into a higher dimensional feature space, and $\boldsymbol{x}$ is the n-dimensional vector of inputs x_j. In Least Squares Support Vector Machines for function approximation, the following optimization problem is formulated,

$$\min_{\boldsymbol{\omega},b,e} J(\omega, e) = \frac{1}{2}\omega^T\omega + \gamma\frac{1}{2}\sum_{i=1}^{N} e_i^2 \tag{3}$$

subject to the equality constraints (inequality constraints in the case of SVMs)

$$e_i = y_i - \hat{y}(\boldsymbol{x}_i), i = 1 \ldots N \tag{4}$$

Solving this optimization problem in dual space, leads to finding the λ_i and b coefficients in the following solution

$$\hat{y} = \sum_{i=1}^{N} \lambda_i K(\boldsymbol{x}, \boldsymbol{x}_i) + b \tag{5}$$

where the function $K(\boldsymbol{x}, \boldsymbol{x}_i)$ is the kernel function defined as the dot product between the $\phi(\boldsymbol{x})$ and $\phi(\boldsymbol{x}_i)$ mappings.

In case we consider Gaussian kernels, the kernel function $K(\boldsymbol{x}, \boldsymbol{x}_i)$ takes the form

$$K(\boldsymbol{x}, \boldsymbol{x}_i) = \exp\left[-\frac{\|\boldsymbol{x} - \boldsymbol{x}_i\|^2}{\sigma^2}\right] \tag{6}$$

where σ is the width of the kernel, that together with the regularization parameter γ, are the hyper-parameters of the problem. Note that in the case in which Gaussian kernels are used, the models obtained resemble Radial Basis Function Networks (RBFN); with the particularities that there is an RBF node per data point, and that overfitting is controlled by a regularization parameter instead of by reducing the number of kernels [6]. In LS-SVM, the hyper-parameters of the model can be optimized by cross-validation. Nevertheless, in order to speed-up the optimization, a more efficient methodology by Lendasse et al. can be found in [3]. A Matlab toolbox for LS-SVMs can be found in [4].

2.1 Incorporating Seasonal Information on LS-SVM Predictors

LS-SVMs present a very good performance for function approximation and time series prediction problems [12]. From the computational complexity point of view, kernel methods in general don't suffer from the curse of dimensionality in the number of input dimensions, but in the number of training data points. This makes easy to include additional information to a kernel-based model, and to check and verify the performance of the modified model.

In case the current series presents seasonality, one or several variables indicating time step within the season of the horizon to be predicted h (for example represented as $s(t+h)$) can be added within the model such that the general modelling equation becomes

$$\hat{x}(t+h) = F\left(s(t+h),\ x\left(t-0\right),\ x\left(t-1\right), \dots\ x\left(t-\tau\right)\right). \tag{7}$$

In this general model, a variable selection approach should be used in order to identify which value of τ is optimum and to identify if the inclusion of $s(t+h)$ improves the performance of the designed model. We note here that there the series of previous series values don't necessarily have the structure $x(t-0)$, $x(t-1), \dots, x(t-\tau)$ but is also taken in many cases as $(x(t-0), x(t-\delta), x(t-2*\delta)$, $\dots, x(t-\tau*\delta))$. A simple wrapper method to perform this variable selection can perform a set of different LS-SVM optimizations, using cross-validation to check which input structure obtained the best performance. In order to decrease the computational demand of the wrapper procedure, the LS-SVM optimizations can be performed using a simple grid-search method using n-fold cross-validation optimization with a low value of n. This simple but effective procedure would assure that a suboptimal subset of variables is obtained.

3 Recursive Prediction

In general, in time series prediction problems, two types of predictions can be needed: short term predictions and long term predictions. Long term time series prediction is a more complex task since the uncertainty increases with the horizon of prediction. In this case, but also valid for short term predictions, two trends can be taken to tackle the problem: direct prediction and recursive prediction [2]. Direct prediction implies the construction of different models, one for each different prediction horizon needed. The performance with this strategy can be expected to be better in prediction accuracy. However, direct prediction has the drawback that too many models might be needed to obtain the long term prediction [7,8], and the time series behaviour understandability vanishes as several different models are needed.

On the other side, recursive prediction only uses one model to predict all the horizons needed. For example, a one-horizon-ahead recursive model would have the structure

$$\hat{x}(t+1) = F\left(s(t+1), \hat{x}\left(t-0\right),\ \hat{x}\left(t-1\right), \ldots\ \hat{x}\left(t-\tau\right)\right) \qquad (8)$$

where for a certain long prediction, the first input values $\hat{x}\left(t-0\right)$, $\hat{x}\left(t-1\right)$, ..., $\hat{x}\left(t-\tau\right)$ are known, and the rest are estimated recursively, using the own model outputs as inputs for further horizons. The performance of the recursive model has to be evaluated, by recursively applying the model to predict all the horizons needed. The drawback of recursive prediction, is that the error committed in nearest horizons can be transmitted to further horizons. However, the interpretability of the predictor improves since a single model is used. It is also important for interpretability of the predictor, to use a low number of input variables in the model. This work considers recursive predictors due to this characteristic. Furthermore, on series presenting seasonality, by using the seasonal information, the number of needed input variables can be highly decreased as it will be shown in the simulations.

4 Time Series Considered

Three well known time series will be used in the simulations section, two natural and one artificial. The evaluation of the performance is done using the mean squared error MSE. First is the ESTSP'2007 conference proposed Benchmark. The data set provided is shown in *fig.* 1. The number of samples in the time series is 875 and it represents the temperature of the pacific ocean along different years (nio phenomenon).

The second example measures the monthly river flow series in cubic feet per second of the Snake river taken from [15]. *Fig.* 2 shows the shape of the series. The series presents a high level of noise, although with a recognizable pattern in the series behavior. One thousand data points are considered from the series.

Third, the Mackey-Glass time series has been considered. It is a well known series, widely used in the literature of time series prediction problems for comparing different methodologies. The time series is described by the following delay differential equation

$$\frac{dx(t)}{dt} = \frac{ax(t-\tau)}{1+x^{10}(t-\tau)} - bx(t) \qquad (9)$$

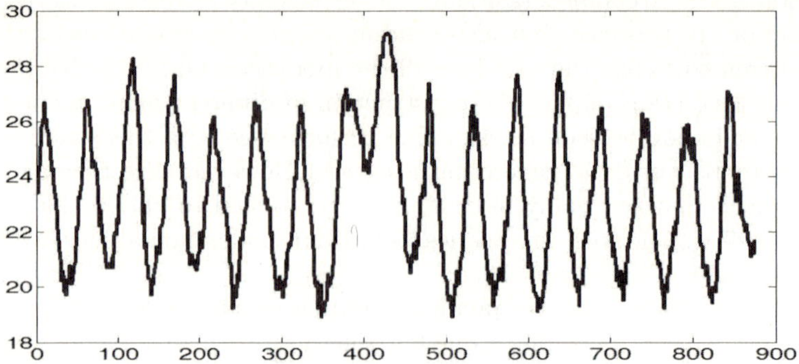

Fig. 1. ESTSP'2007 dataset

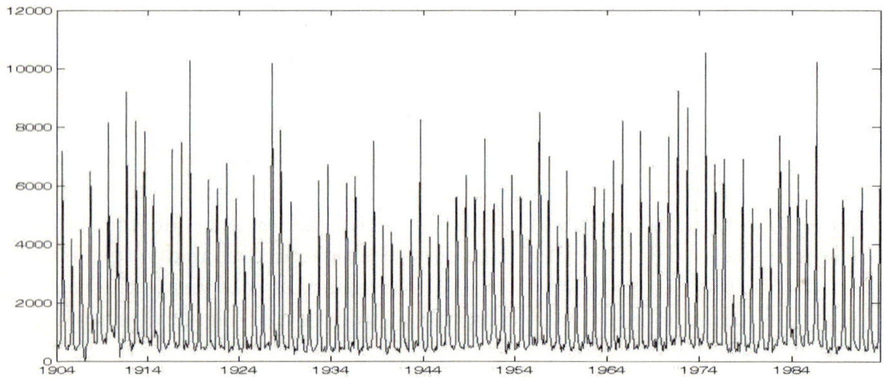

Fig. 2. Monthly river flow series in cubic feet per second of the Snake river, in the period of 1904 until 1994, see [15]

One thousand data points were generated with an initial condition $x(0) = 1.2$ and $\tau = 17$ based on Runge-Kutta method. The plot of the 1000 data points of the series is shown in *fig.* 3.

4.1 Seasonality in Time Series

Seasonality in time series occurs when there is a certain pattern that is repeated each k elements. Both data series in *fig.* 1 and *fig.* 3 show to have a certain seasonality. The period of the seasonality can be obtained by a number of techniques (for example using the autocorrelation function, etc.). In this work, the period has been obtained by evaluating the distances between the supposed-seasons data. The specific period for which the sum of squared distances among the different seasons data was lowest, was taken as a solution.

For the first series, the period found was 52; *fig.* 4 shows the superposition of the different seasons of 52 data samples of the time series. In the LS-SVM

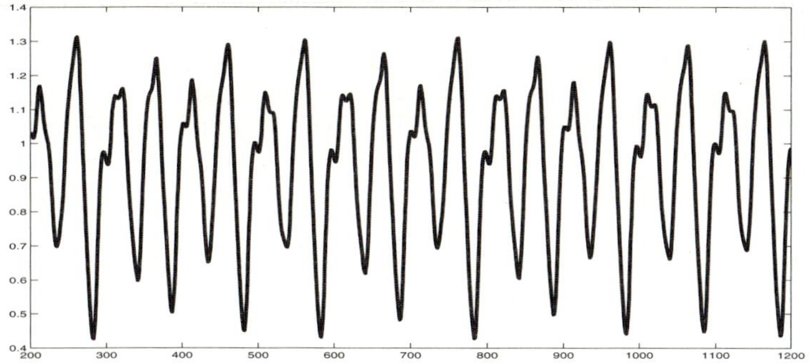

Fig. 3. Mackey-Glass dataset

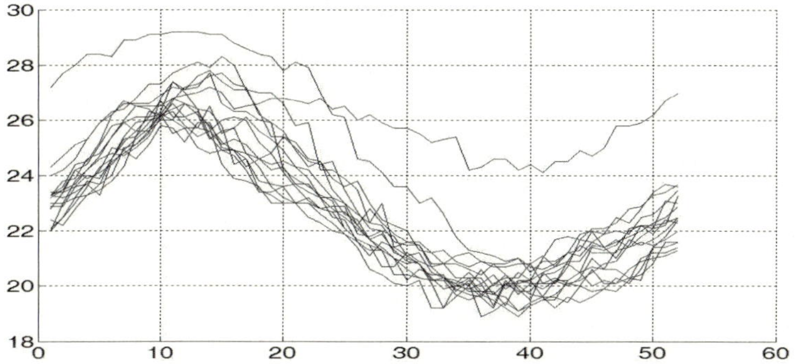

Fig. 4. Different seasons on the original time series

model, as it was explained in the previous section, a variable $s(t+h)$ (indicating the time step within the season of the horizon to be predicted h) is added as a new input variable to improve the prediction accuracy. Therefore this variable will take values from 1..52 in the first series.

For the river flow series, the period found was 12. Again the dummy input variable $s(t+h)$ will take values from 1..12 in the data. For the Mackey-Glass series, the period found was 100. The dummy input variable $s(t+h)$ takes values within 1..100.

5 Simulations

For the ESTSP series, the 875 available data points, were subdivided into 750 for training and 125 for test. The 750 data points of training were subdivided in 5 folds in order to perform cross-validation. Table 1 shows, for the ESTSP series, the 5-fold cross-validation MSE when performing the variable selection procedure for different values of τ, when taking and not taking into account the step in the season $s(t+1)$. The MSE error values were obtained by averaging the

Table 1. MSE for different values of τ in the prediction of the ESTSP series

τ	with $s(t+1)$ CV MSE	without $s(t+1)$ CV MSE	τ	with $s(t+1)$ CV MSE	without $s(t+1)$ CV MSE
$\tau = 0$	1.9200	9.9137	$\tau = 30$	2.7074	2.7937
$\tau = 1$	1.8569	9.1367	$\tau = 31$	3.4971	2.8207
$\tau = 2$	1.8959	7.9905	$\tau = 32$	3.4819	4.0388
$\tau = 3$	1.9212	6.9163	$\tau = 33$	2.8270	4.7843
$\tau = 4$	1.9231	5.8499	$\tau = 34$	2.6153	2.8965
$\tau = 5$	1.9990	4.4584	$\tau = 35$	2.7655	2.8938
$\tau = 6$	1.9877	3.9261	$\tau = 36$	2.9527	2.6832
$\tau = 7$	2.0227	3.2318	$\tau = 37$	4.1308	2.8232
$\tau = 8$	2.0634	3.8275	$\tau = 38$	3.0988	2.7649
$\tau = 9$	2.0954	3.1310	$\tau = 39$	2.8392	3.1840
$\tau = 10$	2.1736	3.5667	$\tau = 40$	2.6673	2.4541
$\tau = 11$	2.1695	3.0814	$\tau = 41$	2.7504	2.5552
$\tau = 12$	2.2897	3.0111	$\tau = 42$	2.8703	2.6058
$\tau = 13$	2.3114	3.1275	$\tau = 43$	2.8337	2.7602
$\tau = 14$	2.3291	3.1848	$\tau = 44$	2.5114	2.7639
$\tau = 15$	2.7397	3.2884	$\tau = 45$	2.7441	2.9026
$\tau = 16$	2.5867	3.2100	$\tau = 46$	2.8401	3.0102
$\tau = 17$	2.8654	3.5485	$\tau = 47$	2.9558	3.1622
$\tau = 18$	2.8517	3.0566	$\tau = 48$	3.0410	3.2489
$\tau = 19$	2.6734	3.0138	$\tau = 49$	3.2325	4.6163
$\tau = 20$	2.7672	3.0245	$\tau = 50$	3.0265	2.9617
$\tau = 21$	3.0418	3.8905	$\tau = 51$	3.2920	3.3118
$\tau = 22$	3.0123	3.1254	$\tau = 52$	2.6466	2.6950
$\tau = 23$	3.0601	3.3506	$\tau = 53$	2.5654	2.7210
$\tau = 24$	3.0054	3.1378	$\tau = 54$	2.9528	2.7376
$\tau = 25$	2.7860	3.4053	$\tau = 55$	2.4548	2.5118
$\tau = 26$	3.0281	3.1866	$\tau = 56$	2.5397	2.5004
$\tau = 27$	3.3303	3.2246	$\tau = 57$	2.6691	4.5925
$\tau = 28$	2.8248	2.9309	$\tau = 58$	2.6442	4.5990
$\tau = 29$	4.0679	2.8582	$\tau = 59$	2.4935	2.4893

evaluation of all possible recursive predictions of 50 values in the validation data sets, for the 5-fold cross-subdivision execution (in total 5 executions$*(n-\tau-50)$ possible recursive evaluations of 50 points).

As it can be seen from the results, the optimal subset of variables selected is given by $\tau = 2$ with $s(t+1)$ as an additional 3rd variable. The results show also a strong improvement by considering the time step within the season $s(t+h)$ when performing the recursive long term prediction. An important aspect too is that the performance of the recursive models that don't include $s(t+h)$ improves as more variables are considered (note however that the CV MSE becomes less reliable as less executions are considered). Therefore we can claim that, for this example, the generalization capability of the model improves by selecting the seasonal variable $s(t+h)$, the number of needed previous regressors is highly reduced in order to obtain a good performance, and therefore the interpretability

Table 2. MSE for different values of τ in the prediction of the river flow series

τ	with $s(t+1)$ CV MSE	without $s(t+1)$ CV MSE	τ	with $s(t+1)$ CV MSE	without $s(t+1)$ CV MSE
$\tau = 0$	6.5685e+005	5.6874e+006	$\tau = 15$	1.0810e+006	2.3535e+006
$\tau = 1$	6.1162e+005	7.1418e+006	$\tau = 16$	1.1014e+006	1.9368e+006
$\tau = 2$	6.4388e+005	4.9638e+006	$\tau = 17$	1.0537e+006	1.4113e+006
$\tau = 3$	6.9242e+005	6.8089e+006	$\tau = 18$	7.6881e+005	7.7071e+005
$\tau = 4$	9.6028e+005	6.2684e+006	$\tau = 19$	3.4948e+006	2.5201e+006
$\tau = 5$	8.7902e+005	3.5305e+006	$\tau = 20$	7.3939e+005	2.0663e+006
$\tau = 6$	7.3270e+005	4.7662e+006	$\tau = 21$	1.1007e+006	9.9358e+005
$\tau = 7$	1.1813e+006	1.4217e+006	$\tau = 22$	9.6651e+005	7.2736e+005
$\tau = 8$	9.3182e+005	2.8930e+006	$\tau = 23$	9.7015e+005	1.4095e+006
$\tau = 9$	2.0916e+006	1.2748e+006	$\tau = 24$	7.1251e+005	7.3901e+005
$\tau = 10$	1.9071e+006	1.3087e+006	$\tau = 25$	7.1103e+005	7.2339e+005
$\tau = 11$	4.2063e+006	1.4718e+006	$\tau = 26$	8.8594e+005	1.0067e+006
$\tau = 12$	1.7122e+006	9.1647e+005	$\tau = 27$	6.8981e+005	7.2266e+005
$\tau = 13$	1.4986e+006	1.1707e+006	$\tau = 28$	7.0264e+005	7.0201e+005
$\tau = 14$	7.3917e+005	8.4034e+005	$\tau = 29$	1.2067e+006	7.9432e+005

of the designed recursive model highly improves too, since only 3 input variables are needed. The average MSE of the test dataset (for all possible evaluations of 50 recursive predictions) for the optimal model is 0.502.

A similar test was performed for the river flow time series. From the 1000 data points, 500 were used for the 5-fold CV procedure, and the rest were left for testing. Table 2 shows, for the river flow time series, the 5-fold cross-validation MSE when performing the variable selection procedure for different values of τ, when taking and not taking into account the step in the season $s(t+1)$. The MSE error values is obtained by averaging the evaluation of all possible recursive predictions of 15 values in the validation data sets, for the 5-fold cross-subdivision execution.

Again from these results we can claim that there is a strong improvement by considering the time step within the season $s(t+h)$ when performing the recursive long term prediction. The optimal subset of variables selected is given again by $\tau = 2$ with $s(t+1)$ as an additional 3rd variable. The improvement in the recursive performance occurs specially for low values of τ, which is more interesting since it provides a better interpretability, apart from the better generalization. The average MSE of the test dataset (for all possible evaluations of 50 recursive predictions) for the optimal model is 3.9434e+007.

For the Mackey-Glass time series, from the 1000 data points, 750 were used for the 5-fold CV procedure, and the rest were left for testing. Table 3 shows, for the Mackey-Glass time series, the 5-fold cross-validation MSE when performing the variable selection procedure for different values of τ, when taking and not taking into account the step in the season $s(t+1)$. Again the MSE error values is obtained by averaging the evaluation of all possible recursive predictions of 50 values in the validation data sets, for the 5-fold cross-subdivision execution.

Table 3. MSE for different values of τ in the prediction of the Mackey-Glass time series

τ	with $s(t+1)$ CV MSE	without $s(t+1)$ CV MSE	τ	with $s(t+1)$ CV MSE	without $s(t+1)$ CV MSE
$\tau = 0$	1.2685e-002	8.4784e-002	$\tau = 10$	5.2343e-004	3.1777e-004
$\tau = 1$	5.6779e-003	2.8835e-002	$\tau = 11$	2.9709e-004	2.7430e-004
$\tau = 2$	3.3021e-003	7.6521e-002	$\tau = 12$	2.1550e-004	1.3528e-004
$\tau = 3$	4.7557e-003	5.5396e-002	$\tau = 13$	1.6134e-004	7.7161e-005
$\tau = 4$	3.4582e-003	2.5383e-002	$\tau = 14$	9.6428e-005	2.6098e-005
$\tau = 5$	3.8077e-003	9.9448e-003	$\tau = 15$	8.5424e-005	2.3221e-005
$\tau = 6$	2.6703e-003	8.0764e-003	$\tau = 16$	6.5617e-005	8.1119e-006
$\tau = 7$	1.9553e-003	3.5603e-003	$\tau = 17$	4.6230e-005	5.6362e-006
$\tau = 8$	1.4060e-003	2.2706e-003	$\tau = 18$	4.0625e-005	3.6452e-006
$\tau = 9$	9.0794e-004	4.5214e-004	$\tau = 19$	2.3944e-005	1.7243e-006

For the Mackey-Glass time series, there is an improvement by considering the time step within the season $s(t+h)$, but only for low values of τ. This can be due to the absence of noise in this artificial time series. From the point of view of the interpretability, that requires a low number of input variables in the model, it is more convenient to include the time step $s(t+1)$ since for low values of τ the performance improves and then less variables are needed. However, from the point of view of the performance, the optimal subset of variables selected is given by $\tau = 24$ without $s(t+1)$ as an additional variable.

6 Conclusion

In time series prediction problems in which the current series presents a certain seasonality, the seasonal information has to be included within the designed predictor. Kernel methods and specifically LS-SVM, allows including the seasonal information by simply adding new variables to the problem, without increasing the computational cost. This work has evaluates the inclusion of the seasonal information of a series presenting seasonality in a recursive model. Three well known time series have been used in the simulations, two natural and one artificial. The results have shown that the performance of the recursive model is upgraded, but also that it is possible to diminish the number of needed input variables needed to perform the modelling. The generalization capabilities of the model increases as less input variables are needed, and therefore the interpretability capability of the recursive model is also improved. However for noiseless series, including additional information can lead to a worse performance. Nevertheless even for those cases, when a low number of input variables is preferred, including the seasonal information can be convenient.

References

1. Weigend, A.S., Gershenfeld, N.A.: Time Series Prediction: Forecasting the Future and Understanding the Past. Addison-Wesley, London, UK (1993)
2. Ji, Y., Hao, J., Reyhani, N., Lendasse, A.: Direct and Recursive Prediction of Time Series Using Mutual Information Selection. IWANN 2005. In: Cabestany, J., Prieto, A.G., Sandoval, F. (eds.) IWANN 2005. LNCS, vol. 3512, pp. 1010–1017. Springer, Heidelberg (2005)
3. Lendasse, A., Ji, Y., Reyhani, N., Verleysen, M.: LS-SVM Hyperparameter Selection with a Nonparametric Noise Estimator. In: Duch, W., Kacprzyk, J., Oja, E., Zadrożny, S. (eds.) ICANN 2005. LNCS, vol. 3697, pp. 625–630. Springer, Heidelberg (2005)
4. LS-SVMlab: a MATLAB/C toolbox for Least Squares Support Vector Machines: http://www.esat.kuleuven.ac.be/sista/lssvmlab
5. Schoelkopf, B., Smola, A.: Learning with Kernels. MIT Press, Cambridge, MA (2002)
6. Suykens, J.A.K., Van Gestel, T., De Brabanter, J., De Moor, B., Vandewalle, J.: Least Squares Support Vector Machines. World Scientific, Singapore (2002)
7. Jung, T., Herrera, L.J., Schoelkopf, B.: Long Term Prediction of Product Quality in a Glass Manufacturing Process Using a Kernel Based Approach. In: Cabestany, J., Prieto, A.G., Sandoval, F. (eds.) IWANN 2005. LNCS, vol. 3512, pp. 960–967. Springer, Heidelberg (2005)
8. EUNITE Competition 2003: Prediction of product quality in glass manufacturing (2003), http://www.eunite.org
9. Tikka, J., Hollmn, J., Lendasse, A.: Input Selection for Long-Term Prediction of Time Series. In: Cabestany, J., Prieto, A.G., Sandoval, F. (eds.) IWANN 2005. LNCS, vol. 3512, pp. 1002–1009. Springer, Heidelberg (2005)
10. Wu, H.-s., Zhang, S.: Power Load Forecasting with Least Squares Support Vector Machines and Chaos Theory. ICNN-B'05 2, 1020–1024 (2005)
11. Espinoza, M., Joye, C., Belmans, R., De Moor, B.: Short term load forecasting, profile identification and customer segmentation: a methodology based on periodic time series. IEEE Trans Power Syst 20(3), 1622–1630 (2005)
12. Herrera, L.J., Pomares, H., Rojas, I., Guillén, A., Prieto, A., Valenzuela, O.: Recursive Prediction for Long Term Time Series Forecasting Using Advanced Models, Neurocomputing, Accepted (2006)
13. Herrera, L.J., Pomares, H., Rojas, I., Guillen, A., Rubio, G.: Incorporating Seasonal Information on Direct and Recursive Predictors Using LS-SVM, 1st ESTSP Conference, Espoo, Finland, pp. 155-164 (2007)
14. Faya, D., Ringwoodb, J.V., Condona, M., Kellyc, M.: 24-h electrical load data-a sequential or partitioned time series? Neurocomputing 55, 469–498 (2003)
15. http://www-personal.buseco.monash.edu.au/hyndman/TSDL/hydrology.html

Can Neural Networks Learn the "Head and Shoulders" Technical Analysis Price Pattern? Towards a Methodology for Testing the Efficient Market Hypothesis

Achilleas Zapranis and Evi Samolada

Department of Accounting and Finance, University of Macedonia of Economic and Social Sciences, P.O. Box 1591, 54006 Thessaloniki, Greece
zapranis@uom.gr, esamo@uom.gr

Abstract. Testing the validity of the Efficient Market Hypothesis (EMH) has been an unsolved argument for the investment community. The EMH states that the current market price incorporates all the information available, which leads to a conclusion that given the information available, no prediction of the future price changes can be made. On the other hand, technical analysis, which is essentially the search for recurrent and predictable patterns in asset prices, attempts to forecast future price changes. To the extend that the total return of a technical trading strategy can be regarded as a measure of predictability, technical analysis can be seen as a test of the EMH and in particular of the independent increments version of random walk. This paper is an initial attempt on creating an automated process, based on a combination of a rule-based system and a neural network, of recognizing one of the most common and reliable patterns in technical analysis, the head and shoulders pattern. The systematic application of this automated process on the identification of the head and shoulders pattern and the subsequent analysis of price behavior, in various markets can in principle work as a test of the EMH.

Keywords: Efficient market hypothesis, technical analysis, head and shoulders price pattern, neural networks.

1 Introduction

Advocates of technical analysis state that there exists predictive power in the price movements of financial assets, and that by analyzing historical prices one can reduce risks and increase returns. Technical analysis is based on the belief that stock time series, volume and other statistics often show similarities that repeat in time and consequently can be used in stock prices prediction and contribute in profits. The tools that technical analysts use are the observation of patterns (head and shoulders, triangles etc), and the calculation of indices (MACD, RSI etc.). One of the most often and reliable pattern in technical analysis is the head and shoulders pattern.

J. Marques de Sá et al. (Eds.): ICANN 2007, Part II, LNCS 4669, pp. 516–526, 2007.

On the other hand, the efficient market hypothesis (EMH) states that security prices incorporate and reflect all the information publicly available and we cannot benefit and make gains by using this information in order to predict stock prices, since it has already been built-in the stock prices. Based on the existence of an efficient market, we can conclude that all prices react and adjust immediately to reflect all information available. Consequently, profits cannot be increased by analyzing historical data, since past data do not influence the outcome of future data. One basic condition of (semi-strong) efficient market hypothesis is the Random Walk 2 (RW2) hypothesis, which states that the natural logarithm of prices $p_t = \ln P_t$ follows a random walk with independent but not identically distributed (INID) increments, i.e.,

$$p_t = \mu + p_{t-1} + \varepsilon_t \, . \tag{1}$$

where μ is the expected price change or drift and ε_t is the random shock. RW2 allows for unconditional heteroskedasticity in the ε_t's, a particularly useful feature given the time-variation in volatility of many financial asset return series.

Technical analysis can be seen as an "economic" test of the RW2 hypothesis. In that view, there several contributions to the relevant literature, however with conflicting results. The overwhelming majority of these contributions concentrate on technical indicators, such as moving averages, oscillators, etc, since it is straightforward to implement the relevant trading systems. Characteristic examples of these studies are, for FX markets are [1] and [2] and for stock markets [3], [4], [5] and [6]. Regarding, the recognition of technical patterns the literature is sparse. To our knowledge, in [7] it is the first time that a systematic and automatic approach to technical pattern recognition is proposed, using nonparametric kernel regression. The authors claimed that several technical indicators do provide incremental information. In discussing those results [8] and [9] warned that data-snooping and survivorship biases can be severe when evaluating technical rules, which can lead researchers to falsely conclude that technical trading strategies can predict future price movements. Finally, in [10] there is the first attempt to quantify a technical pattern with a rule-based system. In particular, the authors specified several explicit criteria that should be met in order to identify the head and shoulders pattern. In [11] the author investigated the profitability of a trading system based on the identification of that pattern in the FX markets.

In this paper we focus on the reliable identification of the head and shoulder pattern in the stock markets. As we demonstrate the rule-based pattern identification system used by [10] and [11] is very sensitive to the values of the parameters used. As a result existing patterns can be overlooked by an automated pattern recognition system. To deal with this problem we incorporate in our two-stage automated pattern recognition system neural models that add robustness to the identification process.

The rest of the paper is organized as follows. In section 2 we discuss the head and shoulders technical pattern and its rule-based automatic recognition. In section 3 we describe how we can add robustness to that process with the inclusion of a post-processing stage based on a neural network. Finally in section 4 we summarize and conclude.

2 Automatic (Rule-Based) Recognition of the Head and Shoulders Pattern

2.1 The Head and Shoulders Price Pattern

In the technical analyst's community, the head and shoulders price pattern is one of the most credible. It could be observed in its normal or inverse (or upside-down) form. An uptrend is formed in the first case as prices make higher-highs (peaks) and higher-lows (troughs) in a stair-step fashion. The end of the uptrend is signified by the formation of the right shoulder.

The line defined by the two troughs is called the neckline, while completion of the pattern and confirmation of a new downtrend occur when the "neckline" is penetrated. After the neckline penetration as a last effort to continue the uptrend and if prices are unable to rise above the neckline, they usually decline rapidly. The same one observes in the reverse head and shoulders pattern, where there are market bottoms and a downtrend is formed in the beginning and it is reversed later on.

Fig. 1. The head and shoulders price pattern for Acher Daniels Midland Co. (ADM) stock traded at NYSE (graph from http://stockcharts.com/education/ChartAnalysis)

In Fig. 1 we can see the head and the two shoulders as well as the two trough of the head and shoulders pattern. The blue line is the neckline. After the formation of the right shoulder the stock price follows a downtrend and after a while it crosses the neckline. This is the point of completion of the pattern.

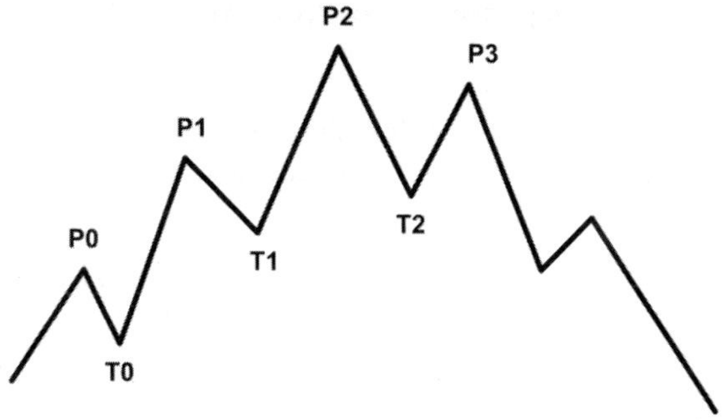

Fig. 2. The head and shoulders price pattern, as defined by the combinations of four consecutive peaks (P0, P1, P2 and P3) and the troughs between them (T0, T1, T2)

The investor should either sell the stock in order to prevent major losses, or short sell it in order to pursue profits. It is worth noticing that the neckline after the completion of the formation is also a strong resistance level for the stock price since the stock price fails to penetrate it for second time.

As we mentioned before the normal head and shoulders pattern signs a trend reversal (downtrend). Besides the graphical part, we should observe that the volume of the stock does not accompany the price, in this case the volume decreases on the head and it is specifically light on the right shoulder as shown on the graph below.

2.2 Rule-Based Definition of the Head and Shoulders Pattern

In automating the recognition of the head and shoulders formation, we use the definition of [10] according to which the following five specific conditions should be met, in order to define the head and shoulders pattern. For simplicity, the four consecutive peaks are named as P0, P1, P2 and P3 and the troughs between them as T0, T1, T2 (see Fig. 2).

SHS1. The head is higher than the shoulders:

$$\mathrm{P2} > \max\left(\mathrm{P1}, \mathrm{P3}\right). \tag{2}$$

SHS2. The pattern is preceded by a generally positive underlying trend:

$$\mathrm{P1} > \mathrm{P0} \;\; \& \;\; \mathrm{T1} > \mathrm{T0}. \tag{3}$$

SHS3 (balance). The left (right) shoulder must be at least as high as the midpoint between right (left) shoulder and its preceding (anteceding) trough:

$$\text{P1} \geq 0.5 \cdot (\text{P3} + \text{T2}) \ \& \ \text{P3} \geq 0.5 \cdot (\text{P1} + \text{T1}). \tag{4}$$

SHS4 (symmetry). The time between left shoulder and head must not be more than 2.5 times the time between head and right shoulder and vice versa:

$$t_{\text{P2}} - t_{\text{P1}} < 2.5 \cdot (t_{\text{P3}} - t_{\text{P2}}) \ \& \ t_{\text{P3}} \cdot t_{\text{P2}} < 2.5 \cdot (t_{\text{P2}} - t_{\text{P1}}). \tag{5}$$

SHS5 (time limit). Let τ denote the time at which the price S_t falls below the neckline:

$$S_{\tau} < \text{T1} + ((t - t_{\text{T1}})/(t_{\text{T2}} - t_{\text{T1}})) \cdot (\text{T1} - \text{T2}). \tag{6}$$

This must not happen too long after the formation of the right shoulder, say:

$$\tau < t_{\text{P3}} + (t_{\text{P3}} - t_{\text{P1}}). \tag{7}$$

The former criteria are crucial because they combine at least 4 technical concepts of smoothed trends, trend reversal, resistance levels and volatility clustering. SHS1 and SHS2 suggest the existence of and upward trend between P0 and P2.

2.3 Automatic Discovery of the Head and Shoulders Pattern in Price Series

The criteria SHS1 to SHS5 were coded in the form of a Matlab script. The script reads through a price series, S_t, and identifies the combination of peaks (P0, P1, P2 and P3) and the troughs between them (T0, T1, T2) that satisfy the above criteria. In doing so, it uses a rolling window of prices of length w, i.e., $\{S_t, S_{t+1}, \ldots, S_{t+w}\}$, for $t = 1, \ldots, n\text{-}w$, where n is the total number of price quotes (S_n is the last available price in the series). The script also computes the neckline intercept and slope and produces a figure with the entire head and shoulders formation and the neckline, with the identified peaks and troughs annotated with the observation number and the corresponding price.

In Table 1 we can see the data generated from the script that corresponds to an identified head and shoulders price pattern, for the stock CNET Networks, Inc. For each peak and through we can read the observation number and the corresponding price, i.e., P0 is identified at $S_{42} = 57.62$, T0 at $S_{66} = 30.31$, P1 at $S_{112} = 60.13$, etc. The length of the rolling window used was $w = 50$ observations. The neckline intercept and slope were also calculated from the script; $a = 38.0530$ and $b = 0.0534$ correspondingly.

In Fig. 3 we can see the graph for CNET generated by the script based on the data given in Table 1 and the calculated intercept and slope. The first five marked points shown (from left) are P1, T1, P2, T2, P3 and they are annotated as: *observation number, price*. The neckline is also plotted. The sixth point (far most in the right), i.e., S_{230} = 53.25, is the neckline and price series bisection point. According to technical analysis when the head and shoulders formation breaks through the neckline a significant decline follows, which is what we observe in Fig.3.

Table 1. Automatic head and shoulders pattern recognition for CNET Networks, Inc. traded at NASDAQ. Rolling window length $w = 50$. Neckline intercept $a = 38.0530$, slope $b = 0.0534$.

Obs.	P0	T0	P1	T1	P2	T2	P3	Price
30	0	0	0	0	0	0	0	40.00
42	**1**	0	0	0	0	0	0	57.62
66	0	**1**	0	0	0	0	0	30.31
112	0	0	**1**	0	0	0	0	60.13
129	0	0	0	**1**	0	0	0	44.94
161	0	0	0	0	**1**	0	0	79.00
191	0	0	0	0	0	**1**	0	48.25
213	0	0	0	0	0	0	**1**	70.56
243	0	0	0	0	0	0	0	24.62
299	0	0	0	0	0	0	0	21.44

In Fig. 4 we can see the graph generated by the Matlab script for ADM for which the technical analysis chart is also given in Fig. 1. As we can see the script has correctly identified the head and shoulders pattern identified by technical analysts in Fig.1. If we were using the script in the context of a technical trading system, at the breakout point $S_{3428} = 39.73$ a short position should be taken.

However, as we can see in those two cases the size of the rolling window is not fixed but it varies (actually, for CNET the rolling window size is $w = 50$, for ADM is $w = 20$). If there does not exist a head and shoulders pattern in the price series, the value of w is irrelevant. But if there exists one, the algorithm will discover the pattern only if w is at least equal to the number of observations needed for the pattern to unfold. A solution to this problem, is to run the algorithm a number of times, with w taking several values, lets say from 20 to 150 with a step of 10.

Moreover, the pattern recognition algorithm is very sensitive to the values of the parameters of the criteria SHS3 (0.5) and SHS4 (2.5) and in many cases, it fails to recognize an existing head and shoulders pattern. For example, it fails to recognize the pattern for EBAY, Inc traded at NASDAQ, but when we change the parameter value for SHS3 from 0.5 to 0.4 it recognizes the pattern successfully as we can see in Fig. 5.

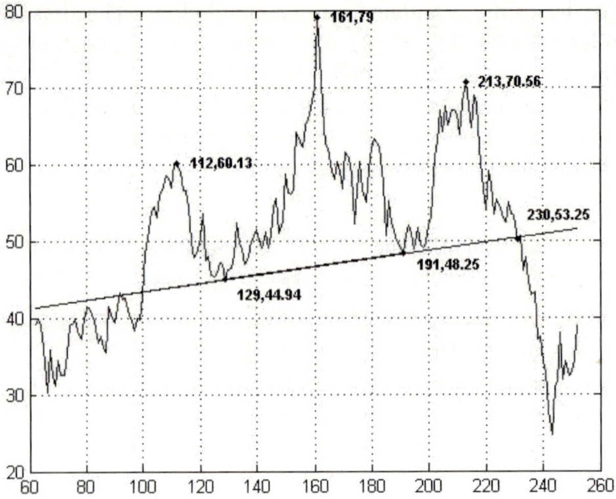

Fig. 3. Automatic head and shoulder price pattern recognition (figure generated from a Matlab script) for CNET Networks, Inc. traded at NASDAQ. Rolling window length $w = 50$. Neckline intercept $a = 38.0530$, slope $b = 0.0534$. Price series used for pattern recognition: 3/5/1999 to 31/8/2000.

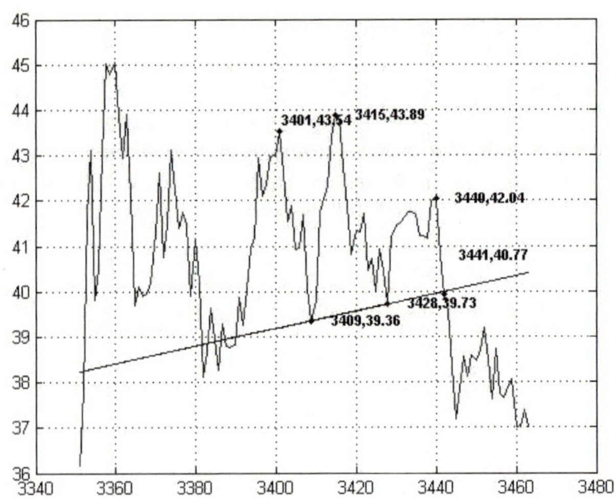

Fig. 4. Automatic head and shoulder price pattern recognition for Acher Daniels Midland Co. (ADM) stock traded at NYSE. Rolling window length $w = 20$. Price series used for pattern recognition: from 2/1/1996 to 31/12/1999.

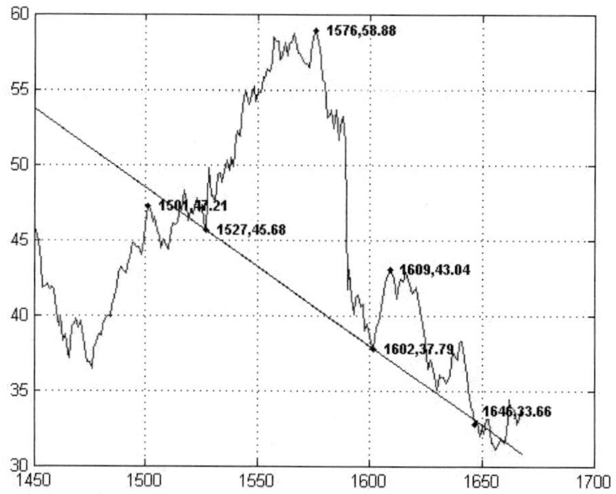

Fig. 5. Automatic head and shoulder price pattern recognition for EBAY, Inc. traded at NASDAQ. The head and shoulders pattern is not recognized, unless the criterion SHS3 is changed to: P1 > 0.4(P3 + T2), P3 > 0.4(P1 + T1).

3 Robust Identification of the Head and Shoulders Pattern with Neural Networks

As we have seen, some of the conditions that should be met in order to identify a head and shoulder pattern impose very strict limits. This is particularly true for conditions SHS3, SHS4 and SHS5. As a result, in many cases (e.g., Fig. 5) existing head and shoulder patterns are not being recognized by the algorithm. This has serious implications on the validity of any study using this algorithm in the context of evaluating the informational content of the particular price formation. Fine-tuning, the criteria SHS3 to SHS5 for every price series under examination, is not a plausible solution either, when hundreds of stocks must be analyzed.

Our approach in dealing with this problem is to build a neural network which will be able to identify robustly the head and shoulders pattern. This will be done by "post-processing" the peaks and troughs recognized by the algorithm, by carefully adding noise to them. The trained network will be then able to identify patterns that otherwise would be missed. We demonstrate this procedure below.

First we have to create our training dataset. A rolling vector with seven inputs, $\mathbf{x} = \{x_1, x_2, \ldots, x_7\}$, reads through the data organized as in Table 1, from right to left and from top to bottom. This is repeated for as many price paths as possible. When P0 = 1 then $x_1 = 1$ and the next inputs x_2 to x_7 are expressed as a percent of their corresponding prices relative to the price at P0. For example, if the price at P0 is 30 and the price at T0 is 40 then $x_1 = 1$, $x_2 = 1.33$, etc. The network target y is 1 when the input vector corresponds to a head and shoulders pattern, otherwise is 0.

For illustrational purposes, we applied the above procedure to three stocks: CNET, MARVEL and ADM. A single training dataset was created with 194 vectors. A part of the training inputs can be seen in Table 2. The line with the bold numbers corresponds to an input identifying a head and shoulders pattern. In total, there were only three lines corresponding to the head and shoulders pattern.

Table 2. Network input vectors for CNET Networks, Inc. traded at NASDAQ, MARVEL Technology Group, Ltd. traded at NASDAQ and Acher Daniels Midland Co. traded at NYSE (194 input vectors)

x_1	x_2	x_3	x_4	x_5	x_6	x_7
…	…	…	…	…	…	…
0.65476936	0.79858067	0.80025047	0.68712169	0.86579002	0.79252766	1.00000000
0.79252766	1.00000000	0.87789606	1.25130453	1.15675224	1.47443123	1.12920058
1.00000000	**0.87789606**	**1.25130453**	**1.15675224**	**1.47443123**	**1.12920058**	**1.27301190**
0.87789606	1.25130453	1.15675224	1.47443123	1.12920058	1.27301190	1.09851805
1.12920058	1.27301190	1.09851805	0.36171989	0.43122521	0.36151117	0.43519098
…	…	…	…	…	…	…

Next, we trained 6 back-propagation neural networks, with one hidden layer, with the number of hidden units, λ, ranging from 3 to 6. The network outputs are given in Table 2. In the first column of the same Table we can see the target output (1 indicates the presence of the head and shoulder pattern).

Table 3. Automatic head and shoulders pattern recognition with neural networks for CNET Networks, Inc. traded at NASDAQ, MARVEL Technology Group, Ltd. traded at NASDAQ and Acher Daniels Midland Co. traded at NYSE (194 input vectors). Number of hidden units λ = 3 - 6. The first column is the target output. Columns 2 to 5 are the network outputs.

	$\lambda = 3$	$\lambda = 4$	$\lambda = 5$	$\lambda = 6$
	…	…	…	…
1	**0.98586000**	**0.99575000**	**0.99493000**	**0.99084000**
0	0.00016041	0.08738100	0.52662000	0.00022027
0	0.00076142	0.00017389	0.00011368	0.00168490
	…	…	…	…
1	**0.96278000**	**0.96752000**	**0.97137000**	**0.96923000**
0	0.00178220	0.01331200	0.02402800	0.00520230
	…	…	…	…
0	0.00017280	0.00016699	0.10528000	0.00018539
1	**0.96932000**	**0.95560000**	**0.93783000**	**0.96248000**
0	0.00019253	0.00018189	0.08615600	0.00025253
	…	…	…	…

As expected, all the neural network models are able to identify the presence of the head and shoulder pattern, although only three such vectors were included in the training dataset. The trained neural networks have learned to recognize the patterns identified by the criteria SH1 to SH5, but they are not still able to identify existing patterns which were wrongly rejected by the algorithm (see Fig. 5 for the stock of EBAY). We rectify this problem, i.e., we make the identification process more robust to small deviations from the identification criteria, by adding noise to the variables of the training dataset. For example, by adding random disturbances from a univariate distribution [0, 0.25] to the inputs x_3, x_4, x_6 and x_7 (see Table 2) we effectively change the criterion SHS3 roughly to: P1 > 0.3(P3 + T2), P3 > 0.3(P1 + T1). Criteria SHS4 and SHS5 are also (indirectly) being affected, since the changes in the values of the peaks and troughs affect the points in time t_{P1}, t_{P2}, t_{P3}, t_{T1} and t_{T2}.

To illustrate this, we added 60 new training vectors generated from the three initial vectors corresponding to head and shoulder price patterns. From each initial vector we generated 20 new training vectors, by adding random disturbances from a univariate distribution [0, 0.25] to the inputs x_3, x_4, x_6 and x_7. Then, we trained a neural network using the new (extended) dataset. Now, the trained network was able to recognize not only the head and shoulder patterns from CNET, MARVEL and ADM, but also from EBAY without making any direct adjustments to the criteria SHS3 to SHS5.

Concluding, our proposed two-step procedure for an automating robust identification of the head and shoulders pattern is as follows:

1. Apply the pattern recognition criteria (in their original form) to several stock price histories and create a dataset in the form of Table 2.
2. Add carefully designed noise to the dataset and then use it to train a neural network.

In the future, in order to identify the head and shoulder pattern for any stock first we have to create the pattern recognition data in the form of Table 2, and then we feed that dataset to the neural network to identify the presence or not of the pattern.

4 Summary and Conclusions

The incentive behind the creation of an automatic recognition tool for the head and shoulders technical pattern is that, it can be used for evaluating the validity of its use as a predictive indicator of the future price behavior. By applying the process of automatic pattern recognition to a large number of stocks across different markets, we can assess the statistical significance of the predictive capacity of the pattern. This will hopefully be a contribution to the continuing debate regarding the hypothesis of efficient markets.

However, first we have to ensure that all (or nearly all) existing patterns are recognized. To that end, here we presented the basic stages of a proposed approach for robust identification of the head and shoulders technical pattern, based on a combination of rule-based data pre-processing and a final step of identification with a neural network.

Rule-based pattern recognition suffers from sensitivity to the parameters of the implemented criteria. Small deviations can lead to wrongly rejecting an existing pattern.

We attempt to rectify this problem by a post-processing stage of the data generated by the rule-based pattern recognition. As we demonstrated, by adding carefully designed noise to that data we effectively transform the cut-off points of the pattern recognition criteria to ranges of acceptable values. A neural network trained with that data, can serve as a robust identification tool of the head and shoulders pattern.

Concluding, here we presented an approach for identifying the head and shoulders pattern which, in principle at least, alleviates the problem of sensitivity to the parameter values of the rule based systems. However, before putting our approach into practice, our immediate future work, is to fine-tune the noise processes we add to the data before training the neural network.

References

1. Osler, C.: Support for resistance: Technical analysis and intraday exchange rates. FRBNY Econ. Pol. Rev., pp. 53–68 (2000)
2. Chang, K., Osler, C.: Methodological madness: technical analysis and the irrationality of exchange rate forecasts. The Econ. J. 109, 636–661 (1999)
3. Neftci, S.: Naïve trading rules in financial markets and Wiener-Kolmogorov prediction theory. A study of technical analysis. J. of Bus 64, 549–571 (1991)
4. Bessembinder, H., Chan, K.: Market efficiency and the returns to technical analysis. Fin. Manag. 27, 5–17 (1998)
5. Sullivan, R., Timmermann, A., White, H.: Data-snooping, technical trading rule performance, and the bootstrap. The J. of Fin. 54, 1647–1691 (1999)
6. Ratner, M., Leal, R.: Tests of technical trading strategies in the emerging equity markets of Latin America and Asia. J. of Bank. and Fin. 23, 1887–1905 (1999)
7. Lo, A., Mamaysky, H., Wang, J.: Foundations of technical analysis: Computational algorithms, statistical inference, and empirical implementation. The J. of Fin. 4, 1705–1765 (2000)
8. Jegadeesh, N.: Foundations of technical analysis: Computational algorithms, statistical inference, and empirical implementation: Discussion. The J. of Fin. 4, 1765–1770 (2000)
9. Sullivan, R., Timmermann, A., White, H.: Data-snooping, technical trading rule performance, and the bootstrap. The J. of Fin. 54, 1647–1691 (1999)
10. Osler, C.L., Chang, P.H.: Head and shoulders not just a flaky pattern. Federal Reserve Bank of New York Staff Report No. 4 (1995)
11. Lucke, B.: Are technical trading rules profitable? Evidence for head-shoulders rules. App. Econ. 35, 33–40 (2003)

Sparse Least Squares Support Vector Regressors Trained in the Reduced Empirical Feature Space

Shigeo Abe and Kenta Onishi

Graduate School of Engineering
Kobe University
Rokkodai, Nada, Kobe, Japan
abe@kobe-u.ac.jp
http://www2.eedept.kobe-u.ac.jp/~abe

Abstract. In this paper we discuss sparse least squares support vector regressors (sparse LS SVRs) defined in the reduced empirical feature space, which is a subspace of mapped training data. Namely, we define an LS SVR in the primal form in the empirical feature space, which results in solving a set of linear equations. The independent components in the empirical feature space are obtained by deleting dependent components during the Cholesky factorization of the kernel matrix. The independent components are associated with support vectors and controlling the threshold of the Cholesky factorization we obtain a sparse LS SVM. For linear kernels the number of support vectors is the number of input variables at most and if we use the input axes as support vectors, the primal and dual forms are equivalent. By computer experiments we show that we can reduce the number of support vectors without deteriorating the generalization ability.

1 Introduction

In a support vector machine (SVM), among training data only support vectors are necessary to represent a solution. However for a difficult classification problem with huge training data, many training data may become support vectors. Since this leads to slow classification, there are several approaches to reduce support vectors [1,2,3]. Keerthi et al. [2] proposed training L2 support vector machines in the primal form. The idea is to select basis vectors by forward selection and for the selected basis vectors train support vector machines by Newton's method. This process is iterated until some stopping condition is satisfied. Wu et al. [3] imposed, as a constraint, the weight vector that is expressed by a fixed number of kernel functions and solved the optimization problem by the steepest descent method.

A least squares support vector machine (LS SVM) [4,5] is a variant of an SVM, in which inequality constraints in an L2 SVM is replaced by equality constraints. This leads to solving a set of linear equations instead of a quadratic programming program. But the disadvantage is that all the training data become support vectors. To solve this problem, in [4,5], support vectors with small absolute values

J. Marques de Sá et al. (Eds.): ICANN 2007, Part II, LNCS 4669, pp. 527–536, 2007.

of the associated dual variables are pruned and an LS SVM is retrained using the reduced training data set. This process is iterated until sufficient sparsity is realized. In [6], LS SVMs are reformulated using the kernel expansion of the square of Euclidian norm of the weight vector in the decision function. But the above pruning method is used to reduce support vectors. Because the training data are reduced during pruning, information for the deleted training data is lost for the trained LS SVM. To overcome this problem, in [7], independent data in the feature space are selected from the training data, and using the selected training data the solution is obtained by least squares method using all the training data. In [8] based on the concept of the empirical feature space proposed in [9], least squares SVMs are formulated as a primal problem and by reducing the dimension of the empirical feature space, sparse LS SVMs are realized

In this paper we extend the sparse LS SVM discussed in [8] to function approximation. Namely, we define the LS support vector regressor (SVR) in the primal form in the empirical feature space. Since the empirical feature space is finite, we can train a primal LS SVM directly by solving a set of linear equations. To generate the mapping function to the empirical feature space, we need to calculate the eigenvalues and eigenvectors of the kernel matrix. Instead, we select the maximum independent components in the kernel matrix by the Cholesky factorization. The independent components are associated with support vectors and reducing the number of independent components we obtain a sparse LS SVM. For linear kernels the number of support vectors is the number of input variables at most and if we use the Euclidean axes as support vectors, the primal and dual forms are equivalent.

In Section 2, we clarify the characteristics of the empirical feature space, and in Section 3 we derive a set of linear equations for training LS SVMs in the empirical feature space and formulate sparse LS SVMs. In Section 4, we show the validity of the proposed method by computer experiments.

2 Empirical Feature Space

In this section, we summarize the characteristics of the empirical feature space.

Let the kernel be $H(\mathbf{x}, \mathbf{x}') = \mathbf{g}^T(\mathbf{x})\,\mathbf{g}(\mathbf{x})$, where $\mathbf{g}(\mathbf{x})$ is the mapping function that maps the m-dimensional vector $\mathbf{x}$ into the l-dimensional space. For the M m-dimensional data $\mathbf{x}_i$, the $M \times M$ kernel matrix $H = \{H_{ij}\}$ $(i, j = 1, \ldots M)$, where $H_{ij} = H(\mathbf{x}_i, \mathbf{x}_j)$, is symmetric and positive semidefinite. Let the rank of H be $N\,(\leq M)$. Then H is expressed by

$$H = U\,S\,U^T, \tag{1}$$

where the column vectors of U are eigenvectors of H and $U\,U^T = U^T U = I_{M\times M}$, $I_{M\times M}$ is the $M \times M$ unit matrix, and $S = \mathrm{diag}\,(\sigma_1, \ldots, \sigma_N, 0, \ldots, 0)$. Here, $\sigma_i\,(> 0)$ are eigenvalues of H, whose eigenvectors correspond to the ith columns of U.

Defining the first N vectors of U as the $M \times N$ matrix P and

$$\Lambda = \mathrm{diag}\,(\sigma_1, \ldots, \sigma_N), \tag{2}$$

we can rewrite (1) as follows:

$$H = P\,\Lambda\,P^T, \tag{3}$$

where $P^T P = I_{N\times N}$ but $P\,P^T \neq I_{M\times M}$.

We must notice that if $N < M$, the determinant of H vanishes. Thus, from $H(\mathbf{x}, \mathbf{x}') = \mathbf{g}^T(\mathbf{x})\,\mathbf{g}(\mathbf{x})$, the following equation holds:

$$\sum_{i=1}^{M} a_i\,\mathbf{g}^T(\mathbf{x}_i) = 0, \tag{4}$$

where $a_i (i = 1, \ldots, M)$ are constant and some of them are nonzero. Namely, if $N < M$, the mapped training data $\mathbf{g}(\mathbf{x}_i)$ are linearly dependent. And if $N = M$, they are linearly independent and there are no non-zero a_i that satisfy (4).

Now we define the mapping function that maps the m-dimensional vector $\mathbf{x}$ into the N-dimensional space called *empirical feature space* [9]:

$$\mathbf{h}(\mathbf{x}) = \Lambda^{-1/2}\,P^T\,(H(\mathbf{x}_1, \mathbf{x}), \ldots, H(\mathbf{x}_M, \mathbf{x}))^T. \tag{5}$$

We define the kernel associated with the empirical feature space by

$$H_\mathrm{e}(\mathbf{x}, \mathbf{x}') = \mathbf{h}^T(\mathbf{x})\,\mathbf{h}(\mathbf{x}'). \tag{6}$$

Clearly, the dimension of the empirical feature space is N. Namely, the empirical feature space is spanned by the linearly independent mapped training data.

We can prove that the kernel for the empirical feature space is equivalent to the kernel for the feature space if they are evaluated using the training data. Namely [9,8],

$$H_\mathrm{e}(\mathbf{x}_i, \mathbf{x}_j) = H(\mathbf{x}_i, \mathbf{x}_j) \qquad \text{for} \quad i, j = 1, \ldots, M. \tag{7}$$

The relation given by (7) is very important because a problem expressed using kernels can be interpreted, without introducing any approximation, as the problem defined in the associated empirical feature space. The dimension of the feature space is sometimes very high or infinite. But the dimension of the empirical feature space is the number of training data at most. Thus, instead of analyzing the feature space, we only need to analyze the associated empirical feature space.

3 Least Squares Support Vector Regressors

3.1 Training in the Empirical Feature Space

The LS SVR in the feature space is trained by minimizing

$$\frac{1}{2}\,\mathbf{w}^T\mathbf{w} + \frac{C}{2}\sum_{i=1}^{M}\xi_i^2 \tag{8}$$

subject to the equality constraints:

$$\mathbf{w}^T \mathbf{g}(\mathbf{x}_i) + b = y_i - \xi_i \quad \text{for} \quad i = 1, \ldots, M, \tag{9}$$

where y_i is the output for input $\mathbf{x}_i$, $\mathbf{w}$ is the l-dimensional vector, b is the bias term, $\mathbf{g}(\mathbf{x})$ is the l-dimensional vector that maps the m-dimensional vector $\mathbf{x}$ into the feature space, ξ_i is the slack variable for $\mathbf{x}_i$, and C is the margin parameter.

Introducing the Lagrange multipliers $\boldsymbol{\alpha} = (\alpha_1, \ldots, \alpha_M)^T$ into (8) and (9), we obtain the dual form as follows:

$$\Omega \boldsymbol{\alpha} + \mathbf{1} b = \mathbf{y}, \tag{10}$$

$$\mathbf{1}^T \boldsymbol{\alpha} = 0, \tag{11}$$

where $\mathbf{1}$ is the M-dimensional vector: $\mathbf{1} = (1, \ldots, 1)^T$ and

$$\Omega_{ij} = \mathbf{g}^T(\mathbf{x}_i)\,\mathbf{g}(\mathbf{x}_j) + \frac{\delta_{ij}}{C}, \quad \delta_{ij} = \begin{cases} 1 & \text{for} \quad i = j, \\ 0 & \text{for} \quad i \neq j, \end{cases} \quad \mathbf{y} = (y_1, \ldots, y_M)^T. \tag{12}$$

Setting $H(\mathbf{x}, \mathbf{x}') = \mathbf{g}^T(\mathbf{x})\,\mathbf{g}(\mathbf{x}')$, we can avoid the explicit treatment of variables in the feature space.

The original minimization problem is solved by solving (10) and (11) for $\boldsymbol{\alpha}$ and b as follows. Because of $1/C\,(> 0)$ in the diagonal elements, Ω is positive definite. Therefore,

$$\boldsymbol{\alpha} = \Omega^{-1}(\mathbf{y} - \mathbf{1}\, b). \tag{13}$$

Substituting (13) into (11), we obtain

$$b = (\mathbf{1}^T \Omega^{-1} \mathbf{1})^{-1} \mathbf{1}^T \Omega^{-1} \mathbf{y}. \tag{14}$$

Thus, substituting (14) into (13), we obtain $\boldsymbol{\alpha}$. We call the LS SVR obtained by solving (13) and (14) *dual LS SVR*.

The LS SVR in the empirical feature space is trained by minimizing

$$Q(\mathbf{v}, \boldsymbol{\xi}, b) = \frac{1}{2}\,\mathbf{v}^T \mathbf{v} + \frac{C}{2} \sum_{i=1}^{M} \xi_i^2 \tag{15}$$

subject to the equality constraints:

$$\mathbf{v}^T \mathbf{h}(\mathbf{x}_i) + b = y_i - \xi_i \quad \text{for} \quad i = 1, \ldots, M, \tag{16}$$

where $\mathbf{v}$ is the N-dimensional vector and $\mathbf{h}(\mathbf{x})$ is the N-dimensional vector that maps the m-dimensional vector $\mathbf{x}$ into the empirical feature space.

Substituting (16) into (15), we obtain

$$Q(\mathbf{v}, \boldsymbol{\xi}, b) = \frac{1}{2}\,\mathbf{v}^T \mathbf{v} + \frac{C}{2} \sum_{i=1}^{M} (y_i - \mathbf{v}^T \mathbf{h}(\mathbf{x}_i) - b)^2. \tag{17}$$

Equation (17) is minimized when the following equations are satisfied:

$$\frac{\partial Q(\mathbf{v}, \boldsymbol{\xi}, b)}{\partial \mathbf{v}} = \mathbf{v} - C \sum_{i=1}^{M} (y_i - \mathbf{v}^T \mathbf{h}(\mathbf{x}_i) - b)\, \mathbf{h}(\mathbf{x}_i) = \mathbf{0} \tag{18}$$

$$\frac{\partial Q(\mathbf{v}, \boldsymbol{\xi}, b)}{\partial b} = -C \sum_{i=1}^{M} (y_i - \mathbf{v}^T \mathbf{h}(\mathbf{x}_i) - b) = 0. \tag{19}$$

From (19)

$$b = \frac{1}{M} \sum_{i=1}^{M} (y_i - \mathbf{v}^T \mathbf{h}(\mathbf{x}_i)). \tag{20}$$

Substituting (20) into (18), we obtain

$$\left(\frac{1}{C} + \sum_{i=1}^{M} \mathbf{h}(\mathbf{x}_i)\, \mathbf{h}^T(\mathbf{x}_i) - \frac{1}{M} \sum_{i,j=1}^{M} \mathbf{h}(\mathbf{x}_i)\, \mathbf{h}^T(\mathbf{x}_j) \right) \mathbf{v} = \sum_{i=1}^{M} y_i\, \mathbf{h}(\mathbf{x}_i) - \frac{1}{M} \sum_{i,j=1}^{M} y_i\, \mathbf{h}(\mathbf{x}_j). \tag{21}$$

Therefore, from (21) and (20) we obtain $\mathbf{v}$ and b. We call the LS SVR obtained by solving (21) and (20) *primal LS SVR*.

3.2 Sparse Least Squares Support Vector Regressors

In training LS SVRs in the empirical feature space we need to transform input variables into the variables in the empirical feature space by (5). But this is time consuming. Thus, instead of using (5), we select independent training data that span the empirical feature space. Let the first M' ($\leq M$) training data be independent in the empirical feature space. Then, instead of (5), we use the following equation:

$$h(\mathbf{x}) = (H(\mathbf{x}_1, \mathbf{x}), \ldots, H(\mathbf{x}_{M'}, \mathbf{x}))^T \tag{22}$$

By this formulation, $\mathbf{x}_1, \ldots, \mathbf{x}_{M'}$ becomes support vectors. Thus, support vectors do not change even if the margin parameter changes. And the number of support vectors is the number of selected independent training data that span the empirical feature space. Thus for linear kernels, the number of support vectors is the number of input variables at most. The selected training data span the empirical feature space but the coordinates are different from those of the empirical feature space, namely those given by (5). Thus, the solution is different from that using (5) because SVRs are not invariant for the linear transformation of input variables [10]. As the computer experiments in the following section show, this is not a problem if we select kernels and the margin parameter properly.

We use the Cholesky factorization in selecting independent vectors [10]. Let H be positive definite. Then H is decomposed by the Cholesky factorization into

$$H = L\,L^T, \tag{23}$$

where L is the regular lower triangular matrix and each element L_{ij} is given by

$$L_{op} = \frac{H_{op} - \sum_{n=1}^{p-1} L_{pn}\,L_{on}}{L_{pp}} \quad \text{for} \quad o = 1,\ldots,M, \quad p = 1,\ldots,o-1, \tag{24}$$

$$L_{aa} = \sqrt{H_{aa} - \sum_{n=1}^{a-1} L_{an}^2} \quad \text{for} \quad a = 1,2,\ldots,M. \tag{25}$$

Here, $H_{ij} = H(\mathbf{x}_i, \mathbf{x}_j)$.

Then during the Cholesky factorization, if the diagonal element is smaller than the prescribed value $\eta\,(> 0)$:

$$H_{aa} - \sum_{n=1}^{a-1} L_{an}^2 \leq \eta, \tag{26}$$

we delete the associated row and column and continue decomposing the matrix. The training data that are not deleted in the Cholesky factorization are independent. If no training data are deleted, the training data are all independent in the feature space.

The above Cholesky factorization can be done incrementally [10,11]. Namely, instead of calculating the full kernel matrix in advance, if (26) is not satisfied, we overwrite the ath column and row with those newly calculated using the previously selected data and $\mathbf{x}_{a+1}$. Thus the dimension of L is the number of selected training data, not the number of training data.

To increase sparsity of LS SVRs, we increase the value of η. The optimal value is determined by cross-validation. We call thus trained LS SVRs *sparse LS SVRs.*

If we use linear kernels we do not need to select independent variables. Instead of (22), we use

$$h(\mathbf{x}) = \mathbf{x}. \tag{27}$$

This is equivalent to using $\mathbf{e}_i$ $(i = 1,\ldots,m)$, where $\mathbf{e}_i$ are the basis vectors in the input space, in which the ith element is 1 and other elements 0. We call the primal LS SVR using (27) *primal LS SVR with orthogonal support vectors (OSV)*, and the primal LS SVR with selected independent training data *LS SVR with non-orthogonal support vectors (NOSV).*

4 Performance Evaluation

We compared the generalization ability of primal, sparse, and dual LS SVRs using the Mackey-Glass [12], water purification [13], orange juice[1], and Boston[2] problems listed in Table 1. For the first three problems, one set of training data set and test data set are given. But the Boston problem is not divided into training and test data sets. As discussed in [14], we used the fifth and the 14th variables as the outputs and call the problems the Boston 5 and 14 problems, respectively. For each problem, we randomly divided the data set into two with almost equal sizes and generated 100 sets of training and test data sets.

For primal LS SVRs we set $\eta = 10^{-9}$ and for sparse LS SVRs, we selected the value of η from $\{10^{-6}, 10^{-5}, 10^{-4}, 10^{-3}, 10^{-2}, 0.05\}$ by fivefold cross-validation.

In all studies, we normalized the input ranges into $[0, 1]$ and used linear and RBF kernels. We determined the parameters C, γ for RBF kernels, and η by fivefold cross-validation; the value of C was selected from among $\{1, 10, 50, 100, 500, 1000, 2000, 3000, 5000, 8000, 10000, 50000, 10^5, \ldots 10^{14}\}$, the value of γ from among $\{0.01, 0.1, 1, 5, 10, 15, 20, 30\}$.

Table 1. Benchmark data sets

Data	Inputs	Train.	Test
Mackey-Glass	4	500	500
Water Purification	10	241	237
Orange Juice	700	150	68
Boston	13	506	—

We determined the margin parameters and kernel parameters by fivefold cross-validation. For RBF kernels we determined the optimal values of γ and C for primal and dual LS SVRs. Then setting the optimal values of γ determined by cross-validation for primal LS SVRs, we determined the optimal values of η and C for sparse LS SVRs.

For the Boston 5 and 14 problems we performed cross-validation using the first five training data sets. For RBF kernels we performed cross-validation for each training data set changing the values of γ and C and selected the value of γ whose associated average of the absolute approximation errors (AAAE) is the smallest. Then we took the median among five γ values as the best value of γ. Then, again we took the median among the best values of C for the best γ associated with the five training data sets. Then, for the best values of γ and C, we trained the SVR for the 100 training data sets and calculated the AAAEs

[1] ftp://ftp.ics.uci.edu/pub/machine-learning-databases/

[2] http://www.cs.toronto.edu/~delve/data/datasets.html

and the standard deviation of the approximation errors for the test data sets. For linear kernels we took the median of the best values of C associated with the first five training data sets.

For sparse LS SVRs, for each value of η we determined the best value of C. We selected the largest value of η whose associated AAAE for the validation data set is comparable with that for $\eta = 10^{-9}$.

Table 2 lists the parameters obtained according to the preceding procedure. The margin parameters for the OSV are the same with the dual LS SVR except for the Mackey-Glass problem. Except for those of OSV, the values of C for the primal LS SVRs are sometimes larger than those for the dual LS SVRs. And the values of γ for water purification, orange juice, and Boston 5 problems are different.

Table 2. Parameter setting for linear and RBF kernels. The parameters were determined by fivefold cross-validation.

Data	Linear			RBF					
	OSV	NOSV	Dual	Primal		Sparse		Dual	
	C	C	C	γ	C	η	C	γ	C
Mackey-Glass	10^{13}	10^5	10^7	30	10^9	10^{-6}	10^9	30	10^9
Water Purification	10	10^6	10	20	50	0.05	50	30	10
Orange Juice	100	10^5	100	10	10^{10}	10^{-4}	10^8	0.01	10^7
Boston 5	1	1	1	10	500	10^{-2}	500	15	50
Boston 14	1	10^4	1	10	3000	10^{-2}	3000	10	50

Table 3 shows AAAEs for the test data sets. For Boston 5 and 14 problems the standard deviations are also shown. For Boston 5 and 14, we statistically analyzed the average and standard deviations with the significance level of 0.05. Numerals in italic show that they are statistically inferior among linear or RBF kernels.

For linear kernels, as theory tells us OSV and dual LS SVR show the same results. The AAAE for the orange juice problem by NOSV is smaller than that of the dual LS SVR (OSV) but the AAAE for the Boston 14 problem by NOSV is statistically inferior. The reason for the Boston 14 problem is that for the dual LS SVR (OSV) the best values of C are the same for the five training data sets but for NOSV the best values ranged from 50 to 10^8. Thus, the median of the best values was not best even for the first five files. For RBF kernels the AAAEs for the Mackey-Glass, water purification, and Boston 5 problems by the primal LS SVR and the sparse LS SVR are worse than by the dual LS SVR but there is not much difference.

Table 3. Comparison of the averages of the absolute approximation errors and the standard deviations of the errors for the linear and RBF kernels

Data	Linear			RBF		
	OSV	NOSV	Dual	Primal	Sparse	Dual
Mackey-G.	0.0804	0.0804	0.0804	0.000508	0.000531	0.000323
Water P.	1.20	1.20	1.20	0.982	0.980	0.962
Orange J.	4.31	4.09	4.31	3.94	4.02	3.99
Boston 5	0.0425	0.0429	0.0425	*0.0290*	*0.0292*	0.0276
	±0.00160	±0.00169	±0.00160	±0.00156	±0.00160	±0.00181
Boston 14	3.41	*3.47*	3.41	2.36	2.38	2.27
	±0.146	±0.148	±0.146	±0.164	±0.158	±0.145

Table 4 lists the number of support vectors for linear and RBF kernels. The numerals in the parentheses show the percentage of the support vectors for the sparse LS SVR against those for the dual LS SVR. For OSV we used all the input variables. Thus, the number of support vectors is the number of input variables. But for NOSV, we selected independent data. For the orange juice problem the support vectors were reduced from 700 to 120. By setting $\eta = 10^{-3}$ and $C = 10^5$ we could still reduce the number to 41 with AAAE of 4.16. Thus even if the number of input variables is larger than that of the training data, we can considerably reduce the number of support vectors by NOSV. If the number of input variables is much smaller than that of the training data, we can reduce support vectors considerably using OSV or NOSV.

For RBF kernels, the number of support vectors for primal solutions is the number of training data at most. By sparse LS SVR the reduction ratio was 42% to 77%.

Table 4. Comparison of support vectors for the linear and RBF kernels

Data	Linear			RBF		
	OSV	NOSV	Dual	Primal	Sparse	Dual
Mackey-G.	4	5 (1)	500	498	384 (77)	500
Water P.	10	10 (4)	241	241	103 (43)	241
Orange J.	700	120 (80)	150	150	63 (42)	150
Boston 5	13±0	13±0.2 (5)	255±12	255±12	134±5 (53)	255±12
Boston 14	13±0	13±0.1 (5)	255±12	255±12	132±5 (52)	255±12

5 Conclusion

In this paper we formulated the primal LS SVR in the empirical feature space and derived the set of linear equations to train the primal LS SVRs. Then we proposed the sparse LS SVR restricting the dimension of the empirical feature space controlling the threshold of the Cholesky factorization. According to the computer experiments, for all the data sets tested, the sparse solutions could realize sparsity while realizing generalization ability comparable with that of primal and dual solutions.

References

1. Burges, C.J.C.: Simplified support vector decision rules. In: Saitta, L. (ed.) Machine Learning, Proceedings of the Thirteenth International Conference (ICML '96), pp. 71–77. Morgan Kaufmann, San Francisco (1996)
2. Keerthi, S.S., Chapelle, O., DeCoste, D.: Building support vector machines with reduced classifier complexity. Journal of Machine Learning Research 7, 1493–1515 (2006)
3. Wu, M., Schölkopf, B., Bakir, G.: A direct method for building sparse kernel learning algorithms. Journal of Machine Learning Research 7, 603–624 (2006)
4. Suykens, J.A.K., Vandewalle, J.: Least squares support vector machine classifiers. Neural Processing Letters 9(3), 293–300 (1999)
5. Suykens, J.A.K., Van Gestel, T., De Brabanter, J., De Moor, B., Vandewalle, J.: Least Squares Support Vector Machines. World Scientific Publishing, Singapore (2002)
6. Cawley, G.C., Talbot, N.L.C.: Improved sparse least-squares support vector machines. Neurocomputing 48, 1025–1031 (2002)
7. Valyon, J., Horvath, G.: A sparse least squares support vector machine classifier. In: Proceedings of International Joint Conference on Neural Networks (IJCNN 2004), Budapest, Hungary, vol. 1, pp. 543–548 (2004)
8. Abe, S.: Sparse least squares support vector training in the reduced empirical feature space. Pattern Analysis and Applications (accepted)
9. Xiong, H., Swamy, M.N.S., Ahmad, M.O.: Optimizing the kernel in the empirical feature space. IEEE Transactions on Neural Networks 16(2), 460–474 (2005)
10. Abe, S.: Support Vector Machines for Pattern Classification. Springer, New York (2005)
11. Kaieda, K., Abe, S.: KPCA-based training of a kernel fuzzy classifier with ellipsoidal regions. International Journal of Approximate Reasoning 37(3), 145–253 (2004)
12. Crowder, R.S.: Predicting the Mackey-Glass time series with cascade-correlation learning. In: Proceedings of 1990 Connectionist Models Summer School, Carnegie Mellon University, pp. 117–123 (1990)
13. Baba, K., Enbutu, I., Yoda, M.: Explicit representation of knowledge acquired from plant historical data using neural network. In: Proceedings of 1990 IJCNN International Joint Conference on Neural Networks, San Diego, vol. 3, pp. 155–160 (1990)
14. Harrison, D., Rubinfeld, D.L.: Hedonic prices and the demand for clean air. Journal of Environmental Economics and Management 5, 81–102 (1978)

Content-Based Image Retrieval by Combining Genetic Algorithm and Support Vector Machine

Kwang-Kyu Seo*

Department of Industrial Information and Systems Engineering, Sangmyung University, San 98-20, Anso-Dong, Chonan, Chungnam 330-720, Korea
Tel.: +81-41-550-5371; Fax: +81-41-550-5185
kwangkyu@smu.ac.kr

Abstract. Content-based image retrieval (CBIR) is an important and widely studied topic since it can have significant impact on multimedia information retrieval. Recently, support vector machine (SVM) has been applied to the problem of CBIR. The SVM-based method has been compared with other methods such as neural network (NN) and logistic regression, and has shown good results. Genetic algorithm (GA) has been increasingly applied in conjunction with other AI techniques. However, few studies have dealt with the combining GA and SVM, though there is a great potential for useful applications in this area. This paper focuses on simultaneously optimizing the parameters and feature subset selection for SVM without degrading the SVM classification accuracy by combining GA for CBIR. In this study, we show that the proposed approach outperforms the image classification problem for CBIR. Compared with NN and pure SVM, the proposed approach significantly improves the classification accuracy and has fewer input features for SVM.

1 Introduction

Content-based image retrieval (CBIR) techniques are becoming increasingly important in multimedia information systems in order to store, manage, and retrieve image data to perform assigned task and make intelligent decisions. CBIR uses an automatic indexing scheme where implicit properties of an image can be included in the query to reduce search time for retrieval from a large database [1].

Features like color, texture, shape, spatial relationship among entities of an image and also their combination are generally being used for the computation of multidimensional feature vector. The features such as color, texture and shape are known as primitive features. Images have always been an essential and effective medium for presenting visual data. With advances in today's computer technologies, it is not surprising that in many applications, much of the data is images. There have been considerable researches done on CBIR using artificial neural networks [1-4].

Recently SVM which was developed by Vapnik [5] is one of the methods that is receiving increasing attention with remarkable results in pattern recognition. SVM classifies data with different class labels by determining a set of support vectors that

* Corresponding author.

J. Marques de Sá et al. (Eds.): ICANN 2007, Part II, LNCS 4669, pp. 537–545, 2007.

are members of the set of training inputs that outline a hyperplane in the feature space. SVM provides a generic mechanism that fits the hyperplane surface to the training data using a kernel function. The user may select a kernel function for the SVM during the training process that selects support vectors along the surface of this function. When using SVM, two problems are confronted such as how to choose the optimal input feature subset for SVM, and how to set the best kernel parameters. These two problems are crucial, because the feature subset choice influences the appropriate kernel parameters and vice versa [6]. Therefore, obtaining the optimal feature subset and SVM parameters must occur simultaneously.

Genetic algorithm (GA) has the potential to generate both the optimal feature subset and SVM parameters at the same time. We aim to optimize the parameters and feature subset simultaneously, without degrading the SVM classification accuracy. The proposed approach performs feature subset selection and parameters setting in an evolutionary way. In the literature, only a few algorithms have been proposed for SVM feature selection [7-9]. Some other GA-based feature selection methods were proposed [10-11]. However, these papers focused on feature selection and did not deal with parameters optimization for the SVM classifier.

This paper focuses on the improvement of the SVM-based model by means of combining GA and SVM in detecting the underlying data pattern for image classification in CBIR using color features based joint HSV (Hue, Saturation and Value) histogram and texture features based on co-occurrence matrix.

2 Research Background

This section presents how to extract image features such as color and texture features [1]. A brief introduction to the SVM and basic GA concepts are also described.

2.1 Image Features

(1) Color

For representing color, we used HSV color model because this model is closely related to human visual perception. Hue is used to distinguish colors (e.g. red, yellow, blue) and to determine the redness or greenness etc. of the light. Saturation is the measure of percentage of white light that is added to a pure color. For example, red is a highly saturated color, whereas pink is less saturated. Value refers to the perceived light intensity. Color quantization is useful for reducing the calculation cost. Furthermore, it provides better performance in image clustering because it can eliminate the detailed color components that can be considered noises. The human visual system is more sensitive to hue than saturation and value so that hue should be quantized finer than saturation and value. In the experiments, we uniformly quantized HSV space into 18 bins for hue (each bin consisting of a range of 20 degree), 3 bins for saturation and 3 bins for value for lower resolution.

In order to represent the local color histogram, we divided image into equal-sized 3×3 rectangular regions and extract HSV joint histogram that has quantized 162 bins for each region. Although these contain local color information, the resulting representation is not compact enough. To obtain compact representation, we extract from each joint histogram the bin that has the maximum peak. Take hue h, saturation s, and value v associated to the bin as representing features in that rectangular region and normalize to be within the same range of [0, 1]. Thus, each image has the $3\times3\times3\,(=27)$ dimensional color vector.

(2) Texture

Texture analysis is an important and useful area of study in computer vision. Most natural images include textures. Scenes containing pictures of wood, grass, etc. can be easily classified based on the texture rather than color or shape. Therefore, it may be useful to extract texture features for image clustering. Like as color feature, we include a texture feature extracted from localized image region.

The co-occurrence matrix is a two-dimensional histogram which estimates the pairwise statistics of gray level. The $(i, j)^{th}$ element of the co-occurrence matrix represents the estimated probability that gray level i co-occurs with gray level j at a specified displacement d and angle θ. By choosing the values of d and θ, a separate co-occurrence matrix is obtained. From each co-occurrence matrix a number of textural features can be extracted. For image clustering, we used entropy, which is mostly used in many applications. Finally, each image has the $3\times3(=9)$ dimensional texture vector.

2.2 Support Vector Machine (SVM)

The goal of SVM is to find optimal hyperplane by minimizing an upper bound of the generalization error through maximizing the distance, margin, between the separating hyperplane and the data. SVM uses the preprocessing strategy in learning by mapping input space, X to a high-dimensional feature space, F. By this mapping, more flexible classifications are obtained. A separating hyperplane is found which maximizes the margin between itself and the nearest training points.

The feature space is very high-dimensional space where linear separation becomes much easier than input space. This is equivalent to applying a fixed non-linear mapping of the data to a feature space, in which a linear function can be used.

A simple description of the SVM algorithm is provided as follows. Consider a pattern classifier, which uses a hyperplane to separate two classes of patterns based on given a training set $S=\{x_i, y_i\}_{i=1}^{n}$ with input vectors $x_i=(x_i^{(1)},\ldots,x_i^{(n)})^T$ and target labels $y_i \in \{-1,+1\}$. Support vector machine (SVM) classifier, according to Vapnik's original formulation, satisfies the following conditions:

$$f(x)=y_i=\langle w\cdot x\rangle+b=\sum_{i=1}^{n} w_i x_i + b \tag{1}$$

where w represents the weight vector and b the bias.

In the non-linear case, we first mapped the data some other Euclidean space F, using a mapping, $x=(x_1,\ldots,x_n)\mapsto \psi(x)=(\psi(x_1),\ldots,\psi(x_n))$. Then instead of the form of dot product, we use kernel function $K(x,y)=\phi(x)\phi(y)$. There are several kernel functions as follows:

– Linear kernel: $K(x_i, x_j) = x_i^T x_j$ (2)

– Polynomial kernel: $K(x_i, x_j) = (\gamma x_i^T x_j + r)^d$, $\gamma > 0$ (3)

– Radial basis function kernel: $K(x_i, x_j) = \exp(-\gamma \|x_i - x_j\|^2)$, $\gamma > 0$ (4)

– Sigmoid kernel: $K(x_i, x_j) = \tanh(\gamma x_i^T x_j + r)$ (5)

Using a dual problem, the quadratic programming problems can be re-written as

$$Q(\alpha) = \sum_{i=1}^{n} \alpha_i - \frac{1}{2} \sum_{i,j=1}^{n} \alpha_i \alpha_j y_i y_j \langle x_i \cdot x_j \rangle \tag{6}$$

subject to

$$0 \le \alpha_i \le C \text{ and } \sum_{i=1}^{n} \alpha_i y_i = 0 \qquad \text{for all } i = 1, 2, \ldots, n \tag{7}$$

with the decision function $f(x) = \text{sgn}(\sum_{i=1}^{n} y_i \alpha_i k(x, x_i) + b)$.

In this paper, we define the image classification problem for CBIR as a nonlinear problem and use the RBF kernel to optimize the hyper plan.

2.3 Genetic Algorithm (GA)

Genetic algorithm (GA) is an artificial intelligence procedure based on the theory of natural selection and evolution. GA uses the idea of survival of the fittest by progressively accepting better solutions to the problems. It is inspired by and named after biological processes of inheritance, mutation, natural selection, and the genetic crossover that occurs when parents mate to produce offspring [12]. GA differs from conventional non-linear optimization techniques in that it searches by maintaining a population of solutions from which better solutions are created rather than making incremental changes to a single solution to the problem. GA simultaneously possesses a large number of candidate solutions to a problem, called a population. The key feature of GA is the manipulation of a population whose individuals are characterized by possessing a chromosome.

Two important issues in GA are the genetic coding used to define the problem and the evaluation function, called the fitness function. Each individual solution in GA is represented by a string called the chromosome. The initial solution population could be generated randomly, which evolves into the next generation by genetic operators such as selection, crossover and mutation. The solutions coded by strings are evaluated by the fitness function. The selection operator allows strings with higher fitness to appear with higher probability in the next generation. Crossover is performed between two selected individuals, called parents, by exchanging parts of their strings, starting from a randomly chosen crossover point. This operator tends to enable to the evolutionary process to move toward promising regions of the search space. Mutation is used to search for further problem space and to avoid local convergence of GA.

GA has been extensively researched and applied to many combinatorial optimization problems. Furthermore GA has been increasingly applied in conjunction with other AI techniques such as NN and CBR. Various problems of neural network design have been optimized using GA. GA has also been used in conjunction with CBR to select relevant input variables and to tune the parameters of CBR. But, few studies have dealt with integration of GA and SVM, though there is a great potential for useful applications in this area.

3 The Combining GA and SVM Approach for CBIR

This study presents approaches for improving the performance of SVM in two aspects: feature subset selection and parameter optimization. GA is used to optimize both the feature subset and parameters of SVM simultaneously for CBIR.

3.1 Parameter Optimization

Feature subset selection is essentially an optimization problem that involves searching the space for possible features to find one that is optimum or near-optimal with respect to a certain performance measures such as accuracy. In a classification problem, the selection of features is important for many reasons: good generalization performance, running time requirements and constraints imposed by the problem itself.

In the literature there are two methods to solve the feature selection problem: The filter method and the wrapper method. The distinction is made depending on whether feature subset selection is done independent of the learning algorithm used to construct the classifier or not. In the filter method, feature selection is performed before applying the classifier to the selected feature subset. The filter method is computationally more efficient than the wrapper method. The wrapper method trains the classifier system with a given feature subset as an input, and it estimates the classification error using a validation set. Although this is a slower procedure, the features selected are usually more optimal for the classifier employed.

Feature subset selection plays an important role in the performance of image classification in CBIR. Furthermore, its importance increases when the number of features is large. This study seeks to improve the SVM based CBIR. The GA-based approach for feature subset selection in the SVM is proposed in this work.

3.2 Feature Subset Selection

One of the important problems in SVM is the selection of the values of parameters that will allow good performance.

Selecting appropriate values for parameters of SVM plays an important role in the performance of SVM. But, it is not known beforehand which values are the best for the problem. Optimizing the parameters of SVM is crucial for the best classification performance.

This paper proposes, GA as the method of optimizing parameters of SVM. In this study, the radial basis function (RBF) is used as the kernel function for CBIR. There are two parameters while using RBF kernels such as C and γ. These two parameters play an important role in the performance of SVMs. In this study, C and γ are encoded as binary strings, and optimized by GA.

3.3 The Proposed Approach

In general, the choice of the feature subset has an influence on the appropriate kernel parameters and vice versa. Therefore the feature subset and parameters of SVM need to be optimized simultaneously for the best classification performance.

The procedure of the proposed approach optimizes both the feature subset and parameters of SVM simultaneously for CBIR. The procedure starts with the randomly selected chromosomes which represent the feature subset and values of parameters of SVM. Each new chromosome is evaluated by sending it to the SVM model. The SVM model uses the feature subset and values of parameters in order to obtain the performance measure. The performance measure is used as the fitness function and is evolved by GA. The chromosomes for the feature subset are encoded as binary strings standing for some subset of the original feature set list. Each bit of the chromosome represents whether the corresponding feature is selected or not. 1 in each bit means the corresponding feature is selected, whereas 0 means it is not selected.

Each of the selected feature subsets and parameters is evaluated using SVM. This process is iterated until the best feature subset and values of parameters are found. The data set is divided into a training set and a validation portion. GA evolves a number of populations. Each population consists of sets of features of a given size and the values of parameters. The fitness of an individual of the population is based on the performance of SVM. SVM is trained on a training set using only the features of the individual and the values of parameters of the individual. The fitness is the average classification accuracy of SVM over a validation set. At each generation new individuals are created and inserted into the population by selecting fit parents which are mutated and recombined. The fitness function is represented as follows:

$$fitness\ \ funtion = W_1 \times SVM\_Accuracy + W_2 \times Nonzeros \tag{8}$$

where W_1 is the weight for SVM classification accuracy, $SVM\_Accuracy$ is SVM classification accuracy; W_2 is the weight for the number of features and $Nonzeros$ is the number of selected features.

During the evolution, the simple crossover operator is used. The mutation operator just flips a specific bit. With the elite survival strategy, we reserve the elite set not only between generations but also in the operation of crossover and mutation so that we can obtain all the benefit of GA operation. The details of the proposed algorithm are explained in Table 1.

Table 1. The proposed algorithm

Step 1 Define the string (or chromosome): features and parameters of SVM are encoded into chromosomes
Step 2 Define population size, probability of crossover and probability of mutation.
Step 3 Generate binary coded initial population randomly.
Step 4 While stopping condition is false, do Step 4–8.
Step 5 Decode *i*th chromosome to obtain the corresponding parameters and feature subsets.
Step 6 Apply the parameters and feature subsets to the SVM model to compute the output.
Step 7 Evaluate fitness
Step 8 Calculate total fitness function of population
Step 9 Reproduction
9.1 Compute the ration of each fitness function over total fitness function
9.2 Calculate cumulative probability
9.3 Generate random number between [0, 1]. If the random variable less than the value gained from 9.1, then select first string. Otherwise, select *i*th string.
Step 10 Generate offspring population by performing crossover and mutation on parent pairs.
Step 11 Stop the iterative step when the stopping condition is reached.

4 Experiments

4.1 Experimental Design

To show the effective classification of the proposed approach, called the combining GA and SVM approach, we checked the classification accuracy. Classification results using color and texture features of real world images will be shown. All experiments were performed on a Pentium IV with 512 Mbytes of main memory and 100 Gbytes of storage. We experimented on 2,000 images where most of them have dimensions of 192×128 pixels. The 2,000 images can be divided into 10 categories with 200 images each such as such as airplane, eagle, horse, lion, polar bear, rose, zebra, tiger, valley and sunset.

The dataset for the combining GA and SVM approach is separated into two parts: the training dataset, and the validation dataset. The ratios are about 0.8 and 0.2. Additionally, to evaluate the effectiveness of the proposed approach, we compare two different models with arbitrarily selected values of parameters and a given all feature subset. The first model uses neural network (NN) and the second model uses pure SVM.

4.2 Experimental Results

In order to evaluate the combining GA and SVM approach, we set the detail parameters for GA as follows: population size 200, crossover rate 0.9, mutation rate 0.1, two-point crossover, roulette wheel selection, and elitism replacement. We set $l = 20$, $m = 20$ and $n = 36$. According to the fitness function of Eq. (8), W_1 and W_2 can influence the experiment result. We defined $W_1 = 0.8$ and $W_2 = 0.2$ for experiments. The termination criteria are that the generation number reaches generation 500 or that the fitness value does not improve during the last 100 generations. The best chromosome is obtained when the termination criteria satisfy.

Table 2 shows both training and validation average success rates that were achieved under the different models.

Table 2. Average image classification accuracy of NN, pure SVM and the combining GA and SVM approach

Classifier	Type of kernel	Training (%)	Validation (%)
NN	–	89.33	86.50
Pure SVM	RBF	92.37	91.15
GA + SVM	RBF	96.75	93.50

As can be seen, the combining GA and SVM approach has consistently given the best performance of the other models as shown 96.75% average success rate on the training dataset and 93.50% on the validation dataset in table 2 and average number of features is 14.

In order to test the superiority of the proposed approach, we perform the McNemar test which is used to examine whether the proposed approach significantly outperforms the other models. This test is a nonparametric test for two related samples using the chi-square distribution. This test may be used with nominal data and is particularly useful with 'before–and-after' measurement of the same subjects. We performed McNemar test to compare the performance for the test data. As a result, the combining GA and SVM approach outperforms NN at the 1% statistical significance level, and pure SVM at the 5% statistical level.

5 Conclusion

In this paper, we presented the GA-based approach to improve the performance of SVM in CBIR. SVM parameters and feature subsets were optimized simultaneously in this study because the selected feature subset has an influence on the appropriate kernel parameters and vice versa.

As far as we know, previous studies have not dealt with the combining GA and SVM approach although there is a great potential for useful applications in CBIR. This paper focuses on the improvement of the image classification accuracy by means of the combining GA and SVM approach for CBIR.

We conducted experiments to evaluate the classification accuracy of the proposed approach with RBF kernel, NN and pure SVM on 2,000 real-world image dataset with 10 image categories. Generally, compared with NN and pure SVM, the proposed approach has good accuracy performance with fewer features.

This study showed experimental results with the RBF kernel of SVM. In future work, we also intend to optimize the kernel function, parameters and feature subset simultaneously. We would also like to expand the proposed approach to apply to instance selection problems.

References

1. Park, S.-S., Seo, K.-K., Jang, D.-S.: Expert system based on artificial neural networks for CBIR. Expert Systems with Applications 29(3), 589–597 (2005)
2. Fournier, J., et al.: Back-propagation Algorithm for Relevance Feedback in Image retrieval. IEEE International Conference in Image Processing (ICIP'01) 1, 686–689 (2001)

3. Koskela, M., Laaksonen, J., Oja, E.: Use of Image Subset Features in Image Retrieval with Self-Organizing Maps. In: Enser, P.G.B., Kompatsiaris, Y., O'Connor, N.E., Smeaton, A.F., Smeulders, A.W.M. (eds.) CIVR 2004. LNCS, vol. 3115, pp. 508–516. Springer, Heidelberg (2004)
4. Pakkanen, J., Iivarinen, J., Oja, E.: The Evolving Tree - a Novel Self-Organizing Network for Data Analysis. Neural Processing Letters 20(3), 199–211 (2004)
5. Vapnik, V.: Statistical learning theory, New York. Springer, Heidelberg (1995)
6. Fröhlich, H., Chapelle, O.: Feature selection for support vector machines by means of genetic algorithms. In: Proceedings of the 15th IEEE international conference on tools with artificial intelligence, pp. 142–148 (2003)
7. Bradley, P.S., Mangasarian, O.L.: Feature selection via concave minimization and support vector machines. In: Proceedings of the 13th international conference on machine learning, pp. 82–90 (1998)
8. Weston, J., et al.: Feature selection for SVM. Advances in neural information processing systems 13, 668–674 (2001)
9. Mao, K.Z.: Feature subset selection for support vector machines through discriminative function pruning analysis. IEEE Transactions on Systems, Man, and Cybernetics 34(1), 60–67 (2004)
10. Raymer, M.L., et al.: Dimensionality reduction using genetic algorithms. IEEE Transactions on Evolutionary Computation 4(2), 164–171 (2000)
11. Yang, J., Honavar, V.: Feature subset selection using a genetic algorithm. IEEE Intelligent Systems 13(2), 44–49 (1998)
12. Goldberg, D.E.: Genetic algorithms in search, optimization and machine learning. Addison-Wesley, New York (1989)

Global and Local Preserving Feature Extraction for Image Categorization

Rongfang Bie*, Xin Jin, Chuan Xu, Chuanliang Chen, Anbang Xu, and Xian Shen

College of Information Science and Technology,
Beijing Normal University, Beijing 100875, P.R. China
rfbie@bnu.edu.cn,
xinjin4@yahoo.com,
abram_win@hotmail.com,
byronccl@126.com,
anbangxu@mail.bnu.edu.cn,
shenxian8307@yahoo.com.cn

Abstract. In this paper, we describe a feature extraction method: Global and Local Preserving Projection (GLPP). GLPP is based on PCA and the recently proposed Locality Preserving Projection (LPP) method. LPP can preserve local information, while GLPP can preserve both global and local information. In this paper we investigate the potential of using GLPP for image categorization. More specifically, we experiment on palmprint images. Palmprint image has been attracting more and more attentions in the image categorization/recognition area in recent years. Experiment is based on benchmark dataset PolyU, using Error Rate as performance measure. Comparison with LPP and traditional algorithms show that GLPP is promising.

1 Introduction

Biometrics has been attracting more and more attentions in recent years. Two possible biometric features, hand geometrical features and palmprint image features, can be extracted from hand. Hand geometrical features such as finger width, length, and the thickness are adopted to represent extracted features, but these features frequently vary due to the wearing of rings in fingers, besides, the width of some fingers may vary during pregnancy or illness. Palmprint image features have several advantages over such physical characteristics [6]: (1) low-resolution imaging; (2) low intrusiveness; (3) stable line feature and (4) high user acceptance.

Palmprint image categorization/recognition approaches can be classified two main classes, namely local feature based methods [2, 5, 7] and feature extraction based methods [1, 3, 4, 6, 8]. Local feature-based methods is mainly centralized on points and lines feature, texture analysis of the original palmprint image while feature extraction based method consider second-order statistics information of the original palmprint image. Not encountering the problem of extracting structure features is the advantage of the latter methods. PCA, ICA, KPCA and KICA are successful subspace-based methods commonly used in palmprint recognition and searching systems.

* Corresponding author.

J. Marques de Sá et al. (Eds.): ICANN 2007, Part II, LNCS 4669, pp. 546–553, 2007.

In this paper we describe a feature extraction method: Global and Local Preserving Projection (GLPP), it is based on PCA and Locality Preserving Projection (LPP) [23] which can preserve local information. GLPP can preserve both global and local information of the data.

The remainder of this paper is organized as follows. Section 2 describes GLPP algorithm and several other methods. Section 3 describes the image recognition procedure. Section 4 presents the experiment results. Conclusions are provided in Section 5.

2 Feature Extraction

In image processing applications, the original data within high-dimension space can not be directly processed because of the course of dimension and need to find a low-dimension representation of data. We perform feature extraction for dimension deduction.

2.1 Global Locality Preserving Projection (GLPP)

We first describe Locality Preserving Projections (LPP). LPP, a linear approximation of the nonlinear Laplacian Eigenmap [9, 10, 23], is a recently proposed method for dimensionality reduction [11]. LPP can make low-dimension representation of the original high-dimension data space with locality information preserving.

The generic problem of linear dimensionality reduction problem is as follows:

1. Given the original data $X = \{x_1, x_2, \cdots, x_m\}$ in high-dimension space R^n.
2. Find a mapping that transforms the original data points into a new set of data points $Y = \{y_1, y_2, \cdots, y_m\}$ in a low-dimension space R^l $(l < n)$.

Because the target transformation is limited to linear transformation, it can be represented as a projection matrix A. These new $y_i = A{\cdot}x_i$ represents the original x_i.

For a high-dimension data X, the distance between x_i indicates the similarity relationship between data. The low-dimension representation is desired to preserve this information. LPP is the algorithm which is equipped with this property [12, 24].

The **LPP** algorithm procedure for data matrix X is as follows [11]:

Step 1. *Constructing the adjacency graph on X*: Let G denotes a graph with m nodes. If x_i and x_j are close, we put an edge between node i and j. Construct the edges by choosing the k nearest neighbors: Nodes i and j are connected by an edge if i is among k nearest neighbors of i or j is among k nearest neighbors of i. The edges can also be constructed by ε-neighborhoods [11].

Step 2. *Choosing the weights on X*: Let W denotes an $m{\times}m$ sparse symmetric matrix and W_{ij} is the weight of the edge between vertices i and j ($W_{ij} = 0$, if there is no edge). Cosine similarity is used to set the weight (other similarity function may also be used).

Step 3. *Eigenmaps on X*: Eigenvectors and eigenvalues are calculated for the generalized eigen-decomposition problem:

$$\mathrm{XLX^T a = \lambda XDX^T a} \tag{1}$$

Let the column vectors $a_1, \cdots, a_l$ be the solutions of Equation (1), ordered according to their eigenvalues, $\lambda_1 < \cdots < \lambda_l$:

$$x_i \rightarrow y_i = A^T x_i,\ A = (a_1, a_2, \cdots, a_l) \tag{2}$$

where y_i is a l-dimensional vector, and A is a $n \times l$ matrix.

LPP can preserve locality information. However, the problem is that when X is in the high dimensional manifold, which is the case for image processing, LPP will fail by the curse of high-dimension. Global Local Preserving Projection (GLPP) can deal with the problem.

GLPP is based on PCA and LPP, it proceeds as follows:

Step 1. *Calculate T as the global information preserving transforming of X.* This step is similar to PCA, by setting a parameter g, we can deduce the original high dimensional space to much lower dimensional space while keeping most global information:

(1) Calculate the data covariance matrix by

$$\mathrm{R_x}(0) = E\{\mathrm{x}(t)\mathrm{x}^{\mathrm{T}}(t)\}. \tag{3}$$

(2) Calculate the SVD of $\mathrm{R_x}(0)$ by

$$\mathrm{R_x}\,(0) = \mathrm{UDV^T} \tag{4}$$

where V is the eigenvector matrix and D is the diagonal matrix whose diagonal elements correspond to the eigenvalues of $\mathrm{R_x}\,(0)$.

(3) Sort the eigenvalues in descending order.

(4) Choose the global preserving parameter g. The value of g varies form 0 to 1. In this paper, we set g to be 0.999 to preserve most global information.

(5) Keep the top k the eigenvalues. The value of k is chosen that

$$\frac{Sum(eigenvalues[i],\ i = 1,...,k)}{total\_eigenvalues} > g \tag{5}$$

where,

$$total\_eigenvalues = Sum(eigenvalues[i],\ i = 1,...,n) \tag{6}$$

where n is X is dimension.

(6) Keep the top k column vectors.

(7) Get T, i.e., from n-dimensional data X to k-dimensional subspace T:

$$\mathrm{T = V^T X}. \tag{7}$$

Step 2. *Constructing the adjacency graph on T.*
Step 3. *Choosing the weights on T.*
Step 4. *Eigenmaps.*

2.2 Algorithms for Comparison

2.2.1 Principal Component Analysis

Principal Component Analysis (PCA) is a classical multivariate data analysis method that is useful in linear feature extraction [18, 19, 25, 26]. PCA is obtained first by subtracting the mean $\overline{x}$ of the n-dimensional data set. The covariance matrix is calculated and its eigenvectors and eigenvalues are found. The eigenvectors corresponding to the m largest eigenvalues are retained, and the input vectors x^n are subsequently projected onto the eigenvectors to give components of the transformed vectors z^n are in the m-dimensional space.

2.2.2 Independent Component Analysis

Independent Component Analysis (ICA) first performs the dimensionality reduction by data sphering (whitening) which project the data onto its subspace as well as normalizing its variance [20, 21, 28]. Data sphering transformation $\mathbf{Q}$ is given by $\mathbf{Q} = \mathbf{D}_s^{-1/2}\mathbf{U}_s^T$, The whitened vector $z \in R^n$ is obtained via $\mathbf{z} = \mathbf{Qx}$. The orthogonal factor V in ICA can be found by minimizing the mutual information in z. Then the ICA transformation W' is given by $y = W'x$, where $W'=VQ$.

2.2.3 Kernel Principal Component Analysis

Kernel Principal Component Analysis (KPCA) generalizes the PCA approach to nonlinear transformations using the kernel trick [33], working in the feature space of a positive semi-definite kernel written implicitly as a dot product in that space. KPCA is to transform the input data into a higher-dimensional feature space. The feature space is constructed such that a nonlinear operation can be applied in the input space by applying a linear operation in the feature space [27].

2.2.4 Kernel Independent Component Analysis

Kernel Independent Component Analysis (KICA) is the kernel counterpart of ICA [13, 14, 15]. Let the inner product be implicitly defined by the kernel function k in F with associated transformation Φ. KICA will extend nonlinearly the centering and whitening of the data.

(1) *Centering* in F: We shift the data $\Phi(x_i)$ $(i=1,\ldots,k)$ with its mean $E(\Phi(x))$, to obtain data

$$\Phi'(x_i) = \Phi(x_i) - E(\Phi(x)) \tag{8}$$

with a mean of 0.

(2) *Whitening* in F: $\hat{\Phi}(x_i) = Q\Phi'(x_i)$ $(i-1,\ldots,k)$ is a unit matrix.

(3) *Transformation* of test vectors:

$$z^* = \hat{\mathbf{W}}\hat{\mathbf{Q}}\Phi(z) = \hat{\mathbf{W}}\mathbf{A}_k(X, z)\,. \tag{9}$$

$\hat{\mathbf{W}}$ denotes the orthogonal transformation matrix, $\hat{\mathbf{Q}}$ is the matrix obtained from kernel centering and whitening.

In this paper, we adopt cosine kernel for both KPCA and KICA.

3 Image Categorization

We use Nearest Neighbors (NN) to perform the image categorization/recognition process. NN is a non-parametric inductive learning paradigm that stores training instances in a memory structure on which predictions of new instances are based [16].

The similarity between the new instance and an example in memory is computed using a distance metric. In this study, we use Euclidian distance.

For palmprint image recognition, NN treats all palmprints as points in the *m*-dimensional space and given an unseen palmprint *p*, the algorithm classifies it by the nearest training palmprint.

4 Experiment Results

The image dataset we use is the benchmark PolyU Palmprint Database, available on [11]. There are 600 palmprint images with 100 classes (each class has 6 images). These images are preprocessed with wavelet [29, 30, 31, 32]. We use the Leave-One-Out method for performance evaluation. For feature extraction, the reduced dimensions vary from 1 to 35. We use Error Rate to estimate recognition performance.

Error Rate: Error Rate is defined by the ratio of the number of correct predictions and the number of all test samples. Error Rate varies from 0 to 1. A perfect classifier should achieve an Error Rate of 0 for each class. The lower the Error Rate, the better the performance.

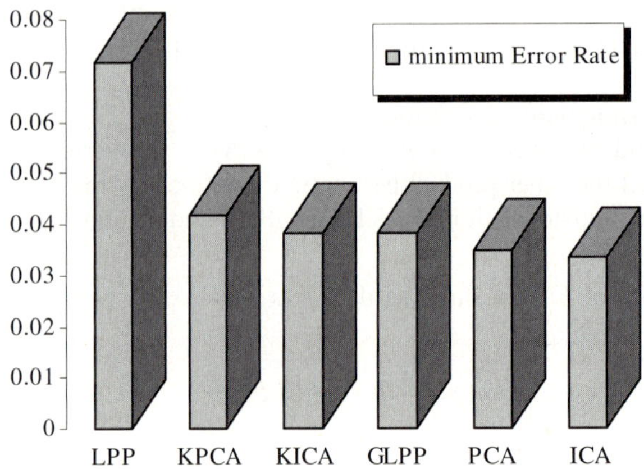

Fig. 1. Minimum Error Rate of GLPP and traditional feature extraction algorithms. Y-axis denotes the Error Rate. X-axis denotes the feature extraction algorithms; their location is arrayed by sorting their performance from the worst to the best (from left to right).

Fig. 1 shows the minimum Error Rate of GLPP, LPP and four traditional feature extraction algorithms: PCA, ICA, KPCA and KICA. The results show that GLPP is much better than LPP, GLPP achieves a minimum Error Rate of 0.038 and LPP achieves 0.072. The results also show that the performance of GLPP is comparable with traditional methods. ICA achieves the best performance with a minimum Error Rate of 0.033, which is just slightly better than the performance of GLPP.

Fig. 2 shows the Error Rate curves under different number of reduced dimensions for GLPP, LPP and the traditional algorithms (PCA, ICA, KPCA and KICA). The results in the figure show that GLPP is uniformly better than LPP. This figure also shows that both GLPP and LPP are much better than the traditional methods when the number of reduced dimensions is less than 15. For reduced dimensions over 25, the performance of GLPP is better than KPCA and KICA, comparable with PCA and ICA. GLPP achieves its best performance (that is, its lowest Error Rate of 0.038) at 24 reduced dimensions. GLPP can still achieve relatively high performance at about just 3 dimensions. This is greatly better than the traditional methods.

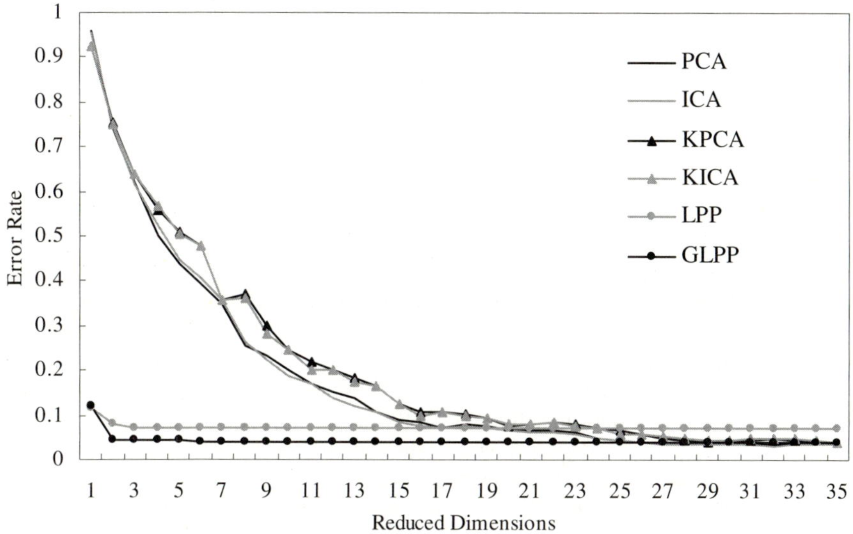

Fig. 2. Error Rate curves of feature extraction methods with different reduced dimensions. The lower is the error rate, the better is the method.

5 Conclusion

In this study, we describe a feature extraction method--Global and Local Preserving Projection (GLPP), which is based on PCA and LPP. While Locality Preserving Projection (LPP) can preserve local information, GLPP can preserve both global and local information. To demonstrate the efficiency of GLPP, we apply it to palmprint image categorization/recognition. In experiment, we use the benchmark palmprint image dataset PolyU, and adopt Error Rate as the performance measure. Comparison is done with LPP and traditional methods: PCA, ICA, KPCA and KICA. The results show that GLPP is better than LPP, KPCA and KICA, and is comparable with PCA and KCA. GLPP can achieve relatively high performance with just very few reduced dimensions, which is a great advantage over traditional methods.

Acknowledgments

This work was supported by the National Science Foundation of China under the Grant No. 60273015 and No. 10001006.

References

1. Feng, G.Y., Hu, D.W., Li, M., Zhou, Z.T.: Palmprint Recognition Based on Unsupervised Subspace Analysis. In: Wang, L., Chen, K., Ong, Y.S. (eds.) ICNC 2005. LNCS, vol. 3610, pp. 675–678. Springer, Heidelberg (2005)
2. Connie, T., Jin, A.T.B, B., K., Ong, M.G., Ling, D.N.C.: An Automated Palmprint Recognition System. Image and Vision Computing 23, 501–515 (2005)
3. Wu, X.Q., Zhang, D., Wang, K.Q.: Fisherpalms Based Palmprint Recognition. Pattern Recognition Letters 24, 2829–2838 (2003)
4. Lu, G.M., Zhang, D., Wang, K.Q.: Palmprint Recognition Using Eigenpalms. Pattern Recognition Letters 24, 1463–1467 (2003)
5. Kumara, A., Zhang, D.: Personal Authentication Using Multiple Palmprint Representation. Pattern Recognition. 38, 1695–1704 (2005)
6. Kong, W.K., Zhang, D., LiW, X.: Palmprint Feature Extraction using 2-D Gabor Filters. Pattern Recognition 36, 2339–2347 (2003)
7. Yang, J., Zhang, D., Yang, J.-Y.: Is ICA Significantly Better than PCA for Face Recognition? ICCV, 198–203 (2005)
8. Wu, X., Wang, K., Zhang, D.: Palmprint Authentication Based on Orientation Code Matching. In: Kanade, T., Jain, A., Ratha, N.K. (eds.) AVBPA 2005. LNCS, vol. 3546, pp. 555–562. Springer, Heidelberg (2005)
9. Bazaraa, M.S., Sherali, H.D., Shetty, C.M.: Nonlinear programming: theory and algorithms. John Wiley &Sons Inc., New York (1993)
10. Belkin, M., Niyogi, P.: Laplacian eigenmaps and spectral techniques for embedding and clustering. In: Dietterich, T.G., Becker, S., Ghahramani, Z. (eds.) Advances in Neural Information Processing Systems, vol. 14, MIT Press, Cambridge, MA (2002)
11. He, X., Niyogi, P.: Locality preserving projection. In: Advances in Neural Information Processing Systems 16 (NIPS 2003), Vancouver, Canada, MIT Press, Cambridge (2003)
12. He, X., Cai, D., Min, W.: Statistical and computational analysis of locality preserving projection. In: International Conference on Machine Learning (ICML), Bonn, Germany (2005)
13. Kocsor, A., Csirik, J.: Fast Independent Component Analysis in Kernel Feature Spaces. In: Pacholski, L., Ružička, P. (eds.) SOFSEM 2001. LNCS, vol. 2234, pp. 271–281. Springer, Heidelberg (2001)
14. Liu, Q.S., Cheng, J., et al.: Modeling Face Appearance with Kernel Independent Component Analysis. Sixth IEEE International Conference on Automatic Face and Gesture Recognition (FGR2004), May 17-19, Seoul, Korea, pp. 761-766 (2004)
15. Francis, R.B., Jordan, M.I.: Kernel independent component analysis. Journal of Machine Learning Research 3, 1–48 (2002)
16. Aha, D., Kibler, D.: Instance-based Learning Algorithms. Machine Learning 6, 37–66 (1991)
17. The PolyU Palmprint Database. http://www.comp.polyu.edu.hk/ biometrics/
18. Zeng, X.-Y., Chen, Y.-W., Nakao, Z., Lu, H.: A New Texture Feature based on PCA Maps and Its Application to Image Retrieval. IEICE Trans. Inf. and Syst. E86-D(5), 929–936 (2003)
19. Diamantaras, K.I., Kung, S.Y.: Principal Component Neural Networks: Theory and Applications. John Wiley & Sons, INC, Chichester (1996)
20. Amari, S.: Natural Gradient for Over- and Under-complete Bases in ICA. Neural Computation 11(8), 1875–1883 (1999)

21. Cardoso, J.F., Laheld, B.H.: Equivariant Adaptive Source Separation. IEEE Trans. Signal Processing 44(12), 3017–3030 (1996)
22. He, X., Niyogi, P.: Locality preserving projections. Technical Report TR-2002-09, University of Chicago Computer Science (October 2002)
23. Hu, Y.Q.: Locality preserving projection for post and expression classification (2006), www.ntu.edu.sg/home/aswduch/Teaching/Assign1-vis/HuYiqun.pdf
24. Ali, G.N., Chiang, P.-J., Mikkilineni, A.K., Chiu, G.T.-C., Allebach, J.P., Delp, E.J.: Application of principal components analysis and Gaussian mixture models to printer identification. In: Proceedings of the IS&T's NIP20: International Conference on Digital Printing Technologies, pp. 301–305 (2004)
25. Addison D., Wermter S., Arevian G.: A Comparison of Feature Extraction and Selection Techniques. In: Proceedings of the International Conference on Artificial Neural Networks, Istanbul, Turkey, pp. 212-215 (June 2003)
26. Jenkins, O.C.: Relative localization from pairwise distance relationships using kernel pca. Technical Report CRES-03-010, Center for Robotics and Embedded Systems, University of Southern California (April 2003)
27. Choi, S., Cichocki, A., Amari, S.: Flexible independent component analysis. Journal of VLSI Signal processing 26, 25–38 (2000)
28. Mallat, S.: A theory for multiresolution signal decomposition: the wavelet representation. IEEE Pattern Anal. and Machine Intell. 11(7), 674–693 (1989)
29. Daubechies, I.: Ten lectures on wavelets, CBMS-NSF conference series in applied mathematics. SIAM Ed. (1992)
30. Pang, Y.-H., Jin, A.T.B., Ling, D.N.C.: Palmprint Authentication System Using Wavelet based Pseudo Zernike Moments Features. International Journal of The Computer, the Internet and Management 13(2), 13–26 (2005)
31. Gan, J., Liang, Y.: Applications of Wavelet Packets Decomposition in Iris Recognition. ICB 2006, 443–449 (2006)
32. Scholkopf, B., Smola, A., Muller, K.-R.: Nonlinear component analysis as a kernel eigenvalue problem. Neural Computation 10, 1299–1319 (1998)

Iris Recognition for Biometric Personal Identification Using Neural Networks

Rahib H. Abiyev and Koray Altunkaya

Near East University, Department of Computer Engineering, Lefkosa, North Cyprus
{rahib,kaltunkaya}@neu.edu.tr

Abstract. This paper presents iris recognition for personal identification using neural networks. Iris recognition system consists of localization of the iris region and generation of data set of iris images and then iris pattern recognition. One of the problems in iris recognition is fast and accurate localization of the iris image. In this paper, fast algorithm is used for the localization of the inner and outer boundaries of the iris region. Located iris is extracted from an eye image, and, after normalization and enhancement it is represented by a data set. Using this data set a neural network is applied for the classification of iris patterns. Results of simulations illustrate the effectiveness of the neural system in personal identification.

1 Introduction

The biometrics technology that uses various physiological characteristics of human plays important role in identification of personals. These physiological characteristics are face, facial thermo grams, fingerprint, iris, retina, hand geometry etc. [1]. Iris recognition is one of the most reliable biometrics that uses iris characteristics of human eyes and plays important role in accurate identification of each individual. Iris region is the part between the pupil and the white sclera. This field sometimes is called iris texture. The iris texture provides many minute characteristics such as freckles, coronas, stripes, furrows, crypts, etc [2-6]. These visible characteristics are unique for each subject. Such unique feature in the anatomical structure of the iris facilitates the differentiation among individuals. The human iris is not changeable and is stable. From one year of age until death, the patterns of the iris are relatively constant over a person's lifetime [1,3]. Because of uniqueness and stability, iris recognition is a reliable human identification.

Iris recognition consists of the iris capturing, pre-processing and recognition of the iris region in a digital eye image. Iris image pre-processing includes iris localization, normalization, and enhancement. Each of these steps uses different algorithms. In iris localization step, the determination of the inner and outer circles of the iris and the determination of the upper and lower bound of the eyelids are performed. The inner circle is located between the iris and pupil boundary, the outer circle is located between the sclera and iris boundary. A variety of techniques have been developed for iris localization. In [3-6], the system with circular edge detector, in [7] a gradient based Hough transform are used for the localizing of the iris. Also circular Hough transform [8], random Hough transform are applied to find the iris circles and

J. Marques de Sá et al. (Eds.): ICANN 2007, Part II, LNCS 4669, pp. 554–563, 2007.

complete the iris localization. In [10] Canny operator is used to locate the pupil boundary. These methods need a long time to locate iris. In this paper a fast iris localization algorithm is proposed.

Various algorithms have been applied for feature extraction and pattern matching processes. These methods use local and global features of the iris. Using phase based approach [3-6], wavelet transform zero crossing approach [8,15], Gabor filtering [10], texture analysis based methods [8,10,11,12] the solving of the iris recognition problem is considered. In [13,14] independent component analysis is proposed for iris recognition.

Daugman [3-6] used multiscale quadrature wavelets to extract texture phase structure information of the iris to generate a 2,048-bit iris code and compared the difference between a pair of iris representations by computing their Hamming distance. Boles and Boashash [9] calculated a zero-crossing representation of 1D wavelet transform at various resolution levels of a concentric circle on an iris image to characterize the texture of the iris. Iris matching was based on two dissimilarity functions. Sanchez-Avila and Sanchez-Reillo [15] further developed the method of Boles and Boashash by using different distance measures (such as Euclidean distance and Hamming distance) for matching. Wildes et al. [8] represented the iris texture with a Laplacian pyramid constructed with four different resolution levels and used the normalized correlation to determine whether the input image and the model image are from the same class.

Today with the development of Artificial Intelligence (AI) algorithms, iris recognition systems may gain speed, hardware simplicity, accuracy and learning ability. In this paper a fast iris segmentation algorithm and also an iris recognition system based on neural networks are proposed.

This paper is organized as follows. In section 2 the iris preprocessing steps that include iris localization, normalization and enhancement are described. In section 3 the neural network which is used for iris pattern recognition is described. Section 4 presents experimental results. Section 4 includes the conclusion of the paper.

2 Iris Recognition

2.1 Structure of Iris Recognition System

Fig.1 shows the architecture of the iris recognition system. The iris recognition system includes two operation modes: training mode and on-line mode. At fist stage using iris images the training of recognition system is carried out. The image recognition system includes iris image acquisition and iris recognition. The iris image acquisition includes the lighting system, the positioning system, and the physical capture system [8]. The iris recognition includes pre-processing and neural networks blocks. During iris acquisition, the iris image in the input sequence must be clear and sharp. Clarity of the iris's minute characteristics and sharpness of the boundary between the pupil and the iris, and the boundary between the iris and the sclera affects the quality of iris image. A high quality image must be selected for iris recognition. In iris preprocessing, the iris is detected and extracted from an eye image and normalized. Normalized image after enhancement is represented by the array that describes greyscale values of the iris image. This array becomes the training data set for the neural network. Neural network is trained for all iris images. After training, in online mode using input digital images of iris, neural network performs classification and recognizes the patterns that belong to a certain person's iris.

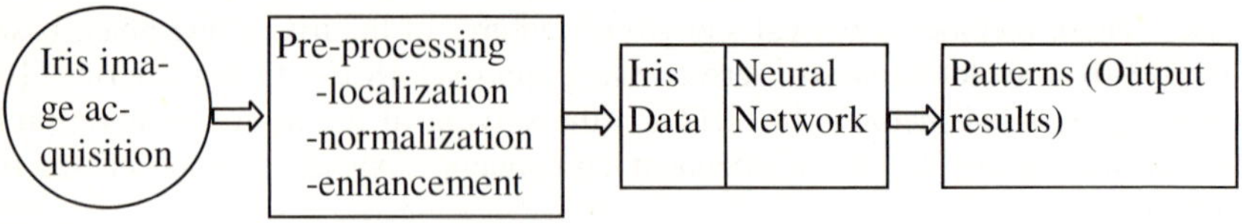

Fig. 1. Steps of iris recognition

2.2 Iris Localization

An eye image contains not only iris region but also some unuseful parts, such as the pupil, eyelids, sclera, and so on. For this reason, at first step, the segmentation will be done to localize and extract the iris region from the eye image. Iris localization is to detect the iris area between pupil and sclera. So we need to detect the upper and lower boundaries of the iris and determine its inner and outer circles. A number of algorithms has been developed for iris localization. One of them is based on the Hough transform. An iris segmentation algorithm based on the circular Hough transform is applied in [7,8]. At first, the canny edge detection algorithm is applied. The eye image is represented using edges by applying two thresholds to bring out the transition from pupil to iris and from iris to sclera. Then circular Hough transform is applied to detect the inner and outer boundaries of the iris (Fig. 2). The circular Hough transform is employed to deduce the radius and centre coordinates of the pupil and iris regions. In this operation, the radius intervals are defined for inner and outer circles. Starting from the upper left corner of iris the circular Hough transform is applied. This algorithm is used for each inner and outer circle separately. The votes are calculated in the Hough space for the parameters of circles passing through each edge point. Here some circle parameters may be found. The parameters that have maximum value are corresponded to the centre coordinates x_c and y_c. After determining centre coordinates,

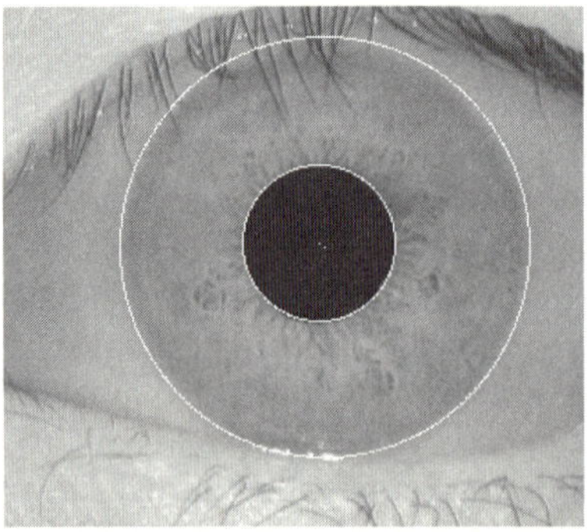

Fig. 2. A localised iris image

the radius r of the inner circle is determined. The same procedure is applied for the outer circle to determine its centre coordinates and radius. Using determined inner and outer radiuses the iris region is detected. The application of the Hough transform needs long time to locate the boundaries of the iris.

2.3 Iris Localization Using Rectangular Areas

In this paper, a fast algorithm for detecting the boundaries between pupil and iris and also sclera and iris has been proposed. To find boundary between pupil and iris we must detect the location (centre coordinates and radius) of pupil. In more case the pupil is not located at the middle of an eye image. The pupil is a dark circular area in an eye image. Besides the pupil eyelids and eyelashes are also characterized by black colour. So it causes difficulties to find right place of pupil by using point by point comparison on the base of threshold technique. In this paper, we are looking for the black rectangular region in an iris image (Fig. 3). Searching starts from the vertical middle point of the iris image and continues to the right side of the image. A threshold value is used to detect the black rectangular area. Starting from the middle vertical point of iris image, the greyscale value of each point is compared with the threshold value. As it is proven by many experiments the greyscale values within the pupil are very small. So a threshold value can be easily chosen. If greyscale values in each point of the iris image are less than threshold value, then the rectangular area will be found. If this condition is not satisfactory for the selected position, then the searching is continued from the next position. This process starts from the left side of the iris, and it continues to the end of the right side of the iris. In case, if the black rectangular area is not detected, the new position in the upper side of the vertical middle point of the image is selected and the search for the black rectangular area is resumed. If the black rectangular area is not found in the upper side of the eye image, then the search is continued in the down side of image. In Fig. 3(a), the searching points are shown by the lines. In Fig. 3(a,b), the black rectangular area is shown in white colour. Choosing

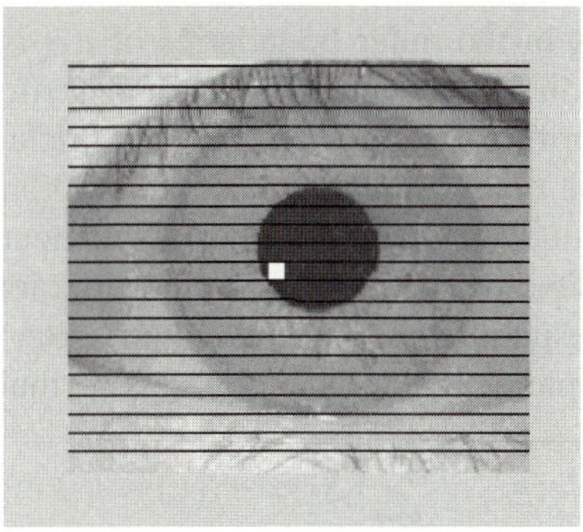

(a)

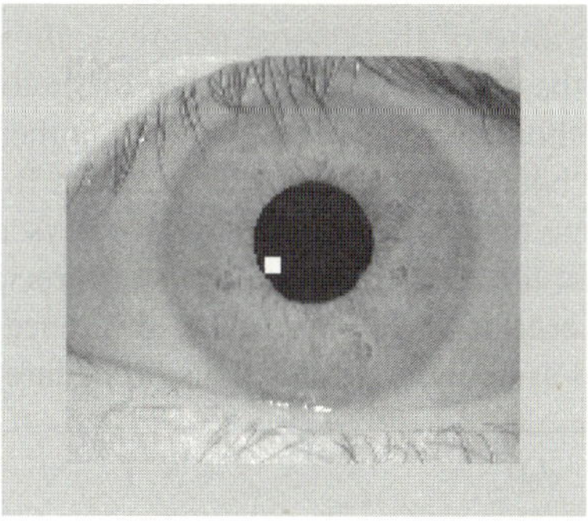

(b)

Fig. 3. Detecting the rectangular area: a) The lines that were drawn to detect rectangular areas b) The result of detecting of rectangular area

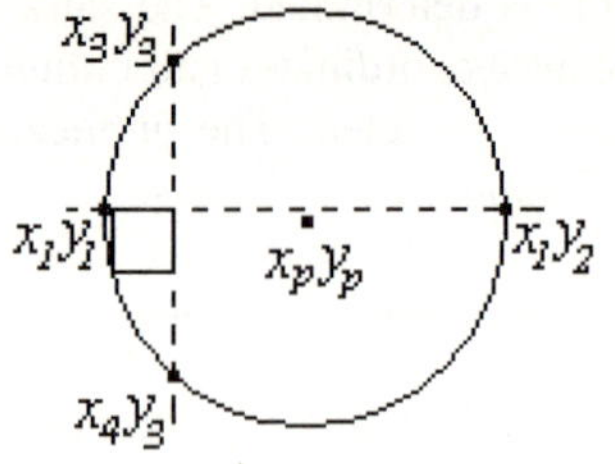

Fig. 4. Finding the centre of the pupil

the size of the black rectangular area is important and it affects the accurate determination of the pupil's position. If we choose small size, then this area can be found in the eyelash region. In this paper a (10, 10) rectangular area is taken to accurately detect the location of the pupil. After finding the black rectangular area, we start to detect the boundary of the pupil and iris. At first step, the points located in the boundary of pupil and iris, in horizontal direction, then the points in the vertical direction are detected (Fig. 4). The border of the pupil and the iris has much a larger greyscale change value. Using a threshold value on the iris image, the algorithm detects the coordinates of the horizontal boundary points of (x_1,y_1) and (x_1,y_2), as shown in Fig. 4. The same procedure is applied to find the coordinates of the vertical boundary points (x_3,y_3) and (x_4,y_3). After finding the horizontal and vertical boundary points between the pupil and the iris, the following formula is used to find the centre coordinates (x_p,y_p) of the pupil.

The same procedure is applied for two different rectangular areas. In case of small differences between coordinates, the same procedure is applied for four and more different rectangular areas in order to detect accurate position of pupil's centre. After finding centre points, the radius of the pupil is determined.

$$r_p = \sqrt{(x_c - x_1)^2 + (y_c - y_1)^2}, \text{ or } r_p = \sqrt{(x_c - x_3)^2 + (y_c - y_3)^2} \tag{1}$$

Because of the change of greyscale values in the outer boundaries of iris is very soft, the current edge detection methods are difficult to implement for detection the outer boundaries. In this paper, another algorithm is applied in order to detect the outer boundaries of the iris. We start from the outer boundaries of the pupil and determine the difference of sum of greyscale values between the first ten elements and second ten elements in horizontal direction. This process is continued in the left and right sectors of the iris. The difference corresponding to the maximum value is selected as boundary point. This procedure is implemented by the following formula.

$$DL_i = \sum_{i=10}^{y_p-(r_p+10)} (S_{i+1} - S_i), \qquad DR_j = \sum_{j=y_p+(r_p+10)}^{right-10} (S_{j+1} - S_j) \tag{2}$$

Here *DL* and *DR* are the differences determined in the left and right sectors of the iris, correspondingly. x_p and y_p are centre coordinates of the pupil, r_p is radius of the pupil, *right* is the right most y coordinate of the iris image. In each point, S is calculated as

$$S_j = \sum_{k=j}^{k+10} I(i,k) \tag{3}$$

where $i=x_p$, for the left sector of iris $j=10,\ldots,y_p-(r_p+10)$, and for the right sector of iris $j=y_p+(r_p+10)$. $I_x(i,k)$ are greyscale values.

The centre of the iris is determined using

$$y_s = (L+R)/2, \quad r_s = (R-L)/2 \tag{4}$$

$L=i$, where i correspond to the value $max(|DL_i|)$, $R=j$, where j correspond to the value $max(|DR_j|)$.

2.4 Iris Normalization

The irises captured from the different people have different sizes. The size of the irises from the same eye may change due to illumination variations, distance from the camera, or other factors. At the same time, the iris and the pupil are non concentric. These factors may affect the result of iris matching. In order to avoid these factors and achieve more accurate recognition, the normalization of iris images is implemented. In normalization, the iris circular region is transformed to a rectangular region with a fixed size. With the boundaries detected, the iris region is normalized from Cartesian coordinates to polar representation. This operation is done using the following operation (Fig. 5).

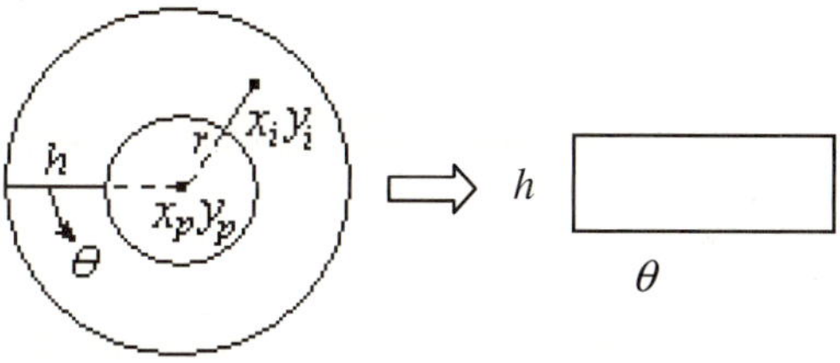

Fig. 5. Normalization of iris

$$\begin{aligned} &\theta \in [0, 2\pi] \quad r \in [R_p, R_L(\theta)] \\ &x_i = x_p + r \times \cos(\theta) \\ &y_i = y_p + r \times \sin(\theta) \end{aligned} \tag{5}$$

Here (x_i, y_i) is the point located between the coordinates of the papillary and limbic boundaries in the direction θ. (x_p, y_p) is the centre coordinate of the pupil, R_p is the radius of the pupil, and $R_L(\theta)$ is the distance between centre of the pupil and the point of limbic boundary.

In the localization step, the eyelid detection is performed. The effect of eyelids is erased from the iris image using the linear Hough transform. After normalization (Fig. 6(a)), the effect of eyelashes is removed from the iris image (Fig. 6(b)). Analysis reveals that eyelashes are quite dark when compared with the rest of the eye image. For isolating eyelashes, a thresholding technique was used. To improve the contrast and brightness of image and obtain a well distributed texture image, an enhancement is applied. Received normalized image using averaging is resized. The mean of each 16x16 small block constitutes a coarse estimate of the background illumination. During enhancement, background illumination (Fig. 6(c)) is subtracted from the normalized image to compensate for a variety of lighting conditions. Then the lighting corrected image (Fig. 6(d)) is enhanced by histogram equalization. Fig. 6(e) demonstrates the preprocessing results of iris image. The texture characteristics of iris image are shown more clearly. Such preprocessing compensates for the nonuniform illumination and improves the contrast of the image.

Normalized iris provides important texture information. This spatial pattern of the iris is characterized by the frequency and orientation information that contains freckles, coronas, strips, furrows, crypts, and so on.

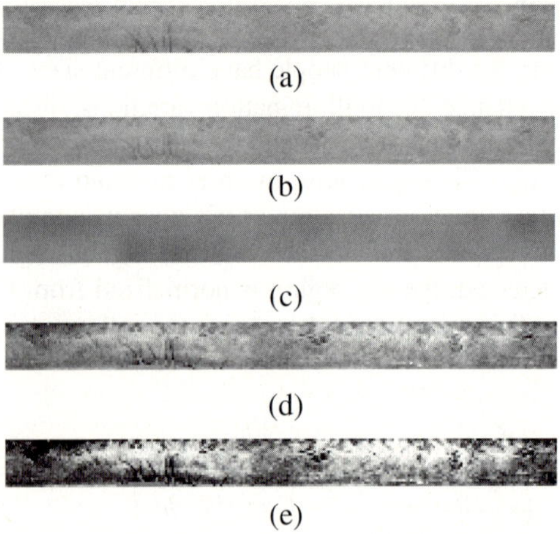

Fig. 6. a) Normalized image, b) Normalized image after removing eyelashes c) Image of non-uniform background illumination, d) Image after subtracting background illumination, d) Enhanced image after histogram equalization

2.5 Neural Network Based Iris Pattern Recognition

In this paper, a Neural Network (NN) is used to recognise the iris patterns. In this approach, the normalized and enhanced iris image is represented by a two-dimensional array. This array contains the greyscale values of the texture of the iris pattern. These values are input signals for the neural network. Neural network structure is given in Fig. 7. Two hidden layers are used in the NN. In this structure, $x_1, x_2, \ldots, x_m$ are greyscale values of input array that characterizes the iris texture information, $P_1, P_2, \ldots, P_n$ are output patterns that characterize the irises.

The k-th output of neural network is determined by the formula

$$P_k = f(\sum_{j=1}^{h2} v_{jk} \cdot f(\sum_{i=1}^{h1} u_{ij} \cdot f(\sum_{l=1}^{m} w_{li} x_l))) \tag{6}$$

where v_{jk} are weights between the hidden and output layers of network, u_{ij} are weights between the hidden layers, w_{il} are weights between the input and hidden layers, f is the activation function that is used in neurons. *m* is the number of input signals, *h1* and *h2* are the number of neurons in hidden layers, *n* is the number of output neurons ($k=1,..,n$).

After activation, the backpropagation learning algorithm is applied for training of NN. The trained network is then used for the iris recognition in online regime.

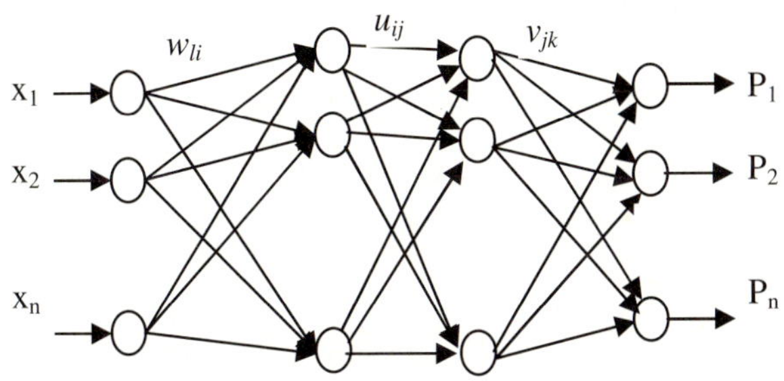

Fig. 7. Neural Network Architecture

3 Experimental Results

In order to evaluate the iris recognition algorithms, the CASIA iris image database is used. Currently this is largest iris database available in the public domain. This image database contains 756 eye images from 108 different persons. Experiments are performed in two stages: iris segmentation and iris recognition. At first stage the above described rectangular area algorithm is applied for the localization of irises. The experiments were performed by using Matlab on Pentium IV PC. The average time for the detection of inner and outer circles of the iris images was 0.14s. The accuracy rate was 98.62%. Also using the same conditions, the computer modelling of the iris localization is carried out by means of Hough transform and Canny edge detection realized by Masek [7] and integrodifferential operator realized by Daugman [3-6]. The average time for iris localization using Hough transform is obtained 85 sec, and 90 sec using integrodifferential operator. Table 1 demonstrates the comparative results of different techniques used for iris localization. The results of Daugman method are difficult for comparison. If we use the algorithm which is given in [16] then the segmentation represents 57.7% of precision. If we take into account the improvements that were done by author then Daugman method presents 100% of precision. The experimental results have shown that the proposed iris localization rectangular area algorithm has better performance. In second stage the iris pattern classification using NN is performed. 50 person's irises are selected from iris database for classification. The detected irises after normalization and enhancement are scaled by using averaging. This help to reduce the size of neural network. Then the images are represented by matrices. These matrices are the input signal for the neural network. The outputs of the neural network are classes of iris patterns. Two hidden layers are used in neural network. The numbers of neurons in first and second hidden layers are 120 and 81, correspondingly. Each class characterizes the certain person's iris. Neural learning algorithm is applied in order to solve iris classification. From each set of iris images, two patterns are used for training and two patterns for testing. After training the remaining images are used for testing. The recognition rate of NN system was 99.25%. The obtained recognition result is compared with the recognition results of other methods that utilize the same iris database. The results of this comparison are given in table 2. As shown in the table, the identification result obtained using the neural network approach illustrates the success of its efficient use in iris recognition.

Table 1. Accuracy rate for iris segmentation

Methodology	Accuracy rate	Average time
Daugman [17]	57.7%	90 s
Wildes [8]	86.49%	110 s
Masek [7]	83.92%	85 s
Proposed	98.62%	0.14 s

Table 2. The recognition performance of comparing with existing method

Methodology	Accuracy rate
Daugman [4]	100%
Boles [9]	92.64%
Li Ma [10]	94.9%
Avila [15]	97.89%
Neural Network	99.25%

4 Conclusion

An iris recognition system for personal identification is presented in this paper. A fast iris localization method is proposed. Using this method, iris segmentation is performed in short time. Average time for iris segmentation is obtained to be 0.14 sec on Pentium IV PC using Matlab. Accuracy rate of iris segmentation 98.62% is achieved. The located iris after pre-processing is represented by a data set. Using this data set as input signal the neural network is used to recognize the iris patterns. The recognition accuracy for trained patterns was 99.25%.

References

1. Jain, A., Bolle, R., Pankanti, S. (eds.): Biometrics: Personal Identification in a Networked Society. Kluwer, Dordrecht (1999)
2. Adler, F.: Physiology of the Eye: Clinical Application, fourth ed. London: The C.V. Mosby Company (1965)

3. Daugman, J.: Biometric Personal Identification System Based on Iris Analysis, United States Patent, no. 5291560 (1994)
4. Daugman, J.: Statistical Richness of Visual Phase Information: Update on Recognizing Persons by Iris Patterns, Int'l J. Computer Vision 45(1), 25–38 (2001)
5. Daugman, J.: Demodulation by Complex-Valued Wavelets for Stochastic Pattern Recognition, Int J. Wavelets, Multiresolution and Information Processing 1(1), 1–17 (2003)
6. Daugman, J.: How Iris Recognition Works, University of Cambridge (2001)
7. Masek, L.: Recognition of Human Iris Patterns for Biometric Identification. School of Computer Science and Soft Engineering, The University of Western Australia (2003)
8. Wildes, R., Asmuth, J., Green, G., Hsu, S., Kolczynski, R., Matey, J., McBride, S.: A Machine-Vision System for Iris Recognition. Machine Vision and Applications 9, 1–8 (1996)
9. Boles, W., Boashash, B.: A Human Identification Technique Using Images of the Iris and Wavelet Transform. IEEE Trans. Signal Processing 46(4), 1185–1188 (1998)
10. Ma, L., Wang, Y.H., Tan, T.N.: Iris recognition based on multichannel Gabor filtering. In: Proceedings of the Fifth Asian Conference on Computer Vision, Australia, pp. 279–283 (2002)
11. Lim, S., Lee, K., Byeon, O., Kim, T.: Efficient Iris Recognition through Improvement of Feature Vector and Classifier. ETRI J. 23(2), 61–70 (2001)
12. Ma, L., Tan, T., Wang, Y., Zhang, D.: Personal Identification Based on Iris Texture Analysis. IEEE Transaction on Pattern Analysis and Machine Intelligence 25(12) (2003)
13. Huang, Y.-P., Luo, S.-W., Chen, E.-Y.: An Efficient Iris Recognition System. In: Proceedings of the First International Conference on Machine Learning and Cybernetics, Beijing (November 2003)
14. Wang, Y., Han, J.-Q.: Iris Recognition Using Independent Component Analysis. In: Proceedings of the Fourth International Conference on Machine Learning and Cybernetics, Guangzhou (2005)
15. Sanchez-Avila, C., Sanchez-Reillo, R.: Iris-Based Biometric Recognition Using Dyadic Wavelet Transform. IEEE Aerospace and Electronic Systems Magazine, 3–6 (2002)
16. Daugman, J., Downing, C.: Recognizing iris texture by phase demodulation. IEEE Colloquium on Image Processing for Biometric Measurement 2, 1–8 (1994)
17. Chavez, R.F.L., Iano, Y., Sablon, V.B.: Process of Recognition of Human Iris: Fast Segmentation of Iris, www.decom.fee.unicamp.br/ rlarico/iris/localizationiris.pdf

No-Reference Quality Assessment of JPEG Images by Using CBP Neural Networks

Paolo Gastaldo, Giovanni Parodi, Judith Redi, and Rodolfo Zunino

Dept. Of Biophysical and Electronic Engineering (DIBE), University of Genoa
Via Opera Pia 11a, 16145, Genoa - Italy
{paolo.gastaldo,giovanni.parodi,judith.redi,
rodolfo.zunino}@unige.it

Abstract. Imaging algorithms often require reliable methods to evaluate the quality effects of the visual artifacts that digital processing brings about. This paper adopts a no-reference objective method for predicting the perceived quality of images in a deterministic fashion. Principal Component Analysis is first used to assemble a set of objective features that best characterize the information in image data. Then a neural network, based on the Circular Back-Propagation (CBP) model, associates the selected features with the corresponding predictions of quality ratings and reproduces the scores process of human assessors. The neural model allows one to decouple the process of feature selection from the task of mapping features into a quality score. Results on a public database for an image-quality experiment involving JPEG compressed-images and comparisons with existing objective methods confirm the approach effectiveness.

Keywords: Image quality assessment, feedforward neural networks, JPEG.

1 Introduction

In most applications, the effectiveness of compression methods for digital images (such as JPEG) depends on their impact on the visual fruition of pictures by consumers. Subjective testing [1] is the conventional approach for quality evaluation; these methods measure perceived quality by asking human assessors to score the overall quality of a set of test images. These tests yield accurate results; nonetheless, they are very difficult to model in a deterministic way.

Objective methods [2] instead, aim to estimate perceived quality, bypassing human assessors. These techniques measure image quality by processing numerical quantities ("objective features") extracted from images. To be effective, objective models must cohere with subjective opinions of quality perception. In most cases, the assessment system has to know the reference (uncompressed) image [2, 3]. By contrast, no-reference (NR) methods assess perceived quality by receiving as input only the compressed image [2, 4-6].

This paper presents a method using neural networks for the objective assessment of JPEG compressed images. A Circular BackPropagation (CBP) feedforward network [7] processes objective features worked out from JPEG images, and returns the

J. Marques de Sá et al. (Eds.): ICANN 2007, Part II, LNCS 4669, pp. 564–572, 2007.

associated quality scores. As the neural-based set-up does not require any information about the original uncompressed image, the overall objective method follows a no-reference approach.

2 Quality Assessment of JPEG Images

Figure 1 shows a schematic representation of the proposed no-reference quality assessment system for JPEG compressed images: a CBP network directly yields the quality assessments associated with input vectors of features worked out form the image. The design of the objective system takes into account that 1) several features characterizing images jointly affect subjective judgments, and 2) non-linear relationships may complicate the modeling process. The effectiveness of the neural-network approach lies in the ability to decouple the problem of feature-selection from the design of an explicit mathematical model; to this purpose, the CBP network provides a paradigm to deal with multidimensional data characterized by complex relationships. The function that maps feature vectors into quality ratings is learned from examples by use of an iterative training algorithm. Hence, the design of the objective metric is not involved in the set-up of the mapping function.

In this research, color correlogram [8] is used to characterize the information in image data. Color correlograms proved to be an effective technique for color texture analysis [8], but they have never been used in objective assessment models. Section 3 will discusses the feature-extraction process and introduces the feature-selection criteria. Indeed, Section 4 will describe the CBP neural network and the design strategy for the neural assessment system.

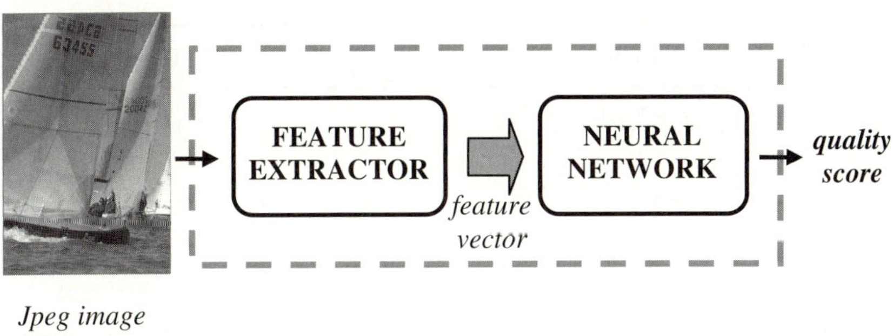

Fig. 1. The proposed no-reference quality assessment system

3 Feature-Based Image Representation: Definition and Selection

3.1 Local-Level Objective Features

The present approach works out objective features on a block-by-block, local basis from pixel values. To this purpose, the image is split into non-overlapping squares of 32x32 pixels, and each block is characterized by the associate set of numerical features; this allows the method to take into account the space-variant nature of perceptual mechanisms.

The picture blocks are characterized by objective features derived from the color correlograms [8]. A color correlogram expresses how the spatial correlation of pairs of colors changes with distance. Let S denote an image subregion that includes $R_S \times C_S$ pixels, and let H_s denote the color correlogram worked out from S. As a result, the matrix element $\mathrm{H}_S^k(i, j)$ counts the pairs of pixels that: 1) have color levels L_i and L_j, respectively, and 2) are separated by k pixels. Formally, if *dist*() denotes the measure of distance between pixels, then $\mathrm{H}_S^k(i, j)$ is defined as follows:

$$\mathrm{H}_S^k(i,j) = \left|\{(m,n), \text{ s.t. } S[m,n] = L_i;\ S[p,q] = L_j; dist(S[m,n], S[p,q]) = k\}\right| \tag{1}$$

In the present work, the operator *dist*() embeds the L_1-norm.

Table 1 presents the set Φ of objective features derived from the correlogram that have been used in this research. The set involves some of the quantities that were formalized in [9] to characterize the co-occurence matrix, an image descriptor similar to the color correlogram. This choice was mostly justified by the established scientific relevance of that fundamental research. In this table, the features *Difference entropy* and *Contrast* exploit the following definition:

$$P(z) = \sum_{\substack{i,j \\ |i-j|=z}} \mathrm{H}_S(i,j) \tag{2}$$

Table 1. Objective features derived from the color correlogram

Feature name	Definition	Feature name	Definition
Energy	$f_1 = \sum_{i,j} [\mathrm{H}_S(i,j)]^2$	*Entropy*	$f_2 = -\sum_{i,j} \mathrm{H}_S(i,j) \log_2 \mathrm{H}_S(i,j)$
Diagonal Energy	$f_3 = \sum_{i} [\mathrm{H}_S(i,i)]^2 / f_1$	*Homogen.*	$f_4 = \sum_{i,j} \mathrm{H}_S(i,j) / \left[1 + (i-j)^2\right]$
Difference Entropy	$f_5 = -\sum_{z} P(z) \log_2 P(z)$	*Contrast*	$f_6 = \sum_{z} z^2 P(z)$

3.2 Global-Level Features

Human assessors are usually asked to provide one overall quality score per image. Thus, from a modeling perspective, one vector must encompass the feature-based image description to be associated with the single score. Toward this end, after local-level feature computation, statistical descriptors unify block-based information to characterize the whole image by using one data vector. From a formal perspective, let $I^{(l)}$ be the l-th original picture, and denote by $\zeta_q(\cdot)$ the JPEG coding algorithm operating at a compression ratio, q; thus, $I^{(l,q)}$ denotes the digital image obtained as $\zeta_q(I^{(l)})$. Then the overall feature-computation algorithm can be outlined as follows.

Inputs: a picture $I^{(l,q)}$, an objective feature f_u;

1. Block-level feature extraction.

   1.a Split $I^{(l,q)}$, into n_b non-overlapping squares to obtain a set of blocks: $\mathrm{B}^{(l,q)} = \{b_j^{(l,q)}; j=1,..,n_b\}$.

   1.b For each block $b_j^{(l,q)} \in \mathrm{B}^{(l,q)}$: compute the feature value $v_{uj}^{(l,q)}$ to obtain

$$\mathrm{F}_u^{(l,q)} = \{v_{uj}^{(l,q)}, j=1,..,n_b\} \tag{3}$$

2. Feature integration into one image descriptor.

   Assemble the global description vector, $\mathbf{x}^{(l,q)}$, for the image $I^{(l,q)}$ as:

$$\mathbf{x}_u^{(l,q)} = \{p_\alpha(F_u^{(l,q)}), \alpha = 0,10,..,100\}. \tag{4}$$

   where $p_\alpha()$ is the percentile of order αth.

To sum up, the feature-extraction process represents the JPEG-compressed image, $I^{(l,q)}$, by a global pattern, $\mathbf{x}^{(l,q)}$, whose dimensionality is d=11.

3.3 A Statistical Approach to Feature Selection

Table 1 presents a list of six objective features derived from the color correlogram. Indeed, a crucial aspect in the proposed approach consists in picking out those descriptors that are most informative; many features will eventually be discarded because they do not carry significant information or because they are mutually correlated.

The present framework tackles that feature-selection procedure empirically; the data set is obtained by applying JPEG compression, $\zeta_q(\cdot)$, to a library of training images, $\Omega=\{I^{(l)}, l=1,..,n_p\}$, at different compression ratios, $q=q_1,..,q_n$. The resulting sample, $\overline{\Omega}=\{I^{(l,q)}, l=1,..,n_p;\ q=q_1,..,q_n\}$ includes $n_s = q_n n_p$ JPEG-compressed images. By applying the feature-extraction process presented in Section 3.2 to each element in $\Phi=\{ f_u, u=1,..,n_f\}$, the vector $\mathbf{x}^{(l,q)}$ characterizing $I^{(l,q)}$ would lie in a huge, intractable space having dimension $d=11n_f$.

In the present research, features have been selected statistically by using a procedure [6] based on Principal Component Analysis (PCA) [10]. The procedure exploits an unsupervised analysis of data and already proved to be effective in quality assessment frameworks [6]. As a result, objective features are selected according to their relevance in characterizing the compressed images, removing any a priori assumption on the relevance of the specific artifacts.

4 CBP for Image Quality Prediction

In the proposed system the feed-forward neural network map feature-based image descriptions into the associated estimates of perceived quality, which, in the present formulation, are represented as scalar values.

Theory proves that feed-forward networks embedding a sigmoidal nonlinearity can support arbitrary mappings [11]. The MultiLayer Perceptron (MLP) model [11] belongs to this class of networks, and has been proved to perform effectively in those problems where the target-mapping function can be attained by a few computing units endowed with global scope. The "Circular Back Propagation" (CBP) network [7] extends the conventional MLP by adding one additional input, which sums the squared values of all the network inputs. By this formulation the properties of the MLP structure remain unaffected. At the same time, CBP theory shows that this additional unit enables the overall network either to adopt the standard, sigmoidal behavior, or to drift smoothly to a bell-shaped radial function. More importantly, the selection between either model is entirely data-driven and stems from the empirical training process, hence model selection does not require any a priori assumption. Such a behavioral adaptiveness makes CBP networks quite interesting for application to perceptual problems, where even the domain structure is often obscure.

The CBP architecture can be formally described as follows. The input layer connects the n_i input values (features) to each neuron of the "hidden layer." The j-th "hidden" neuron performs a non-linear transformation of a weighted combination of the input values, with coefficients $w_{j,i}$ (j=1,…, n_h; i=1,…, n_i):

$$a_j = sigm\left(w_{j,0} + \sum_{i=1}^{n_i} w_{j,i} x_i + w_{j,n_i+1} \sum_{i=1}^{n_i} x_i^{\,2} \right). \tag{5}$$

where $sigm(z)=(1+e^{-z})^{-1}$, and a_j is the neuron activation. Likewise, the output layer provides the actual network responses, y_k, (k = 1,…, n_o):

$$y_k = sigm\left(w_{k,0} + \sum_{j=1}^{n_h} w_{k,j} a_j \right). \tag{6}$$

The structural CBP enhancement still allows one to adopt conventional back-propagation algorithms [11] for weight adjustment. Hence an efficient tool is available for an effective training. A quadratic cost function measures the distortion between the actual NN output and the expected reference score on a sample of training patterns. The cost is expressed as:

$$E = \frac{1}{n_o n_p} \sum_{l=1}^{n_p} \sum_{k=1}^{n_o} \left(t_k^{(l)} - y_k^{(l)} \right)^2 \tag{7}$$

where n_p is the number of training patterns, and t_k are the desired training outputs. In the present application, k=1 and the expected output is given by the quality assessment (score) measured experimentally from a human panel.

4.1 Improving Robustness by Using Ensembles

Statistical fluctuations in the empirical training set give rise to the problem of getting robust estimators. A classic approach to increasing the reliability of the neural approach is to use an "ensemble" [12] of different networks trained on the same problem. Indeed, averaging the predictions of several estimators [12, 13] can reduce the variance, σ, that stems from statistical noise in training data. In principle, an ensemble of N statistically independent estimators can decrease the variance in the estimate up to:

$$\overline{\sigma}^2 = \sigma^2 / N \tag{8}$$

The approach (8) means using several networks in parallel [14, 15]. Of course, building independent estimators is the crucial issue. When few patterns are available as compared with the data dimensionality, an approach based on the theory of receptive fields [11] can apply. This method partitions the high-dimensional input space into several, lower-dimensional subspaces, and provides a specialized neural network for each of them. The overall estimate is obtained by combining (typically, averaging) the contributions of the "local" estimators.

In an ideal ensemble (8), the subspaces should be disjoint from one another. In fact, these assumptions seldom hold in real domains, hence the prediction (8) represents an asymptotic target; in addition, it is not always easy to find a space-partitioning strategy that results in an effective prediction. On the other hand, the space-partitioning method, when applicable, can reduce the dimensionality of input patterns for each neural network in the ensemble, thus enhancing the network's generalization ability.

In the specific problem of quality prediction, a coordinate-partitioning approach indeed allows one to split the input vector, **x**, reasonably into N subvectors of lower dimensionality. These subvectors form the training sets for each network in the ensemble (Fig. 2).

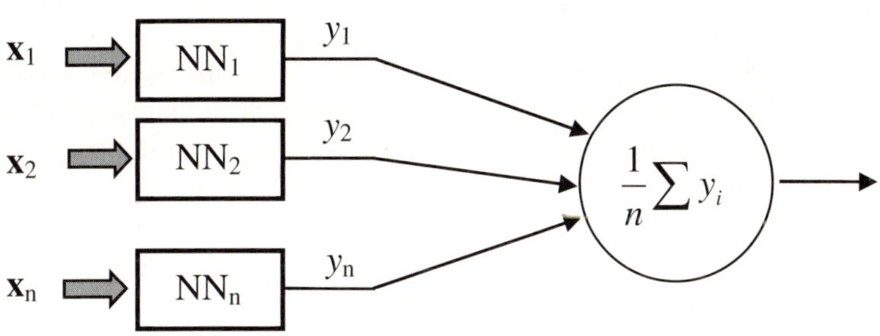

Fig. 2. The data space is partitioned and multiple networks in the ensemble contribute to the final score

5 Experimental Results

The public database [16] of the Laboratory for Image and Video Engineering (LIVE) of the University of Texas at Austin provided a testbed for the proposed method. The database includes 175 test images (including either 480x720 pixels or 768x512 pixels), which had been generated by JPEG-coding 29 original color images at varying compression ratios. The LIVE database associates the compressed images with the quality scores that were prompted by a panel of human assessors that took part in a subjective-quality experiment. People were asked to measure the perceived qualities of the viewed images by using a continuous scale divided into five regions, which was subsequently remapped linearly into the range [1, 100]. The experiments adopted the conventional procedure of working out a Mean Opinion Score (MOS) per image, by averaging all the subjective scores associated with each image; this served the purpose of characterizing pictures with statistically reliable quality judgments. The resulting overall sample included 175 image-MOS pairs.

5.1 System Set-Up

The procedure described in Section 3 for feature selection was first applied to the whole set, Φ, of objective features listed in Table 1. The features derived from the correlogram matrix were worked out in HSV color space, which is supposed to provide better correspondence with human visual perception of color; in particular, the Hue layer was used. Besides, the radius amplitude, k, was set at a fixed value $k = 2$. This choice aimed at minimizing computational complexity and was subsequently supported experimentally, as it was found that larger values of k did not alter the statistical distributions of features significantly.

The unsupervised feature selection showed that, in spite of the large number of possible candidates within Φ, only a limited subset of quantities derived from the color correlogram proved most representative. The selection procedure pointed out three objective measures: *diagonal energy*, *entropy* and *homogeneity*. The relatively high data dimensionality, d=3*11, required the use of an ensemble structure based on the subspace-partitioning strategy. Such an approach inherently matched the definition of the features, which could be naturally partitioned into three groups. Thus, the ensemble included three CBP networks, each covering a single objective feature; as a result, the actual input vector for each network had a dimensionality $d^{(i)} = 11$.

The final system set-up phase consisted in sizing the neural network that constituted each element of the ensemble. The approach proposed by Widrow and Lehr [17] provided an effective and practical method for tackling such a sensitive problem; that method aims to ensure that the available training data will effectively drive the adjustment of the network coefficients. For symmetry, the number of neurons eventually was $n_h = 3$ for all of the involved neural networks. Each CBP network was trained by the accelerated version of BackPropagation [18].

5.2 Evaluating the Run-Time Performance in Quality Assessment

The evaluation of the method's generalization ability adopted a K-fold strategy [19]. Such a procedure was chosen for the present research because it is known to provide reliable results when the available data set is relatively small. Toward that end, the sample of 29 image contents (175 pictures) was repeatedly split into 'folds'; an 'image content' includes one original picture and all pictures derived from that after JPEG coding. Therefore, in each 'experimental run,' 23 image contents made up the training set for the networks in the ensemble, whereas the remaining four image contents provided a test set; the latter ones never entered any step of the training process, and served to assess the system generalization performance empirically.

The overall experimental procedure involved five different experimental runs. For each run, measuring the generalization error required to compare the quality scores, y, predicted by the ensemble with the actual scores, t, collected from human assessors. The discrepancies between these quantities were interpreted by different statistical descriptors:

- Pearson's correlation coefficient, ρ, between y and t;
- the mean prediction error, μ_{err}, between y and t;
- the mean value of the absolute prediction error, $\mu_{|err|}$;
- the root mean square (RMS) error between y and t.

Accordingly, Table 2 report the results obtained by each experimental run. In compliance with cross-validation theory, one might estimate the generalization error by simply averaging over the error results obtained from each run. Individual results from each run, however, were also informative, as empirical evidence showed that the assessment system attained satisfactory performances in all runs. Furthermore, generalization performances were characterized by a very small variance among different runs; this indirectly supports the robustness of the ensemble approach.

Table 2 shows that the system always attained correlation coefficients, ρ, higher than 0.9, and scored an absolute prediction error, $\mu_{|err|}$, smaller than 0.15, corresponding to an estimation accuracy higher than 93%. In this respect, the empirical strategy adopted (K-fold across image contents) proved effective to assess the model's generalization ability, and should be compared with the results obtained by other no-reference approaches. The experiments by Wang et al. [4] and [5], for example, allow fair comparisons, as both involved the images from the LIVE database.

The non-linear quality prediction model proposed by Wang et al. [4] used cross-validation for assessing generalization performance, as the image contents for testing were not included in the training set. The reported results for two runs give RMS errors of 0.76 and 0.89, respectively, on a quality scale in the range [1, 10]. The neural method proposed in this paper improved over that model because, in a larger series of runs, quality predictions (rescaled within the range [1, 10]) scored RMS errors lying in the interval [0.55, 0.84]. Similar conclusions can be drawn when comparing the CBP-based predictions with the results reported in the work by Pan et al. [5]. The adaptive algorithm proposed therein exploited a measure of blocking artifacts and attained a correlation coefficient ρ=0.92 between subjective ratings and estimated quality scores. Table 2 shows that the neural system yielded higher correlation coefficients, with the minor exception of Run #2.

Table 2. Test results on the experiments involved in K-fold cross validation

	Run #1	Run #2	Run #3	Run #4	Run #5
ρ	0.93	0.92	0.97	0.95	0.96
μ_{err}	-0.008	0.01	0.01	-0.05	0.03
$\mu_{\|err\|}$	0.15	0.15	0.09	0.11	0.11
RMS	0.18	0.18	0.11	0.15	0.13

References

1. International Telecommunication Union: Methodology for the subjective assessment of the quality of television pictures. ITU-R BT.500 (1995)
2. Wang, Z., Sheikh, H.R., Bovik, A.C.: Objective video quality assessment. In: Furth, B., Marques, O. (eds.) The Handbook of Video Databases: Design and Applications, CRC Press, Boca Raton, FL (2003)
3. Sheikh, H.R., Sabir, M.F., Bovik, A.C.: A statistical evaluation of recent full reference image quality assessment algorithm. IEEE Trans. Image Processing 15, 3441–3452 (2006)
4. Wang, Z., Sheikh, H.R., Bovik, A.C.: No reference PSNR estimation for compressed pictures. In: Proc. IEEE ICIP, Rochester, New York, vol. 80, pp. 22–25 (September, 2002)
5. Pan, F., Lin, X., Rahardja, S., Lin, W., Ong, E., Yao, S., Lu, Z., Yang, X.: A locally adaptive algorithm for measuring blocking artifacts in images and videos. Signal Processing: Image Communication 19, 499–506 (2004)
6. Gastaldo, P., Zunino, Z.: Neural networks for the no-reference assessment of perceived quality. Journal of Electronic Imaging 14 (2005)
7. Ridella, S., Rovetta, S., Zunino, R.: Circular back-propagation networks for classification. IEEE Trans. on Neural Networks 8, 84–97 (1997)
8. Huang, J., Ravi Kumar, S., Mitra, M., Zhu, W.-J., Zabih, R.: Image indexing using color correlograms. In: Proc. IEEE CVPR '97, pp. 762–768 (1997)
9. Haralick, R.M., Shanmugam, K., Dinstein, I.: Textural Features for Image Classification. IEEE Trans. On Systems, Man and Cybernetics SMC 3, 610–621 (1973)
10. Jolliffe, I.T.: Principal Component Analysis. Springer, New York (1986)
11. Rumelhart, D.E., McClelland, J.L.: Parallel distributed processing. MIT Press, Cambridge, MA (1986)
12. Perrone, M.: Improving regression estimates: Averaging methods for variance reduction with extension to general convex measure optimization. Ph.D. dissertation, Phys. Dep., Brown Univ., Providence, RI (1993)
13. Geman, S., Bienenstock, E., Doursat, R.: Neural networks and the bias/variance dilemma. Neural Computation. 4, 1–48 (1992)
14. Kittler, J., Hatef, M., Duin, R.P.W., Matas, J.: On combining classifiers. IEEE Trans. Pattern Analysis and Machine Intelligence. 20, 226–239 (1998)
15. Rovetta, S., Zunino, R.: A multiprocessor-oriented visual tracking system. IEEE Trans. on Industrial Electronics 46, 842–850 (1999)
16. Sheikh, H.R., Wang, Z., Cormack, L., and Bovik, A.C.: LIVE Image Quality Assessment Database at, http://live.ece.utexas.edu/research/quality
17. Widrow, B., Lehr, M.A.: 30 Years of Adaptive Neural Networks: Perceptron, Madaline and Back Propagation. Proc. IEEE 78, 1415–1442 (1990)
18. Vogl, T.P., Mangis, J.K., Rigler, A.K., Zink, W.T., Alkon, D.L.: Accelerating the convergence of the back propagation method. Biol. Cybern. 59, 257–263 (1998)
19. Hecht-Nielsen, R.: Neurocomputing, Reading, MA. Addison-Wesley, London, UK (1989)

A Bio-inspired Connectionist Approach for Motion Description Through Sequences of Images

Claudio Castellanos-Sánchez

Laboratory of Information Technologies of Centre for Research and Advanced Studies
LTI Cinvestav - Tamaulipas, Ciudad Victoria, Tamaulipas, México
castellanos@cinvestav.mx

Abstract. This paper presents a bio-inspired connectionist approach for motion description through sequences of images. First, this approach is based on the architecture of oriented columns and the strong local and distributed interactions of the neurons in the primary visual cortex (V1). Secondly, in the integration and combination of their responses in the middle temporal area (MT). I propose an architecture in two layers : a causal spatio-temporal filtering (CSTF) of Gabor-like type which captures the oriented contrast and a mechanism of antagonist inhibitions (MAI) which estimates the motion. The first layer estimates the local orientation and speed, the second layer classifies the motion (global response) and both describe the motion and the pursuit trajectory. This architecture has been evaluated on sequences of natural and synthetic images.

1 Introduction

The visual perception of motion helps us to detect the pattern of 3D moving objects, its depth, speed and direction estimation, etc. Out of all of them, the optic flow field is an important source of information about the egomotion of the visual system, for guiding the navigation and in particular for indicating the direction in which the observer is moving.

The research in connectionism is inspired by complexity of neural interactions and their organisation in the brain that can help us to propose a feasible bio-inspired connectionist model. Visual perception of motion has been an active research field for the scientific community since motion is of fundamental relevance for most machine perception tasks [1].

Recent research on computational neuroscience has provided an improved understanding of human brain functionality and bio-inspired models have been proposed to mimic the computational abilities of the brain for motion perception and understanding [2].

Several bio-inspired models exist for visual perception of motion some of them are inspired by the primary visual cortex (V1) with a strong neural cooperative-competitive interactions that converge to a local, distributed and oriented

J. Marques de Sá et al. (Eds.): ICANN 2007, Part II, LNCS 4669, pp. 573–582, 2007.

auto-organisation [3,4,5]. The others are inspired by the middle temporal area (MT) with the cooperative-competitive interactions between V1 and MT and an influence range [6,7]. And some others are inspired by the middle superior area (MST) for the coherent motion and egomotion [8,9]. For more details see [2].

All these models are specialised in each visual cortical area of the brain. These models are based on the local detections by integration of various work directions upon different scales and spaces to end with a global answer. In this paper I present a bio-inspired connectionist approach for motion description through sequences of images that classifies the motion into three types : null motion, motion and egomotion, furthermore it shows the speed and direction of motion and the path pursuit of moving objects in the scene. To begin with, I show the main characteristics of each stage of the proposed approach. Next, I continue with the manipulation of different parameters issued by the mechanism of antagonist inhibitions (MAI). After this, I show three neuromimetic indicators for motion description. Finally, I carried out some experiments on real and synthetic images and end with propositions for future work.

2 Bio-inspired Connectionist Approach

This section broadly describes the mathematical and biological foundations of the proposed bio-inspired model for motion description through sequences of images based on the connectionist approach reported in [2,10].

2.1 General Architecture

The first stage of this bio-inspired connectionist approach is mainly based on the causal spatiotemporal Gabor-like filtering and the second stage is a local and massively distributed processing defined in [2,10], where they have proposed a retinotopically organised model of the following perception principle : the local motion information of a retinal image is extracted by neurons in the primary visual cortex (V1) with local receptive fields restricted to small areas of spatial interactions (first stage : causal spatio-temporal filtering, CSTF); these neurons are densely interconnected for excitatory-inhibitory interactions (second stage : mechanism of antagonist inhibitions, MAI). The figure 1 shows the general architecture for this model.

I will describe in this section these two stages : the causal spatio-temporal filtering (CSTF) integrated by two processes and the mechanism of antagonist inhibitions (MAI). The processes of the first stage are : the spatial processing for modelling the orientation selectiveness of V1 neurons and the temporal processing for modelling the speed selectiveness of MT neurons. Finally the connectionist processing takes advantage of the excitatory-inhibitory local interactions of the cerebral cortex in the human beings and their self-organising mechanisms for coherent motion estimation. The biological foundations and the mathematical details will not be discussed in this paper (for reference see [2,10]). The neuromimetic indicators will be presented in subsection 2.2.

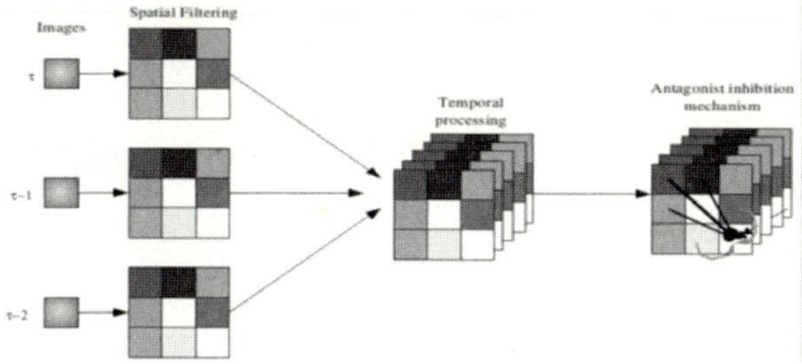

Fig. 1. Architecture of bio-inspired connectionist model (adapted of [10])

Causal Spatio-temporal Filtering (CSTF). The first stage of the model depicted by spatial filtering and temporal processing of figure 1 performs a causal spatio-temporal filtering. It models the magnocellular cells seen as motion sensors that depend on the gradient of image intensity and on its temporal derivatives [11,12,13]. This filtering is performed in two steps (see equation 1) : a Gabor-like spatial filtering and a causal temporal processing [14,10].

$$H_{t,\theta,\boldsymbol{v}}(x,y) = \int S_\theta(x - \hat{v}_1, y - \hat{v}_2)dt \tag{1}$$

$$\hat{v}_1 = \frac{\hat{t}}{\tau - 1} v_1 cos\theta, \quad \hat{v}_2 = \frac{\hat{t}}{\tau - 1} v_2 sin\theta \tag{2}$$

where $S_\theta(\cdot,\cdot)$ is the Gabor-like spatial filtering, $\boldsymbol{v} = (v_1, v_2)$ the speed vector and τ the number of images in the subsequence, and $0 \leq \hat{t}, t < \tau$.

For the spatial filtering, Gabor-like filters are implemented as image convolution kernels in Θ different directions. I usually work with $\Theta = 8$ for simplicity.

Then the causal temporal processing involves the computation of a temporal average of Gabor-like filters for each direction and for a set of search places that correspond to V assumed as different speeds of each pixel (positives and negatives). In other words, for each given assumed direction and speed, these Gabor-like filters reinforce the local motion with the average of the Gabor filters applied to past images on the assumed anterior places. This principle is valid under the strong hypothesis of a very high sampling frequency to ensure a local motion detection and an immediate constant local speed. For more details on this filtering see [2,10].

The computations described in this subsection have been parallelised and implemented on FPGA[1] circuits for real-time embedded motion perception [14].

Mechanism of Antagonist Inhibitions (MAI). The second stage of the model described in [2] (depicted to the right of figure 1) emulates a mechanism of antagonist inhibitions by means of excitatory-inhibitory local interactions in the different oriented cortical columns of V1.

[1] A Field Programmable Gate Array is a semiconductor device containing programmable logic components and programmable interconnects.

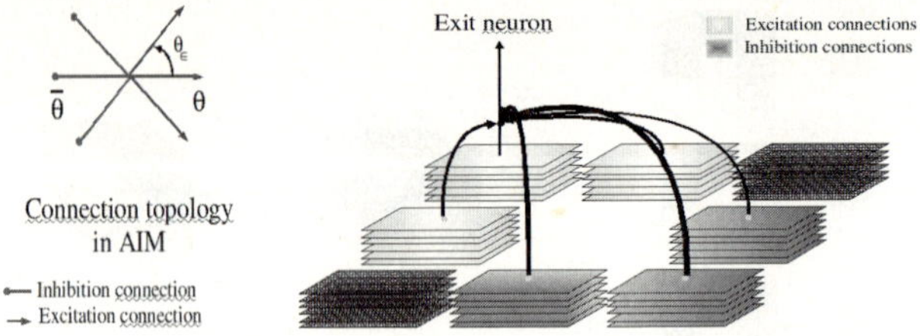

Fig. 2. Topology and interactions in the mechanism of antagonist inhibitions, MAI

In this mechanism each neuron receives both excitation and inhibition signals from neurons in a neighbourhood or influence range to regulate its activity. The figure 2 shows the excitatory and inhibitory local interactions where neurons interact with their close neighbours in this mechanism that change the internal state of neurons and their influence range generates a dynamic adaptive process.

Usually in excitatory-inhibitory neural models, the weighted connections to and from neurons have modulated strength according to the distance from one another. Nevertheless, I call it a mechanism of antagonist inhibitions because the inhibitory connections among neurons regulate downwards the activity of opposing or antagonist neurons, i.e. neurons that do not share a common or similar orientation and speed. On the other hand, excitatory connections increase the neuron activity towards the emergence of coherent responses, i.e. grouping neuron responses to similar orientations and speeds through an interactive process.

Then the updating of the internal state of a neuron is

$$\begin{aligned}\eta\frac{\partial H(x,y,T)}{\partial T} = &-A\cdot H(x,y,T)\\ &+(B-H(x,y,T))\cdot Exc(x,y,T)\\ &-(C+H(x,y,T))\cdot Inh(x,y,T)\end{aligned} \qquad (3)$$

where $-A\cdot H(\cdot)$ is the passive decay, $(B-H(\cdot))\cdot Exc(\cdot)$ the feedback excitation and, $(C+H(\cdot))\cdot Inh(\cdot)$ the feedback inhibition. Each feedback term includes a state-dependent nonlinear signal ($Exc(x,y,T)$ and $Inh(x,y,T)$) and an automatic gain control term ($B-H(\cdot)$ and $C+H(\cdot)$, respectively). $H(x,y,T)$ is the internal state of the neuron localised in (x,y) at time T, $Exc(x,y,T)$ is the activity due to the contribution of excitatory interactions in the neighbourhood $\Omega^{\Omega_E}_{(x,y)}$ and $Inh(x,y,T)$ is the activity due to the contribution of inhibitory interactions in the neighbourhood $\Omega^{\Omega_I}_{(x,y)}$. Both neighbourhoods depend on the activity level of the chosen neuron in each direction. A, B and C are the real constant values and η is the learning rate. For more details on the excitation and inhibition areas see [2,10].

Let ρ be the influence range of neuron (x,y) in this stage. This neuron receives at most ρ^2 excitatory connections from neurons with the same direction and speed and at most $(V\cdot\Theta-1)\cdot\rho^2$ inhibitory connections from other close neurons.

At this level, each pixel correspond to $\Theta\cdot V$ different neurons that encode informations of directions and speeds.

The computations described in this subsection analysing its neural and synaptic parallelism have been implemented on FPGA circuits [15].

2.2 Neuromimetic Indicators

The visual perception of motion is not totally determined in the local responses of the V1 neurons. They are processed to obtain the speed after being collected and combined from V1 and being integrated in MT. It is this combination of signals that resolve the local ambiguity of responses of neurons in V1 [2]. This activity is the inspiration of the last part of figure 1.

Table 1. Experimental ranges for neuromimetic motion indicator (NMI)

Condition	Description
$NMI < 0.10$	Null motion
$NMI < 1.00$	Small moving objects or noise
$NMI < 5.00$	One or two moving objetcs
$NMI < 10.00$	Three to five moving objects
$NMI < 40.00$	Six or more moving objects, or ego-motion
$NMI < 250.00$	Ego-motion or big moving objects
$NMI < 400.00$	Ego-motion
$NMI \geq 400.00$	Not processed

Controlled Generation of Sequences of Real Images. I analysed the active neurons in each direction and speed, the frequencies of active neurons after updating (ANaU) and the negative updating increase (NUI) through m different sequences of real images (about 384×288 pixels per image).

Next, to analyse egomotion, I selected n images of each sequence of real images and for each selected image I generated $\Theta \times V$ controlled subsequences ($\Theta=$ different directions and $V=$ different speeds).

Finally for motion classification, I took a subsequence of each sequence of real images where : a) the motion does not exist, b) one object moves and c) two or more objects move simultaneously. The interpretation of the different obtained values are shown below.

Motion Type. The equation 3 shows the updating rule in the MAI for the active neurons. Let S be an image sequence and let $R \subset S$ be a subsequence of size $Card(R) = \tau$ and let p be the percentage of the neurons to update.

The MAI mechanism updates $p\%$ of active neurons and I obtain in it two frequency percentages : the active neurons after updating ($ANaU$) and negative updating increase, (NUI, see the right side in the equation 3).

The frequencies of the products of $ANaU$ and NUI indicators in all the different controlled subsequences described in section 2.2 inspires us to propose our *neuromimetic motion indicator* : $NMI = ANaU * NUI$. The experimental ranges of NMI are shown in table 1.

Speed and Direction. MT neurons sum the responses of V1 neurons with receptive field positions inside a local spatial neighbourhood that is defined through time and generates a response according to the speed of the visual stimulus [2]. This locality of the MAI mechanism on all the several considered motion directions in V1 bring an emerging answer to the global direction [2,10].

On the other hand, neuro physiological studies roughly indicate that neurons in MT of the visual cortex of primate brains are selective to speed of visual stimuli; which implies that neurons respond strongly to a preferred direction and speed [6].

For each processed subsequence R in the equation 1 I define

$$sat^{+} = max_{t,\theta,\boldsymbol{v}}(H_{t,\theta,\boldsymbol{v}}(x,y)), \quad sat^{-} = min_{t,\theta,\boldsymbol{v}}(H_{t,\theta,\boldsymbol{v}}(x,y)) \tag{4}$$

where sat^{+} and sat^{-} are the positive and negative saturation, respectively.

For each direction and speed of each neuron, I count the neurons with a response greater than at. This parameter is the average of positive and negative saturations. The equation 6 shows its behaviour and the equation 5 computes this frequency in direction θ with speed v.

$$C(\theta,\boldsymbol{v}) = \textstyle\sum_{(x,y)} D(at, H_{t,\theta,\boldsymbol{v}}(x,y))) \tag{5}$$

$$D(at, H_{t,\theta,\boldsymbol{v}}(x,y)) = \begin{cases} 1 \text{ if } H_{t,\theta,\boldsymbol{v}}(x,y) > at \\ 0 \; otherwise \end{cases} \tag{6}$$

where $D(\cdot,\cdot)$ is the threshold of the CSTF filtering.

The collection and combination in MT for direction estimation is:

$$E(\theta,\boldsymbol{v}) = 3\cdot C(\theta,\boldsymbol{v}) + 2\cdot(C(\theta-\phi,\boldsymbol{v}) + C(\theta+\phi,\boldsymbol{v})) + C(\theta-2\phi,\boldsymbol{v}) + C(\theta+2\phi,\boldsymbol{v}) \tag{7}$$

where $\phi = \frac{2\pi}{\Theta}$ is the separation in degrees between each oriented column and $E(\cdot,\cdot)$ is the sum of several oriented responses of V1 that activate a neuron in MT. Finally, I computed the frequencies for negative and positive supposed speeds by the equations:

$$G^{+} = \sum_{\boldsymbol{v}>0,\theta} C(\theta,\boldsymbol{v}), \quad G^{-} = \sum_{\boldsymbol{v}<0,\theta} C(\theta,\boldsymbol{v}) \tag{8}$$

Then I arrange $E(\theta,\boldsymbol{v})$ in direction according to each speed and arranged G^{+} and G^{-} too for processing them to obtain speed and direction indicators. These indicators will be described in the next two paragraphs.

Speed. To obtain the winner speed, I propose the *neuromimetic speed indicator (NSI)* defined by equation 9:

$$NSI = \frac{100 \cdot min(G^{+}, G^{-})}{max(G^{+}, G^{-})} \tag{9}$$

With this indicator I compute the relative speed (rs) that compares the different speed frequencies and their proportion. The table 2 shows my experimental values of $v_i \in \{-2,-1,0,1,2\}$ and $v1, v2$ the frequencies of $|v_i| = 1, |v_i| = 2$, respectively.

Table 2. Experimental ranges for neuromimetic speed indicator (NSI)

Type	Condition	Relative speed	Prototype speed
	$NSI > 70.0$	$rs = (100.0 - NSI)/29.0$	0
Weak if $v1 > v2$	$NSI > 12.0$	$rs = (71 - NSI)/59 + 1$	1
	otherwise	$rs = (12 - NSI) * 0.3529/12 + 2$	2
	$NSI > 22.0$	$rs = (NSI * 0.6470)/22 + 2.3530$	3
Strong if $v1 < v2$	$NSI > 39.0$	$rs = (NSI - 22)/10 + 3$	4
	otherwise	Speed not processed	≥ 5

Direction. Finally, for an interpretation of directions integration for each neuron in MT, I compute $E(\theta, \boldsymbol{v})$ of the equation 7 for each direction and speed.

Next, I arrange their values from major to minor and I take the first three. If these candidates are contiguous, the winner will be at the centre of the three candidates' directions. This is my *neuromimetic direction indicator, NDI*. Finally, if the maximum of the two computed speeds in the equation 8 is the negative one, the winner direction will be its antagonist, ei, $\theta = \theta - 180^\circ$.

3 Experiment Results

I chose four sequences of images : the Yosemite Fly-Through (sequence of synthetic images), the Hamburg Taxi, the BrowseB (issue of video surveillance) and the CatEyes (an experiment with a little camera on cat's head). They include various frames of RGB gray-scaled images (15, 42, 875 and 1334 images) with the sizes of 316×252, 256×191, 384×288, 320×200, respectively.

The figure 3 shows in each row four images of a sequence, their graph of neuromimetic indicators and their path. NMI are between 0 (null motion) and 3000 (egomotion), NSI between 0 and 6 (because I suppose five speeds), and NDI is in $\{1, 2, 3, 4, 5, 6, 7, 8\}$ (0°, 45°,..., 315°, respectively).

The first sequence, the synthetic Yosemite Fly-Through sequence shows an aeroplane flying on the mountains. This sequence presents an egomotion with a speed of five pixels (below the image) that diverge and two pixels for the moving clouds to the right (above the image). The NMI is between 300 and 450, then according to the table 1 it proposes an egomotion with a speed of 2 pixels per image moving at around 45°. The global motion is similar to the Yosemite Fly-Through data. The trajectory describes its evolution.

The real Hamburg Taxi sequence shows three moving cars and a pedestrian. The NMI is between 6 and 18, then according to the table 1 there are about three moving objects and the global speed is 2 pixels per image moving at approximatly 180° in the first half of the sequence and the second half at around 135°. The trajectory shows the evolution of the taxi.

The BrowseB sequence issue of video surveillance in the hall of INRIA laboratory in Grenoble, France, may be split into three parts : (1) a person walks to the centre, stops and returns; (2) null motion; (3) another person walks in, stops and goes farther.

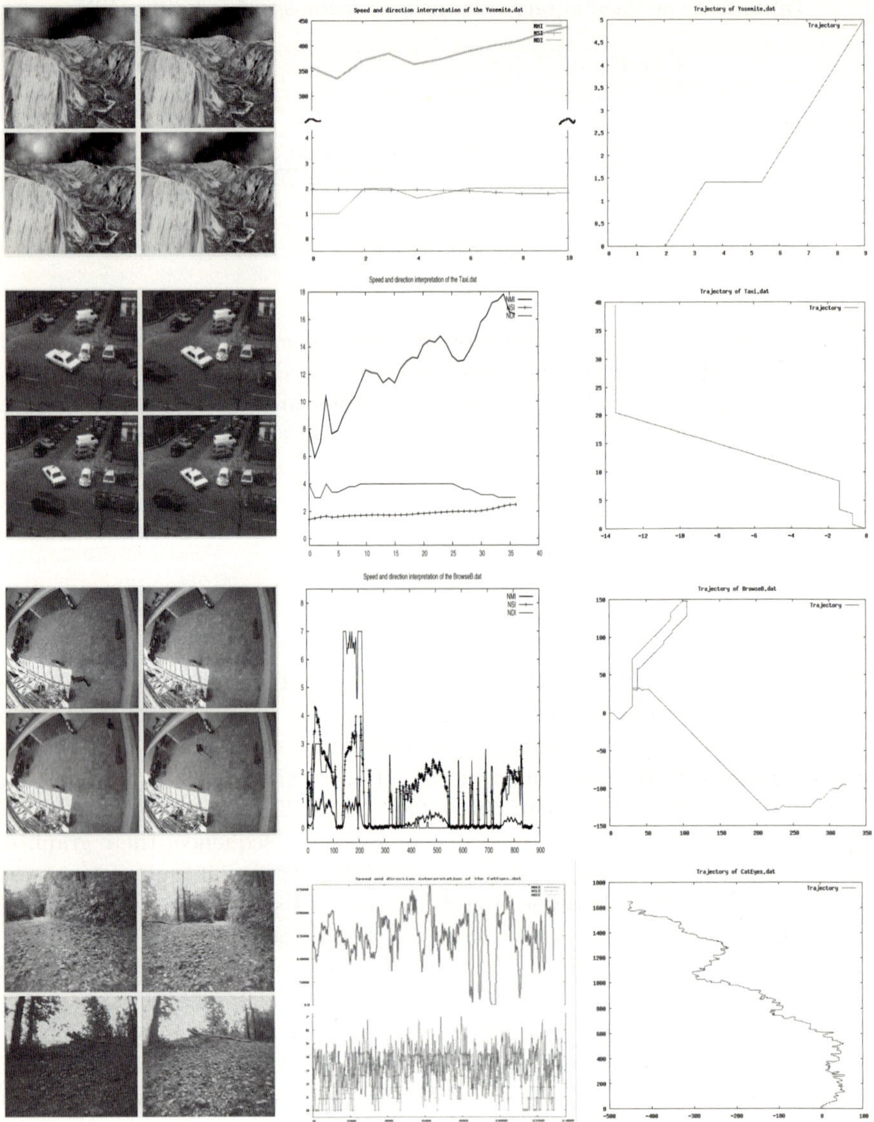

Fig. 3. Sequences of real and synthetic images used in this work: Yosemite, Taxi, BrowseB and CatEyes in each two rows, respectively, and from the top-left to the down-left in clockwise direction, four images of each sequence. In the last two columns on the right their neuromimetic indicators and their trajectory are shown.

The first part in this sequence (images 0 to 220) may be split into three parts according to NMI : two parts with motion and the other part with null motion that correspond to the first person walking between 90° and 135° and with speed of 4 to 2 pixels per image, stops and returns between 270 ° and 315 ° and with speed of 2 to 4 pixels per image, respectively.

In the second part (images 221 to 325) there is null motion. The last part may be split too into three parts according to NMI : (1) motion, (2) generally null motion and (3) motion, respectively to describe this part of the BrowseB sequence. The person walks generally to 0 ° with speed of 1-2 pixels per image. Next, a period of null motion with very weak motions (see pics in the graph between image 550 and 750). Finally, the person moves to about 90 ° with a speed of about 2 pixels per image. The trajectory shows the evolution of the three parts of this sequence.

Finally, the last sequence is the CatEyes sequence which is an issue of an experiment with a little camera on the cat's head that walks in the field. The cat walks toward north-west but it looks to the left, to the right and behind. This sequence has a strong egomotion problem. The NMI is between 10 and 2600. Most of the time it is bigger than 400, then according to the table 1 there is egomotion but it is not processed. On the contrary, the computed speed, direction and trajectory in this model show its real evolution.

4 Conclusion

This work is based on the Castellanos-Sánchez model [10] : a neuromimetic connectionist model for visual perception of motion. A model fully inspired by the visual cortex system, the superior areas and their relations.

In this paper I took advantage of the low-level analysis to detect local motions to obtain the global speed, direction and trajectory. They are determined by the neuromimetic motion indicators issued by MAI mechanism. This is a strong simplification of the model presented in [16] that utilised three layers but here I utilised only the first two layers of this model.

The first experiments show that this model is capable of estimating the null motion, simple motion and egomotion with an estimation of global speed and direction in an environment where other persons or objects move. The estimation of motion is robust in quite complex scenes without any predefined information. Nevertheless, the estimation of NMI is fastidious. The global experimental values are correct for the sequences of real images of ±33% the size of 384 × 288 that is used in this approach.

My current work include experimenting on the other sizes of the images for the generalisation of the NMI, studying the same neuromimetic indicators for the moving fields only, the dynamic multipursuit, the topology optimization of MAI for its implementation in FPGA circuits and its applications in real time for video surveillance and autonomous robotics.

Acknowledgement. "Fondo Mixto Conacyt-Gobierno de Tamaulipas" #51623.

References

1. McCane, B., Novins, K., Grannitch, D., Galvin, B.: On benchmarking optical flow. Computer Vision and Image Undestanding, 126–143 (2001)
2. Castellanos-Sánchez, C.: Neuromimetic connectionist model for embedded visual perception of motion. PhD thesis, Université Henri Poincaré (Nancy I), Nancy, France, Bibliothèque des Sciences et Techniques (2005)
3. Fellez, W.A., Taylor, J.G.: Establishing retinotopy by lateral-inhibition type homogeneous neural fields. Neurocomputing 48, 313–322 (2002)
4. Latham, P.E., Nirenberg, S.: Computing and stability in cortical networks. Neural Computation, 1385–1412 (2004)
5. Moga, S.: Apprendre par imitation: une nouvelle voie d'apprentissage pour les robots autonomes. PhD thesis, Université de Cergy-Pontoise, Cergy-Pontoise, France (2000)
6. Simoncelli, E.P., Heeger, D.J.: A model of neural responses in visual area mt. Visual Research 38(5), 743–761 (1998)
7. Mingolla, E.: Neural models of motion integration and segmentation. Neural Networks 16, 939–945 (2003)
8. Pack, C., Grossberg, S., Mingolla, E.: A neural model of smooth pursuit control and motion perception by cortical area mst. Technical Report CAS/CNR-TR-99-023, Department of Cognitive and Neural Systems and Center for Adaptive Systems, Boston University, 677 Beacon St, Boston, MA 02215 (2000)
9. Zemel, R.S., Sejnowski, T.J.: A model for encoding multiple object motions and self-motion in area mst of primate visual cortex. The Journal of Neurosciences 18(1), 531–547 (1998)
10. Castellanos-Sánchez, C., Girau, B., Alexandre, F.: A connectionist approach for visual perception of motion. In: Smith, L., Hussain, A., Aleksander, I. (eds.) Brain Inspired Cognitive Systems (BICS 2004), vol. BIS3–1, pp. 1–7 (2004)
11. Pollen, D., Ronner, S.: Phase relationships between adjacent simple cells in the visual cortex. Science 212, 1409–1411 (1981)
12. Newsome, W.T., Gizzi, M.S., Movshon, J.A.: Spatial and temporal properties of neurons in macaque mt. Investigative Ophtalmology and Visual Science Supplement 24, 106 (1983)
13. Hammond, P., Reck, J.: Influence of velocity on directional tuning of complex cells in cat striate cortex for texture motion. Neuroscience Letters 19, 309–314 (1981)
14. Torres-Huitzil, C., Girau, B., Castellanos-Sánchez, C.: On-chip visual perception of motion: A bio-inspired connectionist model on fpga. Neural Networks 18(5-6), 557–565 (2005)
15. Torres-Huitzil, C., Girau, B.: Fpga implementation of an excitatory and inhibitory connectionist model for motion perception. In: Brebner, G.J., Chakraborty, S., Wong, W.F. (eds.) IEEE International Conference on Field-Programmable Technology, FPT 2005, pp. 259–266 (2005)
16. Castellanos-Sánchez, C., Girau, B.: Dynamic pursuit with a bio-inspired neural model. In: Blanc.-Talon, J., Philips, W., Popescu, D.C., Scheunders, P. (eds.) ACIVS 2005. LNCS, vol. 3708, pp. 284–291. Springer, Heidelberg (2005)

Color Object Recognition in Real-World Scenes

Alexander Gepperth[1], Britta Mersch[2], Jannik Fritsch[1], and Christian Goerick[1]

[1] Honda Research Institute Europe GmbH, Carl-Legien-Str. 30, Offenbach, Germany
[2] Deutsches Krebsforschungszentrum, Im Neuenheimer Feld 580,
69120 Heidelberg, Germany

Abstract. This work investigates the role of color in object recognition. We approach the problem from a computational perspective by measuring the performance of biologically inspired object recognition methods. As benchmarks, we use image datasets proceeding from a real-world object detection scenario and compare classification performance using color and gray-scale versions of the same datasets. In order to make our results as general as possible, we consider object classes with and without intrinsic color, partitioned into 4 datasets of increasing difficulty and complexity. For the same reason, we use two independent bio-inspired models of object classification which make use of color in different ways. We measure the qualitative dependency of classification performance on classifier type and dataset difficulty (and used color space) and compare to results on gray-scale images. Thus, we are able to draw conclusions about the role and the optimal use of color in classification and find that our results are in good agreement with recent psychophysical results.

1 Introduction

The use of color information in object recognition remains to this day a controversial issue, both from the point of view of psychologists and computer scientists. Although much experimental work has been done on the subject in psychophysical science, the results are sometimes contradicting or inconclusive: early works [1,2] proposed "shape" theories of object recognition, claiming that color is an irrelevant feature for recognition. In contrast, more recent investigations [6,17,13] seem to show that color does improve recognition("shape+surface"), especially when objects have so-called diagnostic, i.e., class-specific intrinsic colors. To our knowledge, however, there are no experiments that investigate the validity of both theories using realistic objects in cluttered real-world scenes.

In computational implementations of object recognition systems, the use of color information is not too common. Instead, many object recognition systems are restricted to the use of shape information (e.g., gradients, local orientation or wavelet representations). Reasons for this are manifold: first of all, the use of color information triplicates the amount of data that needs to be processed. Furthermore, color is an ambiguous cue: its optimal representation should always depend on the task at hand. Hence, little consensus exists about the features that should be extracted from color information, and therefore the use of color always poses quite complex design questions which one would rather avoid if possible.

J. Marques de Sá et al. (Eds.): ICANN 2007, Part II, LNCS 4669, pp. 583–592, 2007.

Lastly, the fact exists that recognition on gray-scale images has been shown to perform successfully in a wide range of domains and applications, so it could be argued that further improvement is not necessary.

In this study, it is investigated whether the use of additional color information improves accuracy in a challenging real-world classification task, and if so, under what circumstances. Obviously, not all outcomes of such an experiment will allow definitive statements about the issue at hand. However, we believe that unambiguously identifying a classification problem where color *does* make a difference would be quite worthwhile in itself and allow to draw meaningful conclusions. Assuming that recognition in the human brain is at least as good as the computational models tested here, one may safely conclude that the human brain could profit still more. In addition to theoretical considerations, this paper should give indications if and how color information can best be used to improve performance in challenging computational classification tasks.

1.1 Related Work in Computational Object Recognition

Interestingly, the number of proposals for object recognition architectures that can use color information is relatively small. Two principal approaches can be tentatively discerned: color histogram and receptive field methods. The color histogram technique was triggered by [12] and followed up by many researchers. Here, color histograms of objects are compared by using dedicated histogram metrics. This approach is powerful and highly invariant to noise and geometric distortions like rotation, occlusion and translation, but does not analyze the spatial structure of objects at all. In contrast, receptive field methods analyze an image by means of spatially localized convolution filters, followed by further processing or direct classification of the obtained information. Convolution filters can directly combine information from different color channels. This approach preserves some of the spatial structure of an object and exhibits invariance to noise and distortions that strongly depends on the convolution filters that are being used. A prominent publication in this direction is [5]. Both approaches, color histogramming and receptive field methods, have also been successfully applied to recognition in gray-valued images. It has been attempted to combine these two techniques theoretically [11] and in a working recognition system [8]. The system presented in [8] is especially interesting since it uses a very large number of visual features including color and, in contrast, a very simple classifier, suggesting that classification works best when combining as many informative features as possible. The classifiers tested in this study use an adaptive receptive field approach since the geometrical structure of objects must be taken into account. We know, of no study that systematically tests the usefulness of color information using real-world classification problems and large datasets of objects.

2 Datasets

The classification problem considered in this study originate from a car classification task within a comprehensive cognitive architecture for advanced driver assistance [9]. The architecture contains modules for (real-time) detection, segmentation, classification and tracking of objects in colored real-world traffic video scenes. Using the architecture, several datasets of increasing difficulty were created, and different steps to encode the color information were performed for each set.

Experiments are conducted for all color representations of each dataset. The goal of classification was to discriminate cars from background objects or object parts (e.g., trees, parts of the horizon, lane markings, guardrails a.s.o.) Since cars do not usually possess a single diagnostic color, and in order to make the classification task still harder, a second object class "signal board" was added. These objects were abundant in some training videos and pose a strong challenge for any classifier since they cannot usually be segmented correctly due to occlusion. In addition, signal boards in Germany have a standardized appearance of diagonal red and white stripes and thus possess unique diagnostic colors, which makes them interesting for this study.

The binary classification problem of car against background is therefore extended to a multi-class problem. This is desirable since the classification task is thus less specialized than a purely binary object-against-background-classification would be. In this way, we expect that the results are more easily generalizable[1]. In the following sections, we will describe steps that were taken in order to increase the generality of the scenario still further.

2.1 Data Generation and Levels of Difficulty

Initially, the architecture described in [9] was used to generate object candidates from several hours of highway and inner-city traffic videos. By visual inspection, datasets of car, signal board and clutter (not belonging to the "car" and "signal board" classes) object images were selected. Object candidates are resized to a common size of 64x64 pixels, and all datasets described below contain images of these dimensions. For the selection of car objects, different criteria were applied to obtain different datasets of object images. For details please consult table 1. Example objects from different datasets are shown in fig. 1.[2]

2.2 Color Representations

By default, the color representation in computer graphics is RGB. Due to the inherently ambiguous nature of color, different color spaces may be used that are tailored for special purposes and circumstances, and indeed a multitude of other color spaces has been proposed. We focus on color spaces that try to match human color perception as closely as possible, like the CIE La^*b^* color space[10] which was designed just for this purpose. We therefore perform the experiments in this study using the RGB, HSV and the polar CIE La^*b^* color spaces concurrently. HSV is a standard computer vision color space which is included for comparison because of its simple and efficient transformation rules. The details of the color space transformations can be found, e.g., in [10].

[1] Although, of course, there is no practical way to prove this.

[2] All datasets are available online from *www.gepperth.net/alexander/downloads.html*

Table 1. Information about the datasets used in this study. Since the criteria are progressively relaxed from dataset I to IV, each preceding dataset is contained in all successors: I ⊂ II ⊂ III ⊂ IV. For all datasets, 4766 non-object(clutter) images and 537 signal board images were used.

dataset	nr of examples	description
I	574	single back-view of a whole car, fills at least 25% of image
II	949	like I, plus front-views
III	1462	single view of a car(back/front), 50% of car must be in image, filling at least 25% of image
IV	1748	not restr. to single view, 25% of car must be in image, filling at least 25% of image

Fig. 1. Typical color object images from datasets I through IV. Top row: car images from dataset I (4 leftmost images) and dataset II (4 rightmost images). Second row: signal board examples, identical in all datasets. Third row: car images from dataset III (4 leftmost images) and dataset IV (4 rightmost images). Bottom row: clutter objects, identical in all datasets. Keep in mind that each dataset contains its predecessors; the shown images illustrate, for each dataset, the kind of objects that are added compared to the preceding dataset. Note that color images are reproduced in gray-scales on paper.

2.3 Error Measures

Since the number of training examples is relatively low, all results are verified by k-fold cross-validation. In k-fold cross-validation, the data is divided into k subsets of equal size. One of the k subsets is then retained as the validation dataset for testing the classifier and the remaining $k-1$ subsets are used as training data. The cross-validation process is then repeated k times, with each of the k subsets used exactly once as validation set to compute the classification error. The k classification results are averaged to produce a single classification error. The classifier is then trained k times, each time leaving out one of the subsets from training and using it to compute the classification error. Note that cross-validation is quite different from "split-sample" or "hold-out" method that are commonly used in machine learning. In the split-sample method, only a single subset (validation set) is used to estimate the generalization error, instead of k different subsets, i. e. there is no crossing. The distinction between cross-validation and split-sample is extremely important because cross-validation is markedly superior for small data sets. This fact is demonstrated in [4]. In this study, a value of $k = 5$ is chosen in order to have a minimum of 100 car and signal board objects in the test set. For each partitioning of a dataset, a receiver-operator characteristic (ROC) is computed and used to obtain an average ROC over 5 partitionings, which is taken to represent the outcome of an experiment for a particular dataset. For reducing a ROC to a single number, we consider the *equal-error condition* where the false positive (non-object examples that are classified as object) and the false negative (object examples that are classified as non-object) rates are identical.

3 Classification Methods

Since it is impossible to test all available classification architectures, we select two models which have been shown to be of value for visual classification tasks: the Visual Hierarchy (VH) [16] and the SCNN [3] classifiers.

Both models differ in the way color is handled: whereas VH extracts form features from a gray-valued version of its input and uses spatially coarse color information only at its last classification stage, SCNN integrates color information from the beginning[3], and no explicit separation between intensity (gray value) and color is made (see fig. 3 for details). Both approaches may be justified or at least made plausible, and one purpose of this paper is to give support to one or the other approach if possible. In this way, hints about the most efficient use of color in computational object classification may be arrived at.

To all intents and purposes, the description of the classification models could stop more or less here, and the less technically inclined reader may skip the rest of this section. In the following, a more detailed account of the working of both models is given.

Both the VH and the SCNN model are convolutional neural network (CNN) models [7] in the sense that they can perform whole-image classifications using block operations, i.e., operations that treat each image pixel independently of its position. The operations are mainly convolutions with filters determined by learning algorithms, but also other operations like subsampling, pooling or competitive mechanisms. Both classification models define unsupervised learning rules for determining well-suited convolution filters. In this way, both models are able to compute a (possibly high-dimensional) feature space which is unique to each classification problem. The final supervised classification takes place in that feature space.

---

[3] SCNN was initially conceived to handle intensity information only, but the extension to color is trivial and is discussed later in this section.

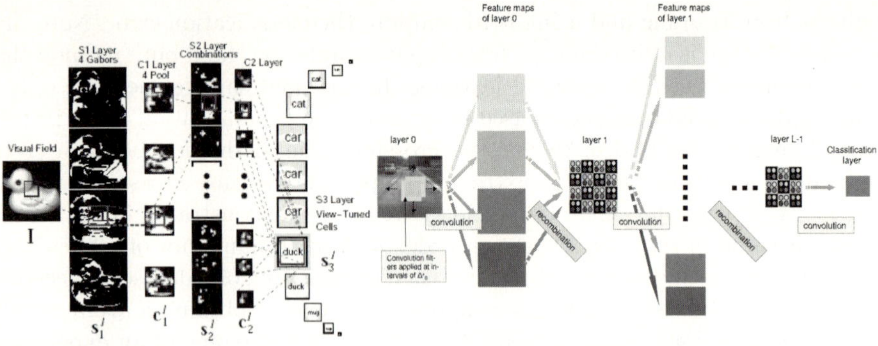

Fig. 2. Classification models used in this study. Left: the VH processing model as described in [16]. The C2 layer calculates 53 features. Right: the SCNN model as described in [3]. 49 filters are applied to the input layer instead of 16 as in the best-performing model given in [3]; this number matches the 53 features of the VH model quite closely. For both models, the input dimension is set to 64x64 pixels. For extensions to both models that are considered here, please see fig. 3.

Both models allow a large number of architectures to be formed by varying layer numbers and sizes, transfer functions, filter sizes a.s.o. Since it is not the goal of this study to perform an in-depth comparison of the two models, they will be taken in the form they are used in recent publications [3,15]. The SCNN model is (trivially) extended to allow the use of color information . In order to reduce computational complexity, and to mimick the pooling stages of the VH classifier, the training examples are resized to a size of 25x25 pixels for use with the SCNN model. In this way, SCNN can be used in the same configuration as in [3]. Fig. 2 shows the computational architecture of both models.

3.1 Extending SCNN in Order to Use Color

In order to apply the SCNN model to vector-valued pixels (as is the case for color images), a simple procedure is applied: each pixel is simply substituted by all vector entries arranged consecutively. In this way, the x-dimension of an image is extended by a factor of N (where N is the dimension of each pixel vector, here $N = 3$) while the y-dimension is unaffected. Care must be taken when choosing the SCNN structure: input layer filters must always come to start on a pixel boundary; this can be ensured by a correct choice of filter sizes and overlaps. Otherwise, the image thus constructed is treated as a gray-valued image, and the normal SCNN training algorithm can be applied. Fig. 3 gives a visual impression of this process.

3.2 Adapting VH to Different Color Representations

In the RGB color representation, VH calculates an intensity value from the RGB data and uses the intensity image for calculating a task-optimized feature space.

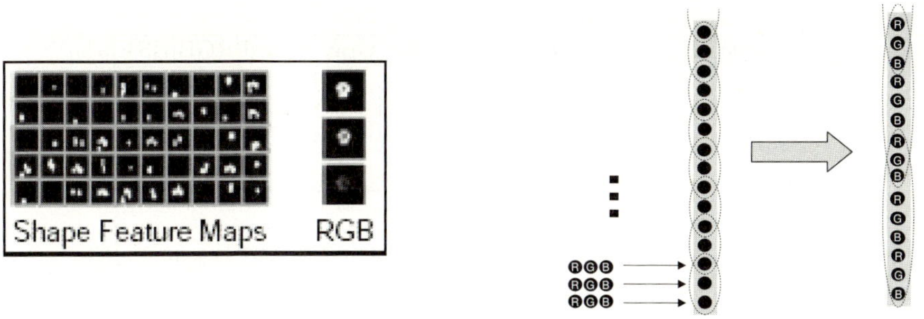

Fig. 3. Right: VH architecture for classification of color objects as described in [15]. The C2 layer is extended to include 3 additional feature maps formed from the downsampled R,G and B color channels. The C2 layer is thus constructed from 53 maps. Left: extension of the SCNN model to handle vector-valued RGB input pixels. As explained in the text, each RGB triplet is represented by 3 pixels extended into the x direction. The input is thus 3 times larger than in the gray-value case. Correspondingly, the x-dimension of filters in the input layer is tripled to 15 pixels. The classification layer consists of 16 feature maps.

When going to the HSV and La^*b^* color spaces, a slightly different approach is used: the Value (V) and the luminance (L) are used for calculating the feature space when using these color representations. Instead of downsampled R, G, and B maps, the downsampled S, V maps in the case of HSV and a^*, b^* maps in the case of La^*b^* are added to the C2 layer. Therefore, the C2 layer comprises only 52 instead of 53 (for RGB) feature maps in these cases.

4 Experiments

Experiments are conducted for the gray-scale, the RGB, the HSV and the La^*b^* representation of datasets I through IV using the VH and the SCNN classification methods summarized in section 3. This gives a total of 16 experiments for each classifier model. Results were obtained according to section 2.3, using datasets described in section 2.

5 Results

As the tables 2 and 3 plainly show, the use of color can improve (VH model) or impair (SCNN model) classification for both object classes. In the rest of this section, we will discuss the improvements obtained by using the VH model.

Table 2. Classification errors for cars. Left table: VH classifier, right table: SCNN classifier. All numerical values are given in percent.

Dataset	Gray-valued	RGB	HSV	pLa^*b^*
I	5.3	4.7	5.3	4.6
II	5.0	4.8	5.8	4.0
III	9.5	6.7	9.5	6.5
IV	11.1	8.0	11.1	7.6

Dataset	Gray-valued	RGB	HSV	pLa^*b^*
I	6.3	8.4	8.1	8.4
II	7.0	9.8	10.0	9.6
III	9.1	12.5	11.2	12.4
IV	11.2	14.0	13.9	14.9

Table 3. Classification errors for signal boards. Left table: VH classifier, right table: SCNN classifier. All numerical values are given in percent.

Dataset	Gray-valued	RGB	HSV	pLa^*b^*
I	11.0	9.3	7.3	10.9
II	10.8	9.8	7.5	10.9
III	11.4	9.8	7.8	11.8
IV	11.1	9.8	7.8	11.4

Dataset	Gray-valued	RGB	HSV	pLa^*b^*
I	11.3	13.3	14.1	12.9
II	10.7	13.9	15.0	14.2
III	11.6	13.3	13.7	13.8
IV	11.2	14.1	14.0	13.4

5.1 Results for Cars

As expected, classification performance deteriorates when going from dataset I to dataset IV. The relative improvement increases, suggesting that color is more useful when the classification task is harder.

5.2 Results for Signal Boards

Since the signal board object class does not differ across datasets, the differences in classification performance are quite small. The differences spring from the fact that a more complex car class can be more easily confused with signal boards. In fact, it is surprising that classification performance is not improved more clearly by the use of color given the fact that signal boards have a clearly defined diagnostic color. This can be easily understood when considering that the main source of confusion are cars and not clutter objects. Preliminary experiments where only signal boards had to be distinguished from clutter indeed showed a far stronger performance difference between gray-scale and color images.

6 Discussion

As a leading remark, we want to state that we have not addressed the difficult issue of color constancy in this article. We are well aware of this fact: the reason we do not believe it plays a role here is that we do not perform object identification but rather categorization with few categories and many objects. As we expected and as was shown, the classifiers are able to generalize sufficiently in order to deal with this problem.

As the results plainly show, the use of color improves classification performance for all datasets when using the VH model. In the case of the SCNN model, results tend to deteriorate when switching to color images. These findings persist, although to different degrees, when treating the problem using different color spaces, suggesting that the color space should always be adapted to the classification task as mentioned in the introduction.

For the signal board class with its clearly defined diagnostic color, the improvements are stronger than for cars but not as strong as one would naively assume. As explained before, this is likely due to confusions with car objects which can have similar colors; when leaving out the car class, the classification of signal boards improves more strongly by using color.

What can be learned from these results? First of all, one can infer conclusions about the preferable way of using color in computational classification. Generally speaking, results are roughly comparable for gray-valued images but get markedly better for color images using the VH model, whereas they deteriorate for the SCNN model. This effect persists over all color spaces and difficulty levels, suggesting that it is systematic: the way color is used in the VH model (see section 3) seems to be more appropriate to the presented task. Although it cannot, from these results, be concluded in all generality that this is a preferable way of using color, it may be concluded that it is a very sensible starting point when going from gray value to color classification.

Secondly, one can use these results to argue against "shape only" theories of object recognition. Based on the classification results, we cautiously argue in the line of [14], where experimental evidence for a "shape+surface" representation in object classification is reviewed. In contrast to many experimental results which suggest "shape only" representations, we believe (based on our results) that color is especially relevant in realistic, cluttered and visually noisy environments. It should be kept in mind that many related experiments were performed under idealized conditions, and that line drawings and images on white backgrounds are not abundant in natural scenes. What is more, recalling the discussion from the previous paragraph, we argue that it is sufficient to represent color as an overall object feature with little spatial structure. Thus, the dimensions for color and shape are well separated: it may be be that color plays some role in the definition of shape, but this study suggests that it is used mainly at a quite abstract level for purposes of overall object class separation.

Acknowledgments

The authors gratefully acknowledge the support of Heiko Wersing, Stephan Hasler and Stephan Kirstein in the use of the VH classifier. Thanks are due to Udo Seiffert for bringing up the subject of using the SCNN model on color images.

References

1. Biederman, I., Ju, G.: Surface versus edge-based determinants of visual recognition. Cognit.Psychol. 20 (1988)
2. Davidoff, J., Ostergaard, A.: The role of color in categorical judgements. Q J Exp Psychol A. 40(3) (1988)

3. Gepperth, A.: Object detection and feature base learning by sparse convolutional neural networks. In: Schwenker, F., Marinai, S. (eds.) ANNPR 2006. LNCS (LNAI), vol. 4087, Springer, Heidelberg (2006)
4. Goutte, C.: Note on free lunches and cross-validation. Neural Computation 9 (1997)
5. Hall, D., de Verdiere, C., Crowley, J.: Object recognition using colored receptive fields. In: Proceedings of the European Conference on Computer Vision (2000)
6. Humphrey, G.: The role of surface information in object recognition: studies of visual form agnosic and normal subjects. Perception 23 (1994)
7. LeCun, Y., Huang, F.-J., Bottou, L.: Learning methods for generic object recognition with invariance to pose and lighting. In: Proceedings of CVPR'04, IEEE Computer Society Press, Los Alamitos (2004)
8. Mel, B.W.: SEEMORE: Combining color, shape and texture histogramming in a neurally-inspired approach to visual object recognition. Neural Computation 9(4) (1997)
9. Michalke, T., Gepperth, A., Schneider, M., Fritsch, J., Goerick, C.: Towards a human-like vision system for resource-constrained intelligent cars. In: The 5th International Conference on Computer Vision Systems Conference Paper (2007)
10. Plataniotis, K.N., Venetsanopoulos, A.N.: Color image processing and applications. Springer, New York (2000)
11. Schiele, B., Crowley, J.L.: Object recognition using multidimensional receptive field histograms. In: Proceedings of the European Conference on Computer Vision (1996)
12. Swain, M., Ballard, D.: Color indexing. International Journal of Computer Vision 7(1) (1991)
13. Tanaka, D., Presnell, L.: Color diagnosticity in object recognition. Percept.Psychophysics 61 (1999)
14. Tanaka, J., Weiskopf, D., Williams, P.: The role of color in high-level vision. Trends in cognitive sciences 5(5) (2001)
15. Wersing, H., Kirstein, S., Götting, M., Brandl, H., Dunn, M., Mikhailova, I., Goerick, C., Steil, J., Ritter, H., Körner, E.: Online learning of objects and faces in an integrated biologically motivated architecture. In: The 5th International Conference on Computer Vision Systems Conference Paper (2007)
16. Wersing, H., Körner, E.: Learning optimized features for hierarchical models of invariant object recognition. Neural Computation 15(7) (2003)
17. Wurm, L.: Color improves object recognition in normal and low vision. L.Exp. Psychol. Human. Perc. Perform. 19 (1993)

Estimation of Pointing Poses on Monocular Images with Neural Techniques - An Experimental Comparison

Frank-Florian Steege, Christian Martin, and Horst-Michael Groß

Department of Neuroinformatics and Cognitive Robotics,
Ilmenau Technical University, Ilmenau, Germany
frank-florian.steege@stud.tu-ilmenau.de, christian.martin@tu-ilmenau.de
http://www.tu-ilmenau.de/neurob

Abstract. Poses and gestures are an important part of the nonverbal inter-human communication. In the last years many different methods for estimating poses and gestures in the field of Human-Machine-Interfaces were developed. In this paper for the first time we present an experimental comparison of several re-implemented Neural Network based approaches for a demanding visual instruction task on a mobile system. For the comparison we used several Neural Networks (Neural Gas, SOM, LLM, PSOM and MLP) and a k-Nearest-Neighbourhood classificator on a common data set of images, which we recorded on our mobile robot HOROS under real world conditions. For feature extraction we use Gaborjets and the features of a special histogram on the image. We also compare the results of the different approaches with the results of human subjects who estimated the target point of a pointing pose. The results obtained demonstrate that a cascade of MLPs is best suited to cope with the task and achieves results equal to human subjects.

1 Introduction and Motivation

In recent years the Human-Machine Interaction has reached a large importance. One of the most important and informative aspects of nonverbal inter-human communication are gestures and poses. In particular, pointing poses can simplify communication by linking speech to objects or locations in the environment in a well-defined way. Therefore, a lot of work has been done in recent years focusing on integrating pointing pose estimation into Human-Machine-Interfaces.

Numerous approaches, which can estimate the target of such a pointing pose have been developed in recent years. Our goal is to provide an approach, which can be used to estimate a pointing pose on a mobile robot by means of low-cost sensors. Therefore, in this paper we refer only to approaches using monocular images to capture the pose of the user. Second, approaches that do not use Neural Networks to estimate the target of the pointing pose like Haasch [1], who used an object-attention system and a skin color map or Nickel [2], who estimated the target by the use of a virtual line through the tracked hand and head of the user, are also not considered in this paper.

J. Marques de Sá et al. (Eds.): ICANN 2007, Part II, LNCS 4669, pp. 593–602, 2007.

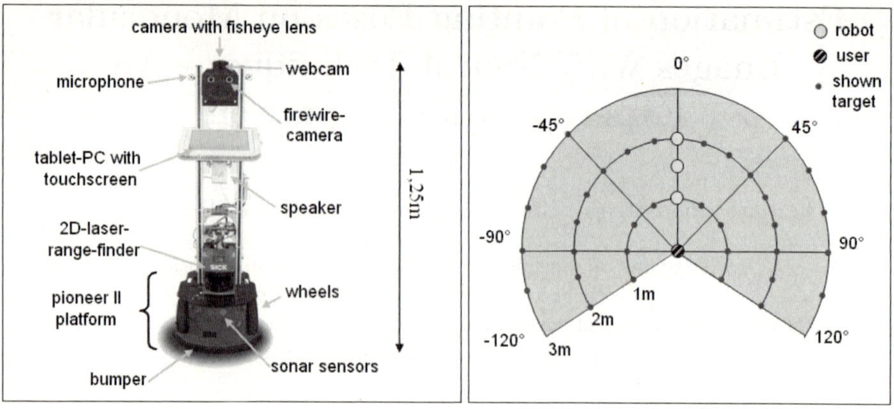

Fig. 1. (left) Our robot HOROS, used for experimental investigation of the pointing pose estimation is shown. The images for the estimation of the pointing target were taken with the firewire camera (located in the right eye). (right) The configuration used for recording the ground truth training and test data. The subject stood in front of the robot and pointed at one of the marked targets on the ground in a distance of 1 to 3 m from the subject. The distance of the robot to the subject varied between 1 m and 2 m.

However there are several approaches that utilize different Neural Networks to estimate the pointing pose. Nölker and Ritter [3] used Gaborfilters in combination with a Local Linear Map (LLM) and a Parametrized Self-Organizing Map (PSOM) to estimate the target of a pointing pose on a screen the user is pointing to. Richarz et al. [4] recently also used Gaborfilters on monocular images and a cascade of Multi-Layer Perceptrons (MLP) as function-approximator to determine the target point of a pointing-pose on the ground. Takahashi [5] suggested to use a special kind of histogram features in combination with a SOM to estimate the pose of a person in an image. Finally, since the head pose is typically also important for a pointing pose, approaches estimating the head pose are also considered in this paper: Krüger and Sommer [6] utilized Gaborfilters and a LLM to estimate the head pose, while Stiefelhagen [7] presented a system that works on edge-filtered images and uses a MLP for head pose estimation.

All these approaches achieved more or less good results for their particular task, but can not be compared with each other, because they use different images captured in different environments and they use different combinations of methods for feature extraction as well as different Neural Networks for approximating the target point or the direction of the pose.

Therefore, for this paper we implemented and compared several selected neural approaches, all trained and tested with the same set of training and test data. In this way we give an overview of the suitability of the different approaches for the task of estimating a pointing pose on a monocular image. The referred approaches suggest different applications for the recognition of a pointing pose. In our comparison we choose an application where a user points at a target on the ground which is similar to the application Richarz [4] suggested. We implemented this approach on our mobile robot HOROS (**HO**me **RO**bot **S**ystem, see Fig. 1 left), making it navigate to the specified targets, thus enabling a user to control the robot only by means of pointing.

The remainder of this paper is organized as follows: First, in Sect. 2 we give an overview of our test environment used to obtain the training and test data for our comparison. In Sect. 3, the preprocessing steps performed on every image and the methods for feature extraction we used in our approach are explained. In Sect. 4, we shortly describe the Neural Network techniques we compare in our approach. Section 5 describes the experimental investigations we conducted and compares the results of the different approaches. We conclude with a summary in Sect. 6 and give a perspective on possible improvements we plan to investigate in the near future.

2 Training-Data and Ground-Truth

We encoded the target points on the floor as (r, φ) coordinates in a user-centered polar coordinate system (see Fig. 1). This requires a transformation of the estimated target into the robot's coordinate system (by simple trigonometry), but the estimation task becomes independent of the distance between user and robot. Moreover, we limited the valid area for targets to the half space in front of the robot with a value range for r from 1 to 3 m and a value range for φ from $-120°$ to $+120°$. The $0°$ direction is defined as user-robot-axis, negative angles are on the user's left side. With respect to a predefined maximum user distance of 2 m, this spans a valid pointing area of approximately 6 by 3 m on the floor in front of the robot in which the indicated target points may lie. Figure 2 shows the configuration we chose for recording the training data and our robot HOROS which was used to record images of the subjects. The subjects stood at distances of 1, 1.5 and 2 m from the robot. Three concentric circles with radii of 1, 2 and 3 m are drawn around the subject, being marked every 15°. Positions outside the specified pointing area are not considered. The subjects were asked to point to the markers on the circles in a defined order and an image was recorded each time. Pointing was performed as a defined pose, with outstretched arm and the user fixating the target point (see Fig. 2).

All captured images are labeled with the distance of the subject and the radius r and angle φ of the target, thus representing the ground truth used for

Fig. 2. Typical examples of images of subjects taken by the frontal camera of the robot in several demanding real world environments with background clutter. The left three images are from the trainig data, the right three images from the test data.

training and also for the comparison with humans as pointing pose estimators (see Section 5). This way, we collected a total of 2,340 images of 26 different interaction partners in demanding real world environments with background clutter. This database was divided into a training subset and a validation subset containing two complete pointing series (i.e two sample sets each containing all possible coordinates (r, φ) present in the training set). The latter was composed from 7 different persons and includes a total of 630 images. This leaves a training set of 19 persons including 1,710 samples.

3 Image Preprocessing and Feature Extraction

Since the interaction partners standing in front of the camera can have different heights and distances, an algorithm had to be developed that can calculate a normalized region of interest (ROI), resulting in similar subimages for subsequent processing. We use an approach suggested by [4] to determine the ROI by using a combination of face-detection (based on the Viola & Jones Detector cascade [8]) and empirical factors. With the help of a multimodal tracker [9] implemented on our robot, the direction and the distance of the robot to the interacting person can be estimated. The cropped ROI is scaled to 160*100 pixels for the body and the arm and 160*120 pixels for the head of the user. Then a histogram equalization is applied. The preprocessing operations used to capture and normalize the image are shown in Fig. 3. Since some of the approaches mentioned in Sect. 1 use a Background Subtraction ([5], [7]) while others do not ([3], [4] and [6]) we optional use a Background Subtraction to test its influence on the pose estimation result.

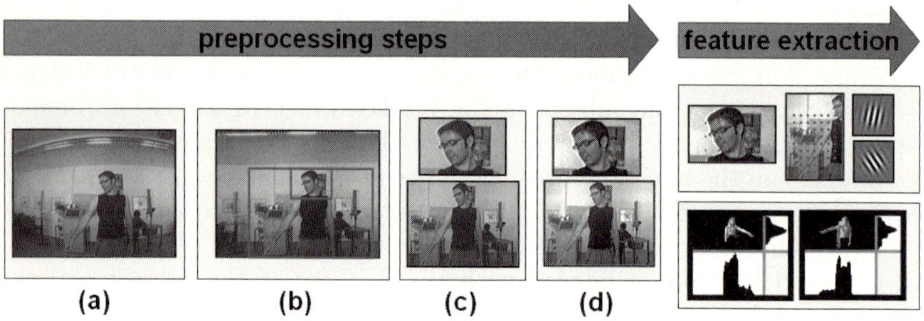

Fig. 3. Steps of preprocessing and feature extraction: the raw distorted image of the camera (a) is transformed into an undistorted image (b) and the face of the user is detected by means of [8]. Based on the height of the face in the picture and the distance of the user, two sections of the image are extracted and transformed into grayscale images (c). On these images a histogram equalization is used (d). Subsequently features are extracted in different ways. First, Gaborfilters placed at defined points of the image (marked as dots in (e)) were used. The second approach is to count how often pixel belonging to a pre-segmented user appear in every row and column of the image (f). A Background Subtraction can optionally be used between steps (d) and (e).

On the normalized image regions we extracted features for the approximation of the target position the user is pointing to. We therefore compared two possible methods. First, we used Gaborfilters of different orientations and frequencies bundled in Gaborjets that we located on several fixed points in the image sections. Gaborjets are also used in the approaches of [3], [4] and [6]. Second, we reimplemented the approach presented by [5]. Based on a background model, in this case, we could subtract the background from the image and count the number of pixels which belong to the user in every row and column. The several steps of preprocessing and feature extraction used in our comparison are shown in Fig. 3.

4 Used Techniques for Approximation of the Target

One objective of our approach is the experimental comparison of selected Neural Network based pointing pose estimators including a simple k-Nearest-Neighbour method well known as reference technique. In the following, the different methods used for comparison are presented:

k-Nearest-Neighbour Classification: The k-Nearest-Neighbour method (k-NN) is based on the comparison of features of a new input with features of a set of known examples from the training data. A distance measure is used to find the k nearest neighbours to the input in the feature space. The label that appears most often at the k neighbours is mapped on the new input. This method allows only classification and not an approximation between the labels of two or more neighbours. Therefore, we slightly modified the method in our approach in a SoftMax-manner where the label for the input $f_k(\mathbf{x})$ is determined as follows:

$$f_k(\mathbf{x}) = \sum_i l_i \cdot \left(\frac{1/d_i}{\sum_j 1/d_j} \right) \tag{1}$$

In this way, the labels l_i of the k nearest neighbours contribute to the output and are weighted by their Euclidian distance d_i to the input.

Neural Gas: A Neural Gas network (NG, [10]) approximates the distribution of the inputs in the feature space by a set of adapting reference vectors (neurons). The weights $\mathbf{w}_i$ of the neurons are adapted independently of any topological arrangement of the neurons within the Neural Net. Instead, the adaptation steps are affected by the topological arrangement of the receptive fields within the input space, which is implicitly given by the set of distortions $D_\mathbf{X} = \{\|\mathbf{x} - \mathbf{w}_i\|, i = 1, \cdots, N\}$ associated with an input signal $\mathbf{x}$. Each time an input signal $\mathbf{x}$ is presented, the ordering of the elements of the set $D_\mathbf{X}$ determines the adjustment of the synaptic weights $\mathbf{w}_i$. In our approach, each neuron also has a label l_i which is adapted to the label of the input signal.

Self-organizing Map: An approach very similar to the NG is the well-known Self-Organizing Map (SOM, [11]). The SOM differs from the NG in the fact that the neurons of the SOM are connected in a fixed topological structure. The neighbours of the best-matching neuron are determined by their relation in this structure and not by their order in the set $D_{\mathbf{X}}$. We modified the SOM so that every neuron has a learned label as we did with the NG above.

Local Linear Map: The Local Linear Map (LLM, [12]) is an extension of the Self-Organizing Map. The LLM overcomes the discrete nature of the SOM by providing a way to approximate values for positions between the nodes. A LLM consists of n nodes which each represent a pair of reference vectors $(\mathbf{w}_i^{in}, \mathbf{w}_i^{out})$ in the in- and output-space and an associated only locally valid linear mapping $\mathbf{A}_i$. The answer $\mathbf{y}_{bm}$ of the best-matching neuron of the LLM to an input $\mathbf{x}$ is calculated as follows:

$$\mathbf{y}_{bm} = \mathbf{w}_{bm}^{out} + \mathbf{A}_{bm}\left(\mathbf{x} - \mathbf{w}_{bm}^{in}\right) \tag{2}$$

The weights $\mathbf{w}_i^{in}, \mathbf{w}_i^{out}$ and the mapping matrix $\mathbf{A}_i$ have to be learned during the training process. See [12] for more details.

Parametrized Self-organizing Map: Like the LLM, a Parametrized Self-Organizing Map (PSOM, [13]) is also an extension of the SOM. While the LLM computes only a linear approximation of the output, a PSOM uses a set of non-linear basis manifolds to construct a mapping through the reference vectors $\mathbf{a}$. A basis function $H(\mathbf{s}, \mathbf{a})$ is associated with each reference vector $\mathbf{a}$. These basis functions realize a smooth interpolation of intermediate positions between the reference vectors. The interpolation is an iterative process starting at the best-matching reference vector. The topological order of the reference vectors has to be provided for the organization of the PSOM. In our approach we use a SOM to obtain this topological order.

Multi-layer Perceptron: For our comparison we used a cascade of several MLPs as decribed in [4]. The (r, φ) coordinates of the target point are estimated by separate MLPs. The radius r is estimated by a single MLP while φ is determined by a cascade of MLPs which first estimate a coarse angle φ' and second the final angle φ depending on r and φ'.

5 Results of Comparing Experimental Investigations

To have a simple reference for the quality of the estimation, 10 subjects were asked to estimate the target point of a pointing pose on the floor. At first, the subjects had to estimate the target on a computer screen where the images of the training data set were displayed. The subject had to click on the screen at the point where they estimated the target. Thus, the subjects were estimating the target on the images having the same conditions as the different estimation systems. Second, we determined the estimation result the subjects achieved under real world circumstances. Here each subject had to point at a target on the ground and a second subject had to estimate the target. The results of the human based reference experiments are included in Fig. 4 and Fig. 5. The label *Human 2D* refers to the experiments on the computer screen and the label *Human 3D* refers to the results under real world conditions.

radius estimation

correct samples in % / mean error in m	k-NN	NG	SOM	LLM	PSOM	MLP
Gaborfilters	48,18 % 0,314 m	33,85 % 0,458 m	42,93 % 0,443 m	54,45 % 0,378 m	29,04 % 0,507 m	70,46 % 0,235 m
Gaborfilters and BG Subtraction (BGS)	64,84 % 0,246 m	65,16 % 0,286 m	65,31 % 0,244 m	77,74 % 0,280 m	58,05 % 0,305 m	**88,21 %** **0,134 m**
Gaborfilters and Discriminant Analysis	60,17 % 0,292 m	48,49 % 0,323 m	56,34 % 0,326 m	64,90 % 0,338 m	47,44 % 0,455 m	74,41 % 0,216 m
Gaborfilters, BGS and Discriminant Analysis	**82,81 %** **0,124 m**	74,16 % 0,208 m	79,27 % 0,186 m	**84,24 %** 0,226 m	72,79 % 0,267 m	**88,40 %** **0,138 m**
Histogram Features and BG Subtraction	64,63 % 0,313 m	51,13 % 0,399 m	45,86 % 0,491 m	70,37 % 0,357 m	26,76 % 0,619 m	77,01 % 0,205 m

Human 3D	Human 2D
84,25 % **0,080 m**	75,00 % 0,350 m

angle estimation

correct samples in % / mean error in °	k-NN	NG	SOM	LLM	PSOM	MLP
Gaborfilters	23,10 % 23,00 °	13,91 % 23,20 °	15,63 % 23,61 °	21,61 % 21,79 °	11,97 % 26,34 °	41,39 % 18,51 °
Gaborfilters and BG Subtraction (BGS)	34,37 % 20,29 °	27,72 % 21,36 °	23,50 % 20,91 °	30,28 % 18,76 °	18,35 % 25,02 °	50,91 % 17,23 °
Gaborfilters and Discriminant Analysis	29,41 % 23,05 °	19,36 % 22,23 °	20,70 % 23,39 °	24,73 % 23,77 °	16,28 % 25,09 °	37,82 % 20,99 °
Gaborfilters, BGS and Discriminant Analysis	41,93 % 17,46 °	30,55 % 20,54 °	29,85 % 20,96 °	37,68 % 19,55 °	20,90 % 23,76 °	57,28 % 15,64 °
Histogram Features and BG Subtraction	35,48 % 18,25 °	28,59 % 17,35 °	23,91 % 19,39 °	40,67 % 15,52 °	15,54 % 24,29 °	51,00 % 15,75 °

Human 3D	Human 2D
74,66 % **4,50 °**	50,00 % 13,76 °

Fig. 4. Results for the estimation of the radius (top) and the angle of the target position (bottom). For each method the percentage of the targets estimated correctly and the mean error is determined. At the right side, the results of the human viewers (2D on computer screen, 3D in reality) are given for comparison. Methods that achieve a result equal to that of the human viewers are marked with a shaded background.

The results of the several neural approaches for estimating the target position are shown in Fig. 4 and Fig. 5. As described in Sect. 2 the ground truth data is a tuple (r, φ) with the target radius r and the target angle φ. The separate results for the estimation of r and φ are shown in Fig. 4. For the correct estimation of the target point, r as well as φ had to be estimated correctly. We defined the estimation result being correct if r differed less than 50 cm from the ground truth radius and φ differed less than 10° from the ground truth angle. Figure 5 shows the results for a correct estimation of both values.

Every of the six selected approaches was trained and tested on the same training data set. For each system, we used five different feature extraction strategies: first only Gaborfilters were utilized, second we combined Gaborfilters with an additional Background Subtraction to reduce the effects of the different cluttered backgrounds in the images. Third, we used only those Gaborfilters that had a high discriminant value extracted by means of a Linear Discriminant Analysis (LDA) executed over all predefined Gaborfilter positions. Fourth, we combined Gaborfilter, Background Subtraction and utilized only the relevant features extracted by the Discriminant Analysis mentioned above. In the last setup, we did not apply the Gaborfilters but the column and row histograms of the presegmented persons in the images as proposed in [5].

target point estimation (correct radius *and* correct angle)						
correct samples in %	k-NN	NG	SOM	LLM	PSOM	MLP
Gaborfilters	11,12%	4,70%	6,70%	11,76%	3,47%	29,16%
Gaborfilters and BG Subtraction (BGS)	22,28%	17,72%	15,34%	23,53%	10,65%	44,90%
Gaborfilters and Discriminant Analysis	17,69%	9,38%	11,66%	16,04%	7,72%	28,14%
Gaborfilters, BGS and Discriminant Analysis	34,72%	22,66%	23,66%	31,74%	15,21%	50,63%
Histogram Features and BG Subtraction	22,93%	14,61%	10,96%	28,61%	4,15%	39,27%

Human 3D	Human 2D
62,90%	37,50%

Fig. 5. The results for the estimation of the target point of the pointing pose. The target point is determined by the radius r and the angle φ. Unlike Fig. 4, showing the separate results for the estimation of r and φ, here the results for the correct estimation of both values are shown. As in Fig. 4 the results of the human viewers (2D on computer screen, 3D in reality) are given for comparison.

computation time						
computation time in ms	k-NN	NG	SOM	LLM	PSOM	MLP
Gaborfilters	285 ms	72 ms	80 ms	63 ms	3308 ms	54 ms
Gaborfilters and BG Subtraction (BGS)	300 ms	87 ms	95 ms	78 ms	3323 ms	63 ms
Gaborfilters and Discriminant Analysis	137 ms	28 ms	33 ms	26 ms	1330 ms	21 ms
Gaborfilters, BGS and Discriminant Analysis	129 ms	38 ms	42 ms	35 ms	995 ms	31 ms
Histogram Features and BG Subtraction	224 ms	33 ms	40 ms	29 ms	1624 ms	21 ms

allowed max.
80 ms

Fig. 6. The computation times of the different methods. A method capable of running with a minimum of 12.5 images per second on our mobile robot has to process one image in less than 80 ms (Athlon 2800, SUSE Linux).

These results demonstrate, that a cascade of several MLPs as proposed in [4] is best suited to estimate the target position of a user's pointing pose on monocular images. A Background Subtraction and the information delivered by a Discriminant Analysis can be used to improve the results. The best system is capable of estimating r as good as humans with their binocular vision system in a real world environment and even better than humans estimating the target on 2D screens. The estimation of φ does not reach equally good values. The system is able to reach a result equally to humans on 2D screens, but it is not able to estimate the angle as good as humans in a real world setting. This is because the estimation of the depth of a target in a monocular image is difficult for both, human and function approximators.

In our experimental comparison, the LLM and the MLP deliver a better result than the SOM and the Neural Gas. We suppose this result is caused by the ability of the MLP and the LLM to better approximate the output function in regions with few examples. The cascade structure of the MLP approach as proposed in [4] makes it possible to estimate φ better than the other approaches. However, since r is estimated by a single MLP and the MLP-result for r is better than that of the other approaches, we believe that a cascade organization of the other Neural Networks would not lead to a better result than that achieved by the MLP cascade. The PSOM delivers a relatively bad result in comparison to the other approaches. This is based on the fact, that only few basis points could be used due to the very long computation time of the PSOM. Figure 6 finally shows the computation times of all methods. Except the k-NN and the PSOM, all methods are able to process more than 12.5 images per second at the robot's on board PC. The k-NN method needs a long running due to the many comparisons which are needed to get the best neighbours to given observations. The computation time of the PSOM is especially high because of the iterative gradient descent along the PSOM structure that is needed to get the best suited output.

6 Conclusion

We presented an experimental comparison of several re-implemented Neural Network based approaches for a demanding visual instruction task on a mobile system. Since our goal is to provide an approach, which copes with the task by means of low-cost sensors, we refered to approaches using monocular images. Of the relevant approaches a cascade of Multi-Layer Perceptrons proved to be best suited for this task. All methods profit from the use of a Background Subtraction and the information delivered by a Discriminant Analysis. The comparison of the different methods had shown, that the usage of Gaborfilters for feature extraction leads to better results than the histogram based features. The best system is able to estimate the radius r of the target point better than human subjects do, but there are still problems in estimating the angle φ of the target due to the use of monocular images. This problem could be reduced by means of a stereo camera, which delivers the lacking depth information. Possibly the angle of the estimated target might not be as important if an other application is chosen, for example, if the user is pointing at certain objects in the surroundings allowing a model-based pointing pose specification instead of a non-specific target point on the ground.

References

1. Haasch, A., Hofemann, N., Fritsch, J., Sagerer, G.: A Multi-Modal Object Attention System for a Mobile Robot. In: Int. Conf. on Intelligent Robots and Systems, pp. 1499–1504 (2005)

2. Nickel, K., Stiefelhagen, R.: Recognition of 3D-Pointing Gestures for Human-Robot-Interaction. In: European Conference on Computer Vision, pp. 28–38 (2004)
3. Nölker, C., Ritter, H.: Illumination Independent Recognition of Deictic Arm Postures. In: Proceedings of the 24th Annual Conference of the IEEE Industrial Electronics Society, Aachen, pp. 2006–2011 (1998)
4. Richarz, J., Martin, C., Scheidig, A., Gross, H-M.: There You Go! - Estimation Pointing Gestures in Monocular Images for Mobile Robot Instruction. In: Int. Symposium on Robot and Human Interactive Communication, pp. 546–551 (2006)
5. Takahashi, K., Tanigawa, T.: Remarks on Real-Time Human Posture Estimation from Silhouette Image using Neural Network. In: Proc. of the IEEE Int. Conf. on Systems, Man and Cybernetics: The Hague, pp. 370–375 (2004)
6. Krüger, V., Sommer, G.: Gabor Wavelet Networks for Efficient Head Pose Estimation. In: Image and Vision Computing 20(9-10), 665–672 (2002)
7. Stiefelhagen, R.: Estimating Head Pose with Neural Networks - Results on the Pointing04 ICPR Workshop Evaluation. In: Pointing04 ICPR, Cambridge, UK (2004)
8. Viola, P., Jones, M.: Rapid Object Detection using a Boosted Cascade of Simple Features. Proc. Conf. of Computer Vision and Patter Recognition 1, 511–518 (2001)
9. Gross, H.-M., Richarz, J., Mller, St., Scheidig, A., Martin, Chr.: Probabilistic Multimodal People Tracker and Monocular Pointing Pose Estimator for Visual Instruction of Mobile Robot Assistants. In: Proc. IEEE World Congress on Computational Intelligence (WCCI 2006), pp. 8325–8333 (2006)
10. Martinetz, T., Schulten, K.: A Neural-Gas Network Learns Topologies. In: Proc. of the ICANN 1991. Helsinki, pp. 397–402 (1991)
11. Kohonen, T.: Self-Organized Formation of Topologically Correct Feature Maps. Biological Cybernetics 43, 59–69 (1982)
12. Ritter, H.: Learning with the Self-Organizing Map. In: Kohonen, T., et al. (eds.) Artifical Neural Networks, pp. 379–384. Elsevier Science Publishers, Amsterdam (1991)
13. Walther, J.A., Ritter, H.: Rapid Learning with Parametrized Self-Organizing Maps. Neurocomputing 12, 131–153 (1996)

Real-Time Foreground-Background Segmentation Using Adaptive Support Vector Machine Algorithm

Zhifeng Hao[2], Wen Wen[1,*], Zhou Liu[3], and Xiaowei Yang[2]

[1] College of Computer Science and Engineering, South China University of Technology, Guangzhou, 510641, China
mathww@126.com
[2] School of Mathematical Science, South China University of Technology, Guangzhou, 510641, China
[3] National Lab of Pattern Recognition, Institute of Automation, Chinese Academy of Sciences, Beijing, 100080, China

Abstract. In this paper, a SVM regression based method is proposed for background estimation and foreground detection. Incoming frames are treated as time series and a fixed-scale working-set selecting strategy is specifically designed for real-time background estimation. Experiments on two representative videos demonstrate the application potential of the proposed algorithm and also reveal some problems underlying it. Both the positive and negative reports from our study offer some useful information for researchers both in the field of image processing and that of machine learning.

Keywords: Support vector machine, regression, foreground-background segmentation.

1 Introduction

Foreground-background segmentation is an essential procedure in real-time video surveillance. It has an important impact on the performance of object recognition and tracking in the successive stage. Fundamentally, foreground- background segmentation is the extraction of the interesting foreground regions or objects, given an incoming video of the scene. There have already been several algorithms to solve this problem, such as algorithms based on background subtraction, color distributions, motion, as well as range and stereo. One category of these algorithms is based on the assumption that the background is known and with very few fluctuations. In this case, a simple algorithm which subtracts the current picture from the stored background picture is used for the foreground detection [1]. However, this method obviously fails when real-world image sequences show a complex structured background and a frequently changing illumination (daylight, clouds, shadows). Therefore a background sequence has to be adapted by the input sequence, which motivates another category of segmentation algorithms. These methods address the problem of segmentation given a dynamic background and model the pixel-wise color distribution of the background via

* Corresponding author.

J. Marques de Sá et al. (Eds.): ICANN 2007, Part II, LNCS 4669, pp. 603–610, 2007.

statistical methods [2, 3]. They have advantages in fast learning and adapting. But they are based on a prior assumption of the distribution of the colors and fluctuations, under which unexpected peak numbers impair their segmentation ability.

To overcome the dilemma of background subtraction model and the statistical methods, researchers proposed to use an estimating-detecting structure for background-foreground segmentation. In this method, the current background is estimated using information from former frames and difference between the estimated value and current value is calculated to determine whether a moving object emerges in the video. One of the popular background estimating methods is Kalman filtering [4]. It is fundamentally a filtering algorithm based on linear time series (though it has non-linear version).

Recently, Support Vector Machine (SVM) has received increasing interest for its remarkable performance in pattern recognition field. It is a kernel based approach, which allows the use of linear, polynomial and RBF kernels and others that satisfy Mercer's condition [5, 6, 7, 8]. Both the theoretical foundation and empirical experiments demonstrate that SVM have good generalization ability for non-linear classification and estimation [6, 9]. Researchers have endeavored to applying SVM to nonlinear noise filters and time series predictions, and encouraging results have been reported [10, 11]. In this paper, SVM is used as a non-linear estimator to capture important information of the dynamic background and a foreground-background segmentation strategy is proposed based on it.

The proposed method is based on least squares support vector machines (LS-SVM), a modified version of standard support vector machines, which employ a sum squared error loss function and just requires solving a quadratic programming problem with equality constraints [12, 13]. LS-SVM has a fundamental relationship with ridge regression. Furthermore, enlightened by the adaptive LS-SVM [14], a specified LS-SVM with fixed-scale working set is designed for the dynamic background modeling.

The rest of this paper is organized as follows: Section 2 provides a brief review to LS-SVM. In Section 3 we propose an estimating-detecting strategy based on LS-SVM regression for foreground-background segmentation. The experimental results are given in Section 4. Finally the discussions and conclusions are presented in Section 5.

2 Brief Reviews to LS-SVM Regression

Given a training set of N samples $\{(x_i, y_i)\}_{i=1}^{N}$ with input features $x_i \in R^d$ and output value $y_i \in R$, the standard LS-SVM regression can be formulated as following optimization problem in primal space:

$$\min J(w,e) = \frac{1}{2} w^T w + \frac{1}{2} \gamma \sum_{i=1}^{N} e_i^2 \qquad (1)$$

$$s.t.\ y_i = w^T \varphi(x_i) + b + e_i, \quad i = 1, \ldots, N$$

In formula (1), $\varphi(\bullet): R^d \to R^{\tilde{d}}$ is a function which maps the input space into a so-called higher dimensional feature space. $w \in R^{\tilde{d}}$ is the weight vector in primal weight space, b is the bias term and $e_i, i = 0,1,\ldots,N$ are error variables.

Using Lagrange multipliers and matrix transformation, optimizing problem (1) can be converted to a set of linear equations in the dual space as formula (2). For more details, see [13]

$$\begin{bmatrix} 0 & 1_v^T \\ 1_v & \Omega + \frac{1}{\gamma} I \end{bmatrix} \begin{bmatrix} b \\ \alpha \end{bmatrix} = \begin{bmatrix} 0 \\ y \end{bmatrix} \tag{2}$$

In (2), $y = [y_1,\ldots,y_N]^T$, $1_v = [1,\ldots,1]^T$, $\alpha = [\alpha_1,\ldots,\alpha_N]^T$ and $\Omega_{ij} = \varphi(x_i)^T \varphi(x_j) = K(x_i, x_j)$, for $i = 1,\ldots N$, $j = 1,\ldots,N$. K is the kernel function, for example, linear kernel, polynomial kernel or RBF kernel. This implies that the training procedure of LS-SVM regression just requires solving a linear system. There have already been some fast training algorithms based on heuristic working-set selecting methods and matrix techniques [14, 16].

In primal weight space one has the model for the estimation of input value x.

$$y(x) = \sum_{i=1}^{N} \alpha_i K(x, x_i) + b \tag{3}$$

3 The Proposed Adaptive LS-SVM for Foreground-Background Segmentation

Literature [13, 15] reveals that LS-SVM has the ability to learn statistical information from the training dataset and for samples generated by the same distribution, it makes good estimations. Based on this property of LS-SVM, an estimating detecting structure is modeled to adapt the dynamic background. Let $\rho_k^{(r,c)}$ be the intensity of pixel (r,c) on frame k. $\{\rho_0^{(r,c)}, \rho_1^{(r,c)}, \ldots, \rho_T^{(r,c)}\}$ consist a series intensities from the incoming frames. Former intensities on each pixel are learned via LS-SVM and the current values are estimated. If the current intensity deviates far away from the estimated value, it is probably caused by a foreground object moving across this pixel. Basically, the procedure can be summarized as Fig. 1.

Two critical problems should be considered in this method. One is how to efficiently learn the statistical information from former frames. The other is what criteria can be used to determine whether the current color represent a pixel of a moving object, given the background itself is changing.

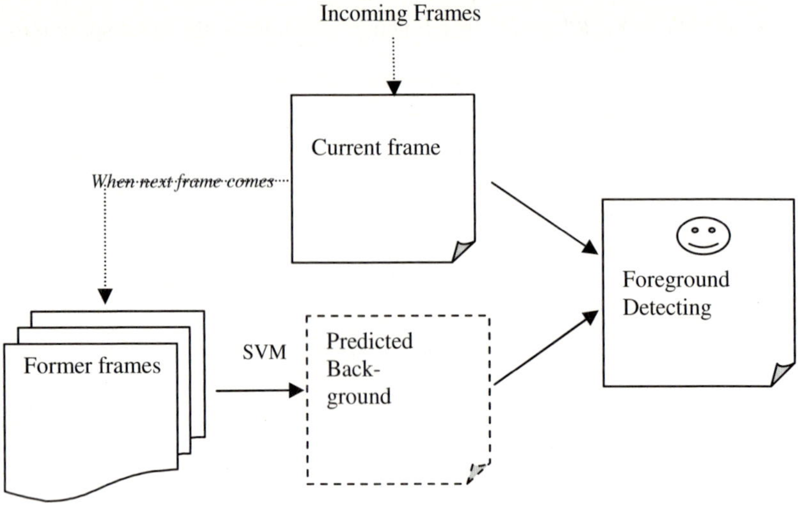

Fig. 1. Estimating-Detecting structure for foreground -background segmentation

A modified training method of LS-SVM is proposed to solve the first problem. For regularly changing cases, given some heuristic strategies, LS-SVM is able to approximate the training dataset via a few support vectors. Enlightened by the adaptive LS-SVM in [14], a few representative incoming frames are selected and added to a *working set* according to given conditions. LS-SVM regressor is trained using these samples. This strategy does not only save the computational time but also adaptively excludes the redundant information remote from the current frame. For the second problem, when a new frame comes, it is firstly estimated using the already trained SVM and then deviations between the predicted and real values are calculated. A dynamic threshold is employed to determine whether the pixel is a background pixel. The threshold is updated according to the errors between the real value and the estimated value.

Pseudocodes of the proposed algorithms are illustrated as Fig. 2. For simplification, ρ_t is used to denote the intensity of pixel (r,c) on the tth frame. In Fig. 2 , S can be viewed as a buffer area which temporarily stores the incoming frames. It is an intermediate strategy for working set selection. W is the number of samples contained in the *working set*. Empirically, it is set 3 to 7. The incoming frames are treated as time series in the proposed algorithm. A time window d is set on them. The input features of a training sample are intensity of pixels on consecutive d frames and the output feature is intensity of that pixel on frame next to them .It is based on the fact that under ordinary conditions, the color of a frame is most related to several frames coming before it. Besides, since the background is dynamic (for example, influenced by regularly changing illumination), an adaptive threshold e_t^* is used in the detecting stage. M in e_t^* is a constant set by the user. In our experiment, M is set 5.0. The updating strategy of the threshold is employed as the last line in Fig. 2.

```
//Initialize.
 S = {s_0, s_1,..., s_{d+W-1}} = {0,0,...,0}; // S is the working set
 t=0;
 //Training and segmentation when a new frame comes.
if(t<W+d){
s_t = ρ_t ;
 }
 Else {
 if( the SVM-training condition is satisfied){
//Incoming frames are treated as time series
//Construct W samples
for(i=0;i<W;i++)
 { x_i = (s_{0+i}, s_{1+i},..., s_{d-1+i}) ;
  y_i = s_{d+i} ;}
Use { (x_i, y_i), i = 0,1,ћ ,W − 1 } to train SVM.
}
 //Background-Foreground Segmentation
 Use SVM to Predict current color ρ̂_t ;
 e_t = |ρ_t − ρ̂_t| ;
 If( e_t < e*_t ) {
 Output that this pixel is a background pixel;}
 Else {
 Output that this pixel is a foreground pixel;}
 if( the updating condition is satisfied){
 for(i=1;i<W+d;i++)
 { s_{i-1} = s_i ; }
  s_i = ρ_t
 }
 t++;
 Adjust e*_t according to e*_t = M • [(t − 1) • e*_{t-1} + e_t]/t ;
 }
```

Fig. 2. Pseudocodes of the adaptive LS-SVM algorithm for real-time foreground-background segmentation

4 Experiments and Results

The proposed algorithm is imbedded in a video-processing framework using Microsoft's Visual C++. It is run on an HP personal computer, utilizing a 3.06GHz Pentium IV processor with a maximum of 512MB memory available. To evaluate the performance of the proposed algorithm, Gaussian mixture model (GMM), a widely used background model, is also implemented in this framework. Besides, RBF kernel with $\sigma = 1.0, \gamma = 100$ is used in the proposed algorithm. (The parameters are empirically selected by our experiments). Two videos are tested both by the proposed algorithm and GMM algorithm.

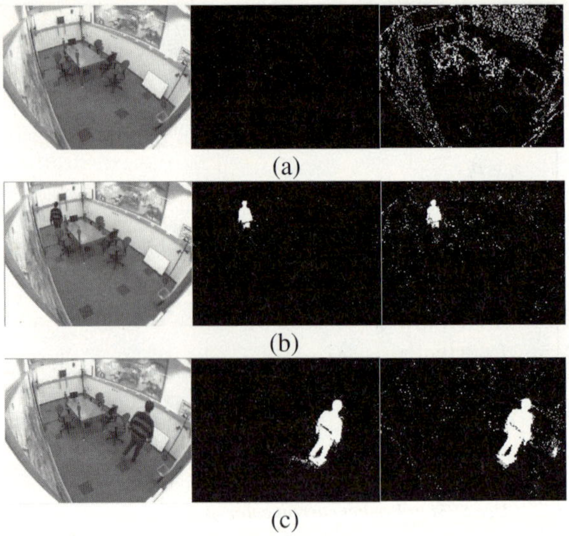

Fig. 3. (a)-(c): Comparisons between the proposed algorithm and GMM on video 1. Images on the left column are grabbed from the original video. Those on the middle column are generated by the proposed algorithm and the right column by GMM algorithm.

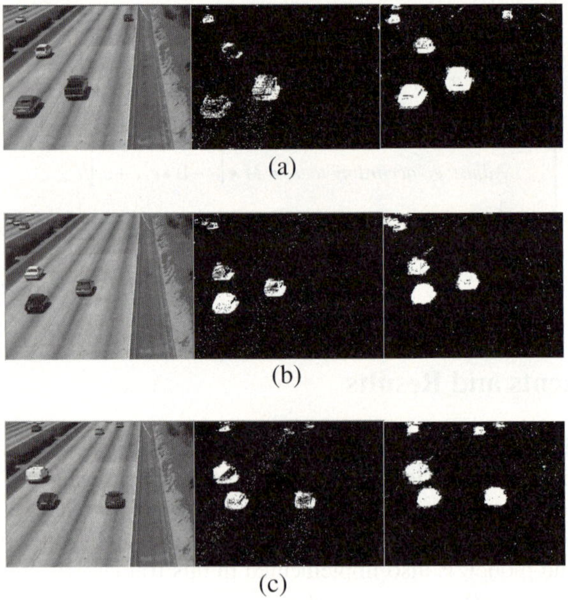

Fig. 4. (a)-(c): Comparisons between the proposed algorithm and GMM on video 2. Images on the left column are grabbed from the original video. Those on the middle column are generated by the proposed algorithm and the right column by GMM algorithm.

Video 1 consists of 300 incoming frames, with 10 frames per second. It is taken in a room for the surveillance of moving objects. Regular changing of illumination make it has a non-static background. At the beginning of this video, there are several frames without any foreground objects. Fig. 3. (a), (b) and (c) illustrate the comparisons between the proposed algorithm and GMM algorithm. It is clear that the proposed algorithm is able to segment the foreground objects from the video while keeping fewer noises than GMM algorithm. This implies that it has good ability to learn the statistical information from the background and in this case the proposed algorithm has a comparable performance. This enlightened us that the proposed algorithm has the potential for background modeling and it once again demonstrates the good statistical learning ability of SVM regression.

However, experiments also exhibit some underlying problems. Video 2 is a video with 500 frames. It is taken on a highway with running cars across the screen from the beginning to the end. Besides, the camera is not fixed but has a regular small vibration, causing the background regular changing. In this case, the proposed algorithm exhibits worse performance than GMM. Two factors cause this. One is that, in video 2, color contrast between the foreground and background is not so sharp as video 1. And unfortunately, the proposed algorithm uses intensity, which makes it lose some color information. The other factor is that, the powerful learning ability of SVM has become a two-edged sword in this case. Since there are moving objects even at the beginning of the video, successive running foreground objects seriously disturb the learning procedure of SVM regression, making it lose the ability for correctly segmenting the foreground objects from background.

Furthermore, experiments show that different values of d in the proposed algorithms have small impact on the effect of the segmentation, namely, results obtained under $d=1$ are almost similar to those under $d=2,3,4,5$. This might be explained by the reason that the current frame is mostly influenced by the preceding frame closest to it. Yet, too large value of W will slow down the running speed of the segmentation procedure. Results presented in Fig. 3 and Fig. 4 are those with the parameters $W=6$ and $d=1$.

5 Discussions and Conclusions

Based on LS-SVM regression, a background-foreground segmentation method is proposed in this paper. It treats successive frames of a video as time series. A fixed-scale working set selecting strategy is specifically designed for the segmenting procedure. It opens a door for the application of SVM regression in dynamic background modeling. Experiments on two videos illustrate the potential of the real-world application of the proposed algorithm. They also demonstrate the good statistical learning ability of LS-SVM regression.

However, studies of this paper also reveal that a lot of work should be done before perfect results are reached. Intensity, instead of complete information of colors, is used in the proposed algorithm for the reason of reducing the computational time. If fast training method, such as the adaptive incremental-decremental algorithm is developed and implemented, full color information can be included. Besides, the working-set updating strategy has to be further improved to suit the application of LS-SVM regression in background modeling. A more self-adaptive strategy should be designed so that the segmentation procedure will be more robust even when there are a lot of moving objects at the beginning of the video. This is what we plan to study in the future.

References

1. Leung, M.K., Yang, Y.H.: Human body motion segmentation in a complex scene. Pattern Recognition 20, 55–64 (1987)
2. Stauffer, C., Grimson, W.E.L.: Adaptive back-ground mixture models for real-time tracking. Proc. of IEEE Conf. on Computer Vision and Pattern Recognition 2, 246–252 (1999)
3. Elgammal, A., Harwood, D., Davis, L.: Non-parametric model for background subtraction. Proc. of European Conference on Computer Vision 2, 751–767 (2000)
4. Ridder, C., Munkelt, O., Kirchner, H.: Adaptive background estimation and foreground detection using Kalman filtering. In: Proc. Of Int. Conf. on Recent Advances in Mechatronics, pp. 193–199 (1995)
5. Vapnik, V.N.: Estimation of Dependencies Based on Empirical Data. Springer, Berlin (1982)
6. Vapnik, V.N.: The Nature of Statistical Learning Theory. Springer, New York (1995)
7. Vapnik, V.N., Golowich, S., Smola, A.: Support vector method for function approximation, regression estimation, and signal processing, Neural Information Processing Systems (eds). MIT Press, Redmond, Washington (1997)
8. Smola, A.J., Scholkopf, B.: A Tutorial on Support vector Regression. Statistics and computing 14, 199–222 (2004)
9. Cortes, C., Vapnik, V.N.: Support vector networks. Machine Learning 20, 273–297 (1995)
10. Wu, C.H., Ho, J.M., Lee, D.T.: Travel-Time Prediction with Support Vector Regression. IEEE Transactions on Intelligent Transportation Systems 5, 276–281 (2004)
11. Cao, L.J., Tay, F.E.H.: Support vector machine with adaptive parameters in financial time series forecasting. IEEE Transactions on Neural Networks 14, 1506–1518 (2003)
12. Suykens, J.A.K., Vandewa, J.: Least Squares Support Vector Machine Classifiers. Neural Processing Letters 9, 200–293 (1999)
13. Suykens, J.A.K., Brabanter, J.D., Lukas, L., Vandewalle, J.: Weighted Least Squares Support Vector Machines: Robustness and Sparse Approximation. Neurocomputing 48, 85–105 (2002)
14. Yu, S., Yang, X.W., Hao, Z.F., Liang, Y.C.: An Adaptive Support Vector Machine Learning Algorithm for Large Classification Problem. In: Wang, J., Yi, Z., Zurada, J.M., Lu, B.-L., Yin, H. (eds.) ISNN 2006. LNCS, vol. 3971, pp. 981–990. Springer, Heidelberg (2006)
15. Suykens, J.A.K., Vandewalle, J.: Recurrent least squares support vector machines. IEEE Transactions on Circuits and Systems I: Fundamental Theory and Applications 47, 1109–1114 (2000)
16. Liu, J.H., Chen, J.P., Jiang, S., Cheng, J.S.: Online LS-SVM for function and classification. Journal of University of Science and Technology, Beijing 10, 73–77 (2003)

Edge-Preserving Bayesian Image Superresolution Based on Compound Markov Random Fields

Atsunori Kanemura, Shin-ichi Maeda, and Shin Ishii

Nara Institute of Science and Technology, Nara 630-0192, Japan
{atsu-kan,ichi,ishii}@is.naist.jp
http://hawaii.naist.jp/

Abstract. This study deals with image superresolution problems simultaneously with accompanying image registration problems. The goal of superresolution is to generate a high resolution image by integrating low-resolution degraded observed images. We propose a Bayesian approach whose prior is modeled as a compound Gaussian Markov random field (MRF). This approach is advantageous in preserving discontinuity in the original image, in comparison to the existing single-layer Gaussian MRF models. Maximum-marginalized-likelihood estimation of the registration parameters is carried out by a variational EM algorithm where hidden variables are marginalized out and the posterior distribution is approximated by a factorized trial distribution. High resolution image estimates are obtained as by-products of the EM algorithm. Experiments show that our Bayesian approach with two-layer compound models exhibits better performance in terms of mean square error and visual quality than the single-layer model.

1 Introduction

Superresolution aims at recovering visual information that has been lost in the imaging process by integrating multiple observed images with low resolution so as to output a higher-resolution (HR) image than defined in the imaging process. An earliest superresolution algorithm was based on a frequency-domain approach by Tsai & Huang [1]. After that, a number of superresolution algorithms have been proposed, as summarized in recent review articles [2,3,4].

Bayesian methods developed originally in neural computation literature [5,6] have recently been applied to image superresolution [7,8] and have yielded fruitful results by successfully avoiding overfitting. However, image models assumed in these studies are not enough because edge preservation is not considered well. Natural images contain edges (discontinuity in pixel values) naturally originating from textures and occlusions of the objects in the scene. Therefore, image models incorporating edge information are desired.

Superresolution often has a problem that the estimation of the HR image cannot be separated from the estimation of registration parameters for the imaging process. If we try to estimate both the parameters and the HR image

J. Marques de Sá et al. (Eds.): ICANN 2007, Part II, LNCS 4669, pp. 611–620, 2007.

simultaneously, we have to search for an optimum point in the very high-dimensional product space composed of possible parameters and HR images. Selecting a single estimate out of a high-dimensional space is faced with the danger of overfitting. A solution to this overfitting problem accompanying the joint optimization is provided by Bayesian marginalization [7,9], where the unknown HR image is integrated out from the search space, which improved the registration accuracy over joint MAP methods [10]. However, since the existing models are based on Gaussian Markov random fields (MRFs) with only a single layer, discontinuity stemming from possible edges is often ruined in the estimated image.

In this article, we propose a Bayesian superresolution scheme for compound MRFs, which have an additional layer of edge representation adaptively inferred from the data. The proposed scheme not only avoids overfitting but also preserves the edge information. Marginalization of this two-layer model is analytically intractable, and then an approximate procedure based on the variational EM algorithm is derived.

The organization of the rest of this article is as follows. In Sect. 2, we briefly review Bayesian superresolution methods. Section 3 describes a linear Gaussian probabilistic forward model for the image formation process. In Sect. 4, we propose compound MRFs used for the prior model and describe the difficulty in marginalizing compound MRFs. In Sect. 5, we present a variational EM algorithm with efficient factorization approximation. Section 6 shows several experimental results and Sect. 7 concludes the article.

2 Bayesian Superresolution

Suppose there are observed T images $\mathcal{D} = \{\mathbf{y}_t\}_{t=1}^{T}$ where each of the observed images $\mathbf{y}_t$ has the size of $M_\mathrm{O} \times N_\mathrm{O}$. The goal of superresolution here is to estimate an $M_\mathrm{H} \times N_\mathrm{H}$ HR image $\mathbf{x}$, where $r = M_\mathrm{H}/M_\mathrm{O} = N_\mathrm{H}/N_\mathrm{H}$ is called a resolution enhancement factor. The numbers of pixels in observed and HR images are denoted by $P_\mathrm{O} = M_\mathrm{O}N_\mathrm{O}$ and $P_\mathrm{H} = M_\mathrm{H}N_\mathrm{H}$, respectively. The images are regarded as lexicographically stacked vectors.

Each observed image is produced from the unknown HR image $\mathbf{x}$ by the imaging process described by unknown parameters $\boldsymbol{\theta}$. We estimate the parameters by maximizing the marginalized likelihood $p(\mathcal{D}|\boldsymbol{\theta})$:

$$\hat{\boldsymbol{\theta}} = \arg\max_{\boldsymbol{\theta}} p(\mathcal{D}|\boldsymbol{\theta}). \tag{1}$$

The marginalized likelihood is derived by integrating the product of the prior (the preference about the HR image) and the likelihood (the imaging process):

$$p(\mathcal{D}|\boldsymbol{\theta}) = \int d\mathbf{x}\ p(\mathbf{x}) \prod_t p(\mathbf{y}_t|\mathbf{x}, \boldsymbol{\theta}). \tag{2}$$

After obtaining the estimate $\hat{\boldsymbol{\theta}}$, we can estimate the HR image based on the posterior distribution $p(\mathbf{x}|\mathcal{D}, \hat{\boldsymbol{\theta}})$, which is again obtained from the prior and the likelihood according to the Bayes rule:

$$p(\mathbf{x}|\mathcal{D}, \hat{\boldsymbol{\theta}}) = \frac{p(\mathbf{x})p(\mathcal{D}|\mathbf{x}, \hat{\boldsymbol{\theta}})}{p(\mathcal{D}|\hat{\boldsymbol{\theta}})}. \tag{3}$$

We have to specify the two fundamental distributions $p(\mathbf{x})$ and $p(\mathbf{y}_t|\mathbf{x}, \boldsymbol{\theta})$. The prior $p(\mathbf{x})$ embodies our a priori desire on the HR image and is usually chosen as an MRF that poses smoothness constraints. The likelihood $p(\mathbf{y}_t|\mathbf{x}, \boldsymbol{\theta})$ is designed according to the image formation process from the HR image $\mathbf{x}$ to an observed image $\mathbf{y}_t$. The parameter vector $\boldsymbol{\theta}$ represents the parameters in the observation process, such as the amount of shift and the angle of rotation.

The prior $p(\mathbf{x})$ and the likelihood $p(\mathbf{y}_t|\mathbf{x}, \boldsymbol{\theta})$ must be set carefully so that they should faithfully represent the physical process and that the integration of (2) should be performed efficiently. Amongst the existing Bayesian superresolution methods [7,9], only the Gaussian distributions have been employed because of their mathematical tractability. However, the Gaussian priors represent merely smoothness constraints between neighboring pixels but do not represent the edges that should be contained in many HR images.

3 Image Formation Process

It is assumed that there is an underlying image formation process from $\mathbf{x}$ to $\mathbf{y}_t$ such that the HR image $\mathbf{x}$ is geometrically warped, blurred with a point spread function (PSF), downscaled, and corrupted by Gaussian noise to form the observed low resolution image $\mathbf{y}_t$:

$$\mathbf{y}_t = W_t\mathbf{x} + \boldsymbol{\varepsilon}_t \quad \text{for } t = 1, \ldots, T, \tag{4}$$

where W_t is the matrix that operates warping, blurring, and downscaling, and $\boldsymbol{\varepsilon}_t$ is the additive noise that obeys iid Gaussian $\mathrm{N}(\mathbf{0}, \beta^{-1}I)$[1], where β is the inverse variance of the isotropic Gaussian. Figure 1 depicts this process. Although this formation process is represented by the linear matrix W_t consisting of a huge number of elements, it is governed by a relatively small number of parameters such as registration parameters $\boldsymbol{\theta}$ and the resolution enhancement factor r.

The image formation process (4) is equivalently represented as the probabilistic forward model

$$p(\mathbf{y}_t|\mathbf{x}, \boldsymbol{\theta}) = \mathrm{N}(\mathbf{y}_t|W_t(\boldsymbol{\theta})\mathbf{x}, \beta^{-1}I) \quad \text{for } t = 1, \ldots, T. \tag{5}$$

The important advantage of such linear Gaussian likelihood is mathematical tractability of the integration in the marginalized likelihood (2) when it is combined with a Gaussian prior for HR images.

[1] $\mathrm{N}(\boldsymbol{\mu}, \Sigma)$ denotes the Gaussian distribution with mean $\boldsymbol{\mu}$ and covariance Σ, and $\mathrm{N}(\mathbf{y}|\boldsymbol{\mu}, \Sigma)$ denotes the Gaussian density function of $\mathbf{y}$.

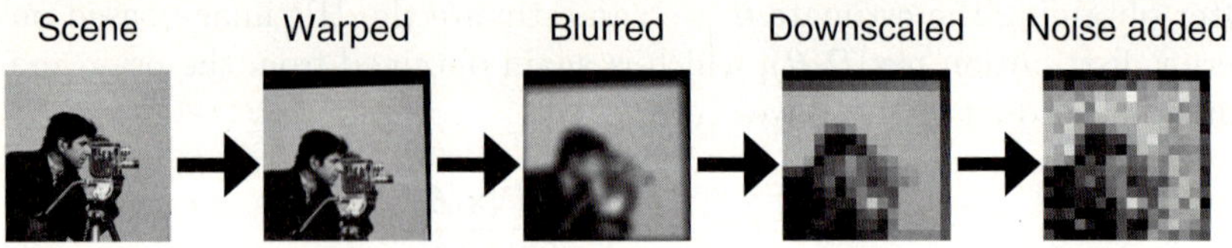

Fig. 1. An illustration of the image forming process

4 Edge-Preserving Compound MRF Prior

A compound MRF is an MRF with an additional latent variable layer called the line process that controls the local correlation among pixel values [11,12]. The introduction of the latent variable enables explicit expression of the possible discontinuity in the HR image. The latent variable is adaptively inferred from the data so that edge preservation should be achieved.

We introduce several notations about neighborhood relations as follows: $\mathcal{N}(i)$ is the set of neighboring pixels of the pixel i, and $i \sim j$ stands for "the pixels i and j are adjacent."

The line process $\boldsymbol{\eta}$ consists of the binary latent variables $\eta_{ij} \in \{0, 1\}$ for the neighboring pixel pairs i and j. Then, the compound MRF is expressed as

$$p(\boldsymbol{\eta}, \mathbf{x}) = \frac{1}{Z} \exp\left\{-\frac{\rho}{2} E(\boldsymbol{\eta}, \mathbf{x}) - \Psi(\boldsymbol{\eta})\right\}, \tag{6}$$

where ρ is an inverse temperature; E is an energy function that defines the characteristics of the probability distribution; and $\Psi(\boldsymbol{\eta}) = \log \int d\mathbf{x} \exp\{-\rho E(\boldsymbol{\eta}, \mathbf{x})/2\}$ and $Z = \sum_{\boldsymbol{\eta}} \exp\{-\rho E(\boldsymbol{\eta}, \mathbf{x})/2 - \Psi(\boldsymbol{\eta})\}$ guarantee that the distribution is normalized. In particular, we use the following energy function[2]:

$$E(\boldsymbol{\eta}, \mathbf{x}) = \sum_{i \sim j} [\eta_{ij} \cdot (x_i - x_j)^2 + (1 - \eta_{ij}) \cdot \lambda], \tag{7}$$

where the summation $\sum_{i \sim j}$ is taken over all pairs of neighboring pixels. The latent variable $\boldsymbol{\eta}$ switches the local characteristics of the prior by indicating whether a pair of pixels take similar values or take independent values. When $\eta_{ij} = 1$, the pixels i and j are strongly smoothed due to the quadratic penalty, whereas there is no smoothing when $\eta_{ij} = 0$. The mixture of the quadratic and constant penalties accomplishes one form of robust estimation; if MAP estimation is employed, this energy function reduces to Hampel's robust cost function [13]. The "smoothing all" case $\boldsymbol{\eta} = \mathbf{1}$ is equivalent to the single-layer Gaussian MRF model in which all pixel pairs are smoothed.

We can rewrite the compound MRF with the line process (6) as

$$p(\boldsymbol{\eta}, \mathbf{x}) = p(\boldsymbol{\eta}) p(\mathbf{x}|\boldsymbol{\eta}), \tag{8}$$

[2] In fact, $\boldsymbol{\eta} = \mathbf{0}$ gives an improper prior, but we can exclude this case by letting $E(\mathbf{0}, \mathbf{x}) = \infty$.

where

$$p(\boldsymbol{\eta}) = \mathrm{Ber}(\boldsymbol{\eta}|\nu) , \qquad p(\mathbf{x}|\boldsymbol{\eta}) = \mathrm{N}(\mathbf{x}|\mathbf{0}, \rho^{-1}A_{\boldsymbol{\eta}}^{-1}). \tag{9}$$

Here, $\nu = \mathrm{sig}(\lambda\rho/2) \equiv 1/(1+\exp(-\lambda\rho/2))$ is the parameter for the Bernoulli distribution $\mathrm{Ber}(\boldsymbol{\eta}|\nu) = \prod_{i\sim j} \nu^{\eta_{ij}}(1-\nu)^{1-\eta_{ij}}$ and the matrix $A_{\boldsymbol{\eta}}$ is defined by

$$[A_{\boldsymbol{\eta}}]_{ij} = \begin{cases} \sum_{k\in\mathcal{N}(i)} \eta_{ik} , & i = j , \\ -\eta_{ij} , & i \sim j , \\ 0 , & \text{otherwise.} \end{cases} \tag{10}$$

To estimate the registration parameters according to the maximum-marginalized-likelihood (MML) rule (1), we need to compute the marginalized likelihood. The marginal prior

$$p(\mathbf{x}) = \sum_{\boldsymbol{\eta}} p(\boldsymbol{\eta}, \mathbf{x}) \tag{11}$$

is used in (2), which gives the marginalized likelihood for the compound model

$$p(\mathcal{D}|\boldsymbol{\theta}) = \sum_{\boldsymbol{\eta}} \int d\mathbf{x}\; p(\boldsymbol{\eta}, \mathbf{x}) \prod_t p(\mathbf{y}_t|\mathbf{x}, \boldsymbol{\theta}). \tag{12}$$

Unfortunately, since this marginalized likelihood has the summation $\sum_{\boldsymbol{\eta}}$ over all $2^{(M_{\mathrm{O}}-1)(N_{\mathrm{O}}-1)}$ configurations of η_{ij}, which requires exponential order of computational complexity, a direct implementation of the above algorithm is intractable. Therefore, some approximation techniques are required for a practical implementation for the compound model. We present a tractable variational EM algorithm in the following section.

5 Variational EM Estimation

5.1 Variational EM Algorithm

First, we review a variational view of the EM algorithm suggested by Neal & Hinton [14], in which the EM algorithm is formulated as minimization of the variational free energy $F(q, \boldsymbol{\theta})$, which is a functional of the trial distribution q and a function of the parameter vector $\boldsymbol{\theta}$. The trial distribution $q(\boldsymbol{\eta}, \mathbf{x})$ is an arbitrary probability distribution for the unknown variables $\boldsymbol{\eta}$ and $\mathbf{x}$. Although the trial distribution can take any form in principle, we assume for the sake of tractability that it possesses a factorized form: $q(\boldsymbol{\eta}, \mathbf{x}) = q(\boldsymbol{\eta})q(\mathbf{x})$. Owing to this assumption, the computational complexity can be relaxed into polynomial order, if log-determinant elimination is performed as will be shown later.

The free energy is given by

$$F(q, \boldsymbol{\theta}) = -\sum_{\boldsymbol{\eta}} \int d\mathbf{x}\; q(\boldsymbol{\eta}, \mathbf{x}) \log \frac{p(\boldsymbol{\eta}, \mathbf{x}, \mathcal{D}|\boldsymbol{\theta})}{q(\boldsymbol{\eta}, \mathbf{x})}. \tag{13}$$

The joint minimum point of the free energy $[\hat{q}, \hat{\boldsymbol{\theta}}] = \arg\min_{q,\boldsymbol{\theta}} F(q, \boldsymbol{\theta})$ gives the MML estimate $\hat{\boldsymbol{\theta}}$ for the registration parameters and the trial distribution $q(\boldsymbol{\eta}, \mathbf{x})$ that optimally approximates a posterior distribution $p(\boldsymbol{\eta}, \mathbf{x}|\mathcal{D}, \boldsymbol{\theta})$ in the Kullback-Leibler divergence sense [14]. Therefore, both of the MML estimate (1) and the posterior distribution (3) are obtained by the free energy minimization; this is Bayesian superresolution.

The EM algorithm minimizes the free energy according to an iterative procedure of a coordinate-descent type; after the initialization of the parameter vector $\boldsymbol{\theta}^0$, the lth ($l \geq 1$) iteration performs the following E and M steps:

$$\text{E-step:} \quad q^l = \arg\min_q F(q, \boldsymbol{\theta}^{l-1}), \tag{14}$$

$$\text{M-step:} \quad \boldsymbol{\theta}^l = \arg\min_{\boldsymbol{\theta}} F(q^l, \boldsymbol{\theta}). \tag{15}$$

These steps are iterated until the convergence is attained. The M-step optimizes the parameters towards the MML direction, whereas the E-step optimizes the trial distribution so as to approach the true posterior distribution $p(\boldsymbol{\eta}, \mathbf{x}|\mathcal{D}, \boldsymbol{\theta})$ [14].

5.2 Free Energy Calculation

Since the joint density $p(\boldsymbol{\eta}, \mathbf{x}, \mathcal{D}|\boldsymbol{\theta})$ in the numerator of the free energy (13) can be decomposed as the product of densities of the three layers, the free energy is calculated as

$$F(q, \boldsymbol{\theta}) = -\langle \ln p(\boldsymbol{\eta}) \rangle - \langle \ln p(\mathbf{x}|\boldsymbol{\eta}) \rangle - \langle \ln p(\mathcal{D}|\mathbf{x}, \boldsymbol{\theta}) \rangle + \langle \ln q(\boldsymbol{\eta}, \mathbf{x}) \rangle, \tag{16}$$

where the angle bracket pair $\langle \cdot \rangle$ denotes the expectation with respect to $q(\boldsymbol{\eta}, \mathbf{x})$. Each term is calculated as

$$-\langle \ln p(\boldsymbol{\eta}) \rangle = -\ln \nu \sum_{i\sim j} \langle \eta_{ij} \rangle - \ln(1-\nu) \sum_{i\sim j} (1 - \langle \eta_{ij} \rangle), \tag{17}$$

$$-\langle \ln p(\mathbf{x}|\boldsymbol{\eta}) \rangle = \frac{P_{\mathrm{H}}}{2} \ln(2\pi/\rho) - \frac{1}{2} \langle \ln |A_{\boldsymbol{\eta}}| \rangle + \frac{\rho}{2} \langle \mathbf{x}^{\mathrm{T}} A_{\boldsymbol{\eta}} \mathbf{x} \rangle, \tag{18}$$

$$-\langle \ln p(\mathcal{D}|\mathbf{x}, \boldsymbol{\theta}) \rangle = \frac{T P_{\mathrm{O}}}{2} \ln(2\pi/\beta) + \frac{\beta}{2} \sum_t \langle \| \mathbf{y}_t - W_t(\boldsymbol{\theta}) \mathbf{x} \|^2 \rangle, \tag{19}$$

$$\langle \ln q(\boldsymbol{\eta}) \rangle = \sum_{i\sim j} [\langle \eta_{ij} \rangle \ln \langle \eta_{ij} \rangle + (1 - \langle \eta_{ij} \rangle) \ln(1 - \langle \eta_{ij} \rangle)], \tag{20}$$

$$\langle \ln q(\mathbf{x}) \rangle = -\frac{P_{\mathrm{H}}}{2} \ln(2\pi e) - \frac{1}{2} \ln |\Sigma|. \tag{21}$$

The log-determinant term $\langle \ln |A_{\boldsymbol{\eta}}| \rangle = \sum_{\boldsymbol{\eta}} q(\boldsymbol{\eta}) \ln |A_{\boldsymbol{\eta}}|$ in (18) is intractable again due to the summation over exponential-order configurations. Here, we simply ignore this term. This type of log-determinant elimination has been implicitly performed in many studies (e.g. [12,15]) and does not seem to bring serious problems.

5.3 E-Step: Optimal Trial Distribution

Under the assumption of the factorized trial posterior and the introduction of log-determinant elimination, the optimal trial distribution is given analytically as

$$q^*(\mathbf{x}) = \mathrm{N}(\mathbf{x}|\boldsymbol{\mu}, \Sigma), \tag{22}$$

$$\text{where } \Sigma = \Big(\rho\langle A_{\boldsymbol{\eta}}\rangle + \beta \sum_{t=1}^{T} W_t^{\mathrm{T}} W_t\Big)^{-1}, \quad \boldsymbol{\mu} = \beta\Sigma\Big(\sum_{t=1}^{T} W_t^{\mathrm{T}} \mathbf{y}_t\Big) \tag{23}$$

and

$$q^*(\boldsymbol{\eta}) = \mathrm{Ber}(\boldsymbol{\eta}|\bar{\boldsymbol{\nu}}) = \prod_{i\sim j} \bar{\nu}_{ij}^{\eta_{ij}} (1-\bar{\nu}_{ij})^{1-\nu_{ij}}, \tag{24}$$

$$\text{where } \bar{\nu}_{ij} = \mathrm{sig}\Big(\frac{\rho}{2}(\lambda - \langle (x_i - x_j)^2\rangle)\Big). \tag{25}$$

Compared to the single layer model in which $q(\boldsymbol{\eta})$ is fixed at $\mathbf{1}$, the compound model can adaptively change the effect of smoothing by balancing λ and $\langle (x_i - x_j)^2\rangle = (\mu_i - \mu_j) + \Sigma_{ii} + \Sigma_{jj} - 2\Sigma_{ij}$. Since the expectation $\langle (x_i - x_j)^2\rangle$ contains variance terms, the uncertainty about pixel values is taken into account, which is beneficial to avoid being too sensitive against noise. And the magnitude of the inverse temperature ρ changes the gradient of the sigmoid function.

5.4 M-Step: Optimizing Parameters

Since the free energy nonlinearly depends on the registration parameters, optimization based on scaled conjugate gradients [16] is employed in the M-step. We can reduce computational efforts by calculating only the term that is dependent on $\boldsymbol{\theta}$, $\langle \|\mathbf{y}_t - W_t(\boldsymbol{\theta})\mathbf{x}\|^2\rangle = \|\mathbf{y}_t - W_t(\boldsymbol{\theta})\boldsymbol{\mu}\|^2 + \mathrm{tr}\,(W_t(\boldsymbol{\theta})^{\mathrm{T}} W_t(\boldsymbol{\theta})\Sigma)$, in which the uncertainty about the HR image is also considered via the posterior covariance.

6 Experiments

We conducted experiments with synthetically generated data sets, and the results by the compound MRF model (CGMRF) were compared with those by the single-layer MRF model (SGMRF).

The data sets were generated by the following procedure. A given original image $\mathbf{x}$ was first transformed by translational and rotational motions, where the amounts of shift and rotation were randomly drawn from the respective uniform distributions $\mathrm{Unif}(-2, 2)$ and $\mathrm{Unif}(-4/(180\pi), 4/(180\pi))$, blurred with a Gaussian PSF with a standard deviation of 2, then downscaled by a factor of $r = 4$, and finally corrupted by Gaussian noise of signal to noise ratio (SNR) 30 dB to generate $T = 16$ observed images $\{\mathbf{y}_t\}$. Figure 2 shows six original images of the size of 40×40; the first five were clipped from public standard

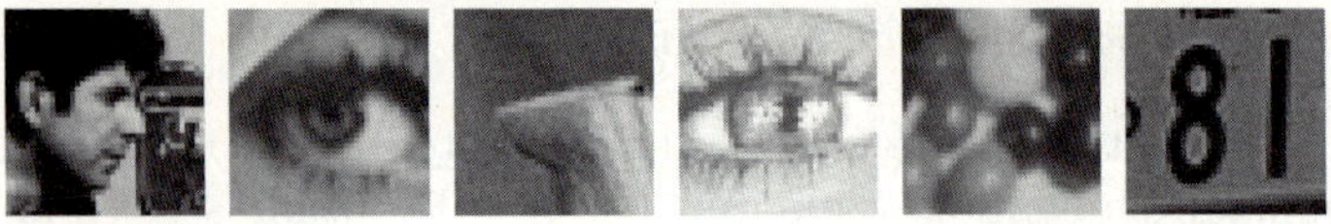

Fig. 2. Six original images used in the experiments. From left to right, Cameraman, Lenna, Peppers, Girl, Beads, License.

Table 1. Mean ISNR with standard deviation in 30 experiments for 6 different images

Model		CGMRF [dB]	SGMRF [dB]
Image	Cameraman	4.92 ± 0.09	4.49 ± 0.06
	Lenna	10.17 ± 0.14	9.64 ± 0.13
	Peppers	6.34 ± 0.13	6.09 ± 0.09
	Girl	7.81 ± 0.05	7.49 ± 0.05
	Beads	10.23 ± 0.19	9.75 ± 0.14
	License	7.66 ± 0.34	7.39 ± 0.09
Total		7.86 ± 1.92	7.47 ± 1.86

images and the sixth License picture was taken by the authors. From each of the original images, 30 different data sets were generated.

The EM algorithm started with initial translational and rotational motion parameters being set at all zero. The EM algorithm was terminated when all of the following three criteria were satisfied: $F^{l+l} - F^l < 10^{-3}$, $\|\boldsymbol{\mu}^{l+1} - \boldsymbol{\mu}^l\|/\|\boldsymbol{\mu}^l\| < 10^{-4}$, and $\|\boldsymbol{\theta}^{l+1} - \boldsymbol{\theta}^l\|/\|\boldsymbol{\theta}^l\| < 10^{-5}$.

For every 180 data sets, both CGMRF and SGMRF superresolution methods of $r = 4$ resolution enhancement were carried out. The quality of the resultant HR images is measured by ISNR (improvement in SNR) defined by

$$ISNR = 10 \log_{10} \frac{\|\tilde{\mathbf{x}} - \mathbf{x}\|^2}{\|\hat{\mathbf{x}} - \mathbf{x}\|^2} \quad [\text{dB}], \tag{26}$$

where $\hat{\mathbf{x}}$ is an estimated HR image, and $\tilde{\mathbf{x}}$ is the bilinearly interpolated image of the first frame $\mathbf{y}_1$. The results are summarized in Table 1. Our edge-adaptive CGMRF model achieved about 0.4 dB improvement on average for all the six images, in comparison to the SGMRF model.

One of the results for Cameraman is shown in Fig. 3, where the HR images estimated by the CGMRF and SGMRF models are displayed together with one of the observed images and an extracted edge field in CGMRF. Figure 4 shows the results for License. Both the CGMRF and SGMRF models exhibit clear resolution enhancement beyond that of the observed image, but the image reconstructed by the CGMRF model yields more enhanced edges.

Throughout the experiments, the hyperparameter values were fixed at $\nu = 0.55$, $\rho = 20$, and $\beta = 3000$, suggesting the robustness of the algorithm against image changes. Although different hyperparameters would prefer different images, we did not focus on hyperparameter optimization in this study.

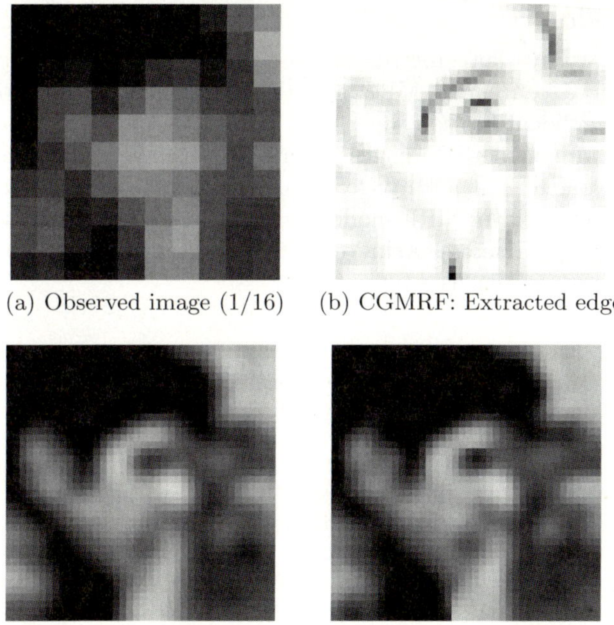

(a) Observed image (1/16) (b) CGMRF: Extracted edge

(c) SGMRF: Estimated image (d) CGMRF: Estimated image

Fig. 3. Superresolution of Cameraman

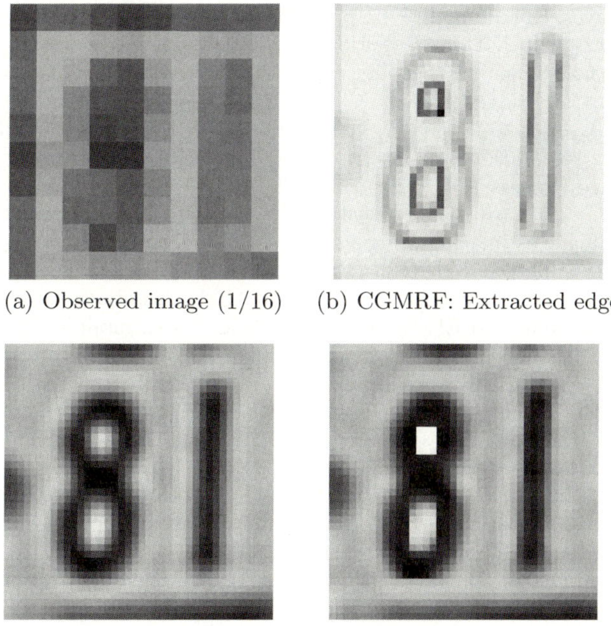

(a) Observed image (1/16) (b) CGMRF: Extracted edge

(c) SGMRF: Estimated image (d) CGMRF: Estimated image

Fig. 4. Superresolution of License

7 Conclusion

We presented a Bayesian superresolution algorithm that explicitly models the edges by using a compound MRF and thus controls the effect of smoothing adaptively. Edge identification was robust and based on the noise-compensated posterior estimate of the HR images. Overfitting of registration parameters was successfully avoided. The advantage of our superresolution algorithm over the non-adaptive, single-layer MRF model was confirmed with various images even with fixed hyperparameters. Achievement of further adaptiveness by training hyperparameters remains as a future study.

References

1. Tsai, R.Y., Huang, T.S.: Multiframe image restoration and registration. In: Advances in Computer Vision and Image Processing, pp. 317–339. JAI Press, Greenwich, CT (1984)
2. Park, S.C., Park, M.K., Kang, M.G.: Super-resolution image reconstruction: A technical overview. IEEE Signal Process. Mag. 20(3), 21–36 (2003)
3. Farsiu, S., Robinson, D., Elad, M., Milanfar, P.: Advances and challenges in super-resolution. Int. J. Imag. Syst. Tech. 14(2), 47–57 (2004)
4. Borman, S., Stevenson, R.L.: Spatial resolution enhancement of low-resolution image sequences: A comprehensive review with directions for future research. Technical report, Dept. of Electrical Enginerring, University of Notre Dame (1998)
5. MacKay, D.J.C.: Bayesian interpolation. Neural Computation 4(3), 415–447 (1992)
6. MacKay, D.J.C.: A practical Bayesian framework for backprop networks. Neural Computation 4(3), 448–472 (1992)
7. Tipping, M.E., Bishop, C.M.: Bayesian image super-resolution. In: Becker, S., Thrun, T., Obermayer, K. (eds.) NIPS 15, pp. 1279–1286. MIT Press, Cambridge (2003)
8. Pickup, L.C., Capel, D.P., Roberts, S.J., Zisserman, A.: Bayesian image super-resolution, continued. In: Schölkopf, B., Platt, J., Hoffman, T. (eds.) NIPS 19, MIT Press, Cambridge (2007)
9. Woods, N.A., Galatsanos, N.P., Katsaggelos, A.K.: Stochastic methods for joint registration, restoration, and interpolation of multiple undersampled images. IEEE Trans. Image Process. 15(1), 201–213 (2006)
10. Hardie, R.C., Barnard, K.J., Armstrong, E.E.: Joint MAP registration and high-resolution image estimation using a sequence of undersamples images. IEEE Trans. Image Process. 6(12), 1621–1633 (1997)
11. Geman, S., Geman, D.: Stochastic relaxation, Gibbs distributions, and the Bayesian restoration of images. IEEE Trans. Pattern Anal. Mach. Intell. PAMI-6(6), 721–741 (1984)
12. Jeng, F.C., Woods, J.W.: Compound Gauss-Markov random fields for image estimation. IEEE Trans. Signal Process. 39(3), 683–697 (1991)
13. Winkler, G.: Image Analysis, Random Fields, and Markov Chain Monte Carlo Methods, 2nd edn. Springer, Heidelberg (2003)
14. Neal, R.M., Hinton, G.E.: A view of the EM algorithm that justifies incremental, sparse, and other variants. In: Jordan, M.I. (ed.) Learning in Graphical Models, Kluwer Academic Publishers, Dordrecht (1998)
15. Molina, R., Mateos, J., Katsaggelos, A.K., Vega, M.: Bayesian multichannel image restoration using compound Gauss-Markov random fields. IEEE Trans. Image Process. 12(12), 1642–1654 (2003)
16. Bishop, C.M.: Neural Networks for Pattern Recognition. Oxford Univ. Press, New York (1995)

A Neurofuzzy Controller for a Single Link Flexible Manipulator

Samaneh Sarraf, Ali Fallah, and T. Seyedena

Amirkabir University of Technology
Hafez Ave., Tehran, Iran
ssaraf@bme.aut.ac.ir, {afallah,seyedena}@aut.ac.ir

Abstract. This paper presents an adaptive neurofuzzy controller for tip position tracking control of a single link flexible manipulator. The controller has a self-organizing fuzzy neural structure in which fuzzy rules are generated during the control process using an online learning algorithm. In order to demonstrate the superior performance of the proposed controller, the results are compared with those obtained by using the proportional-derivative (PD) and neural network controllers. Moreover, since the proposed controller requires no a priori knowledge about the system, it can efficiently cope with the uncertainties such as payload mass variations.

Keywords: Single link flexible manipulator, neural network controller, neurofuzzy controller.

1 Introduction

The control of robotic manipulators has been studied with great interest by many researchers over the past years. Major advantages of flexible manipulators include small mass, fast motion, and large force to mass ratio, which are reflected directly in the reduced energy consumption, increased productivity, and enhanced payload capacity. Flexible manipulators have important applications in space exploration, manufacturing automation, construction, mining, hazardous operations, and many other areas [1].

In general, robot manipulators have to face uncertainties in their dynamics, such as payload mass, friction, and disturbance. Therefore, it is difficult to obtain an accurate model for a flexible link manipulator [2]. Thus, model based control systems may not be easily implemented in flexible link control practice.

There are many different techniques for the control of flexible manipulators such as [3]-[7]. Over the past few decades, intelligent controllers such as neural network, fuzzy and neurofuzzy controllers have attracted considerable attention and interest. This interest resulted from their remarkable model-free characteristic which provides the capability of learning and nonlinear mapping. However, neural network controllers require a predefined structure which may lead to additional time-consuming computations during the control process. The other drawback of neural network controllers is their output dependency on the selection of the initial values of neural network weights. The main problem of fuzzy controllers is the appropriate

J. Marques de Sá et al. (Eds.): ICANN 2007, Part II, LNCS 4669, pp. 621–629, 2007.

definition of fuzzy membership functions and rules. Therefore, in recent years, fuzzy-neural network controllers which combine the human reasoning capabilities of fuzzy systems in dealing with uncertainties with the capabilities of neural networks in learning and generalization have been proposed as a powerful technique for control of flexible link manipulators.

In this paper an adaptive neurofuzzy controller for tip position tracking control of a single link flexible manipulator is presented. In order to show the superior performance of the neurofuzzy controller, the results are compared with the results of the PD and neural network controllers. To achieve this goal, at first we introduce the neural network controller and demonstrate its preference over the PD controller by simulation results. Then, some drawbacks of the neural network controller are mentioned. Finally, to overcome these problems the neurofuzzy controller is proposed and simulation results are discussed.

2 The Flexible Manipulator Dynamics

The flexible link manipulator is shown in Fig. 1, where θ is the hub position and $W(x,t)$ is the elastic deflection along the length x of the link.

The dynamic model of a robot manipulator is derived using a Lagrangian assumed modes method based on Euler- Bernoulli beam theory, and can be expressed as

$$M(q)\begin{bmatrix}\ddot{\theta}\\ \ddot{q}\end{bmatrix}+C(q,\dot{q})\begin{bmatrix}\dot{\theta}\\ \dot{q}\end{bmatrix}+K(\dot{\theta})\begin{bmatrix}\theta\\ q\end{bmatrix}+\begin{bmatrix}Fric(\dot{\theta})\\ 0\end{bmatrix}=\begin{bmatrix}\tau\\ 0\end{bmatrix}. \tag{1}$$

where M, C, K, τ represent the matrices of inertia, Coriolis and centrifugal forces, stiffness effect, and the torque applied to the hub, respectively. $q=\begin{bmatrix}q_1 & q_2 & \cdots & q_N\end{bmatrix}^T$ is the $(N\times 1)$ vector, q_i is the ith elastic mode and N is the number of deflection modes [1, 3]. By incorporating the dynamic model of armature-controlled servomotor into the dynamic model of the flexible link and assuming the armature inductance to be neglected, the model of a robot manipulator including actuator dynamics can be obtained:

$$\hat{M}(q)\begin{bmatrix}\ddot{\theta}\\ \ddot{q}\end{bmatrix}+\hat{C}(q,\dot{q})\begin{bmatrix}\dot{\theta}\\ \dot{q}\end{bmatrix}+K(\dot{\theta})\begin{bmatrix}\theta\\ q\end{bmatrix}+\begin{bmatrix}Fric(\dot{\theta})\\ 0\end{bmatrix}=\begin{bmatrix}v\\ 0\end{bmatrix}. \tag{2}$$

Where v is the control voltage applied to the dc-servomotor.

Since the zero dynamics associated with the tip position are unstable, the system is non-minimum phase and it is difficult to control the tip position [8]. Therefore, to overcome this problem, the redefined output with stable zero dynamics was achieved in [9], choosing the appropriate value for α. The redefined output is defined as

$$y_a=\theta+\alpha\frac{W(h,t)}{h}. \tag{3}$$

where h denotes the length of the link.

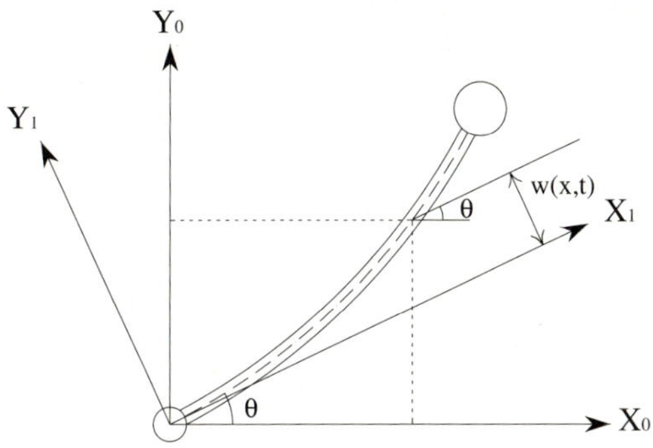

Fig. 1. Model of flexible link manipulator

3 Neural Network Controller

The structure of the proposed control system is depicted in Fig. 2. As can be seen in this figure, the controller is composed of two parts: The conventional proportional-derivative type (PD) controller and the intelligent controller (the neural network or the nerofuzzy controller). The PD controller initially controls the system and maintains the stability of the system during the learning phase of the other controller.

The three-layer neural network controller consists of 6, 15 and 1 neurons in its input, hidden and output layers, respectively. The neural network controller training is accomplished using the back-propagation (BP) algorithm. Since our purpose is to control the tip position of a flexible arm along with suppressing the tip deflection, the terms e , $\dot{e}$ and W should be incorporated in the cost function which is to be minimized in the learning algorithm. The cost function can be expressed as [3]:

$$J = \frac{1}{2}\left(e^T k_1 e + \dot{e}^T k_2 \dot{e} + W^T\left(h,t\right) k_3 W\left(h,t\right)\right). \tag{4}$$

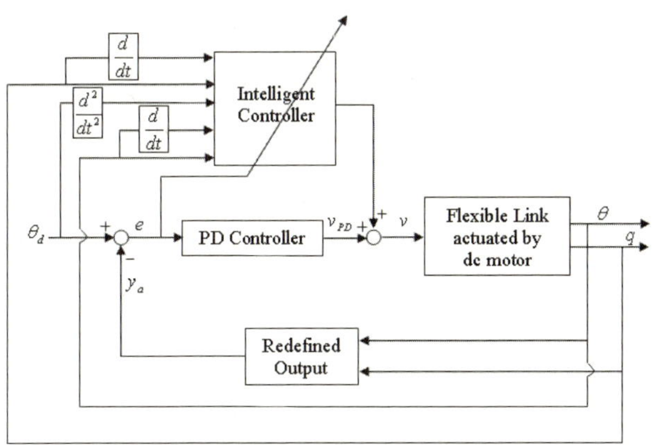

Fig. 2. The proposed control scheme

Simulations are carried out using the link parameters mentioned in [3]. In order to show the effectiveness of the neural network controller, the results achieved by this controller are compared to those obtained from the PD controller.

Figs. 3 and 4 depict the simulation results of the PD controller and the neural network controller for a sequential step desired trajectory, respectively. As shown in Fig. 3, although the PD controller is able to stabilize the system, the tip deflection of the flexible link cannot be effectively suppressed. In contrast, Fig. 4 demonstrates that the control performance is obviously improved and the tip deflection is decreased in the next steps. Fig. 4 also indicates the learning characteristic of the neural network controller.

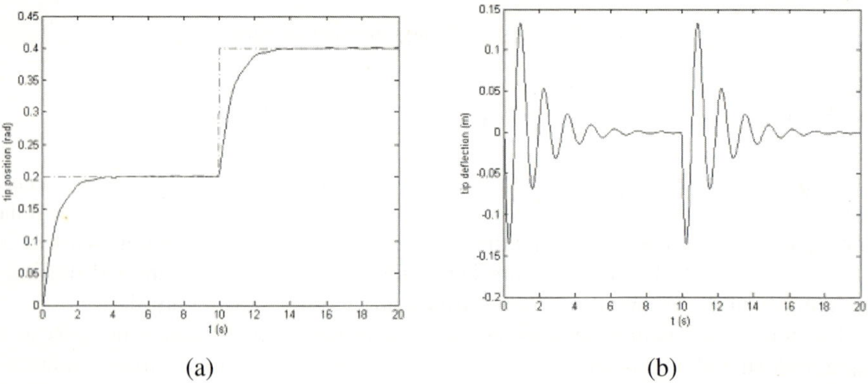

Fig. 3. Simulation results of the PD controller with $M_p = 20gr$. (a) tip position, (b) tip deflection.

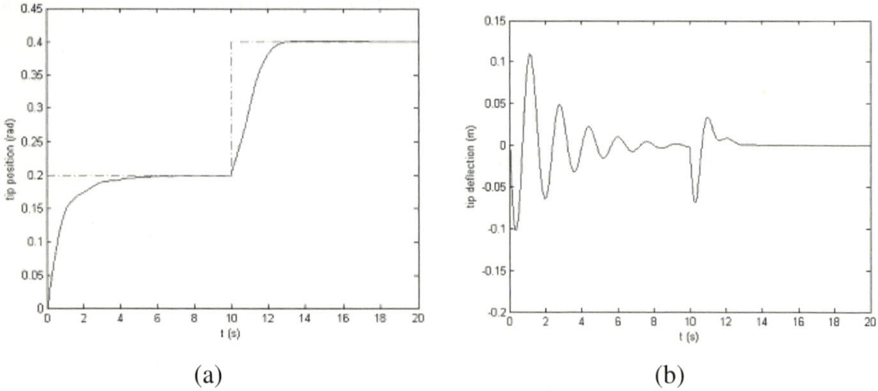

Fig. 4. Simulation results of the neural network controller with $M_p = 20gr$. (a) tip position, (b) tip deflection.

The performance of the neural network controller, having the structure as mentioned in this section, depends on the initial values of its parameters such and the predefined structure. This can be seen in Figs. 4 and 5. In Fig. 4, the neural network weights are initialized to the values between $\left[-0.1 \quad 0.1\right]$. As can be seen in this figure, a satisfactory response of both the tip position and tip deflection are achieved. However, if the initial values of the neural network are assigned in the range of $\left[-0.01 \quad 0.01\right]$, the response becomes deteriorated which is presented in Fig. 5. An examination of Fig. 5 also demonstrates that the learning process is not completed in the second sequence of the step input. It should be noted that although greater initial values of the neural network weights hasten learning process of the neural network, which obviously appears in Fig. 4, the main drawback is that this selection may cause instability for other inputs with different forms and magnitudes. Therefore, a compromise between the selection of neural network initial weights and the learning rate should be performed.

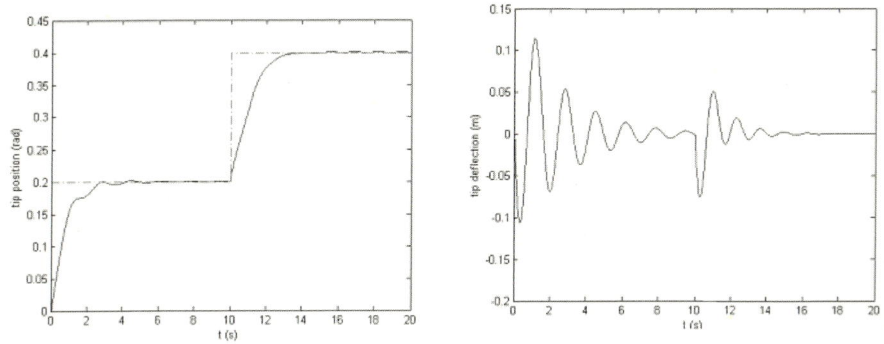

(a) (b)

Fig. 5. Simulation results of the neural network by choosing smaller initial weights with $M_p = 20gr$. (a) tip position, (b) tip deflection.

Moreover, in order to examine the performance of the neural network controller in the presence of uncertainties, such as tip payload variation, we change the tip payload mass from $M_p = 20gr$, which was used in previous simulations, to $M_p = 400gr$ and plot the results. Figs. 6, 7 show the results of the PD and the neural network controller.

By comparing Fig. 6 with Fig. 7, it can be discovered that while the tracking performance obtained by the neural network control strategy remains quite satisfactory with a slightly increased tip deflection and overshoot, results of the PD controller obviously become deteriorated by increasing the tip payload mass. This also confirms the noticeable preference of the neural network controller performance. However, the aforementioned drawbacks of the neural network controller still exist. Therefore, the design of a modified controller such as a neurofuzzy controller receives considerable attention.

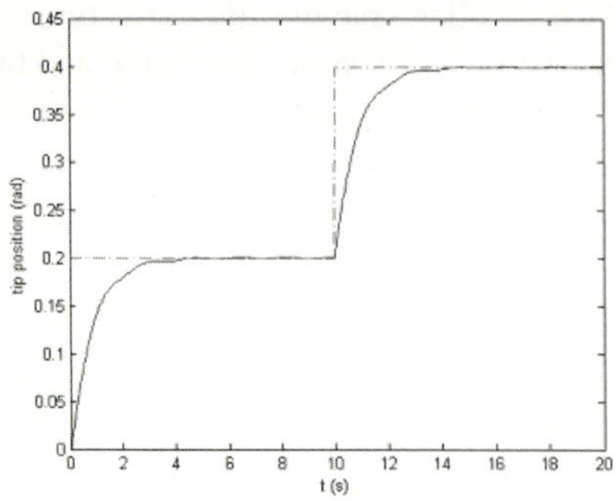

(a)

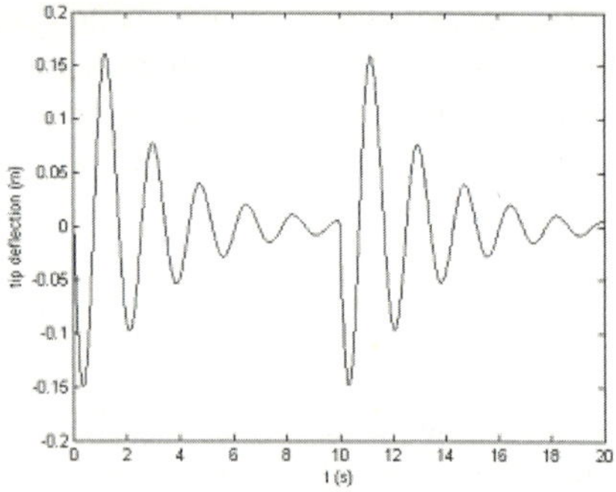

(b)

Fig. 6. Simulation results of the PD controller with $M_p = 400gr$. (a) tip position, (b) tip deflection.

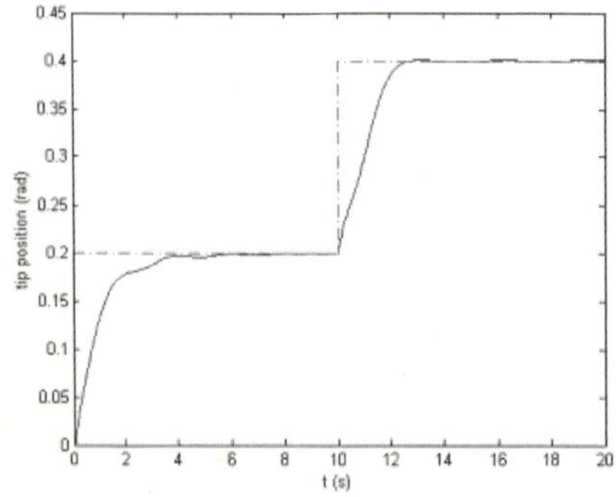

(a)

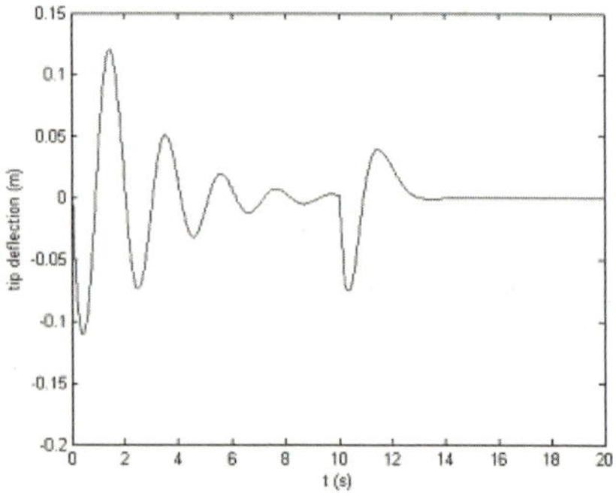

(b)

Fig. 7. Simulation results of the neural network controller with $M_p = 400gr$. (a) tip position, (b) tip deflection.

4 Neurofuzzy Controller

Our proposed neurofuzzy controller is a network based on the radial basis function (RBF) which can be considered a Takagi-Sugeno (TS) fuzzy system by considering each neuron as a fuzzy rule. Fig. 8 shows the schematic diagram of an RBF network.

An RBF network can also be regarded as a special two-layer network in which the hidden layer performs a fixed transformation with no adjustable parameters and it maps the input space onto a new space. The output layer then implements a linear combination on this new space and only adjustable parameters are the weights of this linear combination [10]. In other words, it is anticipated that our neurofuzzy controller (or RBF network) is developed in such a way to obtain the inverse dynamics of the flexible link manipulator and map the input space which consists of $\ddot{\theta}_d, \dot{\theta}, \theta, q, \dot{q}$ onto the control voltage v .

The activation function can be described as

$$\phi_l(X) = \exp\left[-(X - M_l)^T S_l (X - M_l)\right] \qquad l = 1,2,\ldots,L \tag{5}$$

where $X = [x_1 \quad x_2 \quad \cdots \quad x_k]^T$ is the input vector, $M_l = [m_{1l} \quad m_{2l} \quad \cdots \quad m_{kl}]^T$ is the vector of centers, $S_l = diag[1/s_{1l}^2 \quad 1/s_{2l}^2 \quad \cdots \quad 1/s_{kl}^2]$ is the vector of widths, and L is the number of activation function or the number of neurons [4]. The output of the neurofuzzy controller is obtained using (6) which verifies the equivalence between the RBF network and TS fuzzy system by considering the weight vector w as the TS weights.

$$y = [\phi_1 \quad \cdots \quad \phi_L][w_1 \quad \cdots \quad w_L]^T \tag{6}$$

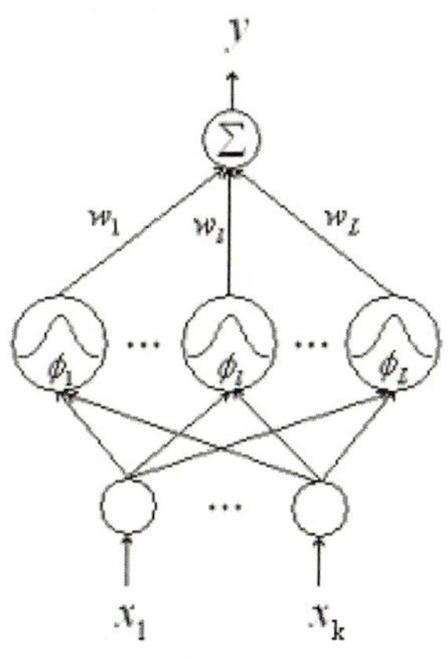

Fig. 8. Structure of the RBF network

The learning algorithm is initially started from a network with no neurons in the hidden layer ($L = 0$). Then the first neuron is added to the network and is initialized to

$$C_1 = X(1), \quad \sigma_1 = \sigma_0 \tag{7}$$

Where σ_0 is a predetermined constant value.

For a given input vector $X(n)$, the Euclidean norm is used to compute the distances of $X(n)$ to all center vectors and to find the minimum distance, $d_{\min}$. If $d_{\min}$ is smaller than a predefined constant threshold, then the adjustable parameters of the RBF network are updated using the conventional gradient descent algorithm that minimizes a modified cost function (4). Otherwise, if $d_{\min}$ is larger than threshold, a new neuron (or a new fuzzy rule) will exist with the parameters defined as

$$C_{L+1} = X(n), \quad \sigma_{L+1} = d_{\min}, \quad w_{L+1}(n) = \frac{1}{L}\sum_{l=1}^{L} w_l(n) \tag{8}$$

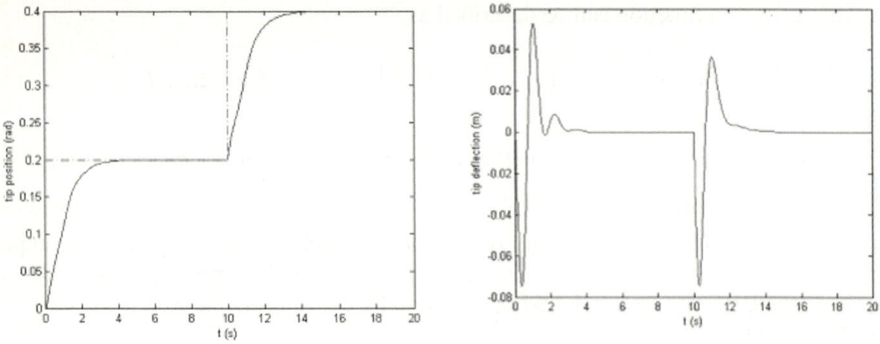

(a) (b)

Fig. 9. Simulation results of the neurofuzzy controller with $M_p = 20gr$. (a) tip position, (b) tip deflection.

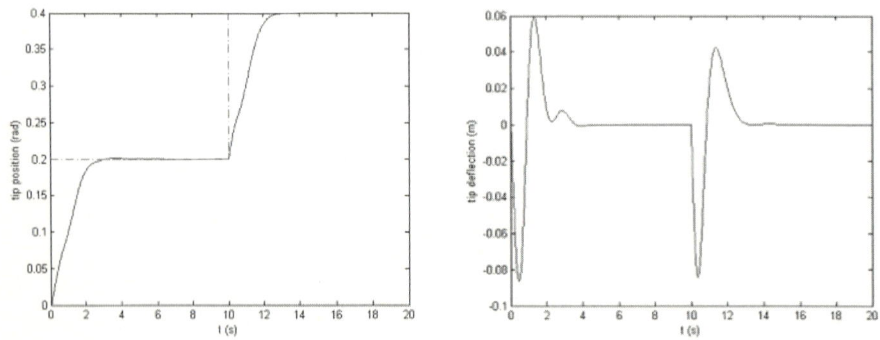

(a) (b)

Fig. 10. Simulation results of the neurofuzzy controller with $M_p = 400gr$. (a) tip position, (b) tip deflection.

Simulation results of the tip position and tip deflection of the flexible link by using the neurofuzzy controller are shown in Fig. 9. As can be clearly seen in this figure, considerable improvement is obtained by using the proposed neurofuzzy controller as compared to the other two controllers shown in Figs. 3 and 4. This is due to the neurofuzzy controller structure which is modified during the online learning algorithm and takes the inverse dynamics of the flexible link. In contrast, the simple-structure linear conventional PD controller exhibit inability to compensate for the nonlinear dynamics of the link. Furthermore, despite rather acceptable performance of the neural network controller, the predefined structure of this controller can not represent the appropriate structure which hinders the controller from getting the inverse dynamics of the system.

To study the effect of an uncertain condition such as payload variation on the performance of the neurofuzzy controller, we change the tip payload mass from $M_p = 20gr$ to $M_p = 400gr$. Fig. 10 shows the results of this situation. It can be seen that the neurofuzzy controller efficiently compensates for the partly changed nonlinear dynamics by modifying its structure and weights and confirms the advantages of the neurofuzzy controller.

5 Conclusion

In this paper, an adaptive neurofuzzy control strategy was proposed and its superiority over the conventional PD controller and the neural network controller are demonstrated. Simulation results also verify the satisfactory performance of the neurofuzzy controller in dealing with uncertainties such as the variation of the tip payload.

References

1. Wang, F.Y., Gao, Y.: Advanced Studies of Flexible Robotic Manipulator. World Scientific, Singapore (2003)
2. Wai, R.J., Chen, P.C.: Intelligent Tracking Control for Robot Manipulator Including Actuator Dynamics via TSK-Type Fuzzy Neural Network. IEEE Trans. Fuzzy Syst. 12(4), 552–559 (2004)
3. Su, Z., Khorasani, K.: A Neural-Network-Based Controller for a Single-Link Flexible Manipulator Using the Inverse Dynamics Approach. IEEE Trans. Ind. Electron. 48(6), 1074–1086 (2001)
4. Er, M.J., Gao, Y.: Robust Adaptive Control of Robot Manipulators Using Generalized Fuzzy Neural Networks. IEEE Trans. Ind. Electron. 50(3), 620–628 (2003)
5. Caswara, F.M., Unbehauen, H.: A Neurofuzzy Approach to the Control of a Flexible-Link Manipulator. IEEE Trans. Robot. Automat. 18(6), 935–944 (2002)
6. Pham, C.M., Khalil, W., Chevallereau, C.: A Nonlinear Model-Based Control of Flexible Robots. Robotica 11, 73–82 (1993)
7. Lin, J.: Flexible Link Robot Arm Control by a Hierarchical Fuzzy Logic Approach. Proc. IEEE Int. Conf. on Fuzzy Systems 2, 1138–1143 (2002)
8. Talebi, H.A., Khorasani, K., Patel, R.V.: Experimental Evaluation of Neural Network Based Controllers for Tracking the Tip Position of a Flexible-Link Manipulator. In: Proc. IEEE Int. Conf. Robotics and Automation, pp. 3300–3305 (1997)
9. Madhavan, S.K., Singh, S.N.: Inverse Trajectory Control and Zero Dynamic Sensitivity of an Elastic Manipulator. Int. J. Robot. Automat. 6(4), 179–192 (1991)
10. Chen, S.C., Cowan, F.N., Grant, P.M.: Orthogonal Least Squares Learning Algorithm for Radial Basis Function Network. IEEE Trans. Neural Networks 2, 302–309 (1991)

Suboptimal Nonlinear Predictive Control with Structured Neural Models

Maciej Ławryńczuk

Institute of Control and Computation Engineering, Warsaw University of Technology
ul. Nowowiejska 15/19, 00-665 Warsaw, Poland
Tel.: +48 22 234-73-97
M.Lawrynczuk@ia.pw.edu.pl

Abstract. This paper details a computationally efficient (suboptimal) nonlinear Model Predictive Control (MPC) algorithm with structured neural models and discusses its application to a polymerisation reactor. Thanks to the nature of the model it is not used recursively, the prediction error is not propagated. The model is used on-line to determine a local linearisation and a nonlinear free trajectory. The algorithm needs solving on-line only a quadratic programming problem. It gives closed-loop control performance similar to that obtained in the fully-fledged nonlinear MPC, which hinges on non-convex optimisation.

1 Introduction

Model Predictive Control (MPC) is recognised as the only advanced control technique which has been very successful in practical applications [8], [15], [17]. MPC algorithms can take into account constraints imposed on both process inputs (manipulated variables) and outputs (controlled variables), which usually decide on quality, economic efficiency and safety. Furthermore, MPC techniques are very efficient in multivariable process control.

Structure of the model and the way it is used decide on accuracy, computational burden and reliability of nonlinear MPC. Fundamental (first-principles) models, although potentially very precise, are usually not suitable for on-line control because they are very complicated and may lead to numerical problems resulting, for example, from ill-conditioning or stiffness. Since neural network models [1] are able to approximate precisely nonlinear behaviour of technological processes [3], [11], have relatively small number of parameters and simple structure, they can be effectively used in MPC algorithms as process models [3], [5], [6], [7], [11], [12], [13], [14], [17], [18]. In light of practical importance linearisation-based MPC techniques, in which only a quadratic programming problem is solved on-line, deserve consideration [2], [6], [7], [10], [11], [17], [18]. Compared to MPC algorithms with full nonlinear optimisation, they are suboptimal, but the accuracy is usually sufficient.

Neural network models are usually trained using the rudimentary backpropagation algorithm which yields one-step ahead predictors. Naturally, they are not suited to be used recursively in MPC for long range prediction since the prediction error is propagated. Recurrent neural models are reasonably less popular.

J. Marques de Sá et al. (Eds.): ICANN 2007, Part II, LNCS 4669, pp. 630–639, 2007.

So as to solve the problem resulting from the inaccuracy of one-step predictors in MPC a multi-model approach has been proposed in [4], [16] for linear processes. For each sampling instant within the prediction horizon one independent linear model is used, the prediction error is not propagated. In this work a different approach to modelling is studied.

The paper describes a computationally efficient suboptimal MPC algorithm with Nonlinear Prediction and Linearisation (MPC-NPL) [6], [7], [17], [18], based on structured neural network models and its application to a polymerisation process. Only one structured model is used, but analogously as the multi-models it has the ability to predict future values of the output without taking into account previous predictions calculated within the prediction horizon. Thanks to this feature the model is not used recursively, the prediction error is not propagated. The algorithm gives good closed-loop performance and, unlike the nonlinear MPC technique, which hinges on nonlinear optimisation, it uses on-line only the numerically reliable quadratic programming approach.

2 Model Predictive Control Algorithms

In the MPC algorithms [8], [17] at each consecutive sampling instant k a set of future control increments is calculated

$$\Delta \boldsymbol{u}(k) = [\Delta u(k|k)\ \Delta u(k+1|k) \ldots \Delta u(k+N_u-1|k)]^T \tag{1}$$

It is assumed that $\Delta u(k+p|k) = 0$ for $p \geq N_u$, where N_u is the control horizon. The objective is to minimise the differences between the predicted values of the output $\hat{y}(k+p|k)$ and the reference trajectory $y^{ref}(k+p|k)$ over the prediction horizon N. The following quadratic cost function is usually used

$$J(k) = \sum_{p=1}^{N}(y^{ref}(k+p|k) - \hat{y}(k+p|k))^2 + \sum_{p=0}^{N_u-1} \lambda_p(\Delta u(k+p|k))^2 \tag{2}$$

where $\lambda_p > 0$ are weighting factors. Typically, $N_u < N$, which decreases the dimensionality of the optimisation problem. Only the first element of the determined sequence (1) is applied to the process, the control law is then

$$u(k) = \Delta u(k|k) + u(k-1) \tag{3}$$

At next sampling instant, $k+1$, the prediction is shifted one step forward and the whole procedure is repeated.

Since the constraints have to be usually taken into account, future control increments are determined as the solution to the following optimisation problem (assuming hard output constraints [8], [17] for simplicity of presentation)

$$\begin{aligned}
&\min_{\Delta u(k|k)\ldots\Delta u(k+N_u-1|k)} \{J(k)\} \\
&\text{subject to} \\
&\quad u^{\min} \leq u(k+p|k) \leq u^{\max}, \quad p = 0, \ldots, N_u - 1 \\
&\quad -\Delta u^{\max} \leq \Delta u(k+p|k) \leq \Delta u^{\max}, \quad p = 0, \ldots, N_u - 1 \\
&\quad y^{\min} \leq \hat{y}(k+p|k) \leq y^{\max}, \quad p = 1, \ldots, N
\end{aligned} \tag{4}$$

The prediction equation for $p = 1, \ldots, N$ is

$$\hat{y}(k+p|k) = y(k+p|k) + d(k) \tag{5}$$

where the quantities $y(k+p|k)$ are calculated from a nonlinear model. The "DMC type" disturbance model is used in which the unmeasured disturbance $d(k)$ is assumed to be constant over the prediction horizon. It is estimated from

$$d(k) = y(k) - y(k|k-1) \tag{6}$$

where $y(k)$ is a measured value while $y(k|k-1)$ is calculated from the model.

Let the Single-Input Single-Output (SISO) process under consideration be described by the following nonlinear discrete-time equation

$$y(k) = f(\boldsymbol{x}(k)) = f(u(k-\tau), \ldots, u(k-n_B), y(k-1), \ldots, y(k-n_A)) \tag{7}$$

where $f : \Re^{n_A+n_B-\tau+1} \longrightarrow \Re$, $\tau \leq n_B$. Using the prediction equation (5) and the model (7), the output predictions for $p = 1, \ldots, N$ are calculated from

$$\begin{aligned}\hat{y}(k+p|k) = f(&\underbrace{u(k-\tau+p|k), \ldots, u(k|k)}_{I_{uf}(p)}, \underbrace{u(k-1), \ldots, u(k-n_B+p)}_{I_u - I_{uf}(p)}, \\ &\underbrace{\hat{y}(k-1+p|k), \ldots, \hat{y}(k+1|k)}_{I_{yp}(p)}, \underbrace{y(k), \ldots, y(k-n_A+p)}_{n_A - I_{yp}(p)}) + d(k)\end{aligned} \tag{8}$$

The predictions $\hat{y}(k+p|k)$ depend on $I_{uf}(p) = \max(\min(p-\tau+1, I_u), 0)$ future values of the control signal (i.e. decision variables of the MPC algorithm), where $I_u = n_B - \tau + 1$, $I_u - I_{uf}(p)$ values of the control signal applied to the plant at previous sampling instants, $I_{yp}(p)$ future output predictions and $n_A - I_{yp}(p)$ plant output signal values measured at previous sampling instants. It is evident that for prediction the model has to be used recursively, because the predictions depend on the predictions calculated for previous sampling instants within the prediction horizon. In spite of the fact that a one-step ahead predictor is given as the result of backpropagation training, it is used for N-step ahead prediction. Since model inaccuracies are unavoidable, the prediction error is propagated.

3 MPC-NPL Algorithm with Structured Neural Models

3.1 Structured Neural Model of the Process

Rewriting the model (7) for sampling instants $k-1, \ldots, k-N+1$ one has

$$\begin{aligned}y(k-1) = f(&u(k-\tau-1), \ldots, u(k-n_B-1), \\ &y(k-2), \ldots, y(k-n_A-1))\end{aligned} \tag{9}$$

$$\vdots$$

$$\begin{aligned}y(k-N+2) = f(&u(k-\tau-N+2), \ldots, u(k-n_B-N+2), \\ &y(k-N+1), \ldots, y(k-n_A-N+2))\end{aligned} \tag{10}$$

$$\begin{aligned}y(k-N+1) = f(&u(k-\tau-N+1), \ldots, u(k-n_B-N+1), \\ &y(k-N), \ldots, y(k-n_A-N+1))\end{aligned} \tag{11}$$

Using (11), the quantity $y(k-N+2)$ given by (10) can be expressed as

$$\begin{aligned} y(k-N+2) = f(&u(k-\tau-N+2),\ldots,u(k-n_B-N+2),\\ &f(u(k-\tau-N+1),\ldots,u(k-n_B-N+1),\\ &\quad y(k-N),\ldots,y(k-n_A-N+1)),\\ &y(k-N),\ldots,y(k-n_A-N+2)) \end{aligned} \tag{12}$$

which can be rewritten as the function

$$\begin{aligned} y(k-N+2) = f_{N-2}(&u(k-\tau-N+2),\ldots,u(k-n_B-N+1),\\ &y(k-N),\ldots,y(k-n_A-N+1)) \end{aligned} \tag{13}$$

Model arguments rearrangement can be repeated for all $y(k-N+2),\ldots,y(k)$, giving the functions $f_{N-2},\ldots,f_0$. Finally, one has

$$\begin{aligned} y(k) = f(&u(k-\tau),\ldots,u(k-n_B),\\ &f_1(u(k-\tau-1),\ldots,u(k-n_B-N+1),\ldots,\\ &\quad y(k-N),\ldots,y(k-n_A-N+1)),\ldots,\\ &f_{n_A}(u(k-\tau-n_A),\ldots,u(k-n_B-N+1),\\ &\quad y(k-N),\ldots,y(k-n_A-N+1))) \end{aligned} \tag{14}$$

which can be rewritten as the function

$$\begin{aligned} y(k) = f_0(&u(k-\tau),\ldots,u(k-n_B-N+1),\\ &y(k-N),\ldots,y(k-n_A-N+1)) \end{aligned} \tag{15}$$

The obtained equation (15) represents the structured model. Using (5) the output predictions for $p=1,\ldots,N$ calculated from the structured model are

$$\begin{aligned} \hat{y}(k+p|k) = f_0(&\underbrace{u(k-\tau+p|k),\ldots,u(k|k)}_{I_{uf}(p)},\underbrace{u(k-1),\ldots,u(k-n_B-N+1+p)}_{I_u-I_{uf}(p)},\\ &\underbrace{y(k-N+p),\ldots,y(k-n_A-N+1+p)}_{n_A})+d(k) \end{aligned} \tag{16}$$

For the structured model $I_u = n_B + N - \tau$. As in the case of the model (7), the predictions $\hat{y}(k+p|k)$ calculated by means of the structured model (15) depend on $I_{uf}(p)$ future values of the control signal, $I_u - I_{uf}(p)$ values of the control signal applied to the plant at previous sampling instants. Unlike the classical predictions (8), they do not depend on the predictions calculated for previous sampling instants within the prediction horizon, but only on n_A values of the plant output signal measured at previous sampling instants. As a result, the structured model is not used recursively, the prediction error is not propagated.

In the sequel it is assumed that a feedforward neural network with one hidden layer and linear output [1] is used as the function f_0 in (15), hence f_0 :

$\Re^{n_A+n_B-\tau+N} \longrightarrow \Re \in C^1$, $\tau \leq n_B + N - 1$. Output of the model can be expressed as

$$y(k) = f_0(\boldsymbol{x}(k)) = w_0^2 + \sum_{i=1}^{K} w_i^2 v_i(k) = w_0^2 + \sum_{i=1}^{K} w_i^2 \varphi(z_i(k)) \tag{17}$$

where $z_i(k)$ and $v_i(k)$ are the sum of inputs and the output of the i-th hidden node, respectively, $\varphi : \Re \longrightarrow \Re$ is the nonlinear transfer function (e.g. hyperbolic tangent), K is the number of hidden nodes. Recalling the prediction of the structured model (16) one has

$$\begin{aligned} z_i(k+p|k) = w_{i,0}^1 + \sum_{j=1}^{I_{uf}(p)} w_{i,j}^1 u(k-\tau+1-j+p|k) + \\ + \sum_{j=I_{uf}(p)+1}^{I_u} w_{i,j}^1 u(k-\tau+1-j+p) + \\ + \sum_{j=1}^{n_A} w_{i,I_u+j}^1 y(k-j-N+1+p) \end{aligned} \tag{18}$$

Weights of the network are denoted by $w_{i,j}^1$, $i = 1, \ldots, K$, $j = 0, \ldots, n_A + n_B - \tau + N$, and w_i^2, $i = 0, \ldots, K$, for the first and the second layer, respectively.

3.2 MPC-NPL Optimisation Problem

In the MPC-NPL algorithm [6], [7], [17], [18] at each sampling instant k the neural model is used on-line twice: to determine the local linearisation and the nonlinear free trajectory. It is assumed that the output prediction can be expressed as the sum of the forced trajectory, which depends only on the future (on future input moves $\Delta \boldsymbol{u}(k)$) and the free trajectory $\boldsymbol{y}^0(k)$, which depends only on the past

$$\hat{\boldsymbol{y}}(k) = \boldsymbol{y}^0(k) + \boldsymbol{G}(k)\Delta\boldsymbol{u}(k) \tag{19}$$

where

$$\hat{\boldsymbol{y}}(k) = [\hat{y}(k+1|k) \ldots \hat{y}(k+N|k)]^T \tag{20}$$

$$\boldsymbol{y}^0(k) = \left[y^0(k+1|k) \ldots y^0(k+N|k)\right]^T \tag{21}$$

The dynamic matrix $\boldsymbol{G}(k)$ of dimensionality $N \times N_u$ is calculated on-line from the linearisation of the nonlinear model taking into account the current state of the plant

$$\boldsymbol{G}(k) = \begin{bmatrix} s_1(k) & 0 & \ldots & 0 \\ s_2(k) & s_1 & \ldots & 0 \\ \vdots & \vdots & \ddots & \vdots \\ s_N(k) & s_{N-1}(k) & \ldots & s_{N-N_u+1}(k) \end{bmatrix} \tag{22}$$

The step-response coefficients of the linearised model are determined from

$$s_j(k) = \sum_{i=1}^{\min(j,n_B+N-1)} b_i(k) - \sum_{i=1}^{\min(j-1,n_A+N-1)} a_i(k)s_{j-i}(k) \tag{23}$$

where $a_i(k)$ and $b_i(k)$ are coefficients of the linearised model. Calculation of these quantities and the nonlinear free trajectory is detailed in the following subsections.

On the one hand, the suboptimal prediction calculated from (19) is different from the optimal one determined from the nonlinear neural model as it is done in MPC algorithms with nonlinear optimisation [17]. On the other hand, thanks to using the superposition principle (19), the optimisation problem (4) becomes the following quadratic programming task

$$\begin{aligned}
&\min_{\boldsymbol{\Delta u}(k)} \left\{ J(k) = \left\| \boldsymbol{y}^{ref}(k) - \boldsymbol{y}^0(k) - \boldsymbol{G}(k)\boldsymbol{\Delta u}(k) \right\|^2 + \left\| \boldsymbol{\Delta u}(k) \right\|^2_{\boldsymbol{\Lambda}} \right\} \\
&\text{subject to} \\
&\quad \boldsymbol{u}^{\min} \leq \boldsymbol{J}\boldsymbol{\Delta u}(k) + \boldsymbol{u}^{k-1} \leq \boldsymbol{u}^{\max} \\
&\quad -\boldsymbol{\Delta u}^{\max} \leq \boldsymbol{\Delta u}(k) \leq \boldsymbol{\Delta u}^{\max} \\
&\quad \boldsymbol{y}^{\min} \leq \hat{\boldsymbol{y}}(k) \leq \boldsymbol{y}^{\max}
\end{aligned} \tag{24}$$

where vectors of length N are

$$\boldsymbol{y}^{ref}(k) = \left[y^{ref}(k+1|k) \ldots y^{ref}(k+N|k)\right]^T \tag{25}$$

$$\boldsymbol{y}^{\min}(k) = \left[y^{\min} \ldots y^{\min}\right]^T \tag{26}$$

$$\boldsymbol{y}^{\max}(k) = \left[y^{\max} \ldots y^{\max}\right]^T \tag{27}$$

vectors of length N_u are

$$\boldsymbol{u}^{\min}(k) = \left[u^{\min} \ldots u^{\min}\right]^T \tag{28}$$

$$\boldsymbol{u}^{\max}(k) = \left[u^{\max} \ldots u^{\max}\right]^T \tag{29}$$

$$\boldsymbol{\Delta u}^{\max}(k) = \left[\Delta u^{\max} \ldots \Delta u^{\max}\right]^T \tag{30}$$

$$\boldsymbol{u}^{k-1}(k) = \left[u(k-1) \ldots u(k-1)\right]^T \tag{31}$$

$\boldsymbol{J}$ is the all ones lower triangular matrix of dimensionality $N_u \times N_u$ and $\boldsymbol{\Lambda} = diag(\lambda_0, \ldots, \lambda_{N_u-1})$.

Structure of the MPC-NPL algorithm is depicted in Fig. 1. At each sampling instant k the following steps are repeated:

1. Linearisation of the structured neural model: obtain the matrix $\boldsymbol{G}(k)$.
2. Find the nonlinear free trajectory $\boldsymbol{y}^0(k)$ using the structured neural model.
3. Solve the quadratic programming problem (24) to determine $\boldsymbol{\Delta u}(k)$.
4. Apply $u(k) = \Delta u(k|k) + u(k-1)$.
5. Set $k := k+1$, go to step 1.

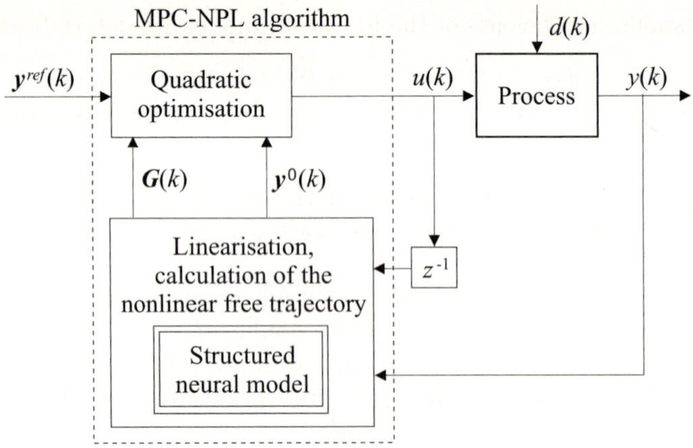

Fig. 1. Structure of the MPC algorithm with Nonlinear Prediction and Linearisation (MPC-NPL)

3.3 On-Line Linearisation of the Structured Neural Model

Defining the linearisation point as a vector composed of past input and output signal values corresponding to the arguments of the structured neural model (15)

$$\bar{\boldsymbol{x}}(k) = [\bar{u}(k-\tau) \ \ldots \ \bar{u}(k-n_B-N+1) \ \bar{y}(k-N) \ \ldots \ \bar{y}(k-n_A-N+1)]^T \tag{32}$$

and using Taylor series expansion at this point, the linear approximation of the model, obtained at a sampling instant k, can be expressed as

$$y(k) = g(\bar{\boldsymbol{x}}(k)) + \sum_{l=\tau}^{n_B+N-1} b_l(\bar{\boldsymbol{x}}(k))(u(k-l) - \bar{u}(k-l)) + \tag{33}$$
$$- \sum_{l=N}^{n_A+N-1} a_l(\bar{\boldsymbol{x}}(k))(y(k-l) - \bar{y}(k-l))$$

where $a_l(\bar{\boldsymbol{x}}(k)) = -\frac{\partial f_0(\bar{\boldsymbol{x}}(k))}{\partial y(k-l)}$, $b_l(\bar{\boldsymbol{x}}(k)) = \frac{\partial f_0(\bar{\boldsymbol{x}}(k))}{\partial u(k-l)}$. Taking into account the structure of the neural model and corresponding equations (17), (18), the coefficients of the linearised model are calculated from

$$a_l(\bar{\boldsymbol{x}}(k)) = \begin{cases} 0 & l = 1, \ldots, N-1 \\ -\sum_{i=1}^{K} w_i^2 \frac{d\varphi(z_i(\bar{\boldsymbol{x}}(k)))}{dz_i(\bar{\boldsymbol{x}}(k))} w_{i,I_u+l-N+1}^1 & l = N, \ldots, n_A+N-1 \end{cases} \tag{34}$$

$$b_l(\bar{\boldsymbol{x}}(k)) = \begin{cases} 0 & l = 1, \ldots, \tau-1 \\ \sum_{i=1}^{K} w_i^2 \frac{d\varphi(z_i(\bar{\boldsymbol{x}}(k)))}{dz_i(\bar{\boldsymbol{x}}(k))} w_{i,l-\tau+1}^1 & l = \tau, \ldots, n_B+N-1 \end{cases} \tag{35}$$

If hyperbolic tangent is used as the function φ, $\frac{d\varphi(z_i(\bar{\boldsymbol{x}}(k)))}{dz_i(\bar{\boldsymbol{x}}(k))} = 1 - \tanh^2(z_i(\bar{\boldsymbol{x}}(k)))$.

3.4 Calculation of the Nonlinear Free Trajectory

The nonlinear free trajectory $y^0(k+p|k)$, $p = 1, \ldots, N$, is calculated on-line recursively from the general prediction equation (5) using the structured neural model defined by (17) and (18)

$$y^0(k+p|k) = w_0^2 + \sum_{i=1}^{K} w_i^2 \varphi(z_i^0(k+p|k)) + d(k) \tag{36}$$

The quantities $z_i^0(k+p|k)$ are determined from (18) assuming no changes in the control signal from a sampling instant k onwards, i.e. $u(k+p|k) := u(k-1)$ for $p \geq 0$. One has

$$z_i^0(k+p|k) = w_{i,0}^1 + \sum_{j=1}^{I_{uf}(p)} w_{i,j}^1 u(k-1) + \sum_{j=I_{uf}(p)+1}^{I_u} w_{i,j}^1 u(k-\tau+1-j+p) +$$
$$+ \sum_{j=1}^{n_A} w_{i,I_u+j}^1 y(k-j-N+1+p) \tag{37}$$

From (6) and (17), the unmeasured disturbance is

$$d(k) = y(k) - \left(w_0^2 + \sum_{i=1}^{K} w_i^2 \varphi(z_i(k)) \right) \tag{38}$$

4 Simulation Results

The process under consideration is a polymerisation reaction taking place in a jacketed continuous stirred tank reactor [9]. The reaction is the free-radical polymerisation of methyl methacrylate with azo-bis-isobutyronitrile as initiator and toluene as solvent. The output $NAMW$ (Number Average Molecular Weight) is controlled by manipulating the inlet initiator flow rate F_I.

Three models of the process are used. The fundamental model [9] is used as the real process during simulations. An identification procedure is carried out, as a result a linear model and a structured neural model are obtained. Three MPC algorithms are compared:

a) the MPC algorithm with the linear model,
b) the MPC algorithm with Nonlinear Optimisation (MPC-NO) with the structured neural model containing 6 hidden nodes ($K = 6$),
c) the MPC-NPL algorithm with the same structured neural model.

The horizons are $N = 5$, $Nu = 3$, the weighting coefficients $\lambda_p = 0.2$. The manipulated variable is constrained, $F_I^{\min} = 0.003$, $F_I^{\max} = 0.06$, the sampling time is 1.8 min.

As the reference trajectories, four set-point changes occurring at $k = 1$ are considered, namely from $NAMW = 20000$ to $NAMW = 25000$, $NAMW = 30000$,

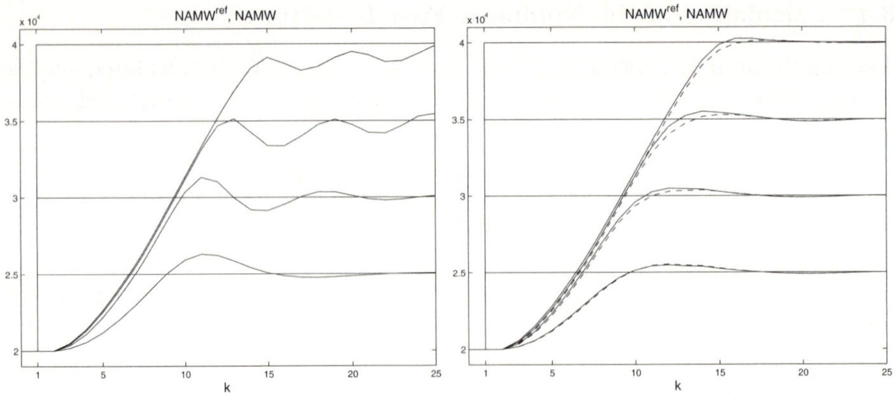

Fig. 2. Left: simulation results of the MPC algorithm with the linear model, right: simulation results of the MPC-NPL (dashed) and MPC-NO (solid) algorithms with the same structured neural network model

$NAMW = 35000$ and $NAMW = 40000$, respectively. Simulation results of the considered MPC algorithms are depicted in Fig. 2. The MPC algorithm with the linear model works satisfactorily for the smallest set-point change, but for bigger ones the system becomes unstable. Both nonlinear algorithms with the structured neural model are stable. Moreover, for four considered set point changes the closed-loop performance obtained in the suboptimal MPC-NPL algorithm with quadratic programming is similar to that obtained in the computationally demanding MPC-NO approach, in which a nonlinear optimisation problem has to be solved on-line at each sampling instant.

5 Conclusion

Reliability, computational efficiency and closed-loop accuracy are the advantages of the presented MPC-NPL algorithm with structured neural network models. The MPC-NPL algorithm uses on-line only the numerically reliable quadratic programming procedure, the necessity of full nonlinear optimisation is avoided. Although suboptimal, in practice the algorithm gives performance comparable to that obtained in MPC schemes with nonlinear optimisation.

The structured neural model predicts future values of the output without taking into account previous predictions calculated within the prediction horizon. The structured model is not used recursively, the prediction error is not propagated. Conceptually, the modelling idea presented in this paper can be regarded as the modification of the linear multi-model approach [4], [16] so as to effectively deal with nonlinear processes. Instead of having a set of separate models, i.e. one model for each sampling instant within the prediction horizon, only one structured neural model is used. The presented idea is general, various structured model types and resulting MPC algorithms can be developed [6].

Acknowledgement

This work was supported by Polish national budget funds 2005-2007 for science as a research project.

References

1. Haykin, S.: Neural networks – a comprehensive foundation. Prentice-Hall, Englewood Cliffs (1999)
2. Henson, M.A.: Nonlinear model predictive control: current status and future directions. Computers and Chemical Engineering. 23, 187–202 (1998)
3. Hussain, M.A.: Review of the applications of neural networks in chemical process control – simulation and online implmementation. Artificial Intelligence in Engineering. 13, 55–68 (1999)
4. Liu, D., Shah, S.L., Fisher, D.G.: Multiple prediction models for long range predictive control. In: Proc. of the IFAC World Congress, Beijing, China (1999)
5. Liu, G.P., Kadirkamanathan, V., Billings, S.A.: Predictive control for non-linear systems using neural networks. Int. Journal of Control. 71, 1119–1132 (1998)
6. Ławryńczuk, M., Tatjewski, P: Suboptimal nonlinear predictive control with structured RBF neural models. In: Proc. of the 13th IEEE Int. Conference on Methods and Models in Automation and Robotics. Międzyzdroje, Poland, Accepted (2007)
7. Ławryńczuk, M., Tatjewski, P.: An efficient nonlinear predictive control algorithm with neural models and its application to a high-purity distillation process. In: Rutkowski, L., Tadeusiewicz, R., Zadeh, L.A., Zurada, J.M. (eds.) ICAISC 2006. LNCS (LNAI), vol. 4029, pp. 76–85. Springer, Heidelberg (2006)
8. Maciejowski, J.M.: Predictive control with constraints, Harlow. Prentice-Hall, Englewood Cliffs (2002)
9. Maner, B.R., Doyle, F.J., Ogunnaike, B.A., Pearson, R.K.: Nonlinear model predictive control of a simulated multivariable polymerization reactor using second-order Volterra models. Automatica. 32, 1285–1301 (1996)
10. Morari, M., Lee, J.H.: Model predictive control: past, present and future. Computers and Chemical Engineering. 23, 667–682 (1999)
11. Nørgaard, M., Ravn, O., Poulsen, N.K., Hansen, L.K.: Neural networks for modelling and control of dynamic systems. Springer, London (2000)
12. Parisini, T., Sanguineti, M., Zoppoli, R.: Nonlinear stabilization by receding-horizon neural regulators. Int. Journal of Control. 70, 341–362 (1998)
13. Piche, S., Sayyar-Rodsari, B., Johnson, D., Gerules, M.: Nonlinear model predictive control using neural networks. IEEE Control Systems Magazine 20, 56–62 (2000)
14. Pottmann, M., Seborg, D.E.: A nonlinear predictive control strategy based on radial basis function models. Computers and Chemical Engineering 21, 965–980 (1997)
15. Qin, S.J., Badgwell, T.A.: A survey of industrial model predictive control technology. Control Engineering Practice 11, 733–764 (2003)
16. Rossiter, J.A., Kouvaritakis, B.: Modelling and implicit modelling for predictive control. Int. Journal of Control 74, 1085–1095 (2001)
17. Tatjewski, P.: Advanced control of industrial processes, Structures and algorithms. Springer, London (2007)
18. Tatjewski, P., Ławryńczuk, M.: Soft computing in model-based predictive control. Int. Journal of Applied Mathematics and Computer Science 16, 101–120 (2006)

Neural Dynamics Based Exploration Algorithm for a Mobile Robot

Jeff Bueckert and Simon X. Yang

Advanced Robotics and Intelligent Systems Lab
School of Engineering
University of Guelph, Guelph, Canada

Abstract. A primary goal for an autonomous mobile robot is to explore and perfrom simultaneous localization and mapping (SLAM). During SLAM, the robot must balance the opposing desires of pose certainty maintenance and information gain. Much of previous research has ignored the need of pose maintenance. This paper provides the first step in developing a neural dynamics based algorithm which considers both information gain and pose maintenance when determining the robot's next pose. Simulation results show that the algorithm is able to provide the robot with an exploration plan to fully explore the tested environments. The next step is to apply the algorithm in a full SLAM environment.

1 Introduction

To build an accurate model of its environment, an autonomous mobile robot must solve three subtasks: localization, mapping and motion control [2,1]. Localization is the task of estimating the robot's position. Mapping is the task of using sensor data to build a representation of the environment. Finally, motion control is the task of steering the robot to the desired location.

In the SLAM problem, motion control is typically not considered. And in most exploration implementations (mapping and motion control), localization is not considered. This is the case for [4,3]. Both search for a new location where the information gain is expected to be the greatest and then use a path planning algorithm to reach the location. Neither algorithm considers pose certainty maintenance. The robot should balance reducing pose uncertainty and information gain when building a map. There is limited research into developing an integrated approach to exploration which satisfies these needs [5].

However, while the work put forward in [5] successfully considers pose uncertainty during exploration, it is based on extended Kalman filter SLAM (EKF SLAM). EKF SLAM requires the presence and recognition of landmarks in the environment; modification to an environment may not be possible in all situations [6,1]. Furthermore, EKF SLAM is limited to sparse maps, has a high update cost; and, if a past data association is found to be incorrect, it is impossible to revise it [6]. The work in [7] combines exploration and pose certainty maintenance by encouraging loop closure. This provides improvement in pose certainty over frontier based exploration [7]. The additional resources required and maintenance

J. Marques de Sá et al. (Eds.): ICANN 2007, Part II, LNCS 4669, pp. 640–649, 2007.

of topological maps are a small increase in computational load [7]. However, the method uses an expensive laser scanner for a sensor payload as compared to relatively inexpensive sonar sensors.

A second flaw with [4] is that it does not update sensor data between frontiers which may be a significant distance apart. As a result the robot does not increase its certainty by revisiting areas, nor is it able to observe changes in these areas. This makes this approach unsuitable for dynamic environments.

Neural dynamics show promise in the field of mobile robotics. The work in [8] uses a shunting model neural network to provide real-time path planning for a point robot in three dimensions and the ability for a point robot to track a moving target in two dimensions. This work is able to cope with a dynamic environment and it is computationally efficient. The shunting model neural network has also been used in [9] to solve the problem of complete area coverage navigation.

The proposed research focuses on using a neural network to provide an exploration plan for an existing grid-based SLAM algorithm. This is an integrated approach as the neural network will consider minimizing pose uncertainty when determining the next pose for the robot. The neural network presented in [9] is the model on which this neural network is based. Unlike the work in [9], the map used by the neural network is an occupancy grid map. Unexplored areas of the map are the excitatory inputs for the network, while the inhibitory inputs are obstacles detected in the environment. A secondary excitatory input is used whose activation will be controlled by pose uncertainty. This input is explored areas; and it will encourage the robot to visit previously explored areas to improve pose and map certainty.

The research is divided into three stages. The initial stage forgoes the SLAM algorithm. Instead the localization procedure is assumed to be perfect. This allows the exploration algorithm to be developed without the additional complications of SLAM. Next the exploration algorithm will be implemented in an environment with SLAM. In this second implementation the algorithm will not use the second excitatory input relegated by pose uncertainty. This allows a baseline to be established for comparison with the final version of the algorithm. Plus, any problems in marrying the exploration algorithm with the grid-based SLAM algorithm may be resolved. The final version of the algorithm implements the second excitatory input. This paper focuses on the first stage of the research.

The problems tested in this paper are four environments of varying complexity. As previously stated, localization is assumed to be perfect. Sensor readings are assumed to be imperfect and corrupted with white Gaussian noise.

2 The Proposed Exploration Algorithm

For mapping the robot maintains an occupancy grid map. Robot movement is restricted to the grid cells adjacent to the robot's location. After moving, the robot re-samples its sensors. This avoids the problem in [4] where the sensors are not sampled between frontier locations that can be a greater than one grid cell apart, making the algorithm suitable for dynamic environments. The proposed algorithm for exploration is composed of 5 steps:

1. Refresh sensor data.
2. Update occupancy grid to reflect sensor data.
3. Fuzzify occupancy grid values using neural input functions.
4. Update neural network using output from neural input functions.
5. Determine next pose for the robot.

The process is repeated until the map is sufficiently revealed.

2.1 Occupancy Grid Updating

Updating the occupancy grid is done as described in [6]. The values used for the parameters in the update process are $l_0 = 0$, $l_{\text{free}} = log\,(0.4/0.6)$ and $l_{\text{occ}} = log\,(0.6/0.4)$. The parameter l_0 denotes prior knowledge of the occupancy of the map; it is chosen to be 0. It is assumed nothing is known apriori to the mapping process; therefore, cells have equal probability of being occupied or free. The information gained by detecting either an obstacle or a free space has the same weight. The magnitude was also chosen to be relatively small so that no single datum indicating an obstacle or a free space can overwhelm the status of the grid cell. Instead, a history of data indicating the same state is needed for the grid cell to converge to a state.

2.2 Neural Input Function

The values of the occupancy grid map are classified before they are used as input for the neural network. Each cell in the occupancy grid has membership in two fuzzy classifications: *unexplored* and *obstacle*. Absolute uncertainty is a value of 0.5 in the occupancy grid; this value generates the maximum excitatory input. A grid cell that is certainly an obstacle, a 1 in the occupancy grid map, generates the maximum inhibitory input. A grid cell that is guaranteed to be empty, a 0 in the occupancy grid map, generates no input at this stage in the research. A single value combining the fuzzy classifications for the single cell is generated by

$$n\left(m_i\right) = f\left(m_i\right) - g\left(m_i\right), \tag{1}$$

where m_i is the ith cell in the occupancy grid map, $f\left(a\right)$ is the *unexplored* membership function, $g\left(a\right)$ is the *obstacle* membership function, and $n\left(a\right)$ is the neural input function response. Five different types of membership functions were used: Gaussian, sigmoidic, step, trapezoidal and triangular Fig.1.

2.3 Neural Network Updating

The neural network in the algorithm employs the shunting model equation,

$$\frac{dx_t}{dt} = -Ax_i + \left(B - x_i\right) S_i^{+} - \left(D + x_i\right) S_i^{-}, \tag{2}$$

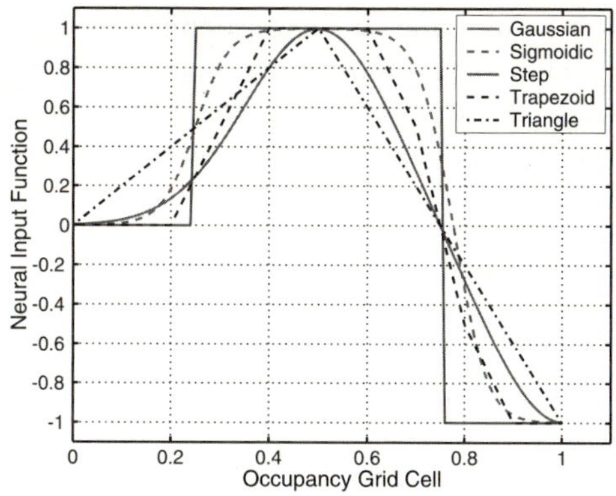

Fig. 1. Neural input functions

where x_i is the neural activity of the ith neuron; A is the passive decay; B and D are the upper and lower bounds; and, S_i^+ and S_i^- are the total excitatory and inhibitory inputs to neuron i respectively. Equation 2 was developed by Grossberg [10] through the work put forth by Hodgkin and Huxley and their neural membrane equation [11].

The neural network used in the algorithm is a two dimensional network where the neurons have a one to one relationship with the occupancy grid cells. Neuron connections only exist between neurons that lie within a radius of $r = 2$ of each other (Fig. 2).

The variables S_i^+ and S_i^- are chosen such that the equation for each neuron's activity is

$$\frac{dx_i}{dt} = -Ax_i + (B - x_i)\left([I_i]^+ + \sum_{j=1}^{M} w_{ij}\,[x_j]^+\right) - (D + x_i)\,[I_i]^- . \quad (3)$$

The passive decay is $A = 50$, and the upper and lower bounds of neural activity, B and D, are 1. The input to a neuron is $I_i = En(m_i)$, where $E = 100$. The

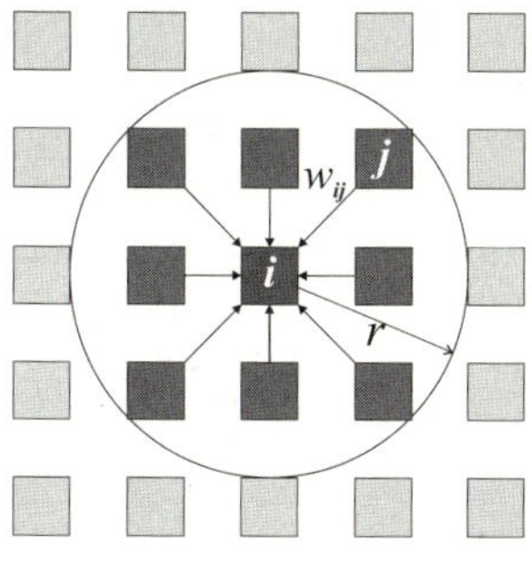

Fig. 2. Adjacent neuron map

two nonlinear functions, $[a]^+$ and $[a]^-$, are defined as $[a]^+ = max\{0, a\}$ and $[a]^- = max\{0, -a\}$. The connection weight between neurons i and j is given by $w_{ij} = f(|m_i - mj|)$, where $|m_i - m_j|$ is the Euclidean distance between m_i and m_j. The function $f(a)$ is $f(a) = \mu/a$ if $0 \leq a < r$; $f(a) = 0$ if $a \geq r$, where $\mu = 2$. This equation allows excitatory signals to propagate throughout the network over time while the inhibitory signals only have a local effect [9].

2.4 Determining Next Pose

The principle factor in determining the robot's next pose is the neural activity of the adjacent neurons. However, there are other costs that we wish to minimize: change in orientation and proximity to obstacles. Changes in orientation should be minimized as they increase the cost of exploration. The robot should not hug obstacles so that greater information is obtained from sensors facing obstacles. Only neurons that have positive neural activity are rewarded for minimizing these secondary costs to avoid neurons with negative activity becoming the best candidate for the next pose.

The work in [9] proposes a reward function for minimizing changes in orientation,

$$y_j = c\left(1 - \frac{|\Delta\theta_j|}{\pi}\right). \tag{4}$$

The parameter c is a positive constant and $|\Delta\theta_j|$ is the absolute change in orientation associated with moving from m_i to m_j; it is bounded on $[0, \pi]$.

A second reward function is developed to encourage the robot to maintain a distance between itself and any identified obstacles,

$$z_j = d \cdot q(r_j). \tag{5}$$

The parameter d is a positive constant and r_j is the Euclidean distance from m_j to the nearest occupancy grid map cell with a value greater than or equal to 0.75. The function $q(r_j)$ is defined as

$$q(r_j) = \begin{cases} r_j/r_{\text{cut-off}} & \text{if } r_j \leq r_{\text{cut-off}} \\ 1 & \text{if } r_j > r_{\text{cut-off}} \end{cases}. \tag{6}$$

The parameter $r_{\text{cut-off}}$ defines at what point extra distance between the robot and the obstacle provides no additional benefit. If $r_{\text{cut-off}}$ is too large, the robot will avoid entering narrow corridors.

The next robot pose is chosen as the adjacent occupancy grid cell m_j that maximizes the sum of the neural activity for the occupancy grid cell, x_j, and the associated rewards from Eq. (4) and Eq. (5):

$$m_{\text{next}} \Leftarrow x_{m_{\text{next}}} = \max\{x_j + y_j + z_j, j = 1, \ldots, k\}, \tag{7}$$

where k is the number of neighboring neurons. In this implementation, $k \leq 8$.

2.5 Computational Complexity

The complexity of the algorithm depends linearly on the area of the occupancy grid that is used [9]. The number of required neurons is $N = N_x \times N_y$, where N_x and N_y are the dimensions of the occupancy grid map [9]. The design of the neural network allows each neuron to have at most 8 neighbors. Therefore, an upper bound on the total number of neural connections is $8N$; giving the algorithm a computational complexity of $O(N)$ [9].

3 Simulation

The simulated robot occupies one grid cell. The simulated robot has 12 sonar sensors that have a maximum range of 10 grid cells and have a beam width of 30°. This configuration was chosen so that no blind spots existed directly adjacent to the robot. If a blind spot did exist, the robot would choose it as its next pose. This is unsafe as there would be no certainty regarding the cell's status as an obstacle.

Four maps are used to test the algorithm; each measures 50 by 50 grid cells. Location $(0, 0)$ is taken as the top left corner; Table 1 gives the starting robot pose for each map. The geometry of the maps is provided in Fig. 3 and they are of increasing complexity. The map M_1 is an environment where there are no corridors; the map M_2 is an environment where one corridor exists in an corner behind a square obstacle; and, the map M_3 is another structured environment that is different by introducing an alcove into a central obstacle. The map M_4 is a map

Table 1. Starting robot pose for test maps

Test map	Starting pose x	y	θ
m_1	34	5	$\pi/2$
m_2	25	25	π
m_3	45	45	π
m_4	39	13	$\pi/2$

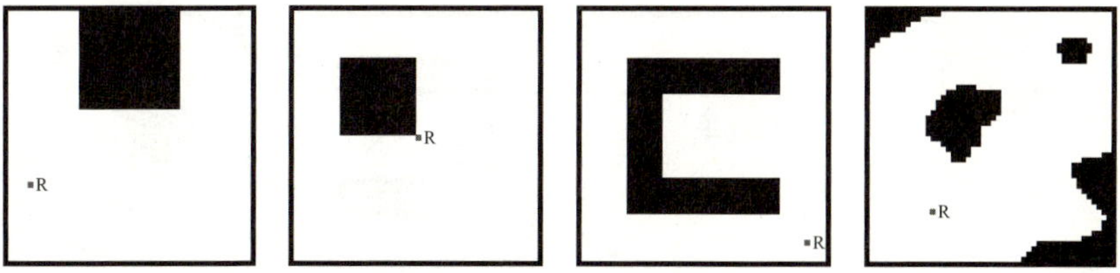

(a) M_1 (b) M_2 (c) M_3 (d) M_4

Fig. 3. Maps used for testing

that represents a less structured environment - one that represents what a mobile robot might encounter in nature. The robot's starting position is denoted by R.

3.1 Simulation Results

Algorithm performance was analysed using two metrics: time and energy. Time is measured as the number of iterations to sufficiently reveal the map, t. Energy is measured as the distance the robot travels to reveal the map, l. The simulation trials were run once with no reward functions and once with the reward functions being used. The values for $c = 2.5 \times 10^{-6}$, $d = 0.04$ and $r_{\text{cut-off}} = 3$ were determined experimentally. In each trial, the five different neural input functions were tested.

No Rewards. With no reward functions, the next pose is determined only through neural activity. There is no extra incentive for the robot to follow a straight path or follow a path that distances itself from obstacles. The results for each map are presented in Table 2; the best paths generated for each map are presented in Fig. 4.

In Fig. 4 two robot behaviors are apparent: wall following, and a tendency to zigzag locally while globally following a straight path. These behaviors are caused by blind spots that occur when the robot is approaching a wall. Once a robot is directly adjacent to the wall, the sensors cannot reliably detect an obstacle diagonal to the robot - leaving the area unexplored and attracting the robot toward it. Free space on the opposite side does not attract the robot as

Table 2. Simulation results with no rewards. *Robot did not completely explore the environment.

Neural input function	M_1 t	M_1 l	M_2 t	M_2 l	M_3 t	M_3 l	M_4 t	M_4 l
Gaussian	328	402.8	364	441.9	352*	410.0*	215*	275.5*
Sigmoidic	332	389.8	345	402.2	338*	363.7*	418*	512.0*
Step	291	377.4	337	405.3	356	394.9	328	402.1
Trapezoidal	294	379.6	324	373.7	358	457.8	343	420.5
Triangular	378	425.0	384	446.1	383	399.6	426	511.7

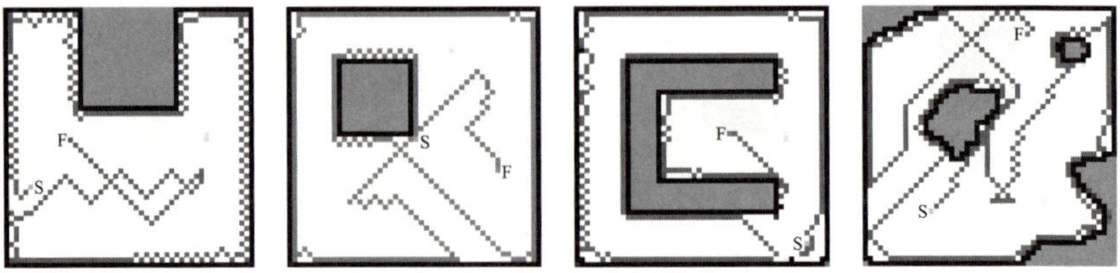

(a) M_1 step (b) M_2 trapezoidal (c) M_3 step (d) M_4 step

Fig. 4. Best paths using no rewards

the unexplored areas do; so, there is nothing to deter circumnavigating the wall. Once the area around the wall is explored, the robot is drawn to the interior, as seen in Fig. 4(a) and Fig. 4(c).

While the robot exhibits these potentially undesirable behaviors, three of the five neural inputs functions were able to provide an exploration plan for each of the maps tested: the step, the trapezoidal and the triangular neural input functions (Table 2).

With Reward Functions. With the inclusion of the reward functions, the mobile robot now attempts to maintain a distance from obstacles and receives a slight reward for minimizing changes in its orientation. Table 3 lists the results for each map while Fig. 5 presents the best paths generated for each map.

Backtracking is more of an issue when using the reward functions, Fig. 5(b) and 5(d). The robot avoids close proximity to obstacles, and while zigzagging is reduced in Fig. 5(a) compared to Fig. 4(a), it could be improved further. The maps M_1 and M_3 show the greatest improvement and present reasonable paths.

No neural input function was able to sufficiently map the environment for the map M_4. All except the triangular neural input function revealed approximately the same map and had the exact same problem area: the top right corner. Due to the narrow corridor that exists in this location, as the robot enters the corridor it detects the walls but it is unsure if the obstacle is directly ahead of it or diagonal to it. Therefore, the robot is deterred from entering the corridor. This problem may be averted by using a greate number of sensors with narrower beam widths.

Table 3. Simulation results with reward functions. *Robot did not completely explore the environment.

Neural input	M_1		M_2		M_3		M_4	
function	t	l	t	l	t	l	t	l
Gaussian	278	358.4	413	508.7	244	310.7	490*	632.9*
Sigmoidic	354	452.2	407	512.6	387	495.1	409*	518.8*
Step	280	358.7	353*	456.1*	263*	328.9*	422*	531.8*
Trapezoidal	335	434.8	459*	554.7*	312	395.7	422*	527.6*
Triangular	499*	623.3*	437*	543.0*	499*	610.0*	499*	623.7*

(a) M_1 Gaussian (b) M_2 sigmoidic (c) M_3 Gaussian (d) M_4 sigmoidic

Fig. 5. Best paths using rewards

4 Conclusion

The research has generated an algorithm that has been shown to be able to provide a successful exploration plan for a mobile robot. The algorithm has been presented with both with and without reward functions.

With no reward functions, the robot tended to maintain a close proximity to obstacles and to exhibit a local zigzagging behavior. The close proximity to obstacles is a concern for sensor efficiency and safety. If a sensor reading is spurious, the robot may run into an obstacle. Different neural input functions were also examined; the trapezoidal and step neural input functions performed the best.

The reward functions had mixed effects on the algorithm's performance. Occasionally the reward functions discouraged the robot from exploring new territory in favor of maintaining the current heading or avoiding a narrow corridor, Fig. 5(d). There were also occasions where the reward functions benefited the algorithm, Fig. 5(a) and Fig. 5(c). The Gaussian neural input function proved to provide the best overall exploration solution.

The algorithm presented is to be used in conjunction with a grid-based SLAM algorithm. This avoids the need for landmark recognition as is the case in [5]. Unlike frontier based approaches in [4, 3], this algorithm allows the path to be changed at any time, making better use of the sensors that are available.

5 Future Work

Further work may yield a better combination of parameters for the reward functions. While providing large advantages in some situations, the reward functions are not as consistent as desired. The problem with pursuing this is a matter of time due to range of values for the paramters. Another avenue of further work is to re-examine the simulation data to find a common problem among the trials and develop a reward function to avoid this fault.

In the larger research framework, the work presented here is a large step in developing a neural dynamics based exploration algorithm for a SLAM environment. What must be done next is to implement this algorithm in a SLAM environment as opposed to an environment where perfect localization is assumed. This will provide a performance baseline for the algorithm. Modifications can then be made with additional excitatory inputs and reward functions. As shown in other works, such as in [7], an exploration algorithm that considers pose maintenance generates maps of higher fidelity. As a result, presumably these new reward functions will be conditioned by pose certainty and may reward a robot who has greater pose uncertainty to choose a new pose that might increase pose certainty. Moreover, the algorithm should be tested in a dynamic environment as the benefits of the safety attained from maintaining a distance from obstacles may become more apparent.

Acknowledgment

This work was supported by Natural Sciences and Engineering Research Council (NSERC), Canada.

References

1. Dudek, G., Jenkin, M.: Computational Principles of Mobile Robotics, United States of America. Cambridge University Press, Cambridge (2000)
2. Siegwart, R., Nourbakhsh, I.R.: Introduction to Autonomous Mobile Robotics, United States of America. MIT Press, Redmond, Washington (2004)
3. Moorehead, S., Simmons, R., Whittakes, W.: Autonomous Exploration Using Multiple Sources of Information. In: Proceedings of IEEE International Conference on Robotics and Automation, Seoul, Korea, pp. 3098–3103 (2001)
4. Yamauchi, B.: Frontier-based Approach for Autonomous Exploration. In: Proceedings of IEEE International Symposium on Computational Intelligence in Robotics and Automation, Monterey, California, USA, pp. 146–151 (1997)
5. Makarenko, A., Williams, S., Bourgault, F., Durrant-Whyte, H.: An Experiment in Integrated Exploration. In: Proceedings of IEEE/RSJ International Conference on Intelligent Robots and Systems, Laussane, Switzerland, pp. 534–539 (2002)
6. Thrun, S., Burgard, W., Fox, D.: Probabilistic Robotics, United States of America. MIT Press, Redmond, Washington (2005)
7. Stachniss, C., Hahnel, D., Burgard, W., Grisetti, G.: On Actively Closing Loops in Grid-based FastSLAM. Advanced Robotics 19(10), 1059–1079 (2005)
8. Yang, S.X., Yuan, X., Meng, M., Yuan, G., Li, H.: Real-time Planning and Control of Robots using Shunting Neural Networks. In: Proceedings of the 2001 IEEE/RSJ International Conference on Intelligent Robots and Systems, Maui, Hawaii, USA, pp. 1590–1595 (2001)
9. Luo, C., Yang, S.X., Meng, M.: Neurodynamics Based Complete Coverage Navigation with Real-time Map Building in Unknown Environments. In: Proceedings of the 2006 IEEE/RSJ International Conference on Intelligent Robots and Systems, Beijing, China, pp. 4228–4233 (2006)
10. Grossberg, S.: Nonlinear Neural Networks: Principles, Mechanisms and Architecture. Neural Networks 1, 17–61 (1988)
11. Hodgkin, A., Huxley, A.: A Quantitative Description of Membrane Current and its Application to Conduction and Excitation in Nerves. Journal of Physiology London 117, 500–544 (1952)

Neural Models in Computationally Efficient Predictive Control Cooperating with Economic Optimisation

Maciej Ławryńczuk

Institute of Control and Computation Engineering, Warsaw University of Technology
ul. Nowowiejska 15/19, 00-665 Warsaw, Poland
Tel.: +48 22 234-73-97
M.Lawrynczuk@ia.pw.edu.pl

Abstract. This paper discusses the problem of cooperation of economic optimisation with Model Predictive Control (MPC) algorithms when the dynamics of disturbances is comparable with the dynamics of the process. A dynamic neural model is used in the suboptimal nonlinear MPC algorithm with Nonlinear Prediction and Linearisation (MPC-NPL), a steady-state neural model is used in approximate economic optimisation which is executed as frequently as the MPC algorithm. The MPC-NPL algorithm requires solving on-line only a quadratic programming problem whereas approximate economic optimisation needs solving a linear programming problem. As a result, the necessity of repeating two nonlinear optimisation problems at each sampling instant is avoided.

1 Introduction

The hierarchical (multilayer) control system structure has been used for years in process control [1], [2], [3], [13]. The regulatory control layer keeps process at given operating points and the optimisation layer calculates these set-points. Model Predictive Control (MPC) algorithms [9], [12], [13] are usually used at the regulatory layer largely due to their unique ability to take into account constraints imposed on process inputs (manipulated variables) and outputs (controlled variables) or state variables, which usually decide on quality, economic efficiency and safety of production and efficiency in multivariable process control.

When the classical multilayer control system structure is used, it is usually assumed that the disturbances are slowly-varying when compared to the dynamics of the process. Thanks to such an assumption, the steady-state nonlinear economic optimisation problem can be solved reasonably less frequently than the MPC controllers execute. Provided that the dynamics of disturbances is much slower than the dynamics of the plant, such an approach gives satisfactory results. In many practical cases, however, the dynamics of the disturbances is comparable with the process dynamics. Very often the disturbances, for example flow rates, properties of feed and energy streams etc., vary significantly and not much slower than the dynamics of the controlled process. In such cases operation

J. Marques de Sá et al. (Eds.): ICANN 2007, Part II, LNCS 4669, pp. 650–659, 2007.

in the classical hierarchical structure with frequency of the economic optimisation much lower than that of MPC can result in a significant loss of economic effectiveness [13]. Ideally, it would be best to perform full nonlinear optimisation but with increased frequency. Obviously, because of high computational burden, such an approach has limited applicability and is rarely implementable on-line.

This paper discusses a computationally efficient approach to the problem of cooperation of economic optimisation and MPC algorithms. An additional approximate economic optimisation problem is solved as frequently as the MPC executes. It recalculates the optimal operating point determined by the nonlinear economic optimisation layer which is activated infrequently. A steady-state neural model is used in the approximate and nonlinear economic optimisation, a dynamic neural model is used in the suboptimal nonlinear MPC algorithm with Nonlinear Prediction and Linearisation (MPC-NPL) [7], [8], [13]. Fundamental models, although potentially very precise, are usually not suitable for on-line control and optimisation because they are very complicated and may lead to numerical problems (e.g. ill-conditioning, stiffness, etc.). Neural models [4], [5], [11] are used because they have excellent approximation abilities, small number of parameters and simple structure. Moreover, problems typical of fundamental models are not encountered because neural models directly describe input-output relations of process variables, complicated systems of algebraic and differential equations do not have to be solved on-line.

2 MPC Cooperating with Economic Optimisation

2.1 Economic Optimisation Problem

The objective of the economic optimisation is to maximise the production profit and satisfy the constraints, which determine safety and quality of production. Typically, the economic optimisation layer solves the following problem (for simplicity of presentation Single-Input Single-Output (SISO) process is assumed)

$$\begin{aligned} &\min_{u^s} \{J_E(k) = c_u u^s - c_y y^s\} \\ &\text{subject to} \\ &\quad u^{\min} \leq u^s \leq u^{\max} \\ &\quad y^{\min} \leq y^s \leq y^{\max} \\ &\quad y^s = f^s(u^s, h^s) \end{aligned} \tag{1}$$

where u is the input of the process (manipulated variable), y is the output (controlled variable) and h is the measured (or estimated) disturbance, the superscript 's' refers to the steady-state. The function $f^s : \Re^2 \longrightarrow \Re \in C^1$ denotes a comprehensive steady-state process model. The quantities c_u, c_y represent prices resulting from economic considerations, $u^{\min}$, $u^{\max}$, $y^{\min}$, $y^{\max}$ denote constraints imposed on input and output variables, respectively. Nonlinear nature of the steady-state model $y^s = f^s(u^s, h^s)$ means that the economic optimisation requires solving on-line the nonlinear optimisation problem (1).

2.2 Model Predictive Control Optimisation Problem

In the MPC algorithms [9], [13] at each consecutive sampling instant k a set of future control increments is calculated

$$\Delta \boldsymbol{u}(k) = [\Delta u(k|k)\ \Delta u(k+1|k) \ldots \Delta u(k+N_u-1|k)]^T \tag{2}$$

Only the first element of the determined sequence (2) is applied to the process, the control law is $u(k) = \Delta u(k|k) + u(k-1)$. At next sampling instant, $k+1$, the prediction is shifted one step forward and the whole procedure is repeated.

Let u^s_{eo} denote the optimal solution to the economic optimisation problem (1). Using the nonlinear steady-state model $y^s = f^s(u^s, h^s)$, the value y^s_{eo} corresponding to u^s_{eo} is calculated. The quantity y^s_{eo} is then passed as the desired set-point to the MPC optimisation problem

$$\begin{aligned}
&\min_{\Delta \boldsymbol{u}(k)} \{J(k) = \sum_{p=1}^{N} (y^s_{eo} - \hat{y}(k+p|k))^2 + \sum_{p=0}^{N_u-1} \lambda_p (\Delta u(k+p|k))^2\} \\
&\text{subject to} \\
&\quad u^{\min} \leq u(k+p|k) \leq u^{\max}, \quad p = 0, \ldots, N_u - 1 \\
&\quad -\Delta u^{\max} \leq \Delta u(k+p|k) \leq \Delta u^{\max}, \quad p = 0, \ldots, N_u - 1 \\
&\quad y^{\min} \leq \hat{y}(k+p|k) \leq y^{\max}, \quad p = 1, \ldots, N
\end{aligned} \tag{3}$$

where N and N_u are the prediction and control horizons, respectively, $\lambda_p > 0$.

3 Neural Models in MPC Cooperating with Economic Optimisation

As increasing the frequency of the economic optimisation layer is limited in practice because of its high computational burden, the MPC layer is often supplemented with additional steady-state target optimisation [6], [12], [13]. Steady-state target optimisation closely cooperates with the MPC layer, the steady-state operating-point determined by the nonlinear economic optimisation layer activated infrequently is recalculated as frequently as the MPC executes. Typically, in steady-state target optimisation a simplified constant linear model is used. Conceptually, it would be better to use a more accurate model.

Structure of the considered control system is depicted in Fig. 1. It consists of nonlinear economic optimisation, approximate steady-state target optimisation and the suboptimal MPC-NPL algorithm. A steady-state neural model is used in economic optimisation and steady-state target optimisation whereas a dynamic neural model is used in the MPC-NPL algorithm. Economic optimisation, in which the nonlinear optimisation problem (1) is solved, is executed infrequently. At each sampling instant steady-state target optimisation and MPC optimisation problems are solved. Steady-state target optimisation recalculates the optimal set-point when economic optimisation is not activated. The MPC-NPL optimisation problem is of quadratic programming type, steady-state target optimisation

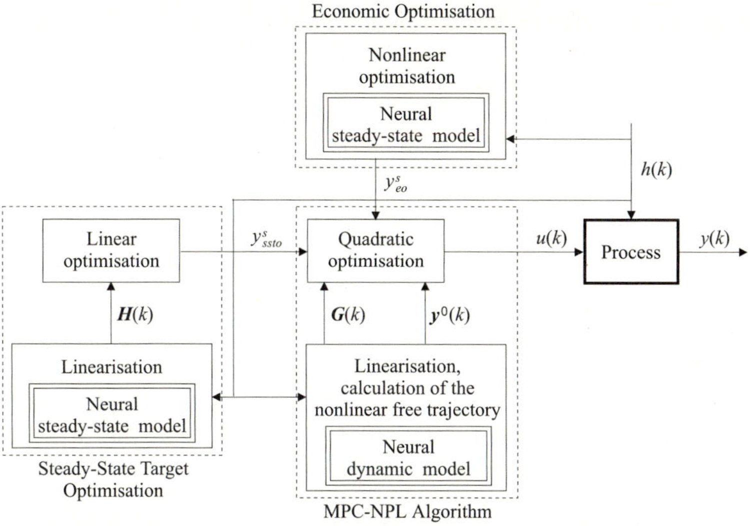

Fig. 1. Structure of the control system with nonlinear economic optimisation, steady-state target optimisation and the MPC-NPL algorithm

is of linear programming type. The measured disturbance h is used in all layers. The optimal operating point determined by nonlinear economic optimisation is denoted by y^s_{eo} whereas the optimal operating point recalculated by steady-state target optimisation is denoted by y^s_{ssto}.

Let the dynamic model of the process under consideration be described by

$$y(k) = g(\boldsymbol{x}(k)) = f(u(k-\tau), \ldots, u(k-n_B), y(k-1), \ldots, y(k-n_A), \\ h(k-\tau_h), \ldots, h(k-n_C)) \tag{4}$$

where $f : \Re^{n_A+n_B+n_C-\tau-\tau_h+2} \longrightarrow \Re \in C^1$, $\tau \leq n_B$, $\tau_h \leq n_C$.

A feedforward neural network with one hidden layer and linear output [4] is used as the function f in (4). Output of the model can be expressed as

$$y(k) = f(\boldsymbol{x}(k)) = w_0^2 + \sum_{i=1}^{K} w_i^2 v_i(k) = w_0^2 + \sum_{i=1}^{K} w_i^2 \varphi(z_i(k)) \tag{5}$$

where $z_i(k)$ and $v_i(k)$ are the sum of inputs and the output of the i-th hidden node, respectively, $\varphi : \Re \longrightarrow \Re$ is the nonlinear transfer function, K is the number of hidden nodes. Recalling input arguments of the nonlinear model (4)

$$z_i(k) = g(\boldsymbol{x}(k)) = w_{i,0}^1 + \sum_{j=1}^{I_u} w_{i,j}^1 u(k-\tau+1-j) + \sum_{j=1}^{n_A} w_{i,I_u+j}^1 y(k-j) \\ + \sum_{j=1}^{I_h} w_{i,I_u+n_A+j}^1 h(k-\tau_h+1-j) \tag{6}$$

Weights of the network are denoted by $w_{i,j}^1$, $i = 1, \ldots, K$, $j = 0, \ldots, n_A + n_B - \tau + 1$, and w_i^2, $i = 0, \ldots, K$, for the first and the second layer, respectively, $I_u = n_B - \tau + 1$, $I_h = n_C - \tau_h + 1$.

Similarly as the dynamic model (4), a feedforward neural network with one hidden layer and linear output is used as the steady-state model $y^s = f^s(u^s, h^s)$

$$y^s = f^s(u^s, h^s) = w_0^{2s} + \sum_{i=1}^{K^s} w_i^{2s} v_i^s = w_0^{2s} + \sum_{i=1}^{K^s} w_i^{2s} \varphi(z_i^s) \tag{7}$$

where

$$z_i^s = w_{i,0}^{1s} + w_{i,1}^{1s} u^s + w_{i,2}^{1s} h^s \tag{8}$$

Weights of the second network are denoted by $w_{i,j}^{1s}$, $i = 1, \ldots, K^s$, $j = 0, 1, 2$, and w_i^{2s}, $i = 0, \ldots, K^s$, for the first and the second layer, respectively. Neural models are trained using input-output data obtained from the real process or from simulation of the fundamental models.

3.1 On-Line Model Approximation in Steady-State Target Optimisation

The steady-state neural model is linearised on-line taking into account the current state of the process determined by $u(k-1)$ and $h(k)$

$$y^s = f^s(u^s, h^s)|_{u^s = u(k-1),\ h^s = h(k)} + \boldsymbol{H}(k)(u^s - u(k-1)) \tag{9}$$

where $\boldsymbol{H}(k)$ is the derivative of the nonlinear steady-state model

$$\boldsymbol{H}(k) = \left. \frac{df^s(u^s, h^s)}{du^s} \right|_{u^s = u(k-1),\ h^s = h(k)} \tag{10}$$

In contrast to the standard steady-state target optimisation approach with a constant linear model [6], [12], [13], the discussed formulation uses a linearised model derived on-line from the steady-state neural model. Taking into account structure of the steady-state neural model described by (7) and (8), one has

$$\boldsymbol{H}(k) = \sum_{i=1}^{K^s} w_i^{2s} \left. \frac{d\varphi(z_i^s)}{dz_i^s} \right|_{u^s = u(k-1),\ h^s = h(k)} w_{i,1}^{1s} \tag{11}$$

Considering the nonlinear economic optimisation problem (1), the equivalent steady-state target optimisation problem is formulated as the linear programming task

$$\begin{aligned}
&\min_{u^s} \{J_E(k) = c_u \Delta u^s - c_y \Delta y^s\} \\
&\text{subject to} \\
&\quad u^{\min} \leq u^s \leq u^{\max} \\
&\quad y^{\min} \leq y^s \leq y^{\max} \\
&\quad y^s = f^s(u^s, h^s)|_{u^s = u(k-1),\ h^s = h(k)} + \Delta y^s \\
&\quad \Delta y^s = \boldsymbol{H}(k) \Delta u^s \\
&\quad \Delta u^s = u^s - u(k-1)
\end{aligned} \tag{12}$$

Let u^s_{ssto} denote the optimal solution to the steady-state target optimisation problem (12). Using the linearised steady-state model the value y^s_{ssto} corresponding to u^s_{ssto} is found. The quantity y^s_{ssto} is then passed as the desired set-point to the predictive controller optimisation problem.

3.2 On-Line Model Approximation in the Suboptimal MPC-NPL Algorithm

In the MPC-NPL algorithm [7], [8], [13] at each sampling instant k the dynamic neural model is used on-line twice: to determine the local linearisation and the nonlinear free trajectory (Fig. 1). It is assumed that the output prediction can be expressed as the sum of the forced trajectory, which depends only on the future (on future input moves $\Delta \boldsymbol{u}(k)$) and the free trajectory $\boldsymbol{y}^0(k)$, which depends only on the past

$$\hat{\boldsymbol{y}}(k) = \boldsymbol{y}^0(k) + \boldsymbol{G}(k)\Delta \boldsymbol{u}(k) \tag{13}$$

where

$$\hat{\boldsymbol{y}}(k) = [\hat{y}(k+1|k)\dots\hat{y}(k+N|k)]^T \tag{14}$$

$$\boldsymbol{y}^0(k) = \left[y^0(k+1|k)\dots y^0(k+N|k)\right]^T \tag{15}$$

The dynamic matrix $\boldsymbol{G}(k)$ of dimensionality $N \times N_u$ is calculated on-line from the nonlinear model taking into account the current state of the plant

$$\boldsymbol{G}(k) = \begin{bmatrix} s_1(k) & 0 & \dots & 0 \\ s_2(k) & s_1 & \dots & 0 \\ \vdots & \vdots & \ddots & \vdots \\ s_N(k) & s_{N-1}(k) & \dots & s_{N-N_u+1}(k) \end{bmatrix} \tag{16}$$

where $s_j(k)$ are step-response coefficients of the linearised model

$$s_j(k) = \sum_{i=1}^{\min(j,n_B)} b_i(k) - \sum_{i=1}^{\min(j-1,n_A)} a_i(k)s_{j-i}(k) \tag{17}$$

Taking into account the structure of the neural model described by the (5) and (6), coefficients of the linearised dynamic model are calculated on-line from

$$a_l(\bar{\boldsymbol{x}}(k)) = -\frac{\partial g(\bar{\boldsymbol{x}}(k))}{\partial y(k-l)} = -\sum_{i=1}^{K} w_i^2 \frac{d\varphi(z_i(\bar{\boldsymbol{x}}(k)))}{dz_i(\bar{\boldsymbol{x}}(k))} w^1_{i,I_u+l} \quad l = 1,\dots,n_A \tag{18}$$

and

$$b_l(\bar{\boldsymbol{x}}(k)) = \begin{cases} 0 & l = 1,\dots,\tau-1 \\ \dfrac{\partial g(\bar{\boldsymbol{x}}(k))}{\partial u(k-l)} = \displaystyle\sum_{i=1}^{K} w_i^2 \frac{d\varphi(z_i(\bar{\boldsymbol{x}}(k)))}{dz_i(\boldsymbol{x}(k))} w^1_{i,l-\tau+1} & l = \tau,\dots,n_B \end{cases} \tag{19}$$

The linearisation point is the vector composed of past input and output signal values corresponding to the arguments of the nonlinear model (4)

$$\bar{\boldsymbol{x}}(k) = \left[\bar{u}(k-\tau)\ldots\bar{u}(k-n_B)\bar{y}(k-1)\ldots\bar{y}(k-n_A)\bar{h}(k-\tau_h)\ldots\bar{h}(k-n_C)\right]^T \tag{20}$$

On the one hand, the suboptimal prediction calculated from (13) is different from the optimal one determined from the nonlinear neural model as it is done in MPC algorithms with nonlinear optimisation [7], [13]. On the other hand, thanks to using the superposition principle (13), the MPC optimisation problem (3) becomes the following quadratic programming task

$$\begin{aligned}
&\min_{\Delta\boldsymbol{u}(k),\ \boldsymbol{\varepsilon}_{\min},\ \boldsymbol{\varepsilon}_{\max}} \left\{J(k) = \left\|\boldsymbol{y}^{ref}(k) - \boldsymbol{y}^0(k) - \boldsymbol{G}(k)\Delta\boldsymbol{u}(k)\right\|^2 + \left\|\Delta\boldsymbol{u}(k)\right\|^2_{\boldsymbol{\Lambda}} + \right.\\
&\qquad\left. +\rho_{\min}\left\|\boldsymbol{\varepsilon}_{\min}\right\|^2 + \rho_{\max}\left\|\boldsymbol{\varepsilon}_{\max}\right\|^2\right\}\\
&\text{subject to}\\
&\boldsymbol{u}^{\min} \leq \boldsymbol{J}\Delta\boldsymbol{u}(k) + \boldsymbol{u}^{k-1} \leq \boldsymbol{u}^{\max}\\
&-\Delta\boldsymbol{u}^{\max} \leq \Delta\boldsymbol{u}(k) \leq \Delta\boldsymbol{u}^{\max}\\
&\boldsymbol{y}^{\min} - \boldsymbol{\varepsilon}_{\min} \leq \hat{\boldsymbol{y}}(k) \leq \boldsymbol{y}^{\max} + \boldsymbol{\varepsilon}_{\max}\\
&\boldsymbol{\varepsilon}_{\min} \geq 0, \quad \boldsymbol{\varepsilon}_{\max} \geq 0
\end{aligned} \tag{21}$$

where vectors of length N are

$$\boldsymbol{y}^{ref}(k) = \left[y^s_{eo}\cdots y^s_{eo}\right]^T \quad \text{or} \quad \boldsymbol{y}^{ref}(k) = \left[y^s_{ssto}\cdots y^s_{ssto}\right]^T \tag{22}$$

$$\boldsymbol{y}^{\min}(k) = \left[y^{\min}\ldots y^{\min}\right]^T \tag{23}$$

$$\boldsymbol{y}^{\max}(k) = \left[y^{\max}\ldots y^{\max}\right]^T \tag{24}$$

vectors of length N_u are

$$\boldsymbol{u}^{\min}(k) = \left[u^{\min}\ldots u^{\min}\right]^T \tag{25}$$

$$\boldsymbol{u}^{\max}(k) = \left[u^{\max}\ldots u^{\max}\right]^T \tag{26}$$

$$\Delta\boldsymbol{u}^{\max}(k) = \left[\Delta u^{\max}\ldots\Delta u^{\max}\right]^T \tag{27}$$

$$\boldsymbol{u}^{k-1}(k) = \left[u(k-1)\ldots u(k-1)\right]^T \tag{28}$$

$\boldsymbol{J}$ is the all ones lower triangular matrix of dimensionality $N_u \times N_u$ and $\boldsymbol{\Lambda} = diag(\lambda_0, \ldots, \lambda_{N_u-1})$. If at the current sampling instant nonlinear economic optimisation is used, $\boldsymbol{y}^{ref}(k) = \left[y^s_{eo}\cdots y^s_{eo}\right]^T$, if approximate steady-state target optimisation is used, $\boldsymbol{y}^{ref}(k) = \left[y^s_{ssto}\cdots y^s_{ssto}\right]^T$. Because the output constraints are present in the MPC optimisation problem, the infeasibility problems may occur. That is why the output constraints are softened by means of slack variables [9], [13] (the vectors $\boldsymbol{\varepsilon}_{\min}$ and $\boldsymbol{\varepsilon}_{\max}$ of length N), $\rho_{\min}$, $\rho_{\max}$ are positive weights. The free trajectory $\boldsymbol{y}^0(k)$ is calculated recursively on-line from the nonlinear neural model (5), (6) assuming no changes in control signals from time instant k (i.e. $u(k+p|k) := u(k-1)$, $p \geq 0$).

4 Simulation Results

The process under consideration is a polymerisation reaction taking place in a jacketed continuous stirred tank reactor [10]. The reaction is the free-radical polymerisation of methyl methacrylate with azo-bis-isobutyronitrile as initiator and toluene as solvent. The output $NAMW$ (Number Average Molecular Weight) is controlled by manipulating the inlet initiator flow rate F_I. Flow rate F of the monomer is the measured disturbance.

Three models of the process are used. The fundamental model [10] is used as the real process during simulations. An identification procedure is carried out, as a result two neural models are obtained, namely a dynamic one ($K = 6$) and a steady-state one ($K^s = 4$). In the MPC-NPL algorithm the horizons are $N = 10$, $Nu = 3$, the weighting coefficients $\lambda_p = 0.2$, the sampling time is 1.8 min.

Since maximum production rate is required, the following performance function is used at the economic optimisation layer

$$J_E(k) = -F_I^s \tag{29}$$

The following constraints are imposed on the manipulated variable

$$F_I^{\min} \leq F_I \leq F_I^{\max} \tag{30}$$

where $F_I^{\min} = 0.0035\ m^3/h$, $F_I^{\max} = 0.033566\ m^3/h$. The product should satisfy some purity criteria, i.e. the output variable is constrained

$$NAMW^{\min} \leq NAMW \tag{31}$$

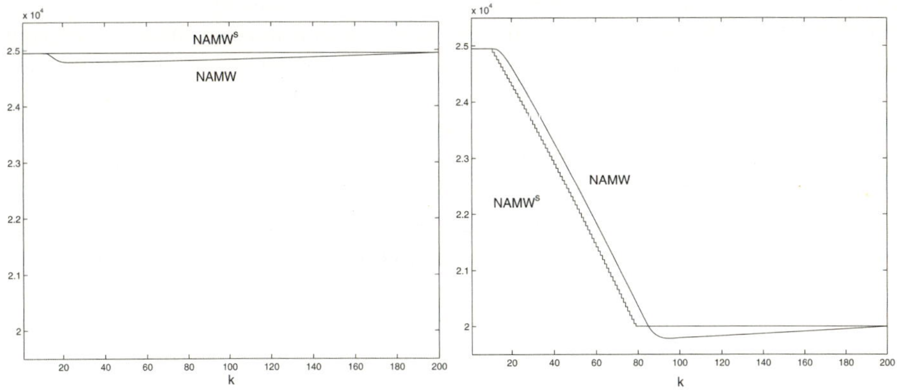

Fig. 2. Simulation results of the polymerisation reactor with nonlinear economic optimisation used once at $k = 3$ (left) and "ideal" nonlinear optimisation repeated as frequently as the MPC controller executes (right), i.e. $T_E = 1$

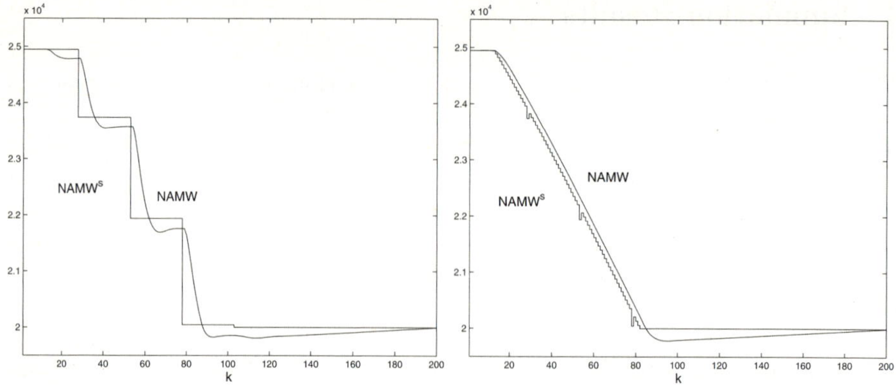

Fig. 3. Simulation results of the polymerisation reactor with nonlinear economic optimisation used 25 times less frequently than the MPC controller executes, i.e. $T_E = 25$: without (left) and with (right) steady-state target optimisation

where $NAMW^{\min} = 20000\ kg/kmol$. The same constraints imposed on manipulated and controlled variables are taken into account in economic optimisation, steady-state target optimisation and MPC-NPL optimisation problems. It is assumed that changes in the disturbance signal can be described by the equation

$$F(k) = 2 - 1.6(\sin(0.008k) - \sin(0.08)) \tag{32}$$

Let T_E denote the intervention period of the nonlinear economic optimisation layer, when it is executed as often as the MPC controller $T_E = 1$. Simulation results of the polymerisation reactor with single economic optimisation (performed at $k = 3$) and with $T_E = 1$ are depicted in Fig. 2. If nonlinear economic optimisation is performed only once, the set-point value is constant, $J_E = -37187.72$ (for comparison J_E is calculated for the whole simulation horizon after completing the simulations). On the contrary, economic optimisation repeated as frequently as the MPC controller executes takes into account changes in the disturbance F, a new optimal steady-state operating point is calculated for each sampling instant, $J_E = -52458.67$. Because the output constraint (31) is implemented as soft in the MPC-NPL algorithm, it is temporarily violated.

Fig. 3 shows simulation results of the polymerisation reactor with nonlinear economic optimisation used 25 times less frequently than the MPC controller executes, i.e. $T_E = 25$ in two cases, namely without ($J_E = -51010.58$) and with steady-state target optimisation ($J_E = -52458.67$). Thanks to the introduction of approximate steady-state target optimisation which needs solving on-line only a linear programming problem, the nonlinear economic optimisation problem does not have to be repeated at each sampling instant. The trajectory of the system is practically the same as in the "ideal" case when nonlinear economic optimisation is repeated at each sampling instant.

5 Conclusion

Cooperation of model predictive control algorithms with economic steady-state optimisation is investigated in the paper. It is particularly important when the dynamics of disturbances is comparable with the dynamics of the process. The emphasis is put on computational efficiency. The MPC-NPL algorithm needs solving on-line only a quadratic programming problem, approximate steady-state target optimisation requires solving only a linear programming problem, the necessity of repeating two nonlinear optimisation problems at each sampling instant is avoided. A dynamic neural model is used in the nonlinear MPC-NPL algorithm, a steady-state neural model is used for economic optimisation.

Acknowledgement

This work was supported by Polish national budget funds 2005-2007 for science as a research project.

References

1. Blevins, T.L., Mcmillan, G.K., Wojsznis, M.W.: Advanced control unleashed. ISA (2003)
2. Brdys, M., Tatjewski, P.: Iterative algorithms for multilayer optimizing control. Imperial College Press, London (2005)
3. Findeisen, W.M., Bailey, F.N., Brdyś, M., Malinowski, K., Tatjewski, P., Woźniak, A.: control and coordination in hierarchical systems, Chichester - New York - Brisbane - Toronto. J. Wiley and Sons, Chichester (1980)
4. Haykin, S.: Neural networks – a comprehensive foundation. Prentice-Hall, Englewood Cliffs (1999)
5. Hussain, M.A.: Review of the applications of neural networks in chemical process control – simulation and online implmementation. Artificial Intelligence in Engineering. 13, 55–68 (1999)
6. Kassmann, D.E., Badgwell, T.A., Hawkins, R.B.: Robust steady-state target calculation for model predictive control. AIChE Journal 46, 1007–1024 (2000)
7. Ławryńczuk, M.: A family of model predictive control algorithms with artificial neural networks. International Journal of Applied Mathematics and Computer Science, accepted for publication (2007)
8. Ławryńczuk, M., Tatjewski, P.: An efficient nonlinear predictive control algorithm with neural models and its application to a high-purity distillation process. In: Rutkowski, L., Tadeusiewicz, R., Zadeh, L.A., Zurada, J.M. (eds.) ICAISC 2006. LNCS (LNAI), vol. 4029, pp. 76–85. Springer, Heidelberg (2006)
9. Maciejowski, J.M.: Predictive control with constraints. Prentice Hall. Harlow, Englewood Cliffs (2002)
10. Maner, B.R., Doyle, F.J., Ogunnaike, B.A., Pearson, R.K.: Nonlinear model predictive control of a simulated multivariable polymerization reactor using second-order Volterra models. Automatica. 32, 1285–1301 (1996)
11. Nørgaard, M., Ravn, O., Poulsen, N.K., Hansen, L.K.: Neural networks for modelling and control of dynamic systems. Springer, London (2000)
12. Qin, S.J., Badgwell, T.A.: A survey of industrial model predictive control technology. Control Engineering Practice 11, 733–764 (2003)
13. Tatjewski, P.: Advanced control of industrial processes, Structures and algorithms. Springer, London (2007)

Event Detection and Localization in Mobile Robot Navigation Using Reservoir Computing

Eric A. Antonelo[1], Benjamin Schrauwen[1], Xavier Dutoit[2], Dirk Stroobandt[1], and Marnix Nuttin[2]

[1] Electronics and Information Systems Department, Ghent University, Belgium
[2] Department of Mechanical Engineering - Katholic University of Leuven, Belgium
Eric.Antonelo@elis.ugent.be

Abstract. Reservoir Computing (RC) uses a randomly created recurrent neural network where only a linear readout layer is trained. In this work, RC is used for detecting complex events in autonomous robot navigation. This can be extended to robot localization based solely on sensory information. The robot thus builds an implicit map of the environment without the use of odometry data. These techniques are demonstrated in simulation on several complex and even dynamic environments.

1 Introduction

Autonomous robot navigation systems have been extensively developed in the literature [1,2,3,4]. Early navigation strategies are either deliberative (generation of robot trajectories based on path planning) or reactive (robot control based on a direct mapping of sensory input to actions). Current state-of-the-art autonomous robot control architectures are *hybrid* [1]: they have an underlying reactive controller which takes care of the real-time critical and simple tasks such as obstacle avoidance; while an upper deliberative control layer steers this reactive part. Information flow in this architecture is both downwards, from abstract deliberative tasks to concrete physical reactive behaviours, and upwards, from physical data to abstract symbols used for deliberation.

This paper investigates two cases of upward information flow: a system for recognizing complex robot events in particular environments (such as detecting if the robot goes through a door, given only sensory input); and a system for determining the current robot location, solely based on sensory information. Both are achieved using the same setup.

These tasks have been shown to be difficult [5]. Traditional algorithms based on the Simultaneous Localization and Mapping (SLAM) concept are expensive to implement due to limited computational efficiency and also hold uncertainties during the calculation of the robot's pose [5].

This work uses an implicit way of forming a representation of the robot's environment that is based on a Recurrent Neural Network (RNN), more specifically using Reservoir Computing (RC). This is a term that groups three similar computing techniques, namely, Echo State Networks [6], Liquid State Machines [7], and BackPropagation DeCorrelation [8]. All three techniques are characterized

J. Marques de Sá et al. (Eds.): ICANN 2007, Part II, LNCS 4669, pp. 660–669, 2007.

by having an RNN that is used as a reservoir of rich dynamics and a linear readout output layer. Only the readout layer is trained by supervised learning, while the recurrent part of the network (the so called reservoir) has fixed weights, but is scaled so that its dynamic regime is at the edge of chaos. Theoretical analysis of reservoir computing methods [9] and a broad range of applications [10] (which often even drastically outperform the current state-of-the-art [11]) show that RC is very powerful and overcomes the problems of tradititonal RNN training such as slow convergence and computational requirements.

The short-term memory, present in these networks, is crucial for solving the event detection and localization tasks. It is not only the instantaneous sensory inputs that are needed to solve the tasks, but also the sensory history [12] and dynamics.

It has already been shown in [13] that RC can be used to detect events in an autonomous robot setting. This work extends these results by also considering dynamic environments for event detection, and goes largely beyond that work by using it to construct implicit maps of the environment for robot localization.

The idea of employing a neural network as a localization model for the robot is also inspired by biological systems. Experiments accomplished with rats show that the hippocampus in their brain forms activation patterns that are associated with locations visited by the rat. These so called *place cells* are the most common evidence for such fact. They fire when an animal is in a particular location in its environment [14].

The data (robot sensors and actuators) are generated using a simulator developed in [3]. It is a completely reactive controller trained by reinforcement learning to explore the environment. The dataset collected from the simulator is used to train an RC system in order to detect events as well as to predict the robot location in particular environments.

2 Reservoir Computing

The current work uses the Echo State Network approach as a learning system for detection of events as well as for robot localisation. The random, recurrent neural network (or reservoir) is composed of sigmoidal neurons and is modelled by the following state update equation:

$$\mathbf{x}(t+1) = f(\mathbf{W}_{\text{in}}\mathbf{u}(t) + \mathbf{W}\mathbf{x}(t)), \tag{1}$$

where: $\mathbf{W}_{\text{in}}$ is the connection matrix from input to reservoir; $\mathbf{W}$ is the weight matrix for the recurrent connections between internal nodes; f is the hyperbolic tangent function; and $\mathbf{u}(t)$ is the input vector at time t. The initial state is $\mathbf{x}(0) = \mathbf{0}$.

The output $\mathbf{y}(t)$ of the network at time t is given by

$$\mathbf{y}(t) = \mathbf{W}_{\text{out}} \begin{bmatrix} \mathbf{x}(t) \\ 1 \end{bmatrix}, \tag{2}$$

where $\mathbf{W}_{\text{out}}$ is the readout matrix.

The matrices $\mathbf{W}_{\text{in}}$ and $\mathbf{W}$ are fixed and randomly created at the beginning. If n_i and n_r denote the number of inputs and neurons inside the reservoir, respectively, then $\mathbf{W}_{\text{in}}$ is a $n_r \times n_i$ matrix. $\mathbf{W}$ is a $n_r \times n_r$ matrix where each element is drawn from a normal distribution with mean 0 and variance 1. It is then rescaled by first dividing the matrix by its spectral radius (the largest absolute eigenvalue) and then multiplying it by 0.95. The spectral radius of the rescaled matrix is thus 0.95 which is close to the edge of stability (around a spectral radius 1). The value of the spectral radius could be further optimised for each experiment, but a quick grid search showed that 0.95 is near optimal for all cases.

The $(n_r+1)\times n_o$ matrix $\mathbf{W}_{\text{out}}$ is the only matrix trained during the experiment (with n_o denoting the number of outputs). Let $\hat{\mathbf{y}}_{\text{fish}}(t)$[1] denote the desired output at time t, then the readout matrix is created by solving (in the mean square sense) the following equation:

$$\mathbf{W}_{\text{out}} \cdot \begin{bmatrix} \mathbf{x}(1) & \mathbf{x}(2) & \dots & \mathbf{x}(n_t) \\ 1 & 1 & \dots & 1 \end{bmatrix} = [\hat{\mathbf{y}}_{\text{fish}}(1)\ \hat{\mathbf{y}}_{\text{fish}}(2)\ \dots\ \hat{\mathbf{y}}_{\text{fish}}(n_t)], \tag{3}$$

where n_t is the total number of time samples.

Prior to training, the desired outputs are relabelled in order to optimise the classification results. Each line of the original target data $\hat{\mathbf{Y}}$ represents one desired output over time, and each output consists of $+1$ and -1. But the number of positive desired outputs can be fairly different from the number of negative desired outputs in each line. In order to get optimal classification through regression (i.e. through solving (3) in the least square sense), each element $\hat{y}^i(t)$ of the i-th line $\hat{\mathbf{y}}^i$ of $\hat{\mathbf{Y}}$ is rescaled so that the whole line $\hat{\mathbf{y}}^i$ sum up to 0:

$$\hat{y}^i_{\text{fish}}(t) = \begin{cases} \frac{n^i_+ + n^i_-}{n^i_+} & \text{if } \hat{y}^i(t) > 0 \\ -\frac{n^i_+ + n^i_-}{n^i_-} & \text{if } \hat{y}^i(t) < 0 \end{cases}, \tag{4}$$

where $n^i_+ = |\{\hat{y}^i(t)|\hat{y}^i(t) > 0\}|$ and $n^i_- = |\{\hat{y}^i(t)|\hat{y}^i(t) < 0\}|$ denote the number of positive and negative required outputs in the i-th line of $\hat{\mathbf{Y}}$, respectively.

3 Robot Model and Controller

The dataset used to train reservoir networks is generated by a simulator used in [3]. Next the environment and robot controller are described briefly. The environment of the robot is composed of repulsive and attractive objects. Each object has a particular color, denoting its respective class. Obstacles are considered repulsive objects while targets are attractive objects [2]. The robot model is shown in Fig. 1. The robot interacts with the environment by distance, color and contact sensors; and by one actuator that controls the adjustment on the movement direction. Sensor positions are distributed homogenously over the front of the robot (from -90° to +90°). Each position holds three sensors (for distance, color and contact perception) [2]. In this work, the robot model has 17 sensor positions, differing from [3]. The velocity of the robot is constant. At each iteration the robot is able to execute a direction adjustment to the left or to the right in the range [0, 15] (degrees).

[1] $\hat{\mathbf{y}}_{\text{fish}}(t)$ refers to the desired output after applying the *fisher* labeling given by (4).

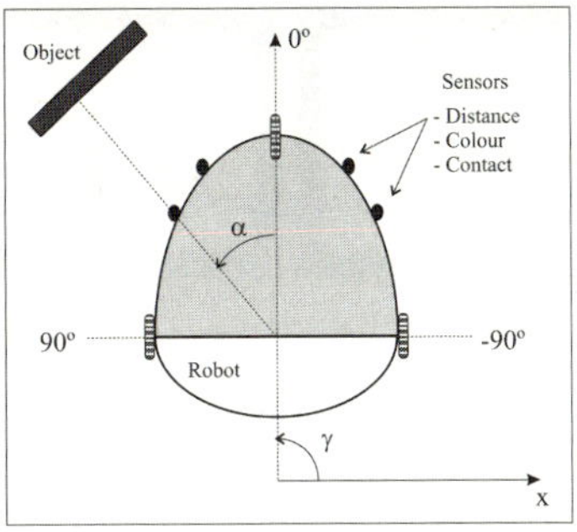

Fig. 1. Robot model

The robot controller (based on [3]) is composed of hierarchical neural networks which are adjusted by classical reinforcement learning mechanisms. The controller constructs its navigation strategy as the robot interacts with the environment. Only already trained robot controllers, which all show very good exploratory behavior after training, are used for generating data. The data (from distance and color sensors, and actuator) collected from the robot simulator are used to train and test reservoir networks in a Matlab environment using the RCT Toolbox[2] [10]. Gaussian noise is added to distance sensors data by the robot simulator.

4 Event Detection in Robot Navigation

Event detection in noisy environments is not a trivial task. There can be very similar scenes from the robot's perspective so that precise event detection becomes very difficult to accomplish [9]. Two different experiments are conducted for the event detection task. The environments used are shown in Fig. 2. They are composed of a large (blue) corridor with a (yellow) target at each end (they appear as dark and light gray objects in black and white format). During simulation, the robot keeps navigating through the corridor and capturing the targets (that are sequentially put back in the same location). There are four possible events of predefined duration and location, which are labeled in Fig. 2. The interpretation should be: when the robot passes through a predefined location, an event should be detected (e.g. entering the corridor corner area, passing through the middle of the corridor). The second environment is the same as the first environment, except for a new blinking object in the middle of the corridor (with random blink interval) which can block the robot's way.

Experiment 1 is accomplished considering the first environment and experiment 2 takes place in the second environment. Both experiments take 120.000 time steps of simulation time. The original dataset is resampled by a factor of

[2] This is an open-source Matlab toolbox for Reservoir Computing which is freely available at http://www.elis.ugent.be/rct

Fig. 2. Environments used for the event detection task. Four events are labeled and shown graphically (by arrows) in the first enviroment. The second environment adds a dynamic obstacle in the middle of the corridor, indicated by an arrow. A typical robot trajectory (after controller learning) is shown in the second environment. Two boxes in the environment are used as targets for the robot.

100, resulting in a smaller dataset of 1.200 observations. The original sampling rate is high (it takes approximately 2000 time steps to go from one side of the simulation environment to the other), which is useful to efficiently control the robot with the implemented controller. But when applying the sensory input to the reservoir it is very important that the internal dynamics and memory of the reservoir are in the same temporal range as the temporal range needed to solve the task. We achieved this by resampling the input, which effectively slows down the dynamics in the reservoir [15]. A grid search of the resampling rate showed that the optimal time range was achieved with a resampling factor of 100.

The inputs to the network are distance and color sensors and a robot actuator (current direction adjustment) suming up 35 inputs which can range from 0 to 1. Parameter configuration is as follows. The reservoir is composed of 400 nodes, scaled to a spectral radius of $|\lambda_{max}| = 0.95$. The readout layer has 4 output units (one for each event detector) which are postprocessed by a winner-take-all function. This function sets the output of the most activated neuron to 1 whereas the others are set to -1. Note that if all the neurons output a negative value, then the winner-take-all function set every output to -1 (this means no event is detected). The input nodes are connected to reservoir nodes by a fraction of 0.3 and are set to -0.15 or 0.15 with equal probabilities. The performance measure considers the number of mispredicted observations and is based on a 3-fold cross-validation method (so, 400 observations, resampled from 40.000 time steps, are selected as test data).

The results are shown in Fig. 3 and summarized in Table 1. Each experiment is evaluated 30 times with different stochastically generated reservoirs and the results are averaged over these 30 runs. It is possible that the robot develops a cyclic and rhythmic trajectory in experiment 1 (see Fig. 3). The trained reservoir is able to detect the 4 events very precisely with a performance of 99 % on test data. One could argue that the reservoir learns to recognize the rhythmic behavior and not the actual event per se. Experiment 2 is devised to test this hypothesis. It shows that a reservoir can still detect the events precisely (performance of 95.8 % on unseen data) even though a dynamic object breaks the rhythmic robot trajectory.

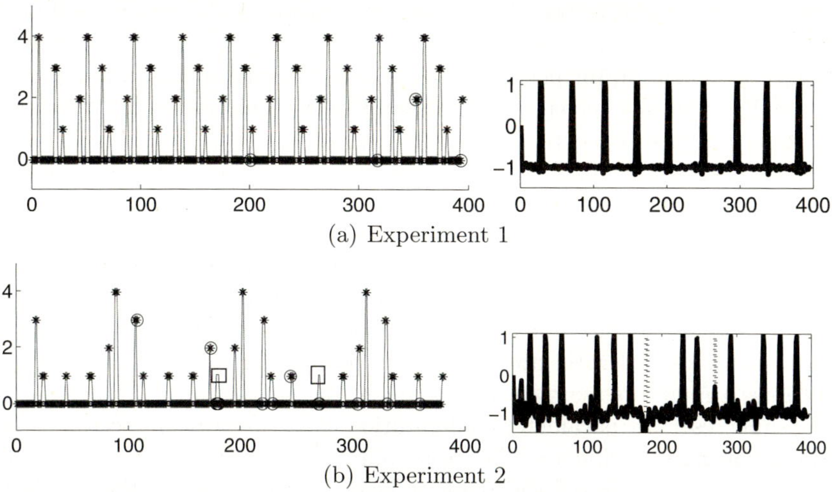

Fig. 3. Left: Events plot during a simulation for experiment 1 (a) and experiment 2 (b). An asterisk represents the predicted event by the reservoir (on test data). The actual events are points connected by lines. The mis-predictions are labeled by an extra circle. Two small rectangles emphasize two events that are not recognized by the reservoir in the bottom plot. Right: Neuron output for detecting event 1 for experiment 1 (a) and experiment 2 (b). The thick line is the actual neuron output whereas the dashed line is the desired outcome (not visible in (a)).

Table 1. Summarized results. For each experiment, thirty (30) runs with the same robot dataset are accomplished (that is resampled by a factor of 100). Every run is based on an 3-fold cross-validation method (experiment 5 uses 6-fold cross-validation). The trainning and test errors are the mean over these 30 runs. The first two experiments are event detection tasks whereas the other three are robot localization tasks. Experiments 2 and 5 consider dynamic objects in the environment.

Experiment	Time steps	Train Error	Test Error
1	1200	0.6 %	1.0 %
2	1200	0.9 %	4.2 %
3	1800	2.9 %	10.3 %
4	1200	1.7 %	12.4 %
5	3600	10.9 %	22.1 %

5 Localization in Robot Navigation

The previous section has shown that an RC network can be used to detect complex events in robot navigation with rather good performance. Now this section extends the experiments to robot localization tasks. Instead of only detecting events, we rather want to predict the current location of the robot based on the same kind of sensory information (giving rise to a more difficult and interesting problem). Localization (or position detection) for mobile robots is usually computationally expensive in terms of space and time requirements [5]. Traditional algorithms are based on explicit maps which must be constructed before robot localization is possible. This section shows how a reservoir can be used for robot localization. Similar work which uses a Long-Short Term Memory RNN for this task is described in [16].

Two maze-like environments are used for the robot localization task (see Fig. 4). The first environment contains 64 predefined locations, that are displayed by small triangles labeled by numbers. Differently from [16], the entire environment is tagged with labels (not only rooms). This feature makes it possible to use a more precise trajectory planner.

The same reservoir parameter configuration as in the previous section is used for the following experiments. The resampling rate for the dataset is also 100. Exceptions are: the size of the readout layer is equivalent to the number of predefined locations in the environment; and the postprocessing function for the readout units is the winner-take-all function which always takes the most activated neuron and set it to 1 (the others are set to -1). So, there is always a predicted location (in contrast to the no event detected situation in previous section). Here also a 3-fold cross-validation is used for performance measure (last experiment is based on a 6-fold cross-validation).

Experiment 3 is accomplished with the first environment from Fig. 4 and lasts 180.000 time steps (before resampling). The resulting robot occupancy grid can be seen in the same figure: it shows that the reservoir is predicting the robot location very well (on test data), with very few mispredictions that are not far located from the actual location (10.3 % is the test error, see Table 1).

Experiment 4, accomplished in the second environment, represents a new challenge for the reservoir-based position detector: the environment has several symmetries and identical areas. For instance, going from position 27 to 26 looks the same for the robot as going from position 22 to 24. The simulation has 120.000 time steps. The resulting occupancy grid in Fig. 4 shows an efficient position detector, featuring a performance of 87.6 % of correct predictions on test data (see Table 1).

Experiment 5 uses the third environment in Fig. 4, that is the same as the first environment, but with additional 11 slow moving obstacles distributed through the environment (moving obstacles are also considered in [16]). These dynamic objects change the robot behavior and also add more noise to sensor readings. The simulation has 360.000 time steps. The respective occupancy grid in Fig. 4 shows that the reservoir is correct in most of the predictions (77.8 %). Some of the mispredictions are located a bit further from the actual position, due to the new source of dynamics and noise.

Experiments only considering distance sensors (removing actuator and color sensor data) result in the same performance reported for the previous

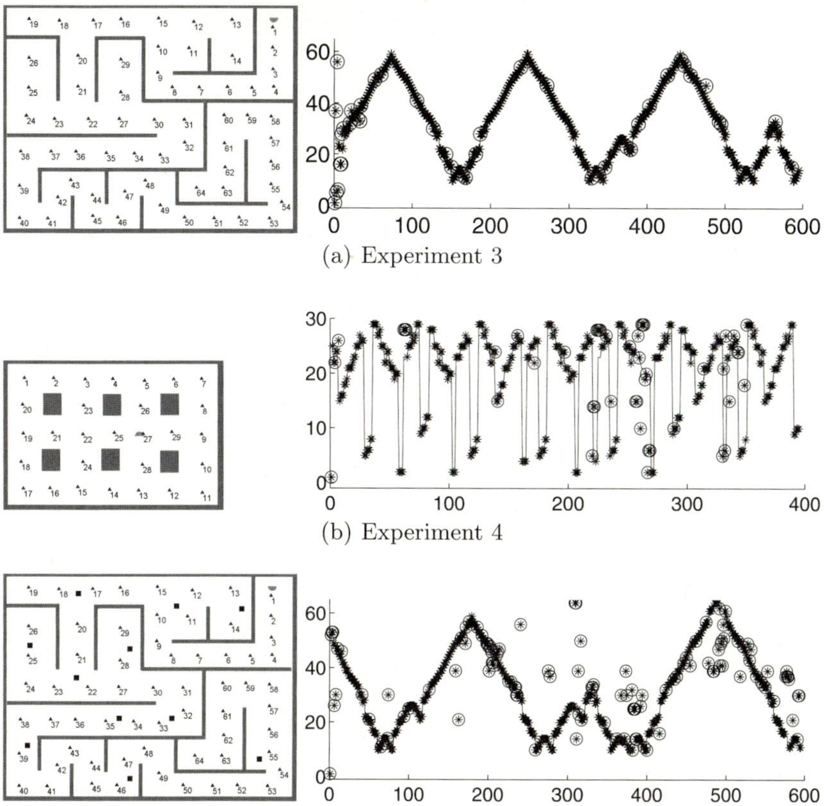

(a) Experiment 3

(b) Experiment 4

(c) Experiment 5

Fig. 4. Environments used for the experiments (left) and respective resulting robot occupancy grids (right). First environment is tagged with 64 labels displayed by small triangles. In the occupancy grid, an asterisk represents the predicted location (on test data, that is 1/3 of the total data) while connected points are the actual robot positions. Mispredicted locations display an additional circle. The second environment has 29 labels distributed through very similar areas. The third environment is the same as the first environment but with additional slow moving obstacles (represented by small rectangles) which add more noise and dynamics to sensor readings and to the robot trajectory, respectively.

experiments in this section. The reservoir network also copes with the kidnapping situation (also reported in [16]). In a new experiment using the environment from experiment 3, the robot is replaced from location 54 to location 27. The network is able to predict sucessfully the robot position after 7 time steps (see Fig. 5). Note that the RC network is not trained with the kidnapping situation.

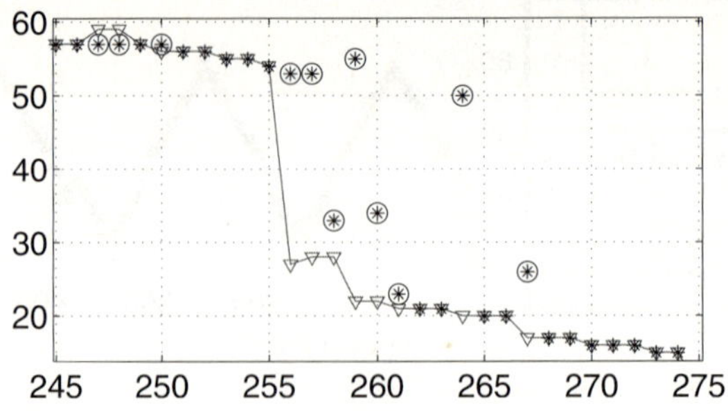

Fig. 5. Occupancy grid after kidnapping the robot in first environment of Fig. 4. An additional triangle is placed at the actual robot position. At time step 255, the robot is moved from position 54 to position 27. The reservoir network takes 7 time steps until it first predict sucessfully the current robot position (at time step 262). The robot visits 4 locations (27, 28, 22 and 21) until the sucessfull prediction.

6 Conclusions and Future Work

In this work we show that it is possible to detect complex events and locate a robot in even dynamic environments with a random dynamic system which is processed by just a single linear readout layer. The proposed system shows very good performance in difficult environments such as mazes or environments which are highly symmetric. To achieve this we only use the dynamics of the sensory information, not the actual behaviors (as in [13]). Besides, no decrease on reservoir performance is reported when experiments on robot localization only consider distance sensors.

This paper only scratched the surface of what could be possible with this technology. As future work we plan to implement it on a real robotic platform, as it is considered the standard and best evaluation method for robotic systems. In this way we can also make comparisons to existing SLAM techniques. Additionaly, a deliberative robotic system can now be constructed so that actual path planning and navigation is accomplished based on the information gathered by the RC-based localization system.

From an RC view, we could improve performance by tuning the reservoir dynamics and time scales for different tasks. For instance, experiments with robots with variable speeds during simulation can be tackled by inducing distinct reservoir dynamics (creating different time scales in the reservoir operation). Future work also includes the unsupervised detection and generation of locations, much resembling actual place cells. Finally, the implicit map stored in the reservoir could be made explicit by using an RC system in a generative setting: given a location as input, the reservoir could start creating expectations (much like 'dreaming') of possible paths and environments.

Acknowledgements. This research is partially funded by FWO Flanders project G.0317.05.

References

1. Arkin, R.: Behavior-based robotics. MIT Press, Cambridge (1998)
2. Antonelo, E.A., Figueiredo, M., Baerlvedt, A.J., Calvo, R.: Intelligent autonomous navigation for mobile robots: spatial concept acquisition and object discrimination. In: Proceedings of the 6th IEEE International Symposium on Computational Intelligence in Robotics and Automation, Helsinki, Finland (2005)
3. Antonelo, E.A., Baerlvedt, A.J., Rognvaldsson, T., Figueiredo, M.: Modular neural network and classical reinforcement learning for autonomous robot navigation: Inhibiting undesirable behaviors. In: Proceedings of IJCNN 2006, Vancouver, Canada (2006)
4. Guivant, J., Nebot, E., Baiker, S.: Autonomous navigation and map building using laser range sensors in outdoor applications. Journal of Robotics Systems 17(10), 565–583 (2000)
5. Bailey, T., Durrant-Whyte, H.: Simultaneous localisation and mapping (SLAM): Part ii state of the art. Robotics and Automation Magazine (2006)
6. Jaeger, H.: The echo state approach to analysing and training recurrent neural networks. Technical Report GMD Report 148, German National Research Center for Information Technology (2001)
7. Maass, W., Natschläger, T., Markram, H.: Real-time computing without stable states: A new framework for neural computation based on perturbations. Neural Computation 14(11), 2531–2560 (2002)
8. Steil, J.J.: Backpropagation-Decorrelation: Online recurrent learning with O(N) complexity. In: Proceedings of IJCNN '04, vol. 1, pp. 843–848 (2004)
9. Jaeger, H.: Short term memory in echo state networks. Technical Report GMD Report 152, German National Research Center for Information Technology (2001)
10. Verstraeten, D., Schrauwen, B., D'Haene, M., Stroobandt, D.: A unifying comparison of reservoir computing methods. Neural Networks 20, 391–403 (2007)
11. Jaeger, H., Haas, H.: Harnessing nonlinearity: predicting chaotic systems and saving energy in wireless telecommunication. Science 308, 78–80 (2004)
12. Schönherr, K., Cistelecan, M., Hertzberg, J., Christaller, T.: Extracting situation facts from activation value histories in behavior-based robots. In: Baader, F., Brewka, G., Eiter, T. (eds.) KI 2001. LNCS (LNAI), vol. 2174, pp. 305–319. Springer, Heidelberg (2001)
13. Hertzberg, J., Jaeger, H., Schönherr, F.: Learning to ground fact symbols in behavior-based robots. In: Proceedings of the 15th European Conference on Artificial Intelligence, pp. 708–712 (2002)
14. O'Keefe, J., Dostrovsky, J.: The hippocampus as a spatial map. Preliminary evidence from unit activity in the freely-moving rat. Brain Research 34, 171–175 (1971)
15. Jaeger, H., Lukosevicius, M., Popovici, D.: Optimization and applications of echo state networks with leaky integrator neurons. Neural Networks 20, 335–352 (2007)
16. Forster, A., Graves, A., Schmidhuber, J.: RNN-based learning of compact maps for efficient robot localization. In: Proceedings of ESANN (2007)

Model Reference Control Using CMAC Neural Networks

Alpaslan Duysak[1], Abdurrahman Unsal[2], and Jeffrey L. Schiano[3]

[1] Computer Engineering, Dumlupinar University, Kutahya, TR
aduysak@dumlupinar.edu.tr
[2] Electrical Engineering, Dumlupinar University, Kutahya, TR
unsal@dumlupinar.edu.tr
[3] Electrical Engineering, The Pennsylvania State University, State College, Pennsylvania, USA
schiano@engr.psu.edu

Abstract. This paper demonstrates the use of CMAC neural networks in real world applications for the system identification and control of nonlinear systems. As a testbed application, the problem of regulating fluid height in a column is considered. A dynamic nonlinear model of the process is obtained using fundamental physical laws and by training a CMAC neural network using experimental input-output data. The CMAC model is used to implement a model reference control system. Successful experimental results are obtained in the presence of disturbances.

1 Introduction

Due to uncertainty and complexity, accurate models of dynamic systems are often difficult to obtain. As a result, it is a challenge to design a controller that causes the system to respond in a desired way. Biological control systems, on the other hand, are successful despite imprecise information and complex situations. Therefore, there has been a great effort to model them. Recent studies in biology, neuroscience, and psychology have led to detailed theories regarding the anatomy and physiology of the cerebellum, which is the part of the brain responsible for learning and voluntary motions. Albus proposed a mathematical description of how the cerebellum computes addresses where control signals are stored, organizes memory, and generates output signals. Based on this description, he proposed a manipulator control system called the cerebellar model articulation controller (CMAC),[1].

CMAC gained more attention after Miller [2] used the CMAC network for real-time control of a full-scale multidegree-of-freedom industrial robot with considerable success. CMAC controllers are widely used in robotic applications. For example, CMAC is utilized to perform feedforward kinematics control of four-legged robot in straight-line walking and step climbing, [3]. Shiraishi [4] used a CMAC controller for fuel-injection system and experimentally evaluated the CMAC performance on a research vehicle in a configuration fully compatible

J. Marques de Sá et al. (Eds.): ICANN 2007, Part II, LNCS 4669, pp. 670–679, 2007.

with production hardware. In order to achieve high-precision position control, an adaptive CMAC model reference control system is proposed [5]. They provided analytical methods based on a discrete-type Lyapunov function to determine the varied learning-rate parameters of the CMAC. Both the modeling and the generalization capabilities are improved in [6], where special kernel network based on the CMAC is proposed.

In this work, we provided detailed analysis on how CMAC can be integrated in model reference control and on how CMAC is trained using experimental data [7].

2 Model Derivation from Physical Laws

The fluid testbed, shown in Fig. 1a, consists of three subsystems. The relationship between the fluid height $y(t)$ in the column and the output flow rate $w_{out}(t)$ is first obtained. The relationship between the armature voltage $V_u(t)$ and the input flow rate $w_{in}(t)$ is then derived. This analysis account for the dynamics of the DC pump. Finally, a pressure sensor is used to determine the column height.

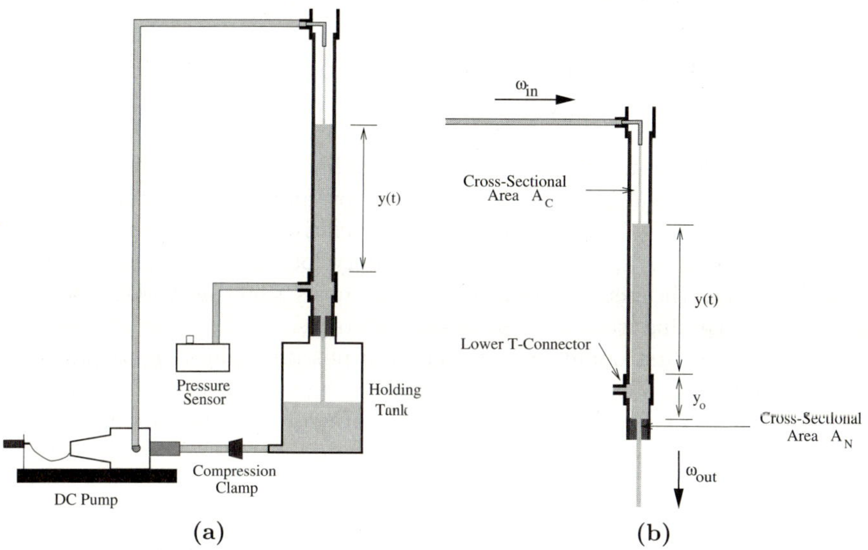

Fig. 1. Experimental setup: (a) Simplified representation of the fluid testbed (b) Schematic representation of the fluid column

A detailed view of the column is shown in Fig. 1b. The pump transfers water from the holding tank to the top of the column with a volume flow rate $w_{in}(t)$ [in^3/sec]. The cross-sectional area of the column is A_C=0.196 [in^2]. At the base of the column water drains back into the holding tank with a flow rate $w_{out}(t)$ [in^3/sec]. The cross-sectional area of the nozzle is A_N=0.00503 [in^2]. The vertical displacement between the top of the nozzle and the T-connector is $y(t) + y_0$, where y_0=1.75 [in]. The fluid dynamics is described by a single-input single-output state-space model whose input is $w_{in}(t)$ and whose output is y(t). Mass conservation leads to the differential equation

$$\dot{y} = \frac{1}{A_c}[w_{in}(t) - w_{out}(t)]. \tag{1}$$

The atmospheric pressure at the top and base of the column are the same and friction is neglected. Torricelli's equation, which provides a relationship between the output flow rate and the fluid height above the nozzle [8], and analyzing the DC pump dynamics yields a first-order nonlinear differential equation relating the input armature voltage $V_u(t)$ to output fluid head height y(t)

$$\dot{y} = -\frac{A_N}{A_C}\sqrt{\frac{2g}{1-\frac{A_N^2}{A_C^2}}}\sqrt{y(t)+y_0} + \frac{1}{A_C}(\alpha + \sqrt{\alpha^2 + \beta V_u}). \tag{2}$$

In the first-order nonlinear model, the only unknown parameters are α and β, which come from DC pump dynamics. The head height $y(t)$ is obtained from the pressure sensor voltage.

3 The Cerebellar Model Articulation Controller (CMAC)

The CMAC network is used to approximate a nonlinear function of the form $y_i = f(x_i)$, where x_i is a multidimensional input vector, and y_i is a multidimensional output vector. The CMAC network is an associative neural network, in that the input x_i serves as an address key to memory locations whose contents are summed to form the network output. Fig. 2 shows that the CMAC network performs the mapping from x_i to y_i in several stages.

The input x_i is first quantized into one of n possible values, q_1 to q_n, which span the input space. The quantization levels can be fixed or variable. The input to the decoder in Fig. 2 is x_i and the output is the quantization level

$$q_k = Q(x_i) = \left\lfloor \frac{x_i - x_{min}}{r} \right\rfloor \tag{3}$$

where $x_i \in [x_{min}, x_{max}]$ and $\lfloor \text{p} \rfloor$ = largest integer less than the real number p. x_{min} and x_{max} are the minimum and the maximum value of the input, respectively. The function Q returns an integer q_k for the input x_i and q_k ranges from 0 to $q_{max} - 1$, which is the total number of available quantization levels. For simplicity, in this work, the quantization resolution is fixed as

$$r = \frac{x_{max} - x_{min}}{q_{max}}. \tag{4}$$

The CMAC neural network contains a conceptual memory whose contents are summed to form the CMAC output. The quantization level q_k maps into

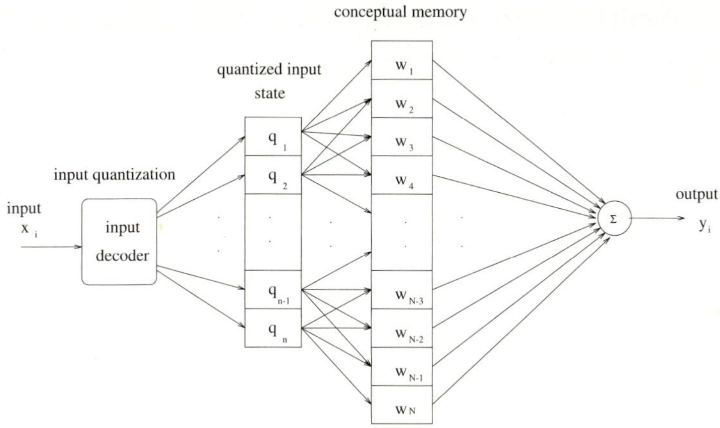

Fig. 2. Diagram of a CMAC neural network

Table 1. The row vector θ_k indicates which cells in the conceptual memory are activated by the quantization level q_k. 1s in the table shows activated memory locations.

q_k	θ_k
0	1 1 1 . . . 1 0 0 0 0 0 0 0 0
1	0 1 1 1 . . . 1 0 0 0 0 0 0 0
2	0 0 1 1 1 . . . 1 0 0 0 0 0 0
.	. . . .
$q_{max} - 1$	0 0 0 0 0 0 0 0 1 1 1 . . . 1

A^* memory cells of the conceptual memory, that is A^* represents the number of memory cells that are activated at any time by a particular quantization level. Any two adjacent quantization levels activate memory cells that overlap by $A^* - 1$. It is convenient to introduce a row vector θ that indicates which memory cells in the conceptual memory are activated by a given quantization level q_k. The number of columns in θ is equal to the size of the conceptual memory. The vector θ_k associated with each quantization levels q_k is shown in table 1.

In this work weights are updated each time an input is applied. The algorithm is

$$w(i,k) = w(i,k-1) + \frac{\gamma}{A*}\theta_k^T[d_k - \theta_k w(i,k-1)] \tag{5}$$

where k=1,2,...,m is the sample number, i represents the iteration, d_k is the desired response and γ is the learning factor. It has been shown that any given input x_k activates A* memory locations in the conceptual memory. The weights stored in these memory locations are summed to produce the output

$$y_k = \theta_k w. \tag{6}$$

4 Identification and Control of the Fluid Testbed Using CMAC Neural Networks

This section shows how the CMAC neural network can be used to obtain a dynamic model of the fluid testbed and to implement a model reference controller for regulating fluid height.

4.1 NARMA Representation of the Fluid Testbed

Based on the nonlinear representation of the fluid testbed model given by equation 2, it is reasonable to represent the nonlinear system with a discrete-time model of the form (NARMA representation [9])

$$y(k+1) = f[y(k)] + g[V_u(k)], \tag{7}$$

where f(·) and g(·) are memoryless and nonlinear functions. This result follows by approximating $\frac{dy}{dt}$ by $(y(k+1)T - y(kT))/T$, where T is the sample period. Using this approximation

$$f[y(k)] = y(k) - T\frac{A_N}{A_C}\sqrt{\frac{2g}{1-\frac{A_N^2}{A_C^2}}}\sqrt{y(k)+y_0},$$

$$g[V_u(k)] = \frac{T}{A_C}\left(\alpha + \sqrt{\alpha^2 + \beta V_u(k)}\right). \tag{8}$$

Given the NARMA representation of the fluid testbed, the task of identifying the functions f(·) and g(·) using CMAC neural networks is considered in the next section.

4.2 System Identification Using CMAC Neural Networks

In system identification, control engineers can obtain a dynamic model by observing the input-output history of the system. Ideally, we desire a model that is valid over a wide operating range. The functions f(·) and g(·) in the NARMA representation of the fluid testbed (equation 8) are approximated using neural networks N_f and N_g, respectively, Fig. 3. It is desired that the fluid testbed output $y(k)$ and the neural network model output $\widehat{y}(k)$ in Fig.3 are identical for the same input $V_u(k)$. Figure 3 suggests that the input for the neural network N_f is the plant output $y(k)$ and the input for the neural network N_g is armature voltage $V_u(k)$. The output of the networks are summed to produce the neural network model output $\widehat{y}(k)$. This response is compared against the plant output. Neural networks N_f and N_g are trained using the error $e(k)$ between the actual plant output $y(k+1)$ and the neural network model output $\widehat{y}(k+1)$ to minimize $J = \sum_{k=1}^{K}[y(k+1) - \widehat{y}(k+1)]^2, = \sum_{k=1}^{K} e(k)$. Experimental data for the system identification is obtained by regulating the fluid height ± 3 [in] about the 11 [in]operating point.

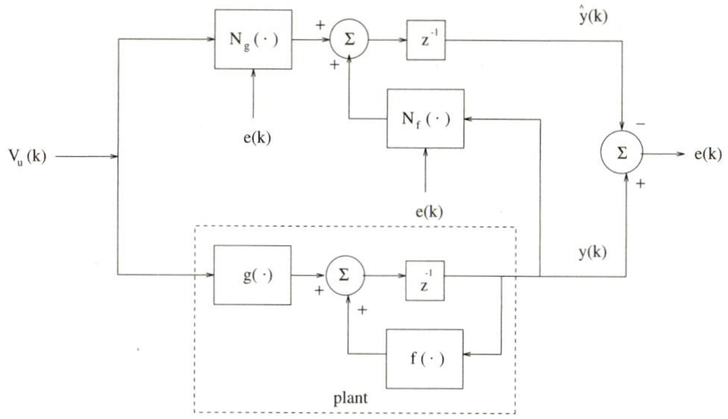

Fig. 3. Identification of the plant. The error between the estimated and actual plant output is used to train the neural networks N_f and N_g.

The output of the neural network model is the sum of the outputs of the two neural networks N_f and N_g as seen in Fig. 3. The output of the neural network N_g for the kth input $V_u(k)$ is

$$N_g[V_u(k)] = \theta_i^g(V_u(k))w_g, \tag{9}$$

where θ_i^g is n dimensional column vector described in section 3, and the weight vector of the neural network N_g is w_g. A similar expression for the output of the neural network N_f is

$$N_f[y(k)] = \theta_i^f(y(k))w_f. \tag{10}$$

It follows that the output of the neural network model is

$$\widehat{y}(k+1) = \theta_i^g w_g + \theta_i^f w_f. \tag{11}$$

Weights of the neural networks N_f and N_g are adjusted using the error between the fluid testbed output and neural network model output using the non-batch training method described in section 3

$$\begin{aligned} w_f(i,k) &= w_f(i,k-1) + \frac{\gamma}{A*}(\theta_k^f)^T e(k) \\ w_g(i,k) &= w_g(i,k-1) + \frac{\gamma}{A*}(\theta_k^g)^T e(k) \end{aligned} \tag{12}$$

For each data point, only A^* number of weights are updated. The training iterations are completed when the magnitude of $e(k)$ fall below a specified value for each data pair (x_k, d_k).

4.3 Model Reference Control

The objective of model reference control is to have the plant output track a reference model output. The design procedure given in [9] is used in this section.

The desired response corresponds to that of a second-order system with settling time, t_s=20 [sec] and peak overshoot, $M_p = 5$ %. In addition a rise time of 8 [sec] is desired. The second-order model transfer function is

$$G_m(s) = \frac{\Delta Y_m(s)}{\Delta R(s)} = \frac{w_n^2}{s^2 + 2\zeta w_n s + w_n^2}, \tag{13}$$

where w_n is the natural frequency, ζ is the dimensionless damping ratio, and $\Delta Y_m(s)$ is the desired response due to an input $\Delta R(s)$. For the given design specifications, w_n and ζ are 0.35 [rad/sec] and 0.7, respectively. The zero-order hold discrete-time equivalent model of the transfer function in equation 13 is

$$\begin{aligned} G_m(z) &= (1 - z^{-1}) Z\{\frac{G_m(s)}{s}\} \\ &= \frac{0.1986z + 0.1002}{z^2 - 1.0729z + 0.3717} \end{aligned} \tag{14}$$

where the sample period T is 2 [sec]. This sample period was chosen because it is a factor of seven smaller than the smallest identified time constant. It is desired that the closed-loop response of the fluid testbed follows that of the reference model

$$y_m(k+1) = \alpha_1 y_m(k) + \alpha_2 y_m(k-1) + \beta_1 r(k) + \beta_2 r(k-1), \tag{15}$$

where α_1=1.0729, α_2=-0.3717, β_1=0.1986, and β_2=0.1002.

By choosing the control input

$$V_u(k) = g^{-1}[-f[y(k)] + \alpha_1 y(k) + \alpha_2 y(k-1) + \beta_1 r(k) + \beta_2 r(k-1)] \tag{16}$$

the closed-loop response is

$$y(k+1) = \alpha_1 y(k) + \alpha_2 y(k-1) + \beta_1 r(k) + \beta_2 r(k-1). \tag{17}$$

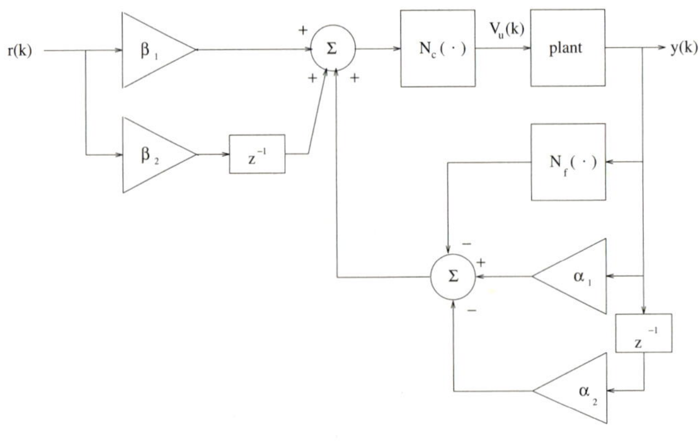

Fig. 4. Block diagram of the model reference control system

This result is obtained by substituting equation 16 into 7. This method is only applicable if the function $g(\cdot)$ is invertible. In the case of the fluid testbed, $g[V_u(k)]$ is given by equation 8 and $V_u(k)$ is positive. For each value of $g[V_u(k)]$, there is a unique value of $V_u(k)$. Also note that this method is only applicable if the functions $f(\cdot)$ and $g(\cdot)$ are known. In the control law specified by equation 16, the function $f(\cdot)$ and $g^{-1}(\cdot)$ are implemented by neural networks N_f and N_c, respectively. A block diagram of the reference model control system is shown in Fig. 4. The previous section described how N_f and N_g were trained. We now consider how N_c is trained.

4.4 Training the Neural Network N_c

In order to implement the control structure given in equation 16, it is necessary to train the neural network N_c so that it approximates the inverse of the function $g(\cdot)$. Because the nonlinear function $g(\cdot)$ is approximated by N_g, the neural network N_c is trained so that $N_c[N_g(V_u(k))] \approx V_u(k)$ as $V_u(k)$ varies over the input range. In other words, we need to find $N_c = N_g^{-1}$. A block diagram of the system used to train N_c is shown in figure 5. The weights of the neural network N_c are adjusted using the error $e(k)$ between the input $V_u(k)$ and the output of the neural network N_c using the non-batch training algorithm

$$w_c(i,k) = w_c(i,k-1) + \frac{\gamma}{A*}(\theta_k^c)^T e(k). \tag{18}$$

Given the training strategies for the neural networks N_f, N_g, and N_c, the model reference control system is implemented as shown in Fig.4.

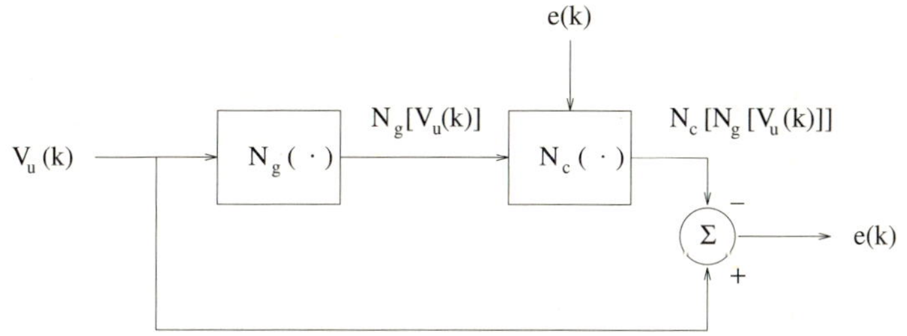

Fig. 5. Block diagram of the system used to train N_c

5 Results and Conclusions

This sections shows the performance of the CMAC-based model reference control system. Four separate experiment were performed. In the first three experiments, the CMAC controller regulated the head height about nominal values of 3, 11, and 19 [in]. In the final experiment, the ability of the CMAC controller to reject a constant disturbance was investigated. Fig. 6a, 6b and 6c show the response

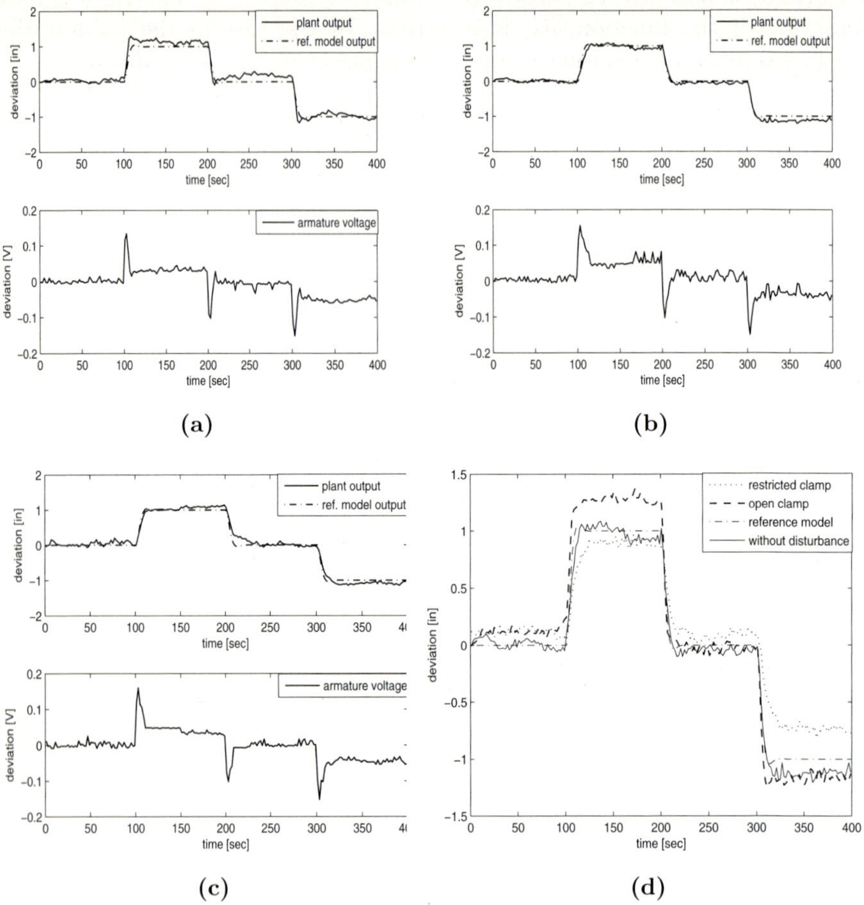

Fig. 6. Experimental response of the closed-loop system using a CMAC controller at operating point of (a) 3, (b) 11 and (c) 19 [in]. Response of the closed-loop system to a disturbance input at 11 [in] (d).

of the reference model, closed-loop system, and the armature input voltage, for the case where the nominal operating point are 3, 11 and 19 [in]. The response of the closed-loop system has a large steady-state error at the 3 [in] operating point. The CMAC controller, however, achieves a small steady-state error at the other two operating points.

The ability of the CMAC controller to disturbance rejection is also examined. The compression clamp located at the pump inlet is restricted at position 1 during system identification and control experiments. To generate a known disturbance, this clamp is set to position 2 and the open position. At position 3, the pipe is completely closed. The plant response to the disturbances is plotted in the Fig. 6d at the operating point of 11 [in].

A model reference control scheme based on the plant differential equation is developed and is experimentally tested. Successful results are obtained even under large disturbances. Future work will include comparisons of the implemented method with well known control techniques.

References

1. Albus, J.S.: A New Approach to Manipulator Control: The Cerebellar Model Articulation Controller (CMAC), Trans. ASME J. Dynamic Syst. Meas. Contr. 97(3), 220–227 (1975)
2. Miller, W.T.: Sensor-Based Control of Robotic manipulators Using a General Learning Algorithm. IEEE Journal of Robotics and Automation ra-3(2), 157–165 (1987)
3. Lin, Y., Song, S.: A CMAC Neural-Network-Based Algorithm for the Kinematic Control of a Walking Machine. Eng. App.Artif. Intell. 5(6), 539–551 (1992)
4. Shiraishi, H., Ipri, S.L., Cho, D.D.: CMAC Neural Network Controller for Fuel-Injection Systems. IEEE Trans. on Control Systems Technology 3(1), 32–36 (1995)
5. Pen, Y.-F., Wai, R.-J., Lin, C.-M.: Adaptive CMAC Model Reference Control System for Linear Piezoelectric Ceramic Motor. In: Proceedings, IEEE hternational Symposium on Computational Intelligence in Robotics and Automation, Kobe, Japan, (July 2003), pp. 30–35 (2003)
6. Horváth, G.: Kernel CMAC: an Efficient Neural Network for Classification and Regression. Acta Polytechnica Hungarica 3(1) (2006)
7. Duysak, A.: CMAC Neural Networks For Identification and Control, Master Thesis, The Pennsylvania State University,USA (1997)
8. Debler, W.R.: Fluid Mechanics Fundamentals, pp. 171–175. Prentice-Hall, Englewood Cliffs (1990)
9. Narendra, K.S., Parthasarathy, K.: Identification and Control of Dynamical Systems Using Neural Networks. IEEE Trans. Neural Networks 1, 4–27 (1990)

A Three-Stage Approach Based on the Self-organizing Map for Satellite Image Classification

Márcio L. Gonçalves[1,2], Márcio L.A. Netto[2], and José A.F. Costa[3]

[1] Department of Computer Science, PUC Minas, Poços de Caldas, MG, Brazil
[2] School of Electrical Engineering, State University of Campinas, Brazil
[3] Department of Electrical Engineering, Federal University of Rio Grande do Norte, Brazil
marcio@pucpcaldas.br, marcio@dca.fee.unicamp.br,
alfredo@dee.ufrn.br

Abstract. This work presents a methodology for the land-cover classification of satellite images based on clustering of the Kohonen's self-organizing map (SOM). The classification task is carried out using a three-stage approach. At the first stage, the SOM is used to quantize and to represent the original patterns of the image in a space of smaller dimension. At the second stage of the method, a filtering process is applied on the SOM prototypes, wherein prototypes associated to input patterns that incorporate more than one land cover class and prototypes that have null activity are excluded in the next stage or simply eliminated of the analysis. At the third and last stage, the SOM prototypes are segmented through a hierarchical clustering method which uses the neighborhood relation of the neurons and incorporates spatial information in its merging criterion. The experimental results show an application example of the proposed methodology on an IKONOS image.

Keywords: Pattern recognition, self-organizing maps, image processing, remote sensing.

1 Introduction

The self-organizing map (SOM), proposed by Kohonen [5], has been widely used in a variety of applications, including areas as pattern recognition, data compression, biological modeling, signal processing, and data mining [6]. Important properties as the input space approximation, topological ordening and density matching, allied with the simplicity of the model and the easiness to implement its learning algorithm justify the success of the SOM and place it as one of the main models of neural nets in the present time.

This work presents a methodology that explores the characteristics and properties of the SOM to perform the unsupervised classification of remotely sensed images. Since the first satellites were launched for the purpose of searching for terrestrial resources, the digital classification of remotely sensed images has acquired a growing importance in the automatic recognition of the land cover patterns [8]. Presently, the enormous quantity of images that are being collected by sensor systems that are more and more sophisticated and modern require the development of innovative classification

J. Marques de Sá et al. (Eds.): ICANN 2007, Part II, LNCS 4669, pp. 680–689, 2007.

methodologies, which allow an automatic and efficient identification of the great volume of information available in the images and at the same time makes the mapping process of terrestrial surfaces less subjective and with greater potential for reuse in subsequent situations.

The classification method proposed here performs a cluster analysis of the image data employing an approach consisting of three processing stages through SOM. Firstly, the SOM is used to map the original patterns of the image to a reduced set of prototypes (neurons) arranged in a two-dimensional grid. Afterwards, a filtering process is applied on the SOM prototypes, wherein prototypes associated to input patterns that incorporate more than one land cover class are excluded in the next stage of the analysis, and prototypes that have null activity are simply discarded. At the last processing stage, an agglomerative hierarchical clustering method is applied over the filtered SOM, generating a dendrogram of clustered prototypes with different degrees of similarity. Each level of dendrogram obtained corresponds to a different clustering configuration of SOM prototypes that can be utilized to represent the classes by which the original image will be classified.

The remainder of the paper is organized in the following form: section 2 describes the SOM and its properties, section 3 presents a brief explanation about the unsupervised classification of satellite images, while section 4 explains the proposed classification methodology. An application example of the proposed approach on an IKONOS image is shown in the section 5, and section 6 gives the conclusions and final considerations pointing out the advantages of the proposed method on conventional classification methods.

2 SOM

SOM is a type of artificial neural net based on competitive and unsupervised learning. The network essentially consists of two layers: an input layer I and an output layer U with neurons generally organized in a 2-dimensional topological array. The input to the net corresponds to a p-dimensional vector, $\boldsymbol{x}$, generally in the space $\Re^p$. All of the p components of the input vector feed each of the neurons on the map. Each neuron i can be represented by a synaptic weight vector $\boldsymbol{w}_i = [w_{i1}, w_{i2}, \ldots, w_{ip}]^T$, also in the p-dimensional space.

For each input pattern $\boldsymbol{x}$ a winner neuron, c, is chosen, using the criterion of greatest similarity:

$$\| \boldsymbol{x} - \boldsymbol{w}_c \| = \min_i \{ \| \boldsymbol{x} - \boldsymbol{w}_i \| \} \tag{1}$$

where ||.|| represents the Euclidian distance. The winner neuron weights, together with the weights of the neighboring neurons, are adjusted according to the following equation:

$$\boldsymbol{w}_i(t+1) = \boldsymbol{w}_i(t) + h_{ci}(t)[\boldsymbol{x}(t) - \boldsymbol{w}_i(t)] \tag{2}$$

where t indicates the iteration of the training process, $\boldsymbol{x}(t)$ is the input pattern and $h_{ci}(t)$ is the nucleus of neighborhood around the winner neuron c.

Once the SOM algorithm has converged, the 2-dimensional array of neurons displays important statistical characteristics of the input space, summarized as follows:

i) *Approximation of the input space:* the basic objective of SOM is to store a large set of input vectors by finding a smaller set of prototypes that provides a good approximation to the original input space.

(ii) *Topological ordering:* the SOM algorithm attempts to preserve as well as possible the topology of the original space, i.e., it tries to make the neighboring neurons in the 2-dimensional array (output space) present weight vectors that represent neighboring patterns in the input space.

(iii) *Density Matching:* the SOM reflects the probability distribution of data in the input space. Regions in the input space where the input patterns $\boldsymbol{x}$ are taken with a high probability of occurrence are mapped onto larger domains of the output space, and thus with better resolution than regions in the input space from which input patterns $\boldsymbol{x}$ are taken with a low probability of occurrence.

In the literature there are some proposed algorithms that seek to automatically (or semi-automatically) interpret and segment the neurons of a trained SOM [1, 2, 9, 10]. As will be presented in section 4, the SOM segmentation strategy employed in this work utilizes an agglomerative hierarchical clustering method that incorporate more information about the data clusters in its merging criterion beyond the usual inter-cluster distance information.

3 Unsupervised Classification of Remotely Sensed Images

The unsupervised classification of remote sensing images is based on the principle that the computational algorithm is capable of identifying by itself the classes of the image. This type of classification is frequently performed through clustering methods.

Although there are a large quantity of different clustering methods in the pattern recognition area [11], the majority of software or computational systems meant to the digital processing of remotely sensed images perform unsupervised classification based on the partitional clustering methods, such as K-means and ISODATA.

Despite being widely used, these partitional clustering methods have various limitations. The objective functions that they used begin with the assumption that the number of classes, K, is known a priori. In the hypothesis that an inadequate K' value has been chosen, the method will impose, through the use of optimization techniques, K' clusters to the data. The user must also manually specify various parameters in order to control the clustering process, among them: the initial centroids of each cluster, the maximum number of iterations, thresholds to perform the division, fusion, or exclusion of clusters. K-means and ISODATA are very sensitive to these parameters, which can generate different partitions when various simulations are done for the same data set. Facing this, the optimal value of these parameters is frequently encountered through trial and error. These needs certainly increase the level of interaction between the user and the computational algorithm, consequently increasing the degree of subjectivity of the categorization process of the image.

Other no less important limitations of partitional algorithms, such as K-means and ISODATA, are: the high computational cost when the data to be analyzed is very large (at each iteration, all of the pixels in the image are compared with all of the clusters centroids) and the existence of suppositions about the cluster forms. Generally only one prototype (centroid) is used to represent a cluster, thus these methods become adequate only for the analysis of clusters that have hyperspherical formats.

Another possible, though uncommon, way of performing unsupervised classification of remotely sensed images is through hierarchical clustering methods. Unlike partitional methods, hierarchical methods do not require the user to specify the number of clusters and other additional parameters beforehand. Another significant advantage of these methods is that they make it possible to visualize the result of the classification by means of a dendrogram which illustrates in a hierarchical form the degree of similarity between the clusters that are formed by fusions (or divisions) at each successive stage of the analysis. However, hierarchical methods have some characteristics that prevent their application in the classification of remotely sensed images: (a) in general they require memory space in the order of $O(N^2)$, in which N is the number of records in the data set; (b) the results can be difficult to interpret, mainly for large data sets; (c) in order to determine the cutoff of the dendrogram (ideal number of clusters) some decision criteria must be applied [11].

4 Proposed Methodology

The key point of the unsupervised classification method of satellites images proposed here is to perform the clusters analysis of the image through a set of SOM prototypes instead of working directly with the original patterns of the scene. For this, an approach consisting of three processing stages through SOM is utilized with the objective to discover representative clusters of prototypes for each land-cover class of interest. The three processing stages, which consist basically of the training, filtering and segmentation of the SOM, are described in the following subsections.

4.1 Training

At the first processing stage a sample set collected from the original image is used to train the SOM. Unlike pixel by pixel approaches that only use the spectral information of individual points to find homogenous regions, the proposed method performs the sampling of the image through pixel windows. The idea is to incorporate in the classification process information about the neighborhood (context) of the pixels, considering that isolated pixels are not able to represent the majority of cover land patterns, especially in the case of images that have higher spatial resolutions. The sample windows are collected uniformly across the entire region of the image, without overlappings and at regular intervals. All of the samples are square and have the same size.

In order to train the SOM, some parameters must be specified to define the structure of the map and to specifically control the stated training. With the objective of guaranteeing good mapping of the original patterns, the proposed methodology defines in a particular way the neural net parameters based on the existing literature, on experimental tests, and some peculiarities of the application of SOM on remotely sensed images. However, since the SOM can be sensitive to choice of its training parameters, other alternatives can also be sought out to obtain good maps [4].

The proposed methodology utilizes the following parameters to train the SOM: linear initialization of weights, batch training mode, gaussian neighborhood function and rectangular shape to organize the two-dimensional array of neurons of the net.

The size of the map is one of the free parameters of SOM that particularly depends on the input image and the objectives of the classification. If the objective is to detect all of the patterns in the image, including those with low probability of occurrence, large-sized maps must be employed in the analysis; in the opposite case, if the interest is concentrated only on the predominant patterns in the scene, a smaller-sized SOM can be utilized. Nevertheless, the performance of the proposed classification methodology is not significantly affected if sufficiently large sizes for the SOM are utilized. Although maps with larger dimensions than are necessary have a larger quantity of inactive neurons, as shows the next processing stage, this event is not prejudicial within the proposed methodology.

4.2 Filtering

The second stage of the proposed approach consists of filtering two types of prototypes that generally appear in the mapping of image patterns through SOM. These prototypes, denominated here as *inactive* and *heterogeneous*, can act as borders (or "interpolation units") in the SOM grid contributing to the separation of clusters.

The inactive prototypes correspond to the neurons that have null activity in the SOM competitive learning process, i.e., they are not associated with any input pattern. These prototypes are simply eliminated of the analysis.

Heterogeneous prototypes are those that have a high degree of spectral-textural heterogeneity and are normally associated with input patterns that incorporate more than one land cover class. Most of the time, these patterns correspond to transition regions between land cover classes present in the image and are captured in consequence of the sampling through pixel windows. Prototypes that are considered heterogeneous are excluded in the third stage of the proposed approach, in which the hierarchical clustering method is applied. The objective of excluding these prototypes is to prevent them (and consequently the input patterns associated with them) from being erroneously attributed to one of the classes that are part of them. Heterogeneous prototypes can be seen as noisy or divergent patterns, and if they are not filtered the hierarchical method can incorporate them in the clusters that will be produced or retain them in separate clusters. And because in hierarchical methods, data or cluster fusion at a determined level can not be corrected in subsequent levels, incorrect interpretations regarding the classes and/or the number of classes of the image can be made. The input patterns associated with these heterogeneous prototypes are particularly classified only at the end of the analysis, considering the neighboring pixels that have already been labelled.

The spectral-textural heterogeneity degree of each prototype of the SOM is computed from Haralick's co-occurrence matrix [3]. Since the weight vectors of the SOM prototypes have the same dimensions as the input patterns (that in this case are pixel windows), it makes it possible to generate an image of each prototype of the net and to calculate the co-occurrence probability of all pairwise combinations of grey levels in each one of them. The *energy* (sometimes called uniformity) was the measure chosen to calculate the spectral-textural heterogeneity of each prototype from co-occurrence matrix. This measure, described through the equation (3), gets values next to 1 when the area of interest presents uniform texture (similar grey levels), and values that tend to zero when the area is not uniform.

$$ENE = \sum_{i}\sum_{j} P(i,j)_{d,\theta} \quad (3)$$

where $P(i,j)_{d,\theta}$ is the co-occurrence probability of two grey levels i and j, separate to a distance d in the direction θ.

The prototypes whose *ENE*'s satisfy the relationship given below are considered heterogeneous and are consequently filtered:

$$ENE > \mu_{ENE} + \frac{1}{2}\sigma_{ENE}\,. \quad (4)$$

Here μ_{ENE} and σ_{ENE} are, respectively the average and the standard deviation of the *ENE*'s of all of the prototypes.

4.3 Segmentation

At the last processing stage of the proposed approach, an agglomerative hierarchical clustering method is applied to the trained and filtered SOM prototypes.

The hierarchical clustering method utilized here has two important characteristics. The first one of them is the imposition of restrictions to the possible SOM prototype fusions. Unlike traditional hierarchical clustering methods, which consist of comparing all of the pairs of objects to decide on a fusion, the approach utilized in this work verifies the possibility of fusions only between adjacent (or neighboring) prototype pairs in the SOM grid. Another important characteristic is that beyond using inter-cluster distance information the employed merging criterion also incorporates space information of the image pixels associated to the SOM prototype clusters.

The distance information (D_{ij}) between two prototype clusters i and j is calculated using the euclidean metric and the nearest neighbor method (or single linkage method). The values of D_{ij} are normalized within the interval [0,1].

The space information is calculated through two indices, denominated *spatial boundary index* and *spatial compactness index*. These indices, developed by Marçal and Castro [7], are computed here from classified image using the SOM prototype clusters (classes) in each level of the dendrogram generated by hierarchic method.

The spatial boundary index (B_{ij}) calculates the boundary length between all class pairs (i,j) considering eight neighbors for each pixel (four adjacent and four oblique). Its formula is given as follows:

$$B_{ij} = 1 - \frac{1}{2}\left(\frac{b_{ij}}{\sum\limits_{k=1(k \neq i)}^{N} b_{ik}} + \frac{b_{ij}}{\sum\limits_{k=1(k \neq j)}^{N} b_{kj}} \right) \tag{5}$$

where b_{ij} is the number of boundary counts for the class pair (i,j) and N is the number of classes in the dendrogram level that is being analyzed. The idea behind this index is that two classes that have a significant common boundary should be more likely to merge than classes with very little or no common boundaries [7].

The spatial compactness index (C_{ij}), defined through the equation (6), is based on the number of self-boundary counts for each class (b_{ii}). This index penalizes the merger of compact classes. In the same way that the index B_{ij}, the index C_{ij} results values within the interval [0,1].

$$C_{ij} = \frac{1}{2}\left(\frac{b_{ii}}{b_{ii} + 6\sum\limits_{k=1(k \neq i)}^{N} b_{ik}} + \frac{b_{jj}}{b_{jj} + 6\sum\limits_{k=1(k \neq j)}^{N} b_{jk}} \right) \tag{6}$$

The merging criterion adopted here establishes that the pair of prototype classes (i,j) that to present the lowest value resultant of the average computed between D_{ij}, B_{ij} and C_{ij} is selected for merger.

At the end of this stage there is a dendrogram that shows in a hierarchical way the similarity levels between the SOM prototypes. Each level of dendrogram obtained corresponds to a different clustering configuration of SOM prototypes that can be utilized to represent the classes by which the original image will be classified.

5 Experimental Results

This section shows an application example of the proposed methodology on a test image. The image used in the experiments (provided by Engesat/Brazil, © Space Imaging) has 385×350 pixels and is composed of three spectral bands of the IKONOS satellite. The study area shows irrigation pivots in the region of Andaraí in the Bahia state, Brazil. Six large land cover classes are present in the scene: sparse vegetation, forest, two types of exposed soil, and two stages of coffee plantation. The Fig. 1(a) shows a color composite of the test image.

In accordance with the procedures described in the subsection 4.1, 1292 sample windows of size 9×9 were collected from test image and used to train a SOM composed of 225 neurons arranged in a 15×15 rectangular grid.

The Fig. 1(b) shows the images of each SOM prototype (neuron) arranged in the rectangular grid after the training. By means of them it is possible to visualize prototype clusters that correspond to the land cover classes present in the original image. It is also possible to observe the *topological ordering* and *density matching* properties in the mapping produced by the neural net. Land cover classes with similar spectral attributes are mapped to neighboring regions of the two-dimensional output grid and those that occupy bigger areas in the original image are mapped to a bigger number of prototypes of the grid.

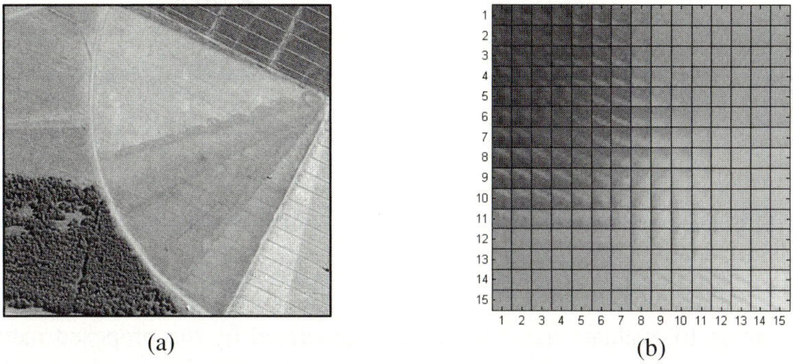

Fig. 1. (a) Color composite of the image used in the tests. (b) Image of the SOM prototypes arranged in the rectangular grid after the training. The images shown in (a) and (b) are in different scales.

After the SOM training, the prototypes filtering process was applied. In this experiment, 10 SOM prototypes presented null activity and 39 presented a high degree of spectral-textural heterogeneity, given that its values of energy (ENE) exceeded the threshold defined in the equation (4). Thus, of the 225 total prototypes, 49 of them were filtered, remaining 176 prototypes to be analyzed in the following stage.

In the last processing stage, the agglomerative hierarchical method (described in the section 4.3) was applied on the filtered SOM prototypes. Consequently, a dendrogram consisting of 176 levels was generated, each level with a different configuration of SOM prototypes clusters.

Since the image presents six large land cover classes (previously cited), the level of the dendrogram composed by six SOM prototype clusters was chosen to effect the classification of the scene. Fig. 2(a) shows the SOM grid segmented in 6 clusters. The squares marked with "o" and "×" are, respectively, the inactive and heterogeneous prototypes that were filtered in the previous stage.

To perform the classification of all pixels of the test image, the image was entirely run through considering 9×9 pixel windows (size equal to the sample windows) and comparing with all of the SOM prototypes. This comparison was performed using distances calculated between considered pixel windows and each one of the prototypes. The central pixel of the pixel window received the prototype label that presented the shortest distance from it. In the sequence, each one of the pixels of the image that were associated to any heterogeneous prototype was reclassified using the neighboring pixel class that has the least (spectral) distance from it. Fig. 2(b) shows the classification results for the test image using the proposed methodology.

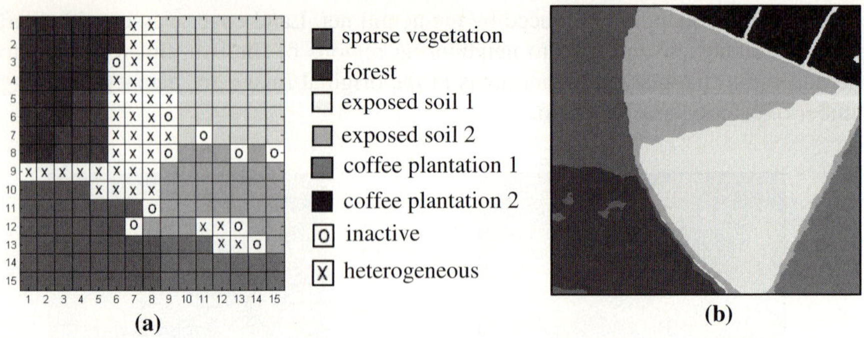

Fig. 2. (a) Filtered and segmented SOM grid. (b) Result of the test image classification by proposed methodology.

Attempting to evaluate the classification generated by the proposed method, and considering the absence of terrestrial truth for the test image, the present work performed the classification of the image in a supervised manner using a multilayers *Perceptrons* (MLP) neural net, and considered these results as a reference (or "true") to calculate the Kappa agreement index (normally used to evaluate the accuracy of image classification). The MLP net was trained with a sample set collected from original image by an image analyst. The Kappa index obtained here was 0.82, which allows to conclude that the classification result of the test image by method presented in this work was very satisfactory.

6 Conclusions and Final Considerations

In this work, an approach consisting of three processing stages through SOM was proposed to perform the unsupervised classification of satellite images. The key point of the proposed method is to perform the clusters analysis of the image through a set of SOM prototypes instead of working directly with the original patterns of the scene. This approach significantly reduces the complexity of the analysis, making it possible to use methods that have not normally been considered viable for the processing of remotely sensed images, such as agglomerative hierarchical methods. Moreover, the proposed method presents advantages that make it as a promising alternative to carry out the classification of remotely sensed images. Among these, we can point out:

(a) The method does not require a previous definition of the number of classes to perform the classification of the image. It does not occur in the majority of the conventional unsupervised classification methods.

(b) The distributed representation of the classes by means of prototype groups gives the method the potential to discover geometrically complex and varied data clusters. Methods such as K-means use a single prototype (centroid) to represent each class and because of this are only capable of adequately detecting clusters that have hyperspherical formats.

(c) The method uses a particular approach to classify pixels situated in transition regions between classes. This procedure contributes to increasing the accuracy of the resulting classification.

(d) The utilization of an agglomerative hierarchical clustering method allows the user to observe the relationships that exist between the land cover patterns existing in the image at different cluster levels. It can be very helpful in applications where the structure of the information present in the image is not clearly known.

(e) The proposed method employs an efficient merge mechanism that incorporates more information about the data in each cluster. Traditional clustering methods use only inter-cluster distance information to decide on the merging of clusters. Integration of spatial characteristics helps to increase the understanding of some classes confusable with others.

In addition to the test image utilized in the experiments shown here, the proposed method has also been applied to other high and medium resolution images, with satisfactory results.

As future works, it is intended to apply modified versions of cluster validation indices, as considered in [2], to automatically determine the cutoff of the dendrogram (ideal number of clusters) for the image. Comparisons of performance with conventional classification methods and sensitivity analysis also must be executed.

References

1. Costa, J.A.F., Netto, M.L.A.: Clustering of Complex Shaped Data Sets via Kohonen Maps and Mathematical Morphology. In: Proceedings of the SPIE Conference on Data Mining and Knowledge Discovery, Orlando, FL, vol. 4384, pp. 16–27 (2001)
2. Gonçalves, M.L., Netto, M.L.A., Costa, J.A.F., Zullo Júnior, J.: Data Clustering using Self-Organizing Maps Segmented by Mathematic Morphology and Simplified Cluster Validity Indexes. Proceedings of IEEE International Joint Conference on Neural Networks 1, 8854–8861 (2006)
3. Haralick, R.M., Shanmugam, K., Dinstein, I.: Textural Features for Image Classification. IEEE Trans. on Systems, Man and Cybernetics 3(6), 610–621 (1973)
4. Kaski, S., Lagus, K.: Comparing self-organizing maps. In: Vorbrüggen, J.C., von Seelen, W., Sendhoff, B. (eds.) Artificial Neural Networks - ICANN 96. LNCS, vol. 1112, pp 809–814. Springer, Heidelberg (1996)
5. Kohonen, T.: Self-Organizing Maps, 2nd edn. Springer, Berlin (1997)
6. Kohonen, T., Oja, E., Simula, O., Visa, A., Kangas, J.: Engineering Applications of the Self-Organizing Map. Proceedings of the IEEE 84(10), 1358–1384 (1996)
7. Marçal, A.R.S., Castro, L.: Hierarchical Clustering of Multispectral Images using Combined Spectral and Spatial Criteria. In: IEEE Geoscience and Remote Sensing Letters 2, 59–63 (2005)
8. Richards, J.A.: Analysis of Remotely Sensed Data: the Formative Decades and the Future. In: IEEE Transactions on Geoscience and Remote Sensing 43, 422–432 (2005)
9. Vesanto, J., Alhoniemi, E.: Clustering of the Self-organizing Map. In: IEEE Transactions on Neural Networks 11, 586–602 (2000)
10. Wu, S., Chow, T.W.S.: Clustering of the Self-organizing Map using a Clustering Validity Index based on Inter-cluster and Intra-cluster Density. Pattern Recognition 37, 175–188 (2004)
11. Xu, R., Wunsch II, D.: Survey of Clustering Algorithms. In: IEEE Transactions on Neural Networks 16, 645–678 (2005)

Performance Analysis of MLP-Based Radar Detectors in Weibull-Distributed Clutter with Respect to Target Doppler Frequency

Raul Vicen-Bueno*, Maria P. Jarabo-Amores, Manuel Rosa-Zurera, Roberto Gil-Pita, and David Mata-Moya

Signal Theory and Communications Department
Escuela Politécnica Superior, Universidad de Alcalá
Ctra. Madrid-Barcelona, km. 33.600, 28805, Alcalá de Henares-Madrid, Spain
{raul.vicen,mpilar.jarabo,manuel.rosa,roberto.gil,david.mata}@uah.es

Abstract. In this paper, a Multilayer Perceptron (MLP) is proposed as a radar detector of known targets in Weibull-distributed clutter. The MLP is trained in a supervised way using the Levenberg-Marquardt back-propagation algorithm to minimize the Mean Square Error, which is able to approximate the Neyman-Pearson detector. Due to the impossibility to find analytical expressions of the optimum detector for this kind of clutter, a suboptimum detector is taken as reference, the Target Sequence Known A Priori (TSKAP) detector. Several sizes of MLP are considered, where even MLPs with very low sizes are able to outperform the TSKAP detector. On the other hand, a sensitivity study with respect to target parameters, as its doppler frequency, is made for different clutter conditions. This study reveals that both detectors work better for high values of target doppler frequency and one-lag correlation coefficient of the clutter. But the most important conclusion is that, for all the cases of the study, the MLP-based detector outperforms the TSKAP one. Moreover, the performance improvement achieved by the MLP-based detector is higher for lower probabilities of false alarm than for higher ones.

1 Introduction

Neural Networks can be applied to detect known targets in coherent Weibull clutter. It is possible because Neural Networks trained in a supervised way can approximate the Neyman-Pearson detector [1], which is usually used in radar systems design. This detector maximizes the probability of detection (Pd) maintaining the probability of false alarm (Pfa) lower than or equal to a given value [2]. The detection of targets in clutter is the main problem in radar detection systems. Many clutter models have been proposed in the literature [3], although one of the commonly accepted models is the Weibull one [4,5].

The research shown in [6] set the optimum detector for target and clutter with arbitrary Probability Density Functions (PDFs). Due to the impossibility

* This work has been supported by Comunidad de Madrid and the University of Alcalá under project CAM-UAH 2005/30.

J. Marques de Sá et al. (Eds.): ICANN 2007, Part II, LNCS 4669, pp. 690–698, 2007.

to obtain analytical expressions for the optimum detector, only suboptimum solutions were proposed. The Target Sequence Known A Priori (TSKAP) detector is one of them and is taken as reference for the experiments. Also, these solutions convey implementation problems, some of which make them non-realizable.

One kind of Neural Network, the MultiLayer Perceptron (MLPs), has been probed to be able to approximate the Neyman-Pearson detector when they are trained in a supervised way to minimize the Mean Square Error (MSE) [7,8]. They have been applied to the detection of known targets in different radar environments [9,10].

In this work, MLPs are trained to approximate the Neyman-Pearson detector for known targets in coherent Weibull clutter and white Gaussian noise. As for designing the MLPs, no assumption is made about the target or the radar environment. So they are expected to outperform the suboptimum solutions, which need to have a priori some knowledge of the environment, as occurs with the TSKAP one. According to it, a study of the MLP size is carried out for typical values of clutter parameters. The work is completed with a sensitivity study with respect to the spectrum spread of the clutter and the doppler frequency of the target, in order to demonstrate the best behavior of the MLP-based detector.

2 Radar Target, Clutter and Noise Models

The radar is assumed to collect N pulses in a scan, so input vectors (z) are composed of N complex samples, which are presented to the detector. Under hypothesis H0 (target absent), z is composed of N samples of clutter and noise. Whereas under hypothesis H1 (target present), a known target characterized by a fixed amplitude (A) and phase (θ) for each of the N pulses is summed up to the clutter and noise samples. Also, a doppler frequency in the target model (f_s) is assumed, where PRF is the Pulse Repetition Frequency of the radar system (the sampling rate of the process).

The noise is modeled as a coherent white Gaussian complex process of unity power, i.e., a power of $\frac{1}{2}$ for the quadrature and phase components. The clutter is modeled as a coherent correlated sequence with Gaussian AutoCorrelation Function (ACF), whose complex samples have a modulus with a Weibull PDF:

$$p(|\mathbf{w}|) = ab^{-a}|\mathbf{w}|^{a-1}e^{-\left(\frac{|\mathbf{w}|}{b}\right)^{a}} \tag{1}$$

where $|\mathbf{w}|$ is the modulus of the coherent Weibull sequence and a and b are the skewness (shape) and scale parameters of a Weibull distribution, respectively.

The NxN autocorrelation matrix of the clutter is given by

$$(\mathbf{M_c})_{h,k} = P_c \rho_c^{|h-k|^2} e^{j(2\pi(h-k)\frac{f_c}{PRF})} \tag{2}$$

where the indexes h and k varies from 1 to N, P_c is the clutter power, ρ_c is the one-lag correlation coefficient and f_c is the doppler frequency of the clutter.

The relationship between the Weibull distribution parameters and P_c is

$$P_c = \frac{2b^2}{a} \Gamma \left(\frac{2}{a} \right) \tag{3}$$

where $\Gamma()$ is the *Gamma function.*

The model used to generate coherent correlated Weibull sequences consists of two blocks in cascade: a correlator filter and a NonLinear MemoryLess Transformation (NLMLT) [4]. To obtain the desired sequence, a coherent white Gaussian sequence is correlated with the filter designed according to (2) and (3). The NLMLT block, according to (1), gives the desired Weibull distribution to the sequence.

Taking into consideration that the complex noise samples are of unity variance (power), the following power relationships are considered for the study:

- Signal to Noise Ratio: SNR $= 10 \cdot log_{10} \left(A^2 \right)$
- Clutter to Noise Ratio: CNR $= 10 \cdot log_{10} \left(P_c \right)$

3 Optimum and Suboptimum Neyman-Pearson Detectors

The problem of optimum radar detection of targets in clutter is explored in [4] when both are time correlated and have arbitrary PDFs. The optimum detector scheme is built around two non-linear estimators of the disturbances in both hypothesis, which minimize the MSE. The study of Gaussian correlated targets detection in Gaussian correlated clutter plus noise is carried out, but for the cases where the hypothesis are non-gaussian distributed, only suboptimum solutions are studied.

The proposed detectors basically consist of two channels. The upper channel is matched to the conditions that the sequence to be detected is the sum of the target plus clutter in presence of noise (hypothesis H1). While the lower one is matched to the detection of clutter in presence of noise (hypothesis H0).

For the detection problem considered in this paper, the suboptimum detection scheme (TSKAP) shown in the fig. 1 is taken. Considering that the CNR is very high (CNR >> 0 dB), the inverse of the NLMLT is assumed to transform the Weibull clutter in Gaussian, so the Linear Prediction Filter (LPF) is a N-1 order linear one. Then, the NLMLT transforms the filter output in a Weibull sequence. Besides being suboptimum, this scheme presents two important drawbacks:

1. The prediction filters have N-1 memory cells that must contain the suitable information to predict correct values for the N samples of each input pattern. So N+(N-1) pulses are necessary to decide if the target is present or not.
2. The target sequence must be subtracted from the input of the H1 channel.

There is no sense in subtracting the target component before deciding if this component is present or not. So, in practical cases, it makes this scheme non-realizable.

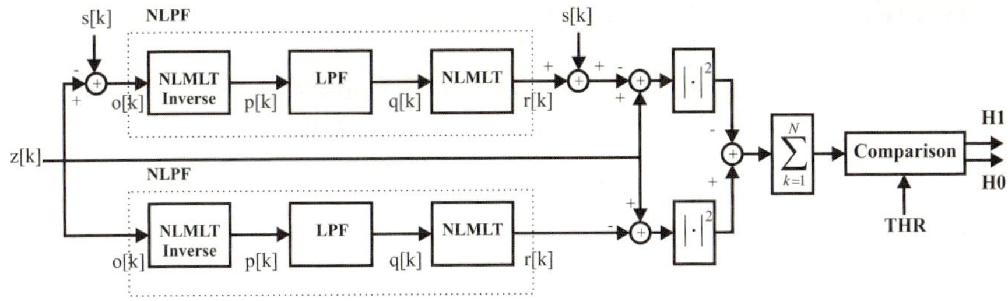

Fig. 1. Target Sequence Known A Priori Detector

4 MLP-Based Detector

A detector based on a MLP with log-sigmoid activation function in its hidden and output neurons with hard limit threshold after its output is proposed. The MLP-based detector tries to overcome the drawbacks of the scheme proposed in the Section 3. Also, as MLPs have been probed to approximate the Neyman-Pearson detector when minimizing the MSE [8], it can be expected that the MLP-based detector outperforms the suboptimum scheme proposed in [4].

In our case of study, MLPs are trained to minimize the MSE using the LM backpropagation algorithm with adaptive parameter [12]. As this algorithm is based on the Newton method, MLPs with few hundred of weights (W) are able to achieve good performances and converge in few epochs with this algorithm.

Three sets of patterns are generated for the training: *train*, *validation* and *test*. *Train*, *validation* sets are used for cross-validation purposes, which avoids overfitting. To improve the generalization of the trained MLPs, the training is stopped if the estimated MSE with the validation set increases during the last ten epochs of the training. Finally, the *test* set is used to obtain the performance of the MLPs trained working as radar detectors, i.e., to obtain the Pfa and Pd estimation by Moltecarlo simulations. All the patterns of the three sets are generated under the same conditions (target parameters: SNR and f_s ; and clutter parameters: CNR, a, f_c and ρ_c) for each case of study.

Before the training process, MLPs are initialized using the Nguyen-Widrow method [14] and, in all cases, the training process is repeated ten times. Once all the MLPs are trained, the best MLP in terms of the estimated MSE with the validation set is selected. With this selection, the problem of keeping in local minima at the end of the training is practically eliminated.

The architecture of the MLP considered for the experiments is $I/H/O$, where I is the number of MLP inputs, H is the number of hidden neurons in its hidden layer and O is the number of MLP outputs. As the MLPs work with real arithmetic, if the input vector (z) is composed of N complex samples, the MLP will have 2N inputs (N in phase and N in quadrature components of the N complex samples). The number of MLP independent elements (weights) to solve the problem is $W = (I+1)\cdot H + (H+1)\cdot O$, considering the bias of each neuron.

5 Results

The performance of the detectors exposed in the previous sections is shown in terms of the Receiver Operating Characteristics (ROC) curves, which relates the Pd with the desired Pfa. The ROC curves are shown with the Pfa axis in a log scale, because in a linear scale it is not able to appreciate correctly the performance differences between the detectors, specially when the study of the MLP size is carried out. The experiments developed are designed for an integration of two pulses ($N = 2$). So, in order to test correctly the TSKAP detector, patterns of length 3 ($N + (N - 1)$) complex samples are generated, due to memory requirements of the TSKAP detector ($N - 1$ pulses).

The MLP architecture used to generate the MLP-based detector is $6/H/1$. The number of MLP outputs (1) is established by the problem (binary detection). The number of hidden neurons (H) is a parameter under study in this work. And the number of MLP inputs (6) is establish because a comparison under the same conditions with a reference radar detector is made. In this case, a total of 3 pulses are considered because the memory requirements of the TSKAP detector.

The a priori probabilities of $H0$ and $H1$ hypothesis is supposed to be the same. Three sets of patterns (*train*, *validation* and *test*) are generated under the same radar conditions for each experiment. The first and the second ones have $5 \cdot 10^3$ patterns, respectively. The third one has $5 \cdot 10^6$ patterns, so the error in the estimations of the Pfa and the Pd is lower than 10% of the estimated values in the wost case (Pfa=10^{-4}). Attending to previous studies of detection of targets in clutter [4,5,6], typical values of the target and Weibull-distributed clutter are taken. The values related with the clutter are: CNR = 30 dB, $f_c = 0$ Hz and $a = 1.2$, whereas the target parameter is: SNR = 20 dB. The coherent white Gaussian noise is considered with unity power. For the case of study, the f_s (target parameter) is modified in order to obtain the behavior of the detectors with respect to target parameter variations. Moreover, the ρ_c (clutter parameter) is modified, as occur in actual radar environments, in order to check if the previous behavior continues or not.

A study of the MLP size is carried out under the conditions exposed above. For this study, the target doppler frequency is fixed to $f_s = 0.5 \cdot PRF$ and the one-lag correlation coefficient of the clutter is established to $\rho_c = 0.90$. The results obtained are shown in fig. 2. As can be observed, a MLP size greater than 6/20/1 gives lower performance improvement and higher computational cost increase than this size. So, a MLP size of 6/20/1 ($W > 121$ weights) is selected for the next experiments because the tradeoff between performance improvement and computational cost. Moreover, greater MLP sizes than 6/30/1 where probed but very low improvements were achieved. The MLP size proposed and lower ones than this are enough to outperform the performance achieved with the TSKAP detector, as it is demonstrated below. Experiments with different f_s and rho_c parameters were carried out and the conclusions obtained about this study were the same.

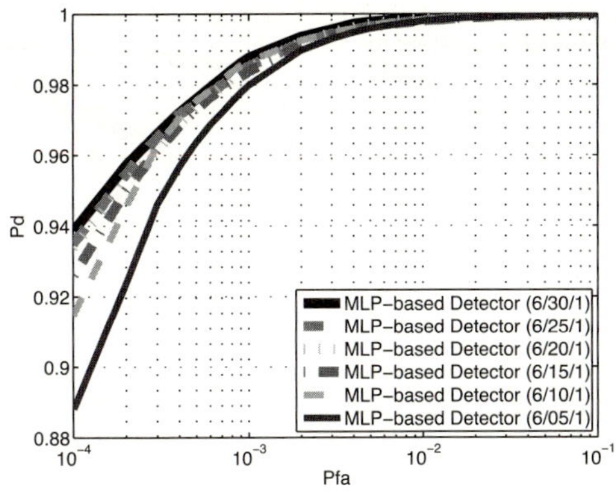

Fig. 2. MLP-based detector performances for different MLP sizes (6/H/1)

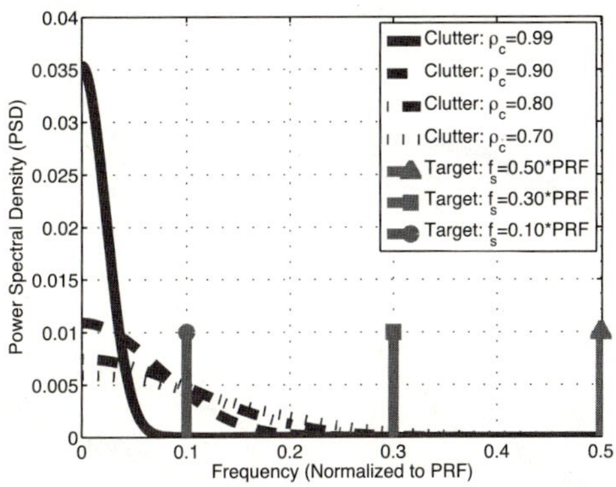

Fig. 3. Power Spectral Densities of the different correlated Weibull-distributed clutters ($a = 1.2$) and different doppler frequencies of the target

The Power Spectral Densties (PSDs) of the Weibull-distributed clutter with unity powers ($P_c = 1$) and target with power P_s are shown in fig. 3. As the target is constant in time, its PSD is a Delta function. The target PSDs for different doppler frequencies are presented. According to the PSD of clutter with different ρ_c, we can observe that the greater is this parameter ($\rho_c \rightarrow 1$), the lower is the spread of the clutter spectrum. So in these situations, the separation of both hypothesis (decision $H0$ or $H1$) will be easier than in situations where the correlation coefficient is far from 1.

Once the MLP-based detector is designed and the target and clutter PSDs are analyzed, the sensitivity of the detectors performances with respect to target parameters is studied. In this case, the detector sensitivity to the f_s is analyzed. Fig. 4 and 5 show the behavior of the MLP-based and TSKAP detectors with the variation of the f_s in different clutter conditions. Analyzing the results, several aspects can be highlighted. First: in all the cases, the MLP-based detector is better than the TSKAP one, independently of the f_s and ρ_c considered in the

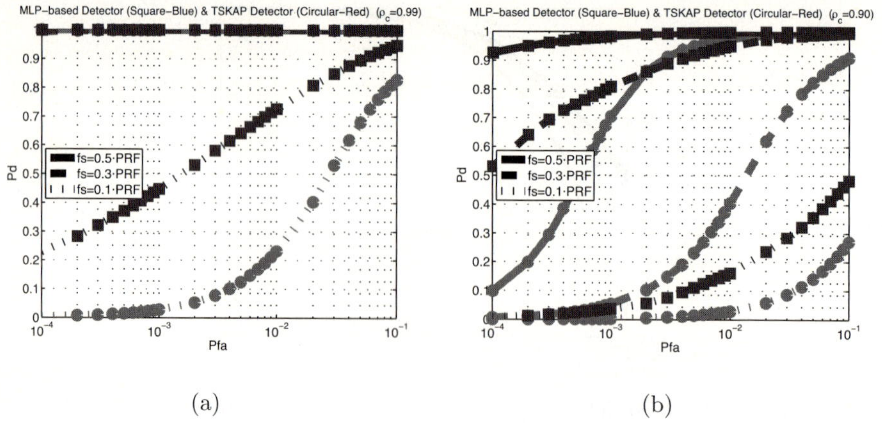

Fig. 4. TSKAP and MLP-based detectors ROC curves for different doppler frequencies of the target and one-lag correlation coefficients: (a) $\rho_c = 0.99$ and (b) $\rho_c = 0.90$

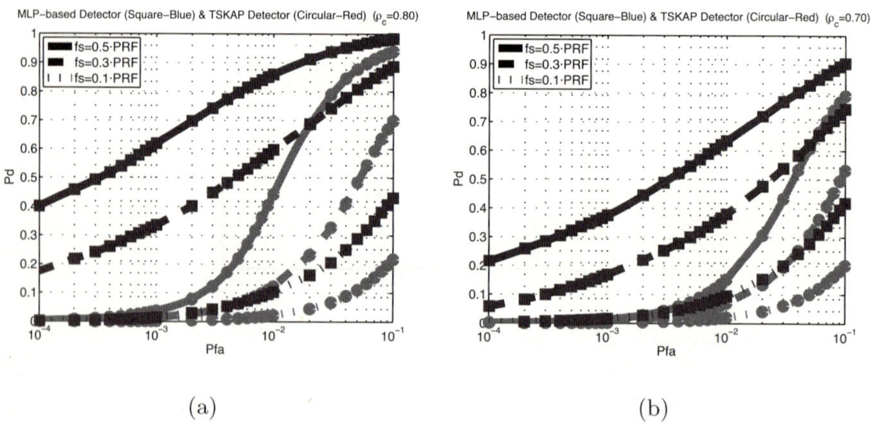

Fig. 5. TSKAP and MLP-based detectors ROC curves for different doppler frequencies of the target and one-lag correlation coefficients: (a) $\rho_c = 0.80$ and (b) $\rho_c = 0.70$ (right)

studies. Second: as was expected with the analisis of the PSDs, the detectors performances are better for high correlated clutter environments ($\rho_c \rightarrow 1$). Third: both detectors are better for targets with high f_s. And four: the performance improvement achieved by the MLP-based detector with respect to the TSKAP one is greater for low Pfa's than for high Pfa's. Finally, because of the difference between performance detectors for all the cases, it is demonstrated that a MLP size lower than 6/20/1 is able outperform the TSKAP one.

6 Conclusion

The influence of the MLP size is studied in order to implement MLP-based radar detectors to detect known targets in a Weibull-distributed plus white Gaussian noise radar environment. This study avoid us to propose a MLP-based detector with a structure of 6/20/1, although a detector with lower MLP size (lower computational cost) than this is able to outperform the detector took as reference, the TSKAP one.

In all the cases under study (variation of f_s for different clutter conditions (ρ_c)), the MLP-based detector outperforms the TSKAP one. Moreover, the performance improvement achieved by the MLP-based detector increases with the decrease of the ρ_c. The obtained results show that the TSKAP detector not only requires that CNR >> 0 dB, but its performance hardly gets worn when ρ_c moves away from 1. On the other hand, the MLP is a non-parametric technique and it doesn't require any a priori information of the inputs, as occurs with the TSKAP detector (parametric technique). In that way, it is able to approximate the Neyman-Pearson detector in all the cases, where the approximation error will only depend on its size (number of weights or freedom degrees of the MLP) and the ability of the training algorithm to find the minimum of the error function.

References

1. De la Mata-Moya, D., et al.: Approximating the Neyman-Pearson Detector for Swerling I Targets with Low Complexity Neural Networks. In: Duch, W., Kacprzyk, J., Oja, E., Zadrożny, S. (eds.) ICANN 2005. LNCS, vol. 3697, pp. 917–922. Springer, Heidelberg (2005)
2. Van Trees, H.L.: Detection, Estimation and Modulation Theory. John Wiley and Sons, New York (1997)
3. Cheikh, K., Faozi, S.: Application of Neural Networks to Radar Signal Detection in K-distributed Clutter. First Int. Symp. on Control, Communications and Signal Processing Workshop Proc., 633–637 (2004)
4. Farina, A., et al.: Theory of Radar Detection in Coherent Weibull Clutter. In: Farina, A. (ed.) Optimised Radar Processors. IEE Radar, Sonar, Navigation and Avionics, Peter Peregrinus Ltd., London, vol. 1, pp. 100–116 (1987)
5. DiFranco, J.V., Rubin, W.L.: Radar Detection, Artech House. U.S.A (1980)
6. Farina, A., et al.: Radar Detection in Coherent Weibull Clutter. IEEE Trans. on Acoustics, Speech and Signal Processing ASSP-35(6), 893–895 (1987)
7. Ruck, D.W., et al.: The Multilayer Perceptron as an Approximation to a Bayes Optimal Discriminant Function. IEEE Trans. on Neural Networks 1(11), 296–298 (1990)
8. Jarabo-Amores, P., et al.: Sufficient Condition for an Adaptive System to Aproximate the Neyman-Pearson Detector. In: Proc. IEEE Workshop on Statistical Signal Processing, pp. 295–300 (2005)
9. Gandhi, P.P., Ramamurti, V.: Neural Networks for Signal Detection in Non-Gaussian Noise. IEEE Trans. on Signal Processing 45(11), 2846–2851 (1997)
10. Andina, D., Sanz-Gonzalez, J.L.: Comparison of a Neural Network Detector Vs Neyman-Pearson Optimal Detector. Proc. of ICASSP-96, 3573–3576 (1996)
11. Haykin, S.: Neural Networks. A Comprehensive Foundation, 2nd edn. Prentice-Hall, London (1999)
12. Bishop, C.M.: Neural networks for pattern recognition. Oxford University Press, New York (1995)

13. Hagan, M.T., Menhaj, M.B.: Training Feedforward Networks with Marquardt Algorithm. IEEE Trans. on Neural Networks 5(6), 989–993 (1994)
14. Nguyen, D., Widrow, B.: Improving the Learning Speed of 2-layer Neural Networks by Choosing Initial Values of the Adaptive Weights. In: Proc. of the Int. Joint Conf. on Neural Networks, pp. 21–26 (1999)

Local Positioning System Based on Artificial Neural Networks

Pedro Claro and Nuno Borges Carvalho

Instituto de Telecomunicações, Aveiro Portugal
nborges@av.it.pt

Abstract. This work describes a complete indoor location system, from its creation, development and deployment. This location system is a capable way of retrieving the position of wireless devices using a simple software solution, no additional hardware is necessary. The positioning engine uses artificial neural networks (ANN) to describe the behaviour of a specific indoor propagation channel. The training of the ANN is assured using a slight variation of the radio frequency fingerprinting technique. Results show that the location system has high accuracy with an average error below two meters.

Keywords: Location, positioning, wifi, wireless, artificial neural networks, backpropagation.

1 Introduction

The location paradigm began years ago when systems like Decca, OMEGA, Alpha and Loran C were developed. Loran C was developed by the United States Navy during World War II. This system main objective was to help US and UK military ships navigation. Several radio beacon towers were deployed along sea coast. Using these radio beacons and their known location, the ships were capable of locating themselves. This was an important tactical advantage and it is still used today, although some modifications were made. Current location systems are more technical advanced, but almost all of them use the same principle of Loran – radio triangulation. This is the case of the well known Global Positioning System (GPS). GPS uses geo-stationary satellites as radio beacons, and provides almost a global coverage real time location system.

1.1 Current State of the Art

Satellite navigation systems, like North American GPS or the future European Galileo, are mainly focused on providing position for outdoor environments. These systems provide a global coverage with a three to five meters average error available for public usage. Although satellite navigation systems have good accuracy they are not suitable to indoor environments where a good clear view to the sky is not available [1]. Other systems are mainly focused on indoor location environments. Systems like Active-Bat [2] developed by AT&T and Cricket [3] use ultrasound time of flight measurements, others like Active-Badge [4] use small infrared tags that

J. Marques de Sá et al. (Eds.): ICANN 2007, Part II, LNCS 4669, pp. 699–708, 2007.

provide symbolic information, like a name of a room. Infrared has limited range hence it has not become very popular. PinPoint 3D [5] uses radio frequency lateration to provide an accuracy of three meters and requires a special developed infrastructure. SpotON [6] project developed a 3D location system using RFID tags. SmartFloor [7] uses a sensor grid installed on a floor and the accuracy depends of the sensor spacing distance. These systems require a special infrastructure that needs to be deployed.

Implementing a location system using WLAN communication infrastructure has been for some time target of intensive research. Various approaches have been proposed, mostly based on the received signal strength. One of the first systems to be introduced was RADAR [8], developed by Microsoft Research Group. RADAR uses received radio strength to map the user current location and it uses an empirical model (K-Nearest Neighbor) as well as a simple signal propagation model. The accuracy of this system is about four meters for 75% of the time and it uses special developed access points.

R. Battiti et al. [9] describe a system capable of deriving the location using neural networks. In this work it is described the training phase, always present in a RF fingerprinting based system, and the neural network architecture used. Despite the accuracy (about 2.3 meters), the results are only compared to test data and information about software developed, radio strength reading method and system architecture is inexistent. M. D. Rodriguez et al. [10] also describe a location system for hospital services based on a neural network. In this work the SNR is used to calculate the user position and all the processing is done in the wireless client. Their results show an error below 4 meters 90% of the time. Bayesian models are also used to calculate a wireless device position, D. Madigan et al. [11] propose a system with an accuracy of four meters. A. Haeberlen et al. [12] describe a probabilistic approach and their location system provides symbolic information, like the office number inside a building.

1.2 Indoor Location System

Indoor location can be important when one thinks of services that can be applied in huge buildings, like shopping centers, office buildings, museums, warehouses, universities, etc. Imagine a warehouse equipped with a location system, one can know exactly the position of a package allowing an increase in the company's efficiency. Think of a museum where tourists can receive in their cell phones important information depending on the place they are or the art work they are seeing. Imagine receiving on your cell phone great price discounts when you walk along a store in your favorite shopping center. This one might not be so great, but the applications and advantages of an accurate and reliable indoor location system are tremendous. This is the aim of this work, to develop a low cost, reliable and accurate location system using today's technology.

2 Technical Description

The indoor location system was developed taking in account two different scenarios. The first, named macro-location, is a large scale implementation and provides symbolic information, like the name of room. This information is retrieved from the access point where the client is associated. The second scenario, named micro-location, gives the position of the wireless client by using spatial coordinates with high accuracy. The macro-location system is a simple solution to retrieve a client location in a symbolic way. Since micro-location solution provides the exact location, with spatial coordinates it was the target of a much more intense research and development. The two systems complement each other.

2.1 Macro-location

In a typical office wireless LAN there are several access points distributed in a defined way in order to offer good signal reception and sufficient bandwidth. Knowing which client is associated with which access point plays a critical role in the macro-location scenario. In fact, the main objective of the macro-location is to retrieve the location of the mobile client using the access point coverage. Despite the existence of several ways of knowing in which access point the client is associated some disadvantages can be found in each one of them. There is the need for a general solution that can be easily deployed to the existing variety of vendors and manufacturers.

The final solution is based on the remote syslog [13] capability of most access points in the market. Remote syslog allows a device to send important messages to a central server that gathers all the data.

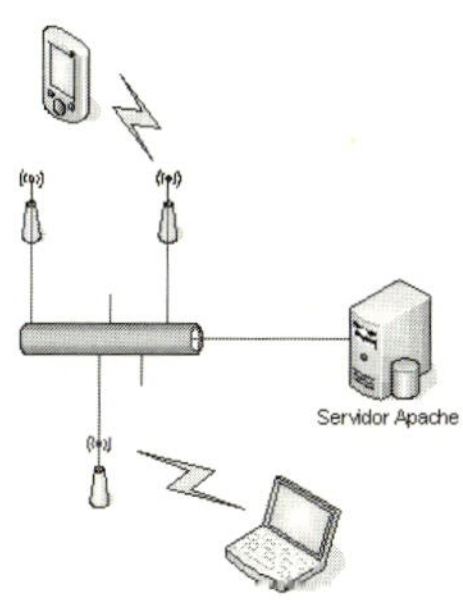

Fig. 1. Macro - location system architecture

With this method when a client associates/disassociates, the access point informs the server of this event by sending a syslog packet. The server has specific developed software in order to treat these messages and collect them into a database. Then it is possible to search for the location of a client or to see which clients are in an area.

2.2 Micro-location

The micro-location system allows the location of a WiFi device in an indoor scenario with an average error below two meters. The system uses a location method, called radio frequency fingerprinting. This technique requires profiling the entire location scenario before the location itself takes place. At each location or point, several measurements are taken and stored. This phase is often denominated calibration. Due to propagation effects, like fading, several measurements are taken to minimize this effect. After profiling the entire area, typically one building with multiple floors, the location system is ready to answer location requests. After the calibration phase is finished there are several RSSI vectors associated to each spatial coordinate. In a typical office floor, the adequate number of RSSI vectors was found to be between 10 and 20. Considering that *n* base stations exist, then the signal strength vector (*R*) is defined as:

$$R_i = (a_{i1}, a_{i2}, .. a_{in}). \tag{1}$$

Where a_{ij} corresponds to the measurement collected from point *i* from base station *j*. The complete RF fingerprinting matrix (*M*) is given by combining spatial information with the RSSI vector:

$$M = \{(x_i, y_i, R_i)\}, i = 1..m. \tag{2}$$

The RSSI vectors become the input of an artificial neural network (ANN)[14]. The number of neurons and of hidden layers of the ANN depends of the application and the size of the input data. The ANN inputs are the RSSI values measured by the wireless client and the output are the spatial coordinates X and Y. In scenarios where a third coordinate is necessary, like multiple floor buildings, the ANN output number increases to three. The multilayer perceptron architecture combined with the nonlinearity of the input and hidden layer activation functions, which are based on the hyperbolic tangent sigmoid function provided the generalization and adaptability needed for a proper and accurate location system.

The ANN training uses the backpropagation algorithm [4] based on a batch approach. The algorithm modifies the weights of each of the neurons to minimize the median square error between the output and the real values. The output layer activation function is linear, since the outputs are spatial coordinates. The number of outputs is typically two, since a two-dimensional plane was used to describe the possible locations of a device in an office floor.

The training process of a neural network must be adequate so that problems like over-fitting should not arise. Over-fitting occurs when too much training is applied to the ANN. This means that the ANN will be fitted exactly to the training data, therefore losing all the generalization capabilities. On the other hand, a poor training makes the ANN not to learn adequately

Another significant parameter in artificial neural networks is the learning rate. It affects the learning capability of the ANN, and a suitable value is required to perform an adequate training. The correct setting of the learning rate is often dependent on the size and type of input data and is typically chosen through experimental testing. Its value can also be adapted during the training phase, therefore becoming time dependent. The value for the learning rate chosen was 0.01. This low value is related to the nature of the input values that were between -1 and 1.

After the training process the ANN is ready to receive data and calculate the wireless client position. The location process follows a client-server architecture. Each time a request is sent by the server, the wireless client that is being located gathers RSSI values and sends them back to the server. The request handling, RSSI measurement and measurement report is assured by special developed software that must be installed on the client. This software does not affect the client processing power and does not require a huge amount of memory usage. After receiving the RSSI measurements, the server normalizes them and uses the trained ANN to calculate the client's position. This position is showed on a web interface special developed for this system. This interface provides a real-time location visualization of the wireless client.

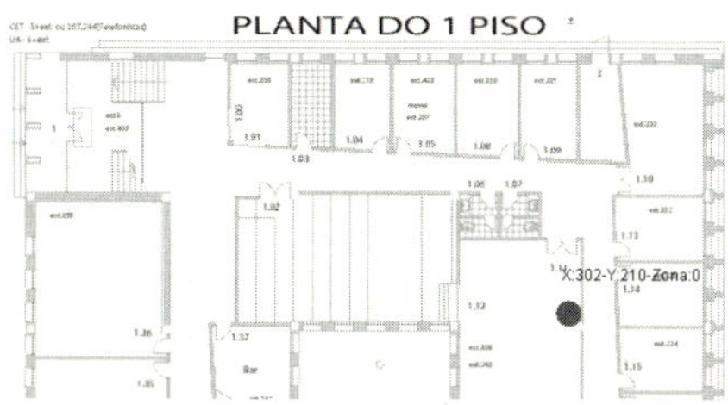

Fig. 2. Wireless client position using a floor plan

Typically, a floor plan is used to help identify the location of the client. Displayed expressions should be numbered for reference. The numbers should be consecutive within each section or within the contribution, with numbers enclosed in parentheses and set on the right margin.

3 Results

3.1 Macro-location

The macro-location system was tested in the *Instituto de Telecomunicações* building using three Cisco access points. These APs were configured to send all the logs to the central server. Various location scenarios were tested, including fast moving clients, turning off one or more APs and also turning off the client. In all cases the location system provided the location information correctly. Increasing the number of access points has little effect in the system. The network performance is unaffected since the log packets are small and not in sufficient number to cause a major impact on typical office LAN.

3.2 Micro-location

Since the location system performance depends on the RSSI values from the various access points, it is important to study the behavior of these signals over time. The standard deviation from the several RSSI values at the same location can not take large values, since it will degrade the location system performance. Normally, the transmitted power in access points in a typical WLAN is constant. The results presented here were collected in a typical deployment scenario – a shopping center.

This test bench had four day duration and consisted of placing a WiFi device collecting 20 measurements at each time in intervals of five minutes. The access points used on this test were the ones that existed on the shopping centre that are from several ISPs. Since the shopping centre entrances are equipped with counting sensors, the RSSI results were crossed with the number of people that were inside the building. This way it might be possible to conclude something about the effect of the constant movement of people and its effect on the radio propagation.

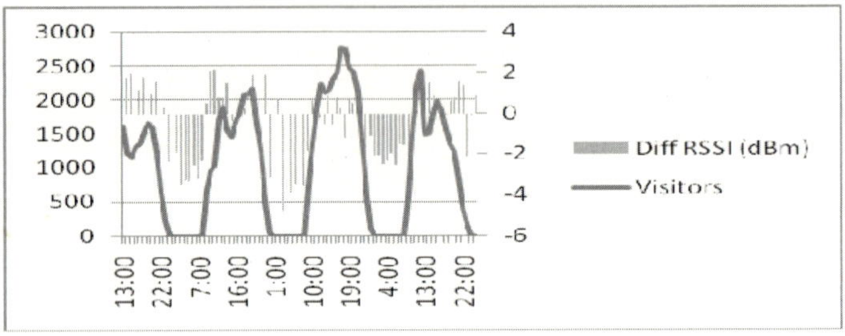

Fig. 3. Shopping center scenario - RSSI difference from its average and visitor number over time -channel 1

Analyzing figure 4, it is clear that the number of people roaming in the shopping centre has an effect of the RSSI value measured, since the difference from the average RSSI has its maximum values (about -4 dB) at night. Most of the time, the RSSI variation is not large enough to have a profound impact on the location system. A signal variation between 0 and 4 dB is a typical small scale fading value, which is observed on the measurements taken at the same point with 300 milliseconds intervals.

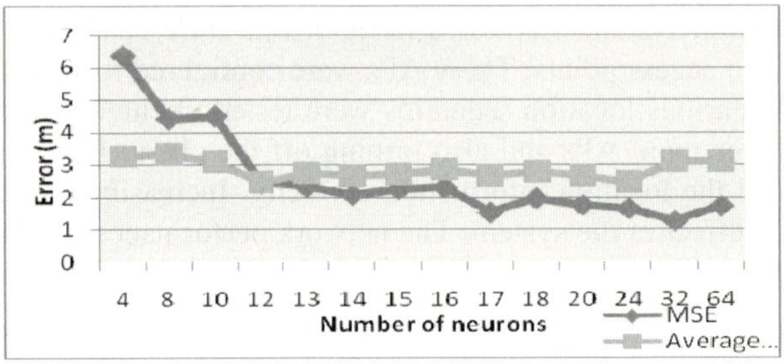

Fig. 4. Average location and training error versus number of neurons

The behavior of the artificial neural network, as mentioned before, has a dependency on the parameters values. The mean square error (MSE) value provides an understanding of the number of neurons that should be used. Crossing the MSE values with the average error gives the optimal number of neurons that, in this case, is near twelve (Figure 5).

In this situation the ideal number of neurons of the hidden layer was found to be twelve. This number makes a good commitment between the training MSE and the average error. Nevertheless, the number of neurons used in the hidden layer must be adapted to each situation and to the size of the training data set. During the location phase the wireless client is asked to perform RSSI measurements. The number of measurements has influence in the accuracy and duration of the location process. Increasing the number of measurements improves accuracy but increases the location duration. A good compromise between location response time and accuracy is a number that goes from five and eight readings, that corresponds to a 2 seconds time interval.

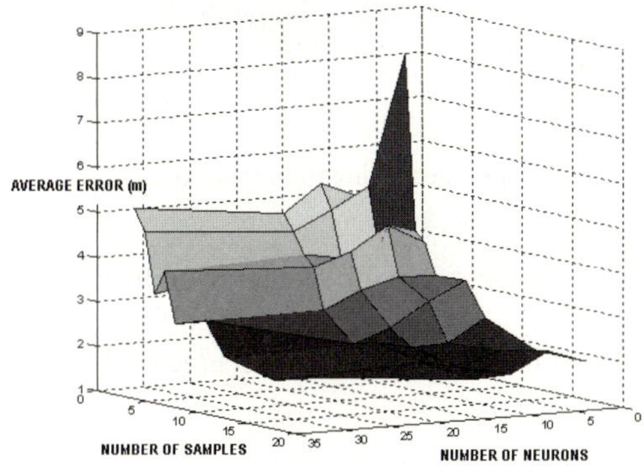

Fig. 5. Average error variation versus number of neurons and sampling size

Analyzing the behavior of the average error when there is a variation in the number of neurons in the ANN hidden layer and the number of samples (readings) used, it is possible to understand that there is an optimal neural network size. According to figure 6, the average error is minimized when the number of neurons of the neural network hidden layer is between 10 and 16. The sampling size also matters, as it is shown on the graphic, a low number of samples, has a negative impact on the location system performance. A low number of samples degrades the final average error values, compromising the system accuracy. This phenomena is explained when one thinks of radio propagation characteristics. Indoor propagation is always affected by effects like small scale fading, multipath, scattering and diffraction. Retrieving just one sample is just too low, since the location algorithm is based on the average RSSI values.

The mean square error is the output value of the neural network training. It is a figure of merit of the NN training and according to its value it is possible to conclude how well the NN has adapted to the input values. As figure 7 shows, MSE has large values when a low number of neurons is used. This is a natural behavior of the neural network, since there is a large set of data, and it is almost impossible to converge to a good final solution with a low number of neurons no matter the number of readings. A low number of readings also degrades the neural network capability of converging. The adequate number of readings should be higher than five. A higher number of readings does not significantly improve the neural network learning performance. The number of neurons used has a high impact on the mean square error of the learning phase.

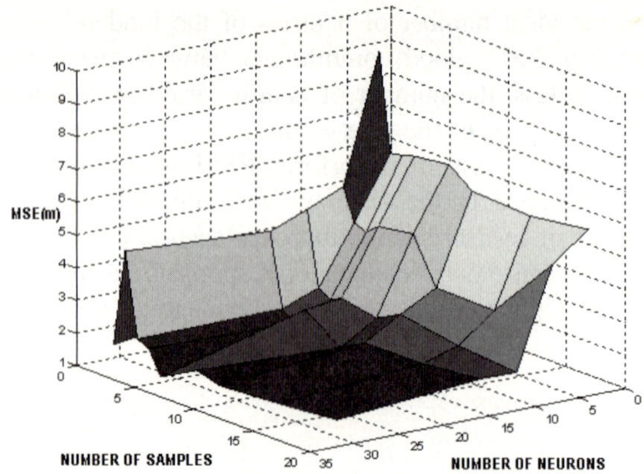

Fig. 6. MSE variation according to number of neurons and samples

It is important to study the behavior of the error in the location system. A low average error is not always a sign of optimal performance. The error histogram provides a clear insight of the location system accuracy.

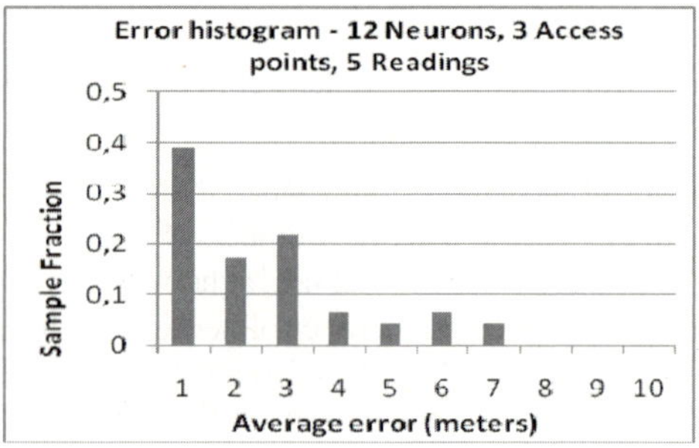

Fig. 7. Error histogram using 12 neurons, 3 access points and 5 readings

Using different settings, like the number of hidden neurons, access points and readings, it is possible to acquire the optimal settings for different scenarios and user requirements. Using a sufficient number of readings for an adequate accuracy (5 readings), the average error value is 2.4 meters (Figure 8). The maximum error value is 7 meters, and for about 80% of the samples, the error is below 3 meters. This result is improved increasing the number of readings to 20 (Figure 9), where the average error drops down to 1.9 meters.

Using additional measurements decreases the maximum error to five meters. According to figure 43 the location system offers an error less than one meter for 50% of the samples. Also for 90% of the samples the error is below three meters.

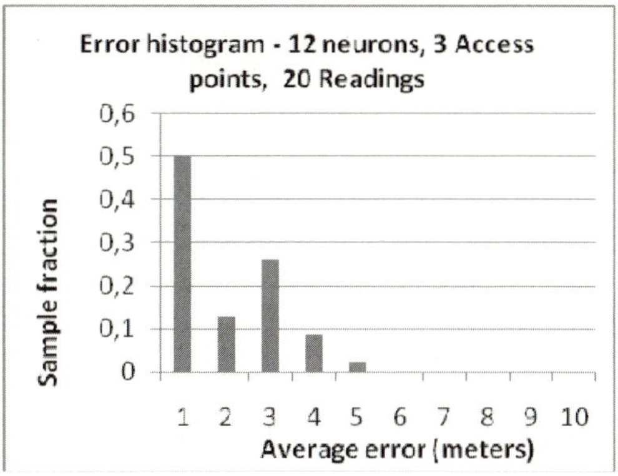

Fig. 8. Error histogram using 12 neurons, 3 access points and 20 readings

4 Conclusion and Future Work

This work provides a description of a wireless location system. Using a simple software solution, with no additional hardware, the location system has a good accuracy that can be improved by increasing the number of readings at the cost of lowering the location response time.

The use of RF fingerprinting is a valid solution to provide accurate positioning techniques. Nevertheless, it can be time consuming gathering measurements and profiling a large indoor scenario. This time was minimized using a simple and fast RSSI measurement and calibration tool, which can be used in an easy way. The usage of radio frequency fingerprinting requires employing mathematical approaches to solve the location problem. Methods like using propagation models and nearest neighbor were evaluated and it was concluded that their performance was inadequate. Artificial neural networks applied to the location paradigm offer sufficient adaptability between different scenarios, contrary to other algorithms used in other location applications. ANN based location algorithm provides enough flexibility and its final accuracy competes directly with the best known location applications.

The idea behind an indoor location system is its capability of providing accuracy while being simple to use. The location system developed can be easily deployed in existing WLAN with minimal cost and difficulty. Additionally it does not require any changes in the existing WLAN infrastructure, only a small software program must be installed in the client's device. This software does not interfere with the client's normal usage of its device. The system robustness is not compromised by minimal environmental changes in the indoor scenario, and it is immune to the multi-path and small scale fading effects, typical encountered on indoor radio propagation.

This location system is divided into two parts, the macro and the micro-location. Although some testing was made to ensure the interaction between those two systems, a final and efficient solution was not achieved. The main goal is to choose the adequate neural network according to the output of the macro-location system. Possible future work includes a seamless integration between the two systems.

Presently, the local system has been deployed in two different scenarios with an average location error below two meters. In the future, one can expect to improve its accuracy and lower the time required to perform the initial calibration.

References

[1] Caffery, J.J.: Wireless Location in CDMA Cellular Radio Systems, Norwell, EUA (1999)
[2] Ward, A., Jones, A., Hopper, A.: A new location technique for the active office. IEEE Personal Communications 4(5), 42–47 (1997)
[3] Priyantha, N.B., et al.: The Cricket Location-Support System. In: Proc. 6th Ann. Int'l Conf. Mobile Computing and Networking (MOBICOM 2000), pp. 32–43. ACM Press, New York (2000)
[4] Want, R., et al.: The Active Badge Location System. ACM Trans. Information Systems, 91–102 (1992)
[5] Werb, J., Lanzl, C.: Designing a positioning system for finding things and people indoors. IEEE Spectrum, 71–78 (1998)
[6] Hightower, J., et al.: SpotON: An indoor 3D location sensing technology based on RF signal strength.Technical Report UW CSE 00-02-02, Department of Computer Science and Engineering, University of Washington, Seattle, WA (February, 2000)
[7] Orr, Robert, et al.: The Smart Floor: a Mechanism for Natural User Identification and Tracking. Human Factors in Computing Systems, Georgia Institute of Technology (2000)
[8] Bahl, P., Padmanabhan, V.N.: RADAR: An in-building RF-based user location and tracking system. In: EEE Infocom 2000, Tel Aviv, Israel (March, 2000)
[9] Battiti, R., et al.: Location-aware computing: a neural network model for determining location in wireless lans. Technical Report DIT-5, Universit'a di Trento, Dipartimento di Informatica e Telecomunicazioni (2002)
[10] Rodriguez, M.D., et al.: Location-Aware Access to Hospital. IEEE Transaction on Information Technology in Biomedicine (December, 2004)
[11] David, M., et al.: Location Estimation in Wireless Networks. A. Rutgers University (2005)
[12] Andreas, H., et al.: Practical Robust Localization Over Large-Scale 802.11. MobiCom'04, 2004, Rice University. Philadelphia: ACM (2004)
[13] BSD Syslog Protocol - RFC 3164. IETF, http://tools.ietf.org/html/rfc3164
[14] Hagan, M.T., et al.: Neural Network Design. Boston, EUA, PWS (1996)
[15] Recurrent neural network design and applications, London: CRC Press (2001)

An Adaptive Neuro-Fuzzy Inference System for Calculation Resonant Frequency and Input Resistance of Microstrip Dipole Antenna

Siddik C. Basaran, Inayet B. Toprak, and Ahmet Yardimci

The Vocational School of Technical Sciences
Akdeniz University, Antalya, Turkey
cbasaran@akdeniz.edu.tr,
ibmutlu@akdeniz.edu.tr,
yardimci@akdeniz.edu.tr

Abstract. The accurate calculation of the resonance frequency and input resistance of microstrip antennas is a key factor to guarantee their correct behavior. In this paper we presented an adaptive neuro-fuzzy inference system (ANFIS) that calculates resonant frequency and input impedance of the microstrip dipole antenna's (MSDAs). Although the MSDAs' resonant frequency greatly depends on the dipole's length, it also depends on the dipole's width, the antenna substrate's permittivity value, and its size (which affects resonant frequency). Input impedance, like resonant frequency, changes with these parameters. According to test results accuracy of ANFIS is calculated 98.91% for resonant frequency while 95.81% for input resistance calculation.

Keywords: Microstrip dipole antenna, ANFIS, resonance frequency, input resistance.

1 Introduction

Technical literature has broadly investigated Microstrip patch antennas (MSPAs). These antennas are lightweight, aerodynamically conformable to aircraft and missile surfaces, compatible with solid-state devices, and simple and inexpensive to construct. Furthermore, by adding loads between the patch and the ground plane (i.e. pins and varactor diodes), one can design adaptive elements containing variable resonant frequency, impedance, polarization, and radiation pattern [1].

Nevertheless, narrow bandwidth is one major drawback of the microstrip antennas. Consequently, printed antennas work efficiently by closely matching their resonant frequency. Therefore, this parameter's accurate evaluation is fundamental in the microstrip antenna design process. A correct evaluation of the resonant frequency and the printed antenna's input resistance requires a rigorous full-wave model [2]. The Finite Element Method (FEM), the Method of Moments (MoM), and Finite Difference Method (FDM) have proved useful for analysis of such antennas by providing rigorous solutions to the present problem [3-5]. However, this technique

J. Marques de Sá et al. (Eds.): ICANN 2007, Part II, LNCS 4669, pp. 709–717, 2007.

requires significant computation and is time-consuming. Recently, studies developed alternative methods for resonant frequency determination using fuzzy logic (FL), neural networks (NNs), and combined adaptive neuro-fuzzy inference systems (ANFIS) [2,6-8].

We developed an adaptive neuro-fuzzy inference system that calculates resonant frequency and the microstrip dipole antenna's (MSDAs) input impedance. Although the MSDAs' resonant frequency greatly depends on the dipole's length, it also depends on the dipole's width, the antenna substrate's permittivity value, and its size (which affects resonant frequency). Input impedance, like resonant frequency, changes with these parameters.

The past two decades have witnessed significant advances in FL and NNs. Some have unsuccessfully used FL and NN separately to find the MSDA's resonant frequency and input impedance. The FL system's difficulty stems from constituting correct membership functions and rule base. Insufficient training sets resulted in NNs producing unstable results. The synergism of FL systems and NN produced a system capable of learning, thinking, and reasoning. This tool determines the imprecisely-defined complex system's behavior. The neuron-fuzzy system's purpose is to apply neural learning techniques to identify and tune the neuro-fuzzy system's parameters and structure. These neuro-fuzzy systems combine the benefits of these two powerful paradigms into a single capsule. Their multi-functionality makes them suitable for a wide range of scientific applications. Their strengths include fast and accurate learning, good generalization capabilities, excellent explanation facilities (formed by semantically meaningful fuzzy rules), and can accommodate both data and existing knowledge about any present problem.

ANFIS can find a model that closely matches the inputs with the target. Fuzzy interface system (FIS) is a knowledge representative where each fuzzy rule describes the system's local behavior. Viewing FIS as a feed forward network structure where the primary inputs and intermediate results are sent to compute the output allows us to apply the same back-propagation principle in the neural networks. The network structure that implements FIS is called ANFIS and employs hybrid learning rules to train a Sugeno-style FIS with linear rule outputs.

Among the various methodology combinations in soft computing, fuzzy logic and neuro-computing are the most common (hence the tem neuro-fuzzy systems). Such systems play an important role in the initiation of rules from observations. It is a powerful tool for quickly and efficiently dealing with imprecision and nonlinearity wherever it occurs. Neuro-adaptive learning techniques work similarly to neural networks. These techniques allow the fuzzy modeling procedure to learn information about a data set that computes the membership function parameters, allowing the associated fuzzy inference system to track the given input/output data. A neural-type structure similar to a neural network that maps inputs through input and output membership functions and associated parameters can be used to interpret the input/output map. This eliminates the normal feed forward multilayer network's disadvantages (difficult to understand or modify). We explain MSDAs and how to use ANFIS to train a fuzzy inference system that calculates both the resonant frequency and input impedance.

2 Microstrip Dipole Antennas

Rectangular microstrip antennas are classified according to their length-to-width ratio. An antenna with a narrow rectangular strip (strip width less than 0.05 λ) is a microstrip dipole. A broad rectangular antenna is called a microstrip patch [9]. Figure 1 shows the microstrip printed dipole antenna containing a conventional half-wave dipole loaded with two open-circuited stubs. The antenna, printed on a PCB substrate, is fed either by cable, surface mount connector, or printed transmission line. Where 'L' represents length, 'W' is dipole width, 'Ds' is distance between the dipole edges and substrate edges, H is the substrate thickness and "ε_r" is the substrate's dielectric's constant value.

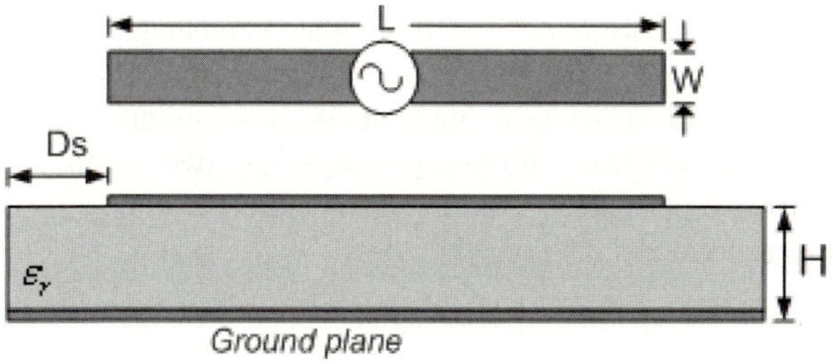

Fig. 1. Geometry of a microstrip dipole antenna

As Figure 1 shows, MSDAs, using the transmission line model, can be designed for the lowest resonant frequency. The formula calculating the value of L and W is shown below. The effective dielectric constant of a microstrip line is:

$$\varepsilon_e = \frac{\varepsilon_r + 1}{2} + \frac{\varepsilon_r - 1}{2}\left(\frac{1}{\sqrt{1 + 12\frac{H}{W}}}\right)$$

The microstrip's line length (L) is:

$$\lambda = \frac{c}{f\sqrt{\varepsilon_e}}$$

Where;

λ :wavelength

c :velocity of light

f :frequency

Thus $L = \frac{\lambda}{2}$.

We fixed the space between the symmetrical metal patches on the substrate at 1mm and the antenna is fed from this space. By changing the antenna parameters L, W, H, ε_r, and Ds, we obtained 61 antenna configurations, using 51 for training and the rest for testing. We used materials used for antenna design such as FR-4, RT/duroid and Rogers TMM. We obtained the antenna substrate's thickness and permittivity values from producer firms' catalogs. We used Ansoft High Frequency Structure Simulator (HFFS) software, based on FEM, to compute and test the data set.

3 Adaptive Neuro-Fuzzy Inference System

3.1 Structure of ANFIS

ANFIS is a multilayer neural network-based fuzzy system [7]. Its topology is shown in Figure 2, and the system has a total of five layers. In this connectionist structure, the input and output nodes represent the descriptors and the activity, respectively, and in the hidden layers, there are nodes functioning as membership functions (MFs) and rules. This eliminates the disadvantage of a normal feedforward multilayer network, which is difficult for an observer to understand or to modify. For simplicity, we assume that the examined fuzzy inference system has two inputs x and y and one output, the activity. To present the ANFIS architecture, two fuzzy if-then rules based on a first order Sugeno model are considered:

Rule 1: If ($x$is A_1)and ($y$is B_1) then ($f_1= p_1x+ q_1y+ r_1$)
Rule 2: If ($x$is A_2)and ($y$is B_2) then ($f_2= p_2x+ q_2y+ r_2$)

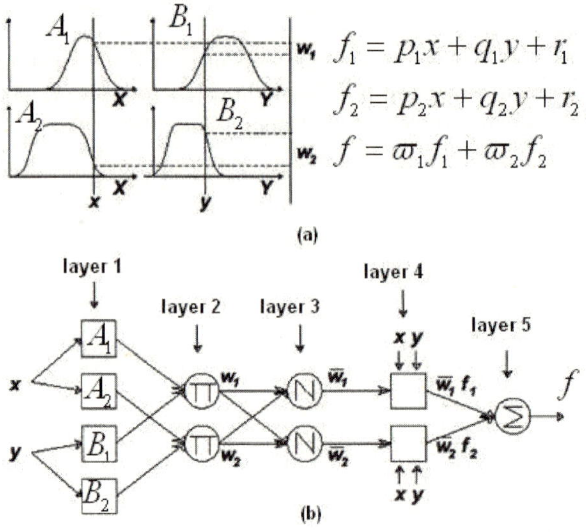

Fig. 2. (a) A two-input first-order Sugeno fuzzy model (b) equivalent ANFIS architecture

where x and y are the inputs, Ai and Bi are the fuzzy sets, $fi$are the outputs within the fuzzy region specified by the fuzzy rule, pi, qi and ri are the design parameters that are determined during the training process. In the first layer, all the nodes are adaptive nodes with anode function:

$$o_{1,i} = \mu_{Ai}(x), \ for\ i = 1,2$$

where x is the input to node i, and A_i is the linguistic label (low, high, etc.) associated with this node function. In other words, $O_{1,i}$ is the membership function of A_i, and it specifies the degree to which the given x satisfies the quantifier A_i. Usually we choose $\mu_{Ai}(x)$ to be bell-shaped with maximum equal to 1 and minimum equal to 0. As the values of the parameters $\{a_i, b_i, c_i\}$ change, the bell-shaped functions vary accordingly, thus exhibiting various forms of membership functions on linguistic label A_i. Parameters in this layer are referred to as premise parameters.

Every nodes in second layer is is a fixed node labeled Π (Figure 2), whose output is the product of all the incoming signals:

$$o_{2,i} = \omega_i = \mu_{A_i}(x) \times \mu_{B_i}(y), \ for\ i = 1,2$$

Each node output represents the firing strength of a rule.

In layer 3 every node is a fixed node labeled N. The ith node calculates the ratio of the ith rule's firing strength to the sum of all rules' firing strengths:

$$o_{3,i} = \varpi_i = \frac{\omega_i}{\omega_1 + \omega_2}, \quad i = 1,2$$

Outputs of this layer are called normalized firing strengths. Every node i in layer 4 is an adaptive node with a node function

$$o_{4,i} = \varpi_i f_i = \varpi_i (p_i x + q_i y + r_i)$$

where ω_i is a normalized firing strength from layer 3 and (p_i ,q_i , r_i) is the parameter set of this node. Parameters in this layer are referred to as consequent parameters. The single node in the last layer is a fixed node labeled Σ, which computes te overall output as the summation of all incoming signals:

$$\text{overall output} = o_{5,i} = \sum_i \varpi_i f_i = \frac{\sum_i \varpi_i f_i}{\sum_i \omega_i}$$

Thus we have constructed an ANFIS system that is functionally equivalent to first-order Sugeno fuzzy model.

3.2 Hybrid Learning Algorithm

From the proposed ANFIS architecture the overall output can be expressed as linear combinations of the consequent parameters. The output f in figure 2 can be written as:

$$f = \frac{\omega_1}{\omega_1 + \omega_2} f_1 + \frac{\omega_2}{\omega_1 + \omega_2} f_2 = \varpi_1 (p_1 x + q_1 y + r_1) + \varpi_2 (p_2 x + q_2 y + r_2)$$
$$= (\varpi_1 x) p_1 + (\varpi_1 y) q_1 + (\varpi_1) r_1 + (\varpi_2 x) p_2 + (\varpi_2 y) q_2 + (\varpi_2) r_2$$

which is linear in the consequent parameters p_1, q_1, r_1, p_2, q_2, and r_2. To train the above ANFIS system, the following error measure will be used:

$$E = \sum_{k=1}^{n} (f_k - {}^{\wedge}f_k)^2$$

where f_k and $^{\wedge}f_k$ are the kth desired and estimated outputs, and n is the total number of pairs (inputs-outputs) of data in the training data set. The learning algorithms of ANFIS consist of the following two parts: (a) the learning of the premise parameters by back-propagation and (b) the learning of the consequence parameters by least-squares estimation [7]. More specifically, in the forward pass of the hybrid learning algorithm, functional signals go forward till layer 4 and the consequent parameters are identified by the least squares estimate. In the backward pass, the error rates propagate backward, and the premise parameters are updated by the gradient descent. During the learning process, the parameters associated with the membership functions will change. The computation of these parameters is facilitated by a gradient vector, which provides a measure of how well the fuzzy inference system is modeling the input/output data for a given set of parameters. It has been proven that this hybrid algorithm is highly efficient in training the ANFIS [13,14]. Therefore, in the present study the proposed ANFIS model was trained with the backpropogation gradient descent method in combination with the least –squares method.

4 Result and Discussion

We presented a new approach based on ANFIS to calculate MSDP's resonant frequency and input impedance. We used two ANFIS classifiers to calculate the resonant frequency and input impedance. Each ANFIS classifier was specially trained for each problem. Sixty-one different antenna parameters were obtained by changing the antenna parameters L, W, H, ε_r, and D_S. The data set was divided into two separate data sets: 51 were used for training while rests were used for testing. Table 1 shows system training and Table 2 shows test data. The data from Table 1 is important to antenna designers because it presents a number of antenna features for different frequencies. All data sets obtained by 3-D simulations used Ansoft High Frequency Structure Simulator (HFFS). All materials that used for this study came from producer firms' catalogs. We used the training data set to train each ANFIS, whereas the testing data set was used to verify the accuracy and effectiveness of each trained ANFIS model.

We used triangular membership functions for resonant frequency calculations and Gaussian membership functions defined below for input resistance:

$$\mu_{A_i} = \frac{1}{1 + \left\{ \left(\frac{x - c_i}{a_i} \right)^2 \right\}^{b_i}}$$

where a_i, b_i and c_i are the membership function's parameters.

Table 1. Part of the training set used for ANFIS

					Frequency [GHz]		Resistance [Ohm]	
ε_r	H [mm]	W [mm]	L [mm]	Ds [mm]	HFSS Results	ANFIS Result	HFSS Results	ANFIS Result 1.0e+003 *
2.1	1.6	3	33	15	5.81	5.8043	1850	1.8500
3.27	1.6	3	33	15	4.87	4.8899	1763	1.7630
4.40	1.6	3	33	15	4.31	4.2996	1682	1.6937
4.90	1.6	3	33	15	4.11	4.1289	1642	1.6420
6.00	1.6	3	33	15	3.77	3.7700	1520	1.5200
9.20	1.6	3	33	15	3.09	3.0900	1306	1.3060
4.40	0.457	3	33	15	4.44	4.4600	730	0.7300
4.40	0.965	3	33	15	4.43	4.3883	1608	1.6080
4.40	3.175	3	33	15	4.09	4.0844	1341	1.3410
4.40	1.6	1	33	15	4.65	4.6500	2889	2.8890
4.40	1.6	2	33	15	4.49	4.4900	2257	2.2570
4.40	1.6	5	33	15	4.23	4.2300	1139	1.1390
4.40	1.6	3	19	15	6.47	6.4710	960	0.9627
4.40	1.6	3	43	15	3.45	3.4431	1945	1.9274
4.40	1.6	3	59	15	2.55	2.5322	2005	1.9926
4.40	1.6	3	67	15	2.25	2.2559	2100	2.1039
4.40	1.6	3	33	5	4.28	4.2798	915	0.9150
3.27	6.35	4	51	15	2.88	2.8800	918	0.9180
9.20	1.91	1.75	15	8	6.01	6.0100	863	0.8630
3.75	3	4.5	37	16	3.85	3.8500	1244	1.2440
2.2	5.08	4.25	43	12	3.92	3.9200	905	0.9050

Each of ANFIS used 51 training data sets in 300 training periods. Error tolerance was 0 and the output membership function was linear. L,W,H, ε_r and Ds were inputs of each ANFIS. The outputs were f, for the first and R_{in} for the second. We used three linguistic terms for each input.

After training, 10 testing data sets were used to validate the ANFIS model's accuracy for the resonant frequency and input resistance's calculation. The testing data consisted of different types of material values and varied in size. Figures 3 and 4 show the comparisons for resonant frequency. While Figure 3 compares the results between the training data set and FIS output, Figure 4 shows the testing set and FIS output. Figures 5 and 6 compare the results of training and test data sets with FIS output for input resistance.

Table 2. Test data set of ANFIS and results

					Frequency [GHz]			Resistance [Ohm]		
ε_r	H [mm]	W [mm]	L [mm]	Ds [mm]	HFSS Results	ANFIS Results	Error (%)	HFSS Results	ANFIS Results	Error (%)
3.75	2.50	3.00	33	15	4.46	4.41	1.07	1490	1612.8	8.23
4.40	3	4.00	33	15	4.00	3.92	1.83	1167	1195.6	2.45
2.10	2.50	4.50	23	20	6.96	6.90	0.83	599	607.7	1.45
4.9	1.91	2.5	29	16	4.68	4.68	0.12	1693	1578.8	6.74
6	1.6	3	47	15	2.73	2.68	1.60	1778	1814.5	2.05

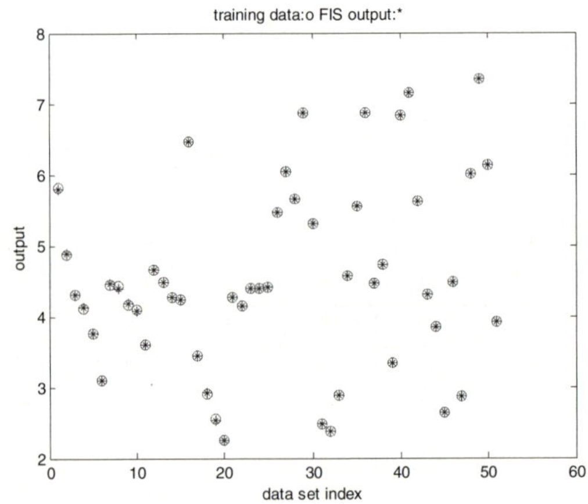

Fig. 3. Comparison of training data set results and FIS results for resonant frequency

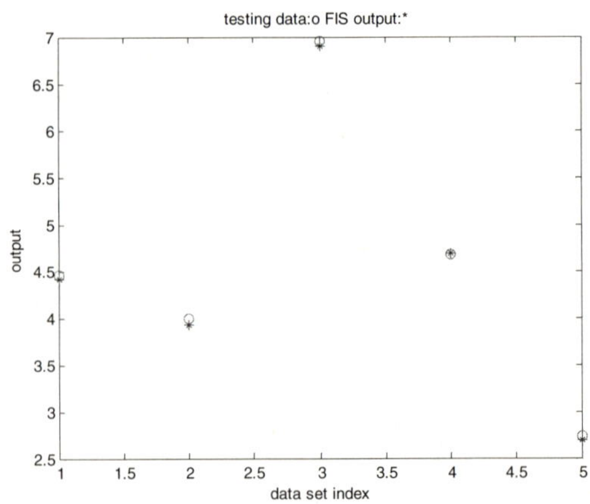

Fig. 4. Comparison of testing data set results and FIS results for resonant frequency

The ANFIS model's test performance was defined after comparing the simulation results and FIS results. In Table 1, the f-labeled column shows the simulator results while the ANFIS-labeled column shows the FIS results. In Table 2, the ANFIS model calculates the resonance frequency with 1.07%, 1.83%, 0.83%, 0.12%, and 1.60% with relative error value, respectively. These results show the ANFIS's mean accuracy at 98.91% with the minimum accuracy as 98.17%. The second ANFIS model's mean accuracy was 95.81%. Our proposed system performed better in calculating resonant frequency than input impedance. These results signify that the proposed ANFIS has the potential in calculating MSDAs' resonant frequency and input resistance.

Acknowledgment. This study was financially supported by Akdeniz University Scientific Research Projects Management Unit.

References

1. Pozar, D.: Microstrip antennas. Fellow, IEEE 80, 79–91 (1992)
2. Angiulli, G., Versaci, M.: Resonant Frequency Evaluation of Microstrip Antennas Using a Neural-Fuzzy Approach. IEEE Transactions on Magnetıcs 39(3) (2003)
3. Volakis, J., Özdemir, T., Gong, J.: Hybrid Finite Element Metodologies for Antennas and Scatterring. IEEE Transactıons on Antennas And Propagatıon 45(3) (1997)
4. Deshpande, M., Shively, D., Cockrell, C.R.: Resonant frequencies of irregularly shaped microstrip antennas using method of moments. NASA Tech. Paper 3386 (1993)
5. Meijer, C.A., Reader, H.C.: Analysıs Of Mıcrostrıp Patch Antennas Usıng A Finite Dıfference Technıque. Electronıcs Letters, (18th, February 1993), vol. 29(4) (1993)
6. Guney, K., Sagiroglu, S., Erler, M.: Generalized neural method to determinate resonant frequencies of various microstrip antennas. International Journal of RF and Microwave Computer-Aided Engineering 12, 131–139 (2002)
7. Jang, J.-S.R.: ANFIS: Adaptive-network based fuzzy inference system. IEEE Trans. Syst. Man Cybern. 23(3), 665–685 (1993)
8. Jang, J.-S.R.: Self-learning fuzzy controllers based on temporal backpropagation. IEEE Trans. Neural Network 3(5), 714–723 (1992)
9. Garg, R., Bhartia, P., Bahl, I.J., Ittipiboon, A.: Microstrip Antenna Design Handbook, Artech House, Dedham, MA (2000)

GARCH Processes with Non-parametric Innovations for Market Risk Estimation*

José Miguel Hernández-Lobato, Daniel Hernández-Lobato, and Alberto Suárez

Escuela Politécnica Superior,
Universidad Autónoma de Madrid,
C/ Francisco Tomás y Valiente, 11, Madrid 28049 Spain
{josemiguel.hernandez,daniel.hernandez,alberto.suarez}@uam.es

Abstract. A procedure to estimate the parameters of GARCH processes with non-parametric innovations is proposed. We also design an improved technique to estimate the density of heavy-tailed distributions with real support from empirical data. The performance of GARCH processes with non-parametric innovations is evaluated in a series of experiments on the daily log-returns of IBM stocks. These experiments demonstrate the capacity of the improved estimator to yield a precise quantification of market risk.

1 Introduction

Market risk is associated with losses in the value of a portfolio that arise from unexpected fluctuations in market prices (security prices) or market rates (interest or exchange rates) [1]. The quantification of market risk is an important tool used by risk analysts to understand, quantify and manage the risk profile of an investment. Risk measures such as VaR and Expected Shortfall are employed to assist in decisions about how much risk an institution is willing to take, to identify risk factors that may be particularly harmful, and for regulatory purposes. The process of measuring risk involves two steps. In the first step, the distribution of future losses and gains for the portfolio value for a time horizon τ in the future (one day, for instance) is modeled. The usual approach consists in assuming that the statistical description for the price changes in the future is similar to that of changes in the recent past. A parametric form for the distribution of the portfolio returns is assumed. The parameters of this model are then selected by performing a fit to historical data. In the second step, measures that quantify the risk associated to extreme losses are computed from this distribution. The most common measure is Value at Risk (VaR), which is a percentile of the distribution at a given probability level p (usually high, e.g. 95% or 99%). [1]. Assuming a time horizon of a day, the value of VaR is the threshold above which one should expect to observe losses $100-p$ days out of a hundred days, on

* This work has been supported by *Consejería de Educación de la Comunidad Autónoma de Madrid*, *European Social Fund*, *Universidad Autónoma de Madrid* and *Direccin General de Investigacin* under project TIN2004-07676-C02-02.

J. Marques de Sá et al. (Eds.): ICANN 2007, Part II, LNCS 4669, pp. 718–727, 2007.

average. Expected Shortfall (ES) is an alternative measure of risk that estimates the expected value of losses conditioned to being above the threshold given by VaR.

The calculation of VaR or ES for a known probability distribution of portfolio returns at horizon τ is relatively straightforward. However, the intrinsic randomness in the behavior of the market, the difficulty in modeling the distribution of extreme events using limited empirical evidence, and the time-dependence of the volatility [2] make it difficult to obtain accurate estimates of the distribution of returns (especially in the tail), and therefore of VaR and ES. In this investigation we use a family of GARCH processes [3,4] with non-parametric innovations whose parameters are estimated in a transformed space to account for the leptokurtosis (heavy tails) of the returns distribution. These processes provide an accurate statistical description of the extreme losses in the tail of the conditional distribution of daily log-returns. Furthermore, they can be used to compute measures of risk that are very precise.

2 Models for Time Series of Financial Asset Prices

Consider the time series of daily prices of a financial asset $\{S_t\}_{t=0}^{T}$. In general, the time-series of financial product prices are non-stationary. For this reason, it is common to model the time series of logarithmic returns

$$r_t = 100 \cdot log\left(\frac{S_t}{S_{t-1}}\right), t = 1, 2, \ldots, T. \tag{1}$$

which, besides being quasi-stationary, has other desirable properties [2].

Typically, the autocorrelations between returns are small and short-lived. By contrast, the time series of financial asset returns are usually heteroskedastic. That is, the volatility or standard deviation of the returns exhibits a time-dependent structure [5]. This is the well-known phenomenon of *volatility clustering*: Large price variations (either positive or negative) are likely to be followed by price variations that are also large. This phenomenon is evidenced by a plot of the autocorrelations in the powers of the absolute values of returns

$$C_\delta = corr(|r_{t+h}|^\delta, |r_t|^\delta), h = 1, 2, \ldots, \tag{2}$$

These correlations are positive for various delays in the range of weeks to months. The highest values are usually achieved for $\delta = 1$ [6]. This behavior, known as the Taylor effect, is displayed in Fig. 1 for the returns of IBM stocks.

GARCH processes are time-series models that have been proposed to account for the time-dependent structure of the volatility in financial time series [3,4]. In this work we consider power GARCH(1,1,δ) processes, which are a generalization of the standard GARCH(1,1) process [6]. A time series $\{r_t\}_{t=1}^{T}$ follows a power GARCH(1,1,δ) process with normal innovations if

$$\begin{aligned} r_t &= \sigma_t \varepsilon_t \\ \sigma_t^\delta &= \gamma + \alpha |r_{t-1}|^\delta + \beta \sigma_{t-1}^\delta , \end{aligned} \tag{3}$$

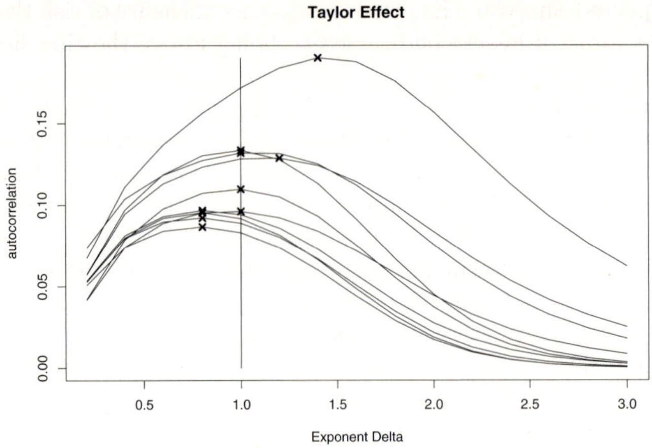

Fig. 1. Plots of the autocorrelation function between $|r_t|^\delta$ and $|r_{t+h}|^\delta$ for $h \in \{1, ..., 10\}$. Each curve corresponds to a different value of $h = 1, 2, \ldots, 10$. The series $\{r_t\}_{t=1}^{T}$, with $T = 9190$ values, corresponds to the standardized returns of IBM stocks from 1962/07/03 to 1998/12/31. The maximum of each function is shown with a cross. All maxima are located close to the value $\delta = 1$.

where $0 < \delta \leq 2$, $\gamma > 0$, $\alpha \geq 0$, $\beta \geq 0$, $\alpha + \beta < 1$ and the innovations $\varepsilon_t \sim \mathcal{N}(0,1)$ are distributed according to a standard normal distribution. The condition $\alpha + \beta < 1$ is sufficient to guarantee the existence of $\mathbb{E}[\sigma_t^\delta]$ and $\mathbb{E}[|r_t|^\delta]$ for any value of δ [6]. Power GARCH processes take into account the correlation between $|r_{t+1}|^\delta$ and $|r_t|^\delta$. The usual choice is $\delta = 2$, which corresponds to the standard GARCH process. The parameters of the model are then estimated by maximizing the model likelihood with standard optimization algorithms that generally employ gradient descent. Nevertheless, an empirical analysis of the data (see Fig. 1) suggests using a value of δ closer to 1. In the experiments carried out in the present investigation, the value $\delta = 1$ is chosen and an optimization method that does not need to compute the gradient is used to maximize the likelihood. The value of $\delta = 1$ is preferred to $\delta = 2$ to allow for the possibility of infinite-variance models for the innovations (for instance, if the innovations are assumed to follow a stable distribution [7], whose second and higher moments do not exist).

If r_t exhibits correlations with r_{t+h} for a range of values of h (usually h small) it is possible to add an autoregressive component to the GARCH process. The simplest one is an AR(1) process

$$\begin{aligned} r_t &= \phi_0 + \phi_1 r_{t-1} + \sigma_t \varepsilon_t \\ \sigma_t &= \gamma + \alpha |r_{t-1} - \phi_0 - \phi_1 r_{t-2}| + \beta \sigma_{t-1}\,, \end{aligned} \tag{4}$$

where $|\phi_1| < 1$ is a necessary condition for stationarity. Assuming Gaussian innovations, the likelihood for the parameters of this GARCH process given a time series $\{r_t\}_{t=1}^{T}$ is

$$\mathcal{L}(\boldsymbol{\theta}|\{r_t\}_{t=1}^T) = \prod_{t=1}^{T} \psi\left(\frac{u_t}{\sigma_t}\right)\frac{1}{\sigma_t}, \tag{5}$$

where $\boldsymbol{\theta} = \{\phi_0, \phi_1, \gamma, \alpha, \beta\}$, $u_t = r_t - \phi_0 - \phi_1 r_{t-1}$ is the empirical autoregressive residual and $\psi(x)$ is the standard Gaussian density function. To evaluate (5) we define $u_0 = 0$, $u_1 = r_1 - \hat{\mu}$ and $\sigma_0 = \hat{\sigma}$ where $\hat{\sigma}$ and $\hat{\mu}$ are the sample standard deviation and sample mean of $\{r_t\}_{t=1}^T$.

If a GARCH process like the one described in (4) is fitted to the daily returns of a financial asset, the model captures the time-dependent volatility quite accurately. However, the empirical innovations of the model $u_t/\sigma_t = (r_t - \phi_0 - \phi_1 r_{t-1})/\sigma_t$ do not follow a standard Gaussian distribution. In particular, the empirical distribution of the residuals has larger kurtosis (i.e., heavier tails) than the Gaussian distribution. This mismatch in the loss tail of the distribution for the innovations reflects the fact that the model is unable to correctly describe extreme events, causing it to severely underestimate measures of market risk such as VaR or Expected Shortfall. We propose to address this shortcoming by estimating the parameters of the GARCH process (4) assuming that the innovations ε_t follow an unknown distribution f which is estimated in a non-parametric way.

3 GARCH Processes with Non-parametric Innovations

The goal is to maximize (5) replacing ψ by f, the (unknown) density function for the innovations $\{\varepsilon_t = u_t/\sigma_t\}_{t=1}^T$. One possible solution to the problem is to use a non-parametric kernel density estimator $\hat{f}$:

$$\hat{f}(x; \{c_i\}_{i=1}^N) = \frac{1}{Nh}\sum_{i=1}^{N} K\left(\frac{x - c_i}{h}\right), \tag{6}$$

where $\{c_i\}_{i=1}^N$ are the centers of N Gaussian kernels denoted by K and h is the smoothing parameter. If $c_1, ..., c_N \sim f$, $h \to 0$ and $Nh \to \infty$ as $N \to \infty$, then $\hat{f}$ can approximate f to any degree of precision [8]. If h is optimally chosen then the Asymptotic Mean Integrated Square Error (AMISE) between f and $\hat{f}$ converges like $O(N^{-4/5})$ [9]. The parameter h can be automatically fixed (under some assumptions) as a function of $\{c_i\}_{i=1}^N$ following Silverman's *rule of thumb* [9]. The function $\hat{f}$ can be seen as a radial basis function network [10] that approximates the actual density. The network uses Gaussian densities as basis functions with standard deviation h and centers $\{c_i\}_{i=1}^N$ (the only free parameters); there is no bias term in the network and the output weights are $1/N$. Alternatively, $\hat{f}$ can be interpreted as a mixture of N Gaussians model.

If ψ is replaced by $\hat{f}$ in (5) then

$$\mathcal{L}(\boldsymbol{\theta}, \{c_i\}_{i=1}^N|\{r_t\}_{t=1}^T) = \prod_{t=1}^{T} \hat{f}\left(\frac{u_t}{\sigma_t}; \{c_i\}_{i=1}^N\right)\frac{1}{\sigma_t}, \tag{7}$$

where the parameters of the model are $\boldsymbol{\theta}$ and $\{c_i\}_{i=1}^N$. Assuming that there are as many Gaussians with centers $\{c_i\}_{i=1}^N$ as training data $\{r_t\}_{t=1}^T$ (N = T) and that h is held fixed, (7) can be maximized by iterating the following steps: First, (7) is maximized with respect to each of the $\{c_t\}_{t=1}^T$ holding $\boldsymbol{\theta}$ fixed. This is accomplished by setting $c_t = u_t/\sigma_t$, $t = 1, 2, \ldots, T$. Second, (7) is maximized with respect to $\boldsymbol{\theta}$ (holding $\{c_t\}_{t=1}^T$ fixed) using a non-linear optimizer. This last step is the same as the one used for calibrating a standard GARCH process. A maximum of (7) is obtained by iterating these steps until the likelihood shows no significant change in two consecutive iterations. Note that the parameter h in the model cannot be determined by maximum likelihood. The reason is that the value of the likelihood increases without bound as h tends to zero. A suitable value of h at the end of the first step is computed using Silverman's *rule of thumb*, as a function of $\{c_t\}_{t=1}^T$. If T is sufficiently large, at convergence, $\hat{f}$ should be an accurate non-parametric estimate of the true distribution f. Note that after the first step in each iteration, $\hat{f}$ can be interpreted as a kernel density estimator of a density g where $u_1/\sigma_1, ..., u_T/\sigma_T \sim g$. This is the basis for the analysis carried out in the next section.

3.1 Density Estimation for Heavy-Tailed Distributions

Density estimators based on Gaussian kernels like $\hat{f}$ do not work well when heavy-tailed distributions are estimated. The difficulties are particularly severe in the modeling of extreme events, which are crucial in the quantification of risk. The origin of this shortcoming is that samples from heavy-tailed distributions usually include very few points in the tails. The estimator designed in the previous section tends to assign very low probability to regions with spare samples.

One possible solution to this problem consists in performing the density estimation in a transformed space where the kernel estimator is believed to perform better [11]. Assuming such a transformation is known, the non-parametric estimator $\hat{f}$, which models the density in the transformed space, is

$$\hat{f}(x; \{c_i\}_{i=1}^N; g_{\boldsymbol{\lambda}}) = |g'_{\boldsymbol{\lambda}}(x)| \frac{1}{Nh} \sum_{i=1}^N K\left(\frac{g_{\boldsymbol{\lambda}}(x) - g_{\boldsymbol{\lambda}}(c_i)}{h}\right), \tag{8}$$

where $g_{\boldsymbol{\lambda}}$ maps the original space to the transformed space and $\boldsymbol{\lambda}$ is a set of parameters that specify $g_{\boldsymbol{\lambda}}$ within a family of monotonic increasing transformations. This process of estimating the density in a transformed space is similar to standard density estimation based on kernels with varying widths [11].

The key aspect that should be considered in the choice of $g_{\boldsymbol{\lambda}}$ is that if $c_i \sim f$ then $g_{\boldsymbol{\lambda}}(c_i)$ should follow a distribution close to normality [11]. A trivial transformation that achieves this is $\Phi^{-1}(F(c_i)) \sim \mathcal{N}(0, 1)$, where F is the accumulated distribution for f and Φ^{-1} is the inverse of the standard Gaussian distribution. The difficulty is that if we knew F then we would also know f as $f = F'$ and the estimation process would not be necessary. Nevertheless, it is possible to approximate F by means of a parametric distribution $\hat{F}_{\boldsymbol{\lambda}}$ that can account for the heavy tails in F. The parameters $\boldsymbol{\lambda}$ of $\hat{F}_{\boldsymbol{\lambda}}$ can be estimated by maximum likelihood to $\{c_i\}_{i=1}^N$. The final transformation is $g_{\boldsymbol{\lambda}}(x) = \Phi^{-1}(\hat{F}_{\boldsymbol{\lambda}}(x))$. In this manner, a hybrid estimation is performed where the parametric model corrects for the tails of the non-parametric estimator.

To incorporate this idea into the algorithm proposed in Section 3 the first step of the algorithm is divided into two parts. In the first one the parameter $\boldsymbol{\lambda}$ of the transformation $g_{\boldsymbol{\lambda}}$ is found by maximizing the likelihood given the values $\{u_t/\sigma_t\}_{t=1}^T$ observed and $\hat{F}_{\boldsymbol{\lambda}}$. In the second part, the density of $\{u_t/\sigma_t\}_{t=1}^T$ is estimated by means of (8) where h is now determined using Silverman's *rule of thumb* in the *transformed space* as a function of $\{g_{\boldsymbol{\lambda}}(u_t/\sigma_t)\}_{t=1}^T$. The second step of the algorithm remains unchanged: $\hat{f}(x\,;\,\{c_i\}_{i=1}^N)$ is replaced in (7) by (8).

One technical problem that remains is that the location and scale of $u_1, ..., u_T$ can be fixed in two different ways. One through the parameters $\boldsymbol{\theta}$ of the GARCH process and another one through $\hat{f}$ and the centers of the kernels $\{c_t\}_{i=1}^N$. This can lead to convergence problems and should be avoided. To address this problem $\hat{f}$ is forced to have always the same location and scale. This is achieved by standardizing $\{u_t/\sigma_t\}_{t=1}^T$ before the first step on each iteration of the algorithm.

Finally, a parametric family of distributions $\hat{F}_{\boldsymbol{\lambda}}$ needs to be selected to fully specify the family of transformations $g_{\boldsymbol{\lambda}}(x) = \Phi^{-1}(\hat{F}_{\boldsymbol{\lambda}}(x))$. The choice used in our investigation is the family of stable distributions [7] which should have enough flexibility to account for the empirical properties of the marginal distribution of returns [5]. This family of distributions is parameterized in terms of a location parameter, a scale parameter, a parameter describing the decay of the tails and, finally, a parameter allowing each tail to have a different behavior. Other possible alternatives would be normal inverse Gaussian distributions [12], hyperbolic distributions [13] or the superclass of these last two families: the generalized hyperbolic distributions [14].

4 GARCH Processes with Stable Innovations

At each iteration of the proposed algorithm we are estimating the density g, where $u_1/\sigma_1, ..., u_T/\sigma_T \sim g$, twice. First, in a parametric way assuming g belongs to the family of stable distributions and, second, in a non-parametric way making use of the former estimate. It is possible that using only the first parametric estimate a better or equivalent solution than the one obtained by means of the non-parametric technique is reached. This would happen if the innovations of the GARCH process in (4) were actually stable. In this case ε_t would follow a stable distribution with 0 as location parameter and 1 as scale parameter: $\varepsilon_t \sim \mathbf{S}(a, b, 1, 0; 0)$, where a is the index of stability, or characteristic exponent, b is the skewness parameter, and the scale and location parameters are held fixed and equal to 1 and 0, respectively. The last parameter with value 0 indicates that the first parameterization proposed by Nolan [7] is used. To account for this situation two alternative models are compared in our experiments. One where the distribution of the innovations is estimated in a non-parametric way as described in Section 3 and another one where the innovations are assumed to be stable. In this way, by comparing the performance of both models it is possible to determine whether implementing the non-parametric estimate represents an actual improvement. The parameters of the GARCH model with stable innovations, $\boldsymbol{\theta}$, a and b are also estimated by maximum likelihood.

5 Model Validation and Results

To assess the capacity of the proposed models to accurately quantify market risk the general backtesting procedure described in [15] is followed. We perform a sliding window experiment on the daily returns of IBM stocks from 1962/07/03 to 1998/12/31, a total of 9190 measurements (see Fig. 3). From the series of returns 8190 overlapping windows with T = 1000 elements each are generated. Each window is equal in length to the previous window but is displaced forward by one time unit (one day). The parameters of the models are estimated by maximum likelihood using the data included in each window. Finally, the performance of the models is tested on the first point to the right of the training window. The testing process consists in calculating the cumulative probability that a model assigns to the first point out of the window and then applying it the inverse of the standard Gaussian distribution. This way, we obtain 8190 *test measurements* for each model that should follow a standard Gaussian distribution under the null hypothesis that the model is accurate. The application of the statistical tests described in [15] to those *test measurements* allows us to validate the models.

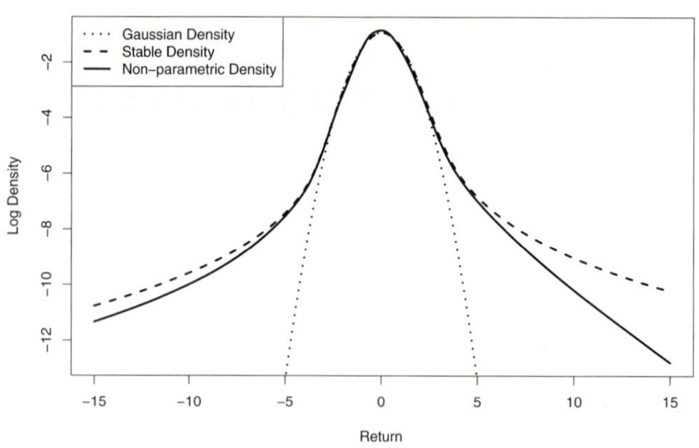

Fig. 2. Plots in logarithmic scale of the different model density functions. The non-parametric density and the stable density have been obtained as averages of the densities estimated in the sliding window experiment. To compute these averages only the densities for one in every fifty windows are used (a total of 163 densities). The stable densities are scaled so that their standard deviation is 1.

The estimation techniques designed are implemented in R [16]. The non-linear optimization method used to maximize the likelihood was the *downhill simplex method* included in the function *constrOptim*. To avoid that the optimization process gets trapped in a local maximum, the algorithm is rerun using different initial random values, and the best solution found is selected. Stable distributions have no general known closed form expressions for their density [7]. To evaluate the stable density we employ the technique described in [17]. The density is computed exactly on a grid of 2^{13} equally spaced points. Outside this grid the density function is calculated by linear interpolation from the values in the grid. To evaluate the accumulated stable distribution we use the R package *fBasics* [18] that implements the method described by Nolan [7]. Finally, the parameter a of the stable densities is restricted to be greater than 1.5, a value which should be sufficiently low to account for the heavy tails in the distribution of returns.

Table 1 displays, for each model, the results of the statistical tests performed over the *test measurements* obtained as described in Section 5. The level of significance considered to reject the null hypothesis on each test is 5%. The GARCH model with stable innovations fails all the tests but the one for Exceedances at the 95% level. The *p-values* obtained by this model for the tests Exc 99% and ES 95% are rather low. In particular, the result for ES 95% reflects the failure of the model to accurately describe the loss tail of the conditional distribution of returns. The results of the GARCH model with non-parametric innovations are remarkable. The model obtains very high *p-values* in all tests except for one. In particular, the test for Expected Shortfall at the 99% level fails. This is the most demanding test and it requires the model to correctly account for extreme events. In our case the test fails because of one single extreme event, a loss of around 5% that took place on January 19, 1970 (see Fig. 3). Immediately prior to that observation, the value estimated for a in the stable distribution that characterizes the transformation g_λ is fairly high 1.97, indicating that the tails are not very heavy and that the local fluctuations of the returns were expected to be small in size (smaller than the loss observed) with a high degree of confidence. If this point is removed, the *p-value* obtained in the test ES 99% is 0.31 and all the other tests give results above the level of significance 0.5. It is also interesting to see that the model can properly account for the 25% loss that took place on October 19, 1987 (Black Monday, see Fig. 3). The difference with the previous instance is that the value estimated for the parameter a right before Black Monday is fairly low (1.88), which signals that the distribution tails are very heavy and that large fluctuations have a non negligible chance of being observed. This low value of a originates in a sequence of large, but not extreme, fluctuations in the period immediately before Black Monday.

Figure 2 displays the average stable density and average non-parametric density obtained in the sliding window experiment. The tails of the stable and non-parametric densities are heavier than those of the standard Gaussian density. Moreover, those two distributions are not symmetric, both densities assign a higher probability to returns close to 5 than to returns close to -5. However, the right tail of the non parametric density displays a fast decay for returns greater

Table 1. *p-values* obtained from the statistical tests performed. From left to right tests for Value at Risk, Exceedances and Expected Shortfall at levels 99% and 95%. See [15] for a description of the tests.

Model	VaR 99%	VaR 95%	Exc 99%	Exc 95%	ES 99%	ES 95%
stable	1.5e-3	0.01	3.7e-5	0.08	0.026	4e-4
non-parametric	0.51	0.66	0.49	0.86	0.042	0.31

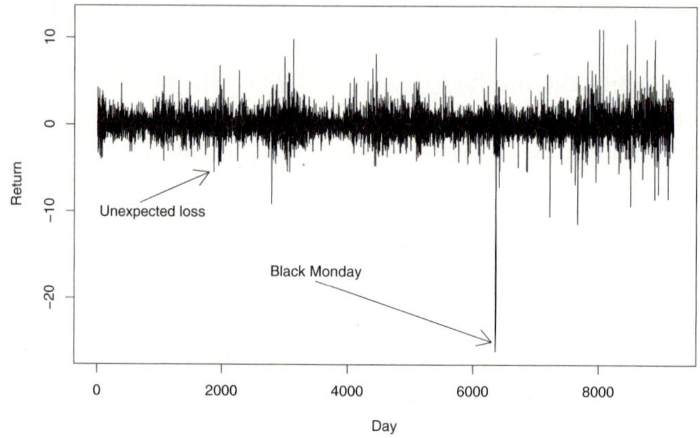

Fig. 3. Returns of IBM stocks from 1962/07/03 to 1998/12/31. The loss on January 19, 1970 has been singled out. The model proposed has difficulties to account for such a loss. The loss on October 19, 1987 known as Black Monday, has also been marked. This loss is correctly modeled.

than 5. This decay does not appear in the right tail of the stable density. This Figure also shows how the tails of the non-parametric density are on average slightly less heavy than those of the stable distribution, specially in the right tail, corresponding to gains. This is probably the reason for the better performance of the non-parametric GARCH process and would agree with previous studies based on extreme value theory, which suggest that the tails of stable densities are probably too heavy to model asset returns [19,20].

6 Conclusion

A GARCH process with non-parametric innovations is proposed. We also describe a procedure to estimate the parameters of the model and the density of the innovations by maximum likelihood in a transformed space. That density is modeled using a non-parametric estimate based on kernels whose width is fixed using Silverman's *rule of thumb*. In the transformed space the non-parametric estimate provides a more accurate model of heavy-tailed densities. A sliding window experiment using the returns of IBM stocks was carried out to evaluate the performance of the model. The experiments show that the new process provides very accurate estimates of risk measures.

References

1. Dowd, K.: Measuring Market Risk. John Wiley & Sons, Chichester (2005)
2. Campbell, J.Y., Lo, A.W., MacKinlay, A.C.: The Econometrics of Financial Markets. Princeton University Press, Princeton (1997)
3. Bollerslev, T.: Generalized autoregressive conditional heteroskedasticity. Journal of Econometrics 31(3), 307–327 (1986)
4. Hamilton, J.D.: Time Series Analysis. Princeton University Press, Princeton University Press (1994)
5. Cont, R.: Empirical properties of asset returns: stylized facts and statistical issues. Journal of Quantitative Finance 1, 223–236 (2001)
6. Ding, Z., C.W.J.G., Engle, R.F.: A long memory property of stock market returns and a new model. Journal of Empirical Finance 1(1), 83–106 (1993)
7. Nolan, J.P.: Stable Distributions. Birkhäuser, Basel (2002)
8. Devroye, L., Gyrfi, L., Lugosi, G.: A Probabilistic Theory of Pattern Recognition (Stochastic Modelling and Applied Probability). Springer, Heidelberg (1997)
9. Silverman, B.W.: Density Estimation for Statistics and Data Analysis. Chapman & Hall/CRC (1986)
10. Bishop, C.M.: Neural Networks for Pattern Recognition. Oxford University Press, Oxford, UK (1995)
11. Wand, M.P., Marron, J.S., Ruppert, D.: Transformations in density estimation. with discussion and a rejoinder by the authors. Journal of the American Statistical Association 86(414), 343–361 (1991)
12. Barndorff-Nielsen, O.E.: Normal inverse gaussian distributions and stochastic volatility modelling. Scandinavian Journal of Statistics 24(1), 1–13 (1997)
13. Eberlein, E., Keller, U.: Hyperbolic distributions in finance. Bernoulli 1(3), 281–299 (1995)
14. Prause, K.: The generalized hyperbolic model: Estimation, financial derivatives and risk measures. PhD Dissertation, University of Freiburg (1999)
15. Kerkhof, J., Melenberg, B.: Backtesting for risk-based regulatory capital. Journal of Banking & Finance 28(8), 1845–1865 (2004)
16. R Development Core Team: R: A language and environment for statistical computing. R Foundation for Statistical Computing, Vienna, Austria ISBN 3-900051-07-0 (2005)
17. Mittnik, S., D.T., D., C.: Computing the probability density function of the stable paretian distribution. Mathematical and Computer Modelling 29(10), 235–240 (1997)
18. Wuertz, D., Others, M.: fBasics: Financial Software Collection - fBasics (2004)
19. Longin, F.M.: The asymptotic distribution of extreme stock market returns. The Journal of Business 69(3), 383–408 (1996)
20. Dennis, W., Jansen, C.G.d.v.: On the frequency of large stock returns: Putting booms and busts into perspective. The Review of Economics and Statistics 73(1), 18–24 (1991)

Forecasting Portugal Global Load with Artificial Neural Networks

J. Nuno Fidalgo and Manuel A. Matos

Power Systems Unit of INESC Porto and Faculty of Engineering of Porto University
jfidalgo@inescporto.pt, mmatos@inescporto.pt

Abstract. This paper describes a research where the main goal was to predict the future values of a time series of the hourly demand of Portugal global electricity consumption in the following day. In a preliminary phase several regression techniques were experimented: K Nearest Neighbors, Multiple Linear Regression, Projection Pursuit Regression, Regression Trees, Multivariate Adaptive Regression Splines and Artificial Neural Networks (ANN). Having the best results been achieved with ANN, this technique was selected as the primary tool for the load forecasting process. The prediction for holidays and days following holidays is analyzed and dealt with. Temperature significance on consumption level is also studied. Results attained support the adopted approach.

Keywords: Artificial neural networks, load forecasting.

1 Introduction

The emergence of the energy markets, demanding higher efficiency levels, together with the establishment of new standards on environment preservation, has been enhancing the load forecasting importance in modern power systems operation. These harder requirements demand more sophisticated tools for power systems operation planning and, therefore, more accurate predictions of future load evolution.

The use of ANN in power systems, particularly, in load forecasting is not new. In fact, load forecasting embodies one of the most successful ANN applications in the power systems area [1]-[23]. This achievement may be explained by ANN top regression abilities, its adapting capacity, its tolerance to noisy data, and because formal system modeling is not necessary.

This paper describes the main results of a research project where the main goal was to forecast the hourly consumption diagram of the Portuguese power system in the following day. This project was developed under the framework of a contract with the Portuguese transmission company.

This paper has the following structure: Section 2 presents the adopted approach. Section 3 synthesizes the preliminary study carry out in order to select the forecasting regression tool. Section 4 describes the used ANN and main results obtained. The problem of holidays is also handled in Section 5. Section 6 describes the analysis of

J. Marques de Sá et al. (Eds.): ICANN 2007, Part II, LNCS 4669, pp. 728–737, 2007.

temperature influence in load consumption in the studied system. Finally, Section 6 presents the main conclusions of this work.

2 Methodology

Computational intelligence methods usually require a groundwork stage of data preparation. In this study, the historical data comprises load evolution, in a 15 min base, from year 2001 to 2005. The first step was to abridge this set to a 60 min base, once the goal is to produce hourly forecasts. In a second step, we divided this set in daylight saving period (summer) and winter, accordingly to Portuguese legislation.

In each training experience, the patterns were normalized to have zero mean and a standard deviation of one. This removes of offset issues and measurement scales. The patterns' set was divided into two sets: one for training and validation (about ¾ of the data) and the other for testing (about ¼).

The research performed subsequently may be divided in the following main phases:

1. Selection of the forecasting regression tool. This study has shown that ANN performed better than competing methods in the actual problem;
2. ANN training (the forecasting tool selected in the previous phase). Goal: to outcome the 24-hour load diagram of the following day. Searching for the best generalization performance by experimenting different architectures and training optimization procedures;
3. Analysis of holidays' effects and establishment of appropriate procedures to deal with these special events;
4. Analysis of temperature effects on load consumption.

Phase 1 consisted of training and performance evaluation of a number of regression tools, in order to verify the most suitable for the current forecasting problem. As shown in the next section, ANN presented the best results, having been selected as the main forecasting tool in this project.

Phase 2 comprised the usual ANN training phase and performance evaluation.

Holidays may be considered as an abnormal event in a load time series. Although their rate of occurrence is relatively small, in relation to the regular days, their effects on performance degradation is substantial. Two types of holidays' effects may be considered: in one hand, forecasts tend to be higher than they should, as the consumption in holidays is comparatively lower; on the other hand, the forecasts that depend on data series values that correspond to holidays have a tendency to be lower than they should. The implementation of procedures to deal with this concern is the main goal of phase 3.

Phase 4 examines the inclusion of temperature effects on the forecasting performance.

3 Selecting the Forecasting Regression Tool

This phase starts with the data preparation. The historical load series involves consumption data from year 2001 to 2005, in a 15 min base. Additionally, there is also information about holidays and dates of daylight saving times (accordingly to Portuguese regulation). The data was arranged in an hourly base, and prepared to be input by the competing regression alternatives: K Nearest Neighbors (KNN), Multiple Linear Regression (MLR), Projection Pursuit Regression (PPR), Regression Trees (RT), Multivariate Adaptive Regression Splines (MARS) and ANN.

For a matter of time economy, these tools were trained to predict the consumption in the following hour, instead of the 24-hour consumption in the following day. This option is based on the assumption that methodologies performance would be ranked similarly in the following hour forecasting and in the following day forecasting.

The performance evaluation was based in the following error measures:

- Mean Squared Error (MSE)
- Root Mean Squared Error (RMSE)
- Mean Absolute Error (MAE)
- Normalized Mean Squared Error (NMSE)
- Normalized Mean Absolute Error (NMAE)
- Correlation coefficient (ρ_r) between actual and predicted values

In order to obtain reliable error estimates, one adopt the "12-fold cross-validation" experimental method, that it consists basically of dividing the set of standards in 12 groups, using 11 of then for training and only 1 for test. The process is repeated 12 times, changing the test group each time. The different algorithms had been adapted and tested with the same training and test sets. Data patterns were constructed both from the original data series and after a normalization process (zero mean and pattern deviation one). The final stage involves the comparison of the average errors attained with each method.

Table 1 presents the best results obtained with each regression tool, showing that ANN performs better than competing methodologies in the present study in all the criteria.

Table 1. Comparison of errors obtained with the competing methods

Method	MSE	RMSE	MAE	NMSE	NMAE	ρ_p
KNN	0.0437	0.2051	0.1559	0.0488	0.1876	0.9784
MLR	0.0865	0.2931	0.2245	0.0985	0.2712	0.9499
PP	0.0153	0.1228	0.0881	0.0173	0.1065	0.9916
RT	0.1061	0.3239	0.2539	0.1205	0.3067	0.9394
MARS	0.0533	0.2294	0.1648	0.0606	0.1991	0.9700
ANN	0.0060	0.0770	0.0553	0.0068	0.0668	0.9966

4 ANN Description

4.1 Training Algorithm

In this work, we used a feedforward type ANN, trained with the Adaptive Backpropagation algorithm (APA) [24]. This algorithm was developed from the classical Backpropagation [25]-[28], but instead of a single fixed learning rate for all weights, it uses a distinct adaptable learning rate for each weight. Each learning rate is incremented or decremented depending of the error surface on that particular weight direction being monotonous or not. This way, the learning rate may increase thousands of times for some weights and decrease until it becomes zero (insignificant) for other weights. The training time required by APA is generally lower than for other training algorithms. The stop training criterion was based on the cross validation principle, in order to fight against overfitting.

4.2 Architecture Definition

The optimization of ANN architecture (number of layers, active connections and number of weights) still remains an open problem. In scientific literature one may find several architecture definition approaches, based on different principles like network pruning, upstart, weight decay, cascade correlation, genetic algorithms and others [29]-[33]. However, there's no recognized methodology able to assure that a given architecture is in fact the optimal one. Still, very good performance results may be obtained with non-optimal ANN architectures, provided that some concerns are taken into consideration during training.

In the present study, the search for an adequate ANN architecture was based on guided trial-and-error experiments that take into account the following:

1. Complex ANN architectures are rarely needed in load forecasting applications. Generally, the function to be mapped is not complicated. Moreover, complex ANN would capture the data noise (load fluctuation);
2. The comparison between training and test errors provides indications about ANN architecture complexity competence. Large errors in both sets suggest a deficient ANN capacity to map the data. In this case, one should increase the number of hidden units. When the test error is considerably higher than training error, this indicates that ANN has captured the training data idiosyncrasies and was not able to generalize it to the testing set. As one uses a validation set for stop training criterion, this is probably caused by an excessive ANN capacity. In this case, the number of hidden units should be reduced.

The application of these ideas led to the ANN architecture presented in Fig. 1. This ANN has 75 inputs:

1 – Holiday (assumes the value 1 in holidays and 0 otherwise)
2, 3 – Weekday information
4-27 – load diagram - 2 days before the forecasting day
28-51- load diagram - 1 week before the forecasting day
52-75- load diagram - 2 weeks before the forecasting day

The use of the cyclic variable *d* (period=7) is advantageous in terms of training, in order to pass that information to the ANN. This can be achieved by defining 2 inputs based in sine and cosine functions. So the weekday is interpreted by a periodic measure, each day type being univocally specified.

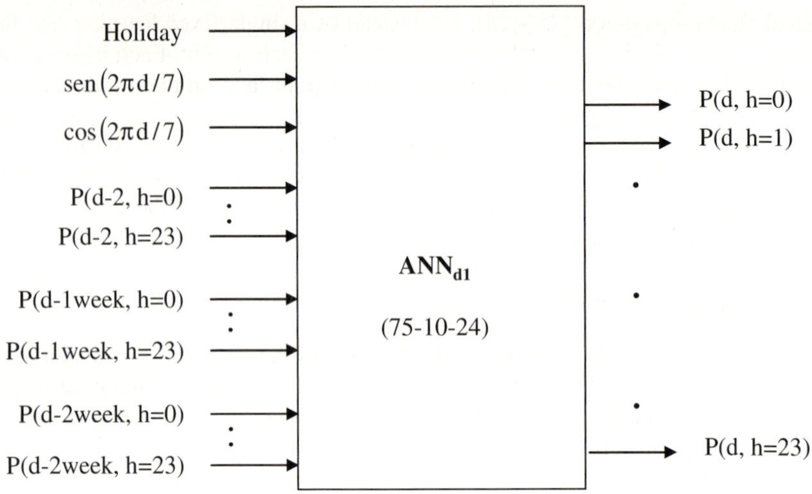

Fig. 1. Architecture of the ANN that produced the best results

The mean absolute percentage error (MAPE) obtained was 2.20% for winter period and 1.99% for summer, corresponding to a global error of 2.07%. Note that these errors refer to the raw performance, *i.e.*, the performance before the holiday treatment described in the next section. Fig. 2 shows two forecasting examples.

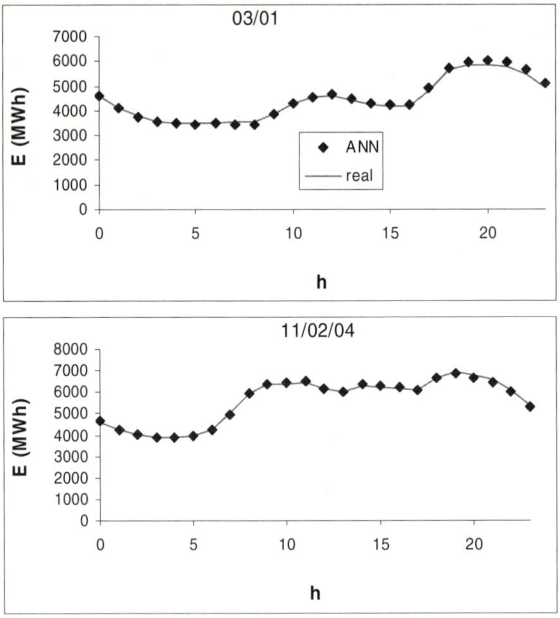

Fig. 2. Next day diagram forecasting examples

5 Holidays Treatment

As mentioned before, holidays occurrences produce two kinds of effects:

1. Holidays' forecasts tend to be higher than they should, as the consumption in holidays is comparatively below a regular day;
2. The forecasts on regular days that depend on holidays' data have a tendency to be lower than they should.

As the ANN described in the previous section has a "holiday" input, the first effect is automatically coped with. Fig. 3 illustrates this outcome, showing the forecasts with (ANN1) and without (ANN2) the holiday variable turned on.

In the second case, if the forecasting day is not weekend (holidays have a minor effect on weekends), input variables corresponding to the holiday are replaced with one of its neighbor workdays. This approach is illustrated in Fig. 4. The last anomalous days type analyzed correspond to holidays' "bridges" – when a holiday occurs in a Thursday or Tuesday, a considerable number of people doesn't work in the day between the weekend and the holiday. The consumption in these days is usually inferior to a regular day.

Table 2 presents the hourly MAPE obtained for the years 2004-05, after the application of the solutions designed to deal with special days. The global average MAPE results in 1.70%.

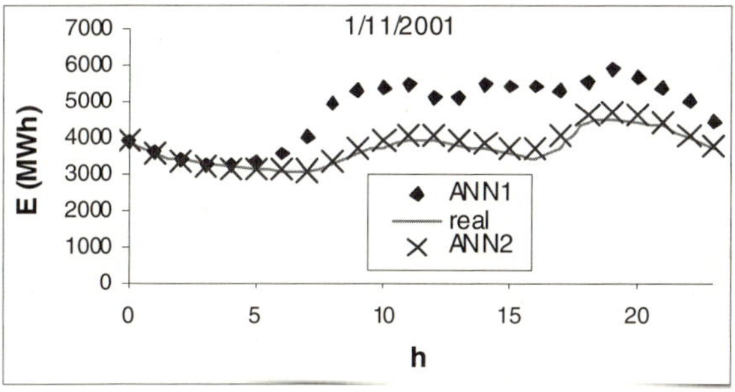

Fig. 3. Illustration of *Holiday* variable effect

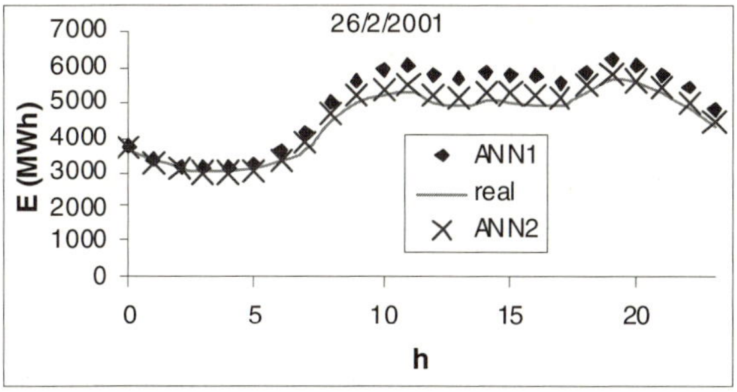

Fig. 4. Forecasting on days that depend on Holidays data

Table 2. Global MAPE obtained for each hour

Hour	MAPE	Hour	MAPE	Hour	MAPE
0	1.40%	8	1.94%	16	2.03%
1	1.39%	9	1.82%	17	1.99%
2	1.40%	10	1.71%	18	1.94%
3	1.38%	11	1.69%	19	1.88%
4	1.38%	12	1.68%	20	1.88%
5	1.35%	13	1.74%	21	1.84%
6	1.38%	14	1.86%	22	1.80%
7	1.63%	15	1.97%	23	1.62%

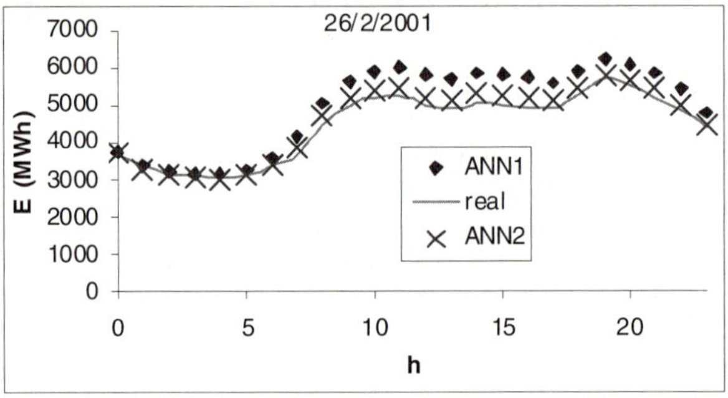

Fig. 5. Dealing with "bridges"

The analysis of the holidays' "bridges" shows that the relation between the consumption in these special days and their neighbor regular days varies a lot, depending on season of the year and on the holiday type. Therefore, it was decided to deal with holidays' "bridges" using the average relation between these days and their neighbors – the consumption in a holiday's "bridge" is about 90% of its ordinary neighbor days. Fig. 5 illustrates the effect of this approach.

6 Temperature Effects

Temperature is frequently perceived as an indispensable variable in the load forecasting process. However, its importance depends, not only on the kind of forecast (long-term, short-term), but also on the type of climate (in temperate climates, like Portugal, temperature effects are not so critical) and on the quality of the temperature forecast (the load depends on the real temperature and not on forecasted one). Besides, the quality of the temperature time series may be also decisive: first, there is the question of representativity (are the temperature values, usually taken in a given spot, descriptive of the all country?); in second place, there is the data quality (absence or scarce occurrence of anomalous temperature variations).

Table 3. Additional temperature inputs considered and performance obtained

ANN name	Additional inputs	MAPE (%)
ANNt1	<T(d)>	2.09
ANNt2	<T(d)>, <T(d-1)>, <T(d-2)>	2.28
ANNt3	<T(d)>, $T_{min}(d)$, $T_{max}(d)$	2.06

Table 4. Additional temperature inputs considered and performance obtained

Case		MAPE (%)
without Temperature (ANN_{d1})		2.07
with Temperature (ANNt3)	Base	2.06
	1	2.13
	2	2.33
	3	2.63

On the other hand, part of the temperature effects are already incorporated in the load data of the previous days.

In this study, the empirical analysis of temperature effects was performed as follows:

1. Inclusion of temperature inputs in the ANN followed by training using real temperature values;
2. Analysis of load forecasting performance sensitivity to errors in temperature forecasting.

In the first step, three ANN were tested. The input/output scheme used was similar to the one presented in Fig. 1 but adding temperature inputs. Table 3 presents the supplementary inputs considered in each tested case and the performance obtained in each case. In this table, <T(d)> stands for average temperature of day d, $T_{max}(d)$ and $T_{min}(d)$ represent the maximum and minimum temperature of day d.

These errors obtained with ANNt1 and ANNt3 are comparable with the one obtained in Section 0 (2.07%). Being so, one may conclude that, the temperature series does not bring significant information gain for the load forecasting problem. Besides, the temperature values used so far were the real ones. In practice, the next day temperature would have to be forecasted and, naturally, subject to inaccuracies. The final step of the present study consists of simulating temperature forecasting deviations and analyzing the subsequent errors changes. For this purpose, we used the best performance case (ANNt3) and introduce small random values to temperature inputs. The random intervals considered were the following:

1. [-1; +1] ° C;
2. [-2; +2] ° C;
3. [-3; +3] ° C.

The results obtained are presented in Table 4. As expected, the error increases with the random limit, which simulates the temperature forecasting error. Moreover, even for the lowest temperature forecasting error considered, the MAPE is larger than in the case without temperature inclusion. This substantiates the conclusion that, in this case, the temperature should not be included in the load forecasting process.

7 Conclusion

This paper describes a load forecasting implementation for the prediction of the hourly load diagram of the following day. Special events like holidays and days following holidays are analyzed and dealt with. The effect of temperature on forecasting performance is also studied, showing that, based on the available data, no advantage is obtained by the inclusion of temperature in the ANN inputs set. A possible explanation for this result may be the moderate Portugal climate, where the temperature variation is smooth and implicitly keyed in the load evolution. A complementary justification is related with the quality of temperature data like the existence of faulted values and the use of local data to represent of the whole country. The results obtained support the adopted approaches.

References

1. Park, D.C., El-Sharkawi, M.A., Marks, R.J., Atlas, I.L.E., Damburg, M.J.: Electric Load Forecasting Using An Artificial Neural Networks. IEEE Transactions On Power Systems 6(2) (May 1991)
2. Lu, C.N., Wu, H.T., Vemuri, S.: Neural Network Based Short Term Load Forescasting. IEEE Transactions on Power Systems 8(1) (February 1993)
3. Kermanshahi, B., Yokoyama, R.: Practical Implementation Of Neural Nets For Weather-Dependent Load Forecasting And Reforecasting At A Power Utility, Power Systems Computer Conference, PSCC96, Dresden, 1996
4. Srinivasan, D., Liew, A.C., Chang, C.S.: Forecasting Daily Load Curves Using a Hybrid Fuzzy.Neural Approach. IEE Proceedings-C, Generation, Transmission and Distribution 141(6) (1994)
5. Sharif, S.A., Taylor, J.H.: J.H, Real-time load forecasting by artificial neural networks. Proc. of the IEEE Power Engineering Society Summer Meeting, 2000, 1, 496–501 (2000)
6. Aboul-Magd, M.A., Ahmed, E.E.-D.E.-S.: An artificial neural network model for electrical daily peak load forecasting with an adjustment for holidays, Power Engineering, 2001. In: Proc. of the LESCOPE '01, Large Engineering Systems Conference, pp. 105–113 (2001)
7. da Silva, A.P.A., Rodrigues, U.P., Reis, A.J.R., Moulin, L.S.: NeuroDem-a neural network based short term demand forecaster, Power Tech Proceedings, 2001 IEEE Porto, vol. 2 (2001)
8. Barghinia, S., Ansarimehr, P., Habibi, H., Vafadar, N.: Short term load forecasting of Iran national power system using artificial neural network, Power Tech Proceedings, 2001 IEEE Porto, vol. 3 (2001)
9. Taylor, J.W., Buizza, R.: Neural network load forecasting with weather ensemble predictions. IEEE Transactions on Power Systems 17(3), 626–632 (2002)
10. Chow, T.W.S, Leung, C.T: Neural network based short-term load forecasting using weather compensation. IEEE Transactions on Power Systems 11(4), 1736–1742 (1996)
11. Chicco, G., Napoli, R., Piglione, F.: Load pattern clustering for Short-Term Load Forecasting of anomalous days, Power Tech Proceedings, 2001 IEEE Porto, vol. 2 (2001)
12. Lamedica, R., Prudenzi, A., Sforna, M., Caciotta, M., Cencellli, V.O.: A neural network based technique for short-term forecasting of anomalous load periods. IEEE Transactions on Power Systems 11(4), 1749–1756 (1996)

13. Piras, A., Germond, A., Buchenel, B., Imhof, K., Jaccard, Y.: Heterogeneous artificial neural network for short term electrical load forecasting. IEEE Transactions on Power Systems 11(1), 397–402 (1996)
14. Khotanzad, A., Afkhami-Rohani, R., Lu, T.-L., Abaye, A., Davis, M., Maratukulam, D.J.: ANNSTLF-a neural-network-based electric load forecasting system. IEEE Transactions on Neural Networks 8(4), 835–846 (1997)
15. Yoo, H., Pimmel, R.L.: Short term load forecasting using a self-supervised adaptive neural network. IEEE Transactions on Power Systems 14(2), 779–784 (1999)
16. Ho, K.L., Hsu, Y.Y., Chen, C.F., Lee, T.E., Liang, C.C., Lai, T.S., Chen, K.K.: Short term load forecasting of Taiwan power system using a knowledge-based expert system. IEEE Trans. Power Systems 5(4), 1214–1221 (1990)
17. Park, J.H., Park, Y.M., Lee, K.Y.: Composite modeling for adaptive short-term load forecasting. IEEE Trans. Power Systems 6(2), 450–457 (1991)
18. Gorr, W.L.: Research prospective on neural network forecasting. Int. J. Forecast. 10, 1–4 (1994)
19. Adya, M., Collopy, F.: How effective are neural networks at forecasting and prediction? A review and evaluation. J. Forecast. 17, 481–495 (1998)
20. Moghram, I., Rahman, S.: Analysis and evaluation of five short-term load forecasting techniques, IEEE Trans. Power Systems 4(4), 1484–1491 (1989)
21. Lu, C.N., Wu, H.T., Vemuri, S.: Neural network based short term load forecasting. IEEE Trans. Power Systems 8(1), 336–342 (1993)
22. Hippert, H.S., Pedreira, C.E., Souza, R.C.: Neural networks for short-term load forecasting: a review and evaluation. IEEE Transactions on Power Systems, 16(1), 44 –55 (2001), Baldonado, M., Chang, C.-C.K., Gravano, L., Paepcke, A.: The Stanford Digital Library Metadata Architecture. Int. J. Digit. Libr. 1, 108–121 (1997)
23. Fidalgo, J.N., Lopes, J.P.: Load Forecasting Performance Enhancement When Facing Anomalous Events. IEEE Transactions on Power Systems 20(1) (2005)
24. Silva, F.M., Almeida, L.B.: Acceleration Techniques For The Backpropagation Algorithm. In: Almeida, L.B., Wellekens, C.J. (eds.) Neural Networks, Springer, Heidelberg (1990)
25. Rumelheart, D.E., et al.: Parallel Distributed Processing. In: Rumelheart, Maccleland (eds.), Mit Press, Cambridge, MA, USA (1986)
26. Haykin, S.: Neural Networks – A Comprehensive Foundation. Prentice-Hall (1999)
27. Bishop, C.M.: Neural Networks for Pattern Recognition. Oxford University Press, Oxford, UK (1995)
28. Zurada, J.M.: Introduction to Artificial Neural Systems. PWS Publishing Company (1992)
29. Frean, M.: The Upstart Algorithm: a method for constructing and training feed-forward neural networks, Technical Report, Department of Physics and Centre for Cognitive Science, Edinburgh University (1990)
30. Fahlman, S.E., Lebiere, C.: The Cascade-Correlation Learning Architecture, Technical Report, School of Computer Science. Carnegie Mellon University (1991)
31. Parekh, R., Yang, J., Honavar, V.: Constructive neural network learning algorithms for multi-category pattern classification. IEEE Tran. Neural Networks 11(2), 436–451 (2000)
32. Burkitt, A.N.: Optimisation of the architecture of feed-forward neural nets with hidden layers by unit elimination, Complex Systems, nr. 5 (1991)
33. Dorsey, R.E., Johnson, J.D.: Evolution of Dynamic Reconfigurable Neural Networks: Energy Surface Optimality Using Genetic Algorithms. In: Levine, D.S., Elsberry, W.R. (eds.) Optimality in Biological and Artificial Networks, Lawrence Erlbaum Associates, Hillsdale, NJ (1997)

Using Genetic Algorithm to Develop a Neural-Network-Based Load Forecasting

Ronaldo R.B. de Aquino[1], Otoni Nóbrega Neto[1] , Milde M.S. Lira[1], Aida A. Ferreira[2], and Katyusco F. Santos[2]

[1] Federal University of Pernambuco, Acadêmico Helio Ramos s/n, Cidade Universitária, Cep: 50.740-530, Recife – PE, Brazil
[2] Federal Federal Center of Technologic Education of Pernambuco, Av. Prof. Luiz Freire, 500 Cidade Universitária, Cep: 50.740-540, Recife – PE, Brazil
rrba@ufpe.br, otoninobrega@hotmail.com, milde@ufpe.br, aidaaf@gmail.com, katyusco@gmail.com

Abstract. This work uses artificial neural networks, whose architecture were developed using genetic algorithm to realize the hourly load forecasting based on the monthly total load consumption registered by the Energy Company of Pernambuco (CELPE). The proposed Hybrid Intelligent System – HIS was able to find the trade-off between forecast errors and network complexity. The load forecasting produces the essence to increase and strengthen in the basic grid, moreover study into program and planning of the system operation. The load forecasting quality contributes substantially to indicating more accurate consuming market, and making electrical system planning and operating more efficient. The forecast models developed comprise the period of 45 and 49 days ahead. Comparisons between the four models were achieved by using historical data from 2005.

Keywords: Genetic Algorithm, Artificial Neural Network, Load Forecasting.

1 Introduction

A task faced every day in the electric utility is the load forecasting, which is accomplished in several periods of time: short-term, mid-term, and long-term forecasts. This data is used as input for several studies, such as planning, operation and electric power market. Traditionally, load forecasting techniques use statistical methods of time series analysis, which include linear regression, exponential damping, and Box Jenkins [1]. Recently, techniques of artificial intelligence such as Artificial Neural Network have been used as a traditional method, presenting better results [2]-[6].

The Brazilian electric sector has gone through constant changes, where organizations such as ANEEL (National Agency of Electric Energy) and ONS (National Power System Operator), which were created in the end of 1990s, have created strict regulation year after year that demands more and more efficiency in electric power supply to the consumer. These requirements are more noticeable in the quality of the supplied energy, in the reduction of the inherent losses of energy and in

J. Marques de Sá et al. (Eds.): ICANN 2007, Part II, LNCS 4669, pp. 738–747, 2007.

the system economy. The appropriate measures to meet the requirements are: the updating of the "Machine of Distribution" and a larger competitiveness between utilities of distribution.

This work presents the description of four developed models in order to mitigate and solve the problem of hourly load forecasting in a period of 45 and 49 days ahead. Two new developed models are added to the two current models of the software PREVER [11]. PREVER is software developed in 2005 by CELPE and the Federal University of Pernambuco - UFPE. The models operate to accomplish the forecast in the period of 45 days (PREVER) and 49 days (proposed models) ahead. The objective of these mid-term forecasts is to inform the ONS about the hourly load in order to formulate the Monthly Program of Operation (PMO).

This paper is organized as follows: section 2 presents the variables of the models and the arrangement of the database; section 3 describes the architecture and the training of the neural networks; section 4 presents the proposed forecast models; section 5 provides experimental results of the analysis and comparisons among the four models; and finally, section 6 concludes with a summary of this paper.

2 Variables and Database Arrangement

Several technical publications in this area [3], [8]-[10] point out the correlation between the electric power consumption and the climatic factors (relative humidity of the air, temperature, depth of rainfall, etc.), considering these climatic variables as important inputs to accomplish the load forecasting. However, these variables present historical series difficult to obtain. Regarding this problem, we decided to use a time-window in the temporal series [2]. Then, a search for characteristics in the series itself was realized. These characteristics were evaluated and used as inputs of the neural networks. The researched inputs were:

- **D':** The hourly consumption of today normalized in the range 0 to 1 to be forecasted in 45 days ahead (D'+45);
- **D'-1:** The hourly consumption of yesterday normalized in the range 0 to 1 to be forecasted in 45 days ahead (D'+45);
- **FSN:** Binary code that indicates whether the day to be forecasted is an anomalous day or not, which is done in the following way:
  - 1 Bit: [0] to no-anomalous day and [1] to anomalous day;
  - 2 Bits: [0 1] to no-anomalous day and [1 0] to anomalous day.
- **CDS:** Binary code from 1 to 7 associated with the days of the week which the day to be forecasted belongs to, for example: Tuesday = [0010000].
- **D:** The hourly consumption of today normalized in the range 0 to 1 to be forecasting 49 days ahead (D+49);
- **M:** Twenty-four inputs containing the average hourly consumption of the current day (e.g. Monday 23 April), the same day of the last week (e.g. Monday 16 April) and the same day of the week before last (e.g. Monday 9 April);
- **Sz:** Two normalized inputs that are formed by the average of the average daily consumptions of the month associated with the days that were used to calculate the variable M; and the average daily consumption of the month associated with the day to be forecasted (D+49);

- **Cm:** Binary code from 1 to 12 associated with the month that the day to be forecast belongs to, for example: February = [010000000000];
- **m:** Normalized number in the range from 0 to 1 representing the month (m) of the day to be forecasted, given by (1) , where m_{max} = 1.1·12, and m_{min} = 0.9·1.

$$\mathbf{m} = \frac{m - m_{min}}{m_{max} - m_{min}} \,. \tag{1}$$

The supplied original data corresponds to the hourly load consumption of the distribution system of CELPE in the period from January 2000 to December 2005 including the period of the rationing responsible for an abrupt decrease in the curve of the load. The hourly load consumptions in the period of the rationing (from May to July 2001) were removed from the original database. Therefore, the database has 2100 days which corresponds to 50.400 hours. The original data supplied by CELPE was preprocessed and used for training and evaluating the performance of the four forecast models developed.

For the forecast model in the period of 45 days ahead (Model of PREVER), the anomalous days were eliminated (holidays), forming the database with 969 training patterns. On the other hand, for the forecast model in the period of 49 days ahead, the anomalous days were included in the training patterns, although they were classified as a weekend day [11] (Saturday or Sunday according to the characteristic of the load curve), resulting a total of 1608 patterns for training. The data of 2005 were used only for evaluating the developed models.

The load values were normalized to fall in the range 0 to 1 according to (2), where L_N is the normalized load value, L is the hourly load value registered by CELPE, L_{max} is the maximum hourly load added to 10% of itself and L_{min} is zero.

$$L_N = \frac{L - L_{min}}{L_{max} - L_{min}} \tag{2}$$

The databases, created to accomplish the load forecasting in the period of 45 and 49 days ahead, have their patterns distributed in the following way: 60% for the training set, 30% for the validation set and 10% for the test set. The patterns of each group were selected in a random way.

The main objective of the models based on ANNs is to learn from patterns of known value and to generalize for new ones. The performance of the system will be measured by the percentage of the mean-square error (MSE) specified in (3), and by the mean absolute percentage error (MAPE) specified in (4):

$$MSE_{\%} = 100 \cdot \frac{L_{max} - L_{min}}{N \cdot P} \sum_{i=1}^{P} \sum_{j=1}^{N} (\hat{L}_{ij} - L_{ij})^2 \,, \tag{3}$$

where $\hat{L}_{ij}$ is the load forecasting value for the pattern i and output j, L_{ij} is the real value of the load for the pattern i and output j, N is the number of output units of the network and P is the total number of patterns.

$$MAPE_{\%} = 100 \cdot \frac{1}{P} \sum_{k=1}^{P} \frac{\left| \hat{L}_k - L_k \right|}{L_k}, \tag{4}$$

where k is the index related to *k-th* hour in the period of analysis, $\hat{L}_k$ and L_k are the load forecasting consumption and real load consumption at hour k, respectively.

3 Architecture and Training

All of the experiments accomplished in this work created ANNs with the MLP architecture, using the resilient backpropagation (RPROP) training algorithm [7]. The RPROP performs a local adaptation of the weight-updates according to the behavior of the error function. The RPROP algorithm operates in the batch mode and falls in a supervised training category.

All of the ANNs used have an input layer, a hidden layer and an output layer. The nodes of the hidden layer use the tan-sigmoid activation function and those of the output layer use the log-sigmoid activation function. The maximum number of iterations for all of the trainings was set to 2500 epochs. The training stops if the early stop method implemented by MATLAB® happens 20 times consecutively, or if the maximum number of epochs is reached, or if the error gradient reaches a minimum, or still if the error goal on the training set is met. The early stop method has the objective of improving generalization of the neural networks.

The networks were created varying number of the nodes in the hidden layer from 30 to 130 with an increment of 1. In each step, ten neural networks were created to mitigate problems of random initialization of weights. The average of the MSE on the validation set was calculated from those ten neural networks created by random weights initializations. Then, the average was used in two different ways to choose the neural network architecture: a) the architecture was chosen by the smallest value of the MSE on the validation set; b) the average was normalized into the range from 0 to 1, then they were given as the input to the fitness function (f_a) of the Genetic Algorithm – GA. This fitness function was defined using a weighted average of the following normalized variables: average of the MSE on the validation set, number of hidden nodes and size of the input vector of the network.

Defined the architecture, the 10-fold cross validation method was used to obtain “the best neural network”. This method has become a standard method in practical terms [12]. Therefore the patterns were divided in ten independent partitions, and each partition has 10% of the data. In each experiment three partitions were used to validate, one, to test and the six remaining partitions were used to train the ANNs.

4 Load Forecasting Models

In this work, the combination of the new variables and the variables, at first, characterized in [11] were also analyzed in the previously section 2. Starting from the combination of those variables, eight trials were carefully analyzed by the procedure described in section 3. Then, 101 networks architectures, for each one of the eight trials, were created because of the varying number of nodes in the hidden layer. The best neural network architecture of each trial was chosen using the smallest average of the MSE on the validation set. The best network among the eight architecture chosen previously was selected using the smallest MSE on the test set, which was the one in the TRIAL 7 (E7) that corresponds to the forecast model in the period of 49 days ahead, as shown Fig. 2(a).

The second model (Fig. 2(b)) was created by the Hybrid Intelligent System – HIS [13], whose architecture was chosen by the GA, developed specially to accomplish this objective. The GA attempts to reduce the architecture size, acting directly in the choice of the number of hidden nodes and in the arrangement of the input variables more appropriate to solve the problem. Two restrictions were imposed to the GA before choosing the best arrangement of the input: either the variables **Cm** or **m** should be present in the network input, not both, since these two variables give the same information (month of the day to be forecast) in a different way; and the variables **M** and **CDS** should be present in all arrangements of the network input.

More than analyzing the trials described in Table 1, the GA can create new networks by combination of possibilities described in Table 2. The final search space by the genetic algorithm is formed using the trials described in Tables 1 and 2, giving a total of 1212 points, where each point is represented by the average of the MSE on the validation set of the 10 networks, the numbers of inputs and hidden nodes. The trials, described in Table 1, give a total of 808 points (8080 networks), where the processing time of the GA was not calculated, and the stop criterion was established by the amount of generations in which no more capable individual could be found.

The GA with the characteristics presented in Table 3 has yielded the architecture of the model shown in Fig. 2 (b).

Table 1. Trials as functions of input

Trials	Input Variables Vector	Number of Inputs	Trials	Input Variables Vector	Number of Inputs
E1	[M DS]	31	E5	[M Sz CDS]	33
E2	[M Cm CDS]	43	E6	[D M Sz Cm CDS]	69
E3	[D M CDS]	55	E7	[D M Sz m CDS]	58
E4	[D M Cm CDS]	67	E8	[M Sz m CDS]	34

Table 2. Combination possibilities for new individuals' creation

Trials	Input Variables Vector	Range of hidden node	Trials	Input Variables Vector	Range of hidden node
GA – E09	[M m CDS]	30-130	GA – E11	[M Sz Cm CDS]	30-130
GA – E10	[D M m CDS]	30-130	GA – E12	[D M Sz CDS]	30-130

Table 3. Select individual to choose the architecture using GA

Trials	Select Individual	Hidden Nodes	Created Individual	Generation	Total of Tested Individual
GA – E08	[M Sz m CDS]	52	140	131	1310

4.1 PREVER Models

Previously, the procedure adopted by CELPE for hourly load forecasts was a mixing of statistical techniques with specialists' knowledge. Then, in order to improve and automate it, the new software denominated PREVER implemented in MATLAB was developed. The new system makes use of a hybrid approach of ANN-based techniques and heuristic rules to adjust the short and mid-term electric load forecasting in the time period of 3, 7, 15, 30, and 45 days. PREVER has all the necessary functions to be considered as a user-friendly computer software: graphical interface, importing, and exporting data from Excel spreadsheets compatible with the spreadsheets used by CELPE, integrated database that allows the validation of the results of the load forecasting by the new system and on-line help function.

Currently, on 15th day of every month a report on the load forecasting of the next month must be sent by CELPE to ONS. The electrical utilities need to accomplish the load forecasting at least 45 days ahead, so PREVER was projected to accomplish the load forecasting in this period. However, it can forecast the electrical load in other time periods, such as 3, 7, 15, and 30 days ahead. The two new models proposed in this paper can operate in the period of 49 days ahead. This specific time of period is useful because the behavior of the seasonal changes in the load can be used to improve the load forecasting.

PREVER makes it possible to forecast for a period of 45 days ahead, accomplished by two models: the first model returns the values of an ANN output and the second returns the values of the combination of an ANN with heuristic rules, whose procedures are described in [11]. The models are shown in Fig. 1 (a) and (b), whose outputs are the 24 normalized hourly load forecasting and inputs are the variables discussed previously in section 2.

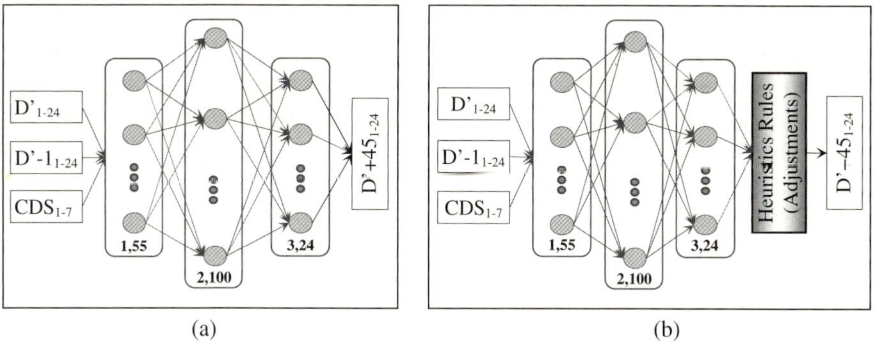

Fig. 1. (a) Neural Model of the PREVER System (b) Hybrid Model of the PREVER System

4.2 Proposed Models

In order to improve performance of the PREVER system, new input variables were investigated. During the search process a Genetic Algorithm was developed to automate the choice of the neural network architecture. This automatic form of creating ANNs constitutes a Hybrid Intelligent System. This hybrid system is necessary to reduce numbers of input variables, without deteriorating the result of the load forecast model. At the end of the search for input variables, we obtained two models: one chosen by the smallest MSE on the validation set [11] and another by the developed HIS. These two models, already discussed in section 4, can be seen in Fig. 2 (a) and (b).

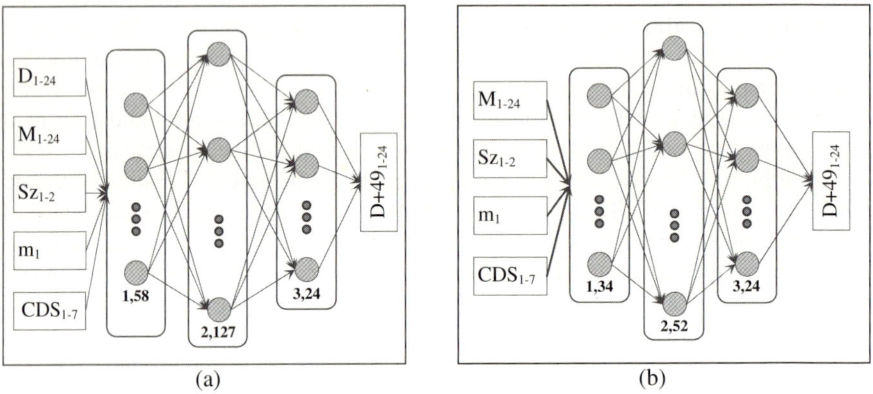

Fig. 2. (a) Model Proposed Created by Traditional Method, (b) Model Proposed Created by Hybrid Intelligent System

5 Comparison of the Models

To compare the four models discussed previously, the hourly load consumption of CELPE in the entire year 2005 was used. In order to show a summery of the results, we decided to divide them into 12 groups corresponding to the months of the year. The forecast error was accomplished in three types: the first, the entire year considering the anomalous days; the second, without considering anomalous days; and the third, just considering the anomalous days. The MAPE (4) was used to evaluate the models.

At first, comparative analysis between the proposed models (E7 and GA-E8) and the PREVER system will be done. After this, analysis of the model E1 with emphases to the holidays will be done. Table 4 presents the monthly average MAPE in 2005. There, we observe that the average error of the proposed models (E7 and GA-E8) and "PREVER-ANN with adjustment" is approximately. The difference between the monthly average error of "PREVER-ANN with Adjustment" and the model E7 is approximately 0.01% which can be considered insignificant. Besides, it can be seen that the developed networks are more stable than "PREVER-ANN without Adjustment", since they present lower standard deviation.

Table 5 describes the behavior of the models without considering the anomalous days. There we can notice that the monthly average MAPE of the model E7 is 2.95% against 2.99% presented by "PREVER-ANN with adjustment " which means that the model E7 has a slight improvement in forecasting electrical load in typical working days. Table 6 presents the MAPE on anomalous days in 2005. The adjustments done in the neural network of PREVER on anomalous days account for its better performance compared with the other models. Although the model GA-E8 presents the best average (7.12%), its standard deviation (4.63) is not only better than the "PREVER-ANN with Adjustment". Observing the data, there are 3 days on which the error was significantly higher than the average, leading to a higher standard deviation. These days are: New Year's Day (01/01/2005), Carnival (07/02/2005) and Christmas (25/12/2005). Obviously these days must be treated in another way, not only as a simple characterization of a Saturday or Sunday behavior.

Analyzing Tables 5 and 6, it can be observed that the errors on holidays or anomalous days are higher than the errors on the others days. In this case, it is very important to observe the model E1, because it was the best as shown Table 5.

In order to consider this potentiality, the two combined models were also analyzed. The first, combining E1 (anomalous days) and E7 (others days) and the second, combining E1 and GA–E8 (others days). The results are presented on Table 4 in column E7/E1 and GA-E8/E1. As expected, the combined models achieved small errors when compared with the models E7 and GA-E8.

Table 4. Monthly average MAPE in 2005

Month	PREVER without Adj.	PREVER With Adj.	E1	E7	GA-E8	E7/ E1	GA-E8/E1
Jan	3.94	3.20	5.19	4.09	4.42	3.95	4.45
Feb	4.11	3.45	4.25	3.49	3.46	3.53	3.62
Mar	6.00	3.55	7.13	4.54	4.34	4.50	4.26
Apr	3.78	3.83	6.09	3.37	3.22	3.34	3.18
May	2.95	2.89	3.05	2.83	3.50	2.82	3.46
Jun	4.12	3.63	3.27	2.97	3.56	2.90	3.54
Jul	2.18	2.32	2.40	1.96	1.92	1.98	1.95
Aug	3.15	2.55	2.65	2.21	2.23	2.21	2.23
Set	5.19	2.55	4.70	2.83	2.90	2.69	2.77
Oct	6.11	2.39	5.43	2.75	2.64	2.60	2.50
Nov	3.88	2.49	5.01	2.78	2.87	2.68	2.70
Dec	3.59	4.63	3.77	3.80	4.26	3.56	3.87
Average	4.08	3.12	4.41	3.13	3.28	3.06	3.21
Std. Dev.	1.18	0.71	1.45	0.75	0.81	0.73	0.79

The comparisons with the "PREVER with adjustment" show that combined model E7/E1 gets small average errors and that the combined model GA-E8/E1 becomes closer. Among all analyses done, it is important to point out that the proposed HIS was able to find the trade-off between errors and network complexity. The complexity of model E7 is 58x127x 24= 10414 connections while the model GA-E8 is 34x52x24=3016 that represents less than 30% of the connections presented by the model E7. On the other hand, the average error (Table 4) is 3.13% to the model E7 against 3.28% presented by the model GA-E8 that represents an increase of less then 5%.

Table 5. Monthly average MAPE in 2005 without anomalous days

Month	PREVER without Adj.	PREVER with Adj.	E1	E7	GA-E8
Jan	2.70	2.51	4.65	3.32	3.86
Feb	3.33	3.04	3.87	3.07	3.17
Mar	6.09	3.51	7.25	4.52	4.28
Apr	3.64	370	6.17	3.34	3.17
May	2.95	2.89	3.02	2.78	3.44
Jun	4.13	3.62	3.23	2.86	3.52
Jul	2.14	2.29	2.34	1.91	1.87
Aug	3.15	3.15	2.65	2.21	2.23
Set	4.74	2.41	4.73	2.65	2.74
Oct	5.75	2.20	5.45	2.53	2.43
Nov	3.39	2.30	5.19	2.70	2.72
Dec	3.27	4.31	3.75	3.52	3.85
Average	3.77	2.99	4.36	2.95	3.11
Std. Dev.	1.20	0.68	1.49	0.68	0.73

Table 6. Daily average MAPE in 2005 of the anomalous days

Month	PREVER without Adj.	PREVER with Adj.	E1	E7	GA-E8
01/01/2005	25.94	16.73	21.73	24.02	20.37
02/01/2005	18.01	11.63	4.27	6.29	4.72
07/02/2005	5.41	11.38	14.28	12.81	11.83
08/02/2005	11.16	3.11	2.97	4.45	2.13
09/02/2005	15.30	6.19	4.83	3.57	3.49
25/03/2005	3.38	4.61	3.72	4.99	6.21
21/04/2005	8.02	7.70	3.57	4.50	4.80
01/05/2005	3.03	3.01	4.05	4.06	5.30
24/06/2005	3.69	3.69	4.18	6.18	4.79
16/07/2005	3.45	3.31	4.11	3.51	3.18
07/09/2005	18.38	6.55	3.67	7.87	7.54
12/10/2005	17.01	8.17	4.64	9.33	9.02
02/11/2005	17.54	4.08	2.07	4.25	5.00
15/11/2005	4.31	6.41	2.76	3.58	5.08
31/12/2005	7.65	9.23	4.28	6.17	7.29
Average	10.68	7.19	5.56	7.19	7.12
Std. Dev.	7.10	3.81	5.09	5.20	4.63

6 Conclusion

The results demonstrated that the developed models presented errors according to the kind of the day. On anomalous days, the model AG-E8 was superior, whereas on days without anomalies the model E7 was better. Finally, the developed models holding much simpler architecture were more precise than the models presented in PREVER.

Another important point to note is that the hybrid intelligent system enables the network architecture to be processed more speedily. It is worth to point out that the HIS really reduce the complexity of the ANN, i.e. decrease numbers of connections in the network. In spite of the fact that the model GA-E8 sometimes presents results inferior to both "PREVER-ANN with adjustment" and E7, it always showed reasonable accuracy compared to them. This was achieved with a more simple architecture, demonstrating strong relationship between the combination of input variables and number of neurons in the hidden layer of the ANN.

The combination of model E1 to forecast the anomalous days has also improved the performance of the new combined models proposed in this paper.

Acknowledgments. The authors would like to thank to CNPq and CAPES-Brazil for financial research support.

References

1. Montgomery, D.C., Johnson, L.A., Gardner, J.S.: Forecasting and time series analysis. Ed. McGraw-Hill International Editions (1990)
2. Bakirtzis, A., Petrldis, V.S., Kiartiz, J., Alexiardis, M.C.: A neural network short term load forecasting model for the Greek power system. IEEE Transactions on Power Systems 11(2), 858–863 (1996)
3. Khotanzad, A., Afkhami-rohani, R., Lu, T., Abaye, A., Davis, M., Maratukulam, D.J.: A Neural-network-based electric load forecasting system. IEEE Transactions on Neural Networks 8(4), 835–845 (1997)
4. Kim, C., Yu, I., Song, Y.H.: Kohonen neural network and wavelet transform based approach to short-term load forecasting. Electric Power Systems Research 63(3), 169–176 (2002)
5. Papadakis, S.E., Theocharis, J.B., Kiartzis, S.J., Bakirtzis, A.G.: A novel approach to short-term load forecasting using fuzzy neural networks. IEEE Transactions on Power Systems 13(2), 480–492 (1998)
6. Silva, A., Moulin, L., Reis, A.J.R.: Feature extraction via multiresolution analysis for short-term load forecasting. IEEE Transactions on Power Systems 20(1), 189–198 (2005)
7. Riedmiller, M., Braun, H.: A direct adaptive method for faster backpropagation learning: The RPROP algorithm. IEEE International Conference on Neural Networks 1, 586–591 (1993)
8. Khotanzad, A., Afkhami-Rohani, R., Maratukulam, D.J.: ANNSTLF – Artificial neural network short-term load forecaster – generation three. IEEE Transactions on Power Systems 13(4), 1413–1422 (1998)
9. Khotanzad, A., Davis, M.H., Abaye, A., Maratukulam, D.J.: An artificial neural network hourly temperature forecaster with applications in load forecasting. IEEE Transactions on Power Systems 11(2), 870–876 (1996)
10. Drinivasan, D., Tan, S.S., Chang, C.S.: Parallel Neural Network-Fuzzy Expert System Strategy For Short-term Load Forecasting: System Implementation And Performance Evaluation. IEEE Transactions on Power Systems 14(3) (1999)
11. Aquino, R.R.B., Ferreira, A.A., Lira, M.M.S., Carvalho, J.M.A., Neto, N.O., Silva, G.B.: Combining artificial neural networks and heuristic rules in a hybrid intelligent load forecast system. In: Kollias, S., Stafylopatis, A., Duch, W., Oja, E. (eds.) ICANN 2006. LNCS, vol. 4132, pp. 757–766. Springer, Heidelberg (2006)
12. Neto, N.O.: Aplicação de redes neurais artificiais e algoritmos genéticos na previsão de carga elétrica em médio prazo. In: Dissertação (Mestrado em Engenharia Elétrica – CTG, UFPE. Recife-PE, In Portuguese (2006)

Separation and Recognition of Multiple Sound Source Using Pulsed Neuron Model

Kaname Iwasa, Hideaki Inoue, Mauricio Kugler, Susumu Kuroyanagi, and Akira Iwata

Nagoya Institute of Technology, Gokiso-cho, Showa-ku, Nagoya, 466-8555, Japan

Abstract. Many applications would emerge from the development of artificial systems able to accurately localize and identify sound sources. However, one of the main difficulties of such kind of system is the natural presence of multiple sound sources in real environments. This paper proposes a pulsed neural network based system for separation and recognition of multiple sound sources based on the difference on time lag of the different sources. The system uses two microphones, extracting the time difference between the two channels with a chain of coincidence detection pulsed neurons. An unsupervised neural network processes the firing information corresponding to each time lag in order to recognize the type of the sound source. Experimental results show that three simultaneous musical instruments' sounds could be successfully separated and recognized.

1 Introduction

By the information provided from the hearing system, the human being can identify any kind of sound (sound recognition) and where it comes from (sound localization) [1]. If this ability could be reproduced by artificial devices, many applications would emerge, from support devices for people with hearing loss to safety devices. With the aim of developing such kind of device, a sound localization and recognition system using Pulsed Neuron (PN) model [2] have been proposed in [3]. PN models deal with input signals on the form of pulse trains, using an internal membrane potential as a reference for generating pulses on its output. PN models can directly deal with temporal data, avoiding unnatural windowing processes, and, due to its simple structure, can be more easily implemented in hardware when compared with the standard artificial neuron model. The system proposed in [3] can locate and recognize the sound source using only two microphones, without requiring large instruments such as microphone arrays [4] or video cameras [5].

However, the accuracy of the system deteriorates when it is used in real environments due to the natural presence of multiple sound sources. Therefore, an important feature of such system is the ability of identifying the presence of multiple sound sources, separating and recognizing each of them. This would enable the system to define an objective sound source type, improving the sound localization performance.

J. Marques de Sá et al. (Eds.): ICANN 2007, Part II, LNCS 4669, pp. 748–757, 2007.

In order to extend the system proposed in [3], this paper proposes a PN based system for separation and recognition of multiple sound sources, using their time lag information difference. Based on the time lags' firing information, the sound sources are recognized by an unsupervised pulsed neural network.

2 Pulsed Neuron Model

When processing time series data (e.g., sound), it is important to consider the time relation and to have computationally inexpensive calculation procedures to enable real-time processing. For these reasons, a PN model is used in this research.

Figure 1 shows the structure of the PN model. When an input pulse $i_k(t)$ reaches the kth synapse, the local membrane potential $p_k(t)$ is increased by the value of the weight w_k. The local membrane potentials decay exponentially with a time constant τ_k across time. The neuron's output $o(t)$ is given by

$$o(t) = H(I(t) - \theta) \qquad I(t) = \sum_{k=1}^{n} p_k(t) \tag{1}$$

where n is the total number of inputs, $I(t)$ is the inner potential, θ is the threshold and $H(\cdot)$ is the unit step function. The PN model also has a refractory period t_{ndti}, during which the neuron is unable to fire, independently of the membrane potential.

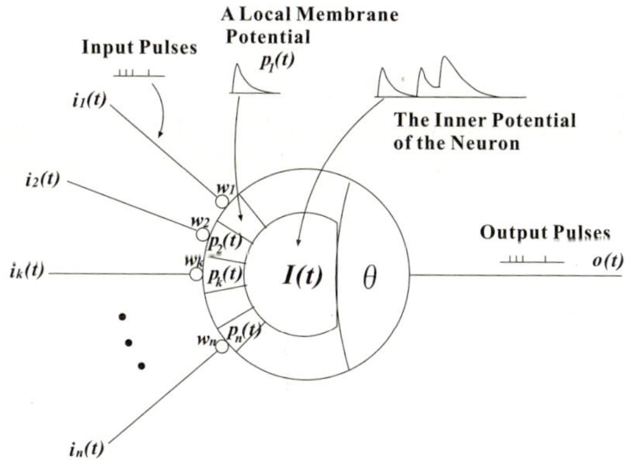

Fig. 1. A pulsed neuron model

3 Proposed System

The basic structure of the proposed system is shown in Fig. 2. This system consists of three main blocks, the frequency-pulse converter, the time difference

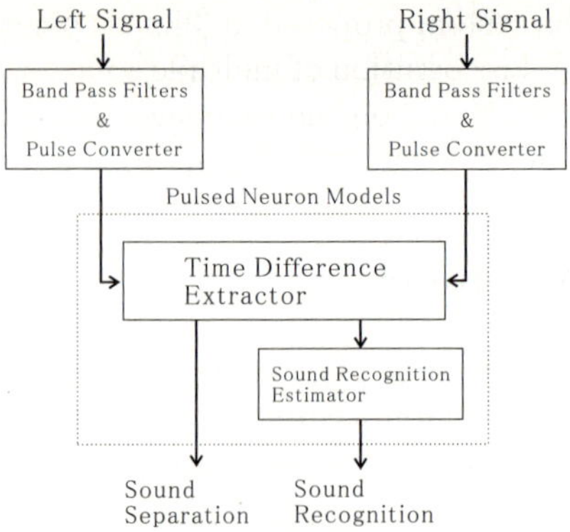

Fig. 2. Basic structure of the proposed system

extractor and the sound recognition estimator. The time difference extractor and sound recognition estimator blocks are based on a PN model.

The left and right signals' time difference information is used to localize the sound source, while the spectrum pattern is used to recognize the type of the source.

3.1 Filtering and Frequency-Pulse Converter

In order to enable pulsed neuron based modules to process the sound data, the analog input signal must be divided on its frequency components and converted to pulses. A bank of band-pass filters decomposes the signal, and each frequency channel is independently converted to a pulse train, which rate is proportional to the amplitude of the correspondent signal. The filters' center frequencies were determined in order to divide the input range (100 Hz to 16 kHz) in 72 channels equally spaced in a logarithm scale.

3.2 Time Difference Extractor

Each pulse train generated at each frequency channel is inputted in an independent time difference extractor. The structure of the extractor is based on Jeffress's model [7], in which the pulsed neurons and the shift operators are organized as shown in Fig. 3. The left and right signals are inputted in opposed sides of the extractor, and the pulses are sequentially shifted at each clock cycle. When a neuron receives two simultaneous pulses, it fires. In this research, the neuron fires when both input's potentials reach the threshold θ_{TDE}. The position of the firing neuron on the chain determines the time difference.

This work uses an improved method, initially proposed in [8], which consists on deleleting the two input pulses when a neuron fires for preventing several false detections due to the matching of pulses of different cycles, as shown in Fig. 4.

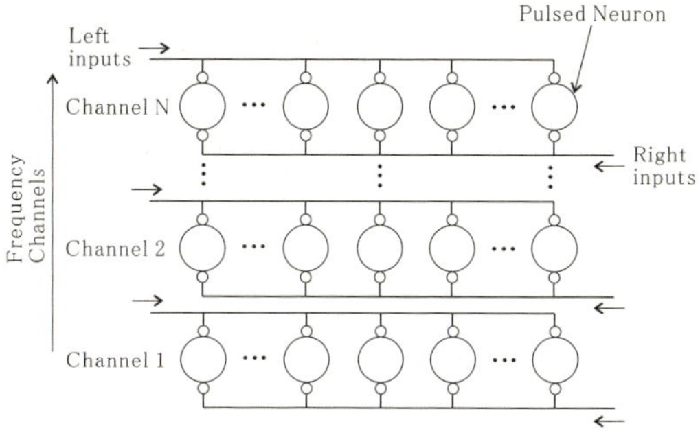

Fig. 3. Time Difference Extractor

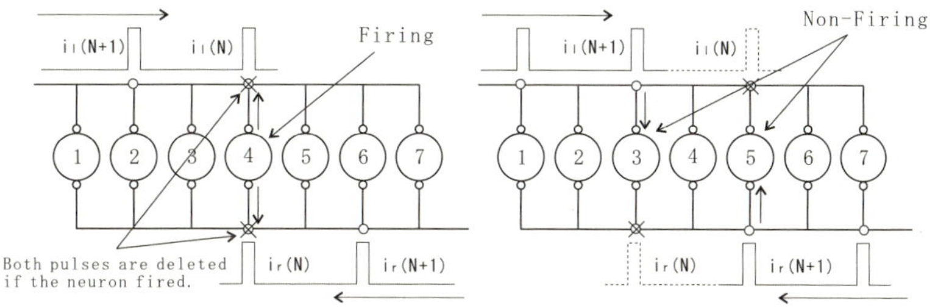

(a) time t (b) time $t+1$

Fig. 4. Pulse deleting algorithm in Time Difference Extractor

3.3 Sound Recognition Estimator

The sound recognition estimator is based on the Competitive Learning Network using Pulsed Neurons (CONP) proposed in [6]. The basic structure of CONP is shown in Fig.5.

In the learning process of CONP, the neuron with the most similar weights to the input (winner neuron) is chosen for learning in order to obtain a topological relation between inputs and outputs. For this, it is necessary to fire only one neuron at a time. However, in the case of two or more neurons firing, it is difficult to decide which one is the winner, as their outputs are only pulses, and not real values. In order to this, CONP has extra external units called control neurons. Based on the output of the Competitive Learning (CL) neurons, the control neurons' outputs increase or decrease the inner potential of all CL neurons, keeping the number of firing neurons equal to one. Controlling the inner potential is equivalent to controlling the threshold. Two types of control neurons are used in this work. The No-Firing Detection (NFD) neuron fires when no CL neuron fires, increasing their inner potential. Complementarily, the Multi-Firing Detection (MFD) neuron fires when two or more CL neurons fire at the same time, decreasing their inner potential.

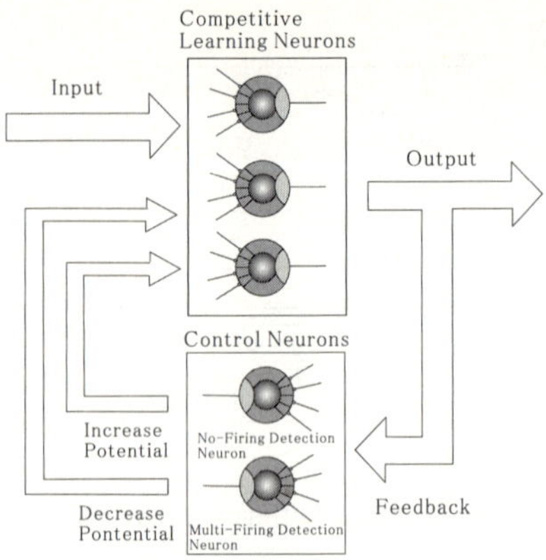

Fig. 5. Competetive Learning Network using Pulsed Neurons (CONP)

The CL neurons are also controlled by another potential, named the input potential $p_{in}(t)$, and a gate threshold θ_{gate}. The input potential is calculated as the sum of the inputs (with unitary weights), representing the frequency of the input pulse train. When $p_{in}(t) < \theta_{gate}$, the CL neurons are not updated by the control neurons and become unable to fire, as the input train has a too small potential for being responsible for an output firing. Furthermore, the inner potential of each CL neuron is decreased by a factor β, in order to follow rapid changes on the inner potential and improving its adjustment.

Considering all the described adjustments on the inner potential of CONP neurons, the output equation (1) of each CL neurons becomes:

$$o(t) = H\left(\sum_{k=1}^{n} p_k(t) - \theta + p_{nfd}(t) - p_{mfd}(t) - \beta \cdot p_{in}(t)\right) \quad (2)$$

where $p_{nfd}(t)$ and $p_{mfd}(t)$ corresponds respectively to the potential generated by NFD and MFD neurons' outputs, $p_{in}(t)$ is the input potential and β ($0 \leq \beta \leq 1$) is a parameter.

4 Experimental Results

In this work, several sound signals generated by computer were used: three single frequency signals (500 Hz, 1 kHz and 2 kHz), and five musical instruments' sounds ("Accordion", "Flute", "Piano", "Drum" and "Violin"). Each of these signals were generated with three different time lags: −0.5 ms, 0.0 ms and +0.5 ms, with no level difference between left and right channels.

4.1 Separation of Multiple Sound Sources

Initially, the time difference information is extracted as described in section 3.2. The used parameters for the signal acquistion, preprocessing and time difference

Table 1. Parameters of each module used on the experiments

Input Sound	
Sampling frequency	48 kHz
Quantization bit	16 bit
Number of frequency channels	72
Time Difference Extractor	
Total number of shift units	121
Number of output neurons	41
Threshold θ_{TDE}	1.0
Time constant	350 μs

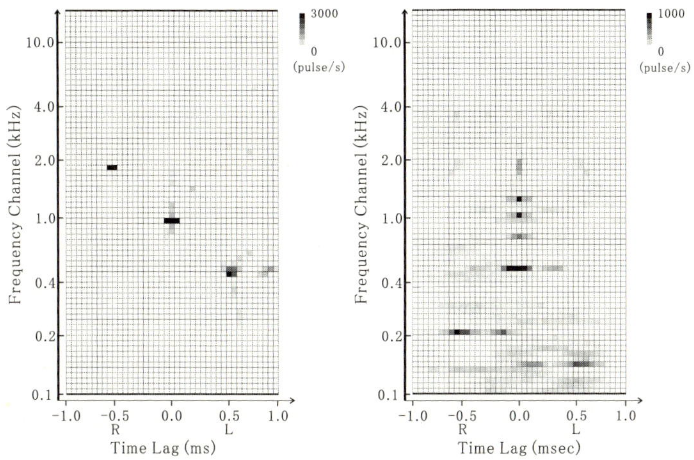

(a) single frequency signals (b) musical instruments

Fig. 6. Output of Time Difference Extractor for three different signals

extraction are shown in Table 1. The 48 kHz sampling frequency causes the pulse train to shift 20.83 μs at each clock cycle (Fig.3), resulting in output time lags of 41.66 μs for each neuron.

Figure 6(a) shows the output of the time difference extractor for an input composed by the 500 Hz single frequency signal (+0.5 ms lag) the 1 kHz signal (0.0 ms lag) and the 2 kHz signal (−0.5ms lag) The x-axis corresponds to the time lag (calculated from the firing neuron in the time difference extractor) and the y-axis corresponds to the channels' frequency. The gray-level intensity represents the rate of the output pulse train. Figure 6(b) shows the output relative to the musical instruments' sounds "Drum" (+0.5 ms lag), "Flute" (0.0 ms lag) and "Violin" (−0.5 ms lag). Again, each time lag shows a different firing pattern in each position.

Figure 7(a) shows the extraction of the firing information for each of the identified instruments in Fig. 6. It can be seen that the frequency components

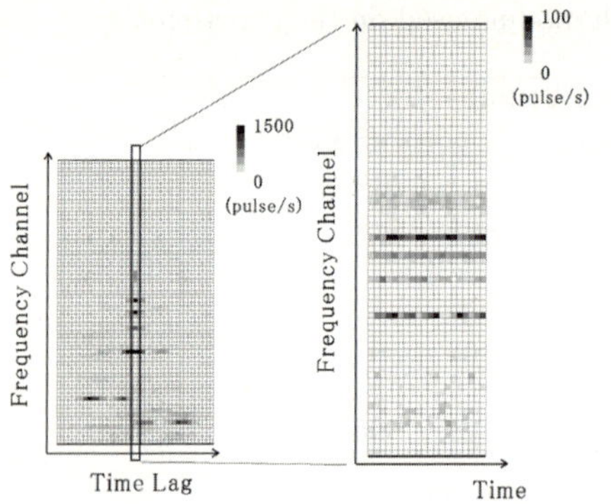

(a) Extraction of a time lag firing information

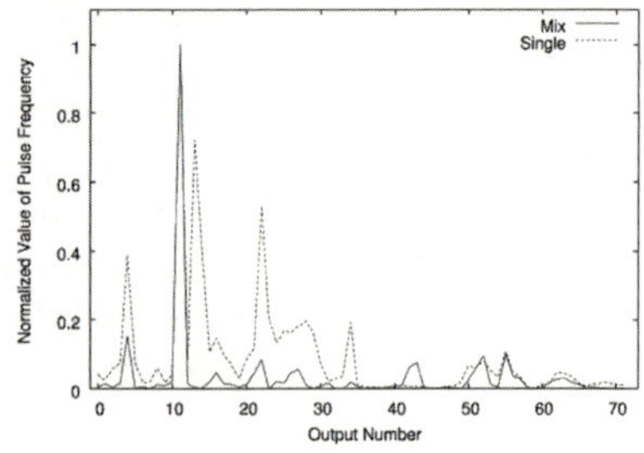

(b) Time Lag = -0.5 ms (Violin)

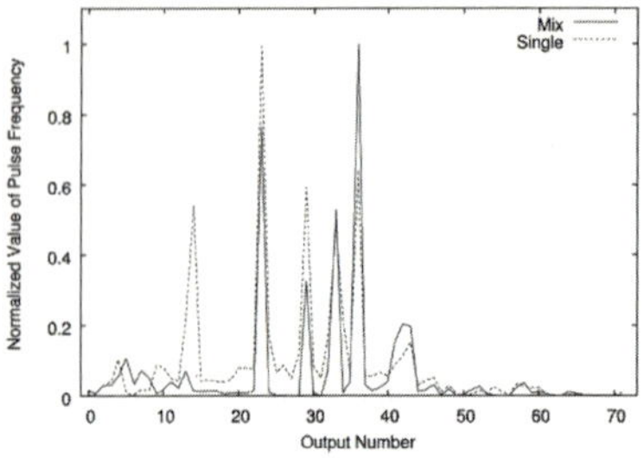

(c) Time Lag = 0.0 ms (Flute)

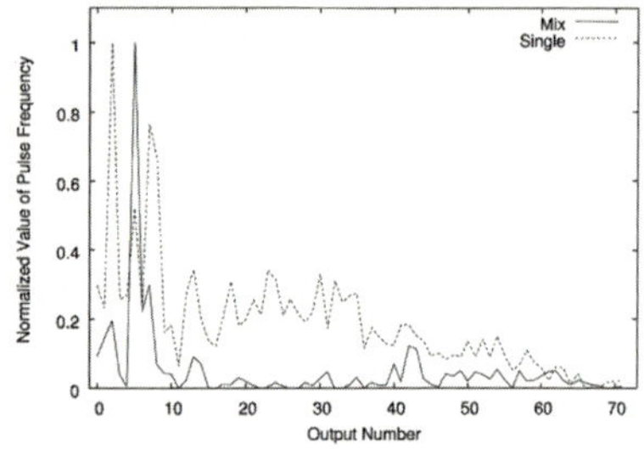

(d) Time Lag = +0.5 ms (Drum)

Fig. 7. Extraction of the independent time lags firing information

are constant along time. Furthermore, Fig 7(b) to (d) show the output firing information of each sound (Mix), together with the original firing information for the independent sounds with no time lag (Single). All data is normalized for comparison, showing that important components are similar. As both results present firing in different frequency components for each time lag, it is possible to recognize the type of sound source for each time difference.

4.2 Recognition of Independent Sound Sources

Each time lag's firing information is recognized by the CONP model described in section 3.3. Initially, the firing information of each type of sound source is extracted with no time lag. This data is used for training CONP, according to the parameters shown in Table 2.

The five musical instruments' sounds were applied to the CONP in all combinations of three simultaneous sounds with the three time lags (60 combinations). Table 3 shows the average accuracy of the CONP model for each instrument in each position. The recognition rate is calculated by the ratio between the number of firings of the neuron corresponding to the correct instrument and the total number of firings.

Table 2. Parameters of CONP used on the experiments

Competitive learning Neuron	
Input Number of CL neurons	72
Number of CL neurons	5[units]
Threshold θ	1.0×10^{-4}
Gating threshold θ_{gate}	100.0
Rate for input pulse frequency β	0.11785
Time constant τ_p	20[msec]
Refractory period t_{ndti}	10[msec]
Learning coefficient α	2.0×10^{-7}
Learning iterations	1000
No-Firing Detection Neuron	
Time constant τ_{NFD}	0.5[msec]
Threshold θ_{NFD}	-1.0×10^{-3}
Connection weight to each CL neurons	0.8
Multi-Firing Detection Neuron	
Time constant τ_{MFD}	1.0[msec]
Threshold θ_{MFD}	2.0
Connection weight from each CL neurons	1.0

Table 3. Results of sound recognition

	Recognition Rate[%]		
Input \ Time Lag	$-0.5ms$	$0.0ms$	$+0.5ms$
Acordion	89.9	88.1	88.8
Flute	92.3	94.4	92.4
Piano	62.5	32.9	64.0
Drum	90.3	89.1	88.6
Violin	79.7	78.4	79.0

In this result, the accuracy of "Piano" was particularly bad at the central position. Figure 8 shows the weights of the neurons corresponding to the sounds of "Accordion", "Flute" and "Piano" after learning. Not only the "Piano" neuron does not present any relevant weight but also some of the highest weights are very similar to the weights of other instruments' corresponding neurons (e.g., inputs 4 and 23). The reason for this pour performance is that the "Piano" sound is not constant, presenting a complex variation along a short period of time. This characteristic makes this kind of sound difficult to be learned by the CONP model. Nevertheless, other instruments' sounds could be correctly identified in all positions with accuracies higher than 78%. This confirms the efficiency of the proposed system on identifying multiple sources based on the time lag information.

Similarly to the human being, the proposed system cannot distinguish between two simultaneous similar sound sources. For instance, the results shown

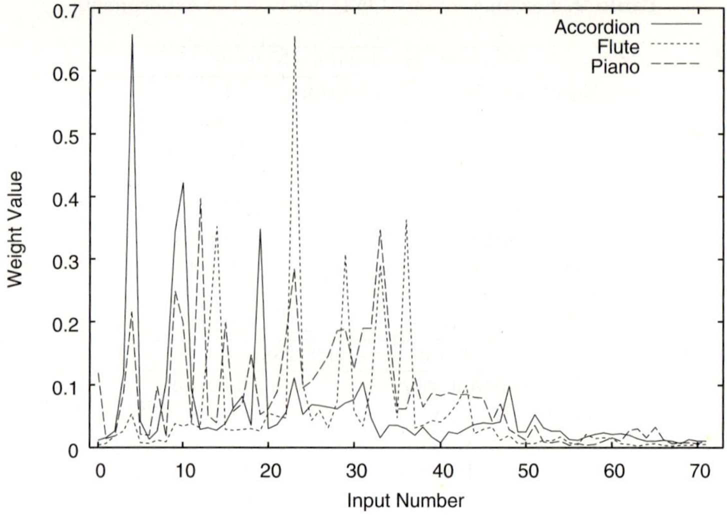

Fig. 8. The weights about three sound source

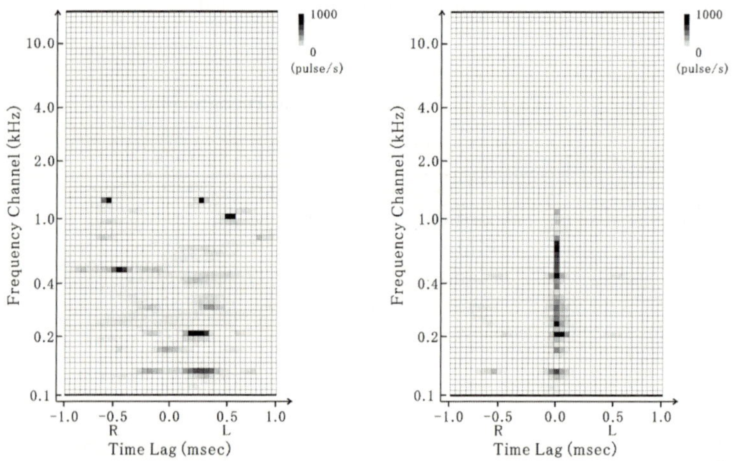

(a) two identical "Violin" signals on the left and central positions and "Flute" signal on the right position

(b) single "Violin" signal on the central position

Fig. 9. Output of Time Difference Extractor for two identical signals

in Fig. 9(a) show the output of the Time Difference Extractor for signal composed by the "Violin" sound coming from the left and central directions (-0.5 ms and 0.0 ms lags) and the "Flute" sound in the right direction (+0.5 ms lag). For reference, Fig. 9(b) shows a single "Violin" signal on the central position. As expected, only two firing patterns can be observed, on corresponding to the "Flute" sound at +0.5 ms and another corresponding to the "Violin" sound at

-0.25 ms. This is, however, an unrealistic situation, as in applications on real environments the occurrence of two identical simultaneous sounds is very improbable, not compromising the applicability of the system.

5 Conclusion

This paper proposes a system for multiple sound source recognition based on a PN model. The system is composed of a time difference extractor, which separates the spectral information of each sound source, and a CONP model which recognizes the sound source type from the firing information of each time lag.

The experimental results confirm that the PN model time difference extractor can successfully separate the spectral components of multiple sound sources. Using the time lag firing information, the sound source type could be correctly identified in almost all cases. The proposed system can separate the multiple sound sources and classify the each sound.

Future works include the application of the proposed system to real sound signals, and also the use of the information of the sound sources type for locating this source with high precision. The implementation of the current system in hardware using an FPGA device is also in progress.

Acknowledgment

This research is supported in part by a grant from the Hori Information Science Promotion Foundation, the Grant-in-Aid for Scientific Research and the Knowledge Clusters (Gifu/Ogaki area), both from the Minister of Education, Culture, Sports, Science and Technology, Government of Japan.

References

1. Pickles, J.O.: An Introduction to the Physiology of Hearing. ACADEMIC PRESS, San Diego (1988)
2. Maass, W., Bishop, C.M.: Pulsed Neural Networks. MIT Press, Cambridge (1998)
3. Kuroyanagi, S., Iwata, A.: Perception of Sound Direction by Auditory Neural Network Model using Pulse Transmission – Extraction of Inter-aural Time and Level Difference. In: Proceedings of IJCNN 1993, pp. 77–80 (1993)
4. Valin, J.M., Michaud, F., Rouat, J., Letourneau, D.: Robust Sound Source Localization Using a Microphone Array on a Mobile Robot. In: Proceedings of IROS 2003, pp. 1227–1233 (2003)
5. Asoh, H., et al.: An Application of a Particle Filter to Bayesian Multiple Sound Source Tracking with Audio and Video Information Fusion. In: Proceedings of The 7th International Conference on Information Fusion, pp. 805–812 (2004)
6. Kuroyanagi, S., Iwata, A.: A Competitive Learning Pulsed Neural Network for Temporal Signals. In: Proceedings of ICONIP 2002, pp. 348–352 (2002)
7. Jeffress, L.A: A place theory of sound localization. J.Comp.Physiol.Psychol. 41, 35–39 (1948)
8. Iwasa, K., et al.: Improvement of Time Difference Detection Network using Pulsed Neuron model, Technical Report of IEICE NC2005-150, pp. 151-156 (2006)

Text-Independent Speaker Authentication with Spiking Neural Networks

Simei Gomes Wysoski, Lubica Benuskova, and Nikola Kasabov

Knowledge Engineering and Discovery Research Institute
Auckland University of Technology, 581-585 Great South Rd, Auckland, New Zealand
{swysoski,lbenusko,nkasabov}@aut.ac.nz
http://www.kedri.info

Abstract. This paper presents a novel system that performs text-independent speaker authentication using new spiking neural network (SNN) architectures. Each speaker is represented by a set of prototype vectors that is trained with standard Hebbian rule and *winner-takes-all* approach. For every speaker there is a separated spiking network that computes normalized similarity scores of MFCC (Mel Frequency Cepstrum Coefficients) features considering speaker and background models. Experiments with the VidTimit dataset show similar performance of the system when compared with a benchmark method based on vector quantization. As the main property, the system enables optimization in terms of performance, speed and energy efficiency. A procedure to create/merge neurons is also presented, which enables adaptive and on-line training in an evolvable way.

Keywords: Spiking Neural Network, Speaker Authentication, Brain-like Pattern Recognition, Similarity Domain Normalization.

1 Introduction

Computer-based speaker authentication presents a number of possible scenarios. Text-dependent, text-independent, long sentences, single words, speaker willing to be recognized, or speaker trying to hide its identity are some examples. For each of these scenarios, different specifically tuned processing techniques seem to be most effective. Of particular interest to our research is the short-sentence text-independent problem, which is typically comprised of input utterances ranging from 3 seconds to 1 minute and to be authenticated, a certain speaker does not necessarily need to present the same word or sentence spoken during the training. Due to the short length of the signal, it is not possible to acquire long-term dependencies of features which could supply additional information to the enhancement of performance. Thus, state machines to detect phonemes, words, and bigrams, can not be setup with full strength.

Based on these properties, during the last years there has been a convergence to the use of, firstly, Vector Quantization (VQ) [1] and, lately Gaussian Mixture Models (GMM) [2][3] to tackle the text-independent speaker authentication problem. These are the methods we use as inspiration for the design of a new spike-based system. VQ is used as benchmark for comparison purposes.

J. Marques de Sá et al. (Eds.): ICANN 2007, Part II, LNCS 4669, pp. 758–767, 2007.

SNNs have been widely used in neuroscience for modelling brain functions. Only recently, attempts to use SNN in machine learning and pattern recognition have been reported [4][5], mainly trying to model vision in a more brain-like way. In speech, SNN has been applied in sound localization [6], and a preliminary study used SNN for recognition of spoken numbers [7][8]. However we are not aware of systems that use SNN to deal specifically with the speaker authentication problem. The use of SNNs for pattern recognition is still in an early stage but it has already being reported that the computation with spikes opens-up some new aspects that are not explicitly present in traditional ways of computing and which could eventually enable further advances in the field. As examples of properties that are not considered in traditional networks we mention:

- Information encoding – Despite it is still not clear how the information encoding effectively happens in the brain, there is strong evidence showing that the spike encoding is optimal in terms of data transmission efficiency (**maximum data transmission**);
- Processing time – Experimental evidence shows that some types of cognitive processes are accomplished in the brain in a very short time (e.g. 150 ms for visual system) and can be improved upon training (**minimum processing time**);
- Energy efficiency – Mammalian brains are known for having more than 10^{10} neurons [9], with neurons operating in a very low spiking rate (1-3 Hz). These numbers suggest that the wiring and connectivity strength are setup is such a way that the processing is done with the minimum energy consumption (**minimum neuronal activity**).

In this work, we present two distinct network architectures that perform classification tasks using spiking neurons. The highlight of the new architectures is the inclusion of two techniques that have already demonstrated to be efficient in traditional methods. They are:

- creation of prototype vectors through unsupervised clustering, and
- adaptive similarity score (*similarity normalization*).

In section 2 we describe the general overview of the speaker authentication system and the speech signal pre-processing stages. Section 3 presents the SNN models and section 4 is devoted to experimental results, which is followed by conclusion and future directions.

2 System Description

Speakers are identified considering physiological (related to the vocal tract) and behavioural characteristics included in a speech signal. The human auditory system extracts these characteristics through the decomposition of the incoming sounds in the frequency domain according to the MEL scale [10], where different frequency bands are processed in parallel pathways. Among numerous models describing such behaviour we use in this work MFCC [10], that are extracted only in the parts of a signal which contains speech. For voice activity detection, we use a simple algorithm based on the energy of a signal after a low-pass Butterworth Filter is applied as described in [11]. Each frame of the signal containing speech fragments generates an MFCC vector that is translated into spikes using Rank Order Coding [4]. A network of spiking neurons is assigned to each speaker, which after proper training decides whether or not a claimant is authenticated.

2.1 Normalizations

Speaker authentication is well known for its high variability between training and test conditions. In order to attenuate this problem, several techniques have been used. The majority of the recent attempts usually normalize the features and/or the way of computing similarity. In the features level (*parameter domain*), we use cepstral mean subtraction of the MFCC features.

In the *similarity domain*, we have embedded within the spiking neural network model a similarity normalization technique, in which the authentication score is calculated not only based on the similarity between a test sample and the speaker model, but on the relative similarity between the test sample and speaker model and test sample and background model. With this procedure, the variations in the testing conditions are taken into account when computing similarity. The normalization on the similarity domain has been already extensively implemented in traditional methods of speaker verification and currently is found in most of state-of-art speaker authentication methods. In our implementation to be described in more details in Section 3, the normalized similarity is computed allocating excitatory connections to neurons representing the claimant model and inhibitory connections to neurons representing the background model.

3 Spiking Neural Network Model

Our design for speaker authentication uses two/three layers feed-forward network of integrate-and-fire neurons where each speaker has its own network. Neurons have a latency of firing that depends upon the order of spikes received and the connections' strengths. The postsynaptic potential (PSP) for a neuron *i* at time *t* is calculated as:

$$PSP(i,t) = \sum \mathrm{mod}^{\,order(j)}\, w_{j,i} \qquad (1)$$

where mod $\in$ (0,1) is the modulation factor, *j* is the index for the incoming connection and $w_{j,i}$ is the corresponding synaptic weight. When *PSP* reaches a given threshold (PSP_{Th}), an output spike is generated and the PSP level is reset. A detailed description of the dynamics of these neurons is given in [4].

3.1 Architecture 1 – Integration of Binary Opinions

Receptive field neurons encode each feature of a frame, typically MFCC, to the time domain (using Rank Order Coding [4]). There is one neuron encoding each feature. The output of the receptive field neurons is a spike time pattern in every frame. Layer 1 (L1) is composed of two neuronal maps. One neuronal map has an ensemble of neurons representing a speaker model (speaker prototypes). Each neuron in the neuronal map is to be trained to respond optimally to different parts of the training utterances. The second neuronal map in L1 is trained to represent the background model. Several ways to represent background models, that can be universal or unique for each speaker, are described and analyzed in [3].

Similarly to L1, L2 has two neuronal maps representing the speaker and the background model. Each L2 neuronal map is composed of a single neuron. L1 and L2 are connected as follows: a) excitatory connections between neurons corresponding to neuronal maps with the same label (L1 speaker to L2 speaker and L1 background to L2 background), and b) inhibitory connections between neurons with differing neuronal map labels (L1 speaker to L2 background and L1 background to L2 speaker). Effectively, L2 neurons accumulate opinions in each frame of being/not being a speaker and being/not being the background. Figure 1 shows the architecture.

The dynamics of the network is described as follows: for each frame of a speech signal, features are generated (currently using MFCC) and encoded into spiking times using the receptive field neurons. The spikes are then propagated to L1 until an L1 neuron emits the first output spike, which is propagated to L2. If a neuron in L2 generates an output spike the simulation is terminated. Otherwise, the next frame is propagated. Before processing the next frame, L1 PSPs are reset to the rest potential, whereas L2 neurons retain their PSPs, which are accumulated over consecutive frames, until an L2 output spike is generated.

The classification is completed when a neuron in L2 generates an output spike or all frames and all spikes in the network have been propagated. If the L2 neuron representing the speaker releases an output spike, the speaker is authenticated. If no spikes occur in L2 after all frames have been processed or an L2 neuron representing the background releases an output spike, the speaker is not authenticated.

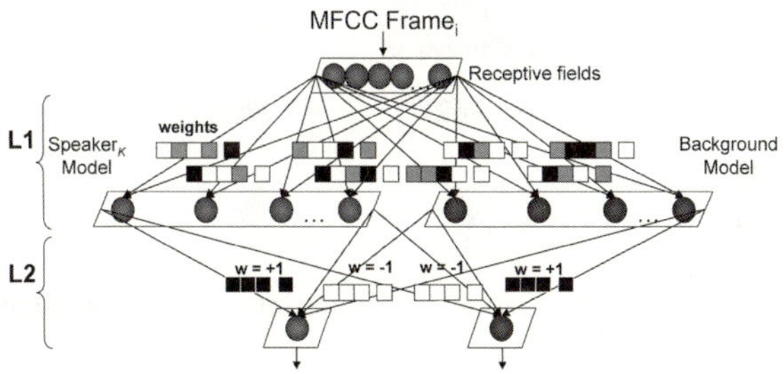

Fig. 1. SNN architecture 1. Frame-by-frame integration of binary opinions.

It is important to notice that in the architecture described, L2 neurons accumulate opinions of being/not being a given speaker collected over several frames. Each frame provides a binary opinion (yes/no), formed based on two criteria: high similarity of the input frame to a certain prototype represented by an L1 neuron, in such way that the similarity causes an L1 neuron to fire (coincidence detector). In addition, the frame needs to be more similar to a speaker prototype than to a prototype representing the background in order to fire first (competition among neurons). The latter property implements the *similarity domain* normalization and enables the network to adapt to variations inherently present in the speaker authentication problem.

The output in each frame, however, does not give a notion of how similar the input frame is from a prototype. In general, traditional methods that apply *similarity domain* normalization, compute the relative distance between the closest prototype of a speaker model and the closest prototype of the background model. To overcome this constraint and to extract the normalized similarity scores in each frame, we propose a second network architecture which requires some additional steps.

3.2 Architecture 2 – Integration of Similarity Scores

In this new and more complex configuration, the network is composed of 3 layers. Similarly to the previous architecture, the encoding of features from each frame into exact spike times in carried out by receptive field neurons. Layer 1 (L1) has two neuronal maps (speaker and background model) where each neuron is trained to respond optimally to a certain input excitation. The neurons in L1 are set to detect the closest prototype in both speaker and background model. Only one neuron is allowed to spike in each L1 map.

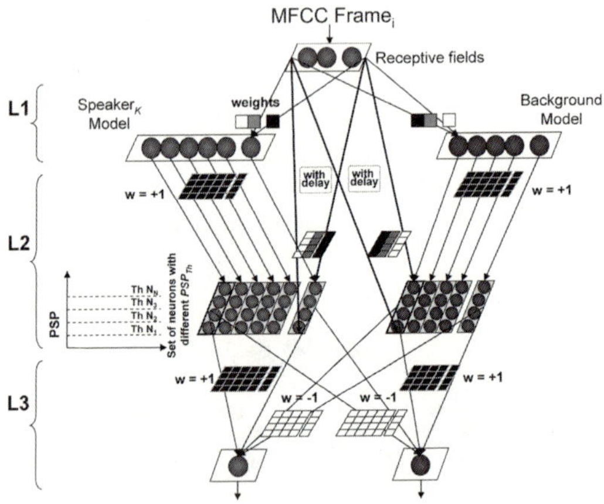

Fig. 2. SNN architecture 2 – Frame-by-frame integration of similarity scores

Each L1 neuron is connected to a set of layer 2 (L2) neurons. The set of L2 neurons are connected to the receptive field neurons with the same connection weights as the corresponding L1 neuron, however they receive the spike train with a certain delay. The delay is set in such a way that L1 output spikes arrive in L2 before the arrival of incoming spikes from the receptive field neurons. L1 output spikes are effectively used to raise the PSP of all neurons in the set to a level where spikes can occur. Thus, in L2 only the neurons belonging to the winner set (closest prototype) become active and can generate output spikes with the arrival of the spikes from the receptive fields. The main characteristic of each set of L2 neurons related to an L1 neuron is that each neuron has the same incoming weight connection from the receptive field neurons, but different PSP thresholds. Therefore the neurons in a set spike at different levels of PSP. Upon the arrival of the input trains of spikes into L2 neurons, several neurons from the winner set are expected to fire. The neurons with lowest PSP_{Th} first, followed by neurons with higher PSP_{Th} levels.

Layer 3 (L3) integrates L2 spikes with excitatory connections between neuronal maps with the same labels and inhibitory connections between neuronal maps with differing labels. Thus, PSP levels in L3 denote the normalized similarity between the most similar speaker prototype and the most similar background prototype.

Similarly to the behaviour of the previous network, PSPs on L1 and L2 are reset every frame, whereas in L3 the PSPs are accumulated over several frames. Figure 2 illustrates the network architecture which implements the normalized similarity between the closest prototypes.

The simulation is terminated when L3 emits an output spike or there are no more frames to be processed. Each frame is processed until all the spikes are propagated or until all L2 neurons representing the speaker or background emit spikes.

3.3 Synaptic Plasticity and Structural Adaptation – SNN Training

The training is done in the synapses connecting receptive field neurons and L1 in a similar fashion for both network architectures. To update weights during training, a simple rule is used [4]:

$$\Delta w_{j,i} = \text{mod}^{order(j)} \tag{2}$$

where $w_{j,i}$ is the weight between the receptive field neuron j and neuron i of the L1, mod $\in$ (0,1) is the modulation factor, *order*(j) is the order of arrival to neuron i of a spike produced by neuron j. For each training sample, we use the *winner-takes-all* approach, where only the neuron that has the highest *PSP* value in L1 has its weights updated.

The postsynaptic threshold (PSP_{Th}) of a neuron is calculated as a proportion $c \in [0, 1]$ of the maximum postsynaptic potential (*PSP*) generated with the propagation of the training sample into the updated weights, such that:

$$PSP_{Th} = c \max(PSP) \tag{3}$$

The procedure for training the network and creating new neurons is adapted from [5] and is summarised with the following pseudo-code:

```
Until weights converge
    For all phrase samples in the training set
        For each frame
            Create a new neuron
            Propagate the frame into the network
            Train the newly created neuron using equations (2) and (3)
            Calculate the similarity between weight vectors of newly created
            neuron and existent neurons within the neuronal map
            If similarity > Threshold
                Merge newly created neuron with the most similar neuron using
                (4) and (5)
```

To merge a newly created neuron with an existing neuron the weights W of the existing neuron n are updated calculating the average as

$$W = \frac{W_{new} + N_{Frames} W}{1 + N_{Frames}} \tag{4}$$

where N_{Frames} is the number of frames previously used to update the respective neuron. Similarly, the average is also computed to update the corresponding PSP_{Th}:

$$PSP_{Th} = \frac{PSP_{Thnew} + N_{Frames} PSP_{Th}}{1 + N_{Frames}} \tag{5}$$

Alternatively, the network structure and the number of desired prototypes (neurons) can be defined *a priori*, using a k-means-like clustering algorithm to update the weights of the winner neuron. In this case, a simple heuristic can be described in two steps:

1. Initialization of neurons' weights

```
For each neuron
   Propagate a random frame of the training set into the network
   Update the neuron's weights using (2) and (3)
```

2. Recursive training

```
Until weights converge
   For all phrase samples in the training set
       For each frame
           Propagate each frame into the network
           Find the maximally activated neuron (the neuron with maximum PSP)
           Create a new neuron and train it using (2) and (3). Update weights of
           the maximally activated neuron merging it to the new neuron (using (4)
           and (5))
```

In our experiments we use the latter method to try to reproduce as close as possible the scenario of the benchmarking algorithm (VQ with k-means clustering).

In SNN architecture 1, L1 neurons are fully connected to neurons in L2. The weights are set in order to accumulate positive or negative opinions of each input frame to each speaker (W = 1 for the links between each L1 neuronal map and its corresponding L2 neuron. W = -1 when the label of L1 neuronal map differs from the label of L2 neuron.

In SNN architecture 2, the connections between L1 neuron and the corresponding set of neurons in L2 are excitatory (W = 1). Neurons in L2 are fully connected to L3 neurons. There are excitatory connections (W = 1) between neurons belonging to the neuronal maps with same label and inhibitory connections (W = -1) otherwise.

4 Experimental Results

We have implemented the spiking network models proposed in the previous sections and used the speech part of the VidTimit dataset [12] for performance evaluation. VidTimit contains 10 utterances of 43 different speakers. In order to reproduce the experiments described in [12], the system is trained to authenticate 35 users using 6 utterances from each user. The remaining 4 utterances of each user have been used for test. In addition, the 4 utterances of the 8 remaining users have been used to simulate impostor access. Thus, the number of true claimants for each individual model is 4 utterances, and the number of impostors that try to break into each model is (35 - 1 remaining users x 4 utterances) + (8 impostors x 4 utterances), which gives a total of 168. For all individual models of the entire dataset, we have (35 users x 4 utterances), totalizing 140 true claimant samples and (35 users x 168 utterances) = 5880 false claimant samples.

The speech signal is sampled at 16 kHz, and features are extracted using standard MFCC with 19 MEL filter sub-bands ranging from 200 Hz to 7 kHz. Each MFCC feature is then encoded into spikes. We have 19 neurons in L1. Encoding was not further optimized despite experiments demonstrate that it plays an important role to the overall performance of the system.

We train a specific background model for each speaker. For the sake of simplicity we use the following procedure. The background model of a speaker *i* is trained using the same amount of utterances used to train the speaker model, with the utterances randomly chosen from the remaining training speakers.

For comparison purposes, we have also implemented a standard vector quantization (VQ) algorithm [1] with k-means clustering. Training was done with the same features (19 MFCCs) and the same strategy for selecting background models was applied. We tested the performance for different number of prototypes. We are only reporting here the best result, which has been obtained with 32 prototypes for speakers and 32 prototypes for the background model. The prototypes were selected without any systematic optimization procedure. The VQ performance can be seen in Figure 3. These results are comparable with the work presented by [12], where with the same dataset authors reported total error TE = false acceptance rate (FAR) + false rejection rate (FRR) = 22 % in slightly different setup conditions using Gaussian Mixture Model.

With respect to the SNN implementation, we have defined *a priori* the number of neurons in L1 neuronal maps for the speaker and background model (80 neurons each). The modulation factor was set to 0.9 for L1 neurons in architecture 1 and L1 and L2 neurons in architecture 2. The other layers are composed of neurons with modulation factor = 1.

In the experiments with SNN architecture 1, PSP_{Th} of L2 neurons are defined as a proportion p of the number of frames used for identification. For instance, if an utterance used for authentication is composed of 40 frames and p is 0.2, the PSP_{Th} used for authentication is 40 x 0.2 = 8. PSP_{Th} of L1 neurons are calculated as a proportion c of the maximum PSP during training according to (3). The performance for $p = 0.2$ and different values of c are shown in Figure 4 (on left).

In the SNN architecture 2, PSP_{Th} of L3 neurons are defined as a proportion p (0.2 was used) of the number of frames used for identification. PSP_{Th} of L1 neurons are calculated as a proportion c of the maximum PSP during training according to (3). A set of L2 neurons have the PSP_{Th} levels ranging from 0 to the maximum PSP_{Th} of their corresponding L1 neuron (equally spaced). Figure 4 (on right) shows a typical performance using 20 PSP_{Th} levels for different c. Notice that, in this scenario for values of c below 0.4 the FRR starts to rise again. This trend is expected when the system reaches an operation point where the set of PSP_{Th} levels in L2 are not acting properly to compute distances.

In our experiments, SNN parameters were optimized by hand. Figure 3 and Figure 4 clearly show that VQ and SNN manifest a similar error trend, with a slightly better performance to VQ when the FAR and FRR curves intersect each other. However, we can not conclude that VQ has a better performance when comparing to SNNs nor that one SNN network configuration is better than another. We can conclude that both network architectures proposed here are able to process frames of speech data using spiking times, can accumulate opinions over many frames and can discern whether they represent or not represent previously trained patterns. The advantages of these networks are that they enable to perform multiple criteria optimization of parameters to reach data transmission efficiency, minimum processing time, and minimum energy consumption.

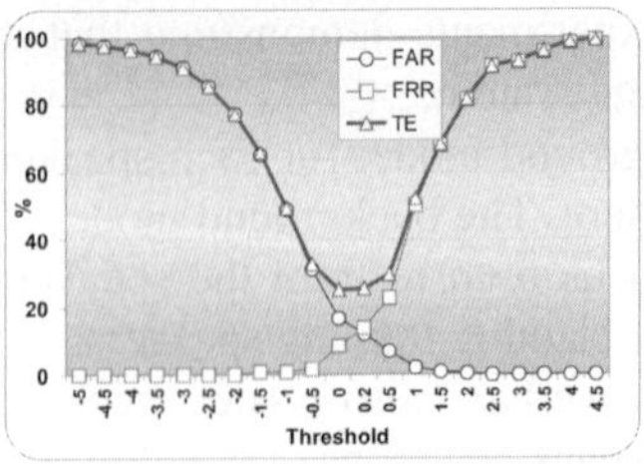

Fig. 3. Vector Quantization (VQ) performance on VidTimit dataset. FAR is the false acceptance rate, FRR is the false rejection rate, and TE is the total error (FAR+FRR).

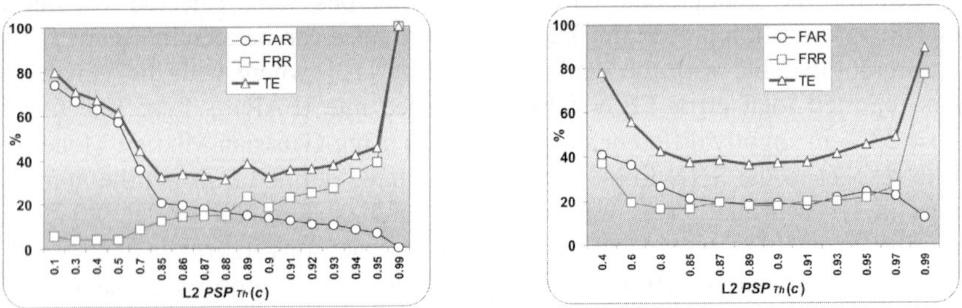

Fig. 4. Typical SNNs performance for different values of c (proportion of the maximum PSP generated by a training sample). On left: SNN Architecture 1. On right: SNN Architecture 2. The same trained weights were used in both architectures.

5 Conclusion and Future Directions

We have presented for the first time two network architectures of spiking neurons for speaker authentication. These networks process streams of speech signals in a frame-based manner. The output layers accumulate positive and negative opinions of being a certain speaker as well as being a background. The main difference is that, in each frame, architecture 1 outputs a binary opinion and architecture 2 gives a notion of similarity between the incoming frame and the closest prototypes.

Connection weights between receptive fields and L1 can be trained to respond to different parts of an utterance, closely corresponding to the usage of k-means algorithm to create codebooks [1], or a set of gaussians in GMM [2][12]. Our models also incorporate the idea of normalized similarity, which demonstrated to be very effective in several classical models [1][2].

The main drawback of the system is that the neuron model implemented in this work [4] reduces its classification efficiency when the features are encoded in small numbers of neurons in a non sparse space. Thus, it may be required to use population of neurons to add sparseness to features encoding to handle more challenging tasks.

The procedures suggested in Section 3.3 based on k-means and network structural adaptation enable continuous and adaptive training. The main properties of these procedures are: a) k-means: needs to define in advance the number of neurons, can present initialization and local minima problem; b) network structural adaptation: an additional parameter for merging neurons needs to be tuned, and a different division of the feature space can be obtained according to the order of the training samples [5].

While the dynamics of the network architectures have proven suitable to perform speaker authentication, further development is needed to take advantage of their main properties, in particular with the inclusion of multi-criteria parameter optimization procedures to reach a better data encoding [13], minimize processing time and reduce the overall number of spikes. In this direction, we plan to optimize parameters in larger scale experiments of speaker authentication.

Acknowledgments

The work has been supported by the Tertiary Education Commission of New Zealand (S.G.W.) and by the NERF grant AUTX02001 funded by FRST (L.B., N.K.).

References

1. Burileanu, C., Moraru, D., Bojan, L., Puchiu, M., Stan, A.: On performance improvement of a speaker verification system using vector quantization, cohorts and hybrid cohort-world models. International Journal of Speech Technology. 5, 247–257 (2002)
2. Reynolds, D.A., Quatieri, T.F., Dunn, R.B.: Speaker verification using adapted Gaussian Mixture Models. Digital Signal Processing. 10, 19–41 (2000)
3. Bimbot, F., et al.: A tutorial on text-independent speaker verification. EURASIP Journal on Applied Signal Processing. 4, 430–451 (2004)
4. Delorme, A., Gautrais, J., van Rullen, R., Thorpe, S.: SpikeNet: a simulator for modeling large networks of integrate and fire neurons. Neurocomputing, 26–27, 986–996 (1999)
5. Wysoski, S.G., Benuskova, L., Kasabov, N.: On-line learning with structural adaptation in a network of spiking neurons for visual pattern recognition. In: Kollias, S., Stafylopatis, A., Duch, W., Oja, E. (eds.) ICANN 2006. LNCS, vol. 4131, pp. 61–70. Springer, Heidelberg (2006)
6. Kuroyanagi, S., Iwata, A.: Auditory pulse neural network model to extract the inter-aural time and level difference for sound localization. Trans. of IEICE. E77-D 4, 466–474 (1994)
7. Loiselle, S., Rouat, J., Pressnitzer, D., Thorpe, S.: Exploration of Rank Order Coding with spiking neural networks for speech recognition. IJCNN, Montreal, pp. 2076–2080 (2005)
8. Rouat, J., Pichevar, R., Loiselle, S.: Perceptive, non-linear speech processing and spiking neural networks. In: Chollet, G., et al. (eds.) Nonlinear Speech Modeling. LNCS(LNAI), vol. 3445, pp. 317–337. Springer, Heidelberg (2005)
9. Gerstner, W., Kistler, W.M.: Spiking Neuron Models. Cambridge Univ. Press, Cambridge MA (2002)
10. Gold, B., Morgan, N.: Speech and Audio Signal Processing. John Wiley & Sons, Chichester (2000)
11. Rabiner, L., Juang, B.: Fundamentals of Speech Recognition. Prentice Hall, New Jersey (1993)
12. Sanderson, C., Paliwal, K.K.: Identity verification using speech and face information. Digital Signal Processing. 14, 449–480 (2004)
13. Bothe, S.M., La Poutre, H.A., Kok, J.N.: Unsupervised clustering with spiking neurons by sparse temporal coding and multi-layer RBF networks. IEEE Trans. Neural Networks 10(2), 426–435 (2002)

Interferences in the Transformation of Reference Frames During a Posture Imitation Task

Eric L. Sauser and Aude G. Billard

Learning Algorithms and Systems Laboratory, LASA
Ecole Polytechnique Fédérale de Lausanne, EPFL, Switzerland
{eric.sauser,aude.billard}@epfl.ch

Abstract. We present a biologically-inspired neural model addressing the problem of transformations across frames of reference in a posture imitation task. Our modeling is based on the hypothesis that imitation is mediated by two concurrent transformations selectively sensitive to spatial and anatomical cues. In contrast to classical approaches, we also assume that separate instances of this pair of transformations are responsible for the control of each side of the body. We also devised an experimental paradigm which allowed us to model the interference patterns caused by the interaction between the anatomical on one hand, and the spatial imitative strategy on the other hand. The results from our simulation studies thus provide predictions of real behavioral responses.

1 Introduction

Although imitation has been extensively addressed in developmental psychology, it has become a current topic in neuroscience and experimental psychology [1, 3, 6, 9, 10, 11, 4]. The starting point of these investigations was the discovery, in monkey and human brain, of *mirror neurons*, which are activated by both the execution and the observation of goal-directed actions [1, 10, 11]. In humans, the mirror circuit, gets also activated during the presentation of intransitive gestures or body postures [10]. In this work, we focus on the process of transformation across frames of reference, required for imitation of arbitrary gestures. In psychology, anatomical and spatial types of imitation are usually considered distinct [5, 6, 9, 10]. On one hand, anatomical imitation considers the observed movements with respect to the observed person's body. On the other hand, spatial imitation considers only the spatial location of the limbs with respect to the observer, regardless of the orientation of the demonstrator. When the imitator and the demonstrator are facing each other, this form of imitation is usually denoted as specular or mirror [4, 6, 5, 10].

We hypothesize that the computations associated with these two forms of imitation are simultaneously computed in the brain. Given the task constraints, a competitive process then selects the correct response [5, 4, 6, 13]. Such a competition usually produces measurable interferences on reaction times [5, 4, 9]. In addition to the previous hypothesis, we suggest that anatomical imitation should

J. Marques de Sá et al. (Eds.): ICANN 2007, Part II, LNCS 4669, pp. 768–778, 2007.

not be considered too strictly. We propose here that an anatomical mapping between contralateral limbs, which mirrors the relationship between the limb joints, also exists. Therefore, our model assumes that distinct pairs of spatial and anatomical transformations are responsible of the control of each side of the body. As a consequence, when an arm posture is presented with either the left or the right arm for example, an imitative response is computed in parallel for both arms of the imitator. Here, we apply a biologically-inspired modeling approach, known as the Dynamic Field Theory [8, 13], to the problem of conflicting transformations across frames of reference. We will first present an experimental paradigm which will help determine the interferences between different imitative strategies during a task requiring the imitation of meaningless body postures. Then, we will briefly describe a neural model, capable of computing both anatomical and spatial imitative transformations. Finally, we will discuss the particular interference patterns predicted by our model and their implications for future research.

2 Model

2.1 Experimental Scenario and Setup

We consider the imitation of body postures where the orientation of the right upper arm is varied. The visual perspective of the demonstrator's body can also vary from side to front view. The task instructions require either a spatial or an anatomical imitative response with either the right arm (the corresponding one) or the left arm (the opposite one). The stimulus variables, shown in Figure 1, are: φ^D, the demonstrator arm elevation, θ^D, its orientation relative to the body in the horizontal plane, and ϕ^D, the orientation of the body with respect to the observer. The response variables are: φ^I_L and φ^I_R, the elevation of the left and right arm of the imitator, and θ^I_L and θ^I_R, their orientations on the horizontal plane. The desired responses are:

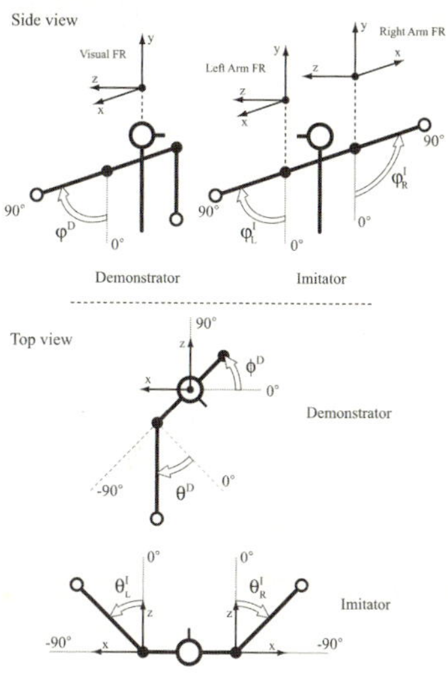

Fig. 1. Experimental variables and considered frames of reference (FR)

$$\begin{aligned} \theta_L^{I,\mathrm{A}} &= \theta_R^{I,\mathrm{A}} = \theta^D \\ \theta_L^{I,\mathrm{S}} &= -\theta_R^{I,\mathrm{S}} = \begin{cases} -180 - (\theta^D + \phi^D) & \theta^D + \phi^D < -90 \\ \theta^D + \phi^D & |\theta^D + \phi^D| \le 90 \\ 180 - (\theta^D + \phi^D) & \theta^D + \phi^D > 90 \end{cases} \end{aligned} \tag{1}$$

where the additional index, A or S, denotes the anatomical or the spatial imitative strategy, respectively. An illustration of these transformations are shown in Figure 2. Let us then define the discrepancy D between the response of both strategies, which is is given by the difference between the response of the instructed strategy and that of the other. Spatial and anatomical transformations are said to be perfectly congruent when the discrepancy $D = 0$. Note that ideal congruency conditions are not equivalent for both arms.

An experimental trial consists first of the presentation of a starting posture which the model is requested to imitate according to the task instructions. Then, the arm posture is abruptly changed, and the subject has to keep imitating as fast as possible. During a single trial, only the arm posture is modified, whereas the body orientation is left unchanged. Experiment 1 investigates the interferences produced by both imitative strategies when their initial responses are congruent and the amplitude of the change of arm posture is kept constant across the trials. The pair of initial and target postures consists of the arm raising from a neutral down position ($\varphi^D = 0°$), where the responses of both transformations are always congruent, to a position on the horizontal plane ($\varphi^D = 90°$). The arm elevation is thus the only degree of freedom which changes during a trial. Complementarily, Experiment 2 investigates the influence of a horizontal postural change, which amount is denoted by $\Delta\theta^D$. Indeed, in such conditions, depending on the stimuli, the discrepancy between the responses of the transformations may vary.

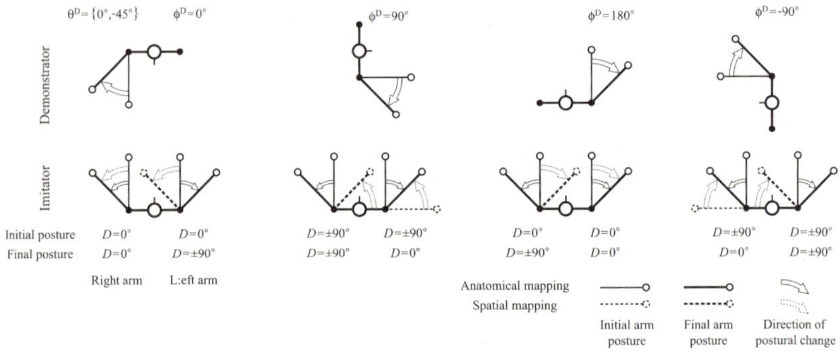

Fig. 2. Examples of anatomical and spatial imitative strategies in various conditions. The discrepancy between the response of the transformations are given for each arm.

2.2 Neural Fields

This section briefly describes our model, which is composed of networks known as neural fields [8,12,13]. Formally, a neural field is composed of a continuous set of neurons, where each of them fires maximally for a specific value $\boldsymbol{r}$ uniformly distributed in the parameter space Γ. Since the modeled variables consist of arm and body orientations, we consider the parameter space as the ensemble of directions in the three dimensional space, i.e, $\Gamma = \{\boldsymbol{r} \in \mathbb{R}^3 |\; \|\boldsymbol{r}\| = 1\}$. Each neuron of the network is fully connected by means of recurrent synaptic weights W^R, exhibiting symmetry, rotational invariance and center-surround characteristics. The neural field dynamics follows

$$\tau\,\dot{u}(\boldsymbol{r},t) = -u(\boldsymbol{r},t) + x(\boldsymbol{r},t) + h(t) + \oint_\Gamma W^R(\boldsymbol{r}',\boldsymbol{r})\, f\big(u(\boldsymbol{r}',t)\big)\, \mathrm{d}\boldsymbol{r}' \qquad (2)$$

where $u(\boldsymbol{r},t) \in \mathbb{R}$ is the membrane potential of the neuron with time constant $\tau \in \mathbb{R}$ and the preferred direction $\boldsymbol{r}$ at time t. $f(y) = \max(0,y)$, $x(\boldsymbol{r},t)$ corresponds to the external input and $h(t)$ to a global modulatory input. The weight linking two neurons, with preferred directions $\boldsymbol{r}'$ and $\boldsymbol{r}$, is given by a periodical Gaussian-like profile defined as

$$W^R(\boldsymbol{r}',\boldsymbol{r}) = \alpha^R \left(g(\boldsymbol{r}',\boldsymbol{r}) - 1\right) \quad \text{where} \quad g(\boldsymbol{r}',\boldsymbol{r}) = \frac{1}{\kappa}\exp\left(\frac{(\boldsymbol{M}\boldsymbol{r}')^T\boldsymbol{r} - 1}{2\sigma^2}\right) \tag{3}$$

$\alpha^R > 0$ and $\sigma > 0$ are, respectively, the amplitude and variance of the weight profile. $\boldsymbol{M}$ refers to a transformation or mapping matrix. In this case $\boldsymbol{M} = \boldsymbol{I}$, i.e., the identity matrix, but different mappings will be described later in the text. $\kappa = 1 - e^{-\frac{1}{\sigma^2}}$ is a normalization factor ensuring that $g(\boldsymbol{r}',\boldsymbol{r}) \in [0,1]$. This type of neural dynamics is known to form an attractor bump on the surface of the neural field (see Fig. 3), through which this class of networks is suggested to convey information. As a read-out mechanism, we consider the population vector $\hat{\boldsymbol{p}}(t) \in \Gamma$. It consists of a weighted summation of the firing rate of each neuron with its preferred direction and is given by

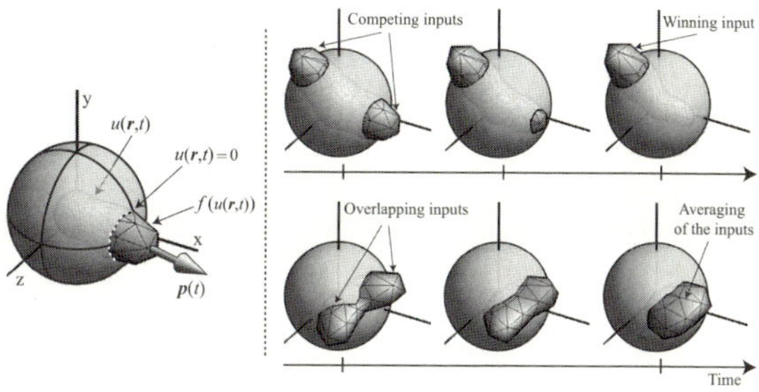

Fig. 3. (Left) Graphical representation of a neural field activity. (Right) Evolution of the neural activity during a selection process involving two competing inputs (top), and between two partially overlapping inputs (bottom).

$$\hat{\boldsymbol{p}}(t) = \boldsymbol{p}(t)/\|\boldsymbol{p}(t)\| \quad \text{where} \quad \boldsymbol{p}(t) = \oint_\Gamma f\big(u(\boldsymbol{r},t)\big)\,\boldsymbol{r}\,\mathrm{d}\boldsymbol{r} \tag{4}$$

Moreover, we define $E(t) = \|\boldsymbol{p}(t)\|$ as the energy of the network response. Since further in the text we consider several neural fields within a large network, we will denote with an index i the network variables corresponding to those of a neural field i. The external input $x^i(\boldsymbol{r},t)$ can be composed of a direct sensory input and of synaptic projections from one or several different neural fields. A sensory input is constrained to represent a variable value $\boldsymbol{s}(t) \in \Gamma$, and projection from a population j to a population i is done through synaptic weights W^{ji}. The external input of a neural field i is then written as

$$x^i(\boldsymbol{r},t) = \beta^i\left(g(\boldsymbol{r},\boldsymbol{s}^i(t)) - \eta\right) + \sum_j \oint_\Gamma W^{ji}(\boldsymbol{r}',\boldsymbol{r})\,f\big(u^j(\boldsymbol{r}',t)\big)\,\mathrm{d}\boldsymbol{r}' \quad \text{where}$$
$$W^{ji}(\boldsymbol{r}',\boldsymbol{r}) = \alpha^{ji}\left(g(\boldsymbol{r}',\boldsymbol{r}) - \eta^{ji}\right) \tag{5}$$

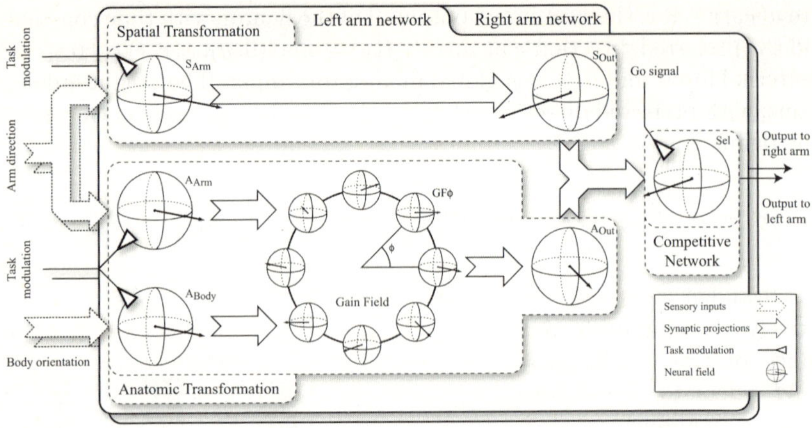

Fig. 4. Architecture of the model: Within each network corresponding to a given arm, two streams compute separately the anatomical and the spatial transformations

where β^i is the strength of the representation of the sensory variable $\boldsymbol{s}^i(t)$, η^{ji} is a normalization term, and $\alpha^{ji} > 0$ is the amplitude of the weights. Similarly, the modulatory input $h^i(t)$ can consist of a constant input or of synaptic projections from other neural populations. In the latter case, we have

$$h^i(t) = \sum_j \oint_\Gamma W^{ji}(\boldsymbol{r}', \boldsymbol{r}^i)\, f\big(u^j(\boldsymbol{r}', t)\big)\, \mathrm{d}\boldsymbol{r}' \qquad (6)$$

where $\boldsymbol{r}^i \in \Gamma$ is constant and W^{ji} corresponds to that defined in Equ. (5).

2.3 Network Architecture

The model architecture, depicted in Figure 4, consists of two networks, one for each arm. Within a single network, two main streams process separately the spatial and the anatomical transformation. Their outputs are projected to a competitive network to select the appropriate response. Since the task instructions specify which arm should be used, the model does not perform the selection of the effector.

As External Inputs, the two streams receive visual input in the form of the arm and body orientation vectors $\boldsymbol{s}^{\mathrm{Arm}}$ and $\boldsymbol{s}^{\mathrm{Body}} \in \Gamma$, relative to the reference frame of the observer (shown in Fig. 1):

$$\begin{aligned} \boldsymbol{s}^{\mathrm{Arm}} &= \big(\sin(\varphi^D)\sin(\theta^D+\phi^D)\,,\ \cos(\varphi^D)\,,\ -\sin(\varphi^D)\cos(\theta^D+\phi^D)\big)^T \\ \boldsymbol{s}^{\mathrm{Body}} &= \big(\sin(\phi^D)\,,\ 0\,,\ -\cos(\phi^D)\big)^T \end{aligned} \qquad (7)$$

These inputs are fed into the input populations of each transformation according to Equ. (5). Note that the spatial transformation does not need the orientation of the demonstrator's body. The input populations also receive an external modulation applied asymmetrically to each stream. According to the task instructions, the inputs of the relevant network receive a positive modulation, whereas those of the other receive inhibition, i.e., h^{Task} and $-(h^{\text{Task}} + \Delta h)$, respectively, where $h^{\text{Task}} > 0$ is a constant, and $\Delta h \in \{0, \Delta h_0\}$. When $\Delta h = \Delta h_0 \gg 0$, the network corresponding to the irrelevant transformation is completely inhibited. This case, where only the relevant transformation is active, is considered as the baseline condition.

The Spatial Transformation consists of mapping the orientation of the demonstrator's arm with the imitator's left and right arm, regardless of the demonstrator's body. For a given arm, two neural populations are required. The former receives the visual input and is connected to the latter through synaptic projections. Using Eqs. (1) and (5), the correct mapping functions for the left and the right arm are given by $\boldsymbol{M} = \boldsymbol{M}^{\text{L,Sp}}$ and $\boldsymbol{M} = \boldsymbol{M}^{\text{R,Sp}}$, respectively, where

$$\boldsymbol{M}^{\text{L,Sp}} = \begin{cases} \mathbf{diag}(1, 1, -1) & \boldsymbol{s}_z^{\text{Arm}} < 0 \\ \boldsymbol{I} & \text{otherwise} \end{cases} \quad \text{and} \quad \boldsymbol{M}^{\text{R,Sp}} = \mathbf{diag}(-1, 1, 1)\boldsymbol{M}^{\text{L,Sp}} \tag{8}$$

Since the frames of references of each arm are symmetric, so are the mapping matrices.

The Anatomical Transformation requires the combination of the orientation of the demonstrator's arm and that of his/her body [12]. Neurophysiological data suggest that such a transformation is performed through gain fields, which are neural populations combining inputs from several external sources [14]. We define a gain field as a continuous set of neural fields denoted by GFϕ, where each of them is preferentially tuned to a specific body orientation ϕ. The population encoding the demonstrator's arm orientation projects to each of them using Equ. (5) with mapping function $\boldsymbol{M} = \boldsymbol{R}_y(-\phi)$, where $\boldsymbol{R}_y(-\phi)$ is the rotation matrix around axis Y with angle $-\phi$. The body orientation is fed to the subfields through their modulatory input $h^{\text{GF}\phi}(t)$ according to Equ. (6), with $\boldsymbol{M} = \boldsymbol{I}$ and $\boldsymbol{r}^{\text{GF}\phi} = (\sin\phi, 0, -\cos\phi)$. The gain field projects to the output population of the transfromation by synaptic projections with $\boldsymbol{M} = \boldsymbol{I}$.

The Response Selection is performed by a neural field receiving projections from the output population of both transformations. The competition arises naturally as an effect of the network recurrent connectivity, producing a winner-take-all type of operation [8, 13]. As illustrated in Figure 3, according to the intrinsic distance metric given by the breadth σ of the recurrent connections, close and overlapping inputs tend to average whereas distant ones compete. In our model, since the output strength of both streams are asymmetrically balanced, the correct response is always selected by the network. The network also receives a go signal by means of its modulatory input. Prior to the presentation of the target posture, $h^{\text{Sel}}(t) = -h^{\text{Go}} \ll 0$ so that the neural field is completely inactive. When the target posture is presented, the network is uninhibited, i.e.,

$h^{\text{Sel}}(t) = 0$, and the selection process begins. The network response is read-out using the population vector (Equ. (4)), which directly represents the selected arm posture in the frame of reference of the imitator.

3 Results

We simulated the two experiments described in Section 2.1. The demonstrator's arm and body postures, were systematically varied across each trial during both experiments[1]. Moreover, in experimental conditions involving the use of the left arm, the subnetwork corresponding to the right arm was not considered, and vice versa.

3.1 Reaction Times and Accuracy

The mean reaction times and the errors resulting from the transformations were measured in both experiments. Reaction times (RT) were defined as the time when the response energy $E(t)$ of the selection network reached a given threshold, whereas the transformation errors (Err) were defined as the angular distance between the population vector response $\hat{\boldsymbol{p}}(t)$ after network convergence, and the correct target position. Moreover, since we do not model the dynamics of arm movements, reaction times should be considered to be times of movement initiation rather than times of movement completion.

Fig. 5. a) Mean reaction times and b) transformation errors observed in both experiments. c) Reaction times during Experiment 2 in the baseline conditions shown according to the amount of postural change $\Delta\theta^D$.

[1] The respective range of arm and body orientations are $\theta^D \in \{k \cdot 22.5°|k \in \{0..8\}\}$ and $\phi^D \in \{k \cdot 22.5°|k \in \{0..15\}\}$. The amplitude of postural change in Exp. 2 is in the range $\Delta\theta^D \in \{k \cdot 22.5°|k \in \{1..8\}\}$. The model parameters are: the amplitudes of the weights, $\alpha^R = 12$, $\alpha^{\text{S}_{\text{Arm}},\text{S}_{\text{Out}}} = \alpha^{\text{A}_{\text{Arm}},\text{GF}\phi} = \alpha^{\text{GF}\phi,\text{A}_{\text{Out}}} = 5.4$, $\alpha^{\text{A}_{\text{Body}},\text{GF}\phi} = 8.0$, and $\alpha^{\text{A}_{\text{Out}},\text{Sel}} = \alpha^{\text{S}_{\text{Out}},\text{Sel}} = 5.0$; the breadth of the weights profiles and their offset, unless specified, $\sigma = 0.5$ and $\eta = \oint_\Gamma g(\boldsymbol{r},\boldsymbol{r}')\,\mathrm{d}\boldsymbol{r}$, then $\sigma^{\text{A}_{\text{Body}},\text{GF}\phi} = \infty$ and $\eta^{\text{A}_{\text{Arm}},\text{GF}\phi} = 1.0$; the amplitude of the inputs, $\beta^{\text{S}_{\text{Arm}}} = \beta^{\text{A}_{\text{Arm}}} = \beta^{\text{A}_{\text{Body}}} = 0.5$. These parameters were chosen so that the response energy of both transformations are equal for an equivalent task modulation. Finally, the task modulatory inputs and go signal are, $h^{\text{Task}} = 0.5$, $\Delta h_0 = 0.75$ and $h^{\text{Go}} = 1.5$.

As reported in Figure 5a, in both experiments, the average reaction times in anatomical conditions were longer than in the spatial task. Indeed, the former transformation requires more computations. In addition, a slight, but not significant increase in average reaction times can be noticed when comparing the normal condition with the baseline. Nevertheless, a difference between these conditions was observed on the transformation errors (see Fig. 5b). Indeed, a competition between the parallel transformations results in larger errors. In Experiment 2, the amplitude of the postural change $\Delta\theta^D$ was different across trials. The reaction times dependency on this experimental variable in the baseline conditions is shown in Figure 5c. For small postural changes, reaction times were longer, but then decreased for larger $\Delta\theta^D$. This effect is caused by the center-surround recurrence in the neural dynamics, resulting in longer convergence times when moving from one attractor state to another, which is sufficiently close.

3.2 Interference Patterns

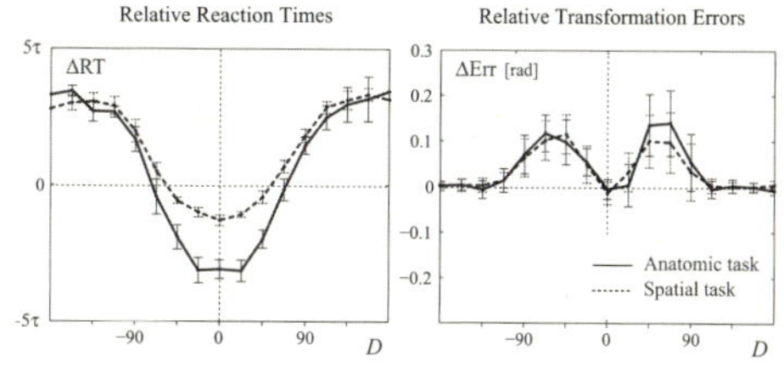

Fig. 6. Results of Experiment 1: Reaction times and transformation errors relative to the baseline condition are shown

Then, we were interested in determining the interference patterns resulting from the competition between the two transformations. The reaction times and transformation errors were considered relative to the baseline conditions. Let us denote them, respectively, by $\Delta\mathrm{RT} = \mathrm{RT} - \mathrm{RT}_0$, and $\Delta\mathrm{Err} = \mathrm{Err} - \mathrm{Err}_0$, where RT_0 and Err_0 correspond to the reaction times and errors measured in the baseline conditions. In Figure 6, data from Experiment 1 are given according to the discrepancy D between the responses of the anatomical on one hand, and the spatial transformation on the other hand. First, since the processing time of the spatial transformation is shorter, it interferes earlier with the anatomical transformation, and conversely. As an effect, the strength of the interferences on reaction times were globally higher in anatomical conditions. Next, the reaction times increased with the discrepancy between the responses, whereas transformation errors behaved slightly differently. The errors did also increase with the discrepancy, but only within a small range. For outermost distances, they decreased until approximately zero. This effect is the result of the averaging of close responses on the neural field. Similar effects were observed in Experiment 2 (see Fig. 7), i.e., the interference patterns were globally more important under anatomical conditions and the error patterns also depended on the discrepancy between the responses. Further, the interference patterns on reaction times exhibited a combination of the effects of both the discrepancy D between the responses and the amount $\Delta\theta^D$ of arm postural change that were shown earlier in Figure 5c and 6. In conditions close to ideal congruency between the transformations, a general facilitatory effect was primarily produced which was even stronger for mid-range distances. In addition, an interaction between both variables on reaction times

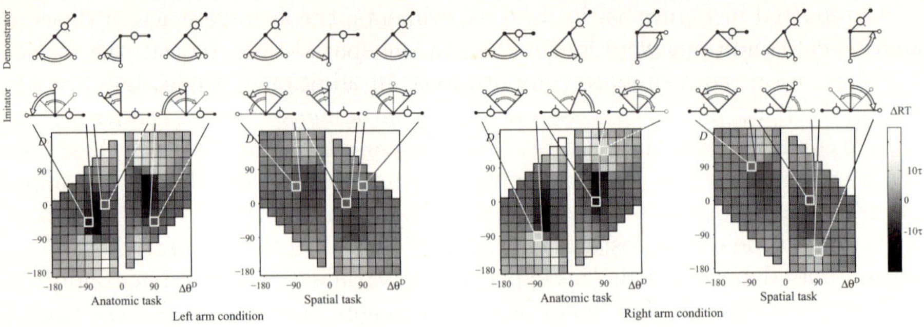

Fig. 7. Results of Experiment 2: Reaction times relative to the baseline condition. On top of each plot, examples of experimental conditions are shown. In each case, the largest arrow corresponds to the response of the relevant transformation.

was observed. It produced a small shift of the interference pattern relative to the discrepancy D, which depended on $\Delta\theta^D$. In anatomical conditions, when the response of the spatial transformation is in the course of that of the anatomical transformation, the facilitory effect is strengthened, whereas when the former is located at a distance, it is weakened. Since this dependency between the responses is primarily caused by the difference in processing times, its effect is reversed in spatial conditions. Finally, because the errors were measured after network convergence, they were not different from those reported in Experiment 1 (see Fig. 6).

4 Discussion

In this paper, we have presented a biologically-inspired neural model addressing the problem of transformations across frames of reference in a posture imitation task. Our modeling is based on the hypothesis that such an imitation process is mediated by two concurrent transformations, corresponding to the spatial and the anatomical imitative strategies [5,4,6,13]. We also devised an experimental paradigm which allowed us to measure the interference patterns that the interaction between the anatomical on one hand, and the spatial imitative strategy on the other hand produced. In addition, we also assumed that separate instances of the pair of transformations are responsible for the control of each side of the body. Since our experiments did not involve the use of both arms simultaneously, this latter hypothesis does not rule out the fact that the processes of each arm may be coupled and located within a single brain region [1,10]. As such, our results provide predictions of real behavioral responses.

Similar to other works which applied the Dynamic Field approach [8,13], our work goes beyond usual binary models, often proposed in experimental psychology [15]. Besides the fact that this framework allows the modeling of continuous stimulus variables and responses, which are more common in imitative behaviors, it is of high biological significance. Neurophysiological studies have shown that, in the superior temporal sulcus, body and arm postures are encoded into neural populations where each neuron exhibits tuning to a specific posture [2]. Similarly distributed representations, and correlates of decisional processes have also been reported in many other sensorimotor brain areas [7, 14]. Together, these findings strengthen our approach by grounding it on a strong biological basis.

Behavioral studies on imitation report greater interferences during tasks where the spatial transformation is irrelevant, as compared to tasks where anatomical imitation has to be avoided [5, 9, 4]. Our model supports this observation, but explains it in terms of the longer processing time required by the anatomical imitative strategy, which needs to process an additional variable. Usual accounts for the greater influence of the spatial transformation consider primarily a stronger linkage with the decisional process [4, 5, 10, 15]. Although both hypotheses are compatible, one may be interested in determining their respective influence, which would need more investigations.

Our modeling study also showed that combining of transformations produces interferences. One may wonder why the nervous system would use a combination of two strategies for solving imitation tasks since they produce interferences. Our simulations show that, in specific conditions, their interaction result in positive effects. For instance, when the imitator and the demonstrator are face to face, mirror imitation is faster, whereas anatomical imitation is more effective when the imitator looks at the back of the demonstrator. From this, we can propose an alternative hypothesis explaining that, in unconstrained conditions and when people are facing each other, mirror imitation is the most usual strategy for copying meaningless gestures [3]. Rather than assuming that mirror imitation has a stronger influence on the selection process [9, 10, 4], we suggest that this strategy is the one which exhibits the maximal congruency between the concurrent transformations. Additional neurophysiological evidence supporting our hypotheses can be found in an fMRI study showing that some of the brain areas activated during the imitation of finger movements are more active during specular than during anatomical imitation [10]. In this experiment, the authors did not consider the hypothesis that an anatomical mapping could exist between contralateral hands. The mirror condition which they showed to produce higher brain activation, corresponds in our approach to a condition where the responses of the parallel strategies are perfectly congruent. Since this case is effectively the one in which our model produces responses with the highest energy, the nervous system may hence be naturally biased toward this strategy.

Acknowledgements. This work was supported by the Swiss National Science Foundation, through grant no 620 – 066127 of the SNF Professorships program.

References

1. Arbib, M., Billard, A., Iacoboni, M., Oztop, E.: Mirror neurons, imitation and (synthetic) brain imaging. Neural Networks 13, 953–973 (2000)
2. Ashbridge, E., Perrett, D.I., Oram, M.W., Jellema, T.: Effect of image orientation and size on object recognition: responses of single units in the macaque monkey temporal cortex. Cognitive Neuropsychology 17, 13–34 (2000)

3. Bekkering, H., Wohlschläger, A., Gattis, M.: Imitation of gestures in children is goal-directed. Quarterly Journal of Experimental Psychology 53, 153–164 (2000)
4. Bertenthal, B.I., Longo, M.R., Kosobud, A.: Imitative response tendencies following observation of intransitive actions. Journal of Experimental Psychology: Human Perception and Performance 32(2), 210–225 (2006)
5. Brass, M., Bekkering, H., Wohlschläger, A., Prinz, W.: Compatibility between observed and executed finger movements: comparing symbolic, spatial and imitative cues. Brain and Cognition 44, 124–143 (2000)
6. Chiavarino, C., Apperly, I.A., Humphreys, G.W.: Exploring the functional and anatomical bases of mirror-image and anatomical imitation: The role of the frontal lobes. Neuropsychologia 45, 784–795 (2007)
7. Cisek, P., Kalaska, J.F.: Neural correlates of reaching decision in dorsal premotor cortex: specification of multiple direction choices and final selection of action. Neuron 45, 801–814 (2005)
8. Erlhagen, W., Schöner, G.: Dynamics field theory of movement preparation. Psychological Review 109(3), 545–572 (2002)
9. Heyes, C., Ray, E.: Spatial S-R compatibility effects in an intentional imitation task. Psychonomic Bulletin & Review 11(4), 703–708 (2004)
10. Koski, L., Iacoboni, M., Dubeau, M.-C., Woods, R.P., Mazziotta, J.C.: Modulation of cortical activity during different imitative behaviors. Journal of Neuophysiology 89, 460–471 (2003)
11. Rizzolatti, G., Fogassi, L., Gallese, V.: Neurophysiological mechanisms underlying the understanding of actions. Nature Reviews Neuroscience 2, 661–670 (2001)
12. Sauser, E.L., Billard, A.G.: Three dimensional frames of references transformations using recurrent populations of neurons. Neurocomputing 64, 5–24 (2005)
13. Sauser, E.L., Billard, A.G.: Parallel and distributed neural models of the ideomotor principle. Neural Networks 19(3), 285–298 (2006)
14. Scherberger, H., Andresen, R.A.: Sensorimotor transformations. In: Chalupa Werner (eds.) The Visual Neurosciences. MIT Press, pp. 1324–1336 (2003)
15. Zhang, H.H., Zhang, J., Kornblum, S.: A parallel distributed processing model of stimulus-stimulus and stimulus-response compatibility. Cognitive Psychology 38, 386–432 (1999)

Combined Artificial Neural Network and Adaptive Neuro-Fuzzy Inference System for Improving a Short-Term Electric Load Forecasting

Ronaldo R.B. de Aquino[1], Geane B. Silva[1], Milde M.S. Lira[1], Aida A. Ferreira[2], Manoel A. Carvalho Jr[1], Otoni Nóbrega Neto[1], and Josinaldo. B. de Oliveira[1]

[1] Federal University of Pernambuco, Acadêmico Helio Ramos s/n, Cidade Universitária, Cep: 50.740-530, Recife – PE, Brazil
[2] Federal Federal Center of Technologic Education of Pernambuco, Av. Prof. Luiz Freire, 500 Cidade Universitária, Cep: 50.740-540, Recife – PE, Brazil
rrba@ufpe.br, geaneufpe@yahoo.com.br, milde@ufpe.br
aidaaf@gmail.com, macj@ufpe.br, otoninobrega@hotmail.com
josinaldoo@ig.com.br

Abstract. The main topic in this work was the development of a hybrid intelligent system for the hourly load forecasting in a time period of 7 days ahead, using a combination of Artificial Neural Network and Adaptive Neuro-Fuzzy Inference System. The hourly load forecasting was accomplished in two steps: in the first one, two ANNs are used to forecast the total load of the day, where one of the networks forecasts the working days (Monday through Friday), and the other forecasts the Saturdays, Sundays and public holidays; in the second step, the ANFIS was used to give the hourly consumption rate of the load. The proposed system presented a better performance as against the system currently used by Energy Company of Pernambuco, named PREVER. The simulation results showed an hourly mean absolute percentage error of 2.81% for the year 2005.

Keywords: Hybrid System, Neuro-Fuzzy System, Artificial Neural Network, Load Forecasting.

1 Introduction

The load forecast is a subject of utmost importance to aid in the planning studies, schedule of operation, enlargements and reinforcements of the basic grid [1], [2].

Recently, short-term load forecasting (7 days ahead) has become extremely important, as this type of forecast is directly linked to the electricity bill forecasting that due to the privatization processes and to the appearance of competition in the Brazilian electricity grid is now a subject of great importance for the agents of the area. Therefore, for the current model of the Brazilian electric system an efficient load forecasting implies making a profit on the commercialization process.

Several researches have been carried out in order to improve planning and operation of these systems. Specifically, that the required load forecasts may be divided into short-term, mid term and long-term forecasts. Traditionally, load forecasting techniques use

J. Marques de Sá et al. (Eds.): ICANN 2007, Part II, LNCS 4669, pp. 779–788, 2007.

statistical methods of time series analysis, which include linear regression, exponential damping and Box Jenkins [3]. In recent years, techniques of artificial intelligence such as artificial neural network (ANN) have been used, obtaining promising results [4]-[7]. Electric load forecasting based on Hybrid systems has also become one of the attractive tools to resolve this problem [8] with quite satisfactory results.

Currently, CELPE uses the software PREVER to accomplish its hourly load forecast, which was developed by the research group of artificial neural network in the Digital Laboratory of Power System - LDSP/UFPE. The PREVER system accomplishes the load forecast using ANN and heuristic rules providing an efficient forecast system with small error [9]. The aim of this work is to improve the hourly load forecast, automating it and incorporating the implicit knowledge of the specialist in the system by the ANFIS. The developed system makes use of a hybrid approach implemented in MATLAB® to accomplish the load forecasting in the period of 7 days ahead.

2 Proposed Scheme for Load Forecasting

The creation of the proposed system was divided in two stages. In the first stage, the daily forecasting consumption system was developed and next, the system that gives the behavior of the hourly load forecasting.

During the accomplishment of the first stage, the behavior of the data represented by the electric load consumption curves was analyzed in order to obtain an effective database. Afterwards, the artificial neural networks were created to provide the forecast of the daily total consumption. From observation of the analyzed data, two architectures of neural networks were created, one to forecast the daily total consumption of the day of the week (Monday through Friday) and other network to forecast the consumption of the holidays and weekend days (Saturday and Sunday). The forecast of the total consumption of one holiday was accomplished as being a Saturday or Sunday, depending on the characteristic of the electric load curve of that specific holiday [9].

In the second stage of the process, a study of the behavior of the load curve relating to the total consumption of the day was accomplished in order to create the database of the neuro-fuzzy system (ANFIS) and its architecture. After the analyses, it was concluded that to obtain a better behavior for the electric load curve of holidays, it would be better to use the coefficients of the hourly consumption of the same holiday in the previous year. However, for the days of the week that are not a holiday, we should use the coefficients generated by the ANFIS system.

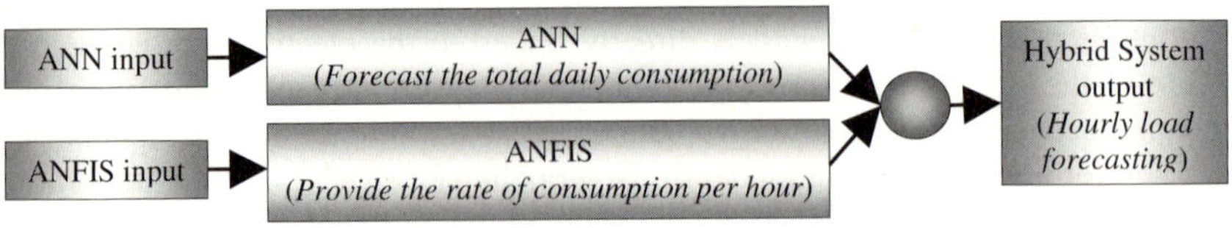

Fig. 1. Diagram of the load forecasting model

An illustration of the flow chart of the load forecasting model is given in Fig. 1. ANN is used to forecast the total daily consumption in the period of 7 days ahead. ANFIS system gives the hourly load forecasting coefficients that multiplied by the total load consumption produce the hourly load forecasting.

3 Neural Network Architecture

3.1 Database Arrangement

The database is of fundamental importance to generate powerful forecasting models, because their outputs are strongly related to the quality and arrangement of the information used in the learning process.

The problem approached in this work is based on the hourly load forecasting for 7 days ahead. The data used in this work were made available by CELPE and they correspond to the hourly load consumption data in the period from January 2000 until December 2005.

All the data were unified in two files, one containing working days and the other the holidays, Saturdays and Sundays. From these data, the pattern was arranged for the information of the year, the day of the week, the month of the year, total load of the day and the day of the week to be forecast using the 1-of-m code.

The value of the hourly load was normalized (L_N) to fall in the range 0 to 1 by using (1):

$$L_N = \frac{L - L_{min}}{L_{max} - L_{min}}. \tag{1}$$

where L_N is the hourly load value registered by the CELPE's system, L_{max} and L_{min} are the maximum and the minimum hourly load value among all the observed values, respectively. In this work L_{min} = 0 and $L_{max} = 1.1 \cdot L_{Amax}$, where L_{Amax} is the maximum value of the actual load data. The objective of factor 1.1 is to turn the values of future loads up to 10% above L_{Amax} into values below the unit after their normalization.

The data to create the working day set was pre-selected using the mean and the standard deviation parameters. Initially, for every month the total consumption of a specific working day (Monday through Friday) was divided by the number of times that specific day appears in the month. Next, the mean $\bar{x}_{ij\_t}$ and the standard deviation σ_{ij_t} were calculated for every day of the week in each month and in each year. Using these values, two limits were created:

$$x_{min\_ij\_t} = \bar{x}_{ij\_t} - \sigma_{ij\_t}. \tag{2}$$

$$x_{max\_ij\_t} = \bar{x}_{ij\_t} + \sigma_{ij\_t}. \tag{3}$$

where, i specifies the month of the year i = 1 to 12, j specifies the day of the week j = 1 to 7 and t specifies the year t = 2000 to 2004 of the training set.

The patterns of the database were limited to lower and upper values given in (2) and (3), respectively. That is, if a given pattern was lower than $x_{min\_ij\_t}$ or greater than $x_{max\_ij\_t}$, it was discarded from the data.

The procedure mentioned above was not used to create the data set of the holidays and weekend days, because this would reduce the amount of data significantly. However, the data regarding the more critical period of the rationing in Brazilian electric power system (May through July 2001) were eliminated from this data set.

In this work, a holiday is considered by the specialist to be as one Saturday or one Sunday, according to [9]. In other words, the specialist indicates beforehand if the load behavior of that specific holiday is more correlated with the load behavior of Saturday or Sunday. Because the load curves of the holidays are either close to the load curves of one Saturday or one Sunday, the hourly load data of holidays were just used in the test set.

The networks are created using the basic principle of dividing the samples into three subsets that are mutually exclusive, defined as: training set used to train the network; validation set used to avoid the overtraining, thus improving generalization; and test set used to compare different models and to plot the test set error during the training process. The idea is that the system performance in the test set represents its performance in the real world. This means that no example of the test set should be available in the training set of the network [10].

The neural network of the working days has eleven inputs: the first three represent values of the electric load demand for 42, 35, 28, 21, 14 and 7 days before the day to be forecast, and the other five define the day of the week to be forecast (Monday through Friday). The output supplies the day to be forecast with total demand.

The neural network of the holidays and weekend days are characterized by the following inputs: the first two represent the value of the electric load for 14 and 7 days before the day to be forecast, and the others define the day of the weekend day to be forecast (Saturday and Sunday). The output supplies the day to be forecast with the total demand.

A total amount of 475 and 374 examples constituted the data that represents the working days and the weekend days, respectively. These data were divided in the following way: 60% for the training set, 30% for the validation set and 10% for the test set. The patterns of each data set were randomly mixed.

The main objective of the load forecasting system based on ANN is to learn from pattern of known values and to generalize for new ones. The performance of the system will be measured by percentage of the mean-square error (MSE) (4) and by the mean absolute percentage error (MAPE) in (5).

$$MSE_{\%} = 100 \times \frac{L_{máx} - L_{mín}}{N \cdot P} \sum_{p=1}^{P} \sum_{i=1}^{N} (L_{pi} - T_{pi})^2 \, , \tag{4}$$

where L_{max} and L_{min} are the maximum and minimum of the hourly load values, in the representation of the problem, respectively; N is the number of output units of the ANN; P is the total number of patterns in database; L_{pi} and T_{pi} are actual and desired target output of the i_{th} neuron in the output layer, respectively.

$$MAPE_{\%} = \frac{1}{P} \sum_{p=1}^{P} \frac{\left| L_p - T_p \right|}{T_p} \times 100 \, , \tag{5}$$

where P is the total number of patterns in database; L_p and T_p are the actual and desired output value for a given input, respectively.

Attempting to achieve an estimated error nearest to the true error, the 10-fold cross validation method [10] was chosen to generate the training, validation and test sets. Therefore, the patterns were divided in ten independent partitions, each partition having 10% of the data. In each experiment, three partitions were used for validation; one, to test; and the six remaining partitions were used to train the ANNs.

3.2 Network Architecture and Training

All of the experiments accomplished in this work created ANNs with the MLP architecture, using Levenberg-Maquardt (LM) training algorithm.

All of the ANNs used have an input, a hidden and an output layer. The maximum number of iterations for all of the trainings was set to 2500 epochs. The training stopped if the *early stopping* implemented by MATLAB® happened 15 times consecutively, or if the maximum number of epochs is reached, or if the error gradient reaches a minimum, or still if the error goal in the training set is met.

To decide on the best configuration of nodes in the hidden layer, ten experiments were carried out with random initialization of weights and with varying number of hidden nodes from 3 to 30 with an increment of 1. The number of hidden nodes in the best neural network for the working days and weekend days are respectively three (Fig. 2a) and five (Fig. 2b).

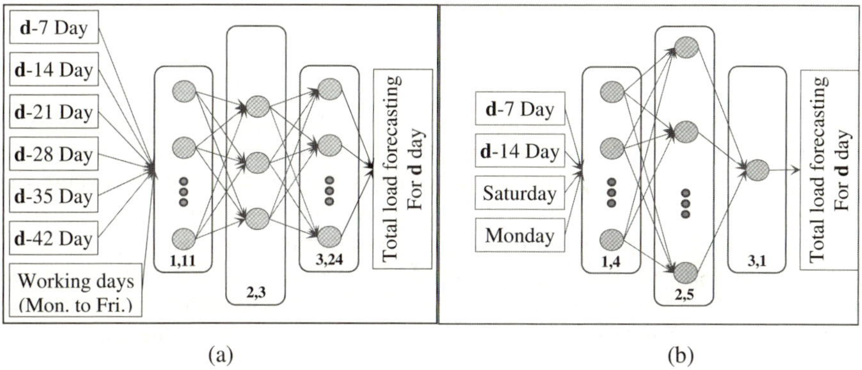

Fig. 2. a) ANN for load forecasting of working days. b) ANN for forecasting of weekend days.

4 ANFIS Architecture

One method of improving the intelligent hybrid systems consists of combining the main characteristics of ANN and fuzzy logic. The ANFIS system creates rules based on any set of input-output data, acquiring the knowledge of the specialist in the form of fuzzy if-then rules. The method for the fuzzy modeling procedure to learn information about a data set works similarly to that of neural networks.

The ANFIS system was developed by Jyh-Shing Roger Jang which combines back-propagation neural network with supervised learning capability and fuzzy inference system. The neural network is basically a multilayer feedforward network in which each node performs a particular function (node function) on incoming signals as well as a set of parameters pertaining to this node. The fuzzy inference system used was the type Takagi-Sugeno [11], where the output of each rule is a linear combination of input variables plus a constant term, and the final output is the weighted average of each rule's output.

The aim of the proposed system is to forecast the hourly load demand in the time period of 7 days ahead. In this way the developed system generates coefficients that represent the hourly proportion of the total load demand of the day to be forecast, which is responsible for supplying the behavior of the load curve to that day.

The aim of the proposed system is to forecast the hourly load demand in the time period of 7 days ahead. In this way the developed system generates coefficients that represent the hourly proportion of the total load demand of the day to be forecast, which is responsible for supplying the behavior of the load curve to that day. The coefficient is called hourly consumption multiplayer m_{ch} , computed by (6).

$$C_h = m_{ch} \cdot C_D \cdot \tag{6}$$

where C_h is the hourly demand and C_D is the total demand of the day.

Analyses of the normalized load curve were accomplished, being verified that to obtain a better behavior of the holiday load curve, it would be better to use the coefficients of the hourly load demand of this same holiday from the previous year. It is important to point out that the proposed system allows the user to use or not this strategy, according to the kind of the holiday.

4.1 ANFIS: Structure and Training

The ANFIS system generates an inference system that supplies the coefficients that represent the hourly load curve behavior; therefore, 24 neuro-inference systems were created, one for every hour of the day. Relating to the holidays, the percentage value for every hour was the percentage value of this same holiday in the previous year, that is, for this kind of day, the neuro-fuzzy system was not used.

In order to create the training set of the ANFIS system, the data of the hourly load demand regarding the period from January 2002 until December 2004 were used except for the load data relating to the holiday. For each hour, the training set is characterized by three inputs and one output:

- Input 1: Relative value of hour *i*, seven days before the day to be forecast.
- Input 2: Day of the week of the day to be forecast corresponding to one value from 1 to 7, denoting in this order the day from Sunday until Saturday.
- Input 3: Month of the day to be forecast corresponding to one value from 1 to 12, denoting in this order the months from January until December.
- Output: Percentage of the hour *i* of the day to be forecast.

The relative value for hour *i* is calculated by the following equation:

$$V_r(i) = \frac{C_h(i)}{C_T}, \qquad i = 1, \ldots, 24 \cdot \tag{7}$$

where, $V_r(i)$ is the relative value of the load demand for hour *i*, $C_h(i)$ is the load demand of hour *i* and C_T is the total load demand of seven days before the day to be forecast.

The training data set forming by the input/targets pairs was properly mixed up. It is worth to pointing out that the data was not normalized, thus the MATLAB® normalized the data automatically using an internal routine.

Each one of the 24 databases has 1089 patterns, which were mixed up and divided into ten independent partitions of 10% to generate the training, validation and test sets. These subsets were distributed in the following way: 70% for the training set, 20% for the validation set and 10% for the test set.

The ANFIS structure that accomplishes the training of the parameters associated to the Fuzzy Inference System (FIS) was chosen, in order to generate the better FIS to represent the problem.

Table 1 represents the ANFIS structure and the number of membership functions of the initial FIS using the subtractive clustering method [12] according to the time interval. In this system, the amount of membership function of the input is equal to the number of cluster centers.

Table 1. Number of membership functions related to the hour of the day and ANFIS structure

Number of Membership Functions	Hour	ANFIS Structure
06	8 h	3-18-6-1
07	7 h	3-21-7-1
08	6, 19, and 24 h	3-24-8-1
09	9, 10, 18, 22 and 23 h	3-27-9-1
10	2 to 5 h, 11 to 17 h and 20 to 21 h	3-30-10-1
11	1 h	3-33-11-1

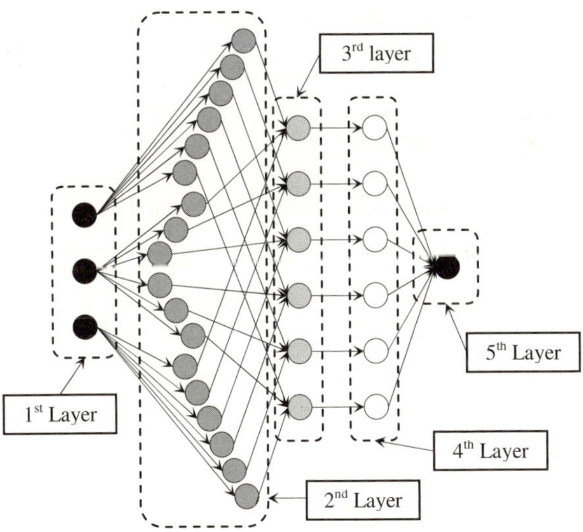

Fig. 3. The ANFIS architecture to supply the hourly coefficients of hour 8

The 24 neuro-fuzzy systems do not necessarily have the same structure. Fig. 3 shows the ANFIS architecture used to model the load forecasting system that supplies the hourly load demand coefficient of hour 8. Note that the figure shows five layers, where the 1st one is the input layer, the 2nd is the layer that contains the membership functions, the 3rd is the rule layer, the 4th is the layer of the output parametric functions and the 5th is the output layer. The figure illustrates a fuzzy inference system that looks like a multilayer feedforward neural network.

At the final training process, the 24 systems were able to represent the knowledge of the specialist based on the information present in the training set. During the training process, the membership functions continually change their parameters, which mean that at the end of the training, the FIS has a different architecture from that given at the beginning.

In order to choose the best FIS, the 10 arrangements of training, validation and test set were trained, that is, 10 systems to each hour were created.

5 Results

In order to verify the performance of the forecasting system on the load data pertaining to CELPE, forecasts were accomplished in the period from January until December 2005.

The results were compared with the PREVER system, software currently used by CELPE to accomplish the load forecasting in short and mid-term [9].

To forecast the hourly load, the coefficients in the time period of 7 days ahead forecasted by the FIS are multiplied by the total load demand in kWh supplied by the neural network. Having the forecasted hourly load, the monthly mean MAPE can be calculated.

Table 2 presents a comparison among the hourly mean MAPE given by the models: "PREVER with and without adjustment" [9] and the developed model, in each month of 2005. As shown on the last line of this table, the mean value of the developed system and the "PREVER with adjustment" presented the same value, showing that the proposed method performed well in representing the heuristic rules implemented in the PREVER system.

Table 2. Hourly mean MAPE in the period of 7 days

Month	Developed model	PREVER without adjustment	PREVER with adjustment
Jan	2.62	4.06	2.21
Feb	2.74	3.80	2.55
Mar	3.59	4.79	4.00
Apr	2.42	3.59	2.53
Mai	3.83	2.59	4.22
Jun	2.84	3.22	2.73
Jul	1.97	2.14	1.87
Aug	1.89	2.53	2.13
Sep	3.26	3.38	3.35
Oct	2.63	3.24	2.48
Nov	2.88	3.53	2.40
Dec	3.01	3.55	3.29
Mean	2.81	3.37	2.81

Table 3. Hourly mean MAPE in the period of 7 days ahead

Month	Developed model	PREVER without adjustment	PREVER with adjustment
Jan	2.62	23.09	4.64
Feb	3.16	7.85	4.26
Mar	3.98	18.23	6.88
Apr	4.03	9.36	5.23
Mai	2.46	2.84	2.55
Jun	4.11	17.08	3.68
Jul	2.30	3.26	2.48
Aug	-	-	-
Sep	1.30	7.62	8.93
Oct	4.73	11.79	4.80
Nov	3.52	5.29	5.50
Dec	6.28	6.70	9.08
Mean	3.50	10.28	5.27

Another analysis comparing the hourly load forecasting error of the holidays given by the developed system and the PREVER is presented in Table 3.

It is important that the proposed model provide a high performance in forecasting the electric load of holidays and anomalous days, which are hardly studied in papers [13].

As can been seen, the developed system presented lower error as against the "PREVER with and without adjustment" in all the months but July. The mean MAPE to the holidays of the year 2005 is 3.50% as against 10.28% and 5.27% presented by the PREVER with and without adjustments, respectively. Consequently, the new system is certainly an improvement on the PREVER system.

6 Conclusion

The results show that the developed system is a trustful system, presenting error inside the acceptable standard [8]. It presents a very simple architecture that uses neural network with small number of nodes, leading to a faster training due to the lower amount of parameter to be updated.

The main advantages of this model are: simple structure, faster training and no necessity of temperature and others climatic variables.

Another advantage of this system is the high performance to model the load curve behavior independent of the type of the day to be forecast, that is, the load curve behavior has the tendency to follow the real load curve, even so the forecast values may be in lower or higher level.

New techniques must be developed to improve the performance of load forecasting, especially for the holidays and anomalous days, as well as, a new approach to modeling load forecasting system for others periods of time then just for only 7 days ahead.

Acknowledgments. The authors would like to thank to CNPq and CAPES-Brazil for financial research support.

References

1. Agência Nacional de Energia Elétrica – Procedimentos de Distribuição de Energia Elétrica no Sistema Elétrico Nacional – PRODIST Módulo 4.W. In: Portuguese
2. Operador Nacional do Sistema Elétrico - Consolidação da Previsão de Carga, Procedimentos de Rede, Submódulo 5.1. In Portuguese
3. Montgomery, D.C., Johnson, L.A., Gardiner, J.S.: Forecasting and Time Series Analysis. In: McGraw-Hill International Editions (1990)
4. Bakirtzis, A., Petrldis, V.S., Kiartiz, J., Alexiardis, M.C.: A Neural Network Short Term Load Forecasting Model for the Greek Power System. In: IEEE Transactions on Power Systems 11(2), 858–863 (1996)
5. Khotanzad, A., Afkhami-rohani, R., Lu, T., Abaye, A., Davis, M., Maratukulam, D.J.: A Neural-Network-Based Electric Load Forecasting System. In: IEEE Transactions on Neural networks, 8(4), 835–845 (1997)
6. Kim, C., Yu, I., Song, Y.H.: Kohonen Neural Network and Wavelet Transform Based Approach to Short-Term Load Forecasting. In: Electric Power Systems Research 63(3), 169–176 (2002)
7. Papadakis, S.E., Theocharis, J.B., Kiartzis, S.J., Bakirtzis, A.G.: A novel approach to Short-term Load Forecasting using Fuzzy neural networks. IEEE Transactions on Power Systems 13(2), 480–492 (1998)
8. Srinivasan, D., Tan, S.S., Chang, C.S., et al.: Parallel Neural Network-Fuzzy Expert System Strategy For Short-term Load Forecasting: System Implementation And Performance Evaluation. In: IEEE Transactions on Power Systems, 14(3) (1999)
9. Aquino, R.R.B., Ferreira, A.A., Lira, M.M.S., Carvalho, J.M.A., Nóbrega Neto, O., Silva, G.B.: Combining artificial neural networks and heuristic rules in a hybrid intelligent load forecast system. In: Kollias, S., Stafylopatis, A., Duch, W., Oja, E. (eds.) ICANN 2006. LNCS, vol. 4132, pp. 757–766. Springer, Heidelberg (2006)
10. Ferreira, A.A.: Comparação de arquiteturas de redes neurais para sistemas de reconhecimento de padrões em narizes artificiais. In: Dissertação de mestrado. UFPE, Recife-PE, In Portuguese (2004)
11. Jang, J.R.: ANFIS: Adaptive-Network-Based Fuzzy Inference System. In: Electric Power Systems Research / IEEE Transactions on systems 23(3), 169–176 (1993)
12. Chiu, S.: Fuzzy Model Identification Based on Cluster Estimation. Journal of Intelligent & Fuzzy Systems 2(3) (1994)
13. Kim, K.H., Youn, H.S., Kang, Y.C., et al.: Short-Term Load Forecasting for Special Days in Anomalous Load Conditions Using Neural Networks. In: IEEE Transactions on Power Systems 15(2) (1993)

A MLP Solver for First and Second Order Partial Differential Equations

Slawomir Golak

Department of Electrotechnology, Silesian University of Technology
Krasinskiego 8, 40-019 Katowice, Poland
slawomir.golak@polsl.pl

Abstract. A universal approximator, such as multilayer perceptron, is a tool that allows mapping of any multidimensional continuous function. The aim of this paper is to discuss a method of perceptron training that would result in its ability to map the functions constituting the solutions of partial differential equations of first and second order. The developed algorithm has been validated by means of equations whose analytical solutions are known.

1 Introduction

Today's science and engineering encounter many problems that can be reduced to finding solutions to partial differential equations. Considering practical applications, the issues such as heat transfer, fluid dynamics, mechanics of materials, and electromagnetic field modeling are among the most important. However, in the majority of real cases there is no possibility of finding analytical solutions to the differential equations connected with the above questions, which has contributed to the development of many numerical methods for differential equations, the most popular being finite boundary method, finite element method or finite volume method (widely used in computational fluid dynamics). Oftentimes they prove to be very efficient; however, not always. As an example 3D models can be mentioned, where the computational complexity of the numerical methods frequently prevents finding the solution within a reasonable time range. Their imperfections can also be noticed while solving the coupled problems, which make allowances for mutual influence of fields, like electromagnetic field and fluid flow field in magnetohydrodynamics problems.

These drawbacks of the classic numerical methods clearly justify the search for alternative algorithms for differential equations solutions. One of the approaches recommends taking advantage of artificial neural networks.

Theoretically, neural networks are universal approximators of any continuous function. Therefore it is assumed that they can map the solution functions of any given differential equation. For these reasons the issue of adopting neural networks to solve differential equations has been taken up by many authors [1-4].

The main justification for the present research is the fact that neural models are characterized by a significantly smaller number of parameters than classic models, which is of crucial importance in case of more complex, high-dimensional problems.

J. Marques de Sá et al. (Eds.): ICANN 2007, Part II, LNCS 4669, pp. 789–797, 2007.

A large majority of literature in this field deals with the application of radial networks (RBF). The networks applied are usually the standard radial networks or the ones with a specially selected structure and radial functions [5-7]. Since it is local approximation of the dependence under consideration that is characteristic of radial networks, the methods incorporating them to solve differential equations must be similar (because of their local approach to the problem) to classic numerical methods like FEM, FDM or FVM, and therefore, inevitably, the same disadvantages will be encountered, and namely the necessity of dense sampling of space as well as potential gross inaccuracies between mesh nodes.

On the contrary, multilayer perceptrons take a global approach, where particular neurons influence the value of the function realized by the network in the whole domain of arguments. All components of the analytical solution of a differential equation also globally and simultaneously influence the value of a depended variable. For this reason the solution obtained by perceptron is significantly closer to the analytical solution of the equation, which can be assumed as the optimal one.

One method of solving a partial differential equation with the help of perceptron involves adopting of a trial solution, which is a function constituted by two components: the analytical component, which ensures that the boundary conditions are met, and the neural network-based component, responsible for differential equation being satisfied [8]. Unfortunately, the application of this method is restricted to the areas with regular, orthogonal boundaries. There have been some attempts to handle the inconvenience by resorting to synergy of two networks: the network of the multilayer perceptron approximating the differential equation and the RBF radial network responsible for satisfying boundary conditions [9]. It must be remembered, however, that application of RBF for boundary conditions modeling involves a partial loss of the advantages that the 'pure' perceptron solution offers. An additional inconvenience of Lagaris method is a limitation to a perceptron with only one hidden layer, which will not always prove the optimal structure.

Delpiano and Zegers have developed a method where a perceptron with any number of layers can be applied to solve differential equations directly [10]. Still, owing to their algorithm only the direct solutions to the first order equations can be obtained in this way, which definitely reduces its applicability.

This paper presents an extension of the above method, which will enable any multilayer perceptron to solve differential equations of both the first and the second order. It will significantly widen the range of applications of the method so as to cover the majority of differential equations employed in most branches of modern science and engineering.

2 Description of the Method

A solution of a partial differential equation aims at finding such a function which satisfies the differential equation within the space of arguments under consideration and simultaneously satisfies the assumed boundary conditions.

In order to solve a differential equation by means of a neural network, an objective function must be defined. Through minimization of the function the neural network will be trained. A differential equation may be reduced to the form of an objective function, which is the sum of squares of deviations from the solution of the given equation at the points sampling the space under consideration, and squares of deviations from the values evaluated by the boundary conditions. The equation (1) is an example of such function for solving PDE with Dirichlet *(DC)* and Naumann *(NC)* boundary conditions. The spatial distribution of the points may be either random or determined by methods analogous to meshing in classic numerical methods.

$$E = W_E \frac{1}{N_E} \sum_{i=1}^{N_E} \left(\frac{\partial^2 y^{(N)}}{\partial x_1^2} + \frac{\partial^2 y^{(N)}}{\partial x_1 \partial x_2} + \frac{\partial^2 y^{(N)}}{\partial x_2^2} + y^{(N)} - F(x_1, x_2) \right)^2 + W_{BC} \left[\frac{1}{N_{DC}} \sum_{1}^{N_D} \left(y^{(N)} - F_{DC}(x_1, x_2) \right)^2 + \frac{1}{N_{NC}} \sum_{1}^{N_N} \left(\frac{\partial y^{(N)}}{\partial x_1} - F_{NC}(x_1, x_2) \right)^2 \right] \tag{1}$$

where: N_E - a number of points probing a differential equation in an analyzed area, N_{DC}, N_{NC} - a number of points probing boundary conditions (Dirichlet, Naumann), $y^{(N)}$ – a dependent variable, a output of the neuron in the last (N-th) layer, which corresponds to the output of network, x_1, x_2 – independent variables, inputs of a neural network.

Another problem concerns weights selection for both performance function component responsible for satisfying of the differential equation (W_E) as well as the component responsible for boundary conditions (W_{BC}). The weights fitted will differ depending on the problem considered, in some cases the focus being on the very differential equation, while other times on the most accurate approach to the boundary conditions.

2.1 Calculation of Partial Derivatives

Perceptron, through its successive layers, transforms the input vector into the ouput vector, realizing the function being approximated.

$$y_i^{(n)} = \begin{cases} f_i^{(n)}\left(s_i^{(n)}\right) & n > 0 \\ x_i & n = 0 \end{cases} ; \quad s_i^{(n)} = \sum_{j=1}^{M^{(n-1)}} w_{ij}^{(n)} y_j^{(n-1)} + b_i^{(n)} \tag{2}$$

where: $y_i^{(n)}$ – an output of the i th neuron in the n-th layer, $f_i^{(n)}$ – an activation function of the i-th neuron in the n-th layer, $s_i^{(n)}$ – a weighted sum of inputs of the neuron, $w_{ij}^{(n)}$ – a weight of the j-th input of the neuron, $b_i^{(n)}$ – a bias of the neuron, $M^{(n)}$ – a number of neurons in n-th layer, x_i – the i-th input of a network.

The application of continuous activation functions (linear, sigmoid) in neurons enables the calculation of partial derivatives (with respect to input vector components) of the functions realized by the network. The method of analytical evaluation of these derivatives resembles calculating of the network output values and consists in the propagation of the given partial derivative through the successive layers of the network in its output direction. Derivatives of the first, second and higher orders can be calculated by means of this technique [11].

The first order derivative of the i-th neuron in the n-th layer with respect to the network input x_a, which corresponds to an independent variable in a differential equation, is evaluated in equation (3).

$$\frac{\partial y_i^{(n)}}{\partial x_a} = \begin{cases} \frac{\partial f_i^{(n)}}{\partial s_i^{(n)}} \sum_j^{M^{(n-1)}} w_{ij}^{(n)} \frac{\partial y_j^{(n-1)}}{\partial x_a} & n > 0 \\ 1 & n = 0, i = a \\ 0 & n = 0, i \neq a \end{cases} \tag{3}$$

The second order derivative with respect to the input x_a is:

$$\frac{\partial^2 y_i^{(n)}}{\partial x_a^2} = \begin{cases} \frac{\partial^2 f_i^{(n)}}{\partial s_i^{(n)2}} \left(\sum_j^{M^{(n-1)}} w_{ij}^{(n)} \frac{\partial y_j^{(n-1)}}{\partial x_a} \right)^2 + \frac{\partial f_i^{(n)}}{\partial s_i^{(n)}} \sum_j^{M^{(n-1)}} w_{ij}^{(n)} \frac{\partial^2 y_j^{(n-1)}}{\partial x_a^2} & n > 0 \\ 0 & n = 0 \end{cases} \tag{4}$$

The mixed derivative with respect to inputs x_a and x_b is:

$$\frac{\partial^2 y_i^{(n)}}{\partial x_a \partial x_b} = \begin{cases} \frac{\partial^2 f_i^{(n)}}{\partial s_i^{(n)2}} \left(\sum_j^{M^{(n-1)}} w_{ij}^{(n)} \frac{\partial y_j^{(n-1)}}{\partial x_a} \right) \left(\sum_j^{M^{(n-1)}} w_{ij}^{(n)} \frac{\partial y_j^{(n-1)}}{\partial x_b} \right) + \frac{\partial f_i^{(n)}}{\partial s_i^{(n)}} \sum_j^{M^{(n-1)}} w_{ij}^{(n)} \frac{\partial^2 y_i^{(n-1)}}{\partial x_a \partial x_b} & n > 0 \\ 0 & n = 0 \end{cases} \tag{5}$$

2.2 Network Learning

Training of the neural network comes down to the optimization of its weights. It also holds true for the neural networks that, besides the function values, also evaluate the values of their partial derivatives with respect to the network inputs. Among the simplest ways of training in case of such a network are evolutionary methods and direct methods [12]. However, it is the gradients methods that prove to be far more efficient a way of network weights optimization.

$$\begin{aligned} \frac{\partial E}{\partial w_{ik}^{(n)}} &= 2W_E \frac{1}{N_E} \sum_{i=1}^{N_E} \left(\frac{\partial^2 y^{(N)}}{\partial x_1^2} + \frac{\partial^2 y^{(N)}}{\partial x_1 \partial x_2} + \frac{\partial^2 y^{(N)}}{\partial x_2^2} + y^{(N)} - F(x_1, x_2) \right) \left(\frac{\partial^3 y^{(N)}}{\partial x_1^2 \partial w_{ik}^{(n)}} + \frac{\partial^3 y^{(N)}}{\partial x_1 \partial x_2 \partial w_{ik}^{(n)}} + \frac{\partial^3 y^{(N)}}{\partial x_2^2 \partial w_{ik}^{(n)}} + \frac{\partial y^{(N)}}{\partial w_{ik}^{(n)}} \right) \\ &+ 2W_{BC} \left[\frac{1}{N_{DC}} \sum_1^{N_D} \left(y^{(N)} - F_{DC}(x_1, x_2) \right) \frac{\partial y^{(N)}}{\partial w_{ik}^{(n)}} + \frac{1}{N_{NC}} \sum_1^{N_N} \left(\frac{\partial y^{(N)}}{\partial x_1} - F_{NC}(x_1, x_2) \right) \frac{\partial^2 y^{(N)}}{\partial x_1 \partial w_{ik}^{(n)}} \right] \end{aligned} \tag{6}$$

In order to apply gradient methods, the gradient of the performance function must be calculated (Eq. 6), and this, when the function incorporates not only the values of the function realized by the network but also its derivatives, necessitates the extension the classic method of error backpropagation:

$$\frac{\partial y^{(N)}}{\partial w_{ik}^{(n)}} = \delta_i^{(n)} y_k^{(n-1)}; \quad \frac{\partial y^{(N)}}{\partial b_i^{(n)}} = \delta_i^{(n)} \tag{7}$$

$$\delta_i^{(n)} = \begin{cases} \frac{\partial f_i^{(n)}}{\partial s_i^{(n)}} \sum_{k=1}^{M^{(n+1)}} w_{ki}^{(n+1)} \delta_k^{(n+1)} & n < N \\ \frac{\partial f_i^{(n)}}{\partial s_i^{(n)}} & n = N \end{cases} \tag{8}$$

Where $y^{(N)}$ is a neuron output in the last (N-th) layer, an output on a neural network, $\delta_i^{(n)}$ is an influence of the i-th neuron in the n-th layer on the network output.

Apart from the need to calculate the derivative of the function being realized by the network with respect to weights, mixture derivatives must be evaluated, both with respect to weights and with respect to one or more network inputs:

$$\frac{\partial^2 y^{(N)}}{\partial x_a \partial w_{ij}^{(n)}} = \frac{\partial \delta_i^{(n)}}{\partial x_a} y_j^{(n-1)} + \delta_i^{(n)} \frac{\partial y_j^{(n-1)}}{\partial x_a} \; ; \quad \frac{\partial^2 y^{(N)}}{\partial x_a \partial b_i^{(n)}} = \frac{\partial \delta_i^{(n)}}{\partial x_a} \tag{9}$$

$$\frac{\partial \delta_i^{(n)}}{\partial x_a} = \begin{cases} \dfrac{\partial^2 f_i^{(n)}}{\partial s_i^{(n)2}} \left(\sum\limits_j^{M^{(n-1)}} w_{ij}^{(n-1)} \dfrac{\partial y_j^{(n-1)}}{\partial x_a} \right) \left(\sum\limits_k^{M^{(n+1)}} w_{ki}^{(n+1)} \delta_k^{(n+1)} \right) + \dfrac{\partial f_i^{(n)}}{\partial s_i^{(n)}} \sum\limits_{k=1}^{M^{(n+1)}} w_{ki}^{(n+1)} \dfrac{\partial \delta_k^{(n+1)}}{\partial x_a} & n < N \\ \dfrac{\partial^2 f_i^{(n)}}{\partial s_i^{(n)2}} \left(\sum\limits_j^{M^{(n-1)}} w_{ij}^{(n-1)} \dfrac{\partial y_j^{(n-1)}}{\partial x_a} \right) & n = N \end{cases} \tag{10}$$

$$\frac{\partial^3 y^{(N)}}{\partial^2 x_a \partial w_{ij}^{(n)}} = \frac{\partial^2 \delta_i^{(n)}}{\partial^2 x_a} y_j^{(n-1)} + 2 \left(\frac{\partial \delta_i^{(n)}}{\partial x_a} \frac{\partial y_j^{(n-1)}}{\partial x_a} \right) + \delta_i^{(n)} \frac{\partial^2 y_j^{(n-1)}}{\partial^2 x_a} \; ; \quad \frac{\partial^3 y^{(N)}}{\partial^2 x_a \partial b_i^{(n)}} = \frac{\partial^2 \delta_i^{(n)}}{\partial^2 x_a} \tag{11}$$

$$\frac{\partial^2 \delta_i^{(n)}}{\partial^2 x_a} = \begin{cases} \left[\dfrac{\partial^3 f_i^{(n)}}{\partial s_i^{(n)3}} \left(\sum\limits_{j=1}^{M^{(n-1)}} w_{ij}^{(n)} \dfrac{\partial y_j^{(n-1)}}{\partial x_a} \right)^2 + \dfrac{\partial^2 f_i^{(n)}}{\partial s_i^{(n)2}} \sum\limits_{j=1}^{M^{(n-1)}} w_{ij}^{(n)} \dfrac{\partial^2 y_j^{(n-1)}}{\partial^2 x_a} \right] \sum\limits_k^{N^{(n+1)}} w_{ki}^{(n+1)} \delta_k^{(n+1)} + & \\ 2 \cdot \dfrac{\partial^2 f_i^{(n)}}{\partial s_i^{(n)2}} \left(\sum\limits_{j=1}^{M^{(n-1)}} w_{ij}^{(n)} \dfrac{\partial y_j^{(n-1)}}{\partial x_a} \right) \left(\sum\limits_k^{M^{(n+1)}} w_{ki}^{(n+1)} \dfrac{\partial \delta_k^{(n+1)}}{\partial x_a} \right) + \dfrac{\partial f_i^{(n)}}{\partial s_i^{(n)}} \sum\limits_k^{M^{(n+1)}} w_{ki}^{(n+1)} \dfrac{\partial^2 \delta_k^{(n+1)}}{\partial^2 x_a} & n < N \\ \dfrac{\partial^2 f_i^{(n)}}{\partial s_i^{(n)2}} \left(\sum\limits_{j=1}^{M^{(n-1)}} w_{ij}^{(n)} \dfrac{\partial y_j^{(n-1)}}{\partial x_a} \right)^2 + \dfrac{\partial^2 f_i^{(n)}}{\partial s_i^{(n)2}} \sum\limits_{j=1}^{M^{(n-1)}} w_{ij}^{(n)} \dfrac{\partial^2 y_j^{(n-1)}}{\partial^2 x_a} & n = N \end{cases} \tag{12}$$

$$\begin{gathered} \frac{\partial^3 y^{(N)}}{\partial x_a \partial x_b \partial w_{ij}^{(n)}} = \frac{\partial^2 \delta_i^{(n)}}{\partial x_a \partial x_b} y_j^{(n-1)} + \frac{\partial \delta_i^{(n)}}{\partial x_a} \frac{\partial y_j^{(n-1)}}{\partial x_b} + \frac{\partial \delta_i^{(n)}}{\partial x_b} \frac{\partial y_j^{(n-1)}}{\partial x_a} + \delta_i^{(n)} \frac{\partial^2 y_j^{(n-1)}}{\partial x_a \partial x_b} \; ; \\ \frac{\partial^3 y^{(N)}}{\partial x_a \partial x_b \partial b_i^{(n)}} = \frac{\partial^2 \delta_i^{(n)}}{\partial x_a \partial x_b} \end{gathered} \tag{13}$$

$$\frac{\partial^2 \delta_i^{(n)}}{\partial x_a \partial x_b} = \begin{cases} \left[\dfrac{\partial^3 f_i^{(n)}}{\partial s_i^{(n)3}} \left(\sum\limits_{j=1}^{M^{(n-1)}} w_{ij}^{(n)} \dfrac{\partial y_j^{(n-1)}}{\partial x_a} \right) \left(\sum\limits_{j=1}^{M^{(n-1)}} w_{ij}^{(n)} \dfrac{\partial y_j^{(n-1)}}{\partial x_b} \right) + \dfrac{\partial^2 f_i^{(n)}}{\partial s_i^{(n)2}} \sum\limits_{j=1}^{M^{(n-1)}} w_{ij}^{(n)} \dfrac{\partial^2 y_j^{(n-1)}}{\partial x_a \partial x_b} \right] \sum\limits_k^{M^{(n+1)}} w_{ki}^{(n+1)} \delta_k^{(n+1)} + & \\ \dfrac{\partial^2 f_i^{(n)}}{\partial s_i^{(n)2}} \left[\left(\sum\limits_{j=1}^{M^{(n-1)}} w_{ij}^{(n)} \dfrac{\partial y_j^{(n-1)}}{\partial x_a} \right) \left(\sum\limits_k^{M^{(n+1)}} w_{ki}^{(n+1)} \dfrac{\partial \delta_k^{(n+1)}}{\partial x_b} \right) + \left(\sum\limits_{j=1}^{M^{(n-1)}} w_{ij}^{(n)} \dfrac{\partial y_j^{(n-1)}}{\partial x_b} \right) \left(\sum\limits_k^{M^{(n+1)}} w_{ki}^{(n+1)} \dfrac{\partial \delta_k^{(n+1)}}{\partial x_a} \right) \right] + & n < N \\ \dfrac{\partial f_i^{(n)}}{\partial s_i^{(n)}} \sum\limits_k^{M^{(n+1)}} w_{ki}^{(n+1)} \dfrac{\partial^2 \delta_k^{(n+1)}}{\partial x_a \partial x_b} & \\ \dfrac{\partial^3 f_i^{(n)}}{\partial s_i^{(n)3}} \left(\sum\limits_{j=1}^{M^{(n-1)}} w_{ij}^{(n)} \dfrac{\partial y_j^{(n-1)}}{\partial x_a} \right) \left(\sum\limits_{j=1}^{M^{(n-1)}} w_{ij}^{(n)} \dfrac{\partial y_j^{(n-1)}}{\partial x_b} \right) + \dfrac{\partial^2 f_i^{(n)}}{\partial s_i^{(n)2}} \sum\limits_{j=1}^{M^{(n-1)}} w_{ij}^{(n)} \dfrac{\partial^2 y_j^{(n-1)}}{\partial x_a \partial x_b} & n = N \end{cases} \tag{14}$$

After the derivatives have been calculated, one of many methods of gradient optimization can be applied. In the research discussed method suggested by Lavenberg-Marquad was used, recognized as the most efficient technique of network training [13].

3 Experiments

As part of this research the method testing was carried out making use of several model examples. For each example the analytical solution was known. The quality of the solution obtained by means of a neural network was measured calculating deviations of the neural network output and the values of the function that constituted the analytical solution.

Neural networks were trained using a mesh of 100 points obtained by considering ten equidistant points in x and y directions. An accuracy of each solution provided by the MLP solver was evaluated at 100 training points and 800 test points (in the domain of the equation) and presented in charts.

The weights of the MLP were initialized with the Nguyen-Widrow rule.

Problem 1: A first order partial differential equation

$$2\frac{\partial z}{\partial x}+3\frac{\partial z}{\partial y}=z \ ; \ x, y\in[-1,1] \text{ with BC: } z(1, y)=y \tag{15}$$

The analytic solution (Fig. 1) is $z(x, y)=(2y-3x+3)/2\cdot e^{(x-1)/2}$.

A multilayer perceptron with two inputs, five sigmoid neurons in one hidden layer and one linear output was used for solving of the equation. Fig. 1b illustrates accuracy of the solution provided by the neural network.

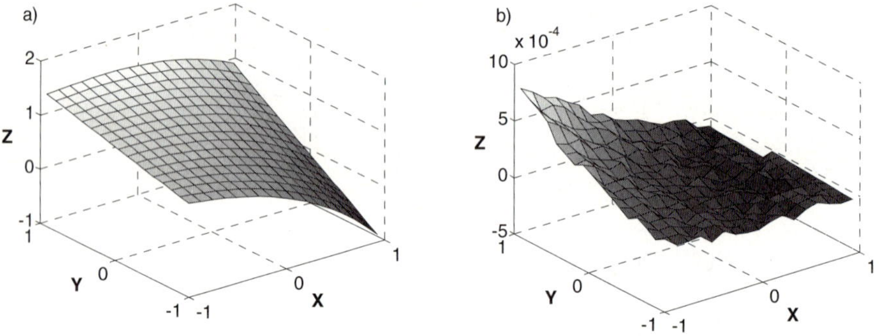

Fig. 1. Problem 1: an exact solution (a) and an accuracy of the computed solution (b)

Problem 2: A second order partial differential equation

$$\frac{\partial^2 z}{\partial x^2}+2\frac{\partial^2 z}{\partial x\partial y}-3\frac{\partial^2 z}{\partial y^2}=0 ; \ x, y\in\left[-\frac{1}{2},\frac{1}{2}\right]$$

$$\text{with BC: } z(x,0)=3x^2 ; \ \frac{\partial z}{\partial y}(x,0)=0 \tag{16}$$

The analytic solution of the equation (Fig. 2a) is $z(x,y)=3x^2+y^2$. The neural network consisted two input units, five hidden and one liner output. Fig. 2b displays an accuracy of the obtained solution.

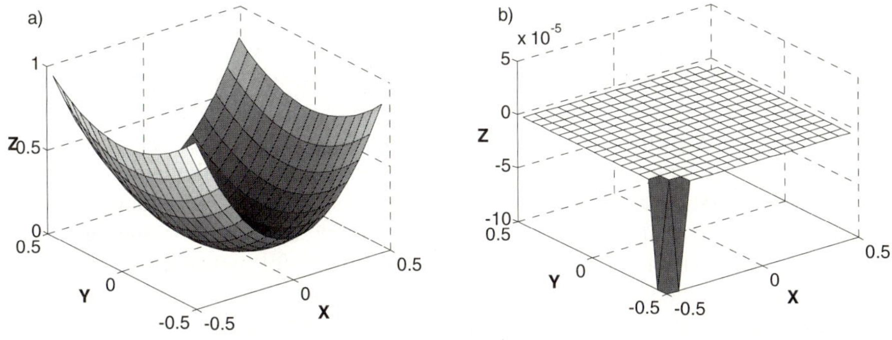

Fig. 2. Problem 2: an exact solution (a) and an accuracy of the computed solution (b)

Problem 3: A second order partial differential equation with irregular boundary conditions

$$\frac{\partial^2 z}{\partial x^2}+2\cos(x)\frac{\partial^2 z}{\partial x \partial y}-\sin^2(x)\frac{\partial^2 z}{\partial y^2}-\sin(x)\frac{\partial z}{\partial y}=0;\; x,y\in[-1,1]$$
$$\text{with BC: } z(x,y)\Big|_{y=\sin(x)}=x+\cos(x);\; \frac{\partial z}{\partial y}(x,y)\Big|_{y=\sin(x)}=\sin(x) \tag{17}$$

The analytic solution is $z(x,y)=x+\cos(x-y+\sin(x))$.

A MLP with following structure was employed: two input neurons, three sigmoid neurons in first hidden layer, eight sigmoid neurons in second hidden layer and one linear output neuron.

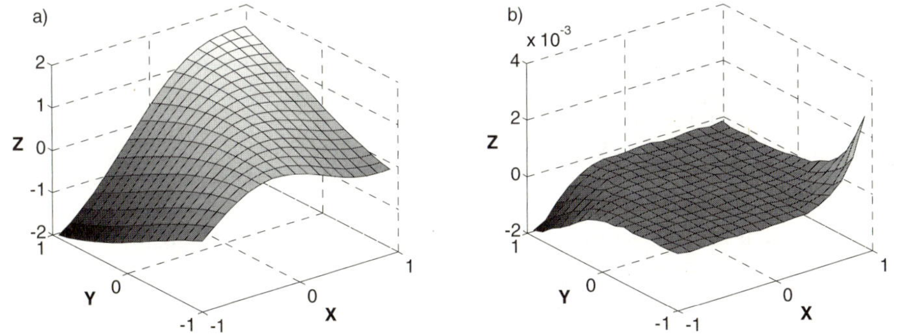

Fig. 3. Problem 3: an exact solution (a) and an accuracy of the computed solution (b)

4 Conclusion

The research has proved that perceptron may be applied as a tool to find solutions to problems represented by partial differential equations of first and second order. The results obtained indicate a good solution convergence.

The method presented is a nondeterministic one, the result obtained being dependent on the random input weights of the network. At the initialization of the neural network no preliminary techniques were applied that would aim at providing for the geometries of the modeled systems or the characteristics of the problem. It can be assumed that a further development of the method in this direction may increase the probability of achieving the equation solution and reducing calculation time.

Since it is assumed that any multilayer perceptron (with continuous activation functions) can be employed in this method, there is room for the application of the already existent, comprehensive methodology for perceptron training and optimization of their structure, which yields huge developmental potential for the improvement of the presented method.

A promising way of achieving a solution faster seems to be the parallelization of the algorithm. The most efficient technique would be to apply the parallelization method through the partition of learning dataset. In case of the algorithm discussed here it would involve a parallel calculation of the performance function gradient in separate areas of the analyzed space, followed by the summing of these values in order to calculate the global gradient used to update the weights. This approach should lead to almost linear increase in performance as a result of a growing number of computer processors in case of high-dimensional problems.

Acknowledgements

The work has been performed within the grant no N508 034 31/1889 (2006 - 2009) financed by Polish Ministry of Science and Higher Education funds.

References

1. Lee, H., Kang, I.: Neural algorithms for solving differential equations. Journal of Computional Physics 91, 110–131 (1990)
2. Dissanayake, M.W.M.G., Phan-Thien, N.: Neural-network-based approximations for solving partial differential equations. Communications in Numerical Methods in Engineering 10, 195–201 (1994)
3. Meade, A.J., Fernadez, A.A.: Solution of nonlinear ordinary differential equations by feedforward neural networks. Mathematical and Computer Modeling 20(9), 19–44 (1994)
4. Ramuhalli, P., Udpa, L., Udpa, S.: Finite-element neural networks for solving differential equations. IEEE Transactions on Neural Networks 16(6), 1381–1392 (2005)
5. Mai-Duy N., Tran-Cong T.: Numerical solution of differential equations using multiquadric radial basis function networks, vol. 14, pp.185–199 (2001)
6. Mai-Duy, N., Tanner, R.I.: Solving high-order partial differential equations with indirect radial basis function networks. International Journal for Numerical Methods in Engineering 63, 1636–1654 (2005)

7. Jianyu, L., Siwei, L., Yingjian, Q., Yaping, H.: Numerial solution of elliptic partial differential equation using radial basis function neural networks. Neural Networks 15, 729–734 (2003)
8. Lagaris, I.E., Likas, A., Fotiadis, I.D.: Artificial neural networks for solving ordinary and partial differentia equation. IEEE Transactions on Neural Networks 9(5), 987–1000 (1998)
9. Lagaris, I.E., Likas, A.C., Papageorgiou, D.G: Neural-network methods for boundary value problems with irregular boundaries. IEEE Transactions on Neural Networks 11(5), 1041–1049 (2000)
10. Delpiano, J., Zagers, P.: Semi-autonomus neural network differential equation solver, International Joint Conference on Neural Networks, Vancouver, Canada, pp. 1863–1869 (2006)
11. Wieczorek, T., Golak, S.: Advance. In: Soft Computing, Proceedings of the International IIS: IIPWM'04 Conference, pp. 470–474 (2004)
12. Aarts, L.P., Veer, P.: Neural network method for solving partial differential equations. Neural Processing Letters 14, 261–271 (2001)
13. Hagan, M.T., Menhaj, M.: Training feedforward networks with the Marquardt algorithm. IEEE Transactions on Neural Networks 5(6), 989–993 (1994)

A Two-Layer ICA-Like Model Estimated by Score Matching

Urs Köster* and Aapo Hyvärinen

University of Helsinki and Helsinki Institute for Information Technology

Abstract. Capturing regularities in high-dimensional data is an important problem in machine learning and signal processing. Here we present a statistical model that learns a nonlinear representation from the data that reflects abstract, invariant properties of the signal without making requirements about the kind of signal that can be processed. The model has a hierarchy of two layers, with the first layer broadly corresponding to Independent Component Analysis (ICA) and a second layer to represent higher order structure. We estimate the model using the mathematical framework of Score Matching (SM), a novel method for the estimation of non-normalized statistical models. The model incorporates a squaring nonlinearity, which we propose to be suitable for forming a higher-order code of invariances. Additionally the squaring can be viewed as modelling subspaces to capture residual dependencies, which linear models cannot capture.

1 Introduction

Unsupervised learning has the goal of discovering the underlying statistical structure of a stream of observed data. This is a difficult problem since most real world data has a complex structure which is hard to capture without prior knowledge. Typically, linear models like Independent Component Analysis (ICA) are utilized. Previous nonlinear extensions of ICA have incorporated prior knowledge on the data [1] [2], so they are not applicable to general data with unknown structure. Therefore we attempt to move towards more general models that can extract complex higher order structure rather than presupposing it. In addition, there is a strong incentive to develop algorithms for the efficient estimation of unsupervised statistical models since recent experiments show they can significantly improve the performance of supervised models [3].

Here we present a model that goes beyond the limitations of ICA without sacrificing generality. It has two layers of weights freely learned from the data, along with a nonlinearity forming a nonlinear representation of the input. The model is specified as a generalization of previous ICA-type models like Topographic ICA (TICA)[4] and Independent Subspace Analysis (ISA)[2]. Since both layers are learned from the data, no prior structure is imposed on the second layer.

* Urs Köster is supported by a scholarship from the *Alfried Krupp von Bohlen und Halbach*-foundation.

J. Marques de Sá et al. (Eds.): ICANN 2007, Part II, LNCS 4669, pp. 798–807, 2007.

Learning in models like this can be done by maximizing the likelihood of the model distribution wrt. observed data. Here one often faces the problem that a model PDF (probability density function) cannot be normalized, and a straightforward estimation of the model is not possible. With Score Matching we present a novel approach to attack this problem. We recently showed [5] that a consistent estimation of the parameters maximizing the likelihood is possible without knowledge of the normalization constant. While other methods based on Monte Carlo methods or approximations have been successfully applied in the past, Score Matching has the advantage that it is a computationally efficient method guaranteeing statistical consistency.

The paper is organized as follows: In section 2, we present the two-layer probabilistic model in more detail, and we explain how it can be estimated using the Score Matching framework. In section 3 we first verify the estimation method by applying the model to artificial data with a known statistical structure. Following this, we present results on real-world data, image patches and natural sounds. The discussion, section 4, puts the new model in perspective with related methods. We highlight the important difference that our model gives rise to sparse connections in the second layer, which is not the case for related work on Contrastive Divergence [6] or modelling "Density Components" [7]. Finally in section 5 we conclude the paper with remarks about the scalability of the model and sketch some possible extensions to other types of data and more than two layers.

2 Model and Estimation

2.1 A Two-Layer Model

While supervised learning methods have often used multiple representation layers, as in multi-layer Perceptrons trained with backpropagation, few unsupervised methods have used such a multi-layer representation. A major problem is that it is usually impossible to obtain the probability distribution of such a model in closed form. For this reason training such models often seems to require a lot of computational resources, because Markov Chain Monte Carlo or similar approximative methods have to be applied.

Still multi-layer models can provide a superior representation for a wide variety of data. We suggest that the lack of suitable estimation principle is a major reason for the poor performance of multilayer models in the past. Using the novel Score Matching approach we show that a very simple and general model can be demonstrated to perform well on a variety of tasks. We propose that our new approach provides a viable alternative to simpler models. Since we formulate it as a generalization of ICA, we find an intuitive way to interpret the results of the model in terms of generalized independent components.

The model that we present here is a bare-bones two layer network with two layers of weights and a scalar nonlinearity acting on the sum of the inputs to each unit.

$$y_i = \mathbf{V}_i g(\mathbf{W}\mathbf{x}) \tag{1}$$

The output of one top-level unit y_i is thus obtained from the data vector $\mathbf{x}$ given the weight matrix $\mathbf{W}$, the row of weights $\mathbf{V}_i$ as well as the nonlinearity $g(\mathbf{u})$. The size of $\mathbf{W}$ and $\mathbf{V}$ is chosen to be equal to the data dimensionality n for simplicity, but the estimation method we propose can also deal with overcompleteness in one or both layers. The weight matrix $\mathbf{V}$ is further constrained to have non-negative elements.

After the first layer of weights $\mathbf{W}$ has performed a linear transform of the data, the scalar nonlinearity $g(\mathbf{u})$ is applied to the outputs. This nonlinearity is the same for all units, and it is fixed in advance rather than learned from the data. We choose to focus on a squaring for the first nonlinearity, i.e. $g(\mathbf{u}) = \mathbf{u}^2$ where the nonlinearity is taken to be element-wise. The second layer $\mathbf{V}$ computes linear combinations of these squared outputs. There are several ways to interpret the squaring nonlinearity that we propose here. Firstly, we would like to point out the connection to our work on Independent Subspace Analysis (ISA) [2], where the components inside a subspace are squared to compute the L_2-norm of the projection onto a subspace. This provides a way to model dependencies of squares that cannot be removed by a simple linear transform. Modelling these dependencies explicitly allows a better fit to the data than linear models could achieve, since high correlations exist between the *activity* of similar features even if they are linearity uncorrelated. [8] The second way to describe the model is to in terms of invariant features. This can provide high selectivity to certain aspects of the data while ignoring aspects that are not relevant to describe the statistical structure of the input. From this point of view the outputs would be features highly invariant under a specific kind of transformation on the input data. A sum of squares, an operation that preserves amplitude but discards the phase of a signal, could perform such an invariant feature extraction [9].

Finally there is an output nonlinearity acting on the second layer outputs. It has the purpose of shaping the overall model PDF to match the statistics of the data. In principle, this could be matched to the optimal distribution for the data under consideration e.g. by an iterative optimization. For simplicity however, we assume the data can be modeled in terms of sparse sources, so we choose an element-wise square root nonlinearity of the form $h(\mathbf{u}) = -\sqrt{\mathbf{u}+1}$. Such a convex choice of h is related to supergaussianity of the PDF.

For learning, the outputs of the second nonlinearity are summed together to define a probability distribution q over the input data.

$$\log q(\mathbf{x}|W, V) = \sum_{i=1}^{n} h\left(\mathbf{V}_i g(\mathbf{W}\mathbf{x})\right) \qquad (2)$$

Intuitively, this model can be thought of as a two layer neural network processing the incoming data vector and computing the probability that the data came from the distribution defined by the model. This immediately provides a means of training the model by adjusting the parameters to maximize the likelihood of the model given the observed training data.

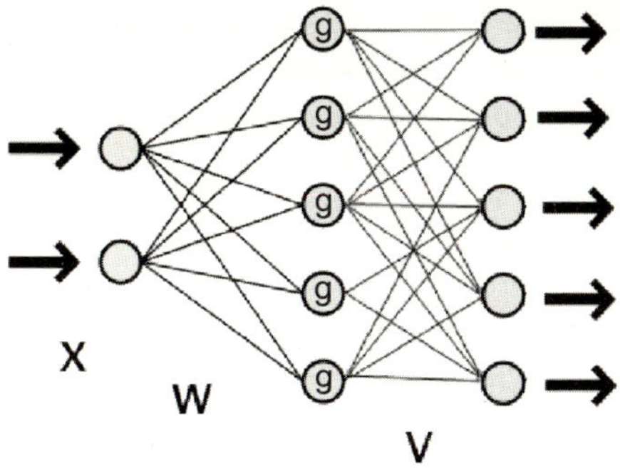

Fig. 1. Graphical representation of the two-layer model

For estimation, we usually need to compute the log-likelihood of the model

$$\log l(\mathbf{W}, \mathbf{V}|\mathbf{x}) = \sum_{i=1}^{n} h\left(\mathbf{V}_i g(\mathbf{W}\mathbf{x})\right) - \log\left(Z(W, V)\right) \tag{3}$$

where Z denotes the normalization constant of the distribution, which is obtained by integrating over all space. It is obvious that the normalization constant cannot be computed in closed form, which makes the estimation of the model impossible with standard methods. Therefore we apply the novel estimation method Score Matching which is described below.

2.2 Score Matching

As we stated above, the probability distribution of the data can in general only be obtained up to a multiplicative constant. This makes it impossible to compute the likelihood of the model, and standard optimization methods like gradient descent on the log-likelihood cannot be used. In the past, Monte Carlo methods such as Contrastive Divergence [10] have been applied to this problem, or approximations of the likelihood were used. Here we circumvent the problem by focusing on the *score function of the density*, $\boldsymbol{\Psi}(\boldsymbol{\eta}; \mathbf{W}, \mathbf{V})$ with respect to $\boldsymbol{\eta}$, where $\boldsymbol{\eta}$ is a variable which replaces the data vector $\mathbf{x}$ for notational unambiguity.

$$\boldsymbol{\Psi}(\boldsymbol{\eta}; \mathbf{W}, \mathbf{V}) = \nabla_{\boldsymbol{\eta}} \log p(\boldsymbol{\eta}; \mathbf{W}, \mathbf{V}) \tag{4}$$

Additionally we can define the *data score function* $\boldsymbol{\Psi}_x(.) = \nabla_\eta \log p_x(.)$ for the distribution of observed data. The model is optimized by matching the data and model score functions (hence the name Score Matching). We can achieve this by minimizing the squared distance

$$J(\mathbf{W}, \mathbf{V}) = \frac{1}{2} \int_{\boldsymbol{\eta}} \|\boldsymbol{\Psi}(\boldsymbol{\eta}; \mathbf{W}, \mathbf{V}) - \boldsymbol{\Psi}_{\mathbf{x}}(\boldsymbol{\eta})\|^2 d\boldsymbol{\eta} \tag{5}$$

This could painstakingly be computed using a nonparametric estimation of the density, but as shown in [5] the expression can be expressed in a much simpler form in terms of derivatives of the data score function:

$$\tilde{J}(\mathbf{W},\mathbf{V}) = \frac{1}{T}\sum_{t=1}^{T}\sum_{i=1}^{n}\left[\frac{\partial}{\partial \eta_i}\boldsymbol{\Psi}_i(\mathbf{x}(t);\mathbf{W},\mathbf{V}) + \frac{1}{2}\boldsymbol{\Psi}_i^2(\mathbf{x}(t);\mathbf{W},\mathbf{V})\right] + C \qquad (6)$$

Here the $\tilde{J}$ indicates a sampled version of the objective function, but in the limit of $T \rightarrow \infty$ and given the existence of a nondegenerate optimum, this estimator is statistically consistent. C is a constant that does not depend on any the parameters. Estimation of the parameters can easily be performed by following the gradient of this function wrt. the parameters.

3 Experiments

3.1 Methods

We performed experiments on a variety of data to show the power and adaptability of the model. The focus was on natural data, i.e. natural image patches and speech recordings, to demonstrate the particular suitability of our model to this very complex and rich kind of data that is poorly modeled by simpler methods. For the natural data we performed preprocessing in the form of whitening (decorrelation), Contrast Gain Control by dividing each data vector by its L_2-norm, and some dimensionality reduction by PCA.

In general we start the optimization by learning the independent components of the data, which is achieved by clamping the second layer weights to the identity matrix. This serves to avoid local minima and speed up the convergence of the algorithm. After this, the second layer connections are learned. It is an important feature of the estimation method that learning for the first layer is not stopped; rather the first layer features start to move away from ICA features to adjust to the second layer as it forms more complex and invariant features.

An additional technical constraint was the use of L_2-normalization on the rows of $\mathbf{V}$, corresponding to the second layer output vectors. This prevents individual units from "dying" and also sets a bound on the maximum activity. We verified that it does not qualitatively change the structure of the outputs. $\mathbf{W}$ was constrained to be orthogonal as it is customary with ICA algorithms. For the optimization we used a stochastic gradient approach with mini batches consisting of 100 data samples. Not only does this significantly increase the speed of convergence, but we found that without stochasticity, local minima hindered the convergence of the second layer weights.

3.2 Artificial Data

As a first test for the model and estimation method we generated data according to the ISA model[2]. This is supergaussian data with dependencies *within*, but not *between* subspaces of the data variables. This data was then mixed with

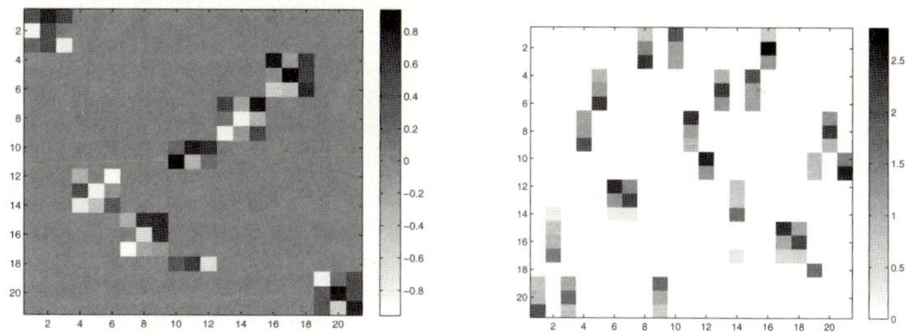

(a) (b)

Fig. 2. The model was tested with ISA data, convergence is fast and finds the global minimum. We show (a) the product of the estimated demixing and known mixing matrix, (b) the learned second layer weights. The rows of the matrices are sorted in ascending order on the columns of **V**. This does not affect the result and is purely for easier visualization.

a random mixing matrix **A**. We used 10,000 samples of 21-dimensional data generated with a subspace size of three.

Figure 2 shows how the first layer weights **W** invert the mixing up to subspace membership, while **V** determines which variables belong together in one subspace. Since the dimensionality of **V** is 21×21, and there are only 7 subspaces, some rows of **V** go to zero and some are duplicated. Contrary to later experiments, both weight layers were initialized randomly and learned simultaneously, and no normalization on the rows **V** was performed.

3.3 Experiments on Natural Images

After confirming the identifiability of the method, we tested the model on natural images which have a particularly rich statistical structure with many higher order dependencies. We use 20,000 image patches of 12×12 pixels, whitened, performed Contrast Gain Control [11] and reduced the data dimensionality to 120 by PCA. We specified the dimensionality of both **W** and **V** to be 120×120. Optimizing **W** first gives familiar ICA features as shown in fig. 3a. In fact variants such as TICA and ISA can easily be performed by setting **V** appropriately. The second layer learns connections between similar first layer features (fig. 3b), giving rise to complex-cell like outputs which are invariant to the spacial phase of the data (fig. 3c). Continued learning on the first layer features increased the similarity of the position and size of filter feeding into the same second layer unit while keeping the phase difference. This result was also confirmed with an overcomplete model.

3.4 Audio Data

In order to demonstrate the general applicability of our model to a variety of data, we also tested it on speech data from the TIMIT database. We sampled

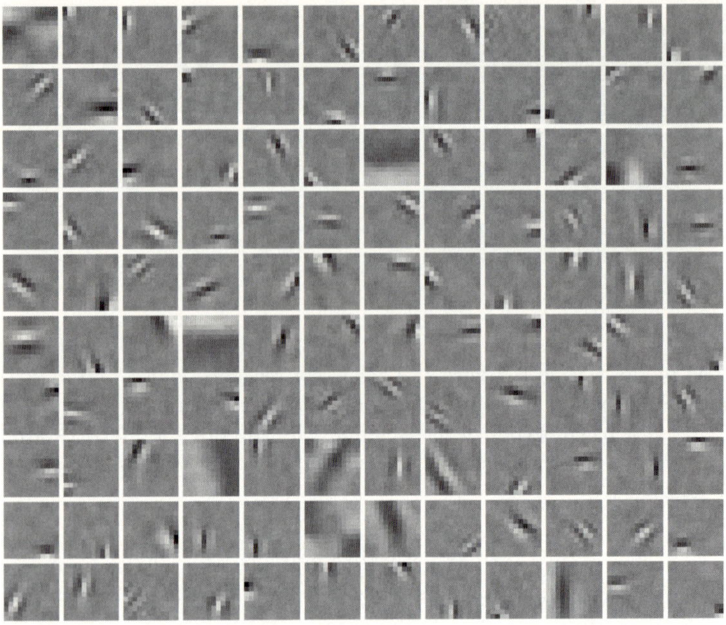

(a) First Layer

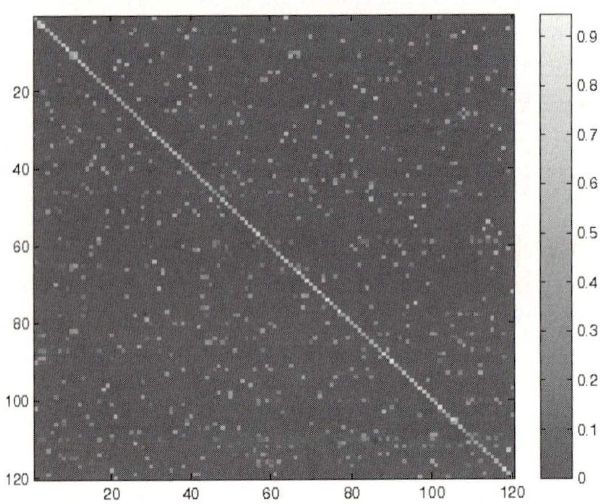

(b) Second Layer

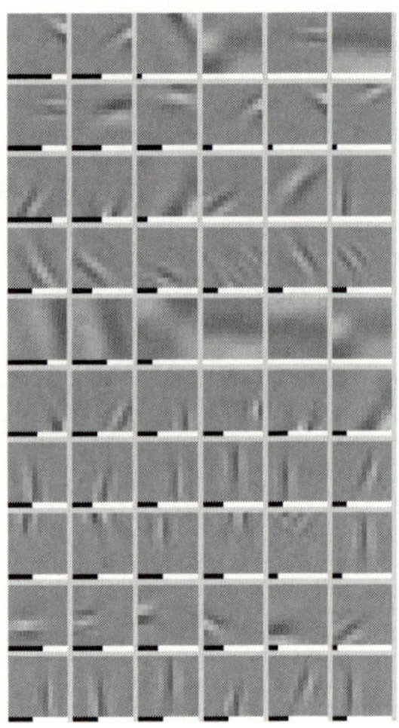

(c) Some Outputs

Fig. 3. a) First layer filters show the classical Simple-Cell type structure. b) Connection in the second layer are sparse, with connections between similar units. c) A random selection of outputs where each row shows the most active contributers to the response with the black bars indicating "synaptic strength", i.e. how strongly the filter contributes to the output.

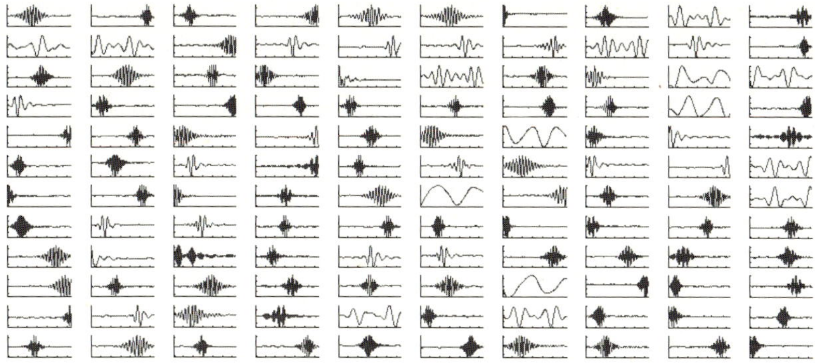

(a) First Layer

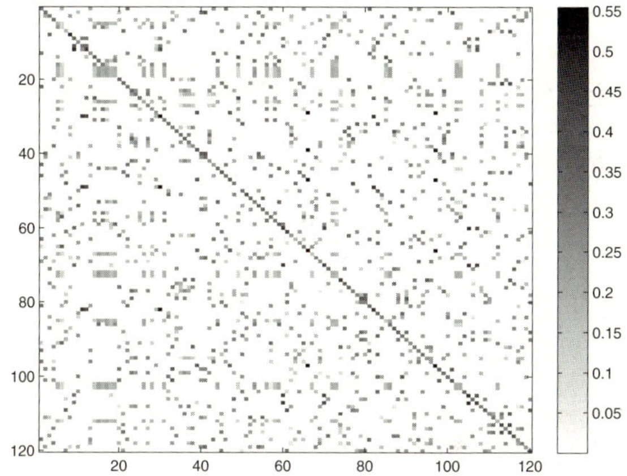

(b) Second Layer

Fig. 4. (a) The first layer gives outputs localized in both frequency and time. (b) The second layer gives connections between features with dependencies of squares.

random rectangular sound windows of 8ms length, and resampled them to 8kHz. We also applied our standard preprocessing consisting of removing the DC component, whitening and contrast gain control. Simultaneously we reduced the dimensionality from 64 to 60 which amounts to low-pass filtering and serves to eliminate artifacts from the windowing procedure. The results are presented in figure 4.

4 Discussion

We have shown that an unsupervised model using two completely flexible layers of weights to learn the statistical structure of its input data can effectively be estimated using Score Matching. While including previous extensions of ICA as special cases, this is far more general that previous models. For example ISA forces the filters to group into subspaces of a constant size and with an equal contribution, and did not allow a single filter to be active in more than one higher order unit. These constraints have been lifted with the new model. Topographic ICA is also included in our model as a special case. If the second layer is fixed to an identity matrix convolved with a kernel (neighborhood function) that leaks activity to off-diagonal elements, a topographic ICA model can be estimated. A more complex topography can be obtained by allowing interactions other than along the main diagonal.

Two models have recently been proposed that have a similar hierarchical structure but are estimated differently. Most close related to our work is the work by Osindero et al. [6]. Instead of using the traditional "independent component" point of view, the model is defined as a "product of experts" model following Student-t distributions. The estimation is performed using contrastive divergence (CD), which was recently shown [12] to be equivalent to Score Matching. The key difference between the models is in the results obtained on natural data. While Osindero et al. report sparse *activation* of second layer units, we also see sparse *connectivity*, which has interesting implications not only because of the striking similarity to biological networks, but also for efficient signal processing.

The second work that we would like to mention is that of Karklin and Lewicki [7]. They present a generative two layer model that performs ICA on the data followed by a variance-modelling stage as in TICA[4]. Contrary to the PoT model of Osindero et al. and our SM model, both layers are estimated separately using the *maximum a posteriori* estimate. The authors observe that in contrast to our model, the first layer units do not change significantly depending on the "density components" modelling the variance of the first layer outputs. Applied to natural stimulus data, this model gives rise to broadly tuned features in the second layer that describe global properties of the data. Again this is in contrast to the sparse connectivity obtained from our model.

5 Conclusion

We have presented a two layer model that that can be used to learn the statistical structure of various kinds of data. By using the novel estimation principle Score Matching, unsupervised learning in this type of model is made faster and more straightforward than with alternatives such as Monte Carlo methods. Contrary to previous linear models, higher order dependencies in the data can be captured to give better models of real world data. Compared to similar models [6] [7], we report the emergence of sparse connectivity in the second layer. Furthermore our model is very general, so it can be overcomplete, and it can be extended to incorporate a third or more layers.

References

1. Cardoso, J.-F.: Multidimensional independent component analysis. In: Proc. ICASSP'98, Seattle (1998)
2. Hyvärinen, A., Hoyer, P.: Emergence of phase and shift invariant features by decomposition of natural images into independent feature subspaces. Neural Computation 12(7), 1705–1720 (2000)
3. Hinton, G.E., Salakhutdinov, R.R.: Reducing the dimensionality of data with neural networks. Science 313(5786), 504–507 (2006)
4. Hyvärinen, A., Hoyer, P., Inki, M.: Topographic independent component analysis. Neural Computation (2001)
5. Hyvärinen, A.: Estimation of non-normalized statistical models using score matching. Journal of Machine Learning Research 6, 695–709 (2005)
6. Osindero, S., Welling, M., Hinton, G.E.: Topographic product models applied to natural scene statistics. Neural Computation 18 (2006)
7. Karklin, Y., Lewicki, M.S.: A hierarchical bayesian model for learning non-linear statistical regularities in non-stationary natural signals. Neural Computation 17(2), 397–423 (2005)
8. Simoncelli, E., Adelson, E.: Noise removal via bayesian wavelet coding. Intl Conf. on Image Processing, 379–382 (1996)
9. Hyvärinen, A., Hurri, J., Väyrynen, J.: Bubbles: a unifying framework for low-level statistical properties of natural image sequences. Journal of the Optical Society of America A 20(7), 1237–1252 (2003)
10. Geoffrey, E.: Training products of experts by minimizing contrastive divergence. Neural Comput. 14(8), 1771–1800 (2002)
11. Köster, U., Hyvärinen, A.: Complex cell pooling and the statistics of natural images. Network: Computation in Neural Systems, in print (2005), available online: `cs.helsinki.fi/u/koster/koster05.pdf`
12. Hyvärinen, A.: Connections between score matching, contrastive divergence, and pseudolikelihood for continuous-valued variables. IEEE Transactions on Neural Networks (in press)

Testing Component Independence Using Data Compressors

Daniil Ryabko

IDSIA, Galleria 2, CH-6928 Manno, Switzerland
daniil@idsia.ch
http://www.idsia.ch/~daniil

Abstract. We propose a new nonparametric test for component independence which is based on application of data compressors to ranked data. For two-component data sample the idea is to break the sample in two parts and permute one of the components in the second part, while leaving the first part intact. The resulting two samples are then jointly ranked and a data compressor is applied to the resulting (binary) data string. The components are deemed independent if the string cannot be compressed. This procedure gives a provably valid test against all possible alternatives (that is, the test is distribution-free) provided the data compressor was ideal.[1]

1 Introduction

In this work we consider a classical problem of mathematical statistics which has important applications in machine learning: component independence. A sample $Z_1, \dots, Z_n$ is given, generated i.i.d. according to some distribution F_Z. Each element Z_i consists of two (or more) components Z_i^1 and Z_i^2. We wish to test whether the components are independent of each other. That is, H_0 is that the marginal distributions are independent whereas H_1 is that there is some dependency. No assumption is made on the distribution F_Z. Type I error of a test occurs when it rejects H_0 while H_0 is actually true, and Type II error occurs when the test accepts H_0 but H_1 is true. Normally one wishes to make both errors as small as possible. In non-parametric statistics a typical approach is to construct a test that has Type I error fixed at some pre-specified level and Type II error tends to 0 when the sample size increases; an asymptotic result of this kind is usually the best one can achieve.

This problem is closely related to the problem of feature selection which is very important for pattern recognition, regression estimation and related machine learning tasks. The problem is to determine which features (components) from a multi-dimensional sample are relevant for estimating (predicting) the value of a distinguished feature (the label). Component independence can be applied to solving this problem either componentwise, that is, testing whether the label is

[1] This research was supported by the Swiss NSF grants 200020-107616 and 200021-113364.

J. Marques de Sá et al. (Eds.): ICANN 2007, Part II, LNCS 4669, pp. 808–815, 2007.

independent of a given component (ignoring al others), or applied to different groups of components. For overview of the problem and issues arising applying these approaches see e.g. [3,7].

Component independence is a well-known problem of mathematical statistics. A classical statistical approach is to model the data with some family of distributions, which leads to constructing parametric tests. However, there are many nonparametric tests also; some of the tests use *ranks* of elements within the joint sample, instead of using the actual samples. Such is, for example, Wilcoxon's test, see [4] for an overview (which also makes some additional assumptions on the distribution, and so is not valid against some alternatives).

In this work we present a simple nonparametric distribution-free rank test for component independence based on data compressors.

The idea to use real-life data compressors for testing classical statistical hypotheses, such as homogeneity, component independence and some others, was suggested in [9,10]. In these works statistical tests based on data compressors are constructed which fall into the classical framework of nonparametric mathematical statistics, in particular, the Type I error is fixed while Type II error goes to 0 under a wide range of alternatives. The hypotheses considered there mostly concern data samples drawn from *discrete* (e.g. finite) spaces. Some tests for continuous spaces are also proposed based on partitioning. Here we extend this approach to *rank* tests, allowing testing component independence without the need of partitioning the sample spaces and making them finite. The idea of using data compressors for tasks other than actual data compression was suggested in [1,2,5], where data compressors are applied to such tasks as classification and clustering. These works were largely inspired by Kolmogorov complexity, which is also an important tool for the present work.

An "ideal" data compressor is the one that compresses its input up to its Kolmogorov complexity. This is intuitively obvious since, informally, Kolmogorov complexity of a string is the length of the shortest program that outputs this string. Such data compressors do not exist; in particular, Kolmogorov complexity itself is incomputable. Real data compressors, however, can be considered as approximations of ideal ones.

In this work we provide a simple empirical procedure for testing component independence with data compressors; we show that for an ideal data compressor this procedure provides a statistical test which is valid against all alternatives (Type II error goes to zero); while Type I error is guaranteed to be below a predefined level (so-called significance level) for all data compressors, not only for ideal ones. It should also be noted that the theoretical assumption underlying data compressors used in real life is that the data to compress is *stationary*. Thus the tests designed in [9,10] are provably valid against any stationary and ergodic alternative, while these tests are based on real data compressors, not only on ideal ones. In our case, the alternative arising in rank test under H_1 is not stationary. Thus we prove theorems only about ideal data compressors, and real data compressors can be used heuristically. However, it can be conjectured that the same results can be proven for some particular real-life data compressors, for example for those which are based on the measure R from [8] or on the LZ algorithm [12].

2 Main Results

Component independence testing is the following task. A sample $Z = Z_1, \ldots, Z_n$ is given where each Z_i consists of r components $Z_i^1, Z_i^2, \ldots, Z_i^r$, $Z_i^j \in \mathbb{R}^{d_j}$. The sample is generated according to some probability distribution F_Z on $\mathbb{R}^d$, where $d := \sum_{j=1}^r d_j$. The goal is to test whether the components are distributed independently. That is, H_0 is that

$$F_Z(Z_1^1 \in T_1, \ldots, Z_1^r \in T_r) = \prod_{j=1}^r F_Z(Z_1^j \in T_j) \tag{1}$$

for all measurable $T_j \subset \mathbb{R}^{d_j}$, $1 \le j \le r$. H_1 is the negation of H_0 (the equality (1) is false for some selection of the sets T_j, $1 \le j \le r$). Again, no assumption is made on the form of the distribution F_Z.

A *code* φ is a function $\varphi : B^* \to B^*$ from the set of all finite words over binary alphabet $B = \{0, 1\}$ to itself, such that φ is an injection (that is, $a \ne b$ implies $\varphi(a) \ne \varphi(b)$ for $a, b \in B^*$). A trivial example of a code is the identity $\varphi_{id}(a) = a$. Less trivial examples that we have in mind are data compressors, such as `zip`, `rar`, `arj`, or others, which take a word and output a "compressed" version of it (which in fact is often longer than the original) from which the original input can always be recovered. We will construct (reasonable) tests for homogeneity from (good) data compressors.

Fix any code φ and construct the test for component independence I_φ in four steps.

Step 1. Break the sample into two halves, leave the first one intact and randomly and independently permute the elements within each component in the second half. More precisely, assume that $n = 2m$ for some m and define the samples X and W as the first and the second half of the sample Z: $X_1 = Z_1, \ldots, X_m = Z_m$ and $W_1 = Z_{m+1}, \ldots, W_m = Z_{2m}$ (if n is odd then make samples X and W of sizes $\lfloor n/2 \rfloor$ and $n - \lfloor n/2 \rfloor$). Construct the sample Y from W by permuting the elements within each component independently: $Y_i^j = W^j_{\pi_j(i)}$, $1 \le i \le m$, $1 \le j \le r$ where π_j are permutations of $\{1 \ldots m\}$, selected at random (with equal probabilities) independently of each other.

Step 2. Make the resulting two samples single-dimensional. Construct samples $\bar{X} = \bar{X}_1, \ldots, \bar{X}_m$ and $\bar{Y} = \bar{Y}_1, \ldots, \bar{Y}_m$ as follows:

$$\bar{X}_t := x_t^{11}, x_t^{21}, \ldots, x_t^{d1}, x_t^{12}, x_t^{22}, \ldots, x_t^{d2}, \ldots$$

where x_t^{ij} is the jth element in the binary expansion of the ith component of X_t (in case the expansion is ambiguous always take the one with more zeros), and analogously for Y. Denote the described function which converts X to $\bar{X}$ (and Y to $\bar{Y}$) by τ.

Step 3. Order two samples jointly and transform them into a binary string, 0 for an element of the first sample and 1 for an element of the second. Let

$$Z_1 \leq Z_2 \leq \cdots \leq Z_n$$

denote the joint sample constructed by ordering jointly two samples $\bar{X}$ and $\bar{Y}$. Construct the word

$$A = A_1 \ldots, A_n$$

as follows: for each i $A_i = 0$ if Z_i is taken from the sample $\bar{X}$ ($Z_i \in \bar{X}$) and $A_i = 1$ if Z_i is from the sample $\bar{Y}$ ($Z_i \in \bar{Y}$) where ties are bracken by randomization: if $Z_j = Z_{j+1} = \cdots = Z_{j'}$ and there are m elements of the sample $\bar{X}$ which are equal to Z_j and k elements of the sample $\bar{Y}$ which are equal to Z_j then the word $A_j \ldots A_{j'}$ is chosen randomly from all $\frac{(m+k)!}{m!k!}$ binary words which have m zeros and k ones, assigning equal probabilities to all words. Let $|K|$ denote the length of a string K.

Step 4. Finally, the actual tests consists in evaluating the length of the binary string A compressed by a code φ.

Definition 1 (Test G_φ). *For any code φ the test for component independence G_φ is constructed as follows. It rejects the hypothesis H_0 (outputs* reject*) at the level of significance α if*

$$|\varphi(A)| \leq \log \alpha N \tag{2}$$

where $N := \frac{n!}{(m!)^2}$ and log *is base 2, and accepts H_0 (outputs* accept*) otherwise.*

The intuition is as follows. Observe that if the components are independent, then permuting them (independently) does not change the distribution of the data. In other words, the two samples constructed in Step 1 (the first one simply a half of the original sample and the second one with permuted components) are distributed according to the same distribution *if and only if* the components are independent.

Further, if we have two samples, order them jointly and construct a binary string as described in Steps 2 and 3, then the resulting binary string is random (more precisely, has equal probability of being any string from the set of all binary strings with m zeros and m ones) if and only if the samples were generated according to the same distribution; that is, if and only if the components were independent.

Thus a data compressor may be able to compress this binary string to about $\log N$ bits, but no code can compress many such strings to less than $\log N - t$ bits ($t > 0$), since there are N such strings and only $2^{-t}N$ binary strings of length $\log N - t$. The next proposition formalizes this property. In other words, it says that for any code the type I error can be made bounded by any pre-specified α.

Proposition 1 (Type I error). *For any code φ and any $\alpha \in [0, 1]$ the Type I error of the test G_φ with level of significance α is not greater than α:*

$$F_Z(G_\varphi(Z) = reject) \leq \alpha \tag{3}$$

for all $F_Z \in H_0$.

Remark 1. The proposition still holds if H_0 is rejected when

$$|\varphi(A)| \leq n + \log \alpha - \log n. \quad (4)$$

Proof. Under hypothesis H_0 the samples X and Y are distributed according to the same distribution. Consequently, the samples $\bar{X}$ and $\bar{Y}$ are distributed according to the same distribution, since they are obtained from X and Y by applying the same function. Thus, under H_0 for every string $a \in B^n$ such that a consists of m zeros and m ones $P(A = a) = 1/N$ (that is, all such strings are equiprobable). Since there are only αN binary strings of length $\log \alpha N$ and φ is an injective function, that is each codeword is assigned to at most one word, we get $F_Z(|\varphi(A)| \leq \log \alpha N) \leq \frac{1}{N} N\alpha = \alpha$ which together with the definition of G_φ implies (3).

The statement of the Remark can be derived from Stirling's expansion for N. □

Remark 2. The term $-\log n$ in (4) is due to the fact that there are only $\frac{n!}{(m!)^2}$ strings with m zeros and m ones (among 2^n all binary strings of this length). So the code φ can specifically assign shorter codewords to these strings. As real data compressors are not designed to favour strings of this particular ratio of zeros and ones, in practice it is recommended to omit the term $-\log n$ in (4).

Obviously, for some codes the test is useless (for example if φ is the identity mapping) and Proposition 1 is only useful when the Type II error goes to zero. Next we will define "ideal" codes (the codes that compress a word up to its Kolmogorov complexity) and show that for them indeed the probability of *accept* goes to zero under any distribution in H_1, that is, the test is valid against all alternatives.

Informally, Kolmogorov complexity of a string A is the length of the shortest program that outputs A (on the empty input). Clearly, the best, "ideal", data compressor can compress any string A up to its Kolmogorov complexity, and not more (except may be for a constant). Next we present a definition of Kolmogorov complexity; for fine details see [11,6]. The complexity of a string $A \in B^*$ with respect to a Turing machine ζ is defined as

$$C_\zeta(A) = \min_p \{l(p) : \zeta(p) = A\},$$

where p ranges over all binary strings (interpreted as programs for ζ; minimum over empty set is defined as ∞). There exists a Turing machine ζ such that $C_\zeta(A) \leq C_{\zeta'}(A) + c_{\zeta'}$ for any A and any Turing machine ζ' (the constant $c_{\zeta'}$ depends on ζ' but not on A). Fix any such ζ and define *Kolmogorov complexity* of a string $A \in \{0,1\}^\infty$ as

$$C(A) := C_\zeta(A).$$

Clearly, $C(A) \leq |A| + b$ for any A and for some b depending only on ζ.

Definition 2 (ideal codes). *Call a code* φ ideal *if some constant* c *the equality* $|\varphi(A)| \leq C(A) + c$ *holds for any binary string* A.

Clearly such codes exist.

Proposition 2 (Type II error: universal validity). *For any ideal code φ Type II error of the test G_φ with any fixed significance level $\alpha > 0$ goes to zero $F_Z(G_\varphi(X,Y) = accept) \to 0$ for any F_Z in H_1.*

Proof. First observe that the function τ that converts d-dimensional samples X and Y to single-dimensional samples $\bar{X}$ and $\bar{Y}$ has the following properties: if X and Y are distributed according to different distributions then $\bar{X}$ and $\bar{Y}$ are also distributed according to different distributions. Indeed, τ is one to one, and transforms cylinder sets, that is sets of the form

$$\{x \in \mathbb{R}^d : x^{i_1 j_1} = b_1, \ldots, x^{i_t j_t} = b_t; b_l \in \{0,1\}, t, i_l, j_l \in \mathbb{N}(1 \le l \le t)\},$$

to cylinder sets. So together with F_X (F_Y) it defines some distribution $F_{\bar{X}}$ ($F_{\bar{Y}}$) on $\mathbb{R}$. If distributions F_X and F_Y are different then they are different on some cylinder set T, but then $F_{\bar{X}}(\tau(T)) \neq F_{\bar{Y}}(\tau(T))$.

We have to show that Kolmogorov complexity $C(A) = |\varphi(A)|$ of the string A is less than $\log \alpha N \ge n + \log \alpha - \log n$ for any fixed α from some n on. To show this, we have to find a sufficiently short description $s(A)$ of the string A; then the Kolmogorov complexity $|\varphi(A)|$ is not greater than $|s(A)| + c$ where c is a constant.

If H_1 is true then $F_X \neq F_Y$ and so there exist some interval $T = (-\infty, t]$ and some $\delta > 0$ such that $|F_X(T) - F_Y(T)| > 2\delta$. Then we will have

$$\frac{1}{m} |\#\{x \in X \cap T\} - \#\{y \in Y \cap T\}| > \delta \tag{5}$$

from some m on with probability 1.

Let A' be the starting part of A that consists of all elements that belong to T and let $m_1 := \#\{x \in X \cap T\}$ and $m_2 := \#\{y \in Y \cap T\}$. A description of A' can be constructed as the index of A' in the set (ordered, say, lexicographically) of all binary strings of length $m_1 + m_2$ that have exactly m_1 zeros and m_2 ones plus the description of m_1 and m_2. Thus the length of such a description is bounded by the sum of

$$\log \frac{(m_1 + m_2)!}{m_1! m_2!} \le (m_1 + m_2) h\left(\frac{m_2}{m_1 + m_2}\right)$$

and

$$\log m_2 + \log m_1 + const,$$

where h is the entropy function

$$h(t) = -t \log t - (1-t) \log(1-t)$$

(the inequality follows from $n! \le n^n$ for all n). Let $\bar{A}$ denote the remaining part of A (that is, what goes after A'). The length of the description of $\bar{A}$ is bounded by

$$(\bar{m}_1 + \bar{m}_2) h\left(\frac{\bar{m}_2}{\bar{m}_1 + \bar{m}_2}\right) + \log \bar{m}_2 + \log \bar{m}_1 + const$$

where $\bar{m}_1 = m - m_1$ and $\bar{m}_2 = m - m_2$. Since h is concave and $1/2$ is between $\frac{m_2}{m_1+m_2}$ and $\frac{\bar{m}_2}{\bar{m}_1+\bar{m}_2}$, from Jensen's inequality we obtain (using $h(1/2) = 1$)

$$1 - \left(\frac{m_1 + m_2}{n} h\left(\frac{m_2}{m_1 + m_2}\right) + \frac{\bar{m}_1 + \bar{m}_2}{n} h\left(\frac{\bar{m}_2}{\bar{m}_1 + \bar{m}_2}\right)\right) > 0.$$

Denote this difference by γ. Clearly, Clearly, γ is positive and depends only on δ. To uniquely describe A we need the description of A' and $\bar{A}$; these have to be encoded in a self-delimiting way; the length of such a description $s(A)$ is bounded by the lengths of description of A', $\bar{A}$ plus $\log n$ and some constant. Thus

$$\begin{aligned} & n + \log\alpha - \log n - |\varphi(A)| \geq \\ & n + \log\alpha - 2\log(k+m) - \frac{m_1 + m_2}{n} h\left(\frac{m_2}{m_1 + m_2}\right) - \frac{\bar{m}_1 + \bar{m}_2}{n} h\left(\frac{m_2}{\bar{m}_1 + \bar{m}_2}\right) - c \\ & \geq n\gamma - 2\log n - c \end{aligned}$$

for some constant c; clearly, this expression is greater than 0 from some k, m on. □

So, as a corollary of Propositions 1 and 2 we get the following statement.

Theorem 1. *For any code φ and any $\alpha \in (0, 1]$ the Type I error of the test G_φ with level of significance α is not greater than α. If, in addition, the code φ is ideal then the Type II of G_φ error tends to 0 as the sample size n approaches infinity.*

3 Discussion

We have presented a theoretical justification for a simple procedure which can be used empirically with any available data compressors. Perhaps the main advantage of the proposed test is that nothing has to be known about the distribution generating the sample: it does not have to be continuous (or discrete), and can have any form.

In particular, in machine learning one is interested in predicting the value of a (real-valued) label based on available features. Some features may be redundant and can hinder the analysis, so one often wishes to exclude such features before applying any machine learning methods for prediction. However, nothing is usually known about the distributions governing the data. In such a setting our test can be applied either to test individual features for independence from the label (applying the test to samples consisting of feature-label pairs) or joint independence of a subset of features from the label.

Let us now consider the requirements on the data compressor that we posed. First of all, to achieve the bound on the Type I error (Proposition 1) the compressor can be arbitrary. However, for the Type II error to go to zero we require the compressor to be ideal. An ideal compressor is a compressor that can notice any regularity in the data (and use it for compression). However, for our purposes, being able to spot only certain type of regularities is sufficient to compress the data. In particular, as we have shown, under H_1 for sufficiently large samples the binary string to be compressed can be bracken into two parts which have different frequencies of 0s and 1s. It can be conjectured that a data compressor which can compress (in asymptotic, up to the entropy), data generated by an arbitrary stationary source, can be used for our purposes also — that is, Type II error should go to zero under H_1 for such compressors too. However, this question is yet open.

A question that we have not addressed explicitly is the speed of convergence in Proposition 2 (the convergence of the probability of Type II error to zero under H_1). This speed of convergence depends on two factors: the data compressor used and the distribution of the sample. Clearly, if the dependence is very slight than a larger sample will be required to detect it. Our analysis (cf. the proof of Proposition 2) suggests that for reasonably behaved data compressors the speed of convergence is exponential in the size of the sample, with constants depending on the actual distribution; this should follow from the fact that the empirical distribution function converges to the true distribution function governing the data with exponential speed. More accurate analysis is, however, a topic for future studies.

References

1. Cilibrasi, R., Vitányi, P.: Clustering by Compression. IEEE Transactions on Information Theory 51(4) (2005)
2. Cilibrasi, R., de Wolf, R., Vitányi, P.: Algorithmic Clustering of Music. Computer Music Journal 28(4), 49–67 (2004)
3. Guyon, I., Elisseeff, A.: An Introduction to Variable and Feature Selection. Journal of Machine Learning Research 3, 1157–1182 (2003)
4. Lehmann, E.: Testing Statistical Hypotheses, 2nd edn. John Wiley & Sons, New York (1986)
5. Li, M., Chen, X., Li, X., Ma, B., Vitányi, P.: The similarity metric, IEEE Trans. Inform. Th. 50(12), 3250–3264 (2004)
6. Li, M., Vitányi, P.: An introduction to Kolmogorov complexity and its applications, 2nd edn. Springer, Heidelberg (1997)
7. Liu, H., Motoda., Hiroshi.: Feature Selection for Knowledge Discovery and Data Mining. Springer, Heidelberg (1998)
8. Ryabko, B.: Prediction of random sequences and universal coding. Problems of Inform. Transmission 24(2), 87–96 (1988)
9. Ryabko, B., Astola, J.: Universal Codes as a Basis for Time Series Testing. Statistical Methodology 3, 375–397 (2006)
10. Ryabko, B., Monarev, V.: Using information theory approach to randomness testing. Journal of Statistical Planning and Inference 133(1), 95–110 (2005)
11. Vereshchagin, N., Shen, A., Uspensky, V.: Lecture Notes on Kolmogorov Complexity, Unpublished (2004), http://lpcs.math.msu.su/~ver/kolm-book
12. Ziv, J., Lempel, A.: Compression of individual sequences via variable-rate coding. IEEE Trans. Inform. Theory IT-24(5), 530–536 (1978)

K-Pages Graph Drawing with Multivalued Neural Networks*

Domingo López-Rodríguez[1], Enrique Mérida-Casermeiro[1], Juan M. Ortíz-de-Lazcano-Lobato[2], and Gloria Galán-Marín[3]

[1] Department of Applied Mathematics, University of Málaga, Málaga, Spain
{dlopez,merida}@ctima.uma.es
[2] Department of Computer Science and Artificial Intelligence, University of Málaga, Málaga, Spain
jmortiz@lcc.uma.es
[3] Department of Electronics and Electromechanical Engineering, University of Extremadura, Badajoz, Spain
gloriagm@unex.es

Abstract. In this paper, the K-pages graph layout problem is solved by a new neural model. This model consists of two neural networks performing jointly in order to minimize the same energy function. The neural technique applied to this problem allows to reduce the energy function by changing outputs from both networks –outputs of first network representing location of nodes in the nodes line, while the outputs of the second one meaning the page where the edges are drawn.

A detailed description of the model is presented, and the technique to minimize an energy function is fully described. It has proved to be a very competitive and efficient algorithm, in terms of quality of solutions and computational time, when compared to the state-of-the-art heuristic methods specifically designed for this problem. Some simulation results are presented in this paper, to show the comparative efficiency of the methods.

1 Introduction

In the last few years, several graph representation problems have been studied in the literature. Most of them are related to the linear graph layout problem, in which the vertices of a graph are placed along a horizontal "node line", or "spine" (where K half-planes or *pages* intersect) and then edges are added to this representation as specified by the adjacency matrix. The objective of this problem is to minimize the total number of crossings (adding over all K pages) produced by such a layout.

Some examples of problems associated to this linear graph layout problem (or K pages crossing number problem) are the bandwidth problem [1], the book thickness problem [2], the pagenumber problem [3,4], the boundary VLSI layout problem [5] and the single-row routing problem [6] and automated graph drawing [7]. Another important application is the design of printed circuit boards [8], since, for the case of non-insulated wires, overlapping wires between electrical components may cause short circuits and thus may be avoided as much as possible.

* This work has been partially supported by Junta de Andalucía project number P06-TIC-01615.

J. Marques de Sá et al. (Eds.): ICANN 2007, Part II, LNCS 4669, pp. 816–825, 2007.

Several authors study a restricted version of this problem in which the vertex order is predetermined and fixed along the node line, and edges are drawn as arcs in one of the pages [9]. Other authors are more interested in the variant in which the node order is not fixed [10]. For this variant, it has been considered necessary to first find an optimal ordering of the vertices in order to compute the layout.

This problem is NP-hard [11,12]. So, many researchers have focused on finding efficient algorithms (some of them specially designed for the case of certain families of graphs) to solve the graph layout problem.

A comparison of several heuristics for this problem is presented in [9,13], including greedy, maximal planar, one-page, bisection and a neural heuristic, among others. Concretely, the neural model developed in [14] (and based on Takefuji and Lee's work [15,16]) was tested and obtained very good results, although authors indicate the possibility of non-convergence of this method. Due to the use of binary neurons, the model needs $2M$ neurons to represent the solution for a graph of M edges. In addition, this model is only intended to solve the 2-pages graph layout problem, and needed to preprocess the graph in order to obtain a good node ordering.

In this work we present a neural model designed to solve this problem. One of the differences of our model with the algorithms developed in literature is that there is no need of assigning a good ordering of the vertices at a preprocessing step. This optimal node order is computed by the model, as well as the relative position of the arcs.

Our model is a variant of the multivalued MREM model which has obtained very good results when applied to other combinatorial optimization problems [17,18,19,20], guaranteeing the convergence to local minima of the energy function.

2 Formal Description of the Problem

Let $G = (V, E)$ be an undirected graph where $V = \{v_i\}$ is the set of vertices and $E = (e_{i,j})$ is a symmetric binary matrix where $e_{i,j} = 1$ if edge (v_i, v_j) exists.

The K-pages book Crossing Number Problem consists in placing graph nodes on a horizontal "node line" in the plane. Every edge can be drawn as an arc in one of the half-planes (pages), which intersect that line, see Fig. 1. The objective is to minimize the number of edge crossings. This problem belongs to the class of NP-hard optimization problems, even if node order is fixed and the number of pages is $K = 2$.

An example of linear embedding of the complete graph K_7 in 4 pages, with 0 crossings, is drawn in Fig. 1. In this figure, we can observe the representation of pages 1 and 2 in the left side and pages 3 and 4 in the right hand side. First and third pages are represented as half-planes above the node line, while the second and fourth pages are under the node line.

Crossings Detection
Let us consider 4 positions in the node line verifying $1 \leq i < k < j < l \leq N$, where i, j, k and l are assigned to nodes v_i, v_j, v_k and v_l. Then, edges (v_i, v_j) and (v_k, v_l) cross each other if, and only if, both are represented (drawn) in the same page.

In Fig. 2, we can observe that edges (v_i, v_j) and (v_k, v_l), represented in the node line and with endpoints $i < k < j < l$, produce a crossing, whereas if $i < j < k < l$ they do not, when both are represented in the same half-plane.

Fig. 1. Optimal linear layout for K_7 in 4 pages

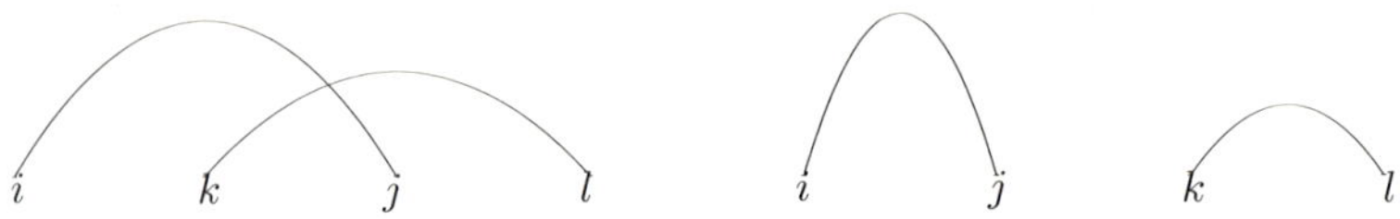

Fig. 2. Crossing condition $i < k < j < l$

It seems reasonable to define $V_{v_a,v_b} = k$ to indicate that the edge (v_a, v_b) will be represented in the k-th page or half-plane. If the edge does not exist, we will denote $V_{v_a,v_b} = 0$ for simplicity.

These definitions allow us to define the number of crossings by means of the cost function:

$$C = \sum_{i} \sum_{k>i} \sum_{j>k} \sum_{l>j} \delta(V_{v_i,v_j}, V_{v_k,v_l})(1 - \delta(V_{v_i,v_j}, 0)) \quad (1)$$

where $\delta(x, y) = 1$ if $x = y$, otherwise it equals 0 (Krönecker delta function).

In Eq. (1), the term $\delta(V_{v_i,v_j}, V_{v_k,v_l})$ expresses that edges (v_i, v_j) and (v_k, v_l) will be drawn in the same page, whereas $(1 - \delta(V_{v_i,v_j}, 0))$ indicates that the edge exists.

3 Previous Heuristics

Cimikowski [9,13] presented a comparison of some heuristic approaches to solve the 2-pages graph drawing problem. All of them, except the neural one developed therein, can be extended to solve the K-pages problem. We make here a brief summary of them:

- *Edge-length heuristic* (e-len): This heuristic initially orders all edges by their length (the length of edge (a, b) is $|b - a|$). Intuitively, longer edges are most likely to produce a big number of crossings than shorter edges and hence should be embedded first in the layout. So, each edge is sequentially added to the page of smallest increase in the number of crossings.
- *One-page heuristic* (1-page): This heuristic initially embeds all edges in the first page. After this, a "local improvement" phase is carried out, in which each edge is moved to the page with the smallest number of new crossings. Edges are selected for movement in order of non-increasing local crossing number, that is, the number of crossings involving an edge.

- *Greedy heuristic* (greedy): In this method, edges are sorted according to the node index of its ends, that is, first, all edges $(1, i)$ (in increasing order of i), then $(2, i)$, etc. Edges are added, in this order, to the page which results in the smallest increase in the number of crossings.

4 The Neural Model MREM

It consists in a series of multivalued neurons, where the state of i-th neuron is characterized by its output (v_i) that can take any value in any finite set $\mathcal{M}$. This set can be a non numerical one, but, in this paper, the neuron outputs only take value in $\mathcal{M} \subset \mathbb{N}$.

The state vector $\boldsymbol{V} = (v_1, v_2, \ldots, v_N) \in \mathcal{M}^N$ describes the network state at any time, where N is the number of neurons in the net. Associated with any state vector, there is an energy function $E : \mathcal{M}^N \to \mathbb{R}$, defined by the expression:

$$E(\boldsymbol{V}) = -\frac{1}{2}\sum_{i=1}^{N}\sum_{j=1}^{N} w_{i,j} f(v_i, v_j) + \sum_{i=1}^{N} \theta_i(v_i) \quad (2)$$

where $W = (w_{i,j})$ is a matrix, $f : \mathcal{M} \times \mathcal{M} \to \mathbb{R}$ is usually a similarity function since it measures the similarity between the outputs of neurons i and j, and $\theta_i : \mathcal{M} \to \mathbb{R}$ is a threshold function. At each step, the state vector will be evolving to decrease the value of the energy function.

The cost function (number of crossings in the graph given by Eq. (1)), must be identified with the energy function of Eq. (2). As a result, we obtain $w_{i,j} = 1$ if $i < j$ and 0 otherwise. The similarity function $f(v_i, v_j)$ and the threshold θ_i can be expressed as:

$$f(v_i, v_j) = -2\sum_{k}\sum_{l>k} \delta(V_{v_i,v_k}, V_{v_j,v_l})(1 - \delta(V_{v_i,v_k}, 0))$$

$$\theta_i(v_i) = -\sum_{j} w_{i,j} \sum_{k\leq j}\sum_{l>k} \delta(V_{v_i,v_k}, V_{v_j,v_l})(1 - \delta(V_{v_i,v_k}, 0))$$

To solve our problem we have considered two MREM neural networks:

- The first network (the 'vertices' net) will be formed by N neurons, being N the number of nodes in the graph. Neurons output (the state vector) indicate the node ordering in the line. Thus, $v_i = k$ will be interpreted as the k-th node being placed in the i-th position in the node line. Hence, the output of each neuron can take value in the set $\mathcal{M}_1 = \{1, 2, \ldots N\}$.
- The second network (the 'edges' net) will be formed by as many neurons as edges in the graph, M. The output of each neuron will belong to the set $\mathcal{M}_2 = \{1, 2, \ldots, K\}$. As mentioned before, for the arc (v_i, v_j), $V_{v_i,v_j} = k$ will indicate that (v_i, v_j) will be embedded in the k-th page. For simplicity, let us denote the absence of edge (v_i, v_j) as $V_{v_i,v_j} = 0$.

Initially, the state of the 'vertices' net is randomly selected as a permutation of $\{1, 2, \ldots, N\}$, and the initial state of the 'edges' net is a random element from $\mathcal{M}_2^M =$

$\{1, 2, \ldots, K\}^M$. At any time, the net is looking for a better solution than the current one, in terms of minimizing the energy function.

In this paper, we study the permutation of two nodes and the change in the location of an edge. These produce the energy increment given in the next subsections. As an additional technique for improvement, we have also considered changes in the position of four edges ('change-4'), since our studies have demonstrated that this dynamics is able to undo some crossings which can not be broken just by changing one edge position.

4.1 Permutation of Two Nodes

When two vertices v_a and v_b permute their order a and b in the node line, we should take into account that the unique edges changing their position (and therefore changing the number of crossings) are those that have exactly one endpoint in $\{v_a, v_b\}$.

Let us study the increase in the number of crossings depending on the relative positions of the endpoints.

Consider a position x in the line and let us see how the number of crossings with the edge (v_x, v_a) is modified when it becomes the edge (v_x, v_b) since nodes a and b permute their positions. Hence, the arc represented with endpoints (x, a) will be drawn, after the update, with endpoints (x, b), and the unique edges modifying the number of crossings due to the change must be in the same page and must have an endpoint v_s represented between a and b ($a < s < b$) and the other, v_t, outside that interval (($t < a$) or ($t > b$)). Some cases, depending on the position of x, are considered:

1. Case $x < a < s < b$: As shown in Fig. 3 (1), if $t < x < a < s < b$ the number of crossings is increased in one unit, since the edge (t_1, s) crosses the arc (x, b), but not (x, a). If $x < t < a < s < b$, a crossing disappears (the arc (t_2, s) cuts (x, a) but not (x, b)) and, at last, if $x < a < s < b < t$, the number of crossings will be increased in 1 unit (analize the arc (s, t_3)).
2. Case $a < x < b$: As shown in Fig. 3 (2), if $t < a < x < s < b$, or $a < x < s < b < t$, a new crossing is introduced (represented by the cuts of arcs (s_2, t_1) and (s_2, t_2) with the new edge (x, b)), whereas if $t < a < s < x < b$ or $a < s < x < b < t$ the number of crossings is reduced since crossings of (s_1, t_1) and (s_1, t_2) with (a, x) disappear.
3. Case $a < s < b < x$: A crossing is introduced if $a < s < b < t < x$ (arc (s, t_2)) and will be erased if $t < a < s < b < x$, or, $a < s < b < x < t$ (arcs (t_1, s) and (s, t_3)), as shown in Fig. 3 (3).

We must also take into account the change in the number of crossings with edges (v_x, v_b). Its study is similar to the already made for (v_x, v_a), it suffices to permute the literals a and b and to change the sense of the inequalities.

Finally, let us consider changes in the number of crossings produced between edges (v_a, v_x) and (v_b, v_y). All possible changes are shown in Fig. 4. There are different cases:

1. Case $a < y < x < b$:
   - Edges (a, x) and (y, b) (Fig. 4 (1)) are transformed into (x, b) and (y, a) (Fig. 4 (2)), vanishing the existing crossing.
   - Edges (a, y) and (x, b) (Fig. 4 (2)) are transformed into (b, y) and (x, a) (Fig. 4 (1)), causing the apparition of a crossing.

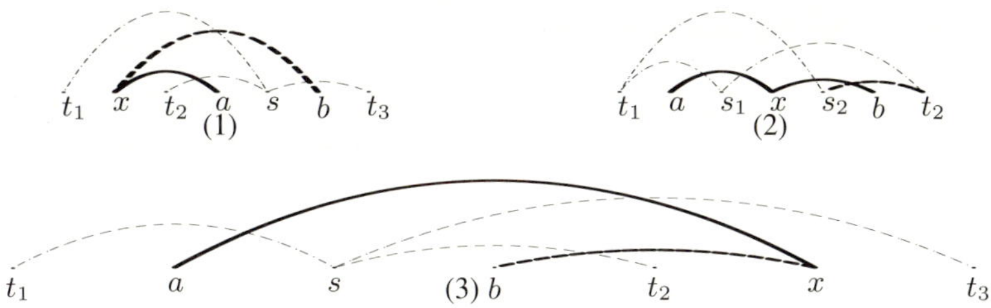

Fig. 3. Changes in the number of crossings when permuting nodes v_a and v_b, represented at positions a and b. An edge represented by the arc (a, x) will be transformed into the arc (b, x).

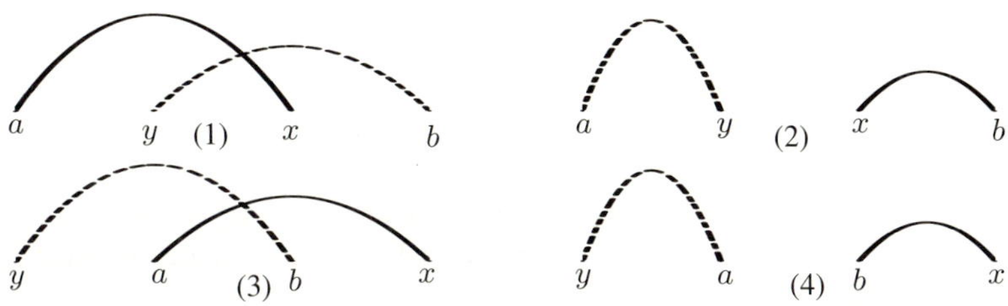

Fig. 4. Changes in the number of crossings when permuting nodes v_a and v_b. Edges represented by arcs (a, x) and (y, b) are transformed into arcs (b, x) and (y, a).

2. Case $y < a < b < x$:
   - When edges (a, x) and (y, b) (Fig. 4 (3)) are transformed into (x, b) and (y, a) (Fig. 4 (4)), a new crossing is formed.
   - Arcs (a, y) and (x, b) (Fig. 4 (4)), are transformed into (b, y) and (x, a) (Fig. 4 (3)), and a crossing is eliminated.

We can derive an explicit formula for the increase of energy related to all these cases, just by considering that Eq. (2) (the number of crossings) can be rewritten as:

$$E = \sum_i \sum_j w_{i,j} \sum_k w_{j,k} \sum_l w_{k,l} \delta(V_{v_i,v_k}, V_{v_j,v_l})(1 - \delta(V_{v_i,v_k}, 0))$$

and by denoting $g(x, y, s, t) = \delta(V_{x,s}, V_{y,t})(1 - \delta(V_{x,s}, 0))$, then the increase of energy caused by the permutation of nodes a and b is given by:

$$\begin{aligned}
\Delta E = & \sum_{i \in \{a,b\}} \sum_j w_{i,j} \sum_k w_{j,k} \sum_l w_{k,l} \left(g(v_i, v_j, v_k, v_l) - g(v'_i, v_j, v_k, v_l)\right) \\
& + \sum_i \sum_{j \in \{a,b\}} w_{i,j} \sum_k w_{j,k} \sum_l w_{k,l} \left(g(v_i, v_j, v_k, v_l) - g(v_i, v'_j, v_k, v_l)\right) \\
& + \sum_i \sum_j w_{i,j} \sum_{k \in \{a,b\}} w_{j,k} \sum_l w_{k,l} \left(g(v_i, v_j, v_k, v_l) - g(v_i, v_j, v'_k, v_l)\right) \\
& + \sum_i \sum_j w_{i,j} \sum_k w_{j,k} \sum_{l \in \{a,b\}} w_{k,l} \left(g(v_i, v_j, v_k, v_l) - g(v_i, v_j, v_k, v'_l)\right)
\end{aligned} \tag{3}$$

where $v'_s = v_a$ if $v_s = v_b$; $v'_s - v_b$, if $v_s = v_a$; otherwise $v'_s = v_s$.

4.2 Change of the Position of an Edge

When the edge with endpoints v_a, v_b is represented in a given page and its location changes from its current page to the k-th page, an increase (or decrease) of the energy function (number of crossings) is produced and is given by

$$\Delta E(k) = (1 - \delta(V_{v_a,v_b}, 0)) \sum_{a<s<b} \sum_{(t<a)\vee(t>b)} (2\delta(V_{v_a,v_b}, V_{v_s,v_t}) - 1) \cdot \\ \cdot (1 - \delta(V_{v_s,v_t}, 0)) \cdot \max\{\delta(V_{v_s,v_t}, V_{v_a,v_b}), \delta(V_{v_s,v_t}, k)\} \quad (4)$$

This expression can be obtained from Eq. (2), by simplifying the difference between the number of crossings before and after the possible change, since edges (v_x, v_y) with both endpoints located between a and b in the node line ($a < x < b$, $a < y < b$), or both placed outside the interval $[a, b]$, i. e., $x, y \notin [a, b]$, do not contribute to modify the number of crossings with (v_a, v_b), which is the changed edge.

In the improvement technique, since four edge positions are to be changed, the procedure to compute the energy increment consists in analysing independently the increase produced by the assignment of each individual edge to each of the K pages. Then, it must be taken into account the possible crossings between the selected edges. This produces a hyper-matrix $\boldsymbol{\Delta E} = (\Delta E(k_1, k_2, k_3, k_4))_{k_1,k_2,k_3,k_4}$ which represents the increase of energy the i-th considered edge is moved to the k_i-th page.

5 Implementation for K-Pages Book Graph Layout Problem

Several dynamics can be used to solve this problem with our model, but we have chosen the following one due to its simplicity and efficiency:

1. Initialization: Given a graph with N nodes and M edges, a random feasible initial configuration $\boldsymbol{V}_0$ is selected for the location of the nodes. This initial state will be a permutation of the set of node indices $\{1, 2, \ldots, N\}$.

   For the edges set, the initial state vector $\boldsymbol{W}_0$ will be a random element of the set $\{1, 2, \ldots, K\}^M$. As usual, the output $V_{v_i,v_j} = k$ means that the arc (i, j) will be embedded in the k-th page.
2. Two positions a and b in the node line are selected in all possible ways.

   Firstly, the net studies the increase of energy when vertices v_a and v_b are permuted by using Eq. (3). If the energy is reduced, the net permutes the vertices: $v_a(t + 1) = v_b(t)$, $v_b(t + 1) = v_a(t)$. Secondly, the net studies to change the page in which the edge (v_a, v_b) is located. To this end, the increase of energy given by Eq. (4) is computed and the change $V_{v_a,v_b}(t + 1) = k_0$ is done, where $\Delta E(k_0) = \min_{1\leq k\leq K} \Delta E(k)$.
3. If all possible combinations of two nodes positions and all possible edge locations have been considered and no change has been made, both networks have converged to a local minimum of the energy function (the cost function) and state vectors represent the obtained solution.

   In this case, in order to break some crossings, we apply the additional improvement technique 'change-4'. For each possible combination of 4 edges, the net studies the hyper-matrix $\boldsymbol{\Delta E}$, computes $\min_{k_1,k_2,k_3,k_4} \Delta E(k_1, k_2, k_3, k_4)$ and the corresponding update is performed.

6 Simulation Results

In this Section we test the performance of our model and compare it with some of the heuristic methods proposed in [9,13] for a test set formed by graphs belonging to well-known graph families. Concretely:

- Complete graphs K_n, where all n nodes are interconnected (no self-connections). the graph K_n has $\frac{n(n+1)}{2}$ edges.
- Circulant graph $C_n(a_1, \ldots, a_k)$, where $0 < a_1 < \ldots < a_k < \frac{n+1}{2}$ is a graph with n vertices such that vertex i is adjacent to vertices $i \pm a_1, \ldots, i \pm a_k(\mathrm{mod}(n))$. The circulant graph $C_n(a_1, \ldots, a_k)$ has $n \cdot k$ edges.

In order to test the efficiency of out method, we use the heuristics given in [9]. Concretely, those named *e-len*, *1-page* and *greedy*, described above. In addition, comparative results of Cimikowski's neural model (CN) are presented in the case $K = 2$.

We must note that CN is a neural model which needs the fine-tuning of some parameters, and our model does not. Also, CN needs of a preprocessing step in which graph nodes are ordered and then remain fixed along the iterations. Our model is able to dynamically obtain a very good ordering.

For every graph, 10 independent executions of our model were performed.

For the case of $K = 2$ pages, results are shown in Table 1. It can be observed that, although CN obtains very good results, our proposal is able to achieve better solutions. Considered heuristics obtain good solutions, but do not perform better than our proposal.

Table 1. Comparative results of our model for the 2-pages graph problem, for the considered test graphs. Numbers between parentheses indicate the average number of crossings in all the 10 executions of our model, if this average differs from the minimum obtained.

Graph	$\lvert V\rvert$	$\lvert E\rvert$	Prop.	CN	e-len	1-page	greedy
K_6	6	21	3	3	3	4	5
K_7	7	28	9	9	11	9	13
K_8	8	36	18	18	18	30	27
K_9	9	45	36	36	42	50	50
K_{10}	10	55	60	60	80	92	84
$C_{20}(1,2)$	20	40	0 (3.8)	2	0	0	0
$C_{20}(1,2,3)$	20	60	19 (24.2)	24	36	48	40
$C_{20}(1,2,3,4)$	20	80	74 (79.3)	74	90	118	108
$C_{22}(1,2,3)$	22	66	22 (26.9)	26	40	54	44
$C_{22}(1,3,5,7)$	22	88	198 (226.5)	200	306	294	286
$C_{24}(1,3)$	24	48	11 (19.4)	14	22	16	22
$C_{26}(1,3)$	26	52	11 (18.8)	16	24	16	24
$C_{28}(1,3,5)$	28	84	80 (98.0)	86	138	138	130
$C_{30}(1,3,5)$	30	90	92 (113)	96	148	150	140

Table 2. Results for $K = 3$ and $K = 4$ pages problems. Numbers between parentheses indicate the average number of crossings in all the 10 executions of our model, if this average differs from the minimum obtained.

	$K = 3$				$K = 4$			
Graph	Prop.	e-len	1-page	greedy	Prop.	e-len	1-page	greedy
K_6	0	1	1	1	0	0	0	0
K_7	2	4	3	4	0	1	1	1
K_8	5	6	8	12	0	4	4	4
K_9	9	15	23	19	3	7	8	8
K_{10}	20	27	49	40	7	12	22	19
$C_{20}(1,2)$	0 (0.9)	0	0	0	0 (0.1)	0	0	0
$C_{20}(1,2,3)$	4 (8.5)	2	10	2	0 (1.4)	1	2	1
$C_{20}(1,2,3,4)$	21 (27.5)	60	71	45	11 (15.3)	6	15	4
$C_{22}(1,2,3)$	5 (9.4)	5	12	4	0 (3.7)	1	1	1
$C_{22}(1,3,5,7)$	96 (106.7)	141	184	166	38 (49.4)	85	101	99
$C_{24}(1,3)$	2 (7.3)	0	12	0	0 (3.6)	0	0	0
$C_{26}(1,3)$	5 (7.7)	2	14	2	0 (2.4)	1	1	1
$C_{28}(1,3,5)$	31 (44.4)	70	71	55	11 (20.8)	29	31	32
$C_{30}(1,3,5)$	43 (53.6)	59	73	61	19 (29.1)	26	30	37

For larger problem sizes, $K = 3$ and $K = 4$, we can observe the same fact as above, since our proposal outperforms the heuristic methods in most cases. This means that our method is a significant improvement for this problem, since these heuristics are well-known to perform very well [13].

Regarding computational time, it must be noted that, although spending more time in obtaining a solution, our method is able to achieve good solutions on average, while the heuristics herein considered always achieve the same solution, no matter the number of times they are executed.

7 Conclusions and Future Work

In this work we have presented a neural model especially designed to solve some kinds of combinatorial optimization problems. This model is a variant of the multivalued model MREM formed by two networks. The dynamics of each of these networks depends on the outputs of the other network, and they are updated alternatively, to reach an equilibrium state corresponding to a local minimum of the common energy function.

We have tested our model with the well-known K-pages graph linear layout problem from graph theory. The proposed model avoids some of the drawbacks of other models in specialized literature, like the absence of convergence guarantees or the fine-tuning of parameters. In addition, it does not need a pre-processing stage to obtain a good node ordering, since it can be achieved dynamically.

By using some test instances, we have observed that our model is, at least, comparable to the other models, and it is able to achieve, in many cases, better solutions.

Future research lines cover aspects such as developing new dynamics for the model which could help to achieve better results. This model is also applicable when the node line is not a straight line, it can be a circle, or another geometry.

References

1. Chinn, P.Z., Chvátalová, L., Dewdney, A.K., Gibbs, N.E.: The bandwidth problem for graphs and matrices – a survey. J. Graph Theory 6, 223–253 (1982)
2. Kainen, P.C.: The book thickness of a graph, ii. Congr. Numer. 71, 127–132 (1990)
3. Chung, F.R.K., Leighton, F.T., Rosenberg, A.L.: Embedding graphs in books: a layout problem with applications to vlsi design. SIAM J. Alg. Disc. Meth. 8, 33–58 (1987)
4. Malitz, S.M.: On the page number of graphs. J. Algorithms 17(1), 71–84 (1994)
5. Ullman, J.D.: Computational Aspects of VLSI. Computer Science Press (1984)
6. Raghavan, R., Sahni, S.: Single row routing. IEEE Trans. Comput C-32(3), 209–220 (1983)
7. Tamassia, R., Di Battista, G., Batini, C.: Automatic graph drawing and readability of diagrams. IEEE Trans. Syst. Man. Cybern. SMC-18, 61–79 (1988)
8. Sinden, F.W.: Topology of thin films circuit. Bell Syst. Tech. Jour. XLV, 1639–1666 (1966)
9. Cimikowski, R.: Algorithms for the fixed linear crossing number problem. Discrete Applied Mathematics 122, 93–115 (2002)
10. Nicholson, T.A.J.: Permutation procedure for minimising the number of crossings in a network. Proc. IEE 115(1), 21–26 (1968)
11. Garey, M.R., Johnson, D.S.: Crossing number is np-complete. SIAM J. Alg. Disc. Methods 4(3), 312–316 (1983)
12. Masuda, S., Nakajima, K., Kashiwabara, T., Fujisawa, T.: Crossing minimization in linear embeddings of graphs. IEEE Trans. Comput. 39(1), 124–127 (1990)
13. Cimikowski, R.: An analysis of some linear graph layout heuristics. J. Heuristics 12, 143–153 (2006)
14. Cimikowski, R., Shope, P.: A neural-network algorithm for a graph layout problem. IEEE Transactions on Neural Networks 7(2), 341–345 (1996)
15. Takefuji, Y.: Design of parallel distributed cauchy machines. In: Proc. Int. Joint Conf. Neural Networks, pp. 529–536 (1989)
16. Takefuji, Y.: Neural Network Parallel Computing. Kluwer Academic Publishers, Dordrecht (1992)
17. Mérida-Casermeiro, E., Galán-Marín, G., MuñozPérez, J.: An efficient multivalued hopfield network for the travelling salesman problem. Neural Processing Letters 14, 203–216 (2001)
18. Mérida-Casermeiro, E., Muñoz Pérez, J., Domínguez-Merino, E.: An n-parallel multivalued network: Applications to the travelling salesman problem. In: Mira, J.M., Álvarez, J.R. (eds.) IWANN 2003. LNCS, vol. 2686, pp. 406–413. Springer, Heidelberg (2003)
19. Mérida-Casermeiro, E., López-Rodríguez, D.: Graph partitioning via recurrent multivalued neural networks. In: Cabestany, J., Prieto, A.G., Sandoval, F. (eds.) IWANN 2005. LNCS, vol. 3512, pp. 1149–1156. Springer, Heidelberg (2005)
20. López-Rodríguez, D., Mérida-Casermeiro, E., Ortiz-de Lazcano-Lobato, J.M., López-Rubio, E.: Image compression by vector quantization with recurrent discrete networks. In: Kollias, S., Stafylopatis, A., Duch, W., Oja, E. (eds.) ICANN 2006. LNCS, vol. 4132, pp. 595–605. Springer, Heidelberg (2006)

Recursive Principal Component Analysis of Graphs

Alessio Micheli[1] and Alessandro Sperduti[2]

[1] Department of Computer Science, University of Pisa, Italy
[2] Department of Pure and Applied Mathematics, University of Padova, Italy

Abstract. Treatment of general structured information by neural networks is an emerging research topic. Here we show how representations for graphs preserving all the information can be devised by Recursive Principal Components Analysis learning. These representations are derived from eigenanalysis of extended vectorial representations of the input graphs. Experimental results performed on a set of chemical compounds represented as undirected graphs show the feasibility and effectiveness of the proposed approach.

1 Introduction

The representation of graphs via numerical vectors is the first necessary step to the application of numerical methods for clustering, classification, and regression. Traditional approaches select apriori a set of structural features of interest and represent each graph via a vector where each component reports how much the associated structural feature is matched. An example of this approach can be found in QSPR/QSAR studies in Chemistry, where topological indexes are used as features.

A different approach has been proposed in the last 10 years in the neural networks field, where Recursive Neural Networks have been proposed and successfully applied in applications involving structured patterns (e.g. see [8,2,4,1,6]). The underpinning idea at the basis of this type of neural networks is the dynamic generation of feed-forward networks (encoding networks) whose topology matches the topology of the input and which exploit shared weights to cope with structures of different sizes. The output of these encoding networks is a numerical representation of the input structure. The advantage of this approach with respect to the former approach is that learning procedures can be exploited to adapt the numerical representations to the classification or regression task at hand. For example, if another neural network (output network) is used to post-process (either for producing a classification or a numerical prediction) the output of the encoding networks, the error obtained by the output network can be back-propagated to the encoding networks and the (shared) weights of the encoding networks adapted to contribute to the minimization of the loss function of interest. Almost all the proposed models in the family of recursive neural networks are only able to directly deal with directed acyclic graphs (and other derivable structures, such as trees and sequences).

An additional approach has been pursued by kernel methods for structured patterns (see [3] for a survey). These methods avoid to explicitly generate numerical vectors representing the input graphs, but directly compute the similarity between two graphs through a kernel function that implicitly projects each graph into a numerical feature space and that returns the dot product between corresponding vectors into the feature

J. Marques de Sá et al. (Eds.): ICANN 2007, Part II, LNCS 4669, pp. 826–835, 2007.

space. One problem with this approach is that almost all the proposed kernels are defined for structures with vertexes annotated by discrete variables, and very often the calculation of the kernel is computationally very demanding. Finally, as in the case of topological indexes, kernels are usually defined apriori, regardless of the specific task of interest and regardless of the available dataset.

In this paper, we address the problem to devise vectorial representations of graphs from a dataset preserving all the information needed to discriminate among them. Specifically, we show how a recently proposed approach to the calculation of Recursive PCA [7] for sequences and trees can be adapted to graphs, either with directed or undirected arcs. The aim is to provide a method to generate informative representations which are amenable to be used into already well known unsupervised and supervised techniques for clustering, classification, and regression. The applicability and effectiveness of the proposed method is evaluated on a dataset of chemical compounds of significant diversity and sizes, involving thousands of atoms and bonds.

2 Principal Components Analysis for Sequences and Trees

In [7] it is shown how Principal Component Analysis can be extended to the direct treatment of sequences and trees. More specifically, given a temporal sequence $\mathbf{x}_1, \mathbf{x}_2, \ldots, \mathbf{x}_t$ of input vectors $\mathbf{x}_i \in \mathbb{R}^k$, where t is a discrete time index, we are interested in modeling the sequence through the following linear dynamical system:

$$\mathbf{y}_t = \mathbf{W}_\mathbf{x}\mathbf{x}_t + \sqrt{\alpha}\mathbf{W}_\mathbf{y}\mathbf{y}_{t-1} \tag{1}$$

where $\mathbf{W}_\mathbf{x} \in \mathbb{R}^{p\times k}$ and $\mathbf{W}_\mathbf{y} \in \mathbb{R}^{p\times p}$ are the matrices of synaptic efficiencies, which correspond to feed-forward and recurrent connections, respectively, $\mathbf{y}_t \in \mathbb{R}^p$ is an output vector, $\alpha \in [0, 1]$ is a gain[1] parameter which modulates the importance of the past history, i.e. $\mathbf{y}_{t-1}$, with respect to the current input $\mathbf{x}_t$. The aim is to define proper synaptic matrices, with *dimension p as small as possible*, such that $\mathbf{y}_t$ can be considered a good "encoding" of the input sequence read till time step t, i.e., the sequence is first encoded using eq. (1), and then, starting from the obtained encoding $\mathbf{y}_t$, it should be possible to reconstruct backwards the original sequence using the transposes of $\mathbf{W}_\mathbf{x}$ and $\mathbf{W}_\mathbf{y}$. This requirement implies that the following equations

$$\mathbf{x}_t = \mathbf{W}_\mathbf{x}^\mathsf{T}\mathbf{y}_t \tag{2}$$
$$\mathbf{y}_{t-1} = \mathbf{W}_\mathbf{x}\mathbf{x}_{t-1} + \mathbf{W}_\mathbf{y}\mathbf{y}_{t-2} = \mathbf{W}_\mathbf{y}^\mathsf{T}\mathbf{y}_t \tag{3}$$

should hold. In fact, the aim of recursive principal component analysis is to find a low-dimensional representation of the input sequence such that the expected reconstruction error, i.e. the sum of the (squared) differences between the vectors generated by equation (2) and the original input vectors for different values of t

$$error(t) = \sum_{i=1}^{t} \|\mathbf{x}_i - \mathbf{W}_\mathbf{x}^\mathsf{T}(\mathbf{W}_\mathbf{y}^\mathsf{T})^{t-i}\mathbf{y}_t\|^2 \tag{4}$$

[1] Here, without loss of generality, we focus on the case where $\alpha = 1$ and $\mathbf{y}_0$ is the null vector.

is as small as possible, i.e. we look for the smallest value of p such that $error(t)$ is minimized.

In [7] it has been shown that, when considering several sequences but the same synaptic matrices, zero error, i.e. an exact solution to the above minimization error, can be obtained by performing eigenanalysis of extended vectorial representations of the input sequences, where a sequence at time t is represented by the vector

$$[\mathbf{x}_t^\top, \ldots, \mathbf{x}_1^\top, \underbrace{\mathbf{0}^\top, \ldots, \mathbf{0}^\top}_{(T-t)}]$$

where T is the maximum length for any input sequence. This representation can be understood as an explicit representation of a stack where a new input vector, e.g. $\mathbf{x}_{t+1}$, is pushed into the stack by shifting to the right the current content by k positions, and inserting (adding) $\mathbf{x}_{t+1}$ into the freed positions:

$$[\mathbf{0}^\top, \mathbf{x}_t^\top, \ldots, \mathbf{x}_1^\top, \underbrace{\mathbf{0}^\top, \ldots, \mathbf{0}^\top}_{(T-t-1)}] + [\mathbf{x}_{t+1}^\top, \underbrace{\mathbf{0}^\top, \ldots, \mathbf{0}^\top}_{(T-1)}] = [\mathbf{x}_{t+1}^\top, \mathbf{x}_t^\top, \ldots, \mathbf{x}_1^\top, \underbrace{\mathbf{0}^\top, \ldots, \mathbf{0}^\top}_{(T-t-1)}]$$

More precisely, let $\mathbf{X}$ be the matrix which collects all the vectors of the above form (for all sequences at any time step). If the input vectors $\mathbf{x}_i \in \mathbb{R}^k$ have zero mean, $s = T \cdot k$, $\mathbf{U}\boldsymbol{\Lambda}\mathbf{U}^\top$ is the eigenvalue decomposition of $\mathbf{X}\mathbf{X}^\top$ and $\widetilde{\mathbf{U}} \in \mathbb{R}^{s \times p^*}$ is the matrix obtained by $\mathbf{U}$ removing all the eigenvectors corresponding to null eigenvalues λ_i, then p^* is the smallest value for which the synaptic matrices defined as:

$$\widetilde{\mathbf{W}}_{\mathbf{x}} \equiv \underbrace{\widetilde{\mathbf{U}}^\top \begin{bmatrix} \mathbf{I}_{k \times k} \\ \mathbf{0}_{(s-k) \times k} \end{bmatrix}}_{\text{adding to the first } k \text{ positions}} \in \mathbb{R}^{p^* \times k}$$

and

$$\widetilde{\mathbf{W}}_{\mathbf{y}} \equiv \underbrace{\widetilde{\mathbf{U}}^\top \begin{bmatrix} \mathbf{0}_{k \times (s-k)} & \mathbf{0}_{k \times k} \\ \mathbf{I}_{(s-k) \times (s-k)} & \mathbf{0}_{(s-k) \times k} \end{bmatrix}}_{\text{shifting to the right of } k \text{ positions}} \widetilde{\mathbf{U}} \in \mathbb{R}^{p^* \times p^*},$$

have $error(T) = 0$. Please, note that smaller synaptic matrices can be obtained by removing from $\widetilde{\mathbf{U}}$ eigenvectors corresponding to smallest eigenvalues, i.e. $p < p^*$. Doing that, however, it is not clear whether the optimal value of $error(T)$ given p is obtained.

A similar, but a bit more elaborated result can be obtained for trees (with maximum outdegree b), where the linear dynamical system considered is

$$\mathbf{y}_u = \mathbf{W}_{\mathbf{x}} \mathbf{x}_u + \sum_{c=0}^{b-1} \mathbf{W}_{\mathbf{c}} \mathbf{y}_{ch_c[u]} \tag{5}$$

where u is a node of the tree, $ch_c[u]$ is the $c+1$-th child of u, and a different matrix $\mathbf{W}_{\mathbf{c}}$ is defined for each child.

3 Graphs and Recursive Principal Component Analysis

The basic idea of standard Principal Component Analysis is to discover orthogonal directions (i.e. principal components) of maximum variance of the data. These directions allow to define the subspace of smallest dimensionality where data is embedded. A nice feature of principal components is that they define a (linear) projection from the original space to the embedding space which can be "inverted" so to reconstruct the "original" vector from its projection into the embedding space. The projection into the embedding space (encoding) and the reconstruction from the embedding space (decoding) are operations which are preserved also into the recursive version of PCA for sequences and trees. When considering the possibility to extend Recursive PCA to graphs either with directed or undirected edges we have to face two problems: *i)* how to deal with cycles during the encoding; *ii)* how to identify the origin and destination of an edge during decoding.

Cycles may be present in directed graphs and are present in undirected graphs by definition[2]. Their presence is problematic when considering the encoding function since it introduces mutual functional dependences among vertexes. In fact, the encoding function is usually defined by induction: the *basis* is applied to vertexes with no out-coming edges, for which there is no functional dependency, and the *induction step* is applied to the remaining vertexes. For example, in the case of rooted trees, the encoding for leaves (*basis*) is given only as function of the label attached to them, while the encoding for internal vertexes is given as function of both the attached label and the encoding of the children. The encoding of a whole tree is obtained by considering the encoding for the root of the tree. If a cycle is present, the above scheme suffers the problem of mutual recursion, which from a mathematical point of view can be translated as the definition of a (linear) dynamical system whom state vector (i.e., the output of our encoding function) may eventually diverge or converge to a single or few attractors. In the first case, no stable encoding can be obtained; in the second case, the same encoding (i.e. attractor) is obtained for different input graphs, which consequently cannot be discriminated.

The second main problem, i.e. the identification of the origin and destination of an edge during decoding, is not present in the case of sequences (which can be seen as linked lists) and trees, since each vertex in this type of data structures is reachable by a single path from the beginning of the list or from the root, respectively. This property implies that it is possible to define the decoding function again by an inductive process: the *basis* is applied to vectors in the embedding space which lie in a designed subspace (e.g. around the origin, or which satisfy a specific "termination" property), i.e. when a vector in the embedding space belongs to the designed subspace or it satisfies the specific termination property, the decoding process is terminated for that vector; the *induction step* is applied when the basis does not apply, i.e. the information about the label is generated as a function of the vector, as well as one (for lists) or several (for trees) further vectors in the embedding space to which the inductive process is recursively applied. This decoding scheme generates trajectories into the origin space which can unambiguously be assigned to paths in a tree (or a single path for a list). When

[2] In fact, edges into undirected graphs can be traveled in both directions, and thus any graph with at least one edge generates a cycle of length 2 if the connected vertexes are different.

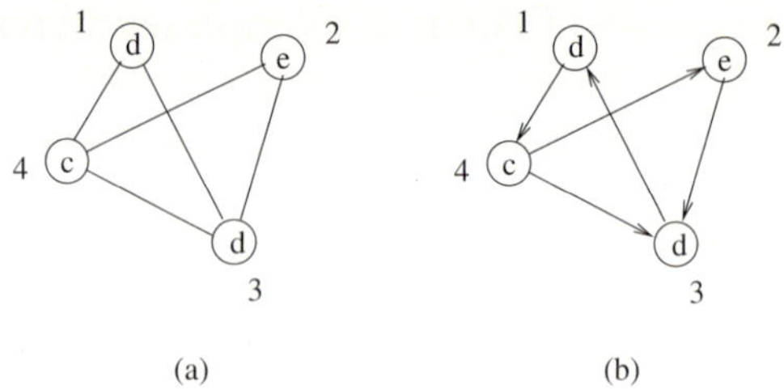

Fig. 1. Examples of undirected (a) and directed (b) graphs with labeled vertexes. The integer number associated to each vertex constitutes the enumeration of the vertex within the same graph.

considering graphs the above scheme cannot work, since in general a vertex can be reached by several paths, and it is not obvious how to assign a "vertex" semantic to each step of the generated trajectories, i.e. if the same label is generated by different trajectories or from the same trajectory at different decoding steps, how can we be sure that its interpretation is that the same vertex of the graph has been generated via different paths or we are facing the generation of different vertexes with the same label ?

Here we propose to solve the above problems with a coding trick. The basic idea is to enumerate the set of vertexes following a given convention and representing a (directed or undirected) graph as an (inverted) ordered list of vertex's labels associated with a list of edges for which the vertex is origin and where the position in the associated list is referring to the destination vertex. The idea is that the list is used by the (linear) neural network during encoding to read one by one the information about each vertex and associated edges, pushing the read information into an internal stack (the encoding space). Decoding is obtained by popping from the internal stack, one by one, the information about vertexes and associated edges. Just to give a concrete example, let consider the two graphs in Figure 1. The enumeration for each vertex is reported as an integer besides each vertex. Assuming that the maximum number of vertexes in the input graph domain is 4, the representation for the undirected graph shown in Figure 1(a) is as follows

$$[(c, [0]), (d, [0, 1]), (e, [0, 1, 1]), (d, [0, 0, 1, 1])]. \tag{6}$$

The representation is used as follows during the encoding process: the first element of the list tells the neural network to push into the internal stack a vertex with associated label c and no edge with itself. Then the neural network reads the second element of the list: a vertex with label d is pushed into the internal stack together with the information that the current vertex has no edge with itself, but shares an edge with the vertex previously pushed into the stack. Subsequently the third element of the list is read and the neural network pushes into the internal stack a vertex with label e and the following information about edges: no edge with itself, shared edge with the vertex previously inserted into the internal stack, and shared edge with the vertex inserted two time steps before, and so on. Please, note that, since the edges are undirected it is sufficient to represent only the upper (or lower) part of the incidence matrix describing the connectivity of the graph.

For the directed graph shown in Figure 1(b) the representation is as follows

$$[(c, [0, 0, 0, 1]), (d, [1, 0, 1, 0]), (e, [1, 0, 0, 0]), (d, [0, 1, 0, 0])].$$

The use of the representation during the encoding is similar to the one described above, with the difference that now the full incidence matrix should be represented in order to retain the information about the direction of the edges. Thus, when considering the first element of the list, the interpretation of the information about the edges, i.e. the list $[0, 0, 0, 1]$, should be understood as follows: the first element of the associated edge list is 0, which means that there is no edge arriving from the vertex pushed as first into the internal stack; the same for the second element and for the third one; the last element of the list is 1, which means that there is an edge arriving from the vertex which will be pushed as fourth into the internal stack.

A linear dynamical system supporting the above idea may be the following

$$\mathbf{y}_i = \mathbf{W}_\mathbf{v}[\mathbf{v}_{label}^\mathsf{T}, \mathbf{v}_{edges}^\mathsf{T}]^\mathsf{T} + \mathbf{W}_\mathbf{y}\mathbf{y}_{i-1} \qquad (7)$$

where i ranges over the enumeration of the vertexes, i.e. positions in the list representing the graph, $\mathbf{v}_{label} \in \mathbb{R}^k$ is the numerical encoding of the current label, $\mathbf{v}_{edges} \in \mathbb{R}^N$ is the vector representing the information about the edges entering the current vertex where N is the maximum number of vertexes that the system can manage for a single input graph, and $\mathbf{y}_0$ is the null vector. Thus $\mathbf{v}^\mathsf{T} = [\mathbf{v}_{label}^\mathsf{T}, \mathbf{v}_{edges}^\mathsf{T}]^\mathsf{T} \in \mathbb{R}^{k+N}$ and the space embedding the explicit representation of the stack is $s = N(k+N)$ since no more than N vertexes can be inserted. It should be noted that this size of the stack is needed only if the input graphs are directed, and the above system is basically equivalent to system (1) for sequences.

However, if undirected graphs are considered, a specific space optimization can be performed. In fact, when inserting the first vertex into the internal stack only the first entry of the vector $\mathbf{v}_{edges}$ may be non null (the one encoding the self-connection), since no other vertex has already been presented to the system. In general, if vertex i is being inserted, only the first i components of $\mathbf{v}_{edges}$ may be non null. Because of that, the shift operator embedded into matrix $\mathbf{W}_\mathbf{y}$ may "forget" the last component of each field into which the internal stack is organized. Just to exemplify this point, let consider the encoding of graph (6). Recall that we assumed that the maximum number of vertexes per graph was 4 and let assume that input symbols are coded via a 10 bits code, so we have for each vertex a coding vector $\mathbf{v}$ of dimension $d = k + N = 14$. Now let consider the organization of the internal stack when all the vertexes of the graph have been read. It can be readily understood that the stack only needs $14+13+12+11$ bits. In fact, the first vertex inserted into the stack has a single edge bit which is non null, the second vertex only 2 bits, and so on. Thus, all the codes for the inserted vertexes can loose the current last bit of the code every time they are shifted to the right because of a push into the stack. Since the first inserted vertex (code) is shifted to the right 3 times, it will "forget" the last three bits of the code, which however are 0s since the first inserted vertex can just have coded an edge which is a self-connection. The second inserted vertex (code) will be shifted to the right 2 times, so it will loose the last 2 bits of the code, which however are 0s since the second inserted vertex can just have coded one edge as self-connection and a second one as a connection with the first inserted vertex, and so on for the other inserted vertexes.

Formally, the shift operator described above can be implemented by the following matrix

$$\mathbf{S} \equiv \begin{bmatrix} \mathbf{0}_{d\times s} & & & \\ \mathbf{I}_{(d-1)\times(d-1)} & \mathbf{0}_{(d-1)\times(s-d+1)} & & \\ \mathbf{0}_{(d-2)\times d} & \mathbf{I}_{(d-2)\times(d-2)} & \mathbf{0}_{(d-2)\times(s-2(d-1))} & \\ \mathbf{0}_{(d-3)\times(2d-1)} & \mathbf{I}_{(d-3)\times(d-3)} & \mathbf{0}_{(d-3)\times(s-3(d-1))} & \\ & \cdots & & \\ & \mathbf{0}_{(k+1)\times(s-k-1)} & \mathbf{I}_{(k+1)\times(k+1)} & \end{bmatrix}$$

and the optimal matrices defined as $\widetilde{\mathbf{W}}_{\mathbf{v}} \equiv \widetilde{\mathbf{U}}^{\mathsf{T}} \begin{bmatrix} \mathbf{I}_{d\times d} \\ \mathbf{0}_{(s-d)\times d} \end{bmatrix}$ and $\widetilde{\mathbf{W}}_{\mathbf{y}} \equiv \widetilde{\mathbf{U}}^{\mathsf{T}}\mathbf{S}\widetilde{\mathbf{U}}$.

4 Experimental Evaluation

The data used for testing our approach is derived from the data set of the PTC (Predictive Toxicology Challenge, [5]) originally provided by the U.S. National Institute for Environmental Health Sciences - US National Toxicology Program (NTP) in the context of carcinogenicity studies. The publicly available dataset (see http://www.predictive-toxicology.org/data/ntp/) is a collection of about four hundred chemical compounds. Figure 2 shows four compounds of the data set using the typical chemical graphical visualization where the vertexes without symbols are carbon atoms (C) and the hydrogens (H) and their bonds (completing the carbon valence) are not shown (hydrogen suppressed graphs). As shown in Figure 2 the data include a range of molecular classes and molecular dimension spanning from small and simple cases to medium size with multi-cycles.

In order to represent these chemical structures and their components, we use for each compound undirected vertex labeled and edge labeled graphs (i.e. a graph with labels associated to vertexes and edges). The vertexes of these graphs correspond to the various atoms and the vertexes labels correspond to the type of atoms. The edges correspond to the bonds between the atoms and the edges labels correspond to the type of bonds. This explicit graph modeling can be obtained through the information directly extracted by standard formats based on connection table representation, limited, in our case, to the information on atoms type (including C and H), bond type (single, double or triple) and their 2D-topology, as implicit in the set of vertexes connections. Here, we do not assume any specific canonical ordering of such information, assuming directly the form provided in the original PTC data set.

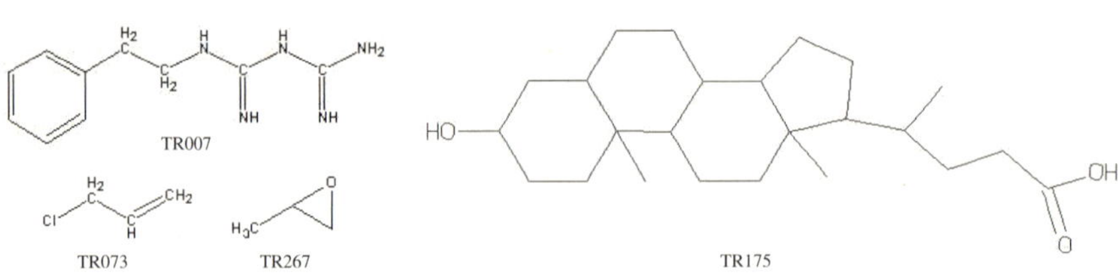

Fig. 2. Four chemical compounds belonging to the used data set

Table 1. Occurrences of atoms symbols in the data set and statistical properties of the dataset and of each split

Chemical Symbol	C	N	O	P	S	F	Cl	Br	H	Na
Frequency	3608	417	766	25	76	11	326	46	4103	22

Dataset Split	# examples	Max. number atoms	Max. number bonds	Avg. number atoms (bonds)	Tot. number items (atoms+bonds)
Training	235	70	73	24.42 (24.76)	11,557
Test	159	67	66	23.03 (23.38)	7,379
Total	394	70	73	23.86 (24.20)	18,936

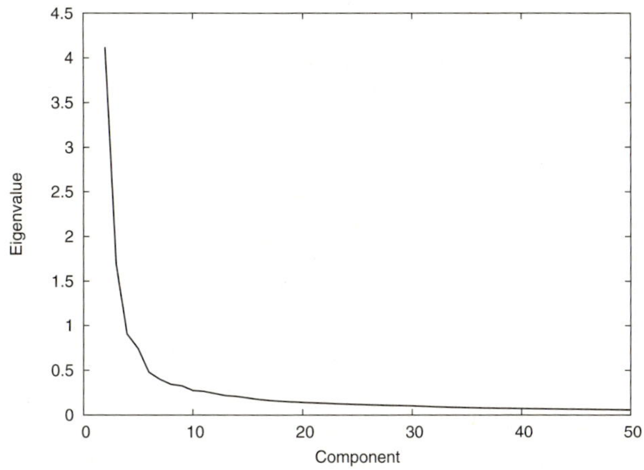

Fig. 3. Eigenvalues from rank 2 to 50 are shown. The most significant eigenvalue, caused by the introduction of $\mathbf{y}_{dummy}$, is not shown since it is very high (32189.69), as well as eigenvalues beyond rank 50. Only 949 eigenvalues over 3115 are non-null, i.e. $p^* = 949$.

For testing our approach, we have considered molecules with atoms occurring at least more than 3 times in the original data set and with a maximum dimension (number of vertexes) of 70. In all, 394 distinct chemical compounds are considered, with the smallest having 4 atoms. 10 distinct atoms occur in the used data set, corresponding to the following chemical symbols: C, N, O, P, S, F, Cl, Br, H, Na. In Table 1 we report the frequencies of such atoms through the compounds. Among the 394 compounds, 235 graphs are selected for training, and the remaining 159 graphs are used for testing the generalization ability of the system, i.e. the ability to successfully decode the components of the input chemical compound starting from the vector encoding the whole compound. In Table 1 we have summarized some general statistics about each split.

Symbols are represented by 10-dimensional vectors (i.e. $k = 10$) following a "one-hot" coding scheme. Bond's type is coded by integers in the set $\{0, 1, 2, 3\}$, where 0 represents the absence of a bond and the other numbers are for single, double and triple

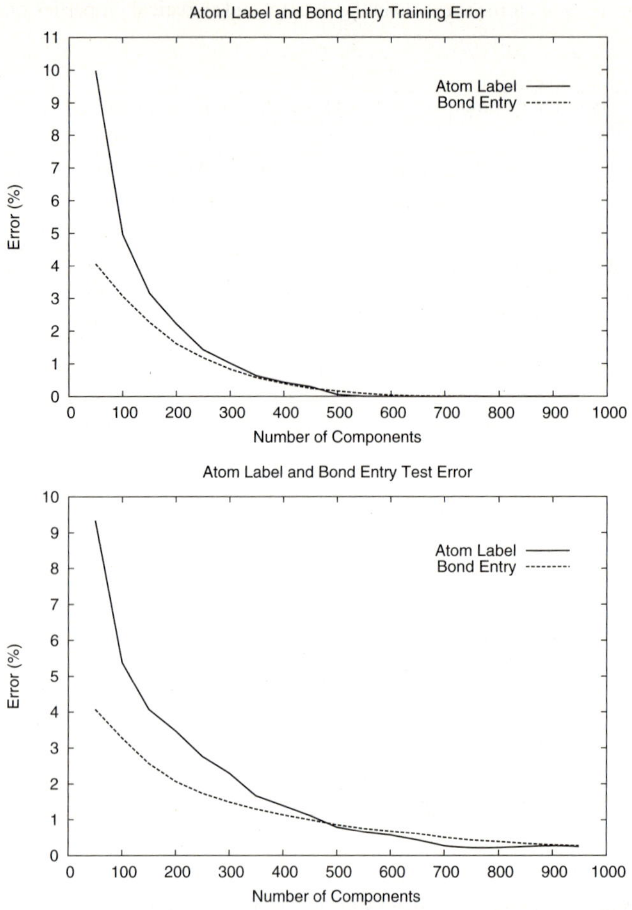

Fig. 4. Plots of the experimental results obtained for the training (top) and test (down) sets

bonds, respectively. Triple bonds occur only 2 times in the training set and 3 times in the test set. Double bonds occur 910 times in the training set and 599 times in the test set. The remaining bonds are single.

Since the graphs do not have self-connections for vertexes, we can avoid to represent the information about self-connections. Thus, because the maximum number of vertexes in the dataset is $N = 70$, we have an input dimension for each vertex which is $d = k+(N-1) = 10+69 = 79$ (recall that we do not consider self-connections) which leads to a stack size of $s = \sum_{i=0}^{N-1}(d-i) = 3115$, since the graphs are undirected. We used the dummy state $\mathbf{y}_{dummy}$ described in [7] to get zero-mean vectors.

The spectral analysis required around 27 cpu/min on an Athlon 1900+ based computer using Scilab. Values for the main eigenvalues are plotted in Figure 3.

In Figure 4 we have reported the training (top) and test (bottom) decoding errors for both label atoms and edges. The error in decoding is computed as follows. Each graph is first fed into the system, so to get the final encoding $\mathbf{y}$ for the graph. Then the final encoding is decoded so to regenerate all the items (atom and bond labels) of the graph. A decoded atom label is considered to be correct if the position of the highest value in the decoded label matches the position of the 1 in the correct label, otherwise a loss of 1 is suffered. A decoded bond entry is considered to be correct if its rounding to the nearest integer matches the target bond entry. If there is a mismatch a loss of 1 is suffered.

The final error is computed as the ratio between the total loss suffered and the total number of items (atom labels and total number of bond entries) in the dataset. For the bond entries we have normalized with respect to the number of bits that have been explicitly decoded by the system.

From the experimental results it is clear that learning is quite successful. In fact, with as few as 350 components it is possible to get a training error below 1% and a test error below 2% for both atoms and edges labels.

5 Conclusion

We have suggested a way to compute recursive principal components for both directed and undirected graphs with labeled vertexes and edges. Feasibility and efficacy of the proposed approach has been demonstrated on a dataset of chemical compound of significant variety and size. The obtained representations are quite informative and can be used as input vectors for any type of classification or regression method, such as Neural Networks and Support Vector Machines.

References

1. Baldi, P., Pollastri, G.: The principled design of large-scale recursive neural network architectures-DAG-RNNs and the protein structure prediction problem. Journal of Machine Learning Research 4, 575–602 (2003)
2. Frasconi, P., Gori, M., Sperduti, A.: A general framework for adaptive processing of data structures. IEEE Trans. Neural Networks 9(5), 768–786 (1998)
3. Gärtner, T.: A survey of kernels for structured data. SIGKDD Explor. New 5(1), 49–58 (2003)
4. Hammer, B.: Learning with Recurrent Neural Networks. Lecture Notes in Control and Information Sciences, vol. 254. Springer, Heidelberg (2000)
5. Helma, C., King, R.D., Kramer, S., Srinivasan, A.: The predictive toxicology challenge 2000-2001. Bioinformatics 17(1), 107–108 (2001)
6. Micheli, A., Sperduti, A., Starita, A., Bianucci, A.M.: Analysis of the internal representations developed by neural networks for structures applied to quantitative structure-activity relationship studies of benzodiazepines. J. of Chem. Inf. and Comp. Sci. 41(1), 202–218 (2001)
7. Sperduti, A.: Exact Solutions for Recursive Principal Components Analysis of Sequences and Trees. In: Kollias, S., Stafylopatis, A., Duch, W., Oja, E. (eds.) ICANN 2006. LNCS, vol. 4131, pp. 349–356. Springer, Heidelberg (2006)
8. Sperduti, A., Starita, A.: Supervised neural networks for the classification of structures. IEEE Transactions on Neural Networks 8, 714–735 (1997)

A Method to Estimate the Graph Structure for a Large MRF Model

Miika Rajala and Risto Ritala

Institute of Measurement and Information Technology,
Tampere University of Technology, Tampere, Finland
firstname.lastname@tut.fi

Abstract. We propose a method to estimate the graph structure from data for a Markov random field (MRF) model. The method is valuable in many practical situations where the true topology is uncertain. First the similarities of the MRF variables are estimated by applying methods from information theory. Then, employing multidimensional scaling on the dissimilarity matrix obtained leads to a 2D topology estimate of the system. Finally, applying uniform thresholding on the node distances in the topology estimate gives the neighbourhood relations of the variables, hence defining the MRF graph estimate. Conditional independence properties of a MRF model are defined by the graph topology estimate thus enabling the estimation of the MRF model parameters e.g. through the pseudolikelihood estimation scheme. The proposed method is demonstrated by identifying MRF model for a telecommunications network, which can be used e.g. in analysing the effects of stochastic disturbances to the network state.

Keywords: Graphical models, Graph structure estimation, Markov random fields (MRF), Multidimensional scaling, Mutual Information.

1 Introduction

Markov random field (MRF) models are used extensively for modelling statistically systems consisting of interacting variables. The structure of MRF model, the form of the joint probability distribution of the variables, can be chosen in many ways to describe best the properties of the underlying system. One such model is a widely used continuous-state linear-Gaussian model [14] with a Gaussian joint probability distribution. Ising model and Potts model [20] are examples of discrete-state models adopted from statistical physics with their joint probability distributions defined through the Boltzmann distribution [15]. Before choosing an appropriate MRF model for the underlying interacting system we need to know which variables of the system interact explicitly with which variables. These interactions are defined by the graph presentation of the MRF model as undirected, symmetric, links between the nodes in the graph, the nodes describing the MRF model variables.

Hence the identification of MRF model is a two-stage process. First the underlying structure of the interacting components, the graph presentation, is identified, and then the parameters for the selected graph structure are identified. In the literature the latter stage has been considered extensively whereas the first stage is usually assumed to be solved through domain knowledge. However, domain knowledge hardly ever is

J. Marques de Sá et al. (Eds.): ICANN 2007, Part II, LNCS 4669, pp. 836–849, 2007.

sufficient in practice for perfect identification of the graph structure. Hence in this paper we will tackle the problem of MRF graph structure estimation.

For Bayesian networks, closely related to MRF models, a Bayesian approach for determining the graph structure has been proposed by [6]. However, this includes exploring the space of topologies which grows exponentially with the number of nodes in the network [2]. Hence the method is computationally challenging in particular for networks having large number of nodes while also suffering some other practical issues [2].

We are considering here mainly large MRF models by which we mean MRF models having a large state space size, usually growing exponentially as the number of network nodes. For example with the binary-state Ising model the number of possible network states grows as 2^M, where M is the number of nodes in the network. The graph structure estimation scheme we are proposing here for such large MRF models is necessary in particular for systems where the node-to-node interactions are at least partially unknown. Also when the topology information available from several sources is inconsistent the method has value in deriving the best graph structure estimate directly from data. An example of inconsistent topology information is a mobile telecommunications network of base stations in which interaction topology is a combination of network connection topology and geographical topology.

The proposed method for estimating the graph structure of a MRF model first estimates the pairwise dependencies of all the MRF variables with a measure based on mutual information (MI), a basic concept in information theory [3]. When identified from real data we will consider the statistical significance that the MI is nonzero; denoted here as SSMI. Pairwise dependencies are interpreted as a similarity matrix. Similarity matrix is then transformed into a dissimilarity matrix and multidimensional scaling (MDS) [5] is applied to construct a 2-dimensional topology estimate, with nodes presenting variables and their distances their maximally preserved similarities. Defining a uniform neighbourhood threshold distance constructs neighbourhood relationships and provides the graph structure estimate.

The rest of the paper is organised as follows. Section 2 contains a brief introduction to MRF models and discusses some typical MRF models. Through Sections 3 and 4 we will introduce the method for estimating the MRF graph structure. Some appropriate dependency measures are presented in Section 3, while in Section 4 we will show how the final graph structure estimate can be derived by applying MDS and thresholding. Section 5 considers a general parameter estimation scheme applicable after having estimated the underlying graph structure. Section 6 contains a case-study by applying binary-state Ising model on a telecommunications network. Finally we conclude in Section 7.

2 Markov Random Field Models

Markov random fields (MRFs) [2] define a set of models which satisfy specified conditional independence properties. MRF models are a natural choice for modelling spatial data consisting of a set of interacting variables, see e.g. [4]. The conditional independence assumptions are represented as a graph, presenting the variables as nodes and the neighbourhood relations as undirected, symmetric links, between the nodes. Based on the graph the MRF model can be defined as a joint probability distribution through a graph property, collection of cliques [2]. After the model structure is chosen and the graph structure known the model parameters can be estimated. Fig. 1 shows an example of MRF graph.

2.1 Some Properties of MRFs

We will denote the states of the nodes in the graph by x_i, where i is the index of a node, and $i = 1,...,M$, where M is the number of nodes (variables) in the model. Conditional independence of states x_i and x_j at nodes i an j, given the states of the rest of the nodes, denoted with $\mathbf{x}_{-ij}$, is defined as [2]

$$p(x_i, x_j \mid \mathbf{x}_{-ij}) = p(x_i \mid \mathbf{x}_{-ij}) p(x_j \mid \mathbf{x}_{-ij}) . \tag{1}$$

For two nodes in MRF to satisfy the conditional independence property, a sufficient and necessary condition is that the nodes are not neighbours to one another in the graph structure, i.e. there does not exist a link between the two nodes in the graph. Hence in MRF only neighbouring nodes are conditionally dependent. This leads to a concept called a Markov blanket [2] defined for each node and consists of nodes in which the node studied is conditionally dependent. For MRF the Markov blanket of node i being the set of its neighbours, $N(i)$, the conditional probability of this node is expressed as

$$p(x_i \mid \mathbf{x}_{-i}) = p(x_i \mid x_{j \in N(i)}) . \tag{2}$$

2.2 MRF Model: The Joint Probability Distribution

The most general form of the joint probability distribution for a MRF, with given conditional independence properties, is defined through the collection of cliques [2]. A clique is a subset of graph nodes which are all neighbours to one another, i.e. they are fully connected in the graph.

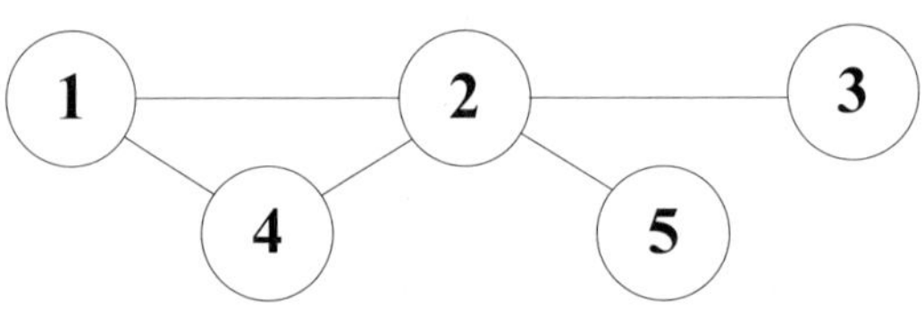

Fig. 1. An example of a MRF graph structure

Hence, for example, each neighbour pair forms a clique. A maximal clique is a clique which may contain other cliques as a subset but is not itself a subset of any other clique. As an example, consider the above Fig. 1 where nodes 1, 2, and 4 form a maximal clique of size 3.

A potential function of a clique is any positive definite function of clique states. The potential function need not have a probabilistic interpretation and thus need not be normalized. The most general form of a joint probability distribution of the node states with given independence properties is a product of potential functions of the maximal cliques. However, in what follows we consider only potential functions of maximal cliques that are products of potential functions of node-pairs within the maximal clique and potential functions of single nodes. When denoting the potential functions as $\psi_C(x_i, x_j)$ and $\psi_D(x_k)$, where C denotes the set of all the node pairs and D the set of all the nodes in the graph, and indices, $(i, j) \in C$ and $k \in D$, the joint probability distribution can be defined as

$$p(\mathbf{x}) = Z^{-1} \prod_{(i,j)\in C} \psi_C(x_i, x_j) \prod_{k\in D} \psi_D(x_k) . \tag{3}$$

Here Z is a normalisation constant, or partition function, and is given by

$$Z = \sum_{\mathbf{x}} \prod_{(i,j)\in C} \psi_C(x_i, x_j) \prod_{k\in D} \psi_D(x_k) . \tag{4}$$

This is a rather general formulation for MRF models into which real models, such as the continuous state Gaussian model, and the discrete state Ising and the Potts models can be reduced to.

2.3 Examples of MRF Models

In this section we will give three examples of typical MRF models, namely the Ising model, the Potts model, and the Gaussian MRF (GMRF). The first two are discrete states models while GMRF is a continuous state model. We will use the notations of Section 2.2, and denote the number of nodes as M.

The most general form of a binary state model is the Ising model [20], origin in statistical physics, but has also many other applications, such as image analysis [18]. When including loading of the nodes to Ising model the joint distribution of the model can be written in a factorised form similar to (3) as

$$\begin{aligned} p(\mathbf{x}) &= Z^{-1} \prod_{(i,j)\in C} \exp(J_{ij} x_i x_j) \prod_{k\in D} \exp(H x_k (h_k - h_0)) \\ &= Z^{-1} \exp\{\sum_{(i,j)\in C} J_{ij} x_i x_j + H \sum_{k\in D} x_k (h_k - h_0)\} \end{aligned} \tag{5}$$

In the first form the exponential factors correspond to the potential functions of single node cliques and node-pair cliques. The product of these factors results into the second form of the joint distribution shown. Here J_{ij}, H, and h_0 define the scalar parameters of the model, and h_i are known loads affecting the states of the nodes. Later in Section 6 we will assume that $J_{ij} = J$, uniform through the MRF structure. Each node x_i can have either state -1, or +1. Due to its properties Ising model can be used for modelling systems that can exhibit coherent behaviour such as discontinuous state transitions and hysteresis [20], hence also applied in the analysis of telecommunications networks [13].

Potts model [20] can be considered as an extension of the Ising model to any integer number, q, of states. The joint distribution of the Potts model factorises similarly to the product of potential functions of single node cliques and node-pair cliques as the Ising model, and can be written in general form as

$$p(\mathbf{x}) = Z^{-1} \prod_{(i,j)\in C} \exp(J_{ij} \delta(x_i, x_j)) \prod_{k\in D} \exp(H(x_k)(h_k - h_0)) . \tag{6}$$

Here δ is the Kronecker delta-function, hence two nodes i and j contribute together to the joint probability only if they are in equal states. Parameters J_{ij} and h_0 are here as with the Ising model, whereas H is a function of the node state x_k. Potts model also has first-order phase transitions ($q > 4$) in two dimensions [20], hence applicable for describing systems exhibiting discontinuous phase transitions.

Continuous-state GMRF can be written as a product of node-pair potentials, hence neglecting the loading terms included in the previous two models. This leads to the usual Gaussian form, and here we assume, without loss of generality, that the expectation values of the joint Gaussian are zero:

$$\begin{aligned} p(\mathbf{x}) &= Z^{-1}\prod\nolimits_{(i,j)\in C} \exp(-J_{ij}x_i x_j) = Z^{-1}\exp\{-\sum\nolimits_{(i,j)\in C} J_{ij}x_i x_j\} \\ &= (2\pi)^{-M/2}|\mathbf{Q}|^{1/2}\exp(-\tfrac{1}{2}\mathbf{x}^T\mathbf{Q}\mathbf{x}) \end{aligned} \quad (7)$$

Here $\mathbf{Q}$ is a MxM precision matrix (inverse covariance matrix) with elements J_{ij}, and Z is a normalisation constant defined by the last form of (7). In GMRFs working with the precision matrices has some advantages to working with the covariance matrices; a value of an element in the precision matrix is non-zero if and only if the two nodes corresponding to that element are neighbours [14]. GMRFs have the property that all the conditional distributions of the nodes are Gaussian, and being probability distributions no explicit normalisation constant is needed.

2.4 Interpretations of the Graph Structure

Graph structure of a MRF only describes direct connections between the variables; the neighbourhood relations. The three MRF models considered above all utilize these neighbourhood relations in the model structures through the products of the states of neighbouring nodes.

We may interpret the graph structure as an estimate of a true topology of a system, in which the node-to-node interactions are more complicated than just binary ones. In the most general concept all the nodes interact continuously with each other and the strength of interaction depends e.g. on the distances between the nodes. If we can obtain an estimate for the true topology of the system first, the graph structure can then simply be defined by thresholding; nodes only within a given threshold distance from one other in the true topology estimate are considered as neighbours, resulting into the binary neighbourhood relations and thus to the graph structure estimate. Hence, in general, there exist an infinite number of true system topologies corresponding to a single graph structure.

In some applications the graph structure may, however, present the true topology perfectly. As an example consider a telecommunications system when neglecting the affect of the loading to the network. Now there only exists logical (binary) connections between the nodes in the network; the graph structure can describe these connections perfectly. In practice the situation is more complicated since the loading of the nodes significantly influences the node states, and the loading between nodes physically close to each other usually correlate strongly. Hence there exist both the binary logical connections and the continuous connections due to physical locations between the nodes, and the graph structure can no longer describe this topology perfectly.

If more complicated node relations need to be considered, in the extreme all the nodes are interpreted as neighbours to each other with their strength of interaction depending on their distance. Although the model has more power to describe more complicated systems, the drawback is in lost simplicity of the model due to lost conditional independence properties. For example, in the case of the continuous state GMRF the precision matrix is full, and the benefits gained in the computation due to sparseness of the precision matrix are lost [14].

3 Measures to Describe Dependencies of MRF Nodes

In this paper we aim at finding an estimate for the graph structure of a MRF model from data. The first step is to model the MRF node (variable) dependencies, which we will consider in this section by defining a dependency measure based on information theory. We will also consider an approximation scheme to this measure, and discuss some alternative measures for describing the node dependencies.

All the measures introduced here are estimated from a finite size data set. Hence the measures are uncertain and the uncertainty depends on the amount of data available. Therefore, instead of using these measures directly as measures of dependency, we will use the probability that – given the data – the null hypothesis of the two nodes being statistically independent must be discarded. We refer this probability as the statistical significance of dependency. Statistical significance is much less sensitive to the amount of data available and hence more appropriate for comparing dependencies of several node pairs or several observation sets of varying size to each other.

3.1 Statistical Significance of Mutual Information

In information theory mutual information (MI) [10] measures the amount of information one random variable contains about another random variable [3]. MI being based on the fundamental concept of entropy is an appropriate measure for describing the node dependencies in a MRF. Let us consider two random variables, *X* and *Y*. The MI of *X* and *Y* is defined as the difference between the entropy of *X* and its conditional entropy given *Y* [3]:

$$I(X;Y) = H(X) - H(X \mid Y) = \sum_{x} \sum_{y} p_{XY}(x,y) \log \left[\frac{p_{XY}(x,y)}{p_X(x) p_Y(y)} \right], \tag{8}$$

where the lower case letters, x and y, refer to the values of the corresponding random variables, *X* and *Y*. The joint probability of *X* and *Y* is denoted as p_{XY}, while p_X and p_Y are the corresponding marginal probabilities. From the definition follows that MI is symmetric: $I(Y; X) = I(X; Y)$.

As the MI estimate is based on a finite size data set we will now consider a more robust dependency estimate, the statistical significance of MI (SSMI). SSMI is defined through a null hypothesis that the two variables, *X* and *Y*, are statistically independent, i.e $p_{XY}^{(0)} = p_X^{(obs)} p_Y^{(obs)}$, where $p_X^{(obs)}$ and $p_Y^{(obs)}$ are the marginal probability distributions of *X* and *Y* derived from the observed joint state distribution, $p_{XY}^{(obs)}$. The MI under the null hypothesis is zero. However, MI estimated from any finite set of observations is positive under null hypothesis with probability one. Estimate for the distribution of MI under null hypothesis and M observations can be obtained by first generating N sets of M state pairs according to $p_{XY}^{(0)}$, and then calculating from each of the M state pairs a total of N mutual information estimates. The probability density of mutual information under null hypothesis and obtained from data of length M, $f^{(0)}(I;M)$, is now estimated as the histogram of the N MI values. The SSMI, σ_{XY}, that when a mutual information $I_{XY}(M)$ has been observed, the null hypothesis must be discarded, is

$$\sigma_{XY} = P^{(0)}\left(I < I_{XY}(M); M\right) = \int_0^{I_{XY}(M)} f^{(0)}\left(I < I_{XY}(M); M\right) dI \,. \tag{9}$$

As SSMI is a probability it always lies between 0 and 1. As SSMI is calculated by simulating random observations under null hypothesis, its computation is not deterministic for any finite M. Consequently, the SSMI fluctuates from estimation to estimation even with identical data.

3.2 Approximation of SSMI: Statistical Significance of Chi-Squared Statistics

By making two approximations, the first in approximating logarithm with the mean value of its lower and upper bounds, and the second in assuming the finite size sample data is χ^2-distributed, we end up in an approximation of SSMI [9]. This approximate measure of the SSMI is here called the statistical significance of chi-squared statistics (SSCSS). When estimated from a sample data set this dependency measure can be derived as follows.

We consider again the two random variables, X and Y, and let M denote the total number of observations. We denote by M_{xy} the number of times X assumes the value x and Y the value y, respectively. By marginalising, we may further derive from M_{xy} the number of times X has a particular value x, M_x, and Y has a value y, M_y. Now, according to the null hypothesis, the expected number of any value pair (x, y), denoted as m_{xy}, is $m_{xy} = (M_x M_y)M^{-1}$. The chi-squared statistics for the two variables is now [11]

$$\chi^2 = \sum\nolimits_{x,y} (M_{xy} - m_{xy})^2 (m_{xy})^{-1} \,. \tag{10}$$

The number of degrees of freedom of test variable (10) is obtained as $D = IJ - I - J + 1$, where I and J are the total number of possible states of variables X and Y [11].

SSCSS is now defined with the chi-square probability distribution (incomplete gamma function) as

$$Q = \frac{1}{\Gamma(D/2)} \int_0^{\chi^2/2} \exp(-t) t^{D/2-1} dt \,, \tag{11}$$

where Γ denotes gamma function [11]. This integral tells the significance of the dependency between the two variables described by the chi-squared statistics, and as SSMI, it lies between zero and one. Not including simulations of random numbers the calculation of SSCSS is more feasible than that of SSMI.

3.3 Dependency Measures Based on Rank Correlation

Spearman rank-order correlation coefficient (SR) and Kendall's tau (KT) are dependency measures of random variables based on rank correlation [11]. In general, the idea in rank correlation is to use the ranks of the values of the data points among all the other sample values, instead of the sample values themselves. In SR the value of each data point is replaced by its rank among all the other sample values, hence employing the absolute ranks of the data points. KT uses relative ranks of the data points by comparing whether a sample value is higher in rank, lower in rank, or equal in rank than the sample value to which it is compared to. Hence, in KT, only magnitudes of the data point values are needed.

Let us first consider the SR in more detail. We will denote the data values of two random variables, X and Y, as x_i and y_i, indices referring to the observation number, and $i = 1,\dots,M$, where M is again the number of observations. When denoting the rank of point x_i as R_i and the rank of y_i as S_i, the SR for the two random variables, X and Y, is defined as the linear correlation coefficient of the ranks

$$r = \{\textstyle\sum_i (R_i - \overline{R})(S_i - \overline{S})\}\{\textstyle\sum_i (R_i - \overline{R})^2\}^{-1/2}\{\textstyle\sum_i (S_i - \overline{S})^2\}^{-1/2}, \qquad (12)$$

where $\overline{R}$ and $\overline{S}$ are the mean values of the R_i's and S_i's [11]. All ties in the ranks are assigned the mean value of the ranks that these values would have if their values were to differ only slightly.

Again we wish to find the significance of a nonzero value of r. This can be done by computing a test value $t = r[(M\text{-}2)/(1\text{-}r^2)]^{-1/2}$, which is approximately distributed according to Student's distribution with M-2 degrees of freedom. This approximation does not depend on the original distribution of the sample values, and hence the approximation is always the same. The statistical significance of SR is now obtained by integrating the corresponding Student's distribution from $-t$ to t [11].

By using the same notations as previously, the calculation of KT is based on the comparison of the ordering of the relative ranks in two consecutive data point values in x to the respective ordering in y. If the relative ordering of the ranks is the same, then the data pair (x_i, y_i) is called concordant, and if opposite, the data pair is called discordant, respectively. If tie occurs only in x's then the pair is called an extra y-pair, and if the tie appears only in y's the pair is called an extra x-pair. In case of both pair ending in tie, the pair is ignored [11]. KT is now

$$\tau = \frac{concordant - discordant}{\sqrt{concordant + discordant + extray}\sqrt{concordant + discordant + extrax}}, \qquad (13)$$

where the labels refer to the number of occurrences of the respective events in the data. Under the null hypothesis of no dependency between the two variables, the statistical significance of KT, can be approximated as a normal distribution with zero mean and variance $(4M+10)/[9M(M\text{-}1)]$ [11].

4 Estimation of the MRF Graph Structure

Dependencies between the variables in a MRF, estimated with any of the measures described in Section 3, can be interpreted as similarities. Transforming these pairwise similarities into dissimilarities they can be presented as a symmetric dissimilarity matrix. The second step in finding an estimate for the graph structure is now to apply multidimensional scaling (MDS) [5] on the dissimilarity matrix. This results in a 2-dimensional topology estimate of the true system topology, with nodes presenting the MRF variables and the distances between the nodes the maximally preserved similarities of the variables. The estimate for the graph structure is now obtained from the topology estimate by considering nodes within a certain threshold distance from one another as neighbours.

4.1 Multidimensional Scaling

MDS transfers dissimilarity measures of variables into a low dimensional distance map with nodes of the map presenting the variables and node inter-distances their similarities. First the similarity measures need to be turned into dissimilarity measures by subtracting similarity value from unity. Both the similarity and the corresponding dissimilarity matrix are symmetric when using any of the dependency measures presented in Section 3.

Let us consider two nodes i and j with symmetric dissimilarity $\delta_{ij} = \delta_{ji}$ [5]. Our goal is to find a topology for the nodes in 2-dimensions in which the inter-distances of the nodes optimally describe their dissimilarities. We will denote the coordinates of nodes i and j as $\mathbf{x}_i$ and $\mathbf{x}_j$ and their Euclidean distance in the mapping as $d_{ij}(\mathbf{x}_i,\mathbf{x}_j)$. We apply non-metric multidimensional scaling (e.g. [17]) in which the optimal coordinate values in 2-dimensions are searched by minimising Kruskal's stress-1 criterion [7], [8]

$$S_1 = \{\sum_{i,j} \{d_{ij}(\mathbf{x}_i,\mathbf{x}_j) - \hat{d}_{ij}\}^2 \Big/ \sum_{i,j} d_{ij}(\mathbf{x}_i,\mathbf{x}_j)^2\}^{1/2}. \tag{14}$$

Here $\hat{d}_{ij} = f(\delta_{ij})$ are called the target distances, or disparities, and are monotonically related to the observed similarities δ_{ij}: $f(\delta_{ij}) < f(\delta_{kl}) \Leftrightarrow \delta_{ij} < \delta_{kl}$. The stress-1 criterion can be minimised e.g. by using the Shepard-Kruskal scaling algorithm, see e.g. [7], [8].

4.2 MRF Graph Structure Estimate as Applying MDS on Node Similarities

Applying MDS on the similarities estimated for the MRF variables, the topology estimate that results approximates continuously the similarities of the variables in two dimensions. Hence we have obtained a topology estimate in two dimensions for the system variables. The graph structure estimate for the MRF can be derived by defining a threshold for node distances, uniform through the estimated topology, and then interpreting nodes within this distance from one another as neighbours. The graph structure estimate is now defined as undirected links between the neighbouring nodes. The graph structure estimate defines the conditional independence properties of the MRF model, and hence enables the estimation of the parameters of the joint probability distribution of the MRF model.

An obvious difficulty related to the estimation of the node neighbourhoods is in choosing an appropriate threshold value. One option is to specify the average number of node neighbours on the basis of the properties of the system considered, and then define the threshold value accordingly. If no information on the preferred average number of neighbours exists the only way is to vary the threshold value and analyze the uncertainty of the model parameters in the parameter identification stage.

One has to keep in mind that since the graph estimate is based on the threshold value, this choice may have drastic effects on the qualitative properties of the model. For example, when selecting too small a threshold value for a system behaving coherently the resulting network is not fully connected and the coherence may be lost. Similarly, choosing a threshold ending in too large a neighbourhood size for a system consisting of nearly independent variables may result in false coherent behaviour.

The method for estimating the graph structure is valuable in many practical situations where the graph structure may not be completely known in advance by the domain knowledge. In cases there exists some partial prior knowledge about the neighbourhoods of the variables these can be easily implemented to the graph obtained by simply adding or removing links between the nodes.

4.3 Identifiability of the Graph Structure

The estimate for the graph structure was obtained by thresholding the node distances in the network topology estimate resulted when applying MDS. There are two obvious situations where the topology estimate can not be directly obtained with MDS. The first one is when the system variables to be modelled are behaving coherently, i.e. when all the variables are constantly in equal states through all the observations. In this situation the similarity lies in unity through all the variable pairs, and no MDS map can be obtained. We may interpret all the nodes interacting explicitly with each other, and hence define all the nodes as neighbours to one another.

The second, opposite, situation arises when the variables are independent on each other, thus their states do not correlate. Since the similarity equals zero through all the variable pairs, again no solution is obtained with MDS. In this case the system can be modelled as a set of independent variables.

5 Estimation of the MRF Model Parameters

Successful identification of the graph structure enables the second part of the MRF model identification; the estimation of the MRF model parameters. How to estimate the parameters may depend on the model type chosen. A general approach for finding model parameter estimates is the maximum likelihood (ML) method [16]. To apply the ML method the partition function (normalisation term) of the joint distribution is needed. For large MRF models this is extremely difficult to calculate [19].

Pseudolikelihood estimation scheme [1], closely related to the ML method has been applied successfully in the literature [4]. In this procedure the model parameters are estimated from the product of the conditional probability distributions of the nodes (full conditionals). This is highly advantageous since now the partition function of the joint probability distribution need not to be calculated. By further utilising the conditional independence assumptions of (2), we find that each of these full conditionals only depends on the set of neighbouring nodes (Markov blanket) of a node, defined by the MRF graph structure estimate. By using the same notations as in Section 2, and by denoting the model parameters with $\boldsymbol{\theta}$, the pseudolikelihood estimation can be formulated as

$$\hat{\boldsymbol{\theta}} = \arg\max_{\boldsymbol{\theta}} \prod_{i=1}^{M} p(x_i \mid \{x_j\}_{j\in N(i)}, \boldsymbol{\theta}) = \arg\max_{\boldsymbol{\theta}} \sum_{i=1}^{M} \log p(x_i \mid \{x_j\}_{j\in N(i)}, \boldsymbol{\theta}), \quad (15)$$

where the second equality is obtained by using the monotonicity of log-function. The full conditionals needed in the estimation can be easily estimated as fixing the states of the neighbouring nodes of a node studied in the MRF joint probability distribution. Hence this estimation scheme is easily derived and it is effective in many situations, in particular when the state space of the MRF model is large.

6 Case Study: Applying Ising Model on Telecommunications Network Data

To illustrate the methods proposed in this paper, we will consider here the identification of a MRF model for a telecommunications network. Telecommunications networks are examples of complex networks, and since exhibiting diverse phenomena such as coherence in state, discontinuous phase transitions, and hysteresis, we apply here the Ising model appropriate to describe such phenomena. In particular we aim here at modelling the behaviour of a set of base transceiver stations (BTS) which are the basic elements of which a telecommunications network, such as a GSM network, consists of.

6.1 Synthetic Network Data

Before using the MRF graph structure estimation scheme for real telecommunications networks, we will first test that the graph estimation scheme works with a synthetic network by comparing the topology recovered from the generated data with SSMI similarity measure and MDS to the true topology of the network. In [12] we have tested this reconstruction scheme with few parameterisations of model (5) and found that this scheme works quite well except the special cases discussed in Section 4.3.

To demonstrate the method, the Ising model parameters (J, H, h_0) were given values (0.1, 0.55, 0.3) and the external loads of the nodes were drawn independently from uniform distribution with interval [0, 1]. The synthetic network contained 30 nodes and we generated 270 network state observations by using Markov Chain Monte Carlo methods with Gibbs sampling [10].

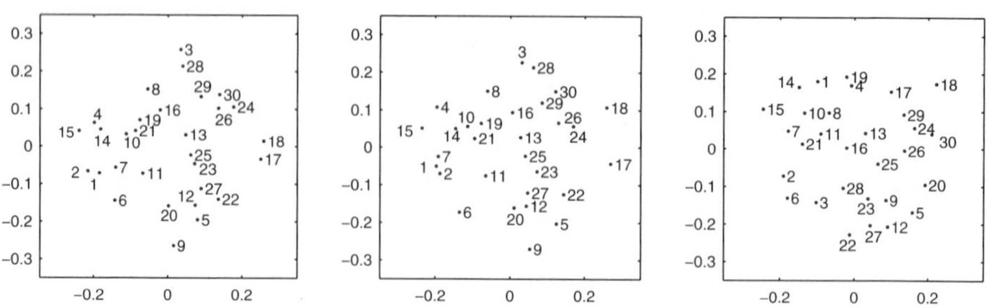

Fig. 2. The geographical (left), the logical (middle) and the estimated topology (right) of the synthetic network

As explained in Section 2.4, in general there exists two inconsistent topologies for a telecommunications network; the first, the logical topology, resulting from binary logical connections of the nodes and the second, the geographical topology, resulting from continuous connections due to physical locations between the nodes. Here the logical topology is defined through thresholding the node distances in the geographical topology and then applying MDS. We specified the average number of neighbours to be 8.8, and the value for the threshold was selected accordingly. Since the logical and geographical topologies are here consistent, the estimated topology is quite similar to them, as Fig. 2 shows.

6.2 Real Network Data

In real networks as state data for a BTS we use key performance indicators (KPIs), which describe qualitative properties of BTSs. The state data is obtained as unifying four KPI variables, each scaled into interval [0, 1], by calculating the Euclidean distance of each observation consisting of the four KPI values from an optimal value, 1. Then defining a threshold distance from 1 the unified state data is discretised into

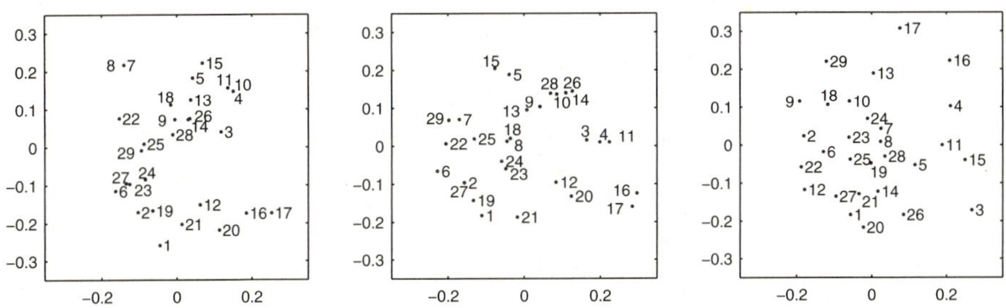

Fig. 3. The geographical (left), the logical (middle) and the estimated topology (right) of the real network

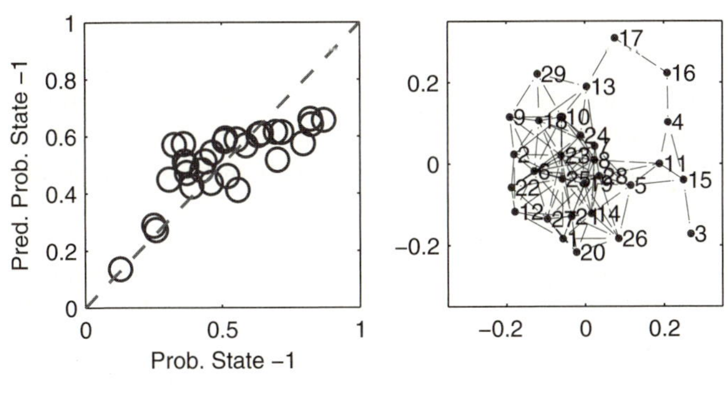

Fig. 4. On left the predicted probabilities for state -1 ('o') as a function of corresponding data based probabilities. Optimal predictions are presented with dashed line. The estimated MRF graph structure is shown on right.

binary values (-1, +1) – here 51.3% of states had value -1. As the load data for the nodes we use another KPI variable. The network contains 29 nodes with 269 network state observations, giving 269*29=7801 node state observations. More information on the real data set can be found in [12].

Fig. 3 shows the geographical (left) and logical topologies (middle) of the real network and the topology estimate based on SSMI and MDS. The logical topology is obtained by applying MDS on the domain-based logical node connections. Fig. 3 shows that many of the neighbour relations in the estimated topology are the same as in the geographical and logical topologies. Recall from Section 2 that now both the physical locations of the nodes and the logical connections affect the state behaviour of the nodes. Since the two topologies based on the domain-knowledge are inconsistent the estimated topology is now expected to be somewhat different to them. Further, since both the domain-based topologies affect the behaviour of the nodes, the best topology estimate is the one derived directly from the data, hence combining the effects of the two topologies.

Now the estimate for the graph structure, shown on right of Fig. 4, is defined according to the system properties by selecting the threshold value as having on the average of 8.34 neighbours per node, corresponding to the average amount of neighbours among the true logical node relations. Now the graph structure estimate can be compared e.g. to the graph structure according to the logical topology by comparing the average logical distance between the neighbouring nodes in the graph structure estimate, here 1.55, to the average logical distance among all the nodes, here 1.85. Since the first number is lower than the second the estimated graph structure captures some of the true logical neighbours.

The model prediction results, after estimating the model parameters (J, H, h_0) with the pseudolikelihood method of (15), are presented on left of Fig. 4, showing the average predicted probabilities for state -1 for each node as a function of the corresponding state probabilities calculated from data. Hence the results show that the proposed graph estimation scheme works here leading to decent results.

7 Conclusion

Identification of a Markov random field (MRF) model can be considered in two parts: first the identification of the graph structure and second the selection of the model structure and the identification of the model parameters. The latter is considered extensively in the literature and the first one is usually bypassed by assuming the topology information known e.g. through the domain knowledge.

In many practical applications, however, the domain knowledge is often imperfect and the topology is either partly or completely unknown, or there may exist several inconsistent topology information. Hence in this paper we proposed a method for estimating the graph structure for a MRF model from data. The method is particularly valuable for large MRF models where the graph structure information is uncertain. We demonstrated the method by applying the topology estimation scheme on a telecommunications network, where the method was found to give decent results in spite of using a binary approximation scheme for continuous state variables.

References

1. Besag, J.E.: Statistical analysis of non-lattice data. The Statistician 24(3), 179–195 (1975)
2. Bishop, C.M.: Pattern Recognition and Machine Learning, pp. 383–393. Springer, Heidelberg (2006)
3. Cover, T.M., Thomas, J.A.: Elements of Information Theory, pp. 5–21. John Wiley & Sons, Chichester (1991)
4. Cressie, N.A.C.: Statistics for Spatial Data, pp. 383–573. John Wiley & Sons, Inc, Chichester (1993)
5. Everitt, B., Rabe-Hesketh, S.: Kendall's Library of Statistics 4. The Analysis of Proximity Data. Arnold, 11–68 (1997)
6. Friedman, N., Koller, D.: Being Bayesian about Network Structure: A Bayesian Approach to Structure Discovery in Bayesian Networks. Machine Learning 50, 95–126 (2003)
7. Kruskal, J.B.: Multidimensional scaling by optimizing goodness of fit to a nonmetric hypothesis. Psychometrika 29, 1–27 (1964)
8. Kruskal, J.B.: Nonmetric multidimensional scaling: A numerical method. Psychometrika 29, 115–129 (1964)
9. Kullback, S.: Information Theory and Statistics. John Wiley (1959); reprinted by Dover Publications, Inc. pp. 155–188 (1997)
10. MacKay, D.: Information Theory, Inference and Learning Algorithms, pp. 357–421. Cambridge Univ. Press, Cambridge (2003)
11. Press, W., et al.: Numerical Recipes: The Art of Scientific Computation, pp. 476–481. Cambridge Univ. Press, Cambridge (1986)
12. Rajala, M., Ritala, R.: Mutual Information and Multidimensional Scaling as Means to Reconstruct Network Topology. In: Proc. SICE-ICCAS pp. 1398–1403 (2006)
13. Rajala, M., Ritala, R.: Statistical Model Describing Networked Systems Phenomena. In: Proc. IEEE Symp. on Comp. and Comm, pp. 647–654 (2006)
14. Rue, H., Held, L.: Gaussian Markov Random Fields: Theory and Applications, pp. 15–83. Chapman&Hall/CRC (2005)
15. Schroeder, D.V.: An Introduction to Thermal Physics, pp. 220–256. Addison-Wesley, London, UK (1999)
16. Stuart, A., Ord, K., Arnold, S.: Kendall's Advanced Theory of Statistics. In: Classical Inference & the Linear Model, 6th edn., vol. 2A, pp. 46–116. Oxford Univ. Press, Oxford, UK (1999)
17. Steyvers, M.: Multidimensional Scaling. In: Encyclopedia of Cognitive Science. Macmillan, London (2002)
18. Winkler, G.: Image Analysis, Random Fields and Markov Chain Monte Carlo Methods. Springer, Heidelberg (2003)
19. Yang, C.N.: The Spontaneous Magnetization of a Two-Dimensional Ising Model. Physical Review 85(5), 808–816 (1952)
20. Yeomans, J.M.: Statistical Mechanics of Phase Transitions. Oxford University Press, Oxford, UK (1992)

Neural Substructures for Appraisal in Emotion: Self-esteem and Depression

Nienke Korsten, Nickolaos Fragopanagos, and John G. Taylor

Department of Mathematics, King's College, Strand, London WC2R2LS, UK
{nienke.korsten,nickolaos.fragopanagos,john.g.taylor}@kcl.ac.uk

Abstract. In an attempt to bridge the gap between appraisal theory and the neuroscience of emotions, we have created a computational neural model in which a discrepancy between the internal value of global self-esteem and a more temporary, stimulus-inspired current self-esteem initiates an ongiong emotional response. We assign possible neural correlates to the nodes in this model, amongst which the orbitofrontal cortex and cingulate gyrus. We propose disruptions of the model analogous to states of depression.

Keywords: Self-esteem, Depression, Computational model, Anger, Neural simulation.

1 Introduction

The goal of the present paper is to find a common ground between appraisal theory, a framework firmly grounded in decades of psychological research on the emergence of emotions, and neuroscientific literature on the structures in the human brain that may underlie this process. According to appraisal theory, emotions are continuously created as responses to a recurrent, permanent process of assessment as to whether ongoing events are (potentially) beneficial or damaging, be it directly or indirectly. Different emotions represent differential outcomes of this assessment [1].

A recent and detailed appraisal based framework is the Component Process Model (CPM) [1], in which several Stimulus Evaluation Checks (SEC's) are identified. These SEC's are organised in terms of four appraisal objectives: relevance, implication, coping potential and normative significance. These form the main headers under which SEC's such as novelty and discrepancy-from-expectation are divided. In the CPM, it is claimed that these SEC's will always be applied in the same order.

Although appraisal theory, and the associated CPM to an even greater extent, point to several aspects of emotion of which neural correlates could potentially be found, precious little effort has been made to make the connection to the neurosciences. This area of science has made enormous progress in recent years, and many structures have been identified that have a role in emotional function, both on the side of emotion perception as on the production and regulation of affective states [2]. The structures involved form closely intertwined networks, and it has been proposed that there is a ventral system for rapid production of affective states, centered around the amygdala, insula, ventral striatum and ventral regions of anterior cingulate gyrus and prefrontal

J. Marques de Sá et al. (Eds.): ICANN 2007, Part II, LNCS 4669, pp. 850–858, 2007.

cortex, as well as a dorsal system for regulation of these states, encompassing hippocampus and dorsal regions of cingulate gyrus and prefrontal cortex [2]. Due to the necessarily non-intrusive nature of research techniques and distributed character of emotional brain correlates, however, it remains difficult to pinpoint the exact function of an area, and thereby the exact mechanism underlying particular behaviours.

2 Self-esteem

In the current paper, we attempt to bridge the gap between these two research traditions through a neurally based computational model of the appraisal of self-esteem. Self-esteem has long been known to be involved in the development of feelings of anger. It has long been thought that it was low self-esteem per se that caused anger and aggression [3] but more recently it has been suggested that an instability of [4] or/combined with threat to self-esteem [5] creates violent responses. Baumeister et al (1996) [3] also point out that it is not low self-esteem per se, but a discrepancy between a global high self-esteem and a more temporary perception of a disagreement with this self-assessment somewhere in the outside world that gives rise to anger. This global self-esteem is generally viewed as a trait that is determined in youth and adolescence, and that retains similar values throughout life [6]. There is a distinction between this and self-evaluations, which refers to the way people evaluate their various abilities and characteristics separately. Then there is also a distinction between global self-esteem and more temporary emotional states, arising from positive or negative feedback. These are referred to as feelings of self-worth or as state self-esteem.

Thus, we can view self-esteem as a specific appraisal that consists of the assessment of a discrepancy between two values of self-esteem, one (semi)permanent and the other temporary. This appraisal could result in an anger response or one of embarrassment, possibly dependent on a later appraisal of coping potential. In case of a highly appraised coping potential and an anger response, the ensuing (aggressive) behaviour can be seen as an attempt to restore self-esteem by standing up to whatever it was that caused the reduction of self-esteem in the first place, either in reality or imagination.

3 Depression

A neuroscientific notion that has often been connected to self-esteem (or a lack thereof) is depression. We can find numerous sources of a connection between the two [4] but here we must also be careful to take into account the more finely tuned distinctions involved. [7] find that it is a discrepancy between implicit and explicit self-esteem that is a correlate of depression, (again) not low self-esteem per se. In any case, it is not unlikely that if there is a neurally distinct mechanism for the maintenance of self-esteem, its disruption would play a part in the causes of depression.

On the neuroscientific side, numerous findings from postmortem as well as structural and functional neuroimaging methods have implied involvement of the subgenual cingulate gyrus, orbitofrontal cortex, dorsolateral and ventrolateral prefrontal cortex and amygdala in depression [8].

Recently, there have been advances in the treatment of patients with depression resistant to other treatments such as medication and psychotherapy, by application of chronic deep brain stimulation (DBS) to the subgenual cingulate region (Brodmann area 25) [9]. This area has been shown to be hyperactive in treatment resistant depression, and DBS was shown to counteract this hyperactivity in this structure as well as reduce activity in several other areas of the brain (e.g. orbitofrontal cortex, medial frontal cortex) and increase activity in several others (e.g. dorsal cingulate) in four out of six patients. The treatment had both direct and long-term effects in terms of sensations reported by the patients and measured improvements in interest, motor speed, activity level and depression test scores. Depression has also been treated through lesioning of the anterior cingulate region [8].

Mayberg et al. [9] distinguish between several regions of the cingulate that appear to have different roles in depression. The subgenual cingulate (Brodmann area 25), a relatively dorsal region, is stimulated in DBS, and is hyperactive in depression. More anterior regions (BA 24) they find to be *hypo*active in depression. In a different paper, Seminowicz et al. [10] present a path modeling analysis of the areas involved, using PET data and Structural Equation Modeling.

4 Architecture

We have created a neural model (see figure 1 for the basic structure of the model) consisting of a small number of graded nodes, each containing a single sigmoid response function, that produces an emotional response when a stimulus reducing the current value of self-esteem is perceived. This output then continues until the conflict between the remembered, internal value of global self-esteem and the more variable, current value is resolved by the perception of a stimulus that increases the value of self-esteem. We thereby propose that an emotional response is a (primitive) behaviour with as its goal the maintenance of ones value of self-esteem.

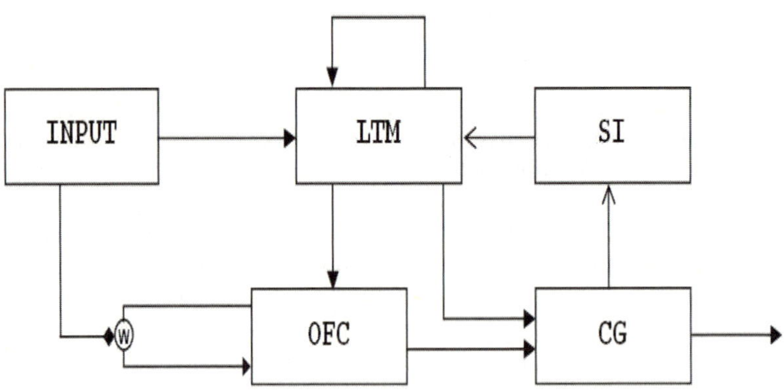

Fig. 1. Schematic outline of the model. Closed arrows represent excitatory connections, open arrows inhibitory ones. The diamond shaped node represents modulation of w.

In this model, we have assumed the existence of a representation of the current value of self-esteem, encoded in the orbitofrontal cortex (OFC) analogous to state self-esteem as described above. The OFC has been associated with representations of value in a multitude of ways: OFC activation has been shown to represent the value of predicted reward/punishment in primates (see [11] as well as [12] for a 'minireview'), and similar results have been found in the human brain through neuroimaging of reward learning [13]. The OFC has close anatomical interconnections with the amygdala, and is often activated in a similar if subtly different manner [14]. Therefore, the amygdala may also play a role in the representation of current self-esteem.

We have also assumed the existence of a more permanent value of self-esteem encoded in long-term memory (LTM), analogous to global self-esteem. This could consist of some representation of episodic self-related memories, where we could think of separate nodes for memories related to specific characteristics of the self together representing the global self-image. This representation could reside in the medial prefrontal cortex, based on results obtained by [15], who found differential activation for self-referential judgement of trait adjectives when compared to other-referential judgments. However, the neural location of this function might also lie in the medial temporal lobe, which is more traditionally regarded as the locus of episodic memory[16]. It is also possible that this representation of global self-esteem is not directly related to episodic memory but rather to a value attached to the concept of self, in which case it might reside in the OFC, or that this representation is distributed over several areas. Because of this uncertainty as to the neural correlate of global self-esteem, we will refer to this node as LTM rather than assign an analogous area in the brain.

In addition to this, the model contains a node representing the cingulate gyrus (CG), monitoring the difference between these two values of self-esteem. Cingulate cortex activation has been found to correlate with monitoring functions [17], more specifically in Error Related Negativity, the ERP signal following error trials of almost any kind, and the cingulate is known to play a role in decision making as well as emotional processing. As mentioned above, this area is also highly involved in depression [9], as well as the voluntary suppression of negative affect [18].

The fourth node in the simulation spontaneously inhibits the LTM node. We could propose regions of the cingulate as a neural correlate for this node, but this is highly speculative. We will elaborate on this in the discussion.

When a self-esteem related stimulus comes in, this will activate the LTM. The input reaching the LTM is not valenced, meaning that it is excitatory in response to any stimulus pertaining to self-esteem, regardless of whether this is negative (e.g. insult) or positive (e.g. compliment).

The LTM in turn activates the OFC. A comparison between the two values takes place in the CG where, if the new value is lower than the old one (by a certain threshold), an emotional response will be initiated and the spontaneous inhibition node will be inhibited, thereby allowing the LTM to remain activated through its recurrent connectivity. If the new self-esteem value is not lower than the remembered value, the CG will not be activated and the spontaneous inhibition will continue to deactivate the LTM. So we have a system where any self-esteem related input to LTM will provoke a comparison (in CG) between the current value of self-esteem (represented by OFC activation) and its permanent, remembered value (LTM). If the new value is lower, disinhibition of the LTM (via SI) will then cause all three modules (LTM, OFC and CG) to remain active until this difference in self-esteem values is changed.

In order for this difference in self-esteem values to arise (or be resolved), and the recurrent activity in the system to start or stop, the value of the OFC will have to be changed. This occurs through changes in the recurrent weight of the OFC (w), caused by the valence of the self-esteem related input. So, if the input is negative, the recurrent activity of the OFC node will be reduced, thereby creating a difference in the CG and producing an emotional response and disinhibiting the LTM. This activation (and thus the emotional response) will continue until the OFC weight is increased again. This could occur either by a positive input from the outside world, or possibly by the creation of a coping strategy and thereby an internal re-evaluation.

This model was created using Matlab Simulink, with a fixed step size of 0.01 and appying the Euler method of integration. In table 1, we describe the internal structure of each node in more detail, as well as weights and other parameters. Each neuron in the model performs a simple linear addition of all incoming signals and passes the summed activity x through a sigmoid response function y, where $y(x) = 1/(1+e^{-x}) - 0.5$. The recurrent weight in the OFC node, w, consists of a continuous integration of the valenced inputs, with a maximum output of 1.05 and a minimum of 0.7, then multiplied by a constant (4).

Table 1. Detailed description of nodes and parameters

Node	Function	Weights	Other parameters
Input	Produce 50 ms pulses, all positive (to LTM) or positive/negative dependent on valence (to w).	to LTM: 1	
LTM	Integrate excitatory input pulses with recurrent excitation and SI inhibition	recurrent: 5 to OFC: 1.1 to CG: 1	
OFC	Recurrent weight w is changed where OFC = y (OFC*w*c + LTM) where y is sigmoidal and c = (see also text)	recurrent: w to CG: 0.8	
CG	Ongoing comparison between OFC and LTM activation where CG = y (LTM – OFC – threshold) if LTM – OFC – threshold > 0.	to SI: 4	
SI	Spontaneous inhibition of LTM, unless inhibited by CG so SI = y (c – CG)	to LTM: 1	spontaneous exc. (c): 0.8

Summarizing the functionality of this architecture: when a stimulus is perceived that is both relevant to self-esteem (input to LTM) and negative (input to w), the representation of the current value of self-esteem in OFC is reduced, leading to a mismatch between the value of global self-esteem (in LTM) and of current self-esteem, represented in the CG. This mismatch causes an emotional response, and this response continues to be produced until the discrepancy is resolved through a positive input, restoring the value of current self-esteem. This process of mismatch analysis could be viewed as an appraisal, since it is a continuous monitoring, resulting in an emotional response.

5 Results

As can be seen in figure 2, a negative input results in expected activations (CG, OFC, LTM) and deactivations (SI) of the nodes in the model. These nodes remain active / deactivated until a positive stimulus comes in to restore the value of self-esteem. OFC has a baseline activation (low stable state) and a higher activation while the LTM is active, reflecting the self-esteem value.

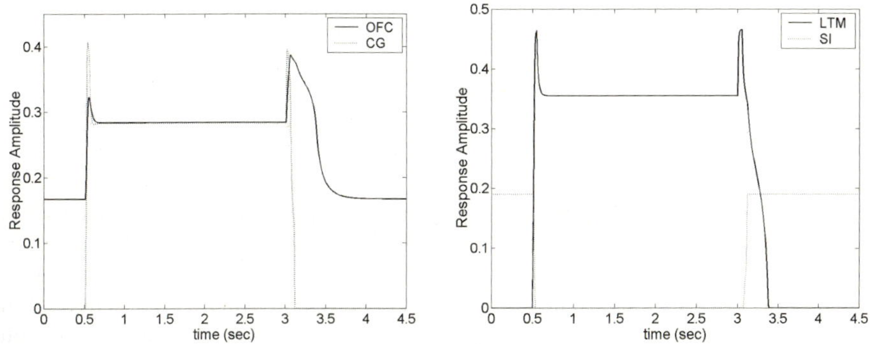

Fig. 2. Activations for OFC, CG (left), LTM and SI (right) nodes with a negative (at time 0.5) and a positive (at time 3) 50 ms input pulse

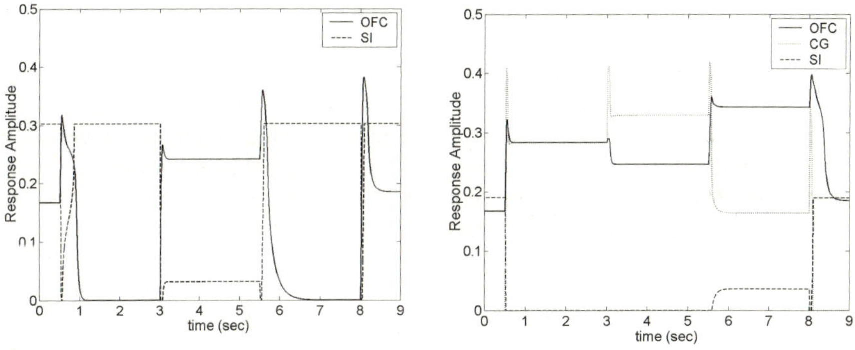

Fig. 3. Activations in the model when spontaneous activation constant of the SI node is increased to 1.4 (left) or at its normal value of 0.8 (right), with two negative (t = 0.5 and 3) and two positive (t = 5.5 and 8) 50 ms input pulses. In the right hand figure, the negative input at t = 0.5 causes activation of CG as well as increased activation of OFC, where a reduction of OFC activation causes an increase in CG activation. In the left hand figure, the negative input at t = 0.5 causes a deactivation of OFC, and a second negative input is needed at t = 3 to boost the system into activation. CG shows activity from t = 3 to 5.5. These activations were omitted for purposes of clarity.

Thus, we have proposed a mechanism by which an ongoing emotional response is produced when a negatively valenced stimulus pertaining to self-esteem is perceived, and a positive input is needed to stop this response. We assume that, in depression, this system is disregulated in some way. We hypothesize that this occurs through a hyperactivity of the SI node, which represents the subgenual cingulate, shown to be hyperactive in depression [9]. Figure 3 shows that this hyperactivity causes a failure to activate the CG and LTM as well as a deactivation of the OFC to the extent that it can be turned off completely, instead of retaining its normal low stable state. In a 'healthy' version of the model, negative input causes an increase in activation of the CG and LTM, instead of a decrease, which puts the OFC into its high activity state. Behavioural symptoms of depression seem to be in agreement with this pattern of activations and deactivations. A reduction in activity can be explained by the lack of CG output and feelings of worthlessness are in agreement with the reduced levels of both global (LTM) and current (OFC) self-esteem. This fits in with the data concerning DBS of the subgenual cingulate (CG 24) as a treatment for depression, if we assume that this region is the neural correlate of the SI node. A deactivation of this area by DBS would reduce its hyperactivity.

In order to determine if the SI module is the only structure in the model that could be analogous to the hyperactive subgenual cingulate in depression, we have looked at the effects of increased activation of the other modules as well. However, for CG and LTM this results in an increased CG output, which would result in increased responses to stimuli rather than the reduced emotionality that is found in depression. Changes in the weight between the CG and SI node could also decrease CG output initially, but a value of w low enough to activate the CG can still be reached with a large amount of negative input.

6 Discussion

We have presented a model in which a self-esteem related stimulus evokes an emotional response through a neurally based architecture. The assessment in the CG node of the model of the discrepancy between global and current self-esteem could be viewed as an appraisal of implication. Since the information that the stimulus is relevant is already avaiable in the input, we can take the appraisal of relevance to have gone before. The appraisal of coping potential would logically follow, where an adjustment might be made as to the kind of emotional response: with a high coping potential, an anger response could be justified, whereas with a low coping potential sadness or embarrassment may be more appropriate. In this way, we incorporate aspects of appraisal, the neuroscience of emotion and the psychological and neuroscientific aspects of depression in one simple model.

However, this simplicity is also a weakness. The view presented in this paper may be somewhat oversimplified, in the sense that emotional processes in the brain form a highly intricate system, in which various different components continuously interact, and the processes after our cingulate output still have an influence on what goes on inside the model. Having just the one output module may seem like a bit of a stretch.

We do feel that this can partly be justified because the cingulate gyrus does appear to play a large and deciding role in producing and regulating affective states [2] and action generation [19]. Projections of this area include amygdala and ventral striatum, known to be important areas in the generation of emotions. Phan et al [18] find that activity within the dorsal anterior cingulate is inversely related to intensity of negative effect in a task where the subjects were instructed to suppress this affect. This may provide support for our tentative proposition of the subgenual cingulate as an inhibitory area, analogous to the SI in our model. We intend to extend the model to incorporate more of this functionality in the future. Another function we intent to add in future simulations is that of positive emotion when self-esteem is increased.

Acknowledgements

This work has been partially supported by the HUMAINE EC Network of Excellence and the Biotechnology and Biological Sciences Research Council (BBSRC), UK; JGT would like to thank the EC for financial support under the GNOSYS cognitive robot project (FP6-IST-2-003835) to allow him to pursue this work.

References

1. Sander, D., Grandjean, D., Scherer, K.R.: A Systems Approach to Appraisal Mechanisms in Emotion. Neural Networks 18, 317–352 (2005)
2. Phillips, M.L., Drevets, W.C., Rauch, S.L., Lane, R.: Neurobiology of Emotion Perception I: the Neural Basis of Normal Emotion Perception. Biological Psychiatry 54, 504–514 (2003)
3. Baumeister, R.F., Smart, L., Boden, J.M.: Relation of Threatened Egotism to Violence and Aggression: The Dark Side of High Self-Esteem. Psychological Review 103, 5–33 (1996)
4. Franck, E., De Raedt, R.: Self-Esteem Reconsidered: Unstable Self-Esteem Outperforms Level of Self-Esteem As Vulnerability Marker for Depression. Behaviour Research and Therap. In: Press, Corrected Proof
5. Kuppens, P., van Mechelen, I.: Interactional Appraisal Models for the Anger Appraisals of Threatened Self-Esteem, Other-Blame and Frustration. Cognition and Emotion 21, 56–77 (2007)
6. Brown, J.D., Dutton, K.A., Cook, K.E.: From the Top Down: Self-Esteem and Self-Evaluation. Cognition and Emotion 15, 615–631 (2001)
7. Franck, E., De Raedt, R., Dereu, M., Van den Abbeele, D.: Implicit and Explicit Self-Esteem in Currently Depressed Individuals With and Without Suicidal Ideation. Journal of Behavior Therapy and Experimental Psychiatry 38, 75–85 (2007)
8. Phillips, M.L., Drevets, W.C., Rauch, S.L., Lane, R.: Neurobiology of Emotion Perception II: Implications for Major Psychiatric Disorders. Biological Psychiatry 54, 515–528 (2003)
9. Mayberg, H.S., Lozano, A.M., Voon, V., McNeely, H.E., Seminowicz, D., Hamani, C., Schwalb, J.M., Kennedy, S.H.: Deep Brain Stimulation for Treatment-Resistant Depression. Neuron 45, 651–660 (2005)
10. Seminowicz, D.A., Mayberg, H.S., McIntosh, A.R., Goldapple, K., Kennedy, S., Segal, Z., Rafi-Tari, S.: Limbic-Frontal Circuitry in Major Depression: a Path Modeling Metanalysis. NeuroImage 22, 409–418 (2004)
11. Roesch, M.R., Olson, C.R.: Neuronal Activity in Primate Orbitofrontal Cortex Reflects the Value of Time. J Neurophysiol 94, 2457–2471 (2005)
12. Schoenbaum, G., Roesch, M.: Orbitofrontal Cortex, Associative Learning, and Expectancies. Neuron 47, 633–636 (2005)
13. O'Doherty, J.P.: Reward Representations and Reward-Related Learning in the Human Brain. Insights From Neuroimaging. Curr. Opin. Neurobiol 14, 769–776 (2004)

14. Rolls, E.T.: The Functions of the Orbitofrontal Cortex. Brain Cogn 55, 11–29 (2004)
15. Kelley, W.M., Macrae, C.N., Wyland, C.L., Caglar, S., Inati, S., Heatherton, T.F.: Finding the Self? An Event-Related FMRI Study. J Cogn Neurosci. 14, 785–794 (2002)
16. Eichenbaum, H.: Hippocampus: Cognitive Processes and Neural Representations That Underlie Declarative Memory. Neuron 44, 109–120 (2004)
17. van Veen, V., Carter, C.S.: The Anterior Cingulate As a Conflict Monitor: FMRI and ERP Studies. Physiology & Behavior 77, 477–482 (2002)
18. Phan, K.L., Fitzgerald, D.A., Nathan, P.J., Moore, G.J., Uhde, T.W., Tancer, M.E.: Neural Substrates for Voluntary Suppression of Negative Affect: A Functional Magnetic Resonance Imaging Stud. Biological Psychiatry 57, 210–219 (2005)
19. Rushworth, M.F.S., Behrens, T.E.J., Rudebeck, P.H., Walton, M.E.: Contrasting Roles for Cingulate and Orbitofrontal Cortex in Decisions and Social Behaviour. Trends in Cognitive Sciences 11, 168–176 (2007)

The Link Between Temporal Attention and Emotion: A Playground for Psychology, Neuroscience, and Plausible Artificial Neural Networks

Etienne B. Roesch[1], David Sander[1,2], and Klaus R. Scherer[1,2]

[1] Geneva Emotion Research Group, University of Geneva, Switzerland
Etienne.Roesch@pse.unige.ch
[2] Swiss Center for Affective Sciences, Geneva, Switzerland

Abstract. In this paper, we will address the endeavors of three disciplines, Psychology, Neuroscience, and Artificial Neural Network (ANN) modeling, in explaining how the mind perceives and attends information. More precisely, we will shed some light on the efforts to understand the allocation of attentional resources to the processing of emotional stimuli. This review aims at informing the three disciplines about converging points of their research and to provide a starting point for discussion.

Keywords: Attention, Emotion, Cognitive Science.

1 Introduction

In this paper, we address the endeavors of three disciplines, Psychology, Neuroscience, and Artificial Neural Network (ANN) modeling, in explaining how the mind perceives and attends information. More precisely, we address the efforts to understand the allocation of attentional resources to the processing of emotional stimuli. By bringing the three disciplines together, we aim at informing researchers about some of the recent advances in the other disciplines: whereas temporal attention and emotion are often studied separately, and with very disciplinary approaches, we argue that advances in one domain can help to refine the others. We further argue that the interplay between Psychology, Neuroscience, and ANN modeling lies in the constraints that each discipline can impose on the others, offering converging evidence towards one common goal. To focus our enterprise, we will address results from studies investigating the modulation of *temporal attention* by emotional stimuli. Temporal attention contrasts with other types of attention, like spatial attention, by setting the focus on the unfolding allocation of attentional resources to the processing of stimuli over time, and the underlying processing dynamics. Studies addressing temporal attention use experimental paradigms like Rapid Serial Visual Presentations (RSVP), for which it has been shown that emotional stimuli elicit particular patterns of response. Our paper is structured as follows. In the next section, we present the recent theoretical advances in emotion psychology with regards to

J. Marques de Sá et al. (Eds.): ICANN 2007, Part II, LNCS 4669, pp. 859–868, 2007.

the modulation of temporal attention by emotion. The third section describes the brain mechanisms highlighted by neuroscience, pointing out the key areas of the brain involved in the cognitive functions of attention and emotion, and the brain mechanisms underlying their interaction. In the fourth section, we will briefly present some of the ANN modeling proposed to account for the modulation of temporal attention. We will conclude by highlighting some of the issues an interaction between Psychology, Neuroscience, and ANN modeling can help to resolve.

2 Psychological Perspectives on Pre-attentive Processes and Emotion

Psychology plays a major role in preparing the stage for the interplay between the disciplines. In this section, we will first describe the modulation of temporal attention by emotion, along with some of the behavioral results issued from experimental psychology. We will then introduce two sets of theories that have a special interest in describing the unfolding dynamics underlying the allocation of attentional resources to the processing of emotional stimuli.

2.1 The Modulation of Temporal Attention by Emotion

In a typical RSVP experiment, participants are presented with rapidly flowing images (presented at a frequency ± 10 Hz), one replacing the other at the same spatial location on the screen. Participants are asked to spot and perform tasks on one or more targets embedded within distracting images. Varying the time interval between two targets renders it possible to indirectly measure the amount of resources that is allocated to the processing of targets: results in a typical dual task experiment indeed show that the perception and processing of a first target (T1) hinders the perception and processing of a second target (T2) if it appears within 200-400 milliseconds after T1 (Figure 1). This phenomena has been rhetorically named an "Attentional Blink" (AB) [13]. Interestingly, emotional targets seem to benefit from a processing bias, alleviating the blink. If extensive research has been done on temporal attention, surprisingly little has been devoted to the modulation of temporal attention by emotion.

In an early study, Anderson and Phelps [1] showed that, not only did negative words alleviate the typical blink response compared to neutral words, but also that the amygdala was critical to benefit from the emotional significance of the words. These authors concluded that a critical function of the human amygdala is to enhance the perception of stimuli that have emotional significance. As we will discuss in the next section of the paper, the amygdala seems to play an important role in the interaction between attention and emotion. The modulation of the blink by the intrinsic significance of the targets has also been reported in a study showing that participants did not experience an AB for their own names but did for either other names or nouns [18]. Equivalent results have been reported when T2 targets were familiar faces compared to unfamiliar faces, the former

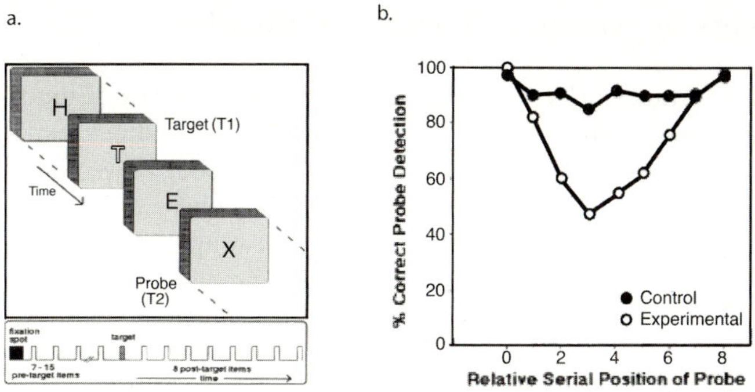

Fig. 1. *Panel a.* Illustration of the RSVP, T1 was a white letter, T2 was the black letter "X". *Panel b.* Group mean % of trials in which T2 was correctly detected, after correct identification of T1, plotted as a function of the relative serial position after T1 [13], with permission, copyright American Psychology Association 2007.

alleviating the blink [8]. The latter results have been shown to be sensitive to trait personality differences, like trait anxiety.

In the remain of this section, we will provide the reader with accounts from current theoretical frameworks of emotion psychology. In the landscape of emotion theories, two sets of theories have a particular interest in describing the unfolding dynamics underlying the genesis of emotions, and the allocation of attentional and cognitive resources to the processing of emotional stimuli.

2.2 Accounts by "Basic Emotions" Theories

The concept of basic emotions refers to the postulate that there is a small number of emotions, fundamentally distinct from one another [5, e.g.]. As a result of this perspective, some emotions have been studied more thoroughly than others, as in the case of fear for which several models are proposed. In this theoretical thread, Öhman and colleagues [12] proposed an evolved module of fear and fear learning. Shaped by evolutionary pressure, this so-called *fear module* would have become specialized in the solving of potentially harmful situations for the species, like snakes, spiders or particular social encounters. The authors further argue that a dysfunction of this module would explain the selective anxiety disorders that are commonly described, like snake phobias, spider phobias, or social phobias, respectively [11]. The authors describe this module as being selective, automatic, encapsulated, and implemented in a dedicated cerebral circuit centered on the amygdala. This module is related to the research on fear conditioning in rats by LeDoux [10], who showed evidence for two routes involved in the processing of fearful stimuli, emphasizing the role of the amygdala in early stages of the processing. Interestingly, this last model has been computationally modeled using plausible ANN [2], offering converging evidence that responses to fear conditioning could occur even without the impulse from primary auditory cortices.

2.3 Accounts by Appraisal Theories of Emotions

The theoretical framework offered by the appraisal theories of emotions posits that emotions result from cognitive appraisals that individuals make about occurring events. Unlike the theories previously described, these theories suggest common mechanisms to the genesis of all emotions. These mechanisms can take the form of rapid, automatic, unconscious, cognitive appraisals evaluating stimuli against particular criteria. In other words, the process causally linking a stimulus to an emotional response is divided into multiple cognitive processes, which are common to every emotions, and that, in turn, evaluate the stimulus against a finite number of criteria. The result of this appraisal process yields to the genesis of a particular emotion. Several appraisal theories are available [16, for a review], out of which we selected Scherer's Component Process Model [15] for being sufficiently detailed to allow precise predictions, and sufficiently general to encompass a fair number of phenomena.

The Component Process Model defines an emotion as an episode of interrelated, synchronized changes in the states of all or most of the five subsystems classically described in the emotion literature, in response to the evaluation of an external or internal stimulus event as relevant to major concerns of the organism. More precisely, Scherer defines the nature and the functions of the cognitive evaluations yielding to the genesis of an emotion. These evaluations are described in terms of cognitive appraisals, named Stimulus Evaluation Checks, allowing the genesis of differentiate emotions. One can see checks as devoted processes evaluating the stimulus in regards to a specific criteria. The concept of *relevance* is at the core of the theory, being the first step in the sequence of appraisals [15]. It is also of particular importance in our endeavor as this first appraisal process is believed to determine the amount of cognitive resources to be allocated to the further processing of the perceived stimulus. This mechanism is evolutionarily justified in that it provides the organism with the economy of available resources, only allocating processing resources to important stimuli.

In general, any stimulus that could potentially influence the goals, or maintain the individual, in a sustained level of well-being is considered relevant. From this first appraisal unfold the orienting of the attention towards the stimulus event, the allocation of cognitive resources to its further processing, and the preparation of the organism to a behavioral response. A facial expression of fear or anger, for instance, will both represent a relevant information for the individual, signaling the occurrence of a negative event, obstructing the goals of the individual, or a potential danger. The degree to which the individual will process this information, allocating more or less attentional resources to its processing, attributing to it a particular emotional value, and adequately choose a line of reaction will depend on his goals, his needs, or the context in which the stimulus appeared. As will be discussed in the next section, the amygdala may be a potential candidate for implementing some kind of relevance detector [14].

To summarize, both sets of theories emphasize the role of pre-attentive processing mechanisms in the unfolding of temporal attention and allocation of resources to cognitive processing. Whereas the predictions from the "basic emotions" theories are mainly focused on threat-related stimuli, the predictions from the appraisal theories of emotion extend to any stimulus which is relevant to the organism, for any of many different reasons. Finally, both sets of theories emphasize the role of subcortical brain areas, like the amygdala, in subserving these mechanisms. Therefore, we suggest that the influence of emotion over temporal attention can occur in a pre-attentive scenario, as soon as the perceived, relevant stimuli enters the thalamo-amygdala circuitry.

3 Pre-attentive Processes and Emotion in the Light of Neuroscience

The knowledge gained from Neuroscience is of two kinds. Using very different technologies, researchers either describe the topography of the brain networks involved in specific tasks, or the temporal dynamics at play in these networks of structures. In this section, we will set a particular emphasis on the description of the dynamics involved in the modulation of both attention, and temporal attention, by emotion.

3.1 Brain Mechanisms Underlying the Modulation of Attention by Emotion

Endogenous modulation of attention (e.g., spatial attention) by emotion, is often described in terms of top-down bias effects from one region of the brain in favor of other, lower level, regions of the brain [20]. Many neuroscience studies have indeed shown enhanced responses to emotional stimuli relative to neutral stimuli, and researchers suggest that direct top-down signals might be emitted not only from fronto-parietal regions (e.g., Pre-Frontal Cortices, PFC), but also from subcortical regions like the amygdala. As discussed in the previous section, the amygdala, in particular, is known to be crucial in fear processing and fear learning [10]. Its position in the processing stream of perceptual information makes it a perfect candidate to potentially influence many cortical and subcortical regions. However, if two routes have reliably been identified in rats, there is still some debate about the precise circuitry involving the amygdala in humans. Regardless of the hypothesis advanced, researchers agree nonetheless to attribute an initial appraisal of emotional significance to this structure, based on limited information, early in the processing stream. This influence could take the form of direct feedback to sensory cortices, but also as indirect modulation of parietal and frontal regions (e.g., PFC). This latter signals would then produce a cascade of events which would signal emotional significance.

3.2 Brain Mechanisms Underlying the Modulation of Temporal Attention by Emotion

Dehaene and colleagues used recordings of event-related potentials (ERP) to compare the temporal dynamics of seen and unseen (blinked) words in a typical AB experiment [4,17]. Describing the cortical activations of unseen words, the authors report a drop in the waveform of components peaking around 300 milliseconds, which correlated with behavioral visibility ratings. Whereas ERP methodology cannot be used to make unambiguous inferences about brain localizations, estimations techniques allow to roughly determine the sources of cortical electrical activity. Using this approach, the authors report that seen words, compared to unseen words, initiated an intense spread of activation within left temporal and inferior frontal regions (about 300 milliseconds after stimulus onset), which would then spread to lateral prefrontal and anterior cingulate cortices (about 440 milliseconds), before extending in more posterior regions (about 580 milliseconds). In their interpretation, the authors introduce the concept of a *global workspace*, which refers to the notion that the different high-level, specialized, brain areas involved in the processing of visual stimuli interconnect to each other, to form a global workspace processing the stimuli into a unitary assembly supporting conscious reportability. Perceived stimuli would thus compete to recruit this global workspace that, once activated, only affords exclusive access, yielding to the inability to process subsequent stimuli for a transient period of time. Areas in the global workspace theoretically map onto the description of the processing streams involved in visual perception, from perceptual areas to higher associative areas of temporal, parietal, frontal, and cingulate cortex [4]. Which is to say that the bottleneck described in AB studies would therefore lay in such top-down influence of higher-level areas, like the PFC or more parietal areas, over the lower-level areas involved in the visual streams.

A second scenario has been proposed by Hommel and colleagues in an article reviewing evidence from numerous imaging techniques [7]. In this article, the authors present a neurocognitive model of the AB, which situates the processing bottleneck reported in behavioral results in the rendering of an intrinsically parallel system into effectively serial dynamics: after nonselective processing in specialized perceptual cortices, stimuli are fed to object-specific temporal areas, where they are matched against long-term knowledge and, consequently, identified. Identified objects are then maintained in frontal working memory, and receive support by means of the synchronization of the relevant structures in frontal and parietal cortices. By closing a perceptual window, this synchronization stabilizes the representation maintained in working memory, increasing the likelihood that the target be reported, and preventing other stimuli from entering further processing.

If these two scenarios have received considerable interest from the research community and, being somehow complementary, do provide consistent explanations about the dynamics underlying temporal attention, and its modulation by top-down processes, none explicitly takes into account the modulation of temporal attention by emotion, and no better account is being offered as of today. To summarize, both representative scenarios start from the assumption that the processing of perceived stimuli is initiated in perceptual cortices, which then feed into fronto-parietal systems processed and aggregated feature-representations. These systems then complete the processing of the stimuli, whilst preventing other stimuli to interfere with the processing by one means or another. We therefore suggest that emotion may modulate temporal attention by means of top-down enhancing signals emitted very early in the processing stream, mediated by the amygdala, for instance.

4 ANN Modeling of Temporal Attention and Emotion

Because attention in general, and the AB in particular, represent fairly constrained and well described phenomena, a reasonable number of ANN models are available, compared to other phenomena, each emphasizing a different aspect of the findings. Out of the few models that exist, we selected three models for representing the range of focus ANN models can entail. This range of focus is best appreciated at the light of the abstraction level emphasized by each ANN model. In that sense, the first model we describe focuses on the high-level interpretation of the findings of behavioral experiments, whereas the second and third models set a special emphasis on the low-level neurobiological plausibility of the network and the dynamics involved. We discuss the appropriate balance between these levels in the conclusion of this manuscript.

In a recent effort to converge the different findings of the literature into a unifying theory of the AB, Bowman & Wyble [3] presented a model of temporal attention and working memory encapsulating 5 principles that represent the main effects described in psychology literature. This model, called the Simultaneous Type, Serial Token (ST^2) model, is modeled using several layers and explicit mechanisms closely mapping the effects that are described in the literature. Whereas results obtained with this model provide a good fit with the results reported in the literature, the authors acknowledge that their ambition was to provide a "cognitive-level explanation of the AB, and temporal attention in general", rather than a plausible implementation of the mechanisms described in Neuroscience studies. Therefore, we argue that it is unclear to what extent this data-driven implementation can fully explain the dynamics underlying the interaction between temporal attention and emotion, even though some of the principles this model rely on do refer to emotional significance, at least semantically.

A competing ANN model is provided by Dehaene and colleagues [4]. Unlike the previous ANN model, the authors based their model on the neurobiology of neural pathways from early sensory regions (Areas A and B) to higher association areas (Areas C and D). The global workspace, as described in the previous section, lays in the interconnections of the nodes in higher areas C and D (Figure 2, panel a). In addition, the basic brick from which they built the model closely models a thalamo-cortical column, reproducing the laminar distribution of projections between excitatory and inhibitory spiking neurons. In doing so, they reproduced the basic computational unit that can be found in neural pathways, and managed to explicitly compare results from their model with actual neuroimaging data, providing converging evidence for a global workspace hypothesis. However, if their implementation closely models some aspects of neurobiology, it neglects other aspects that have been highlighted in Neuroscience research, and described in the previous section: one of which being the close interaction between subcortical and cortical structures in the processing of visual stimuli. Furthermore, their model does not provide any mechanism from which a modulation of temporal attention by emotion could arise.

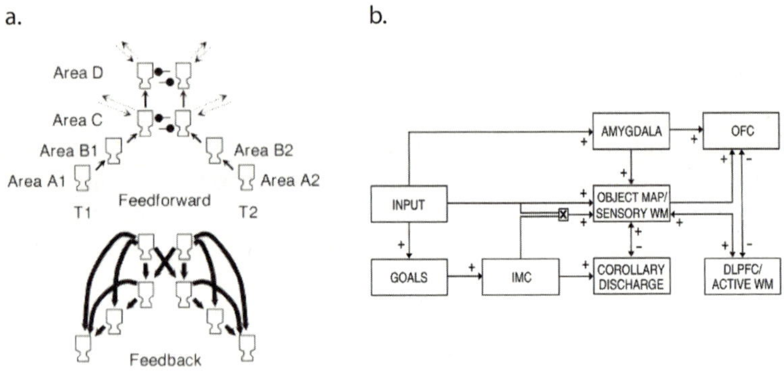

Fig. 2. *Panel a.* Schematic architecture of the global workspace model. Areas A and B represent the perceptual pathways leading to the global workspace, areas C and D implement the global workspace including fronto-parietal regions [4]. Adapted with permission, copyright National Academy of Science of U. S. A. 2007. *Panel b.* CODAM architecture extended by Amygdala and Orbito-Frontal Cortex to address emotional influences on attention [9]. Adapted with permission, copyright Elsevier 2007.

A third attempt at modeling temporal attention has been made by Taylor and colleagues [20,6] in the COrollary Discharge of Attention Movement (CODAM) model. This model was developed by analogy to models of motor controls applied to attention [20,19,9] in which the creation of the attention modulation signal is emitted from a controller structure onto separate modules where activations are to be modulated (Figure 2, panel b). The CODAM model is based on the descriptions provided by Neuroscience, both in terms of the structure of the network and of the dynamics implemented. As such, it contains input pathways leading to a working memory that can be influenced by the conjunction of several signals, both exogenous and endogenous. Critical to this approach is the Inverse Model Controller (IMC), which boosts the representation of perceived stimuli. This attentional boost is required for stimuli to reach working memory, and thus be reported. In conjunction with a conflict monitor, the IMC interferes with subsequent stimuli if the first stimuli has not yet reached the working memory. As a result, the monitor will suppress the second stimulus in the IMC, and thus withhold its attentional boost, hindering its successful encoding in working memory (i.e., yielding to an AB). This model also contains modules representing subcortical structures, like the amygdala. Input to the amygdala comes from crude, early signals directly from the input module, representing posterior sites in the brain. It also interacts with the OFC module, which encodes the value of stimuli and can influence attention via top-down activations.

5 Conclusion

In this paper, we addressed the perspectives of three disciplines in explaining temporal attention, and its modulation by emotion. We highlighted concepts issued from Psychology that are now being rediscovered by Neuroscience. The concept of *relevance*, for instance, central in the appraisal theories of emotion [15] seems to play a major role in the definition of the functional domain of the amygdala [14]. We then introduced several scenarios proposed by neuroscientists to account for the unfolding of temporal attention. Finally, we described examples of ANN models accounting for the AB, and some of the mechanisms implemented to underlie a modulation by emotional stimuli. However, ANN approaches are very different, depending on the abstraction level that modelers choose to pursue. The ST^2 model, for instance, only semantically represents the computations that could be implemented in the brain and, even though it provides a good fit with behavioral results, it does not, however, offer plausible converging evidence as to how emotional stimuli modulate temporal attention in the brain. The CODAM model, on the other hand, is structured on the basis of what has been described in the Neuroscience literature. By doing so, the authors had to fill in the blanks by making a number of assumptions. For instance, they introduced the notion of corollary discharge mediated by an Inverse Model Controller for which there is only indirect evidence. This offers new tracks to explore to both psychologists and neuroscientists.

In the introduction of this paper, we proposed that advances in one discipline could help to refine the other disciplines. We further argued that the interplay between the disciplines lied in the *constraints* that each discipline can impose on the others. These constraints can be expressed in the form of the *goodness-of-fit* between the models proposed by each discipline. In other words, by providing the three disciplines with a common goal, i.e. the modulation of temporal attention by emotion, we argue that the findings in one discipline should be able to address the findings in the others. Taking this interdisciplinary view, we raise a number of questions:

- Most of what is known about the modulation of temporal attention by emotion has been investigated using threat-material. What is the effect of positive *relevant* stimuli over the unfolding process of attention?
- If relevance is subjectively determined by the appraised propensity of stimuli to affect the goals, the needs of the individual, how do inter-individual or personality factors modulate the unfolding process of attention? How could this be accounted for in Neuroscience, and in plausible ANN models?
- If mechanisms like the corollary discharge have proven useful in modeling the modulation of temporal attention by emotion [6], how do these mechanisms relate to recent findings in Neuroscience? In what way can we model the perceptual window described by Hommel et al. [7]?

References

1. Anderson, A.K., Phelps, E.A.: Lesions of the human amygdala impair enhanced perception of emotionally salient events. Nature 411(6835), 305–309 (2001)
2. Armony, J.L., Servan-Schreiber, D., Cohen, J.D., LeDoux, J.: Computational modeling of emotion: explorations through the anatomy and physiology of fear conditioning. Trends in the Cognitive Sciences 1(1), 28–34 (1997)
3. Bowman, H., Wyble, B.: The simultaneous type, serial token model of temporal attention and working memory. Psychol Rev 114(1), 38–70 (2007)
4. Dehaene, S., Sergent, C., Changeux, J.-P.: A neuronal network model linking subjective reports and objective physiological data during conscious perception. Proc Natl Acad Sci U S A 100(14), 8520–8525 (2003)
5. Ekman, P.: Are there basic emotions? Psychological Review 99, 550–553 (1992)
6. Fragopanagos, N., Kockelkoren, S., Taylor, J.G.: A neurodynamic model of the attentional blink. Cognitive Brain Research 24, 568–586 (2005)
7. Hommel, B., Kessler, K., Schmitz, F.: How the brain blinks: towards a neurocognitive model of the attentional blink. Psychological Research 70(6) (2005)
8. Jackson, M.C., Raymond, J.E.: The role of attention and familiarity in face identification. Percept Psychophys 68(4), 543–557 (2006)
9. Korsten, N.J.H., Fragopanagos, N.F., Hartley, M., Taylor, N., Taylor, J.G.: Attention as a controller. Neural Netw 19(9), 1408–1421 (2006)
10. LeDoux, J.: The emotional brain: the mysterious underpinnings of emotional life. Simon & Schuster, New York, NY (1996)
11. Öhman, A.: Psychology. conditioned fear of a face: a prelude to ethnic enmity? Science 309(5735), 711–713 (2005)
12. Öhman, A., Mineka, S.: Fears, phobias, and preparedness: toward an evolved module of fear and fear learning. Psychological Review 108(3), 483–522 (2001)
13. Raymond, J.E., Shapiro, K.L., Arnell, K.M.: Temporary suppression of visual processing in an rsvp task: an attentional blink? J Exp Psychol Hum Percept Perform 18(3), 849–860 (1992)
14. Sander, D., Grafman, J., Zalla, T.: The human amygdala: An evolved system for relevance detection. Reviews in the Neurosciences 14, 303–316 (2003)
15. Scherer, K.R.: Appraisal considered as a process of multilevel sequential checking. In: Scherer, K.R., Schorr, A., stone, T.J. (eds.) Appraisal processes in emotion: Theory, methods, research, pp. 92–120. Oxford University Press, New York (2001)
16. Scherer, K.R., Schorr, A., Johnstone, T.: Appraisal processes in emotion: theory, methods, research, volume Series in affective science. Oxford University Press, Oxford, New York (2001)
17. Sergent, C., Baillet, S., Dehaene, S.: Timing of the brain events underlying access to consciousness during the attentional blink. Nat Neurosci 8(10), 1391–1400 (2005)
18. Shapiro, K.L., Caldwell, J., Sorensen, R.E.: Personal names and the attentional blink: a visual "cocktail party" effect. J Exp Psychol Hum Percept Perform 23(2), 504–514 (1997)
19. Taylor, J.G., Fragopanagos, N.F.: The interaction of attention and emotion. Neural Netw 18(4), 353–369 (2005)
20. Taylor, J.G., Rogers, M.: A control model of the movement of attention. Neural Networks 115, 309–326 (2002)

Inferring Cognition from fMRI Brain Images

Diego Sona[1], Sriharsha Veeramachaneni[2], Emanuele Olivetti[1], and Paolo Avesani[1]

[1] FBK-irst, Trento, Italy
[2] Thomson R&D, MN, USA

Abstract. Over the last few years, functional Magnetic Resonance Imaging (fMRI) has emerged as a new and powerful method to map the cognitive states of a human subject to specific functional areas of the subject brain. Although fMRI has been widely used to determine average activation in different brain regions, the problem of automatically decoding the cognitive state from instantaneous brain activations has received little attention. In this paper, we study this prediction problem on a complex time-series dataset that relates fMRI data (brain images) with the corresponding cognitive states of the subjects while watching three 20 minute movies. This work describes the process we used to reduce the extremely high-dimensional feature space and a comparison of the models used for prediction. To solve the prediction task we explored a standard linear model frequently used by neuroscientists, as well as a *k-nearest neighbor* model, that now are the state-of-art in this area. Finally, we provide experimental evidence that non-linear models such as *multi-layer perceptron* and especially *recurrent neural networks* are significantly better.

1 Introduction

Thanks to the advent of functional Magnetic Resonance Imaging (fMRI), neuroscientists received impressive help in studying the functionalities of the human brain. This fMRI technology enables detailed analysis of neural activity by providing the means to collect brain activation data at high spatial and temporal resolution. Thanks to this, many studies identify regions of brain activated when humans perform specific cognitive tasks. Although traditionally fMRI scans have been used by neuroscientists to identify brain regions correlated with external conditions such as sense stimuli, there is burgeoning interest in adopting the reverse view, i.e., to use pattern recognition techniques and machine learning to predict external stimuli based on fMRI data [1].

Preliminary studies have shown that it is possible to decode visual perception cognition [2] looking at the brain state image acquired through an fMRI scan. The same approach is currently being extended to infer further high level cognitive functions [3]. The challenge of decoding mental states has been defined as a learning problem. The goal is to train a classifier that given an fMRI brain image, the mental state, predicts the associated cognitive state [4].

J. Marques de Sá et al. (Eds.): ICANN 2007, Part II, LNCS 4669, pp. 869–878, 2007.

The machine learning community addressed the brain interpretation mainly using linear models and achieving controversial results. Previous works [5] indicate that *support vector machines* (*SVMs*) outperform *Gaussian naïve Bayes* and a k-nearest neighbor classifiers (*k-NN*). On the other hand, more recently [6], it has been provided empirical evidence where a *k-NN* classifier outperformed a linear kernel model.

The above empirical analyses have been performed on datasets determined with cognitive experiments designed to address a single stimulus, i.e., the interpretation issue are just discriminative tasks between two alternative cognitive states. While this way of proceeding is effective from the point of view of neuroscientists that are looking for brain mapping, it affects the generality of empirical analysis on learning models since the evaluation is restricted to a single cognitive function.

In 2006, a team of neuroscientists organized a competition on decoding of mental states, the Pittsburgh Brain Activity Interpretation Competition (PBAIC)[1]. They provided fMRI data collected from three subjects while they were watching three 20 minutes movie segments from a television show. The subjects themselves later annotated the movies with respect to several ratings (e.g., language, attention, amusement etc.). The competition consisted in predicting the ratings of the third movie for all three subjects from the functional data, using the annotated ratings for the first two movies as training.

The contribution of this work is a report of our winning entry in the above competition, and an investigation of the use of non-linear learning model as feed forward neural networks and recurrent neural networks for the task of brain image interpretation. We provide empirical results on the dataset of PBAIC competition that provides the labeling of many different cognitive functions on the same brain scans.

In Section 2 we describe fMRI and feature ratings data in detail and define the prediction task. Section 3 is devoted to the description of our approach to preprocessing. In Section 4 we present the models used to predict the cognitive state of the subject from the brain images. Finally, Section 5 presents the result and in Section 6 conclusions are drawn.

2 Description of the Task

The task consists on the analysis of fMRI brain data of human subjects watching movie segments of a tv-series. The data was produced by the neuroscience group at the University of Pittsburgh for a competition held in 2006. Each movie segment is rated by the subjects themselves with multi-valued categories, such as the presence of faces, tools, sadness, arousal, individual actors, language, music, etc[2]. The challenge is to interpret the brain activity of the human subject allowing predicting what he/she is experiencing.

[1] http://www.ebc.pitt.edu/2006/competition.html

[2] The subject cognitive experience is made of 13 feature ratings, 3 actor presence ratings and 3 location ratings.

In more detail, the same segments of movie were shown to three subjects (2 male and 1 female, with mean age of 26 years) that afterwards rated the movies with their personal impressions[3]. Both fMRI data and features ratings were sampled at a frame-rate of one frame every 1.75 seconds.

Each frame of the fMRI data is a 3-dimensional image made of 64x64x34 voxels. The intensity of the voxel represents the amount of blood arriving at the particular area of the brain measured using *blood-oxygen-level dependant fMRI contrast*. This is an indirect measure of the brain activity in the corresponding area. The image sequences were preprocessed for motion correction, slice time correction, linear trend removal, and spatial normalization. Each brain image has an associated vector of ratings provided by the human subjects (the target of our task). These ratings were temporally convolved with a hemodynamic filter that makes these features real-valued[4]. The features scored in the competition are of two types:

Content: body parts, environmental sounds, faces, food, language, laughter, motion, music, and tools;
Reaction: amusement, attention, arousal, and sadness;

but there were also other optional scores which were, actors and locations. We refer the reader to [7] for a more detailed description of the task and the data.

For the remaining part of the paper, we evaluated the proposed models using a subset of the available features: Amusement, Body parts, Faces, Language, and Motion. The reason is that for these ratings the evaluation of the quality of the used models is more robust. All the discarded ratings are characterized by the absence of a significant number of positive samples, hence the model evaluation are not as robust as for those ratings well sampled that we selected.

3 Image Processing and Dimensionality Reduction

The fMRI data and the corresponding ratings that we used for the experiments in the present paper are made of two movies collected in a total time span of 40 minutes at the frame-rate of 1.75 seconds. Hence, the sum of the temporal sequences is made of about 10^3 brain images and rating samples. From these sequences, we have to select training and testing sets. In addition, each of these 3-D brain images are extremely noisy and high dimensional (10^5 voxels). This suggests that these images need to be reduced in their dimensionality both from a computational standpoint as well as to alleviate the loss of prediction accuracy due to the *curse of dimensionality*. In the remainder of this section, we describe the process of image feature attribute generation. The main idea is to select

[3] The ratings given by the subjects resulted to have high discrepancies, i.e., the correlations between the ratings of different subjects were sometimes very low.

[4] The reason for this convolution is the need for temporal realignment of the fMRI sequences with the events in the movies. The blood flow increases in the interested brain area few seconds after the area activity. This delay is well studied and the hemodynamic filter compensates this delay.

the most informative voxels in the brain and then to cluster the voxels with similar behavior in time. Each cluster then is used to extract one image feature attribute.

3.1 Noise Removal

We observed that the variation of intensity of a voxel over time has much higher frequency components than the feature ratings. Therefore, the first step in the preprocessing phase consists on the filtering of high frequency noise for all voxels in the brain images by a low pass filter, i.e., a moving average window. The dimension of the moving window was 5 time steps. The reason for the choice of this window size is that it preserves the important part of the signal without losing those frequencies that also appear on the ratings (see Fig. 1 for an example of signal smoothing).

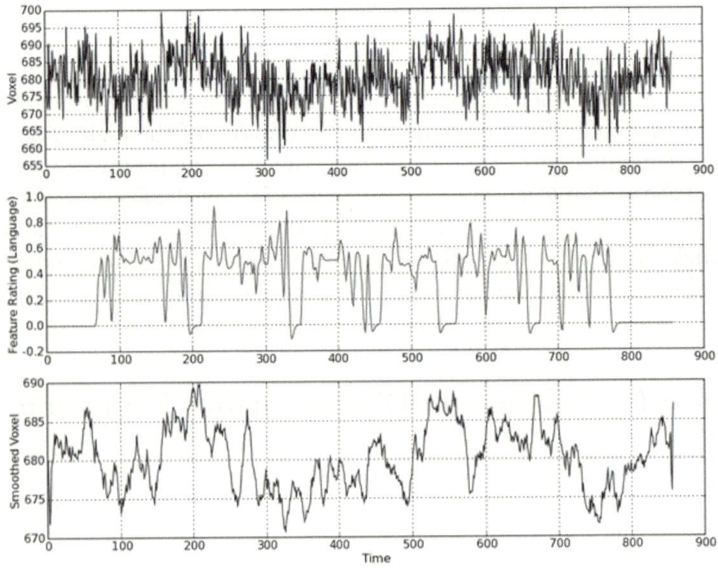

Fig. 1. The top graph describe the temporal behavior of a voxel randomly selected in the brain. The middle graph describes the temporal trend of the Language feature rating. The bottom graph is the result of low-pass filtering applied to the top graph.

3.2 Feature Selection

The second preprocessing step is the selection of the most informative voxels in order to discard all the voxels not informative for the task. We adopted the mutual information measure to evaluate the informativeness of a voxel with respect to the desired target. The feature extraction was conducted as follows. For each feature rating, we found its mutual information to the value at every voxel in

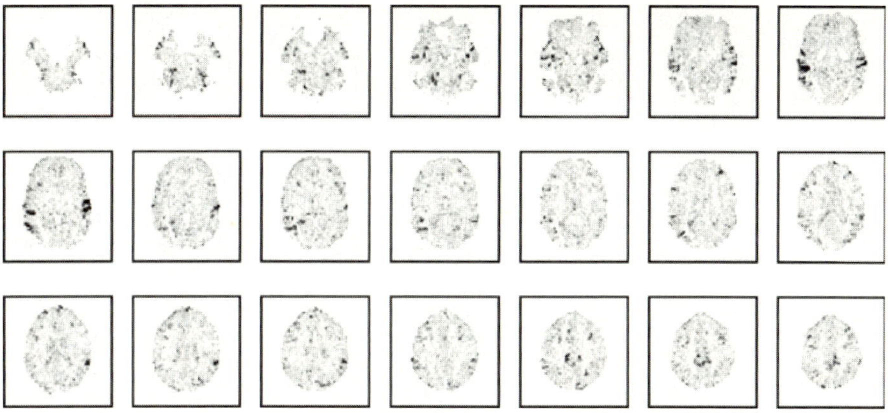

Fig. 2. Mutual information of some slices of brain image of subject 1 against Language feature ratings. The scale goes from 0.0 (bright voxels) to 0.4 (dark voxels). Notice the evident highlight of the hearing center in the temporal lobe of both left and right hemispheres

the image[5], separately for every subject. For each subject and feature rating, we ranked the voxels according to their mutual information and we selected the best 10% voxels obtaining a subject-feature mask of approximately 10^4 voxels. Figure 2 shows an example of the values of mutual information computed for the voxels in some slices of a brain image.

3.3 Features Extraction with Voxel Clustering

The third step in the preprocessing phase consisted in a further reduction of the dimension of the brain representation from 10^4 to 200 attributes. This reduction is obtained by grouping the voxels into a smaller set of representatives. To obtain these representatives we clustered the voxels with a simple *k-means* algorithm. However, in the neuroscience community there is still a debate about whether the cognitive processes are centered in a few specific and well-organized areas of the brain, or are distributed in many smaller and sparse agglomerates of neurons. Hence, we decided to adopt a measure of similarity taking into account these two possibilities. Our clustering algorithm therefore is designed to group voxels which are both near in space and similar in the temporal trend. This is obtained adopting a distance measure able to combine these two kinds of information:

$$d = d_{spatial}^{\alpha}(1 - r_{temporal})^{1-\alpha}, \tag{1}$$

[5] To compute mutual information between the time-course of each voxel and a given feature rating we quantized both signals (50 steps for voxels and 16 steps for feature rating) and estimated probabilities with the help of Laplace smoothing. The choice of the quantization grids has been motivated by two opposite factors: representing signals without losing relevant information and reducing estimation problems.

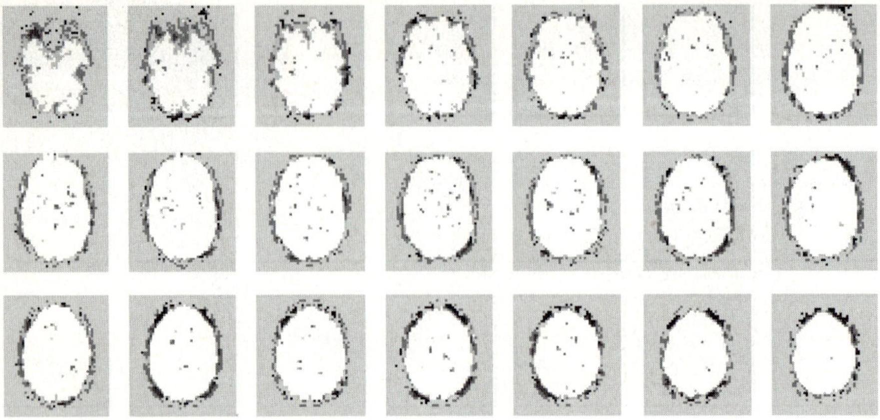

Fig. 3. Example of an important cluster (with approximately 300 voxels). The voxels are organized into sparse small agglomerates.

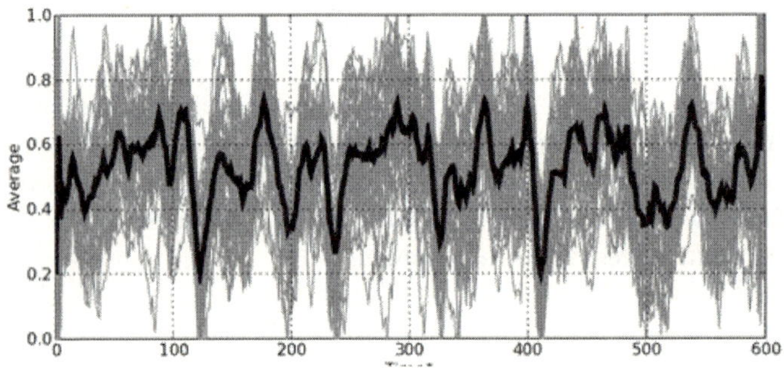

Fig. 4. The dark black line is the average behavior of all the voxels belonging to a cluster

which is a geometric mean of a standard Euclidean distance (d_{spatial}) that uses the 3D coordinates in the brain, and the Pearson's correlation over time (r_{temporal}). The weighting factor α can be used to give priority to space or correlation. With $\alpha = 1$ only the Euclidean distance is used, hence a spatial clustering is performed. On the contrary, with $\alpha = 0$ only correlation is used, performing in this way a sort of temporal clustering. In our experiments we gave the same importance to spatial distance and correlation (i.e., $\alpha = 0.5$).

In both movies there are some scenes lasting for a few seconds each, where nothing was projected on the screen[6]. Since the location of these blank sections was known, we decided to remove these parts from the brain image sequences while computing the correlation. This was done to eliminate noise due to the possibly random states of the brain during these blank sections.

[6] These "blank" parts were used during data collection to refine the calibration of the MRI machine.

Each cluster is therefore a spatio-temporally proximal set of "informative" voxels. Figure 3 shows an example of a cluster of voxels determined with the above processing. The example shows that the voxels in the cluster are sparse agglomerates in the brain. At the end, the representation of the brain at a given time instance, i.e., an fMRI volume, was reduced to 200 features. These features were computed as the average intensity of all the voxels in the image associated to the corresponding cluster (Figure 4 shows an example of how the voxels in a cluster participate to the creation of a unique average feature). We constructed the feature attribute datasets for each subject and feature rating.

After preprocessing, the data was normalized to improve the quality of the models. We decided to perform a linear scaling of the feature ratings to the interval [0, 1]. The brain data was, instead, normalized according to mean and variance[7]. For each of the 200 features, the mean was forced to 0 and the variance to 1.

4 The Prediction Models

As previously mentioned our intention was to let the "data do the talking", i.e., to let the data determine whether it was possible to find useful dependencies between brain activation and cognitive states. Therefore, excluding the spatio-temporal assumption made for "data compression" during clustering, we made no any other particular assumption on the distribution of data or on the dependencies between brain activation and stimuli. Starting from this hypothesis, we did not know whether there exists any linear or non-linear dependence between the brain activations and the desired output. For this reason, we preferred to select a non-linear model. Clearly, the drawback was that a bad choice of the model complexity (i.e. the number of free parameters) can cause a severe over-fitting of the model on the given data if the data is not as complex as the model. Our choice went to Neural Networks, even considering that these models cannot help to identify the regions of interest in the brain (regions devoted to specific cognition tasks).

Moreover, in the data there are at least two kinds of temporal dependencies inherent in the above brain activation sequences. The first kind of dependence is what we like to call "latency" (or "inertia" of the scenes). Since the subjects are watching a video, there are few drastic changes in the scene from one frame to the next. We can assume that the features appearing in a scene of the movie (music, a face, a location, some food, etc.) have some persistence, i.e., they will last for a certain time in the movie. Hence, the "instantaneous" forecasting given the current input can be reinforced by the hypothesis made in the past scenes.

The second kind of dependence encoded within the sequence can be referred to as "adaptivity" of the brain. When a stimulus or a combination of stimuli arrives to a subject, his/her brain is activated in certain locations according to the cognitive process. These locations and the strength of activity, however, change in time according to changes in the cognitive process. For example, the

[7] All 200 features were normalized with $x' = \frac{x-\bar{x}}{\sigma}$.

first time that a subject sees an actor in a movie his/her brain may be involved in solving several cognitive problems (identifying who is the actor, comparing the appearance with respect to his/her memory of past movies, memorizing the face if unknown, and many other specific unconscious activities). As far the actor continuously appears in the movie, the cognitive efforts change in time, maybe reducing to just recognizing the character. The idea is that the brain activation can slowly change in time for the same stimulus.

Both the above-described varieties of dependence motivate a temporal analysis of the data. Hence, we believe *recurrent neural networks* (*RNN*) to be appropriate models both because of the ease with which time dependence can be modeled as well as their ease of learning or estimation.

As comparison versus *RNN* we also tested other linear and non-linear models. The first obvious choice was to compare the *RNN* versus the *multi-layer perceptron* (*MLP*). This was necessary to see whether the hypotheses on temporal dependencies were sustained by experiments. Then we decided to evaluate also a very simple *general linear model* (*GLM*) described by linear equations:

$$Y = XA \tag{2}$$

where the parameters A are determined with an ordinary least squares estimation. The reason for this evaluation is that this is a reference model in the neuroscience community; hence, it was quite natural to consider it as a baseline.

The other model we tested is a *k-NN* that, together with *SVMs*, is considered the state-of-art for the current application. Many authors showed that for the current task *k-NN* frequently gives better results than *SVMs*.

4.1 Experimental Setting

We used the data of the two movies both for training and for testing adopting a cross-validation approach. Since each movie is composed of segments separated by intervals of blank video, we used these blank intervals to split the two movies into 12 consistent segments (6 for each movie). Each of these segments is made of approximately 100 temporally ordered samples. Then we performed the leave-one-segment-out training on all possible combinations of 11 segments, iterating the test set over all the 12 segments. Each feature rating was predicted separately with preprocessing of data and training of the model performed separately for that rating.

Independently of the model one of the major problems of the neural networks is the choice of the network topology (i.e., the number of free parameters). During our experiments we observed that very few hidden units were usually enough to create overfitting problems both for *RNN* and *MLP*. Hence, to avoid overfitting we decide to adopt cross-validation as stopping criterion. From the 11 training segments, we were holding-out 2 randomly selected segments used to stop the back-prop training algorithm on the remaining 9 segments. In particular *RNNs* were trained with a standard *back-propagation through time*. Both networks were using the hyperbolic tangent output function for hidden units and the logistic output function for the output unit. The recurrences in *RNN*s were only on the hidden layer. Weights were randomly initialized in the interval [-0.05, 0.05] and updated with a dynamic learning parameter and with moment.

The evaluation of the results was done using the Pearson's correlation between estimated and real feature ratings as suggested by the competition guidelines. The reason for this measure of correctness is that we have sequences of real values to compare. In this case, there is not any kind of correctness but just similarity; hence, it was not possible to determine a standard precision/error evaluation. The *NN*s models, due to their indeterministic behavior, were experimented with 5 trials and the results were then averaged.

5 Results and Discussion

Both *MLP* and *RNN* were tested with 4 hidden units, still having severe problems of overfitting. Regarding *k-NN*, since this is a regression problem, we have seen that the best solution was to average the ratings of the most similar k brain activations. Aalmost all studies of *k-NN* applied to this task are designed on controlled experiments, were the subjects are repeatedly presented with a controlled set of stimuli (usually positive and negative). In this task, however, there is not control on the sequence and the combination of stimuli, because they are consequence of a real experience (watching a movie). Hence, a lot of noise appears in the brain signals. For this reason, the best solution was to smooth the noise averaging the best k ranked elements. The average was weighted with values inversely proportional to the corresponding Euclidean distances. In particular, we discovered that for all features the best results were with $115 \leq K \leq 120$. We chosen $K = 118$.

In Tab. 1 are shown the average results of the different models for the 5 different feature ratings. Apparently, non-linear models are always better than linear models. Moreover, the exploitation of temporal autocorrelation, as expected, gives a little improvement in the quality of results.

Table 1. The results of the 4 models over all feature rating and their average

	Amusement	Body parts	Faces	Language	Motion	Average
GLM	0.209	0.327	0.311	0.426	0.381	0.331
k-NN	0.087	0.334	0.394	0.439	0.446	0.340
MLP	0.285	0.432	0.468	0.605	0.506	0.459
RNN	0.306	0.446	0.480	0.621	0.543	0.479

6 Conclusion

The PBAIC team provided an extremely rich data set that includes complex spatial and temporal dependencies among brain voxels and cognitive states. For the first time it was possible to evaluate the performance of a learning model across many and cognitive functions. Our preliminary study shows that non-linear models as feed-forward and recurrent neural networks are effective in dealing with the complex task of decoding brain states as acquired by fMRI scan. The behavior has been shown quite stable with respect to different classes of cognitive tasks.

The slight enhancement introduced by the recurrent neural networks suggests that the extraction of relational knowledge, both at temporal and spatial level, remains an open challenge. Anyway, it is still not clear whether the temporal response of the brain is different with respect to concurrent stimuli rather than a single stimulus.

References

1. Editorial, B.: What's on your mind. Nature Neuroscience 9(8) (2006)
2. Kamitani, Y., Tong, F.: Decoding the visual and subjective contents of the human brain. Nature Neuroscience 8(5), 679–685 (2005)
3. Haynes, J.D., Rees, G.: Decoding mental states from brain activity in humans. Nature Neuroscience 7(7) (2006)
4. Mitchell, T.M., Hutchinson, R., Niculescu, R.S., Pereira, F., Wang, X., Just, M., Newman, S.: Learning to decode cognitive states from brain images. Machine Learning 9(8) (2004)
5. Wang, X., Hutchinson, R., Mitchell, T.M.: Training fmri classifiers to detect cognitive states across multiple human subjects. In: International Conference on Neural Information Processing Systems Foundation (2003)
6. Vishwajeet Singh, K.P., Miyapuram, R.S.B.: Detection of cognitive states from fmri data using machine learning techniques. In: Proceedings of Twentieh International Conference on Artificial Intelligence, pp. 587–592 (2007)
7. Schneider, W., Siegle, G.: Pittsburgh brain activity interpretation competition guidebook (2006), http://www.ebc.pitt.edu/2006/competition.html

Modelling the N2pc and Its Interaction with Value

John G. Taylor and Nickolaos Fragopanagos

Department of Mathematics, King's College, Strand, London WC2R2LS, UK
{john.g.taylor,nickolaos.fragopanagos}@kcl.ac.uk

Abstract. Attention and emotion are closely interlinked and recent results have shown some of the neuro-physiological details of the effects of attention on emotion through the distractor devaluation (DD) effect. We develop a possible neural attention control architecture to explain the DD effect, and show by specific simulation how the N2pc (an early component of attention movement) can be encoded to produce encoding of devaluation of distractors.

Keywords: Attention, Emotion, Competition, Feedback, Devaluation by Inhibition, Orbito-Frontal Cortex, Face Processing.

1 Introduction

Attention is now appreciated as acting as a filter on complex environments so as to reduce the effects of distractors. The manner in which such distractors are handled under attention has been explored recently through analyses of the early signals associated with the N200, an indicator of specific brain activations at about 180-300msecs after a stimulus input. For stimulus inputs which are roughly symmetrical across the midline, a distinct increase in inhibition has been detected contralateral to the target or alternatively to the distractors (if salient enough) as part of the N200 signal [1, 2, 3]. The difference between the contralateral and ipsilateral activations is termed the N2pc (standing for 'N2 posterior contralateral'), and has been suggested to be an indicator of the initial focussing of attention [4, 5]. As such this signal is of great interest in understanding in more detail the dynamics of attention, either as arising through amplification of the target activity, or by means of distractor inhibition or from both forms of activity.

In association with the distractor inhibition interpretation of the N2pc there has been discovered a change in emotional valuation of the distractor stimulus [6, 7]. Subjects were shown a pair of faces, one of which had to be selected as being of a particular gender and then its colour reported on. The trustworthiness of the distractor face was found, on a subsequent test, to be reduced compared to the target. Numerous experiments have duplicated this effect (termed distractor devaluation or DD for short), although some of its characteristics still need to be uncovered.

Recently data on the N2pc were analysed in relation to such devaluation [8], so as to help develop a model of the underlying mechanism of the devaluation process. The data involved not only observations of the N2pc arising from processing of a target (as a face of a specific gender) and a distractor face, but also of the correlation of the size of the N2pc with later evaluation of trustworthiness of the distractor face: the

J. Marques de Sá et al. (Eds.): ICANN 2007, Part II, LNCS 4669, pp. 879–888, 2007.

larger the N2pc, the larger the devaluation of the distractor. This result thereby provides an important experimental clue as to how the devaluation process might be occurring in the brain.

In this paper we will develop a simple neural architecture to begin to model the N2pc itself, as well as propose its extension to be able to handle the more delicate question of encoding the level of distractor inhibition in appropriate parts of the brain so as to produce a devaluing effect of the encoded inhibition. At the same time we are interested in the mere exposure ((ME) effect [9]), for which there is very recent relevant data in [10]. Several conjectures have to be made here, beyond simplifications assumed in the visual architecture, in order to model the N2pc and the associated distractor devaluation. These conjectures lead to questions/predictions concerning the manner in which limbic system value codes can be manipulated by use of the inhibition activated during attention search. We can obtain a rough fit to the empirical data of [8] but considerably more experimental work must be done to answer some of our questions and allow the model to be suitably refined.

2 Methods

2.1 General Architecture

The modules involved in attention control are known to be in parietal and pre-frontal sites. Brain imaging results indicate the goal-nature of the pre-frontal cortical activity, with the attention control signal being generated, for spatial attention movement, in the parietal cortex, very likely in the superior parietal lobe (SPL) [11, 12, 13]. For feature/object based attention nearby sites, such as the temporo-parietal junction (TPJ) have been observed active under feature-based attention control [14]. The stimulus representations are well known to be created in occipital and temporal cortices. At the same time trustworthiness evaluation is expected to involve limbic sites in orbito-frontal cortex and amygdala, as well as components such as insula and superior temporal lobe [15].

The simplest attention control architecture is thus that of ballistic control, in which there is a goal region (in prefrontal cortex) biasing stimulus input (in a spatially independent manner (associated with face representations) so as to generate (in superior parietal lobe, fed by the spatial maps of the face inputs) feedback to cause attention amplification/inhibition of posterior cortical activity of stimulus representations in occipital and temporal cortices in a spatially dependent manner. The information flow in this ballistic model is:

$$\text{Goal module} \rightarrow \text{Attention signal generator} \rightarrow \text{Posterior cortex (stimulus activity)} \quad (1)$$

In this flow diagram the goal module receives its activation as arising endogenously from earlier task conditions, so that suitable goal nodes will have been activated prior to the stimulus input so as to prepare the overall system for suitable target inputs. In the face input case, there will be a node for the suitable face gender active, so biasing, by the feedback in (1), the input to lower level cortex of stimulus input. Since the side on which the target face is to be presented is only specified once the input enters, there will also be bottom-up control of the spatial focus of attention, as studied, for example, in [16]. This will activate top-down attention control in the posterior parietal cortex (PPC) [17, 12] biased by the spatial positions of the faces, so as to provide an attention feedback which generates hemispheric asymmetry in the processing at the level of posterior cortex (which we take here to be V4). Finally we include value maps that are negatively influenced by the endogenously driven attention to the distractor nodes in V4 through attention feedback, as shown in (2) and discussed in more detail later. Thus the more complete architecture for this mixed endogenous/exogenous attention paradigm is as in the flow diagram (2) below:

$$\text{Object goal module}\ \langle \begin{array}{lcl} \to & \text{Value maps} & \\ \to & \text{Object attention control} \to & \text{Feature maps} \\ & \text{Spatial attention control} \leftrightarrow & \text{Feature maps} \end{array} \qquad (2)$$

2.2 Experimental Paradigm

The detailed paradigm to which we are applying the above architecture is that of [8]: a sequence of face pairs is presented to each subject for 200 msecs while they are fixating centrally, with each face on either side of the central fixation cross. The subject is required to attend to the faces so as to determine which face has a particular pre-assigned gender (M or F), and then to report the colour of the chosen face. At a later stage, 1200 msecs after the gender response, they are shown a copy of a face (either M or F) and asked to rate its level of trustworthiness for them, on a scale from 1 to 5). It was observed in [8] that there was an N2pc in relation to the distractor faces, which was well correlated to the level of trustworthiness, with a low value (more untrustworthy) corresponded to a larger N2pc.

This result could be interpreted, as noted by [8], to there being a larger inhibition created by the more untrustworthy (distractor) faces, as evidenced by the size of the N2pc. The N2pc was thereby acting as an indicator of the inhibition needed to prevent distractor interference, and this inhibition resulting in reducing the value attached to the distractor face.

2.3 Specific Architecture

The overall architecture of the model is based on substantial volume of research on the perceptual and attention systems and a reasonable amount of research on the emotion system. The links between the perception/attention system and the emotion system in the model are also broadly based on the known anatomical connectivity from both primate and human neuroanatomy. However, we have made a number of assumptions regarding the more detailed characteristics of these links that were motivated mainly by the extant literature on the DD effect that, itself, provides primarily behavioural results. As such, the model goes beyond a mere reproduction of the extant results from the DD effect studies by proposing a range of predictions that can be tested experimentally both behaviourally and neuroscientifically.

The attention control system needs to be influenced by the stimulus input so as to achieve the goal. This latter is endogenous: 'to detect the appropriate gendered face and then report its colour.' We assume this detection proceeds by using bottom-up guidance of spatial attention biased by a top-down influence from the prefrontal cortex (PFC) (as schematise in (2) above), and given in more detail along the lines of figure 1.

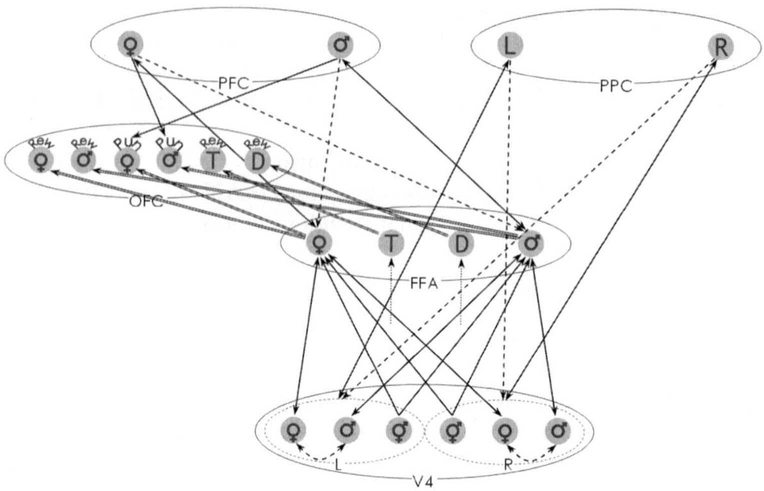

Fig. 1. The overall architecture of the DD model. Solid lines correspond to excitatory connections while dashed lines correspond to inhibitory connections. Double lines indicate connections with weight adaptation according to modified hebbian described in main text. The blobs with biological gender-symbols correspond to face-gender-feature-sensitive neurons (V4) or face-gender-sensitive neurons (FFA, PFC, OFC). Blobs with letter L(R) correspond to left(right)-hemisphere-sensitive neurons while blobs with T(D) letters correspond to target(distractor)-coding neurons. OFC blobs with Rew(Pun) arched above them are coding for reward(punishment) of blob object. V4 module has left and right hemispheric components indicated by L and R dotted-line-defined groups. For details of nature of coding in various modules see main text above.

The various levels of encoding in the various modules of figure 1 are as follows:

1) The visual area 4 (V4) module consists of a left-hemisphere and a right-hemisphere part that each process the contralateral face (ilpsilateral activations can be neglected for simplicity). Each hemispheric part contains three neurons, one coding for male features, one for female features and one for non-gender-specific or ambiguous features. In figure 1 the latter is represented by a superposition of the male and female biological symbols whereas the male-feature–coding neurons are represented by the male biological symbol and the female-feature-coding neurons by the female one. We choose here only one node for each of these, but note that these nodes represent a cluster of living neurons. Such single node representations should therefore act as graded neurons and not spiking neurons, since we are averaging over the responses of thousands of neurons at a time. A second set of nodes also code more generally for faces without any gender specificity (the hermaphrodites); these are also included in the module.

2) The neurons in the fusiform face area (FFA) module are coding for gender only, as a cluster responding to male of female faces (so denoted by the male and female biological symbols respectively). They have receptive field of the whole display (across both hemispheres). They feedback additively to the gender-sensitive nodes in V4. This feedback is both excitatory to same-gender sensitive nodes in V4 and inhibitory to all other nodes in V4 of any sort, with greatest inhibition to those nodes representing the opposite gender.

3) The neurons at PFC level code in a similar way to those in the face fusiform area (FFA) only for gender; there are thus only two nodes (also denoted by the male and female biological symbols respectively) in each module PFC, FFA, with no spatial specificity. These feed back additively to the gender-sensitive nodes of the same gender in FFA, together with inhibitory feedback to all other nodes of any sort.

4) The PPC module for the attention feedback signal, codes only for spatial position, so is composed only of the two nodes denoted L and R.

5) The orbitofrontal cortex (OFC) module has two neurons for each gender, one coding for reward and the other for punishment [18], although it could still be that there is only one node for each gender coding for its overall value, and that these lie only in one or other of the regions distinguished by [18] as coding for reward or punishment separately. We discuss this coding in more detail later.

We note that there is lateral feedback inhibition both from the PFC nodes to those coding for the complimentary gender from PFC to FFA and from a given hemisphere to the opposite one from PPC to V4 with spatial topography. There is also lateral inhibition in V4 between nodes coding for the features of a given gender. In real brains, this would be achieved by local inhibitory interneurons activated by the neurons coding for the opposite gender features. In our model, there is simply an inhibitory input to any node of the network that can be driven by other nodes either within or across modules. There may also exist lateral inhibition between the left and right nodes of the PPC module; however, we haven't included this in our model at the time, as the contralateral feedback inhibition generated by these nodes was sufficient to achieve target-distractor discrimination. Finally, we must note that suitable delays have been added in the model between the input and the V4 and FFA modules to simulate the known ventral visual path latencies reported in the relevant literature. Further delays could be added between the other modules as well, however, the graded response neurons we used for these simulations are not particularly sensitive to small inter-modular delays so they were left out at this point. If the model was implemented using more detailed spiking neurons, more care would have to be taken with the signal delays across the various modules and subtle differences such as the ipsilateral versus contralateral (interhemispheric) signal delays should be considered and accounted for.

2.4 Interpreting the N2pc

Due to ambiguities of uncovering the underlying neural activity from surface-measured ERPs (related to vertical flows in multi-layered cortical sheets as well as variations in the cortical surface with respect to the skull), we therefore do not attempt to model the signs of the appropriate ERPs, but only their magnitude. The mechanism for the N2pc is as follows. The target gender goal in PFC sends spatially independent feedback to lower levels, and most specifically to V4 (either directly or through FFA). In the scenario of [8] distractor shapes could be all possible faces or orientation conjunctions of importance in face detection for the opposite gender.

A further important contribution to the competition arises from PPC to direct the spatial attention focus, in an exogenous manner, to the site of nodes in V4 coding for the target face, together with inhibition to the opposite hemisphere; this contribution will be spatially asymmetric, whereas the feedback from PFC/FFA is spatially symmetric. The resulting competition on V4, spatially biased by feedback from PPC and gender-wise from PFC/FFA, will involve the outputs from the target-contributing node, in either hemisphere, being sent from V4 to PPC (and back to V4 by recurrence). The target-contributing nodes in V4 will be boosted by the PPC attention feedback, the others in V4 on the opposite hemisphere being inhibited.

The N2pc will therefore be created through the excess of activation in V4 in the target hemisphere as compared to that on the opposite hemisphere, as brought about by the target-representing nodes on the target side winning the overall competition in their modules, being boosted recurrently by the nodes in PPC coding the spatial co-ordinates of the target face, with larger inhibition to the hemisphere containing the distractor face (which is not helped by the endogenous excitatory bias from PFC/FFA).

To conclude, the hemispheric activity difference suggested as being at the base of the N2pc is that brought about by the excess of feedback inhibition over excitation from PPC to V4 in a given hemisphere, through the greater extent of the feedback inhibition from the target side of PPC as compared to the inhibition on V4 from the distractor side of PPC to the target hemisphere. This asymmetry is boosted by feedback excitation to V4 from the target face goal nodes in PFC/FFA; this is spatially symmetric, but helps bias the excitatory feedback from PPC to the target side of V4.

2.5 Detailed Mechanism of Devaluation

To develop a devaluation mechanism we must note that the evaluation phase takes place under quite different attention conditions as compared to the initial target search phase. In the evaluation phase only the distractor is presented as a stimulus, so that there will not be any lateral competition on PPC. Nor is the stimulus to be valued in the same spatial position, it now being at fixation as compared to its previously being to left or right of fixation. Thus valuation encoding of inhibition to the distractor face representation must have occurred in a spatially-independent or spatially-spread manner. It would thus be expected to arise from ventral inhibition, say at PFC or FFA level, but be encoded, say in OFC, where spatially invariant values of stimuli are stored. Such inhibition was in any case supposed to be the source of the N2pc, so it is the level of that inhibition which must somehow drive the devaluation process, and explain the correlation between the size of the N2pc and the DD level.

We take the relevant inhibition to arise from PFC, with inhibition onto the distractor node in FFA or V4, say, from the target node in PFC. These will determine the level of relative bias given to the target and distractor nodes in V4, hence the level of the N2pc. At the same time this inhibition is encoded as a modification of value (so trustworthiness) in OFC.

We assume that the value of a stimulus is encoded in OFC by stimulus input to an excitatory node, whose output is determined by the predicted reward available form the stimulus. Associated with the excitatory node is an inhibitory one, which can reduce the level of output of the excitatory node as appropriate from later experience.

Using this pair of nodes coding for reward, it is possible to achieve distractor devaluation by either an 'active' mechanism or an 'adaptive' one. The former uses the continued activity of the inhibitory inter-neuron in OFC so that it feeds continually to its relevant excitatory stimulus node so as to lower its value; on a later trustworthiness evaluation the value reported, being modulated by the inhibitory input, will thereby be reduced. On the other hand for the adaptive mechanism it is assumed there is some longer term modification (increase) of synaptic weights to the inhibitory input, so producing a devaluation of trustworthiness in later testing. These two mechanisms are quite different at cellular level, and also could, we expect, be differentiated by testing the time course of the DD effect as the lag between the attention task and the valuation task is increased from the initial 1200 msecs in [8], say to several hours or days (when the continued inhibitory node activity will be expected to have abated). We present results below for only the synaptic learning mechanism.

We assume these long range axons from PFC to FFA or V4, which have collaterals going to an inhibitory interneuron accessing the M or F node in V4, also send their collaterals to the similarly coded nodes in OFC. This could arise by correlated signals in V4 and OFC occurring when either the M or F face is presented, so helping modify the connections onto OFC in the present paradigm. Thus when the representation of the distractor suffers inhibition in V4 a similar learning of this reduced reward will occur by learnt increase of the synaptic strength from the PFC target node onto that inhibiting the distractor-coded node in OFC; this will devalue the distractor, as required.

Details of the longer-term adaptation mechanism assumed to be at the basis of the DD effect are by the information flows:

Long range axon collateral from FFA → Inhibitory node in OFC → excitatory output node in OFC (3)

We denote by w the connection strength of the input from FFA to the inhibitory node in OFC. Then the long-term synaptic modification equation is taken to be of simple Hebbian form:

$$\tau \, dw/dt = -w + \text{Hebbian term} \quad (4)$$

where the Hebbian term is proportional to product of the input to the inhibitory neuron and its output. The resulting increase in w due to the inhibition of the distractor will cause an increase in w (by equation (4)) and hence bring about DD by greater inhibition of the relevant OFC excitatory output node.

The model can be extended to take into account the trial-to-trial variations caused by individual stimuli by addition of noise to the input nodes.

2.6 Detailed Mechanism of the ME

We have also extended the model to incorporate the mere exposure effect (ME). This is achieved by use of a similar Hebbian learning to (4) for the connection weights of activity arriving from given inputs onto the FFA. These will be biased to be a larger increase under attention for the target and a smaller on for the distractor, so will lead to larger inputs for targets than distractors; this will be expected to lead to easier processing and hence an increased familiarity effect [9].

3 Results

We are able to reproduce N2pc size variation leading to DD size variation as in [8]. In figure 2 are presented the dynamical flows of activity in V4 for the two different values of the inhibitory feedback strength of 0.2 and 0.8; in table 1 we give the N2pc levels (the middle line) for the inhibitory strengths of 0.2, 0.4, 0.6 and 0.8. We see that in each of figure 2 there is a faster and higher peak of such activity on the contralateral side to the target, as reported in [8].

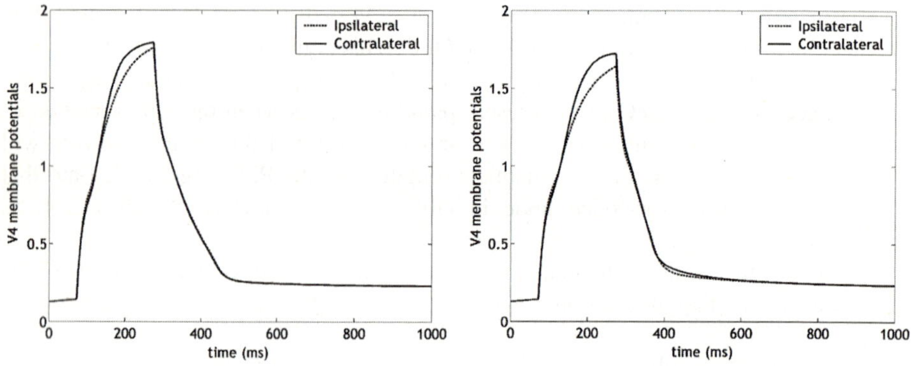

Fig. 2. N2pc for ventral attention inhibition strength 0.2 (left) and 0.8 (right)

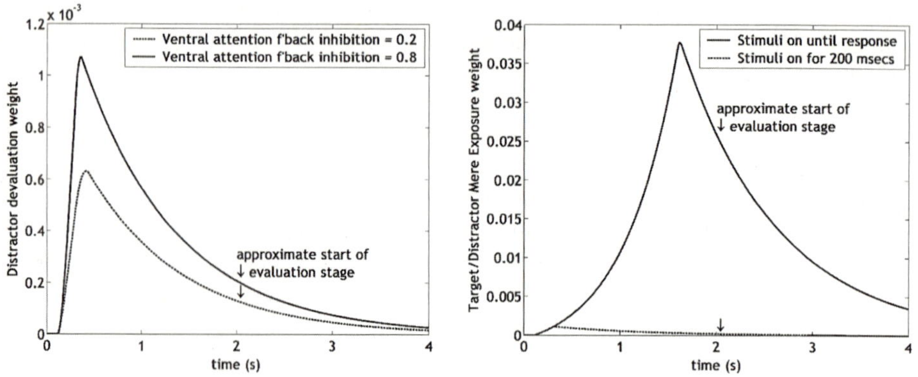

Fig. 3. Weight adaptation for distractor devaluation (left) and mere exposure (right)

In figure 3 (left) is plotted the temporal development of the inhibitory weights onto the relevant OFC nodes for the two extreme feedback inhibition strengths to V4 used in figure 2. In figure 3 (right) we present the weight adaptation for mere exposure for the two cases of stimulus only on for 200 msecs or on until response. Finally the main results are presented in tabular form in Table 1, with a clear correlation between the N2pc and the DD for the range of inhibitory feedback values. We assume that these strengths vary across the subjects, so fitting with the variations observed in [8].We note that the results presented here depend heavily not only on the general architecture but also on parameters chosen for the simulations. More experimental data is needed to be able to refine these choices.

Table 1. Simulation results for different values of ventral attention feedback inhibition strength

Ventral Attention Feedback Inhibition Strength	*0.2*	*0.4*	*0.6*	*0.8*
Contralateral – Ipsilateral activations	0.035	0.045	0.060	0.085
Distractor Devaluation Weight at Evaluation	1.309	1.338	1.555	2.075

4 Conclusion

We have presented a simulation architecture enabling a recent important result on the relation between attention and emotion to be simulated. The relation of distractor devaluation is robust, and has been tested in a variety of paradigms. We have had to make several assumptions.

Acknowledgements

This work has been partially supported by the Biotechnology and Biological Sciences Research Council (BBSRC), UK.

References

1. Woodman, G.F., Luck, S.J.: Electrophysiological measurement of rapid shifts of attention during visual search. Nature 400, 867–869 (1999)
2. Conci, M., Gramann, K., Muller, H.J., Elliott, M.A.: Electrophysiological correlates of similarity-based interference during detection of visual forms. J.Cogn Neurosci. 18, 880–888 (2006)
3. Hickey, C., McDonald, J.J., Theeuwes, J.: Electrophysiological evidence of the capture of visual attention. J.Cogn Neurosci. 18, 604–613 (2006)
4. Luck, S.J., Hillyard, S.A.: Spatial filtering during visual search: evidence from human electrophysiology. J.Exp.Psychol.Hum.Percept.Perform. 20, 1000–1014 (1994)
5. Eimer, M.: The N2pc component as an indicator of attentional selectivity. Electroencephalogr.Clin.Neurophysiol. 99, 225–234 (1996)
6. Raymond, J.E., Fenske, M.J., Tavassoli, N.T.: Selective attention determines emotional responses to novel visual stimuli. Psychol.Sci. 14, 537–542 (2003)

7. Raymond, J.E., Fenske, M.J., Westoby, N.: Emotional devaluation of distracting patterns and faces: a consequence of attentional inhibition during visual search? J.Exp.Psychol.Hum.Percept.Perform 31, 1404–1415 (2005)
8. Kiss, M., Goolsby, B., Raymond, J.E., Shapiro, K.L., Silvert, L., Nobre, A.C., Fragopanagos, N., Taylor, J.G, Eimer, M.: Efficient attentional selection predicts distractor devaluation: ERP evidence for a direct link between attention and emotion. Journal of Cognitive Neuroscience (in press)
9. Kunst-Wilson, W.R., Zajonc, R.B.: Affective discrimination of stimuli that cannot be recognized. Science 207, 557–558 (1980)
10. Goolsby, B., Raymond, J.E., Silvert, L., Kiss, M., Fragopanagos, N., Taylor, J.G, Eimer, M., Nobre, A.C., Shapiro, K.L.: Affective Consequences of Effortful Selection of Faces. Visual Cognition. Submitted
11. Corbetta, M., Shulman, G.L.: Control of goal-directed and stimulus-driven attention in the brain. Nat.Rev.Neurosci. 3, 201–215 (2002)
12. Corbetta, M., Tansy, A.P., Stanley, C.M., Astafiev, S.V., Snyder, A.Z., Shulman, G.L.: A functional MRI study of preparatory signals for spatial location and objects. Neuropsychologia 43, 2041–2056 (2005)
13. Kincade, J.M., Abrams, R.A., Astafiev, S.V., Shulman, G.L., Corbetta, M.: An event-related functional magnetic resonance imaging study of voluntary and stimulus-driven orienting of attention. J.Neurosci. 25, 4593–4604 (2005)
14. Giesbrecht, B., Woldorff, M.G., Song, A.W., Mangun, G.R.: Neural mechanisms of top-down control during spatial and feature attention. Neuroimage. 19, 496–512 (2003)
15. Winston, J.S., Strange, B.A., O'Doherty, J., Dolan, R.J.: Automatic and intentional brain responses during evaluation of trustworthiness of faces. Nat.Neurosci. 5, 277–283 (2002)
16. Hopf, J.M., Boelmans, K., Schoenfeld, M.A., Luck, S.J., Heinze, H.J.: Attention to features precedes attention to locations in visual search: evidence from electromagnetic brain responses in humans. J.Neurosci. 24, 1822–1832 (2004)
17. Yantis, S., Schwarzbach, J., Serences, J.T., Carlson, R.L., Steinmetz, M.A., Pekar, J.J., Courtney, S.M.: Transient neural activity in human parietal cortex during spatial attention shifts. Nat.Neurosci. 5, 995–1002 (2002)
18. O'Doherty, J., Kringelbach, M.L., Rolls, E.T., Hornak, J., Andrews, C.: Abstract reward and punishment representations in the human orbitofrontal cortex. Nat.Neurosci. 4, 95–102 (2001)

Biasing Neural Networks Towards Exploration or Exploitation Using Neuromodulation

Karla Parussel and Lola Cañamero

Adaptive Systems Research Group, School of Computer Science,
University of Hertfordshire, College Lane, Hatfield, Herts, AL10 9AB, U.K.
{K.M.Parussel,L.Canamero}@herts.ac.uk

Abstract. Taking neuromodulation as a mechanism underlying emotions, this paper investigates how such a mechanism can bias an artificial neural network towards exploration of new courses of action, as seems to be the case in positive emotions, or exploitation of known possibilities, as in negative emotions such as predatory fear. We use neural networks of spiking leaky integrate-and-fire neurons acting as minimal disturbance systems, and test them with continuous actions. The networks have to balance the activations of all their output neurons concurrently. We have found that having the middle layer modulate the output layer helps balance the activations of the output neurons. A second discovery is that when the network is modulated in this way, it performs better at tasks requiring the exploitation of actions that are found to be rewarding. This is complementary to previous findings where having the input layer modulate the middle layer biases the network towards exploration of alternative actions. We conclude that a network can be biased towards either exploration of exploitation depending on which layers are being modulated.

1 Introduction

In the brain, different levels of neuro-active substances modulate the sensitivity-to-input of neurons that have receptors for them [1, page 94]. Fellous [2] proposes that emotion can be seen as continuous patterns of neuromodulation of certain brain structures. Kelley [3] argues that in their broadest possible sense, emotions are required for any organism or species to survive. They allow animals to satisfy needs and act more effectively within their environment. She argues that emotions are derived from neurochemically coded systems. These systems have been present in one form or another throughout our evolutionary history. Emotions can be influenced by altering the levels of these neuromodulators in the nervous system.

Emotions also help the reasoning process [4]. Evans puts this idea in a game-theoretical framework in his search hypothesis [5], according to which, in a rational agent confronted to an open-ended and partially unknown environment, emotions constrain the range of outcomes to be considered and subjectively applies a utility to each. The search hypothesis can be seen as an example of an

J. Marques de Sá et al. (Eds.): ICANN 2007, Part II, LNCS 4669, pp. 889–898, 2007.

agent moving from exploration of possible outcomes to an exploitation of the action providing the current expected highest expected utility. However, the best course of action does not need to be learnt through experience. Nesse [6] defines emotions as specialised states of operation that give an evolutionary advantage to an agent in particular situations. LeDoux [7] describes a distinguishing characteristic of cognitive processing as flexibility of response to the environment. Emotions provide a counter-balance to this by narrowing the response of an agent in ways that have a greater evolutionary fitness. As an example, predator avoidance driven by fear is an ideal behaviour to be selected for and optimised by evolution. It is a behaviour that needs to be maintained until the prey reaches assured safety regardless of whether it is able to continually sense the predator or not [8]. Nor will the prey benefit from being distracted by less important sensory input while it is still in danger. Successful fleeing behaviour might not require exploration of different actions when instead, exploitation of known strategies for a successful escape should be given priority. On the contrary, positive emotional states are thought to promote openness to the world and exploration of new courses of actions [9].

2 The Agent

We have used the simplest possible agent to test the effect of neuromodulation when applied to an artificial neural network, an agent that cannot directly sense its external environment. It can only sense two critical resources of its simulated body which it must maximise. These resources are referred to here as 'energy' and 'water'. The agent can execute a set of actions that either increase or decrease by a given amount the energy or water level in the body, plus two neutral actions. Neutral actions are useful because if they are used differently to each other then it throws doubt on how well the agent is adapting. The 'inactive' action is used by default when an agent does not choose for itself. This can happen if no activation reaches the output neurons of its neural network. It results in each resource of the agent being reduced by the maximum cost. The effect of this is more costly to the agent than if it deliberately chose the most costly action available to it as that would only result in a reduction of one resource.

2.1 The Neural Network

The agent adapts using a feed forward neural network of spiking leaky integrate-and-fire neurons based on the model described in [10] and [1, page 339]. The network learns which outputs should be most frequently and strongly fired to minimise the subsequent level of input signal in the next turn. Each neural network is made up of three distinct layers; input, middle and output layer. The network is iterated over a fixed number of times within a single turn.

For each resource, the input layer has two neurons that output to the middle layer. One neuron signals the need for the resource and the other neuron signals the satisfaction of that need. There are situations in which an effective behaviour for an agent may be to decrease a need but not satisfy it. Alternatively there may be situations in which an agent needs to store more resources than it is used to doing. In these experiments the agent is tasked only with maximising its resources.

There is one output neuron per action. The action performed by the agent directly and immediately alters the level of a resource. This consequently determines the strength of the corresponding input signal fed to the network in the next turn. This is fed via the input neurons corresponding to the resource effected by the action. In this way the network acts as a minimal disturbance system [11] as it settles upon actions that reduce its total input activation.

2.2 The Neuron

Spiking neurons were used in the neural network, each one acting as a capacitor to integrate and contain the charge delivered by synaptic input. This charge slowly leaks away over time. The neurons have a fixed voltage threshold and base leakage which are genetically determined.

The neurons also have an adaptive leakage to account for how frequently they have recently spiked. If a neuron spikes then its leakage is increased by a genetically determined amount. If the neuron does not spike then the leakage is decreased by that same amount. Leakage is constrained within the range [0, 1]. The spiking threshold is the same for all neurons in the network and is constant. The neurons are stochastic so that once the spiking threshold has been reached, there is a random chance that a spike will be transmitted along the output weights; either way the cell loses its activation. The neurons send out a stereotypical spike. This is implemented as a binary output. The weights connecting the neurons are constrained within the range [0, 1]. The learning rule employed uses spike timing-dependent plasticity (STDP). The rule used here is implemented using a two-coincidence-detector model [12] Each neuron has its own post-synaptic recording function that is incremented when the neuron spikes and which decays over time in-between spikes. This is compared to the pre-synaptic recording function of the neuron that has transmitted the activation. Each layer of neurons has its own increment and decay rates determined prior to testing via automated parameter optimisation.

2.3 Modulators

Several variants of the network were created; either modulating or non-modulating. Used here, a modulator is a global signal that can influence the behaviour of a neuron if that neuron has receptors for it. The signal decays over time, specified by the re-uptake rate, and can be increased by firing neurons that have secretors for it.

Neurons that are to be modulated are given a random number of receptors. These can be modulated by neurons in other layers that have secretors for those modulators. The receptors modulate either the neuron's sensitivity to input or probability of firing. The effect of this modulation is determined by the level of the associated modulator and whether the receptor is inhibitory or excitatory.

Neurons can also have secretors. These increase the level of an associated modulator. The modulator re-uptake rate, the modulation rate of the receptors and the increment rate of the secretors is determined by artificial evolution along with many other parameters of the neural network before the model is tested.

2.4 Parameter Optimisation

The parameters of the networks were initially optimised using artificial evolution so as to make a fair comparison. Once these constrained evolutionary runs were finished the parameters were hard-coded and tested as a population of 450 agents in order to determine the average performance of the neural network. An average fitness is required because the mapping from genotype to phenotype is stochastic. This is due to the randomisation of weights and the connectivity between neurons. The fitness function used during parameter optimisation was $Energy + Water + Age - absolute(Energy - Water)$.

The absolute difference between the energy and water resource was subtracted from the fitness as both resources were essential for the agent to stay alive. The age was only used for the fitness function during the evolutionary runs and not used afterwards when comparing the average performance of agents with the optimised architectures. This is because agents would generally only die at the beginning of an evolutionary run before the architecture had been optimised.

3 Discrete and Continuous Actions

Modulating and non-modulating versions of the network were implemented and compared in [13]. In all the networks a winner-takes-all selection scheme was used. A single action was chosen each turn by determining the output neuron that had the strongest average activation over multiple iterations of the network. The difference in activation strength between the winning output neuron and the losing neurons was of no consequence. Nor did it matter how strongly the losing output neurons were activated.

If the network is to be used to drive the motors of a robot, or to provide input signals to other neural networks, then it needs to be able to balance the activations of all of its output neurons concurrently.

The previous experiments have used actions that each have one single discrete effect. In the experiments described here, the networks are provided with continuous actions whose effect depend upon the level of activation of the corresponding output neuron. The stronger the activation the greater the effect provided by the continuous action.

In a robot, discrete actions would be the equivalent of motors that either ran at full speed or were switched off. Continuous actions would be the equivalent of motors that ran at a speed determined by the level of the activation they received. The networks have to learn to provide the correct activation to all of the output neurons concurrently rather than only be concerned about which neuron is more strongly activated than all the others.

3.1 Exploratory Two-Modulator Network Optimised for Use with Discrete Actions

The modulating network analysed in [14] and [13] had two modulators, one to signal hunger and another to signal thirst. The neurons in the input layer each had a secretor for the modulator that corresponded to the resource the input neuron pertained to. The neurons in the middle layer had a random number of excitatory or inhibitory receptors for these modulators, see Fig.1a).

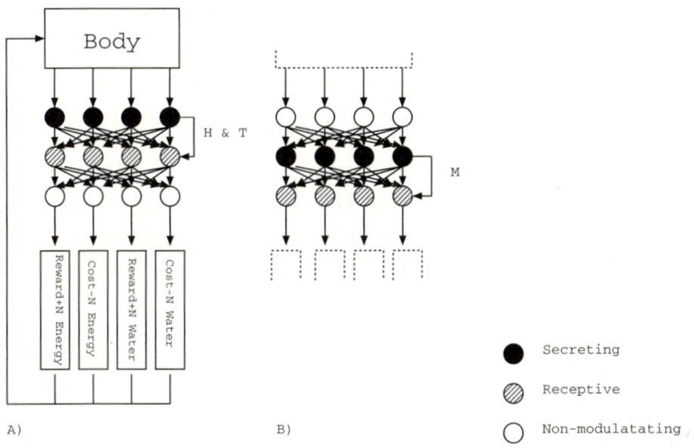

Fig. 1. The agent consists of a body that contains water and energy levels. A) Two-modulator agent: Hunger (and thirst) neurons increase the strength of the hunger (or thirst) modulator when they fire. Neurons in the middle layer have a random number of inhibitory receptors for these modulators. B) Single-modulator agent: Neurons in the middle layer increase the strength of a single modulator when they fire. Neurons in the output layer have a random number of excitatory receptors for this modulator.

Having the input layer modulate the middle layer was shown to increase exploration. As a consequence of this the performance of the modulating agent was slightly below that of the non-modulating network. Actions that were costly or neutral were less likely to be ignored throughout the evaluation period. But conversely, the modulating network was more able to adapt when the effect of actions changed.

3.2 Networks Optimised for Use with Continuous Actions

Many different variants of the network were implemented and tested. The aim was to find the best way of modulating a minimal disturbance network for use with continuous actions. Permutations included having the input layer modulate the output layer, using between one and four modulators and having layers modulate themselves. The parameter sets were optimised for use with continuous actions using artificial evolution. If the architecture performed particularly well then the parameters were hard-coded and tested more thoroughly.

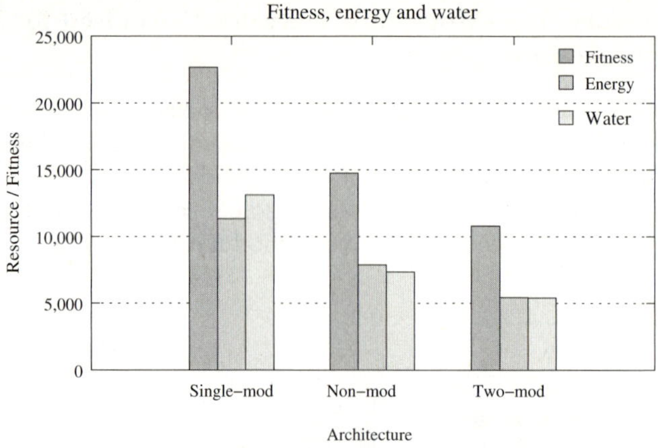

Fig. 2. Fitness, energy and water levels of the different architectures. In order of performance: single modulator (middle to output layer), non-modulating and two modulators (input to middle layer).

The best performing design used a single modulator secreted by the middle layer to modulate neurons in the output layer, see Fig.1b). The non-modulating and two-modulator architectures, originally optimised for use with discrete actions, were re-optimised for use with continuous actions. The parameters of all three architectures were hard-coded and tested using a population of 450 agents. The average fitness, energy and water levels for each architecture can be seen in Fig.2. The explorative behaviour of the two-modulator architecture carries a cost in performance when used in relatively stable environments as the agent tries other actions that have not necessarily proven successful in the past.

4 Adaptive Performance of the Networks

The synaptic weights between the input and the middle layer of the network can be thought of as providing 'activity diffraction' to allow the input signals to filter through the system at different speeds. The synaptic weights between the middle layer and the output layer can be thought of as providing 'activity integration', integrating those signals back into combinations that allow particular output neurons to fire more frequently than others.

Because activity filters through the network at different speeds, some output neurons will fire earlier than others. If an action is rewarding and subsequently reduces the input signal to the network, synaptic activity will be reduced for the other neurons and therefore will be less likely to fire. If an action is not rewarding, the input signal is not reduced, other neurons will eventually fire and other actions will be tried instead.

4.1 Input to Middle Layer Modulation

The hunger and thirst modulators of the two-modulator agent optimised for use with discrete actions inhibit the neurons in the middle layer. The strongest firing neurons have more activation to lose when being inhibited. These are also the neurons more likely to be firing the output neurons that lead to actions that reduce total input activity into the network. So by inhibiting the neurons in the middle layer the 'diffraction' of activation throughout the network is reduced and other actions have a greater chance of being performed. This increases exploratory behaviour.

4.2 Middle to Output Layer Modulation

The most successful network optimised for use with continuous actions has the middle layer modulating the output layer. The receptors of the output layer for the single-modulator network have all evolved to be excitatory. This suggests that modulation is used to excite output neurons that lead to rewarding actions. In other words, modulation is used to balance the outputs of the neural network.

Evidence for this comes from using the single-modulator network with discrete actions even though it has been optimised for use with continuous actions. It performs better than a non-modulating network optimised for use with discrete actions. Not only does the single-modulating network achieve greater average energy and water resource levels (energy=853, water=853) than the non-modulating network (energy=790, water=757), it also manages to avoid having more of one resource than the other.

With the non-modulating network, the more rewarding an action, the stronger the activation of the corresponding output neuron. In contrast, the single-modulating network only fires the outputs leading to rewarding actions and ignores the neutral ones even when there is no need to do so, (see Table 1).

Table 1. The average frequency of discrete actions chosen by a population of 450 agents. Two architectures are compared, the non-modulating architecture optimised for use with discrete actions, and the single-modulator architecture optimised for use with continuous actions.

Action	Amt	Resource	Non-mod freq.	Single-mod freq.
Inactive	-2&-2	E&W	0.0131111%	0.0948889%
Cost	-2	E	1.32978%	0.944%
Cost	-1	E	1.30733%	0.960889%
Neutral	0	E	2.42911%	0.994889%
Reward	+1	E	7.15533%	5.92333%
Reward	+2	E	37.9384%	41.2469%
Cost	-2	W	1.53378%	0.922667%
Cost	-1	W	1.652%	0.956222%
Neutral	0	W	2.60533%	1.00356%
Reward	+1	W	7.57089%	5.59422%
Reward	+2	W	36.4649%	41.3584%

This suggests that the single-modulating network provides reduced activations for all of its output neurons by default and uses modulation to excite the output neurons which are rewarding.

5 Exploitation vs. Exploration

Having the middle layer modulate the output layer helps the agent exploit the actions that are found to be the most rewarding whilst ignoring those actions that are neutral or costly. To further demonstrate this, the networks were tested using discrete cost / reward actions modified to work on the principle of 'use-it-or-lose-it'. This gives exploitative agents an advantage. The actions work as follows:

- If an action is performed for the first time then it provides its maximum effect.
- If the agent continues to perform that action then the it will continue to provide its maximum effect.
- If another action is performed then the potential effect of the original action will decrease each turn until it reaches a minimum regardless of whether the agent uses it or not. The minimum potential effect is anything less than 1 resource point. After the action reaches this minimum it will return to providing its maximum effect when used.

If an agent explores other actions and returns to the original action found to be the most rewarding so far, the effect of that action will be reduced for each turn that the agent performed other actions. If the agent continues to use that action thereafter, the effect will continue to be reduced each turn until it reaches a minimum. At this point the action returns to providing its maximum effect again.

Each network was tested using a population of 450 agents. They were tested 102 times; for each evaluation the ratio of the action's previous effect being retained was incremented by 0.01. For example, at a ratio of 0.5 the potential effect that an action can provide is halved each round once the agent stops exploiting it continuously. The actions are discrete so the agents can only pick one action per turn. This is the action whose corresponding output neuron has the strongest average activation.

The performance of the three architectures can be compared in Fig.3. It can be seen that the performance of each architecture declines as the ratio reaches 0.99. This is because once the agent stops using an action, it takes longer for the potential effect of using that action to reduce to the minimum before returning to its maximum level again. When the ratio reaches 1 the performance of all three architectures reverts to the same level as at 0. It is not plotted here for the sake of clarity.

The single-modulator architecture is the best performer with each agent in the population increasing their energy and water levels by the highest average amount each time. At a ratio of 0.99, the single-modulator architecture performs as well as the non-modulating architecture but the performance increases as

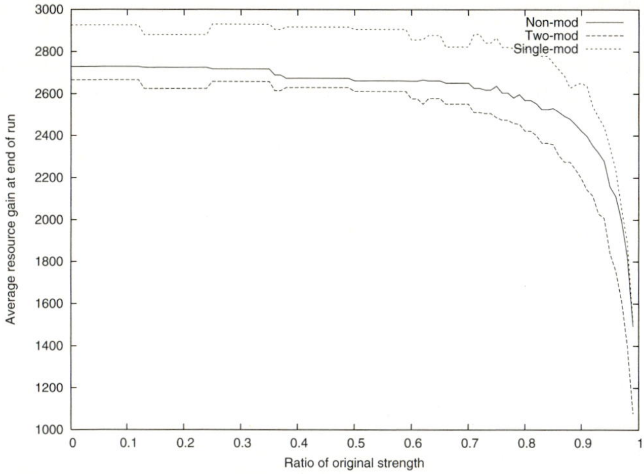

Fig. 3. Testing the networks using discrete use-it-or-lose-it actions show how well they cope with tasks benefitting from exploitative behaviour. The single-modulator network performs significantly better. The two-modulator network, previously shown to perform better at tasks benefitting from explorative behaviour, performs worst of all.

the ratio decreases. The two-modulator architecture performs worst of all. Its performance at a ratio of 0.99 is significantly below that of the other two.

6 Conclusion

Taking modulation as a mechanism underlying emotions, we have investigated how such a mechanism can bias an artificial neural network towards exploration of new courses of action, as seems to be the case in positive emotions, or exploitation of known possibilities, as in negative emotions such as predatory fear. Modulation can be used to both concurrently provide the correct activation to each neuron in the output layer, and to bias a network towards either exploration or exploitation.

If an emotion is merely a particular subset of neural functions found by evolution to provide the optimal behaviour for an agent given a certain environmental or bodily state, then those neural substrates need to be activated concurrently. Each neural function may also require a different degree of activation. This means that we may need a single neural network to find the optimal balance of activation for each of its output neurons so that it can later be used to drive other neural networks.

Further work is required to determine whether exploration and exploitation networks should be driven by a third, arbitrating neural network, and whether the correct network can be selected using neuromodulators. It may also be the case that a single neural network can be biased towards either exploitation or exploration at runtime, as in [9], by modulating the re-uptake rate.

References

1. Koch, C.: Biophysics of Computation. Oxford University Press, Oxford (1999)
2. Fellous, J.M.: The neuromodulatory basis of emotion. The neuroscientist 5(5), 283–294 (1999)
3. Kelley, A.E.: 3. In: Who needs emotions? The brain meets the robot, pp. 29–77. Oxford University Press, Oxford (2005)
4. Damasio, A.: Descartes' Error: Emotion, Reason, and the Human Brain. Quill (1994)
5. Evans, D.: The search hypothesis of emotion. British Journal for the Philosophy of Science 53(4), 497–509 (2002)
6. Nesse, R.: Evolutionary explanations of emotion. Human Nature 1(30), 261–289 (1990)
7. LeDoux, J.E.: The Emotional Brain. Simon & Schuster (1998)
8. Avila-García, O., Cañamero, L.: Hormonal modulation of perception in motivation-based action selection architectures. In: Avila-García, O. (ed.) Proceedings of the Symposium on Agents that Want and Like: Motivational and Emotional roots of Cognition and Action at the AISB-05 conference, The society for the study of artificial intelligence and the simulation of behaviour, pp. 9–16 (2005)
9. Blanchard, A., Cañamero, L.: Developing affect-modulated behaviors: Stability, exploration, exploitation, or imitation? In: Kaplan, F. (ed.) Proc. 6th Intl. Workshop on Epigenetic Robotics, vol. 128, Lund University Cognitive Studies (2006)
10. Wehmeier, U., Dong, D., Koch, C., van Essen, D.: Modeling the mammalian visual system. In: Koch, C., Segev, I. (eds.) Methods in Neuronal Modeling: From synapses to networks, pp. 335–360. MIT Press, Cambridge (1989)
11. Wörgötter, F., Porr, B.: Temporal sequence learning, prediction and control - a review of different models and their relation to biological mechanisms. Neural Computation 17, 1–75 (2004)
12. Karmarkar, U.R., Najariana, M.T., Buonomano, D.V.: Mechanisms and significance of spike-timing dependent synaptic plasticity. Biological Cybernetics 87, 373–382 (2002)
13. Parussel, K.M.: A bottom-up approach to emulating emotions using neuromodulation in agents. PhD thesis, University of Stirling (2006)
14. Parussel, K., Smith, L.: Cost minimisation and reward maximisation. a neuromodulating minimal disturbance system using anti-hebbian spike timing-dependent plasticity. In: Proceedings of the Symposium on Agents that Want and Like: Motivational and Emotional roots of Cognition and Action at the AISB-05 conference, The society for the study of artificial intelligence and the simulation of behaviour, pp. 98–101 (2005)

A Simple Model of Cortical Activations During Both Observation and Execution of Reach-to-Grasp Movements

Matthew Hartley and John G. Taylor

King's College London, The Strand, London, WC2R 2LS
mhartley@mth.kcl.ac.uk, John.G.Taylor@kcl.ac.uk

Abstract. We discuss evidence for the existence of mirror systems in the brain, including recent experimental results that demonstrate the use of shared pathways for the observation and execution of reaching and grasping actions. We then describe a brain based model of observational learning that explains the similarities and differences in levels of activation of brain regions during observation and execution of actions. We simulate a very simple paradigm whereby an actor performs an action which is observed and then repeated by the simulated animal. We discuss the implications and possible extensions of our model.

1 Background

1.1 Mirror Systems in the Brain

There is considerable evidence for the existence of "mirror systems" in the primate brain - areas activated both in the production of actions and observation of those actions in others [5,12]. These were first discovered in the F5 region of the monkey cortex (known to be involved in formation of grasping movements) and inferior parietal and thought to be confined to those areas.

More recent studies [11] (and unpublished data) have shown that mirror activity in the brain is widespread, and substantial parts of the action execution pathway are activated during observation tasks. In particular, the primary motor and somatosensory cortices (M1 and S1) are activated during observation of movements. This suggests that our understanding of the motor actions of others requires us to "mentally simulate" those actions using parts of cortical (and possibly sub-cortical) circuitry that would be used for the production of these actions [13].

While single cell recording studies examine monkeys, there is evidence from imaging studies that similar mirror systems exist in the human brain - for example, substantial evidence that viewing hand movements activates sensory cortex [1] and that viewing of speech activates speech production regions [15].

1.2 Existing Models of Imitation

There are a number of existing models of imitation/observational learning. Broadly, we can divide these models into two categories - those involving

J. Marques de Sá et al. (Eds.): ICANN 2007, Part II, LNCS 4669, pp. 899–911, 2007.

machine learning approaches and those that involve coupled forward and inverse models for motor control. The models often address mirror neurons specifically, rather than a particular focus on observational learning, but provide valuable insight nonetheless.

Of the machine learning variety, examples include [14] which uses a dynamical systems approach to organise observation and execution of dynamical actions and examines how these can share components, and [2], which uses evolutionary algorithms to develop behaviour of agents which react to teaching agents by generating outputs. For more details of computational mirror neuron models, see [10] for a recent summary.

The coupled inverse/forward model approach is popular since it extends existing attempts to understand motor learning - see [16], [4], [8], [3] for details of how brain based motor control might operate. For examples of this type of model see [6], the Mosaic model which is a well developed motor control model extended to mirror neurons/imitation and [9] a model of infant grasp learning that makes use of motor control circuitry. The coordinate systems between which these models transform are often not entirely clear. Some models suggest a transformation from Euclidean space coordinates to joint angle coordinates taking place in premotor/parietal areas, however experimental evidence for this is lacking, and most data seem to suggest that coding in premotor cortices is related to direction of action in Euclidean space [7].

Another interpretation of the concept of internal models is that, rather than always necessitating a conversion from physical to muscle/joint space coordinates, they are involved with the transformation of goal and current state to the action necessary to achieve that goal. In this case, some of the role of premotor/parietal regions in the transformation of current hand/arm state and desired affordances on an object into direction vectors for movement could be considered to be part of an inverse model system.

1.3 Problems to Address

The biggest questions to be answered by a model of neural activation during observation are:

Why is there so much activation of brain regions associated with the execution of movement during observation of those same movements?

In particular, how is S1, the primary somatosensory cortex, activated during observation (since this area involves sensorimotor feedback, which is obviously not present during observation)?

It seems reasonable to assume that part of the reason for extensive co-activation of motor pathways during action and observation is that attempting to understand the movement observed requires use of the same circuitry for executing movements. This may also explain why activation of motor regions is not always seen during observation of reaching movements, since unless there is sufficient reason to attempt to understand the movements, the brain circuitry is not recruited.

The activation of S1 is in some sense more puzzling, since it is considered to be activated primarily by proprioceptive feedback from muscles, which does not occur without execution of movements. One possible suggestion is that the region is involved with preservation of "sense of self" during observation of movement and general attribution of agency, but this does not explain how it is activated. For there to be S1 activation during observation of movement, something must be providing a substitute for proprioception, since muscles are not active. The neural projection from M1 to S1 is a candidate for filling this role, and we suggest this as a reason for the activation, as we will see below.

2 Model Details

Our model comprises several modules, here we describe their individual function. A diagram of the model can be seen in Figure 1. Some of these modules correspond to clearly identifiable brain regions (the visual system, M1 and S1 for example), others, such as sensorimotor integration and extraction of affordances correspond to several related areas.

Visual system. We assume that considerable low level visual processing takes place before input to our system, such that inputs occur as processed spatial coordinates. The visual system then allows observation of actions, determination of goal end points and observation of instructions to perform actions.

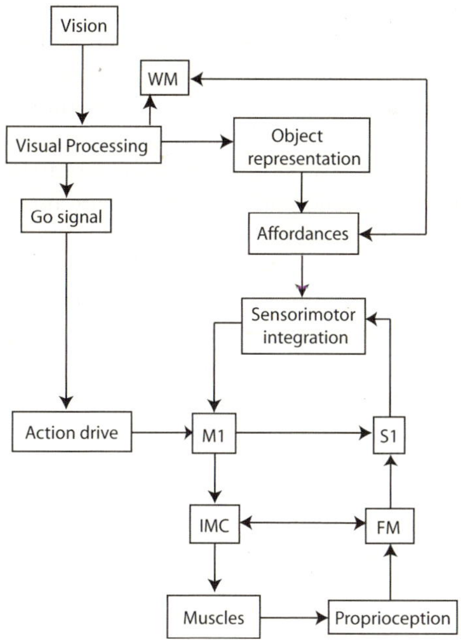

Fig. 1. Model structure details showing the connectivity of the regions used in the model

Object representations. Input from the visual system activates an object representation in this module, each possible object occurring as a dedicated single node (graded neuron).

Working memory. The WM serves to store the coordinates observed contact points in physical space during observation of actions.

Action drive. The action drive provides the reason for which the simulated animal performs an action. The drive arises from some presumed association of goal end points with food rewards (based on previous training). The action drive is modelled as a single graded neuron that is activated by the go signal.

Affordances. This module is involved in extracting affordances from objects, based on the way the object is to be used, and the observed actions on the objects. In this simple simulation, we assume that affordances take the form of a contact point on the button to be pressed.

Sensorimotor integration. Here the target of movement (the contact point generated by the affordances module) is integrated with sensory feedback about the current position of the hand to form a motor plan to reach the target. This output takes the form of a direction vector for movement to M1.

M1. The primary motor cortex receives a movement plan from the IMC and activates the IMC which generates suitable muscle movements from the plan. We model it using a small population of graded neurons that can encode a direction vector.

S1. Primary sensory cortex. It receives input from proprioceptive feedback (via the forward model) to update the position of the arm based on known movements. It also receives an efferent copy of the motor plan from M1. Like M1, it is modelled using a population of graded neurons.

IMC. In the Inverse Model Controller a target direction vector in physical space is converted into the muscle space movements necessary to carry out the planned movement. This involves a conversion from the direction vector input from M1 (coded in Euclidean space) to a code involving the joint angles necessary to perform the movement.

FM. The forward model (FM) takes proprioceptive feedback from the muscles and calculates how this updates the position of the arm. This information is passed to the sensorimotor integration module via S1 so that it can update the motor plan, involving a transformation from joint angle coordinates to the consequent movement of the arm in physical space.

We can use the activation of the parts of the system modelled as graded neurons to compare to experimental results (specifically M1, S1, the object representations, action drive and the affordances extraction network). It is more difficult to make comparisons with non-neural simulated components (the sensorimotor integration module, working memory and inverse and forward models), but we can record when these components are used and generate activation levels based on this usage.

2.1 Identification of Brain Regions Modelled

We can consider which brain regions might correspond to some of the components of our model. This allows us to compare to imaging data which include activations, and also to make predictions. We can see these allocations in Table 1.

Table 1. Identification of model components with brain regions

Model area	Brain region
Working memory	Prefrontal cortex (PFC), parietal lobe
Object representations	Superior Temporal Sulcus (STS), temporal lobe (TL)
Affordances	Parietal lobe
Sensorimotor integration	F5 and other premotor areas
IMC and FM	Brainstem (possibly also cerebellar/striatal network)

3 Simulation Paradigm

We consider initially a very simple simulation paradigm. The environment consists of an actor which contributes only by performing actions which can be observed by the simulated learner. Both actor and learner have a simple rod-like arm with a single joint of variable angle. The environment also contains a button object which can be reached by both actor and learner by arm movement. We assume that this object has some previously associated value (such as association with a food reward) which causes it to be of interest to the learner. When this joint reaches a certain angle θ, the arm is in contact with the button object and is considered to have reached it.

The simulation then consists of two stages:

Stage 1: The actor makes an arm movement terminating at the location of the button. This action is observed by the learner, and we record the activation of the various model regions. The learner also registers the location of the button and the point of contact used.

Stage 2: The learner repeats the previously observed movement, terminating at the button. Again we record activation of the model regions which we can then compare to the observation phase.

The simulation sees the teacher's actions from the point of view of the learner. We assume that because our very simple model consists of a single arm only, the issue of translating the actor's movement into an egocentric perspective is absorbed into some stage of visual processing.

These stages can be seen in Figure 2.

3.1 The Simulation During Observation

During observation of action, the visual system passes on the observed object to the object representation module, and the contact points to the working

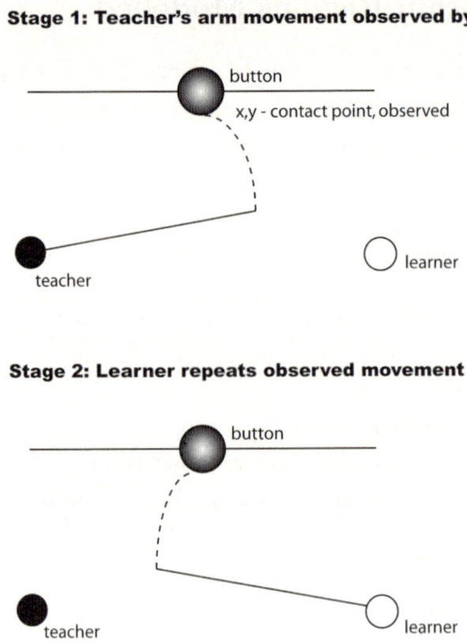

Fig. 2. Explanation of paradigm stages showing the simple actor and learner components and the action involved

memory. The contact points from the working memory are integrated with those generated by previous association of the object, and the result passes to the sensorimotor integration module. This output from this is fed both to M1 (where it is insufficient to actually cause muscle activation) which also activates S1.

3.2 The Simulation During Execution

During execution, the observed go signal is processed by the visual system and then used to activate both the action drive and the goal module. The visual system also activates the object representation module. The sensorimotor integration module combines information about the spatial location of the contact points with sensory feedback about the current position of the arm and produces a direction vector for movement (by performing a target - hand position calculation). This is passed to M1 and onwards to the IMC which converts it to a suitable set of impulses to muscles. Sensory feedback from muscles returns to the forward model, which calculates the change in arm position implied by the proprioception, and passes this information to the sensorimotor feedback module so the motor plan can be updated.

4 Results

To gain some idea of how well the model replicates experimental data, we can examine activation levels of both M1 and S1 during observation and execution phases. Since these are modelled as graded neurons, we can examine their activation levels to give an indication of that region's output. Each of these has a short "rise time", where the initial inactivity in the model M1 and S1 neurons is raised by the periodic input from the sensorimotor feedback module, whenever it recalculates the motor plan.

We are interested in comparing our results to those showing activations of M1 and S1 under observation/execution, which can be found in [11] and are reproduced here in Figure 3.

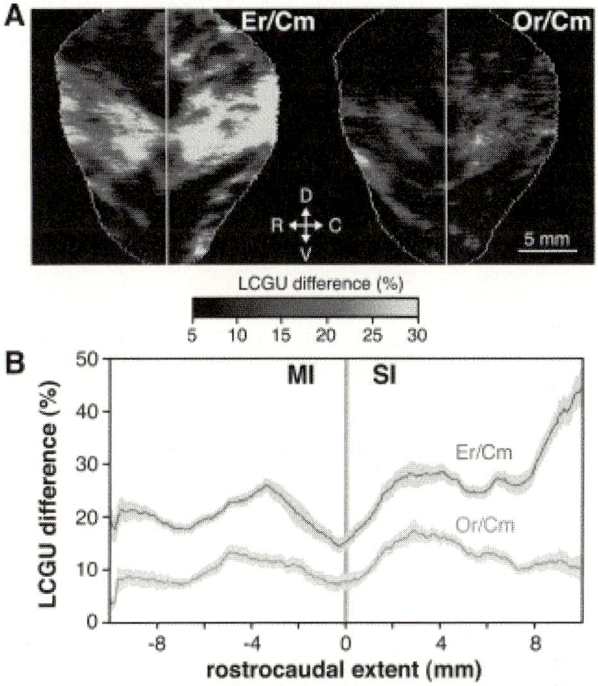

Fig. 3. Cortical activity of S1 and M1 during both observation and execution (showing percentage differences in activation) of reach-to-grasp movements, taken from [11]. A: Maps of activations in M1 (left) and S1 (right) showing execution of grasping (Er) and observation of grasping (Or) compared to biological motion control (Cm). B: Percent differences in activation over the anterior-posterior extent of M1 and S1 where -10 is the anterior crown, +10 the posterior crown and 0 the fundus. Data use local cerebral glucose utilisation (LGCU) values, and were recorded with the quantitative 14C-deoxyglucose method.

4.1 M1

In Figure 4 we see how activation of M1 during execution of the action is higher than activation during observation. Each 250ms, the motor plan is recalculated and the direction vectors for movement are fed to M1 causing activation, this being smoothed by the neuron's response time. The higher activation during execution occurs as a result of the action signal fed to M1.

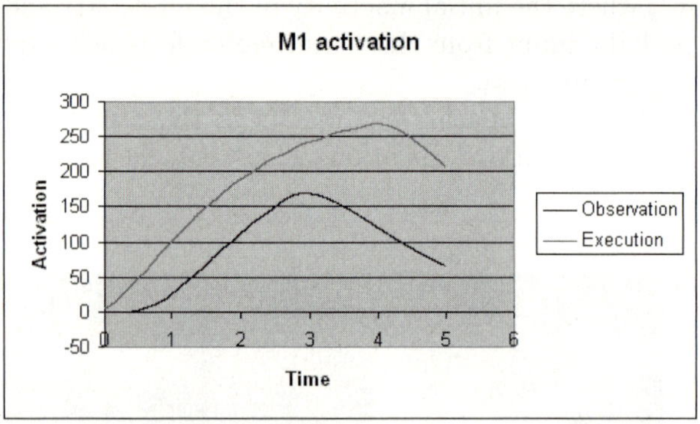

Fig. 4. Activation of our simulated M1 during both action observation and action execution showing activity against time

4.2 S1

Figure 5 shows a similar pattern of activation for S1. S1 receives proprioceptive feedback from muscles (via the forward model), which is why it is activated more strongly during execution of actions.

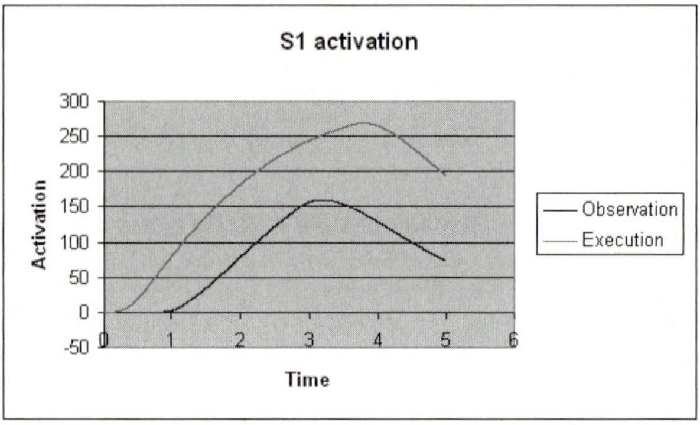

Fig. 5. Activation of S1 during observation and execution phases showing activity against time

4.3 Sensorimotor Integration

We can consider what activation we expect from our sensorimotor region, by considering its activation to increase when it performs a calculation, then to decay exponentially over time between these events. We can see the results of this process in Figure 6 - the peak activation level is similar during execution to integration, although it begins earlier (due to the timing of movement versus observation of that movement), and persists for longer.

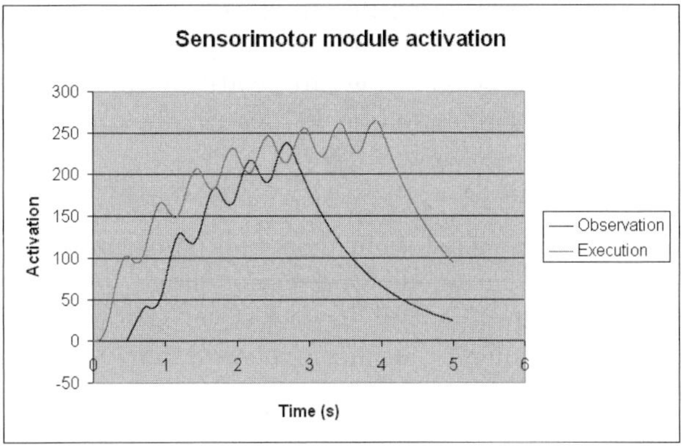

Fig. 6. Activation of sensorimotor region during observation and execution phases showing activity against time

4.4 Summary

The results demonstrate activation of both M1 and S1 during the execution phase of the simulation (as would be expected). They also show activations of both M1 and S1 (at lower levels to the execution case) during the observation phase, a more surprising result. These activations arise from the nature of the model's connectivity (which in turn provides predictions for experiment), as we shall now discuss.

5 Discussion

An important component of the model is the capacity to simulate the sensory consequences of an action that is being observed. This is the reason why activation of S1 occurs in our model during observation (a non-obvious result!), and why there is greater activation during execution, since the input from simulation of the action is added to proprioceptive feedback from muscles.

5.1 Comparison with Experiment - What Our Results Show

The primary areas we use for comparison here are the M1 and S1 regions. As discussed previously, experimental results on monkeys show partial activation of both M1 and S1 during observation of reaching and grasping movements. In these results, M1 and S1 both show significant (above threshold) activity during observation, as we can see in Figure 3. When we compare these to Figures 4 and 5 we see that our model also demonstrates similar lower (but above threshold) activations for M1 and S1 during observation.

It is important to consider what the nontrivial aspects of the results are. Activation of both M1 and S1 during observation of movement is an inherently surprising result and one replicated in our model. Although the reasons these occur (the projection from M1 to S1 to activate S1, and the activation of the sensorimotor system during observation) are relatively simple, they provide predictions for experimental verification (as we describe below).

5.2 Model Predictions

Since one of the key components of our model is the use of the projection from M1 to S1 to activate S1 during the observation stage, we can consider what might happen if that connection were interfered with. In particular, we predict that no activation of S1 would occur during observation.

Another prediction given by our model is that the action drive signal is crucial for the difference between activation of muscles during actual movement and activation that does not cause movement during observation. This is our current answer to the question of why no movement occurs during observation of actions (since motor cortex is activated) - without the action drive there is insufficient. Another possible explanation is that actual movement is inhibited in some way (although since something must be causing this inhibition, these explanations are similar in basic concept, although differ in mechanism).

5.3 Unanswered Questions and Extensions

Some data show that activation of M1 and S1 does not always occur for pure reaching movements. Why this is the case is unclear. One possible explanation is that, unless the movement is of interest (perhaps important to observe since repetition will aid in obtaining a food reward), no attention is paid to the moment. This provokes an interesting question as to how attention interacts with the system, which is an interesting direction for further study.

As mentioned above, we suggest the action drive as an explanation for cause of actual movement. This explanation raises questions as to what controls this action drive and to whether is operates directly by providing stimulation to muscle controllers, or indirectly by releasing inhibition on muscles.

The existing model is very simple and open to several avenues of extension, in particular more neurophysiologically realistic modelling of the sensorimotor integration module. Currently this performs non-neural calculations based on deriving the target vector from the target contact points and information about current hand position. It may be possible to use a trainable network here, although that would add a large degree of complexity. The F5 region of the brain is thought to be involved (possibly along with other premotor areas) in the target - hand calculation, and it may be possible to examine further biological data here to suggest suitable mechanisms.

Another possibility for extension is to a more complicated paradigm - a simple reach to a button position provides a basic demonstration of the model's operation. However, there is some question as to the difference in motor/sensory cortex activation during the observation of reaching and grasping (vs. purely reaching) movements. A paradigm involving full reaching and grasping (as separate movements) would allow us to cover this situation.

6 Conclusion

Our approach is that of "functional modelling", in that we assume certain functional modules are needed for the overall process, and then relate to brain science results to locate these functional modules in connected sets of brain modules. The set of modules are then simulated and compared to further details of experiment.

In our model, the activation of M1 and S1 during both action and observation arises because the same cortical circuitry is activated in both situations from visual input coupled with a need to understand the nature of the observed movement. S1 is activated during observation because of the projection from M1 (which is active as part of the process of observing movement in a manner intended to extract information. Actual motor action occurs because an action drive signal is necessary to provide sufficient motor cortex activity (and possibly also to release inhibition of actual movement).

Our model extends existing research by providing an explanation for the unexpected activation of motor and sensory cortices during observation of reaching and grasping movements. We also provide a very simple experimentally testable mechanism for the production of the activation of S1 during observation of movements.

References

1. Avikainen, S., Forss, N., Hari, R.: Modulated activation of the human SI and SII cortices during observation of hand actions. Neuroimage 15, 640–646 (2002)
2. Borenstein, E., Ruppin, E.: The evolution of imitation and mirror neurons in adaptive agents. Cognitive Systems Research 6 (2005)
3. Davidson, P.R., Wolpert, D.M.: Widespread access to predictive models in the motor system: a short review. J Neural Eng 2, 313–319 (2005)
4. Desmurget, M., Grafton, S.: Forward modeling allowed feedback control for fast reaching movements. Trends Cogn Sci 4, 423–431 (2000)
5. Gallese, V., Fadiga, L., Fogassi, L., Rizzolatti, G.: Action recognition in the premotor cortex. Brain 119, 583–609 (1996)
6. Haruno, W., Wolpert, D.M., Kawato, M.: Mosaic model for sensorimotor and learning control. Neural Computation 13, 2201–2220 (2001)
7. Kakei, S., Hoffman, D.S., Strick, P.L.: Direction of action is represented in the ventral premotor cortex. Nat Neurosci 4, 1020–1025 (2001)
8. Mehta, B., Schaal, S.: Forward models in visuomotor control. J. Neurophysiol 88, 942–953 (2002)
9. Oztop, E., Bradley, N.S., Arbib, M.A.: Infant grasp learning: a computational model. Exp Brain Res 158, 480–503 (2004)
10. Oztop, E., Kawato, M., Arbib, M.: Mirror neurons and imitation: A computationally guided review. Neural Netw 19, 254–271 (2006)
11. Raos, V., Evangeliou, M.N., Savaki, H.E.: Observation of action: grasping with the mind's hand. Neuroimage 23, 193–201 (2004)
12. Rizzolatti, G., Fadiga, L., Gallese, V., Fogassi, L.: Premotor cortex and the recognition of motor actions. Brain Res Cognit 3, 131–142 (1996)

13. Rizzolatti, G., Fogassi, L., Gallese, V.: Neurophysiological mechanisms underlying the understanding and imitation of action. Nat Rev Neurosci 2, 661–670 (2001)
14. Tani, J., Ito, M., Sugita, Y.: Self-organization of distributedly representated multiple behavious schemata in a mirror system: Reviews of robot experiements using RNNPB. Neural Networks 17, 1273–1289 (2004)
15. Watkins, K.E., Strafella, A.P., Paus, T.: Seeing and hearing speech excites the motor system involved in speech production. Neuropsyhcologia 41, 989–994 (2003)
16. Wolpert, D.M., Ghaharamani, Z., Flanagan, J.R.: Perspectives and problems in motor learning. Trends Cog Sci 5, 487–494 (2001)

A Appendix

A.1 Model Details

Parameters involved are given fully in Table 2.

The simulated M1 and S1 consist of simple graded neurons with membrane potential obeying the equation:

$$C\frac{dV}{dt} = g_{leak}(V - V_{leak}) + I, \tag{1}$$

where I is the neuron's input current. Their output is sigmoidal in form:

$$\frac{1}{1 + exp(-V/V_{scale})} \tag{2}$$

where V_{scale} scales the rate of change of the sigmoid function.

The sensorimotor integration module is represented by a calculating machine that provides a constant activation level when performing a calculation. When it integrates the goal contact points with current sensory information, its activation level is set to Act_{smi}. It performs this transformation every t_{smi} ms, and when it does so, its output becomes SMI_{out} for t_{calc} ms.

The action drive provides a constant current output I_{action} when active. When performing a movement, the muscles have a proprioceptive feedback signal that translates to a current equal to $I_{proprio}$ which is fed to S1.

A.2 Table of Parameters

Table 2 shows the parameters used in the model with units where applicable.

Table 2. Values of constants

Variable name	Value	units
V_{leak}	70	mV
C	25	nF
g_{leak}	0.025	μS
I_{action}	10	nA
$I_{proprio}$	15	nA
t_{smi}	100	ms
t_{calc}	50	ms
SMI_{out}	5	nA
Act_{smi}	1	n/a

A Cognitive Model That Describes the Influence of Prior Knowledge on Concept Learning

Toshihiko Matsuka and Yasuaki Sakamoto

Stevens Institute of Technology, Hoboken, NJ 07030, USA
{tmatsuka,ysakamot}@stevens.edu

Abstract. It is well known that our prior knowledge and experiences affect how we learn new concepts. Although several formal modeling attempts have been made to quantitatively describe the mechanisms about how prior knowledge influences concept learning behaviors, the underlying cognitive mechanisms that give rise to the prior knowledge effects remains unclear. In this paper, we introduce a computational cognitive modeling framework that is intended to describe how prior knowledge and experiences influence learning behaviors. In particular, we assume that it is not simply the prior knowledge stored in our memory trace influencing our behaviors, but it is also the learning strategies acquired through previous learning experiences that affect our learning behaviors. Two simulation studies were conducted and the results showed promising outcomes.

1 Introduction

A body of empirical studies in human concept acquisition indicates that our learning behaviors are strongly influenced by our own prior knowledge and experiences (e.g.[1]). Several formal modeling attempts have been made to quantitatively evaluate hypotheses about how our prior knowledge influences our cognitive behaviors associated with concept learning (e.g. [2]). Yet, most of them take the form of a retrospective model only allowing post-hoc data fitting, a model that does not offer a priori predictions about the observed effects of prior knowledge (but see [3]). Furthermore, these models of prior knowledge tend to emphasize the forward processes (i.e., how stored information is used) without specifying any learning algorithms (i.e., how information is updated) and thus cannot describe the underlying cognitive mechanisms that yield prior knowledge effects.

In the paper, we introduce a computational cognitive model with a learning algorithm that is intended to describe how prior knowledge and experiences influence learning behaviors. One unique contribution of our work is that we apply a hybrid meta-heuristic optimization method to describe how prior knowledge affects biases in acquired knowledge as well as biases in learning strategies (e.g. biases in hypothesis generation). Our new cognitive model, PIKCLE (PrIor Knowledge on Concept LEarning), integrates the effect of prior knowledge on concept learning. A novel characteristic of PICKLE is its assumption that it is not simply the prior knowledge stored in our memory trace influencing our learning behaviors, but it is also the learning strategies acquired through previous learning experiences that affects our learning behaviors. Another significant aspect of PICKLE is that it may generate insightful a priori predictions about the role

J. Marques de Sá et al. (Eds.): ICANN 2007, Part II, LNCS 4669, pp. 912–921, 2007.

of prior learning experiences on subsequent learning, as we demonstrate in the current simulation studies.

Note that the terms "category learning" and "concept learning" will be used interchangeably throughout this paper, because conceptual knowledge is considered to be categorically organized.

2 Categorization Algorithm

Rather than introducing new forward algorithms we chose to apply PIKCLE's learning processes to SUPERSET's [4] [5] categorization processes, because our main objective is to introduce a learning model that explains the effects of prior knowledge on concept acquisition. SUPERSET is a computational model of category learning based on the assumption that humans have a very flexible knowledge representation system, capable of adapting a situationally "optimal" representation scheme (e.g. rules, prototypes, or exemplars).

Unlike previous modeling approaches to category learning research which modify a single notion (i.e., a single complete set of coefficients, $\boldsymbol{\theta}^n$: $\{\mathbf{w}^n, \mathbf{D}^n, \mathbf{C}^n\} \in \boldsymbol{\theta}^n$), SUPERSET maintains, modifies, and combines a set of notions. The idea of having a population of notions (as opposed to having a single notion) is important because it allows not only the selection and concept combination in learning, but also the creation of diverse notions, making learning more robust. Thus, unlike previous categorization models, SUPERSET assumes that humans have the potential to maintain a range of notions and are able to apply a notion that is most suitable for a particular set of situational characteristics.

The SUPERSET framework, like several other models of the human categorization process, assumes that humans hold memorized internal reference points (i.e., rules, prototypes, or exemplars) and utilize similarities or conformities between the input stimulus and the reference points as evidence to probabilistically assign the input stimulus to an appropriate category. The similarity (i.e., s_j) between input x and j-th reference point (i.e., R_j) in SUPERSET are formulated as:

$$s_j^n(x) = \exp\left[-\beta\left[\sum_{i=1}^{I}\frac{(R_{ji}^n - x_i)^2}{1+\exp\left(-D_{ji}^n\right)} + \sum_{i=1}^{I-1}\sum_{m=i+1}^{I} 2C_{jim}^n[R_{ji}^n - x_i][R_{jm}^n - x_m]\right]\right] \quad (1)$$

where D_j and C_j are R_j's dimensional and correlational selective attention, respectively. A free parameter β defines an overall similarity gradient. Superscript *n* is an index for different notions. Subscripts *i* and *m* indicate feature dimensions, and *I* is the number of feature dimensions. Note that it is assumed that $C_{jim} = C_{jmi}$, $C_{jim}^2 \le \left(1+\exp\left(-D_{ji}^n\right)\right)^{-1} \cdot \left(1+\exp\left(-D_{jm}^n\right)\right)^{-1}$. The correlational attention weights can take a negative value, where its signum indicates direction of attention field while its magnitude indicates the strength of attention.

The psychological similarities are fed forward to the output category nodes using the following function:

$$O_k^n(x) = \sum_j w_{kj}^n s_j^n(x) \quad (2)$$

where w_{kj} is a learnable association weight between reference point j and category node k. The probability that x being classified as category A is calculated using the following choice rule:

$$P(A|x) = \frac{\exp(\phi \cdot O_A^o(x))}{\sum_k \exp(\phi \cdot O_k^o(x))} \quad (3)$$

where ϕ controls decisiveness of classification response, and superscript o indicates the notion adopted to make a categorization response (i.e., the most useful notion).

3 PIKCLE

3.1 Qualitative Descriptions of Learning Mechanisms

PIKCLE assumes that human learning involves the consideration of multiple notions according to their usefulness in a given task. Each of these notions determines which aspects of the categories are psychologically salient and which can be ignored. PICKLE assumes that the definition of the learning objective is not based solely on the accuracy of knowledge, but also on the subjectively and contextually determined utility of knowledge being acquired.

PIKCLE's learning algorithm is built on the basis of the Evolution Strategy (ES) optimization method. PIKCLE, as in a typical ES application, assumes three key processes in learning: *crossover*, *mutation*, and (survivor) *selection*. In the *crossover* process, the randomly selected notions (i.e., one complete set of SUPERSET coefficients indicated by superscript *n*) form a pair and exchange elements (i.e. coefficients) of notions, creating a new notion. In human cognition, the crossover process can be interpreted as conceptual combination, where new notions are created based on merging ideas from existing effective notions. In the *mutation* process, each element of notion (i.e., coefficient) is pseudo randomly altered (see section below for detail). A mutation can be considered as a modification of a notion by randomly generating a new hypothesis. Note that each element of a notion (i.e., SUPERSET's coefficient) is associated with a unique dynamically altering random hypothesis generator function, allowing it to be sensitive to the topology of the hypersurface of objective function (e.g., searching within a smaller area if coefficients are close to an optimum). In the *selection* process, a certain number of notions are deterministically selected on the basis of their fitness in relation to the environment for survival. Those selected notions (i.e., a set of coefficient) will be kept in PIKCLE's memory trace, while non-selected notions become obsolete or are forgotten.

However, unlike typical ES application, PIKCLE also assume a systematic bias caused by prior knowledge and experiences in concept modification. That is, while there is still a stochastic process involved in the knowledge *mutation* process, there is a force or momentum created by prior experiences that systematically influences the likelihood of particular patterns of notions (coefficient configurations) to emerge in the process.

3.2 PIKCLE Learning Algorithm

For the sake of simplicity, we use the following notation $\{\mathbf{w}^n, \mathbf{D}^n, \mathbf{C}^n\} \in \boldsymbol{\theta}^n$.

Knowledge Combinations. In PIKCLE, randomly selected pairs of notions exchange information to combine notions. In particular, PIKCLE utilizes the discrete recombination of SUPESET coefficients. Thus, parent notions $\boldsymbol{\theta}^{p1}$ and $\boldsymbol{\theta}^{p2}$ produce a child notion $\boldsymbol{\theta}^{c}$ or

$$\theta_l^c = \begin{cases} \theta_l^{p1} & \text{if } UNI \leq 0.5 \\ \theta_l^{p2} & \text{otherwise} \end{cases} \tag{4}$$

where UNI is a random number drawn from the Uniform distribution. This combination process continues until the number of children notions produced reaches the memory capacity of PIKCLE.

In PICKLE, self-adapting strategy coefficients denoted by σ_l^n, which define the widths of random search areas for SUPERSET's coefficients, are also combined during this stage. Unlike $\boldsymbol{\theta}^n$, the intermediate crossover method is used for the self-adapting strategy coefficients. Thus, $\sigma_l^c = 0.5 \cdot (\sigma_l^{p1} + \sigma_l^{p2})$.

Knowledge Modifications. Two separate hypothesis generation mechanisms are involved in PIKCLE' knowledge modification phase. One is a traditional stochastic modification process, modifying each coefficient with a random number drawn from the Gaussian distribution (i.e., last term in Eq.5). The other is systematic hypothesis generation (i.e. middle term in Eq. 5) that takes relationships among different types of knowledge in its memory trace into account (e.g. interaction between attention weighs allocated to feature dimensions 1 and 2, i.e., $D_{j1} \& D_{j2}$). Specifically, one element of notion θ_l^n at time $t+1$ is updated as follows:

$$\theta_l^n(t+1) = (1-\eta)\theta_l^n(t) + \eta \mathrm{A}(\mathbf{v}, \mathbf{u}, \boldsymbol{\theta}^n(t)) + N(0, \sigma_l^n(t+1)) \tag{5}$$

where *t* indicates time, A is a simple autoassociative network with one hidden layer causing a systematic bias in knowledge modification, a free parameter η controls the relative influence of the systematic bias, and $N(0, \sigma_l)$ is a random number drawn from the Gaussian distribution with the corresponding parameters. The random search width for notion element l is given as:

$$\sigma_l^n(t+1) = \sigma_l^n(t) \cdot \exp\left(N_G(0, \gamma_G)) + N_l(0, \gamma_l)\right) \tag{6}$$

where $N_G(0, \gamma_G)$ is a random number drawn from the Gaussian distribution with the corresponding parameters that is applicable to all knowledge elements within a notion (G stands for Global) where a free parameter γ_G is fixed for all knowledge elements within and across notions. $N_l(0, \gamma_l)$ is another random number drawn from the Gaussian distribution that is only applicable to notion element l. Unlike γ_G, γ_l varies across knowledge elements, and thus each knowledge element (i.e., θ_l) has its own unique tendency of thoroughness in a hypothesis search process. Specifically, since we assume that useful knowledge elements would usually take non-zero values across a variety of knowledge types, we define γ_l to be a mean the absolute value of corresponding coefficients amalgamating all prior experiences. For example, if a person experienced that information on feature dimension "color" has been more useful than feature dimension "size" in classifying several different categories in the past, then the person is more likely to have thorough knowledge modification on the basis of the applicability of "color" than "size" of input stimuli in new categorization tasks.

The other important element of notion modification (Eq. 5) is a systematic bias given below:

$$\mathrm{A}_l(\mathbf{v}, \mathbf{u}, \boldsymbol{\theta}^n(t)) = \sum_h v_{lh} \left[1 + \exp\left(-\sum_l u_{hl} \theta_l^n(t)\right)\right]^{-1} \qquad (7)$$

Equation 7 is a simple autoassociative network (AAN) with one nonlinear hidden layer. The fundamental idea behind integrating AAN in the notion modification phase is that even with some stochastic changes in notions resulting from the random knowledge recombination process (i.e., Eq. 4) and the stochastic notion modification process (i.e., Eq. 6), PIKCLE has a tendency to update notions such that the configural pattern of a new notion set is similar to that of previously acquired and applied notions; it is influenced by successful prior experiences. Note that AAN can be considered as a form of data reduction system (e.g. PCA), and thus, through this ANN, PIKCLE might acquire "eigen-notions" that are applicable to various type of concepts and categoriesto some extent.

The coefficients within ANN (i.e., $\mathbf{v}, \mathbf{u}$) are learned using a standard online version of gradient descent. We incorporate a *pseudorehearsal* method [6] for learning in ANN, assuming that this ANN serves as long term memory units in the PIKCLE framework. This allows ANN to avoid catastrophic forgetting (e.g. [7] [8]).

In summary, two separate notion modification processes are involved in the knowledge modification phase in PIKCLE. Those two processes are influenced by prior knowledge and experiences acquired in previous concept learning tasks; but the ways in which prior knowledge affect the two processes are different. One type of notion modification is stochastic updating in which previously useful notion elements (i.e. model coefficients) are more thoroughly searched in a current learning task. The other type is systematic updating that–with the influence of previously acquired knowledge–causes PIKCLE to generate new notion sets that resembles previously acquired effective notions.

Selection of Surviving Knowledge. After creating new sets of notions, PIKCLE selects a limited number of notions to be maintained in its memory. In PIKCLE, the survivor selection is achieved deterministically, selecting best notions on the basis of estimated utility of concepts or knowledge. The function defining utility of knowledge is described in the next section.

3.3 Estimating Utility

The utility of each notion or a set of coefficients determines the selection process in PIKCLE, which occurs twice. During categorization, PIKCLE selects a single notion with the highest predicted utility to make a categorization response (referred to as concept utility for response or UR hereafter). During learning, PIKCLE selects best fit notions to update its knowledge (utility for learning or UL hereafter). In both selection processes, the notion utility is subjectively and contextually defined, and a general function is as follows:

$$U(\boldsymbol{\theta}^n) = \Upsilon\left(E(\boldsymbol{\theta}^n), Q_1(\boldsymbol{\theta}^n), ..., Q_L(\boldsymbol{\theta}^n)\right) \qquad (8)$$

where Υ is a function that takes concept inaccuracy (i.e., E) and L contextual factors (i.e., Q) and returns an estimated utility value for a corresponding notion (Note that PIKCLE's learning is framed as a minimization problem). There are virtually infinite contextual functions appropriately defined for Eq. 8 (e.g. concept abstraction, domain expertise and knowledge commonality). For example, in ordinary situations, humans prefer simpler notions (e.g. requiring a smaller amount of diagnostic information to be processed) to complex ones, as long as both are sufficiently accurate, whereas in highly critical tasks (e.g. medical diagnosis), many might choose a notion with the highest accuracy disregarding complexity.

Note that functions for UR and UL do not have to be the same. For example, domain experts often know multiple approaches to categorize objects. This ability appears to be a very important characteristic and thus be a part of their UL. However, "knowledge diversity" is only relevant for selecting a population of notions (for survival), but not for selection of a particular notion to make a categorization response. Thus, knowledge diversity should not be considered for UR.

In PIKCLE, the predicted (in)accuracy of a notion during categorization is estimated based on a retrospective verification function [9], which assumes that humans estimate the accuracies of the notions by applying the current notions to previously encountered instances with a memory decay mechanism. Thus,

$$E\left(\boldsymbol{\theta}^{n}\right)=\sum_{g=1}^{G}\sum_{k=1}^{K}\left(\frac{\sum\limits_{\forall i|x^{(i)}=x^{(g)}}\left(\tau^{(i)}+1\right)^{-\delta}}{\sum\limits_{\forall i|x^{(i)}=x^{(0)}}\left(\tau^{(i)}+1\right)^{-\delta}}\right)\left(d_k^{(g)}-O_k^{(n)}\left(x^{(g)}\right)\right)^2 \tag{9}$$

where g indicates particular training exemplars, G is the number of unique training exemplars, the last term is the sum of squared error with d being the desired output, and the middle term within a parenthesis is the (training) exemplar retention function defining the strength of the retaining training exemplar $x^{(g)}$ [10]. Memory decay parameter, δ, in the exemplar retention function controls speed of memory decay, and τ indicates how many instances were presented since $x^{(g)}$ appeared, with the current training being represented with "0." Thus, $\tau = 1$ indicates $x^{(g)}$ appeared one instance before the current trial. The denominator in the exemplar retaining function normalizes retention strengths, and thus it controls the relative effect of training exemplar, $x^{(g)}$, in evaluating the accuracy of knowledge or concept. $E(\boldsymbol{\theta})$ is strongly influenced by more recently encountered training exemplars in early training trials, but it evenly accounts for various exemplars in later training trials, simultaneously accounting for the Power Law of Forgetting and the Power Law of Learning [10].

4 Simulations

In order to investigate the capability of PIKCLE to exhibit cognitive behaviors that resemble real people, two simulation studies were conducted. We created two category structures shown in Table I. There are three binary feature dimensions in those categories.

Table 1. Schematic representation of stimulus set used in Simulation Study

Stimulus Set						
Dim1	Dim2	Dim3	Label - Set 1A	Label - Set 1B	Label - Set 2A	Label - Set 2B
1	1	1	A	C	E	G
1	1	2	A	C	E	H
1	2	1	A	C	F	H
1	2	2	A	D	F	G
2	1	1	B	C	F	G
2	1	2	B	C	F	H
2	2	1	B	D	E	H
2	2	2	B	D	E	G

4.1 Simulation 1

Previous empirical studies indicate that prior knowledge can strongly interfere with the acquisition of new concepts[11] [12]. For example, learning linearly separable categories can become a difficult task when people are informed that categories that are about to learn can be potentially linearly non-separable [12]. In Simulation 1, we sequentially organized two hypothetical concept learning tasks. Our simulated subjects learned Set1A where information on Dim 1 provides sufficient evidence of perfect classification, followed by learning of Set1B. Set 1B would require our simulated subjects to pay attention to all feature dimensions in order to classify all exemplars correctly.

Methods: There were three types of PIKCLE learners involved in the present simulation study, namely BP who has a bias in generating new hypotheses on the basis of prior knowledge and experience in concept learning ($\eta = 0.2$ in Eq. 5) and maintains acquired knowledge; FP who does not have the bias ($\eta = 0.0$) but still possesses previously acquired knowledge in their memory trace; and FF who does not have the bias and forgets acquired notions. For the sake of simplicity, we omitted correlational attention weights from SUPERSET (see Eq. 1). In addition, we also assumed that $D_{ji} = D_{li}, \forall j\&l$. Furthermore, as in typical SUPERSET implementation [4] [5], we defined the knowledge utility function for BP, FP, and FF to be:

$$U(\boldsymbol{\theta}^n) = E(\boldsymbol{\theta}^n) + \lambda_w \sum_k \sum_j (w_{kj}^n)^2 + \lambda_\alpha \sum_i \left[1 + (\alpha_i^n)^{-2} \cdot \sum_l^I (\alpha_l^n)^2 \right]^{-1} \quad (10)$$

where $E(\boldsymbol{\theta}^n)$ is as given in Eq.9, and $\alpha_i^n = (1 + \exp(-D_i^n))^{-1}$. The second term (i.e., a weight decay function regularizing *w*) and the third term (i.e., the relative attention elimination function reducing the number of dimensions attended) regularize SUPERSET's knowledge complexity. λ's are constant free parameters weighting different regularizing factors.

All BP, FP, and FF were run in a simulated training procedure to learn the correct classification responses for the stimuli with corrective feedback. For each category set, there were a total of 10 training blocks, each of which was organized as a random presentation of the eight unique exemplars. The model parameters were selected arbitrarily;

Table 2. Results of Simulation 1

	Set 1A				Set 1B			
Types	Acc	Attn D1	Attn D2	Attn D3	Acc	Attn D1	Attn D2	Attn D3
BP	0.968	0.709	0.133	0.158	0.743	0.531	0.271	0.198
FP	0.969	0.686	0.149	0.165	0.895	0.402	0.413	0.185
FF	0.970	0.712	0.140	0.147	0.951	0.237	0.550	0.184

$c = 5$, $\phi = 5$, $\delta = 1$,$\gamma_G = 0.2$, $\lambda_w = 0.1$, $\lambda_\alpha = 1$. Note that the same parameter values were used for all three types of learners, except η. The identical parameter configuration was applied to learning of Set1A and Set1B. The memory sizes for the two models were 10 (i.e., possessing 10 notions at a time). There were a total of 100 simulated subjects for both models.

Predictions: We predict that all three type of learners will be able to learn Set1A very accurately in a similar manner. However, we predict that BP is more likely to be fixated at Dim 1 in the Set1B task, managing only mediocre accuracy performance. FP and FF, on the other hand, would shift his/her attention to Dim 2, thus resulting in a relatively highly accuracy rate.

Results: The results of Simulation 1 are shown in Table 2. As we predicted, all three types of learners resulted in similar accuracies and attention allocation patterns for the Set1A task. For the Set1B task, BP performed the worst, followed by FP then FF. And, as predicted, BP, because of the interference by prior knowledge, paid the most attention to Dim1 which was less diagnostic than Dim2. FP paid more attention to Dim 2 which was imperfect (but the most diagnostic dimension), but it also paid a similar amount of attention to Dim 1. FF paid the most attention to Dim2 and equally less attention to Dims 1 and 3, allowing it to be the best performer for the Set1B task. Note that although FF performed the best in Set1B, this does not necessarily mean that it is the best model in terms of the objective of the present research. Rather, it is probably the worst model, unable to exhibit the interference effect of prior knowledge [11] [12]. Instead, BP's prediction is most consistent with the results of previous empirical studies.

4.2 Simulation 2

Previous empirical studies indicate that expectation and context affect our concept learning behaviors [11] [12] [13]. In Simulation 2, we let PIKCLE learn hypothetical learning tasks with two hypothetical contextual elements and then observed a knowledge generalization pattern.

Methods: Table 1 shows a schematic representation of category structure used in Simulation 2. There are two category types, Set 2A and Set 2B, both of which are defined by three a feature dimension plus one additional dimension (Dim 4, not shown) representing situational characteristics. Dim 4 was fixed at "1" for the Set2A task and "2" for the Set2B task. Note that both Set2A and Set2B are simple XOR-logic type categories (Dim1&Dim2 for Set2A and Dim2&Dim3 for Set2B). After successful learning, we gave PIKCLE a generalization tasks to see how situational characteristic influence the choice of notion to be manifested. In particular, we estimated the amount of attention allocated to correlations between the key feature dimensions (Dim1&Dim2 for Set2A and Dim2&Dim3 for Set2B).

There were two types of PIKCLE learners involved in the present simulation study, namely BP and FF. The basic simulation method follows that of Simulation 1, except that there were a total of 50 training blocks; and that we applied a none-restrictive version of SUPERSET. There were a total of four conditions - BP learning Set2A first followed by Set 2B (i.e., BPA condition); BP learning Set2B first (BPB); FF learning Set2A first (FFA); FF learning Set2B first (FFB).

Results & Discussion. For both FFA and FFB, there was a very strong recency effect in that this particular PIKCLE implementation was not sensitive to situational characteristics and was "too adaptive." A recency effect was also observed in BPA (0.11) and BPB (0.09) conditions, but the magnitudes were much smaller. BP was shown to be affected by prior knowledge and experiences and sensitive to situational characteristics (i.e., in PIKCLE, BP paid attention to the covariation between Dim1 & Dim2 when the situational characteristic or Dim4 is "1", while it paid attention to the covariation between Dim2 & Dim3 when Dim4 is "2"). The results also indicate that BP is capable of maintaining diverse effective notions.

Although this simulation result seems easily anticipated and perhaps too simplistic given that PIKCLE contains an autoassociative network, it is worth noting that most computational models of concept learning take a form of radial basis function network, which usually does not allow different types of notional elements to directly interact with each other (e.g. predicting association weight between basis unit j and output unit k using association weight between l and m). PIKCLE, on the other hand, allows direct interaction among elements in a notion by integrating the effect of prior knowledge on concept learning.

5 Conclusion

It is well known that our prior knowledge and experiences affect how we learn new concepts (e.g.[1] [2]). However, the cognitive mechanisms that give rise to the prior knowledge effects remains unclear. In the present paper, we introduced a computational cognitive modeling framework that is intended to describe how prior knowledge and experiences influence learning behaviors. Our new model's main contribution is that it is not simply the prior knowledge stored in our memory trace that influences our learning behaviors, but it is also the learning strategies acquired through previous learning experiences that affect our learning behaviors.

In addition, our new model PIKCLE can acquire general knowledge called "eigen-notions" that are applicable to various type of concepts and categories along with sensitivity to situational characteristics. There is an interesting psychological phenomenon called ad-hoc categories [14], i.e., people can instantaneously and spontaneously create a category, typically in the service of some goal [15]. The cognitive mechanism underlying the creation of ad-hoc categories may be described by the "eigen-notions."

Acknowledgment

This research was supported by the Office of Naval Research, grant #N00014-05-1-00632.

References

1. Murphy, G., Medin, D.: The role of theories in conceptual coherence. Psychological Review 92, 289–316 (1985)
2. Heit, E., Bott, L.: Knowledge selection in category learning. In: Medin, D.L. (ed.) Psychology of Learning and Motivation (1999)
3. Rehder, B., Murphy, G.L.: A Knowledge-Resonance (KRES) model of category learning. Psychonomic Bulletin & Review 10, 759–784 (2003)
4. Matsuka, T., Sakamoto, Y.: A Model of Concept Formation with a Flexible Representation System. In: Advances in Neural Networks, ISNN07. Lecture Notes on Computer Science, vol. 4491, pp. 1139–1147. Springer, Heidelberg (2007)
5. Matsuka, T., Sakamoto, Y.: Integrating a Flexible Representation Machinery in a Model of Human Concept Learning. In: Proc. of IJCNN2007. Forthcoming (2007)
6. Robins, A.: Catastrophic forgetting, rehearsal and pseudorehearsal. Connection Science 7, 123–146 (1995)
7. French, R.M.: Catastrophic forgetting in connectionist network. Trends in Cognitive Sciences 3, 128–135 (1999)
8. Grossberg, S.: Competitive learning: from interactive activation to adaptive resonance. Cognitive Science 11, 23–63 (1987)
9. Matsuka, T., Chouchourelou, A.: A Model of Human Category Learning with Dynamic Multi-Objective Hypotheses Testing with Retrospective Verifications. In: Proc. IJCNN 2006, pp. 3648–3656 (2006)
10. Anderson, J.R., Bothell, R., Lebiere, C., Matessa, M.: An integrated theory of list memory. Journal of Memory and Language 38, 341–380 (1998)
11. Pazzani, M.: The influence of prior knowledge on concept acquisition: Experimental and computational results. Journal of Experimental Psychology: Learning, Memory & Cognition 17, 416–432 (1991)
12. Wattenmaker, W.D., Dewey, G.L., Murphy, T.D., Medin, D.L.: Linear separability and concept learning: Context, relational properties, and concept naturalness. Cognitive Psychology 18, 158–194 (1986)
13. Barrett, S.E., Abdi, H., Murphy, G.L., McCarthy Gallagher, J.: Theory-based correlations and their role in children's concepts. Child Development 64, 1595–1616 (1993)
14. Barsalou, L.W.: Ad Hoc Categories. Memory and Cognition 11, 211–227 (1983)
15. Medin, D.L., Ross, B.H., Markman, A.B.: Cognitive Psychology (4th Ed.) Hoboken, NJ: Wiley (2005)

Developing Concept Representations

Stathis Kasderidis

Foundation for Research and Technology Hellas,
Institute of Computer Science, Vassilika Vouton, 71110 Heraklion, Greece
stathis@ics.forth.gr

Abstract. The paper discusses a novel model for concept learning and representation. Two levels of representation are used: exemplars and (generalized concepts) prototypes. The internal structure of the model is based on a semantic network using spreading activation. Categorisation and addition operations are supported in parallel. Forgetting of learned concepts is used in order to track dynamic and novel environments. The model is inspired by the corresponding psychological theories of exemplars and prototypes. Simulation results support the formulation of the model.

1 Introduction

According to [1] there are three levels of representation: the associationist, the conceptual and the symbolic. Concepts belong to the sub-symbolic level and they represent a good comprise between compression of information, so as to avoid representing every possible stimulus with a unique class, and resolution of information so as to allow differentiating between different things in the world.

The purpose of this paper is to present a new concept system model, which combines seamlessly aspects from two main classes of psychological theories of concepts. The first class is called the exemplar-view and it assumes that concept classification and learning is based on storing (learning) and using individual objects as representations of the general class. The second class of theories is called prototype-view and assumes that people somehow build more general representations than that of individual concrete members of the class. Thus the category in this case represents a summary representation of all known examples of the class. There is finally a third-view, which assumes that concepts are more complex knowledge structures, this view is known as the theory-theory or knowledge-view, with the belief that there are general relations that hold between the attributes of a concept. Declarative knowledge about the physical world of the form "all material bodies have volume" is an example of the latter view and makes hard to devise concept representations that are easily fused with the representations of the other two views. Psychological evidence suggests that each of the aforementioned class of theories has support, advantages and disadvantages in particular areas of psychological interest, e.g. concept formation, categorisation, concept combination, language & semantics, etc. For further discussions see [2].

Our approach is based on the exemplar and prototype theories. We develop a computational model for use in autonomous agents, such as in robotics or software agents. The structure of the paper is organised as follows: In section 2 we provide a

J. Marques de Sá et al. (Eds.): ICANN 2007, Part II, LNCS 4669, pp. 922–933, 2007.

brief discussion on the basic ideas of the aforementioned theories. In section 3 we develop a computational model based on the previous ideas. In section 4 we provide simulation results that support the current model formulation. In section 5 we conclude with a summary of its main characteristics and a discussion about future extensions of the current model.

2 Background on the Psychology of Concepts

2.1 Concept Properties and Formation

Assume that a sensory signal $\mathbf{s(t)}=(s_1, s_2, \ldots, s_N) \in R^N$ is given which carries information from a number, N, of input channels at a time instance t. The input channels represent external and internal sources of information. We might call this signal the *native state*. We also assume the presence of a *perceptual* system, which is responsible for extracting *features* out of the incoming sensory signal. Thus the following transformation takes place in general:

$$\mathbf{f}(t) = \mathbf{P}(\mathbf{s}(t)) \tag{1}$$

where, $\mathbf{f}(t)$ represents the derived feature vector which in general is in a space R^M. $\mathbf{P}$ corresponds to the perceptual system transformation.

The feature vector $\mathbf{f(t)}$ corresponds then to an *instance* of a class and also represents a concrete object in the world. The components of feature vectors correspond to *attributes*. Now we can state the general concept formation problem: *Given feature vectors of concrete objects in the world develop an internal representation that groups these objects according to a similarity criterion and preserve the relation of such objects with other objects that belong to the same or to a different class in a manner such that the corresponding concepts continue to have a relation that is isomorphic to the one observed in the instances.*

We provide here a number of properties of concepts found by psychological research:

- O1. Concepts form incrementally;
- O2. There are typicality effects present;
- O3. Concepts depend on context, i.e. their essential features change depending on the context of use;
- O4. A given instance may belong at the same time to a given concept and a more abstract one but also not have a lot of commonalities with other members of the abstract class (goal-derived concepts);
- O5. Transitive inference might not always hold;
- O6. Concepts combine to produce new concepts with emergent properties that might not be predicted by the knowledge of properties of the constituent concepts.

Due to lack of space we will not provide any examples and explanations of the above properties here. The interested reader should consult [2].

2.2 Review of Exemplar Theory

According to Exemplar theory concept formation takes place by memorising observed instances. Consequently instance classification takes place by comparing the new instance to the set of stored instances. In practice there are many Exemplar theories, as each one specifies different rules as to which instances are stored, how many are retained, and what is the set of instances that the new input compares against. Deficiencies of the exemplar theory include the lack of prototypes and thus difficulty in constructing abstractions and concept combinations. It has been found experimentally that there are people that use an exemplar-based strategy in classification tasks.

2.3 Review of Prototype Theory

Prototype theory suggests the existence of *summary representations*, which encode the general information about the class, which is not necessary present in any given instance. The theory assumes that a schema structure is often in place. *Schema* is a data structure with a number of slots (called frames); each one of them corresponding to a feature of the schema. The feature values of the instances are stored in the corresponding frame of the schema. Usually information regarding the probability distribution out of which the values are drawn is stored as well in each frame.

This type of theory is strong in explaining abstractions and at least some cases of concept combination effects. However, the schema approach cannot easily incorporate constrains and correlations among the frame values, nor it can easily provide a set of prototypes when multiple values are common for each schema feature. See [2], chapter 2 for further information.

2.4 Typicality and Context Effects

It has been observed experimentally that there are instances that are somehow more typical than others as representatives of a class, while others seem to be almost border cases between two classes. To account for these phenomena psychologists have postulated the existence of suitable metric spaces (which we will call conceptual spaces) where the notion of a distance function between two instances can be represented. Assuming for the moment that the feature vectors consist only from numerical[1] variables, one can easily use a Euclidean distance function as a possible distance measure:

$$D(i,j) = \sqrt{\sum_k (x_{ik} - x_{jk})^2} \tag{2}$$

where, x_{ik} and x_{jk} are the components of the features vectors $\mathbf{x}_i$ and $\mathbf{x}_j$ corresponding respectively to instances i and j. The feature vectors $\mathbf{x}_i$ and $\mathbf{x}_j$ are derived by use of (1) and correspond to dinstinct values of the perceptual transformation **P**. Contextual effects can be explained by assuming a suitable weighting for the component domains, i.e. for the Euclidean metric:

[1] I.e. not categorical variables.

$$D(i,j) = \sqrt{\sum_k w_k * (x_{ik} - x_{jk})^2} \quad (3)$$

where the coefficients w_k are called *attention coefficients* (in the psychological literature) and they encode the relative importance of the various components. Different sets of the attention coefficients represent different contexts. This naturally leads us to consider the notion of *similarity* between any two instances. It has been proposed that similarity is a function of distance of any two instances. Usually it is assumed to be a Gaussian or (negative) exponential function as in (4) and (5):

$$s_{ij} = \exp(-c * D(i,j)^2) \quad (4)$$

or

$$s_{ij} = \exp(-c * D(i,j)) \quad (5)$$

where c is a variable, called *specificity,* that controls the effective size of the neighbourhood in which any two instances are considered similar enough.

It is now easy to explain how typicality effects arise. Assume that somehow we have some instances (or prototypes) that are considered representative of a class. As closer as an instance is to a prototype the more easily it will be recalled. For example, when one recalls birds usually consider first (as a typical prototype) a robin rather than a penguin.

3 Computational Details

3.1 Exemplar Representation

Let us start with the issue of exemplar representation. The idea that we use is related to the work of J. McClelland [3]. Figure 1 presents a simplified version of the representation of an exemplar.

Let us consider objects with three features, such as Colour, Shape and Size information. The set of all features is called Feature Space. A node inside any feature set represents an actual measurement (we assume a mixture of numerical and nominal variables, e.g. Colour and Shape information are of nominal type while Size is of numerical type). All nodes inside a feature set are linked with bi-directional links so as to inhibit each other; in the simplest case we use a common weight, I, for these links. Assuming that we have observed Object 1 {Colour="BLUE", Shape="Stick", Size=20} we see that there are links from the corresponding feature value nodes to the exemplar node of Object 1. The value "Stick" for the Shape feature can be treated as a linguistic label but this is not necessary; for example it could be a vector of activations in a suitable neural network model for vision. Each one of the four semantic nodes (3 for the feature values, 1 for the exemplar) has associated variables of Activation, Input and Effect. With the arrival of a new feature vector the activations of all nodes change.

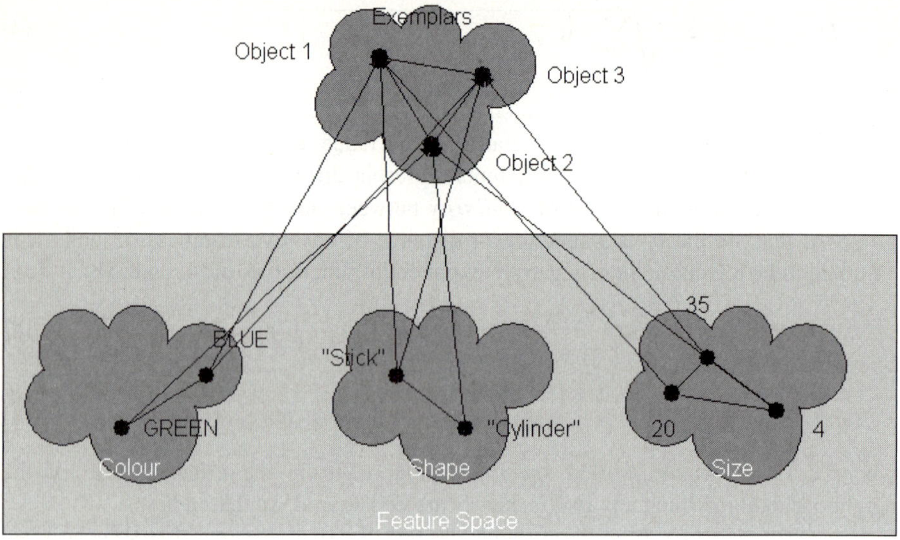

Fig. 1. An instantiation of the Concept System. View of the Exemplar level of representation.

The activation is calculated as follows:

$$Input_i(t) = \Pr obe_i(t) + E * \sum_j e_{ij} - I * \sum_j i_{ij} \tag{6}$$

$$Effect(t) = (M - Act_i(t)) * Input_i(t), Input_i(t) \geq 0 \tag{7a}$$

or

$$Effect_i(t) = (Act_i(t) - m) * Input_i(t), Input_i(t) < 0 \tag{7b}$$

and

$$\frac{dAct_i(t)}{dt} = Effect_i(t) - D * (Act_i(t) - R) \tag{8}$$

where in the above relations $Input_i(t)$, $Effect_i(t)$ and $Act_i(t)$ is the net input, the effect and the activation of each semantic node respectively. $Probe_i(t)$ represents a contribution arising by incoming percepts. This term in effect calculates a similarity function that is defined in analogy with (4) as:

$$\Pr obe_i(t) = A * \exp(-c * Dist(i, p(t))) \tag{9}$$

where in (9) "attention weights" are included in the distance function. A is the magnitude of activation for the similarity function with value of A=1.0 and we use a negative exponential for the Distance function between the node (i) and the probe p at time t. Typical values of the "attention weights" and "specificity"are w_i=0.3 and c=10 respectively. The distance function in (9) is defined both for nominal and numerical values. It takes values in [0,1] and it is given by (10):

$$Dist_{nominal}(i,j) = 1, i \neq j \quad \text{or} \tag{10a}$$
$$Dist_{nominal}(i,j) = 0, i = j$$

and

$$Dist_{numerical}(i,j) = \frac{|i-j|}{range} \tag{10b}$$

where we have defined in (10a) the distance function for nominal and in (10b) for numerical variables respectively. In (10b) the range is defined as range=max-min for all encountered instances. For further details for these and other heterogeneous distance function definitions see [4].

Formulae (6)-(8) define a spreading activation mechanism, which is used to calculate the activation of exemplar nodes from presentation of a given percept. The parameters which appear have the following values E=0.05 (excitatory weight), I=0.03 (inhibitory weight), M=1 (maximum activation), m=-0.2 (minimum activation), D=0.1 (decay coefficient) and R=0 (resting level). e_{ij} and i_{ij} correspond to excitatory and inhibitory activations from other nodes j respectively. The activation is called excitatory if it is greater or equal to zero and the activating node is in another set. All nodes that coexist in the same set and have activation greater than zero provide inhibitory excitation to the node in question.

On the presentation of a new percept we collect its feature vector and we use as probes its components to the corresponding feature set. Using (9) we calculate the input due to the probe in all values inside a set. Next, using (6)-(8) we calculate the activation of all nodes in each feature set and the exemplar set. Before any new activation calculation we initialise the activations of all nodes in random values in the interval [m, 0]. We allow the activation of a probe to persist in each iteration. We allow a number of iterations of equations (6)-(8) until the activations reach an equilibrium state. Typically 1000 iterations suffice to reach equilibrium.

3.2 Prototype Representation

The same spreading activation mechanism, described in section 3.1, is also in place in the Prototype Representation. Thus the same form of representation is used. The same parameter values are used as well. In this level we apply the probes to the (summary) feature values, calculate the effect on each (summary) feature value, which in turn influences the activation of the concept nodes. At the end some concept nodes are activated more than others and assuming a decision threshold one returns a set of activated concept nodes as the reply of the system to the incoming percept. Figure 2 shows an instantiation of the Prototype level corresponding to that of figure 1. On the reception of a new percept a classification process takes place. Classification starts at the Prototype Level using the mechanism of equations (6)-(8) to calculate the activations of the concept nodes. If these activations are less than a threshold, say T1=0.7, then a second classification takes place in the Exemplar Level in the same manner. If the classification succeeds, then a concept update process takes place. If the activations of Exemplars nodes are less than a second threshold, e.g. T2=0.56, then we enter in the phase of concept formation which is described in section 3.3. In case that the activations of Exemplar nodes are in [T3, T1], where T3 is a third threshold, say T3=0.6, we assume that the percept describes an existing exemplar that needs to be modified slightly or new information to be added to it, such as a new feature. This will be discussed in section 3.4 together with the update process of the Prototype Level.

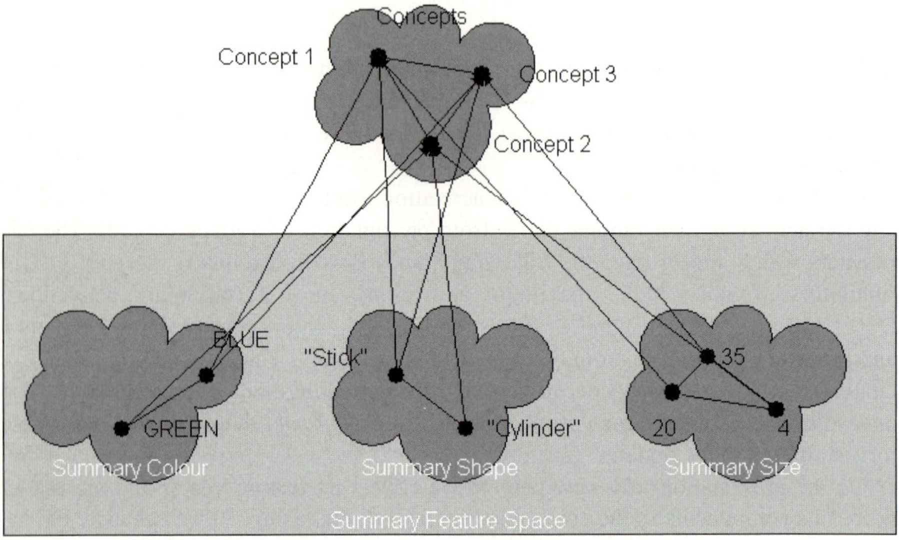

Fig. 2. Prototype level representation

3.3 Adding a New Concept

If our two-stage classification procedure fails to return any activated nodes we proceed on adding the new percept to the concept system as a new concept. The main idea is that we use a dual representation using both exemplars and prototype nodes. Some percepts will be represented as exemplars; others will only influence the (summary) feature values for the features of a prototype and they will not have explicit nodes created for them. The set of exemplars that is linked to a given prototype node corresponds to the *supporting set* of the concept, which is represented by the prototype node. The whole process consists of four steps. These are described below:

Step1 (Exemplar Insertion): In this step we add in the Exemplar and Feature Spaces nodes that represent the corresponding feature components of the percept as well its related exemplar node. The exemplar node forms links with the newly created feature value nodes assuming a weight of 1 for each link. The exemplar and the feature nodes also form links with other exemplar or feature nodes in their including sets. These links also carry a weight of 1. If a component exists in the percept that does not correspond to an existing feature set in the Feature Space, then a new feature set is created where the component value is entered as its first value node. When an exemplar node is inserted in the Exemplars set an internal variable called LifeTime is initialised as in (11):

$$LifeTime_i(0) = \frac{R - Value_i}{\sum_j R - Value_j} \tag{11}$$

i.e. it is the relative R-value of the exemplar inside the supporting set. R-Value is the reinforcement value of an item. The lifetime variable is used in order to control the time that the exemplar will be stored in the concept system. Every time that an Exemplar classification takes place, the winning Exemplar increases its lifetime according to (12a) while all other loosing exemplars reduce theirs according to (12b):

$$\Delta LifeTime_i(t) = D_1 * LifeTime_i(0), Act_i(t) \geq T_2 \tag{12a}$$

or

$$\Delta LifeTime_i(t) = -D_2 * LifeTime_i(0), Act_i(t) < T_2 \tag{12b}$$

Typical values for D_1 and D_2 are $D_1 = 0.1$ and for $D_2=0.01$.

Step2 (Concept Insertion): In this step we also add nodes in the concept (or prototype) set and the corresponding summary feature sets. If a new feature set is required then it is created as needed. Links between the concept node and its corresponding summary feature value nodes are created. Finally competitive links of the new nodes with other nodes in their including sets are also formed.

Step3 (Deriving the Summary Values): Having formed already nodes for the features of an exemplar we need to specify how these nodes, as well as the corresponding summary value nodes, are initialised. In the Exemplar Representation, the feature value nodes take the value of the corresponding component in the percept. However, in the case of the corresponding summary node, its value is the average of the percept component value and the "centre of gravity" value of the other values present already in the summary feature set. In the case of nominal variables, the above-described numerical procedure does not obviously work. In this case the summary values are the same with the corresponding feature values.

Setp4 (Linking Exemplars to Concepts): In the final step links are formed between the newly created exemplar and the concept nodes present in the concept space. These links use weights that are given by a Hebbian-like rule, more specifically of ARTMAP style (see [5] for details) (13):

$$\frac{dW_{ec}(t)}{dt} = \gamma * Act_e(t) * (-\delta * W_{ec}(t) + Act_c(t)) \tag{13}$$

This rule builds correlations between the exemplar node (e) and the concept node (c) for all concept nodes. In (13) W_{ec} is the weight of the link connecting the exemplar to the concept node. Act_e and Act_c are the corresponding activations. The parameters γ and δ have the values of 0.1 and 0.02 respectively. All weights start with zero value and by use of (13) evolve over time to stable values (assuming a stationary environment).

3.4 Incremental Change of a Concept

In this section we discuss the related problem of how an existing concept can change given new percepts. There are two principal ways in which this can be accomplished. The first is by change of the summary feature values corresponding to a concept node. The second relates to the "amendment" of an exemplar node

Change of Summary Feature Values: If a new percept is successfully classified in the first stage of Prototype Representation then a change process in the (summary) feature values of all activated concepts (above T1) takes place. This process in effect creates the "average" values for the features of the concept. We use an incremental SOM algorithm to achieve this, even though there are other ways to achieve the same effect. However, we believe that a SOM-based approach is both more biologically plausible as well as it provides nice topology preserving properties for the maps of the various feature sets.

Change of Exemplars: It is also possible that a new percept will be successfully recognised in the Exemplar stage and the activations of the resulting exemplars will be in the interval of [T3, T1]. In this case we assume that the activated exemplars closely resemble the percept, so the percept should be a "noisy" encoding of the activated exemplar(s). In this case, some adjustment takes place for the feature values of the exemplar(s) based on the newly observed features. Section 3.3 – Step1 described the process, covering also the case of novel features. For features that are of numerical nature we assume that the "amended" value is the mean of the two values (of the exemplar's and the percept's corresponding values), while for features of nominal nature we use, between the two values, the one which has the higher probability.

3.5 Controlling the Capacity of the Supporting Sets

Each supporting set has a capacity, which is determined dynamically. Capacity is defined as the number of exemplars that are retained in the set. We assume initially that there is a fixed maximum number of exemplars that can be stored dynamically inside the system. Let us call this number K. Assuming that in the system currently exist L < K exemplars and that these are partitioned in C classes then the *capacity* for the supporting set of concept c is given by (14):

$$Capacity_c = K * \frac{Utility_c}{\sum_c Utility_c} \tag{14a}$$

$$Utility_c = \sum_j R-Value_j^c - IE(P_c) \tag{14b}$$

i.e. it is the ratio of the set's Utility, against all other set Utilities. For cases where an exemplar might be linked to more than one concept we count its contribution to the concept for which its link has the highest weight. R-Value$_j^c$ is the reinforcement value of the exemplar j which belongs to concept c. IE is the information entropy of the distribution of exemplars belonging to class c, see equation (15) below:

$$IE(P) = -\sum_j P_j * \log_2(P_j) \quad (15)$$

Equation (15) holds strictly for a discrete distribution, P_j, but a straightforward generalisation to an integral can be made for continuous distributions. The usefulness of (15) lies in the fact that it can be applied to both numerical and nominal distributions, while use of standard deviation only works on numerical distributions.

4 Simulation Results

In this section we will explain the operation of the system using two test cases, which also provide simulation evidence for the model. We assume that we have initialised the system with the five exemplars present in table 1, (Ex1-5). We test the system using unobserved instances (Pr1), and missing values from the feature vector of a previously known exemplar (Pr2). Due to lack of space we discuss only the exemplar level classification process. Elsewhere we will discuss more fully the operation in all other modes, such as concept formation, classification in the prototype level, etc. Table 2 provides activations of the five exemplars when two different probes are used.

Table 1. Exemplars used for initialisation of the system. Probes are used for testing.

Exemplar Features									
ID	**Tail**	**Legs**	**Arms**	**Fins**	**Peak**	**Color**	**Weight**	**Size**	**Class**
Ex1	1	2		2	1	Black	20	80	PENGUIN
Ex2		2	2			Black	75	190	HUMAN
Ex3	1	4				Grey	180	150	LION
Ex4	1	3				Brown	40	75	DOG
Ex5	1	4				White	5	40	CAT
Pr1		2	1			Yellow	60	170	HUMAN
Pr2		3				Brown	40		DOG

Table 2. Activation of exemplars due to presentation of probes

Exemplar Activations					
	Ex1	**Ex2**	**Ex3**	**Ex4**	**Ex5**
Pr1	0.38	0.67	0.22	0.29	0.11
Pr2	0.20	0.18	0.21	0.61	0.24

Note the fact that in the first probe we have an one-armed man, while in the second one two out of five attributes are missing. The last column in table 1 provides the hypothetical class of the object in question and it is not used in the calculation. It serves only to provide context.

4.1 Testing with Novel Values

The first test uses probe 1 that describes a human, which is somewhat smaller in size and weight from that of exemplar 2, and he has only one arm. As it is expected the highest activation takes place for exemplar 2. The value is somewhat lower than the maximum one due to the lack of matching in other shared attributes. The closer animal is a penguin, which due to sharing two legs and "medium" distance on the height and weight attributes comes with reasonable activation.

4.2 Testing with Missing Values

In the next testing task, we used as a probe an existing exemplar but with missing values in some of the attributes (that of exemplar 4). As it is also expected the best match took place with the correct exemplar but at a reduced activation due to lack of presence of values for two of the original attributes. The other animals in the exemplar set had rather low-level activations.

4.3 Details of Simulation

The simulations took place using the similarity function (9) while using functions (10) for calculation of distance per component depending on its type. The attention weights in (9) were kept fixed at value 0.125 for each component that entered in the calculations. We allowed 500 steps for reaching equilibrium of activations and the initial activation values were set randomly with a uniform distribution in the interval [-0.2, 0]. Ten simulation experiments took place in each case using different initial conditions for activations. Table 2 shows the average activations that were calculated on the basis of the individual experiments. The specificity value used was c=10 and the values for all parameters were set to the corresponding numbers that were given in previous sections. None of the above parameter values are critical for the proper operation of the concept system. The values of the parameters were determined initially by using guidance from the literature [1, 2] and by modifying them afterwards through trial and error. No misclassifications took place with this set of parameter values.

5 Conclusion

In this paper we have presented our proposal for a concept system that closely follows ideas from human concept research and provides enhanced flexibility. The system does not only allow the insertion of new concepts, with variable feature information between observations, but also the incremental update of the created concepts. It also allows for forgetting of rarely used or unimportant concepts. In this way it can track dynamic environments and transient situations. We have also presented a scheme that allows one to form "summary" level representations for feature values of more abstract (general) entities from the corresponding values of more concrete entities. The concept system uses the idea of similarity for classification of new precepts to existing concepts and uses a spreading activation mechanism for calculating the classification answer. It allows in essence a multi-concept answer supporting thus a fuzzy interpretation of the results. It also distinguishes between "typical" cases, that are used to update a concept's (summary) feature values and which are consequently discarded, from "boundary" cases (exemplars), which are maintained inside the system. The boundary cases might belong to different concepts albeit with different degrees of membership. In this paper we have demonstrated with simulations results, that in classification tasks the system can return answers even in cases where feature information is missing, but other provided features are similar enough to existing concepts. In this case, it also supplies back "default" values for the missing components of a probe. Due to lack of space we could not provide further simulation results on other aspects of the system. This evidence will be presented elsewhere.

The system borrows heavily from human concept research and it shows promise for supporting and developing flexible enough concept representations bringing together ideas from two out of the three dominant psychological theories of concepts. The disadvantage of the system is its relative complexity, but this is a fact that should be expected for increased flexibility. Obviously the future effort will concentrate on introducing further simplifications, while increasing the classification and concept learning capability. Paramount to this work is to explore in an extensive manner the complex relations of the various system parameters and devise ways in which these can be specified in a semi-automatic setting. An obvious idea is to link the classification thresholds to a cost function for dynamic adjustment of the thresholds. An important part of the future work will be to devise action representations so as to allow the linking of object concepts with affordances. Another future research direction includes basic linguistic capabilities, as it is expected that the more direct use of language will speed up the process of developing (summary) feature values for concept nodes, as in the case of joint attention between infant and caretaker. Finally a thorough investigation of the influence of the various parameters' values on the success of classification tasks will take place.

Acknowledgments. We would like to thank the EU CogSys GNOSYS project (FP6-003835) for its support to our work. Its contribution is greatly acknowledged.

References

1. Gardenfors, P.: Conceptual Spaces: The Geometry of Thought. MIT Press, Cambridge (2000)
2. Murray, G.: The big book of concepts. MIT Press, Cambridge (2004)
3. McClelland, J.: Retrieving General and Specific Information from Stored Knowledge of Specifics. In: Proceedings of the 3rd Annual Conference of the Cognitive Science Society, Berkeley, California, pp. 19–21 (August 1981)
4. Wilson, R., Martinez, T.: Improved Heterogeneous Distance Functions. Journal of Artificial Intelligence Research 6, 1–34 (1997)
5. Bullock, D., Grossberg, S., Gunther, F.: A self-organising neural model of motor equivalent reaching and tool use by a multijoint arm. Journal of Cognitive Neuroscience 5(4), 408–435 (1993)

Self-perturbation and Homeostasis in Embodied Recurrent Neural Networks: A Meta-model and Some Explorations with Mechanisms for Sensorimotor Coordination

Jorge Simão

DCC-Faculty of Sciences-University of Porto & LIACC
jsimao@dcc.fc.up.pt

Abstract. We present a model of a recurrent neural network, embodied in a minimalist articulated agent with a single link and joint. The configuration of the agent defined by one angle (degree of freedom), is determined by the activation state of the neural network. This is done by contracting a muscle with many muscular fibers, whose contraction state needs to be coordinated to generate high amplitude link displacements. In networks without homeostasic (self-regulatory) mechanism the neural state dynamics and the configuration state dynamics converges to a fixed point. Introduction of random noise, shows that fixed points are metastable. When neural units are endowed with homeostasic mechanisms in the form of threshold adjustment, the dynamics of the configuration angle and neural state becomes aperiodic. Learning mechanisms foster functional and structural cluster formation, and modifies the distribution of the kinetic energy of the network. We also present a meta-model of embodied neural agents, that identifies self-perturbation as a mechanism for neural development without a teacher.

1 Introduction

Motivated by concepts and ideas from autopoetic philosophy, ecological psychology, complex systems theory, and situated artificial intelligence research, we present a recurrent neural network model [1] to study how a minimalist embodied agent can be made to develop basic sensori-motor coordination skills. The agent consists of a link and a rotational joint in a 2D plane. The agent configuration is fully determined by a single degree of freedom — the joint angle. Attached to the link is a muscle whose contraction/distension produces an angular displacement of the link. The activity of the neural units determine the level of contraction of the muscle. The units in our neural model are endowed with homeostasis modeled as an adaptive threshold adjustment. Our goal is to study what neural and behavioral mechanisms are required for the emergence of effective movement of body limbs in humans and high-order animals. In particular, we want to explore if self-perturbation — perceptual input generate by the agents own action, is an effective mechanism to guide neural development. In

J. Marques de Sá et al. (Eds.): ICANN 2007, Part II, LNCS 4669, pp. 934–942, 2007.

this article we present preliminary results related to the dynamics of the neural controller, looking both at neural state and resulting configuration state.

Below, we present first a meta-model for embodied neural agents (section 2). Next, we describe the particular embodied neural agent model studied in this article (section 3), and present the results of several computational experiments using the model (section 4). Section 5 presents related work and discusses simulation results.

2 A Meta-model for Embodied Neural Agents

To study self-organization in embodied neural agents and the development of sensori-motor skills, we have made an abstract characterization of this type of agent models. [See [2] for another characterization of embodied developmental agents].

Agents are characterized at two levels: the macro-level and the micro-level. The macro-level is defined by the configuration state — a formal description of the agents body posture in space, as seen by an external observer or as made apparent to the agent through self-perception. A small number of degrees of freedom is often required to describe an agent at this level. The micro-level is a characterization of the state of its neural controller. This includes the activation level of neural units (e.g. mean firing rate), units thresholds, and neural connections weights. Usually, the micro-level requires a much higher number of degrees of freedom to be fully described than the macro-level, since an agent with few links and joints may have a controller with many neural units. Interfacing the micro and macro-levels, agent descriptions include the way the neural controller is connected to the agents body — both in muscular connections (efferent) and in the way sensation-perception cells/inputs impinge in the neural controller.

Agents are often situated in some environment, in such a way that its behavior and interaction with the environment may be observed by some external observer. In fig. 1**left)**, we make a sketch representation of the relation between the agent, its environment, the external observer, and the two levels of description.

A key aspect of natural agents, is that the mapping from the (micro) neural level and the (macro) configuration level is not one-to-one. The coordinated action of a large number of neurons and muscular cells is usually required to generate strong and high-amplitude body movements. Additionally, many different neural states may mandate the same body configuration.

Because agents have units sensitive to environmental and body state (the sensation-perception inputs), agents can sense the effects of their own actions. Thus the micro and macro level are connected in a two-way causality loop. The state of the micro-level determines the body configuration (apart from external mechanical perturbations of the agent body, such as gravity and social manipulation), and the body configuration perturbates the internal dynamics of the neural controller. Fig. 1**right)**, represents the causality loops in agent behavior according to our meta-model. **X** represent the state of the neural controller of

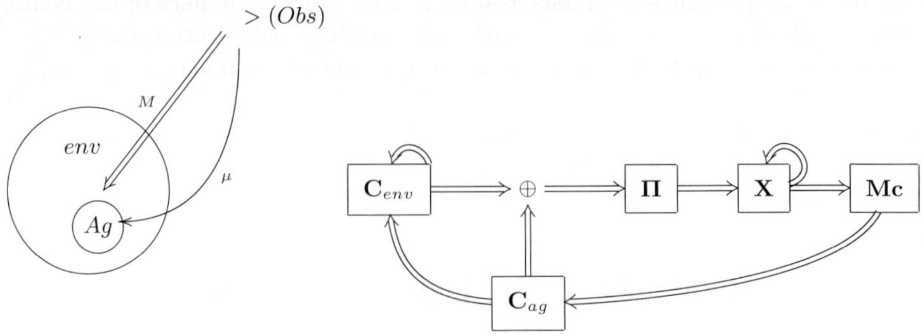

Fig. 1. Meta-Model of Embodied Neural Agents: **left)**: conceptual diagram of the agent, the external observer, the environment, the macro-level and the micro-level descriptions; **right)**: schematic block diagram of agent-agent and agent-environment functional dependencies

the agent (part of micro-level or internal state), and $\mathbf{C}_{ag}$ represents the body configuration of the agent (the macro-level or external state).

3 A Model of a Minimalist Embodied Neural Agent

We model an embodied agent with a single link and a single joint. The joint angle ψ fully defines the body configuration of the agent. The joint angle is determined by the contraction of a simplified muscle that works like a mechanical lever. The muscle has a large number of muscular units m_i. The contraction/extension of a muscular unit m_i produces a spatial displacement Δs_i, and the summation of all displacements determines the joint angle. Formally, $\psi = f(\sum_i \Delta s_i)$, where f is a function of the detailed geometry of the agent.We assume that the contraction of a single neural units produces a relatively small link displacement. In particular, the simultaneous contraction of a large proportion of muscular units is required to generate maximum displacement of the link. Moreover, the joint angle ψ is always constrained to lie within a maximum amplitude interval $[-\frac{pi}{2}, \frac{pi}{2}]$. In figure 2, we show the abstract design of the agent (left), and the graphical design as visualized in our simulator (right).

Muscle contraction (and thus body configuration) is controlled by a neural population with N_m units, whose activation/excitation state we represent by the vector $\mathbf{X_m} \equiv [x_1, \ldots, x_i, \ldots, x_{N_m}]$. We make a simple attachment between this motor control neural population and the muscle units, by making the number of muscular units equal to the number of neural units, and connecting them one-to-one (unidirectional). When all units are in a rest/natural activation value ϕ takes value 0 (the link is horizontal).

Neural units are connected in a network/graph as a fully recurrent neural network (all units connect to all) [1]. Connection strengths are represented with a connectivity matrix $\mathbf{M}$, where element c_{ij} represents the connection strength

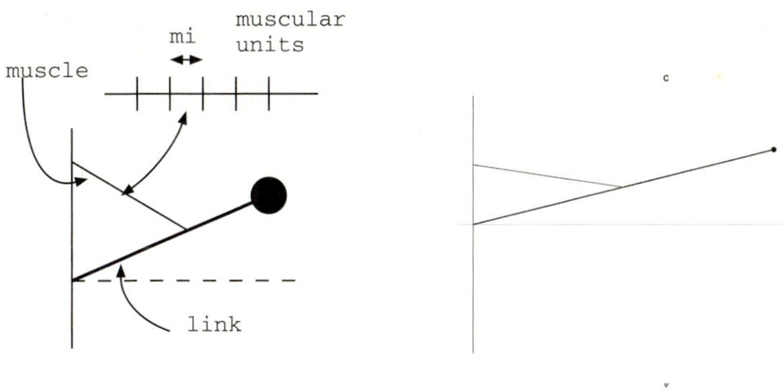

Fig. 2. Body Configuration of Minimalist Articulated Agent (one degree of freedom, one set of muscular units): **(left)**: abstract design; **(right)**: visualization in neural simulator

or weight between unit i and j. In the simulation results presented below we experiment both with fixed connection weights, and with time varying connection weights that change continuously as the system runs. There is no separation between learning phase and performance/test phase (as is often the case in traditional connectionist models [3]).

Neural unit are assumed to be initially connected in random weights, using a normal distribution with mean value 0 and variance $\sigma^2(M)$. When learning is used, weights are modified in a single direction. Positive/excitatory weights can only be increased, and negative/inhibitory weights can only be decreased. Learning is modeled as an hebbian-like learning rule for positive connections, and an anti-hebbian-like learning rule for negative connections.

Neural units have an adaptive threshold that is used to maintain units in a sensitive state. This is equivalent to cellular homeostasis mechanisms in biological neural networks [4]. For unit i we represent its threshold as θ_i. When a unit's activation is very high, a slow adaptation process takes place that gradually moves the activation value to a rest or natural activation value x_0. Likewise, when units activation value is low the same adaptation process takes place to raise activation level to x_0.

The operation of units is formally defined using two ordinary first-order differential equations [approximate by the Euler method in the simulations below]. The first equation below describes the (fast) dynamics of individual units activation. The second equation describes the (slower) dynamics of homeostasis. The learning dynamics is modelled by the third equation presented below.

$$\begin{cases} \tau_1 \dot{x}_i = -x_i + x_0 + f(\sum_j c_{ji} x_i + c_i \pi_i - \theta_i) + \xi \\ \tau_2 \dot{\theta}_i = x_i - x_0 \\ \tau_3 \dot{c}_{ji} = sign(c_{ji}) \cdot g(x_{\max} - x_i) g(x_{\max} - x_j), \end{cases}$$

above τ_1, τ_2, and tau_3, with $\tau_1 << \tau_2$, are constants for the characteristic times of the neural processes modelled. x_0 is the resting or natural activation of units. f is a activation gain function. We use a constant gain G. Units activation x_i is constrained to be in the interval $[x_{\min}, x_{\max}]$, where $x_{\max}$ is the saturation value and $x_{\min}$ is the lowest/depression value. ξ is a normally distributed random noise value. π_i represent the current value of the self-perturbation input for unit i and c_i is a fixed input gain. Only for units in $\mathbf{X}_p$ is π_i non-zero. g is an auxiliary function that produces higher values when the argument is close to 0. Its used to implement the (anti-)hebbian learning function. *sign* is the standard sign function.

Solving for equilibrium in the first equation shows that at rest $x_i = x_0 + f(\sum_j c_{ji} x_i + c_i \pi_i - \theta_i) + \xi$, which is a fast quiescent/rest state. Solving for equilibrium for the second equation show that at rest $x_i = x_0$, which is a slow quiescent/rest state (since $\tau_1 << \tau_2$).

4 Experimental Results

Embodied recurrent neural networks in articulated agents have receive modest attention in the research community [5,6]. Thus, we have made some preliminary explorations in the dynamics of our model, before moving to complex model setting. We present the simulation results in increasing level of complexity, starting with a stripped/simplified version of the neural model and progressively increasing the number of model elements at work.

The simplest operation regime for the system is when neural units are not connected, that is, the connection matrix $\mathbf{M}$ is null. In this case, network dynamics is governed by the internal noise variables ξ_i and, if units homeostasis is turned off, the body configuration angle ψ takes values normally distributed around 0. [The dynamics of the proprio-perceptive population $\mathbf{X}_p$ in not considered here.] Fig.3 show the distribution of angle ϕ and the time-series of ϕ in one particular simulation run (with 500 steps) with a high value of $\sigma^2(\xi)$. The distribution is clearly bell-shaped (given some sampling error). This is the expected result since ξ_i is normally distributed, and link displacement is proportional to units activation. All simulation runs without connections produce similar results. The standard deviation of ϕ is a monotonic increasing function of the standard deviation of ξ since units activation values are statistically independent. If units are set to have homeostasis, fluctuation around $\phi = 0$ can be made arbitrarily low by reducing $\sigma^2(\xi)$ and reducing the characteristic time for homeostasis τ_2.

When units are connected according to some random weight matrix $\mathbf{M}$, with mean 0 and variance $\sigma^2(M)$, and weights are fixed (no learning), the behavior of the system changes. If noise is removed and homeostasis is also removed, the activation state and the configuration angle converges in most simulation runs (with different random matrices) to a **fixed point**. In the fixed point most units are either fully saturated or fully depressed. Different random matrices produce different fixed points. Fig.4**left)** shows the evolution of the neural state over 100 time steps. High activation of units is color coded as red, low activation as blue, and values near x_0 as green[1]. Also in fig.4**left)** we show the evolution of the

[1] We use this color code since red can be associated with heat produced by cells' activity, and blue with cold.

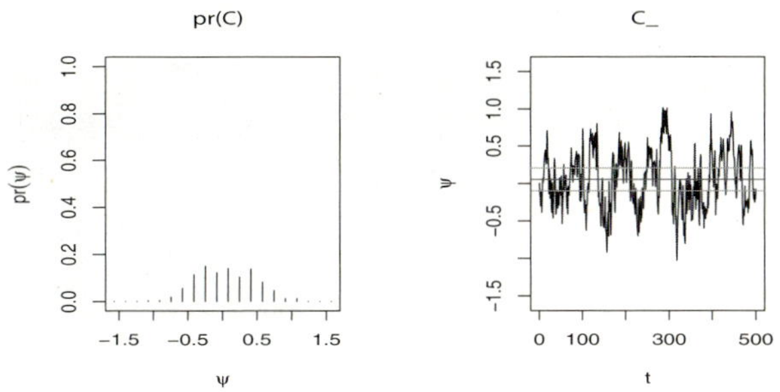

Fig. 3. Distribution and time-series of ψ when units are not connected and units are not homeostasic ($\sigma^2(\xi) = 10, N = 16$)

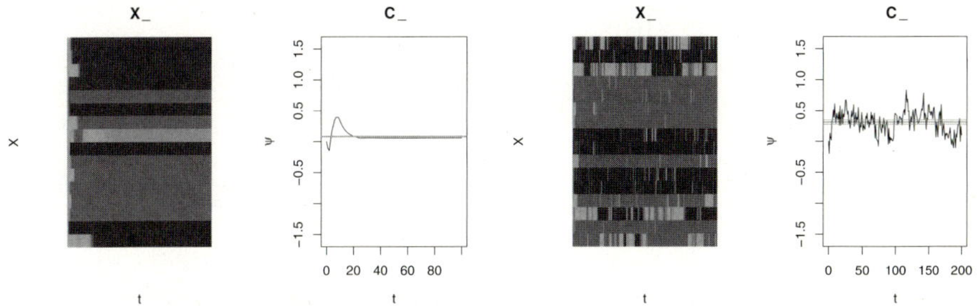

Fig. 4. Typical system's behavior with random matrix ($\sigma^2(M) = 1$) and no homeostasis ($\tau_2 = \infty$): **left)** no noise ($\sigma^2(\xi) = 0$): Convergence to a fixed point with most units saturated or depressed ($N = 16$); **right)** high noise ($\sigma^2(\xi) = 5$): Convergence to a small region of the state state (meta-stable)

configuration state for the same simulation run. In a set of 10 consecutive runs with the same settings, the results obtained where qualitatively the same. Analysis of the Hopfield energy function of neural states, $E_t(X) = -\frac{1}{2}\sum_{i,j} c_{ij}x_i x_j$, showed that energy values do not always reach a global minimum, which is expected because connections are not symmetric [1].

Fig.4**right)** show the system evolution when noise is added to the system of a typical simulation run. The results show that the neural and configuration state converges to a small region of state space, and remain in that region (the region is meta-stable). [The results were qualitatively the same for a set of 10 consecutive runs.] This suggests that fixed points in the previous case (no noise) are Lyapunov stable (neural states tend to stay within a small distance of a fixed point when perturbed). Most units remain for most of the time close to saturation or depressive points (red and blue colors). Fluctuations of ψ are small. Fluctuations of ψ around a mid-point can not be assumed to be normally distributed since units are interconnected.

When units use homeostasis, the behavior of the system changes considerably. The proportion of time units are not saturated or depressed increases, as inspection of the differential equation for the threshold above would suggest. However,

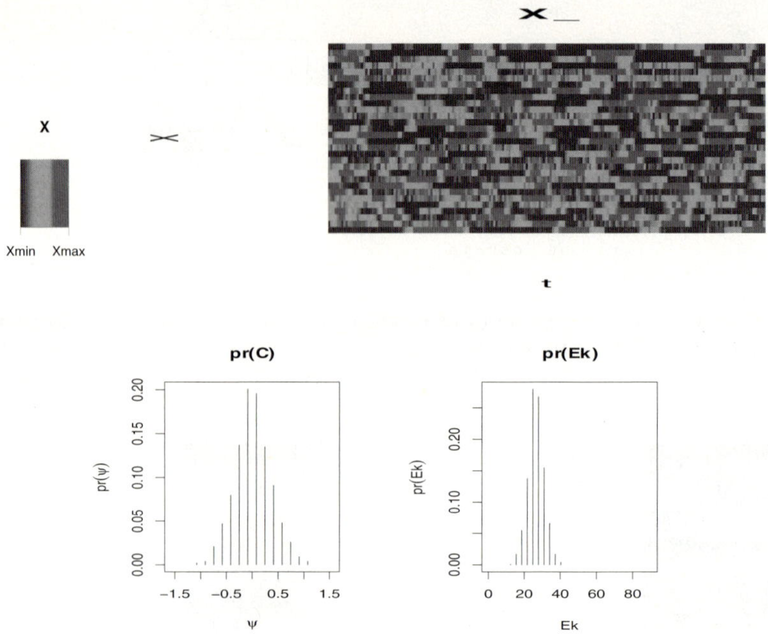

Fig. 5. A qualitatively typical system's behavior with random matrix ($\sigma^2(M) = 1$), no learning, moderate noise ($\sigma^2(\xi) = 0.2$), and homeostasis ($\tau_2 = \frac{\tau_1}{5}$): The system exhibits non-periodic behavior ($N = 30$). **top)**evolution of neural state; **bottom)**Distribution of configuration angle ψ, and kinetic energy.

most units do not remain with an activation value near x_0 all the time since they are taken away from homeostasis due to interconnection with other units. Fig. 5 shows a qualitatively typical simulation run (from a set of 10 consecutive runs). [Noise was set to a moderate level ($\sigma^2(\xi) = 0.2$, when compared with the set maximum activation level ($x_{\max} = 3$).] The system state does not converge to any attractor, but exhibits **non-periodic behavior** due to threshold adjustments. The distribution of the configuration angle and the kinetic energy of the units is better fitted by a power-law $pr(\psi) \propto \psi^{-\lambda}$ (one for each side of the distribution), than by a Gaussian curve since values near $x0$ are more likely than would be expected in a Gaussian curve[2]. This shows that threshold adjustment is a valuable mechanism for generating complex patterns in neural controllers' activity, preventing the system to become locked in a fixed-point.

When the learning rule is used to modify connection weights, the system behavior remains non-periodic but now units form high-connectivity clusters that produce correlated activity. We measure this by using a structural clustering

[2] We are currently working on further simulations to obtain reliable statistics for the λ exponents. In this particular run, the sum of squared errors for the best fitting parameters was one order of magnitude lower for the power-law than for the Gaussian.

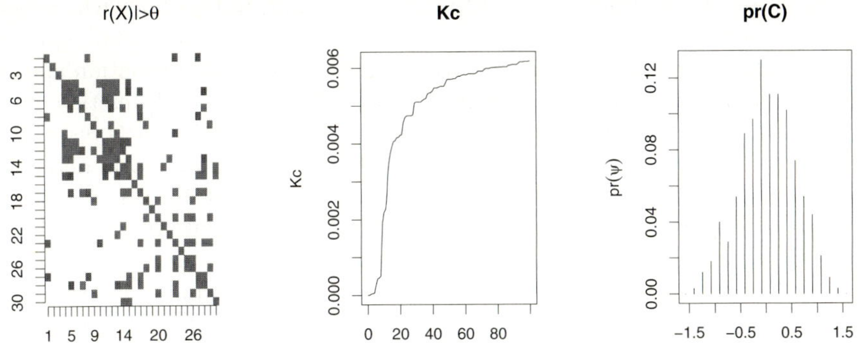

Fig. 6. Learning generates clusters of structural connectivity and functional correlations. **left)** functional correlation matrix; **middle)** evolution of structural clustering index; **right)** distribution of configuration angle ψ

index defined as $Kc = \frac{1}{NM_{\max}} \cdot \sum_{ijk} c_{ij} c_{ik} c_{jk}$ (only positive connections considered). Fig. 6 shows that Kc and the absolute value of connections increases till the learning capability of the network is reached. The clusters formed have the effect of increasing the average kinetic energy, and breaking the symmetry of the configuration angle distribution. Low angle values (corresponding to many depressed units) are less likely than higher values (corresponding to many saturated units). Configuration angles are also more likely to take extreme values (thus changing the λ exponent in the power-law). This is the result of simulatenous activity of units is the formed clusters. Making connections weights change to be in one direction only helps the learning process in the initial phase, but is a contributing factor in limiting the network ability for continuous learning. We are currently investigating ways to extend networks' ability for continuous learning.

5 Summary Discussion, Related Work, and Conclusions

We have presented a model of a recurrent neural network with homeostasic units, for an embodied agent with a single degree of freedom. Activity of neural cells generates muscular contraction, and determines the configuration angle of the agent. The neural model is motivated by a meta-model of embodied neural agents, whose goal is to inspire the design of agents that learn without a teacher using the mechanism of self-perturbation — that is, perceptual input generated by the agents own actions. Simulation work realize so far enabled us to arrive at some preparatory and preliminary conclusions. Homeostasis in neural cells allow neural controller to "escape" fixed points and allow the agent to explore its configuration space. Homeostasic mechanisms have been identified in the biological neural networks literature [4],and its behavioral relevance is starting to be explored [7]. The presented results are along this direction. Connections between neural units produces aperiodic behavior in our neural model. This as been previously advanced in the neural networks literature [8] and fits known empirical data about animal and human brain activity [9]. Our simulation results replicate these findings. Learning mechanisms foster functional clusters formation and increases average kinetic energy of units. Configuration angle and kinetic energy distribution is better characterized by a power-law rather than a Gaussian — because homeostasis brings connections to rest activity, and clusters allow aggregates of units to move far from the rest activation level. Similar activation distribution has been reported in empirical data about activity of human brain [10]. We are currently working to improve the statistical robustness of our results, using multiple simulation runs and networks with a higher number of units. The role of self-perception as a driving force in neural and behavior development is another focus of our current experimental work. In particular, we are studying how neural dynamics and learning are affected by self-perturbation. Theoretical and experimental advances in robotics research have identified complexity theory as a promising tool to understand how neural agents can self-organize to produce adaptive behavior [11]. Recent work on developmental robotics has been proposing candidate mechanisms for understanding the development of embodied agents [2]. Our work in self-perturbation is a further step in this direction.

References

1. Hopfield, J.J.: Neural networks and physical systems with emergent collective computational abilities. Proc. Natl. Acad. Sci. USA 70, 2554–2558 (1982)
2. Weng, J.: Developmental robots: theory and experiments. International Journal of Humanoid Robotics 1(2), 199–236 (2004)
3. McClelland, J.L., Rumelhart, D.E.: Parallel Distributed Processing: Explorations in the Microstructure of Cognition. MIT Press, Cambridge (1986)
4. Turrigiano, G., Nelson, S.B.: Homeostatic plasticity in the developing nervous system. Nature Reviews Neuroscience, 97–101 (2004)
5. Vogels, T.P., Rajan, K., Abbott, L.F.: Neural network dynamics. Annual Review of Neuroscience 28(1), 357–376 (2005)
6. Cessac, B., Samuelides, M.: From neuron to neural network dynamics. In Topics in Dynamical Neural Networks : From Large Scale Neural Networks to Motor Control and Vision (EPJ special issue). Forthcoming.
7. Di Paolo, E.A.: Organismically-inspired robotics: Homeostatic adaptation and natural teleology beyond the closed sensorimotor loop. In Dynamical Systems Approach to Embodiment and Sociality, Advanced Knowledge International, International Series on Advanced Intelligence, Magill, Australia: Advanced Knowledge International Press, pp. 19–42 (2003)
8. Harter, D., Kozma, R.: Aperiodic dynamics and the self-organization of cognitive maps in autonomous agents. International Jounral on Artificial Intelligence Tools 21(9), 955–971 (2005)
9. Freeman, W.J.: How Brains Make Up Their Minds. Columbia University Press (2001)
10. Victor M. Eguìluz, Dante R. Chialvo, Guillermo A. Cecchi, Marwan Baliki, and A. Vania Apkarian. Scale-free brain functional networks. *Physical Review Letters*, 94(1), 018102.1–4 (2005)
11. Nolfi, S.: Behaviour as a complex adaptive system: On the role of self-organization in the development of individual and collective behaviour. ComplexUs, 2(3–4), 195–203 (2004/2005)

An Oscillatory Model for Multimodal Processing of Short Language Instructions

Christo Panchev

School of Computing and Technology, University of Sunderland
Sunderland SR2 7QB, United Kingdom
christo.panchev@sunderland.ac.uk
http://www.his.sunderland.ac.uk/christo

Abstract. Language skills are dominantly implemented in one hemisphere (usually the left), with the pre-frontal areas playing a critical part (the inferior frontal area of Broca and the superior temporal area of Wernicke), but a network of additional regions in the brain, including some from the non-dominant hemisphere, are necessary for complete language functionality. This paper presents a neural architecture built on spiking neurons which implements a mechanism of associating representations of concepts in different modalities; as well as integrating sequential language input into a coherent representation/interpretation of an instruction. It follows the paradigm of temporal binding, namely synchronisation and phase locking of distributed representations in nested gamma-theta oscillations. The functionality of the architecture is presented in a set of experiments of language instructions given to a real robot.

1 Introduction

Sound and vision are processed in different parts of the brain. Nevertheless, a spoken word describing an object and an object present in the visual field would raise activity in the brain which leads to a coherent representation of the object associating the interpretation of the sound and vision, as well as features related to other modalities, e.g. affordances related to possible actions (Shtyrov et al., 2004; Arbib, 2005; Arbib and Bota, 2003). In the context of language processing, the cortex can be seen as a multimodal information-merging computational device which uses neurons representing and processing information from various sensory and motor modalities. Similar to the brain, where language processing is related to actual interaction with the environment and is performed in conjunction with internal constraints, computational language systems should be able to associate objects and actions with their perceptions and affordances. Consequently, a growing number of computational models of language processing are being based on multi-modal associations (Burns et al., 2003; Oates, 2001; Siskind, 2000).

The development of the neural architectures presented here follow the above arguments for language representations and processing in the brain. It is based on the view that concepts are represented by distributed cell assemblies across

J. Marques de Sá et al. (Eds.): ICANN 2007, Part II, LNCS 4669, pp. 943–952, 2007.

multiple areas of different modalities, with objects activating neurons in both auditory and visual areas, and actions activating neurons in auditory and motor areas.

Different mechanisms have been considered for the task of associating/binding distributed representation (features) into one coherent pattern of activity. The temporal correlation hypothesis, first proposed by von der Malsburg (1981), postulates that features are bound into complex representations based on the temporal correlation of the activity patterns of the neurons representing them. It is based on neurophysiological studies showing that neurons driven by a single stimulus respond synchronously with oscillations in the 30-70 Hz range. Such synchronous oscillation have also been observed between different sensory and motor regions (Roelfsema et al., 1997; Bressler et al., 1993). Temporal binding for a language processing task has been discussed in a number of experimental and modelling studies, including the slot-filler model presented by Sougné (1999) which was evaluated and found to have similar to human performance in a range of reasoning tasks: such as in short-term memory span, serial effects of STM, similarity effects of STM, double dissociation between STM and LTM, in the negation effects of conditional reasoning, and the effects related to multiple instantiation. Although, in most models temporal binding has been based on synchronous activity, the principles of the idea can also be applied in a more general form where the condition is a phase-locked activity of the neurons representing the individual features. Such an approach is taken in the development of part of the model presented here. It introduces binding across different modalities via associative connections trained using Spike-Timing Dependent Plasticity (STDP), implements a working memory model, and in further work (not covered by this paper) incorporates it into an embodiment platform for evaluation in a real world environment and tasks.

2 Overall Model

The overall objective of the work, part of which is presented in this paper, is to build a spiking neural network architecture for robot control using language instructions. The robot moves in an environment with objects of different shape and colour. A spiking neural network controls the robot in avoiding any obstacles while navigating around the environment within a suitably constrained scenario of finding and picking objects. The network also implements the recognition and execution of language instructions for: direct navigation, finding and recognising objects, and moving objects in the environment. The lexicon includes the words: *go, stop, turn, left, right, back* for navigation; *find, grab, drop* for object manipulation; and *red, green, blue, ball, box, cone* for describing the 9 objects on the scene. The set of instructions that can be given to the robot includes short (1-3 word) phrases, such as *go*, *turn left* or *find red ball.*

The work involves modelling of primary sensory areas - auditory, visual and tactile, higher cognitive functions areas - language processing/understanding, working memory, and motor control areas (Panchev and Wermter, 2006; Panchev, 2006).

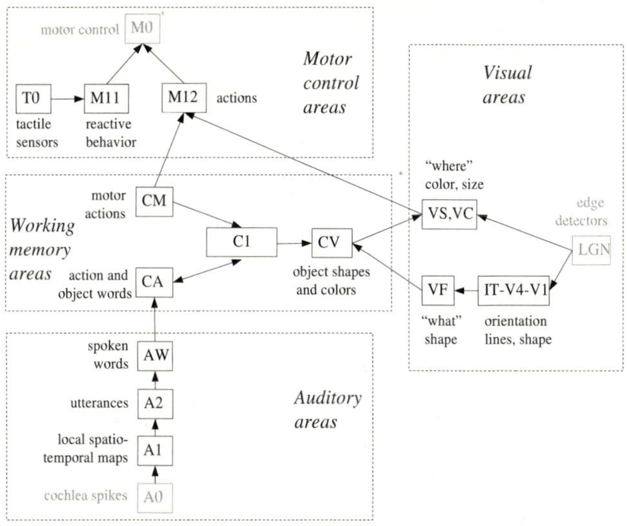

Fig. 1. Multi-modal ADDS spiking neural network architecture for robot control

The overall architecture is presented in figure 1. It consists of four main interconnected modules: the auditory (A0, A1, A2 and AW), the visual (LGN, V1, V4, IT, VF, VC and VS), the motor control (T0, M0, M11 and M12) and the central associative and working memory modules (CA, CV, CM and C1). All areas are implemented using (Active Dendrites and Dynamic Synapses (ADDS) spiking neurons (Panchev, 2005). The rest of this paper will concentrate on the detailed description of the architecture and functionality of the (central) working memory areas which implement the multimodal integration of language input in an neural circuit with nested gamma-theta oscillations.

3 Multi-modal Sensory-Motor Integration and Working Memory

3.1 Architecture

The Working Memory (Central) modules (figure 2) implement three main functions: (1) The networks in the Central module of the model (the AW-CA-C circuit in particular) implement the simple grammar of the robot's instruction set and the integration of a sequence of words into a phrase (object and/or instruction); (2) The neural circuits from the Working Memory (CA-C1a and CA-C1o) maintain the instructions given to the robots for the period necessary for their execution; and (3) Neurons from the central modules (CA-C1a-CM, CA-C1o-C1 and CA-C1o-C2) are the points of integration and transfer of information across the different modalities.

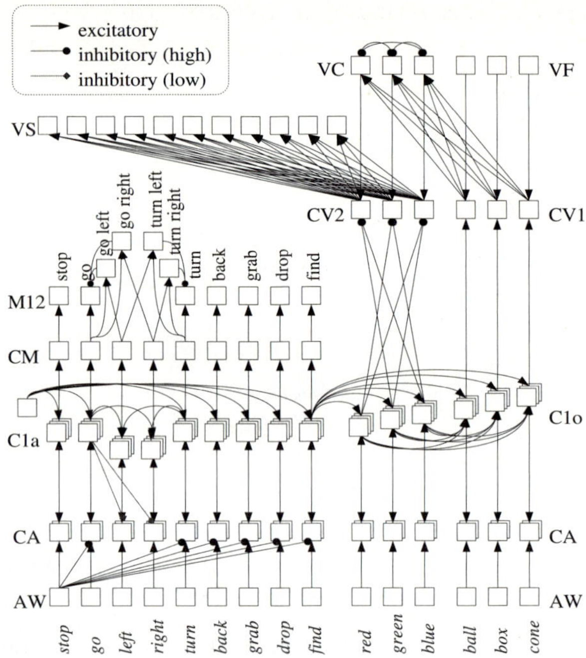

Fig. 2. Architecture of the Central (Working Memory) module. For clarity only some inhibitory connections from AW to CA and from C1a,o to CA are shown.

The central areas (CA, C1a, C1o, CV1, CV2 and CM) include small clusters of neurons representing each of the words in the robot's lexicon. C1a is a sub-area for action words and C1o is a sub-area for object words. Lateral and feed-back connections implement the relationship between the words in an instruction. The inhibitory connections are setup so that high inhibition prevents the neurons from firing whereas low inhibition delays their firing. The strong inhibition from AW to CA connects neurons representing words which have the same order position in the instructions. The weak inhibition from C to CA connects neurons representing words in successive position in the instruction phrases. The central layers (C1a and C1o) interact with the other modules via neurons in area CA for auditory, areas CV1,2 for the visual and area CM for the motor areas. The connectivity pattern of the WM module was hard-wired with the strength of the inhibitory synapses fixed and the weights of the excitatory synapses trained using the STDP-type synaptic plasticity protocol presented in (Panchev et al., 2002; Panchev, 2007).

Each concept recognised by the robot is represented by distributed cell assemblies across different modalities. For example the representation of *go* includes neurons from the AW, CA, C1a, CM and M12 areas, and entities such as *box* activate neurons from AW, CA, C1o, CV and VF (and possibly for colour and spatial location also including the VC and VS) areas.

3.2 Language Understanding: Recognising Sequences of Words

The main constraints considered during the design and implementation of the computational architecture for language processing presented here are: (1) The words forming an instruction would arrive in a sequence, and the mechanism should support gradual build up of the semantic information contained in the sequence under a set of syntactic and semantic constraints; (2) The temporal structure of a spoken sequence of words contains relatively high levels of noise, more specifically, the interval between two consecutive words can range from a few hundred milliseconds to a few seconds. The proposed neural mechanism should cope with such fluctuations; (3) The mechanism should allow for insertions such as adjectives. For example, the instructions *find box* and *find red box* should lead to the same behaviour of the robot if there is a red box in front of it.

Part of the architecture presented here is influenced by a model presented by Lisman and Jensen (Lisman and Idiart, 1995; Idiart and Lisman, 1995; Jensen and Lisman, 2005). It is implemented with the CA-C1a,o circuit running two nested oscillations. The main difference here is the implementation of the neuronal mechanisms supporting the oscillations. While in the original model it relies on activity-dependent changes of membrane excitability (mainly the membrane after-depolarisation effect), the current model uses long range circuits (implemented by the recurrent CA-C1 connection) and lateral excitation (implemented by the lateral C1-C1 connections) forming reverberatory cell assembly activity as proposed by Hebb (1949). The oscillation representing a phrase runs at 10 Hz. Within each cycle of this theta oscillation, the cell assemblies representing each of the currently active concepts spike in a sequence forming a 30 Hz oscillation (figure 3).

Each cycle of the theta oscillation can include one phrase (instruction). The start of the cycle is marked by a Central Pattern Generator (CPG) neuron in

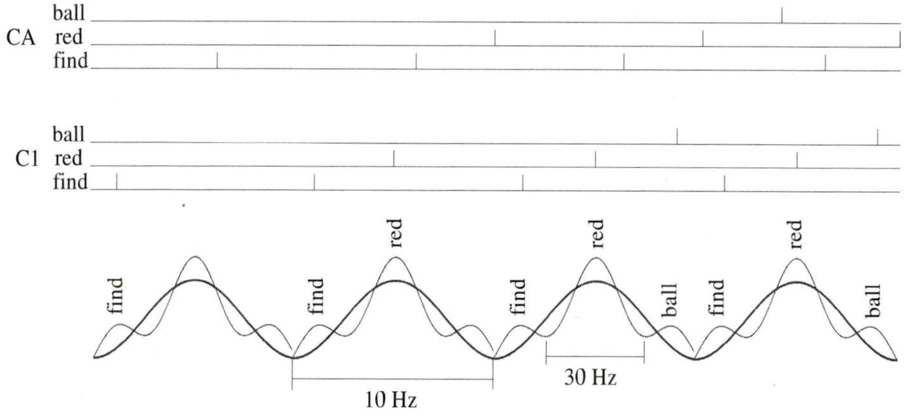

Fig. 3. CA-C1a,o circuit implements nested oscillations where each word in the instruction is represented within a subsequent cycle of a high-frequency oscillation. The instruction is repeated in each cycle of a low frequency oscillation.

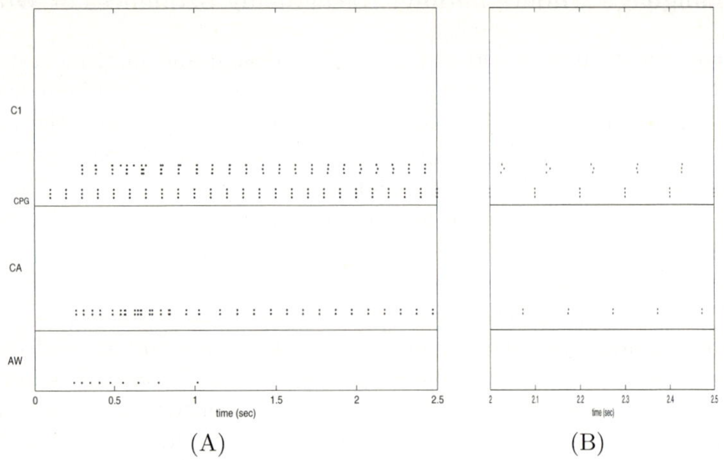

Fig. 4. Working memory activity in processing the instruction *go* given at time 0. (A) At approximately 250 milliseconds the word is recognised and activates the AW neuron representing *go*. The spike burst from AW activates the neurons in CA and thereby the working memory oscillation C1a-CA. (B) Zoom in the oscillation after the activation pattern has stabilised.

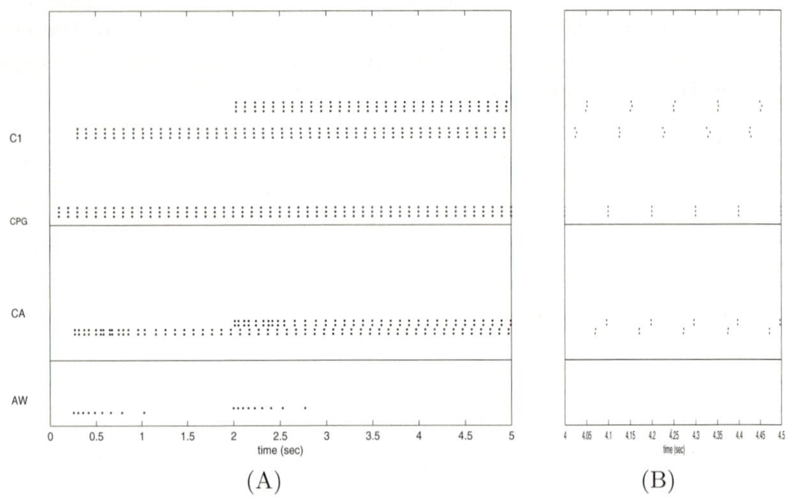

Fig. 5. Working memory activity in processing the instruction *turn left* given as a sequence of the words *turn* at time 0 sec and *left* at 1.66 sec. (A) At approximately 250 milliseconds the first word is recognised and activates the AW neuron representing *turn*. The spike burst from AW activates the neurons in CA and thereby the working memory oscillation C1a-CA. At approximately 2 sec the word *left* is recognised an enters the working memory oscillation after *turn* (B) Zoom in the oscillation after the activation pattern has stabilised.

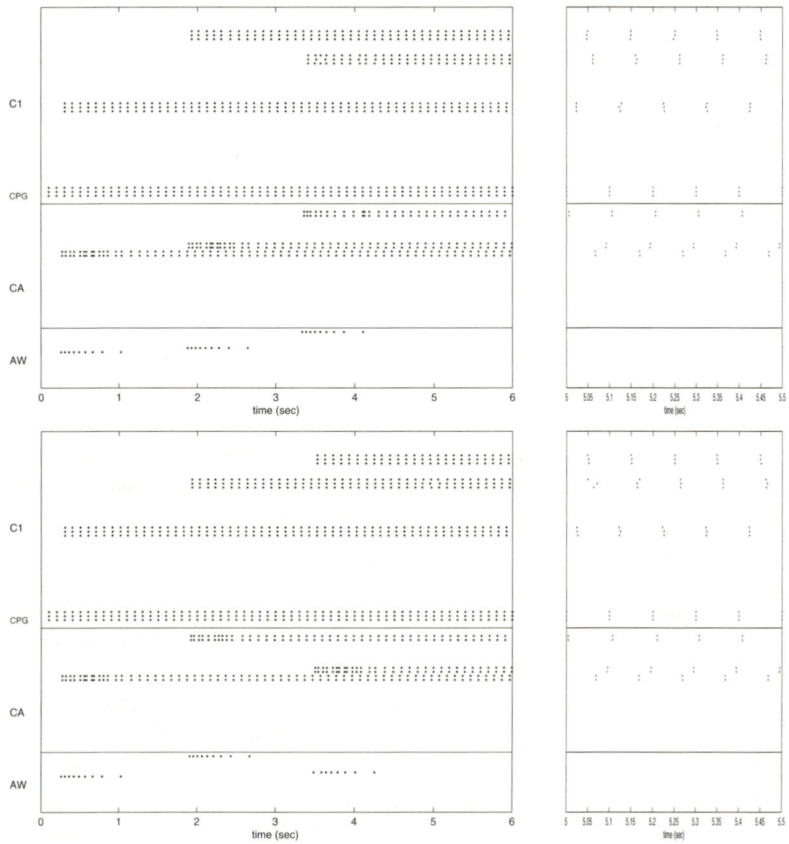

Fig. 6. Working memory activity in processing the instruction *find red box* (top) and *find box red* (bottom). In both cases the pattern of the final stable oscillation represents the sequence *find red box.*

area C1 that spikes at 10 Hz and sends signals to all C1 neurons representing words which can appear at the beginning of a phrase. These signals generate subthreshold membrane potentials at the C1 neurons and alone are not sufficient to activate the assemblies. Additional input from the neurons in CA is required for these neurons to spike. The CA and C1 neurons in a cell assembly representing a particular concept have recurrent connections and formulate a reverberating oscillatory activity between the two areas. Thus, activation of a single word constituting an instruction (or being the first word of an instruction), e.g. *go*, would be as follows (figure 4): upon recognition of the auditory input stream as the word *go* the AW neurons for that word will respond with a decaying spike burst causing several spikes in the CA neurons for *go*. In parallel the inhibition from AW to CA will shut down any oscillatory activity of a word which can take the same place as *go*, e.g. *stop* or *find*. Being the first word, this will remove any instruction currently held in CA-C. The combined input from CPG and CA

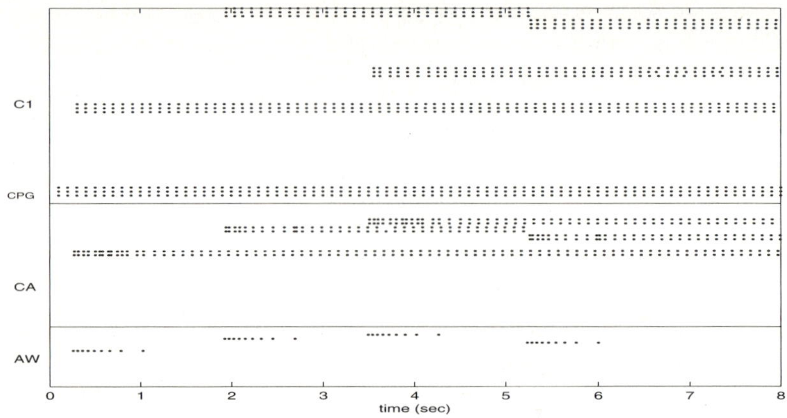

Fig. 7. Correcting the target in the instruction: Working memory activity in processing the instruction *find blue ball* followed by the word *green.* The final activation pattern of the working memory is the sequence *find green ball.*

neurons will activate the neurons representing *go* in C1a. In return, the neurons from C1a will activate (with some delay) the CA again as well as motor control neurons in CM and M12. Following the propagation of activity, the CA-C1a neurons for the word *go* will oscillate with a precise frequency led by the spikes from CPG and maintain this activity while subsequent words come in and/or the instruction is being executed.

The processing of words taking second and third position in the instruction phrases follows a similar activation patter (figure 5). The main difference is that instead of receiving inputs from CPG, the neurons representing such words in area C1 receive lateral inputs from the C1 neurons representing words which can precede them in a valid instruction. For example the C1a neurons representing *left* receive lateral excitatory connections from the C1a neurons representing *go* and *turn.* In addition, the CA neurons for the word *left* will also receive low strength fast inhibition from the C1a neurons of *go* and *turn.* This inhibition is not sufficient to prevent the CA neurons from firing but rather delays their firing and facilitates the order of activation in the CA-C1 oscillations. Critically, the weak inhibition from C1 to CA ensures that when new words come as input, they enter the CA-C1 oscillation at the appropriate place, i.e. after the word which should precede them. This is in contrast with some earlier working memory models based on oscillation, where the new items join the oscillation at the front, e.g. (Lisman and Idiart, 1995).

This architecture supports a gradual build up of the current context which allows a wide range of fluctuations in the intervals between the consecutive words in the input stream, including insertions. The CA-C1 oscillation maintains the current context until the next word arrives. Upon arriving of a word from the auditory stream (that is activation in AW), the new entity is included at the appropriate place in the current context in accordance with the semantic and syntactic constraints in the robot's dictionary. For example, if the current context is *find red* and the new input word is *box*, the new oscillation will be a sequential firing of the assemblies for *find red box* (figure 6 top), whereas if the current context is *find box*, a subsequent input word *red* will be included in just before *box* and again lead to the representation of *find red box* (figure 6 bottom).

4 Conclusion

The paper presented a neural architecture implementing association of object descriptions and actions represented in different perceptual and motor modalities and constructing a short language instruction. The architecture follows the temporal binding paradigm, and implemented a synchrony binding of multimodal distributed representations of the same concept and phase locking for binding language entities from an instruction - both running in nested gamma and theta oscillations respectively. Furthermore, the language instruction was maintained in the working memory during its execution by the robot. The architecture was implemented using ADDS spiking neurons and the excitatory connections were trained using STDP learning algorithm. The model was able to correctly recognise the instructions given to the robot and in further studies was shown to achieve good performance in executing the required actions. Although implemented with a simple grammar and in a relatively constrained environment, the results from the model showed that the architecture and mechanisms on which it is based have the potential to play a role in the higher level cognitive processes in the brain as well as in computational models integrating such processes. Further modelling and experimental work will accommodate higher number of words, more complex grammar and robot's environment, and will allow further evaluation of the mechanisms and principles employed in this architecture.

References

Anderson, J., Rosenfeld, E. (eds.): Neurocomputing: Foundations of Research. MIT Press, Cambridge (1988)

Arbib, M.: From monkey-like action recognition to human language: An evolutionary framework for neurolinguistics. Behavioural and Brain Sciences (submitted) (2005)

Arbib, M., Bota, M.: Language evolution: neural homologies and neuroinformatics. Neural Networks 16(9), 1237–1260 (2003)

Bressler, S.L., Coppola, R., Nakamura, R.: Episodic multi-regional cortical coherence at multiple frequencies during visual task performance. Nature 366, 153–156 (1993)

Burns, B., Sutton, C., Morrison, C., Cohen, P.: Information theory and representation in associative word learning. In: Prince, C., Berthouze, L., Kozima, H., Bullock, D., Stojanov, G., Balkenius, C. (eds.) Proceedings Third International Workshop on Epigenetic Robotics: Modelling Cognitive Development in Robotic Systems, Boston, MA, USA, pp. 65–71 (2003)

Hebb, D.: The Organization of Behavior. Wiley, New York. Partially reprinted in Anderson and Rosenfeld (1949)

Idiart, M.A.P., Lisman, J.: Short-term memory as a single cell phenomenon. In: Bower, J.M., editor, The Neurobiology of Computation: Proceedings of the third annual computational and neural systems conference, chapter 14. Kluwer Academic Publishers (1995)

Jensen, O., Lisman, J.E.: Hippocampal sequence-encoding driven by a cortical multi-item working memory buffer. Trends in Neurosciences 28(2), 67–72 (2005)

Lisman, J., Idiart, M.A.P.: Storage of 7+/-2 short-term memories in oscillatory subcycles. Science 267, 1512–1515 (1995)

Oates, T.: Grounding Knowledge in Sensors: Unsupervised Learning for Language and Planning. PhD thesis, University of Massachusetts, Amherst (2001)

Panchev, C.: Spatio-Temporal and Multimodal Processing in a Spiking Neural Mind of a Robot. PhD thesis, University of Sunderland (2005)

Panchev, C.: Temporal processing in a spiking model of the visual system. In: Proceedings of the International Conference on Artificial Neural Networks. LNCS, Springer, Heidelberg (2006)

Panchev, C.: Computing with active dendrites. Neurocomputing. in press (2007)

Panchev, C., Wermter, S.: Temporal sequence detection with spiking neurons: towards recognizing robot language instructions. Connection Science 18(1), 1–22 (2006)

Panchev, C., Wermter, S., Chen, H.: Spike-timing dependent competitive learning of integrate-and-fire neurons with active dendrites. In: Dorronsoro, J.R. (ed.) ICANN 2002. LNCS, vol. 2415, pp. 896–901. Springer, Heidelberg (2002)

Roelfsema, P.R., Engel, A.K., König, P., Singer, W.: Visio-motor integration is associated with zero time-lag synchronization among cortical areas. Nature 385, 157–161 (1997)

Shtyrov, Y., Hauk, O., Pulvermüller, F.: Distributed neuronal networks for encoding catergory-specific semantic information: The mismatch negative to action words. European Journal of Neuroscience 19, 1–10 (2004)

Siskind, J.M.: Learning word-to-meaning mappings. In: Broeder, P., Murre, J. (eds.) Models of Language Acquisition: Inductive and Deductive Approaches, pp. 121–153. Oxford University Press, Oxford (2000)

Sougné, J.P.: INFERNET: A Neurocomputational Model of Binding and Inference. PhD thesis, Université de Liège (1999)

von der Malsburg, C.: The correlation theory of brain function. Technical report, Max-Planck-Institute for Biophysical Chemistry. Internal Report 81-2 (1981)

Towards Understanding of Natural Language: Neurocognitive Inspirations

Włodzisław Duch[1], Paweł Matykiewicz[1,2], and John Pestian[2]

[1] Dept. of Informatics, Nicolaus Copernicus University, Toruń, Poland
[2] Department of Biomedical Informatics, University of Cincinnati,
Cincinnati Children's Hospital Medical Center, OH, USA
{Pawel.Matykiewicz,John.Pestian}@cchmc.org

Abstract. Neurocognitive processes responsible for representation of meaning and understanding of words are investigated. First a review of current knowledge about word representation, recent experiments linking it to associative memory and to right hemisphere synchronous activity is presented. Various conjectures on how meaning arises and how reasoning and problem solving is done are presented. These inspirations are used to make systematic approximation to spreading activation in semantic memory networks. Using hierarchical ontologies representations of short texts are enhanced and it is shown that high-dimensional vector models may be treated as a snapshot approximation of the neural activity. Clustering short medical texts into different categories is greatly enhanced by this process, thus facilitating understanding of the text.

1 Introduction

Low-level cognitive functions involving perception and motor control have reasonable neural models at different level of complexity, from sophisticated spiking neuron biophysical models to quite approximate Hopfield-like and self-organized networks that provide qualitative ideas rather than detailed explanations. Unfortunately, despite great progress in neuroscience, the higher cognitive functions: language, thinking, reasoning, planning, problem solving, creativity, understanding of visual scenes are all poorly understood and lack good working models. Great progress in neuroimaging has not elucidated the precise mechanisms of high-level cognitive functions, because they depend on synchronization of processes at a single neuron or a microcircuit level. Attempts to elucidate such processes at present must be speculative. Even if they prove ultimately too simplistic they may still be fruitful by helping to formulate neurocognitive models of various higher cognitive functions.

In this paper neurolinguistic insights are used to elucidate the process of text understanding and to find useful approximations to the spreading of brain activity during text comprehension. The connectionist approach to natural language has been introduced already in [1], where it was used to explain qualitatively a few linguistic phenomena. The only known system that can deal with linguistic structures is the human brain. The neurocognitive approach to linguistics "is an attempt to understand the linguistic system of the human brain, the system that makes it possible for us to speak and write, to understand speech and writing, to think using language ..." [2].

J. Marques de Sá et al. (Eds.): ICANN 2007, Part II, LNCS 4669, pp. 953–962, 2007.

Although this approach has been quite fruitful for understanding neuropsychological, language-related problems, it is relatively unknown in the natural language processing (NLP) community; no practical algorithms for large-scale text analysis have been derived from it.

The basic assumption of neurocognitive computing is that words activate microfeature-based associative networks and that the activation spreads to other parts of the network, which increases the probability of priming dynamic activations of states that facilitate semantic interpretation of words, concepts, sentences and episodes. Basic words and concepts label the action-perception subnetworks, acquiring the meaning directly through references to actions in the environment [3-4]. Constrained spreading activation techniques have recently been applied in information retrieval [5], semantic search techniques [6] and word sense disambiguation [7], although their application is still quite limited.

A brief introduction to the putative neurocognitive processes behind higher cognitive functions is presented in the next section. The section focuses on the use of words and symbols, analysis of priming experiments with pairwise word associations, and recent observations of insight states in the brain. Various approximations of the spreading activation processes in brain networks are discussed and related to the methods used in natural language processing. The challenge is to create approximations that could be used in large-scale, practical NLP projects. An example of how hierarchical ontologies can enhance the representation of short medical texts (summary discharges) illustrates the usefulness of simple approximations. Discussion of the results and their wider implications closes this paper.

2 Representation of Words and Meanings

Linguists have employed symbol manipulation, grammars and parsing techniques, trying to understand languages in conceptual terms. Progress in understanding languages in this way has been rather slow, which has led to the use of statistical techniques to study patterns of language use in large corpora [8]. Although language is based on symbols, logical linguistic analysis may provide only an awkward approximation of the spreading activation and associative processes in the brain. The neurocognitive approach to language draws its inspiration from brain research in trying to understand the processes that make language understanding and production possible.

Sensory systems transform incoming stimuli by extracting from auditory and visual streams such basic quantized elements as phonemes in speech or edges with high contrast in vision. These elementary building blocks form larger patterns, building discrete representations of words and shapes, and in a hierarchical way filling the working memory with information about whole scenes and complex objects, some of them abstract and not even directly related to activation of sensory cortices [9]. The cortex has a layered, modular structure, with columns of about 10^5 densely interconnected neurons, which communicate with other cortical columns in the neighborhood and sometimes also in quite distant areas across the brain, including the opposite hemisphere. Each column contains thousands of microcircuits with different properties (due to the different type of neurons, neurotransmitters and neuromodulators), acting as local resonators that may respond to sensory signals, converting them into intricate patterns of excitations.

Hearing words activates a strongly linked subnetwork of microcircuits that bind articulatory and acoustic representations of a spoken word. Such patterns of activation are localized in most brains in the left temporal cortex, with different word categories coded in the anterior and posterior parts [8-10]. Psycholinguistic experiments show that acoustic speech input is quickly changed into categorical, phonological representation. A small set of phonemes, quantized building blocks of phonological representations are linked together in an ordered string by a resonant state representing word form, and extended to include other microcircuits defining the semantic concept. From the N200 feature of auditory event-related potentials, it has been conjectured that phonological processing precedes semantic activations by about 90 ms [4]. Words seem to be organized in a lexicon, with similar phonological forms activating adjacent resonant microcircuits. Upon hearing a word, a string of connected resonators is activated, creating representation of a series of phonemes that is categorized as a word. Spoken language has a number of syllables and longer chunks of sounds (morphemes) that are strongly associated with each other. They are easily activated when only part of the word is heard, creating the illusion that the whole word has been heard. Categorical auditory perception enables understanding of speaker-independent speech and has clear advantages in a noisy environment, providing speaker-independent speech representation. Strong associations sometimes lead to activation of wrong representations. For example, when only a part of some personal name is heard, often a more common name is substituted.

Phonological representations of words activate an extended network that binds symbols with related perceptions and actions, grounding the meaning of each word in a perception/action network. Various neuroimaging techniques confirm the existence of semantically extended phonological networks, which lends this model of word representation strong experimental support [3,4,10,11]. Symbols in the brain are thus composed of several representations: their sound patterns, pronunciation (vocal motor programs), and their visual and motor associations. This does not resemble the traditional idea of a representation. Learning new concepts prompts minimal changes (convergence) of neural connections that assure unique dynamical states that have the correct relational properties. Hearing a word activates a string of phonemes, increasing the activity (priming) of all candidate words and non-word combinations. A polysemic word probably has a single phonological representation that differs only in its semantic extensions. This encoding automatically ensures that many similarity relations between words, phonological as well as semantic, may automatically be retrieved. Meanings are stored as activations of associative subnetworks that may be categorized and processed further by other areas of the brain. Context priming selects an extended subnetwork corresponding to a unique word meaning, while competition and inhibition in the winner-takes-all processes leaves only the most active candidate networks. The meanings of concepts listed in thesauri or dictionaries are only approximations, because the actual meaning is always modified by the context. Overlapping patterns of brain activations for subnetworks coding word representations lead to strong transition probabilities between the words, and thus to semantic and phonological associations that easily "come to mind".

During text comprehension, background knowledge stored in the semantic memory is activated, resulting in brain states that contain unique interpretations. Two approaches to knowledge representation prompted by semantic memory are Collins/Loftus spreading activation model [12], and Collins/Quillian's hierarchical semantic memory model [13]. The first has been used in connectionist models of language [1]; the second is the basis for various ontologies. No large-scale semantic networks capturing commonsense knowledge have been built for practical applications, although considerable theoretical work has been done in this area [14,15]. Collecting knowledge for semantic networks that would approximate associative processes in the brain has proved to be quite difficult, since lexical resources such as Wordnet [16] do not contain structural descriptions of concepts. Statistical approaches to context analysis are insufficient in this area because most common sense knowledge is acquired through embodiment and perception, and is so obvious that it is never written down. Recent attempts to analyze machine-readable sources for the creation of large-scale semantic memories have been examined in [17], and the use of word games and active dialogues to extend and correct such knowledge is promising [18]. Ontologies, on the other hand, though they offer taxonomies of concepts [19] that are useful for experts, do not reflect common sense knowledge and lateral associations.

3 Words and Creative Processes

Understanding of words can be regarded as a simple version of problem solving. Recent experiments using the EEG and functional MRI techniques on the "Aha!" insight experience that accompanies some solutions have contrasted insight with analytical problem solving that does not require insight [20,21]. An increased activity in the right hemisphere anterior superior temporal gyrus (RH-aSTG) has been observed during initial solving efforts and during insights. This area is probably involved in higher-level abstractions that can facilitate indirect associations. About 300 ms before insight, a burst of gamma activity was observed. This has been interpreted as "making connections across distantly related information during comprehension (…) that allow them to see connections that previously eluded them" [21]. Bowden *et al.* [20] performed a series of experiments that confirmed the EEG results using fMRI techniques. It is probable that the initial impasse in problem solving is due to the inability of the processes in the left hemisphere, focused on the precise representation of the problem, to make progress. This deadlock is removed when less-focused right hemisphere projects back relevant activations, allowing new dynamical associations to be formed. An emotional component is needed to increase the plasticity of the brain and remember these associations. The "Aha!" experience may thus result from the activation of larger left hemisphere areas by the right hemisphere, with a gamma burst winning the competition for working memory access and thus reaching consciousness. This process occurs more often when the activation of the left hemisphere decreases (giving up conscious efforts to solve the problem), perhaps leading to a short period of knowing that the solution has been found although it has not yet been formulated in symbolic terms. This last step requires synchronization between states in the left hemisphere, defining the transition from the start to the goal through intermediate states.

Such observations may be used as inspirations for neurocognitive models. The LH network codes phonological and visual representations in the visual word form area (VWFA) in the left unimodal occipitotemporal sulcus area. The adjacent lateral inferotemporal multimodal area (LIMA) reacts to both auditory and visual stimulation, and has cross-modal phonemic and lexical links [22]. Extended representations reach to the sensory, motor and premotor cortices [3-4]. Distal connections between the left and right hemispheres require long projections, and therefore neurons in the right hemisphere may generalize over similar concepts and their relations. Most of these RH activations do not have phonological components; the activations result from diverse associations, temporal dependencies and statistical correlations that create certain expectations. For example, hearing the word "left lung" may activate several RH cortical areas that react to all concepts related to lungs and the left side of the upper part of the body, including the heart; hearing "left nose" or "left head" creates a strange feeling. It is not clear what brain mechanism is behind the signaling of this lack of familiarity, but one can assume that interpretation of text is greatly enhanced by "large receptive fields" in the RH, which can constrain possible interpretations, help in the disambiguation of concepts and provide ample stereotypes and prototypes that generate various expectations.

Distributed activations in the right hemisphere also form configurations that activate larger regions of the left hemisphere. High-activity gamma bursts projected to the LH prime its subnetworks with sufficient strength to allow for synchronization of groups of neurons that create distant associations. In problem solving, this synchronization links the initial description D with partial or final solutions S. Such solutions may initially be difficult to justify, they become clear only when all intermediate states T_k between D and S are transversed. If each step from T_k to T_{k+1} is an easy association, a series of such steps is accepted as an explanation. An RH gamma burst activates emotions, increasing the plasticity of the cortex and facilitating the formation of new associations between initially distal states. The same neural processes should be involved in sentence understanding, problem solving and creative thinking.

According to these ideas, approximation of the spreading activation in the brain during language processing should require at least two networks activating each other. Given the word $w = (w_f, w_s)$ with phonological/visual component w_f and extended semantic representation w_s, and the context *Cont*, the meaning of the word results from spreading activation in the left semantic network *LH* coupled with the right semantic network *RH*, establishing a global state $\Psi(w,Cont)$. This state changes with each new word received in sequence, with quasi-stationary states formed after each sentence is understood. It is quite difficult to decompose the $\Psi(w,Cont)$ state into components, because the semantic representation w_s is strongly modified by the context. The state $\Psi(w,Cont)$ may be regarded as a quasi-stationary wave, with its core component centered on the phonological/visual brain activations w_f and with quite variable extended representation w_s. As a result the same word in a different sentence creates quite different states of activation, and the lexicographical meaning of the word may be only an approximation of an almost continuous process. To relate states $\Psi(w,Cont)$ to lexicographical meanings, one can clusterize all such states using dendrograms and use different cutoffs to define prototypes for different meanings.

4 Approximations to Brain States

The high-dimensional vector model of language is a very crude approximation that does not reflect essential properties of the perception-action-naming activity of the brain [3-4]. The process of understanding words (spoken or read) starts from activation of the phonological or grapheme representations that stimulate networks containing prior knowledge used for disambiguation of meanings. This continuous process may be approximated through a series of snapshots of microcircuit activations $\phi_i(w,Cont)$ that may be treated as basis functions for the expansion of the state $\Psi(w,Cont) = \Sigma_i\, \alpha_i\, \phi_i(w,Cont)$, where the summation extends over all microcircuits that show significant activity resulting from presentation of the word w. The high-dimensional vector model used in NLP measures only the co-occurrence of words $\mathbf{V}_{ij} = \langle \mathbf{V}(w_i),\mathbf{V}(w_j)\rangle$ in some window, averaged over all contexts. A better approximation of the brain processes involved in understanding words should be based on the overlap between waves $\langle\Psi(w_1,Cont) \mid \Psi(w_2,Cont)\rangle = \Sigma_{ij}\, \alpha_i\, \alpha_j\, \langle\phi_i(w_1,Cont) \mid \phi_j(w_2,Cont)\rangle$ that depends on time. Systematic study of transformations between the two bases: activation of microcircuits ϕ_i and activation of complex patterns $\mathbf{V}(w_i)$, has not yet been done. The use of waves to describe states makes this formalism similar to that used in quantum mechanics, although no real quantum effects are implied here.

Spreading activation in semantic networks should provide enhanced representations that involve concepts not found directly in the text. Approximations of this process are of great practical and theoretical interest. The model should reflect activations of various concepts in the brain of an expert reading such texts. A few crude approximations to this process may be defined. First, semantic networks that capture many types of relations among different meanings of words and expressions may provide space on which words are projected and activation spread. Each node w in the semantic network represents the whole state $\Psi(w,Cont)$ with various contexts clusterized, leading to a collection of links that capture the particular meaning of the concept. Usually only the main differences among the meanings of the words with the same phonological representation are represented in semantic networks (meanings listed in thesauruses), but the fine granularity of the meanings resulting from different contexts may be captured in the clusterization process and can be related to the weights of connections in semantic networks. The spreading activation process should involve excitation and inhibition, and "the winner takes most" processes. Current models of semantic networks used in NLP are only vaguely inspired by the associative processes in the brain and do not capture such details [14,15].

Quite crude approximation to the spreading activation processes leads to enhancement of the initial text being analyzed by adding new concepts linked by semantic or hierarchical ontological relations. Inhibition between concepts arising from the same phonological word forms should then lead to formation of graphs of consistent concepts, applied recently to disambiguate concepts in medical domain [23]. The enhanced representations are very useful in document clusterization and categorization, as is illustrated using short medical texts in the next section. Vector models may be related to semantic networks by looking at snapshots of the activation of nodes after several steps of spreading the initial activations through the network. In view of the remarks about the role of the right hemisphere, larger "receptive fields" in the linguistic domain should be defined and used to enhance text representations. This is much more difficult because many of these processes have no phonological component and thus have representations that are less constrained and have no directly identifiable meaning. Internal representations formed by neural networks are also not meaningful to us, as only the final result of information processing or decision making can be interpreted in symbolic terms. Defining prototypes for different categories of texts, clusterizing topics or adding prototypes that capture some *a priori* knowledge useful in document categorization [24], is a process that goes in the same direction.

Relationships between creativity and associative memory processes have been noticed long ago [25]. Further experimental support for the ideas described above may be found in pairwise word association experiments using different priming conditions. In [26] puzzling results from using nonsensical words were observed for people with high compared to those with low creativity levels. Analysis of these experiments provided in [27] reinforces the idea that creativity relies on associative memory, and in particular on the ability to link distant concepts together. Adding neural noise by presenting nonsensical words in priming leads to activation of more brain circuits and facilitates in a stochastic resonance-like way a formation of distal connections for not obvious associations. This is possible only if weak connections through chains involving several synaptic links exist, as is presumably the case in creative brains. For simple associations the opposite effect is expected, with strong local activations requiring longer times for the inhibitory processes to form consistent interpretations. Such experiments show that some effects cannot be captured at the symbolic level. It is thus quite likely that language comprehension and creative processes both require subsymbolic models of neural processes realized in the space of neural activities, reflecting relations in some experiential domain, and therefore cannot be modeled using semantic networks with nodes representing whole concepts. Recent results on creation of novel words [27] give hope that some of this process can be approximated by statistical techniques at the morphological level.

5 Visualization of Semantic Similarity

The time-dependent state of the brain $\Psi(w_i,Cont)$ that arises after reading or hearing texts that are understood by the experts should show high similarity for documents of the same category and should be different if documents from other categories are processed. Documents have usually quite sparse representation; for example, hospital discharge summaries by different specialties, but for the same disease, may use completely different vocabularies. Therefore, agglomerative hierarchical clustering methods will show a poor performance in document clustering. The simplest extension is to replace single words (terms) by associations based on synonyms, for example by using the Wordnet synsets [16]. This simulates some of the spreading activation processes in the brain increasing the similarity of documents that use different words to describe the same topic. However, synsets are not useful for very specific concepts that have no synonyms, such as medical concepts used in discharge summaries. To avoid problems with shared common words, only specific concepts that belong to selected semantic types may be used – the process presumably facilitated by the RH.

A better approximation to spreading activation in brain networks is afforded by soft evaluation of the similarity of different terms. Distributional hypothesis assumes that similarity of terms results from similar linguistic contexts [8]. However, in the medical domain and other specialized areas it may be quite difficult to estimate similarity reliably on the basis of co-occurrence, because there are so many specific concepts that there will never be sufficient data to do that. Statistical approaches cannot replace systematic, structured knowledge describing medical concepts. To illustrate that process, two-steps of spreading activation have been made in a network built from ontological relations found in the Systematized Nomenclature of Medicine –Clinical Terms (SNOMED CT) section of the National Library of Medicine's Unified Medical Language System (UMLS) [19]. Discharge summaries for 10 initial diagnoses are represented by a carefully selected semantic feature space (described in [24]). Figs. 1-3 show Multidimensional Scaling (MDS) visualization of records from three strongly overlapping classes only to improve legibility: pneumonia (class 1, 609 records), juvenile rheumatoid arthritis (class 6, 41 records) and otitis media (class 9, 493 records).

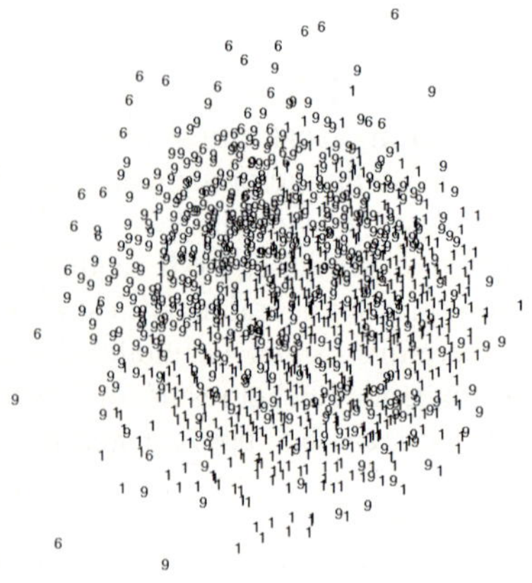

Fig. 1. MDS for original data

Fig. 2. MDS after first enhancement

Initial feature space is composed from 488 SNOMED CT concepts with high feature-class correlation coefficient (CC>0.5). Visualization of these documents using multi-dimensional scaling (Fig. 1) shows great mixing of documents. A single step of spreading activation through the network, followed by feature selection based on CC > 0.27 extends the feature space to 761 concepts. MDS in this space (Fig. 2) already shows a clear cluster structure. The second iteration with CC>0.5 increases the space to 1138 features and shows even more detailed and fine-grained structure, identifying different subclusters within each category (Fig. 3). For example, bacterial infections may come from Yersinia, Salmonella, Streptococcal and other infections, increasing similarity of all diseases caused by bacteria. In the extended spaces accuracy of classification is also greatly improved – for the 3 classes presented here from about 81% to 87% and 88±4% in crossvalidation tests using linear SVM (for the 10-class case the improvement is on more than 20%). Even quite simple approximations of the spreading of neural activation leads to a significantly improved accuracy in classification.

Fig. 3. MDS on medical discharge summaries after two enhancement steps

6 Conclusion

Although linguistic processes are not yet completely understood, following neurolinguistic inspirations may be quite fruitful, allowing one to formulate some crude models of the processes that are responsible for text understanding in real brains. Various approximations to the putative brain processes responsible for language comprehension have been considered, leading to useful algorithms for text analysis. Vector representations of concepts may be regarded as a snapshot of activity patterns, defining connections with other concepts. Relations between spreading activation in neural and in semantic networks, and the vector model of concepts have been elucidated. The role of the right hemisphere, which constrains and guides the spreading activation processes by providing "large receptive fields" for concepts, has been discussed.

It is perhaps surprising that even a crude approximation using two steps of spreading activation with feedback loops leads to such good clusterization and to great improvement in classification on a very difficult problem of summary discharge categorization [24]. Background knowledge has been derived here from synsets, statistical co-occurrences and ontologies. In [24] prototypes of concepts representing *a prio*ri medical knowledge were used, providing crude approximation of the activity of neural cell assemblies in the brain of a medical expert who thinks about a particular disease. Creating numerical representations of various concepts that may be useful in large-scale NLP applications is an interesting challenge. Neurocognitive inspirations lead here to many ideas that will be explored in future work.

Acknowledgment. Support by the Polish Committee for Scientific Research, research grant 2005-2007, and Ohio's Third Frontier is gratefully acknowledged.

References

1. Rumelhart, D.E., McClelland, J.L. (eds.): Parallel Distributed Processing: Explorations in the Microstructure of Cognition, vol. 1 & 2. MIT Press, Cambridge (1986)
2. Lamb, S.: Pathways of the Brain: The Neurocognitive Basis of Language. J. Benjamins Publishing, Amsterdam & Philadelphia (1999)

3. Pulvermüller, F.: The Neuroscience of Language. On Brain Circuits of Words and Serial Order. Cambridge University Press, Cambridge, UK (2003)
4. Pulvermüller, F., Shtyrov, Y., Ilmoniemi, R.: Brain signatures of meaning access in action word recognition. Journal of Cognitive Neuroscience 17, 884–892 (2005)
5. Crestani, F.: Application of Spreading Activation Techniques in Information Retrieval. Artifical Intelligence Review 11, 453–482 (1997)
6. Crestani, F., Lee, P.L.: Searching the web by constrained spreading activation. Information Processing & Management 36, 585–605 (2000)
7. Tsatsaronis, G., Vazirgiannis, M., Androutsopoulos, I.: Word Sense Disambiguation with Spreading Activation Networks Generated from Thesauri, 20th Int. Joint Conf. in Artificial Intelligence, Hyderabad, India, pp. 1725–1730 (2007)
8. Manning, C.D, Schütze, H.: Foundations of Statistical Natural Language Processing. MIT Press, Cambridge, MA (1999)
9. Caplan, D., Waters, G.S.: Verbal working memory and sentence comprehension. Behavioral and Brain Sciences 22, 77–94 (1999)
10. Damasio, H., Grabowski, T.J., Tranel, D., Hichwa, R.D., Damasio, A.R.: A neural basis for lexical retrieval. Nature 380, 499–505 (1996)
11. Martin, A., Wiggs, C.L, Ungerleider, L.G, Haxby, J.V: Neural correlates of category-specific knowledge. Nature 379, 649–652 (1996)
12. Collins, A.M., Loftus, E.F.: A spreading-activation theory of semantic processing. Psychological Reviews 82, 407–428 (1975)
13. Collins, A.M., Quillian, M.R.: Retrieval time from semantic memory. Journal of Verbal Learning and Verbal Behavior 8, 240–247 (1969)
14. Sowa, J.F. (ed.): Principles of Semantic Networks: Explorations in the Representation of Knowledge. Morgan Kaufmann Publishers, San Mateo, CA (1991)
15. Lehmann, F.(eds.): Semantic Networks in Artificial Intelligence. Pergamon, Oxford (1992)
16. See http://wordnet.princeton.edu
17. Szymanski, J., Sarnatowicz, T., Duch, W.: Towards Avatars with Artificial Minds: Role of Semantic Memory. Journal of Ubiquitous Computing and Intelligence (in press)
18. Szymanski, J., Duch, W. (eds.): Semantic Memory Knowledge Acquisition Through Active Dialogues. Int. Joint Conf. on Neural Networks (in press). IEEE Press, Orlando (2007)
19. UMLS Knowledge Sources, available at http://www.nlm.nih.gov/research/umls
20. Bowden, E.M., Jung-Beeman, M., Fleck, J., Kounios, J.: New approaches to demystifying insight. Trends in Cognitive Science 9, 322–328 (2005)
21. Jung-Beeman, M., Bowden, E.M., Haberman, J., Frymiare, J.L., Arambel-Liu, S., Greenblatt, R., Reber, P.J.: Neural activity when people solve verbal problems with insight. PLoS Biology 2, 500–510 (2004)
22. Gaillard, R., Naccache, L., Pinel, P., Clémenceau, S., Volle, E., Hasboun, D., Dupont, S., Baulac, M., Dehaene, S., Adam, C., Cohen, L.: Direct intracranial, FMRI, and lesion evidence for the causal role of left inferotemporal cortex in reading. Neuron 50, 19–204 (2006)
23. Matykiewicz, P., Duch, W., Pestian, J.: Nonambiguous Concept Mapping in Medical Domain, Lecture Notes in Artificial Intelligence. In: Rutkowski, L., Tadeusiewicz, R., Zadeh, L.A., Zurada, J.M. (eds.) ICAISC 2006. LNCS (LNAI), vol. 4029, pp. 941–950. Springer, Heidelberg (2006)
24. Itert, L., Duch, W., Pestian, J.: Medical document categorization using a priori knowledge, Lecture Notes in Computer Science. In: Duch, W., Kacprzyk, J., Oja, E., Zadrożny, S. (eds.) ICANN 2005. LNCS, vol. 3696, pp. 641–646. Springer, Heidelberg (2005)
25. Mednick, S.A.: The associative basis of the creative process. Psychological Review 69, 220–232 (1962)
26. Gruszka, A., Nęcka, E.: Priming and acceptance of close and remote associations by creative and less creative people. Creativity Research Journal 14, 193–205 (2002)
27. Duch, W., Pilichowski, M.: Experiments with computational creativity. Neural Information Processing – Letters and Reviews 11, in press (2007)

A Computational Model of Metaphor Understanding Consisting of Two Processes

Asuka Terai and Masanori Nakagawa

Tokyo Institute of Technology,
2-12-1 O-okayama, Meguro-ku, Tokyo, 152-8552, Japan

Abstract. The purpose of this study is to construct a computational model of the metaphor understanding process. This study assumes that metaphor understanding consists of two processes. The first is a categorization process; a target is assigned to an ad hoc category of which the vehicle is a prototypical member. The second is a dynamic interaction process; the target assigned to the ad hoc category is influenced by dynamic interaction among features. Feature emergence is extracted through this dynamic interaction. In this study, a model of metaphor understanding is constructed based on this assumption by applying a statistical analysis of large-scale corpus. Further a psychological experiment is conducted in order to verify the psychological validity of the constructed model of metaphor understanding. Reflecting the fact that the constructed model represents more appropriate features of a metaphor than a model incorporating only the categorization process, the experimental results support its validity.

1 Introduction

The purpose of this study is to construct a computational model of the metaphor understanding process. Metaphorical expressions are frequently used in daily conversation. But, generally it is difficult for people who do not share the language and culture of a metaphor to fully understand the appropriate meaning of the metaphor. It is particularly noteworthy that even when the meaning of each word within a metaphor is explained, it is not easy for the strangers to that language and culture to understand the metaphor. As Kusumi [1] observes, this highlights the linguistically and culturally bound nature of the knowledge structure required for metaphor understanding. It is, therefore, quite hard for Japanese language learners or children who have grown up outside of Japanese culture to understand some Japanese metaphors, due to differences in terms of knowledge structures. For example, the Japanese language has the metaphorical expression "cheeks like apples", which is quite difficult for Europeans to understand. For Europeans, apples are typically objects with the features of green and small. On the other hand, Japanese apples are generally red and big and round in shape. Thus, the Japanese metaphorical expression "cheeks like apples" means red colored and round shaped cheeks. This example clearly underscores the need to construct a system which can provide detailed meanings for metaphorical

J. Marques de Sá et al. (Eds.): ICANN 2007, Part II, LNCS 4669, pp. 963–972, 2007.

expressions. In other words, a system that is able to decompose metaphorical expressions into their elements of meaning would be highly useful for both language learners and children. This paper constructs a computational model which realizes the understanding process of metaphorical expression, represented in the form of "A like B", as a first step toward such a system.

There are two theories that seek to account for the understanding process of similes ("A is like B") and metaphors ("A is B") in psychology. One is the comparison theory, which holds that metaphor understanding is realized by aligning similar elements between the target and the vehicle with each other [2]. For example, in comprehending the metaphor "Socrates is like a midwife", the understanding process is realized when similar elements relating to "helpfulness" are identified in both "Socrates" and "midwife" and are mutually aligned. In other words, this metaphor is comprehensible when one notices that Socrates was someone who "helped" his students to grasp certain ideas and that a midwife is someone who "helps" a pregnant woman in giving birth to a child. However, this theory faces difficulties in distinguishing between targets and vehicles. The second theory is the categorization theory. Here, metaphor understanding is explained in terms of class-inclusion statements, where a target is regarded as a member of an ad hoc category of which the vehicle is a prototypical member [3]. For example, in comprehending the metaphor of "Socrates is like midwife", the target of "Socrates" is considered as belonging to a "helpful" category which could be typically represented by a vehicle like "midwife".

A computational model of metaphor understanding based on these theories requires a numerically-represented knowledge structure. Some models using knowledge structures obtained from psychological experiments have been developed [4][5]. However, it is not practically feasible to collect sufficient data to cover enough concepts by such psychological methods alone, because participants cannot rate all of the vast range of concepts that are commonly used in metaphorical expressions within limited time frames. Accordingly, a model based only on psychological experimentation cannot be extended to computational systems (e.g., a search engine).

There are some computational models of metaphor understanding based on a knowledge structure for concepts obtained from language corpora [6][7]. Kintsch's model[6], based on the categorization theory, employs a knowledge structure estimated by Latent Semantic Analysis (LSA) (Deerwester, et al. [8]). Also utilizing LSA, Utsumi[7] constructs two models; one based on categorization theory, with an identical algorithm to Kintsch's, and one based on comparison theory. Even though the meanings of concepts are represented by vectors in LSA, each dimension of the vectors does not have a meaning in itself. Therefore, the meaning of a metaphor represented by a particular vector must be generally defined in terms of the cosines of angles existing between other vectors according to the LSA method. This aspect of LSA makes it quite difficult to interpret metaphors represented by vectors.

Terai & Nakagawa [9] constructed a model using a knowledge structure estimated from the statistical language analysis developed by Kameya & Sato [10].

The meanings of concepts estimated from the statistical language analysis are represented by conditional probabilities of concepts given features. A concept is represented by a vector, that is, a set of the conditinal probabilities of the concept given features. In this case, each dimension of the vector has its own meaning as a feature. This makes it easier to determine the estimated meaning of a metaphor than with the LSA approach. In this context, it is worth noting that some studies have reported on low-salient features of a target and a vehicle being emphasized in the process of metaphor understanding; a phenomenon referred to as feature emergence [11][12]. Terai & Nakagawa's model implements the phenomenon of feature emergence by using a recurrent neural network to represent the dynamic interaction among features with the metaphor understanding process. However, the model suffers somewhat in its inability to distinguish targets and vehicles because it is based on the comparison theory.

In order to overcome these problematic aspects with previous models, this study assumes that metaphor understanding is realized through two processes. The first is a categorization process; a target is assigned to an ad hoc category of which the vehicle is a prototypical member. The second is a dynamic interaction process; the target assigned to the ad hoc category is influenced by dynamic interaction among features. Feature emergence is realized through this dynamic interaction. Based on this assumption, the procedure for constructing the model is as follows:

- Step 1: Knowledge structure of concepts is estimated using statistical language analysis [10].
- Step 2: The assigned meaning of a target to an ad-hoc category of a vehicle is computed within the knowledge structure.
- Step 3: The recurrent neural network model, representing the dynamic interaction among features, estimates the meaning of the metaphor based on the assigned meaning of the target in Step 2.
- Step 4: A psychological experiment is conducted in order to verify the psychological validity of the constructed model.

2 Statistical Language Analysis

Knowledge structure has been estimated using LSA in previous models [6][7]. However, the LSA approach has some problems in addition to the central problem of defining meaning. In particular, LSA is susceptible to noise due to function words and to the data sparseness problem. The Tf-idf method and stop-word lists are often used with LSA in order to avoid the noise of function words. However, such avoidance strategies are based on procedures that must be manually implemented. In order to solve these problems, this study applies a statistical method developed by Kameya & Sato [10], using extracted frequency data for adjective-noun modifications. The noise problem associated with function words is effectively eliminated by this method.

The statistical method assumes that the co-occurrence probabilities of a term n_i(noun) and a term a_j(adjective), $P(n_i,a_j)$ can be computed using the following formula (1):

$$P(n_i, a_j) = \sum_k P(n_i|c_k)P(a_j|c_k)P(c_k), \quad (1)$$

where c_k indicates a latent semantic class assumed in the method. When θ represents the parameter vector for $P(n_i|c_k)$, $P(a_j|c_k)$ and $P(c_k)$, and D denotes the data, the likelihood of the data is represented using formula (2):

$$P(D|\theta) = \sum_{i,j} F(n_i, a_j) log P(n_i, a_j), \quad (2)$$

where $F(n_i,a_j)$ is the co-occurrence frequency of the term n_i and the term a_j. The parameter vector θ is estimated as the value that maximizes the likelihood of co-occurrence data using the EM algorithm. In order to avoid the data sparseness problem, the prior distribution of θ is assumed to be a Dirichlet distribution.

$$P(\theta) = \gamma \prod_k P(c_k)^{\alpha_1} (\prod_i P(n_i|c_k)^{\alpha_2}) (\prod_j P(a_j|c_k)^{\alpha_3}), \quad (3)$$

where, γ indicates a normalization constant, and α_1, α_2, α_3 are the hyper-parameters of the Dirichlet distribution.

In order to estimate the meaning of a concept, the conditional probability of an adjective given a particular noun, $P(a_j|n_i)$ is computed. The probabilities are computed using function (4) for $P(n_i|c_k)$, $P(a_j|c_k)$ and $P(c_k)$ based on Bayes theory:

$$P(a_j|n_i) = \frac{\sum_k P(n_i|c_k)P(a_j|c_k)P(c_k)}{\sum_k P(n_i|c_k)P(c_k)}. \quad (4)$$

Extracted frequency data for adjective-noun modifications was used for this analysis. The data was extracted from the Japanese newspaper "MAINICHI SHINBUN" for the period 1993-2002 using a modification analysis tool called "Cabocha" [13]. The number of semantic classes in the statistical analysis was fixed at 70. The relationship between the hyper-parameters and latent classes is such that if the hyper-parameter values are larger, then fewer latent classes are estimated. If one assumes that $\alpha_1 = \alpha_2 = \alpha_3$, then the largest value for them will be 0.12 in the case of 70 latent classes.

These conditional probabilities are computed based on the newspaper corpus reflecting the characteristics of newspaper writing. One concern here is that a noun's most basic meaning may only rarely be expressed in a newspaper. For example, even though ice is naturally cold, the expression "cold ice" is rather unlikely to appear in a newspaper, because newspaper articles would not normally state such natural and obvious facts. Thus, $P(cold|ice)$ has a low value. In order to overcome this problem, the conditional probabilities of the basic meaning given the noun are revised based on a dictionary [14] as follows. If an adjective (a_j) appears more than once in the dictionary explanation of a noun (n_i), it is assumed that the adjective represents a basic meaning of the noun (n_i). The set of these adjectives is represented by $Dic(n_i)$. The conditional probabilities

Table 1. Meaning vectors computed from the language statistical analysis for "work", "mountain", "teacher", and "demon". The component values of the vectors are shown in parentheses.

	work like a mountain		a teacher like a demon	
	$V_j(work)$	$V_j(mountain)$	$V_j(teacher)$	$V_j(demon)$
1	important (0.0659)	high (1.0714)	young (0.0968)	awful (0.8814)
2	new (0.0628)	deep (0.0403)	good (0.0318)	mysterious (0.0465)
3	main (0.0373)	low (0.0400)	favorite (0.0292)	white (0.0400)
4	good (0.0282)	wide (0.0337)	strict (0.0269)	high (0.0255)
5	bad (0.0214)	beautiful (0.0330)	weak (0.0204)	enormos (0.0252)
6	strict (0.0202)	near (0.0305)	bad (0.0202)	black (0.2119)
7	favorite (0.0194)	bright (0.0212)	inconvenient (0.0141)	red (0.0209)
8	able (0.0190)	narrow (0.0209)	happy (0.0132)	strange (0.0194)
9	various (0.0186)	white (0.0137)	famous (0.0128)	beautiful (0.0186)
10	novel (0.0174)	many (0.0136)	strong (0.0122)	blue (0.0156)

are then replaced by the derived value and the meaning vector of the concept ($V(n_i)$) is computed from the following formula (5):

$$V_j(n_i) = \begin{cases} P(a_j|n_i) + max_{i,j}(P(a_j|n_i)) \text{ if } a_j \in Dic(n_i) \\ P(a_j|n_i) \text{ else,} \end{cases} \tag{5}$$

where $V_j(n_i)$ indicates the jth component of the vector which means the concept n_i. Dimensions of the vectors represent features (adjectives). Knowledge structure is constructed based on estimations of the meaning vectors for a concept.

In this study, "work like a mountain" in Japanese, meaning "a mountainous load of work", and "a teacher like a demon" are used as examples. The meaning vectors for "work", "mountain", "teacher", "demon" are shown in Table 1. The listed features are ordered according to their respective values for $P(a_j|work)$, $P(a_j|mountain)$, $P(a_j|teacher)$, and $P(a_j|demon)$.

3 The Metaphor Understanding Model

The model consists of two kinds of process. One is the categorization process and the other is the dynamic interaction process.

3.1 The Categorization Process

A vector, representing an assigned target as a member of an ad hoc category of a vehicle, is estimated based on categorization theory using the meaning vectors of concepts. The algorithm for the categorization process is as follows.

First, the semantic neighborhood ($N(n_i)$) of a vehicle of size s_1 is computed on the basis of similarity to the vehicle, which is represented by the cosine of angles existing between meaning vectors using the following formula (6):

$$sim(n_i, n_h) = \frac{V(n_i) \cdot V(n_h)}{\|V(n_i)\| \|V(n_h)\|}, \tag{6}$$

where $sim(n_i, n_h)$ indicates the similarity between concept n_i and concept n_h. Next, L concepts are selected from the semantic neighborhood ($N(n_i)$) of the vehicle on the basis of similarity to the target. Finally, a vector ($V(M)$) is computed as the centroid of the meaning vectors of the target, the vehicle and the selected L concepts (n'_l: $l = 1, 2, .., L$). The computed vector ($V(M)$) indicates the assigned meaning of the target as a member of the ad-hoc category of the vehicle concerning the metaphor M. $V(M)$ is computed using the formula (7):

$$V(M) = \frac{\sum_l V(n'_l) + V(target) + V(vehicle)}{L+2}, \tag{7}$$

where n'_l indicates the lth selected concepts and L denotes the number of the selected concepts.

This algorithm is the same as Kintsch's [6] algorithm and that of Utsumifs categorization model [7]. The category consisting of the vehicle and the selected k concepts is considered to be an ad hoc category of which the vehicle is a prototypical member.

3.2 The Dynamic Interaction Process

The meaning of the metaphor is computed using the vector estimated by the categorization process model ($V(M)$) by applying the dynamic interaction process model. The algorithm for the dynamic interaction process is as follows.

First, features are selected if $V_j(M)$ exceeds the threshold ζ. These selected features are related to metaphor understanding. Next, the recurrent neural network model is constructed using the selected features (Fig. 1). Each node corresponds to the selected feature. These nodes have both inputs and outputs.

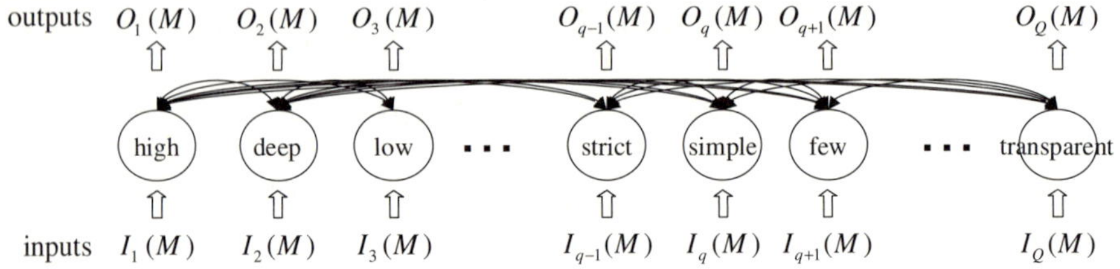

Fig. 1. Architecture of the model for "work like a mountain" (M="work like a mountain"). The nodes represent the selected features. These are both input and output nodes.

The dynamics of the network are based on the following set of simultaneous differential equations (8):

$$\frac{dx_q(t)}{dt} = -x_q(t) + f(\beta \sum_r w_{qr} x_r(t) + I_q(M)), \tag{8}$$

where $x_q(t)$ represents the activation strength of the qth node at time t and where the function f is a logistic function. The range is between -1 and 1. When $dx_q/dt = 0$, the node outputs $O_q(M) = x_q(t)$. The vector $(O(M))$, which is a set of $O_q(M)$, represents the meaning of the metaphor M. $I_q(M)$ represents the input value of the qth node concerning the metaphor M. A normalization of the $V_{j(q)}(M)$ is used as input value $I_q(M)$, because the domain of the function f include minus value although each value of $V_{j(q)}(M)$ is positive. $I_q(M)$ is computed using formula (9):

$$I_q(M) = \frac{V_{j(q)}(M) - \overline{V'(M)}}{SD'(M)},$$
$$\overline{V'(M)} = \frac{\sum_q V_{j(q)}(M)}{Q}, \tag{9}$$
$$SD'(M) = \sqrt{\frac{\sum_q \left(V_{j(q)}(M) - \overline{V'(M)}\right)^2}{Q-1}},$$

where $V_{j(q)}(M)$ indicates the jth components of $V(M)$, the meaning of the jth dimension corresponds to the meaning of the qth node, and Q means the number of the selected features. In the formula (8), β denotes the influences of the dynamic interaction among features. w_{qr} denotes the weight of the connection from the rth to the qth node and is the correlation coefficient among the qth and rth features related to the sibling concepts of the target and the vehicle. A sibling neighborhood ($N^s(vehicle)$) for a vehicle of size s_2 and a sibling neighborhood ($N^s(target)$) for a vehicle of size s_2 are computed on the basis of similarity. The concepts included in $N^s(vehicle)$ and $N^s(target)$ are regarded as sibling concepts.

Thus, the mutual and symmetric connections among nodes (w_{qr}) represent interaction among features in the metaphor understanding. If the metaphor is changed, then the weights of connection between the same pair of features may change. For example, in the case of "a dog like a cloud", "white" and "puffy" should be connected strongly. On the other hand, in the case of "skin like snow", "white" and "puffy" should only be weakly connected. Therefore, each weight for the mutual connections between nodes is estimated using the correlation coefficient between the two features.

3.3 Model Simulation

In this study, the model is simulated using the parameters $s_1 = 250$, $L = 5$, $s_2 = 100$, $\zeta = 0.0029$ ($= 10/the\ number\ of\ adjectives$)), $\beta = 0.5$. The model simulation results for the metaphors of "work like a mountain" and "a teacher like a demon" are shown in Table 2. The results of the categorization process model ($V_j(M)$) and the results of the dynamic interaction process model ($O_q(M)$) are shown. Features with relatively strong values for $V_j(M)$ and $O_q(M)$ respectively can be regarded as meanings of the vectors.

Reflecting the influence of revision based on the dictionary, the first values in the categorization model are very high. The results of both models (CPM and

Table 2. Metaphor meaning computed by the categorization process model and the two-process model ("work like a mountain", "a teacher like a demon"). The output values are shown in parentheses.(CPM:categorization process model, TPM:two-process model).

	M = work like a mountain		M = a teacher like a demon	
	CPM (V_j(M))	TPM (O_q(M))	CPM (V_j(M))	TPM (O_q(M))
1	high (0.2290)	high (0.9761)	awful (0.1310)	awful (0.9954)
2	low (0.0409)	many (0.9196)	mysterious (0.0382)	young (0.9950)
3	many (0.0229)	strict (0.7148)	young (0.0239)	natural (0.9943)
4	near (0.0197)	near (0.6776)	strange (0.0158)	horrible (0.9869)
5	new (0.0189)	wonderful (0.5901)	good (0.0154)	good (0.9793)
6	good (0.0180)	new (0.5676)	high (0.0143)	powerful (0.9766)
7	important (0.0164)	important (0.5033)	horrible (0.0099)	strong (0.9766)
8	strict (0.0150)	possible (0.4724)	favorite (0.0095)	special (0.9695)
9	main (0.0101)	various (0.4707)	natural (0.0095)	beautiful (0.9677)
10	bad (0.0096)	hard (0.4678)	bad (0.0092)	black (0.9570)

TPM) seem to be appropriate. However, in order to verify which model is more appropriate and to examine the validity of the two-process model, the following psychological experimentation is needed.

4 Psychological Experiment

In order to examine the validity of the model, a psychological experiment was conducted.

The participants were 31 undergraduates. The metaphorical expressions used in the psychological experiment were "work like a mountain" and "a teacher like a demon". First, the target ("work" or "teacher"), the vehicle ("mountain" or "demon") and the metaphor ("work like a mountain" or "a teacher like a demon") were presented to the participants. Next, they were asked to respond with appropriate features of the target, of the vehicle and of the metaphor in the form of adjectives.

Table 3 lists features that were given by two or more participants when the metaphor ("work like a mountain" or "a teacher like a demon"), the vehicle ("mountain" or "demon") and the target ("work" or "teacher") were presented.

Features of "work like a mountain", such as "many", "hard" and "strict", which were given by two or more participants in the experiment, were successfully estimated by the two-process model although "hard" can not be estimated by the categorization process model. The results of the experiment indicate the existence of emergent features, especially in the cases of "many", "hard", "challenging", "never ending", "troublesome" and "strict". The constructed model also emphasizes the emergent features of "many" and "hard", which were given by more than two participants, in the dynamic interaction process. Similarly, features of "a teacher like a demon", such as "horrible" and "strong", which were

Table 3. The results of the psychological experiment ("work like a mountain", "a teacher like a demon"). The numbers of the participants who responded with a particular feature are shown in parentheses.

work like a mountain			a teacher like a demon		
work	mountain	work like a mountain	teacher	demon	a teacher like a demon
bitter (10)	big (9)	many (20)	gentle (10)	horrible (19)	horrible (20)
happy (9)	high (8)	busy (8)	awful (8)	strong (12)	strict (16)
busy (7)	beautiful (4)	bitter (8)	strict (5)	big (9)	strong (7)
troublesome (5)	cold (3)	hard (7)	irritable (3)	awful(6)	irritable (3)
difficult (3)	wide (2)	difficult (3)	great (2)	red (4)	violent (2)
stiff (2)		challenging (3)	wise (2)	muscular (2)	big (2)
important (2)		never ending (2)	beautiful (2)	blue (2)	
toilsome (2)		troublesome (2)	wonderful (2)		
		strict (2)			

given by multiple participants in the experiment, were successfully estimated by the model although "strong" can not be estimated by the categorization process model. The results of the experiment support that the two-process model is more appropriate than the categorization process model and suggests the psychological validity of the constructed model.

5 Discussion

In this study, it is assumed that a target assigned as a member of the ad-hoc category of a vehicle in the categorization process is influenced by the dynamic interaction among features. The present model is constructed based on this assumption. First, the knowledge structure of concepts is estimated using a statistical language analysis [10]. Next, the meaning of a target as a member of the ad-hoc category of a vehicle is computed within the knowledge structure by applying Kintsch's algorithm. Third, the recurrent neural network model, which represents the dynamic interaction among features, estimates the meaning of the metaphor using the assigned meaning of the target. Finally, the validity of the model was examined by a psychological experiment. The results of the model were found to be more appropriate than the results for the categorization process model which uses the same algorithm as Kintsch's model. The present model also implements the dynamic interaction process that leads to the emergence of features in the metaphor. The emergent features estimated by this model are also validated by the psychological expriment.

In order to avoid the biases inherent in newspaper corpora, the current study revised the knowledge structure by supplementing the statistical analysis with dictionary data. While this undoubtedly enhanced the linguistic analysis, this procedure alone is still insufficient to realize an ideal knowledge structure, because even dictionaries do not contain very obvious information beyond the simplest definitions (e.g., "yellow tiger"). In order to realize even more sophisticated knowledge structures, statistical analysis for very large-scale corpora need to be carried out (e.g., literature and school books). Furthermore, in oreder to examine more clealy the psychological validity of the model, more detailed psychological experiment have to be conducted.

Acknowledgement. This research is supported by the Tokyo Institute of Technology 21COE Program, "Framework for Systematization and Application of Large-scale Knowledge Resources".

References

1. Kusumi, T.: Hiyu no Syori Katei to Imikozo. Kazama Syobo. (1995)
2. Gentner, D., Wolff, P.: Alignment in the processing of metaphor. Journal of memory and language 37, 331–355 (1997)
3. Glucksberg, S., Keysar, B.: Understanding Metaphorica Comparisons: Beyond Similarity. Psychological Review. 97(1), 3–18 (1990)
4. Nakagawa, M., Terai, A., Hirose, S.: A Neural Network Model of Metaphor Understanding. In: Proc. of Eighth International Conference on Cognitive and Neural Systems, vol. 32 (2004)
5. Utsumi, A.: Hiyu no ninchi / Keisan Moderu. Computer Today. 96(3), 34–39 (2000)
6. Kintsch, W.: Metaphor comprehension: A computational theory. Psychonomic Bulletin & Review 7(2), 257–266 (2000)
7. Utsumi, A.: Computational exploration of metaphor comprehension processes. In: Proc. of the 28th Annual Meeting of the Cognitive Science Society, pp. 2281–2286 (2006)
8. Deerwester, S., Dumais, S., Furnas, G., Landauer, T., Harshman, R.: Indexing by Latent Semantic Analysis. Journal of the Society for Information Science 41(6), 391–407 (1990)
9. Terai, A., Nakagawa, M.: A Neural Network Model of Metaphor Understanding with Dynamic Interaction based on a Statistical Language Analysis. In: Kollias, S., Stafylopatis, A., Duch, W., Oja, E. (eds.) ICANN 2006. LNCS, vol. 4131, pp. 495–504. Springer, Heidelberg (2006)
10. Kameya, Y., Sato, T.: Computation of probabilistic relationship between concepts and their attributes using a statistical analysis of Japanese corpora. In: Proc. of Symposium on Large-scale Knowledge Resources: LKR2005, pp. 65–68 (2005)
11. Nueckles, M., Janetzko, D.: The role of semantic similarity in the comprehension of metaphor. In: Proc. of the 19th Annual Meeting of the Cognitive Science Society, pp. 578–583 (1997)
12. Gineste, M., Indurkhya, B., Scart, V.: Emergence of features in metaphor comprehension. Metaphor and Symbol. 15(3), 117–135 (2000)
13. Kudoh, T., Matsumoto, Y.: Japanese Dependency Analysis using Cascaded Chunking. In: Proc. of the 6th Conference on Natural Language Learning, pp. 63–69 (2002)
14. Kindaichi, H., Ikeda, Y.: KOKUGO DAIJITEN 2nd Eddition, GAKKEN (1988)

A Novel Novelty Detector

Neill R. Taylor and John G. Taylor

King's College London, Department of Mathematics, The Strand,
London, WC2R 2LS, U.K.
{neill.taylor,john.g.taylor}@kcl.ac.uk

Abstract. We develop a model of a set of novelty and familiarity detectors in the hippocampus which possess unique properties, and have been recently reported in [1]. The model uses both inhibition and disinhibition, together with a suitable output function of prefrontal object representations, to create the separate novelty and familiarity detectors with the observed properties. We conclude the paper with a discussion of the relation of this novelty system with that presented by numerous other techniques.

Keywords: Familiarity, hippocampus, prefrontal cortex, object representations, inhibition, disinhibition.

1 Introduction

Novelty detection is an important faculty for any information system required to venture efficiently into new environments and improve its repertoire of responses. In order to do that in the presence of novel objects affordances for these objects must be constructed so as to enable them to be manipulated most efficiently. These affordances involve novelty detection and then gradual learning of the associated visual and motor responses to an object, such as the grasps that can be made to it [2].

There have been numerous novelty detectors created through use of various neural network architectures. Methods include: multi-layer perceptrons [3], support vector machines [4], radial basis functions [5], auto-associator networks [6], Hopfield networks [7], self-organising maps (SOMs) [8], and adaptive resonance theory (ART) [9]. The paper of [1] has shown that in the hippocampus (HC) there exist, amongst others, two sorts of memory-based neurons: novelty detectors (ND) and familiarity detectors (FD). The ND are turned on by novel stimuli but not by familiar ones. The FD are turned on by familiar stimuli, but not by novel ones. Moreover the FD are not stimulus specific, but respond to any of a large category of input stimuli that had been seen half an hour previously. It is these two classes that we proposed to simulate.

We discuss the data in more detail in the next section, and present a model architecture in the following one. Results of simulations are presented in section 4. The paper finishes with a conclusion.

J. Marques de Sá et al. (Eds.): ICANN 2007, Part II, LNCS 4669, pp. 973–983, 2007.

2 Discussion of Rutishauser et al. Data [1]

Rutishauser and colleagues [1] recorded single cell activity from neurons in the hippocampal-amygdala complex in the medial temporal lobe (MTL). Subjects were presented with familiar and novel objects. In the learning phase all stimuli were novel, the learning set was composed of 12 objects presented on a computer screen only once in a random quadrant, it is not clear whether a central fixation point was present. Subjects needed to remember both the object and location. The recognition phase occurred ~30 minutes later, where the data set included objects that were familiar (i.e. had been included in the learning phase) or novel. Subjects had to indicate if the stimulus was novel or familiar and for familiar objects indicate which quadrant the object had previously been presented; in all cases the recognition data set was presented at the centre of the computer screen. Novel and familiar objects were identified with high accuracy 88.5% ± 2.8%, whilst the location for familiar stimuli was correctly identified 49.5% ± 8.0%. Hence correct identification of stimuli as familiar or novel did not depend on correct recall of spatial location. If a central fixation point is not present this might account for the lower accuracy in recalling the location information, since the subjects will quickly fixate on the presented stimulus. They were able to classify neurons as either NDs or FDs, depending on how their activity altered in comparing responses in novel and familiar trials. NDs showed an increase in activation for those trials involving novel stimuli versus those of familiar objects. It was found that one-shot learning could occur, with firing rate changes: if during the recognition phase a novel object was presented then a node classified as ND would be active, but at the next presentation of that same stimulus it was a node classified as FD that was active and the ND node was now silent. Fig. 1 shows the response of a single hippocampal neuron (from [1]) during the recognition phase, this node however did not respond in a significant manner during the initial presentation of the objects.

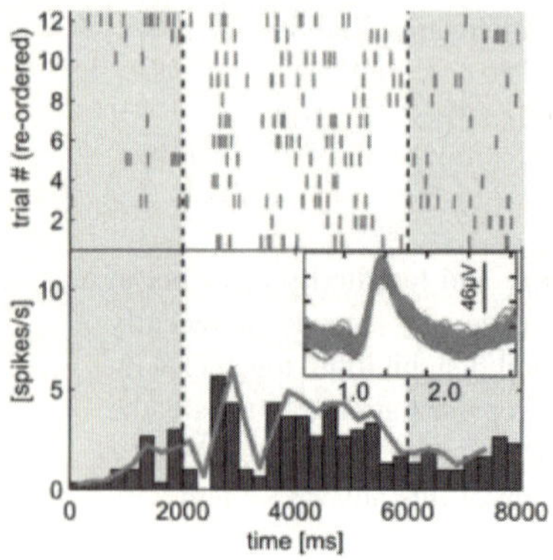

Fig. 1. Taken from [1] their fig. 2e. Response of a hippocampal neuron neuron during presentation of a familiar stimulus. Stimulus presentation from 2000-6000ms. Raster plot for individual presentation is at top; bottom: binned histograms are across all trials and inset is the spike waveform.

3 Architecture

To simplify the architecture we do not model the ventral stream of the visual pathway. Object representations are based on simulated TE responses to different objects previously reported elsewhere [10], and input to the inferior frontal gyrus (IFG) in the pre-frontal cortex (PFC). The IFG is a known target of the ventral visual stream, and the model can easily be extended with the addition of a suitable visual model (such as [10]). We propose the architecture of fig. 2 as a possible manner that object representations, with FD and ND neurons could interact to generate similar results to [1]. Here the IFG is composed of four nodes, three of which are preferentially responsive to one of three familiar objects (such as: square, triangle and circle), the fourth will be used for learning a novel object. The FD and ND regions are modelled as a single node each, since the experimental results show that these neurons respond in a general manner to the group of familiar objects and novel objects, respectively, rather than to a specific object (known or unknown). A further region that is inhibitory (INHIB) acts upon the ND node. Reciprocal inhibitory connections between the FD and ND generally produce a 'winner-take-all' response, though this can vary due to the noise in the system. The FD to ND inhibitory weight is important during the presentation of a familiar stimulus, with the reverse connection useful during learning. Weight values are indicated in table I, these can vary substantially.

Table 1. Simulation connection strengths

Connection	Strength
IFG →FD	$0.8*10^{-10}$
ND→FD	$-5*10^{-12}$
INHIB→ND	$-40*10^{-12}$
IFG→INHIB	$-2.4*10^{-12}$
FD→ND	$-2.5*10^{-12}$

The basis of the model is based on assessment of the IFG activity brought about by a given input stimulus. The basis of the model is in terms of:

1) If a familiar stimulus is input to the system one or more of the IFG goal nodes are activated by the input to a good level (exceeding some threshold).
2) If a novel stimulus is input there will be low-level activation of many (if not all) of the IFG goal nodes, but none of these activations will be close to that caused by a familiar input stimulus.

The difference of responses of IFG neurons to the two different classes of stimuli is most clearly given by the maximum operator MAX applied to the IFG nodes (though only for an excitatory projection to the FD):

1) For a familiar stimulus: MAX(IFG activity) $\geq$ threshold=>IFG output ~ 1
2) A novel stimulus: IFG < threshold, and hence MAX(IFG) ~ 0, zero output to FD.

A familiarity detector FD can be created most simply by:

MAX(IFG activity) > threshold → Excitatory input onto FD neuron.

A novelty detector ND can be created simply by:

(IFG (activity) < threshold) → Inhibitory input onto a spontaneously active neuron (INHIB) → inhibitory input onto ND neuron (also spontaneously active, when inhibition is released). However, for this structure any IFG activity that is sufficiently strong will release the ND from inhibition whether the stimulus is novel or familiar. Adding an inhibitory connection from the FD to ND will reduce any ND activation in the case where the stimulus is familiar. Hence the following states apply:

1) For a familiar input => MAX(IFG) ~ 1 => FD ~ 1 and ND ~ 0
2) For a novel input => MAX(IFG) ~ 0 => FD ~ 0 and ND ~ 1, as required for ND.

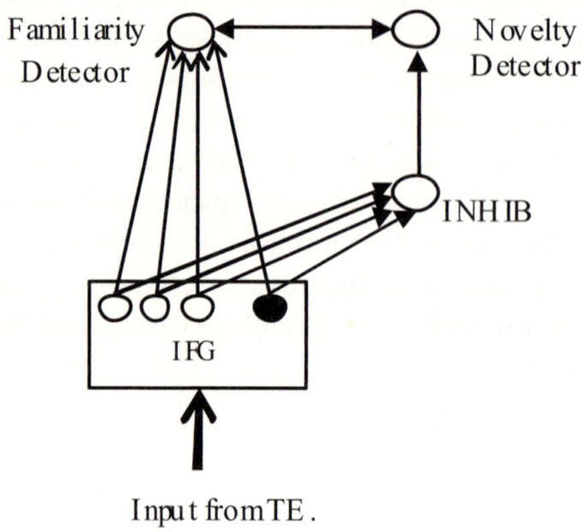

Fig. 2. Model architecture. Open arrow-heads indicate excitatory weights and closed arrow-heads are inhibitory. In the IFG those nodes that are preferentially responsive to an object are unfilled whilst a node that currently is not responsive to any of the current known objects is filled in black.

However it is necessary to prevent the spontaneously active ND from firing when no stimuli are present hence the need for the INHIB input that must be present, under contextual control. This results in the final architecture of fig. 2.

The experimental data of [1] results from tests performed by adults. In these subjects whilst the total form of the input stimulus is novel the component parts are not. There will be responses at V2 to angles formed by pairs of bars, V4 will respond to more complex combinations of 3 or more lines, similarly TEO, TE and IFG will have firing rates increased from spontaneous rates. Indeed a simulation of occlusion, [10] which combined 2 known (square and triangle) objects into a composite novel object, showed firing rates within the IFG nodes of up to 40Hz for the new input. This was reduced from the responses of the nodes to their preferred input (which were >80Hz), but increased form spontaneous levels. The connectivity from IFG to the INHIB region is such that a single modelled IFG node firing at ~10Hz will generate a response from the ND node, the FD neuron will remain at spontaneous levels since all IFG nodes have firing rates below the threshold of the MAX function.

All nodes are leaky-integrate-and-fire neurons with potentials defined by:

$$C_m \frac{dV}{dt} = -g_{leak}(V - V_{leak}) + (V_m - V)I_{excit} + (V - V_{shunt})I_{GABA} + I_{irtrinsic} + noise \quad (1)$$

where $V(t)$ is the potential, C_m is the neuron capacitance set to $5*10^{-10}$ F, V_{leak} is the resting potential of –70mV, g_{leak} =$2.5*10^{-8}$ S, I_{excit} is the total excitatory input, I_{GABA} is the total inhibitory input, V_{shunt} is the shunting potential of –80mV, V_m is set to 0mV. The spiking threshold is set to –52mV, $I_{intrinsic}$ is an intrinsic current present for INHIB and ND to generate spontaneous firing rates of 30Hz and 10Hz, respectively.

Output from IFG to the FD node is controlled by a MAX function, as indicated by our earlier discussion. To cause a non-zero response the MAX detector requires that the IFG nodes fire at 80 Hz or greater within a moving time-window of 50ms (4 spikes or more within the time-window). Hence the first inputs to the FD node occur after a delay of ~50ms, before such spikes are integrated at the FD synapse. Hence only when an IFG node is strongly activated by the presence of a familiar input will the FD node become active.

Noise is added to the IFG and INHIB nodes in the form of inputs from 800 nodes firing with a Poisson spike-train of 3Hz. Increasing the population size of each region to give distributed representations as well as extending the noise to the ND and FD nodes should give the model the variability seen in the experimental results [1], where incorrect decisions are made, i.e. a novel input is misclassified as a familiar one and with a more complex IFG by including inhibitory inter-neurons causing misclassifications of familiar inputs. We hope to consider this elsewhere.

Learning a novel stimulus uses the unclassified IFG node. Learning is initiated by the activation of the ND, which allows the model to learn the connectivities from TE to the unclassified node, whose weights are initialized as non-zero. To prevent learning to the other IFG nodes these can be inhibited, or learning can be directed to a specific node by the presence of some neuro-transmitter. A simple causal Hebbian learning rule is used with a time-window of 50ms, and high learning rate to accomplish one-shot learning. We have assumed that a novel representation is learnt during a single presentation, since the experimental regime [1] includes a distracting cognitively demanding task during the 30 minute delay phase between learning and recognition phases that is designed to prevent rehearsal. The experimental results do not show a change of response during a learning trial from an ND to an FD. This is where the inhibitory synapse from the modelled ND to FD is important, by preventing FD response even when learning has reached a level to satisfy the MAX function for IFG output to the FD, since the experimental results [1] do not show a change in responses from ND to FD during learning of a novel input. For this reason the connection strength from ND to FD is double that of the reverse weight, this results from spike timing issues which are only present in reduced models, if the FD and ND were modelled as populations the weights could be equal. The MAX function is, as previously, set to 80Hz, though could be set at a lower threshold and learnt as the new IFG representation improves with learning.

4 Simulation Results

The simulations are run over 5000ms, with the inputs (novel and familiar) being presented for 4000ms with a stimulus onset time of 500ms. For the familiar objects (square, triangle and circle from [10]) one IFG node fires at ~85Hz (this is the node that is preferentially responsive to the current input) whilst the two other nodes (which prefer the two non-presented objects) have firing rates of ~70 and 45 Hz. Table 2 shows the mean number of spikes of the IFG nodes, within the moving 50ms time-window along with the standard deviations (SD), that respond to the known inputs. The unclassified IFG node fires at ~8Hz. Hence only the IFG node preferentially responsive to the input has a sufficient firing rate to turn-on an IFG to FD synapse, in this case IFG node 1 the mean number of spikes throughout the presentation of the input is > 4, this satisfies the MAX function which requires a firing rate of 80Hz hence 4 spikes per 50ms bin. The results for a familiar object are shown in fig. 3 (only the first 1000ms of the trial is shown), where only the IFG neuron shown in fig. 3a has sufficient activity to excite the FD node.

The familiar input leads to the activation of the FD node, as well as inhibiting the INHIB node. In the absence of any input (either familiar or novel) only the INHIB node of the disinhibitory route from IFG → INHIB → ND is active. Whilst the INHIB and ND nodes both have intrinsic currents the greater firing rate of the INHIB node dominates the ND, keeping the latter silent. In the presence of a familiar or novel input the INHIB region is silenced by IFG output, releasing the ND node. For the case of a familiar object the FD node is active and this takes over the inhibition of the ND from the INHIB region. Hence only the FD node becomes active, firing at 16 Hz (fig. 3e) whilst the ND remains silent (fig. 3f), inhibited by the INHIB node in the absence of any input initially and later by the FD node as it becomes active in response to the familiar object. The responses of the FD and ND agree with the results of [1] shown in their fig. 2, where for a familiar stimulus only the FD has a response that increases from the spontaneous levels, whilst the ND response remains at the spontaneous levels.

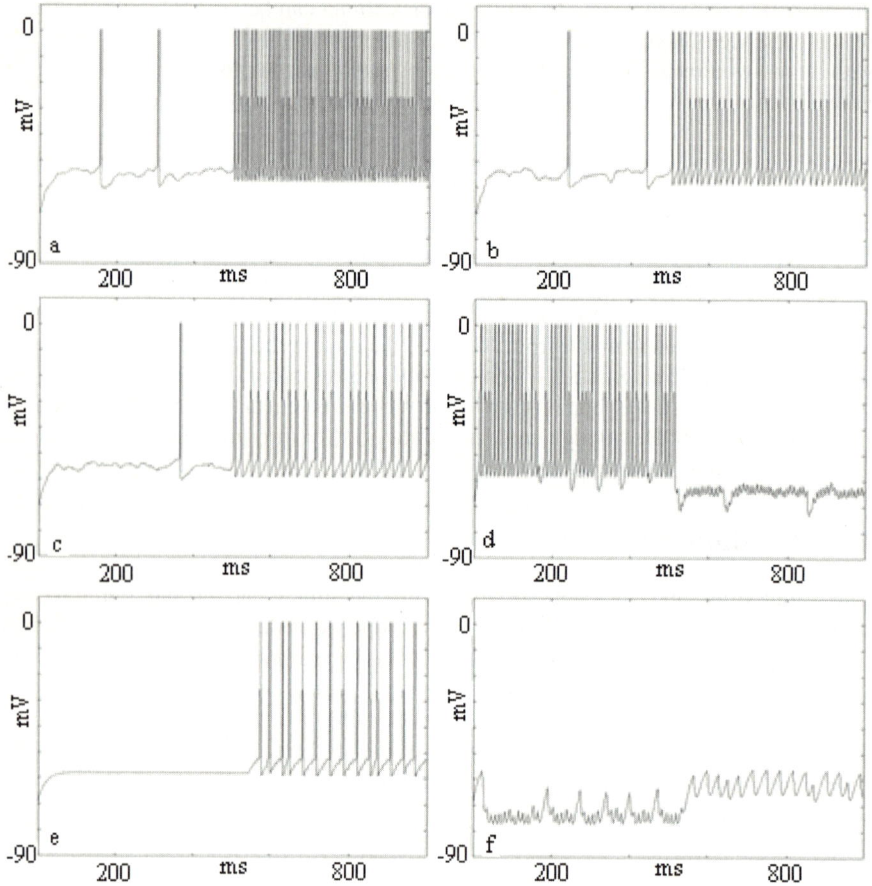

Fig. 3. Responses of the model neurons to a familiar stimulus (the first 1000ms of a trial is shown). Plots a), b) and c) show the potentials of the 3 IFG nodes only the first fires at >80Hz to the familiar input, d) INHIB node, e) FD node, f) ND node.

Presentation of a novel object generates IFG firing rates of: ~70, 70 and 45 Hz for the classified IFG nodes and again 8Hz for the unclassified node. Hence all IFG nodes have firing rates below the threshold to activate the IFG to FD synapses. The neuron

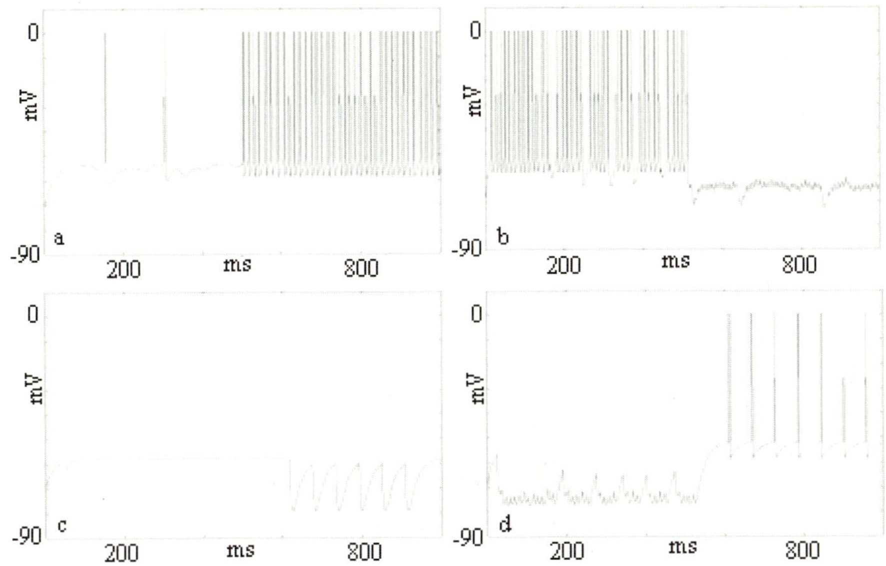

Fig. 4. Plots of neuronal potentials for a novel input. Plot a) IFG neuron that previously spiked at >80Hz; all 3 IFG nodes fire below 80Hz (the other two IFG nodes have firing rates and patterns as shown in fig. 3 b and c. Plot b) is the INHIB node, c) the FD and d) the ND.

potential plots for a novel input are shown in fig. 4 (again only the first 1000ms); for this particular case the learning has been turned off.

Table 2. IFG mean number of spikes in the 50ms time-window (figures in brackets are mean firing rates calculated over 4 seconds of presentation)

IFG node 1	IFG node 2	IFG node 3
4.110±0.019 (85Hz)	3.457±0.021 (70Hz)	2.169±0.021 (45Hz)

Since all three IFG nodes with a preference have firing rates below 80Hz (only one shown in fig. 4 a, the other two have the same response as shown in fig 3 b and c), the FD is not activated (fig. 4c) and as before the INHIB node (fig 4 b) is inhibited by the IFG activations (only one of the IFG nodes actually needs be active to achieve this inhibition). Turning off the INHIB node releases the ND node (fig 4d) from its inhibited state and since in this case the FD is silent (due to the MAX function not being relevant), the ND begins firing at a rate of 16Hz.

It should be noted that the INHIB node is not useful in determining whether a given input is novel or familiar since it has similar firing patterns for both input types, as it is always inhibited by the IFG for both novel and familiar stimuli (figs. 3d, 4b).

We show the results for a novel stimulus (the same as used for fig. 4) with the learning (of Hebbian form, and from the temporal lobe input onto the novel node in IFG) now turned on in fig. 5. The initial response of the new IFG node (the first 1000ms of the trial) is shown in fig 5a. The responses during this period of the

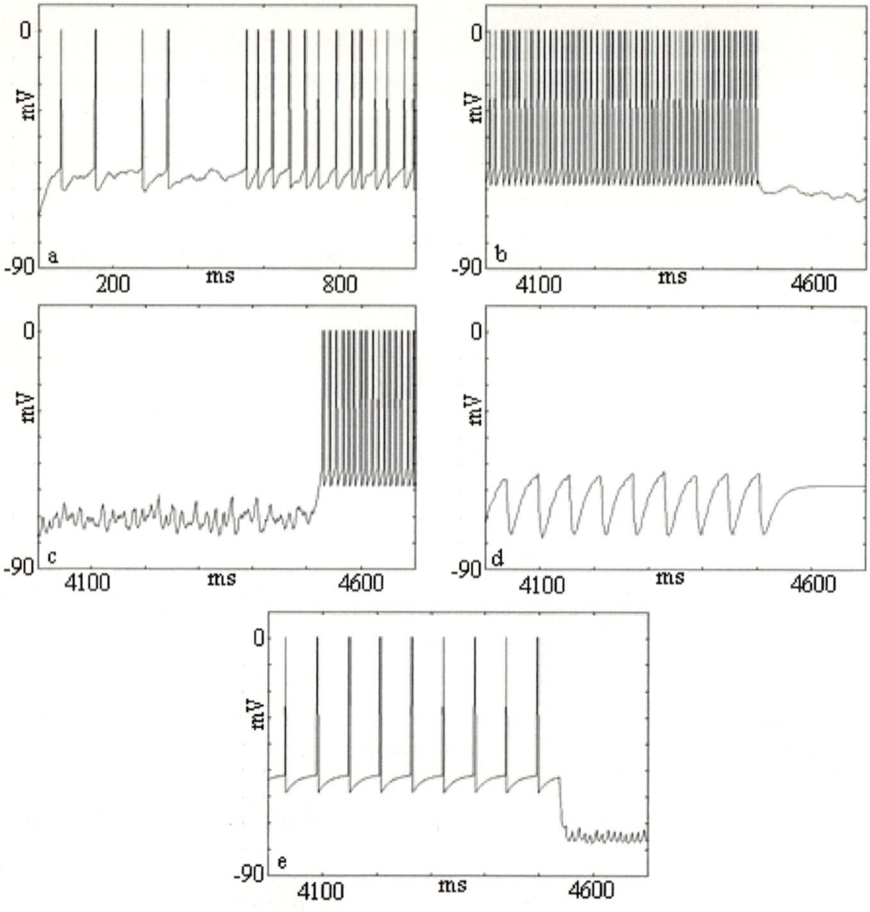

Fig. 5. Responses of the new IFG node during the early (a) and late (b) period of stimulus presentation of a novel object with learning. The later period for the INHIB node (c), the FD node (d) and ND node (e) are shown. Note that although the firing rate of the IFG node exceeds 80 Hz around 4300ms, and leads to an excitatory input to the FD the inhibition from the active ND prevents any activation.

INHIB, FD and ND nodes are as in fig. 4 b, c and d, respectively. We also show the final 700ms starting at 4000ms for new IFG node, INHIB, FD and ND in fig 4 b, c, d and e, respectively. The firing rate of the new IFG node has increased substantially during the presentation of the novel stimulus, the learning being sufficiently fast that the firing rate exceeds 80Hz at ~ 4300ms of trial time. This allows output from this IFG node via the MAX function to the FD. However since the ND is active inhibition to the FD prevents it firing. The next presentation of this same stimulus is shown in fig. 6. In this case the new IFG node (fig 6a) satisfies the MAX function (with the other three IFG nodes all being sub-threshold) and the FD node (fig 6b) is now activated which inhibits any possible spiking of the ND. Hence a representation of the

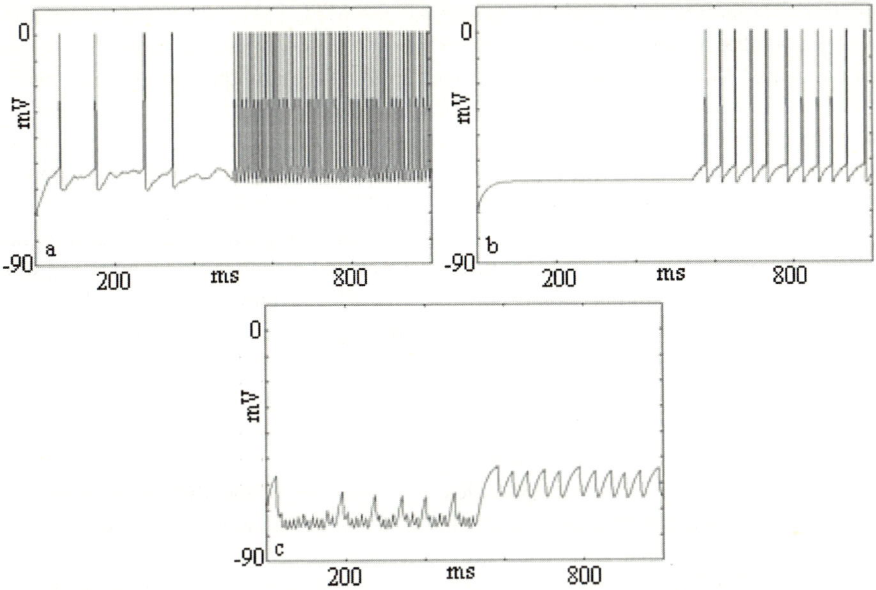

Fig. 6. Presentation of the same stimulus used in training for fig. 5. The new IFG node (a) now responds strongly to the input easily exceeding the 80Hz threshold for excitatory output to the FD (b), which becomes active. The FD keeps the ND (c) inhibited. Hence on the second presentation of a previously novel input it is now defined as familiar.

novel stimulus was learnt during its first presentation (fig. 5) when it was defined as novel allowing learning to be turned on by the activation of the ND node and at the subsequent presentation it is now recognized as familiar.

5 Conclusion

We have shown that a simple structure can be used to produce simulated results that are similar to experimental single cell recordings [1], in that distinct groups of neurons that define whether a stimulus is novel or familiar can be hard-wired, where in the latter case a familiar stimulus activates a FD node but not a ND node. By including learning, of a causal Hebbian one-shot form, an object seen for the first time and so novel activates the ND but then on subsequent presentations the stimulus now activates FD, also replicating the experimental results of [1]. From the data of [1] it appears that with one presentation of a novel stimulus, 30 minutes later there is a stored representation that now indicates that the object is familiar. This is what is to be expected in a HC with one-(or a few)-shot learning. The addition of a dorsal visual stream (the 'where' route) would allow for further investigation of the experimental results of [1], and the authors' suggestion that contextual information is not necessarily required for novelty and familiarity distinction, it is unclear whether a fixation point was included in the experimental paradigm which could affect the single-cell recording results. Such an architecture could be based on the proposed functional organisation of memory system of the MTL [11], where the information from the ventral visual stream enters the HC via the perirhinal and lateral entorhinal cortex, whilst the path of the dorsal stream is via the parahippocampal and medial entorhinal cortex and finally the HC. Hence the pathways are kept separate and are only integrated at the HC. The addition of the lower cortical regions of the ventral stream, perhaps using the model of [10], will also require learning to be extended certainly as far as forward projections to TE and TEO from V2 and V4.

Using the architecture described above, lesions affecting the MAX operator would potentially lead to familiar stimuli being mis-identified as novel by the NDs and FDs. The thresholded IFG output would not reach the FD, leaving it silent, whilst, as we have seen, the INHIB node is always inhibited (fig. 3 d and fig. 4 b) which releases the ND node which without inhibition from an active FD will begin to spike. Hence the object could be identified correctly within higher regions of the ventral visual pathway and also IFG but then defined as being novel. It would be interesting to know whether such results occur within any patient groups.

We note that there are several theories as to the role of the PFC. One is that the PFC does not store information long-term, but has a more 'write and erase' adaptive structure [12]. Neuronal responses are then dependent on what is pertinent for completion of the current task. These tasks are often trained for long periods of time, indeed the subjects may be over-trained. Other results especially with recognition of faces have shown that PFC does have memory sites [13].

Finally, to compare the model presented in the paper with that of others, we can consider the visual input to IFG as giving a distance or confidence measure as to how the current input compares with the learnt and hence familiar object representations. There is the structure on top of this including the MAX function to further help interpret the IFG firing rates as familiar or novel. For the specific models mentioned in section 1: a number have a non-neural basis or require post-processing that is non-neural only the SOM (using a distance measure) and ART (comparison measure) approaches are similar. There is a certain level of similarity in a number of previous schemes for novelty detection, but few have separate FD & ND nodes as in fig. 2.

Acknowledgments. NRT would like to thank EPSRC, JGT would like to thank the EC, under the GNOSYS (FP6-003835) project.

References

1. Rutishauser, U., Mamelak, A.N., Schuman, E.M.: Single-Trial Learning of Novel Stimuli by Individual Neurons of the Human Hippocampus-Amygdala Complex. Neuron 49, 805–813 (2006)
2. Raos, V., Umilta, M.-A., Murata, A., Fogassi, L., Gallese, V.: Functional Properties of Grasping-Related Neurons in the Ventral Premotor Area F5 of the Macaque Monkey. J Neurophysiol 95, 709–729 (2006)
3. Moya, M.R., Koch, M.W., Hostetler, L.D.: One-class classifier networks for target recognition applications. In: Proc. WCNN, INNS, pp. 797–801 (1993)
4. Tax, D.M.J., Duin, R.P.W.: Support vector domain description. Pattern Recognition Letters 20, 1191–1199 (1999)

5. Bishop, C.: Neural networks for pattern recognition. Oxford University Press, Oxford (1995)
6. Sohn, H., Worden, K., Farrar, C.R.: Novelty detection under changing environmental conditions. In: Proc. 8th Annual SPIE Inter. Symp. Smart Structures & Mat (2001)
7. Jagota, A.: Novelty detection on a very large number of memories stored in a Hopfield-style network. Proc. IJCNN-91 2, 905 (1991)
8. Kohonen, T.: Self-organisation and associative memory. Springer-Verlag, Berlin (1988)
9. Granger, E., Grossberg, S., Rubin, M.A., Streilein, W.W.: Familiarity discrimination of radar pulses. Advances in NIPS 11, 875–881 (1999)
10. Taylor, N.R., Panchev, C., Hartley, M., Kasderidis, S., Taylor, J.G.: Occlusion, Attention and Object Representations. In: Kollias, S., Stafylopatis, A., Duch, W., Oja, E. (eds.) ICANN 2006. LNCS, vol. 4131, pp. 592–601. Springer, Heidelberg (2006)
11. Eichenbaum, H.: Remembering: Functional organization of the declarative memory system. Curr. Biol. 16, 643–645 (2006)
12. Duncan, J.: An adaptive coding mode of neural function in the prefrontal cortex. Nat. Rev. Neurosci 2, 820–829 (2001)
13. Scalaidhe, S.P., Wilson, F.A., Goldman-Rakic, P.S.: Face-selective neurons during passive viewing and working memory performance of rhesus monkeys: evidence for intrinsic specialization of neuronal coding. Cereb. Cortex 9, 459–475 (1999)

Author Index

Printing: Mercedes-Druck, Berlin
Binding: Stein+Lehmann, Berlin